Optical Fiber Telecommunications VII

Optical Fiber Telecommunications VII

Edited by

ALAN E. WILLNER
University of Southern California, Los Angeles, CA, United States

Academic Press is an imprint of Elsevier
125 London Wall, London EC2Y 5AS, United Kingdom
525 B Street, Suite 1650, San Diego, CA 92101, United States
50 Hampshire Street, 5th Floor, Cambridge, MA 02139, United States
The Boulevard, Langford Lane, Kidlington, Oxford OX5 1GB, United Kingdom

British Library Cataloguing-in-Publication Data
A catalogue record for this book is available from the British Library

Library of Congress Cataloging-in-Publication Data
A catalog record for this book is available from the Library of Congress

ISBN: 978-0-12-816502-7

For Information on all Academic Press publications
visit our website at https://www.elsevier.com/books-and-journals

Publisher: Mara Conner
Acquisition Editor: Tim Pitts
Editorial Project Manager: Joshua Mearns
Production Project Manager: Anitha Sivaraj
Cover Designer: Greg Harris

Typeset by MPS Limited, Chennai, India

Dedication

In memory of Ivan P. Kaminow and Tingye Li
They were pioneers, mentors, and champions of our field. They guided this *Optical Fiber Telecommunications* book series through many editions, and their impact lives on.

Dr. Ivan P. Kaminow (March 3, 1930–Dec. 18, 2013), ז"ל
缅怀 Dr. Tingye Li (July 7, 1931–Dec. 27, 2012)

For Florence Kaminow and Edith Li
For Michelle, Moshe, Asher, Ari, and Yaakov Willner

Contents

List of contributors

Kazi S. Abedin
OFS Laboratories, Somerset, NJ, United States

Shaif-ul Alam
Optoelectronics Research Centre (ORC), University of Southampton, Southampton, United Kingdom

Miles H. Anderson
Laboratory of Photonics and Quantum Measurements (LPQM), École Polytechnique Fédérale de Lausanne (EPFL), Lausanne, Switzerland

Cristian Antonelli
Department of Physical and Chemical Sciences, University of L'Aquila, L'Aquila, Italy

Moritz Baier
Fraunhofer Heinrich Hertz Institute (FhG-HHI), Berlin, Germany

Polina Bayvel
Optical Networks Group, University College London, London, United Kingdom

Neal S. Bergano
SubCom, Eatontown, NJ, United States

Keren Bergman
Columbia University in the city of New York, New York, NY, United States

John Bowers
University of California Santa Barbara, Santa Barbara, CA, United States

Jin-Xing Cai
SubCom, Eatontown, NJ, United States

You-Chia Chang
Department of Photonics and Institute of Electro-Optical Engineering, National Chiao Tung University, Hsinchu, Taiwan, ROC

Di Che
Nokia Bell Labs, Holmdel, NJ, United States; Department of Electrical and Electronics Engineering, The University of Melbourne, Parkville, VIC, Australia

Xi Chen
Nokia Bell Labs, Holmdel, NJ, United States

Qixiang Cheng
Columbia University in the city of New York, New York, NY, United States

Francesco Da Ros
DTU Fotonik, Technical University of Denmark, Lyngby, Denmark

Edson Porto da Silva
Federal University of Campina Grande, Campina Grande, Paraíba, Brazil

Ning Deng
Huawei Technologies, Shenzhen, P.R. China

Nicholas M. Fahrenkopf
AIM Photonics, SUNY Polytechnic Institute, New York, NY, United States

John Fakidis
Li-Fi Research and Development Center and Institute for Digital Communications, School of Engineering, The University of Edinburgh, Edinburgh, United Kingdom

Qirui Fan
Photonics Research Centre, Department of Electrical Engineering, The Hong Kong Polytechnic University, Hong Kong

Wolfgang Freude
Institute of Photonics and Quantum Electronics (IPQ), Karlsruhe Institute of Technology (KIT), Karlsruhe, Germany

Michael Galili
DTU Fotonik, Technical University of Denmark, Lyngby, Denmark

Peter David Girouard
DTU Fotonik, Technical University of Denmark, Lyngby, Denmark

Richard Gladhill
Toppan, Tokyo, Japan

Madeleine Glick
Columbia University in the city of New York, New York, NY, United States

Rich Goldman
Lumerical Inc., Vancouver, BC, Canada

Pengyu Guan
DTU Fotonik, Technical University of Denmark, Lyngby, Denmark

Harald Haas
Li-Fi Research and Development Center and Institute for Digital Communications, School of Engineering, The University of Edinburgh, Edinburgh, United Kingdom

Tetsuya Hayashi
Sumitomo Electric Industries, Ltd., Yokohama, Japan

Gloria Hoefler
Infinera, Sunnyvale, CA, United States

Yongmin Jung
Optoelectronics Research Centre (ORC), University of Southampton, Southampton, United Kingdom

Pawel Marcin Kaminski
DTU Fotonik, Technical University of Denmark, Lyngby, Denmark

Maxim Karpov
Laboratory of Photonics and Quantum Measurements (LPQM), École Polytechnique Fédérale de Lausanne (EPFL), Lausanne, Switzerland

Juned N. Kemal
Institute of Photonics and Quantum Electronics (IPQ), Karlsruhe Institute of Technology (KIT), Karlsruhe, Germany

Faisal Nadeem Khan
Photonics Research Centre, Department of Electrical Engineering, The Hong Kong Polytechnic University, Hong Kong

Tobias J. Kippenberg
Laboratory of Photonics and Quantum Measurements (LPQM), École Polytechnique Fédérale de Lausanne (EPFL), Lausanne, Switzerland

Fred Kish
Analog Photonics, Boston, MA, United States

Frederik Klejs
DTU Fotonik, Technical University of Denmark, Lyngby, Denmark

Christian Koos
Institute of Photonics and Quantum Electronics (IPQ), Karlsruhe Institute of Technology (KIT), Karlsruhe, Germany; Institute of Microstructure Technology (IMT), Karlsruhe Institute of Technology (KIT), Karlsruhe, Germany

Saurabh Kumar
Amazon, Seattle, WA, United States

Cedric F. Lam
Google Fiber, Mountain View, CA, United States

Gilles S.C. Lamant
Cadence Design Systems, Inc., San Jose, CA, United States

Alan Pak Tao Lau
Photonics Research Centre, Department of Electrical Engineering, The Hong Kong Polytechnic University, Hong Kong

Gerd Leuchs
Max Planck Institute for the Science of Light, Erlangen, Germany

Ming-Jun Li
Corning Incorporated, Corning, NY, United States

Xinying Li
Georgia Institute of Technology, Atlanta, GA, United States

Michael Liehr
AIM Photonics, SUNY Polytechnic Institute, New York, NY, United States

Gabriele Liga
Signal Processing Systems Group, Department of Electrical Engineering, Eindhoven University of Technology, Eindhoven, The Netherlands

Mads Lillieholm
DTU Fotonik, Technical University of Denmark, Lyngby, Denmark

Michal Lipson
Department of Electrical Engineering, Columbia University, New York, NY, United States

Xiang Liu
New Jersey Research Center, Futurewei Technologies, Bridgewater, NJ, United States

Chao Lu
Photonics Research Centre, Department of Electronic and Information Engineering, The Hong Kong Polytechnic University, Hong Kong

Pablo Marin-Palomo
Institute of Photonics and Quantum Electronics (IPQ), Karlsruhe Institute of Technology (KIT), Karlsruhe, Germany

Dan M. Marom
The Hebrew University of Jerusalem, Jerusalem, Israel

Christoph Marquardt
Max Planck Institute for the Science of Light, Erlangen, Germany

Hanaa Marshoud
Li-Fi Research and Development Center and Institute for Digital Communications, School of Engineering, The University of Edinburgh, Edinburgh, United Kingdom

Antonio Mecozzi
Department of Physical and Chemical Sciences, University of L'Aquila, L'quila, Italy

Georg Mohs
SubCom, Eatontown, NJ, United States

David T. Neilson
Nokia Bell Labs, Holmdel, NJ, United States

Leif Katsuo Oxenløwe
DTU Fotonik, Technical University of Denmark, Lyngby, Denmark

Peter O'Brien
Tyndall Institute, University College Cork, College Cork, Ireland

Xiaodan Pang
KTH Royal Institute of Technology, Stockholm, Sweden

George Papen
University of California, San Diego, CA, United States

Loukas Paraschis
Systems Engineering, Cloud Transport, Infinera, Sunnyvale, CA, United States

James Pond
Lumerical Inc., Vancouver, BC, Canada

Kannan Raj
Infrastructure and Region Build, Oracle Cloud Infrastructure, San Diego, CA, United States

Siddharth Ramachandran
Electrical and Computer Engineering Department, Boston University, Boston, MA, United States

David J. Richardson
Optoelectronics Research Centre (ORC), University of Southampton, Southampton, United Kingdom

Roland Ryf
Nokia Bell Labs, Holmdel, NJ, United States

Luis L. Sánchez-Soto
Max Planck Institute for the Science of Light, Erlangen, Germany; Departamento de Óptica, Facultad de Física, Universidad Complutense, Madrid, Spain

Elham Sarbazi
Li-Fi Research and Development Center and Institute for Digital Communications, School of Engineering, The University of Edinburgh, Edinburgh, United Kingdom

Katharine Schmidtke
Facebook, Menlo Park, CA, United States

Marco Secondini
Institute of Communication, Information, and Perception Technologies, Scuola Superiore Sant'Anna, Pisa, Italy

William Shieh
Department of Electrical and Electronics Engineering, The University of Melbourne, Parkville, VIC, Australia

Dmitry V. Strekalov
Max Planck Institute for the Science of Light, Erlangen, Germany; Jet Propulsion Laboratory, Pasadena, CA, United States

Zhan Su
Analog Photonics, Boston, MA, United States

Erman Timurdogan
Analog Photonics, Boston, MA, United States

Peter J. Winzer
Nokia Bell Labs, Holmdel, NJ, United States

Chongjin Xie
Alibaba Group, Sunnyvale, CA, United States

Tianhua Xu
School of Engineering, University of Warwick, Coventry, United Kingdom; Optical Networks Group, University College London, London, United Kingdom

Metodi Plamenov Yankov
DTU Fotonik, Technical University of Denmark, Lyngby, Denmark

Shuang Yin
Google Fiber, Mountain View, CA, United States

Jianjun Yu
ZTE TX Inc., Morristown, NJ, United States

Preface: Overview of Optical Fiber Telecommunications VII

Introduction

Optical Fiber Telecommunications VII (OFT VII) is the seventh installment of the *OFT* series, and each edition reflects the current state-of-the-art at the time. Now 40 years old, the series is a compilation by the research and development community of progress in optical fiber communications. Each edition starts with a clean slate, and chapters and authors of *OFT VII* have been selected to elucidate topics that have evolved since *OFT VI* or that have now emerged as promising areas of research and development.

This book incorporates completely new content from *OFT VI* (2013) and presents the latest advances in optical fiber communication components, subsystems, systems, and networks. The chapters are written by leading authorities from academia and industry, and each chapter gives a self-contained overview of a specific technology, covering the state-of-the-art and future research opportunities.

This book is intended to be an ideal reference on the latest advances in the key technologies for future fiber optic communications, suitable for university and industry researchers, graduate students, optical systems implementers, network operators, managers, and investors.

Seven editions

Installments of the series have been published roughly every 5–8 years and chronicle the natural evolution of the field:

- In the late 1970s, the original *OFT* (Miller & Chenoweth, 1979) was concerned with enabling a simple optical link, in which reliable fibers, connectors, lasers, and detectors played the major roles.
- In the late 1980s, *OFT II* (Miller & Kaminow, 1988) was published after the first field trials and deployments of simple optical links. By this time, the advantages of multi-user optical networking had captured the imagination of the community and were highlighted in the book.
- *OFT III* (Kaminow & Koch, 1997) explored the explosion in transmission capacity in the early to mid-1990s, made possible by the erbium-doped fiber amplifier, wavelength-division multiplexing (WDM), and dispersion management.
- By 2002, *OFT IV* (Kaminow & Li, 2002) dealt with extending the distance and capacity envelope of transmission systems. Subtle nonlinear and dispersive effects,

requiring mitigation or compensation in the optical and electrical domains, were explored.

- *OFT V* (Kaminow, Li & Willner, 2008) moved the series into the realm of network management and services, as well as employing optical communications for ever-shorter distances. Using the high bandwidth capacity in a cost-effective manner for customer applications started to take center stage.
- *OFT VI* (Kaminow, Li & Willner, 2013) explored photonic integrated circuits (PICs), higher capacity transmission systems, and flexible network architectures. Highlighted areas included advanced coherent technologies, higher-order modulation formats, and space-division multiplexing.
- The present edition, *OFT VII*, continues the trend of multidisciplinary topics, including aspects of foundry technologies, data center interconnections, and electronics-based solutions to optical networking problems. Areas that have gained much interest in increasing performance include advanced components to enable space-division multiplexing, software tools for PICs, nanophotonic technologies, and free-space classical and quantum communications. In addition, many of the topics from earlier editions are brought up to date and new areas of research that show promise are featured.

Although each edition has added new topics, it is also true that each edition has addressed new emerging challenges as they relate to older topics. For example, certain devices may have adequately solved transmission problems for the systems of that era. However, as systems become more complex, critical device technologies that might have been considered a "solved problem" would now have new requirements placed upon them and need a fresh technical treatment.

Finally, each edition has dealt with the issue of cost effectiveness in the consideration of optical fiber telecommunications solutions, for which cost and performance are critical drivers of research and impact our entire field.

Ivan P. Kaminow and Tingye Li

Ivan Kaminow and Tingye Li contributed to the science and engineering of light wave technology and have helped revolutionize telecommunications. Ivan's and Tingye's contributions include the *Optical Fiber Telecommunications* Series, in which Ivan co-edited *OFT II-VI* and Tingye co-edited *OFT IV-VI*. They passed away as *OFT VI* was being published, and they are greatly missed by the many people who knew and were impacted by them. Moreover, their contributions to our field are evident in the ongoing activities of the researchers, engineers, and companies working in this field today.

Working on the *Optical Fiber Telecommunications* Series books brought Ivan and Tingye great joy, enabling them to use their love and skill of writing, help explore the most impactful issues, and interact with the best people in our field. The book series itself is a legacy to their efforts, leadership, wisdom, and insight.

Ivan and Tingye were luminaries in the photonics community. They had been researchers and friends at Bell Laboratories for over 40 years, where they explored several key aspects of optical telecommunication technologies. Ivan and Tingye received many distinguished awards, and both received the Frederic Ives Medal from the Optical Society and the Edison Medal and Photonics Award from the IEEE. They are remembered by friends and colleagues for their intelligence, honesty, inquisitiveness, vision, leadership, and gentleness.

Their impact on the optical fiber communications community and technical field has been monumental and will not be forgotten.

Perspective of the past 6 years

OFT VI was published in 2013. During the preceding few years, our field had emerged from the unprecedented "bubble-and-bust" upheaval circa 2000, at which time worldwide telecom traffic ceased being dominated by the slow-growing voice traffic and was overtaken by the rapidly growing internet traffic. Today, our field continues to gain strength, demand for bandwidth continues to grow at a very healthy rate, and optical fiber telecommunications is firmly entrenched as part of the global information infrastructure.

It is important to emphasize that our field is critical to the way society functions. By way of example, consider that: (1) there would be no internet as we know it if not for optical communications, (2) modern data centers may have more than 1 million lasers to help interconnect boards and machines with high bandwidth and low cost, and (3) smart phones wouldn't be so smart without the optical fiber backbone.

A key question for the immediate future is how deeply will optical communications penetrate and complement other forms of communications, such as wireless access, on-premises networks, interconnects, and satellites. *OFT VII* examines the opportunities for future optical fiber technology by presenting the latest advances on key topics such as:

- Fiber and 5G wireless access networks;
- Inter- and intra- data center communications;
- Quantum communications;
- Free-space optical links.

Another key issue today is the use of advanced photonics manufacturing and electronic signal processing to lower the cost of services and increase system performance. To address this, *OFT VII* covers:

- Foundry capabilities for widespread user access;
- Software tools for designing PICs;

- Nano- and micro-photonic components;
- Advanced and nonconventional data modulation formats.

The traditional emphasis of achieving higher data rates and longer transmission distances are also addressed through chapters on space-division multiplexing, undersea cable systems, and efficient reconfigurable networking.

The odds are that optics will continue to play a significant role in assisting nearly all types of future communications. This is in stark contrast to the voice-based future envisioned by the original *OFT*, published in 1979, which occurred before the first commercial intercontinental or transatlantic cable systems deployed in the 1980s.

In this edition, *OFT VII*, the various authors have attempted to capture the rich and varied technical advances that have occurred in our field. Innovations continue to abound, and we hope our readers learn and enjoy from all the chapters.

Acknowledgments

The authors sincerely thank Tim Pitts, Joshua Mearns, and Anitha Sivaraj of Elsevier for their gracious and invaluable support throughout the publishing process. We are also deeply grateful to all the authors for their laudable efforts in submitting their scholarly works of distinction. Finally, we wish to thank the many people whose insightful suggestions were of great assistance.

Chapter highlights

Below are brief highlights of the different chapters in the book:

Optical Fiber Telecommunications VII: Chapter titles, authors, and abstracts

Chapter 1 Advances in low-loss, large-area, and multicore fibers

Ming-Jun Li and Tetsuya Hayashi

In this chapter, recent advances are discussed in single-core and multicore optical fibers for increasing capacity for transmission systems. For single-core fibers used in long-haul transmission, impairments such as chromatic dispersion and polarization-mode dispersion can be compensated by digital signal processing; therefore, the fiber parameters that can be optimized further are fiber attenuation and effective area. System figure of merit and design trade-offs are discussed for these two parameters, and recent results on ultralow loss and large effective area fibers are presented. In terms of next-generation fibers, multicore fibers for space-division multiplexing have the potential to increase the capacity by an order of magnitude. Design considerations of and recent progress on multicore fibers are also presented.

Chapter 2 Chip-based frequency combs for wavelength-division multiplexing applications

Juned N. Kemal, Pablo Marin-Palomo, Maxim Karpov, Miles H. Anderson, Wolfgang Freude, Tobias J. Kippenberg, and Christian Koos

Optical frequency combs have the potential to become key building blocks of optical communication subsystems. In general, a frequency comb consists of a multitude of narrowband spectral lines that are strictly equidistant in frequency and that can serve both as carriers for massively parallel data transmission and as local oscillator tones for coherent reception. Recent experiments have demonstrated the viability of various comb generator concepts for communication applications. These comb generators must offer low phase noise and line spacings of several tens of gigahertz while being amenable to chip-scale integration into compact transceiver assemblies. Among the various approaches, so-called Kerr frequency combs stand out as a particularly promising option. Kerr comb generators exploit broadband parametric gain in Kerr nonlinear microresonators and allow the providing of tens or even hundreds of tones from a single device. This chapter describes advances regarding different types of chip-scale frequency comb sources and their use in optical communications with a special emphasis on high-performance Kerr frequency combs.

Chapter 3 Nanophotonic devices for power-efficient communications

You-Chia Chang and Michal Lipson

Power-efficient high-speed nanophotonic silicon devices, particularly silicon modulators, are key components for optical communications. Design of active silicon photonic devices involves trade-offs between different performance metrics. Here these trade-offs and current state-of-the-art active silicon photonics devices are reviewed. Emerging approaches are also introduced for improving the performance via device design, including the use of vertical phase noise junctions for improving modulation strength, as well as novel device designs that allow for fabrication tolerance and temperature fluctuation. Finally, material integration approaches are described that expand the available modulation mechanisms and open new possibilities for high-performance active devices. The set of materials includes Ge, GeSi, III-V materials, organic and inorganic $\chi^{(2)}$ materials, and two-dimensional materials.

Chapter 4 Foundry capabilities for photonic integrated circuits

Michael Liehr, Moritz Baier, Gloria Hoefler, Nicholas M. Fahrenkopf, John Bowers, Richard Gladhill, Peter O'Brien, Erman Timurdogan, Zhan Su, and Fred Kish

Thanks to the successful application of Si-based photonic integrated circuits (PICs) to data communications, demand for PICs has increased dramatically. As a result, integrated device manufacturers, as well as foundries, have provided much improved capability and capacity since 2010. PIC foundries, in particular, offer capability that is accessible to users around the world and in a variety of technology platforms. This chapter is meant to teach the community what has advanced in the past decade to enable a suite of processes for different types of PICs, typically dedicated to a particular market demand. The chapter is not meant to describe the operation of a specific foundry, but rather a vision of PIC foundries with examples from various institutions. After reading the chapter, the reader should have a better understanding of the advances that have enabled PIC foundry capabilities and the background to be able to interact with a PIC foundry.

Chapter 5 Software tools for integrated photonics

James Pond, Gilles S.C. Lamant, and Rich Goldman

Driven by the need for low-cost, high-speed, and power-efficient data connections, integrated photonics is becoming a reality for communications applications (e.g., transceivers for data centers). Integrated photonic devices can be single chip (i.e., monolithic) but are increasingly incorporating the three-dimensional assembly of multiple chips (i.e., hybrid) to create integrated electrical-and-optical systems. However, designing with and for light is not the same as designing for pure electronics, and producing systems that utilize both photons and electrons as their information carrier brings new challenges to the design automation area. One such challenge is the extreme scale differences; from simulation to layout, designers need to manage accuracy, performance, and memory usage for quantities that coexist but have very different time and size scales. In this chapter, existing techniques are reviewed that are used to enable codesign of electronic-photonic systems, covering a schematic driven methodology with system-level simulation, layout generation, and layout parameter back-annotation. Also discussed are design automation challenges and the need for the electronic-photonic package to connect to the outside world. Finally, current development areas are highlighted that are driven by the designer base of commercial design automation providers.

Chapter 6 Optical processing and manipulation of wavelength division multiplexed signals

Leif Katsuo Oxenløwe, Frederik Klejs, Mads Lillieholm, Pengyu Guan, Francesco Da Ros, Pawel Marcin Kaminski, Metodi Plamenov Yankov, Edson Porto da Silva, Peter David Girouard, and Michael Galili

This chapter describes optical processing concepts that allow for simultaneous manipulation of multiple wavelength channels in a single or few optical processing units. This offers a potential for collective sharing, among the channels, of the energy associated with the processing, thus lowering the required processing energy per channel. Optical processing allows for ultra-broadband processing, thus increasing the potential energy savings, and could play a role in flexible networks by, for example, converting wavelength grids, modulation, or signal formats. This chapter describes the means to regenerate multiple wavelength channels for improved transmission performance, compress or magnify the wavelength grid for better bandwidth utilization, and complement the optical signal processing with its digital cousin. In particular, this chapter describes optical time lenses and phase-sensitive amplifiers, as well as optical phase conjugation paired with digital probabilistic shaping. The chapter also gives an overview of efficient nonlinear materials that could support these advanced optical signal processing schemes.

Chapter 7 Multicore and multimode optical amplifiers for space division multiplexing

Yongmin Jung, Shaif-ul Alam, David J. Richardson, Siddharth Ramachandran, and Kazi S. Abedin

Space-division multiplexing (SDM) has attracted considerable attention within the fiber optics communication community as a very promising approach to significantly increase the transmission capacity of a single optical fiber and to reduce the overall cost per transmitted bit of information. Various SDM transmission fibers (e.g., multicore fibers and multimode fibers) have been introduced, and a more than 100-fold capacity increase (>10 Pbit/s) relative to conventional single-mode fiber systems has successfully been demonstrated. Also, a wide range of new SDM components and SDM amplifiers has accordingly been developed, with most now realized in a fully fiberized format. These fully integrated devices and subsystems are one of the key requirements for the deployment of practical SDM in future networks due to their potential for cost-, energy-, and space-saving benefits. In this chapter, the state of the art in optical amplifiers is reviewed for the various SDM approaches under investigation, with particular focus on multicore and multimode devices.

Chapter 8 Transmission system capacity scaling through space-division multiplexing: a techno-economic perspective

Peter J. Winzer

This chapter presents a unified view of the possible optical network capacity scalability options known today, with a focus on their techno-economics. Generalized Shannon capacity scaling considerations are combined with practical engineering principles, showing that space-division multiplexing seems to provide the only viable long-term capacity scaling solution, from chip-to-chip interconnects to submarine ultra-long-haul networks.

Chapter 9 High-order modulation formats, constellation design, and digital signal processing for high-speed transmission systems

Tianhua Xu, Gabriele Liga, and Polina Bayvel

The achievable capacity of optical communication networks is currently limited by the Kerr effect inherent to transmission using optical fibers. The signal degradations due to the nonlinear distortions become more significant in systems using larger transmission bandwidths, closer channel spacing, and higher order modulation formats, and optical fiber nonlinearities are seen as the major bottleneck to the performance of optical transmission networks. This chapter describes the theory and experimental investigations for a series of techniques developed to unlock the capacity of optical communications to overcome the capacity barriers in transmission over nonlinear fiber channels. This chapter covers four key areas seen as effective in combatting optical fiber nonlinearities: (1) nonlinearity in optical communication systems with higher order modulation formats; (2) electronic nonlinearity compensation techniques such as digital backpropagation and Volterra equalization; (3) performance of nonlinearity compensation considering other physical impacts (e.g., polarization mode dispersion and laser phase noise); (4) signal processing techniques that make use of coded modulation and probabilistic constellation shaping. This chapter reviews and quantifies different examples of the joint application of digital signal processing-based nonlinearity compensation and further possible increases in the achievable capacity and transmission distances.

Chapter 10 High-capacity direct-detection systems

Xi Chen, Cristian Antonelli, Antonio Mecozzi, Di Che, and William Shieh

A direct detection (DD) system is a conventional light communication system in which the power (i.e., intensity) of the light is modulated and detected. Conventional DD systems refer to systems that use a single photodiode as a receiver. In contrast to modern coherent receivers that have more complex structure and can detect four dimensions (i.e., amplitude and phase of both polarizations) of an optical field, the conventional DD receivers detect power, which is only one dimension of the light. The corresponding transmitter modulates optical power/intensity. The DD systems are often referred as intensity-modulated and directly detected (IM-DD) systems. This chapter describes recent advances in increasing performance and reducing complexity in IM-DD optical communication systems.

Chapter 11 Visible-light communications and light fidelity

Harald Haas, Elham Sarbazi, Hanaa Marshoud, and John Fakidis

There is significantly increased interest in visible light communications (VLC) and light fidelity (LiFi) during the last 10 years. This chapter describes many aspects as such systems, including a taxonomy of the various optical wireless communications technologies, their key discriminating features, typical applications, important features of the optical wireless propagation channel, and essential differences with radio-frequency-based wireless communications. Various source and receiver technologies are introduced, as well as recent advances in digital modulation techniques for intensity modulation/direct detection for single-input single-output channels. The chapter also discusses multichannel data transmission by considering spatial and wavelength domains, as well as the ability to serve randomly moving mobile terminals. Finally, the chapter introduces key functions, such as nonorthogonal multiuser access, cochannel interference mitigation when multiple light sources transmit different signals to different users, and seamless services when users roam through a room or inside a building.

Chapter 12 R&D advances for quantum communication systems

Gerd Leuchs, Christoph Marquardt, Luis L. Sánchez-Soto, and Dmitry V. Strekalov

Understanding the nature of light leads to the question of how the principles of quantum physics can be harnessed in practical optical communications. A deeper understanding of fundamental physics has always advanced technology. However, the quantum principles certainly have a distinctly limiting character from an engineering point of view. A particle cannot have well-defined momentum and position at the same time. An informative measurement will unpredictably alter the state of a quantum object. One cannot reliably clone an arbitrary quantum state. These and a number of other similar principles give rise to what is commonly known as the quantum "no-go theorems"—a disconcerting term when it comes to building something practical. And yet a search for novel principles of communication enabled by quantum physics began already in its early days and has only intensified since. On this path physicists are faced with a remarkable challenge: to turn a series of negative statements into new technological recipes. This chapter deals with the path to answering this challenge by describing recent advances in quantum communication systems.

Chapter 13 Ultralong-distance undersea transmission systems

Jin-Xing Cai, Georg Mohs, and Neal S. Bergano

Ultralong-distance undersea transmission systems have gone through revolutionary changes since the widespread introduction of coherent transponder technology. In the "dry" plant, transponders using higher order modulation formats and digital coherent technology are being routinely deployed with 100–400 Gbit/s per line cards. In the "wet" plant, many undersea systems optimized for coherent transmission have been deployed. Today's undersea cable operators require "open cable systems" with flexible capacity and improved cost per bit. In this chapter, the most recent technology evolutions and industry trends are reviewed. The chapter introduces the Gaussian noise model, discusses generalized optical signal-to-noise ratio, and presents the "open cable" concept. Also highlighted are technologies to combat fiber nonlinearities and achieve high capacity in single-mode fiber, including symbol rate optimization, nonlinearity compensation techniques, variable spectral efficiency, and nonlinear system optimization. The chapter describes several different optical amplification technologies to increase optical bandwidth and enhance nonlinear tolerance. Furthermore, the chapter reviews space-division multiplexing as a means to achieve better power efficiency and higher overall system capacity. Finally, the chapter discusses system value improvements using "wet" wavelength selective switch–based reconfigurable optical add-drop multiplexers and new cable types.

Chapter 14 Intra-data center interconnects, networking, and architectures

Saurabh Kumar, George Papen, Katharine Schmidtke, and Chongjin Xie

This chapter describes the interconnect technologies applied in a data center network (DCN) architecture along with those used in the modern network systems. The chapter starts with a discussion on the characteristics of a DCN and shows how these characteristics drive the development of the network topology, architecture, and network cabling in data centers. Subsequently, the chapter discusses different characteristics and applications of interconnect technologies used inside data centers. These technologies include direct attach cables, active optical cables, and optical transceivers. Pluggable optical technologies for 40, 100, and 400G networks are presented in detail, including multimode and single-mode technologies. Also shown is how optical interconnect technologies have evolved as the DCN bandwidth has increased from one generation to the next. This leads to a discussion on the advantages and limitations of different technologies in data centers. Finally, perspectives are presented on future development for intra-data center interconnects and networks, including coherent detection, mid-board optics, copackaging of electronic and optical circuits, and optical switching technologies.

Chapter 15 Innovations in DCI transport networks

Loukas Paraschis and Kannan Raj

The capacity of the transport networks interconnecting data centers (DCI) has grown more than any other traffic type due to the proliferation of "cloud" services. Consequently, DCI has motivated the evolution of dedicated DCI networks and DCI-optimized transport systems. This chapter reviews important current and emerging innovations and synergies in technology, systems, and networks, as well as the related research, development, and standards efforts, that collectively have facilitated the DCI evolution to its current multi-Tbit/s global infrastructure. This infrastructure employs some of the most spectrally efficient deployed fiber networks. Purpose-built DCI transport systems have been optimized for the data center operational requirements, simpler routing, and state-of-the-art coherent wavelength-division multiplexing (WDM) transmission that has already exceeded 6 bit/s/Hz. Moreover, in WDM transport, DCI has pioneered the extensive adoption of software-defined networking innovations in programmability, automation, management abstraction, and control-plane disaggregation to simplify operations and enable "open" transport architectures.

Chapter 16 Networking and routing in space-division multiplexed systems

Dan M. Marom, Roland Ryf, and David T. Neilson

Optical networks serve as the cornerstone of our connected society. As the number of users and data services increase, the network technology and architecture must adapt for it to continue to efficiently and economically support the larger traffic loads. Currently, these optical networks consist of optical transceivers of different wavelengths whose signals are wavelength-division multiplexed (WDM), and the paths these signals traverse can be selected by using reconfigurable optical add-drop multiplexers at network nodes. This chapter addresses current architectures of WDM networks and how they may evolve in the future to support even greater capacities through the use of additional spatial paths, an approach referred to as space-division multiplexing. While the addition of SDM to optical transport simply adds linearly to the number of channels in an unused degree of freedom, the challenges in switching are more significant. Since the spatial degrees of freedom are already being used in the optical switches to enable WDM and switching to multiple ports, adding SDM compromises some aspect of the switching performance. To manage increased numbers of spatial channels, it may be necessary to compromise on the spectral resolution and therefore the channel count. It seems that the most likely approach to SDM, at least initially, will be to use increasing numbers of uncoupled spatial channels. It also seems likely that as the number of these spatial channels increases, there will be a corresponding reduction in the number of wavelength channels to be switched, ultimately leading to a purely spatial switching network with significant changes in architecture and control.

Chapter 17 Emerging optical communication technologies for 5G
Xiang Liu and Ning Deng

To address the new requirements on optical networks imposed by the upcoming fifth-generation wireless (5G) (e.g., high bandwidth, low latency, accurate synchronization, high reliability, and flexible application-specific network slicing), a new generation of optical networks that are optimized for 5G is in great demand. This chapter presents enabling technologies for such 5G-oriented optical networks, including: 5G wireless trends and technologies such as cloud radio access networks, massive multiple-input and multiple-output, and coordinated multiple-point; recent advances on the common public radio interface (CPRI) and the Ethernet-based CPRI; issues related to point-to-point and point-to-multiple point fronthaul architectures. Also discussed are emerging optical communication technologies, such as low-cost intensity-modulation and direct detection schemes, 400-Gbit/s coherent modulation and detection, next-generation reconfigurable optical add/drop multiplexers and optical cross-connects, and techniques to achieve low-latency wireless fronthaul and backhaul networks. Finally, the chapter describes 5G-oriented optical network architectures of various optical transport and access networks to better support the upcoming 5G wireless.

Chapter 18 Optical interconnection networks for high-performance systems
Qixiang Cheng, Madeleine Glick, and Keren Bergman

Large-scale high-performance computing systems in the form of supercomputers and warehouse-scale data centers permeate nearly every corner of modern life from applications in scientific research, medical diagnostics, and national security to film and fashion recommendations. Vast volumes of data are being processed at the same time that the relatively long-term progress of Moore's Law is slowing advances in transistor density. Data-intensive computations are putting more stress on the interconnection network, especially those feeding massive data sets into machine learning algorithms. High bandwidth interconnects, essential for maintaining computation performance, are representing an increasing portion of the total energy and cost budgets. Photonic interconnection networks are often cited as ways to break through the energy-bandwidth limitations of conventional electrical wires to solve bottlenecks and improve interconnect performance. This chapter presents an overview of the recent trends and potential solutions to this challenge.

Chapter 19 Evolution of fiber access networks

Cedric F. Lam and Shuang Yin

This chapter discusses commercial fiber access technologies and their applications in deployed production networks today, as well as the drivers leading to next-generation fiber access technologies. Currently, passive optical network (PON) is used almost synonymously to represent a fiber access network, although it is only one and probably the most important one of the many architecture options. For this reason, this chapter will be mostly devoted to the discussion of the evolution of future PON technologies, covering the development of new PON standards, their performance targets, and implementation challenges. Traditional PON systems have adopted the time-division multiplexing (TDM) approach to share a common transmission medium with multiple users, but scaling with TDM alone becomes quite difficult as the baud rate increases. The latest PON standards adopt combinations of TDM and wavelength-division multiplexing (WDM) techniques to overcome some of these transmission challenges. WDM brings both benefits and new challenges in fiber access networks, which are discussed in this chapter. TDM and WDM together will enable new levels of access network scaling and different network economy afforded by the so-called "super-PON" networks, which is now within the reach of commercial applications. As the baud rate of fiber access networks increases, optical components with demanding performance specifications are also needed. Since the cost of optical transceivers accounts for the majority of fiber access networks, new modulation schemes and electronic processing techniques that can offset the demanding requirements of optical components in future fiber access networks are also being explored.

Chapter 20 Information capacity of optical channels

Marco Secondini

This chapter describes some basic concepts of information theory and the notion of channel capacity. Moreover, the chapter focuses on some issues that are particularly relevant for the optical fiber channel but often only briefly touched on in classical textbooks on information theory: how to deal with waveform channels, with memory, and with the unavailability of an exact channel model. The optical fiber channel is described by presenting the main equations governing the propagation of light in optical fibers and by discussing a few different approximated channel models that can be deployed for an information theoretical analysis, providing different trade-offs between accuracy and complexity. Finally, the chapter explores the capacity of the optical fiber channel, providing both easy-to-compute capacity bounds and more accurate but complex bounding techniques, and considering different scenarios and link configurations. Future perspectives and open problems are also discussed.

Chapter 21 Machine learning methods for optical communication systems and networks

Faisal Nadeem Khan, Qirui Fan, Chao Lu, and Alan Pak Tao Lau

Machine learning (ML) is being hailed as a new direction of innovation to transform future optical communication systems. Signal processing paradigms based on ML are being considered to solve certain critical problems in optical communications that cannot be easily tackled using conventional approaches. Recent applications of ML in various aspects of optical communications and networking such as nonlinear transmission systems, network planning and performance prediction, cross-layer network optimizations for software-defined networks, and autonomous and reliable network operations have shown promising results. However, to comprehend true potential of ML in optical communications, a basic understanding of the nature of ML concepts is indispensable. This chapter describes mathematical foundations of several key ML methods from communication theory and signal processing perspectives, and it highlights the types of problems in optical communications and networking where ML can be useful. The chapter also provides an overview of existing ML applications in optical communication systems with an emphasis on physical layer.

Chapter 22 Broadband radio-over-fiber technologies for next-generation wireless systems

Jianjun Yu, Xinying Li, and Xiaodan Pang

The ever-increasing bandwidth demand has motivated the exploration of radio-over-fiber (RoF) for future broadband 5G cellular communication networks. The integration of RoF networks makes full use of the huge bandwidth offered by fiber links and the mobility feature presented via wireless links. Therefore RoF can satisfy the various demands of the access network on capacity and mobility enhancement, as well as power consumption and cost reduction. It is expected that millimeter-wave (mm-wave) bands in future optical wireless access networks will be utilized to solve the problem of frequency congestion and to meet the demand for higher signal bandwidth. In this chapter, several key enabling technologies for very high throughput RoF networks are reviewed, including simple and cost-effective broadband optical mm-wave signal generation and transmission, multidimensional multiplexing techniques to improve the transmission capacity, radio-frequency-transparent photonic demodulation technique applied for novel RoF network architecture, and low-complexity high-efficiency digital signal processing for RoF systems. The chapter also summarizes recent progress on RoF systems, including field trials of high-speed and long-distance delivery using these enabling techniques. The results show that the integrated systems are practical solutions for offering very high throughput wireless to end users in optically enabled RoF systems.

Alan E. Willner
University of Southern California, Los Angeles, CA, United States

PART I

Devices/Subsystems Technologies

CHAPTER 1

Advances in low-loss, large-area, and multicore fibers

Ming-Jun Li[1] and Tetsuya Hayashi[2]
[1]Corning Incorporated, Corning, NY, United States
[2]Sumitomo Electric Industries, Ltd., Yokohama, Japan

1.1 Introduction

Low-loss optical fiber has revolutionized the telecommunication industry in the last nearly five decades. Since the first low-loss optical fiber with less than 20 dB/km at 632.8 nm in 1970 [1], optical fiber loss has continued to evolve toward lower levels. Fig. 1.1 shows the record of fiber loss evolution over the past 48 years [1–10]. Fiber loss decreased very quickly at beginning of optical fiber development. Three years after the first low-loss fiber, the loss was reduced to 5 dB/km at 850 nm [2], and in the following year, fiber loss of 2.5 dB/km at 1060 nm was reported. In 1976, fiber with loss of 0.47 dB/km at 1200 nm was demonstrated [3]. Within three years, the fiber attenuation reached 0.2 dB/km at 1550 nm [5], which is the typical attenuation value of today's standard single-mode fiber products using germanium-doped core. After that, the rate of fiber loss change slowed down, but researchers continued to look for glass compositions for lower fiber loss. In 1986, fiber loss of 0.154 dB/km was reported using pure silica core [6]. It took another 16 years for the loss to break the barrier of 0.15 dB/km in 2002 [7]. Below 0.15 dB/km, the loss reduction became even harder because the loss is so close to the fundamental limit. However, research efforts have continued to push the fiber loss closer to its limit. It took another 11 years for the fiber loss to reach a new record. Three new results were reported after 2013 with loss reduction in the order of a few thousandths of dB/km [8–10]. Today, the lowest loss has reached 0.142 dB/km at 1560 nm [10].

To benefit from the lower loss at longer wavelength, the long-haul transmission technology evolved from 850 nm multimode fiber systems [11] to 1310 nm single-mode fiber systems [12–14], and then to 1550 nm single-mode fiber systems [15–17]. The low loss of single-mode fiber and erbium-doped fiber amplifiers (EDFA) in the 1550 nm window has enabled multiple wavelength channel transmission via wavelength-division multiplexing (WDM), [18] which increases the system capacity.

Optical fiber designs have also evolved to adapt system technologies. Standard single-mode fiber was designed for 1310 nm transmission with nearly zero chromatic

Optical Fiber Telecommunications V11
DOI: https://doi.org/10.1016/B978-0-12-816502-7.00001-4

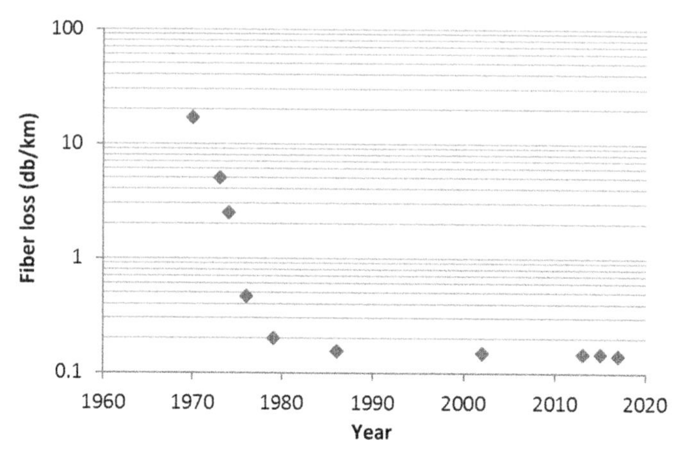

Figure 1.1 Optical fiber loss evolution.

dispersion [12–14]. When transmission systems moved to 1550 nm window where the fiber loss is lower, the high chromatic dispersion in the 1550 nm window (~17 ps/km nm) became a limitation for high data rate systems. To reduce chromatic dispersion at the 1550 nm window, dispersion-shifted optical fiber (DSF) was proposed [15–17]. The chromatic dispersion of DSF was designed to be around zero at the 1550 nm wavelength, which is optimal for single-channel transmission at 1550 nm. However, this fiber is not suitable for WDM transmission because the crosstalk penalty due to the nonlinear effect of four-wave mixing is the strongest when the dispersion is zero [19]. To overcome the four-wave mixing effect, nonzero dispersion-shifted fiber (NZDSF) was proposed [20–23]. An NZDSF has dispersion of 3–5 ps/nm km, which is high enough to reduce the four-wave mixing effect but small enough to minimize the dispersion penalty.

In parallel to WDM development, the data rate in each wavelength channel has also improved to meet the increasing bandwidth demand. With the advances in electronics, the channel data rate has increased from 2.5 to 10 Gb/seconds and then to 40 Gb/seconds using intensity modulation and direct detection. With the continuous increase in channel rate, coherent technologies have attracted a large interest in recent years [24]. Coherent detection allows information to be encoded with two degrees of freedom, which increases the amount of information per channel. It also allows integrating digital signal processing (DSP) function into a coherent receiver. With coherent detection techniques, advanced modulation formats [25] such as BPSK, QPSK, 8PSK, 16QAM, and higher levels of modulation formats have been proposed, which have pushed the channel capacity to 100 Gb/seconds and beyond. Coherent detection in conjunction with DSP can compensate transmission impairments from chromatic dispersion and polarization mode dispersion. This new system technology has shifted the fiber design direction toward fibers with lower loss and larger effective area that mitigates nonlinear transmission impairments.

While continuing improvements in conventional fiber-optic technologies will increase the system capacity further in a short term, recent studies show that the transmission capacity over single-mode optical fibers is rapidly approaching its fundamental Shannon limit [26]. To overcome this limit, new technologies using space division multiplexing (SDM) will be needed to provide a solution to the future capacity growth [27]. There are two approaches for space division multiplexing. One is to use multicore fibers (MCFs) and the other is to use few-mode fibers (FMFs). MCFs can be divided into two types: uncoupled and coupled MCFs. For SDM transmission systems using coupled MCFs and FMFs, digital signal processing (DSP) using multiple-input and multiple-output (MIMO) is necessary to deal with mode coupling effects, which increase system complexity. For uncoupled MCF-based SDM transmission systems, MIMO DSP is not needed because each core behaves like an isolated individual fiber. Therefore, uncoupled MCFs are straightforward to implement in SDM systems. However, randomly coupled MCFs have been confirmed to outperform the equivalent SMF bundle and are expected to be one of the strong candidates for the next-generation ultra-long-haul submarine transmission fiber because of their ultralow-loss and ultralow modal dispersion characteristics.

In this chapter, we discuss recent advances in single-core and multicore optical fibers for increasing capacity for transmission systems. This chapter is organized as follows. In Section 1.2, we focus on single-mode fibers with ultralow loss and large effective area. We discuss first the fiber system's figure of merit related to the two parameters of loss and effective area in Section 1.2.1. Then we examine loss mechanisms and present approaches for lowering the fiber loss in Section 1.2.2, and we discuss fiber designs for large, effective area fibers in Section 1.2.3. In Section 1.2.4, we review recent progress on low-loss and large effective area fibers and present system testing results using these fibers. Section 1.3 is devoted to multicore fibers for space division multiplexing. In Section 1.3.1, we introduce important design parameters related to different types of multicore fibers. Then in Section 1.3.2, we analyze multicore fibers using coupled power and coupled mode theories and describe characteristics of uncoupled and coupled multicore fibers. In Section 1.3.3, we present different multicore fiber designs that have been proposed and review recent progress toward practical realizations of multicore fibers.

1.2 Low-loss and large effective area fibers

1.2.1 Figure of merit of fiber loss and effective area on transmission systems

For coherent transmission systems, transmission impairments from fiber chromatic dispersion and polarization mode dispersion effects can be compensated using digital signal processing. This greatly simplifies not only optical system designs but also optical fiber designs. For optical fiber designs, the most important fiber parameters for high-capacity long-haul transmission become the fiber attenuation and effective area.

The fiber effective area and attenuation affect the optical signal-to-noise ratio (OSNR) of a transmission system. The main factors that determine the OSNR for a given system link at a given distance are the channel launch power into each span, the noise figure of the optical amplifiers, the loss per span, and the total number of spans in the link. Of these, the factors that are directly related to optical fiber parameters are channel launch power and span loss. The span loss includes both fiber propagation loss and the total splice loss. Using the current splice technology, the splice loss is minor contribution to the span loss; therefore, the dominant factor is fiber propagation loss. For a linear optical transmission system, the channel launch power is limited only by the amount of power available from a laser. Unfortunately, optical fiber is a nonlinear transmission medium. The channel launch power is limited by the fiber nonlinear effects such as four- wave mixing, self-phase modulation, cross-phase modulation, and Brillouin and Raman scattering. The fiber nonlinearity is described by its nonlinear coefficient:

$$\gamma = \frac{2\pi}{\lambda}\frac{n_2}{A_{\text{eff}}} \tag{1.1}$$

where A_{eff} is the fiber effective area and n_2 is the nonlinear index of refraction. Because the nonlinear index of refraction does not change much for silica-based optical fibers, the only way to reduce the fiber nonlinear coefficient is to increase the fiber effective area.

To quantify the performance of fiber designs for coherent transmission systems, a system model has been proposed to analyze power spectral density of nonlinear interference as a function of fiber parameters based on the Gaussian noise model theory [28]. By treating the nonlinear interference as an additional noise factor, a generalized OSNR formula is derived to predict the optimal signal power that maximizes OSNR. Based on the model, a fiber figure of merit (FOM) is proposed as a means of comparing different fiber designs and their performance. The FOM has been validated with excellent agreement to the experimental results [29–31]. To quantify the benefits of large effective area and low loss on OSNR, a simplified version of the FOM is given in Eq. (1.2), in which the FOM is defined relative to a reference fiber:

$$\begin{aligned} \text{FOM(dB)} = {} & \frac{2}{3}10\log\left(\frac{A_{\text{eff}}\,n_{2,\text{ref}}}{A_{\text{eff,ref}}\,n_2}\right) - \frac{2}{3}(\alpha - \alpha_{\text{ref}})L \\ & - \frac{1}{3}10\log\left(\frac{L_{\text{eff}}}{L_{\text{eff,ref}}}\right) + \frac{1}{3}10\log\left(\frac{D}{D_{\text{ref}}}\right) \end{aligned} \tag{1.2}$$

where A_{eff}, α, n_2, L_{eff}, and D are, respectively, the effective area, attenuation coefficient, nonlinear refractive index, effective length, and chromatic dispersion of the fiber under consideration; $A_{eff,ref}$, α_{ref}, $n_{2,ref}$, and $L_{eff,ref}$ are, the effective area, attenuation coefficient, nonlinear refractive index, effective length, and chromatic dispersion of the reference fiber; and L is the span length. In Eq. (1.2), we have neglected the splice because the value is normally very small compared to other factors.

Fig. 1.2 plots the fiber FOM changes with effective area for fibers with different attenuation coefficients for a fiber span length of 100 km. The reference fiber parameters used are for a typical standard single-mode fiber with an attenuation coefficient of 0.2 dB/km and an effective area of 82 μm^2. We assume that the fibers with attenuation above 0.175 dB/km are made of germanium-doped core with an n_2 value of 2.3×10^{-20} m^2/W, and the fibers with attenuation below 0.175 dB/km are made of pure silica core with a lower n_2 value of 2.1×10^{-20} m^2/W. Fig. 1.2 shows that the fiber FOM can be increased by either increasing the fiber effective area or decreasing the fiber attenuation. The impact of effective area on FOM is almost identical for fibers with different attenuation coefficients. For an effective area increase from 82 to 150 μm^2, which is considered to be the current limit for effective area, the FOM is increased by about 1.8 dB. On the other hand, the attenuation has more impact on FOM. For a given effective area, the FOM is improved by about 3.2 dB when attenuation is decreased from 0.2 to 0.15 dB/km. By combining both the benefits of low attenuation and a large effective area, an FOM increase of more than 5 dB can be obtained by moving to a fiber with attenuation of 0.15 dB/km and an effective area of 150 μm^2. For longer spans or unrepeated long links, the ultralow loss and large effective area become even more important. In the next two subsections, we discuss how to reduce fiber loss and design fibers with large effective areas.

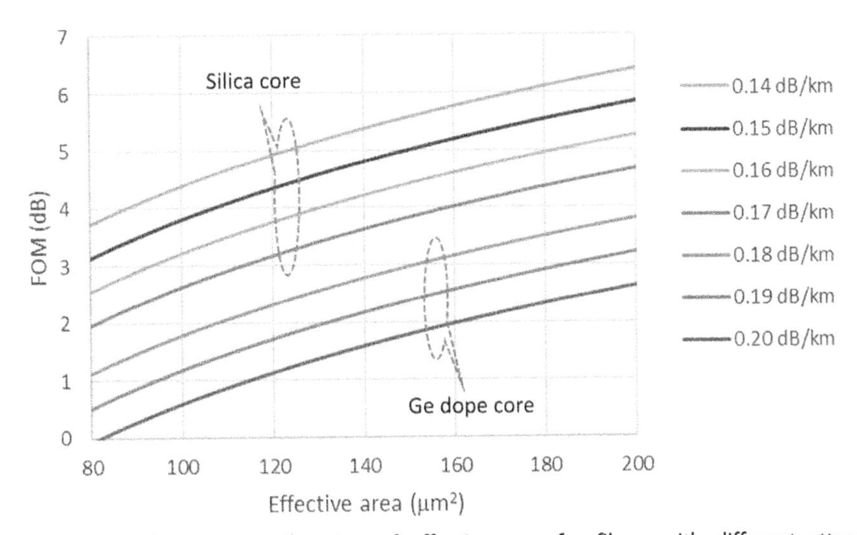

Figure 1.2 Figure of merit as a function of effective area for fibers with different attenuation coefficients.

1.2.2 Fiber loss mechanism and approaches for lowering fiber loss

Low-loss optical fibers are made of silica-based glass materials. For silica glass—based optical fibers, the total attenuation of an optical fiber is the addition of intrinsic and extrinsic loss factors, as shown in Fig. 1.3. Intrinsic loss is due to fundamental properties of glass materials used to construct the fiber core and cladding, including Rayleigh scattering, infrared absorption, and ultraviolet absorption. Extrinsic factors include absorptions due to OH ions and transition metals, scattering due to waveguide imperfections, and loss due to fiber bending effects.

For the extrinsic factors, the absorptions due to contaminants such as transition metals and OH ions can be largely eliminated in the current fiber manufacturing processes using chemical vapor deposition techniques such as OVD, VAD, MCVD, and PCVD. Waveguide imperfection loss is due to the geometry fluctuation at the core and cladding boundary. The fluctuation causes scattering of light in the forward direction. Because the scattering is confined to an angle of less than 10 degrees, the loss due to waveguide imperfection is also referred to as small angle scattering (SAS) [32,33]. The boundary fluctuation is mainly due to the residual stress that is induced during the manufacturing process. The residual stress depends on the magnitude of the viscosity difference between the core and cladding and the fiber draw tension. The stress can be reduced by matching the viscosity of the core and cladding [34]. To design a fiber with viscosity matching between the core and cladding, two dopants are needed. If the relative refractive index change due to dopant A is Δ_A in the core, and the relative refractive index change due to dopant B is Δ_B in the cladding, then the core relative refractive index change to the cladding is

$$\Delta = \Delta_A - \Delta_B \tag{1.3}$$

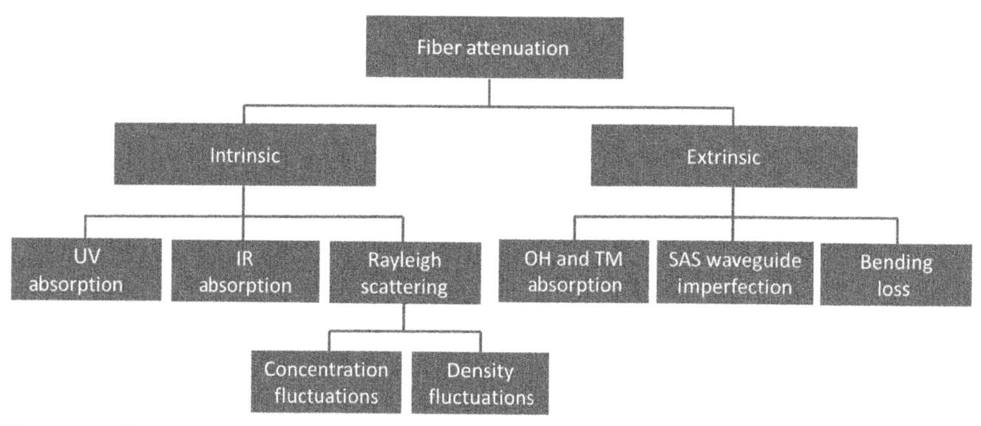

Figure 1.3 Fiber attenuation components.

We denote the changes in the viscosity logarithm per unit of relative refractive index change induced by dopants A and B as K_A and K_B:

$$\frac{d\log\eta_A}{d\Delta_A} = K_A \tag{1.4a}$$

$$\frac{d\log\eta_B}{d\Delta_B} = K_B \tag{1.4b}$$

To match the viscosity, $K_A\Delta_A = K_B\Delta_B$, we get

$$\Delta_A = \frac{\Delta}{1 - \left(\frac{K_A}{K_B}\right)} \tag{1.5a}$$

$$\Delta_B = \frac{\Delta}{\left(\frac{K_B}{K_A}\right) - 1} \tag{1.5b}$$

Using Eqs. (1.5a) and (1.5b), if the design target Δ is set, we can determine the relative refractive index change contributions Δ_A *and* Δ_B from dopants A and B to match the viscosity between the core and cladding based on the viscosity change constants K_A and K_B.

Matching viscosity to reduce waveguide imperfection loss was demonstrated experimentally using germania-doped core and fluorine-doped cladding [35]. A reduction of 0.025 dB/km in waveguide imperfection loss was reported.

The viscosity matching approach requires two dopants, which constrain the fiber design and increase process complexity. Another approach to reducing waveguide imperfection loss is to use graded index profile designs instead of step index profile designs [36]. One type of graded index profile can be described by an alpha profile:

$$\Delta = \Delta_0\left[1 - \left(\frac{r}{r_0}\right)^{\alpha}\right], \tag{1.6}$$

where Δ_0 is the maximal relative refractive index change in the center of the core, r_0 is the core radius, and α is the profile shape parameter. A profile with $\alpha = 1$ is a triangular profile, and a profile with $\alpha = 2$ is a parabolic profile. For a step index profile, α is greater than 10. It has been reported that SAS can be reduced by from 50% to more than 80% by using alpha profiles with an alpha value of less than 2.5.

Another extrinsic loss factor is the bending losses. Bending losses include both macro-bending loss and micro-bending loss. The macro-bending loss is related to fiber profile designs. The microbending loss depends on not only fiber profile designs but also coating material designs. We will discuss the bending losses in more detail in Section 1.2.3.

For the intrinsic factors, UV and IR absorption tails in the transmission windows are determined mostly by the silica glass material. The dopants used for making optical fibers have very minor effects on the UV and IR absorption tails. The most important intrinsic factor is the Rayleigh scattering loss. Fig. 1.4 shows typical fiber attenuation components for a generic standard single-mode fiber in the 1300 and 1550 nm windows. From this figure, it is clear that the largest opportunity for reducing the total fiber attenuation is reduction of the Rayleigh scattering component.

The Rayleigh scattering loss α_{RS} has two components: scattering due to density fluctuation α_ρ and scattering due to composition fluctuation α_c [37]:

$$\alpha_{RS} = \alpha_\rho + \alpha_c. \tag{1.7}$$

The scattering coefficient due to density fluctuation depends on the fictive temperature, T_f, which is defined as the temperature where the glass structure is the same as that of the supercooled liquid:

$$\alpha_\rho = \frac{8\pi^3}{3\lambda^4} n^8 p^2 \beta_T k_B T_f, \tag{1.8}$$

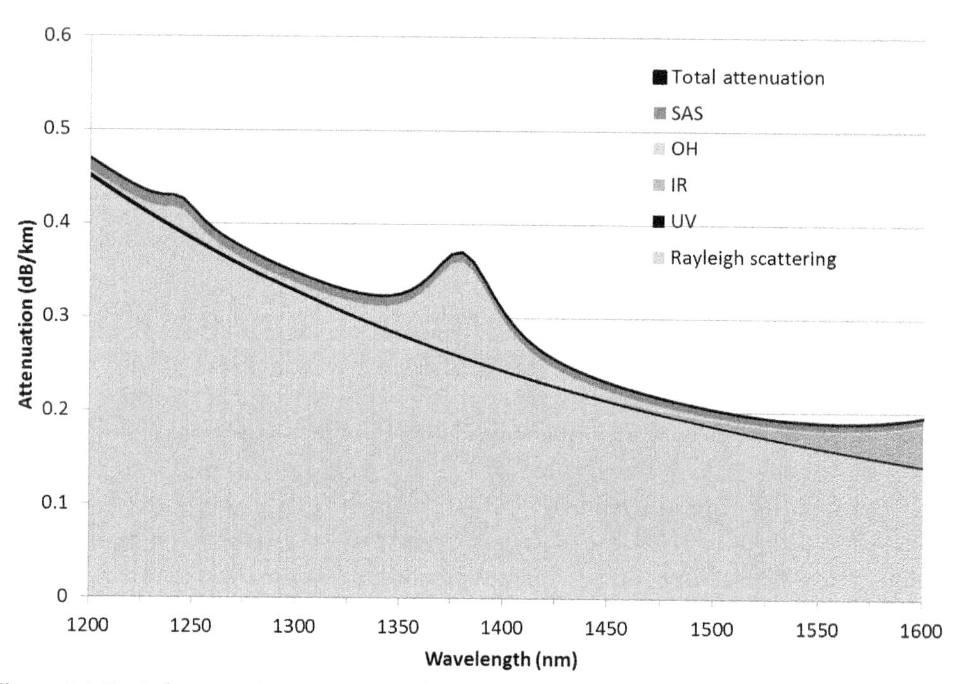

Figure 1.4 Typical attenuation spectrum and contributions from different loss components.

where λ is the wavelength of incident light, p the photoelastic coefficient, n the refractive index, k_B the Boltzmann constant, and β_T the isothermal compressibility. The composition fluctuation is proportional to:

$$\alpha_c \sim \left(\frac{\partial n}{\partial C}\right)^2 \langle \Delta C^2 \rangle T_f, \tag{1.9}$$

where $< C^2 >$ is the mean square fluctuations of concentration. Because the Rayleigh scattering is mainly caused by frozen-in density fluctuation, and is proportional to the fictive temperature, T_f, it is important to reduce the fictive temperature in order to reduce the Rayleigh scattering coefficient. When a silica glass is doped, the fictive temperature is normally reduced because most dopants reduce the glass viscosity and lower the glass melting temperature. However, with the increase of dopant concentration, the scattering due to composition concentration fluctuation increases. Therefore, the Rayleigh scattering change with dopant concentration normally follows a V-shaped curve, as shown schematically in Fig. 1.5. To minimize the Rayleigh scattering coefficient, it is critical to choose a dopant that reduces the fictive temperature without increasing the concentration fluctuation.

To reduce the concentration fluctuation, it is advantageous to reduce GeO_2 dopant level in the core because GeO_2 increases the Rayleigh scattering coefficient related to concentration fluctuation. In principle, pure silica without any dopants would eliminate the scattering component due to dopant concentration fluctuation. Theoretical prediction of scattering coefficient of pure silica is about 0.11 dB/km. Therefore, all the ultralow-loss fibers use essentially pure silica core designs without doping with GeO_2. Fig. 1.6 compares the loss components of Ge-doped core fiber and silica core fiber. The Rayleigh scattering coefficient is reduced by a factor of more than 17% for silica core fiber.

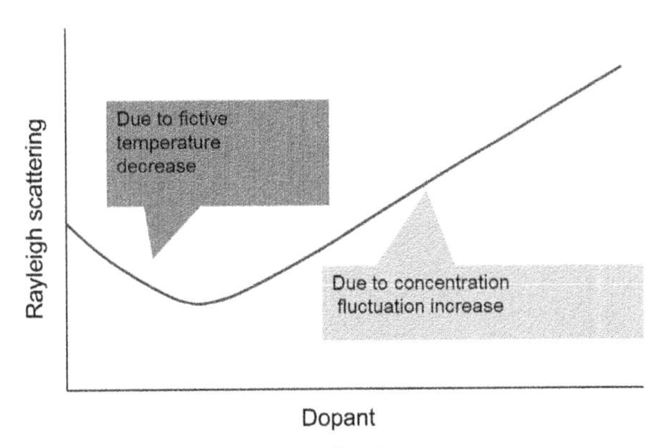

Figure 1.5 Schematic of effects of dopant on Rayleigh scattering.

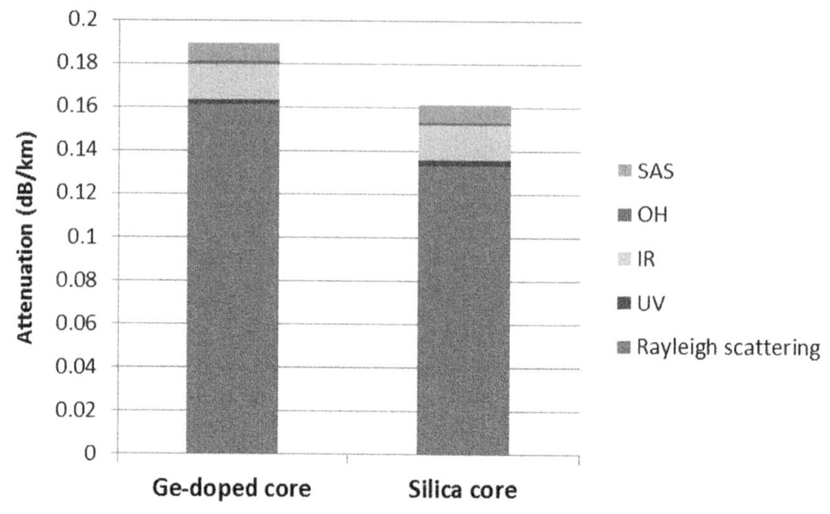

Figure 1.6 Comparison of loss components of Ge-doped and silica core fibers.

For pure silica core fibers, the main cause of transmission loss is Rayleigh scattering from glass density fluctuation due to irregularity of Si and O atom arrangements. This irregularity originates from thermal vibrations when optical fibers are drawn at approximately 2000°C. The irregularity can be reduced by glass structural relaxation during the cooling process. Slow cooling of fiber during the fiber draw process promotes the structural relaxation, resulting in less density fluctuation and lower scattering loss. The density fluctuation can also be reduced by trace amount of dopants such as alkali metals of potassium and sodium [38], and halogen elements of chlorine [39] and fluorine [40].

1.2.3 Fiber design for large effective area

Fiber bending losses including both macro- and microbending losses are important factors for designing low-loss optical fibers. They become even more critical for large effective area fibers. Another fiber property that needs to be considered for large effective area fiber designs is the cable cutoff wavelength, above which the fiber can operate in a single-mode condition. For single-mode operation within the C and L bands, the cable cutoff needs to be below 1530 nm. To increase the effective area while keeping good bending performance for single-mode fiber, the core refractive index profile needs to be carefully designed.

Fig. 1.7 shows a schematic of typical refractive index profile design for a low-loss and large effective area fiber. The profile consists of a core, an inner cladding, a trench, and an outer cladding. The core can be a step-index profile or a graded-index profile. The simplest profile design is step-index profile design with matched cladding. The profile has only two parameters, core Δ_1 and core r_1 ($\Delta_2 = \Delta_3 = \Delta_4 = 0$). To increase

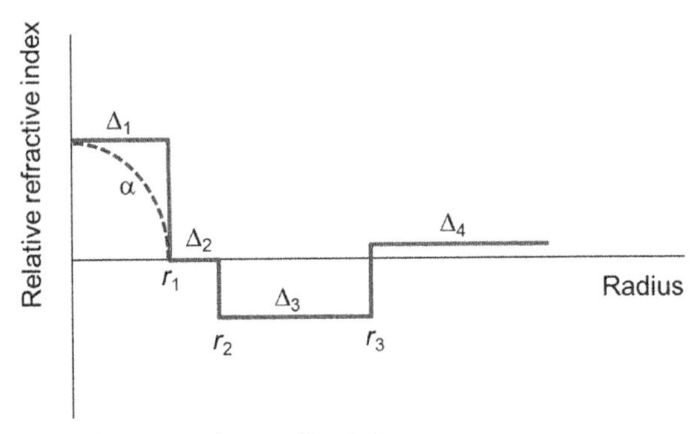

Figure 1.7 Schematic of refractive index profile design.

the effective area, the core radius needs to be increased but core delta needs to be reduced to keep the cable cutoff wavelength below 1530 nm. Due to the cable cutoff wavelength limitation, bending losses will increase when the effective area gets larger. Fig. 1.8A shows the modeled bending losses at 1550 nm as a function of effective area for step index profiles with matched cladding at three bend diameters: 20, 30, and 40 mm. For 40 mm bend diameter, the bending loss remains low for effective area up to 150 μm^2. Bending loss increases for 30 mm bend diameter but remains below 1 dB/turn for the effective area up to 150 μm^2. At a bend diameter of 20 mm, bending loss increases more rapidly with effective area. At this bend diameter, the bending loss is 1 dB/turn at effective area of 130 μm^2 and 6 dB/turn at 150 μm^2.

A depressed cladding layer or a low index trench can be added to reduce macrobending losses while keeping the cable cutoff wavelength below 1530 nm. For a depressed cladding design, $r_1 = r_2$ and $\Delta_2 = \Delta_3 < 0$. For a trench profile design, there is an offset between the core and the low index layer. Fig. 1.8B shows the bending losses as a function of effective area at 1550 nm for step-index core profiles with depressed cladding designs. Compared to matched cladding designs, depressed cladding designs reduce the bending loss of large effective area fibers significantly. The bending loss is below 0.5 dB turn even for bend diameter of 20 mm and effective area of 150 μm^2.

Effective area is limited not only by macrobending loss but also by microbending loss [41]. Microbending loss is an attenuation increase caused by high-frequency longitudinal perturbations to the core due to a set of very small radius bends of the fiber. These perturbations result in power coupling from the light guided in the core to higher order modes in the cladding, leading to absorption by the coating materials. Microbending depends on both profile designs and coating materials. A phenomenological model was proposed by Olshansky. In his mode the microbending loss coefficient can be estimated using the following formula [42]:

$$\gamma = N\langle h^2 \rangle \frac{a^4}{b^6 \Delta^3} \left(\frac{E}{E_f} \right)^{3/2} \tag{1.10}$$

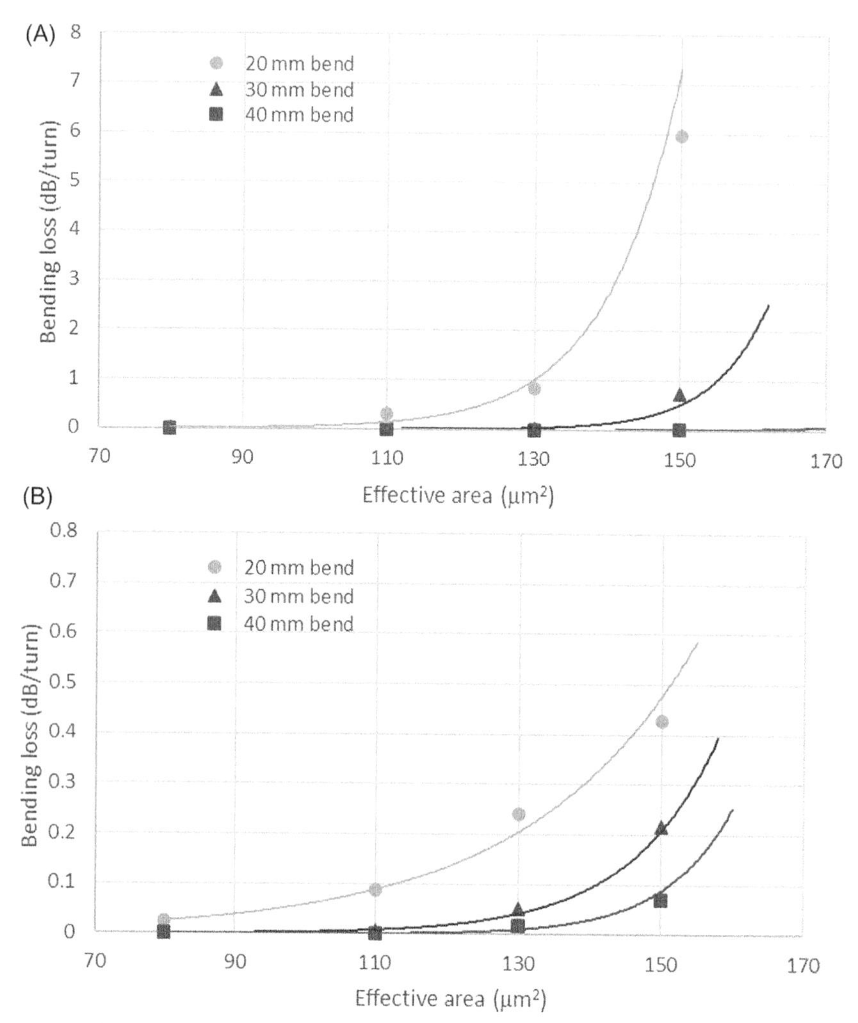

Figure 1.8 Bending loss as a function effective area for (A) matched cladding and (B) depressed cladding profile designs.

where N is the number of bumps of average height h per unit length, a is the core radius, b is the cladding radius, Δ is the relative refractive index of the core, E_f is the elastic modulus of fiber, and E is the elastic modulus of the primary coating layer that surrounds the glass. This model captures the importance of treating the glass and coating as a composite system for reducing the microbending loss. We have shown that the depressed cladding profile design reduces the macrobending loss sensitivity. It can also improve the microbending performance based on Eq. (1.10), because it allows increasing the core Δ. But the increase in core Δ is limited by the cable cutoff wavelength. The inner primary coating modulus is the most important factor for reducing

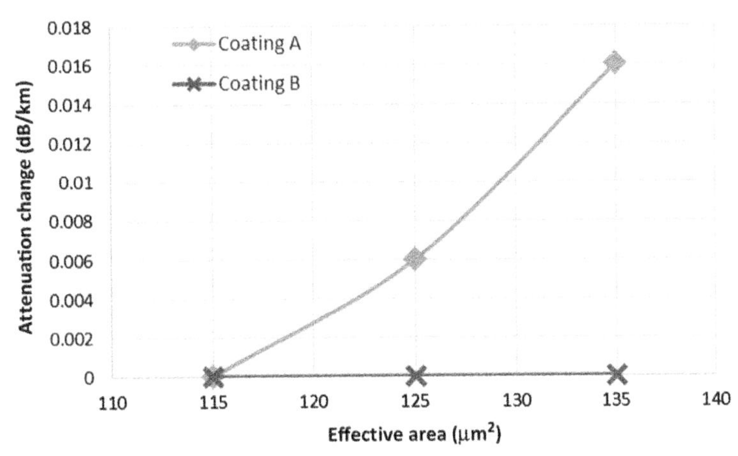

Figure 1.9 Attenuation changes of fibers with different effective areas with different coatings.

microbending sensitivity in large effective fibers. Making the inner primary coating softer will help cushion the glass from external perturbations and improve the microbending performance. This is shown in Fig. 1.9, where the measured fiber attenuation changes are plotted as a function of effective area relative to a reference fiber with an 80 μm^2 effective area for two primary coatings materials in a conventional test condition designed to apply microbending forces to the fiber. Coating A has a higher modulus, while Coating B has a lower modulus. For effective area of 115 μm^2, there is no attenuation change for the fibers with the two coatings. This shows that intrinsic attenuation can be realized with effective area less than 115 μm^2. For effective area larger than 115 μm^2, the microbending loss starts to increase with Coating A, while Coating B keeps the microbending loss to nearly zero for effective area up to 135 μm^2. It is evident that to derive the benefits of ultralow attenuation, a fiber with very large effective area requires a coating with an optimized inner primary modulus to protect against microbending. With an optimized profile design and optimized primary coating, effective area of about 150 μm^2 has been realized. The 150 μm^2 effective area is considered to be the current upper limit for single-mode fiber.

Fiber microbending loss can be reduced by increasing the glass fiber diameter because the microbending sensitivity is inversely proportional to the sixth power of glass fiber diameter, as suggested by Eq. (1.10). It has been reported that by increasing the glass diameter to 170 μm and coating diameter to 270 μm, fiber with effective area of 211 μm^2 and loss of 0.159 dB/km is achievable [43].

As discussed earlier, the fiber bending performance is limited by the cable cutoff wavelength for single-mode fiber with large effective area. If the cable cutoff wavelength is allowed to move to longer wavelengths, the bending performance can be improved and the effective area of the fundamental LP_{01} mode can be increased further. In this case, the fiber is not a single-mode fiber anymore but becomes a few-mode fiber. A few-mode fiber can be used for single-mode transmission by launching the light into the fundamental mode. In this case, the fiber is referred as a quasi-single-mode (QSM) fiber [44,45]. For QSM fiber designs, it has been shown that the effective area of LP_{01} mode can be increased to over 200 μm^2 [46,47]. The large-effective-area nature of the LP_{01} mode is expected to extend the transmission distance considerably. However, mode coupling can happen at the splice points and along the fiber during propagation for QSM fibers, causing multipath interference (MPI). MPI produces modal noise due to power fluctuations that limit the transmission distance. MPI must be mitigated before the full benefit of large effective area of QSM fibers can be realized. Because MPI is a linear impairment, in principle it can be compensated using coherent transmission technology. It has been shown that MPI can be effectively compensated for by using digital signal processing in a coherent receiver [46]. It has also been found that signal modulation formats with multiple electrical subcarriers may be more tolerant to MPI than single carrier formats. This is due to the lower baud rate of the subcarriers compared to a single carrier format with the same overall data rate [47]. Another way to reduce MPI is the use of hybrid fiber span designs in which each span is composed of a combination of QSM fiber and single-mode fiber. In this approach, an optimized length of QSM fiber is deployed in the first part of each span, followed by single-mode fiber in the latter part of the span [48]. This design uses QSM fiber in the portion of the span where the channel powers are highest to reduce nonlinear impairments. At the same time, MPI is reduced because the QSM fiber length is shorter than the full span length and the single-mode fiber after QSM fiber serves as mode filter.

1.2.4 Recent progress on low-loss and large effective area fiber and system results

Significant progress has been made in ultralow-loss fiber over the last few years using pure silica core fiber technology. Pure silica core fiber was proposed for make low-loss fiber in 1986, and a record loss of 0.154 dB at 1550 nm was demonstrated [6]. Since then, the research effort has continued to improve the attenuation of silica core fiber. Fig. 1.10 shows the attenuation improvement of silica core fiber since 1986. Notice that the attenuation improvement was slow from 1986 till 2013, and then accelerated afterward. The acceleration is due to the fact that attenuation is one of the two important fiber parameters for coherent transmission systems, as discussed previously. In 2013 fiber attenuation broke the 0.15 dB/km limit to 0.149 dB/km at 1550 nm [8]. Two years later in 2015 the attenuation reached 0.146 dB/km at 1550 nm [9]. The lowest loss was demonstrated in 2017 with attenuation of 0.142 dB/km at 1550 nm [10].

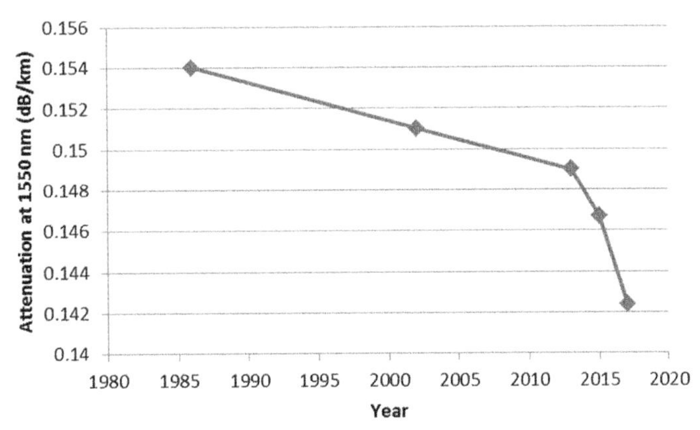

Figure 1.10 Record attenuation of silica core fiber.

Table 1.1 Commercial products of single-mode fibers with low loss.

ITU standard description	G.652 low loss	G.652 ultralow loss	G.654.E ultralow loss	G.654.B ultralow loss	G.654.D ultralow loss
Attenuation (dB/km)	0.18	0.16	0.17	0.15	0.15
Core dopant	GeO_2	Silica	Silica	Silica	Silica
Effective area (μm^2)	82	82	125	112	150
Dispersion at 1550 nm (ps/nm/km)	17	17	21	21	21

In parallel to low-loss fiber research efforts, commercial products have also improved both in fiber attenuation and effective area. Table 1.1 summarizes typical commercial product offerings of low-loss fibers according to ITU standards. For silica core fibers, typical attenuation specifications are between 0.15 and 0.17 dB/km, and effective areas range from 82 to 150 μm^2. For Ge-doped G.652 standard single fiber, the attenuation has been reduced to 0.18 dB/km for low-loss fiber.

Optical fibers with ultralow loss and large effective area enable transmission systems with high data rates and long reach distances. These benefits have been experimentally demonstrated by many recent transmission experiments. Table 1.2 summarizes record transmission experiments using low-loss and large effective area fibers. In these experiments, single optical fibers used have attenuation in the range of 0.15−0.18 dB/km, and effective area of 80−152 μm^2. For quasi-single-mode fibers, the effective area is as large as 200 μm^2.

The first 10 transmission experiments in Table 1.2 are unrepeated transmission systems with span lengths from 160 to 633 km. Ref. [49] reported 100 Gb/seconds PAM-4 transmission experiments over optical fibers with ultralow attenuation,

Table 1.2 Transmission experiments using low-loss and large effective area fibers.

Fiber		System								
α (dB/km)	A_{eff} (μm^2)	Span (km)	Amplifier	Data rate (Gb/s)	Modulation format	No. of channels	Channel spacing (GHz)	Reach length (km)	Total capacity (Tb/s)	Refs.
0.162	82	160	None	100	PAM-4	39	50	160	3.9	[49]
0.168	125	160	None	100	PAM-4	39	50	160	3.9	[49]
0.163	76–128	365	EDFA/Raman	112	PM-QPSK	40	50	365	4	[50]
0.163	115	440	Raman/ROPA	43	PDM-RZ-BPSK	64	50	440	2.6	[51]
0.158	110	304	Raman	256	PM-16QAM	40	50	304	8	[52]
0.154	152									
0.158	112	409.6	Raman/ROPA	120	PM-QPSK	150	150	409.6	15	[53]
0.158	110	401.1	Raman/ROPA	256	DP-16QAM	108	33.3	401.1	20.74	[54]
0.157	200									
0.155	125	515	Raman/ROPA	128	PDM-QPSK	8	100	515	0.8	[55]
0.17	78	333.4	Raman/ROPA	120	PM-QPSK	150	50	633	15	[56]
		298.6								
0.158	110	62.3	Raman/ROPA	200	DP-16QAM	8	50	458.8	1.6	[57]
0.154	152	212.4								
0.157	200	116.4								
0.17	80	125	EDFA	112	PDM-QPSK CRZ-	1	50	3000	0.26	[58]
				42.8	DQPSK	4				
0.162	134	100	EDFA	112	PM-QPSK	16	50	7200	1.6	[59]
0.165	85,134	200	EDFA/Raman	112	PM-QPSK	8	50	5400	0.8	[60]
						32		6000	3.2	
0.161	112	50	EDFA	112	PDM-QPSK	80	50	9000	8	[61]
0.160	112, 146	121	EDFA	120.5	PDM-8QAM-OFDM	115	25	10181	11.5	[62]
0.160	112, 146	121	EDFA	85.74	DP-QPSK	350	25	10181	30	[63]
0.162	112	100	Raman	112	PM-QPSK	40	50	10200	4	[64]
0.166	112	112.3	EDFA	256	PM-16QAM	20	50	2359	4	[65]
								14491	2	
				128	PM-QPSK	40		7414	4	
	143				PM-QPSK					
0.18	152	55	EDFA	162	PM-16QAM	81	33.3	9100	49.3	[66]
				180		201				
SM: 0.153	151	50.3	EDFA	214.5	SCM-16QAM	111	33	6600	22.2	[47]
QSM: 0.157	200	51.3								
SM: 0.153	150	30	EDFA	207.7	64APSK	168	25	6375	34.9	[67]
QSM: 0.163	176	26.1								

demonstrating reach lengths of up to 160 km. This transmission system can be attractive for low-cost long-reach interconnects between data centers.

Refs. [50–57] demonstrated high-capacity long unrepeated transmission systems in the range of 300 to over 600 km distances with total capacity of about 1–20 Tb/seconds. High-capacity unrepeated systems such as these can be used for submarine networks connecting islands to the mainland, one island to another, and mainland points to each other in a festoon arrangement. Long-reach unrepeated systems have the potential to reduce system complexity and installation costs.

The other experiments in Table 1.2 aimed to increase the capacity and distances for long-haul and submarine networks. In these experiments, the span length ranged from 50 to 200 km, the reach distance was from 3000 to over 14400 km, the channel data rate was from 40 to 250 Gb/seconds, and the capacity was from 0.26 to 49.3 Tb/seconds [47], [58–67]. Effective area management was used in Refs. [62,63] and [65] to reduce nonlinearity by using a larger effective area fiber at beginning of each span and to increase Raman efficiency by using a smaller effective area at the end of each span. The long reach lengths demonstrated in the experiments can cover system link distances for both terrestrial and submarine applications. The fiber spans used in the experiment of Refs. [47] and [67] are hybrid spans consisting of a single-mode fiber and a few-mode fiber configured for QSM transmission. These experiments and results highlight the long reach lengths and good system performance enabled by the fiber low-loss and large effective area attributes.

1.3 Multicore fibers

An MCF is an optical fiber that includes multiple cores in one common cladding. MCFs offer more degrees of freedom in fiber parameters than single-core fibers, which implies that the MCFs are more complex to design.

In this section, we describe the MCF design parameters and mode/power coupling characteristics based on [68–70], and review the progress on MCF developments.

1.3.1 Design parameters and types of multicore fibers

As summarized in Fig. 1.11, an MCF has more design freedom in core structure, core count, core layout, outer cladding thickness (OCT: the minimum distance between the outmost core center and the cladding–coating interface), and cladding/coating diameters. These parameters, of course, affect optical/mechanical characteristics, as discussed in the following, and therefore have to be optimized in order to realize preferable optical/mechanical properties, depending on various applications.

MCFs can be divided into two types, uncoupled and coupled, based on coupling characteristics that will be discussed in detail in the next section. Table 1.3 briefly

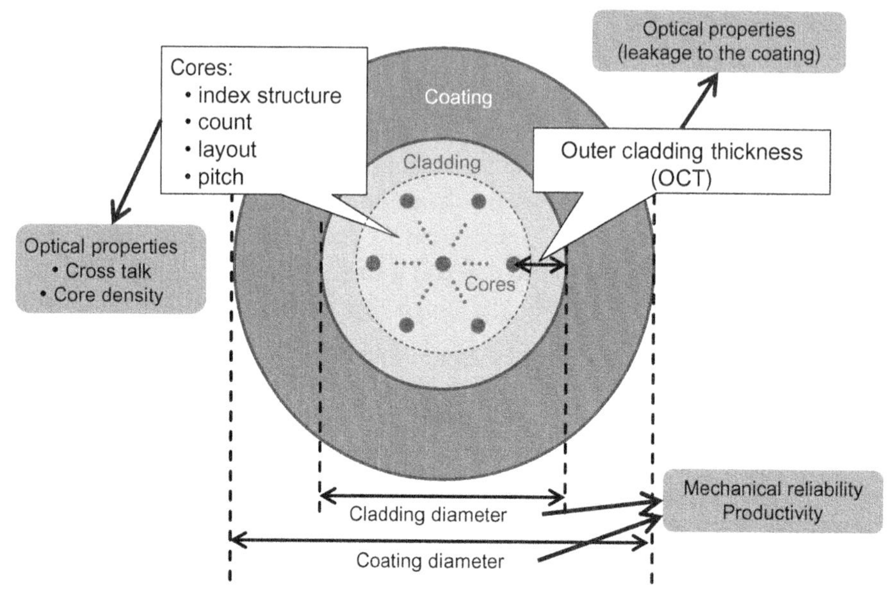

Figure 1.11 MCF-particular design parameters and their effects on the optical/mechanical properties.

Table 1.3 Characteristics of multicore fibers.

Fiber type	Coupled MCF		UC-MCF
	SC-MCF	RC-MCF	
Core pitch	Small	Medium	Large
Coupling between core modes	Strong		Weak
	Systematic	Random	
Coupling between eigenmodes[a]	Weak	Strong	
	Random		Systematic
Dominant source of DGD	DGD between eigenmodes	Both may affect	DGD between core modes
DGD accumulation	Linear proportion to fiber length	Square root proportion to fiber length	Linear proportion to fiber length
MDL accumulation	Linear proportion to fiber length	Square root proportion to fiber length	Linear proportion to fiber length

[a]The eigenmodes of the whole multicore waveguide system.

summarizes the classification of MCFs in view of coupling characteristics. Uncoupled MCF (UC-MCF) is the most representative type of the MCFs, in which the coupling between the cores (core-to-core crosstalk: XT) is suppressed and each core is used as an individual spatial channel. Therefore, the UC-MCF is basically compatible with conventional transceivers for single-mode fibers. The coupled MCF is subdivided into systematically coupled MCF (SC-MCF) and randomly coupled MCF (RC-MCF), depending on the coupling strength between the cores and the strength of longitudinal perturbations. The RC-MCF is now considered one of the strong candidates for the next-generation ultralong-haul submarine transmission fiber and has been studied intensively in recent years. It can simultaneously realize a high core density, ultralow transmission loss, and strong mode mixing, which can greatly suppress the accumulations of the differential group delay (DGD) between modes, and the low mode-dependent loss (MDL). The SC-MCF has been theoretically and numerically studied very well, but there have been a limited number of experimental demonstrations. Further experimental studies are expected.

1.3.1.1 Core pitch

Core pitch is, of course, one of the most important parameters of MCF to design its coupling or crosstalk characteristics, which will be discussed in Section 1.3.2. Besides coupling or crosstalk, cutoff wavelength is also affected by the core pitch. When each core is surrounded by an index trench layer and the core pitch is too small, a core (surrounded core) that is surrounded by other cores (surrounding cores) is surrounded by the trenches of the surrounding cores; therefore, the cutoff wavelength of the surrounded core can be elongated [71].

For coupled MCFs, by decreasing the core pitch, supermode effective indices deviate from the core mode effective indices, and the effective indices of higher order supermodes could approach to the refractive index of the cladding. Therefore, the macrobend/microbend loss and cutoff wavelength of such higher-order supermodes should be taken into account, especially for the systematically coupled case.

1.3.1.2 Outer cladding thickness

The minimum distance between the center of the outmost core and the cladding—coating interface is often referred to as "outer cladding thickness (OCT)." When the OCT is too small, some power in the outermost core couples to the coating and the transmission loss of the core increases [72]. A high-confinement core design such as trench-assisted index profile can help to simultaneously achieve the small OCT and the coating-leakage loss suppression [73]. Operating wavelength band optimization also helps to reduce the OCT and increase core density [74].

1.3.1.3 Cladding diameter

A thick fiber can pack more cores in its cladding, and core density can also be improved by reducing the fraction of the unutilized area in the outer cladding. Ultrahigh-capacity transmission experiments have been conducted using various MCFs with a cladding diameter more than 200 μm [75–77]. However, thicker cladding degrades fiber productivity and the mechanical reliability of a bent fiber. Technically, mechanical failures due to microdefects can be improved by increasing proof test stress even for thick-cladding fibers, but a large tensile force would be a limiting factor of the screening, because the tensile force required for the proof test is proportional to the cube of the cladding diameter (the required proof strain or stress is proportional to the cladding diameter, and the required force is proportional to the required stress and the cladding area). In addition, failures due to intrinsic glass strength cannot be suppressed by the proof test; therefore, there is a fundamental trade-off between the cladding diameter and the minimum acceptable bending radius. Recently, MCFs with the standard 125-μm-diameter cladding have been proposed by various groups [74,78–80], to avoid these challenges in the mechanical reliability and the productivity, and thus to promote the practical realization of the MCFs in communication systems.

1.3.2 Coupling characteristics of propagating modes

1.3.2.1 Uncoupled multicore fibers

XT of an MCF is affected by various longitudinal perturbations on the propagation constants; thus it is difficult to express the electric field of XT in a closed-form expression like the coupled-mode equation for directional coupler. However, the power coupling between the cores can be well predicted using closed-form expressions [69], [73,81]. In addition, the discrete coupling model can calculate/estimate the XT behavior in the electric field in a bent and twisted MCF and has been used for investigating the characteristics of the XT [73,82–86]. Based on these theories and models, this section describes coupling characteristics of MCFs.

1.3.2.2 Coupled-power theory for uncoupled MCF

Since UC-MCF is a weakly coupled multimode system with longitudinal random perturbations/deformations, XT can be considered using the coupled power theory [87]. XT, as a power coupling, is described using coupled-power equations:

$$\frac{dP_n}{dz} = \sum_{m \neq n} h_{nm}[P_m - P_n], \tag{1.11}$$

where P_n is the average power in Core n, z is the longitudinal position of the MCF, and h_{nm} is the power coupling coefficient (PCC) from Core m to Core n. The PCC is expressed as [87]:

$$h_{nm} = \kappa_{nm}^2 \left\langle \left| F(\beta_n - \beta_m) \right|^2 \right\rangle, \tag{1.12}$$

where κ_{nm} is the mode coupling coefficient from Core m and Core n, the bracket $\langle \cdot \rangle$ represents the ensemble average, $\left\langle \left| F(\beta_n - \beta_m) \right|^2 \right\rangle$ is the power spectrum density (PSD) of [87]:

$$f(\beta_n - \beta_m) = \exp\left[-j \int_0^z (\beta_n - \beta_m) d\zeta \right], \tag{1.13}$$

with respect to z, and β_n is the propagation constant of Core n.

In the UC-MCF, the propagation constant of each core can be longitudinally perturbed by bends, twists, and structure fluctuations. Now, we can equivalently express β_n as

$$\beta_{\mathrm{eq},n} = \beta_{\mathrm{c},n} + \beta_{\mathrm{v},n}, \quad \beta_{\mathrm{v},n} = \beta_{\mathrm{b},n} + \beta_{\mathrm{hf},n}, \tag{1.14}$$

where β_{c} and β_{v} are the constant and variable components of β, respectively, β_{b} represents the low-spatial-frequency perturbations induced by the macrobends and twists [88,89], and β_{hf} represents the high spatial frequency perturbations induced by the structure fluctuation [90] and microbends [69].

Fig. 1.12 shows the schematics of perturbations on β_{eq}—expressed in the effective refractive index $n_{\mathrm{eff}} = \beta/(2\pi/\lambda)$. As shown in Figs. 1.12A and B, microbend and structure fluctuation can induce a slight change in β_{v} in one core, which can be the cause of β_{hf}. Between two cores, as shown in Fig. 1.12C, macrobend can induce relatively large β_{v} in a core by taking the other core as a reference, which can be the cause of β_{b}.

The effect of β_{hf} on the PSD component in Eq. (1.12) was empirically derived as [91]

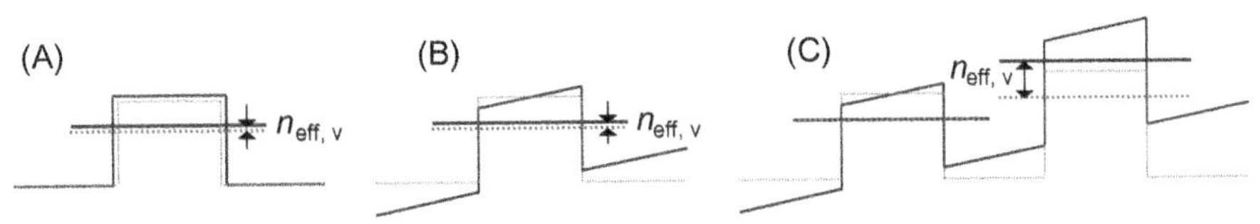

Figure 1.12 (A) Slight fluctuations in the refractive index contrast and/or core diameter can induce n_{eff} fluctuation (corresponding to β_{hf}). (B, C) Fiber bends can induce the perturbation on the refractive index profiles (RIPs), which can be regarded as a tilt of the RIP. The RIP tilt can induce $n_{\mathrm{eff,eq}}$ change (B) within a core and (C) between cores. In MCFs, β change between cores is a dominant perturbation from the bends. Narrow and thick lines represent RIP and effective index $n_{\mathrm{eff,eq}} = \beta_{\mathrm{eq}}/(2\pi/\lambda)$.

$$\left\langle \left| F(\Delta\beta_c + \Delta\beta_b + \Delta\beta_{hf}) \right|^2 \right\rangle = S_{hf}(\Delta\beta_c + \Delta\beta_b) = \frac{2l_{cor}}{1 + (\Delta\beta_c + \Delta\beta_b)^2 l_{cor}^2}, \quad (1.15)$$

where l_{cor} is the correlation length. Eq. (1.15) is the Lorentzian PSD and represents the effects of high spatial frequency perturbations induced by structure fluctuations [90] and microbends [69].

The effect of β_b can be analytically considered. By focusing on β_b (ignoring β_{hf}), we can describe a bent fiber with the refractive index profile (RIP) $n(r,\theta)$ as a corresponding straight fiber with the equivalent refractive index [92]:

$$n_{eq}(r, \theta) \approx n(r, \theta)\left(1 + \frac{r \cos \theta}{R_b}\right), \quad (1.16)$$

where (r, θ) is the local polar coordinate in the cross section of the fiber, $\theta = 0$ in radial direction of the bend, and R_b is the bending radius of the MCF. One can replace R_b with the effective bend radius $R_{b,eff}$ to account for photoelastic effect due to bend-induced stress. According to Eq. (1.16), as shown in Fig. 1.13, the equivalent effective index $n_{eff,eq,n}$ of Core n can be represented as

$$n_{eff,eq,n} \approx n_{eff,c,n}\left(1 + \frac{r_n \cos \theta_n}{R_b}\right), \quad (1.17)$$

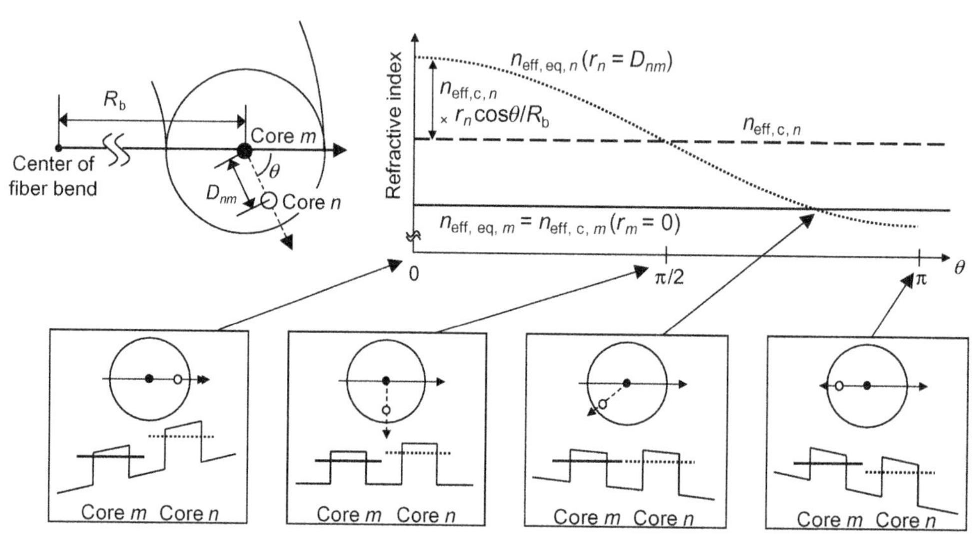

Figure 1.13 Fiber bends and twists induce the variation of equivalent effective refractive index ($n_{eff,eq}$) from intrinsic (constant) effective index ($n_{eff,c}$). The center of Core m is taken as the origin of the local coordinate for simple description.

and $\beta_{\mathrm{eq},n}$ and $\beta_{\mathrm{b},n}$ can be represented as

$$\beta_{\mathrm{eq},n} \approx \beta_{\mathrm{c},n} + \beta_{\mathrm{b},n}, \quad \beta_{\mathrm{b},n} \approx \beta_{\mathrm{c},n} \frac{r_n \cos \theta_n}{R_{\mathrm{b}}}. \tag{1.18}$$

Since $\beta_{\mathrm{eq},n}$ can become equal to $\beta_{\mathrm{eq},m}$, the phase matching can occur when

$$\left|\beta_{\mathrm{c},n} - \beta_{\mathrm{c},m}\right| \leq \frac{\beta_{\mathrm{c},n} D_{nm}}{R_{\mathrm{b}}}, \tag{1.19}$$

where D_{nm} is the distance between Core n and Core m, the bend-induced phase matching can be suppressed when R_{b} is larger than the threshold bending radius R_{pk} [89]:

$$R_{\mathrm{pk}} = \frac{D_{nm} \beta_{\mathrm{c},n}}{\left|\beta_{\mathrm{c},n} - \beta_{\mathrm{c},m}\right|}. \tag{1.20}$$

If we assume the twist is sufficiently random and gradual, the probability density of $\Delta\beta_{\mathrm{b}}$ can be expressed as the arcsine distribution:

$$S_{\mathrm{b}}(\Delta\beta_{\mathrm{b}}) = \begin{cases} \dfrac{1}{\pi\sqrt{\left(\beta_{\mathrm{c}} D/R_{\mathrm{b}}\right)^2 - \Delta\beta_{\mathrm{b}}^{\,2}}}, & \left|\Delta\beta_{\mathrm{b}}\right| \leq \left|\beta_{\mathrm{c}} D/R_{\mathrm{b}}\right|, \\ 0, & \text{otherwise,} \end{cases} \tag{1.21}$$

and the PCC averaged over longitudinal positions or twists can be expressed as the convolution of the Lorentzian and the arcsine distribution [68,69]:

$$\bar{h}_{nm} = \kappa^2 (S_{\mathrm{b}} \times S_{\mathrm{hf}})(\Delta\beta_{\mathrm{c}}) = \kappa^2 \int S_{\mathrm{b}}(\Delta\beta_{\mathrm{b}}) S_{\mathrm{hf}}(\Delta\beta_{\mathrm{c}} + \Delta\beta_{\mathrm{b}}) d(\Delta\beta_{\mathrm{b}}). \tag{1.22}$$

Fig. 1.14 shows the schematic illustration of the relationship between $\bar{h}$, the intrinsic propagation constant mismatch $\Delta\beta_{\mathrm{c}}$, and the fiber bend curvature $(1/R_{\mathrm{b}})$, where $\Delta\beta_{\mathrm{b}}^{\mathrm{dev}} = \beta_{\mathrm{c},n} D_{nm}/R_{\mathrm{b}}$ is the maximum deviation of $\Delta\beta_{\mathrm{b}}$.

This average PCC can be expressed in the closed-form expressions as [81]

$$\bar{h}_{nm} = \kappa_{nm}^2 \sqrt{2} l_{\mathrm{cor}} \left[\frac{1}{\sqrt{a\left(b + \sqrt{ac}\right)}} + \frac{1}{\sqrt{c\left(b + \sqrt{ac}\right)}} \right], \tag{1.23}$$

$$a = 1 + l_{\mathrm{cor}}^2 \left(\Delta\beta_{\mathrm{c},nm} - \frac{B_{nm}}{R_{\mathrm{b}}} \right)^2 \approx 1 + l_{\mathrm{cor}}^2 \left(\Delta\beta_{\mathrm{c},nm} - \frac{\beta_{\mathrm{c},n} D_{nm}}{R_{\mathrm{b}}} \right)^2, \tag{1.24}$$

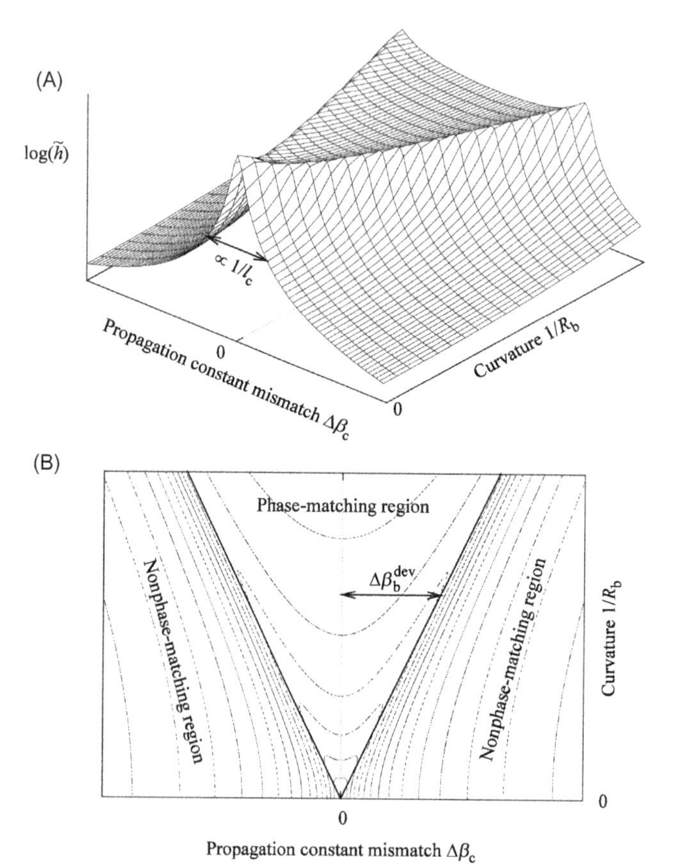

Figure 1.14 A schematic example of the power coupling coefficient $\bar{h}$ averaged over the rotation direction, as a function of the propagation constant mismatch $\Delta\beta_c$ and the curvature $1/R_b$, in case that twist of an MCF is gradual and random enough. (A) A 3-dimensional plot. (B) A contour map of $\log(\bar{h})$. Thick solid lines in (B) are the thresholds between the phase-matching region and the nonphase-matching region.

$$b = 1 + l_{cor}^2\left[\Delta\beta_{c,nm}^2 - \left(\frac{B_{nm}}{R_b}\right)^2\right] \approx 1 + l_{cor}^2\left[\Delta\beta_{c,nm}^2 - \left(\frac{\beta_{c,n}D_{nm}}{R_b}\right)^2\right], \tag{1.25}$$

$$c = 1 + l_{cor}^2\left(\Delta\beta_{c,nm} + \frac{B_{nm}}{R_b}\right)^2 \approx 1 + l_{cor}^2\left(\Delta\beta_{c,nm} + \frac{\beta_{c,n}D_{nm}}{R_b}\right)^2, \tag{1.26}$$

$$B_{nm} = \sqrt{\left(\beta_{c,n}x_n - \beta_{c,m}x_m\right)^2 + \left(\beta_{c,n}y_n - \beta_{c,m}y_m\right)^2}, \tag{1.27}$$

where B_{nm} can be approximated as $\beta_{c,n}D_{nm}$ if $\beta_{c,m}/\beta_{c,n} \simeq 1$.

For some special cases, $\overline{h}$ can be expressed in very simple expressions. When the cores are identical and $l_{\text{cor}} << R_{\text{b}}/B_{\text{nm}} \sim R_{\text{b}}/(\beta_{\text{c},n} D_{nm})$, $\overline{h}$ is expressed as [73,81]

$$\overline{h}_{nm} \approx \kappa_{nm}^2 \frac{2R_{\text{b}}}{\beta_{\text{c},n} D_{nm}}. \tag{1.28}$$

When the cores are dissimilar,

$$\overline{h}_{nm} \approx \kappa_{nm}^2 \frac{2R_{\text{b}}}{\beta_{\text{c},n} D_{nm}}. \tag{1.29}$$

can be a good approximation, but Eq. (1.18) is still precise enough (<0.1 dB error) when $\Delta\beta_{\text{c}} < 0.21\beta_{\text{c}} D/R_{\text{b}}$. When R_{b} is sufficiently larger than R_{pk}, h is expressed as [81]

$$\overline{h}_{nm} \approx \kappa_{nm}^2 \frac{2l_{\text{cor}}}{1 + (\Delta\beta_{\text{c}} l_{\text{cor}})^2}. \tag{1.30}$$

When R_{b} is at around R_{pk}, $\overline{h}$ has a maximum value and $\overline{h}$ at R_{pk} is expressed as [81]

$$\overline{h}_{nm} \approx \kappa_{nm}^2 \sqrt{2} l_{\text{cor}} \sqrt{\frac{1 + \sqrt{1 + (2\Delta\beta_{\text{c}} l_{\text{cor}})^2}}{1 + (2\Delta\beta_{\text{c}} l_{\text{cor}})^2}}. \tag{1.31}$$

1.3.2.2.1 Discrete coupling model and statistical distribution of the cross talk

The average power coupling coefficient is sufficient for predicting the intensity XT (i.e., the ensemble/wavelength/time average of the XT), but the information on the stochastic behavior of the XT is important for studying the impact of the XT on the transmission quality. The discrete coupling model is one of the simple models that can calculate/estimate the XT behavior in the electric field [73,82–86]. Fig. 1.15 shows an example of the mode coupling in a bent and twisted MCF, which was calculated using Eq. 1.18 and the coupled mode equation with perturbed propagation constants:

$$\frac{dA_m}{dz} = -j \sum_{n \neq m} \kappa_{mn} \exp\left[j \int_0^z \left(\beta_{\text{eq},m} - \beta_{\text{eq},n} \right) dz \right] A_n. \tag{1.32}$$

In Fig. 1.15, discrete-like dominant XT changes (bend-induced resonant couplings) are observed at every phase-matching point (PMP) where the difference in β_{eq} between the cores equals zero, and the XT changes in the other positions can be regarded just as local fluctuations, as shown. The dominant changes appear random, because the phase differences between Cores m and n are different for every PMP.

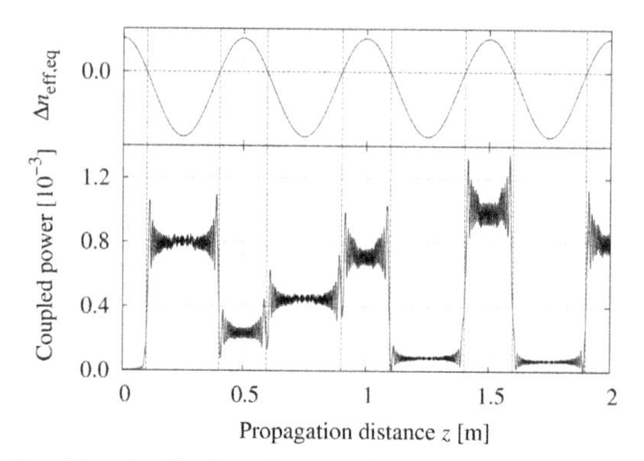

Figure 1.15 An example of longitudinal evolution of coupled power in a bent and twisted MCF. Source: *Replotted from the data in Ref. T. Hayashi, T. Nagashima, O. Shimakawa, T. Sasaki, E. Sasaoka, Crosstalk variation of multi-core fibre due to fibre bend, in: Eur. Conf. Opt. Commun. (ECOC), Torino, 2010, p. We.8.F.6, [89].*

The phase differences can easily fluctuate in practice by slight variations of the perturbations. Therefore, the XT can be understood as a practically stochastic parameter.

When PMPs are discretized by the bends and twists, mode coupling from Core m to Core n can be approximated by discrete changes [73]:

$$A_n(N_{\mathrm{PM}}) \approx A_n(0) - j\sum_{l=1}^{N_{\mathrm{PM}}} \chi_{nm}(l)\exp\left[-j\varphi_{\mathrm{rnd}}(l)\right]A_m(l-1), \tag{1.33}$$

where $A_n(N_{\mathrm{PM}})$ represents the complex amplitude of Core n after N_{PM}-th PMP, $\phi_{\mathrm{rnd}}(N_{\mathrm{PM}})$ is the phase difference between Cores m and n at N_{PM}-th PMP, and χ_{nm} is the coefficient for the discrete changes caused by the coupling from Core m to Core n. ϕ_{rnd} can be regarded as a random sequence, and ϕ_{rnd} varies with wavelength and with time variations of the bend, twist, and other perturbations. In the special case of the homogeneous MCF with constant bend and twist, χ_{nm} can be approximated as [73]

$$\chi_{nm} \cong \sqrt{\frac{\kappa_{nm}^2}{\beta_{\mathrm{c}}}\frac{R_{\mathrm{b}}}{D_{nm}}\frac{2\pi}{\omega_{\mathrm{twist}}}}\exp\left[-j\left(\frac{\beta_{\mathrm{c}}D_{nm}}{\omega_{\mathrm{twist}}R_{\mathrm{b}}} - \frac{\pi}{4}\right)\right]. \tag{1.34}$$

When the XT is adequately low ($|A_n(N_{\mathrm{PM}})| << 1$) and $A_m(0) = 1$, Eq. (1.33) is simplified as

$$A_n(N_{\mathrm{PM}}) \approx -j\sum_{l=1}^{N_{\mathrm{PM}}} \chi_{nm}\exp\left(-j\varphi_{\mathrm{rnd},l}\right), \tag{1.35}$$

and the XT X_{nm} in intensity can be approximated as $|A_n(N_{\mathrm{PM}})|^2$.

When the MCF is sufficiently long and has sufficient PMPs, the probability density function (pdf) of the I–Q components converge to Gaussian distributions with the standard deviation $\sigma_{4\mathrm{df},nm}$, based on the central limit theorem. When assuming random polarization mode coupling, the XT X can be represented as a sum of powers of $\Re A_n(N_{\mathrm{PM}})$'s and $\Im A_n(N_{\mathrm{PM}})$'s of two polarization modes; therefore, the pdf of X_{nm} is a scaled chi-square distribution with 4 degrees of freedom [73]:

$$f_{X,4\mathrm{df}}(X_{nm}) = \frac{X_{nm}}{4\sigma^4_{4\mathrm{df},nm}} \exp\left(-\frac{X_{nm}}{2\sigma^2_{4\mathrm{df},nm}}\right), \tag{1.36}$$

and the statistical average $\mu_{X,nm}$ of X_{nm} is [73]:

$$\mu_{X,nm} = 4\sigma^2_{4\mathrm{df},nm} = N_{\mathrm{PM}}\left|X_{nm}\right|^2. \tag{1.37}$$

In actual UC-MCFs, the phase difference ϕ_{rnd} between cores varies with wavelength and time; therefore the XT varies with wavelength and time, as shown in Fig. 1.16. The fringes in the XT spectrum are the interference fringes of coupled lights from many PMPs; therefore, the inverse Fourier transform can extract the information of the skew (differential group delay between cores) accumulated between PMPs—the fringe period becomes shorter when the skew becomes larger.

Now, XT for a modulated signal can be considered based on the XT behavior, as previously discussed and summarized in Fig. 1.16. From the Gaussian distribution on the in-phase and quadrature (I–Q) plane, XT may behave as additive Gaussian-distributed noise when the bandwidth of the signal light is adequately broad, since

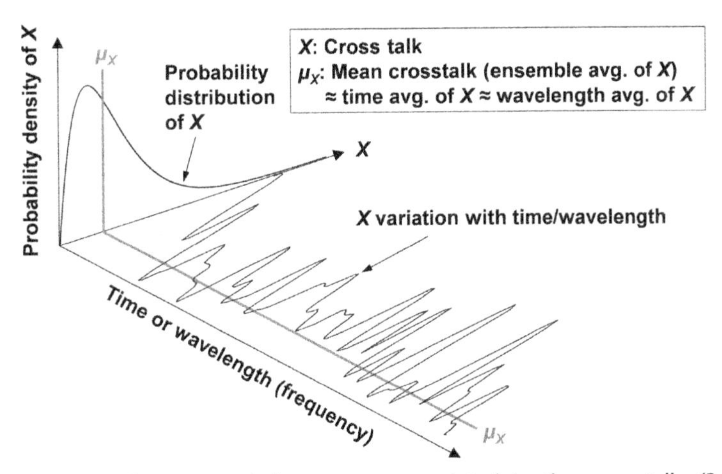

Figure 1.16 Schematic illustration of the parameters related to the cross talk. *(Reused from Y. Kokubun, M. Koshiba, Novel multi-core fibers for mode division multiplexing: proposal and design principle, IEICE Electron. Express 6 (8) (2009) 522–528, [93] ©2014 IEICE).*

the instantaneous frequency of the modulated signal light may rapidly vary with time over the broad bandwidth of the XT spectrum. In this case, the short-term average XT (STAXT: X averaged over a short term) equals the statistical average of the XT, because the modulated signal light averages the wavelength dependence of X.

When the bandwidth becomes narrower, the XT distribution on the $I-Q$ plane can become more distorted from the Gaussian distribution. Moreover, if the bandwidth is adequately narrow, the XT may behave like a static coupling, since the XT spectrum variation is very gradual compared to the symbol rate on the order of picoseconds to nanoseconds. However, this static coupling may gradually vary with time, and the STAXT varies with time and obeys the chi-square distribution in power or the normal distribution in the $I-Q$ plane, as shown in Fig. 1.16.

Rademacher et al. investigated such an XT behavior from the relationship between the variance ($\mathrm{Var}_{\mathrm{XT}}$) of the STAXT and the modulated signal bandwidth for the on–off keying (OOK) case and the quadrature phase-shift keying (QPSK) case, as shown in Fig. 1.17, by numerically solving the extension of Eq. (1.35) for a skew and dual polarizations [82]:

$$A_n(L,\omega) \approx -j\sum_{l=1}^{N_{\mathrm{PM}}} \chi_{nm,l}\mathbf{R}_l \begin{bmatrix} \exp\left[-j\varphi_{\mathrm{rnd},l} - j\tau_{nm,l}\omega\right] \\ \exp\left[-j\varphi_{\mathrm{rnd},l} - j\tau_{nm,l}\omega\right] \end{bmatrix}. \tag{1.38}$$

where $\mathbf{R}_l$ is a random unitary 2×2 matrix simulating random polarization rotations between the PMPs, and τ_{nm} is the differential group delay between Core n and Core m at l-th PMP. In [82], τ_{nm} was assumed to be proportional to the fiber longitudinal position, and the linear birefringence was ignored. According to Fig. 1.17, the variance of the STAXT is high when the skew and symbol rate product is low, and low when the skew and symbol rate product is high. In the case of QPSK modulation, $\mathrm{Var}_{\mathrm{XT}}$ converges to zero when the skew and symbol rate product is high. However, in the case of OOK modulation, $\mathrm{Var}_{\mathrm{XT}}$ never converge to zero even when the skew and symbol rate product is very high. This can be understood from the spectrum of the modulated signal lights, shown in the bottom of Fig. 1.17. The spectrum of the QPSK signal light has a flat-top-like shape; therefore, the XT of the signal light is well averaged over the signal band. On the other hand, the spectrum of the OOK signal light has a strong and sharp peak at the center of the spectrum, which results from the carrier frequency of the OOK signal; therefore, the XT of the signal light is not well averaged over the signal band. Due to the STAXT variation, the modulation formats with a strong carrier component such as OOK and pulse amplitude modulation (PAM) require some margin from the statistical average XT for the XT suppression. These XT behaviors were also experimentally confirmed [82].

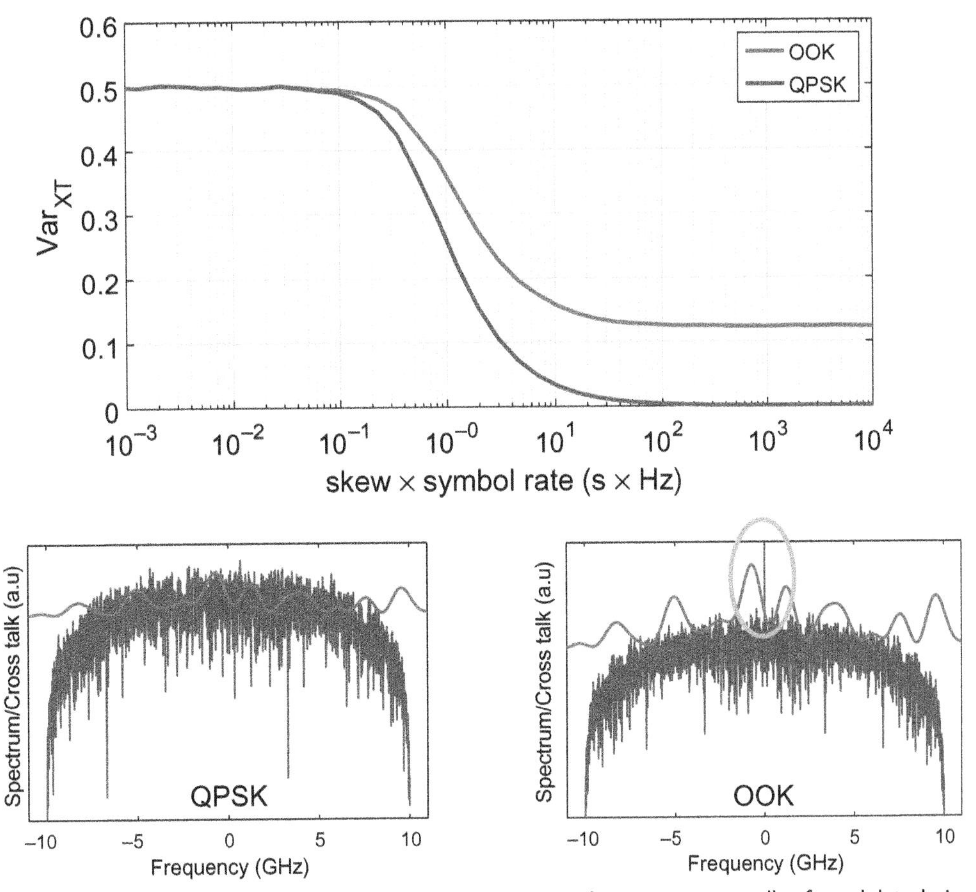

Figure 1.17 (Top) The variance of the short-term averaged intercore crosstalk of modulated signals, and (bottom) schematic spectra of signal and crosstalk for the OOK and QPSK modulations Source: *(Top) Reused from G. Rademacher, R.S. Luís, B.J. Puttnam, Y. Awaji, N. Wada, Crosstalk dynamics in multi-core fibers, Opt. Express 25 (10) (2017) 12020–12028, [82] ©OSA 2017. (Bottom) Reused from the presentation material of G. Rademacher, R.S. Luis, B.J. Puttnam, Y. Awaji, and N. Wada, Crosstalk fluctuations in homogeneous multi-core fibers, in: Photonic Networks and Devices, New Orleans, 2017, p. NeTu2B.4, [83] by courtesy of the authors.*

1.3.2.3 Coupled multicore fibers

In the coupled MCF (also referred to as strongly coupled MCF, coupled-core fiber, coupled-core MCF), the coupling between the cores is strong and the one set of multiple coupled cores guides multiple spatial modes; therefore, each core cannot be used as an independent spatial channel. The coupled MCF can be further subdivided to SC-MCF and RC-MCF, depending on the coupling strength. These two subtypes of the coupled MCFs are often simply called "coupled MCF" or the like without distinction in much literature.

1.3.2.3.1 Systematically coupled multicore fiber

When coupling between the cores is strong, the eigenmodes of the whole multicore system can be supermodes—that is, the superpositions of core modes—and stably propagate in the MCF without coupling. Thus, the coupling between the core modes is very systematic and deterministic. Such supermodes can be discussed with core modes using the nonorthogonal coupled mode theory. We express the coupled mode equation of the core modes as

$$\frac{d}{dz}|E\rangle = -j\mathbf{H}|E\rangle, \tag{1.39}$$

where $\mathbf{H}$ is the transfer matrix and Hermitian for power conservation. Any Hermitian matrix can be diagonalized as

$$\mathbf{H} = \mathbf{U}\boldsymbol{\Lambda}\mathbf{U}^*, \tag{1.40}$$

where $\mathbf{U}$ is a unitary matrix, the superscript * of a matrix represents the conjugate transpose, and $\boldsymbol{\Lambda}$ is a diagonal matrix. The diagonal elements of $\boldsymbol{\Lambda}$ are the eigenvalues of $\mathbf{H}$ and are the propagation constants of the eigenmodes. $\mathbf{U}$ is composed of the orthonormal eigenvectors of $\mathbf{H}$ written as its column, and converts the eigenmodes $|E_{\text{eig}}\rangle$ and the core modes $|E\rangle$ as

$$\begin{cases} |E\rangle = \mathbf{U}|E_{\text{eig}}\rangle, \\ |E_{\text{eig}}\rangle = \mathbf{U}^*|E\rangle. \end{cases} \tag{1.41}$$

The orthonormal eigenvectors are the eigenmodes represented as the superpositions of the core modes.

Using Eqs. (1.40) and (1.41), we can reduce the coupled mode Eq. (1.39) to

$$\frac{d}{dz}|E_{\text{eig}}\rangle = -j\boldsymbol{\Lambda}|E_{\text{eig}}\rangle, \tag{1.42}$$

which is a simple propagation equation without coupling. Therefore, in ideal cases, there are no couplings between the supermodes. In actual cases, various longitudinal perturbations can induce mode coupling between the supermodes, but such coupling can be suppressed by increasing the propagation constant mismatch between the supermodes.

To explain these equations in detail and for the sake of simplicity, we will consider a two-core case and ignore polarizations,. Eq. (1.39) can be rewritten as

$$\mathbf{H} = \begin{bmatrix} \beta_1 & \kappa \\ \kappa^* & \beta_2 \end{bmatrix}, \tag{1.43}$$

where $\beta_{1/2}$ is the propagation constants of Core 1/2, the superscript * of a scalar represents the complex conjugate, and κ is the mode coupling coefficient between Core 1 and Core 2. **H** can be eigendecomposed with the eingenvalues and the corresponding orthonormal eingenvectors:

$$\Lambda_1 = \beta_{\text{avg}} + \sqrt{\Delta^2 + |\kappa|^2}, \quad |u_1\rangle = \frac{1}{\sqrt{\left(\Delta+\sqrt{\Delta^2+|\kappa|^2}\right)^2 + |\kappa|^2}} \begin{bmatrix} \Delta + \sqrt{\Delta^2 + |\kappa|^2} \\ \kappa^* \end{bmatrix}, \tag{1.44}$$

$$\Lambda_2 = \beta_{\text{avg}} - \sqrt{\Delta^2 + |\kappa|^2}, \quad |u_2\rangle = \frac{1}{\sqrt{\left(\Delta+\sqrt{\Delta^2+|\kappa|^2}\right)^2 + |\kappa|^2}} \begin{bmatrix} -\kappa \\ \Delta + \sqrt{\Delta^2 + |\kappa|^2} \end{bmatrix}, \tag{1.45}$$

where $\beta_{\text{avg}} = (\beta_1 + \beta_2)/2$ and $\Delta = (\beta_1 - \beta_2)/2$. Here, the diagonal matrix $\mathbf{\Lambda}$ and the unitary matrix **U** can be chosen as

$$\mathbf{\Lambda} = \begin{bmatrix} \Lambda_1 & 0 \\ 0 & \Lambda_2 \end{bmatrix}, \quad \mathbf{U} = [|u_1\rangle, |u_2\rangle] \quad . \tag{1.46}$$

According to Eqs. (1.44) and (1.45), when $\Delta \neq 0$, the eigenmodes are the superpositions of unequally excited core modes. When $\Delta = 0$ (identical cores, $\beta = \beta_1 = \beta_2$), Eqs. (1.44) and (1.45) are reduced to

$$\Lambda_1 = \beta + |\kappa|, \quad |u_1\rangle = \frac{1}{\sqrt{\kappa + \kappa^*}} \begin{bmatrix} \sqrt{\kappa} \\ \sqrt{\kappa^*} \end{bmatrix}, \tag{1.47}$$

$$\Lambda_2 = \beta - |\kappa|, \quad |u_2\rangle = \frac{1}{\sqrt{\kappa + \kappa^*}} \begin{bmatrix} -\sqrt{\kappa} \\ \sqrt{\kappa^*} \end{bmatrix}. \tag{1.48}$$

Here, each supermode is the superposition of the two core modes with equal amplitude where the even mode (Eq. (1.47)) consists of the core modes with the same phase and the odd mode (Eq. (1.48)) consists of the core modes with the opposite phases. The propagation constant mismatch $\Delta\Lambda$ between the supermodes is

$$\Delta\Lambda_{12} = \Lambda_1 - \Lambda_2 = 2|\kappa|, \tag{1.49}$$

and increases when core-to-core coupling becomes strong. In general cases of more than two identical cores, Λ and $\Delta\Lambda$ can be represented as

$$\Lambda_i = \beta + \sum_n a_{i,n} |\kappa_n|, \tag{1.50}$$

$$\Delta\Lambda_{ij} = \sum_n \left(a_{i,n} - a_{j,n}\right) |\kappa_n| \tag{1.51}$$

where $a_{n,i}$ represents the coefficient of the coupling coefficient κ_n. for i-th supermode.

By taking advantage of the orthogonality of the supermodes, Kokubun and Koshiba proposed uncoupled SDM transmission over the supermodes of linearly arranged $(1 \times n)$ coupled cores [94]. In the linear layout, all the supermodes have different propagation constants and there are no degenerated modes. So, when κ is large enough, the coupling between the supermodes can be suppressed even with small longitudinal perturbations and all the spatial modes can be transmitted without mode mixing. In the transmission experiments, modal XT suppression at mode multiplexing/demultiplexing and at fiber splicing/connection would be the challenge for the transmission without modal XT compensation, similar to the weakly coupled few-mode fiber transmissions.

1.3.2.3.2 Randomly coupled multicore fiber

When core-to-core coupling is weaker than the SC-MCF but not negligible like the UC-MCF, strong and random mode mixing can occur in the MCF caused by the longitudinal perturbations on the MCF [95]. This type of the MCF is referred to as randomly coupled MCF (RC-MCF). The RC-MCF requires the MIMO DSP to undo the random mode mixing, but this random mixing provides some preferable features for the transmission [96]. Strong random mode mixing can suppress the accumulations of DGD, mode dependent loss/gain (MDL/MDG), and nonlinear impairments. As mentioned, DGD suppression is a critical factor for reducing the calculation complexity in the MIMO DSP for the XT compensation. RC-MCF can simultaneously realize low DGD and ultralow loss comparable to the lowest loss realized in SMF, because the RC-MCF can suppress DGD with simple step-index-type pure-silica cores thanks to the random coupling. In contrast, the single-core few-mode/multimode fiber (FMF/MMF) needs a GeO_2-doped precisely-controlled graded-index-type core for the DGD suppression, and mode coupling in the FMF/MMF is weak.

In this section, the mechanism of the random coupling is described by using the two-core case for the sake of simplicity. In addition, the DGD and MDL characteristics are also briefly reviewed.

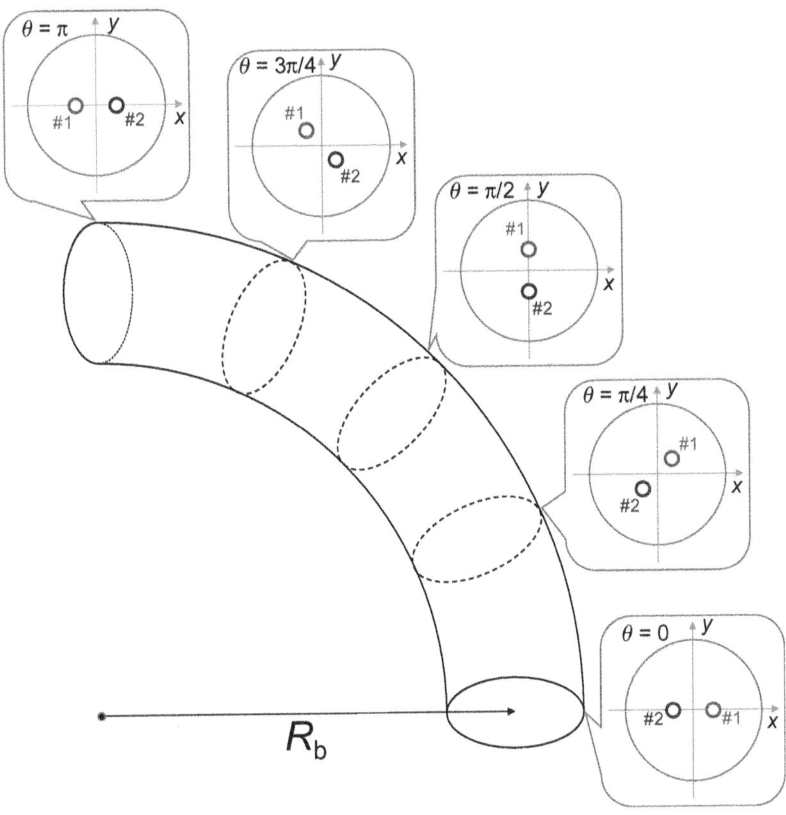

Figure 1.18 Schematic illustration showing the coordinates relating to the fiber bend and twist. θ is the fiber rotation angle that is defined such that $\theta = 0$ when the direction from Core 2 to Core 1 is the radial direction of the bend. The direction of the x axis of the local coordinates on a fiber cross section is chosen to always be the radial direction of the fiber bend.

Mechanism of random mode coupling We now consider coupled identical two cores perturbed by bends and twists. When longitudinal perturbation is dominantly induced by the bends and twists, Δ in Eqs. (1.44) and (1.45) can be expressed as

$$\Delta = \sqrt{\left(\frac{\beta_c D}{2R_b} \cos\theta\right)^2 + |\kappa|^2}. \tag{1.52}$$

where the coordinates are defined as shown in Fig. 1.18. Therefore, the changes of R_b and/or θ can vary the excitation ratio ρ of the core modes in the eigenmode. By assuming κ is real for simplicity, and defining a and b such that $|u_1\rangle$ and $|u_2\rangle$ in Eqs. (1.44) and (1.45) can be expressed as

$$|u_1\rangle = \begin{bmatrix} a \\ b \end{bmatrix}, \quad |u_2\rangle = \begin{bmatrix} -b \\ a \end{bmatrix}, \tag{1.53}$$

we may define ρ as [70]:

$$\begin{aligned}\rho &\equiv a/b \\ &= \Delta/\kappa + \sqrt{(\Delta/\kappa)^2 + 1} \\ &= M\cos\theta + \sqrt{M^2\cos^2\theta + 1},\end{aligned} \tag{1.54}$$

where M is the maximum ratio of the bend-induced perturbation to the core-coupling-induced perturbation on the propagation constant:

$$M = \frac{\beta_c D/R_b}{2\kappa}. \tag{1.55}$$

Now, we write the coupled mode equation for the bend-perturbed eigenmodes in the coupled MCF as

$$\frac{dA_m}{dz} = -j\sum_{n\neq m} Q_{mn}\exp\left[j\int_0^z (\Lambda_m - \Lambda_n)dz\right]A_n, \tag{1.56}$$

where the subscripts (m, n) represent the eigenmode identification numbers. [70,97]. Again, to explain these equations in detail and for the sake of simplicity, we will consider a two-core case. By modeling the coupled MCF as concatenated infinitesimal sections and assuming that the coupling between the eigenmodes is induced by the field profile mismatch between the sections, the coupling coefficient between the eigenmodes $|u_1\rangle$ (even mode) and $|u_2\rangle$ (odd) can be derived as [70,97]

$$Q_{12} \approx \frac{-j}{1+\rho^2}\frac{\partial\rho}{\partial z}, \quad Q_{21} \approx Q_{12}^*. \tag{1.57}$$

When $Q >> 0$ and $\Lambda_1 \sim \Lambda_2$, the mode propagation becomes nonadiabatic and nonnegligible coupling between the eigenmodes can occur [97].

Here, the variations of the core excitation ratio ($|a|^2$ and $|b|^2$) of the eigenmodes are shown in Fig. 1.19, and the mode coupling coefficient between the eigenmodes is shown in Fig. 1.20, both for various M values. At $\theta = \pi/2$, the slopes of the changes of $|a|^2$ and $|b|^2$ are steepest, and $|Q|/(\partial\theta/\partial z)$ has the highest value of $M/2$. The shapes of $|Q|/(\partial\theta/\partial z)$ are similar among the cases of $M = 0.01$, 0.1, and 1, because the variations of a and b along θ are similar to sinusoidal variations whose amplitudes depend on M. When M is 10 or 100, the variations of a and b become more localized around $\theta = \pi/2$. Therefore, the coupling coefficient not around $\theta = \pi/2$ becomes more suppressed.

Fig. 1.21 shows the intensity evolutions—or mode couplings—of eigenmodes and core modes in a constantly bent and twisted homogeneous two-core fiber. Only one

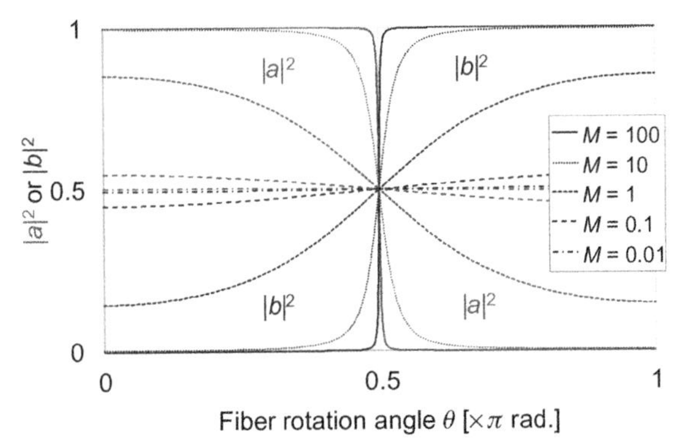

Figure 1.19 The variations of the core excitation ratio of the eigenmodes of a homogeneous two-core fiber for various *M*. Source: *Reused from T. Hayashi, Multi-core fibers for space division multiplexing, in: G.-D. Peng (Ed.), Handbook of Optical Fibers, Springer, Singapore, 2018, pp. 1–46, [70] ©Springer 2018.*

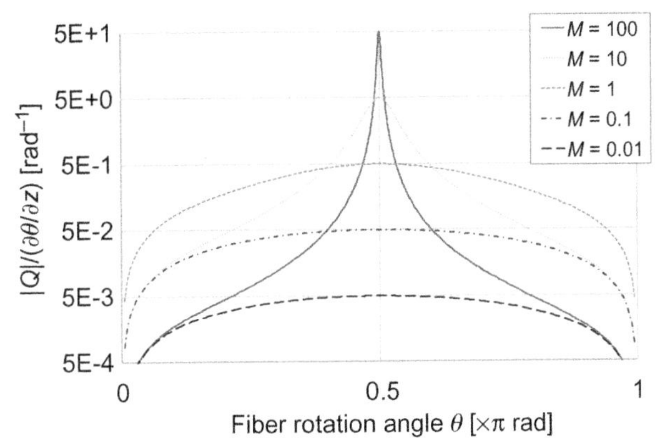

Figure 1.20 The variations of the mode coupling coefficient between the eigenmodes of a homogeneous two-core fiber for various *M*. The coefficient *Q* represents the coupling per unit length, and $Q/(\partial\theta/\partial z)$ the coupling per unit rotation angle. Source: *Reused from T. Hayashi, Multi-core fibers for space division multiplexing, in: G.-D. Peng (Ed.), Handbook of Optical Fibers, Springer, Singapore, 2018, pp. 1–46 ©Springer 2018.*

eigenmode (even mode) or core mode (Core 1, see Fig. 1.18) is excited at $\theta = 0$. The intensity evolutions for the core pitches of 15, 20, and 30 μm were calculated. The core diameter was 9 μm, the refractive index contrast was 0.35%, the bend radius was 8 cm, and the twist rate was 2π rad/m. Here the *M* value becomes larger when the core pitch becomes larger. When the core pitch was 15 μm, the eigenmodes were almost uncoupled and no significant intensity variation was observed. In contrast, the core modes coupled significantly but it was a systematic coupling. The high-frequency oscillation of the intensity was due to the beating between the eigenmodes.

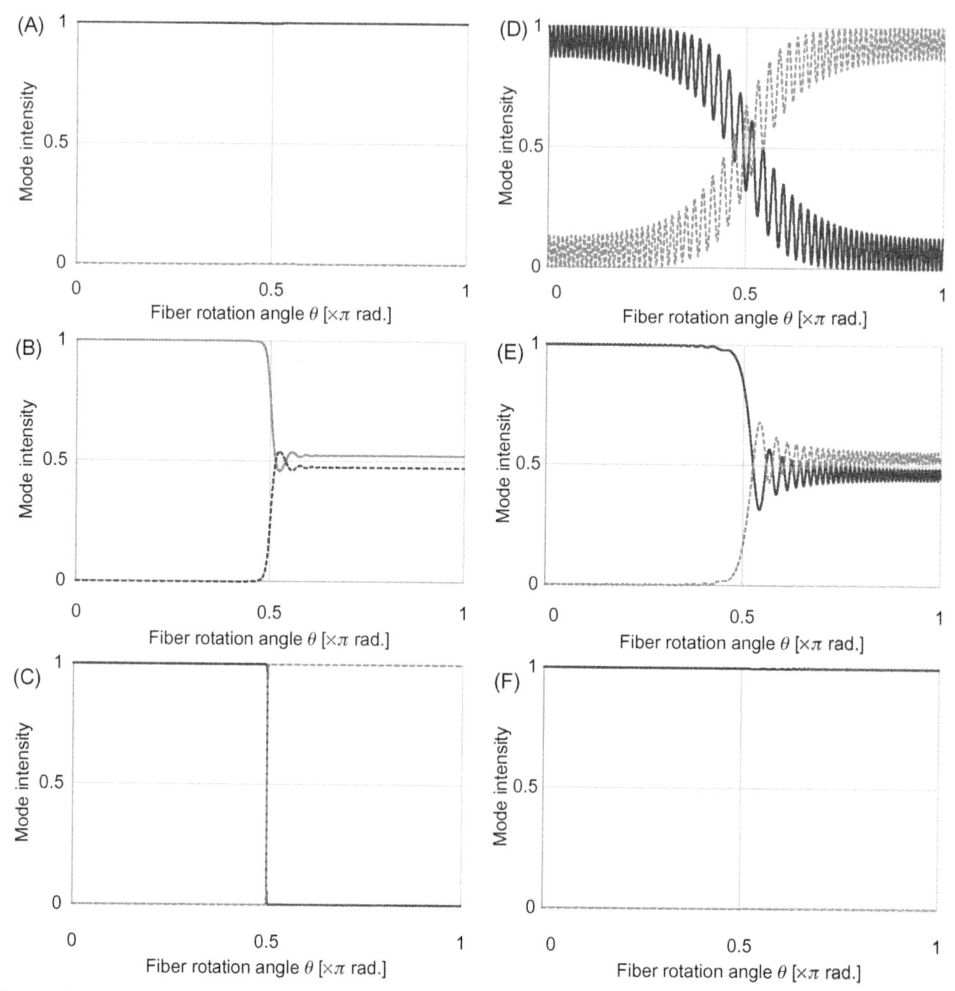

Figure 1.21 Intensity evolutions of (A–C) eigenmodes and (D–F) core modes in a homogeneous two-core fiber with a core pitch of (A, D) 15 μm, (B, E) 20 μm, or (C, F) 30 μm, when only one (A–C) eigenmode or (D–F) core mode is excited at $\theta = 0$ Source: *Reused from T. Hayashi, Multi-core fibers for space division multiplexing, in: G.-D. Peng (Ed.), Handbook of Optical Fibers, Springer, Singapore, 2018, pp. 1–46, [70] ©Springer 2018. The core diameter is 9 μm, the refractive index contrast is 0.35%, the bend radius is 8 cm, and the twist rate is 2π radians.*

The low-frequency intensity variations represent the transfer of the power from the excited mode to the other mode, because most of the power of the even mode is always guided in a core that is located at the outer side of the bend. When the core pitch was 20 μm, both the eigenmodes and the core modes experienced a significant coupling around $\theta = \pi/2$, which can be regarded as a PMP. So, every time the lights pass a PMP, random power mixing occurs. When the core pitch was 30 μm, the core modes were almost uncoupled and no significant intensity variation was observed. On the other hand, the eigenmodes exhibited a full power transfer from the even mode to the odd mode, and the power transfer was greatly localized at $\theta = \pi/2$. This can be interpreted as follows:

1. At $\theta = 0$, the even mode is excited. The even mode is almost equivalent to the core mode of Core 1, because when $\theta < \pi/2$, Δ is very large and the even mode is localized in Core 1 (outside the bend) and the odd mode in Core 2 (inside the bend).
2. At $\theta = \pi/2$, Δ becomes zero and the eigenmodes suddenly spread over the two cores. Therefore, the light in the outside core mode in $\theta < \pi/2$ couples to both even and odd modes at $\theta = \pi/2$. However, all power is still guided in Core 1, because the phases of the even and odd modes are the same in Core 1 and opposite in Core 2.
3. When $\theta > \pi/2$, Δ becomes very large. Therefore, each eigenmode is localized to each core again. However, the even mode is localized in Core 2 and the odd mode in Core 1, because Core 2 is outside the bend and Core 1 is inside the bend now.
4. Since all the power of the two eigenmodes is guided only in Core 1 at $\theta = \pi/2$, the light in Core 1 at $\theta = \pi/2$ couples to the eigenmode localized in Core 1 in $\theta > \pi/2$, which is the odd mode.

Group delay spread The preceding simplification cannot hold when the bend-induced DGD is not dominant or the core count is more than two. In this case, the DGD has to be numerically calculated. In the same manner as the GDs of randomly coupled polarizations are defined [98], the GDs of randomly coupled modes can be defined [99], which is reviewed in this subsection.

When a lossless system is assumed for the sake of simplicity, an input state $|E_{\text{in}}\rangle$ and an output state $|E_{\text{out}}\rangle$ of the modes are related by the unitary transfer matrix $\mathbf{T}_{\text{tot}}$ of an RC-MCF as

$$|E_{\text{out}}\rangle = \mathbf{T}_{\text{tot}}|E_{\text{in}}\rangle. \tag{1.58}$$

The generalized (unitary) Jones matrix $\mathbf{J}_{\text{tot}}$ corresponding to $\mathbf{T}_{\text{tot}}$ can be defined as

$$\mathbf{T}_{\text{tot}} = \exp(-j\phi_0)\mathbf{J}_{\text{tot}}, \tag{1.59}$$

where ϕ_0 is the common phase. The mean τ_0 of the GDs of the modes can be expressed as

$$\tau_0 = \frac{\partial \phi_0}{\partial \omega}. \tag{1.60}$$

Since $\mathbf{J}_{\text{tot}}$ is unitary, the following relationship holds:

$$\mathbf{J}_{\text{tot}}\mathbf{J}_{\text{tot}}^{*} = \mathbf{I}, \tag{1.61}$$

where $\mathbf{I}$ is the identity matrix.

By using Eqs. (1.58−1.61), we can express the derivative of $|E_{\text{out}}\rangle$ with respect to ω as

$$\frac{\partial|E_{\text{out}}\rangle}{\partial\omega} = -j(\tau_0\mathbf{I} + \mathbf{G})|E_{\text{out}}\rangle, \tag{1.62}$$

$$\mathbf{G} = j\frac{\partial\mathbf{J}_{\text{tot}}}{\partial\omega}\mathbf{J}_{\text{tot}}^{*}, \tag{1.63}$$

where the operator $\mathbf{G}$ is referred to as "group-delay operator." All the eigenvalues of $\mathbf{G}$ are real, because $\mathbf{G}$ is Hermitian ($\mathbf{G} = \mathbf{G}^*$), which can be proved by differentiating Eq. (1.61) with respect to ω.

Now, substituting an eigenvector of $\mathbf{G}$ to $|E_{\text{out}}\rangle$ in Eq. (1.62) yields

$$\frac{\partial|E_{\text{out}}\rangle_n}{\partial\omega} = -j(\tau_0\mathbf{I} + \mathbf{G})|E_{\text{out}}\rangle_n = -j(\tau_0 + \tau_n)|E_{\text{out}}\rangle_n, \tag{1.64}$$

where $|E_{\text{out}}\rangle_n$ and τ_n are an eigenvector and its corresponding eigenvalue of the group-delay operator $\mathbf{G}$, respectively. From Eq. (1.64), $\tau_0 + \tau_n$ is the GD that is independent of ω (zero dispersion) to first order when the output state of the modes is an eigenvector $|E_{\text{out}}\rangle_n$. Therefore, every eigenvector of the group-delay operator $\mathbf{G}$ can be understood as an output "principal mode" [99] in analogy to "principal state of polarization."

The GD spread of an RC-MCF can be evaluated by the standard deviation (STD) of the GDs of the principal modes. If a RC-MCF (or MMF in general) can be modeled as concatenated K uncoupled sections with equal length and group delay variance $\sigma_\tau{}^2$, and strong mode mixing occurs between each section, the overall GD spread can be expressed as [100]

$$\sigma_{\tau_{\text{g,tot}}} = \sqrt{K}\sigma_\tau. \tag{1.65}$$

Equivalently, the GD spread is square root proportional to the fiber length. Thus, the GD spread of RC-MCFs are in square root proportion to the fiber length, as observed in various transmission experiments and fiber characterizations [79,97,101,102].

Mode-dependent loss Mode-dependent loss in an RC-MCF (or MMF in general) is described by the singular values of the overall transfer matrix of the fiber. Using singular value decomposition, the overall transfer matrix $\mathbf{M}$ can be decomposed as

$$\mathbf{M} = \mathbf{V}\mathbf{\Lambda}\mathbf{U}*, \tag{1.66}$$

where **U** and **V** are the input and output mode coupling matrices, and **Λ** is the diagonal matrix with the singular values (Λ_i) in its diagonal elements. Overall MDL can be defined as the STD of the log powers of the singular values ($\ln\Lambda_i^2$) [103], which can be expressed as

$$\sigma_{\mathrm{mdl}} = \sqrt{E\left[\left(\ln \Lambda^2 - E\left[\ln \Lambda^2\right]\right)^2\right]} = \sqrt{\frac{1}{D}\sum_{i=1}^{D}\left(\ln \Lambda_i^2 - \frac{1}{D}\sum_{j=1}^{D}\ln \Lambda_j^2\right)^2}, \tag{1.67}$$

where E[x] represents the expected value of x, and D is the number of the modes including both polarization and spatial modes. σ_{mdl} in decibel can be obtained by multiplying $10/\ln 10 \sim 4.34$ to σ_{mdl} in log power. Ho and Kahn investigated the statistics of MDL in the random coupling (strong coupling) regime and revealed that when the overall MDL is small, the STD of the overall MDL σ_{mdl} can be expressed as [103]

$$\sigma_{\mathrm{mdl}} = \xi\sqrt{1 + \frac{\xi^2}{12(1 - D^{-2})}}, \tag{1.68}$$

where ξ is the square root of the accumulated MDL variance, defined such that ξ^2 is the sum of the MDL variances $\sigma_{\mathrm{mdl,uc}}{}^2$ of uncoupled fiber sections. So, if an RC-MCF (or MMF in general) can be modeled as concatenated K uncoupled sections with equal MDL variance $\sigma_T{}^2$, and the strong mode mixing occurs between each section, ξ can be expressed as [103]

$$\xi = \sqrt{K}\sigma_{\mathrm{mdl,uc}}. \tag{1.69}$$

Overall MDL of the RC-MCF only depends on ξ and D via Eq. (1.58). When σ_{mdl} is lower than 10 dB and $D \geq 4$ (2 or more spatial modes with two polarizations), Eq. (1.58) can be approximated by its limit for $D \rightarrow \infty$ as

$$\sigma_{\mathrm{mdl}} \simeq \xi\sqrt{1 + \xi^2/12}, \tag{1.70}$$

with a sufficient accuracy of $\leq 1\%$ error. Since the square root term in Eqs. (1.58) and (1.60) can be approximated as

$$\sqrt{1 + \frac{\xi^2}{12(1 - D^{-2})}} \approx \begin{cases} 1, & \text{for } \xi^2 \ll 12\left(1 - D^{-2}\right), \\ \dfrac{\xi}{\sqrt{12(1 - D^{-2})}}, & \text{for } \xi^2 \gg 12\left(1 - D^{-2}\right), \end{cases} \tag{1.71}$$

the overall MDL σ_{mdl} may be in square root proportion to the fiber length when the MDL is sufficiently low (1–2 dB or lower), and in linear proportion to the fiber length when the MDL is sufficiently high (1000 s dB or higher). However, within the range of practical interest (MDL < a few 10 s dB even for few-mode fibers [104]), the overall MDL proportionality to the fiber length is near square root and well below linear.

1.3.3 Various MCFs proposed for communications and progress toward practical realization

Along with the elucidation of the MCF characteristics, a variety of MCFs have been proposed and demonstrated. Some of the representative examples of actually demonstrated MCFs are summarized in Fig. 1.22. Note that due to limited space only limited examples are shown and many other characteristic MCFs have been proposed.

As shown in the SC-MCF section, few MCFs have been actually demonstrated for communications so far [105–107], even though some theoretical works have proposed and investigated the SC-MCF in the very early stage of the SDM fiber research [94,108]. This might be because most of the fiber manufacturers have focused on other types of the SDM fibers, such as UC-MCFs, RC-MCFs, and non-MCF few-mode fibers, which are easier to fabricate and use. The lack of low-XT mode multiplexers/demultiplexers (MUX/DEMUX) for eigenmodes also presents a challenge for the fiber research. Though some MUX/DEMUX methods have been proposed [94], no low-XT MUX/DEMUX have actually been demonstrated. Once the low-XT MUX/DEMUX for eigenmodes are demonstrated, the research for the SC-MCF might be accelerated.

In the RC-MCF section, many fibers with different core counts are shown but these fibers can be grouped into three types: (1) MCFs with a single group of coupled multiple single-mode cores [79,101,102,109], (2) MCFs with multiple uncoupled groups of coupled multiple single-mode cores [110], and (3) MCFs with a single group of coupled multiple few-mode cores [111].

Type-1 RC-MCF is the major type of the RC-MCFs, and various long-haul or real-time MIMO DSP transmission experiments have been conducted using these RC-MCFs [96,101,102,112]. Among the SDM fibers, the lowest loss of 0.158 dB/km at 1550 nm and the lowest SMD of 3.14 ps/$\sqrt{}$km have been realized by an RC-MCF with four Ge-free silica cores with enlarged effective core-mode areas of ~112 μm^2 at 1550 nm [79]. This fiber was confirmed to outperform the equivalent low-loss large-effective-area SMF in nonlinearity tolerance and transmission performance [96]. Up to 12 cores can be packed into the standard 125-μm-cladding with suppressed SMD [109]. The calculation complexity of the MIMO DSP would be an issue if the total spatial channel count increases in Type-1 RC-MCFs, but dividing the spatial channels into multiple uncoupled groups like Type-2 RC-MCF helps the MIMO complexity suppression. In Type-3 RC-MCF, coupled 3-mode (2-LP-mode) 7-core fiber was proposed for further increasing the spatial density in the standard-diameter cladding

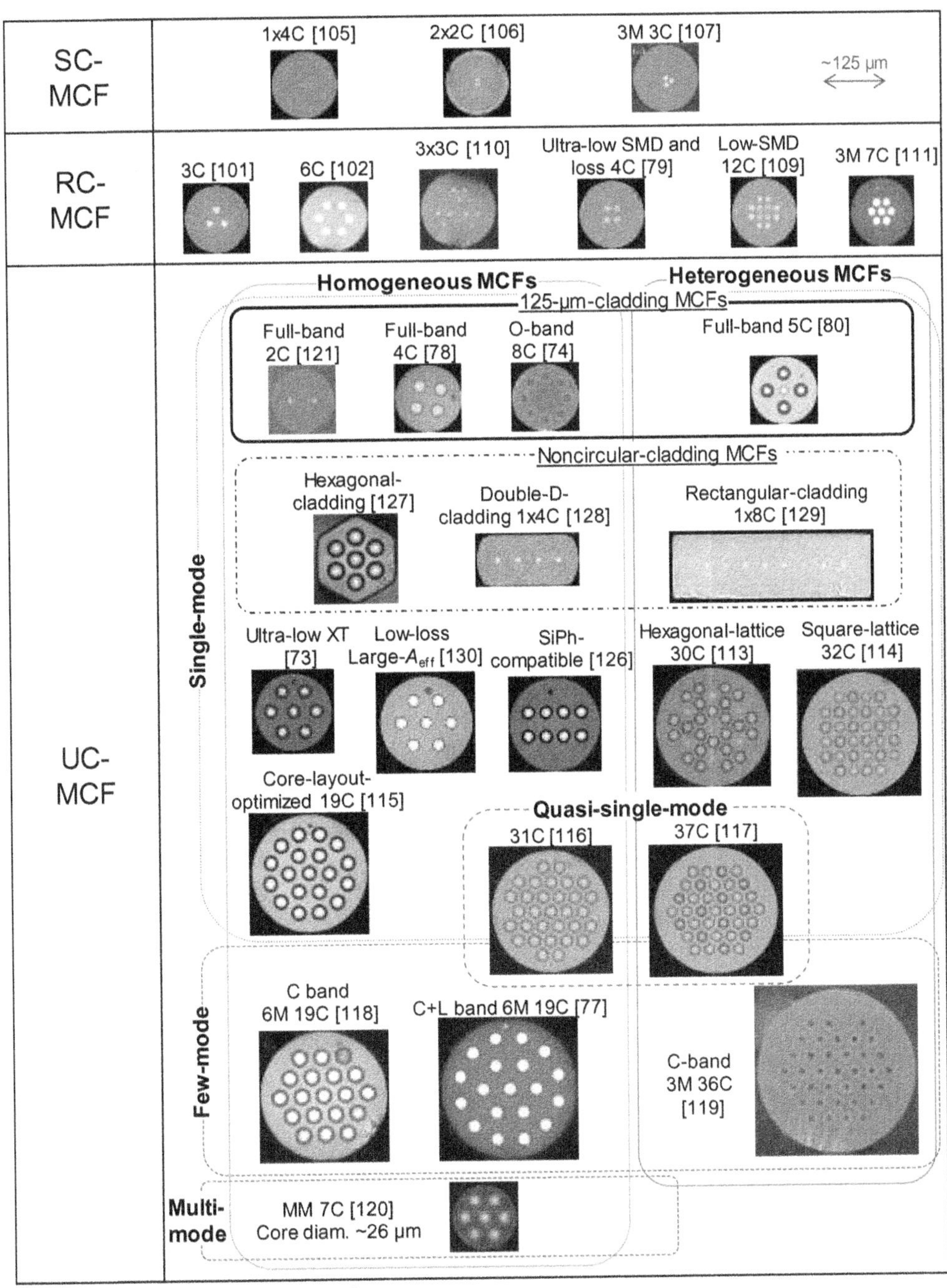

Figure 1.22 Representative examples of various types of actually demonstrated MCFs. Source: *The figures in the table are adapted from [73,74], [77–80], [101,102], [105–107], [109–111], [113–121], [126–130].*

[111]. Although the intracore coupling between LP_{01}-based modes and LP_{11}-based modes were rather suppressed, the intercore coupling within the LP_{01}-based modes or within the LP_{11}-based modes well suppressed intercore skew within each (LP_{01}-based or LP_{11}-based) mode group.

In the UC-MCF section, we see a variety of MCFs. The homogeneous MCFs with identical cores are preferable in optical characteristics homogeneity among the cores, and heterogeneous MCFs with dissimilar cores are preferable in XT suppression. MCFs with single-mode cores are more practical, thanks to the compatibility to the conventional SMF transceivers, and MCFs with few-mode cores are suited for achieving extremely high capacity using ultrahigh number of spatial channels. Some MCFs employ the quasi-single-mode (QSM) cores (i.e., few-mode cores where only the fundamental mode is used for transmission) for improving the light confinement to the cores and suppressing the XT.

From the difference in the research directions, these uncoupled MCFs can be categorized into two groups:

A. High-spatial-channel-count MCFs for extremely high capacity
B. MCFs proposed for near-term practical realization

Type-A MCFs are basically designed to have more than 10 cores and cladding diameter thicker than 200 μm [77,113–119], for achieving ultrahigh transmission capacity per fiber. For example, the highest fiber transmission capacity of 10.16 Pb/seconds was also achieved using an FM-MCF (6-mode 19-core fiber) over C + L band (1530–1625 nm) [77], but there are still many technical/economic challenges for their practical realization, such as DGD/MDL suppression in the FM-MCFs.

Type-B MCFs can be further divided into three types by their strategies for practical realization. The first type is MCFs with the standard 125-μm-diameter cladding [74,78,80,120,121]. The standard 125-μm-diameter cladding achieves the field-proven mechanical reliability and can suppress production cost increase because a longer fiber can be drawn from a preform compared with a thicker fiber. The 125-μm-cladding MCFs are also compatible with the conventional drawing and cabling technologies [74,122], and connector ferrules [123]. The optical properties, including the XT, have been reported to show no degradation due to cabling [122,124]. Interoperability demonstration achieved 118.5 Tb/seconds transmission over 316 km using 125-μm-cladding 4-core fibers and MCF connectors fabricated by three fiber manufacturers with common specifications, and multicore erbium-doped amplifiers fabricated by three companies [125]. The second type is MCFs with a core layout compatible to transceivers, which enables end-to-end MCF transmission links. For example, the 2 × 4-core-layout MCF was codesigned with Si Photonics die so that the core array can match the grating coupler array [126]. The arrays of vertical cavity surface emitting lasers (VCSELs) and photo diodes were developed for 125-μm-cladding multimode 7-core fiber so that the outer 6 cores can match the VCSEL and PD arrays [120]. The third type is MCFs with a noncircular cladding for easy rotational alignment [127–129]. Noncircular cladding realizes passive rotational alignment, but the manufacturing tolerances in core positions and cladding dimensions limit the alignment accuracy. In addition, the drawing speed is limited for avoiding cladding shape deformation due to the surface tension

References

[1] F.P. Kapron, D.B. Keck, R.D. Maurer, Radiation losses in glass optical waveguides, Appl. Phys. Lett. 17 (10) (1970) 423–425.
[2] D.B. Keck, R.D. Maurer, P.C. Schultz, On the ultimate lower limit of attenuation in glass optical waveguides, Appl. Phys. Lett. 22 (7) (1973) 307–309.
[3] J.B. MacChesney, D.B. O'Connor, F.V. Dimarcello, J.B. Simpson, P.D. Lazay, Preparation of low-loss optical fibers using simultaneous vapor deposition and fusion, in: Proc. 10th Int. Congr. on Glass, Kyoto, Japan, 1974, vol. 6, pp. 40–44.
[4] M. Horiguchi, H. Osanai, Spectral losses of low-OH-content optical fibres, Electron. Lett. 12 (12) (1976) 310–312.
[5] T. Miya, Y. Terunuma, T. Hosaka, T. Miyashita, Ultimate low-loss single-mode fibre at 1.55 μm, Electron. Lett. 15 (4) (1979) 106–108.
[6] H. Kanamori, et al., Transmission characteristics and reliability of pure-silica-core single-mode fibers, J. Lightw. Technol. 4 (8) (1986) 1144–1150.
[7] K. Nagayama, M. Kakui, M. Matsui, I. Saitoh, Y. Chigusa, Ultra-low-loss (0.1484 dB/km) pure silica core fibre and extension of transmission distance, Electron. Lett. 38 (20) (2002) 1168–1169.
[8] M. Hirano, et al., Record low loss, record high FOM optical fiber with manufacturable process, in: Opt. Fiber Commun. Conf. (OFC), 2013, p. PDP5A.7.
[9] S. Makovejs, et al., Record-low (0.1460 dB/km) attenuation ultra-large Aeff optical fiber for submarine applications, Opt. Fiber Commun. Conf. (OFC), 2015, p. Th5A.2.
[10] Y. Tamura, et al., Lowest-ever 0.1419-dB/km loss optical fiber, in: Opt. Fiber Commun. Conf. (OFC), 2017, p. Th5D.1.
[11] R.J. Sanferrare, Terrestrial lightwave systems, AT&T Tech. J. 66 (1) (1987) 95–107.
[12] B. Ainslie, K. Beales, C. Day, J. Rush, The design and fabrication of monomode optical fiber, IEEE J. Quantum Electron. 18 (4) (1982) 514–523.
[13] P. Lazay, A. Pearson, Developments in single-mode fiber design, materials, and performance at Bell Laboratories, IEEE J. Quantum Electron. 18 (4) (1982) 504–510.
[14] T. Li (Ed.), Optical Fiber Communications, Vol. 1: Fiber Fabrication, Academic, New York, 1985.
[15] L.G. Cohen, C. Lin, W.G. French, Tailoring zero chromatic dispersion into the 1.5-1.6 μm low-loss spectral region of single-mode fibres, Electron. Lett. 15 (12) (1979) 334–335.
[16] P. Francois, Zero dispersion in attenuation optimized doubly clad fibers, J. Lightw. Technol 1 (1) (1983) 26–37.
[17] V.A. Bhagavatula, M.S. Spotz, D.E. Quinn, Uniform waveguide dispersion segmented-core designs for dispersion-shifted single-mode fibers, in: Opt. Fiber Commun. Conf. (OFC), 1984, p. MG2.
[18] H. Taga, S. Yamamoto, K. Mochizuki, H. Wakabayashi, 2.4 Gbit/s 1.55 μm WDM/bidirectional optical fiber transmission experiments, Trans. IEICE E71 (10) (1988) 940–942.
[19] R.W. Tkach, A.R. Chraplyvy, F. Forghieri, A.H. Gnauck, R.M. Derosier, Four-photon mixing and high-speed WDM systems, J. Lightw. Technol. 13 (5) (1995) 841–849.
[20] P. Nouchi, et al., Low-loss single-mode fiber with high nonlinear effective area, in: Opt. Fiber Commun. Conf. (OFC), 1995, p. ThH2.
[21] V.L.D. Silva, Y. Liu, D.Q. Chowdhury, M. Li, A.J. Antos, A.F. Evans, Error free WDM transmission of 8 × 10 Gbit/s over km of LEAF™ optical fiber, in: Eur. Conf. Opt. Commun. (ECOC), 1997, pp. 154–158.

[22] D.W. Peckham, A.F. Judy, R.B. Kummer, Reduced dispersion slope, non-zero dispersion fiber, in: Eur. Conf. Opt. Commun. (ECOC), 1998, pp. 139–140.
[23] Y. Liu, A.J. Antos, Dispersion-shifted large-effective-area fiber for amplified high-capacity long-distance systems, in: Opt. Fiber Commun. Conf. (OFC), 1997, p. TuN5.
[24] J.M. Kahn, K.-P. Ho, Spectral efficiency limits and modulation/detection techniques for DWDM systems, IEEE J. Sel. Top. Quantum Electron. 10 (2) (2004) 259–272.
[25] S. Tsukamoto, K. Katoh, K. Kikuchi, Coherent demodulation of optical multilevel phase-shift-keying signals using homodyne detection and digital signal processing, IEEE Photon. Technol. Lett. 18 (10) (2006) 1131–1133.
[26] R.-J. Essiambre, R.W. Tkach, Capacity trends and limits of optical communication networks, Proc. IEEE 100 (5) (2012) 1035–1055.
[27] D.J. Richardson, J.M. Fini, L.E. Nelson, Space-division multiplexing in optical fibres, Nat. Photon. 7 (2013) 354–362.
[28] P. Poggiolini, The GN model of non-linear propagation in uncompensated coherent optical systems, J. Lightw. Technol. 30 (24) (2012) 3857–3879.
[29] M. Hirano, Y. Yamamoto, V. Sleiffer, T. Sasaki, Analytical OSNR formulation validated with 100G-WDM experiments and optimal subsea fiber proposal, in: Opt. Fiber Commun. Conf. (OFC), 2013, p. OTu2B.6.
[30] Y. Yamamoto, M. Hirano, V. a. J. M. Sleiffer, T. Sasaki, Analytical OSNR formulation considering nonlinear compensation, in: OptoElectron. Commun. Conf./Photonics in Switching (OECC/PS), 2013, p. WR4-3.
[31] V. Curri, et al., Fiber figure of merit based on maximum reach, in: Opt. Fiber Commun. Conf. (OFC), 2013, p. OTh3G.2.
[32] E.G. Rawson, Measurement of the angular distribution of light scattered from a glass fiber optical waveguide, Appl. Opt. 11 (11) (1972) 2477–2481.
[33] E.G. Rawson, Analysis of scattering from fiber waveguides with irregular core surfaces, Appl. Opt. 13 (10) (1974) 2370–2377.
[34] M. Tateda, M. Ohashi, K. Tajima, K. Shiraki, Design of viscosity-matched optical fibers, IEEE Photon. Technol. Lett. 4 (9) (1992) 1023–1025.
[35] M. Ohashi, M. Tateda, K. Shiraki, K. Tajima, Imperfection loss reduction in viscosity-matched optical fibers, IEEE Photon. Technol. Lett. 5 (7) (1993) 812–814.
[36] W.B. Mattingly, III, S.K. Mishra, L. Zhang, Low attenuation optical fiber, US7171090B2, Jan 30, 2007.
[37] L. Maksimov, A. Anan'ev, V. Bogdanov, T. Markova, V. Rusan, O. Yanush, Inhomogeneous structure of inorganic glasses studied by Rayleigh, Mandel'shtam-Brillouin, Raman scattering spectroscopy, and acoustic methods, IOP Conf. Ser.: Mater. Sci. Eng 25 (1) (2011) 012010.
[38] M.E. Lines, Can the minimum attenuation of fused silica be significantly reduced by small compositional variations? I. Alkali metal dopants, J. Non-Cryst Solids 171 (3) (1994) 209–218.
[39] H. Kakiuchida, E.H. Sekiya, K. Saito, A.J. Ikushima, Effect of chlorine on Rayleigh scattering reduction in silica glass, Jpn. J. Appl. Phys. 42 (12B) (2003) L1526.
[40] Y. Tamura, et al., The first 0.14-dB/km loss optical fiber and its impact on submarine transmission, J. Lightw. Technol. 36 (1) (2018) 44–49.
[41] S.R. Bickham, Ultimate limits of effective area and attenuation for high data rate fibers, in: Opt. Fiber Commun. Conf. (OFC), 2011, p. OWA5.
[42] R. Olshansky, Distortion losses in cabled optical fibers, Appl. Opt. 14 (1) (1975) 20.
[43] M. Tsukitani, M. Matsui, K. Nagayama, E. Sasaoka, Ultra low nonlinearity pure-silica-core fiber with an effective area of 211 μm^2 and transmission loss of 0.159 dB/km, in: Eur. Conf. Opt. Commun. (ECOC), Copenhagen, 2002, p. 3.2.2.
[44] F. Yaman, N. Bai, B. Zhu, T. Wang, G. Li, Long distance transmission in few-mode fibers, Opt. Express 18 (12) (2010) 13250–13257.
[45] F. Yaman, et al., 10x112Gb/s PDM-QPSK transmission over 5032 km in few-mode fibers, Opt. Express 18 (20) (2010) 21342–21349.
[46] Q. Sui, et al., Long-haul quasi-single-mode transmissions using few-mode fiber in presence of multi-path interference, Opt. Express 23 (3) (2015) 3156–3169.

[47] F. Yaman, et al., First quasi-single-mode transmission over transoceanic distance using few-mode fibers, in: Opt. Fiber Commun. Conf. (OFC), 2015, p. Th5C.7.
[48] J.D. Downie, et al., Quasi-single-mode fiber transmission for optical communications, IEEE J. Sel. Top. Quantum Electron. 23 (3) (2017) 31−42.
[49] J.D. Downie, J. Hurley, R. Nagarajan, T. Maj, H. Dong, S. Makovejs, 100 Gb/s wavelength division multiplexing four-level pulse amplitude modulated transmission over 160 km using advanced optical fibres, Electron. Lett 54 (11) (2018) 699−701.
[50] J.D. Downie, et al., 40 × 112 Gb/s transmission over an unrepeatered 365 km effective area-managed span comprised of ultra-low loss optical fibre, in: Eur. Conf. Opt. Commun. (ECOC), Torino, Italy, 2010, p. We.7.C.5.
[51] P. Bousselet, H. Bissessur, J. Lestrade, M. Salsi, L. Pierre, D. Mongardien, High capacity (64x 43Gb/s) unrepeatered transmission over 440 km, in: Opt. Fiber Commun. Conf. (OFC), 2011, p. OMI2.
[52] J.D. Downie, J. Hurley, I. Roudas, D. Pikula, J.A. Garza-Alanis, Unrepeatered 256 Gb/s PM-16QAM transmission over up to 304 km with simple system configurations, Opt. Express 22 (9) (2014) 10256−10261.
[53] D. Chang, et al., 150 x 120 Gb/s unrepeatered transmission over 409.6 km of large effective area fiber with commercial Raman DWDM system, Opt. Express, OE 22 (25) (2014) 31057−31062.
[54] Y. Huang, et al., 20.7-Tb/s repeater-less transmission over 401.1-km using QSM fiber and XPM compensation via transmitter-side DBP, in: OptoElectron. Commun. Conf./Photonics in Switching (OECC/PS), 2016, p. PD1-4.
[55] B. Zhu, et al., 800Gb/s (8x128Gb/s) unrepeatered transmission over 515-km large-area ultra-low-loss fiber using 2nd-order Raman pumping, Opt. Express, OE 24 (22) (2016) 25291−25297.
[56] D. Chang, W. Pelouch, S. Burtsev, P. Perrier, H. Fevrier, High capacity 150 × 120 Gb/s transmission over a cascade of two spans with a total loss of 118 dB, in: Opt. Fiber Commun. Conf. (OFC), 2016, p. Th1B.6.
[57] Y.-K. Huang, et al., Real-time 8 × 200-Gb/s 16-QAM unrepeatered transmission over 458.8 km using concatenated receiver-side ROPAs, in: Opt. Fiber Commun. Conf. (OFC), 2017, p. Th2A.59.
[58] C. Zhang, et al., 112 Gb/s PDM-QPSK transmission over 3000 km of G.652 ultra-low-loss fiber with 125 km EDFA amplified spans and coherent detection, in: 16th Opto-Electronics and Communications Conference, 2011, pp. 210−211.
[59] J. Downie, J.E. Hurley, J. Cartledge, S.R. Bickham, S. Mishra, Transmission of 112 Gb/s PM-QPSK signals over 7200 km of optical fiber with very large effective area and ultra-low loss in 100 km spans with EDFAs only, in: Opt. Fiber Commun. Conf. (OFC), 2011, p. OMI6.
[60] J.D. Downie, J. Hurley, J. Cartledge, S. Bickham, S. Mishra, 112 Gb/s PM-QPSK transmission up to 6000 km with 200 km amplifier spacing and a hybrid fiber span configuration, Opt. Express 19 (26) (2011) B96−B101.
[61] M. Salsi, C. Koebele, P. Tran, H. Mardoyan, S. Bigo, G. Charlet, 80 × 100-Gbit/s transmission over 9,000 km using erbium-doped fibre repeaters only, in: Eur. Conf. Opt. Commun. (ECOC), Torino, Italy, 2010, p. We.7.C.3.
[62] D. Qian, et al., Transmission of 115 × 100G PDM-8QAM-OFDM channels with 4 bits/s/Hz spectral efficiency over 10,181 km, in: Eur. Conf. Opt. Commun. (ECOC), Geneva, 2011, p. Th.13. K.3.
[63] D. Qian, et al., 30 Tb/s C- and L-bands bidirectional transmission over 10,181 km with 121 km span length, Opt. Express 21 (12) (2013) 14244−14250.
[64] J.D. Downie, 112 Gb/s PM-QPSK transmission systems with reach lengths enabled by optical fibers with ultra-low loss and very large effective area, in: Proc. SPIE, 2012, vol. 8284, p. 828403.
[65] J.D. Downie, J. Hurley, D. Pikula, Transmission of 256 Gb/s PM-16QAM and 128 Gb/s PM-QPSK signals over long-haul and submarine systems with span lengths greater than 100 km, in: Eur. Conf. Opt. Commun. (ECOC), 2013, p. Tu.1.D.3.
[66] J.-X. Cai, et al., Transmission over 9,100 km with a capacity of 49.3 Tb/s using variable spectral efficiency 16 QAM based coded modulation, in: Opt. Fiber Commun. Conf. (OFC), 2014, p. Th5B.4.

[67] S. Zhang et al., Capacity-approaching transmission over 6375 km at spectral efficiency of 8.3 bit/s/Hz, in: Opt. Fiber Commun. Conf. (OFC), 2016, p. Th5C.2.
[68] T. Hayashi, Multi-core optical fibers, in: A.I.P. Kaminow, T. Li, A.E. Willner (Eds.), Optical Fiber Telecommunications, sixth ed, Academic Press, New York, 2013, pp. 321–352.
[69] T. Hayashi, T. Sasaki, E. Sasaoka, K. Saitoh, M. Koshiba, Physical interpretation of intercore crosstalk in multicore fiber: effects of macrobend, structure fluctuation, and microbend, Opt. Express 21 (5) (2013) 5401–5412.
[70] T. Hayashi, Multi-core fibers for space division multiplexing, in: G.-D. Peng (Ed.), Handbook of Optical Fibers, Springer, Singapore, 2018, pp. 1–46.
[71] K. Takenaga, et al., Reduction of crosstalk by trench-assisted multi-core fiber, in: Opt. Fiber Commun. Conf. (OFC), 2011, p. OWJ4.
[72] B. Zhu, T.F. Taunay, M.F. Yan, J.M. Fini, M. Fishteyn, E.M. Monberg, Seven-core multicore fiber transmissions for passive optical network, Opt. Express 18 (11) (2010) 11117–11122.
[73] T. Hayashi, T. Taru, O. Shimakawa, T. Sasaki, E. Sasaoka, Design and fabrication of ultra-low crosstalk and low-loss multi-core fiber, Opt. Express 19 (17) (2011) 16576–16592.
[74] T. Hayashi, et al., 125-μm-cladding eight-core multi-core fiber realizing ultra-high-density cable suitable for O-band short-reach optical interconnects, J. Lightw. Technol. 34 (1) (2016) 85–92.
[75] B.J. Puttnam, et al., 2.15 Pb/s transmission using a 22 core homogeneous single-mode multi-core fiber and wideband optical comb, in: Eur. Conf. Opt. Commun. (ECOC), 2015, p. PDP.3.1.
[76] T. Kobayashi, et al., 1-Pb/s (32 SDM/46 WDM/768 Gb/s) C-band dense SDM transmission over 205.6-km of single-mode heterogeneous multi-core fiber using 96-gbaud PDM-16QAM channels, in: Opt. Fiber Commun. Conf. (OFC), 2017, p. Th5B.1.
[77] D. Soma, et al., 10.16 Peta-bit/s Dense SDM/WDM transmission over Low-DMD 6-Mode 19-Core Fibre across C + L Band, in: Eur. Conf. Opt. Commun. (ECOC), Gothenburg, 2017, p. Th. PDP.A.1.
[78] T. Matsui, et al., Design of 125 μm cladding multi-core fiber with full-band compatibility to conventional single-mode fiber, in: Eur. Conf. Opt. Commun. (ECOC), Valencia, 2015, p. We.1.4.5.
[79] T. Hayashi, Y. Tamura, T. Hasegawa, T. Taru, 125-μm-cladding coupled multi-core fiber with ultra-low loss of 0.158 dB/km and record-low spatial mode dispersion of 6.1 ps/km$^{1/2}$, in: Opt. Fiber Commun. Conf. (OFC), 2016, p. Th5A.1.
[80] T. Gonda, K. Imamura, R. Sugizaki, Y. Kawaguchi, T. Tsuritani, 125 μm 5-core fibre with heterogeneous design suitable for migration from single-core system to multi-core system, in:, Eur. Conf. Opt. Commun. (ECOC), Düsseldorf (2016) 547–549.
[81] M. Koshiba, K. Saitoh, K. Takenaga, S. Matsuo, Analytical expression of average power-coupling coefficients for estimating intercore crosstalk in multicore fibers, IEEE Photon. J. 4 (5) (2012) 1987–1995.
[82] G. Rademacher, R.S. Luís, B.J. Puttnam, Y. Awaji, N. Wada, Crosstalk dynamics in multi-core fibers, Opt. Express 25 (10) (2017) 12020–12028.
[83] G. Rademacher, R.S. Luis, B.J. Puttnam, Y. Awaji, N. Wada, Crosstalk fluctuations in homogeneous multi-core fibers, in: Photonic Networks and Devices, New Orleans, 2017, p. NeTu2B.4.
[84] R.S. Luis, et al., Time and modulation frequency dependence of crosstalk in homogeneous multi-core fibers, J. Lightw. Technol. 34 (2) (2016) 441–447.
[85] A.V.T. Cartaxo, T.M.F. Alves, Discrete changes model of inter-core crosstalk of real homogeneous multi-core fibers, J. Lightw. Technol. 35 (12) (2017) 2398–2408.
[86] G. Rademacher, B.J. Puttnam, R.S. Luis, Y. Awaji, N. Wada, Time-dependent crosstalk from multiple cores in a homogeneous multi-core fiber, in: Opt. Fiber Commun. Conf. (OFC), 2017, p. Th1H.3.
[87] D. Marcuse, "Coupled power theory," in Theory of Dielectric Optical Waveguides, second ed, Academic Press, San Diego, 1991.
[88] J.M. Fini, B. Zhu, T.F. Taunay, M.F. Yan, Statistics of crosstalk in bent multicore fibers, Opt. Express 18 (14) (2010) 15122–15129.
[89] T. Hayashi, T. Nagashima, O. Shimakawa, T. Sasaki, E. Sasaoka, Crosstalk variation of multi-core fibre due to fibre bend, in: Eur. Conf. Opt. Commun. (ECOC), Torino, 2010, p. We.8.F.6.

[90] K. Takenaga, et al., An investigation on crosstalk in multi-core fibers by introducing random fluctuation along longitudinal direction, IEICE Trans. Commun. E94.B (2) (2011) 409–416.
[91] M. Koshiba, K. Saitoh, K. Takenaga, S. Matsuo, Multi-core fiber design and analysis: coupled-mode theory and coupled-power theory, Opt. Express 19 (26) (2011) B102–B111.
[92] D. Marcuse, Influence of curvature on the losses of doubly clad fibers, Appl. Opt. 21 (23) (1982) 4208–4213.
[93] T. Hayashi, T. Sasaki, E. Sasaoka, Behavior of inter-core crosstalk as a noise and its effect on Q-Factor in multi-core fiber, IEICE Trans. Commun. E97.B (5) (2014) 936–944.
[94] Y. Kokubun, M. Koshiba, Novel multi-core fibers for mode division multiplexing: proposal and design principle, IEICE Electron. Express 6 (8) (2009) 522–528.
[95] T. Hayashi, Coupled multicore fiber for space-division multiplexed transmission, in: Proc. SPIE, Next-Generation Optical Communication: Components, Sub-Systems, and Systems VI, San Francisco, 2017, vol. 10130, p. 1013003.
[96] R. Ryf, et al., Long-haul transmission over multi-core fibers with coupled cores, in: Eur. Conf. Opt. Commun. (ECOC), Gothenburg, 2017, p. M.2.E.1.
[97] T. Sakamoto, T. Mori, M. Wada, T. Yamamoto, F. Yamamoto, K. Nakajima, Fiber twisting and bending induced adiabatic/nonadiabatic super-mode transition in coupled multi-core fiber, J. Lightw. Technol. 34 (4) (2016) 1228–1237.
[98] J.P. Gordon, H. Kogelnik, PMD fundamentals: polarization mode dispersion in optical fibers, PNAS 97 (9) (2000) 4541–4550.
[99] S. Fan, J.M. Kahn, Principal modes in multimode waveguides, Opt. Lett. 30 (2) (2005) 135–137.
[100] K.P. Ho, J.M. Kahn, Statistics of group delays in multimode fiber with strong mode coupling, J. Lightw. Technol. 29 (21) (2011) 3119–3128.
[101] R. Ryf, et al., Space-division multiplexed transmission over 4200 km 3-core microstructured fiber, in: Opt. Fiber Commun. Conf. (OFC), 2012, p. PDP5C.2.
[102] R. Ryf, et al., 1705-km transmission over coupled-core fibre supporting 6 spatial modes, in: Eur. Conf. Opt. Commun. (ECOC), Cannes, 2014, p. PD.3.2.
[103] K.-P. Ho, J.M. Kahn, Mode-dependent loss and gain: statistics and effect on mode-division multiplexing, Opt. Express 19 (17) (2011) 16612–16635.
[104] G. Rademacher, et al., 159 Tbit/s C + L band transmission over 1045 km 3-mode graded-index few-mode fiber, in: Opt. Fiber Commun. Conf. (OFC), 2018, p. Th4C.4.
[105] Y. Kokubun, T. Komo, K. Takenaga, S. Tanigawa, S. Matsuo, Selective mode excitation and discrimination of four-core homogeneous coupled multi-core fiber, Optics express 19 (26) (2011) B905–B914.
[106] S. Saitoh, K. Takenaga, K. Aikawa, Demonstration of a rectangularly-arranged strongly-coupled multi-core fiber, in: Opt. Fiber Commun. Conf. (OFC), 2018, p. Th2A.22.
[107] C. Xia, et al., Supermodes in coupled multi-core waveguide structures, IEEE J. Sel. Top. Quantum Electron. 22 (2) (2016) 4401212.
[108] C. Xia, N. Bai, I. Ozdur, X. Zhou, G. Li, Supermodes for optical transmission, Opt. Express 19 (17) (2011) 16653–16664.
[109] T. Sakamoto, et al., Randomly-coupled single-mode 12-core fiber with highest core density, in: Opt. Fiber Commun. Conf. (OFC), 2017, p. Th1H.1.
[110] R. Ryf, et al., Space-division multiplexed transmission over 3 × 3 coupled-core multicore fiber, in: Opt. Fiber Commun. Conf. (OFC), San Francisco, 2014, p. Tu2J.4.
[111] T. Sakamoto, T. Mori, M. Wada, T. Yamamoto, F. Yamamoto, K. Nakajima, Coupled few-mode multi-core fibre for ultra-high spatial density space division multiplexing, in: Eur. Conf. Opt. Commun. (ECOC), Düsseldorf, 2016, pp. 553–555.
[112] S. Randel, et al., First real-time coherent MIMO-DSP for six coupled mode transmission, in: IEEE Photon. Conf. (IPC), Reston, 2015, pp. 1–2.
[113] Y. Amma, et al., High-density multicore fiber with heterogeneous core arrangement, in: Opt. Fiber Commun. Conf. (OFC), Los Angeles, 2015, p. Th4C.4.
[114] Y. Sasaki et al., Crosstalk-managed heterogeneous single-mode 32-core fibre, in: Eur. Conf. Opt. Commun. (ECOC), 2016, pp. 550–552.

[115] K. Imamura, T. Gonda, R. Sugizaki, "19-core fiber with new core arrangement to realize low crosstalk," in OptoElectron. Commun. Conf./Australian Conf. Opt. Fibre Technol. (OECC/ACOFT), Melbourne, Australia, 2014, p. TU5C-3.
[116] Y. Sasaki et al., Quasi-single-mode homogeneous 31-core fibre, in: Eur. Conf. Opt. Commun. (ECOC), 2015, pp. 1–3.
[117] Y. Sasaki, K. Takenaga, K. Aikawa, Y. Miyamoto, T. Morioka, Single-mode 37-core fiber with a cladding diameter of 248 μm, in: Opt. Fiber Commun. Conf. (OFC), 2017, p. Th1H.2.
[118] T. Sakamoto, et al., Low-loss and low-DMD few-mode multi-core fiber with highest core multiplicity factor, in: Opt. Fiber Commun. Conf. (OFC), Anaheim, 2016, p. Th5A.2.
[119] J. Sakaguchi, et al., Realizing a 36-core, 3-mode fiber with 108 spatial channels, in: Opt. Fiber Commun. Conf. (OFC), 2015, p. Th5C.2.
[120] B.G. Lee, et al., End-to-end multicore multimode fiber optic link operating up to 120 Gb/s, J. Lightw. Technol. 30 (6) (2012) 886–892.
[121] Y. Geng, et al., High-speed, bi-directional dual-core fiber transmission system for high-density, short-reach optical interconnects, in: Proc. SPIE, next-generation optical networks for data centers and short-reach Links II, San Francisco, 2015, vol. 9390, p. 939009.
[122] M. Tsukamoto, T. Miura, H. Yutaka, T. Gonda, K. Imamura, R. Sugizaki, Ultra-high density optical fiber cable with rollable multicore fiber ribbon, in: Int. Wire Cable Symp. (IWCS), Providence, RI, 2016, pp. 597–599.
[123] T. Morishima, et al., MCF-enabled ultra-high-density 256-core MT connector and 96-core physical-contact MPO connector, in: Opt. Fiber Commun. Conf. (OFC), 2017, p. Th5D.4.
[124] T. Hayashi, T. Nakanishi, F. Sato, T. Taru, T. Sasaki, Characterization of interconnect multi-core fiber cable: mechanical/thermal characteristics and inter-core crosstalk of the straightened cable, in: IEEE Optical Interconnects Conference, San Diego, 2016, p. WB4.
[125] T. Matsui, et al., 118.5 Tbit/s transmission over 316 km-long multi-core fiber with standard cladding diameter, in: 2017 Opto-Electronics and Communications Conference (OECC) and Photonics Global Conference (PGC), Singapore, 2017, p. 10.1109/OECC.2017.8115049.
[126] T. Hayashi, et al., End-to-end multi-core fibre transmission link enabled by silicon photonics transceiver with grating coupler array, in: Eur. Conf. Opt. Commun. (ECOC), Gothenburg, 2017, p. Th.2.A.4.
[127] M. Tanaka, M. Hachiwaka, Y. Fujimaki, S. Kusunoki, H. Taniguchi, Study of multi-core fiber suitable for connectivity (in Japanese with English abstract), IEICE Tech. Rep. 112 (5) (2012) 5–8. OFT2012-2.
[128] T. Nagashima, H. Sakuma, S. Toyokawa, T. Hayashi, T. Nakanishi, Multi-core fibre with concaved double-D shape cross section, in: Eur. Conf. Opt. Commun. (ECOC), Gothenburg, 2017, p. M.2.B.5.
[129] O.N. Egorova, et al., Multicore fiber with rectangular cross-section, Opt. Lett., OL 39 (7) (2014) 2168–2170.
[130] T. Hayashi, T. Taru, O. Shimakawa, T. Sasaki, E. Sasaoka, Uncoupled multi-core fiber enhancing signal-to-noise ratio, Opt. Express 20 (26) (2012) B94–B103.

CHAPTER 2

Chip-based frequency combs for wavelength-division multiplexing applications

Juned N. Kemal[1], Pablo Marin-Palomo[1], Maxim Karpov[2], Miles H. Anderson[2], Wolfgang Freude[1], Tobias J. Kippenberg[2] and Christian Koos[1,3,*]
[1]Institute of Photonics and Quantum Electronics (IPQ), Karlsruhe Institute of Technology (KIT), Karlsruhe, Germany
[2]Laboratory of Photonics and Quantum Measurements (LPQM), École Polytechnique Fédérale de Lausanne (EPFL), Lausanne, Switzerland
[3]Institute of Microstructure Technology (IMT), Karlsruhe Institute of Technology (KIT), Karlsruhe, Germany

2.1 Wavelength-division multiplexing using optical frequency combs

Wavelength-division multiplexing (WDM) has been used for longhaul optical fiber links over distances of hundreds or even thousands of kilometers since the 1990s [1–3]. For such links, the fiber infrastructure represents the most expensive asset, while the costs of transceiver components are comparatively low. However, with the explosive growth of data rates across all network levels, WDM is now becoming increasingly important also for shorter links, which are deployed in much larger quantities and which are much more sensitive to cost and size of the transceiver assemblies. This evolution is, for example, witnessed by a strong market growth in the field of so-called data-center interconnects, which link two or more data centers or parts thereof across a metropolitan area with typical distances of less than 100 km. In the future, even shorter WDM links are likely to emerge, for example, in the context of high-throughput campus area networks, which connect data center buildings over distances of up to a few kilometers. At present, these networks still rely on parallel transmission using spatially separated channels in thousands of single-mode fibers (SMF), each of which is operated at a comparatively low data rate of at most a few hundred Gbit/s, see Fig. 2.1. It is foreseeable that the concept of plain spatial parallelization will soon reach its scalability limits and that it must be complemented by spectral parallelization of data streams in each fiber to sustain further increases in data rates. This may unlock an entirely new application space for WDM techniques, in which utmost scalability of channel counts and data rates is of the highest importance.

* Christian Koos retains copyright for images/figures.

Optical Fiber Telecommunications V11
DOI: https://doi.org/10.1016/B978-0-12-816502-7.00002-6

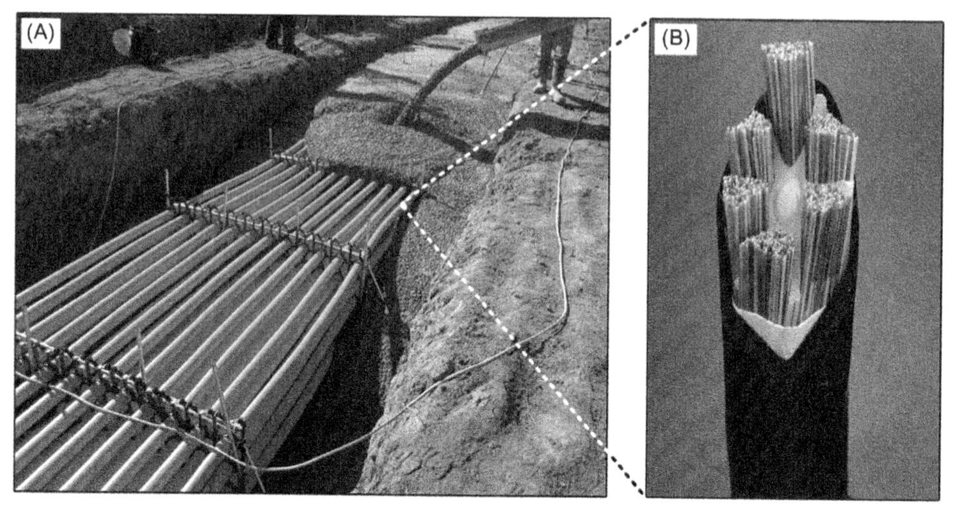

Figure 2.1 Capacity scaling in campus area networks through spatially parallel transmission. (A) Buried fiber conduit connecting two facilities on a data center campus. At present, such links might comprise tens of thousands of individual fibers, thus unveiling the fundamental scalability limitations. To sustain the unabated growth of data traffic, spatially parallel transmission must be complemented by spectral parallelization using highly scalable wavelength-division multiplexing (WDM) transmission. (B) Fiber cable, combining more than 1000 individual single-mode fibers (SMF). Source: *Images courtesy Google and Sumitomo Electric Industries, 2018.*

In this context, optical frequency comb generators (FCGs) may play a key role as compact and robust multi-wavelength light sources that can provide large numbers of well-defined optical carriers. A particularly important advantage of frequency combs is the fact that comb lines are inherently equidistant in frequency, hence relaxing the requirements for interchannel guard bands [4] and avoiding frequency control of individual lines as needed in conventional schemes that combine arrays of independent distributed feedback (DFB) lasers. Note that these advantages do not apply only to the WDM transmitter but also to the receiver, where an array of discrete local oscillators (LO) may be replaced by a single comb generator [5–7]. Using an LO comb further facilitates joint digital signal processing of the WDM channels, which may reduce receiver complexity and increase phase noise tolerance [8]. Moreover, parallel coherent reception using an LO comb with phase-locked tones might even allow the reconstruction the time-domain waveform of the overall WDM signal and thus permit compensation of impairments caused by optical nonlinearities of the transmission fiber [9].

Besides these conceptual advantages of comb-based transmission, compact footprint and cost-efficient mass production are key for future WDM transceivers. Among the various comb generator concepts, chip-scale devices are therefore of particular interest [6,7,10–20]. When combined with highly scalable photonic integrated circuits for modulation, multiplexing, routing, and reception of data signals, such devices may

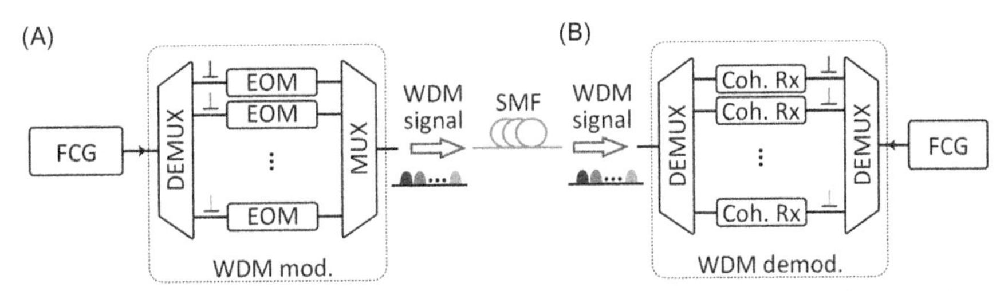

Figure 2.2 Concept of comb-based wavelength-division multiplexing (WDM) transmission and coherent reception. (A) WDM transmitter using a multitude of carriers from a first frequency comb generator (FCG). Information is encoded onto the various carriers via electro-optic modulators (EOM) using advanced quadrature amplitude modulation (QAM) formats. *DEMUX*, Demultiplexer; *MUX*, multiplexer. (B) Coherent WDM receiver using a multitude of local oscillator (LO) tones that are derived from a second FCG.

become the key to compact and energy-efficient WDM transceivers that can be fabricated in large quantities at low cost and that can offer transmission capacities of tens of Tbit/s per fiber.

The basic setup of a WDM transmitter with an optical FCG as a multi-wavelength source is depicted in Fig. 2.2A. The comb lines are first separated in a demultiplexer (DEMUX) and then fed to electro-optic modulators (EOM). For best spectral efficiency (SE), the schemes usually rely on advanced quadrature amplitude modulation (QAM) formats [21,22]. After recombining the data channels in a multiplexer (MUX), the WDM signal is transmitted through a single-mode fiber (SMF). Fig. 2.2B shows the corresponding WDM receiver (WDM Rx), exploiting LO tones from a second FCG for multi-wavelength coherent detection. The channels of the incoming WDM signal are separated by a DEMUX and then fed to an array of coherent receivers (Coh. Rx). The demultiplexed tones of the LO comb serve as phase reference for each coherent receiver. The performance of such WDM links clearly depends strongly on the underlying comb generators, in particular on the optical linewidth and the optical power per comb line. A more detailed discussion will be provided in the next section.

2.2 Properties of optical frequency combs

In this section the most important performance metrics for optical frequency combs in optical communications are introduced. These metrics will help to select the most suitable comb generator for a specific communication application and will build the base of a comparative discussion of different comb generator concepts in Section 2.3.

The schematic of a frequency comb spectrum is shown in Fig. 2.3. The comb consists of a set of narrowband spectral lines numbered by an integer index m, evenly spaced in frequency around a central line. The comb line frequencies $f_m = f_c + mf_r$ are fully determined by the center frequency f_c and by the so-called free spectral range

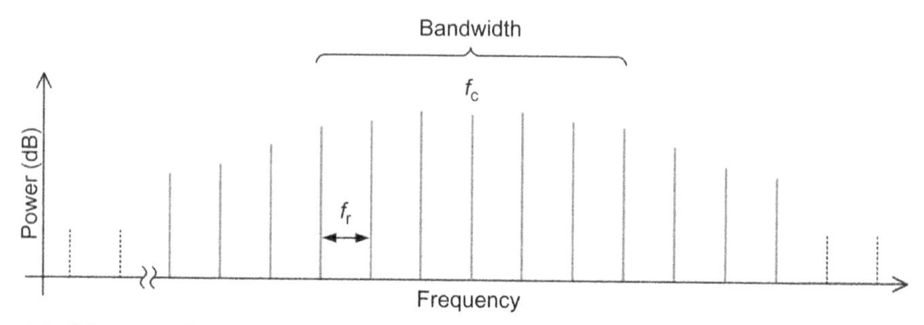

Figure 2.3 Schematic illustration of an optical frequency comb spectrum. The comb consists of a set of narrowband lines evenly spaced by the repetition frequency f_r of the time domain optical pulses. The comb is centered at frequency f_c in the middle of the frequency range associated with the comb's usable spectral bandwidth.

(FSR) f_r, which is equivalent to the repetition frequency of the time domain optical pulses from the frequency comb source. The center frequency is usually defined as the frequency of the comb line at the center of the used spectral bandwidth of the comb, see Fig. 2.3.

2.2.1 Center frequency, line spacing, and line count of frequency combs

The choice of the center frequency depends essentially on the link length and on the data rate. If chromatic dispersion of standard SMF is an issue, light sources operating close to the zero-dispersion wavelength 1.3 μm are desirable. However, if minimum fiber attenuation and the availability of widely used telecommunication devices like erbium-doped fiber amplifiers are important, a center wavelength near 1.55 μm is preferable.

The frequency spacing of the comb lines should be tens of gigahertz and adhere to the International Telecommunication Union (ITU) standard for WDM transmission specifying line spacings of 12.5, 25, 50, or 100 GHz and integer multiples thereof. For isolated point-to-point links, for example, in proprietary data center networks, the line spacing could also deviate from the ITU standards. In general, comb generators with tunable line spacing provide highest flexibility.

Finally, the number of lines is a crucial parameter, dictating the number of WDM channels that can be transmitted or detected in parallel. Note that there is no generally accepted metric for quantifying the line count of a frequency comb—publications may refer to the number of lines within the 3-dB bandwidth [23,24] or the 10-dB bandwidth [25] of the comb or to the number of lines that were usable for data transmission [4,16]. As the total comb power is generally limited, the individual lines will have lower power as the number of lines increases. For large line counts, optical amplifiers are needed to boost the optical power prior to modulation, and this will have direct impact on the achievable optical signal-to-noise power ratio (OSNR) of the WDM transmitter, see Section 2.2.3. If the comb is amplified as a whole, the optical amplifier bandwidth may additionally limit the number of usable tones.

2.2.2 Optical linewidth and relative intensity noise

The performance of comb-based WDM systems depends not only on the number and the spacing of the carriers but also on their power as well as on the phase and the amplitude noise. In optical communication systems that rely on advanced modulation formats, the phase of the optical carrier is also used to transmit information, and phase noise may have detrimental effects on the signal quality. Phase noise is usually quantified by the linewidth Δf of the carrier as an easy-to-handle figure of merit. Note that this parameter does not reveal the complete statistical characteristics of the phase noise that can, for example, be obtained from the so-called FM noise spectrum [26]. Strictly speaking, the linewidth Δf represents the spectral width of a Lorentzian spectral line that is associated with a white frequency-independent FM noise spectrum. It does not account for an increase of the FM noise spectrum towards lower frequencies.

The impact of phase noise on the data transmission performance depends on the random change of the carrier phase during one optical symbol. Phase noise hence becomes less problematic at shorter symbol durations T_s, that is, higher symbol rates. For a given modulation format and a chosen symbol rate, the linewidth requirements can therefore be expressed by the product of the carrier linewidth Δf and the symbol duration T_s. To quantify these requirements, the linewidth-induced penalty of the OSNR for a reference bit error ratio (BER) of 10^{-3} is considered [27]. In general, the OSNR penalty is specified by the ratio of the OSNR that is actually required to achieve a certain reference BER and the OSNR that would theoretically be required in an ideal transmission setup without phase noise. Table 2.1 specifies the $\Delta f\, T_s$ product corresponding to a 1 dB OSNR penalty [27]. As expected, higher order modulation formats are less tolerant to phase noise as a higher number of discrete phase states need to be discriminated for correct detection of a symbol.

Table 2.1 Maximum product $\Delta f\, T_s$ of linewidth and symbol duration for different modulation formats, leading to a 1dB optical signal-to-noise power ratio (OSNR) penalty at a reference bit error ratio (BER) of 10^{-3} [27].

Modulation format	$\Delta f\, T_s$
QPSK	4.1×10^{-4}
16QAM	1.4×10^{-4}
64QAM	4.0×10^{-5}

QPSK, Quadrature phase-shift keying; *16QAM*, 16-state quadrature amplitude modulation; *64QAM*, 64-state quadrature amplitude modulation.

Besides phase noise, relative intensity noise (RIN) of the comb tones may additionally reduce the transmission performance. Note that RIN does not usually play an important role in coherent communications, since balanced receivers suppress the RIN influence of both the carrier and the LO. If RIN is to be considered, however, then the RIN of the individual comb lines is more important than the RIN of the comb as a whole. As an example, mode partition noise in quantum-dash mode-locked laser diodes (QD-MLLD) results in spectral lines having a large RIN, whereas the total comb power shows a much smaller RIN [28,29]. This can be attributed to the fact that all optical tones are fed by the same gain medium and their powers are hence correlated such that a power increase in one tone is compensated by a power decrease of other tones.

2.2.3 Comb line power and optical carrier-to-noise power ratio

Besides phase noise and RIN of an individual comb line, additive noise arising from spontaneous emission in the comb source or in the amplifiers along the transmission link may impair the performance of a WDM system. The level of the additive noise background is generally quantified by the OSNR, which is defined as the power ratio of the data signal to the noise power within a certain reference bandwidth B_{ref}. In the following, the OSNR is specified with respect to a constant reference bandwidth of $B_{\mathrm{ref}} = 12.5$ GHz, corresponding to a reference wavelength span of 0.1 nm at a center wavelength of 1.55 μm, see Appendix A.2 for a more detailed explanation. The OSNR within the reference bandwidth can be translated into the signal-to-noise power ratio OSNR_{sig} that refers to the actual bandwidth B of a specific data signal by means of Eq. (2.14) in Appendix A.2.

The performance of a WDM system is eventually quantified by the optical signal-to-noise power ratio at the receiver ($\mathrm{OSNR}_{\mathrm{Rx}}$), which depends both on the comb line power P_ℓ and on the optical carrier-to-noise power ratio (OCNR) of the comb line OCNR_ℓ. In analogy to the OSNR, the OCNR relates the comb line power to the power of the background noise, measured again within a reference bandwidth $B_{\mathrm{ref}} = 12.5$ GHz centered at the comb line frequency. Comb line power and OCNR determine the $\mathrm{OSNR}_{\mathrm{Rx}}$ of a WDM channel and thus limit the link reach, restrict the choice of the modulation format, and set the maximum symbol rate. For a quantitative analysis, consider the WDM link depicted in Fig. 2.4A consisting of a WDM transmitter, a link with $\nu_{\max}$ fiber spans, and a WDM receiver. Note that the comb line powers P_ℓ of most chip-scale comb generators are usually much weaker than the power levels emitted by state-of-the-art continuous-wave (CW) laser diodes as used in conventional WDM systems. It is therefore necessary to amplify the comb tones prior to modulation by sending them through a dedicated comb amplifier with gain G_0. Data is encoded onto the various carriers by a WDM modulator unit (WDM mod.),

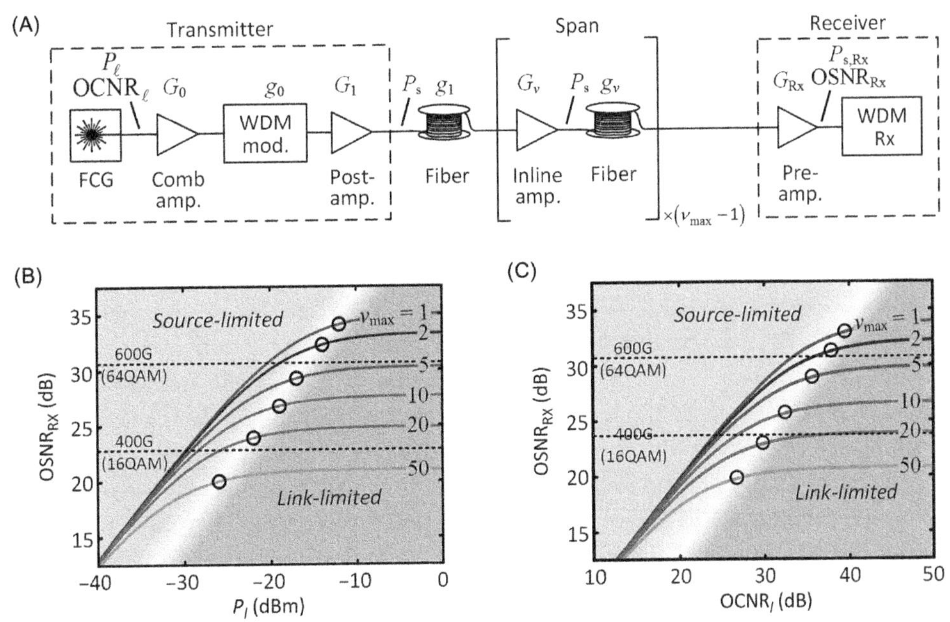

Figure 2.4 Influence of the comb line power P_ℓ and of the carrier-to-noise power ratio OCNR_ℓ on the achievable optical signal-to-noise power ratio OSNR_{Rx} at the receiver for different numbers ν_{max} of spans. (A) Schematic of a wavelength-division multiplexing (WDM) transmission and reception system. The comb lines are first amplified by a comb amplifier ("Comb amp.") before data are encoded on each line in the WDM modulation unit ("WDM mod."). The WDM signal is then boosted by a post-amplifier ("Post-amp.") and transmitted through the fiber link consisting of at least one span. Each of the $\nu_{max} - 1$ additional spans contains an in-line amplifier ("Inline amp.") which compensates the loss of the corresponding fiber section. The data from each WDM channel are recovered at the WDM receiver, which contains an optical pre-amplifier ("Pre-amp"). (B) OSNR_{Rx} as a function of P_ℓ in the limit of high OCNR_ℓ. The plot reveals two regimes: For source-limited transmission at low line powers, the OSNR_{Rx} is dominated by the contribution of the comb amplifier, whereas the contributions of the various amplifiers along the link dominates for link-limited transmission at high line powers. The transition between both regimes is indicated by black circles which mark the points where the OSNR_{Rx} has decreased by 1 dB in comparison to its limit at high comb line powers P_ℓ. The dashed horizontal lines indicate the minimum OSNR_{Rx} required for transmission of a net data rate of 400 Gbit/s using 16QAM and of 600 Gbit/s using 64QAM as a modulation format. (C) OSNR_{Rx} as a function of OCNR_ℓ for comb line powers P_ℓ that correspond to the transition points marked by circles in subfigure (B). For low OCNR_ℓ, OSNR_{Rx} is dominated by the noise of the comb source, and the link performance is source-limited. For high OCNR_ℓ, the OSNR_{Rx} is independent of the OCNR_ℓ, and the transmission performance is link-limited. Source: *Adapted from P. Marin-Palomo, J. N. Kemal, W. Freude, S. Randel, and C. Koos. "OSNR limitations of chip-based optical frequency comb sources for WDM coherent communications" arXiv preprint arXiv: 1907.01042 (2019).*

comprising a WDM DEMUX, an array of dual-polarization in-phase/quadrature I/Q modulators, and a WDM MUX, see Fig. 2.2A. The power transmission factor of the WDM modulator unit is denoted as g_0 and accounts for the insertion losses of the MUX and the DEMUX as well as for the insertion and modulation losses of the I/Q modulators. The WDM signal is then boosted by a post-amplifier with gain G_1 and transmitted through a fiber link of at least one span, where each of the $(\nu_{max} - 1)$ additional spans contains an in-line amplifier and a fiber section. Each of the fiber spans attenuates the signal power by a factor $g_\nu < 1$, which is compensated by the gain $G_\nu = 1/g_\nu$ of the post-amplifier or the corresponding in-line amplifier, $\nu = 1 \ldots \nu_{max}$. For simplicity, the signal power P_s of a single WDM channel is assumed to be 0 dBm at the beginning of each fiber span, offering a good compromise between nonlinear impairments of the data channels and OSNR [30,31]. All spans are assumed to be identical, $G_\nu = G$, $g_\nu = g$, $Gg = 1$, and all amplifiers have the same noise figure $F_\nu = F$. At the receiver, a pre-amplifier with gain G_{Rx} and noise figure F_{Rx} amplifies the received signal. A WDM Rx, comprising a WDM demodulator unit (WDM demod.) and a FCG, is used to recover the transmitted data, see Fig. 2.2B. For simplicity, the gain and the noise figure of the receiver pre-amplifier is assumed to be identical to that of the in-line amplifiers, leading to a signal power of 0 dBm per wavelength channel at the input of the receiver DEMUX. Note that for a high number of spans, the noise of the receiver pre-amplifier does not play a role any more, since the noise background is dominated by the contributions from the various amplifiers along the link. Note also that the additive noise background of the LO comb tones can be neglected with respect to that of the received signal. The LO comb is hence not considered further in the subsequent analysis.

To obtain the OSNR_{Rx} for a single WDM channel at the receiver, the ratio of the received signal power and the noise power in the reference bandwidth B_{ref} needs to be calculated, see Appendix A.1 for a detailed mathematical description. For a given line power P_ℓ and OCNR_ℓ, the OSNR_{Rx} can be expressed as [32]

$$\mathrm{OSNR}_{Rx} = \frac{g_0 G_0 P_\ell}{\underbrace{\frac{g_0 G_0 P_\ell}{\mathrm{OCNR}_\ell}}_{\text{ampl. FCG noise}} + \underbrace{hf\,B\left(g_0(G_0 - 1)F_0 + g\nu_{max}(G-1)F + \frac{(G_{Rx} - 1)}{G_{Rx}}F_{Rx}\right)}_{\text{amplifier noise}}}. \tag{2.1}$$

In this relation, the quantity hf is the photon energy of the comb line under consideration. Eq. (2.1) is the foundation of Fig. 2.4B, which shows the OSNR_{Rx} as a function of the line power P_ℓ and the number ν_{max} of fiber spans for a realistic comb-based WDM system. For this plot, the OCNR_ℓ of the comb lines is assumed to be infinite, that is, the FCG itself does not introduce any practically relevant additive noise background. This can, for example, be accomplished by Kerr comb generators. The various parameters used for this study are specified in Table 2.2. For the WDM modulator unit, an overall insertion loss of 25 dB is assumed, comprising 3.5 dB of insertion loss

Table 2.2 Model parameters used for generating Fig. 2.4A and B from Eq. (2.1).

Variable	Description	Value	
B_{ref}	Reference bandwidth for optical signal-to-noise power ratio (OSNR) and optical carrier-to-noise power ratio (OCNR) calculation	12.5 GHz	
P_s	Signal power in a single wavelength-division multiplexing (WDM) channel at the input of each fiber span	1 mW	0 dBm
g_0	Power transmission factor of WDM modulator unit	3.2×10^{-3}	−25 dB
G_0	Power gain factor of comb amplifier	$P_s/(P_\ell g_0 G)$	
F_0	Noise figure of comb amplifier	3.2	5 dB
$g_\nu = g$	Power transmission factor of fiber section	3.2×10^{-2}	−15 dB
$G_\nu = G = g^{-1}$	Power gain factor of post and in-line amplifier	32	15 dB
$F_\nu = F$	Noise figure of post and in-line amplifier	3.2	5 dB
G_{Rx}	Power gain factor of pre-amplifier	32	15 dB
F_{Rx}	Noise figure of pre-amplifier	3.2	5 dB

The fiber attenuation g_ν, amplifier gain G_ν, and noise figures F_ν are assumed to be the same for all ν_{max} fiber links ($\nu = 1 \ldots \nu_{max}$).

each for the WDM DEMUX and the MUX [33], 13 dB of loss for the dual-polarization I/Q modulators [34], and an additional 5 dB of modulation loss, which are assumed to be independent of the modulation format for simplicity. Note that the results shown in Fig. 2.4B do not change significantly when varying these losses by a few dB. The fiber spans are assumed to feature a power loss of 15 dB each, corresponding to 75 km of SMF with a propagation loss of 0.2 dB/km. The power loss of each span is exactly compensated by the 15 dB gain of the corresponding post-amplifier or in-line amplifier such that a signal launch power of $P_s = 0$ dBm is maintained for all spans. The gain G_0 of the comb amplifier is adjusted such that this launch power is reached after the post-amplifier.

For low line powers P_ℓ, Fig. 2.4B shows that the noise level at the receiver is dominated by the contribution of the comb amplifier. As a consequence, the OSNR_{Rx} increases in proportion to P_ℓ and is essentially independent of the span count ν_{max}. In this regime, transmission performance is limited by the comb source and its associated amplifier ("source-limited"). For high comb line powers P_ℓ, in contrast, the noise level at the receiver is dominated by the contributions of the various amplifiers along the link ("link-limited"). In this regime, the OSNR_{Rx} is essentially independent of P_ℓ and decreases with each additional span. The transition between both regimes is indicated by black circles in Fig. 2.4B, indicating the points where the OSNR_{Rx} has decreased by 1 dB in comparison to its limit at high line power P_ℓ. When using frequency combs in WDM systems, operation in the link-limited regime is preferred. Depending on the number of spans, this requires a minimum comb line power between −25 and −15 dBm. Note that, in principle, the gain G_1 of the post-amplifier could be decreased by increasing the gain G_0 of the comb amplifier. For source-limited transmission, the overall impact would be small since the OSNR_{Rx} is mainly dictated by the comparatively low line power P_ℓ that enters the comb amplifier. For link-limited transmission over a small number of spans, the OSNR_{Rx} can be slightly improved by approximately 2 dB by decreasing the gain G_1 and increasing the gain G_0 while maintaining realistic power levels at the output of the comb amplifier.

Fig. 2.4B can be used as a guide for estimating the performance requirements for comb sources in WDM applications. As a reference, the plot indicates the minimum OSNR_{Rx} required for transmission of a net data rate of 400 and 600 Gbit/s per WDM channel, using 16QAM and 64QAM as a modulation format, respectively. In both cases, advanced forward-error correction (FEC) schemes with 11% overhead and BER thresholds of 1.2×10^{-2} [35] are assumed, requiring a symbol rate of 56 GBd to provide the specified net data rates.

For many practically relevant comb sources, the OCNR_ℓ is finite, which will further decrease the transmission performance. Fig. 2.4C shows the OSNR_{Rx} as a function of OCNR_ℓ for various span counts ν_{max}. In this plot, the comb line power P_ℓ was assumed to correspond to the minimum value required for link-limited transmission, as indicated by the corresponding transition points marked by circles in Fig. 2.4B. For low OCNR_ℓ, the noise level at the receiver is dominated by the noise background of the comb source, and the OCNR_ℓ and OSNR_{Rx} are essentially identical. In this case the link performance is again source-limited. For high OCNR_ℓ, the OSNR_{Rx} is independent of the OCNR_ℓ, and the transmission performance is link-limited. Depending on the number of spans, the minimum OCNR_ℓ values needed for link-limited transmission range between 25 and 40 dB. The transition between both regimes is again marked by black circles in Fig. 2.4C, indicating the points where the OSNR_{Rx} has decreased by 1 dB in comparison to its respective limit at high OCNR_ℓ.

2.3 Chip-scale optical frequency comb generators

Besides the performance parameters discussed in the previous section, comb generators have to provide a compact footprint and be amenable to cost-efficient mass production to qualify as light sources in highly scalable WDM transceivers. In this context, chip-scale devices or assemblies are of particular interest. This section provides an overview and a comparative discussion of different realization concepts for chip-scale comb generators.

2.3.1 Mode-locked laser diodes

The phenomenon of mode-locking was first observed [36–38] and theoretically explained [39–41] in the 1960s and subsequently became the foundation of pulsed femtosecond solid-state lasers [42–44]. A mode-locked laser emits a regular train of pulses that are spaced by the repetition period T_r in the time domain, corresponding to a broad frequency comb with FSR $f_r = T_r^{-1}$ in the frequency domain. In general, mode-locking requires an active medium with an inhomogeneously broadened gain spectrum that allows different longitudinal modes to oscillate simultaneously in the laser cavity, in combination with an additional mechanism that establishes a fixed phase relationship between these modes. The line spacing of the comb is then dictated by the FSR of the cavity, which is the inverse of the cavity round-trip time. For line spacings of tens of GHz as used in WDM applications, so-called passive mode-locking techniques are particularly important [45], offering stable device operation at manageable technical complexity. In general, passive mode-locking requires a nonlinear optical element that allows the various optical modes to interact.

Regarding mode-locked lasers as multi-wavelength light sources for WDM transmission, experimental demonstrations were mainly based on devices that exploit low-dimensional structures of III–V semiconductors such as quantum dashes and quantum dots as active media. QD-MLLD rely on four-wave mixing (FWM) or cross-gain modulation as passive mode-locking mechanisms [46,47]. The devices can be operated by a simple direct current (DC), Fig. 2.5A, and provide a relatively flat comb spectrum that may span 3-dB bandwidths in excess of 1 THz, see Fig. 2.5B. QD-MLLD have been realized with repetition rates from 10 GHz [48,49] to more than 300 GHz [50], dictated by the cavity length of the respective device. The simple operation and the compact footprint of QD-MLLD in combination with the comparatively broad and smooth spectral envelope of the associated combs make these devices promising candidates for particularly compact WDM transceivers. In communication experiments, QD-MLLD have been shown to allow for WDM transmission of more than 50

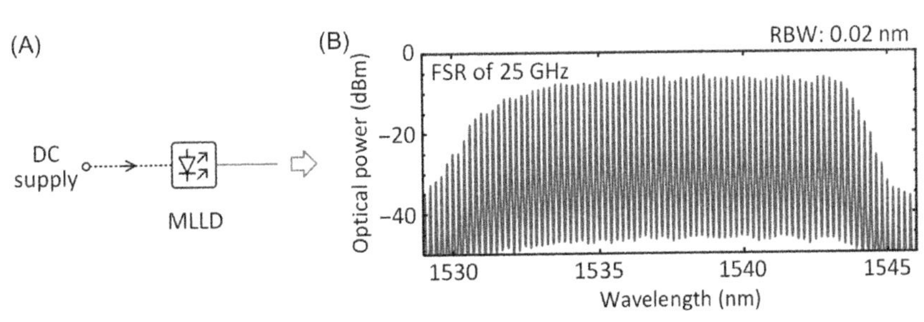

Figure 2.5 Concept of frequency comb generation in mode-locked laser diodes that rely on InAs/InGaAsP quantum dashes as an active material. (A) Quantum-dash mode-locked laser diodes (QD-MLLD) stand out due to easy operation—a frequency comb can simply be generated by driving the device with a direct current (DC). (B) QD-MLLD can generate combs with broad and smooth spectral envelopes, featuring 3-dB bandwidths well in excess of 1 THz, corresponding to a wavelength span of approximately 10 nm at a center wavelength of 1.55 μm. The line power amounts to approximately −10 dBm.

channels and data rata rates in excess of 10 Tbit/s [15]. These experiments also revealed strong low-frequency phase noise of the QD-MLLD as one of the main limitations of transmission performance, see Section 2.3.6 for more details. If this problem can be overcome, comb line powers of the order of −10 dBm [14,15,51], Fig. 2.5B, and OCNR values in excess of 40 dB would allow for link-limited transmission using QD-MLLD in realistic WDM systems, see Fig. 2.4B and C.

2.3.2 Electro-optic modulators for comb generation

Frequency combs may also be generated by periodic modulation of a CW laser tone, using high-speed electro-optic modulators (EOM), see Fig. 2.6A for an illustration of the concept. The output of a CW laser is fed into a single EOM or a series thereof. The EOM are usually driven with a sinusoidal electrical signal from a radio frequency (RF) oscillator. This leads to modulation sidebands at the output of the EOM, which are spaced by the fundamental frequency of the RF drive signal. Over the last years, various approaches have been explored to improve the spectral bandwidth and flatness of EOM-based combs [52], which are generally limited by the modulation depth that can be achieved in the various EOM. These approaches have relied on asymmetrically driven Mach–Zehnder modulators [53] or cascades of intensity and phase modulators [54,55]. Fig. 2.6B shows a comb spectrum generated in such a cascade, featuring approximately 30 lines within a 3 dB bandwidth of approximately 900 GHz, corresponding to a wavelength span of 8 nm [54]. Note, however, that these demonstrations were still obtained with conventional lithium niobate ($LiNbO_3$) modulators in

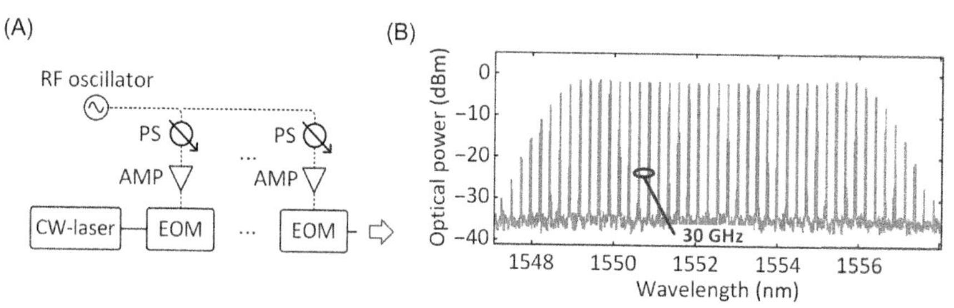

Figure 2.6 Optical frequency comb generation using single or cascaded electro-optic modulators (EOM). (A) Schematic of the comb generator. The output of a continuous-wave (CW) laser is sent through a single EOM or through a cascade of EOM. The EOM are driven by sinusoidal signals that are derived from a common radio frequency (RF) oscillator and that are carefully adjusted in amplitude and phase. This allows for generation of comparatively broadband combs with flat spectral envelopes; *AMP*, RF amplifier; *PS*, RF phase shifter. (B) Example spectrum of a frequency comb generated by cascading a lithium niobate ($LiNbO_3$) phase and intensity modulator in a benchtop type setup [54]. EOM-based comb generators are amenable to chip-scale integration [16,17,56,58,59]. Source: *Adapted from C. Chen, C. Zhang, W. Zhang, W. Jin, K. Qiu, Scalable and reconfigurable generation of flat optical comb for WDM-based next-generation broadband optical access networks, Opt. Commun. 321 (2014) 16–22. © 2014 Elsevier.*

benchtop-type laboratory setups. While the concept is clearly amenable to chip-scale integration, the performance of integrated modulator-based comb generators is still inferior to that of the discrete counterparts. Integrated modulator-based comb generators have been demonstrated on the indium phosphide (InP) platform [56–58] and on silicon photonics [16,17,59]. Using an integrated silicon-organic hybrid (SOH) modulator, combs with up to 7 lines in a 3-dB bandwidth of 280 GHz and WDM data rates of up to 1 Tbit/s were demonstrated [16]. Assuming that the performance of integrated modulator-based comb generators may be improved to match that of their discrete-element counterparts in the future, line powers in excess of −10 dBm, Fig. 2.6B [54], might make link-limited transmission a realistic scenario provided that the corresponding OCNR requirements according to Fig. 2.4C can be met.

Modulator-based comb generators stand out due to their flexibility: The center frequency can be adjusted by simply tuning the emission wavelength of the CW laser, and the FSR can be set via the RF modulation frequency. However, they suffer from the limited number of lines as well as from the comparatively high complexity that is associated with driving a multitude of modulators with RF signals of precisely defined amplitude and phase. Modulator-based comb generators might therefore be a viable option for transmission of WDM superchannels with a limited number of lines and flexible line spacing.

2.3.3 Gain-switched laser diodes

Another technique for generating frequency combs with tunable line spacing relies on gain switching of a semiconductor laser diode that is injection-locked by a CW master laser. Fig. 2.7A shows the concept of comb generation in a gain-switched laser diode

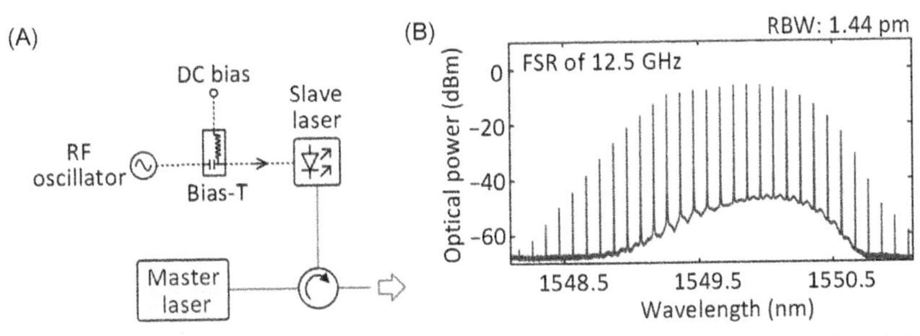

Figure 2.7 Optical frequency comb generation by gain-switched laser diodes (GSLD). (A) Schematic setup. A semiconductor laser, biased close to threshold, is driven by a sinusoidal signal from an radio frequency (RF) oscillator. This leads to a train of chirped output pulses with a repetition rate determined by the frequency of the RF drive signal. External injection of continuous-wave (CW) light from a master laser establishes phase coherence between the pulses and hence allows for frequency comb formation. (B) Spectrum of the frequency comb from a gain-switched distributed feedback (DFB) slave laser. Taking into account the 1.44 pm resolution bandwidth of the optical spectrum analyzer (OSA), an optical carrier-to-noise power ratio (OCNR) of approximately 30 dB can be estimated from the spectrum—a typical value for GSLD-based combs.

(GSLD). A semiconductor slave laser is biased close to its threshold and driven by a sinusoidal signal from an RF oscillator. This periodically brings the carrier concentration in the active zone of the slave laser from below to above the lasing threshold, thus generating a train of strongly chirped optical pulses with a repetition rate corresponding to the RF modulation frequency. External injection of CW light emitted from a master laser is used to establish phase coherence between the generated pulses, thus leading to a frequency comb with discrete narrowband lines [60], see Fig. 2.7B. The bandwidth of the comb is dictated by the intrinsic relaxation dynamics of photons and electrons in the slave laser. The injection also increases the relaxation frequency of the semiconductor lasers, enabling the generation of frequency combs with a larger FSR as compared to no external injection [61]. In this scheme, the linewidth and RIN characteristics of the carrier of the master laser are transferred to each of the comb lines [62]. This allows for generation of low-noise frequency combs by using a master laser with low linewidth and small RIN.

For frequency combs from GSLD, both the center frequency and the FSR can be adjusted by tuning the emission wavelength of the master laser and by adjusting the RF modulation frequency [63]. GSLD-based frequency combs typically cover 3-dB bandwidths of approximately 200 GHz and provide line spacings of up to 20 GHz along with typical line powers of the order of −10 dBm [11,60]. Using GSLD-based combs, WDM transmission on 24 channels with data rates of up to 2 Tbit/s was reported [11]. Note that these demonstrations were still performed with benchtop-type GSLD setups built from discrete devices. Integration of the slave and the master laser on a common InP chip has been demonstrated [64], but the performance is currently still inferior to that of a discrete-element setup. Note also that the overall transmission performance of GSLD-based combs in WDM is not only limited by the line spacing and the number of lines but also by the OCNR values that typically amount to approximately 30 dB [6,11]. According to Fig. 2.4C, transmission performance might then be limited by the source rather than by the transmission link itself.

2.3.4 Kerr-nonlinear waveguides for spectral broadening

One of the most stringent limitations of the comb sources discussed in the previous sections is the limited number of carriers. A potential solution to this problem is spectral broadening of narrowband seed combs using waveguides with strong third-order nonlinearities such as self-phase modulation (SPM), cross-phase modulation, and FWM. This concept has been exploited for WDM communications using benchtop-type comb generators in combination with highly nonlinear fibers (HNLF) for spectral broadening [65,66]. However, for generating flat comb spectra, these schemes often rely on delicate schemes for dispersion management [65] or spectral stitching [66], which usually require intermediate fiber amplifiers and are hence not amenable to chip-scale integration. This can be overcome by using Kerr-nonlinear integrated optical waveguides for spectral broadening instead of HNLF [18,67,68]. The viability of integrated Kerr-nonlinear waveguides (KNWG) for spectral broadening has been demonstrated by using low-loss aluminum-gallium-arsenide-on-insulator (AlGaAsOI) waveguides that feature high nonlinearity parameters of up to 660 $(\mathrm{Wm})^{-1}$. These devices allow for generation of more than 300 densely spaced carriers within the telecommunication C-band (1530 nm ... 1565 nm), offering a WDM transmission capacity of 22 Tbit/s [18]. The underlying comb generation scheme is summarized in Fig. 2.8A, and the corresponding spectra are shown in Fig. 2.8B. Line powers of more than −15 dBm make this approach promising for link-limited transmission, see Fig. 2.4B. Since spectral broadening of combs in KNWG works independently of the seed-comb line spacing and—within limitations—independently of the center wavelength, the approach could be particularly interesting for broadband comb generation with tunable FSR and center frequency. Note that in the experiment published so far [18], the seed comb was still generated by a solid-state mode-locked laser with a repetition rate of 10 GHz followed by an erbium-doped fiber amplifier to achieve sufficient launch power into the integrated waveguide. Moreover, spectral stitching was used to achieve a flat and high-quality frequency comb for the data transmission demonstration. Transferring this concept to a chip-scale assembly would require cointegration of all these elements.

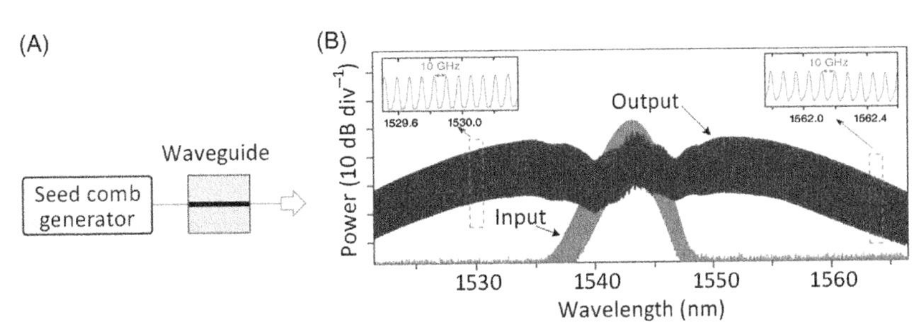

Figure 2.8 Integrated Kerr-nonlinear waveguides (KNWG) for spectral broadening of narrowband seed combs. (A) Schematic setup. A narrowband seed comb is coupled to an integrated optical waveguide with high Kerr nonlinearity. Self-phase modulation (SPM) and cascaded four-wave mixing (FWM) leads to a broadband comb. (B) Spectra of seed frequency comb at the input and of the broadened comb at the output of an aluminum-gallium-arsenide-on-insulator (AlGaAsOI) waveguide, pumped by a solid-state mode-locked laser [18]. Source: *Adapted from H. Hu, F. Da Ros, M. Pu, F. Ye, K. Ingerslev, E. Porto da Silva, et al., Single-source chip-based frequency comb enabling extreme parallel data transmission, Nat. Photonics 12 (2018) 469–473. © 2018 Springer.*

2.3.5 Microresonator-based Kerr-comb generators

A particularly promising option to generate frequency combs in chip-scale devices relies on resonantly enhancing Kerr-nonlinear interaction in integrated optical waveguides. This leads to the concept of Kerr comb generation in microring resonators [69,70], which is illustrated in Fig. 2.9A. A high-Q resonator is resonantly pumped by a CW laser, thus producing a strong intracavity power. Under appropriate conditions, degenerate and nondegenerate FWM leads to formation of new spectral lines by converting pairs of pump photons into pairs of photons that are upshifted and downshifted in frequency, see Section 2.4 for a more detailed explanation of the underlying mechanisms. The broadband nature of parametric gain can be used to convert the CW pump into octave-spanning combs, without any additional spectral broadening [71]. Over the previous years, comb generation has been demonstrated in a variety of different Kerr-nonlinear microresonators, comprising silica toroidal cavities [69], crystalline whispering-gallery mode (WGM) microresonators [72], and silicon nitride (Si_3N_4) ring resonators [70,73], which are particularly well suited for wafer-level mass production [70].

Of particular interest for WDM applications are so-called single dissipative Kerr soliton (DKS) states, which consist of only one ultrashort pulse circulating around the microresonator, see Section 2.4 for details. This leads to broadband combs with particularly smooth spectral envelopes [74,75]. An example of a DKS comb spectrum is shown in Fig. 2.9B. The comb offers more than 100 carriers spaced by approximately

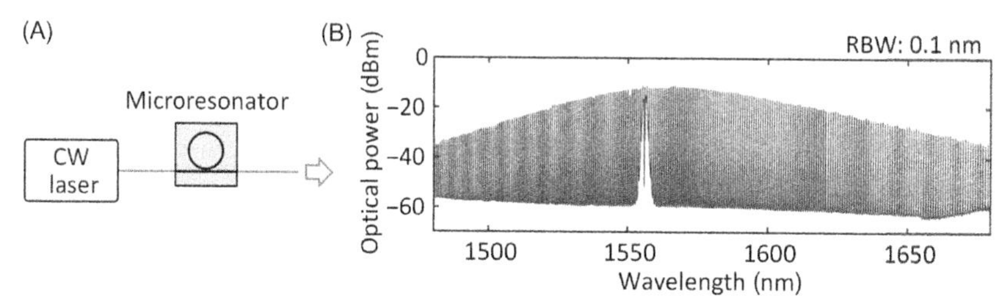

Figure 2.9 Kerr comb generation in high-Q microresonators. (A) Principle of Kerr comb generation. A high-Q microring resonator is pumped by a continuous-wave (CW) laser. Under appropriate conditions, degenerate and nondegenerate four-wave mixing (FWM) leads to the formation of new spectral lines, see Section 2.4 for details. Of particular interest for optical communications are so-called single dissipative Kerr soliton (DKS) states, for which the superposition of the phase-locked optical tones forms an ultrashort soliton pulse circulating in the cavity. This leads to a comb spectrum with a broadband smooth envelope. (B) Spectrum of a DKS frequency comb generated in a Si_3N_4 microresonator [7]. The comb offers more than 100 carriers in the telecommunication C and L-band (1530 nm ... 1610 nm), spaced by approximately 100 GHz. In the center of the spectrum, the comb line power amounts to −11 dBm, and the optical carrier-to-noise power ratio (OCNR) is approximately 48 dB. Source: *Adapted from P. Marin-Palomo, J.N. Kemal, M. Karpov, A. Kordts, J. Pfeifle, M.H.P. Pfeiffer, et al., Microresonator-based solitons for massively parallel coherent optical communications, Nature 546 (2017) 274–279.*

100 GHz within the telecommunication C and L-band (1530 nm ... 1610 nm) [7]. In the center of the comb, the line powers reach—11 dBm and the OCNR amounts to approximately 48 dB [7]. This would safely permit link-limited transmission in realistic WDM systems. Toward the edges of the C and L-band, the line powers still drop to slightly lower levels, requiring further device improvements.

For the WDM transmission experiments demonstrated so far, DKS comb generation has still relied on benchtop-type setups with high-power EDFA that provide CW pump powers of more than 30 dBm at the input of the microresonator chip [7]. For chip-scale integration of WDM transceiver modules, these power levels must be reduced. In this context, improvements of microresonator fabrication are key, allowing the increase of the quality-factors (Q-factors) of the cavities and thus enabling single-soliton comb formation at pump powers down to a few milliwatts [76–78], which can be provided by a cointegrated light source [78]. Even though the output power levels of these comb generators is still below the threshold of link-limited transmission, the experiments show the great potential of reducing the pump-power requirements of Kerr comb sources by improved design and fabrication techniques. A more comprehensive discussion of Kerr comb generation covering basic principles, technological aspects, and application demonstrations of massively parallel WDM transmission is provided in Section 2.4.

2.3.6 Comparative discussion

Chip-scale comb sources have been used in a variety of WDM transmission experiments, relying on different comb generation approaches and covering a wide range of data rates, channel counts, and line spacings. Table 2.3 gives an overview of selected examples of such demonstrations, specifying the type of comb source along with the number of lines, the maximum and minimum power of the native comb lines, as well as the spacing f_r and the intrinsic optical linewidth Δf of the comb lines used for transmission, see Column 1 ... 5 of Table 2.3. Note that not all techniques of measuring the optical linewidth actually allow the direct extraction of the intrinsic linewidth that is associated with the spectrally white background of the FM noise spectrum [26]. Some of the values provided in Column 5 of Table 2.3 and in the associated references have therefore to be understood as an upper limit of the corresponding intrinsic linewidth. Columns 6 ... 9 of Table 2.3 finally give the symbol rate, the modulation format, the aggregate net data rate, and the net SE for dual-polarization WDM transmission. In this context, the net data rate refers to the usable rate for transmission of payload data after subtracting the overhead associated with the respective FEC scheme. Note that the line powers in Column 3 refer to the native comb lines prior to any spectral flattening that is usually applied to achieve a more uniform spectrum of carriers. For link-limited transmission over a single-span WDM connection with

Table 2.3 Selected experimental demonstrations of wavelength-division multiplexing (WDM) transmission using different chip-scale frequency comb generators (FCGs) as multi-wavelength light sources.

Comb source	# lines	PI max/min	f_r [GHz]	Δf [MHz]	Sym. rate [GBd]	Mod. format	Net data rate [Tbit/s]	SE $\left[\frac{bit}{s \cdot Hz}\right]$	References
Quantum-dash mode-locked laser diodes (QD-MLLD)	36	N.A.	34.5	<4	12.5	QPSK	1.89	1.5	[13]
QD-MLLD	52	$-5/-20$	42.0	<2	40.0	QPSK	7.74	3.5	[14]
QD-MLLD	60	$-10/-20$	25.0	<0.1	20.0	32QAM	11.22	7.5	[15]
EOM	5	$5/-5$	20.0	<0.1	16.0	16QAM	0.75	7.5	[17]
EOM	9	$-17/-27$	40.0	<0.1	28.0	QPSK	0.94	2.6	[16]
GSLD	6	$-4/-7$	12.5	0.3	12.0	QPSK	0.24	3.2	[12]
GSLD	13	$-5/-20$	12.5	<0.1	12.0	16QAM/QPSK	1.03[a]	5.2	[6]
GSLD	24	$-10/-30$	12.5	<0.1	12.0	16QAM/QPSK	1.87	6.2	[11]
KNWG	320	$-10/-30$	10.0	<0.1	40.0	16QAM	22.03[b]	6.9	[18]
Kerr comb	6	$10/-10$	690	<0.1	20.0	16QAM	0.90	0.22	[20]
Kerr comb	20	$0/-20$	25.0	N.A.	18.0	QPSK	1.35	2.7	[19]
Kerr comb	20	$10/-20$	230	<0.1	20.0	64QAM	4.04	0.88	[10]
Kerr comb	94	$-10/-20$	95.8	<0.1	40.0	16QAM	28.0	3.1	[7]
Kerr comb	179	$-18/-28$	95.8	<0.1	40.0	16QAM/QPSK	50.2[c]	5.2	[7]
Kerr comb	93	$-10/-20$	95.8	<0.1	50.0	16QAM	34.6[a]	3.9	[7]

The third column specifies the maximum and the minimum power of the native comb lines, prior to any spectral flattening. The net data rate refers to the aggregate rate that is usable for transmission of payload data after subtracting the forward-error correction (FEC) overhead.

EOM, Electro-optic modulator; Section 2.3.2; *GSLD*, gain switched laser diode, Section 2.3.3; *Kerr comb*, comb generation in Kerr-nonlinear microresonators, Section 2.3.5; *KNWG*, Kerr-nonlinear waveguides for spectral broadening of narrowband seed comb, Section 2.3.4; *N.A.*, values are not available; *QD-MLLD*, quantum-dash mode-locked laser diode, Section 2.3.1.

[a]These experiments rely on frequency combs both as multi-wavelength light source at the transmitter and as multi-wavelength LO at the receiver.

[b]The indicated data rate refers to the WDM capacity of a single-mode fiber (SMF) core. In the experiment, WDM was combined with space-division multiplexing on 30 parallel fiber cores, leading to an overall data of 661 Tbit/s [18].

[c]The transmission experiment relies on interleaved frequency combs that were obtained from a pair of separate Kerr comb generators.

realistic technical parameters, line powers should not be less than approximately −12 dBm, see Fig. 2.4. In some of the experiments [6,7], chip-scale comb generators were used both as multi-wavelength light source at the transmitter and as multi-wavelength LO at the receiver—the corresponding data rates are marked with a star in the second column. Note that all experiments listed in Table 2.3 were performed at center wavelengths around 1550 nm, relying on the telecommunication C and L bands that are most commonly used for WDM transmission. However, the concepts may be also transferred to other wavelength ranges, for example, around 1300 nm.

Among the various comb generation concepts, EOM-based approaches, GSLD, and spectral broadening in KNWG allow for frequency comb generation with tunable line spacing. The same is true for spectral broadening in KNWG, provided that the tunability is supported by the seed comb generator. For these devices, the line spacings are typically below 50 GHz. In contrast to that, the line spacing of combs generated in QD-MLLD or Kerr-nonlinear microresonators is determined by the cavity length and can hence not be tuned over a significant range. However, these devices also allow the access of large line spacings, which may range up to hundreds of gigahertz for very small cavities [79].

Comparing the demonstrated WDM transmission performances, EOM and GSLD are useful for aggregate WDM data rates up to a few Tbit/s at high SE. The transmission capacity is mainly limited by the available number of lines and by the line spacing, which restricts the usable symbol rate. The number of lines can be increased by using QD-MLLD as light sources, leading to aggregate WDM data rates in excess of 10 Tbit/s [15]. These light sources, however, may still suffer from comparatively strong phase noise, both in terms of intrinsic linewidth and with respect to a strong increase of the FM noise power towards small frequencies [14,80,81]. The strong phase noise has initially limited transmission demonstrations with QD-MLLD to rather simple modulation formats such as QPSK. Later experiments showed that these limitations may be overcome by dedicated phase-noise reduction schemes, relying, for example, on feed-forward phase-noise compensation of the carrier prior to modulation at the transmitter [82], advanced digital phase tracking [14], or self-injection locking, which allows for reducing the optical linewidths from a few MHz to less than 100 kHz [15]. Note that the viability of these schemes has been demonstrated using benchtop-type experimental setups, and that chip-scale integration has not yet been addressed. Note also that frequency combs from QD-MLLD exhibit RIN due to mode partition noise [28,29], but the impact on the performance of coherent transmission schemes is usually small, unless the comb is used as a multi-wavelength LO [51].

Regarding scalability to large WDM channel counts, Kerr combs and spectral broadening in KNWG appear to be the most promising concepts. Both approaches were shown to permit generation of combs that span more than an octave of frequencies [68,83], which would in principle allow to cover all optical telecommunication bands in the near infrared by a single device. Note, however, that spectral broadening in KNWG usually results in a comb with an irregular spectral envelope and may require spectral slicing techniques for high-performance WDM transmission [18]. In contrast to that, single DKS states lead to combs with smooth spectral envelopes that can be readily used for WDM transmission. Using a combination of two interleaved Kerr comb generators, WDM transmission with aggregate net data rates of more than 50 Tbit/s could be demonstrated, see Section 2.4.6 for a more detailed description.

Another important aspect of the various comb sources is the power consumption and the wall-plug efficiency, defined by, the ratio of the optical output power of the comb and the associated electrical input power. In this respect, quantitative considerations are only available for few exemplary cases [11,84,85] such that a broad comparison of different comb generation schemes based on commercial devices is not possible. Qualitatively, QD-MLLD seem attractive not only due to their simple operation scheme but also in terms of energy efficiency, converting a DC laser pump current directly into a frequency comb at the output of the device. In comparison to GSLD or EOM-based schemes, this avoids additional RF sources and amplifiers that may easily feature power consumptions of a few watts that come on top of the power consumption of the laser source [11,58]. Moreover, QD-MLLD do not contain any lossy optical coupling interfaces that would decrease the overall efficiency. Regarding KNWG, a very good optical power conversion efficiency of 66% from the seed comb to the broadened comb at the device output is reported [18]. Note, however, that this does not account for generation of the seed comb and that the overall efficiency of KNWG-based comb generators is always worse than that of the underlying seed comb source [84]. For Kerr combs, the main hurdle towards power-efficient operation is the optical conversion efficiency from the CW pump to the comb tones. Reported conversion efficiencies range from less than 1% up to 30% for bright and dark soliton combs, respectively [7,86]. Also here, this figure does not account for the wall-plug efficiency of the CW pump laser, which might need to be complemented by an optical amplifier. Nevertheless, despite the rather low efficiency of the nonlinear conversion process, Kerr comb generators or devices based on spectral broadening in KNWG can reduce the overall power consumption of a WDM transceiver by eliminating the need for individual frequency control of a multitude of CW lasers [84].

2.4 Kerr comb generators and their use in wavelength-division multiplexing

Based on the comparative analysis presented in Section 2.3, Kerr combs offer an attractive route toward highly scalable WDM transceivers with hundreds of channels and transmission capacities of tens of Tbit/s. Since Kerr comb generation is a particularly active field of ongoing research, such systems are likely to benefit from further scientific and technological progress regarding both design and implementation of the comb source. The following section gives an overview on the current state of the technology, explaining the basic principles and technological aspects of Kerr comb generation as well as the associated application demonstrations in the field of WDM transmission.

Kerr comb generation, as a field of research in its own right, was established in 2007 after a seminal paper by Del'Haye et al. [69], where a novel type of optical frequency comb generated from a CW-driven, high-Q optical microcavity was first experimentally demonstrated. Due to their extreme compactness and large range of available FSRs, spanning from microwave to THz, which are not generally easily accessible with other types of optical frequency combs (like fiber-based or Ti:Sapphire optical combs), these systems have quickly become the focus of intensive research effort. The field has evolved significantly over the past decade and a deep understanding of the fundamental aspects of Kerr comb formation and dynamics has been developed, and it has already reached the stage of notable experimental demonstration of powerful applications in spectroscopy, coherent communications, light detection and ranging (LIDAR), optical frequency synthesis, and others.

2.4.1 Principles and applications of microresonator comb generators

The key component, and historically the first resonator system, where the formation of Kerr combs were observed are optical WGM microresonators. Commonly these take the form of mm- or μm-size optical ring-cavities operating on the basis of total internal reflection, which is enabled by their typical microring, microtoroid, or microsphere shapes. WGM microresonators are able to reach ultrahigh Q-factors (up to 10^{11} [87]) and, depending on the choice of material and fabrication approach, can possess FSRs from below 10 to more than 1000 GHz with a high degree of light confinement.

The ultrahigh Q-factor and high finesse of such microresonators significantly facilitates access to optical nonlinear effects, even with moderate powers of laser light coupled to the cavity. Since most microresonator materials do not have inversion symmetry, the elementary nonlinear optical interactions in the system are typically presented by the third-order (Kerr) nonlinearity, which can be described with an intensity-dependent refractive index $n = n_0 + n_2 I....$, where I is the optical intensity, and n_2 is the nonlinear refractive index. Due to the intracavity intensity buildup and high confinement (which can reach up to several MW/cm^2 in a microresonator), even a comparably weak Kerr nonlinearity can lead to frequency conversion processes. The most important among such process is that of FWM (or "hyperparametric oscillations"), where two initial pump photons with frequency ω_p are parametrically converted into two photons with different frequencies (signal ω_s and idler ω_i) such that, according to energy conservation, $\hbar\omega_{p1} + \hbar\omega_{p2} = \hbar\omega_s + \hbar\omega_i$. The threshold for such oscillations in a microcavity scales as Q^{-2}, highlighting the strong advantage of these ultrahigh Q-factors.

Another important feature impacting the process of Kerr comb formation is the dispersion, or frequency-dependent refractive index, of the microresonator. The dispersion generally defines the bandwidth and envelope of the generated optical comb, as it induces the variation of the cavity FSR with wavelength. For the comb lines oscillating far from the pump, the dispersion can induce a large detuning from the corresponding cavity resonances, thus reducing their power. In order to describe the microresonator dispersion, it is often convenient to use the so-called "integrated dispersion," which is introduced as the frequency deviation of the comb lines from an equidistant frequency grid:

$$D_{\text{int}} = \omega_\mu - (\omega_0 + D_1\mu) = \sum_{k\geq 2} D_k \frac{\mu^k}{k!}, \tag{2.2}$$

where $\omega_\mu = \omega_0 + \sum_{k\geq 1} D_k\mu^k/k!$ is the Taylor series expansion of the frequency of the longitudinal modes of the microresonator, as a function of the mode index number μ, around the pump resonance at ω_0. The frequency spacing of the Kerr comb is introduced as $D_1 = 2\pi f_r$, and the following order term $D_2 = -cD_1^2\beta_2/n_0$ describes the cavity modal dispersion (in rad/s) where c is the speed of light, n_0 is the effective group refractive index, and β_2 is the waveguide group-velocity dispersion (GVD) (in s^2/m). Depending on the sign of D_2, the microresonator dispersion is called anomalous ($D_2 > 0$, higher frequencies are faster than the lower ones), or normal ($D_2 < 0$, lower frequencies are faster). The general requirement for Kerr comb formation in a microresonator is *anomalous* GVD, as the deviating microresonator modes are able to be compensated by the nonlinear frequency pulling of the generated FWM sidebands, allowing for their resonant enhancement. This effect is not possible under the conditions of normal GVD. In the absence of free carriers, the two main contributions to the microresonator dispersion are the material and geometrical dispersions. Material dispersion is defined by the dispersion of the microresonator medium (e.g., Si, Si_3N_4, MgF_2, etc.) and can usually be well approximated empirically with the Sellmeier equation. Geometrical dispersion depends on the geometrical parameters of the resonator including the profile of the resonator circumference (where WGM mode propagates), bending radius, and resonator waveguide dimensions. While the material contribution can hardly be tuned for a device of a given material, the geometrical dispersion provides a convenient tool to tune the dispersion within a broad range of wavelengths in order to access Kerr combs in different spectral windows and achieve desirable bandwidth and frequency comb envelope. This is particularly the case for photonic waveguides where the dimensions of the cross section can be finely controlled [83,88,89].

To date, the formation of Kerr combs has been already demonstrated in a large variety of nonlinear platforms, starting from SiO_2 microtoroids [69], Si_3N_4 [70], hydex [90], AlN [91] and $LiNbO_3$ microrings, MgF_2, CaF_2 and SrF_2 crystalline resonators and silica spheres [92], and also more recently in photonic waveguide materials such as AlGaAs [93,94], GaP [94]. The latter waveguide materials hold particular promise for their extremely high Kerr nonlinearity, enabling comb generation at very low threshold powers. They also possess, unlike in Si or Si_3N_4, a significant second-order nonlinearity, thereby enabling three-wave mixing processes such as sum- and difference-frequency generation and the electro-optic effect to influence frequency comb generation.

The advantages and disadvantages of different platforms are defined by the strength of the nonlinearity and ability to confine light in small mode volume, as well as available for the fabrication FSR and ability to controllably engineer dispersion and mode spacing of the resulting comb. In general, crystalline-based microresonator systems are able to overcome quality factors of microfabricated devices by several orders of magnitude but possess smaller nonlinearity and are more difficult and tedious in fabrication. On the other hand, the recently developed microfabrication processes are able to deliver devices in large volumes and provide precise control over microresonator FSR as well as large possibilities for dispersion engineering, which has enabled access to octave-spanning frequency combs [95].

2.4.2 Microresonator fabrication

The fabrication approach can vary significantly depending on the microresonator platform and materials used. Here, a brief review of the fabrication process for two currently widely used microresonator platforms is given.

The crystalline microresonators, such as MgF_2, CaF_2, and SrF_2, are typically fabricated from the crystalline preforms (rods), using a process that can be divided in two steps, as shown in Fig. 2.10. In the first step, the crystalline rod is shaped to have a small, axis-symmetric protrusion around its circumference. For this, the preform is fixed on the rotating spindle and diamond-turning machine is used to define the resonator protrusion. After this step, the roughness of the fabricated resonators can be rather high, and the second step, polishing, is needed. In the polishing step, the resonator is polished with diamond slurries in several stages using decreasing size of the diamond nanoparticles, down to tenth of nanometers. Thus, the resonator surface roughness can be brought below one nanometer rms, allowing the access to ultrahigh Q-factors above 10^9 (up to 10^{11}). Despite their low nonlinear refractive index and relatively large effective mode area, these microresonators can still have microwatt-level thresholds for hyperparametric oscillations resulting from exceptionally high-quality factors. Light can be evanescently coupled to such microresonators via coupling prism or tapered fiber.

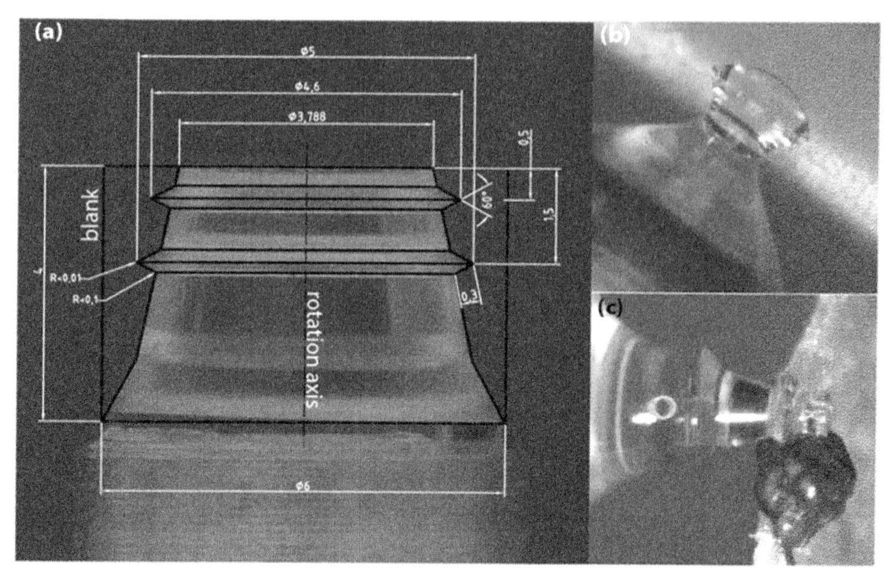

Figure 2.10 Microresonator fabrication. (A) MgF_2 resonator before the final polishing step. The sample holds two resonators. Superimposed is the technical drawing (length in units of mm) used for the diamond turning process. The original cylindrical magnesium fluoride blank is indicated by the black dash-dotted line. (B) Manually shaped resonator and abrasive diamond film (in the background). (C) Diamond turned resonator rotating in diamond particle paste to remove residual cracks originating from the diamond turning process.

On the other hand, the vast majority of on-chip microresonators, which are currently used for Kerr comb experiments, are produced using microfabrication techniques. The most advanced among these material platforms in terms of DKS generation is that of Si_3N_4. Si_3N_4 was introduced as a waveguide platform for Kerr comb generation by Levy et al. [70] as a suitable alternative to Si, which exhibited two-photon and free-carrier absorption at telecom wavelengths [96]. As Si_3N_4 has its bandgap located near the ultraviolet domain [97], it does not suffer from this problem, limiting the major sources of losses to impurities and waveguide roughness [98], and so is able to exhibit a high nonlinearity-to-loss ratio. There have been two major methods for the fabrication of low-loss Si_3N_4 waveguides: subtractive and Damascene, the process flows depicted in Fig. 2.11A and B, respectively.

The subtractive process is the most conventional process and has been used since the beginning of Si_3N_4 based microresonator production, and it is still widespread in the fabrication of waveguides of all materials. This process involves widely depositing a film of Si_3N_4 across the SiO_2 and Si substrate before etching into it through a patterned resist. The Si_3N_4 is commonly deposited using low-temperature chemical vapor deposition (LPCVD), and then a high-temperature annealing step is usually required before further material cladding in order to remove hydrogen-bond impurities left

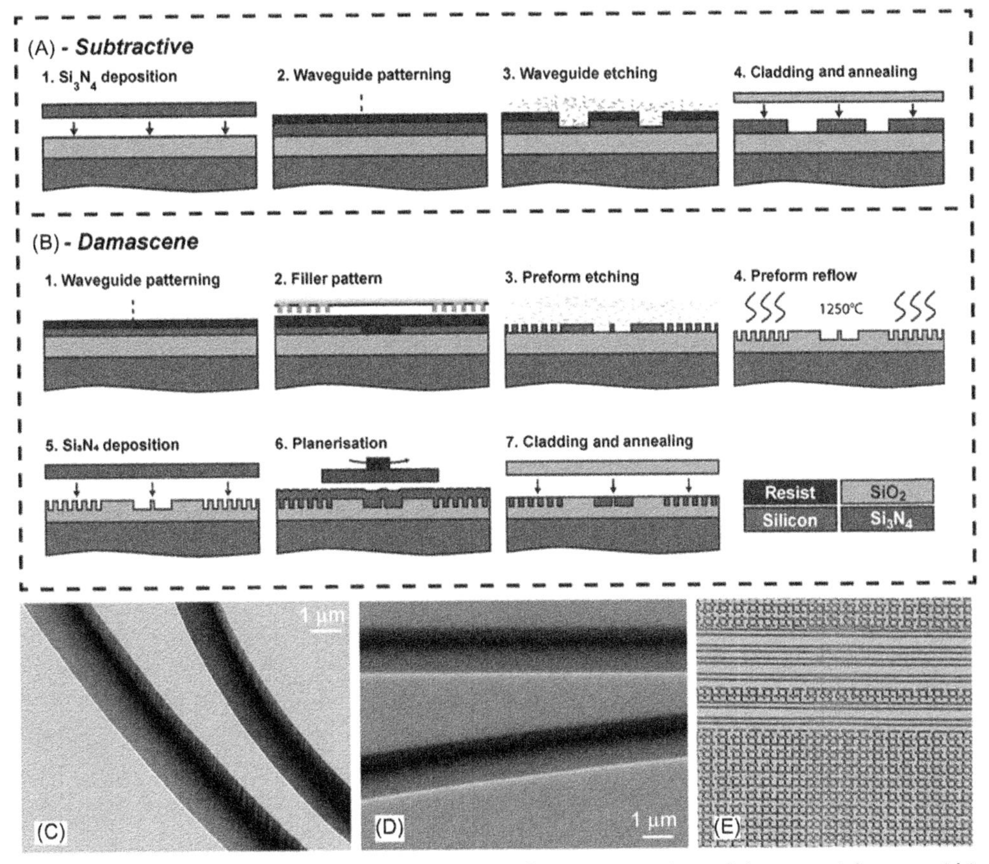

Figure 2.11 Microresonator fabrication steps. (A) Schematic overview of the essential steps within a conventional subtractive process. (B) Schematic process flow of the photonic Damascene process. (C) Scanning electron micrograph (SEM) of the etched SiO_2 preform before reflow. (D) Preform after reflow, made smooth. (E) Dense stress-release pattern inhabiting the space surrounding waveguides, preventing crack propagation.

inside the Si_3N_4 that cause significant absorption in the near-infrared [99]. The simple form of this process tends to leave significant roughness on the waveguide sidewall, but in the most current implementations of the method, this roughness has been drastically reduced with the use of hard masks, clearing polymer residue, and improving the stability of the e-beam lithography [100]. As a result, maximum Q-factors of microrings fabricated using the subtractive process have improved over the last few years from the order of 10^5 to now $>10^7$.

An alternative fabrication method, the *photonic Damascene* process, was developed recently [101]. In this method, a preform is etched into the SiO_2 cladding layer before Si_3N_4 deposition, such that the waveguides are formed in an additive way directly into the mold of the cladding. The excess Si_3N_4 deposited over the top of the SiO_2 layer is then removed by way of chemical mechanical polishing, leaving a highly smooth top surface. For creating smooth waveguide sidewalls, the preform can be put under a *reflow* stage where the SiO_2 is gradually heated slightly above its glass transition temperature in order to melt away fine roughness. With these steps implemented, microring Q-factors fabricated with the Damascene process have increased to a consistent average of above 10^7, allowing full frequency combs to be generated with comb line spacings less than 100 GHz for as little as 30 mW coupled on to the chip [77]. A further advantage, when accompanied with a dense interlocking "filler pattern" deposited between the waveguides, this method has demonstrated a high resilience to crack propagation, a severe problem for LPCVD Si_3N_4 [102]. In this way, waveguides thick as 2.3 μm have been fabricated for the purpose of nonlinear photonics, beyond the conventional maximum of 1.0 μm [103].

2.4.3 Application overview of Kerr combs and their spectral coverage

Due to the number of unique properties, such as high compactness, access to wide range of repetition rates from microwave to THz domains and compatibility with standard microfabrication processes, Kerr comb generators have quickly become potential candidates for various applications. A particular boost in the interest has happened, when a new operating regime of Kerr combs, associated with the formation of self-sustained intracavity pulses—DKS—has been demonstrated. The new *soliton* Kerr combs were shown to be fully coherent and feature low-noise performance. Their formation has been demonstrated in several microresonator platforms including crystalline microresonators with ultrahigh Q-factors such as MgF_2 [74], and in chip-integrated platforms such as SiO_2 microdiscs [104], or in Si_3N_4 [105], Si [106], high-index doped SiO_2 [107], and AlN microring resonators [108].

Since the discovery of soliton Kerr combs in 2014, the majority of research efforts have been mostly focused on the telecom band, determined by the wide range of available laser sources, amplifiers, and fiber components at around 1.55 μm. In recent years, however, DKS formation has also been demonstrated using driving lasers outside of this spectral region, including CW lasers operating at 2.80 μm [106], 1.30 μm [83], 1.06 μm [109,110], and 0.78 μm [109]. Two important spectral regions, whose investigation is strongly driven by the promising potential for applications, have been already covered by DKS-based combs. The first is the near-infrared domain from 0.7 to 1.3 μm, which is of particular interest for a large class of biomedical applications due to its high penetration depth in biological tissues. Another reason is the ability to interface with commonly used atomic transitions such as Rb or Cs, which can be used for the stabilization of chip-scale DKS sources for building compact atomic clocks. The second spectral region of high interest is the mid-infrared spectral band from 2.4 to

4.3 μm, mainly due to the broad range of molecular spectroscopy applications [111] (molecular fingerprinting) for medical, sensing, and military purposes. DKS in both spectral regions are being actively developed, with a further interest in shifting the operating region further toward the visible or infrared spectral regions.

After the discovery and experimental demonstration of DKS in optical microcavities, the Kerr combs have started to prove themselves in applications outside of research. Due to the formation of fully coherent optical combs, featuring broad bandwidth and low linewidth with large spacing of individual comb lines, they became viable for various typical applications of fiber-based optical combs and beyond. Over the last four years, we have witnessed an explosion of various experimental proof-of-concepts involving the soliton-based Kerr combs, which showed their advantages in a wide range of applications.

These applications include counting of the cycles of light by establishing a direct microwave-to-optical link via self-referencing assisted with external-fiber-based nonlinear broadening [112], or by exploiting its own dispersion-engineered broadband spectrum [113]; high-precision integrated optical frequency synthesis able to generate frequency-stable light from a microwave reference [79]; ultrafast distance measurements (LIDAR) with submicrometer resolution [114,115]; dual-comb spectroscopy for high-speed acquisition of linear absorption spectra [116–118]; low-noise microwave generation [119,120]; astrophysical spectrometer calibration [121,122]. Fig. 2.12 summarizes a number of applications of DKS microcombs in which either compactness is required or the high repetition rates of DKS microcombs are advantageous.

2.4.4 The physics of Kerr comb generation

The Kerr comb formation process is complex and typically involves several stages of evolution that depend on the pump power of the driving laser and its detuning from the pumped resonance.

A simple method to access the different stages, which is also used in a majority of the experiments for Kerr comb generation, is to tune the CW laser, at a fixed pump power, into a microcavity resonance from the blue-detuned side, see Fig. 2.13. This enables a gradual increase of the intracavity power, and once it reaches the threshold for hyperparametric oscillations, the system starts to develop the first (primary) parametric sidebands when the cavity decay-rate is exceeded by the parametric gain from the FWM process. The spectral position of the sidebands is defined by the GVD term D_2, and in the first-order approximation, they appear $\mu_{th} \approx \sqrt{\kappa/D_2}$ FSRs away from the pump. These first sidebands appear as a result of the degenerate FWM process, where two initial pump photons have the same frequency. As the power in these first two sidebands increases, cascaded nondegenerate FWM takes place, which start to fill the envelope of the comb with optical lines spaced by $\Delta = \mu_{th} \cdot f_r$, forming the so-

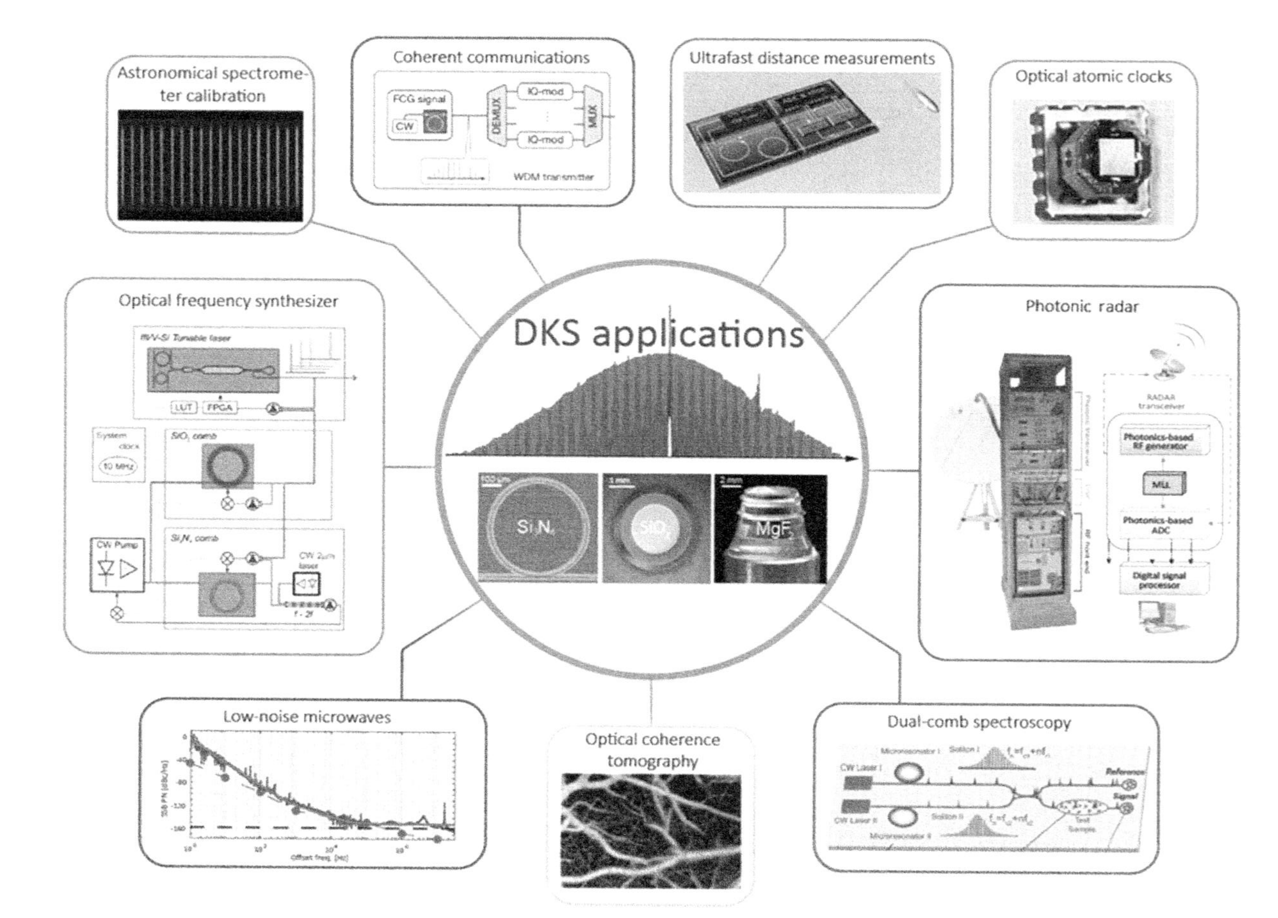

Figure 2.12 Applications of dissipative Kerr solitons (DKS).

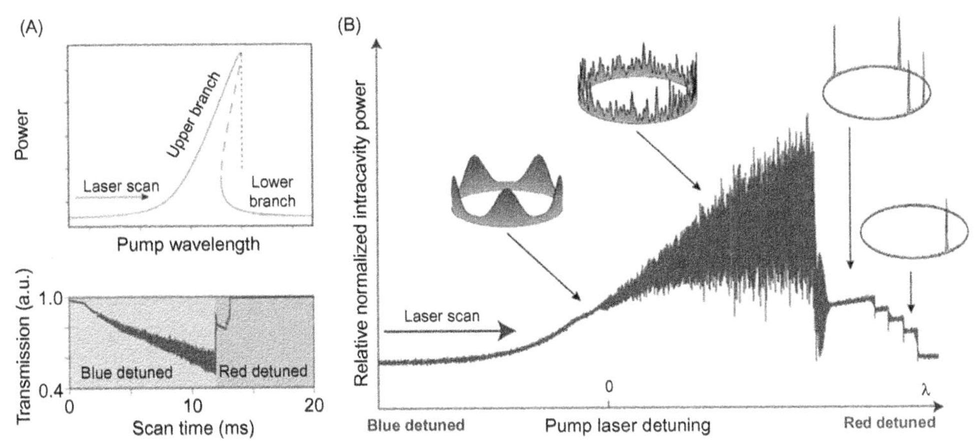

Figure 2.13 Kerr comb generation and transition to dissipative soliton states. (A) Top: sketch of the intracavity power evolution as the pump wavelength is scanned over the resonance in the nonlinear microresonator. Due to the nonlinearity, the system becomes *bistable* over a certain range. Here is where solitons are able to coexist stably with a continuous-wave (CW) background. Bottom: experimentally measured transmission and detuning of the pump as it scans over the resonance of a MgF_2 microresonator. (B) Generation of Kerr combs and dissipative Kerr soliton (DKS) states. Figure shows the evolution of the total intracavity power as the pump laser is scanned over the cavity resonance. Insets show intracavity waveforms in each of the stages. From left to right: primary comb, merged secondary comb, multisoliton state (consisting of four DKS pulses and CW background), and single-soliton state.

called "primary comb." Depending on the dispersion, μ_{th} can be 1, leading to the formation of the fully populated comb (natively mode-spaced), which, however, has a strongly reduced bandwidth, limiting the number of available optical lines for applications. Such primary combs, despite having a large comb spacing, feature high coherence and low linewidth for the generated comb lines. Due to this fact, this stage of the Kerr comb formation can already be used for certain applications, and particularly for telecom applications, such as direct data transmission [123].

The second stage of Kerr comb formation starts when the intracavity power grows further by reducing the detuning of the driving laser from the pumped resonance. The large bandwidth of the FWM gain lobes can initiate parametric oscillations in the resonances between the primary lines, which is also referred to as subcomb formation. The growing subcombs can merge and form the gap-free, fully populated optical comb spectrum. In the general case, the mode spacing of the subcombs (δ) does not necessary commensurate with the primary comb spacing (Δ), leading to the situation where multiple optical lines may oscillate in the same cavity resonance. This results in structured and broadband comb lines (as well as their RF beatnotes) and very poor coherence of the generated Kerr optical comb. Such states are often referred to as noisy Kerr combs or chaotic combs, highlighting the fact that the intracavity

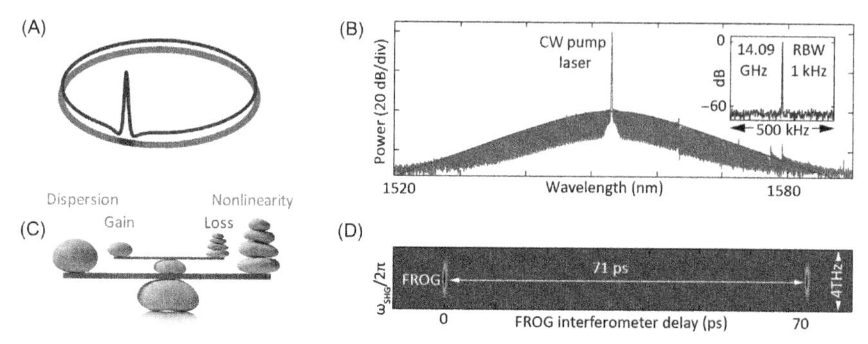

Figure 2.14 (A) Intracavity waveform of a single dissipative Kerr soliton (DKS) state in the microring resonator. (B) Experimentally realized single soliton state in a 14-GHz-FSR MgF_2 crystalline microresonator. The inset shows the low-noise cavity repetition-rate beatnote. (C) Double balance of the dispersion, nonlinearity gain and losses in the DKS state. (D) Frequency-resolved optical gating (FROG) trace of the DKS state shown in (B).

waveform is chaotic in the time domain. Despite having a large bandwidth and populating every cavity resonance, these combs can hardly be used for most applications, including telecommunications and spectroscopy, due to their broad linewidth. Several works have reported that chaotic combs can also operate in low-noise regimes, as in $\delta - \Delta$ matching (when Δ corresponds to an integer number of δ) [124] or when using the parametric seeding technique [125]. But in all cases, they have either lacked the clear and robust mechanism to access such states, or imposed limitations on the available FSRs of the Kerr comb.

Notwithstanding, several applications of Kerr combs (mostly in such low-noise regimes) were still demonstrated at this early stage of the field evolution, which included pulse detection and shaping [126], optical coherent telecommunications with complex modulation formats, microresonator-based optical clocks [127], and optical multicasting [128].

2.4.5 Dissipative Kerr solitons

A significant step forward in the field of Kerr combs was made in 2014, when the low-coherence problem had been addressed through the formation of intracavity optical solitons [74]. Earlier, several theoretical works had predicted that a CW-driven Kerr-nonlinear microcavity could support the formation of a certain type of spatially localized intracavity waveforms—optical solitons—leading to the mutual phase-locking of the comb lines and drastically reducing their linewidth. At the same time, several experiments with CW-driven fiber cavities had also demonstrated that similar optical soliton pulses could be synthesized in fiber resonators [129].

In 2014, Herr et al. experimentally demonstrated that this phenomenon could also be observed in a crystalline MgF_2 WGM microresonator. It was shown that under CW driving, the system could spontaneously synthesize single or multiple optical pulses circulating inside the cavity—DKS. See Fig. 2.14. Despite being dissipative structures, these solitons are continuously supplied with the energy from the CW drive (pump), whose photons are redistributed among other frequencies through the FWM processes. Stable operation of the Kerr-nonlinear microcavity in the soliton regime is enabled by a dynamical double balance between the dispersion of the cavity and nonlinearity, which counteract to maintain the pulse duration during its circulation inside the cavity, as well as continuous driving and cavity losses. In the frequency domain, soliton pulse formation manifests as a fully coherent frequency comb spectrum with a well-defined envelope.

Analytically, the system can be described very well with the damped-driven nonlinear Schrodinger equation, which for the simplest case of the dispersion limited to the second-order term (D_2) is equivalent to the so-called Lugiato–Lefever equation (LLE):

$$\frac{\partial A}{\partial t} - i\frac{1}{2}D_2\frac{\partial^2 A}{\partial \phi^2} - i\gamma|A|^2A = -\left(\frac{\kappa}{2} + i\big(\omega_0 - \omega_{\mathrm{p}}\big)\right)A + \sqrt{\frac{\kappa_{\mathrm{ex}}P_{\mathrm{in}}}{\hbar\omega_0}}, \tag{2.3}$$

where $A(\phi, t)$ is the slowly varying intracavity field amplitude, $\phi = \varphi - D_1 t$ is the corotating angular coordinate, and φ is the regular polar angle of the resonator. Input power of the laser is denoted as P_{in}, κ_{ex} is the energy coupling rate to the cavity, and $\kappa = \kappa_{ex} + \kappa_0$ is the total cavity decay rate with internal losses κ_0. The nonlinear coupling coefficient

$$\gamma = \frac{\hbar\omega_0^2 c n_2}{n_0^2 V_{\mathrm{eff}}} \tag{2.4}$$

describes the Kerr nonlinearity, with nonlinear refractive index n_2, and effective mode volume V_{eff}.

An approximate analytical solution for the field amplitude of the DKS pulse has a characteristic sech^2-shape profile:

$$A(\phi, t) \approx \sqrt{\frac{\kappa(\omega_0 - \omega_{\mathrm{p}})}{\gamma}}\,\mathrm{sech}\left(\sqrt{\frac{(\omega_0 - \omega_{\mathrm{p}})}{D_2}}\phi\right), \tag{2.5}$$

where one can see that the pulse duration scales inversely (and correspondingly, the resulting comb bandwidth) with the pump-cavity detuning and is inversely proportional to the cavity-mode dispersion term. When using the maximum detuning value available for soliton generation (i.e., critical detuning), it is possible to compute the maximum power per line for the optical comb exiting the resonator:

$$P(\mu) \approx \frac{\kappa_{ex} D_2 \hbar \omega_0}{4\gamma} \mathrm{sech}^2 \left(\frac{\mu \kappa}{2} \sqrt{\frac{D_2 \hbar \omega_0}{\gamma \kappa_{ex} P_{\mathrm{in}}}} \right). \tag{2.6}$$

In order to excite the DKS state, it is necessary to have a seed pulse (or pulses) inside the cavity, while maintaining two key parameters of the system—pump power and detuning—in a certain region where the soliton state can exist stably, that is, the soliton existence range. It happens that the chaotic intracavity waveform obtained in the modulation instability state (the noisy Kerr comb) is a very good source of such seed pulses. Thus, the typical procedure for DKS excitation, which historically first appeared for DKS in microresonators, is a one-way tuning of the CW pump, at fixed power, over the cavity resonance from its blue-detuned side to its red-detuned side. During this approach, the system consecutively passes all the stages of Ker comb formation, resulting in the chaotic intracavity waveform consisting of many transient optical pulses. Once the pump passes over the resonance and enters the region of bistability on the red-detuned side (the soliton existence range), the system transitions into the DKS state with single or multiple solitons forming out of the high-energy pulses of the preceding chaotic field, while the remainder of the cavity field collapses to the low energy CW state. In order to facilitate this process, as well as minimize the impact of additional thermal instabilities, the detuning speed needs to be carefully chosen. After the pump sweep is stopped in the soliton existence range and is kept within, the system can maintain the DKS operating regime.

Two main challenges that, however, are often faced during the process of soliton generation are (1) strong thermal effects, which due to the microresonator thermal nonlinearity can destabilize the pump-cavity detuning in the soliton existence range, prohibiting direct access to the soliton states from the preceding high-energy state; and (2) access to a *single* soliton state, when there is only one pulse inside the cavity, which forms a smooth and predictable sech^2 envelope on the comb spectrum. In most cases, especially for chip-scale devices, the number of initially generated solitons is large. Both challenges have been successfully addressed using more sophisticated approaches for pump tuning. Overcoming thermal effects is usually possible by using fast-sweeping of the pump with EO modulators (DQPSK or intensity modulators), where the solitons are able to be formed at a greater speed than the cavity thermal relaxation time [88]. Additionally, locking techniques allow the cavity detuning to be fixed, even in the cases when the cavity resonance shifts due to a change in the cavity temperature. For deterministic access to a single soliton state, a backward tuning technique [130] was developed, which allows on-demand access to DKS states with any soliton number via soliton "switching," and which was successfully employed in multiple applications for accessing single-soliton states.

The majority of the first works with the DKS have been mainly focused on the fundamental aspects of their excitation and dynamics. Similar to pulse propagation in silica optical fibers, a DKS propagating in a microresonator experiences a compounded impact of the dispersive, nonlinear, and thermal properties of the cavity medium. The resulting effect on soliton dynamics, however, can deviate significantly from the case of fiber solitons. A good example of such a difference is in the effect of stimulated Raman scattering, and in particular, intrapulse Raman scattering, which in the fiber case leads to the formation of a copropagating Raman soliton pulse whose spectrum is continuously shifting towards longer wavelengths (towards the red). In contrast, the Raman effect in microresonators induces the formation of stable Raman-locked solitons [131–133], with a fixed red-shifted spectrum, which essentially correspond to a new DKS state with a triple-energy-balance between pumping, cavity losses and the Raman scattering. Similarly distinct behavior is also observed for soliton-induced dispersive waves [105,134], which in the case of DKS are locked to the soliton and provide a coherent extension of the DKS spectrum in the regions of normal GVD. The discovery of this fact has enabled significant broadening of the DKS comb sources and has essentially paved the way to frequency metrology applications, where a broad bandwidth (and in particular, octave-spanning bandwidth) is necessary for self-referencing of the comb sources [79,113].

Despite having essentially only two control parameters—pump power and detuning—the system possesses a high level of complexity due to its nonlinear dissipative nature, resulting in rich physics, and a plethora of nonequilibrium states. Apart from the stable stationary regimes of single- and multiple-soliton states, which correspond to one or several randomly distributed pulses inside the cavity, the system can also form stable nonstationary states, where the DKS pulses experience a periodic evolution in terms of their amplitude and duration, that is, breathing [135–137]. It was shown that such phenomena bear resemblance to a long-standing puzzle of nonlinear dynamics, called Fermi–Pasta–Ulam recurrence [138]. DKS breathing begins with a stable, continuous periodicity, but it has also been shown to transition into a fully chaotic cycle [139]. This breathing can be induced naturally as part of the fundamental Kerr cavity dynamics, but also as a result of higher-order perturbations such as stimulated Raman scattering [140] or interspatial mode interactions [141]. Coupling between spatial modes in particular has been shown to induce dispersive waves that serve to both stabilize and destabilize multiple soliton states in different cases [142,143]. One dramatic example of this is where a high-number soliton state organizes itself into a periodic or near-periodic lattice, otherwise known as a soliton crystal [144], forming a highly regular spectral pattern depending on the varieties of defects appearing in the temporal lattice. These complex dynamics together have a marked impact on the profile of the frequency comb, as well as its long-term stability, and therefore may have to be taken into account for applications.

Besides bright DKS states, the microresonator system has also been shown to support another type of spatially localized structures—platicons—that can provide an interesting alternative to the solitons. Platicons are flat-top pulses resulting from the interlocking of two switching waves [145,146]. Platicons can occupy a small or large part of the cavity, correspondingly forming short bright or dark pulses in the microresonator. Platicons tend to have a much higher conversion efficiency from the pump to the generated comb lines and, in contrast to bright DKS, are formed in the regions of normal dispersion [147]. Despite these advantages, the platicons are not yet able to replace DKS, and have not been as widely used for applications. Platicons have a non-deterministic procedure for their excitation, which requires the presence of avoided modal crossings between the transverse mode families of the microcavity. The first major application based on platicons has been demonstrated recently for high-order coherent telecommunications [10].

One of the strongest features of DKS-based combs is their ability to access a broad range of comb mode spacings, which is particularly well developed for the on-chip devices. To date, a variety of mode spacing in chip-scale DKS combs have been demonstrated ranging from the GHz level (which is typical for more established systems such as Ti:Sapphire systems) up to as high as several THz, which has been hardly explored previously. Another advantage is that the geometrical dispersion can be precisely engineered via strict control of the rectangular dimensions of the waveguide cross section, enabled by the microfabrication processes. Dispersion can be controlled in this way commonly up to the fourth order, where the ultimate achievable comb bandwidth will be limited, possibly at an octave [83,88]. Even higher order dispersion engineering has been shown to be possible in longitudinally modulated waveguides and in waveguides with more exotic cross sections [89,148], potentially increasing the bandwidth or flattening the comb profile further. Moreover, photonic integration of soliton Kerr comb sources allows them to be mass-produced in a compact form, and integrated with additional optical components such as lasers and modulators on the same chip.

For these reasons, DKS-based combs in integrated devices make a suitable and relatively cheap multi-wavelength light source for massively parallel WDM telecommunications, which can serve as a set of coherent carriers and LO, which have low linewidth and can be adapted to any industrial frequency grids.

2.4.6 Massively parallel wavelength-division multiplexing transmission using dissipative Kerr soliton comb

The viability of DKS comb generators for WDM applications has been demonstrated in multiple experiments that are described in more detail in the subsequent sections [7]. In the first experiment, data was transmitted on 94 carriers that span the entire telecommunication C and L bands with a line spacing of about 100 GHz. Using

16QAM as a modulation format, an aggregate line rate of 30.1 Tbit/s was achieved. In the second experiment, two DKS combs were interleaved to double the number of carriers and to improve the SE. This led to a total of 179 carriers and an aggregate line rate of 55.0 Tbit/s transmitted over a distance of 75 km. In the third experiment, coherent transmission with DKS comb generators both at the transmitter and the receiver was demonstrated. This experiment relied on 93 carriers and led to an aggregated line rate of 37.2 Tbit/s. Note that these line rates include the overhead required for FEC and that they are therefore slightly higher than the net data rates specified in the last three rows of Table 2.3.

2.4.6.1 Data transmission with single and interleaved soliton

The general concept of massively parallel data transmission using a DKS frequency comb as a multi-wavelength light source is depicted in Fig. 2.15A. In laboratory experiments, massively parallel WDM transmission is usually emulated by using simplified schemes that rely, for example, on having only two independent data streams on neighboring WDM channels rather than on a dedicated WDM transmitter for each channel [66,84,149]. In the first transmission experiment [7], a line spacing of approximately 100 GHz and a symbol rate of 40 GBd were used in combination with band-limited Nyquist pulses that have approximately rectangular power spectra, see Fig. 2.15B. As a modulation format, 16QAM was chosen. At the receiver, each channel was individually characterized using a CW laser as local oscillator in combination with an optical modulation analyzer, which extracts signal quality parameters such as the error-vector magnitude and the BER. The BER values of the first transmission experiment are depicted as triangles in Fig. 2.16, with different BER thresholds indicated as horizontal dashed lines. Of the 101 carriers derived from the comb in the C and L bands, 94 channels were used for data transmission, resulting in a total line rate of 30.1 Tbit/s. In this experiment, the transmission capacity was restricted by the fact that the line spacing of about 100 GHz substantially exceeds the signal bandwidth of about 40 GHz, leading to unused frequency bands between neighboring channels, as shown in Fig. 2.15B, and hence to a rather low SE of 2.8 bit/s/Hz.

These restrictions could be overcome in a second experiment by using interleaved frequency combs, as shown in Fig. 2.15C. This scheme used a pair of DKS combs with practically identical line spacing, which are shifted with respect to each other by half the line spacing. At the receiver, the demonstration experiment still relied on individual CW lasers as LO for coherent detection. The line spacing of approximately 50 GHz enabled dense packing of the 40-GBd data channels in the spectrum, as shown in Fig. 2.15D. The BER results of the second transmission experiment are depicted as diamonds in Fig. 2.16, with different thresholds indicated as horizontal *dashed lines*. For a given FEC scheme, these thresholds define the maximum BER of

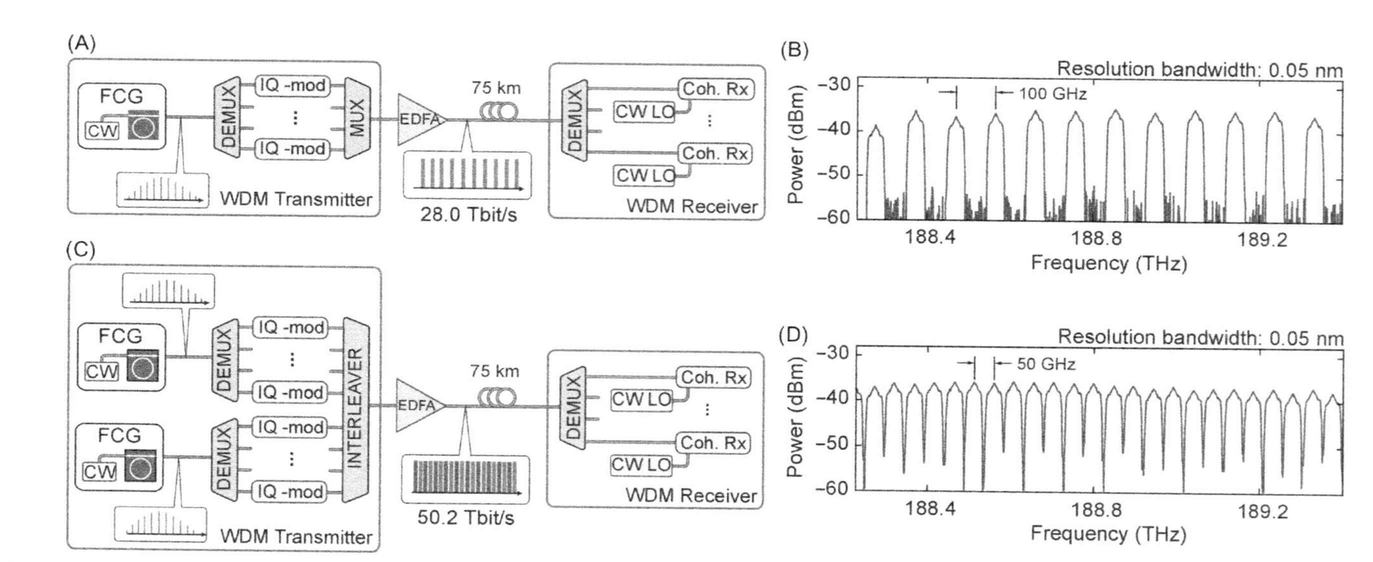

Figure 2.15 Coherent data transmission concept using dissipative Kerr soliton (DKS) comb sources at the transmitter. (A) Principle of data transmission with a single DKS source, similar to the scheme shown in Fig. 2.2. *CW LO*, Continuous-wave local oscillator laser; *Coh. Rx*, coherent receiver; *DEMUX*, demultiplexer; *EDFA*, erbium-doped fiber amplifier; *FCG*, frequency comb generator; *IQ-mod*, dual-polarization in-phase/quadrature modulator; *MUX*, multiplexer. (B) Section of the optical spectrum of the wavelength-division multiplexing (WDM) data stream recorded at the WDM transmitter output. (C) Principle of data transmission with two interleaved DKS combs at the transmitter. (D) Section of the optical spectrum of the WDM data stream recorded at the WDM transmitter output using interleaved DKS combs at the transmitter.

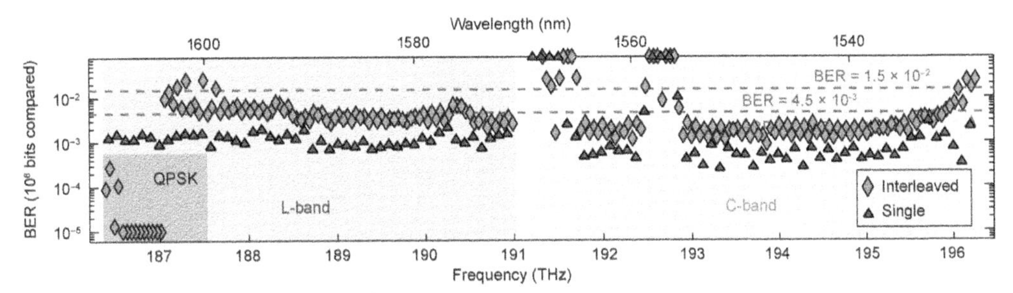

Figure 2.16 Results of transmission experiments using dissipative Kerr soliton (DKS) combs for massively parallel wavelength-division multiplexing (WDM). The triangles (diamonds) show the BER values of the transmitted channels for the single-comb (interleaved) experiment. The dashed lines show the bit error ratio (BER) thresholds for error-free propagation when applying FEC schemes with 7% overhead (4.5×10^{-3}, *lower dashed line, orange*) and 20% overhead (1.5×10^{-2}, *upper dashed line, blue*).

the raw data channel that can still be corrected to a BER of less than 10^{-15}, which is considered error-free [150]. In the experiment, 179 out of a total of 204 carriers in the C and L bands could be used for data transmission. Using a combination of 16QAM and QPSK as modulation formats, a total line rate of 55.0 Tbit/s and a net data rate of 50.2 Tbit/s were achieved, leading to a net SE of 5.2 bit/s/Hz. These data rates were not limited by the performance of the DKS comb source, but by the underlying experimental transmission setup [84].

As a benchmark, the transmission performance of a single comb line was compared to that of a reference carrier derived from a high-quality benchtop-type external-cavity laser (ECL). The ECL features an optical linewidth of approximately 10 kHz, an optical output power of 15 dBm, and an OCNR in excess of 60 dB. As a metric for the comparison, the OSNR penalty at a reference BER of 4.5×10^{-3} was used, which corresponds to the threshold for FEC with 7% overhead [150]. The results for 40-GBd 16QAM transmission are shown in Fig. 2.17A for three different comb lines and for ECL reference transmission experiments at the corresponding comb line frequencies. For all cases, an OSNR penalty of 2.6 dB with respect to the theoretically required value (black line in Fig. 2.17A) is observed for a BER of 4.5×10^{-3}, that is, there is no additional penalty that can be attributed to the comb source. DKS-based comb sources can hence markedly improve the scalability of WDM systems without impairing the signal quality. The error floor in Fig. 2.17A is attributed to transmitter nonlinearities and electronic receiver noise of the experimental setup. Fig. 2.17B shows the measured constellation diagrams for the ECL and the comb line at 193.56 THz, both taken at the same OSNR of 35 dB.

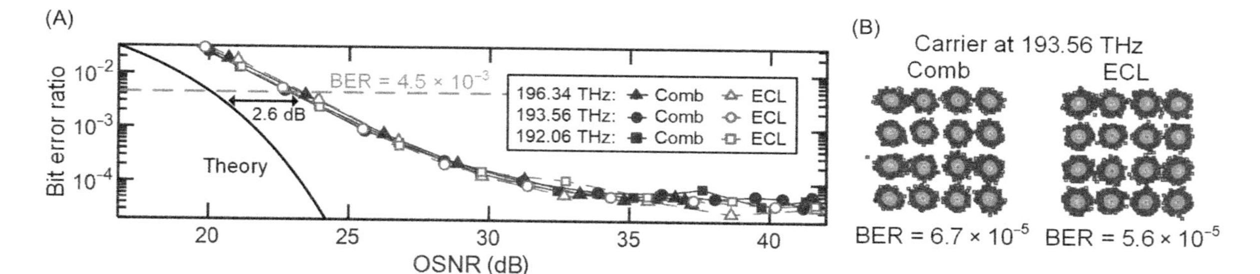

Figure 2.17 Transmission performance of single comb lines compared to that of reference carriers derived from high-quality benchtop-type external-cavity laser (ECL) (A) Measured bit error ratio (BER) versus optical signal-to-noise power ratio (OSNR) of three different channels, centered at frequencies of 196.34, 193.56, and 192.06 THz and derived from a dissipative Kerr soliton (DKS) frequency comb (*open symbols, blue*) and a high-quality ECL (*filled symbols, red*). All experiments were performed with 16QAM signaling at 40 GBd. The comb lines do not show any additional penalty compared to the ECL tones. The black solid line ("Theory") indicates the theoretical dependence of the BER on the OSNR for an ideal transmission system. (B) Constellation diagrams obtained for an ECL and DKS comb tone at a carrier frequency of 193.56 THz.

2.4.6.2 Data transmission with solitons both at the transmitter and at the receiver

DKS frequency combs may also be used as multi-wavelength LO to realize highly scalable WDM receivers. A demonstration experiment is schematically shown in Fig. 2.18A [7]. At the transmitter, a first DKS comb generator with a FSR of about 100 GHz was used as a multi-wavelength source. At the receiver, a second DKS comb source with approximately the same FSR was used to generate the corresponding LO tones for intradyne detection. The LO tones feature an optical linewidth of less than 100 kHz [84]. In this experiment, an aggregate data rate of 34.6 Tbit/s was transmitted on 99 WDM channels. As a reference, the same experiment was repeated using a high-quality ECL with a 10 kHz linewidth as LO for channel-by-channel demodulation. The resulting BER values for both types of LO are shown in Fig. 2.18B. Overall, no considerable penalty is observed that could be systematically attributed to using the DKS comb as a local oscillator.

2.4.6.3 Progress toward integrated wavelength-division multiplexing transceiver modules

DKS combs have a vast potential as multi-wavelength light sources in a variety of applications, see Fig. 2.12. To leverage this potential, integration of DKS comb generators into chip-scale photonic modules is of crucial importance. A first important step in this direction was the reduction of the required pump power levels [76–78], which has enabled operation of the devices by cointegrated CW light sources [78], see Section 2.3.5 for a more detailed discussion. This opens a path toward more complex modules that combine DKS comb generators with other photonic integrated circuits for processing and detection of optical signals. A vision of a chip-scale WDM transmitter is illustrated in Fig. 2.19 [19]. The overall concept corresponds to the architecture shown in Fig. 2.2A, combining a WDM DEMUX for separating the comb lines, an array of variable optical attenuators and I/Q modulators for spectral flattening and for encoding of data onto the lines, and a MUX that combines the data streams into a single-mode output fiber. The implementation illustrated in Fig. 2.19 exploits the concept of photonic multi-chip integration, which allows to combine individually optimized and tested dies into a powerful chip-scale system. Chip-chip and fiber-chip connections are realized by so-called photonic wire bonds [151–153], which offer high flexibility with respect to the devices that are to be connected and allow for fully automated module assembly. In the illustrated system, the CW light source relies on highly efficient III–V lasers while the Kerr-nonlinear microresonator for DKS comb generation exploits low-loss Si_3N_4 waveguides. Demultiplexing, flattening, and modulation of the carriers and multiplexing of the WDM signals may be accomplished by a single silicon photonic transmitter chip, as illustrated in Fig. 2.19. Alternatively, MUX and DEMUX can be realized on a separate die, allowing, for example, to exploit an

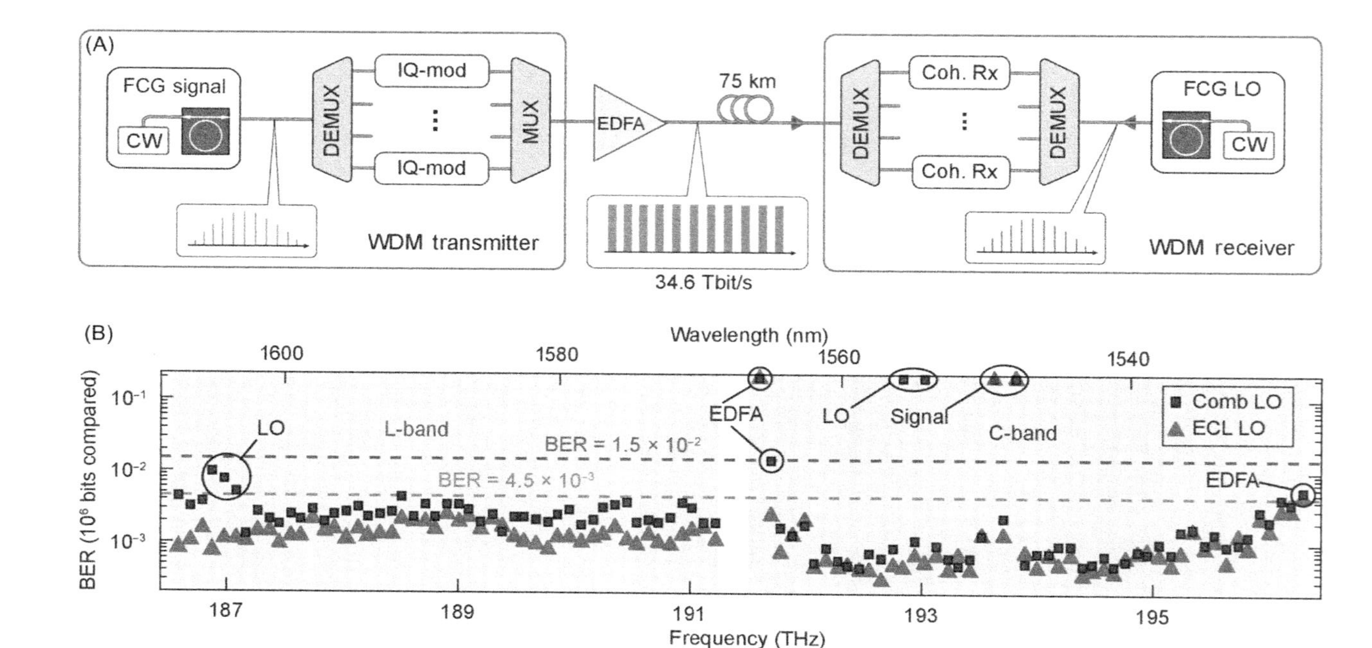

Figure 2.18 Coherent data transmission using dissipative Kerr soliton (DKS) combs both at the transmitter and at the receiver. (A) Principle of massively parallel wavelength-division multiplexing (WDM) with DKS comb generators used as multi-wavelength sources at the transmitter and the receiver. (B) Measured bit error ratio (BER) for each WDM channel. Squares indicate the results obtained when using a DKS comb as a multi-wavelength local oscillator (LO), and triangles show the results of a reference measurement using a high-quality external-cavity laser (ECL) as LO. Dashed lines mark the BER thresholds of 4.5×10^{-3} (1.5×10^{-2}) for hard-decision (soft-decision) FEC with 7% (20%) overhead. Black circles show the channels with BER above the threshold for 7% FEC. The associated impairments are caused by the specific implementation of the transmission experiment and do not represent a fundamental problem of the transmission scheme [7].

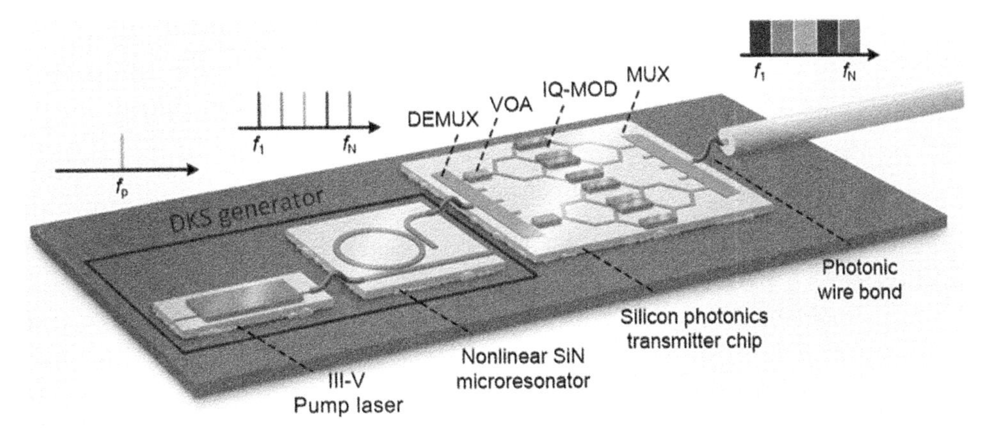

Figure 2.19 Illustration of a future chip-scale terabit-per-second transmitter, leveraging a Kerr frequency comb generator (FCG) as optical source. The basic concept corresponds to the architecture illustrated in Fig. 2.2A. Chip-chip and fiber-chip connections are realized by so-called photonic wire bonds [151–153]. *DEMUX*, Demultiplexer; *IQ-Mod*, *IQ*-modulator; *MUX*, multiplexer; *VOA*, variable optical attenuator. Source: *Adapted from J. Pfeifle, V. Brasch, M. Lauermann, Y. Yu, D. Wegner, T. Herr, et al., Coherent terabit communications with microresonator Kerr frequency combs, Nat. Photonics. 8, 2014, 375–380. https://doi.org/10.1038/nphoton.2014.57.*

integration platform that is optimized towards athermal behavior. The WDM transmitter system illustrated in Fig. 2.19 is only one example of the tremendous opportunities that might be unlocked by combining DKS comb sources with high-performance photonic circuits in a common multi-chip module. The concept may be transferred to other application fields such as metrology and sensing [114].

2.5 Conclusions

Chip-scale optical FCGs are likely to become key elements of future WDM transceivers, enabling efficient scaling of channel counts and transmission capacity. The devices can act as multi-wavelength light sources that produce a multitude of optical carriers at the WDM transmitter or as a multi-wavelength local oscillator for massively parallel coherent reception at the receiver. In both cases, the comb generators must offer low phase noise along with large line spacings of several tens of gigahertz while being amenable to integration into chip-scale transceiver assemblies. This chapter provides an overview of different comb generator concepts and introduces key performance metrics that allow their evaluation and benchmarking with respect to WDM applications. Among the various approaches, comb generators based on dissipative Kerr solitons (DKS) stand out as particularly promising options. DKS combs feature broadband and smooth spectral envelopes with tens or even hundreds of narrowband tones, thereby enabling massively parallel WDM transmission with aggregate data rates of tens of Tbit/s. The chapter further provides an overview on the current state of DKS comb generator technology, covering basic device principles, technological aspects, and application demonstrations in the field of WDM transmission.

Appendix A

Appendix A.1 details the derivation of Eq. (2.1), which expresses the OSNR at the receiver, $\mathrm{OSNR}_{\mathrm{Rx}}$, as a function of the power per line P_ℓ of the frequency comb for different link distances [32]. The equation is derived considering the transmitter setup and channel structure depicted in Fig. 2.4A. Appendix A.2 gives the expression of the required OSNR at the receiver for a desired target BER, $\mathrm{OSNR}_{\mathrm{sig}}$, for a given modulation format and symbol rate; see dashed lines in Fig. 2.4B.

A.1 Calculation of optical signal-to-noise power ratio (OSNR) at the receiver

For WDM applications, the power and OCNR per line of a frequency comb will dictate the maximum achievable OSNR of the WDM channel at the receiver, $\mathrm{OSNR}_{\mathrm{Rx}}$. To estimate the $\mathrm{OSNR}_{\mathrm{Rx}}$, the setup depicted in Fig. 2.4A is considered. The noise power at the output of the FCG is given by

$$P_n = S_n B_{\mathrm{ref}} = \frac{P_\ell}{\mathrm{OCNR}_\ell}. \tag{2.7}$$

where S_n is the co-polarized noise power density, and $B_{\mathrm{ref}} = 12.5$ GHz is the reference noise measurement bandwidth. An optical amplifier with gain $G_0 > 1$ increases the comb line power P_ℓ and the noise power $P_{\mathrm{n},\ell}$ to $P_{\ell,0}$ and $P_{\mathrm{n},0}$, respectively, and adds some co-polarized noise of its own,

$$P_{\ell,0} = G_0 P_\ell, \qquad P_{n,0} = \underbrace{S_n G_0 B_{\mathrm{ref}}}_{\text{amplified OFC noise}} + \underbrace{F_0 hf(G_0 - 1) B_{\mathrm{ref}}}_{\text{ASE noise of comb amplifier}}. \tag{2.8}$$

In these relations, the quantity F_0 is the noise figure of the amplifier after the comb source, and hf is the photon energy. The noise power $P_{n,0}$ is reduced by a factor $g_0 < 1$ through the loss in the WDM modulator, comprising a DEMUX, an array of electro-optic modulators, and a MUX, Fig. 2.2A. A post-amplifier with gain $G_1 > 1$ increases the signal, but also contributes noise,

$$P_s = G_1 g_0 P_{\ell,0}, \quad P_{n,1} = G_1 g_0 P_{n,0} + F_1 hf(G_1 - 1) B_{\mathrm{ref}}. \tag{2.9}$$

This power is then launched into the first fiber section, which attenuates the power by $g_1 < 1$ such that $G_1 g_1 = 1$. This fiber span might be followed by $(\nu_{\max} - 1)$ additional fiber sections, each attenuating the power by $g_\nu < 1$ and by the same number of additional amplifiers with a gain G_ν such that $G_\nu g_\nu = 1$. Assuming that all $\nu_{\max}$ links are identical ($G_\nu = G$, $g_\nu = g$, $Gg = 1$, $F_\nu = F$ for $\nu = 1 \ldots \nu_{\max}$), the signal power and the noise power after the transmission link are given by

$$P_{s,\nu_{\max}} = g\,P_s, \quad P_{n,\nu_{\max}} = g\,P_{n,1} + (\nu_{\max} - 1)gFhf(G-1)B_{\mathrm{ref}}. \tag{2.10}$$

The signal power $P_{\mathrm{s,Rx}}$ and the noise power $P_{\mathrm{n,Rx}}$ in the reference bandwidth B_{ref} at the output of the pre-amplifier are given by

$$P_{s,\mathrm{Rx}} = G_{\mathrm{Rx}} P_{s,\nu_{max}}, \quad P_{n,\mathrm{Rx}} = G_{\mathrm{Rx}} P_{n,\nu_{\max}} + F_{\mathrm{Rx}}\,hf(G_{\mathrm{Rx}} - 1)B_{\mathrm{ref}}. \tag{2.11}$$

The $\mathrm{OSNR}_{\mathrm{Rx}}$ at the input of the receiver reads

$$\mathrm{OSNR}_{\mathrm{Rx}} = \frac{G_{\mathrm{Rx}} g_0 G_0 P_\ell}{P_{n,\mathrm{Rx}}}. \tag{2.12}$$

Introducing Eqs. (2.7–2.11) into Eq. (2.12) leads to Eq. (2.1). In typical WDM links, an optical channel power of 0 dBm is launched into each fiber span for reducing the nonlinear impairments while maintaining good OSNR [30,31]. This leads to the curves shown in Fig. 2.4.

A.2 Required receiver optical signal-to-noise power ratio

For a given modulation format and bandwidth B, a required $\mathrm{OSNR}_{\mathrm{sig}}$ at the receiver can be calculated for the desired target BER. In the case of non-data-aided reception of polarization multiplexed signals impaired by additive white Gaussian noise, the relation between BER and $\mathrm{OSNR}_{\mathrm{sig}}$ for quadratic M-ary QAM constellations is given by [154]

$$\mathrm{BER} = \frac{\sqrt{M} - 1}{\sqrt{M}\log_2\sqrt{M}}\,\mathrm{erfc}\sqrt{\frac{3}{2(M-1)}\mathrm{OSNR}_{\mathrm{sig}}}. \tag{2.13}$$

$$\mathrm{OSNR}_{\mathrm{sig}} = \frac{2B_{\mathrm{ref}}}{pB}\mathrm{OSNR}_{\mathrm{ref}}. \tag{2.14}$$

In these relations, the $\mathrm{OSNR}_{\mathrm{sig}}$ refers to the power ratio of the data signal to the noise power within the true signal bandwidth B, which differs from the reference bandwidth $B_{\mathrm{ref}} = 12.5$ GHz that was used to specify the $\mathrm{OSNR}_{\mathrm{Rx}}$ in the previous sections and that corresponds to a reference wavelength span of 0.1 nm at $\lambda = 1.55$ μm. The $\mathrm{OSNR}_{\mathrm{ref}}$ is determined by measuring the unpolarized noise power density in x and y-polarization within B_{ref} using with an optical spectrum analyzer, $S_{n,x} + S_{n,y} = 2S_{n,x}$. If the signal comprises only one polarization that carries a power P_x, an

$OSNR_{\text{ref}} = P_x/(2S_{n,x}B_{\text{ref}})$ results. If the signal power is composed of two orthogonally polarized contributions with equal powers, and overall signal power of $P_x + P_y = 2P_x$ is measured, and $OSNR_{\text{ref}} = 2P_x/(2S_{n,x}B_{\text{ref}})$ holds. In contrast to that, the signal-relevant OSNR_{sig} of the polarized signal power has to be referred to the copolarized noise power density only. This explains the factor p in Eq. (2.14), which is $p = 1$ for single-polarization and $p = 2$ for dual-polarization signals.

References

[1] E. Lowe, Current european WDM deployment trends, IEEE Commun. Mag. 36 (1998) 46–50. Available from: https://doi.org/10.1109/35.648756.

[2] J.P. Ryan, W.D.M. North, American deployment trends, IEEE Commun. Mag. 36 (1998) 40–44. Available from: https://doi.org/10.1109/35.648755.

[3] G.E. Keiser, A review of WDM technology and applications, Opt. Fiber Technol. 5 (1999) 3–39. Available from: https://doi.org/10.1006/ofte.1998.0275.

[4] B.J. Puttnam, R.S. Luis, W. Klaus, J. Sakaguchi, J.-M. Delgado Mendinueta, Y. Awaji, et al., 2.15 Pb/s transmission using a 22 core homogeneous single-mode multi-core fiber and wideband optical comb, in: IEEE 2015 Eur. Conf. Opt. Commun., 2015, pp. 1–3. doi:10.1109/ECOC.2015.7341685.

[5] N.K. Fontaine, G. Raybon, B. Guan, A. Adamiecki, P.J. Winzer, R. Ryf, et al., 228-GHz coherent receiver using digital optical bandwidth interleaving and reception of 214-GBd (856-Gb/s) PDM-QPSK, in: Eur. Conf. Exhib. Opt. Commun., OSA, 2012, p. Th.3.A.1. doi:10.1364/ECEOC.2012.Th.3.A.1.

[6] J.N. Kemal, J. Pfeifle, P. Marin-Palomo, M.D.G. Pascual, S. Wolf, F. Smyth, et al., Multi-wavelength coherent transmission using an optical frequency comb as a local oscillator, Opt. Express 24 (2016) 25432. Available from: https://doi.org/10.1364/OE.24.025432.

[7] P. Marin-Palomo, J.N. Kemal, M. Karpov, A. Kordts, J. Pfeifle, M.H.P. Pfeiffer, et al., Microresonator-based solitons for massively parallel coherent optical communications, Nature 546 (2017) 274–279. Available from: https://doi.org/10.1038/nature22387.

[8] L. Lundberg, M. Mazur, A. Lorences-Riesgo, M. Karlsson, P.A. Andrekson, Joint carrier recovery for DSP complexity reduction in frequency comb-based superchannel transceivers, in: 2017 IEEE Eur. Conf. Opt. Commun., 2017, pp. 1–3. doi:10.1109/ECOC.2017.8346044.

[9] E. Temprana, E. Myslivets, B.P.-P. Kuo, L. Liu, V. Ataie, N. Alic, et al., Overcoming Kerr-induced capacity limit in optical fiber transmission, Science (80–) 348 (2015) 1445–1448. Available from: https://doi.org/10.1126/science.aab1781.

[10] A. Fülöp, M. Mazur, A. Lorences-Riesgo, Ó.B. Helgason, P.-H. Wang, Y. Xuan, et al., High-order coherent communications using mode-locked dark-pulse Kerr combs from microresonators, Nat. Commun. 9 (2018) 1598. Available from: https://doi.org/10.1038/s41467-018-04046-6.

[11] J. Pfeifle, V. Vujicic, R.T. Watts, P.C. Schindler, C. Weimann, R. Zhou, et al., Flexible terabit/s Nyquist-WDM super-channels using a gain-switched comb source, Opt. Express 23 (2015) 724–738. Available from: https://doi.org/10.1364/OE.23.000724.

[12] T. Shao, R. Zhou, V. Vujicic, M.D. Gutierrez Pascual, P.M. Anandarajah, L.P. Barry, 100 km Coherent nyquist ultradense wavelength division multiplexed passive optical network using a tunable gain-switched comb source, J. Opt. Commun. Netw. 8 (2016) 112–117. Available from: https://doi.org/10.1364/JOCN.8.000112.

[13] V. Vujicic, A. Anthur, V. Panapakkam, R. Zhou, Q. Gaimard, K. Merghem, et al., Tbit/s optical interconnects based on low linewidth quantum-dash lasers and coherent detection, in: Conf. Lasers Electro-Optics, OSA, Washington, D.C., 2016, p. SF2F.4. doi:10.1364/CLEO_SI.2016.SF2F.4.

[14] P. Marin-Palomo, J.N. Kemal, P. Trocha, S. Wolf, K. Merghem, F. Lelarge, et al., Comb-based WDM transmission at 10 Tbit/s using a DC-driven quantum-dash mode-locked laser diode, Opt. Express (2019). accepted.

[15] J.N. Kemal, P. Marin-Palomo, K. Merghem, G. Aubin, C. Calo, R. Brenot, et al., 32QAM WDM transmission using a quantum-dash passively mode-locked laser with resonant feedback, in: Opt. Fiber Commun. Conf. Postdeadline Pap., OSA, Washington, D.C., 2017, p. Th5C.3. doi:10.1364/OFC.2017.Th5C.3.
[16] C. Weimann, P.C. Schindler, R. Palmer, S. Wolf, D. Bekele, D. Korn, et al., Silicon-organic hybrid (SOH) frequency comb sources for terabit/s data transmission, Opt. Express 22 (2014) 3629–3637. Available from: https://doi.org/10.1364/OE.22.003629.
[17] J. Lin, H. Sepehrian, Y. Xu, L.A. Rusch, W. Shi, Frequency comb generation using a CMOS compatible SiP DD-MZM for flexible networks, IEEE Photonics Technol. Lett. 30 (2018) 1495–1498. Available from: https://doi.org/10.1109/LPT.2018.2856767.
[18] H. Hu, F. Da Ros, M. Pu, F. Ye, K. Ingerslev, E. Porto da Silva, et al., Single-source chip-based frequency comb enabling extreme parallel data transmission, Nat. Photonics. 12 (2018) 469–473. Available from: https://doi.org/10.1038/s41566-018-0205-5.
[19] J. Pfeifle, V. Brasch, M. Lauermann, Y. Yu, D. Wegner, T. Herr, et al., Coherent terabit communications with microresonator Kerr frequency combs, Nat. Photonics. 8 (2014) 375–380. Available from: https://doi.org/10.1038/nphoton.2014.57.
[20] A. Fülöp, M. Mazur, A. Lorences-Riesgo, T.A. Eriksson, P.-H. Wang, Y. Xuan, et al., Long-haul coherent communications using microresonator-based frequency combs, Opt. Express 25 (2017) 26678. Available from: https://doi.org/10.1364/OE.25.026678.
[21] P.J. Winzer, High-spectral-efficiency optical modulation formats, J. Light. Technol. 30 (2012) 3824–3835. Available from: https://doi.org/10.1109/JLT.2012.2212180.
[22] E. Agrell, M. Karlsson, A.R. Chraplyvy, D.J. Richardson, P.M. Krummrich, P. Winzer, et al., Roadmap of optical communications, J. Opt. 18 (2016) 063002. Available from: https://doi.org/10.1088/2040-8978/18/6/063002.
[23] R. Maher, P.M. Anandarajah, S.K. Ibrahim, L.P. Barry, A.D. Ellis, P. Perry, et al., Low cost comb source in a coherent wavelength division multiplexed system, in: 36th IEEE Eur. Conf. Exhib. Opt. Commun., 2010, pp. 1–3. doi:10.1109/ECOC.2010.5621515.
[24] L. Tao, J. Yu, N. Chi, Generation of flat and stable multi-carriers based on only integrated IQ modulator and its implementation for 112Gb/s PM-QPSK transmitter, in: Natl. Fiber Opt. Eng. Conf., OSA, Washington, D.C., 2012, p. JW2A.86. doi:10.1364/NFOEC.2012.JW2A.86.
[25] D.O. Otuya, K. Kasai, M. Yoshida, T. Hirooka, M. Nakazawa, Single-channel 192 Tbit/s, Pol-Mux-64 QAM coherent Nyquist pulse transmission over 150 km with a spectral efficiency of 75 bit/s/Hz, Opt. Express 22 (2014) 23776. Available from: https://doi.org/10.1364/OE.22.023776.
[26] K. Kikuchi, Characterization of semiconductor-laser phase noise and estimation of bit-error rate performance with low-speed offline digital coherent receivers, Opt. Express 20 (2012) 5291. Available from: https://doi.org/10.1364/OE.20.005291.
[27] T. Pfau, S. Hoffmann, R. Noe, Hardware-efficient coherent digital receiver concept with feedforward carrier recovery for M-QAM constellations, J. Light. Technol. 27 (2009) 989–999. Available from: https://doi.org/10.1109/JLT.2008.2010511.
[28] G.P. Agrawal, N.K. Dutta, Semiconductor Lasers, second ed., Van Nostrand Reinhold, New York, 1993.
[29] Y. Ben M'Sallem, Q.T. Le, L. Bramerie, Q.-T. Nguyen, E. Borgne, P. Besnard, et al., Quantum-dash mode-locked laser as a source for 56-Gb/s DQPSK modulation in WDM multicast applications, IEEE Photonics Technol. Lett. 23 (2011) 453–455. Available from: https://doi.org/10.1109/LPT.2011.2106116.
[30] P.S. Cho, V.S. Grigoryan, Y.A. Godin, A. Salamon, Y. Achiam, Transmission of 25-Gb/s RZ-DQPSK signals with 25-GHz channel spacing over 1000 km of SMF-28 fiber, IEEE Photonics Technol. Lett. 15 (2003) 473–475. Available from: https://doi.org/10.1109/LPT.2002.807934.
[31] G. Raybon, P.J. Winzer, C.R. Doerr, 1-Tb/s (10 × 107 Gb/s) Electronically multiplexed optical signal generation and WDM transmission, J. Light. Technol. 25 (2007) 233–238. Available from: https://doi.org/10.1109/JLT.2006.886723.
[32] P. Marin-Palomo, J.N. Kemal, W. Freude, S. Randel, C. Koos, OSNR limitations of chip-based optical frequency comb sources for WDM coherent communications, arXiv preprint arXiv: 1907.01042, 2019.

[33] Auxora, CATV Module, 2018. Available from: <http://www.auxora.com/product/info_2.aspx?itemid = 171&lcid = 45&ppid = 7&pid = 23> (accessed 07.01.19).

[34] Fujitsu, DP-QPSK 100G LN Modulator, 2014. Available from: <http://www.fujitsu.com/downloads/JP/archive/imgjp/group/foc/services/100gln/ln100gdpqpsk-e-141105.pdf> (accessed 07.01.19).

[35] Y. Cai, W. Wang, W. Qian, J. Xing, K. Tao, J. Yin, et al., FPGA investigation on error-floor performance of a concatenated staircase and hamming code for 400G-ZR forward error correction, in: Opt. Fiber Commun. Conf. Postdeadline Pap., OSA, Washington, D.C., 2018, p. Th4C.2. doi:10.1364/OFC.2018.Th4C.2.

[36] K. Gürs, R. Müller, Breitband-modulation durch steuerung der emission eines optischen masers (Auskoppel modulation), Phys. Lett. 5 (1963) 179–181. Available from: https://doi.org/10.1016/S0375-9601(63)96191-7.

[37] L.E. Hargrove, R.L. Fork, M.A. Pollack, Locking of He-Ne laser modes induced by synchronous intracavity modulation, Appl. Phys. Lett. 5 (1964) 4–5. Available from: https://doi.org/10.1063/1.1754025.

[38] A.J. DeMaria, D.A. Stetser, H. Heynau, SELF mode-locking of lasers with saturable absorbers, Appl. Phys. Lett. 8 (1966) 174–176. Available from: https://doi.org/10.1063/1.1754541.

[39] M. DiDomenico, Small-signal analysis of internal (coupling-type) modulation of lasers, J. Appl. Phys. 35 (1964) 2870–2876. Available from: https://doi.org/10.1063/1.1713121.

[40] M. Crowell, Characteristics of mode-coupled lasers, IEEE J. Quantum Electron. 1 (1965) 12–20. Available from: https://doi.org/10.1109/JQE.1965.1072174.

[41] A. Yariv, Internal modulation in multimode laser oscillators, J. Appl. Phys. 36 (1965) 388–391. Available from: https://doi.org/10.1063/1.1713999.

[42] D.E. Spence, P.N. Kean, W. Sibbett, 60-fsec pulse generation from a self-mode-locked Ti:sapphire laser, Opt. Lett. 16 (1991) 42. Available from: https://doi.org/10.1364/OL.16.000042.

[43] C. Spielmann, P.F. Curley, T. Brabec, F. Krausz, Ultrabroadband femtosecond lasers, IEEE J. Quantum Electron. 30 (1994) 1100–1114. Available from: https://doi.org/10.1109/3.291379.

[44] U. Keller, Recent developments in compact ultrafast lasers, Nature 424 (2003) 831–838. Available from: https://doi.org/10.1038/nature01938.

[45] X. Huang, A. Stintz, H. Li, L.F. Lester, J. Cheng, K.J. Malloy, Passive mode-locking in 1.3 μm two-section InAs quantum dot lasers, Appl. Phys. Lett. 78 (2001) 2825–2827. Available from: https://doi.org/10.1063/1.1371244.

[46] F. Lelarge, B. Dagens, J. Renaudier, R. Brenot, A. Accard, F. van Dijk, et al., Recent advances on InAs/InP Quantum dash based semiconductor lasers and optical amplifiers operating at 1.55 μm, IEEE J. Sel. Top. Quantum Electron. 13 (2007) 111–124. Available from: https://doi.org/10.1109/JSTQE.2006.887154.

[47] Z.G. Lu, J.R. Liu, S. Raymond, P.J. Poole, P.J. Barrios, D. Poitras, 312-fs Pulse generation from a passive C-band InAs/InP quantum dot mode-locked laser, Opt. Express 16 (2008) 10835. Available from: https://doi.org/10.1364/OE.16.010835.

[48] A. Shen, J.-G. Provost, A. Akrout, et al., Low confinement factor quantum dash (QD) mode-locked Fabry-Perot (FP) laser diode for tunable pulse generation, in: 2008 IEEE Conf. Opt. Fiber Commun. Fiber Opt. Eng. Conf., OFC/NFOEC 2008, 2008, pp. 1–3. doi:10.1109/OFC.2008.4528481.

[49] A. Akrout, A. Shen, A. Enard, G.-H. Duan, F. Lelarge, A. Ramdane, Low phase noise all-optical oscillator using quantum dash modelocked laser, Electron. Lett. 46 (2010) 73. Available from: https://doi.org/10.1049/el.2010.2886.

[50] K. Merghem, A. Akrout, A. Martinez, G. Aubin, A. Ramdane, F. Lelarge, et al., Pulse generation at 346 GHz using a passively mode locked quantum-dash-based laser at 1.55 μm, Appl. Phys. Lett. 94 (2009) 021107. Available from: https://doi.org/10.1063/1.3070544.

[51] J.N. Kemal, P. Marin-Palomo, V. Panapakkam, P. Trocha, S. Wolf, K. Merghem, et al., Coherent WDM transmission using quantum-dash mode-locked laser diodes as multi-wavelength source and local oscillator, Opt. Express (2019).

[52] V. Torres-Company, A.M. Weiner, Optical frequency comb technology for ultra-broadband radio-frequency photonics, Laser Photon. Rev. 8 (2014) 368–393. Available from: https://doi.org/10.1002/lpor.201300126.

[53] T. Sakamoto, T. Kawanishi, M. Izutsu, Widely wavelength-tunable ultra-flat frequency comb generation using conventional dual-drive Mach–Zehnder modulator, Electron. Lett. 43 (2007) 1039. Available from: https://doi.org/10.1049/el:20071267.
[54] C. Chen, C. Zhang, W. Zhang, W. Jin, K. Qiu, Scalable and reconfigurable generation of flat optical comb for WDM-based next-generation broadband optical access networks, Opt. Commun. 321 (2014) 16–22. Available from: https://doi.org/10.1016/j.optcom.2014.01.059.
[55] R. Wu, V.R. Supradeepa, C.M. Long, D.E. Leaird, A.M. Weiner, Generation of very flat optical frequency combs from continuous-wave lasers using cascaded intensity and phase modulators driven by tailored radio frequency waveforms, Opt. Lett. 35 (2010) 3234. Available from: https://doi.org/10.1364/OL.35.003234.
[56] N. Dupuis, C.R. Doerr, L. Zhang, L. Chen, N.J. Sauer, P. Dong, et al., InP-based comb generator for optical OFDM, J. Light. Technol. 30 (2012) 466–472. Available from: https://doi.org/10.1109/JLT.2011.2173463.
[57] T. Yamamoto, K. Hitomi, W. Kobayashi, H. Yasaka, Optical frequency comb block generation by using semiconductor Mach–Zehnder modulator, IEEE Photonics Technol. Lett. 25 (2013) 40–42. Available from: https://doi.org/10.1109/LPT.2012.2227696.
[58] R. Slavik, S.G. Farwell, M.J. Wale, D.J. Richardson, Compact optical comb generator using Inp tunable laser and push-pull modulator, IEEE Photonics Technol. Lett. 27 (2015) 217–220. Available from: https://doi.org/10.1109/LPT.2014.2365259.
[59] Y. Xu, J. Lin, R. Dubé-Demers, S. LaRochelle, L. Rusch, W. Shi, Integrated flexible-grid WDM transmitter using an optical frequency comb in microring modulators, Opt. Lett. 43 (2018) 1554. Available from: https://doi.org/10.1364/OL.43.001554.
[60] P.M. Anandarajah, R. Maher, Y.Q. Xu, S. Latkowski, J. O'Carroll, S.G. Murdoch, et al., Generation of coherent multicarrier signals by gain switching of discrete mode lasers, IEEE Photonics J. 3 (2011) 112–122. Available from: https://doi.org/10.1109/JPHOT.2011.2105861.
[61] L.P. Barry, P. Anandarajah, A. Kaszubowska, Optical pulse generation at frequencies up to 20 GHz using external-injection seeding of a gain-switched commercial Fabry-Perot laser, IEEE Photonics Technol. Lett. 13 (2001) 1014–1016. Available from: https://doi.org/10.1109/68.942678.
[62] X. Jin, S.L. Chuang, Relative intensity noise characteristics of injection-locked semiconductor lasers, Appl. Phys. Lett. 77 (2000) 1250–1252. Available from: https://doi.org/10.1063/1.1290140.
[63] R. Zhou, S. Latkowski, J. O'Carroll, R. Phelan, L.P. Barry, P. Anandarajah, 40nm wavelength tunable gain-switched optical comb source, Opt. Express 19 (2011) B415. Available from: https://doi.org/10.1364/OE.19.00B415.
[64] M.D.G. Pascual, V. Vujicic, J. Braddell, F. Smyth, P. Anandarajah, L. Barry, Photonic integrated gain switched optical frequency comb for spectrally efficient optical transmission systems, IEEE Photonics J. 9 (2017) 1–8. Available from: https://doi.org/10.1109/JPHOT.2017.2678478.
[65] V. Ataie, E. Temprana, L. Liu, E. Myslivets, B.P.-P. Kuo, N. Alic, et al., Ultrahigh count coherent WDM channels transmission using optical parametric comb-based frequency synthesizer, J. Light. Technol. 33 (2015) 694–699. Available from: https://doi.org/10.1109/JLT.2015.2388579.
[66] D. Hillerkuss, R. Schmogrow, M. Meyer, S. Wolf, M. Jordan, P. Kleinow, et al., Single-laser 32.5 {Tbit/s Nyquist WDM} transmission, J. Opt. Commun. Netw. 4 (2012) 715–723. Available from: https://doi.org/10.1364/JOCN.4.000715.
[67] B. Kuyken, X. Liu, R.M. Osgood Jr, R. Baets, G. Roelkens, W.M.J. Green, Mid-infrared to telecom-band supercontinuum generation in highly nonlinear silicon-on-insulator wire waveguides, Opt. Express 19 (2011) 20172. Available from: https://doi.org/10.1364/OE.19.020172.
[68] R. Halir, Y. Okawachi, J.S. Levy, M.A. Foster, M. Lipson, A.L. Gaeta, Ultrabroadband supercontinuum generation in a CMOS-compatible platform, Opt. Lett. 37 (2012) 1685. Available from: https://doi.org/10.1364/OL.37.001685.
[69] P. Del'Haye, A. Schliesser, O. Arcizet, T. Wilken, R. Holzwarth, T.J. Kippenberg, Optical frequency comb generation from a monolithic microresonator, Nature 450 (2007) 1214–1217. Available from: https://doi.org/10.1038/nature06401.

[70] J.S. Levy, A. Gondarenko, M.A. Foster, A.C. Turner-Foster, A.L. Gaeta, M. Lipson, CMOS-compatible multiple-wavelength oscillator for on-chip optical interconnects, Nat. Photonics 4 (2010) 37–40. Available from: https://doi.org/10.1038/nphoton.2009.259.
[71] P. Del'Haye, T. Herr, E. Gavartin, M.L. Gorodetsky, R. Holzwarth, T.J. Kippenberg, Octave spanning tunable frequency comb from a microresonator, Phys. Rev. Lett. 107 (2011) 063901. Available from: https://doi.org/10.1103/PhysRevLett.107.063901.
[72] A.A. Savchenkov, A.B. Matsko, V.S. Ilchenko, I. Solomatine, D. Seidel, L. Maleki, Tunable optical frequency comb with a crystalline whispering gallery mode resonator, Phys. Rev. Lett. 101 (2008) 093902. Available from: https://doi.org/10.1103/PhysRevLett.101.093902.
[73] M.A. Foster, J.S. Levy, O. Kuzucu, K. Saha, M. Lipson, A.L. Gaeta, Silicon-based monolithic optical frequency comb source, Opt. Express 19 (2011) 14233. Available from: https://doi.org/10.1364/OE.19.014233.
[74] T. Herr, V. Brasch, J.D. Jost, C.Y. Wang, N.M. Kondratiev, M.L. Gorodetsky, et al., Temporal solitons in optical microresonators, Nat. Photonics. 8 (2014) 145–152. Available from: https://doi.org/10.1038/nphoton.2013.343.
[75] X. Yi, Q.-F. Yang, K.Y. Yang, M.-G. Suh, K. Vahala, Soliton frequency comb at microwave rates in a high-Q silica microresonator, Optica 2 (2015) 1078. Available from: https://doi.org/10.1364/OPTICA.2.001078.
[76] N. Volet, X. Yi, Q.-F. Yang, E.J. Stanton, P.A. Morton, K.Y. Yang, et al., Micro-resonator soliton generated directly with a diode laser, Laser Photon. Rev. 12 (2018) 1700307. Available from: https://doi.org/10.1002/lpor.201700307.
[77] J. Liu, A.S. Raja, M. Karpov, B. Ghadiani, M.H.P. Pfeiffer, B. Du, et al., Ultralow-power chip-based soliton microcombs for photonic integration, Optica 5 (2018) 1347. Available from: https://doi.org/10.1364/OPTICA.5.001347.
[78] B. Stern, X. Ji, Y. Okawachi, A.L. Gaeta, M. Lipson, Battery-operated integrated frequency comb generator, Nature 562 (2018) 401–405. Available from: https://doi.org/10.1038/s41586-018-0598-9.
[79] D.T. Spencer, T. Drake, T.C. Briles, J. Stone, L.C. Sinclair, C. Fredrick, et al., An optical-frequency synthesizer using integrated photonics, Nature 557 (2018) 81–85. Available from: https://doi.org/10.1038/s41586-018-0065-7.
[80] V. Panapakkam, A.P. Anthur, V. Vujicic, R. Zhou, Q. Gaimard, K. Merghem, et al., Amplitude and phase noise of frequency combs generated by single-section InAs/InP quantum-dash-based passively and actively mode-locked lasers, IEEE J. Quantum Electron. 52 (2016) 1–7. Available from: https://doi.org/10.1109/JQE.2016.2608800.
[81] M.O. Sahni, S. Trebaol, L. Bramerie, M. Joindot, S.P.Ó. Dúill, S.G. Murdoch, et al., Frequency noise reduction performance of a feed-forward heterodyne technique: application to an actively mode-locked laser diode, Opt. Lett. 42 (2017) 4000. Available from: https://doi.org/10.1364/OL.42.004000.
[82] J. Pfeifle, R. Watts, I. Shkarban, S. Wolf, V. Vujicic, P. Landais, et al., simultaneous phase noise reduction of 30 comb lines from a quantum-dash mode-locked laser diode enabling coherent Tbit/s data transmission, in: Opt. Fiber Commun. Conf., OSA, Washington, D.C., 2015, p. Tu3I.5. doi:10.1364/OFC.2015.Tu3I.5.
[83] M.H.P. Pfeiffer, C. Herkommer, J. Liu, H. Guo, M. Karpov, E. Lucas, et al., Octave-spanning dissipative Kerr soliton frequency combs in Si_3N_4 microresonators, Optica 4 (2017) 684. Available from: https://doi.org/10.1364/OPTICA.4.000684.
[84] P. Marin-Palomo, J.N. Kemal, M. Karpov, A. Kordts, J. Pfeifle, M.H.P. Pfeiffer, et al., Supplementary information—microresonator-based solitons for massively parallel coherent optical communications, Nature 546 (2017) 274–279. Available from: https://doi.org/10.1038/nature22387.
[85] H. Hu, F. Da Ros, M. Pu, F. Ye, K. Ingerslev, E. Porto da Silva, et al., Supplementary information—single-source chip-based frequency comb enabling extreme parallel data transmission, Nat. Photonics 12 (2018) 469–473. Available from: https://doi.org/10.1038/s41566-018-0205-5.
[86] X. Xue, P.-H. Wang, Y. Xuan, M. Qi, A.M. Weiner, Microresonator Kerr frequency combs with high conversion efficiency, Laser Photonics Rev. 11 (2017) 1600276. Available from: https://doi.org/10.1002/lpor.201600276.

[87] A.A. Savchenkov, A.B. Matsko, V.S. Ilchenko, L. Maleki, Optical resonators with ten million finesse, Opt. Express 15 (2007) 6768. Available from: https://doi.org/10.1364/OE.15.006768.
[88] Q. Li, T.C. Briles, D.A. Westly, T.E. Drake, J.R. Stone, B.R. Ilic, et al., Stably accessing octave-spanning microresonator frequency combs in the soliton regime, Optica 4 (2017) 193. Available from: https://doi.org/10.1364/OPTICA.4.000193.
[89] S.-W. Huang, A.K. Vinod, J. Yang, M. Yu, D.-L. Kwong, C.W. Wong, Quasi-phase-matched multispectral Kerr frequency comb, Opt. Lett. 42 (2017) 2110. Available from: https://doi.org/10.1364/OL.42.002110.
[90] L. Razzari, D. Duchesne, M. Ferrera, R. Morandotti, S. Chu, B.E. Little, et al., CMOS-compatible integrated optical hyper-parametric oscillator, Nat. Photonics. 4 (2010) 41–45. Available from: https://doi.org/10.1038/nphoton.2009.236.
[91] H. Jung, C. Xiong, K.Y. Fong, X. Zhang, H.X. Tang, Optical frequency comb generation from aluminum nitride microring resonator, Opt. Lett. 38 (2013) 2810. Available from: https://doi.org/10.1364/OL.38.002810.
[92] K.E. Webb, M. Erkintalo, S. Coen, S.G. Murdoch, Experimental observation of coherent cavity soliton frequency combs in silica microspheres, Opt. Lett. 41 (2016) 4613. Available from: https://doi.org/10.1364/OL.41.004613.
[93] M. Pu, L. Ottaviano, E. Semenova, K. Yvind, Efficient frequency comb generation in AlGaAs-on-insulator, Optica 3 (2016) 823. Available from: https://doi.org/10.1364/OPTICA.3.000823.
[94] D.J. Wilson, K. Schneider, S. Hoenl, M. Anderson, T.J. Kippenberg, P. Seidler, Integrated gallium phosphide nonlinear photonics, 2018. Available from: <http://arxiv.org/abs/1808.03554>.
[95] Y. Okawachi, K. Saha, J.S. Levy, Y.H. Wen, M. Lipson, A.L. Gaeta, Octave-spanning frequency comb generation in a silicon nitride chip, Opt. Lett. 36 (2011) 3398. Available from: https://doi.org/10.1364/OL.36.003398.
[96] M.A. Foster, A.C. Turner, J.E. Sharping, B.S. Schmidt, M. Lipson, A.L. Gaeta, Broad-band optical parametric gain on a silicon photonic chip, Nature 441 (2006) 960–963. Available from: https://doi.org/10.1038/nature04932.
[97] C.J. Krückel, A. Fülöp, Z. Ye, P.A. Andrekson, V. Torres-Company, Optical bandgap engineering in nonlinear silicon nitride waveguides, Opt. Express 25 (2017) 15370. Available from: https://doi.org/10.1364/OE.25.015370.
[98] M.H.P. Pfeiffer, J. Liu, A.S. Raja, T. Morais, B. Ghadiani, T.J. Kippenberg, Ultra-smooth silicon nitride waveguides based on the damascene reflow process: fabrication and loss origins, Optica 5 (2018) 884. Available from: https://doi.org/10.1364/OPTICA.5.000884.
[99] C.H. Henry, R.F. Kazarinov, H.J. Lee, K.J. Orlowsky, L.E. Katz, Low loss $Si_3N_4-SiO_2$ optical waveguides on Si, Appl. Opt. 26 (1987) 2621. Available from: https://doi.org/10.1364/AO.26.002621.
[100] X. Ji, F.A.S. Barbosa, S.P. Roberts, A. Dutt, J. Cardenas, Y. Okawachi, et al., Ultra-low-loss on-chip resonators with sub-milliwatt parametric oscillation threshold, Optica 4 (2017) 619. Available from: https://doi.org/10.1364/OPTICA.4.000619.
[101] M.H.P. Pfeiffer, C. Herkommer, J. Liu, T. Morais, M. Zervas, M. Geiselmann, et al., Photonic damascene process for low-loss, high-confinement silicon nitride waveguides, IEEE J. Sel. Top. Quantum Electron. 24 (2018) 1–11. Available from: https://doi.org/10.1109/JSTQE.2018.2808258.
[102] K.H. Nam, I.H. Park, S.H. Ko, Patterning by controlled cracking, Nature 485 (2012) 221–224. Available from: https://doi.org/10.1038/nature11002.
[103] H. Guo, C. Herkommer, A. Billat, D. Grassani, C. Zhang, M.H.P. Pfeiffer, et al., Mid-infrared frequency comb via coherent dispersive wave generation in silicon nitride nanophotonic waveguides, Nat. Photonics 12 (2018) 330–335. Available from: https://doi.org/10.1038/s41566-018-0144-1.
[104] Q.-F. Yang, X. Yi, K.Y. Yang, K. Vahala, Spatial-mode-interaction-induced dispersive waves and their active tuning in microresonators, Optica 3 (2016) 1132. Available from: https://doi.org/10.1364/OPTICA.3.001132.
[105] V. Brasch, M. Geiselmann, T. Herr, G. Lihachev, M.H.P. Pfeiffer, M.L. Gorodetsky, et al., Photonic chip-based optical frequency comb using soliton Cherenkov radiation, Science (80–) 351 (2016) 357–360. Available from: https://doi.org/10.1126/science.aad4811.

[106] M. Yu, Y. Okawachi, A.G. Griffith, M. Lipson, A.L. Gaeta, Mode-locked mid-infrared frequency combs in a silicon microresonator, Optica 3 (2016) 854. Available from: https://doi.org/10.1364/OPTICA.3.000854.
[107] Z. Lu, W. Wang, W. Zhang, M. Liu, L. Wang, S.T. Chu, et al., Raman self-frequency-shift of soliton crystal in a high index doped silica micro-ring resonator [Invited], Opt. Mater. Express 8 (2018) 2662. Available from: https://doi.org/10.1364/OME.8.002662.
[108] Z. Gong, A. Bruch, M. Shen, X. Guo, H. Jung, L. Fan, et al., High-fidelity cavity soliton generation in crystalline AlN micro-ring resonators, Opt. Lett. 43 (2018) 4366. Available from: https://doi.org/10.1364/OL.43.004366.
[109] S.H. Lee, D.Y. Oh, Q.-F. Yang, B. Shen, H. Wang, K.Y. Yang, et al., Towards visible soliton microcomb generation, Nat. Commun. 8 (2017) 1295. Available from: https://doi.org/10.1038/s41467-017-01473-9.
[110] M. Karpov, M.H.P. Pfeiffer, J. Liu, A. Lukashchuk, T.J. Kippenberg, Photonic chip-based soliton frequency combs covering the biological imaging window, Nat. Commun. 9 (2018) 1146. Available from: https://doi.org/10.1038/s41467-018-03471-x.
[111] M. Yu, Y. Okawachi, A.G. Griffith, N. Picqué, M. Lipson, A.L. Gaeta, Silicon-chip-based mid-infrared dual-comb spectroscopy, Nat. Commun. 9 (2018) 1869. Available from: https://doi.org/10.1038/s41467-018-04350-1.
[112] J.D. Jost, T. Herr, C. Lecaplain, V. Brasch, M.H.P. Pfeiffer, T.J. Kippenberg, Counting the cycles of light using a self-referenced optical microresonator, Optica 2 (2015) 706. Available from: https://doi.org/10.1364/OPTICA.2.000706.
[113] V. Brasch, E. Lucas, J.D. Jost, M. Geiselmann, T.J. Kippenberg, Self-referenced photonic chip soliton Kerr frequency comb, Light Sci. Appl. 6 (2017). Available from: https://doi.org/10.1038/lsa.2016.202. e16202–e16202.
[114] P. Trocha, M. Karpov, D. Ganin, M.H.P. Pfeiffer, A. Kordts, S. Wolf, et al., Ultrafast optical ranging using microresonator soliton frequency combs, Science (80–) 359 (2018) 887–891. Available from: https://doi.org/10.1126/science.aao3924.
[115] M.-G. Suh, K.J. Vahala, Soliton microcomb range measurement, Science (80–) 359 (2018) 884–887. Available from: https://doi.org/10.1126/science.aao1968.
[116] M.-G. Suh, Q.-F. Yang, K.Y. Yang, X. Yi, K.J. Vahala, Microresonator soliton dual-comb spectroscopy, Science (80–) 354 (2016) 600–603. Available from: https://doi.org/10.1126/science.aah6516.
[117] N.G. Pavlov, G. Lihachev, S. Koptyaev, E. Lucas, M. Karpov, N.M. Kondratiev, et al., Soliton dual frequency combs in crystalline microresonators, Opt. Lett. 42 (2017) 514. Available from: https://doi.org/10.1364/OL.42.000514.
[118] A. Dutt, C. Joshi, X. Ji, J. Cardenas, Y. Okawachi, K. Luke, et al., On-chip dual-comb source for spectroscopy, Sci. Adv. 4 (2018) e1701858. Available from: https://doi.org/10.1126/sciadv.1701858.
[119] W. Liang, D. Eliyahu, V.S. Ilchenko, A.A. Savchenkov, A.B. Matsko, D. Seidel, et al., High spectral purity Kerr frequency comb radio frequency photonic oscillator, Nat. Commun. 6 (2015) 7957. Available from: https://doi.org/10.1038/ncomms8957.
[120] E. Lucas, J.D. Jost, K. Beha, M. Lezius, R. Holzwarth, T.J. Kippenberg, Low-noise microwave generation with optical microresonators, in: IEEE 2017 Conf. Lasers Electro-Optics Eur. Eur. Quantum Electron. Conf., 2017: pp. 1–1. doi:10.1109/CLEOE-EQEC.2017.8087466.
[121] E. Obrzud, M. Rainer, A. Harutyunyan, M.H. Anderson, M. Geiselmann, B. Chazelas, et al., A microphotonic astrocomb, 2017. Available from: <arXiv e-prints>.
[122] M.-G. Suh, X. Yi, Y.-H. Lai, S. Leifer, I.S. Grudinin, G. Vasisht, et al., Searching for exoplanets using a microresonator astrocomb, 2018. Available from: <arXiv e-prints>.
[123] J. Pfeifle, A. Coillet, R. Henriet, K. Saleh, P. Schindler, C. Weimann, et al., Optimally coherent kerr combs generated with crystalline whispering gallery mode resonators for ultrahigh capacity fiber communications, Phys. Rev. Lett. 114 (2015) 093902. Available from: https://doi.org/10.1103/PhysRevLett.114.093902.
[124] T. Herr, K. Hartinger, J. Riemensberger, C.Y. Wang, E. Gavartin, R. Holzwarth, et al., Universal formation dynamics and noise of Kerr-frequency combs in microresonators, Nat. Photonics 6 (2012) 480–487. Available from: https://doi.org/10.1038/nphoton.2012.127.

[125] S.B. Papp, P. Del'Haye, S.A. Diddams, Parametric seeding of a microresonator optical frequency comb, Opt. Express 21 (2013) 17615. Available from: https://doi.org/10.1364/OE.21.017615.
[126] F. Ferdous, H. Miao, D.E. Leaird, K. Srinivasan, J. Wang, L. Chen, et al., Spectral line-by-line pulse shaping of on-chip microresonator frequency combs, Nat. Photonics 5 (2011) 770–776. Available from: https://doi.org/10.1038/nphoton.2011.255.
[127] S.B. Papp, K. Beha, P. Del'Haye, F. Quinlan, H. Lee, K.J. Vahala, et al., Microresonator frequency comb optical clock, Optica 1 (2014) 10. Available from: https://doi.org/10.1364/OPTICA.1.000010.
[128] C. Bao, P. Liao, A. Kordts, M. Karpov, M.H.P. Pfeiffer, L. Zhang, et al., Demonstration of optical multicasting using Kerr frequency comb lines, Opt. Lett. 41 (2016) 3876. Available from: https://doi.org/10.1364/OL.41.003876.
[129] F. Leo, S. Coen, P. Kockaert, S.-P. Gorza, P. Emplit, M. Haelterman, Temporal cavity solitons in one-dimensional Kerr media as bits in an all-optical buffer, Nat. Photonics 4 (2010) 471–476. Available from: https://doi.org/10.1038/nphoton.2010.120.
[130] H. Guo, M. Karpov, E. Lucas, A. Kordts, M.H.P. Pfeiffer, V. Brasch, et al., Universal dynamics and deterministic switching of dissipative Kerr solitons in optical microresonators, Nat. Phys. 13 (2016) 94–102. Available from: https://doi.org/10.1038/nphys3893.
[131] M. Karpov, H. Guo, A. Kordts, V. Brasch, M.H.P. Pfeiffer, M. Zervas, et al., Raman self-frequency shift of dissipative Kerr solitons in an optical microresonator, Phys. Rev. Lett. 116 (2016) 103902. Available from: https://doi.org/10.1103/PhysRevLett.116.103902.
[132] C. Milián, A.V. Gorbach, M. Taki, A.V. Yulin, D.V. Skryabin, Solitons and frequency combs in silica microring resonators: interplay of the Raman and higher-order dispersion effects, Phys. Rev. A. 92 (2015) 033851. Available from: https://doi.org/10.1103/PhysRevA.92.033851.
[133] X. Yi, Q.-F. Yang, K.Y. Yang, K. Vahala, Theory and measurement of the soliton self-frequency shift and efficiency in optical microcavities, Opt. Lett. 41 (2016) 3419. Available from: https://doi.org/10.1364/OL.41.003419.
[134] C. Milián, D.V. Skryabin, Soliton families and resonant radiation in a micro-ring resonator near zero group-velocity dispersion, Opt. Express 22 (2014) 3732. Available from: https://doi.org/10.1364/OE.22.003732.
[135] F. Leo, L. Gelens, P. Emplit, M. Haelterman, S. Coen, Dynamics of one-dimensional Kerr cavity solitons, Opt. Express 21 (2013) 9180. Available from: https://doi.org/10.1364/OE.21.009180.
[136] M. Yu, J.K. Jang, Y. Okawachi, A.G. Griffith, K. Luke, S.A. Miller, et al., Breather soliton dynamics in microresonators, Nat. Commun. 8 (2017) 14569. Available from: https://doi.org/10.1038/ncomms14569.
[137] E. Lucas, M. Karpov, H. Guo, M.L. Gorodetsky, T.J. Kippenberg, Breathing dissipative solitons in optical microresonators, Nat. Commun. 8 (2017) 736. Available from: https://doi.org/10.1038/s41467-017-00719-w.
[138] C. Bao, J.A. Jaramillo-Villegas, Y. Xuan, D.E. Leaird, M. Qi, A.M. Weiner, Observation of fermi-pasta-ulam recurrence induced by breather solitons in an optical microresonator, Phys. Rev. Lett. 117 (2016) 163901. Available from: https://doi.org/10.1103/PhysRevLett.117.163901.
[139] M. Anderson, F. Leo, S. Coen, M. Erkintalo, S.G. Murdoch, Observations of spatiotemporal instabilities of temporal cavity solitons, Optica 3 (2016) 1071. Available from: https://doi.org/10.1364/OPTICA.3.001071.
[140] W. Chen, B. Garbin, A.U. Nielsen, S. Coen, S.G. Murdoch, M. Erkintalo, Experimental observations of breathing Kerr temporal cavity solitons at large detunings, Opt. Lett. 43 (2018) 3674. Available from: https://doi.org/10.1364/OL.43.003674.
[141] H. Guo, E. Lucas, M.H.P. Pfeiffer, M. Karpov, M. Anderson, J. Liu, et al., Intermode breather solitons in optical microresonators, Phys. Rev. X. 7 (2017) 041055. Available from: https://doi.org/10.1103/PhysRevX.7.041055.
[142] Y. Wang, F. Leo, J. Fatome, M. Erkintalo, S.G. Murdoch, S. Coen, Universal mechanism for the binding of temporal cavity solitons, Optica 4 (2017) 855. Available from: https://doi.org/10.1364/OPTICA.4.000855.

[143] C. Bao, Y. Xuan, D.E. Leaird, S. Wabnitz, M. Qi, A.M. Weiner, Spatial mode-interaction induced single soliton generation in microresonators, Optica 4 (2017) 1011. Available from: https://doi.org/10.1364/OPTICA.4.001011.

[144] D.C. Cole, E.S. Lamb, P. Del'Haye, S.A. Diddams, S.B. Papp, Soliton crystals in Kerr resonators, Nat. Photonics 11 (2017) 671–676. Available from: https://doi.org/10.1038/s41566-017-0009-z.

[145] C. Godey, I.V. Balakireva, A. Coillet, Y.K. Chembo, Stability analysis of the spatiotemporal Lugiato–Lefever model for Kerr optical frequency combs in the anomalous and normal dispersion regimes, Phys. Rev. A. 89 (2014) 063814. Available from: https://doi.org/10.1103/PhysRevA.89.063814.

[146] V.E. Lobanov, G. Lihachev, T.J. Kippenberg, M.L. Gorodetsky, Frequency combs and platicons in optical microresonators with normal GVD, Opt. Express 23 (2015) 7713. Available from: https://doi.org/10.1364/OE.23.007713.

[147] X. Xue, Y. Xuan, Y. Liu, P.-H. Wang, S. Chen, J. Wang, et al., Mode-locked dark pulse Kerr combs in normal-dispersion microresonators, Nat. Photonics 9 (2015) 594–600. Available from: https://doi.org/10.1038/nphoton.2015.137.

[148] G. Moille, Q. Li, S. Kim, D. Westly, K. Srinivasan, Phased-locked two-color single soliton microcombs in dispersion-engineered Si_3N_4 resonators, Opt. Lett. 43 (2018) 2772. Available from: https://doi.org/10.1364/OL.43.002772.

[149] J. Cai, M. Nissov, C. Davidson, A. Pilipetskii, Measurement techniques for high-speed WDM experiments, in: IEEE Opt. Fiber Commun. Conf., OFC 2003, vol. 2, pp. 575–577. doi:10.1109/OFC.2003.315946.

[150] F. Chang, K. Onohara, T. Mizuochi, Forward error correction for 100 G transport networks, IEEE Commun. Mag. 48 (2010) S48–S55. Available from: https://doi.org/10.1109/MCOM.2010.5434378.

[151] N. Lindenmann, G. Balthasar, D. Hillerkuss, R. Schmogrow, M. Jordan, J. Leuthold, et al., Photonic wire bonding: a novel concept for chip-scale interconnects, Opt. Express 20 (2012) 17667. Available from: https://doi.org/10.1364/OE.20.017667.

[152] N. Lindenmann, S. Dottermusch, M.L. Goedecke, T. Hoose, M.R. Billah, T.P. Onanuga, et al., Connecting silicon photonic circuits to multicore fibers by photonic wire bonding, J. Light. Technol. 33 (2015) 755–760. Available from: https://doi.org/10.1109/JLT.2014.2373051.

[153] M.R. Billah, M. Blaicher, T. Hoose, P.-I. Dietrich, P. Marin-Palomo, N. Lindenmann, et al., Hybrid integration of silicon photonics circuits and InP lasers by photonic wire bonding, Optica 5 (2018) 876. Available from: https://doi.org/10.1364/OPTICA.5.000876.

[154] R. Schmogrow, B. Nebendahl, M. Winter, A. Josten, D. Hillerkuss, S. Koenig, et al., Error vector magnitude as a performance measure for advanced modulation formats, IEEE Photonics Technol. Lett. 24 (2012) 61–63, Erratum: *ibid.* **24**(23), 2198 (2012). Available from: https://doi.org/10.1109/LPT.2011.2172405.

CHAPTER 3

Nanophotonic devices for power-efficient communications

You-Chia Chang[1] and Michal Lipson[2]
[1]Department of Photonics and Institute of Electro-Optical Engineering, National Chiao Tung University, Hsinchu, Taiwan, ROC
[2]Department of Electrical Engineering, Columbia University, New York, NY, United States

3.1 Current state-of-the-art low-power GHz silicon photonic devices

Power-efficient high-speed modulators in silicon are the workhorse that converts electrical signals to the optical domain for optical communications [1–7]. Silicon modulators rely on mechanisms that electrically tune the optical properties of silicon. In this section, we provide an introduction to the operation principle of silicon modulators and review the state of the art of current low-power silicon modulators.

The two mechanisms available in silicon for tuning its optical properties are the thermo-optic effect and the plasma dispersion effect. The thermo-optic effect can create a strong refractive index change of 10^{-2} to 10^{-3} under a typical temperature change ($dn/dT = 1.86 \times 10^{-4}\,\mathrm{K}^{-1}$). However, it is too slow ($\sim 10\,\mu$s time scale) to fulfill the need of high-speed communications [8–10]. The plasma dispersion effect is not as strong as the thermo-optical effect but allows for high-speed tuning (~ 10 s GHz). So far the most mature technologies that realize silicon modulators have been based on the plasma dispersion effect. Modulators based on the plasma dispersion effect are compatible to complementary metal-oxide-semiconductor (CMOS) process and require no integration of other exotic materials. Comprehensive review of plasma-dispersion-based silicon modulators can be found in Refs. [3–5,11].

The plasma dispersion effect arises from light interaction with free electrons and holes in silicon, which can be described by the Drude model [12]. By manipulating the free carrier concentration in the waveguide, one can modify the refractive index (electro-refraction) and the absorption (electro-absorption) of silicon. The plasma dispersion effect is a broadband mechanism, unlike, for example, the Franz-Keldysh effect that appears only near the band edge [13,14]. The degree of tuning with the plasma dispersion effect has been carefully characterized in the literature [12]. At the

Optical Fiber Telecommunications V11
DOI: https://doi.org/10.1016/B978-0-12-816502-7.00003-8

wavelength of 1.55 μm, the tuning of the refractive index Δn and the absorption $\Delta\alpha$ in silicon is given by

$$\begin{aligned}\Delta n &= -\left[8.8\times10^{-22}\times\Delta n_e + 8.5\times10^{-18}\times\Delta n_h^{0.8}\right],\\ \Delta\alpha &= 8.5\times10^{-18}\times\Delta n_e + 6.0\times10^{-18}\times\Delta n_h.\end{aligned} \tag{3.1}$$

Here Δn_e and Δn_h are the change of electron and hole density in cm^{-3}, respectively. $\Delta\alpha$ is the absorption coefficient in cm^{-1}. The plasma dispersion effect is slightly weaker at the wavelength of 1.3 μm, given by

$$\begin{aligned}\Delta n &= -\left[6.2\times10^{-22}\times\Delta n_e + 6.0\times10^{-18}\times\Delta n_h^{0.8}\right],\\ \Delta\alpha &= 6.0\times10^{-18}\times\Delta n_e + 4.0\times10^{-18}\times\Delta n_h.\end{aligned} \tag{3.2}$$

Note that free holes are more effective in tuning the refractive index and produce less absorption modulation than free electrons [12], as indicated by Eqs. (3.1) and (3.2).

The plasma dispersion effect is usually used to provide phase tuning. Phase tuning requires sufficient length, especially considering the small Δn from the plasma dispersion effect. For example, a typical carrier concentration change of 10^{17} cm^{-3} only gives rise to Δn of -4.26×10^{-4} and $\Delta\alpha$ of 1.45 cm^{-1}. Creating a π phase shift would require a length of $\lambda/(2\Delta n)$. Therefore, devices based on the plasma dispersion effect often require lengths on the order of millimeter, which increases the insertion loss and the real estate. The insertion loss can be as high as 2–6 dB [15]. The real estate can be as high as $\sim10^4$ μm^2 for a phase shift modulation but can be decreased by resonant enhancement [3], as will be discussed in Section 3.1.2.2. The phase tuning with the plasma dispersion effect is also accompanied by nonnegligible absorption modulation.

In order to evaluate modulators, one often uses a performance metric that quantifies the effectiveness of a phase tuning mechanism with respect to an applied voltage, namely the product of voltage and length for creating a π phase shift (denoted by $V_\pi L$). A smaller value is desired, as it indicates lower voltage and/or more compact footprint. This definition also implies that one can trade length for lower driving voltage, and vice versa. Small $V_\pi L$ usually indicates lower power consumption. This is because the power consumption of an active device is proportional to V^2. A tuning mechanism with smaller $V_\pi L$ allows more room to design for low-voltage, and therefore low-power, devices. We will discuss the values of $V_\pi L$ for different tuning mechanisms in the following sections.

3.1.1 Carrier manipulation mechanisms

To create modulation with the plasma dispersion effect, one needs to electrically manipulate the carrier concentration in the silicon waveguide, that is, either decrease or increase it in specific regions. Carrier manipulation can be achieved with carrier depletion, carrier injection, and carrier accumulation, as shown in Fig. 3.1 [3]. Carrier depletion has been the most commonly used mechanism for communications due to its high operation speed [15–19]. Carrier depletion devices usually consist of a PN junction operated in the reverse bias, as shown in Fig. 3.1C. Electrical signals modify the width of the depletion region and thus change the carrier concentration near the junction. These devices consume only the switch energy for charging and discharging the junction capacitance, with very little static power from the leakage current. The typical voltage-length product $V_{\pi}L$ of carrier depletion devices is $\sim 1-2$ V · cm [15,18–20], which remains fairly constant with frequency typically up to ~ 30 GHz, as shown in Fig. 3.2 [20]. Carrier injection devices are usually realized with a PIN diode structure [20–23], as shown in Fig. 3.1B. When operating in the forward bias, electrons and holes are injected into the intrinsic silicon waveguide core and change the refractive index. The speed of carrier injection devices is limited by the carrier recombination time (~ 1 ns) [23]. As a result, although carrier injection devices produce much stronger tuning compared to carrier depletion devices at DC ($V_{\pi}L \sim 0.02$ V · cm), the $V_{\pi}L$ at 10 s of GHz for these devices is worse than for the carrier depletion devices [20,23]. This is demonstrated by the strongly frequency-dependent $V_{\pi}L$ shown in Fig. 3.2 [20]. In order to mitigate the frequency limitation, carrier injection devices operating up to 50 Gb/s have been shown using preemphasis signals [20]. These PIN devices

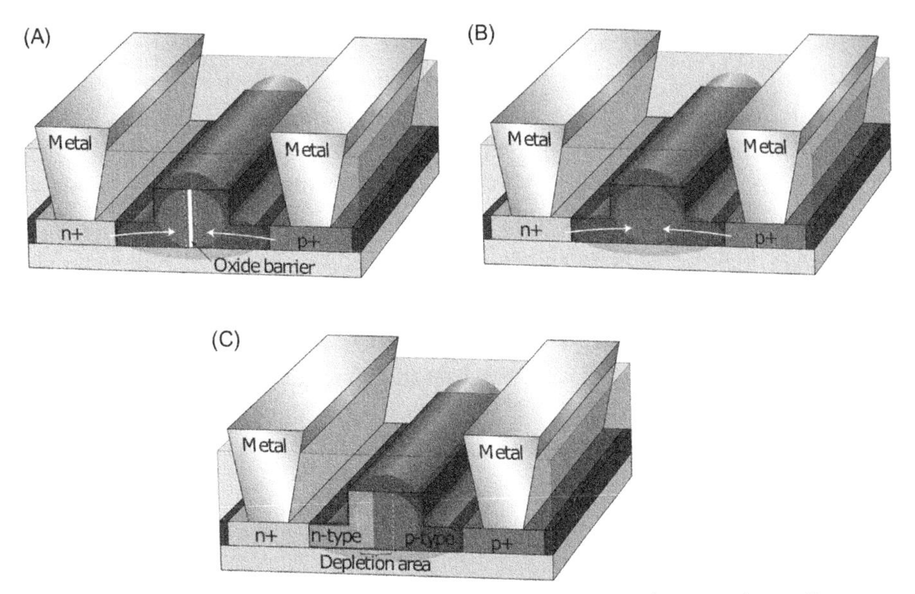

Figure 3.1 Different device structures for carrier manipulation. (A) silicon-insulator-silicon capacitor structure (SISCAP) for carrier accumulation. (B) PIN structure for carrier injection. (C) PN structure for carrier depletion. *Reprinted with permission from G.T. Reed, G. Mashanovich, F.Y. Gardes, D.J. Thomson, Silicon optical modulators, Nat. Photonics. 4 (2010) 518–526.*

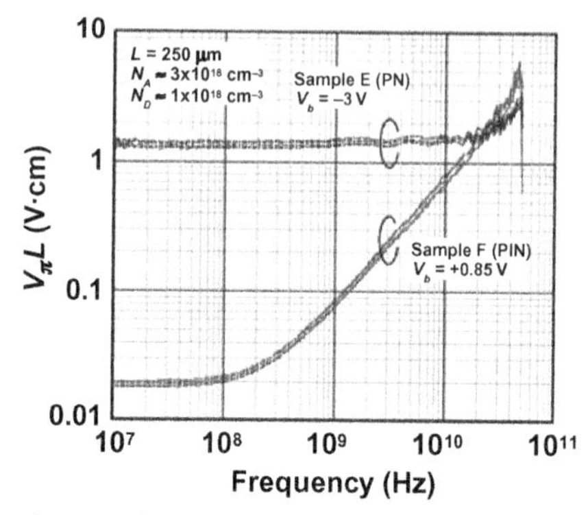

Figure 3.2 Frequency dependence of $V_\pi L$ for carrier depletion PN device (blue curve, Sample E) and carrier injection PIN device (red curve, Sample F). *Reprinted with permission from S. Akiyama, M. Imai, T. Baba, T. Akagawa, N. Hirayama, Y. Noguchi, et al., Compact PIN-diode-based silicon modulator using side-wall-grating waveguide, IEEE J. Sel. Top. Quantum Electron. 19 (2013) 74–84.*

consume static power due to the current generated by the forward bias of the diode, in addition to the switch energy of charging the capacitor [20]. Carrier injection devices may not be competitive for high-speed modulators, but the much smaller $V_\pi L$ at low frequencies may find applications in other emerging fields such as compact phase shifters for optical phased arrays [24,25]. Carrier accumulation devices are usually realized with a silicon-insulator-silicon capacitor (SISCAP) structure, as shown in Fig. 3.1A [26–28]. This structure is also called metal-oxide-semiconductor (MOS) capacitor structure in the literature. The capacitor structure induces accumulation of opposite charges on the two sides of the insulator when a bias voltage is applied, which changes the carrier concentration in the waveguide. Carrier accumulation devices can provide a smaller $V_\pi L$ (~ 0.2 V · cm) compared to carrier depletion devices [27] but require more complex fabrication process and introduce more optical loss. We will discuss more about SISCAP in Section 3.2.1.

3.1.2 Photonic designs of silicon modulators

The photonic designs of the plasma-dispersion-based modulators can generally be classified as Mach-Zehnder modulators and resonant modulators, as shown in Fig. 3.3. In the following section we describe the current state of the art of these two photonic designs and discuss their difference in performance.

3.1.2.1 Mach-Zehnder modulators

In a Mach-Zehnder modulator, as shown in Fig. 3.4, an active PN, PIN, or SISCAP junction is placed in one or both of the interferometer arms to create a phase shift. The interference between the two arms converts the phase shift to amplitude

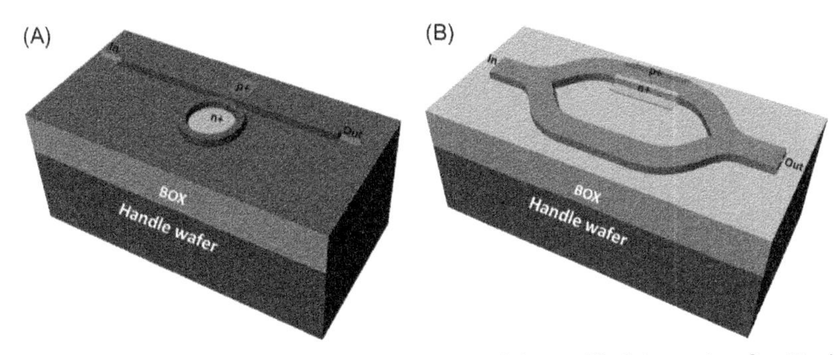

Figure 3.3 (A) Schematic of a resonant microring modulator. (B) Schematic of a Mach-Zehnder modulator. The PIN doping is shown for illustration purposes only. *Reprinted with permission from H. Subbaraman, X. Xu, A. Hosseini, X. Zhang, Y. Zhang, D. Kwong, et al., Recent advances in silicon-based passive and active optical interconnects, Opt. Express. 23 (2015) 2487–2511.*

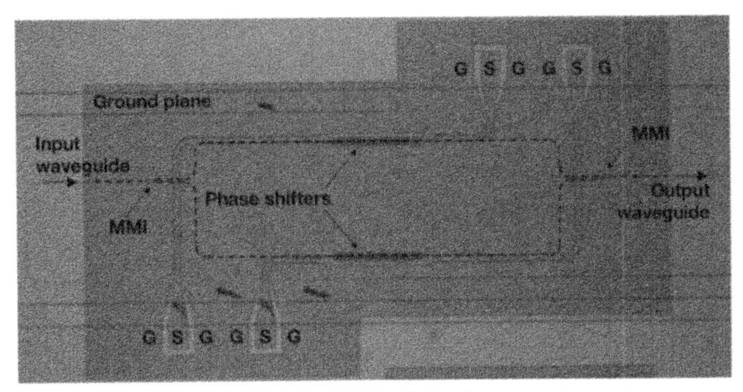

Figure 3.4 Optical microscope image of the Mach-Zehnder modulator demonstrated in Ref. [33]. *Reprinted with permission from D.J. Thomson, F.Y. Gardes, Y. Hu, G. Mashanovich, M. Fournier, P. Grosse, et al., High contrast 40Gbit/s optical modulation in silicon, Opt. Express. 19 (2011) 11507–11516.*

modulation to carry information. Compared to resonant modulators, Mach-Zehnder modulators have the advantage of robustness to fabrication errors and environment temperature, a direct consequence of the broad optical bandwidth. Mach-Zehnder modulators, however, consume more power and require larger footprints because of the lack of resonant enhancement. A typical depletion-type Mach-Zehnder modulator ($V_\pi L \sim 1-2$ V · cm [15,18–20]) requires a length on the order of millimeter. Because this length exceeds the radio-frequency (RF) wavelength, Mach-Zehnder modulators cannot be treated as lumped elements. Mach-Zehnder modulators therefore usually use traveling-wave electrodes, which critically determine the device performance. Traveling-wave electrode design requires considerations in the velocity match between RF and optical waves, the RF loss, and the impedance match [30,31]. All of these are crucial for the power consumption and the operation speed.

We summarize here the performance metrics of some state-of-the-art low-power Mach-Zehnder modulators. The power consumption P of traveling-wave modulators is dictated by the drive voltage, as $P \propto V^2/Z$, where Z is the impedance [4]. Earlier reports of depletion-type Mach-Zehnder modulators described devices that consume $\sim$ 2–5 pJ/bit at 40–50 Gb/s [32,33]. Recent progress of traveling-wave electrode design enables long devices with low driving voltage while maintaining high operational speed, leading to Mach-Zehnder modulators with orders of magnitude reduction in power consumption. Baehr-Jones et al. demonstrated 200 fJ/bit at 20 Gb/s [18], and Ding et al. demonstrated 32 fJ/bit at 40 Gb/s [19]. Using multi-electrode design, Samani et al. demonstrated a 128 Gb/s modulator using PAM-4 (Pulse Amplitude Modulation) with a power consumption of $\sim$ 400 fJ/bit [34].

3.1.2.2 Resonant modulators

Since the first introduction of high-speed resonant modulator in 2005 [21], resonators have been employed widely to enable power-efficient compact devices. In resonant modulators, the active junctions are embedded in the cavities. These active junctions modify the refractive index and shift the cavity resonance to create amplitude modulation. With enhanced light-matter interaction, resonant modulators enable smaller footprint and less power consumption when compared to nonresonant Mach-Zehnder modulators. The enhanced interaction originates from allowing light to circulate multiple round trips within resonators [35], which increases the effective interaction length. One can quantify the enhancement with the finesse of a resonator—a resonator with a finesse F reduces the power consumption and the length by approximately F/π times [6,36]. Compared to the millimeter lengths of Mach-Zehnder modulators, resonant modulators typically have footprints on the order of $\sim 10^2$ μm^2 or less, and power consumption as low as few fJ or even sub fJ [17]. Resonant modulators are lumped elements of which the power consumption, proportional to CV^2, is dictated by the driving voltage [4]. Therefore, design of power-efficient resonant modulators requires reducing the driving voltage with least increase of the capacitance.

The enhanced tuning from resonant structures, however, comes with trade-offs. First, the optical bandwidth is dramatically reduced to $1/F$ of the free spectral range. This is a consequence of the requirement for all the circulating waves in the resonator to constructively interfere with each other. The narrow optical bandwidth results in tight fabrication and temperature tolerance. As a result, resonant modulators usually operate with additional heaters to actively tune the narrowband resonance to the laser wavelength. This active thermal tuning, which consumes extra power and adds control complexities, are often considered the biggest price of using resonant modulators. In addition, the photon cavity lifetime in resonators imposes a limit on the operation speed of devices [16,36]. The 3-dB modulation bandwidth is thus limited to $\sqrt{\sqrt{2}-1} \times f_{\mathrm{FWHM}}$, where f_{FWHM} is the full-width-half-maximum of the resonance [17,37]. Furthermore, encoding advanced modulation formats is challenging with resonant modulators when compared to Mach-Zehnder modulators due to the coupled response between the amplitude and the phase modulation [38].

Since the first introduction of high-speed silicon microring modulators by Xu et al. in 2005 [21], as shown in Fig. 3.5A, the power efficiency of resonant modulators has shown great progress over the past decade. For example, early demonstration of depletion-type microring modulators in 2009 by Dong et al. consumes 50 fJ/bit and has an electro-optic 3 dB bandwidth of 11 GHz [39]. In 2011, Li et al. demonstrated a lateral-junction depletion-type microring modulator, as shown in Fig. 3.5B, exhibiting a power consumption of 7 fJ/bit and an electro-optic 3 dB bandwidth of 16.3 GHz [40]. In 2014, Timurdogan et al. demonstrated one of the most efficient silicon modulators with a microdisk resonator [17], which consumes 0.9 fJ/bit. The device structure is shown in Fig. 3.7A. This modulator employed vertical PN junction design to increase the overlap between the junction and the optical mode, which will be discussed in Section 3.2.1. The electro-optic 3 dB bandwidth was ~ 21 GHz, limited by the photon lifetime, and operation up to 25 Gb/s was demonstrated. In 2018 resonant modulators with high data rate were

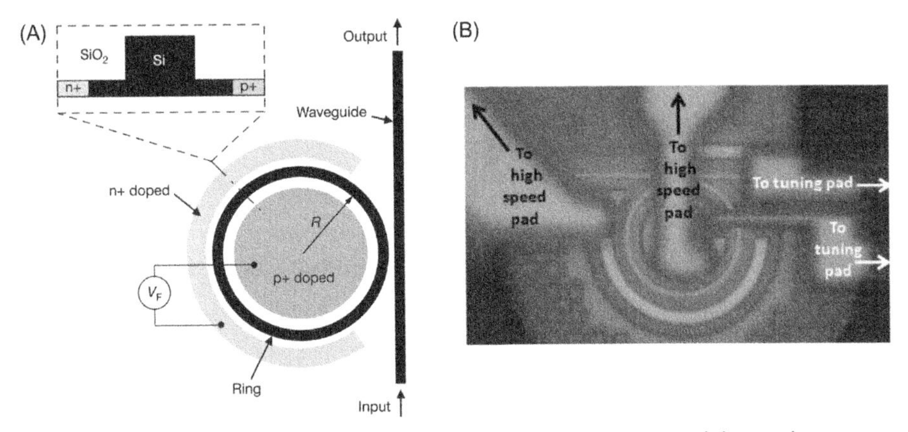

Figure 3.5 (A) Schematic of the first injection-type silicon microring modulators demonstrated in Ref. [21]. (B) Optical microscope image of the lateral-junction depletion-type microring modulator demonstrated in Ref. [40]. *Reprinted with permission from Q. Xu, B. Schmidt, S. Pradhan, M. Lipson, Micrometre-scale silicon electro-optic modulator, Nature 435 (2005) 325–327 and G. Li, X. Zheng, J. Yao, H. Thacker, I. Shubin, Y. Luo, et al., 25Gb/s 1V-driving CMOS ring modulator with integrated thermal tuning, Opt. Express. 19 (2011) 20435–20443.*

also demonstrated. For example, Sun et al. demonstrated a high-speed microring modulator that can send 128 Gb/s PAM4 data rate [16], and the electro-optic 3 dB bandwidth reached $\sim$ 50 GHz.

3.2 Emerging approaches for improving performance via device design

Reducing the power consumption of active photonic devices is crucial for next-generation communications and information processing [6,41]. The key to enable power reduction is to enhance the light interaction with the material of which the refractive index is tuned by the applied electrical signal. In this section we introduce emerging approaches of novel device designs that aim to enhance this interaction. One important approach is to design novel devices that have better mode overlap with the junctions [16,17,27,31]. This allows light to see more tuning as the refractive index change is maximal near the junctions. In Section 3.2.1 we discuss vertical junction designs, which have better mode overlap than conventional lateral junctions. Another common approach for enhancing the interaction is by resonant structures. Resonant structures, however, are extremely constrained in optical bandwidth, which results in susceptibility to fabrication and temperature variations. In Section 3.2.2 we introduce emerging device designs that relax the constraints of resonant structures, enabling larger fabrication tolerance and operational temperature range [42–45]. We also introduce in Section 3.2.3 a novel approach called resonance-free light recycling [46–48]. It allows light to circulate multiple passes in the device for enhancing the interaction, similar to the resonator approach. In this approach, light is converted to a different orthogonal spatial mode after each pass, which bypasses the problem of optical bandwidth reduction in the resonator approach.

3.2.1 Novel junction design for improving mode overlap

Junction design is crucial to plasma-dispersion-based devices, as it determines how efficiently one can manipulate the carrier concentration in a waveguide. This is particularly important for carrier depletion and carrier accumulation mechanisms, in which the modulation of carrier concentration is maximized near the junctions. Examples of different junction designs, including lateral, vertical, and interleaved junctions [17], are shown in Fig. 3.6A. While lateral junction design benefits from its fabrication simplicity, the emerging approach of vertical junction design enables more efficient modulation because of the better overlap between the junction and the optical mode. Vertical junctions have better mode overlap simply because most of the waveguides employed in silicon photonic circuits have wide and thin cross sections. Fig. 3.6B shows a simulated example that compares the tuning of a resonant modulator with different junction designs [17]. The vertical junction (30.9 GHz/V)

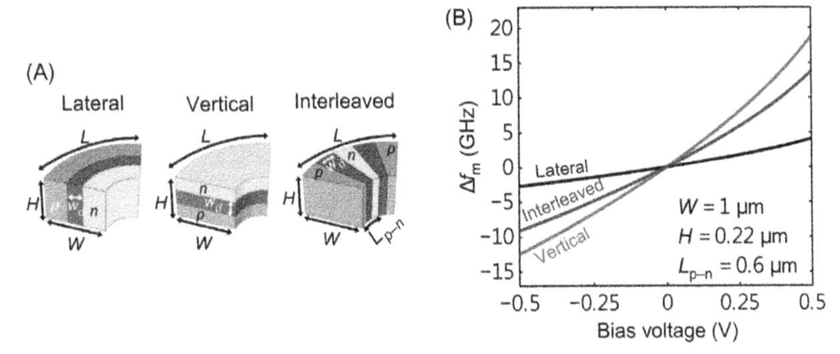

Figure 3.6 (A) Schematic of different junction designs. (B) Simulated tuning of the resonance frequency of a microdisk modulator with different junction designs. *Reprinted with permission from E. Timurdogan, C.M. Sorace-Agaskar, J. Sun, E. Shah Hosseini, A. Biberman, M.R. Watts, An ultralow power athermal silicon modulator, Nat. Commun. 5 (2014) 4008.*

enables a tuning of the resonant frequency ~ 4.5 times stronger than the lateral junction (6.9 GHz/V). This translates to 4.5 times lower driving voltage and roughly 20 times lower power consumption (switching energy $\propto V^2$). One quantitative way to discuss the mode-junction overlap is via the junction capacitance per unit length C/L seen by the waveguide mode [17,31]. Vertical junction designs have larger C/L than lateral junctions due to larger junction area. Because the change in carrier number per length $\Delta Q/L$ equals $(C/L)\Delta V$, larger C/L allows more efficient carrier number manipulation and thus stronger refractive index change. The use of vertical junctions has its trade-offs, however. The enhancement comes at an expense of a more complex fabrication process (due to complicated dopant implantation), slower speed (due to increased RC), and higher optical propagation loss (due to typically higher dopant concentration) [31].

Vertical junction design has been employed by Timurdogan et al. to demonstrate one of the most efficient silicon microdisk modulators based on the plasma dispersion effect, which consumes only 0.9 fJ/bit and operates up to an electro-optic 3 dB bandwidth of ~ 21 GHz [17]. Fig. 3.7A shows the device structure of this depletion-type microdisk modulator with a vertical PN junction. Although a larger capacitance per length C/L enables stronger tuning, it is desirable to keep total capacitance as low as possible for smaller switching energy ($\propto CV^2$) and smaller RC. Timurdogan et al. showed that vertical junction also allows a more compact microdisk to keep the total capacitance low. This is because it eliminates the need of the ridge waveguide, which is used in lateral junction designs to make P + and N + contacts. As shown by Fig. 3.7B, microdisks with hard outer walls can maintain a high quality factor even with small radii.

Other variations of vertical PN junction design have also been explored to achieve a good overlap between the junction and the optical mode. For example,

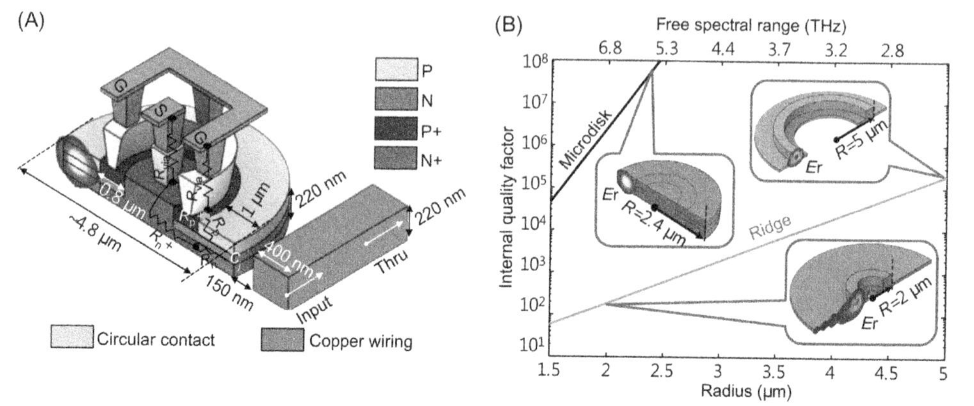

Figure 3.7 (A) Schematic of a microdisk modulator with vertical junction design. (B) Simulated internal quality factor as a function of the radius for a microdisk and a ridge resonator. *Reprinted with permission from E. Timurdogan, C.M. Sorace-Agaskar, J. Sun, E. Shah Hosseini, A. Biberman, M.R. Watts, An ultralow power athermal silicon modulator, Nat. Commun. 5 (2014) 4008.*

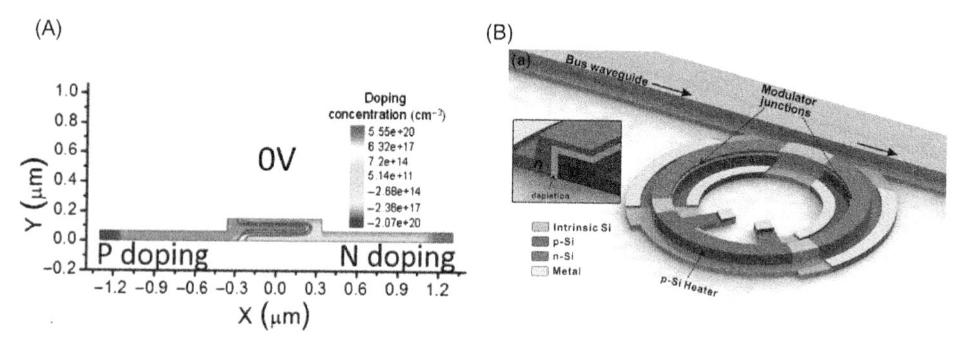

Figure 3.8 (A) Doping profile of a U-shape vertical PN junction. *Reprinted with permission from Z. Yong, W.D. Sacher, Y. Huang, J.C. Mikkelsen, Y. Yang, X. Luo, et al., U-shaped PN junctions for efficient silicon Mach-Zehnder and microring modulators in the O-band, Opt. Express. 25 (2017) 8425–8439.* (B) A microring modulator with a junction of both vertical and horizontal segments. *Reprinted with permission from J. Sun, M. Sakib, J. Driscoll, R. Kumar, H. Jayatilleka, Y. Chetrit, et al., A 128 Gb/s PAM4 silicon microring modulator, in: Opt. Fiber Commun. Conf. Expo. San Diego, 2018, p. Th4A.7.*

Yong et al. demonstrated a U-shape vertical PN junction, as shown in Fig. 3.8A [31]. The larger junction surface area of the U shape allowed high junction capacitance per length C/L, without requiring high dopant concentration. This reduced the optical loss penalty of the vertical junction design. The authors demonstrated a voltage-length product $V_{\pi}L$ of 0.46 V · cm, which was ~ 2 to 4 times smaller than typical lateral junction design ($V_{\pi}L$ ~ 1–2 V · cm) [15,18–20]. The propagation loss was 12.5 dB/cm, which was only 1 ~ 2 times higher than the lateral junction design. The demonstrated electro-optic 3 dB bandwidth was ~ 13 GHz. Sun et al. demonstrated a microring modulator in which a junction with both

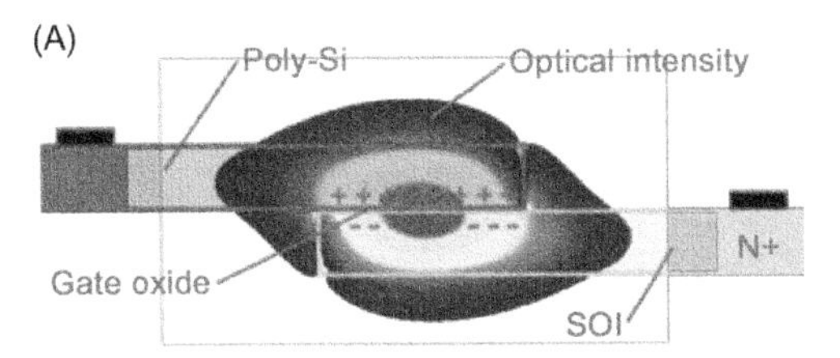

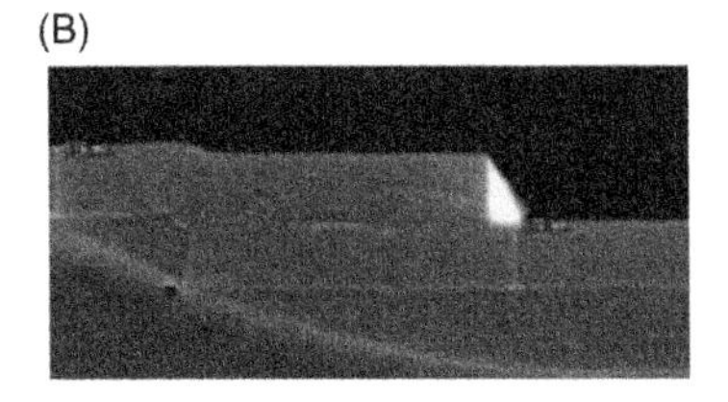

Figure 3.9 (A) Schematic of the SISCAP structure and the optical mode. (B) SEM picture of the SISCAP structure. *Reprinted with permission from M. Webster, P. Gothoskar, V. Patel, D. Piede, S. Anderson, R. Tummidi, et al., An efficient MOS-capacitor based silicon modulator and CMOS drivers for optical transmitters, in: 11th Int. Conf. Group IV Photonics, Paris, 2014, p. W.B1.*

vertical and horizontal directions was created inside a ridged waveguide, as shown in Fig. 3.8B [16]. The authors demonstrated a voltage-length product $V_\pi L$ of 0.52 V · cm and an electro-optic bandwidth of 50 GHz.

Another approach to achieve efficient modulation is to use a vertical junction with SISCAP embedded within the waveguide mode [27]. The capacitor formed by the polysilicon, gate oxide, and crystalline silicon allows high capacitance per length for charge accumulation, as shown in Fig. 3.9. Webster et al. demonstrated a SISCAP with $V_\pi L$ of 0.2 V · cm [27], which is 5 ~ 10 times smaller than the typical lateral junction design [15,18–20]. The demonstrated data rate is up to 40 Gb/s. The major price to pay for gaining the enhanced tuning is the high propagation loss of 65 dB/cm, mainly due to the p-doped polysilicon.

3.2.2 Novel resonator design

The strong sensitivity of resonators remains a major challenge for broadening the applications of resonators. Here we introduce emerging approaches that tackle this challenge by designing resonators insensitive to fabrication and temperature variations.

We briefly explain the origin of the enhanced interaction and the increased sensitivity in resonators. Resonators are fundamentally narrowband devices. They provide a dramatic reduction of power consumption and footprint by circulating light many round trips to increase the effective interaction length. The requirement of constructive interference between all circulating waves, however, restricts the optical bandwidth. Here we analyze the sensitivity to fabrication and temperature variations for typical resonators. A resonator with a finesse F reduces the power consumption and the device length by approximately F/π times [6,36]. The optical bandwidth, however, reduces to only $1/F$ of the free spectral range, that is,

$$\text{Optical bandwidth} = \frac{FSR}{F} = \frac{\lambda_{\text{res}}}{Q}, \tag{3.3}$$

where FSR is the free spectral range, given by $\lambda^2/(n_g L)$. n_g is the group index. L is the round-trip length of the resonator. λ_{res} is the resonant wavelength, and Q is the quality factor [35]. For example, a typical resonator with a $Q = 10{,}000$ operating at λ_{res} of 1550 nm has an optical bandwidth of only 0.155 nm. This subnanometer optical bandwidth imposes challenges to maintain the resonance under variations. We can calculate the shift in the resonance wavelength $\Delta\lambda_{\text{res}}$.

$$\frac{\Delta\lambda_{\text{res}}}{\lambda_{\text{res}}} = \frac{\Delta n_{\text{eff}}}{n_g}, \tag{3.4}$$

where n_{eff} is the effective refractive index of the waveguide mode [35]. Δn_{eff} is commonly caused by the variation in waveguide width (w), waveguide height (h), and temperature (T).

$$\Delta n_{\text{eff}} = \frac{\partial n_{\text{eff}}}{\partial w}\Delta w + \frac{\partial n_{\text{eff}}}{\partial h}\Delta h + \frac{\partial n_{\text{eff}}}{\partial T}\Delta T \tag{3.5}$$

Fig. 3.10 shows the sensitivity of n_{eff} to the width variation in a wire waveguide. For a typical SiO_2-cladded wire waveguide with a 450 nm width and a 220 nm height, the sensitivity of λ_{res} to width, height, and temperature variation is $\approx$ 0.8 nm/nm, 1.4 nm/nm, and 0.1 nm/K, respectively [49]. Consider a resonator of $Q \sim 10{,}000$. The sensitivity translates to a very challenging subnanometer fabrication tolerance. The temperature needs to be controlled actively within $\sim$ 1°C. The preceding analysis shows the strong sensitivity to dimensions and temperature in resonators. In Section 3.2.2.1 and 3.2.2.2 we introduce novel device designs to decrease the sensitivity.

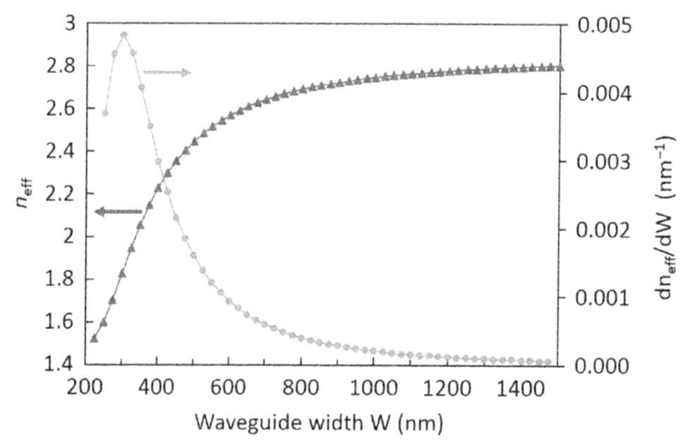

Figure 3.10 The effective refractive index (green, left axis) and the sensitivity of the effective refractive index with respect to the width (pink, right axis) for a TE_0 wire waveguide at the wavelength of 1550 nm. The waveguide height is 220 nm. The waveguide is cladded with SiO_2.

3.2.2.1 Robust resonators to fabrication variations

To fabricate resonators with a yield for which their resonance frequencies align with the laser frequencies requires subnanometer control of waveguide dimensions, as given by the analysis with Eqs. (3.3–3.5). This is challenging, if not impossible. Even though the height is more controllable by the silicon-on-insulator (SOI) wafer thickness, the width variation can be caused by normal fluctuations in the lithography and etching [49]. The uncertainty of the resonance frequencies of fabricated resonators is the reason why practical resonant modulators require thermal tuning to compensate for the fabrication errors.

One approach to design resonators robust to fabrication errors is to propagate only in the fundamental mode within a wide multimode waveguide. This can be understood from Fig. 3.10, which shows that the sensitivity $\partial n_{\text{eff}}/\partial W$ decreases drastically if a wide multimode waveguide is used. The insensitivity of a fundamental mode in a wide waveguide is a result of confining most of the mode inside the silicon core. The effective index is closer to the refractive index of silicon and therefore more insensitive to the dimensions. In this approach, the coupling between the bus waveguide and the resonator requires careful design to excite only the fundamental mode while maintaining appropriate coupling strength. Luo et al. applied this approach to demonstrate microring modulators based on the plasma dispersion effect with improved fabrication tolerance [42]. As shown by Fig. 3.11, the 3.3-μm radius microring has a multimode waveguide width of 600 nm. Fabricated in a 130 nm CMOS

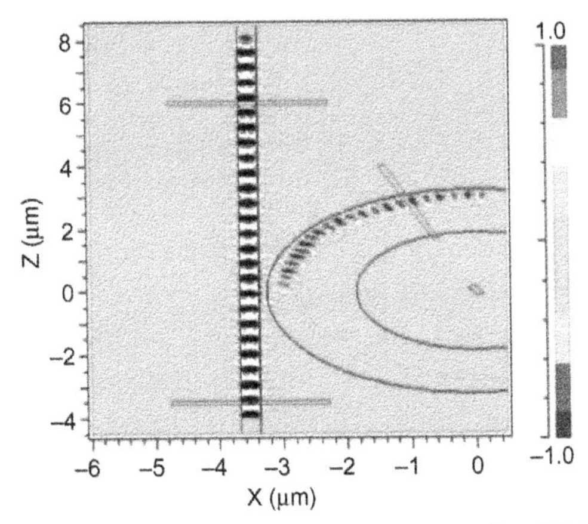

Figure 3.11 Finite-difference time-domain (FDTD) simulation of a fabrication-tolerant microring that uses the fundamental mode in a wide multimode waveguide. *Reprinted with permission from Y. Luo, X. Zheng, S. Lin, J. Yao, H. Thacker, I. Shubin, et al., Modulator based on multi-mode waveguides, IEEE Photonics Technol. Lett. 28 (2016) 1391–1394.*

foundry on 300-nm-thick SOI wafer, these microrings exhibited a standard deviation of ~ 1 nm of their resonance wavelengths over wafer scale, which is 3 times smaller compared to other microrings that use a single-mode 390 nm waveguide width. The variation in the resonance wavelength $\Delta\lambda_{res}$, however, still exceeds the optical bandwidth (~0.8 nm). This approach can potentially be extended to even wider and thicker multimode waveguides to achieve better robustness.

3.2.2.2 Athermal resonators

Resonant silicon modulators provide better energy efficiency and more compact footprint, but the susceptibility to temperature variations can easily render these devices inoperable [50]. The sensitivity of the resonance wavelength to temperature in typical silicon microrings is ~ 0.1 nm/K, as calculated by Eqs. (3.4) and (3.5) [43]. This poses a challenge to maintain the resonance of these narrow-bandwidth devices under temperature fluctuations. For example, a typical resonator with a Q of ~15,000 goes off-resonance even when the temperature changes by only 1°C. Temperature fluctuation can come from the joule heat of driving active photonic devices, heat transfer from nearby electronics, optical absorption in the doped silicon waveguide, and the ambient temperature changes. Two categories of solutions that take completely opposite approaches have been employed: (1) solutions that reduce the thermal sensitivity to ensure insensitivity to variations of temperature and (2) solutions that increase the local thermal sensitivity and actively tune the resonance [51]. Here we focus on the former approach because it is desirable for low-power devices to operate without the need for active control. As shown by Eqs. (3.4) and (3.5), the temperature sensitivity of the resonance wavelength comes from the temperature dependence of n_{eff}, which is given by

$$\frac{\partial n_{\mathrm{eff}}}{\partial T} = \Gamma_{\mathrm{core}} \frac{\partial n_{\mathrm{core}}}{\partial T} + \Gamma_{\mathrm{clad}} \frac{\partial n_{\mathrm{clad}}}{\partial T} + \Gamma_{\mathrm{sub}} \frac{\partial n_{\mathrm{sub}}}{\partial T}, \tag{3.6}$$

where n_{core}, n_{clad}, and n_{sub} are the refractive indices of the silicon core, cladding, and substrate, respectively. Γ_{core}, Γ_{clad}, and Γ_{sub} represent the confinement factors in these three materials. Silicon has a thermo-optic coefficient $\partial n_{\mathrm{core}}/\partial T$ of $1.86 \times 10^{-4}\,\mathrm{K}^{-1}$ [8]. Silicon dioxide, the common cladding and substrate material, has a lower thermo-optic coefficient of $9.5 \times 10^{-6}\,\mathrm{K}^{-1}$ [52].

One approach to realize athermal resonators via decrease of thermal sensitivity is to use a cladding material with a negative thermo-optic coefficient [43,53–56]. By designing the confinement factors to balance the positive and negative thermo--optic coefficients of different materials, one can achieve significant decrease in thermal sensitivity. TiO_2, for example, has a negative thermo-optic coefficient of ~ $-1 \times 10^{-4}\,\mathrm{K}^{-1}$, comparable to the positive thermo-optic coefficient of silicon [43]. It is available in standard CMOS process for gate dielectrics. Guha et al.

demonstrated athermal microring resonators by tailoring the silicon waveguide width to delocalize the mode and designing an appropriate TiO_2 cladding thickness [43], as shown in Fig. 3.12. Alipour et al. demonstrated athermal microdisk resonators using polymer cladding [56].

An alternative approach to decrease the temperature sensitivity is to combine a resonator with another photonic device that has the opposite temperature dependence. Guha et al. demonstrated athermal modulators by embedding a silicon microring in a Mach-Zehnder interferometer [44,45], as shown in Fig. 3.13. A balanced Mach-Zehnder interferometer with different waveguide widths in the two arms was used. Because the mode in the narrower waveguide is more delocalized into the cladding, it has smaller $\partial n_{\text{eff}}/\partial T$ compared to the other arm. This allows engineering the sign of temperature-dependent phase shift between the two arms. The microring is overcoupled to the Mach-Zehnder interferometer arm with the narrower waveguide to serve as a phase shifter. As the temperature increases, the microring gives a positive phase shift, which is compensated by the negative phase shift of the Mach-Zehnder interferometer. A PIN silicon modulator based on this approach shows an open eye diagram over 35°C of temperature change, as shown in Fig. 3.13B.

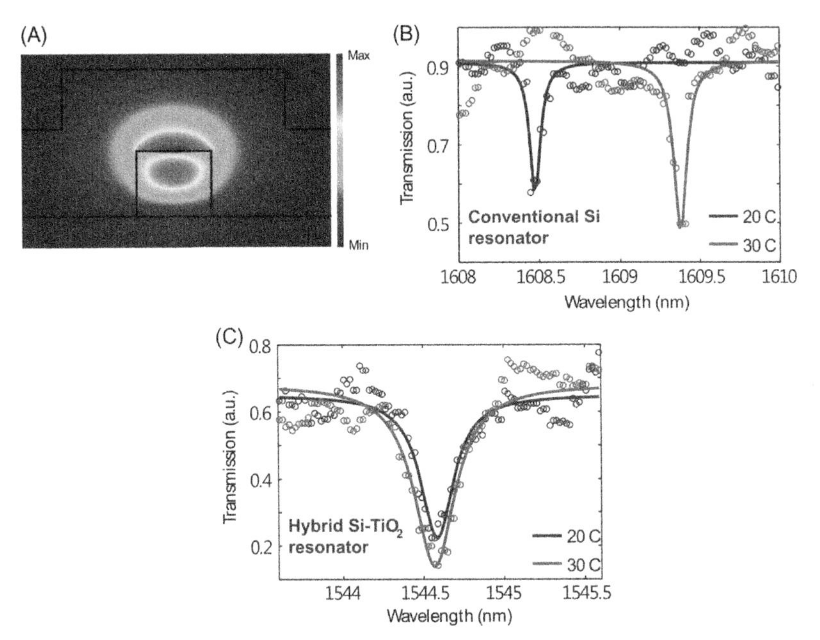

Figure 3.12 (A) Athermal waveguide consisting of a silicon core and a TiO_2 over cladding (TM mode). Transmission spectra of a conventional Si resonator (B) and the athermal hybrid Si-TiO_2 resonator (C) under temperature variations. *Reprinted with permission from B. Guha, J. Cardenas, M. Lipson, Athermal silicon microring resonators with titanium oxide cladding, Opt. Express. 21 (2013) 26557–26563.*

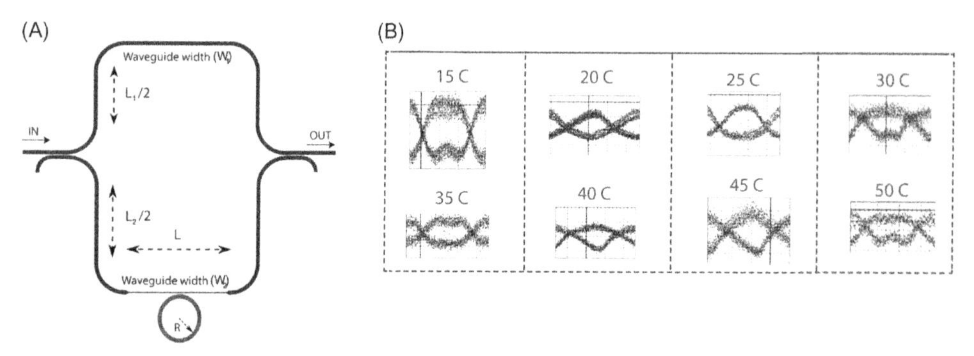

Figure 3.13 (A) Schematic of a resonator embedded in a Mach-Zehnder interferometer for athermal operation. *Reprinted with permission from B. Guha, B.B.C. Kyotoku, M. Lipson, CMOS-compatible athermal silicon microring resonators, Opt. Express. 18 (2010) 3487–3493.* (B) Eye diagrams at different temperatures for a PIN silicon microring modulator embedded in a Mach-Zehnder interferometer. *Reprinted with permission from B. Guha, K. Preston, M. Lipson, Athermal silicon microring electro-optic modulator, Opt. Lett. 37 (2012) 2253–2255.*

3.2.3 Resonance-free light recycling

The future scalability of silicon photonics involving systems with a large number of devices requires an approach for reducing the power consumption of active devices. Although resonators can reduce the power consumption by circulating light by multiple round trips within the active devices, they are inherently narrowband. The narrow optical bandwidth is a fundamental consequence of requiring all the circulating waves inside the resonators to interfere constructively. Any deviation from the resonance wavelength impairs the constructive interference. The narrow optical bandwidth results in high sensitivities to fabrication and temperature variation, as discussed in Sections 3.2.2.1 and 3.2.2.2. This causes great difficulties for scaling the number of resonators in a photonic integrated circuit due to power-hungry active tuning of all resonators.

We have demonstrated a scalable, resonance-free light recycling approach for power-efficient active devices [46,47]. This approach exploits the spatial modes of a multimode waveguide. Similar to resonators, this approach allows light to circulate inside the device multiple passes to enhance the power efficiency. However, the mode in the waveguide is converted to a different order after each pass [57–59]. Because of the orthogonality between different modes, circulating waves do not interfere with each other. This avoids the optical bandwidth reduction associated with the requirement of constructive interference in resonators.

A recycling structure consists of a multimode bus waveguide and multiple passive mode converters. Fig. 3.14A and B show the schematics of the recycling structure and the recycling passes with associated spatial modes. Light enters the recycling structure with TE_0 mode. After light passes through the structure the first time, it is converted

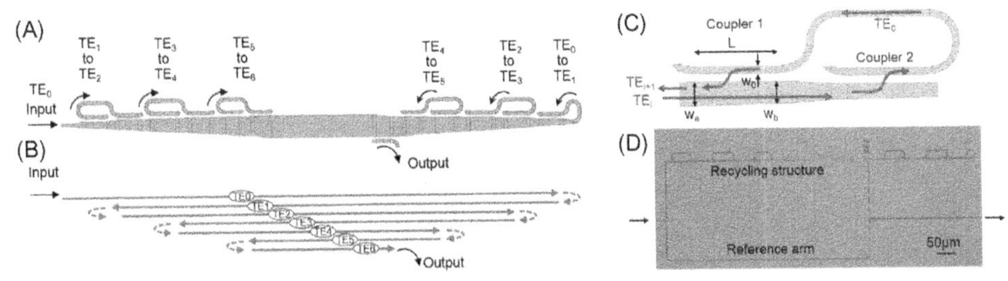

Figure 3.14 (A) Schematic of the recycling structure that enables 7-pass light recycling. (B) Schematic of the recycling passes and the associated spatial modes. (C) The mode converter that raises TE_i mode to TE_{i+1} mode. (D) Optical microscope image of a fabricated recycling structure embedded in a Mach-Zehnder interferometer. *Reprinted with permission from Y.-C. Chang, S.P. Roberts, B. Stern, I. Datta, M. Lipson, Resonance–free light recycling in waveguides, in: Conf. Lasers Electro-Opt. San Jose, 2017, p. SF1J.5.*

into TE_1 mode via a passive mode converter and sent back in the opposite direction. In the second pass, light travels through the bus waveguide, and upon reaching the other end of the structure, it is converted to TE_2 mode and sent back. Light is therefore recycled back and forth as the mode is promoted consecutively. Eventually the mode order reaches highest order and exits the bus waveguide as TE_0 mode. The passive mode converter consists of two directional couplers, as shown in Fig. 3.14C. It is designed to pick only one particular mode, raise the mode order by one, and reverse the propagation direction, while keeping all other lower-order modes unperturbed. These directional couplers are designed to be mode-selective by engineering the effective refractive index via the waveguide width. They are designed to be broadband and robust to fabrication variation with adiabatic coupling [47].

Resonance-free light recycling is a general approach that can potentially enhance a variety of active and sensing devices, including thermo-optic phase shifters, electro-optic modulators, chemical sensors, and so on. An active photonic device or sensing material can be placed in the middle multimode section to gain enhancement by the multiple passes. Using 7-pass recycling, the demonstrated thermo-optic phase shifter requires only 1.7 mW per π, representing more than an eightfold enhancement, as shown in Fig. 3.15A. This approach exhibits a broad optical bandwidth exceeding 100 nm, as shown in Fig. 3.15B. The 7-pass recycling has an insertion loss of 4.6 dB. The insertion loss decreases to 2.2 dB if 5-pass recycling is used. The simulated fabrication tolerance is ± 15 nm [47], which is in stark contrast to the nonscalable subnanometer tolerance in typical silicon resonators.

Miller et al. demonstrated the scalability of this light-recycling approach with a 512-element active optical phased array, showing record-low array power consumption [48]. The optical images of the optical phased array are shown in Fig. 3.16. The enhanced tuning and large fabrication tolerance of the light-recycling approach

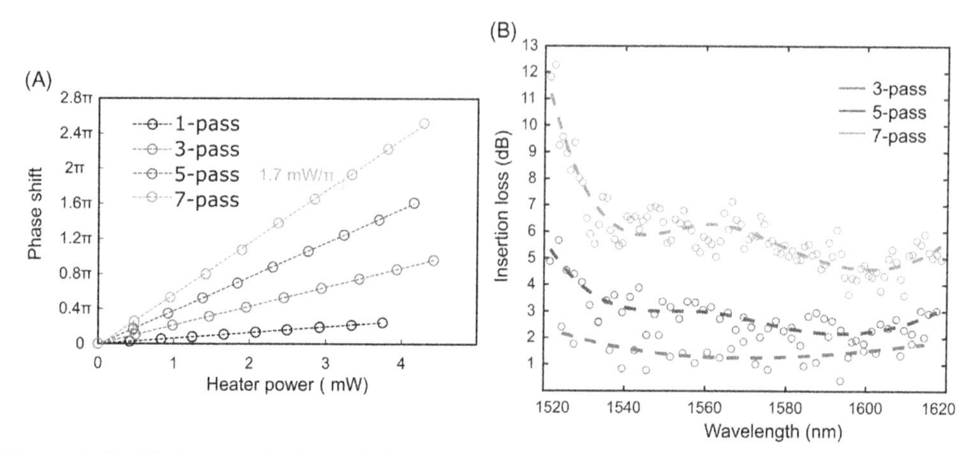

Figure 3.15 (A) Measured phase shift in the recycling-enhanced phase shifters. (B) Measured insertion losses of the 3-pass, 5-pass, and 7-pass recycling structures.

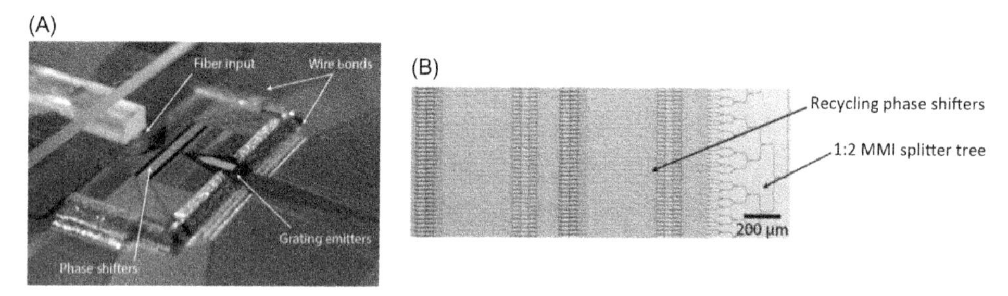

Figure 3.16 (A) Optical image of the 512-element silicon optical phased array with recycling-enhanced phase shifters. (B) Optical microscope image of the recycling-enhanced phase shifters and the splitter tree. *Reprinted with permission from S.A. Miller, C.T. Phare, Y.-C. Chang, X. Ji, O. Jimenez, A. Mohanty, et al., 512-element actively steered silicon phased array for low-power LIDAR, in: Conf. Lasers Electro-Opt. San Jose, 2018, p. JTh5C–2.*

enable low power consumption and scalability, while the broad optical bandwidth also allows large-range 2D beam steering with one axis steered via wavelength tuning [24,60]. The authors implemented 5-pass light recycling structure on each of the 512 thermo-optic phase shifters. These phase shifters consume only ~1.9 W in total when accessing any steering direction. This power consumption is at least an order of magnitude lower compared to other demonstrated large-scale active phased arrays. For example, the 1024-element phased array demonstrated by Chung et al. consumes 55 W [60], while a smaller phased array with 128 elements demonstrated by Hutchison et al. consumes ~10 W [61].

In the light-recycling approach, the broadband enhancement is achieved at the expense of the insertion loss and the footprint. Both are determined by the coupling efficiency and mode selectivity of each adiabatic directional coupler. In the current demonstration, adiabatic directional couplers with linear taper geometry are used [46–48]. In order to enable this approach to further lower the power consumption, one potential direction is to implement more sophisticated geometries for the adiabatic directional couplers that ensure adiabaticity along the whole coupler length [62,63]. This can lead to lower insertion loss, more high-order modes, and/or smaller footprint for the light-recycling approach.

3.3 Emerging approaches for improving performance via material integration

Material integration can open up new tuning mechanisms that are not available in silicon. This can enable active devices that significantly outperform conventional silicon devices. The performance of active photonic devices has been limited by the two available tuning mechanisms in silicon: the plasma dispersion effect and the thermo-optic effect. The plasma dispersion effect provides only a weak change to the refractive index of silicon, typically on the order of 10^{-4} (see Section 3.1) [12]. The thermo-optic effect can provide a larger change to the refractive index ($dn/dT = 1.86 \times 10^{-4}\ \mathrm{K}^{-1}$), but the time scale of the effect is too slow to carry data for optical communications [8–10]. The limited choice of tuning mechanisms places a fundamental constraint on the performance of active photonic devices.

Here we introduce examples of novel active photonic devices that hybridize various materials with silicon. Integration of germanium and germanium-silicon alloys makes use of the Franz-Keldysh effect and the quantum-confined Stark effect for index tuning [13,14,64,65]. These effects modify the band edge absorption and enable compact, high-speed electro-absorption modulators. Integration of III-V semiconductor such as InGaAsP provides stronger plasma dispersion effect with less free-carrier absorption, thanks to the smaller electron effective mass and higher electron mobility in these materials [66–68]. Integration of $\chi^{(2)}$ nonlinear materials enables the Pockels effect for index tuning that is otherwise absent in silicon [69]. Integration of novel two-dimensional materials brings new tuning mechanisms that rely on gating, such as band-filling (i.e., Pauli blocking) in graphene [70].

3.3.1 Materials with strong electro-absorption

3.3.1.1 Germanium and germanium-silicon alloys

Integration of germanium (Ge) and the germanium-silicon (GeSi) alloys has been possible in several CMOS foundries. This integration enables electro-absorption modulators and photodetectors [5]. The direct bandgaps of Ge and GeSi are close to the photon energy of telecommunication wavelengths. This enables electro-absorption modulation based on the Franz-Keldysh effect and the quantum-confined Stark effect, neither of which is present in silicon at these wavelengths [13,14,64,65]. While the energy bandgap of Ge corresponds to wavelengths in the *L* band, one can engineer the bandgap energy of GeSi for the *C* band by controlling the composition of silicon and germanium [71]. The ability to epitaxially grow Ge and GeSi on selective regions of silicon allows monolithic integration of these materials in the CMOS process flow [72].

Both the Franz-Keldysh effect and the quantum-confined Stark effect are electro-absorption mechanisms occurring near the band edge. The Franz-Keldysh effect is observed in bulk semiconductors, in which an external electric field modifies the band edge absorption, inducing a "tail" in the absorption spectrum inside the bandgap, as shown in Fig. 3.17A. The quantum-confined Stark effect is observed in quantum wells when the external electric field is applied perpendicular to the layers. The absorption spectra at the band edge are steplike due to quantum confinement, as shown in Fig. 3.17B, which can be tuned by the applied electric field. The Franz-Keldysh effect and the quantum-confined Stark effect can provide a strong change of absorption coefficient from $\sim 100\ cm^{-1}$ to $\sim 1000\ cm^{-1}$ [6], which enables electro-absorption modulators with very small footprint. Compared to Franz-Keldysh effect, the stronger modulation of quantum-confined Stark effect allows lower applied voltage and better energy efficiency, with a price of process complexity.

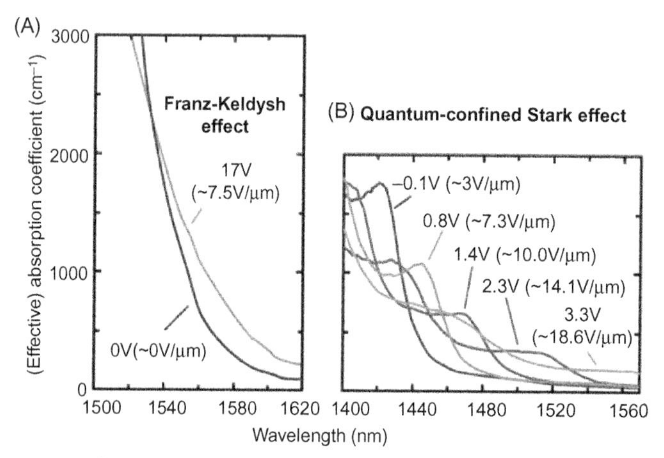

Figure 3.17 (A) Measured Franz-Keldysh effect in $Si_{0.006}Ge_{0.994}$ alloy. The absorption coefficient near the band edge inside the bandgap is modified by the applied electric field. (B) Measured quantum-confined Stark effect in 5 Ge quantum wells (14 nm thick) separated by 18 nm $Si_{0.19}Ge_{0.81}$ barriers. Curves with different colors correspond to the measurements obtained at different biased voltages. *Reprinted with permission from D.A.B. Miller, Attojoule optoelectronics for low-energy information processing and communications, J. Light. Technol. 35 (2017) 346–396.*

Ge and GeSi electro-absorption modulators monolithically integrated on silicon platform have been demonstrated by different groups [71,73–75]. Srinivasan et al. [73] demonstrated a Ge electro-absorption modulator based on Franz-Keldysh effect, as shown in Fig. 3.18A. Ge is epitaxially grown on recessed Si region to form a Ge waveguide. The Ge waveguide is further doped to form a PIN diode, which allows creating an electric field under reverse bias. The process is done on an SOI wafer in a CMOS foundry. The modulator in Ref. [73] exhibits high speed (3 dB electro-optic bandwidth >50 GHz), great energy efficiency (12.8 fJ/bit), and a very small footprint (40 × 10 μm). It has an insertion loss of 4.8 dB and consumes a static power of 1.2 mW, mainly due to photocurrent generation from the absorbed photons [73,76]. The operation wavelength is in the *L* band (1615 nm), as determined by the bandgap energy of Ge. The same research group also demonstrated a similar modulator with a GeSi waveguide [71], in which the operation wavelength shifts to the *C* band with 0.8% Si incorporation. This GeSi modulator shows a similar device performance as the Ge modulator demonstrated by the same group [71]. Recently real-time 100 Gb/s NRZ-OOK transmission over 500 m of a standard single-mode fiber and 2 km of a dispersion-shifted fiber has been demonstrated with this GeSi modulator [77]. Ren et al. demonstrated an electro-absorption modulator based on quantum-confined Stark effect with Ge/SiGe quantum wells epitaxially grown on SOI wafer, as shown in Fig. 3.18B [74]. It operates in the C band, showing a speed of ~ 3.5 GHz, energy efficiency of 0.75 fJ/bit [76], and a footprint of 0.8 × 10 μm.

Electro-absorption modulators based on monolithic integration of Ge and GeSi on silicon have shown competitive performance particularly in footprint, energy consumption, and speed. More research is required to reduce the static power from

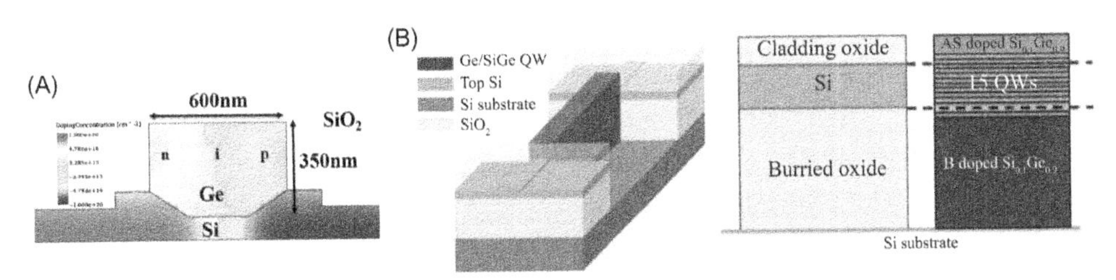

Figure 3.18 (A) Ge waveguide electro-absorption modulator based on Franz-Keldysh effect. *Reprinted with permission from S.A. Srinivasan, M. Pantouvaki, S. Gupta, H.T. Chen, P. Verheyen, G. Lepage, et al., 56 Gb/s germanium waveguide electro-absorption modulator, J. Light. Technol. 34 (2016) 419–424.* (B) Ge/SiGe quantum well electro-absorption modulator based on quantum-confined Stark effect. *Reprinted with permission from S. Ren, Y. Rong, S. Claussen, R.K. Schaevitz, T.I. Kamins, J. Harris, et al., Ge/SiGe quantum well waveguide modulator monolithically integrated with SOI waveguides, in: IEEE Photonics Technol. Lett. 24 (2012) 461–463.*

photocurrent generation [76] and improve the extinction ratio. As Franz-Keldysh effect and quantum-confined Stark effect only occur near the band edge, the optical bandwidth is only about 20–40 nm [71,73]. This is in contrast to the broadband plasma dispersion effect, which has only a weak frequency dependence. The difficulty to produce independent amplitude and phase modulation with electro-absorption modulators also limits the application to advanced modulation formats and coherent communications.

3.3.2 Materials with improved plasma dispersion effect

3.3.2.1 III-V semiconductors

The plasma dispersion effect in silicon offers a relatively weak modulation in the refractive index, accompanied by nonnegligible absorption modulation. Integrating III-V semiconductors on silicon enables stronger and almost phase-only plasma dispersion effect [66–68]. The superior plasma dispersion effect is a result of the smaller electron effective mass and higher electron mobility in III-V semiconductors. The integration is achieved by direct wafer bonding [78,79], which has been intensively developed for InP laser integration. The underlying physics can be understood with the Drude model. The change of the refractive index Δn by the plasma dispersion effect scales as $\Delta N_{e,h}/m^*_{e,h}$, while the change of absorption Δk scales as $\Delta N_{e,h}/\left(m^{*2}_{e,h}\mu_{e,h}\right)$, where $\Delta N_{e,h}$ are the change in the electron and hole concentration, $m^*_{e,h}$ are the electron and hole effective masses, and $\mu_{e,h}$ are the electron and hole mobilities [12,66,68]. The Drude model indicates that an ideal material should have a small effective mass to enable a large $-\Delta n$ for strong tunability and a high mobility to enable a large $-\Delta n/\Delta k$ ratio for phase-only tuning. As shown by Fig. 3.19B, hole-doped silicon has a relatively large $-\Delta n/\Delta k$ ratio. Electron-doped silicon, however, has a $-\Delta n/\Delta k$ ratio of less than 10. To achieve pure phase modulation, it is therefore more desirable to tune silicon with only hole doping. In contrast to silicon, III-V semiconductors, in particular InGaAsP, are good electron-doped materials due to their smaller electron effective masses and higher electron mobilities. As Fig. 3.19A and C show, $-\Delta n$ and $-\Delta n/\Delta k$ of electron-doped InGaAsP are significantly larger than the values of silicon. One can therefore achieve strong and pure phase modulation with material integration by combining hole doping in silicon and electron doping in InGaAsP.

Recently Han et al. [66] and Hiraki et al. [67] demonstrated modulators based on InGaAsP-silicon integration. In Ref. [66], InGaAsP/Al_2O_3/Si forms a MOS capacitor, as shown by Fig. 3.20. This allows electron accumulation on the InGaAsP side and hole accumulation on the Si side. Al_2O_3 serves both as the gate dielectric and as the wafer-bonding interface. This Mach-Zehnder modulator shows a voltage-length product $V_\pi L$ of 0.047 $V \cdot cm$, about 4 times smaller than a Si MOS modulator [27]

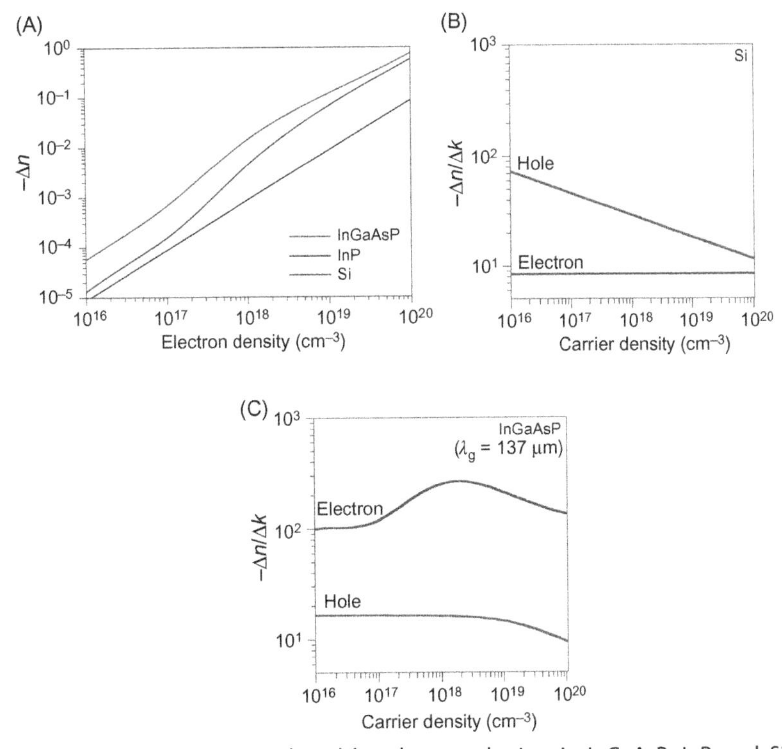

Figure 3.19 (A) Theoretical $-\Delta n$ induced by electron doping in InGaAsP, InP and Si, showing the stronger tunability in InGaAsP. (B) Theoretical $-\Delta n/\Delta k$ induced by electron and hole doping in Si. (C) Theoretical $-\Delta n/\Delta k$ induced by electron and hole doping in InGaAsP. Here Δn and Δk denote the change of the real and imaginary part of the complex refractive index, respectively. *Reprinted with permission from J.-H. Han, F. Boeuf, J. Fujikata, S. Takahashi, S. Takagi, M. Takenaka, Efficient low-loss InGaAsP/Si hybrid MOS optical modulator, Nat. Photonics. 11 (2017) 486–490, J. Witzens, Silicon photonics: modulators make efficiency leap, Nat. Photonics. 11 (2017) 459.*

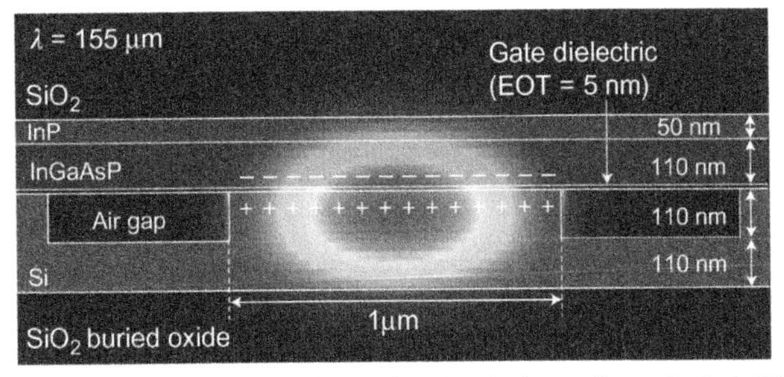

Figure 3.20 Device structure and optical mode simulation of an InGaAsP/Si MOS capacitor. *Reprinted with permission from J.-H. Han, F. Boeuf, J. Fujikata, S. Takahashi, S. Takagi, M. Takenaka, Efficient low-loss InGaAsP/Si hybrid MOS optical modulator, Nat. Photonics. 11 (2017) 486–490.*

and 20 ~ 40 times smaller than the more common Si depletion-type modulator [15,18–20]. The induced loss to reach a π phase shift is only 0.23 dB, about 10 times smaller than a Si MOS modulator [66]. This represents an almost pure phase modulation. The length of the phase shifter in this InGaAsP-silicon modulator is 500 μm.

The highest demonstrated speed of these InGaAsP/Si hybrid modulators is only 2.2 GHz [67], which is substantially lower than the state-of-the-art silicon modulators. The demonstrated speed is limited by the high contact resistance resulting from process issues, while the theoretical speed can reach 37 GHz. Low power consumption is expected in these modulators because of the small $V_{\pi}L$, but the energy per bit of these devices has yet to be reported. The phase-only modulation of these devices is attractive for phased arrays and coherent communications with complex modulation formats.

3.3.3 Materials with $\chi^{(2)}$ nonlinearity

The success of lithium niobate ($LiNbO_3$) Pockels modulators in fiber communications has motivated the integration of $\chi^{(2)}$ materials with silicon. Pockels effect is absent in silicon due to its centrosymmetric crystal structure [69]. In spite of the great progress of the plasma-dispersion-effect-based modulators over the past decades, as discussed in Section 3.1, their speed is fundamentally limited to ~ 60 GHz due to carrier transport [80]. In addition, the phase modulation with the plasma dispersion effect is always accompanied by amplitude modulation from free-carrier absorption. Integration of $\chi^{(2)}$ materials paves a way to break these limitations and realize modulators that outperform the state-of-the-art modulators in speed and pure phase modulation. Here we describe novel Pockels modulators that hybridize silicon with two types of nonlinear materials—organic and inorganic. Organic nonlinear materials have large electro-optic coefficients and can fill in small gaps where electric field is strongly confined. Organic nonlinear materials therefore enable Pockels modulators with very high performance in almost all aspects—power consumption, footprint, and speed. However, the long-term stability of these materials remains an issue that needs to be addressed. On the other hand, integration with inorganic nonlinear materials enables Pockels effect in silicon with less stability concern; however, the performance of these modulators is limited by the weaker electro-optic coefficients and the smaller overlap between the nonlinear materials and the optical mode.

3.3.3.1 Organic nonlinear materials

Organic nonlinear materials are promising particularly because of their large electro-optic coefficients and straightforward fabrication process. Because of the electro-refractive nature of the Pockels effect, this approach can realize pure phase shifters without amplitude modulation. This is important for applications such as optical phased arrays and coherent communications [81]. Organic nonlinear materials, such as chromophore, possess electro-optic coefficient r_{33} as large as 359 pm/V after poled at an elevated temperature [82,83]. This is about 10 times higher than that of crystalline lithium niobate [69]. Organic materials can be spin-coated on silicon waveguides and serve as the cladding. Silicon slot waveguides [84] allow fully exploiting the Pockels effect of nonlinear organic materials via strong confinement of both the optical and the RF fields [85]. As shown in Fig. 3.21A, the optical field is strongly confined and enhanced in the slot, typically 120 nm to 160 nm in width. The field enhancement in the lower-index nonlinear organic material is attributed to the large E field discontinuity in order to satisfy the continuity of the normal component of the D field [84]. The silicon rails and slabs are n-doped such that the voltage applied between the electrodes drops almost entirely across the slot, as shown in Fig. 3.21B. The enhanced optical and RF field overlaps inside the slot filled with nonlinear organic material, enabling efficient Pockels effect.

The approach of hybridizing organic materials with silicon has shown excellent performance in speed and power consumption. By embedding highly confined silicon slot waveguides with organic nonlinear materials [84], one can realize modulators with modulation speed up to 100 GHz [86], voltage-length product $V_{\pi}L$ of only 0.05 V · cm [83], and power consumption as low as 0.7 fJ/bit [87]. The key challenge lies in the long-term stability and power handling of the organic nonlinear materials. The poling of the organic materials degrades with time by thermally activated reorientation [87]. For example, the device lifetime of the 0.7 fJ/bit modulator in Ref. [87] is only a month in ambient conditions. Further material research on thermally stable nonlinear organic materials and hermetic sealing are of key importance to bring this approach to practical applications [88].

Recently plasmonic metal-insulator-metal (MIM) structures with embedded organic $\chi^{(2)}$ materials have attracted a lot of attention [89–91]. They show promises in realizing extremely small, fast, and efficient modulators. Plasmonic MIM structures

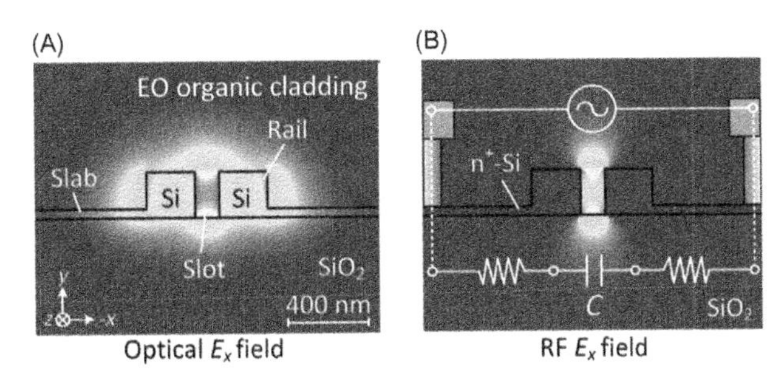

Figure 3.21 Simulated optical field (A) and RF field (B) of a silicon slot waveguide embedded with an organic nonlinear material. *Reprinted with permission from C. Koos, J. Leuthold, W. Freude, M. Kohl, L. Dalton, W. Bogaerts, et al., Silicon-organic hybrid (SOH) and plasmonic-organic hybrid (POH) integration, J. Light. Technol. 34 (2016) 256–268.*

allow extreme confinement in the insulator region. The enhancement can be much stronger than in silicon slot waveguides due to the existence of tightly-confined surface plasmons at the metal-insulator interfaces [92]. As shown by Fig. 3.22A, the confinement in plasmonic MIM structure enables stronger tuning in the effective refractive index at the expense of more propagation loss. Different configurations of plasmonics-assisted Pockels modulators, resonant and nonresonant, have been demonstrated [89–91,93,94]. In 2014, Haffner et al. demonstrated a plasmonic Mach-

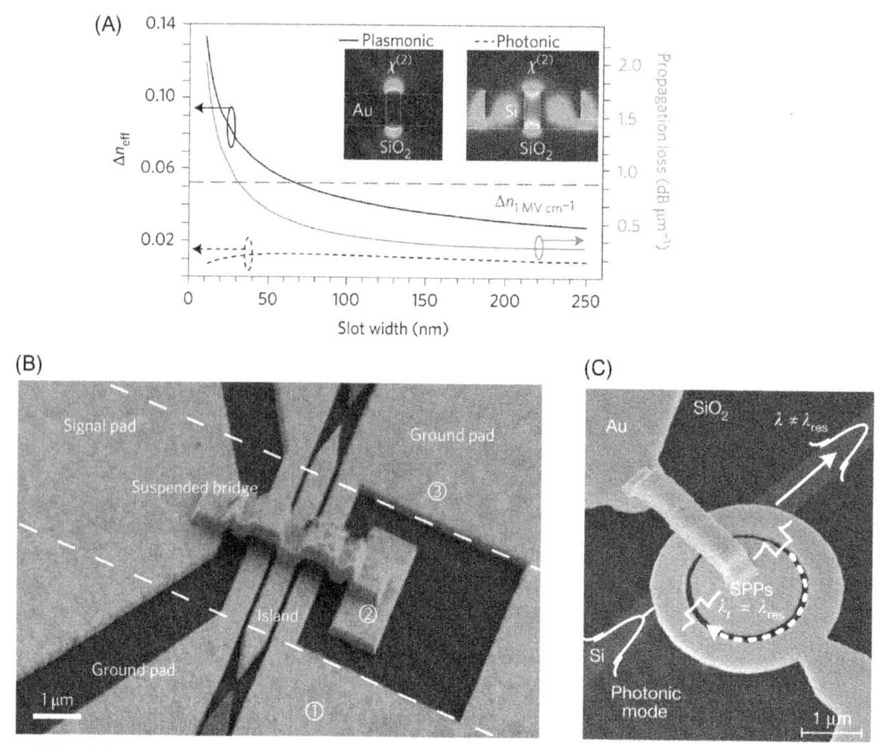

Figure 3.22 (A) Comparison of the change of effective refractive index Δn_{eff} achieved by the plasmonic MIM structure (black solid line, left axis) and the silicon slot waveguide (black dashed line, left axis). The inset shows the optical field in both structures. (B) False-colored SEM image of the plasmonic Mach-Zehnder modulator (silicon: purple; gold: yellow). *(A,B) Reprinted with permission from C. Haffner, W. Heni, Y. Fedoryshyn, J. Niegemann, A. Melikyan, D.L. Elder, et al., All-plasmonic Mach–Zehnder modulator enabling optical high-speed communication at the microscale, Nat. Photonics. 9 (2015) 525–528.* (C) False-colored SEM image of the plasmonic microring modulator. *Reprinted with permission from C. Haffner, D. Chelladurai, Y. Fedoryshyn, A. Josten, B. Baeuerle, W. Heni, et al., Low-loss plasmon-assisted electro-optic modulator, Nature 556 (2018) 483–486.*

Zehnder modulator [89], as shown by Fig. 3.22B. A single-mode silicon waveguide is tapered to couple to surface plasmons in the MIM structure. The gap in the MIM structure is filled with DLD-164, an organic $\chi^{(2)}$ material with an electro-optic coefficient of 180 pm/V. This modulator has a length of 10 μm and a voltage-length product $V_{\pi}L$ of only 60 V · μm, about one order of magnitude better than modulators based on hybrid silicon-organic slot waveguides [83]. The electro-optic bandwidth exceeds 70 GHz. The power consumption is 25 fJ/bit. This modulator shows a slightly higher insertion loss of 8 dB due to ohmic losses in metal, when compared to the typical 3–9 dB in plasma-dispersion-based silicon modulators [15,18,32,34]. In 2018, the same research group improved the insertion loss by designing a low-Q resonant modulator based on a plasmonic MIM microring coupled to a silicon waveguide [91], as shown in Fig. 3.22C. The authors bypassed the insertion loss from ohmic loss in the metal using "resonant switching." In their design, light couples to lossy plasmonic microring only in the off state (on resonance) of the modulator in which attenuation is desired. Light passes through without seeing the plasmonic microring in the on state (off resonance). This modulator shows an insertion loss of only 2.5 dB, power consumption of 12 fJ/bit, and electro-optic bandwidth exceeding 100 GHz. Although promising in almost all performance metrics, plasmon-assisted modulators ultimately rely on the Pockels effect in nonlinear organic materials that have challenges in long-term stability [87]. The use of noble metals such as gold also raises some questions for CMOS compatibility.

3.3.3.2 Inorganic nonlinear materials

Integration of inorganic $\chi^{(2)}$ nonlinear materials such as $LiNbO_3$ [95–97] and lead zirconate titanate (PZT) [98,99] close to silicon or silicon nitride (SiN) waveguides can enable Pockels effect on the integrated platform. Using inorganic nonlinear materials avoids the issues of thermal stability, aging, and power handling in the organic counterparts. However, the voltage-length product $V_{\pi}L$ is much higher due to the order-of-magnitude smaller electro-optic coefficients and weaker confinement in both optical mode and RF field. For example, the electro-optic coefficients r_{33} and r_{13} of $LiNbO_3$ are 31 and 8 pm/V, respectively [100], while r_{33} of binary chromophore organic glass is as large as 230 pm/V [83]. The confinement approaches employed in organic materials, such as slot waveguides and plasmon waveguides discussed in Section 3.3.3.1, are not applicable to inorganic nonlinear materials due to the difficulty in filling nonlinear materials in small gaps.

For decades the Pockels effect in bulk $LiNbO_3$ has been the workhorse for fiber communications [101]. There have been efforts in bonding crystalline $LiNbO_3$ with silicon waveguides to enable Pockels effect on the integrated photonic platform. Chen et al. demonstrated ring modulators based on hybrid silicon-$LiNbO_3$ waveguides [95],

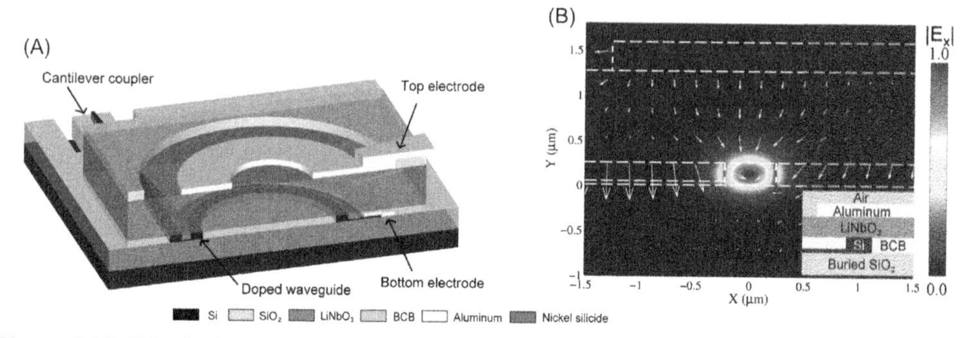

Figure 3.23 Hybrid silicon-$LiNbO_3$ ring modulator (A) schematic of the device structure, (B) the optical mode and the DC voltage-induced electric field. *Reprinted with permission from L. Chen, Q. Xu, M.G. Wood, R.M. Reano, Hybrid silicon and lithium niobate electro-optical ring modulator, Optica 1 (2014) 112–118.*

as shown by Fig. 3.23. The ring modulator shows a voltage-length product $V_\pi L$ of 9.1 V · cm. The 3-dB electro-optic bandwidth is RC-limited to ~ 5 GHz. The power consumption is 4.4 pJ/bit. Weigel et al. demonstrated a hybrid silicon-$LiNbO_3$ Mach-Zehnder modulator [97], also fabricated with direct bonding. This modulator shows a $V_\pi L$ of 8.2 V · cm and an electro-optic bandwidth greater than 6 GHz. Note that $V_\pi L$ demonstrated in this hybrid silicon-$LiNbO_3$ platform remains higher than the values in typical plasma-dispersion-based modulators (1–2 V · cm, as in Refs. [15,18–20]), due to small fraction of optical mode power in $LiNbO_3$. There are also efforts at another platform of making $LiNbO_3$ nanophotonic waveguides without hybridizing silicon [102–104]. Modulators in this $LiNbO_3$ platform show better $V_\pi L$ (such as 1.8 V · cm in Ref. [103]) and higher speed (such as 100 GHz electro-optic bandwidth in Ref. [104]). However, this $LiNbO_3$ platform lacks the mature building blocks and mass production capabilities in the CMOS-compatible silicon platform.

Recently PZT has been also investigated as a nonlinear inorganic material to enable Pockels modulators on the SiN platform. SiN is an important CMOS-compatible material that has enabled many applications including high-Q resonators and frequency combs [105,106], but only very few active photonic devices have been demonstrated on this platform [107,108]. Alexander et al. deposited PZT via chemical solution deposition and then poled it to provide $\chi^{(2)}$ nonlinearity [98,99]. The device structure of microring Pockels modulators on this PZT-SiN platform is shown in Fig. 3.24. The optical mode in SiN is less confined compared to typical modes on the Si platform. This enables better mode interaction with nonlinear PZT. These modulators show a voltage-length product $V_\pi L$ of 3.2 V · cm and electro-optic bandwidth of 33 GHz [99]. The hybrid PZT-SiN waveguides also exhibit low propagation loss of ~ 1 dB/cm.

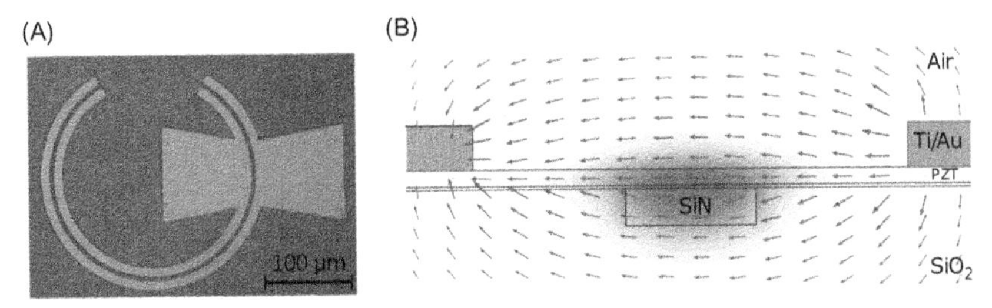

Figure 3.24 Hybrid PZT-SiN ring modulators based on Pockels effect. (A) Optical microscope image. (B) The cross section, optical mode, and the DC voltage-induced electric field. *Reprinted with permission from K. Alexander, J.P. George, J. Verbist, K. Neyts, B. Kuyken, D.V. Thourhout, et al., Nanophotonic Pockels modulators on a silicon nitride platform, Nat. Commun. 9 (2018) 3444.*

3.3.4 Two-dimensional materials

Two-dimensional (2D) materials have attracted great attention as a new class of building blocks for photonics [70]. Two-dimensional materials can be transferred onto arbitrary substrates and adhere to the substrates via the Van der Waals force. These materials provide a versatile approach to add a variety of high-quality crystalline materials to any layer of devices without being limited by epitaxial growth. They also enable active devices on an otherwise passive platform such as silicon nitride. These 2D materials include semimetal (graphene) [109], semiconductors with different bandgaps (transition metal dichalcogenide and black phosphorus), and insulators (hexagonal boron nitride) [110,111].

Graphene provides a modulation mechanism via band-filling (also called Pauli-blocking in the literature) [112]. Because of the low density of states of graphene, electrical gating can efficiently tune the Fermi level and provide an electro-absorption mechanism [113–115], as shown in Fig. 3.25A. Compared to the Franz-Keldysh effect in Ge and GeSi [13,14,64,65], electro-absorption in graphene is not limited by the bandgap of the material and can be designed to operate over a broad optical bandwidth. In addition, when operated in the regime where the interband transition is fully Pauli-blocked (i.e., Fermi energy > twice the photon energy), graphene can also be electro-refractive to realize a phase modulator, as shown in Fig. 3.25B. The high mobility of graphene enables these devices to exhibit high speeds.

Various graphene-on-silicon and graphene-on-silicon nitride modulators have been demonstrated [107,116–119]. As shown in Fig. 3.26A, Liu et al. placed a double-layer graphene capacitor on a silicon waveguide and demonstrated a 1 GHz electro-absorption modulator with a power consumption of 1 pJ/bit [116]. Phare et al. integrated a double-layer graphene capacitor on a silicon nitride microring [107] and used the electro-absorption of graphene to tune the microring between under-coupled and critical coupled regimes, as shown in Fig. 3.26B. The authors obtained 30 GHz electro-optic bandwidth and 800 fJ/bit power consumption. Datta et al. and

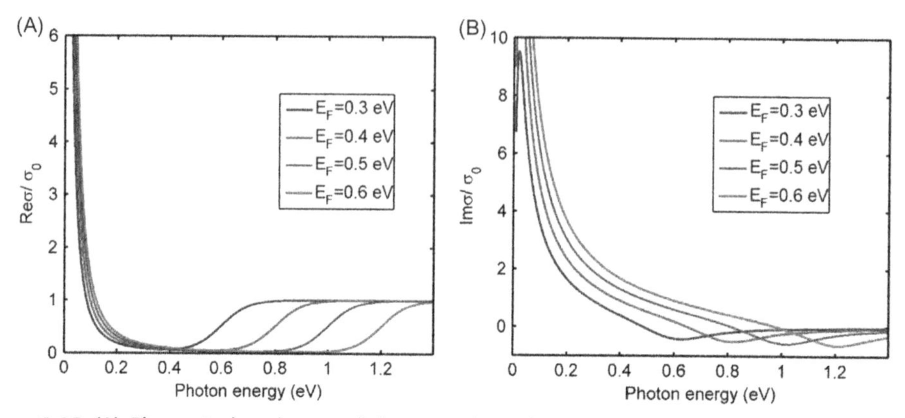

Figure 3.25 (A) Theoretical real part of the optical conductivity (electro-absorption) of graphene at different Fermi energies. (B) Theoretical imaginary part of the optical conductivity (electro-refraction) of graphene at different Fermi energies. We plot the figure using random-phase approximation [113,114]. Intraband scattering rate is taken as 40 meV. The optical conductivity is normalized to the universal conductivity of graphene.

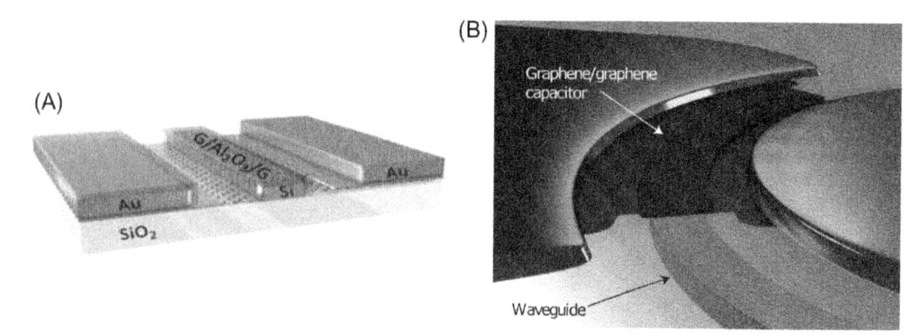

Figure 3.26 (A) Graphene-on-silicon electro-absorption modulator demonstrated in Ref. [116]. *Reprinted with permission from M. Liu, X. Yin, X. Zhang, Double-layer graphene optical modulator, Nano Lett. 12 (2012) 1482–1485.* (B) Graphene-on-silicon nitride microring modulator demonstrated in Ref. [107]. *Reprinted with permission from C.T. Phare, Y.-H.D. Lee, J. Cardenas, M. Lipson, Graphene electro-optic modulator with 30 GHz bandwidth, Nat. Photonics. 9 (2015) 511–514.*

Sorianello et al. demonstrated electro-refraction modulators by gating graphene into the Pauli-blocking regime to eliminate the amplitude modulation [117,118].

Two-dimensional transition metal dichalcogenides (TMDs) provide another material platform. Recently Datta et al. demonstrated a phase-only modulator with an ITO-WS_2 capacitor on a silicon nitride waveguide [120], as shown in Fig. 3.27. This modulator operates in the near-infrared spectral range, well-detuned from the WS_2 exciton absorption around 2 eV [121], which ensures a pure electro-refractive modulation. The authors demonstrated a refractive index change in WS_2 of up to 3%, accompanied by an absorption change of only 0.2%. This result shows that the integration of 2D TMD materials enables pure phase modulation in integrated

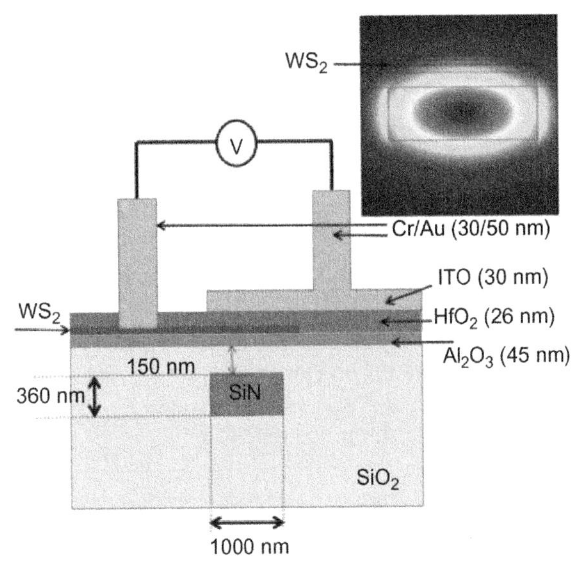

Figure 3.27 Phase-only modulator that integrates WS_2 with silicon nitride waveguides. *Reprinted with permission from I. Datta, S.H. Chae, G.R. Bhatt, B. Li, Y. Yu, L. Cao, et al., Giant electro-refractive modulation of monolayer WS2 embedded in photonic structures, in: Conf. Lasers Electro-Opt. San Jose, 2018, p. STu4N.7.*

photonic devices. The integration of 2D materials with silicon and silicon nitride provides versatility for making active devices on otherwise passive platforms. The design of the 2D material-based devices requires optimization of the optical mode overlap with the monolayer 2D materials. Currently, the speed of these 2D material-based devices is RC-limited, particularly due to the high contact resistance between the 2D material and the metal. Ongoing research on contact approaches such as making edge contact or producing cleaner surfaces could potentially push the limit [122]. To apply 2D material-based active devices to system-level applications, further research on large-area chemical-vapor-deposited growth and high-quality, reliable transfer are required [123–128].

3.4 Concluding remarks

We have discussed various approaches, including novel photonic device designs and material integration, to enable power-efficient nanophotonic devices. We summarize here the most important performance metrics to reveal the pros and cons as well as the potential of each approach.

We compare in Fig. 3.28 the voltage-length product and the electro-optic 3-dB bandwidth of the representative devices discussed in this chapter. Voltage-length product quantifies the effectiveness of a mechanism on tuning the optical properties of

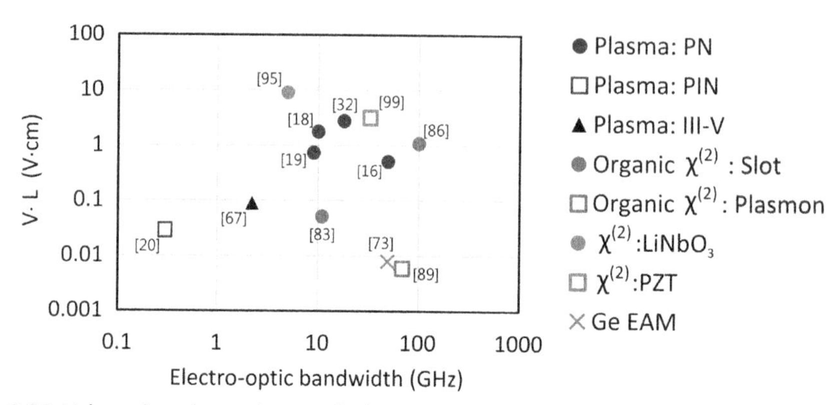

Figure 3.28 Voltage-length product and electro-optic bandwidth of the representative devices discussed in this chapter. *EAM*, electro-absorption modulator.

waveguiding materials. This performance metric is directly related to how much change of the effective refractive index Δn_{eff} or the absorption $\Delta\alpha$ is seen by a waveguide mode at a given voltage. Strong tuning (i.e., small voltage-length product) can enable small device footprint and better power efficiency. For electro-refraction modulators, we plot the commonly used $V_\pi L$, the voltage-length product to produce a phase tuning of π. For Ge electro-absorption modulators, we plot the product of the swing voltage and device length [73]. We use the PN depletion-type modulators (blue solid circles) as the benchmark, because these plasma-dispersion-based modulators represent the most mature state of the art. We compare other technologies with this benchmark to reveal their potential. Among different PN modulators [16,18,19,32], vertical junction design [16] shows improved $V_\pi L$, thanks to the better overlap between the optical mode and the junction. PIN injection-type modulators (blue empty square) [20], also based on the plasma dispersion effect, show orders of magnitude stronger tuning, but the speed is slow due to the recombination time of the injected carriers [23]. PIN modulators therefore may find applications such as phased arrays where the speed is not as crucial as in communications. Integration of III-V materials with silicon shows promising improvement of $V_\pi L$ (black solid triangle) [67]. The speed of the initial experimental demonstration remains slow, but analysis predicts a potential speed of 37 GHz. Integration of organic $\chi^{(2)}$ materials with silicon slot waveguides (red solid circles) shows promising strong tuning [83] and high speed [86]. A similar approach of integrating organic $\chi^{(2)}$ materials with plasmonic waveguides (red empty square) achieves the best $V_\pi L$ and almost the highest speed in the figure [89]. However, the longevity of organic $\chi^{(2)}$ materials remains to be addressed. The integration of inorganic $\chi^{(2)}$ materials such as $LiNbO_3$ (green solid circle) [95] and PZT (green empty square) [99] does not reveal competitive metrics in tunability and speed. This approach may not be promising in making power-efficient high-speed

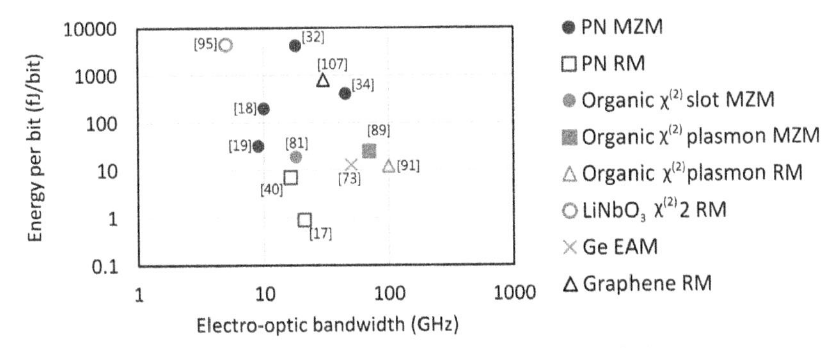

Figure 3.29 Power consumption and electro-optic bandwidth of the representative devices discussed in this chapter. *MZM*, Mach-Zehnder modulator; *RM*, Resonant modulator.

devices, but it may find applications in optical phased arrays and coherent communications because of the purity in phase tuning. Ge and GeSi electro-absorption modulators (pink cross) [73] are now technologically mature and exhibit competitive performance metrics at the bottom right corner of Fig. 3.28. Compared to the benchmark—PN depletion-type modulators, encoding advanced modulation formats with electro-absorption modulators is not as straightforward as with electro-refraction modulators.

We summarize in Fig. 3.29 the energy per bit and the electro-optic 3-dB bandwidth of the representative devices. Energy per bit is the most direct performance metric to quantify the power efficiency of an active photonic device used for communications. This performance metric presents the lumped result of all the design aspects of a device, including the physical tuning mechanism, the photonic design, and the driving electrode design. Energy per bit also depends on whether an enhancement such as resonance (see Section 3.2.2) or nonresonant mode multiplexing (see Section 3.2.3) is employed. We again use PN depletion-type modulator as the benchmark. We distinguish in this figure whether the photonic design is based on a nonresonant Mach-Zehnder interferometer or a resonant structure. One can clearly see that resonance helps reduce the power consumption, as indicated by the comparison between PN Mach-Zehnder modulators (blue solid circle) [18,19,32,34] and PN resonant modulators (blue empty square) [17,40]. The state-of-the-art PN resonant modulators represent the most energy-efficient devices in the figure. However, their speeds are limited by the cavity photon lifetime and thus slower than their nonresonant counterpart. Also, resonant enhancement comes at a cost of fabrication and temperature sensitivity, as we discussed in Section 3.2.2. Integration of organic $\chi^{(2)}$ materials with silicon slot waveguides (red solid circle) [81] and plasmonic waveguides (red solid square and red empty triangle) [89,91] shows encouraging results in moving to higher speed while maintaining low power consumption. These results are achieved without employing high-Q resonators and therefore avoid the challenges of cavity-lifetime-limited bandwidth, fabrication intolerance, and temperature sensitivity. The longevity issue of organic $\chi^{(2)}$ materials remains to be addressed. Integrating $LiNbO_3$ (green empty circle) [95] and graphene (black empty triangle) [107] to silicon does not seem to help the power consumption. Integrating Ge and GeSi with silicon (pink cross) [73] to make electro-absorption modulators shows competitive metrics in energy consumption and speed.

Nanophotonic devices on the silicon photonic platform have proved their uniqueness in communications, especially for the optical interconnects in data centers and high-performance computing systems [7,11]. There is a pressing need for reducing the power consumption and increasing the data bandwidth for optical interconnects [6,41]. With the rapid progress in the past few years, we expect to see more novel designs and material integration that continuously push the frontier of the silicon photonic technology for power-efficient communications.

References

[1] D.A.B. Miller, Rationale and challenges for optical interconnects to electronic chips, Proc. IEEE. 88 (2000) 728–749.
[2] D.A.B. Miller, Device requirements for optical interconnects to silicon chips, Proc. IEEE. 97 (2009) 1166–1185.
[3] G.T. Reed, G. Mashanovich, F.Y. Gardes, D.J. Thomson, Silicon optical modulators, Nat. Photonics. 4 (2010) 518–526.
[4] G.T. Reed, G.Z. Mashanovich, F.Y. Gardes, M. Nedeljkovic, Y. Hu, D.J. Thomson, et al., Recent breakthroughs in carrier depletion based silicon optical modulators, Nanophotonics. 3 (2014) 229–245.
[5] D. Thomson, A. Zilkie, J.E. Bowers, T. Komljenovic, G.T. Reed, L. Vivien, et al., Roadmap on silicon photonics, J. Opt. 18 (2016) 073003.
[6] D.A.B. Miller, Attojoule optoelectronics for low-energy information processing and communications, J. Light. Technol. 35 (2017) 346–396.
[7] A. Biberman, K. Bergman, Optical interconnection networks for high-performance computing systems, Rep. Prog. Phys. 75 (2012) 046402.
[8] G. Cocorullo, I. Rendina, Thermo-optical modulation at 1.5 mu m in silicon etalon, Electron. Lett. 28 (1992) 83–85.
[9] R.L. Espinola, M.C. Tsai, J.T. Yardley, R.M. Osgood, Fast and low-power thermooptic switch on thin silicon-on-insulator, IEEE Photonics Technol. Lett. 15 (2003) 1366–1368.
[10] M.R. Watts, J. Sun, C. DeRose, D.C. Trotter, R.W. Young, G.N. Nielson, Adiabatic thermo-optic Mach–Zehnder switch, Opt. Lett. 38 (2013) 733–735.
[11] Z. Zhou, R. Chen, X. Li, T. Li, Development trends in silicon photonics for data centers, Opt. Fiber Technol. 44 (2018) 13–23.
[12] R. Soref, B. Bennett, Electrooptical effects in silicon, IEEE J. Quantum Electron. 23 (1987) 123–129.
[13] M. Fox, Optical Properties of Solids, Oxford University Press, Oxford, 2010.
[14] C. Pollock, M. Lipson, Integrated Photonics, Springer Science & Business Media, 2013.
[15] X. Xiao, H. Xu, X. Li, Z. Li, T. Chu, Y. Yu, et al., High-speed, low-loss silicon Mach–Zehnder modulators with doping optimization, Opt. Express. 21 (2013) 4116–4125.
[16] J. Sun, M. Sakib, J. Driscoll, R. Kumar, H. Jayatilleka, Y. Chetrit, et al., A 128 Gb/s PAM4 silicon microring modulator, in: Opt. Fiber Commun. Conf. Expo. San Diego, 2018, p. Th4A.7.
[17] E. Timurdogan, C.M. Sorace-Agaskar, J. Sun, E. Shah Hosseini, A. Biberman, M.R. Watts, An ultralow power athermal silicon modulator, Nat. Commun. 5 (2014) 4008.

[18] T. Baehr-Jones, R. Ding, Y. Liu, A. Ayazi, T. Pinguet, N.C. Harris, et al., Ultralow drive voltage silicon traveling-wave modulator, Opt. Express. 20 (2012) 12014–12020.
[19] J. Ding, R. Ji, L. Zhang, L. Yang, Electro-optical response analysis of a 40 Gb/s silicon Mach-Zehnder optical modulator, J. Light. Technol. 31 (2013) 2434–2440.
[20] S. Akiyama, M. Imai, T. Baba, T. Akagawa, N. Hirayama, Y. Noguchi, et al., Compact PIN-diode-based silicon modulator using side-wall-grating waveguide, IEEE J. Sel. Top. Quantum Electron. 19 (2013) 74–84.
[21] Q. Xu, B. Schmidt, S. Pradhan, M. Lipson, Micrometre-scale silicon electro-optic modulator, Nature. 435 (2005) 325–327.
[22] W.M. Green, M.J. Rooks, L. Sekaric, Y.A. Vlasov, Ultra-compact, low RF power, 10 Gb/s silicon Mach-Zehnder modulator, Opt. Express. 15 (2007) 17106–17113.
[23] G. Zhou, M.W. Geis, S.J. Spector, F. Gan, M.E. Grein, R.T. Schulein, et al., Effect of carrier lifetime on forward-biased silicon Mach-Zehnder modulators, Opt. Express. 16 (2008) 5218–5226.
[24] J.C. Hulme, J.K. Doylend, M.J.R. Heck, J.D. Peters, M.L. Davenport, J.T. Bovington, et al., Fully integrated hybrid silicon two dimensional beam scanner, Opt. Express. 23 (2015) 5861–5874.
[25] F. Aflatouni, B. Abiri, A. Rekhi, A. Hajimiri, Nanophotonic projection system, Opt. Express. 23 (2015) 21012–21022.
[26] A. Liu, R. Jones, L. Liao, D. Samara-Rubio, D. Rubin, O. Cohen, et al., A high-speed silicon optical modulator based on a metal–oxide–semiconductor capacitor, Nature. 427 (2004) 615–618.
[27] M. Webster, P. Gothoskar, V. Patel, D. Piede, S. Anderson, R. Tummidi, et al., An efficient MOS-capacitor based silicon modulator and CMOS drivers for optical transmitters, in: 11th Int. Conf. Group IV Photonics, Paris, 2014, p. W.B1.
[28] J. Fujikata, M. Takahashi, S. Takahashi, T. Horikawa, T. Nakamura, High-speed and high-efficiency Si optical modulator with MOS junction, using solid-phase crystallization of polycrystalline silicon, Jpn. J. Appl. Phys. 55 (2016) 042202.
[29] H. Subbaraman, X. Xu, A. Hosseini, X. Zhang, Y. Zhang, D. Kwong, et al., Recent advances in silicon-based passive and active optical interconnects, Opt. Express. 23 (2015) 2487–2511.
[30] H. Yu, W. Bogaerts, An equivalent circuit model of the traveling wave electrode for carrier-depletion-based silicon optical modulators, J. Light. Technol. 30 (2012) 1602–1609.
[31] Z. Yong, W.D. Sacher, Y. Huang, J.C. Mikkelsen, Y. Yang, X. Luo, et al., U-shaped PN junctions for efficient silicon Mach-Zehnder and microring modulators in the O-band, Opt. Express. 25 (2017) 8425–8439.
[32] D.J. Thomson, F.Y. Gardes, J.-M. Fedeli, S. Zlatanovic, Y. Hu, B.P.P. Kuo, et al., 50-Gb/s silicon optical modulator, IEEE Photonics Technol. Lett. 24 (2012) 234–236.
[33] D.J. Thomson, F.Y. Gardes, Y. Hu, G. Mashanovich, M. Fournier, P. Grosse, et al., High contrast 40 Gbit/s optical modulation in silicon, Opt. Express. 19 (2011) 11507–11516.
[34] A. Samani, D. Patel, M. Chagnon, E. El-Fiky, R. Li, M. Jacques, et al., Experimental parametric study of 128 Gb/s PAM-4 transmission system using a multi-electrode silicon photonic Mach Zehnder modulator, Opt. Express. 25 (2017) 13252–13262.
[35] W. Bogaerts, P. De Heyn, T. Van Vaerenbergh, K. De Vos, S. Kumar Selvaraja, T. Claes, et al., Silicon microring resonators, Laser Photonics Rev. 6 (2012) 47–73.
[36] J. Müller, F. Merget, S.S. Azadeh, J. Hauck, S.R. García, B. Shen, et al., Optical peaking enhancement in high-speed ring modulators, Sci. Rep. 4 (2015) 6310.
[37] I.-L. Gheorma, R.M. Osgood, Fundamental limitations of optical resonator based high-speed EO modulators, IEEE Photonics Technol. Lett. 14 (2002) 795–797.
[38] P. Dong, C. Xie, L. Chen, N.K. Fontaine, Y. Chen, Experimental demonstration of microring quadrature phase-shift keying modulators, Opt. Lett. 37 (2012) 1178–1180.
[39] P. Dong, S. Liao, D. Feng, H. Liang, D. Zheng, R. Shafiiha, et al., Low V_{pp}, ultralow-energy, compact, high-speed silicon electro-optic modulator, Opt. Express. 17 (2009) 22484–22490.
[40] G. Li, X. Zheng, J. Yao, H. Thacker, I. Shubin, Y. Luo, et al., 25Gb/s 1V-driving CMOS ring modulator with integrated thermal tuning, Opt. Express. 19 (2011) 20435–20443.
[41] Z. Zhou, B. Yin, Q. Deng, X. Li, J. Cui, Lowering the energy consumption in silicon photonic devices and systems, Photonics Res. 3 (2015) B28–B46.

[42] Y. Luo, X. Zheng, S. Lin, J. Yao, H. Thacker, I. Shubin, et al., Modulator based on multi-mode waveguides, IEEE Photonics Technol. Lett. 28 (2016) 1391–1394.
[43] B. Guha, J. Cardenas, M. Lipson, Athermal silicon microring resonators with titanium oxide cladding, Opt. Express. 21 (2013) 26557–26563.
[44] B. Guha, K. Preston, M. Lipson, Athermal silicon microring electro-optic modulator, Opt. Lett. 37 (2012) 2253–2255.
[45] B. Guha, B.B.C. Kyotoku, M. Lipson, CMOS-compatible athermal silicon microring resonators, Opt. Express. 18 (2010) 3487–3493.
[46] Y.-C. Chang, S.P. Roberts, B. Stern, I. Datta, M. Lipson, Resonance–free light recycling in waveguides, in: Conf. Lasers Electro-Opt. San Jose, 2017, p. SF1J.5.
[47] Y.-C. Chang, S.P. Roberts, B. Stern, M. Lipson, Resonance-free light recycling, in: arXiv:1710.02891 [physics], 2017. http://arxiv.org/abs/1710.02891 (accessed May 28, 2018).
[48] S.A. Miller, C.T. Phare, Y.-C. Chang, X. Ji, O. Jimenez, A. Mohanty, et al., 512-Element actively steered silicon phased array for low-power LIDAR, in: Conf. Lasers Electro-Opt. San Jose, 2018, p. JTh5C–2.
[49] S.K. Selvaraja, W. Bogaerts, P. Dumon, D. Van Thourhout, R. Baets, Subnanometer linewidth uniformity in silicon nanophotonic waveguide devices using CMOS fabrication technology, IEEE J. Sel. Top. Quantum Electron. 16 (2010) 316–324.
[50] K. Padmaraju, K. Bergman, Resolving the thermal challenges for silicon microring resonator devices, Nanophotonics 3 (2013) 269–281.
[51] P. Dong, W. Qian, H. Liang, R. Shafiiha, D. Feng, G. Li, et al., Thermally tunable silicon racetrack resonators with ultralow tuning power, Opt. Express. 18 (2010) 20298–20304.
[52] A. Arbabi, L.L. Goddard, Measurements of the refractive indices and thermo-optic coefficients of Si_3N_4 and SiO_x using microring resonances, Opt. Lett. 38 (2013) 3878–3881.
[53] S.S. Djordjevic, K. Shang, B. Guan, S.T. Cheung, L. Liao, J. Basak, et al., CMOS-compatible, athermal silicon ring modulators clad with titanium dioxide, Opt. Express. 21 (2013) 13958–13968.
[54] J. Teng, P. Dumon, W. Bogaerts, H. Zhang, X. Jian, X. Han, et al., Athermal silicon-on-insulator ring resonators by overlaying a polymer cladding on narrowed waveguides, Opt. Express. 17 (2009) 14627–14633.
[55] M. Han, A. Wang, Temperature compensation of optical microresonators using a surface layer with negative thermo-optic coefficient, Opt. Lett. 32 (2007) 1800–1802.
[56] P. Alipour, E.S. Hosseini, A.A. Eftekhar, B. Momeni, A. Adibi, Temperature-insensitive silicon microdisk resonators using polymeric cladding layers, in: Conf. Lasers Electro-Opt. Baltimore, 2009, p. CMAA4.
[57] L.-W. Luo, N. Ophir, C.P. Chen, L.H. Gabrielli, C.B. Poitras, K. Bergmen, et al., WDM-compatible mode-division multiplexing on a silicon chip, Nat. Commun. 5 (2014) 3069.
[58] J. Wang, S. He, D. Dai, On-chip silicon 8-channel hybrid (de)multiplexer enabling simultaneous mode- and polarization-division-multiplexing: on-chip Si 8-channel hybrid (de)multiplexer for mode-/polarization-division-multiplexing, Laser Photonics Rev. 8 (2014) L18–L22.
[59] B. Stern, X. Zhu, C.P. Chen, L.D. Tzuang, J. Cardenas, K. Bergman, et al., On-chip mode-division multiplexing switch, Optica. 2 (2015) 530–535.
[60] S. Chung, H. Abediasl, H. Hashemi, A monolithically integrated large-scale optical phased array in silicon-on-insulator CMOS, IEEE J. Solid-State Circuits. 53 (2018) 275–296.
[61] D.N. Hutchison, J. Sun, J.K. Doylend, R. Kumar, J. Heck, W. Kim, et al., High-resolution aliasing-free optical beam steering, Optica. 3 (2016) 887–890.
[62] X. Sun, H.-C. Liu, A. Yariv, Adiabaticity criterion and the shortest adiabatic mode transformer in a coupled-waveguide system, Opt. Lett. 34 (2009) 280–282.
[63] S.-Y. Tseng, Counterdiabatic mode-evolution based coupled-waveguide devices, Opt. Express. 21 (2013) 21224–21235.
[64] D.A.B. Miller, D.S. Chemla, T.C. Damen, A.C. Gossard, W. Wiegmann, T.H. Wood, et al., Band-edge electroabsorption in quantum well structures: the quantum-confined Stark effect, Phys. Rev. Lett. 53 (1984) 2173–2176.
[65] Y.-H. Kuo, Y.K. Lee, Y. Ge, S. Ren, J.E. Roth, T.I. Kamins, et al., Strong quantum-confined Stark effect in germanium quantum-well structures on silicon, Nature. 437 (2005) 1334–1336.

[66] J.-H. Han, F. Boeuf, J. Fujikata, S. Takahashi, S. Takagi, M. Takenaka, Efficient low-loss InGaAsP/Si hybrid MOS optical modulator, Nat. Photonics. 11 (2017) 486–490.
[67] T. Hiraki, T. Aihara, K. Hasebe, K. Takeda, T. Fujii, T. Kakitsuka, et al., Heterogeneously integrated III–V/Si MOS capacitor Mach–Zehnder modulator, Nat. Photonics. 11 (2017) 482–485.
[68] J. Witzens, Silicon photonics: modulators make efficiency leap, Nat. Photonics. 11 (2017) 459.
[69] R.W. Boyd, Nonlinear Optics, Academic Press, London, 2003.
[70] Z. Sun, A. Martinez, F. Wang, Optical modulators with 2D layered materials, Nat. Photonics. 10 (2016) 227–238.
[71] S.A. Srinivasan, P. Verheyen, R. Loo, I. De Wolf, M. Pantouvaki, G. Lepage, et al., 50Gb/s C-band GeSi waveguide electro-absorption modulator, in: Opt. Fiber Commun. Conf. Exhib. Anaheim, 2016, p. Tu3D.7.
[72] G. Wang, F.E. Leys, L. Souriau, R. Loo, M. Caymax, D.P. Brunco, et al., Selective epitaxial growth of germanium on Si wafers with shallow trench isolation: an approach for Ge virtual substrates, ECS Trans. 16 (2008) 829–836.
[73] S.A. Srinivasan, M. Pantouvaki, S. Gupta, H.T. Chen, P. Verheyen, G. Lepage, et al., 56 Gb/s germanium waveguide electro-absorption modulator, J. Light. Technol. 34 (2016) 419–424.
[74] S. Ren, Y. Rong, S. Claussen, R.K. Schaevitz, T.I. Kamins, J. Harris, D.A.B. Miller, Ge/SiGe quantum well waveguide modulator monolithically integrated with SOI waveguides, *IEEE Photonics Technol. Lett.* 24, 2012, 461–463.
[75] J. Liu, M. Beals, A. Pomerene, S. Bernardis, R. Sun, J. Cheng, et al., Waveguide-integrated, ultralow-energy GeSi electro-absorption modulators, Nat. Photonics. 2 (2008) 433–437.
[76] D.A.B. Miller, Energy consumption in optical modulators for interconnects, Opt. Express. 20 (2012) A293–A308.
[77] J. Verbist, M. Verplaetse, S.A. Srivinasan, P. De Heyn, T.D. Keulenaer, R. Pierco, et al., First real-time 100-Gb/s NRZ-OOK transmission over 2 km with a silicon photonic electro-absorption modulator, in: Opt. Fiber Commun. Conf. Exhib. Los Angeles, 2017, p. Th5C.4.
[78] T. Komljenovic, M. Davenport, J. Hulme, A.Y. Liu, C.T. Santis, A. Spott, et al., Heterogeneous silicon photonic integrated circuits, J. Light. Technol. 34 (2016) 20–35.
[79] C. Zhang, P.A. Morton, J.B. Khurgin, J.D. Peters, J.E. Bowers, Ultralinear heterogeneously integrated ring-assisted Mach–Zehnder interferometer modulator on silicon, Optica. 3 (2016) 1483–1488.
[80] F.Y. Gardes, G.T. Reed, N.G. Emerson, C.E. Png, A sub-micron depletion-type photonic modulator in Silicon On Insulator, Opt. Express. 13 (2005) 8845–8854.
[81] M. Lauermann, R. Palmer, S. Koeber, P.C. Schindler, D. Korn, T. Wahlbrink, et al., Low-power silicon-organic hybrid (SOH) modulators for advanced modulation formats, Opt. Express. 22 (2014) 29927–29936.
[82] C. Kieninger, Y. Kutuvantavida, H. Zwickel, S. Wolf, M. Lauermann, D.L. Elder, et al., Record-high in-device electro-optic coefficient of 359 pm/V in a silicon-organic hybrid (SOH) modulator, in: Conf. Lasers Electro-Opt. San Jose, 2017, p. STu3N.2.
[83] R. Palmer, S. Koeber, D.L. Elder, M. Woessner, W. Heni, D. Korn, et al., High-speed, low drive--voltage silicon-organic hybrid modulator based on a binary-chromophore electro-optic material, J. Light. Technol. 32 (2014) 2726–2734.
[84] V.R. Almeida, Q. Xu, C.A. Barrios, M. Lipson, Guiding and confining light in void nanostructure, Opt. Lett. 29 (2004) 1209–1211.
[85] C. Koos, J. Leuthold, W. Freude, M. Kohl, L. Dalton, W. Bogaerts, et al., Silicon-organic hybrid (SOH) and plasmonic-organic hybrid (POH) integration, J. Light. Technol. 34 (2016) 256–268.
[86] L. Alloatti, R. Palmer, S. Diebold, K.P. Pahl, B. Chen, R. Dinu, et al., 100 GHz silicon–organic hybrid modulator, Light Sci. Appl. 3 (2014) e173.
[87] S. Koeber, R. Palmer, M. Lauermann, W. Heni, D.L. Elder, D. Korn, et al., Femtojoule electro-optic modulation using a silicon–organic hybrid device, Light Sci. Appl. 4 (2015) e255.
[88] D. Jin, H. Chen, A. Barklund, J. Mallari, G. Yu, E. Miller, et al., EO polymer modulators reliability study, in: Org. Photonic Mater. Devices XII, San Francisco, 2010, p. 75990H.

[89] C. Haffner, W. Heni, Y. Fedoryshyn, J. Niegemann, A. Melikyan, D.L. Elder, et al., All-plasmonic Mach–Zehnder modulator enabling optical high-speed communication at the microscale, Nat. Photonics. 9 (2015) 525–528.
[90] C. Hössbacher, A. Josten, B. Bäuerle, Y. Fedoryshyn, H. Hettrich, Y. Salamin, et al., Plasmonic modulator with >170 GHz bandwidth demonstrated at 100 GBd NRZ, Opt. Express 25 (2017) 1762–1768.
[91] C. Haffner, D. Chelladurai, Y. Fedoryshyn, A. Josten, B. Baeuerle, W. Heni, et al., Low-loss plasmon-assisted electro-optic modulator, Nature. 556 (2018) 483–486.
[92] S.A. Maier, Plasmonics: Fundamentals and Applications, Springer Science & Business Media, New York, 2007.
[93] J.A. Dionne, K. Diest, L.A. Sweatlock, et al., PlasMOStor: a metal – oxide – Si field effect plasmonic modulator, Nano Lett. 9 (2009) 897–902.
[94] V.J. Sorger, N.D. Lanzillotti-Kimura, R.-M. Ma, X. Zhang, Ultra-compact silicon nanophotonic modulator with broadband response, Nanophotonics. 1 (2012) 17–22.
[95] L. Chen, Q. Xu, M.G. Wood, R.M. Reano, Hybrid silicon and lithium niobate electro-optical ring modulator, Optica. 1 (2014) 112–118.
[96] S. Jin, L. Xu, H. Zhang, Y. Li, $LiNbO_3$ thin-film modulators using silicon nitride surface ridge waveguides, IEEE Photonics Technol. Lett. 28 (2016) 736–739.
[97] P.O. Weigel, J. Zhao, D. Trotter, D. Hood, J. Mudrick, C. Dallo, et al., Foundry-compatible hybrid silicon / lithium niobate electro-optic modulator, in: Conf. Lasers Electro-Opt. San Jose, 2018, p. SF2I.4.
[98] K. Alexander, J.P. George, B. Kuyken, J. Beeckman, D. Van Thourhout, Broadband electro-optic modulation using low-loss PZT-on-silicon nitride integrated waveguides, in: Conf. Lasers Electro-Opt. San Jose, 2017, p. JTh5C.7.
[99] K. Alexander, J.P. George, J. Verbist, K. Neyts, B. Kuyken, D.V. Thourhout, et al., Nanophotonic Pockels modulators on a silicon nitride platform, Nat. Commun. 9 (2018) 3444.
[100] K.K. Wong (Ed.), Properties of Lithium Niobate, INSPEC, The Institution of Electrical Engineers, London, 2002.
[101] E.L. Wooten, K.M. Kissa, A. Yi-Yan, E.J. Murphy, D. Maack, D.V. Attanasio, et al., A review of lithium niobate modulators for fiber-optic communications systems, IEEE J. Sel. Top. Quantum Electron. 6 (2000) 69–82.
[102] A.J. Mercante, P. Yao, S. Shi, G. Schneider, J. Murakowski, D.W. Prather, 110 GHz CMOS compatible thin film LiNbO3 modulator on silicon, Opt. Express. 24 (2016) 15590–15595.
[103] C. Wang, M. Zhang, B. Stern, M. Lipson, M. Lončar, Nanophotonic lithium niobate electro-optic modulators, Opt. Express. 26 (2018) 1547–1555.
[104] C. Wang, M. Zhang, X. Chen, M. Bertrand, A. Shams-Ansari, S. Chandrasekhar, P. Winzer, M. Lončar, 100-GHz low voltage integrated lithium niobate modulators, in: Conf. Lasers Electro-Opt. San Jose, 2018, p. SM3B.4.
[105] W. Xie, Y. Zhu, T. Aubert, S. Verstuyft, Z. Hens, D. Van Thourhout, Low-loss silicon nitride waveguide hybridly integrated with colloidal quantum dots, Opt. Express. 23 (2015) 12152–12160.
[106] A. Dutt, C. Joshi, X. Ji, J. Cardenas, Y. Okawachi, K. Luke, et al., On-chip dual-comb source for spectroscopy, Sci. Adv. 4 (2018) e1701858.
[107] C.T. Phare, Y.-H.D. Lee, J. Cardenas, M. Lipson, Graphene electro-optic modulator with 30 GHz bandwidth, Nat. Photonics. 9 (2015) 511–514.
[108] A. Mohanty, Q. Li, M.A. Tadayon, G. Bhatt, E. Shim, X. Ji, et al., A reconfigurable nanophotonics platform for sub-millisecond, deep brain neural stimulation, in: arXiv:1805.11663 [physics.app-ph], 2018. http://arxiv.org/abs/1805.11663 (accessed August 18, 2018).
[109] K.S. Novoselov, A.K. Geim, S.V. Morozov, D. Jiang, Y. Zhang, S.V. Dubonos, et al., Electric field effect in atomically thin carbon films, Science. 306 (2004) 666–669.
[110] A.K. Geim, I.V. Grigorieva, Van der Waals heterostructures, Nature. 499 (2013) 419–425.
[111] K.S. Novoselov, A. Mishchenko, A. Carvalho, A.H.C. Neto, 2D materials and van der Waals heterostructures, Science. 353 (2016) aac9439.

[112] F. Wang, Y. Zhang, C. Tian, C. Girit, A. Zettl, M. Crommie, et al., Gate-variable optical transitions in graphene, Science. 320 (2008) 206–209.

[113] L.A. Falkovsky, A.A. Varlamov, Space-time dispersion of graphene conductivity, Eur. Phys. J. B 56 (2007) 281–284.

[114] L.A. Falkovsky, S.S. Pershoguba, Optical far-infrared properties of a graphene monolayer and multilayer, Phys. Rev. B. 76 (2007) 153410.

[115] Y.-C. Chang, C.-H. Liu, C.-H. Liu, Z. Zhong, T.B. Norris, Extracting the complex optical conductivity of mono- and bilayer graphene by ellipsometry, Appl. Phys. Lett. 104 (2014) 261909.

[116] M. Liu, X. Yin, X. Zhang, Double-layer graphene optical modulator, Nano Lett. 12 (2012) 1482–1485.

[117] I. Datta, C.T. Phare, C.T. Phare, A. Dutt, A. Dutt, A. Mohanty, et al., Integrated graphene electro-optic phase modulator, in: Conf. Lasers Electro-Opt. San Jose, 2017, p. STu3N.5.

[118] V. Sorianello, M. Midrio, G. Contestabile, I. Asselberghs, J.V. Campenhout, C. Huyghebaert, et al., Graphene–silicon phase modulators with gigahertz bandwidth, Nat. Photonics. 12 (2018) 40–44.

[119] M. Liu, X. Yin, E. Ulin-Avila, B. Geng, T. Zentgraf, L. Ju, et al., A graphene-based broadband optical modulator, Nature. 474 (2011) 64–67.

[120] I. Datta, S.H. Chae, G.R. Bhatt, B. Li, Y. Yu, L. Cao, et al., Giant electro-refractive modulation of monolayer WS2 embedded in photonic structures, in: Conf. Lasers Electro-Opt. San Jose, 2018, p. STu4N.7.

[121] Y. Li, A. Chernikov, X. Zhang, A. Rigosi, H.M. Hill, A.M. van der Zande, et al., Measurement of the optical dielectric function of monolayer transition-metal dichalcogenides: MoS_2, WS_2, and WSe_2, Phys. Rev. B. 90 (2014) 205422.

[122] L. Wang, I. Meric, P.Y. Huang, Q. Gao, Y. Gao, H. Tran, et al., One-dimensional electrical contact to a two-dimensional material, Science. 342 (2013) 614–617.

[123] Y. Hao, M.S. Bharathi, L. Wang, Y. Liu, H. Chen, S. Nie, et al., The role of surface oxygen in the growth of large single-crystal graphene on copper, Science. 342 (2013) 720–723.

[124] L. Banszerus, M. Schmitz, S. Engels, J. Dauber, M. Oellers, F. Haupt, et al., Ultrahigh-mobility graphene devices from chemical vapor deposition on reusable copper, Sci. Adv. 1, 2015, p. e1500222.

[125] L.A. Shiramin, A. Bazin, S. Verstuyft, S. Lycke, P. Vandenabeele, G. Roelkens, et al., Transfer printing of micron-size graphene for photonic integrated circuits and devices, ECS J. Solid State Sci. Technol. 6 (2017) P435–P439.

[126] Y. Gao, Z. Liu, D.-M. Sun, L. Huang, L.-P. Ma, L.-C. Yin, et al., Large-area synthesis of high--quality and uniform monolayer WS_2 on reusable Au foils, Nat. Commun. 6 (2015) 8569.

[127] K. Kang, S. Xie, L. Huang, Y. Han, P.Y. Huang, K.F. Mak, et al., High-mobility three-atom-thick semiconducting films with wafer-scale homogeneity, Nature. 520 (2015) 656–660.

[128] K.M. McCreary, A.T. Hanbicki, G.G. Jernigan, J.C. Culbertson, B.T. Jonker, Synthesis of large-area WS_2 monolayers with exceptional photoluminescence, Sci. Rep. 6 (2016) 19159.

CHAPTER 4

Foundry capabilities for photonic integrated circuits

Michael Liehr[1], Moritz Baier[2], Gloria Hoefler[3], Nicholas M. Fahrenkopf[4], John Bowers[5], Richard Gladhill[6], Peter O'Brien[7], Erman Timurdogan[8], Zhan Su[8] and Fred Kish[3]

[1]AIM Photonics, SUNY Polytechnic Institute, New York, NY, United States
[2]Fraunhofer Heinrich Hertz Institute (FhG-HHI), Berlin, Germany
[3]Infinera, Sunnyvale, CA, United States
[4]AIM Photonics, SUNY Polytechnic Institute, New York, NY, United States
[5]University of California Santa Barbara, Santa Barbara, CA, United States
[6]Toppan, Tokyo, Japan
[7]Tyndall Institute, University College Cork, College Cork, Ireland
[8]Analog Photonics, Boston, MA, United States

4.1 Outline of the ecosystem

The photonic integrated circuit (PIC) foundry industry ecosystem can generally be divided into three major categories: wafer foundry or wafer R&D facilities; packaging, test, and assembly facilities; and the supporting electronic-photonic design automation (EPDA) companies. The last of these categories is the subject of the subsequent chapter, while the first two will be discussed in this chapter.

Wafer foundries have sprung up typically to address specific market demands that a chosen materials system is particularly well suited to address via performance and/or price characteristics.

For years, indium phosphide-based PICs have played a role in long-haul communication links. The key advantages of indium phosphide (InP) are obvious: it is a direct bandgap material and hence allows easy integration of the light source, and allows amplification as well as high-efficiency detectors. The key disadvantages of the material are its cost, the lack of a mature equipment support base, and its sensitivity to defects.

Si photonics has been developed much more recently by primarily applying the know-how and the infrastructure of the Si semiconductor industry to integrate optical elements, such as waveguides, detectors, modulators, and laser sources. While all but the last of these can be built using relatively standard Si technology processing, the integration of the laser represents a substantial challenge from a cost and complexity perspective and will be subject of a section of this chapter.

Niche foundry technologies, such as all passive PICs, take advantage of the material's intrinsic ability to provide waveguides at very low loss, something that is more difficult

Optical Fiber Telecommunications V11
DOI: https://doi.org/10.1016/B978-0-12-816502-7.00004-X

to achieve in a more comprehensive Si PIC foundry due to process limitations. Such PICs tend to be combined with active chips, such as InP PICs, to form a system. This trend has been enabled by the emergence of interposer technologies based on 2.5D and 3D die or wafer stacking that permit a novel, high-performance first-level packaging approach.

An orthogonal differentiation has crystallized because of a market forces: research and development facilities with a high degree of flexibility at the leading edge, which focus on both leading-edge process development but also cater to a customer base that explores future or niche markets (e.g., sensors, defense applications) versus large-scale foundries that aim to satisfy the growing demand of larger end users primarily in the data communication sector where cost is the major driver. Some of the R&D centers have found success in markets such as laser manufacturing, a current demand bottleneck, and a subset of what an InP foundry can offer. This chapter will discuss all these aspects.

Of all the processes typically performed in a foundry, lithography is the most critical and most costly. Additionally, photonic systems impose radically different constraints given the curvilinear nature of photonic devices, as contrasted to the "Manhattan" nature of complementary metal-oxide-semiconductor (CMOS) features. This problem ripples significantly into photonic mask making and design rule checking (DRC), as the CMOS approach is wasteful when applied to round shapes. A section of the chapter will discuss the unique challenges of the mask process.

Packaging costs represent an estimated 80% of a PIC, a mirror image of standard CMOS devices. Primary challenges for PIC packaging are the steps that are not readily automated, as they require active alignment: fiber attach and light source incorporation. The section about PIC packaging and test will detail the challenges and progress made in this area.

Designers typically do not access foundries at the very basic level of a design manual that specifies physical ground rules that cannot be violated and a set of device models. Instead, as practiced by the electronics industry since the 1980s, a designer ideally is sheltered from having to know the physical details of the process by being provided with a process design kit (PDK) that encodes those details. Little development has taken place in the integrated photonics industry to define preconfigured blocks of design intellectual property (IP), a shortcoming that the industry is addressing currently. This will be subject of the last section of this chapter.

4.2 InP pure play foundries

Classically, InP technologies are developed with a specific product in mind; for example, a tunable laser. In contrast, the PIC technologies discussed here are typically referred to as "generic"; that is, they are designed to allow for the widest range of products possible. This section gives an overview over modern generic InP technologies as they are offered by pure play foundries today. Due to a number of European Union-funded research projects [1–3], the landscape of true pure play InP foundries was dominated until recently by Europe. Just like Europractice [4] operates as a technology broker for application-specific integrated circuits (ASIC) and application-specific photonic integrated circuits (ASPIC) services in silicon-on-insulator (SOI) technology, the organization Jeppix [1] was established to serve as an InP broker. The American Institute for Manufacturing Integrated Photonics (AIM Photonics, aka "AIM") initiative aims to provide this service in the United States.

Historically, integrated photonics emerged from the development of semiconductor lasers, detectors, and modulators. At least until around 2000, integrated photonics was dominated by III-V technologies, such as GaAs and InP. InP has prevailed until this day as the material system of choice for a wide range of applications. InP is appealing for a number of reasons:

- Direct bandgap, enabling monolithic integration of lasers and optical amplifiers
- Mature epitaxy, enabling control over the bandgap from 1 to 1.7 μm
- Several fast electro-optic effects (Pockels, quantum-confined Stark)
- High carrier mobilities

Rather than just being a single material, InGaAsP/InP (and InGaAlAs/InP) is actually a material system. The core functionalities of InP devices are enabled not by InP itself, but by the quaternary $In_{1-x}Ga_xAs_yP_{1-y}$ materials class. Having to fulfill the condition of lattice match to InP (~5.87 Å), one remaining degree of freedom is tuning the bandgap from just below 1 μm to above 1.7 μm (see Fig. 4.1). Therefore, the performance of InP devices is closely related to the stoichiometric control over four or more constituents. Due to this range of achievable bandgaps, the general integration strategy is clear: large bandgaps can be used to implement waveguides and passive optical devices like couplers, filters, and phase modulators. Small bandgaps can be used for absorbers and photodetectors. Here, "large" and "small" are in comparison with the photon wavelength. Bandgaps that are close to the photon energy can be used for semiconductor optical amplifiers (SOAs) and lasers. Finally, electro-absorption modulators (EAMs) can also be realized in this regime.

4.2.1 InP-specific manufacturing challenges

These considerations make it clear that the key challenge of photonic integration in InP is the integration of a number of different bandgaps on one substrate. Technologically, two principal approaches are known to tackle this problem: vertical and lateral integration. As an example, both techniques are used in the integration schemes offered by Fraunhofer HHI (Fhg-HHI). Waveguide-integrated photodiodes and offset quantum wells (QWs) both fall under vertical integration (see Fig. 4.2). Vertical integration was widely generalized by the Canadian company OneChip

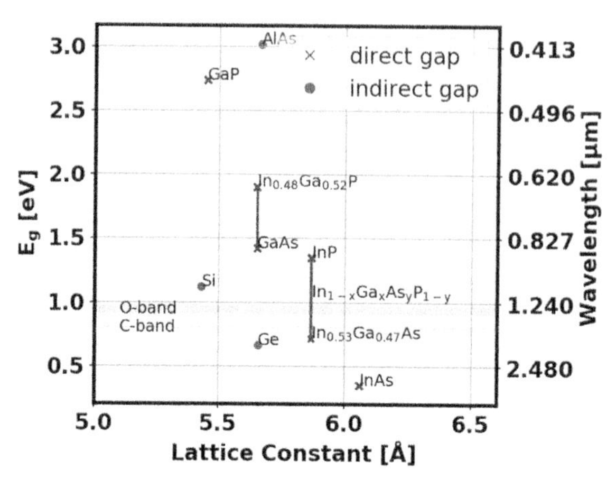

Figure 4.1 Bandgap E_g and equivalent wavelength versus lattice constant of the InGaAsP/InP material system and various other materials. The bandgap of InGaAsP lattice matched to InP can be tuned to any wavelength in the O- and C-bands. The longest wavelength that is still lattice matched to InP is around 1.72 μm, obtained by In_0.53 Ga_0.47 As. This material is commonly referred to as InGaAs/InP or just InGaAs. Source: *Credit: FhG-HHI. The parameters for the III-V materials are from ONECHIP PHOTONICS. <http://www.onechip-photonics.com/>, 2018 (accessed 15.05.18); the group IV parameters are from PIC Magazine. GCS, Intengent, VLC to offer PIC Fab Services. <https://picmagazine.net/article/103667/GCS_Intengent_VLC_to_offer_PIC_Fab_Services>, 2018 (accessed 15.05.18).*

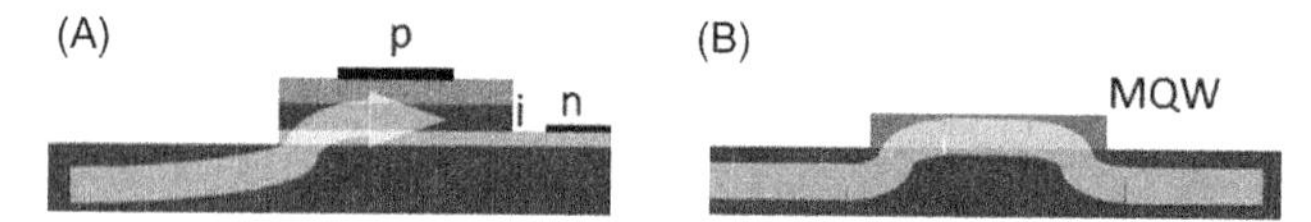

Figure 4.2 (A) Schematic of a waveguide-integrated pin-photodiode. The intrinsic absorber (*green*) has a larger refractive index than the underlying waveguide, giving rise to evanescent coupling. (B) shows a multiquantum well (MQW) stack on top of a waveguide. This structure can be used for lasers, semiconductor optical amplifiers, or electro-absorption modulators. Source: *Credit: FhG-HHI.*

Photonics [5] and is now being commercialized by Global Communication Semiconductors LLC [6]. This technology avoids any regrowth steps and relies on lateral tapers to couple between the different guiding layers.

While epitaxial InP technology is challenging from a deposition perspective, the demands on lithography are rather relaxed compared to SOI technology. Simply put, the buried oxide (BOX) layer in SOI technology, or the cladding material for nitride waveguides, increases the index contrast by around one order of magnitude compared to InP, making it possible to build the waveguides one order of magnitude smaller. So, while silicon waveguides are typically around 200 nm wide, single mode waveguides in InP tend to be around 2 μm wide. The critical feature sizes scale accordingly. Because topography can be quite significant and the critical dimensions are often large enough, contact lithography is still employed in some processes. In this case, the wafer effectively gets flattened out by the pressure from the mask. For example, the integration platform at FhG-HHI still uses contact lithography, even though stepper lithography is on the roadmap [7].

4.2.2 State-of-the-art generic InP photonic integrated circuit technologies

4.2.2.1 Example 1: Fraunhofer HHI

The current generic integration platform at the FhG-HHI is an extension to a receiver (Rx) platform that is based on 40 GHz waveguide-integrated photodiodes as in Fig. 4.2A. Much of the extension was carried out as part of reference [2] and led to what is referred to an integrated transceiver-receiver (TxRx) platform today [8, 9].

An overview of the process flow is shown in Fig. 4.3. It consists of three epitaxial growth steps, one electron beam lithography step for grating definition, and about 25 mask layers for optical lithography. The integration platform relies on Fe-doped substrates and passive waveguide layers, ensuring electrical insulation between the individual building blocks. So, unlike in other technologies, all components can be disconnected electrically not only on the *p*-side but also the *n*-side. The wafer size is currently 3 in and is planned to migrate to 4 in by 2022 [9]. Fabricated wafers undergo back end processing to yield diced chips with antireflection coated facets. Chip sizes can vary between 4×6, 8×6, and 12×6 mm^2.

An overview of the available building blocks and their respective performance is given in Fig. 4.4.

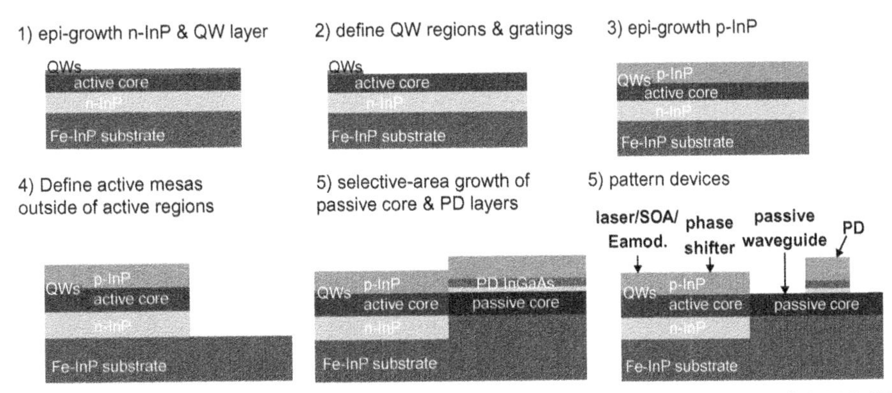

Figure 4.3 Overview of the process flow of the generic integration technology at FhG-HHI. (1) The base epitaxy is grown on 3 in semiinsulating substrate. (2) The quantum-well- (QW) based gain material is patterned and optical gratings are written by e-beam lithography. (3) A *p*-cladding is grown to form the diodes of the active components. (4) Mesas are etched, which will eventually form the active components. (5) Layers for passive waveguides and photodetectors are grown selectively, forming a butt-joint at the interface to the active mesas. (6) Waveguides are patterned and detector mesas are formed. Source: *Credit: FhG-HHI.*

Component	Specification	Value
Lasers and Amplifiers		
SOA	Gain	92 cm^{-1} @7kA/cm2
	Saturation Power	3 dBm
DBR grating	Tuning range	4 nm
DFB laser	Tuning range	4 nm
	Output power	3 mW @ 150 mA
DBR laser	Tuning range	4 nm
	Output power	3.5 mW @ 150 mA
Electro-absorption modulator	dynamic extinction	10 dB @2V$_{pp}$
	f_{3dB}	15 GHz
Polarization devices		
Polarization splitter	Loss	<4 dB
	Max polar ratio	25 dB
Polarization converter	Loss	<3 dB
	Conversion efficiency	>10 dB
PIN photodiode		
	3 dB bandwidth	>35 GHz
	Dark current	<100 nA @ -2 dV
	sensitivity	0.8 A/W
Modulators		
Thermo-optic phase modulator	Loss	2 dB/cm
	I(PI) x L	20 mA x mm
Current injection phase modulator	Loss	2 dB for 100-200 μm
	I(PI) x L	20 mA x mm
Spot size converter		
	Coupling to SSMF	<2 dB
Passive waveguides		
Straight Waveguide	Loss	2 dB/cm
Arc waveguide	Minimal radius	150 μm;
Tapered waveguide	Loss	2 dB/cm
Couplers:		
1x2 MMI coupler	Loss	<1 dB
2x2 MMI coupler	Loss	<1 dB

Figure 4.4 Overview of available building blocks and their characteristics in the FhG-HHI photonic integration technology. Source: *Credit: FhG-HHI.*

Figure 4.5 Optical photograph of a finished MPW on 3-in. wafer. Source: *Credit: FhG-HHI.*

4.2.3 Multiproject wafer runs

Pure-play foundries typically offer multiproject wafer runs (MPWs) with a fixed tapeout schedule. Both SMART Photonics and FhG-HHI have four tapeout dates per year with a turnaround time of about 6 months. Using typical tile sizes, up to 50 different users can participate in one MPW run. Fig. 4.5 shows an example of a finished MPW on a 3 in wafer. Examples for PICs that have been fabricated using those MPWs include all-optical memory [10], microwave PICs [11], Bragg-sensor interrogators [12], quantum number generators [13], dual-polarization transmitters [14] or tunable wavelength filters [15], and many more. There are initiatives like PICs4All [3] to help interested parties arrange feasibility studies for their respective application and find the right foundry.

Designs for MPWs are usually submitted in GDS format with placeholders for the foundry proprietary design blocks (IP) that are available to customers as black boxes. To ensure a smooth submission process, each foundry has a PDK with all the necessary layout information and design rules. After receipt of the design data, the foundry or the broker swaps the black box content into the placeholders (e.g., layout for lasers) and assembles the mask reticle content prior to submission to a mask house.

4.3 Turn-key InP foundry

Over the past few decades, the exponential demand to increase transmission capacity in optical communication networks, as well as the need to mitigate transmission impairment, has led to breakthrough developments in photonic integration. Thus far, optical network solutions with InP technologies have contributed to more than an order of magnitude increase in capacity every decade. Fig. 4.6 illustrates the trend started with electro-absorption-modulated lasers [16] and continued through with the

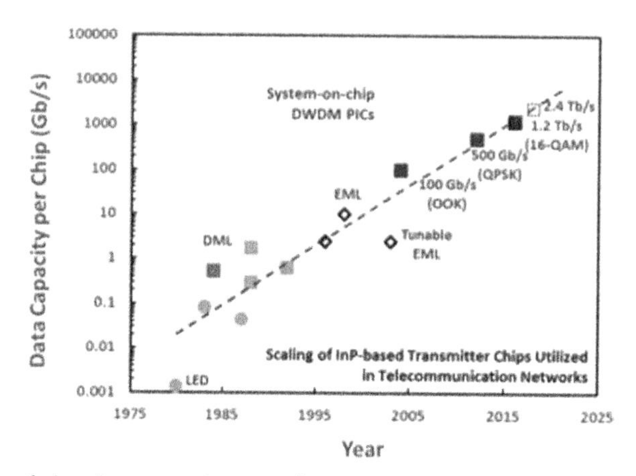

Figure 4.6 Scaling of the data capacity per chip in commercial optical networks for InP-based transmitter chips. The data capacity per chip has doubled an average of every ~2.2 years (fit to the data) for the last 30+ years, enabled by the employment of photonic ICs. Since 2004, SOC dense wavelength division multiplexing (DWDM) PICs have been essential to this scaling (*dark blue squares*). The open square illustrates a research demonstration of a SOC DWDM PIC capable of 2.4 Tb/s capacity. Source: *Credit: Infinera.*

commercial introduction of the first fully integrated multichannel optical system-on-chip (SOC) 100 Gb/s transmitter and receiver PICs in 2004 [17].

Each generation of commercial SOC InP-based PIC transmitters and receivers has increased integration and functionality density, resulting in significant benefits to the economics, power consumption, density, and reliability of high-capacity dense wavelength division multiplexing (DWDM) systems. The transition in scaling of data rate from discrete devices to integrated devices in commercial optical networks has been driven by the presence of performance impairments in the network that arise at higher bit rates, making discrete scaling less cost-effective than that of integrated devices [18]. In 2014 over 1700 optical functions were monolithically integrated to realize a 2.25 Tb/s polarization multiplexed-quadrature phase shift keying (PM-QPSK) transmitter with a 1 THz optical bandwidth [19].

The deployment of multichannel InP-DWDM PICs utilizing optimal, yet scalable, optical functions demonstrates a capability that can simultaneously achieve both optimal performance and high yield.

4.3.1 State-of-the-art photonic integrated circuit product examples enabled by InP technologies

Several generations of commercial SOC InP-based PIC transmitters (Tx) and receivers (Rx) have been deployed based on compact PIC architectures that leverage coherent modulation with unique scalability, efficiency, and security [20, 21]. Fig. 4.7A shows a schematic of a monolithically integrated multichannel Tx PIC. Each channel includes

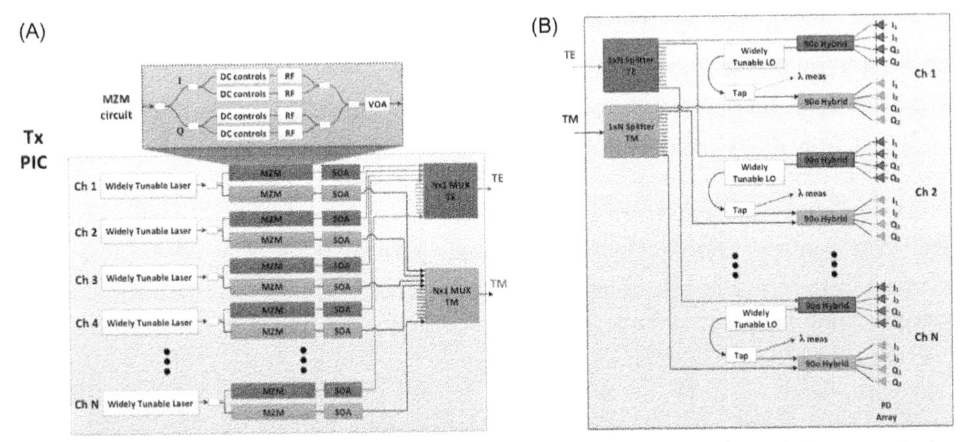

Figure 4.7 (A) The multichannel widely tunable coherent transmitter photonic integrated circuit (Tx PIC) architecture schematic, integrating widely tunable lasers, IQ-MZMs, and per-polarization VOAs and SOAs; (B) multichannel Tx PIC architecture integrating 1 × 6 MMI-based power splitters for the two orthogonal polarization data streams, per-channel widely tunable lasers with taps as local oscillators, 90 degrees optical hybrids, and photodetector arrays for coherent detection. Source: *Credit: Infinera.*

an independently controlled, widely tunable laser (WTL) providing continuous tuning over the entire extended C-band (4.8 THz). The output from each laser is split into two nested in phase-quadrature Mach-Zehnder modulators (IQ-MZMs). At the output of the MZM, a variable optical attenuator (VOA) and a SOA is used to provide additional power balancing and amplification of the output power from the Tx PIC. DC control elements in each of the four arms of the nested MZM are used for power balancing and biasing the modulator to the required phase condition [20].

The corresponding Rx PIC architecture is shown in Fig. 4.7B [18, 21]. The monolithically integrated InP receiver PIC accepts two inputs polarized to be transverse electric (TE) on chip from free space optics that rotate one signal path polarization from transverse magnetic (TM) to TE. Each polarization signal path is split with a multimode interference (MMI) coupler-based 1 × N power splitter to feed N 90 degrees optical hybrids. The optical hybrids mix the incoming signals with one of the independently controlled, integrated widely tunable local oscillator laser outputs. MMI-based optical taps on the laser output paths are used to couple light out from the chip to measure wavelength over the extended C-band (> 40 nm).

A 6-channel implementation with C-band tunable 1.2 Tb/s transceiver modules operating at 33 Gbaud, 16-QAM was transitioned into manufacturing in 2016 [20, 21]. Next-generation Tx PIC platform is 1 Tb/s per wave capable operating at 100 GBd × 32QAM and supports modulation formats with variable spectral efficiency [22,23]. Moreover, 100 GBd × 16QAM signals generated with an integrated Tx PIC assembly were transmitted over a distance of 1400 km using a reference receiver [24].

4.3.2 InP photonic integrated circuit packaging

The success of the PIC SOC scaling is critically dependent on the codesign and optimization of the key technologies critical to the PIC SOC DWDM optical module: optical interconnect, high-speed electrical interconnect, DC control/bias interconnect, thermal management, analog driver/amplifier ICs, and control ICs. The complexity of the electrical interconnect inside a 500 Gb/s coherent SOC DWDM transmitter module is shown in Fig. 4.8A [22]. A total of >17 ft of Au wire bonding is utilized to interconnect the elements within the module. The widely tunable transmitter and receiver PICs are copackaged in a hermetic high-temperature cofired ceramic (HTCC) stacked-ceramic land grid array (LGA) package to yield a 1.2 Tb/s coherent transceiver module. The Tx (Rx) PICs are assembled with MZM driver array ASICs integrating 24 high-speed bit streams in SiGe BiCMOS technology, TE coolers, and DC stacked control ASICs drivers (for the Tx PIC). The transceiver also integrates free-space optics for polarization multiplexing/de-multiplexing the polarization multiplexed signals [20]. Looking ahead, the continued demand for higher baud rates poses challenges scaling the wire bond interconnect assembly technology while maintaining the requisite signal integrity and crosstalk. Recently, employing flip chip assembly technology, Infinera has developed the direct bonding of the MZM and InP-PIC to a common interposer, maintaining high interconnect density to minimize size (cost) penalty on the InP-based photonic IC and removing all wire bonds. Fig. 4.8B shows a cross-sectional micrograph of the high-quality bond achieved between an InP PIC and interposer. The MZM driver (MZMD) and interposer show comparable bond quality (cross section not shown). This manufacturable, high-fidelity interconnection between MZM and MZMD was implemented to facilitate 1 Tb/s per wavelength capability in the next generation of SOC coherent Tx PICs [24].

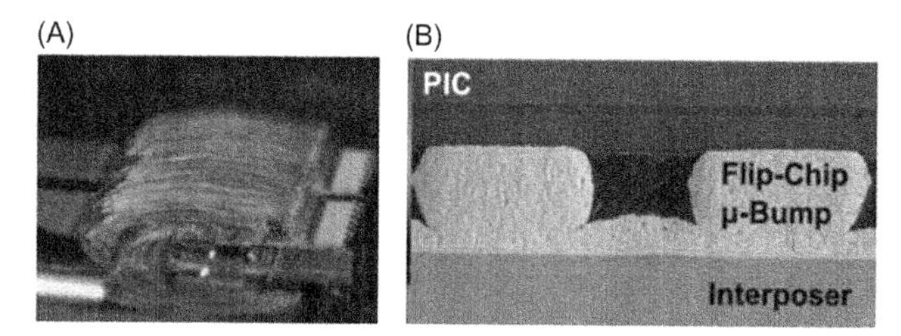

Figure 4.8 (A) Image of a 14-channel photonic integrated circuit (PIC) attached and wire bonded to a ceramic carrier for testing. Seven layers of stacked wire bonds for DC control are utilized to achieve a dense electrical interconnection to the PIC. Radio frequency interconnects to a high-speed application-specific integrated circuits are not shown; (B) cross section of the high-quality flip chip bond between the PIC and interposer. Source: *Credit: Infinera.*

4.3.3 InP photonic integrated circuit manufacturing challenges

Essential to the commercial success of any integrated circuit (IC) is that the economic value derived from the integrating component must outweigh the cost of the integration itself. To realize manufacturable PICs with a high density of monolithically integrated functions and channels, at least two requirements need to be met for sufficient performance and yield in each PIC channel to meet all requisite system metrics and the ability to deliver this performance across multiple channels without incurring a substantial yield penalty. These requirements impose a minimum constraint on design capability of the PIC as well as the manufacturing process.

Advances in silicon and III-V semiconductor fabrication technologies have enabled a highly capable fabrication platform for InP-based PIC manufacturing. The wafer fabrication yields that can be achieved in a state-of-the-art InP PIC fab (e.g., Infinera Corporation) are now equivalent to that of Si CMOS circa mid-1990s [25, 26].

Fig. 4.9 compares the line yield (normalized per 10 mask layers) for a Si CMOS fab circa 1990s and multiple generations of transmitter SOC InP-based photonic ICs. The line yield (LY) is calculated as

$$LY = \frac{\mathrm{WO}}{\mathrm{WO} + \mathrm{SC}}, \tag{4.1}$$

where WO is the number of wafers completed during the period and SC is the number of wafers scrapped during the same period. To compare disparate fabrication processes, the line yield is normalized to yield per 10 mask levels by the relation

$$LY_{10} = LY^{\left(\frac{10}{ML}\right)}, \tag{4.2}$$

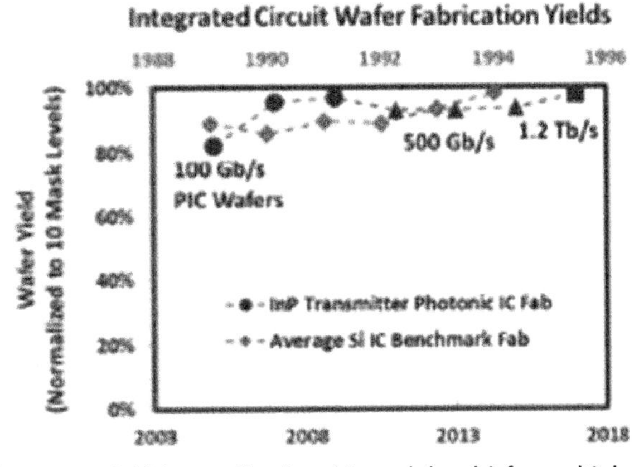

Figure 4.9 Wafer fabrication yield (normalized to 10 mask levels) for multiple generations of transmitter PIC wafers compared to a benchmark Si CMOS fab (1989–94). Source: *Credit: Infinera.*

where LY is the overall line yield, ML is the number of masking layers, and LY_{10} is the calculated line yield per 10 layers. In order to attain the photonic IC manufacturing capability shown in Fig. 4.9, it was necessary to adopt a methodology for InP PICs similar to electronic ICs, where designers are given a fixed (albeit limited) tool set to design within, resulting in a manufacturable (cost effective) manufacturing process. Thus, both highly capable fabrication capability and device performance have been simultaneously achieved.

Furthermore, for direct comparison purposes, the line yield excludes the crystal growth steps. Currently, the yields of the PIC crystal growth processes are comparable to the cumulative wafer fabrication yields. Like those for Si electronic ICs, the data shows that line yields for four consecutive generations of InP-based PIC products are in the high 90% (per 10 mask levels).

State-of-the-art PIC fabrication technology have reduced the incidents of "killer defects," which has been essential in enabling sufficiently low defect densities to realize fully integrated SOC PICs [27]. "Killer defects" are catastrophic defects that result in electrical zero yield of device or are optically zero yielding by the shortening of metal lines or creating topography disruptions at current or subsequent fabrication steps. A recent comparison of random killer defect densities versus time was demonstrated for Si CMOS ICs and four generations of InP-based SOC transmitter PICs [18]. The defect density numbers for InP-based PICs in maturity have a rate of decrease (defect reduction learning) similar to what the Si industry achieved in 1975–95. The demonstrated manufacturing capability of InP-based photonic IC manufacturing has matured to realize economically viable sophisticated SOC architectures with integration levels approaching 500 optical functions for commercial devices. Such capability enables the development of even higher functionality and integration levels.

For in-line manufacturability and final known good die (KGD) reasons, it is imperative to perform extensive PIC testing during manufacturing and prior to shipment. Typically, over 100 parametric performance parameters are tested on a given PIC. Fig. 4.10 presents yield statistics for one of the most challenging requirements of the tunable laser threshold current and continuous tuning range for the worst channel tested on each PIC. The data was gathered from all fourth-generation transmitter production PICs produced from 2016 to March 2017. The figure plots the cumulative probability of the measurement and demonstrates better than 99% yield (on a per PIC basis) to the threshold current specification (left) and the continuous tuning range (right) [18].

These manufacturing results demonstrate the availability of a highly capable design environment for the widely tunable lasers integrated into the coherent PIC platform and the development of robust designs and tightly controlled fabrication processes. Similarly high specification yields have been reported across all components in multiple generations of PICs, indicating a systematic improvement in capability capitalized

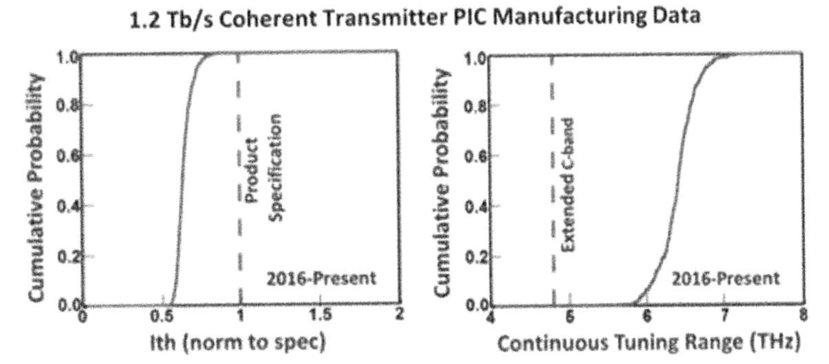

Figure 4.10 Cumulative probability distribution plots for the worst channel performance for 1.2 Tb/s transmitter photonic ICs manufactured from 2016 to present. Data for threshold current (normalized to specification, left) and continuous tuning range (right) show very high-performance yields (> 99% for each specification). Source: *Credit: Infinera.*

from the semiconductor learnings derived from each successive deployment of the next generation of PICs [18,26,28,29]. An InP-based foundry approach [30] derived from a commercially deployed integration platform capability can support the realization of a wide range of market applications with path to commercialization.

4.3.4 Turn-key photonic integrated circuit foundry

The successful implementation of InP-PICs using a foundry kit is largely determined by the maturity of the technology, the integration platform capability, and the predictable performance of all integrated components. A turn-key foundry kit has been developed at Infinera Corporation to enable the realization of InP-PICs for a variety of applications. This foundry kit development leverages Infinera's fabrication capability with optimized designs and yield improvements gained through multiple generations of commercial PIC deployment. The approach permits the design of photonic circuits in a schematic form using optical functions realized using *p*-cells.

The information necessary to realize a PIC from its schematic is contained within the PDK library. The PDK library consists of a collection of fully characterized *p*-cells used by the designer to create a PIC schematic to realize the desired circuit functionality and performance metrics [31]. Each *p*-cell has input and output ports that define electrical connectivity and or optical routing. A multitude of *p*-cells can be used to create a composite building block with its defined input and output ports. These building blocks can be instantiated in a hierarchal fashion to create the final PIC schematic. Furthermore, each *p*-cell and composite building block are described by functional models that can used to perform circuit simulations. Each *p*-cell is DRC compliant and fully described by design parameters and a physical layout consistent with the specific foundry process flow.

Similar to the design of electronic integrated circuits, turn-key photonic ICs require multiple design and verification steps supported by EPDA tools. However, photonic circuits pose unique challenges for current automated design environments typically used in the electronics industry, as the optical connections between components do not behave like electrical wires. Specifically, EPDA tools must support the PIC layout while accounting for such photonic-unique items as optical phase and radiation losses. The EPDA generated PIC layout must also meet the intended performance metrics, such as optical power budget and signal-to-noise requirements in addition to satisfying layout versus schematic (LVS). Furthermore, the physical PIC layout must also meet the foundry's DRC specific to the integration platform and the fabrication process [31,32].

It is important to note that while LVS ensures that the PIC layout meets the circuit schematic connectivity, it does not verify the intended functionality. Therefore, a fully bidirectional EPDA interface between a PIC schematic and its layout is needed to ensure that changes performed at the physical layout level are back annotated to the intended performance metrics of the PIC. Consequently, EPDA tools need further customization with additional constrains and capability beyond commercially available plug-in EPDA tools. Once the physical layout meets the designers' intent, it is used to generate the final GDSII files.

Fig. 4.11 illustrates a foundry process flow, automated design (EPDA), fabrication, and test steps. The "Process Specifications" encompasses a set of foundry-defined specifications tailored to each integration platform, including the epitaxial target layers, device cross sections, process flow, and GDS layer list.

A successful foundry run requires that all the *p*-cells and building blocks defined in the PDK library meet their performance specifications. To this end at the foundry,

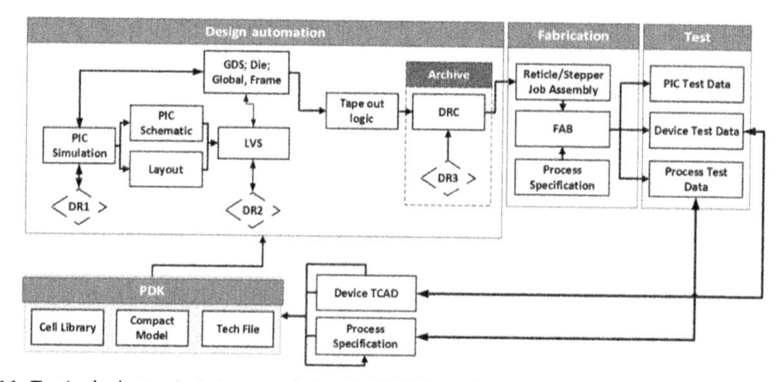

Figure 4.11 Typical photonic integrated circuit (PIC) foundry process flow, showing process design kit (PDK) inputs and multiple design review (DR) steps consisting of DR1: PIC feasibility review; DR2: layout design review; DR3: tapeout review; process waiver. The PDK library is encompassed in the shaded gray area. Source: *Credit: Infinera.*

special test structures are used to monitor and characterize the PIC fabrication and *p*-cell performance. Each lot has a set of verification dies that contain process control monitors (PCM), electrical-electrical (E-E), electrical-optical (E-O), and optical-optical (O-O) test structures. The PCM test structures provide monitoring and feedback on the fabrication process uniformity, while the O-O, E-E, and E-O test structures verify that each stand-alone p-cell or building block meets its specified performance metrics. The process and device tests results may be used as feedback to update the foundry process specifications or *p*-cell designs as shown in Fig. 4.11. These tests can be performed at wafer-level (on chip measurements) or at die-level. However, from a practical standpoint, many optical tests require measurements in die form. For this purpose, wafers are first cleaved into die/chips and facet coated, and then chips are attached to a ceramic carrier and wire bonded for electrical connectivity. Mounting the die on a ceramic carrier as a chip-on-carrier (CoC) ensures a controlled thermal environment and minimal handling damage during the testing. An example of ceramic carrier design is shown in Fig. 4.12. Alignment fiducials are placed on the PIC and carrier for assembly and test automation. Chips on ceramic carrier can be directly attached to electronic boards with the respective drivers to complete their characterization.

One of the challenges with fabricating a CoC is making provisions for high-speed connections without detrimentally altering radio frequency (RF) performance due to additional parasitic contributions from the wire bond and the ceramic bond-pad capacitance. The carrier shown in Fig. 4.12 comprises a "ceramic pedestal" of similar thickness as the PIC to minimize the length of wire bonds. The pedestal is further designed such that it can include an impedance matching network to further enhance the RF performance.

Probe cards similar to the electronic industry are needed to test DC and RF measurements of PICs in wafer and die form. A custom-designed interface is used to

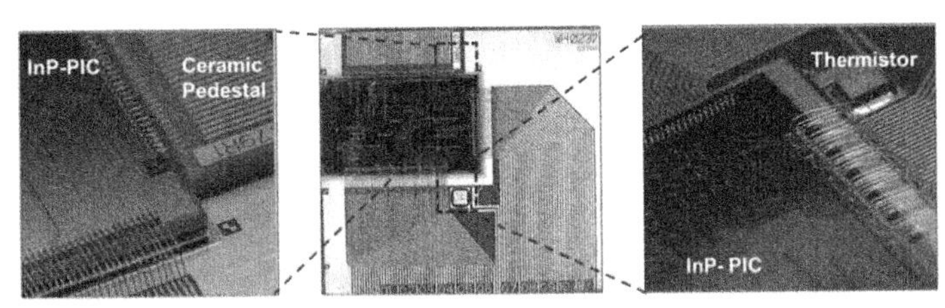

Figure 4.12 Micrograph chip-on-carrier using a custom-designed ceramic carrier to enable a standard foundry photonic integrated circuit (PIC) input/output port assignment. The center image shows the image of the carrier with identification (ID) and alignment fiducials to assist with automated assembly tools. The insets depict (left) a ceramic pedestal attached to the carrier to reduce radio frequency wire bond lengths and (right) a thermistor used for temperature control. Source: *Credit: Infinera.*

Figure 4.13 Photograph of the realization of a monolithically integrated colliding-pulse mode-locked laser with a coupled optoelectronic oscillator, modulator, and optical amplifiers. *Source: Credit: Infinera.*

provide electrical connectivity between the probe card or handler and the test system. Multiple DC and RF tip probes are required to simultaneously land on the PIC or carrier, without causing damage to the PIC die or the metallization pads during touch-down. Consequently, probe cards are highly customized to meet the PIC bond-pad location and pitch and in accordance with the card technology for electrical and mechanical stability. Thus, by confining the PIC physical layout to fit within a die frame with fixed input and output optical ports as well as fixed bond-pad (DC and RF) locations and geometry, the need for multiple probe card and carrier designs is eliminated, thereby simplifying testing infrastructure.

Fig. 4.13 shows a photograph of the implementation of a monolithically integrated racetrack colliding-pulse mode-locked laser (CP-MLL) [33] stabilized using a coupled optoelectronic oscillator (COEO) injection-locking scheme [34]. The entire PIC is realized using the building blocks defined in the foundry's PDK library. The COEO consists of a tunable laser and cascaded single sideband (SSB) and a nested MZM. The CP-MLL architecture implements the functionality of CP-MLL with an integrated pulse-picking MZM and SOAs. To mitigate degradation of the RF performance of the SOA and MZM, the PIC was wire bonded directly to an RF compatible FR4 board to enable hybrid integration of electronic drivers.

4.4 Si photonics development

While the transmission of data using optical methods has been in place for many decades, the PIC, especially when based on silicon, is much younger—at most, 20 years old. It has been noted [35] that PICs are about 40 years behind their electronic counterparts and, indeed, Jack Kilby demonstrated the IC in 1958 [36], which is more than 40 years before Luxtera unveiled their PIC [37]. Before the IC, electronic circuits consisted of discrete components (capacitors, transistors, resistors, inductors) connected by wires, and before the photonics IC, optical circuits consisted of discrete components (modulators, lasers, detectors) connected by fibers. Today, these photonic components can be cofabricated on a common substrate, and that common substrate is often the same that popularized electronic ICs: silicon.

Silicon is, of course, not the only material system used for ICs (electronic or photonic) and, in some cases, is not the best material system [38]. Other systems, like SiGe, SiC, or InP, sometimes provide significant device performance advantages over silicon that are critical for specialized applications. However, the proliferation of silicon-based electronic ICs means that there is an unsurpassed infrastructure for silicon-based ICs and PICs that leverage the fabs, tooling, and supply chains already in place. This gives a significant advantage to silicon photonics in terms of commercialization, mass production, and integration with electronic components. Fig. 4.14 illustrates the considerable cost differential between 300 mm Si wafers (per cm^2) and smaller wafers of other materials (e.g., 150 mm GaAs and 100 mm InP wafers).

Using established high-volume silicon factories comes with a different set of challenges in that the most advanced fabrication facilities (fabs) are designed, optimized, and focused on silicon ICs. Until Si PICs drive volumes comparable to the capacity of a full Si fab, this leads to less flexibility in tool and process changes due to risk aversion

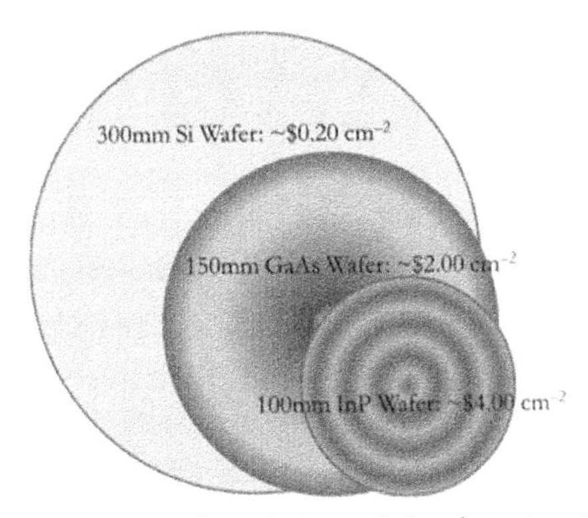

Figure 4.14 While PICs can be fabricated in GaAs or InP substrates, these materials systems are more expensive per unit area and are limited to smaller wafer sizes. Source: *Credit: Data from A. Seeds, S. Chen, J. Wu, M. Tang, H. Liu, Monolithic integration of quantum dot lasers on silicon substrates, Semicon West, <http://www.semiconwest.org/sites/semiconwest.org/files/data15/docs/5_AlwynSeeds_UniversityCollegeLondon.pdf>, 2016 (accessed 07.03.18) [39], M. Liehr, AIM Photonics: manufacturing challenges for photonic integrated circuits, Semicon West, <http://www.semi.org/eu/sites/semi.org/files/events/presentations/02_MichaelLiehr_SUNY.pdf>, 2017 (accessed 07.03.18) [40].*

to the core electronics technologies that a fab supports. In addition, while PICs can use the same materials, equipment, and facilities as electronic ICs, the process development and device integration know-how is typically not directly transferrable [41]. The physical dimensions alone are incongruent: leading-edge CMOS fabs are focused on 14, 10, or 7 nm node devices, while leading-edge PIC technologies currently do not and may never need structures much below 100 nm [42]. Instead, a typical challenge is fabricating photonic devices that are "large" (micron sized) using tools that are designed for nanoscale components. These challenges are not insurmountable, and therefore, with a modest allocation of resources to focus on photonics technology, they can be mitigated and the advantages of a silicon-based facility can be leveraged for PIC technology.

Building a PIC on silicon requires a number of basic components: passive devices to move light in, out, and around the chip (waveguides); modulators to convert electrical signals to optical ones; detectors to convert optical signals to electrical ones; and a light source (which currently is often off-chip). More complex circuits might include electronic circuits, integrated lasers, optical fiber input/output (I/O), or other, more exotic components, such as gain material or optical isolators. Fig. 4.15 illustrates typical active and passive optic components for PICs.

In theory, an on-chip waveguide is straightforward: the core of a waveguide should have a refractive index greater than the refractive index of the cladding material which leads to total internal reflection. As mentioned earlier in the chapter, the larger the refractive index difference, the more efficient the waveguiding. Therefore, silicon has an inherent advantage over other integrated photonics platforms; for example, InP. Silicon itself has a refractive index of 3.46, while silicon dioxide has a refractive index of 1.45, and these materials are very common in a silicon fab. Other materials commonly used in silicon processing that can also be repurposed for waveguiding are

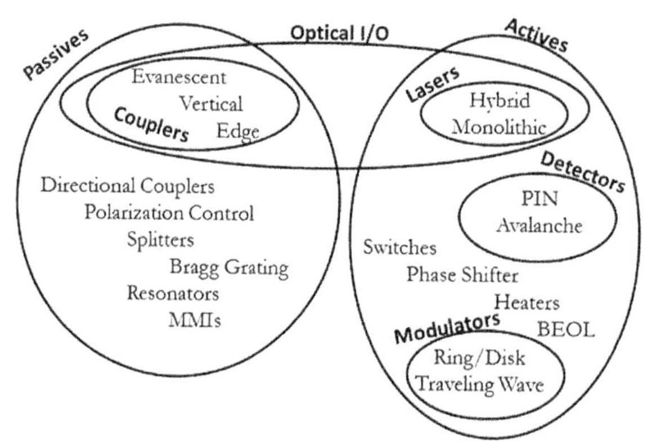

Figure 4.15 Graphical representation of typical active and passive optic components used by photonic integrated circuits. Source: *Credit: AIM Photonics.*

silicon oxynitride (SiON, $n \approx 1.6$) and silicon nitride (SiN, $n \approx 1.9$). In practice, however, material and structural properties of the waveguide and cladding films used for CMOS applications need to be tuned to reduce or eliminate light absorption or scattering. The patterning of the waveguiding layer (usually with photolithography and reactive-ion etching, RIE) needs to be optimized for pattern fidelity and sidewall roughness in order for the passive devices to function as designed and to minimize light loss through sidewall scattering [38]. As mentioned previously, while these materials and process steps are common in an electronics fab, they are not tailored for photonics. For example, SiN deposited via low- pressure chemical vapor deposition (LPCVD) yields a high-quality SiN film, but the process thermal cycle may not be compatible with the other photonic devices previously created on the wafer (discussed below) [43]. Alternatively, SiN deposited with lower temperature plasma-enhanced CVD (PECVD) usually suffers from excess hydrogen in the film, which absorbs light around 1520 nm. This is very close to the functional wavelength peak of the C-band (1550 nm) and therefore causes excess waveguide losses. Finally, any SiN deposition needs to be tuned for the thicknesses required for waveguiding. Thin films in CMOS manufacturing are on the order of tens of angstroms or thinner, while SiN must be deposited at hundreds, if not thousands, of angstroms to be useful for photonic applications.

Passive photonic components are fabricated in the same manner and in the same process step as the waveguide. Bends must be designed with a large enough radius of curvature to minimize light loss. The minimum bend radius is a function of the material system, cross-sectional dimensions of the waveguide, and the supported wavelength of light. Light can be moved from one waveguide to another if they are close enough together that the modal fields overlap and the coupling length and spacing is designed appropriately [38]. Similarly, other passive components like splitters, ring resonators, and multimodal interference couplers must be precisely designed and require high pattern fidelity on the wafer in order to function properly.

Passive photonics devices will function when light enters them without any additional inputs. Active devices require an electrical signal (in DC or RF) in order to operate and therefore change the flow of light around the PIC. Light can be modulated by thermal or electromagnetic fields, and both of these methods can be facilitated with patterned structures of the silicon that have been doped via ion implantation [44–48]. Ion implantation of silicon is a core segment of an electronics flow to define transistors, resistors, and diodes, and so those processes can be directly applied to silicon photonics in order to define modulators, switches, heaters, and filters. Modulators are one of the most important active devices that encode an electric signal into the optical domain by switching the intensity of incident light high or low. For applications like datacom, these modulators must be able to function at very high speeds (56 GHz and beyond) to be competitive. There is a significant amount of research and development into their design, as well as process optimization in the fab, in order to increase modulation speed [49].

The other major active device is the photodetector, which complements the modulator by encoding an incident optical signal into an electric one through the photoelectric effect. While there are some exceptions, most photodetectors in silicon photonics applications are P-I-N diodes of epitaxially grown germanium [50–54]. This, too, takes advantage of the modern silicon fab's experience with heteroepitaxy, and, like the modulator, much research and development focuses on new designs and process development to create very high speed detectors with high responsivity and low leakage current.

After the creation of active devices in the silicon with ion implantation and Ge epitaxy, these devices are buried in silicon dioxide, contacted and connected using standard middle-of-the-line (MOL) and back-end-of-the-line (BEOL) fab processes [42]. The MOL contacts the Ge and Si, and the BEOL connections are typically are very similar or identical to what is used for electronics interconnects.

Terminating the silicon photonics chip in Al for wire bonding applications or Cu for under ball metallurgy (UBM)/solder ball applications allows for straightforward electronic packaging of the PICs. However, many PICs, including datacom PICs such as transceivers, also need optical packaging to get light in and out of the chip [55]. Optical I/O can be accomplished using vertical coupling, edge coupling, or evanescent coupling. In vertical coupling, a passive photonic device called a grating coupler redirects light from a waveguide up to a fiber oriented normal to the surface of the PIC [38]. Likewise, light from a fiber that shines on a grating coupler will direct that light into the waveguide. Alternatively, a fiber can be positioned parallel to the waveguide, provided that a deep trench is fabricated to provide a smooth surface for the fiber to butt up against the end of the waveguide [42]. In these applications, a passive edge coupler device serves to convert the optical mode of the fiber to the waveguide and vice versa. Finally, an evanescent coupler is similar to on-chip passive devices where the fiber is brought very close to a waveguide through a deep trench so that the modes of the fiber and the waveguide overlap, allowing for efficient coupling [56,57].

For certain applications of silicon photonics, an on-chip laser is required and, for others, an on-chip laser is highly desired in order to minimize the size and/or the cost of the system (see Fig. 4.16) [55,58]. The current state of the art of silicon photonics is to attach lasers grown on a different material system, either through wafer scale bonding or chip-attach [57]. One drawback to wafer scale bonding is that typically III-V lasers are grown on much smaller substrates than a silicon PIC is fabricated on, so multiple III-V wafers are needed to cover a significant amount of the silicon wafer before being patterned into lasers. This is expensive, and, as with any wafer scale bonding work, yield loss is multiplied (a bad silicon photonics die or a bad III-V laser

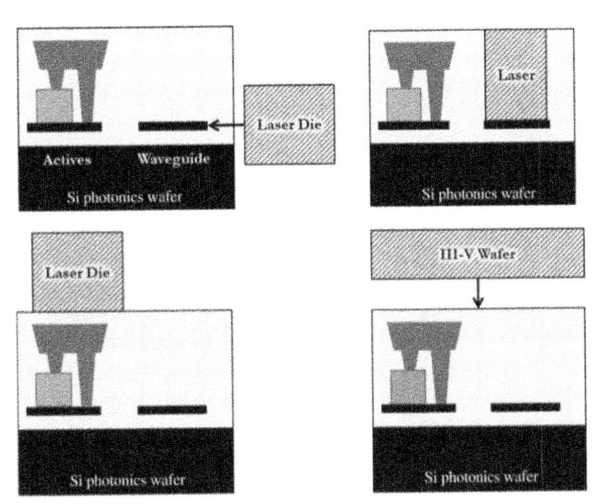

Figure 4.16 Integration of a laser source for silicon photonics can be done in different ways, but all methods have challenges (clockwise from top right): monolithically integrated laser where the lasing material has been grown on the silicon is currently under development; wafer scale bonding of a III-V wafer to a silicon wafer is inefficient due to size differences between III-V and silicon wafers; chip attach of III-V laser to the surface of the silicon photonics wafer is a slow process; or assembly of a laser die next to the photonic integrated circuit die on a chip carrier or other packaging substrate. Source: *Credit: AIM Photonics.*

die will yield a bad bonded die). The advantage is if the lasers are patterned after the III-V material is bonded, the lasers can be aligned to the waveguides with lithographic precision, decreasing coupling losses. With die-attach, there is the "known good die" advantage: only good III-V lasers are used, and only get attached to good silicon photonics die. This process is less wasteful since III-V material is only being placed where it is needed; however, it is very time consuming (i.e., costly) and can lead to higher coupling losses into the waveguide depending on the placement accuracy of the chip-attach tool.

More recently, there has been a push to develop a method to grow a III-V laser directly on the silicon PIC, which would be fast and accurate like the bonding approach, but easier to integrate than bonding multiple III-V wafers [55]. The primary challenges to this approach are the lattice mismatch between Si and III-V materials, the size of the grown laser in comparison to the rest of the PIC stack, thermal budget from III-V growth, and the cost of a large scale III-V reactor. However, for many years, the electronics industry has shown interest in III-V on larger silicon wafers so, again, that investment and experience can be leveraged. A more in-depth discussion on laser integration options and trends is presented later in this chapter.

As described, the basic building blocks of a silicon PIC include passive waveguide devices, modulators, filters, switches, and detectors; BEOL wiring; and optical I/O via

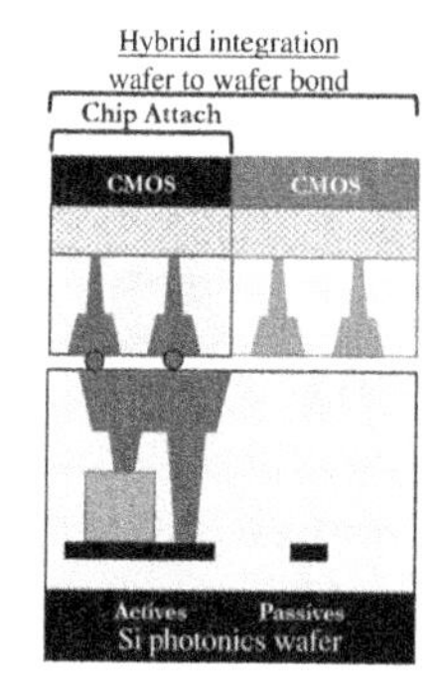

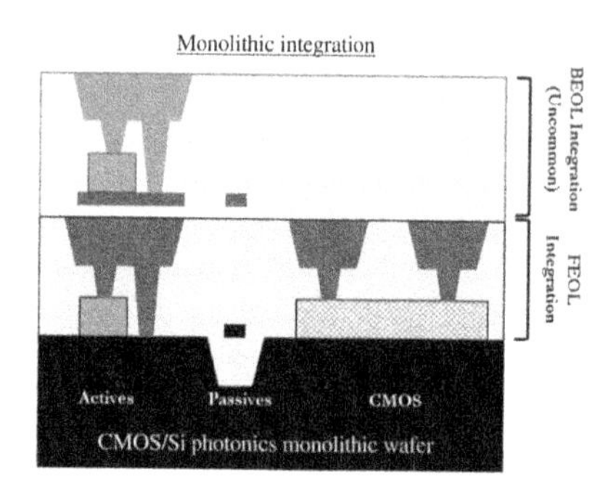

Figure 4.17 Left: Most silicon photonics MPW runs and research originating from academia use silicon-on-insulator (SOI) wafers to build the PICs. Driver circuits are either connected with wire bonds in a package, chip bonded, or wafer scale bonded to the surface of the PIC. This allows for separate optimization of the CMOS and PIC process flows. Right: Most semiconductor fabs introduced silicon photonic elements into standard CMOS process flows alongside the front-end-of-line (FEOL) or, less commonly, BEOL. This monolithic integration requires careful cooptimization of the electronics and photonics, which can present significant problems. Source: *Credit: AIM Photonics.*

fibers or perhaps an on-chip laser. They can be integrated onto a silicon substrate in two major ways: using SOI wafers or bulk silicon wafers (see Fig. 4.17) [42,55,59,60]. SOI wafers provide the best substrate for silicon waveguide fabrication, since the SOI is very high-quality single-crystal silicon, as long as the BOX is thick enough to isolate the propagating light from the handle wafer. Building PICs on bulk silicon wafers requires creating a deep trench to serve as the bottom cladding layer for the waveguides (since there is no BOX) and requires depositing not only SiN or SiON waveguides, but also the Si waveguide (as polycrystalline or amorphous silicon). While this can present additional challenges to PIC fabrication, the primary advantage is that the PIC can be co-fabricated with CMOS circuits to drive or control the photonic elements. For SOI PIC wafers, the CMOS must be flip chip or wafer scale bonded, which carries their own set of process and yield challenges. Considering cost, bulk wafers are much less expensive than SOI photonics wafers and have a much more established supply chain. Specifically, the thick BOX required for photonics applications makes wafer procurement even more difficult and expensive than CMOS-grade SOI wafers, which have thinner SOI and BOX layers.

While silicon photonics products are being commercialized at a rapid rate, especially for datacom applications, most product manufacturers are fabless [38,41,42,55,57,61]. Even large companies like Cisco, Juniper, and Huawei do not own or operate their own PIC factories, partially as a result of the consolidation of cutting-edge CMOS fabs. Instead, they leverage the technology offered by high volume 300 mm facilities, such as Intel, GlobalFoundries, or TSMC (Intel being the exception to "fabless," in that they design and build their own electronic and photonic ICs). There are also niche fabs that run more modest volumes, such as TowerJazz, STMicroelectronics, and Skorpios. Finally, there are numerous research facilities with an interest in silicon photonics, from small laboratory operations consisting of electron beam lithography writers and tabletop RIE tools to government funded fab-like facilities, such as AMF (formerly IME A*STAR) in Singapore, CEA-Leti in France, IMEC in Belgium, and the SUNY Polytechnic Institute in the United States (see Fig. 4.18). The high-volume fabs, and even the smaller commercial foundries, might have steep entry costs, while these research clusters are often specifically charged by their funding agencies with providing lower price points for small-to-medium-sized enterprises (SMEs), including academic researchers.

Besides collaborations and private/custom development for those that can afford it, each of the four research centers mentioned (plus commercial TowerJazz) offer MPW services and PDKs. A fab's PDK typically contains predefined device designs with known performance specifications to allow for designers to lay out functional chips on the first try, which helps reduce the number of cycles of learning required. A more detailed discussion of PDKs and how they are developed is included later in this chapter. In a Si MPW run, the cost for ordering reticles and processing wafers in a fab is shared by multiple (up to dozens) of customers or "riders" who split up the reticle field. A rider might be able to secure a small chip for a fee on the order of $10,000, whereas the cost of reticles alone could approach $100,000. The MPW programs are an extremely affordable way for a designer to test out an idea, a design, and a fab's capabilities. After prototyping on MPW runs, a designer could opt to start scaling to dedicated wafer runs in the same fab or eventually scale up to a production run in a foundry or high-volume fab. Alternatively, other riders, such as academic researchers, may opt to continue working on MPW runs and publish their results while keeping their costs low.

Figure 4.18 Publicly available silicon photonics multiproject wafer run (MPW) services (clockwise from top left): CEA-Leti and IMEC MPW runs are organized through the ePIXfab consortium; SUNY Polytechnic Institute's MPW runs are facilitated by AIM Photonics; AMF (formerly IME/A*STAR); and, most recently, TowerJazz. Source: *Credit: AIM Photonics.*

Besides cost, MPW runs allow SMEs, including academic researchers, the ability to sample industrial-grade PICs. While a typical university-based fabrication facility might be able to create simple passive and even active components for scientific inquiry, these types of facilities typically are not suited to create prototypes for commercial products, or for demonstrating a developers' novel approach that leads to a product. The larger research centers, foundries, and high-volume fabs operate with wafers of at least 200 mm in diameter. SUNY Poly's MPW program, and some of IMEC research and CEA-Leti's programs, take advantage of industry-leading 300 mm wafers. Using the larger wafers reduces the cost of a chip since fewer wafers are needed in order to yield a specific number of chips (assuming the program is leveraging an existing fab and tool set, which is more expensive to begin with). In addition to the use of automated material handling systems that improve contamination control, the equipment used to fabricate 300 mm wafers is often the tool vendors' best-of-breed platform, which benefits process stability, uniformity, and repeatability. Finally, 300 mm fabs are typically newer facilities where wafers are never handled by workers or ever exposed to the air outside of the tools themselves. In general, 300 mm fabs and 300 mm equipment have been developed and installed for truly leading-edge CMOS efforts (65 nm down to 5 nm or below process nodes), which have remarkably stringent specifications. A PIC technology operating on these kinds of tool sets has a built-in advantage in quality.

4.5 Future device integration

There now exist several different platforms for photonic integration, each with its own tradeoffs in performance and cost. As PICs become more sophisticated and complex, it becomes increasingly necessary to integrate the light source along with the rest of the PIC. In the case of communication PICs, this is necessitated by the need for multiple light sources at different wavelengths for DWDM. For other applications, the need for lower- cost packaging (reduced or no external fiber connections), lower-cost testing (using conventional electronic testers and complete self-test without external optical input or output), and more compact devices drive the integration of the light source. Lastly, an integrated light source allows SOAs. Here we consider the primary options for photonic integration of the light source. For GaAs or InP substrate PICs, the laser is easily integrated; that topic is covered in the section on InP PICs. For silicon substrates, the situation is more difficult since silicon, unlike GaAs or InP, is indirect bandgap and does not efficiently emit light. The options are (1) silicon- or Ge-based light sources or (2) III-V- or II-VI-based light sources on silicon. These are discussed in order.

4.5.1 Si- or Ge-based lasers on Si

Silicon lasers using Raman or Brillouin gain work well but still require an optical pump [62]. Here we confine our discussion to electrically pumped sources. Germanium, Ge-Si, and Ge-Si-Sn combinations have been proposed as possible light sources for monolithic silicon photonics [63]. Ge-Si lasers monolithically grown on silicon have been demonstrated using strained SiGe on Si to reduce the energy difference between direct and indirect emission, and also n-doping to increase the number of electrons in the gamma valley, which is advantageous for direct emission [64]. Presently this approach is not technically viable for commercial use, given that the required threshold current density is typically $>100\ \mathrm{kA/cm^2}$ [64,65], which is more than 1000 times higher than III-V-based lasers on silicon [66]. Low operating power is a key requirement for most applications, which precludes the use of indirect gap light emitters [67].

4.5.2 III-V-based lasers on Silicon

4.5.2.1 Heterogeneous bonded lasers

Heterogeneous integration is the process of combining III-V functionality with monolithic silicon photonics via wafer bonding of the two materials together (see Fig. 4.19). With this approach, one can take advantage of the advanced silicon-based processing in a monolithic silicon photonics process, and also utilize III-V materials for on-chip active optical functionalities such as gain, photodetection, and phase or amplitude modulation [68]. As an added benefit, integration of multiple diverse materials on a single PIC connected by a common silicon waveguide is possible. A number of R&D foundries, such as CEA-LETI and IMEC, are currently exploring this process. Intel has recently released products formed by this technology [69], while Juniper Networks [70] and HPE [71] are actively investing in the technology.

Heterogeneous integration presents opportunities for the realization of novel device architectures with enhanced performance that are otherwise unattainable using

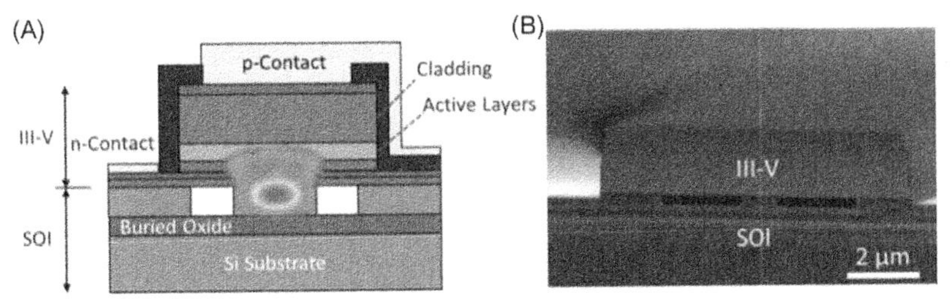

Figure 4.19 (A) A schematic diagram of heterogeneously integrated III-V laser on Si including the evanescent optical mode. (B) Micrograph of an etched III-V laser ridge bonded to a patterned silicon-on-insulator (SOI) substrate. Source: *Credit: Reference J.C. Norman, D. Jung, Y. Wan, J.E. Bowers, The future of quantum dot photonic integrated circuits, APL Photonics 3 (2018) 030901.*

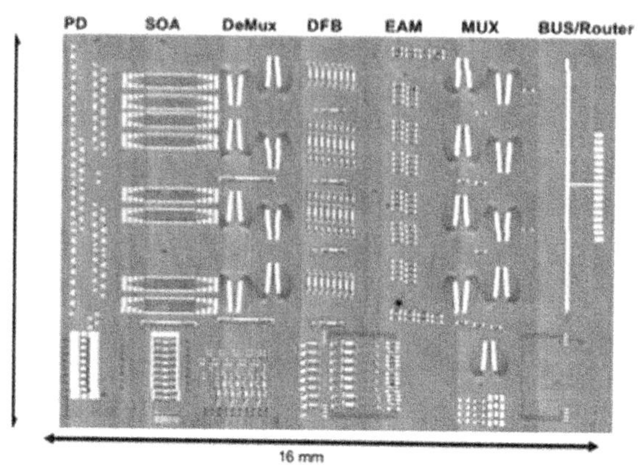

Figure 4.20 Optical photo of a 2.56 Tbps optical network on chip. Source: *Credit: Reference C. Zhang, S. Zhang, J.D. Peters, J.E. Bowers, 8 × 8 × 40 Gbps fully-integrated silicon photonic network-on-chip, Optica 3 (7) (2016) 785–786.*

a pure monolithic platform alone. Some prominent examples of this are the achievement of ultranarrow linewidths from heterogeneously integrated III-V laser coupled to a high-Q silicon cavity [72] and integrated waveguide photodiodes with record output power and bandwidth by engineering the hybrid III-V silicon mode to smooth out the absorption profile [73]. Compared to monolithic III-V PICs or lasers, heterogeneously integrated active components have no significant incident photon density on exposed III-V facets or mirrors, minimizing III-V facet related degradation mechanisms and lending to improved reliability of the active devices. The most complex heterogeneously integrated PIC demonstrated is an $8 \times 8 \times 40$ Gbps network-on-chip with 2.56 Tbps of data capacity shown in Fig. 4.20 [74]. A recent review of heterogeneous integration can be found in [75].

4.5.2.2 Epitaxially grown lasers

Epitaxial growth of III-V compound semiconductors such as GaAs and InP on silicon for the fabrication of photonic devices is a research area that has been pursued for many decades, motivated by the potential to integrate traditional CMOS logic with the superior RF and optical properties of III-Vs, and, more recently, by silicon photonics. This approach has also been referred to as "heteroepitaxy" or "direct (heteroepitaxial) growth on silicon" (Fig. 4.21). There are two potential variations of this approach for the fabrication of PICs, which can be distinguished by the choice of passive circuitry to be used. One option is to couple light generated in a III-V section to passive photonic circuitry in silicon on an SOI substrate. The other approach is to fabricate the entire PIC from epitaxially deposited III-V layers on silicon. In this case, III-V layers will be used for both active devices as well as passive circuitry (instead of silicon as in the former case). This would be analogous to the case of photonic

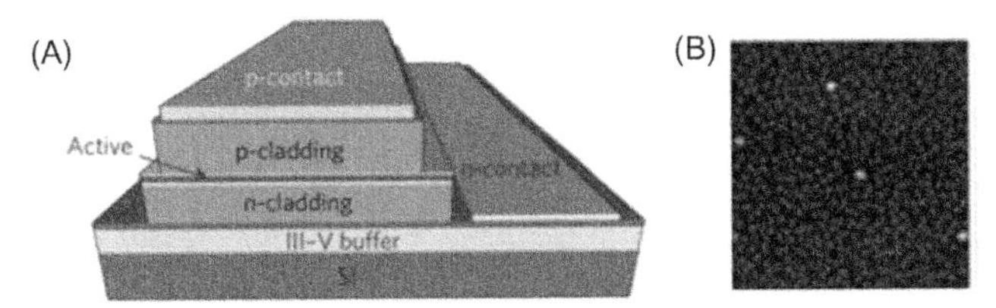

Figure 4.21 (A) Schematic diagram of quantum dot laser on silicon; (B) AFM image of quantum dot active region. Source: *Credit: Reference J.C. Norman, D. Jung, Y. Wan, J.E. Bowers, The future of quantum dot photonic integrated circuits, APL Photonics 3 (2018) 030901.*

integration on InP; however, the difference is that the III-V layers are grown on a much cheaper and larger silicon substrate. In this case, the purpose of the silicon substrate is only to serve as a low-cost substrate to facilitate scalable manufacturing with more modern and more advanced processing, packaging, and testing equipment. Multiple bandgaps may be obtained from selective area growth, regrowth steps, or intermixing, as is done in the case of photonic integration with InP. Integrating III-V materials on silicon by epitaxial growth is an active area of research both within academia as well as commercially by AIM, IMEC, IBM, and NTT, among others. Performance and reliability issues had previously relegated this field to the realm of academic research focused on the continual improvement of individual device performance. However, a number of recent breakthroughs have significantly de-risked the technology from a performance standpoint. These recent developments include using quantum dot (QD) active regions to realize significant improvement in device performance and reliability of the light source, as well as demonstration of other optoelectronic devices from epitaxial III-V on silicon material, such as modulators and photodetectors.

There are various approaches to monolithically grow light sources on silicon; however, by far the most successful approach has been using III-V QD lasers directly grown on silicon [66,67].

4.5.2.3 Quantum dot (QD) lasers on silicon

There are seven reasons that QD lasers have advantages over QW lasers grown on silicon. This are discussed in turn.

1. Better reliability and less sensitivity to defects
2. Lower diffusion length
3. Lower threshold
4. Higher temperature operation (220°C)
5. Lower linewidth enhancement factor
6. Narrower linewidth lasers
7. Lower reflection sensitivity

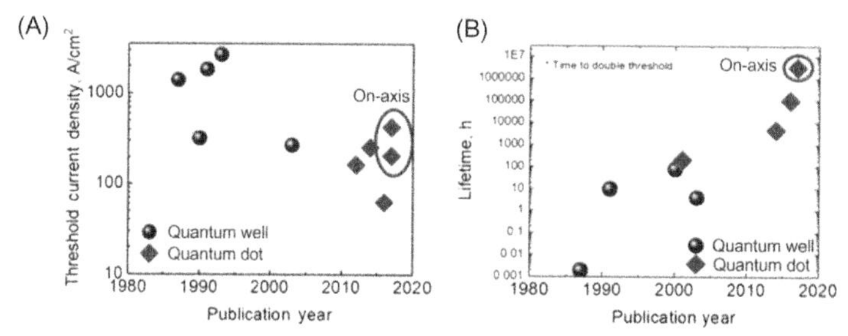

Figure 4.22 (A) Threshold current density and (B) device lifetime (either extrapolated or measured) for lasers on Si operating in the continuous wave mode. The distinction is noted between historical results on miscut Si substrates and recent results on CMOS compatible on-axis (001) Si. Source: *Credit: Reference J.C. Norman, D. Jung, Y. Wan, J.E. Bowers, The future of quantum dot photonic integrated circuits, APL Photonics 3 (2018) 030901.*

These advantages are related. For example, the shorter diffusion length results in a lower sensitivity to defects. If the threading dislocation density is $10^8/cm^2$, then a QW with a diffusion length of 1 μm is very sensitive to defects, while a QD layer with a particle density of $10^{10}/cm^2$ or higher and a diffusion length of 0.1 μm is much less sensitive to defects. This is illustrated in Fig. 4.22, where it can be seen that the lifetimes of QD lasers on Si are 10,000 longer than the best QW lasers on Si. Furthermore, the threshold current densities are now lower (Fig. 4.22A). The lower threshold allows lower current operation, which is also important for better reliability. The higher temperature operation is important for integration with electronics, where the substrate temperatures may be quite high. Presently the highest temperature lasers are QD lasers operating at 220°C [76], considerably higher than QW lasers, operating up to 140°C [76]. The lower linewidth enhancement factor [77] results in lower linewidth lasers, as well as lower chirp and lower reflection sensitivity [78].

One past problem with heteroepitaxially integrated lasers was that the process relied on the use of special silicon substrates that employed a slight miscut angle from the (001) orientation. The original reason for the use of miscut substrates is that the double-atomic steps induced from the miscut angle are favorable for the suppression of the formation of antiphase domains resulting from the growth of the polar III-V material on the non-polar silicon surface. As traditional CMOS fabs all use on-axis (001) oriented wafers, use of the miscut silicon wafers was deemed to be incompatible (mostly due to cost reasons) with traditional CMOS processes. Today, this is a non-issue, as the field has migrated to the use of CMOS-compatible standard on-axis silicon substrates using novel methods to solve the antiphase domain problem, obviating the need for miscut substrates (see circled data points in Fig. 4.22). These approaches range from direct nucleation using conventional two-step growth methods, the use of a thin GaP buffer layer, and patterned growth on exposed [111] v-groove facets of silicon [79]. The most recent results from on-axis devices—employing an GaP interlayer—

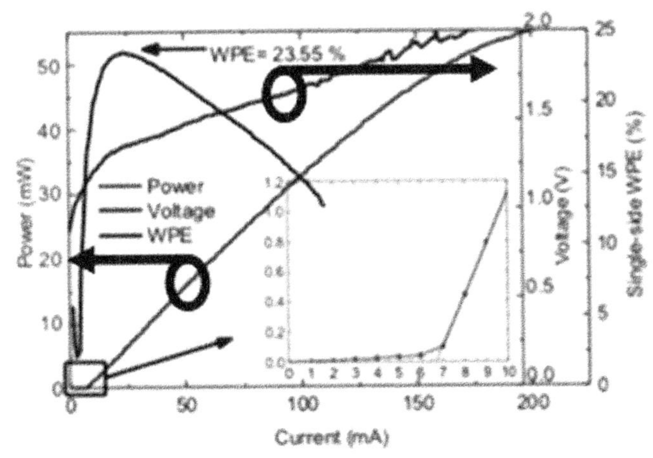

Figure 4.23 Light-current curve for quantum dot (QD) lasers along with voltage-current and wall-plug efficiency. Source: *Credit: Reference D. Jung, Z. Zhang, J. Norman, R. Herrick, M.J. Kennedy, P. Patel, et al., Highly reliable low threshold InAs quantum dot lasers on on-axis (001) Si with 87% injection efficiency, ACS Photonics 5 (3) (2017) 1094–1100.*

now show room temperature continuous-wave Fabry Perot threshold currents as low as 9.5 mA, single-facet output powers of 175 mW, ground state lasing up to 80ºC, and wall-plug efficiencies as high as 38.4%, all obtained with as-cleaved facets [66]. A typical example is shown in Fig. 4.23. Ultimately, this technology is agnostic to what the initial buffer approach is, and thus whichever approach yields the best material quality should be used.

These results are for epitaxial lasers on silicon. An important issue is how to integrate lasers with silicon waveguides on SOI substrates. Fig. 4.24 shows a number of approaches to integrate epitaxial QD material with silicon waveguides. The optimum approach depends on the application and on the results of further process development. The last approach (Fig. 4.24E) shows the use of all III-V devices on silicon, which is similar to a conventional InP PIC except that it is on a silicon substrate. In addition to the rapid progress for lasers on silicon, there is rapid progress being made in epitaxial InGaAs photodetectors on silicon [80], as well as InGaAsP modulators on silicon [81], so this approach may be viable as well.

4.6 Photonics mask making

Photomask pattern fidelity and manufacturability for curvilinear structures microfabrication techniques, used widely for the construction of ICs, are being used in an increasing variety of non-IC applications, such as the production of photonics and microelectromechanical systems (MEMS) devices. Unlike standard ICs with critical features composed from rectilinear structures, photonics and MEMs devices require

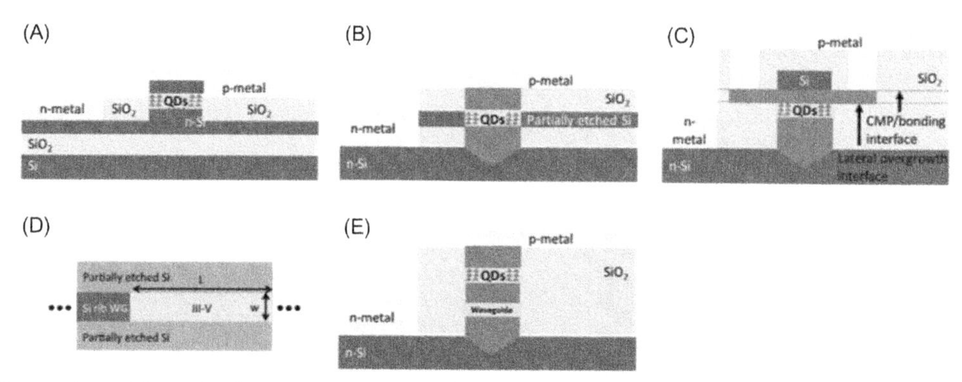

Figure 4.24 Schematic illustrations of potential III-V/Si integration schemes including embodiments using a silicon waveguide: (A) direct growth on silicon-on-insulator (SOI), (B) growth on patterned SOI from the handle wafer with butt-coupling to a Si device layer waveguide, and (C) growth on Si with a bonded Si waveguide on top of the III-V epi. (D) Top-down schematic of structures (A–C) showing III-V and Si waveguides. (E) An all III-V integration scheme where a separate waveguide layer is grown in the III-V layers for evanescent coupling. Source: *Credit: Reference J.C. Norman, D. Jung, Y. Wan, J.E. Bowers, The future of quantum dot photonic integrated circuits, APL Photonics 3 (2018) 030901.*

critical features with curved structures. Maintaining the fidelity of these curved patterns while simultaneously achieving manufacturability through the design, layout, photomask data preparation, and manufacturing process steps requires special care that takes into account the current characteristics of computer-aided design (CAD) software and manufacturing hardware that has been designed to produce rectilinear features.

The design of a curved feature may start with a mathematical formula of infinite accuracy but is rendered by a CAD layout tool in a format of finite accuracy. Two of the most often used formats, graphic database system (GDS) and open artwork system interchange standard (OASIS), share some common characteristics. Both of these formats use a fixed database unit on an orthogonal grid to define the coordinate space in which figures can be described. Coordinates are given as integer multiples of the database unit. The GDS format uses 32-bit integers for coordinates, which limits the database unit spanning a 200 mm substrate to approximately 0.05 nm, or about the Bohr radius of the hydrogen atom. The OASIS format can make use of 64-bit integers for coordinates, thereby decreasing the size of the minimum database unit spanning a 200 mm substrate to about 10^{-20} m, which is about five orders of magnitude smaller than the charge radius of the proton. Since the print address of photomask writers is much larger than either of these database unit limits, the limitations of the database unit size of the layout is not a practical limit for pattern fidelity.

The two basic types of figures in these layout formats are a path, which is a list of coordinates describing the center line of a fixed width line, and a polygon, which is a list of vertices for a closed boundary. Curved figures must be approximated by a series of straightedge segments, with coordinates defined on the fixed orthogonal grid based upon the database unit. The smaller the database unit, the closer the representation of the curved feature can be to the ideal curve; however, the more vertices used to describe the curved feature, the larger the layout database, which can have negative effects upon subsequent processing steps. To reduce the layout database size, curved features can be represented by a series of angled edge segments, rather than a stair-step arrangement of vertices at every possible database grid point. Ultimately, the design will be transferred to a photomask that will be used in an optical projection system. Optical image simulation techniques of this system can be used to find a reasonable balance between image fidelity and edge segment length. Ideally, this limit on the edge segment length should be imposed late in the process of transferring the design to the photomask, but current characteristics of available software tools require that this limit be imposed during the layout phase of the process.

The next step in the photomask production process requires the transformation of the design layout into formats compatible with mask writing tools, a process termed fracturing because the multisided paths and polygons are broken down into four-sided trapezoids. The fracture address unit is equal to or larger than the layout database unit, since the mask writing tools employ printing grids that are typically larger than the layout database unit. This fracture address unit describes an orthogonal array of points that are integer multiples of the fracture address. The vertices of the layout figures are mapped onto this fracture grid; a vertex not located at an integer multiple of the fracture address unit is moved to a nearby fracture grid point, potentially distorting the image from its desired shape.

It is also fairly common during the mask data preparation process to adjust figures to compensate for processing effects, either in the processing of the mask itself or in the subsequent processing of the image on the final substrate. To a first level of approximation, these process effects are often assumed to be isotropic and independent of local image context, in which case a simple algorithm to move all edges by the same amount is sufficient. However, to achieve the highest levels of image fidelity, more computationally intensive mask process compensation (MPC) techniques may have to be employed. It may be possible to reduce the computational load of MPC if the layout database contained the true curved design intent rather than an approximation of a series of angled edge segments. The OASIS format contains the extensible data element XGEOMETRY that could be used to describe curved features, but making use of this capability would also necessitate significant modifications to the CAD tools currently used in mask data preparation.

The writing of the photomask is done with either a raster scan or vector scan tool. On a raster scan laser beam tool, the fractured figures are rasterized into an orthogonal array of modulated pixels, with angled edges approximated by stair steps of exposed pixels. Interlacing the pixel arrays and modulating the beam exposure can achieve an effective print grid that is less than a single pixel size. Patterns with high figure counts can take longer to write, since the tool has more figures to rasterize. On a vector scan electron beam (VSB) tool, the fractured figures are partitioned into rectangular shots formed by illuminating a rectangular aperture with a beam of electrons, with angled edges approximated by stair steps of rectangular shots. To achieve the smooth edge required for many photonics applications, the size of the stair stepped shots must be small relative to the size of the feature, resulting in more shots and longer write times than purely orthogonal designs. For VSB tools, a model-based approach to fracture could result in an arrangement of shots that reproduces the desired feature outline while using fewer shots than the current stair-step approach. For raster scan laser tools, an approach to fracture that implements a minimum edge segment length for curved or angled edges could help reduce figure counts, potentially resulting in shorter photomask write times.

4.7 Photonic packaging

There have been significant developments to realize cost-effective PIC fabrication processes, but there now exists a packaging bottleneck that is impeding the growth of these markets. Key challenges to be overcome include low-cost optical fiber and microoptical packaging processes that provide high coupling efficiencies (e.g., <1 dB loss per optical interface), the ability to package large numbers of optical channels per chip (e.g., >20 channels per optical interface), the integration of different PIC platforms (e.g., Si, InP and SiN), and the hybrid and heterogeneous integration of photonic and electronic devices in a common package. Challenges also remain in providing high-speed electrical packaging as required for communication systems with bandwidth requirements of >100 G. In addition, the drive to develop highly integrated photonic and electronic components in a single package adds to the difficulty of efficiently managing thermal loads. Critically, the technologies developed to overcome all these challenges must be implemented in high-volume manufacturing environments, using cost-effective materials and packaged using equipment that, where possible, operates using automated machine vision (passive) alignment processes.

Existing photonic packaging processes typically rely on component-level packaging, where optical and electrical connections are assembled after the photonic device has already been placed in a mechanical package. This serial type process flow has limited throughput, with scale-up in manufacturing depending directly on the number of packaging machines. Historically, such serial processes have been acceptable for low-volume high-value applications, such as fiber-optic telecommunication, that require extremely high reliability and long lifetimes. The typical packaging for photonic devices involve (1) the photonic device assembled in the package, (2) electrical connection using gold ribbon wire bonding, (3) active alignment of the optical fiber

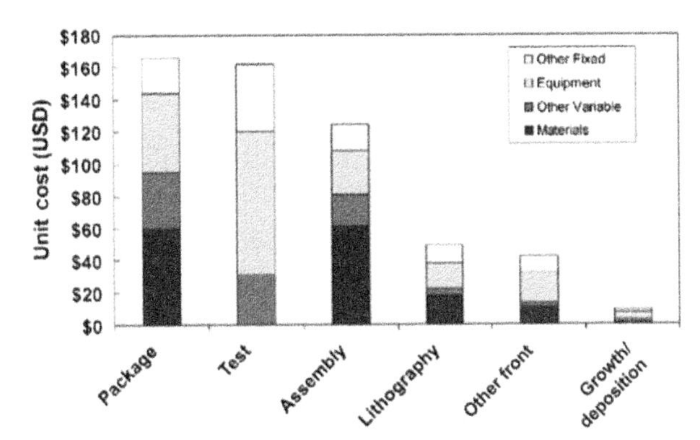

Figure 4.25 Cost breakdown for an integrated InP photonic device, showing the majority of product cost can be attributed to assembly, packaging, and test. Source: *Credit: Reference E.R.H. Fuchs, E. J. Bruce, R.J. Ram, R.E. Kirchain, Process-based cost modeling of photonics manufacture: the cost competitiveness of monolithic integration of a 1550-nm DFB laser and an electroabsorptive modulator on an InP platform, IEEE J. Lightwave Technol. 24 (8) (2006) 3175–3186* .

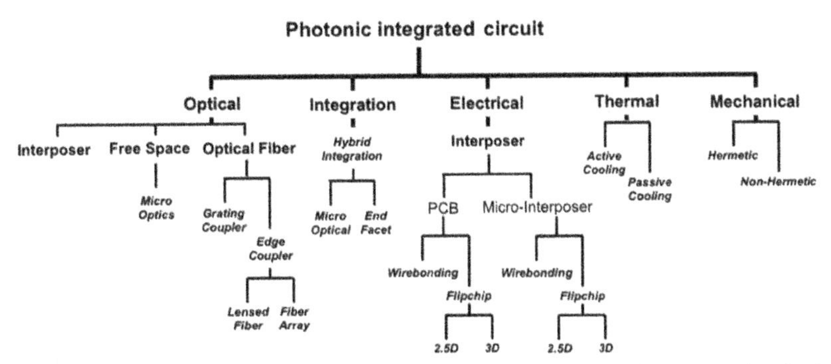

Figure 4.26 Organization chart providing a breakdown of the key packaging technologies, including optical, electrical, thermal, and mechanical aspects. Source: *Credit: Tyndall Institute.*

to minimize insertion loss, and (4) hermetic sealing of the mechanical housing in an inert atmosphere—although non-hermetic packaging is becoming more acceptable for many applications, such as data centers. However, this serial packaging approach cannot meet the all of the cost and volume demands of emerging mass markets in photonics. These packaging processes can account for the majority of the product manufacturing cost [82], making photonic device packaging prohibitively expensive for many emerging low-cost applications (see Fig. 4.25).

4.7.1 Key photonic integrated circuit packaging technologies

PIC packaging requires a wide variety of technologies, which can result in a complex manufacturing process. Fig. 4.26 shows a high-level organization chart that provides a breakdown of the key PIC packaging technologies.

The following sections provide a summary of the critical features for each of the key PIC packaging technologies.

4.7.1.1 Optical packaging

Optical packaging typically involves bonding of optical fibers to the PIC device, where single or fiber arrays can be used. The conventional packaging process uses an active alignment procedure where the coupled optical power is continuously measured and maximized, and the fiber is fixed in position using a laser welding or UV epoxy curing process; sub-micro alignment tolerances are required. This process is usually performed at the individual component or package level and can suffer from throughput limitations, with cycle times on the order of minutes to tens of minutes per package, depending on the complexity of the photonic device to the packaged.

The main challenge associated with coupling light from an optical fiber to a PIC waveguide is the large mismatch between the mode sizes and refractive indices of the two systems. In a single-mode fiber (e.g., SMF28), the mean field diameter (MFD) for 1550 nm light is approximately 10 μm, while the cross section of the corresponding strip SOI-waveguide is usually just 220 nm × 450 nm. There are two distinct approaches to fiber coupling: (1) edge coupling and (2) grating coupling. As its name suggests, edge coupling involves the transfer of light between the edge of the Si-PIC and the optical fiber. Edge coupling is already an industry standard for the optical packaging of many discrete III-V laser devices [83]. The alternative to edge coupling is grating coupling, where a sub-wavelength periodic structure is etched into the SOI platform, creating a condition that diffractively couples light from a near-normally incident fiber-mode into the SOI-waveguide [84]. Ultimately, the best fiber coupling solution for a given PIC is strongly application- and cost-dependent.

4.7.1.2 Fiber edge coupling

The most convenient means of coupling light from an edge-emitting photonic device into an optical fiber is edge coupling and is a well-established approach for the commercial packaging of laser chips. However, in the field of Si-photonics, edge coupling has been less widely adopted, despite being able to deliver broadband, polarization-agnostic insertion losses of better than −1 dB, after careful alignment [85]. A typical edge-coupler on a Si-PIC consists of a relatively long (typically 300 μm) inverted taper, on which a polymer- or nitride-based mode adaptor has been added using a post-processing step [86]. Introducing an edge coupler onto a Si-PIC tends to increase the device processing costs, as it usually calls for the accurate etching and dicing of the PIC facets. Typical MFD for edge couplers on Si-PICs are on the order of 3 μm, although larger MFDs are currently in development to better match the MFD of standard optical fibers. Edge coupling schemes, in which grooves are etched directly into the SI-PIC to passively align the fibers with respect to the SOI-waveguides, have

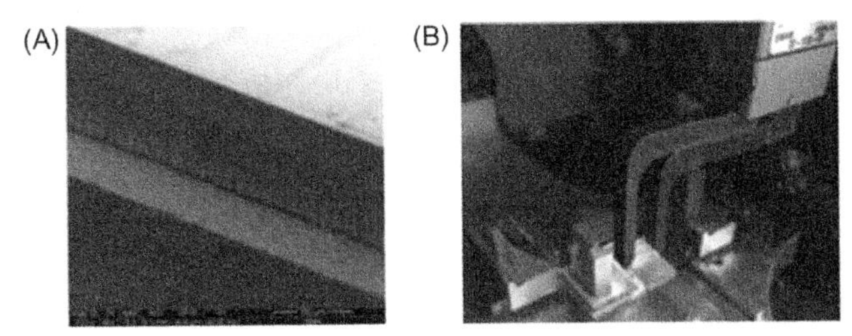

Figure 4.27 (A) Etched facet of silicon photonic devices with diced ledge to enable edge coupling to a silicon photonic device; (B) coupling of fiber array to silicon photonic device using an active alignment process. Source: *Credit: (A) AIM Photonics; (B) Tyndall Institute.*

been demonstrated (Fig. 4.27A) [87]. However, this approach also adds to the fabrication costs and has the additional drawback of populating a significant fraction of the usable Si-PIC footprint with V-grooves. To date, edge coupling is usually made to/from a lensed fiber, with a focused spot size of the order of 3 μm, to reduce insertion loses by improving the modal overlap with the mode adapter. The 1 dB alignment tolerance for a typical edge coupling is sub-micron (approximately ± 500 nm), and so requires careful active-alignment. Such tight alignment tolerances make edge coupling to a fiber-array extremely challenging, and so the majority of optical-packaging for edge couplers have been restricted to single-fiber channel applications (Fig. 4.27B). Laser welding is the preferred means of securing the lensed fiber (mounted in a metallic ferrule) with respect to the Si-PIC after alignment. Laser welds do not suffer from the small alignment drift that sometimes occurs in epoxy bonds due to age or environmental effects. These drifts are not relevant in grating coupler schemes because of their more relaxed alignment tolerances, but they cannot be ignored with the ± 500 nm 1 dB alignment tolerance of edge couplers. For similar reasons, it is recommended that modules with edge couplers are mounted in a Kovar package, to minimize the effects of thermal expansion and contraction on the optical alignment. A recent development in the area of silicon photonics has been the introduction of high numerical aperture (NA) optical fibers with reduced MFD (e.g., 3–4 μm) [88], matching the MFD of the Si-PIC waveguide. A key benefit of these high NA fibers is the ability to produce arrays, similar to standard fiber arrays that can be attached to the Si-PIC using UV curable epoxies. These specialty fibers can be spliced to standard fibers with low insertion loss [89].

4.7.1.3 Fiber grating coupling

For optimum coupling efficiency, the fiber mode should be near-normally incident on the grating structure. However, if the fiber itself is optically packaged at near-normal incidence—that is, in the "pigtail" geometry—then the overall device becomes bulky and delicate. To address this issue, a "quasi-planar" approach has been developed, in which the fiber lies on the surface of the PIC, with a 40 degrees polished facet providing a total internal reflection (TIR) condition that directs the fiber mode onto the grating coupler at the correct angle of 10 degrees [90]. After active alignment, index-matching UV-curable epoxy is used to bond the fiber to the surface of the Si-PIC, exploiting capillary action and surface tension effects to ensure an even application of the adhesive. To ensure the TIR condition at the polished facet is not disturbed, the epoxy cannot be allowed to flow onto the facet. The quasi-planar approach has an easily manageable 1 dB "roll" tolerance of ± 2.5 degrees, in addition to the ± 2.5 μm 1 dB lateral alignment tolerance, and offers an insertion loss that is on a par with that of pigtail coupling.

4.7.1.4 Microoptical coupling

Wafer-level packaging has the ability to overcome the throughput limitations of serial packaging processes, where parallel alignment and bonding processes can achieve significantly higher throughputs: seconds per device as opposed to minutes. Eliminating the need to bond fibers to the photonic chip through the use of microoptics assembled on-wafer offers one solution. This style of optical interconnect is ideal for low-cost photonic packages, such as pluggable transceivers for data centers or disposable biosensors for medical diagnostics [91]. This type of high-volume and low-cost interconnect is expected become a viable alternative to the dominant fixed-fiber design.

4.7.1.5 Evanescent coupling

In this design, light is coupled between the optical fiber and PIC waveguide via an optical interposer, much like how an electrical interposer is used to interface between adjacent electronic devices. Structuring the PIC at the wafer level to include an evanescent coupling element is relatively straightforward, and when bonded on top of the optical interposer via a flip chip process, enables light to efficiently transfer between the PIC and interposer waveguides [92]. An added advantage of the evanescent technique is the ability to form an optical connection at any point on the PIC surface, rather than at the PIC edge or facet, as with standard fiber coupling. Optical interposers can be fabricated in glass, with embedded waveguides interfacing between the optical fibers and PIC [93]. Although at a relatively early stage of development, both microoptics and evanescent coupling indicate the benefits of using wafer-level processes to ease the burden of optical packaging.

4.7.2 Electrical packaging

Electrical packaging for photonics can be the largest contributor to material costs within the package. As the demand for higher operating frequencies rises, there is a need to integrate the electronics closer to the PIC. In some cases, electronic and photonic functionality can be integrated in the same device. However, hybrid

Figure 4.28 Planar fiber array coupled to gratings on a silicon photonic device. The image also shows flip chip of an electronic integrated circuit on the photonic device. Source: *Credit: Reference M. Rakowski, M. Pantouvaki, P. De Heyn, P. Verheyen, M. Ingels, C. Hongtao, et al., A 4x20Gb/s WDM ring-based hybrid CMOS silicon photonics transceiver, ISSCC Conference, 2015. pp. 408–409 [94].*

integration using 2.5 and 3D integration processes tend to be the preferred technology, as photonic and electronic devices can be fabricated from different technology nodes, resulting in a more cost-effective solution.

2.5D integration involves placing both photonic and electronic devices side by side on a common electrical carrier or interposer. The interposer should be selected from a material that provides good electrical insulation capable of supporting high-frequency electrical signal transmission lines and offering high thermal conductivity. Ceramics, such as AlN, meet all these requirements and are commonly used electrical interposers. To further improve high-frequency signal performance, both photonic and electrical devices can be flip chip assembled on the interposer using stud bumps, solder spheres, or plated copper pillar capped with solder, such as AgSnCu or just Sn. 3D integration involves stacking the photonic and electronic devices on top of each other (see Fig. 4.28). The main advantages of this approach are further improvements in high-frequency performance and a smaller package footprint. However, depending on the power requirements of both devices, thermal issues can arise, especially from the electronic devices, which can impact on the performance of the photonic device, affecting sensitive functions such as switches and multiplexer and demultiplexer elements. A recent development in interposer technology is the combination of both electrical and optical signal routing in a common electro-optic interposer. For example, if a glass interposer is used, optical waveguides can be defined in the glass substrate using processes such as femtosecond laser inscription or ion diffusion to modify the local refractive index [93]. This type of dual functioning interposer is under development by a number of academic and industrial research groups and is expected to become a commonly used technology for PICs that have complex electrical and optical functionality.

4.7.2.1 Thermomechanical packaging

Photonic elements, such as integrated semiconductor optical amplifiers and microring resonators, exhibit a strong temperature dependence. It is not uncommon for a temperature drift of 10°C to drive a PIC outside of its operational tolerances. For example, the channel spacing in DWDM systems is approximately 1 nm, and the tuning coefficient of a typical μRing resonator on a Si-PIC is 0.1 nm/K, which means that a temperature change of just 10°C can result in channel hopping. Consequently, there is generally a need for thermal control of the PIC [95]. This can be achieved actively using a thermoelectric cooler (TEC), or passively by ensuring a relatively large thermal mass is in contact with the PIC, providing an effective heat dissipation path. TECs are highly inefficient, typically exhibiting 10%–15% efficiency, and consume large amounts of energy, thus making them unsuitable for low-cost power-efficient applications, such as photonic transceiver modules. The alternative is to assemble the PIC on a thermally conductive substrate, such as AlN, and mount this subassembly on or within a metal housing. Future improvements to passive cooling depend on optimization of module designs (modeling and simulation) and access to new housing and bonding materials (e.g., plastic enclosures and graphene-based epoxies) to reduce the cost of the packaged photonic modules while increasing their thermal performance. Active cooling involves inserting a TEC between the PIC and module housing (which brings an unwanted increase the thermal resistance) and adding both a thermistor and control circuit to power the TEC, which then allows the PIC temperature to be set independently of the module housing. Future improvements to active cooling depend on improving the coefficient of performance (CoP) of the TEC through better footprint-matching to the PIC, and optimized module design and adopting new materials, as well as hybridization with passive-cooling strategies to reduce the power budget for driving the TEC.

4.7.3 Photonic integrated circuit packaging design rules and standards

The need to create standardized design rules for photonic packaging is slowly gaining recognition in the PIC community, driven by the dual advantages of removing a design burden from researchers/engineers and streamlining the design-to-device process. Developers of photonic packaging solutions are working with industry partners to establish packaging/processing design kits (PDKs) and rules (PDRs) to help users ensure that their Si-PICs and photonic devices are compatible with reliable and cost-effective best practices in photonic packaging. The implementation of these packaging design rules will be especially beneficial to new entrants to Si-photonics space, by helping avoid costly design mistakes that could result in a newly designed Si-PIC being impossible to fully package. Critical PDR parameters include the pitch of grating arrays on the Si-PIC, the pitch of fibers in the fiber-arrays, the pitch of DC and RF bond pads, the metallization type of bond pads, and the location of coupling elements (grating couplers and edge couples) with respect to the electrical interconnects [96]. Although currently at an early stage of development, it is expected that these packaging design rules will become more widely adopted, especially at the layout phase of the chip design process. To increase their ease of use, the PDKs are being implemented in software layout tools and templates. This will allow designers to easily access and implement the packaging rules, in conjunction with the photonic elements available from different Si-photonic foundries.

Integration of packaging software design tools is also required in the future. Current design process flows typically involve separate optical, electrical, thermal, and mechanical phases. Future developments will require the seamless integration of these different tools, enabling designers to view how changes in one aspect of package affect others. Finally, as packaging typically consumes the majority of PIC module costs, accurate cost modeling software needs to be developed. This software tool should also be linked with design tools, providing designers with the ability to assess the cost implications of various package design changes.

4.8 Silicon photonics integrated circuit process design kit

Integrated electronics enable rapid innovation in a wide variety of application areas and replaced discrete solutions by leveraging CMOS processes on silicon wafers. Standardized CMOS processes and material choice lead to development of consistent transistors that can be lithographically defined over an entire reticle. A single reticle can host billions of transistors, which will take years to design and lay out without the aid of PDKs and electronic design automation (EDA) flows. The PDKs abstract processes provide building blocks that represent circuit functionality and include rules to lay out large-scale integrated circuits with the help of EDA tools. The abstraction and rules enable the separation of circuit design from underlying complicated semiconductor physics and process development that typically requires specific expertise and education. This reduced barrier to entry for design allows for an increased number of large-scale IC designs. The design reuse and hierarchical design proliferate semiconductor intellectual property (IP) market and applications to achieve mass production. The mass production is enabled by stamping the design reticle across the wafer and increasing wafer diameter to produce large quantities of chips. The revenue gained out of the sales of these chips support the research and development of advanced CMOS processes, resulting in a cycle that provides cost, size, and power savings.

Integrated optics have the same promise with integrated electronics. However, the major CMOS foundries did not provide a silicon photonics design and fabrication service until recently. The omission of the silicon photonics service did not stem from the lack of commercial interest to large-scale integrated optics. For instance, the communications industry was being served by alternative integrated optics solutions for years. The main reason for this omission was the lack of separation of PIC design from the process development. Although silicon photonics have a late start in CMOS foundries, fabrication of photonics uses the mature CMOS processes and materials and share some design rules. Today, this enables foundries, design houses, and EPDA tools to adapt and develop silicon photonics PDKs (e.g., design guide and component library) at a rapid pace [97–99]. Silicon photonics PDKs combined with EPDA tools enable accurate modeling and predictable performance of large-scale integrated photonics. The reuse of verified component libraries from these PDKs enables hierarchical designs, which leads toward a photonics semiconductor IP market. This market will enable quick turn-around commercial applications and mass production. The mass production similar to integrated electronics industry will provide the necessary revenue for research and development of improved PDKs and silicon photonic processes.

4.8.1 Silicon photonics process design kit

4.8.1.1 Process design kit hierarchy

A typical silicon photonics PDK provides primitive elements and component libraries that are essential to the end user. The primitive elements provide sufficient information about the foundry process to design, lay out, and tape out. These elements, shown in Fig. 4.29, are the design guide, technology file (tech file), and DRC deck. The design guide can include mask layer names, types (negative or positive), opacities, thicknesses, purposes, stress gradients, and tolerances. The technology file provides a layer map that transfers the design intent layers to fabrication-ready layer numbers and shows interactions between these layers. This layer map is used to convert the designs to a graphic database system (GDSII) or OASIS file formats that are recognized by the DRC deck. After passing the DRC, designs will be ready to be submitted to a broker or a foundry for mask generation. This submission process is typically referred to as the tapeout.

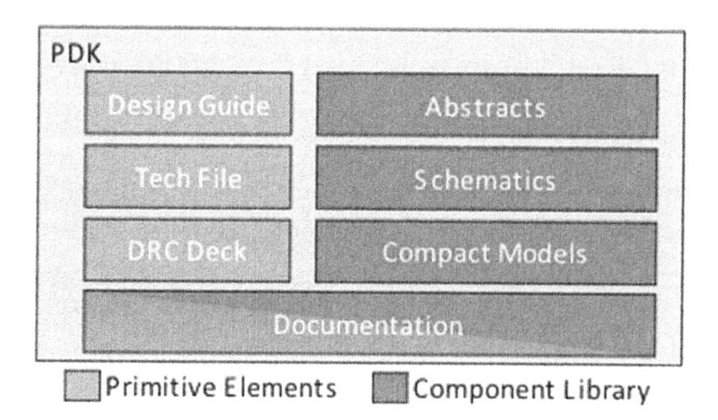

Figure 4.29 Hierarchy of a typical process design kit primitive elements and component library. Source: *Credit: Analog Photonics Analog Photonics, Process development kit, <http://www.analog-photonics.com/pdk/>, 2017 (accessed 07.08.18).*

Although primitive elements are sufficient for a custom-design tapeout, the custom design requires expertise in process, design, test, and layout. The custom design typically takes multiple fabrication runs (iterations) to meet the specifications, which increases the cost of development and slows down the development of a product. The added cost and development time can lead to a loss of competitive advantages that came with the custom design. Instead, a PDK component library aided design is preferred to accurately predict PIC behavior and iterate at a rapid pace, which reduces the risk and time to market. In addition, the overall cost of licensing a component library can be less than the custom development of those components.

The PDK component library, shown in Fig. 4.29, typically includes the abstracts, schematics, and compact models. The abstracts allow the PDK developer to obstruct parts of true geometries that contain design IP while enabling the user to interact with the designated optical and electrical ports of components. The abstracts are replaced with true geometries for fabrication-ready files at a broker or a foundry. The schematics and compact models of these components are used to predict the performance of the design, including the custom designed elements. If PDK component library provides state-of-the-art verifiable performance and high yield, it will attract customers for design in that silicon photonics foundry services.

A quality PDK also includes documentations of both primitive elements and component library. These documentations guide and inform the designers about PDK hierarchy, changes, design methodologies, supported EPDA tools, component library performances, application notes, and instructions (installation, licensing, and tapeout of the PDK).

4.8.1.2 Development cycle of a process design kit component library

A typical development cycle for PDK component library is shown in Fig. 4.30. Initially, the focus of component library should be determined. For instance, electronic-photonic functionalities that address the needs of sensors and communications are the focus of Analog Photonics-SUNY Poly (APSUNY) PDK [97]. The next step is to select design geometries to achieve these functionalities. There are typically multiple design geometries to achieve the same functionality. The design geometries that provide high performance within process tolerances and variations are selected. These designs are then turned into true (e.g., actual) geometries. After consideration of layout, testing, and routing, the true geometries are transformed into abstracts and schematics with electrical and optical ports (Fig. 4.30 inset, abstract). The true geometries and variations of the true geometries for optimization are placed in a layout and taped out for fabrication at the foundry (Fig. 4.30 inset, layout). Upon receiving the fabricated test sites, the components are measured and compared against the simulations. If there is good agreement between the simulations and experimental results of a component, that component is considered to be verified (see Fig. 4.30-inset, verification). Each

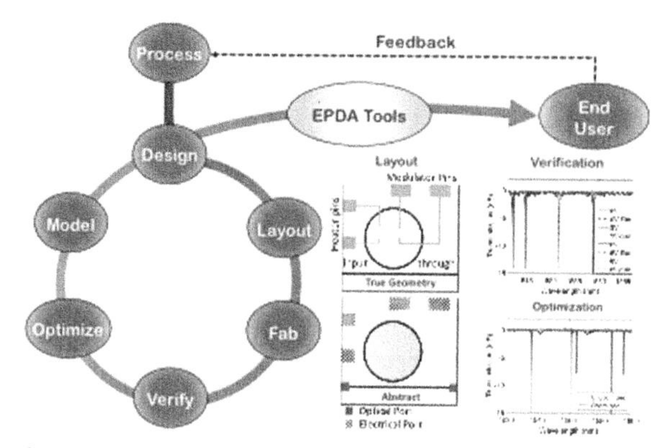

Figure 4.30 A typical process design kit component library development cycle. Source: *Credit: Analog Photonics Analog Photonics, process development kit, <http://www.analogphotonics.com/pdk/>, 2017 (accessed 07.08.18).*

component in the library typically needs to go through the development cycle several times to be verified. The verified results will be feedback to simulations to achieve not only verified performance but also optimized performance specifications (see Fig. 4.30-inset, optimization). After this optimization stage, the components are converted to compact models. The compact models, schematics, and abstracts of the component library are released to EPDA partners. After PDK implementation of EPDA partners, the component library is released to the end user.

Throughout the development cycle, end user feedback is crucial, which can point out inconsistencies, tailor the focus and component specifications, and update the process for the next development cycle. For example, APSUNY PDK component library currently provides over 25 active and 25 passive components and is updated four times a year, expanding functionalities while maturing performance since 2016 [97].

4.8.1.3 Organization of a process design kit component library

Fig. 4.31A shows a typical photonic PDK organization that consists of interface, waveguiding, passive, and active component libraries. The optical, E-O, or optical-electric (O-E) component testing requires interfaces where an external optical power supply (i.e., laser) and electrical instrumentation are connected to optical I/O couplers and electrical pads. The couplers facilitate a bridge between on-chip waveguides and fibers, laser chips, or other PICs and components. However, there are a couple important differences between optics and electronics interfaces. First, the couplers require tight alignment tolerances to efficiently transfer light from the fibers (see Fig. 4.31B). Second, the couplers, and especially on-chip single mode waveguides, support two orthogonal polarization that have different propagation constants. This polarization dependence stems from the fact the on-chip waveguides are wide and thin.

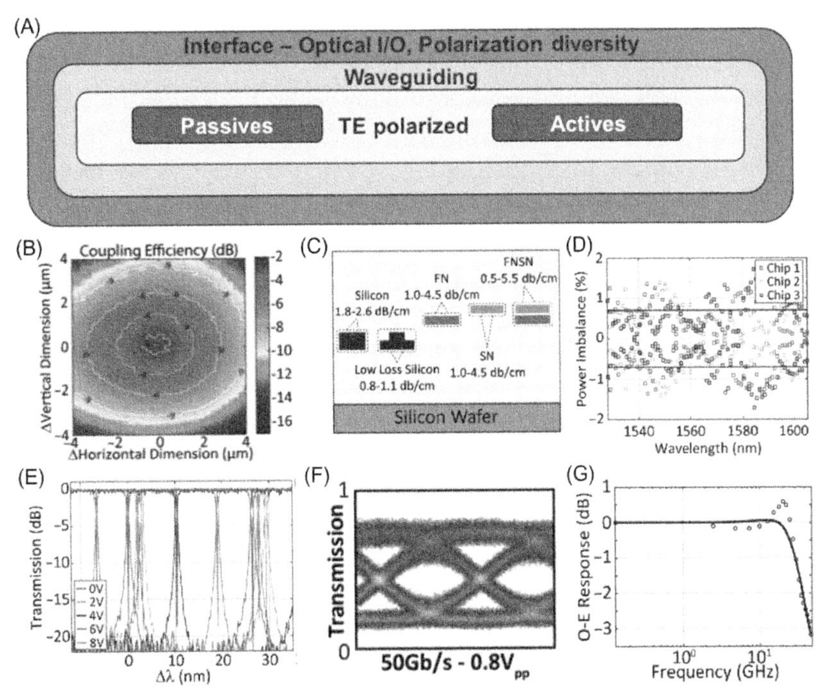

Figure 4.31 (A) A typical photonic PDK organization; (B) APSUNY PDK SMF-28e fiber to edge coupler coupling efficiency and alignment tolerance from [97]; (C) APSUNY PDK waveguide geometries and propagation losses from [97]; (D) power imbalance of APSUNY PDK 3 dB splitter ports; (E) reconfigurable response of the APSUNY PDK filter; (F) 50 Gbps non-return to zero (NRZ) on off keying transmission eye diagrams from the APSUNY PDK MZM; (G) O-E response of APSUNY PDK photodetector with a 50 GHz bandwidth. Source: *Credit: Analog Photonics E. Timurdogan, Z. Su, C.V. Poulton, M.J. Byrd, S. Xin, R. Shiue, et al., AIM process design kit (AIMPDKv2.0): silicon photonics passive and active component libraries on a 300mm wafer, in: Opt. Fiber Comm. Conf. OSA Technical Digest, 2018, paper M3F.1.*

The waveguide geometry provides different boundary conditions for TE and TM polarization modes. In a wide waveguide, TE polarized light is highly confined within silicon and favored due to low crosstalk and bend losses. However, the polarization coming out of an input fiber is typically scrambled and not always aligned with on-chip TE polarization mode. Therefore, a polarization diversity scheme [100] where input TE and TM polarized modes are split and TM polarized mode is rotated to TE polarized mode, enables both polarization to be represented in the chip. At the output interface, the TE and TM modes are combined back together with a polarization splitter and rotator and passed to the fiber with a coupler. The couplers, polarization diversity components (i.e., polarization multiplexers, polarization splitters, and rotators), and electrical pads form the interface library of a photonic PDK. The interface library ensures all the modes are converted to TE modes inside the chip, which enables waveguides and passive and active component libraries to be designed and tested for a single-mode operation, saving testing time and providing maximum performance.

Optical waveguides are utilized for routing light from the interface to other components. In a typical silicon photonics foundry process, silicon (i.e., crystalline, poly, amorphous), nitride, and germanium waveguides in an oxide background are common. The reason for many different waveguide materials is the difference between refractive index, electrical conductivity, roughness, power handling, and optical transparency. Fig. 4.31C shows the waveguide library in APSUNY PDK that consists of high confinement and low loss silicon waveguides, and first, second, and dual silicon nitride waveguides [97]. In addition to different waveguide geometries, seamless transitions between these layers are essential to combine advantages of silicon and silicon nitride waveguides.

The passive and active silicon photonic component libraries are connected to waveguides inside the chip to provide single or multiple wavelength manipulation of light using wavelength division multiplexed (WDM) systems. Passive libraries provide complex routing with layer transitions and waveguide crossings, and enable optical phase and power distribution with splitters and wavelength division multiplexers. A key parameter for passive splitters is to maintain the desired split ratio across a broad wavelength range. Fig. 4.31D shows the deviation from the desired split ratio of a 50%/50% (3 dB) splitter in APSUNY PDK. Active libraries use thermo-optic and E-O effects to manipulate light. This library imprints electrical signals to optical signals with switches and modulators (Fig. 4.31E), converts optical signals to electrical signals with detectors (Fig. 4.31F), and removes unwanted signals using reconfigurable filters (Fig. 4.31G). A high-performance modulator is required to maintain low-voltage drive while achieving low optical losses and high electrical bandwidths. For instance, modulators in APSUNY PDK now support data rates up to 50 Gbps with less than 1 Vpp drive, minimizing power consumption.

4.8.1.4 Verification of a process design kit component library

The verification of specifications and compatibility of these specs to existing or emerging standards are critical for the PDK component library. Although performance of the interface library can be measured in isolation, the rest of the component libraries require the addition of interfaces, waveguides, passive terminations or splitters, and active detectors or phase shifters for characterization. Therefore, characterization of these test structures require de-embedding of all the additional parts in the measurement to extract isolated PDK performance. Although this process is similar to electronics testing, there are three major difficulties. First, the deembedding macros can occupy a large footprint, since the size of PICs is quite large compared to transistors. This increases the fabrication and design cost of development. Second, all responses are analog rather than digital in photonics and phase and amplitude of the light is strongly dependent to fabrication and environmental variations (width, height, implant conditions, temperature). Due to this factor, the sensitivity of de-embedding parameters to these variations is large compared to electronics. Third, component response is polarization dependent. If the goal is to measure polarization-dependent response, then de-embedding parameters need to be measured for each polarization. Due to these three reasons, parameter extractions are typically performed with more than one measurement technique and results of these different techniques are compared. Each measurement technique comes with certain assumptions and requires different setup and conditions, leading to further uncertainty of the measurement results.

The following paragraph summarizes current measurement techniques. To characterize broadband devices, a wavelength-dependent loss (WDL), insertion loss (IL), and return loss (RL) test is performed. For phase-dependent devices, free spectral range is measured to extract group index and estimate the refractive index. For polarization-dependent devices, a polarization-dependent loss (PDL) measurement is performed using optical vector analyzer. For devices with multiple ports, WDL is performed on multiple ports to extract optical crosstalk. For extracting split ratio, extinction ratio (ER) of a Mach-Zehnder interferometer is measured. For detectors, the dark current, responsivity, S11 (reflection), impedance, $S12_{OE}$ (bandwidth), and saturation power are measured. For modulators and switches, the E-O phase shift ($V_{\pi}L$), loss, rise and fall times, S11 (reflection), and $S12_{EO}$ (bandwidth) are measured. To verify communication signal integrity, extinction ratio, insertion loss, transmission penalty, and pattern dependence are measured. To measure spurious free dynamic range (SFDR) and link gain of analog links, compression point and linearity of the RF to optical and optical to RF conversions are tested by monitoring harmonic distortion. Fortunately, PDK component library developers perform all these experiments and de-embed all components except the device under test (DUT) and compare with compact models for verification. Analog Photonics performs these verifications prior to any PDK release and then provides accurate models for every new component [99]. These models remove the guesswork and reduce risk and cost of developing a PIC for the end user.

In this section, the need for silicon photonics PDKs was addressed. The reuse of CMOS processes and photonic design is accelerating the adoption of silicon photonics while applications spaces, such as sensors and communications, are increasing the demand for increased bandwidth. The hierarchy of a photonic PDK was explained, as well as the silicon photonic component library development cycle and organization and verification processes. The development steps, organic component growth, and added value of the PDK component library were discussed. The APSUNY PDK component library lowers the barrier to entry for PIC design and provides accurate models. PDKs like this one can lead to a large photonic IP market as well as proliferate new applications that leverage photonic-electronic codesign using the maturing EPDA tools.

4.9 Conclusions

In this chapter, the advancements that enabled a suite of processes for different types of PICs were discussed. Moreover, the components of the PIC foundry industry ecosystem were described, including wafer foundry and R&D facilities, the enabling PDK, mask challenges, and finally the currently costly packaging, test, and assembly. Specifically, this chapter discussed InP-based research and development foundries, and an InP turn-key PIC foundry capable of Tb/s class SOC PICs. InP is both advantaged by its intrinsic materials features (direct bandgap) as well as disadvantaged by them (defectivity, compressive strength). By contrast, silicon allows straightforward integration with CMOS but lacks cheap, easy integration of optically active materials. Promising developments for integration were described, with the most compelling being QD lasers on Si. These demonstrations point to a future where silicon-based systems approach the best results for III-V lasers on native substrates, and point to the possibility for a revolution in photonics, where PICs are made on silicon at high volume and low cost.

4.10 Disclosure

Some of this material is based on research sponsored by Air Force Research Laboratory throughout the American Institute for Manufacturing Integrated Photonics ("AIM Photonics") under agreement number FA8650-15-2−5220. The US Government is authorized to reproduce and distribute reprints for Governmental purposes notwithstanding any copyright notation thereon. The views and conclusions contained herein are those of the authors and should not be interpreted as necessarily representing the official policies or endorsements, either expressed or implied, of Air Force Research Laboratory or the US Government.

Acknowledgments

The authors thank Hiroko Sueyoshi and Nicole Neu-Baker, MPH for assistance with reviewing and editing this chapter.

References

[1] EuroPIC. European manufacturing platform for photonic integrated circuits. <http://europic.jeppix.eu/>, 2013 (accessed 24.02.14).
[2] PARADIGM. <http://paradigm.jeppix.eu/>, 2014 (accessed 07.08.18).
[3] pics4all. <http://pics4all.jeppix.eu/>, 2016 (accessed 15.05.18).
[4] Europractice. <http://www.europractice-ic.com/index.php>, 2018 (accessed 15.05.18).
[5] ONECHIP PHOTONICS. <http://www.onechip-photonics.com/>, 2018 (accessed 15.05.18).
[6] PIC Magazine. GCS, Intengent, VLC to offer PIC Fab Services. <https://picmagazine.net/article/103667/GCS_Intengent_VLC_to_offer_PIC_Fab_Services>, 2018 (accessed 15.05.18).

[7] JePPIX, JePPIX Roadmap 2018. OFC 2018, San Diego, Mar-2018. Downloadable at <http://www.jeppix.eu/vision/>, 2018 (accessed 07.08.18).
[8] F.M. Soares, M. Baier, T. Gaertner, M. Feyer, M. Möhrle, N. Grote, et al., High-performance InP PIC technology development based on a generic photonic integration foundry, in Optical Fiber Comm. Conf. OSA Technical Digest (online) Optical Society of America (2018), paper M3F.3.
[9] V. Dolores-Calzadilla, F.M. Soares, M. Baier, T. Gaertner, M. Feyer, M. Möhrle, et al., InP-based Photonic integration platform: status and prospects, presented at the European Conference on Integrated Optics, Warsaw (2016).
[10] G. Mourgias-Alexandris, C. Vagionas, A. Tsakyridis, P. Maniotis, N. Pleros, All-optical 10Gb/s ternary-CAM cell for routing look-up table applications, Opt. Express 26 (6) (2018) 7555–7562.
[11] M.P. Chang, E.C. Blow, J.J. Sun, M.Z. Lu, P.R. Prucnal, Integrated microwave photonic circuit for self-interference cancellation, IEEE. Trans. Microw. Theory. Tech. 65 (11) (2017) 4493–4501.
[12] S.K. Ibrahim, M. Farnan, D.M. Karabacak, Design of a photonic integrated based optical interrogator, SPIE Proc. 10110 (2017) 101100U.
[13] C. Abellan, W. Amaya, D. Domenech, P. Muñoz, J. Capmany, S. Longhi, et al., Quantum entropy source on an InP photonic integrated circuit for random number generation, Optica 3 (9) (2016) 989–994.
[14] M. Baier, F.M. Soares, T. Gaertner, A. Schoenau, M. Moehrle, M. Schell, New polarization multiplexed externally modulated laser PIC, 2018 European Conference on Optical Communication (ECOC), 2018, pp. 1–3.
[15] K. Rylander, R. Broeke, R. Stoffer, D. Melati, A. Melloni, A. Bakker, Design of integrated, tuneable filters for telecom application, in: Proc. 18th European Conference Integrated Optics, 2016.
[16] Y. Kawamura, K. Wakita, Y. Itaya, Y. Yoshikuni, H. Asahi, Monolithic integration of InGaAs/InP DFB lasers and InGaAs/InAlAs MQW optical modulators, Electron. Lett. 22 (5) (1986) 242.
[17] R. Nagarajan, C.H. Joyner, R.P. Schneider, J.S. Bostak, T. Butrie, A.G. Dentai, et al., Large-scale photonic integrated circuits, IEEE J. Sel. Top. Quantum Electron. 11 (1) (2005) 50–65.
[18] F. Kish, V. Lal, P. Evans, S.W. Corzine, M. Ziari, T. Butrie, et al., System-on-chip photonic integrated circuits, IEEE J. Sel. Top. Quantum Electron. 24 (1) (2018) 6100120.
[19] J. Summers, 40 channels x 57 Gb/s monolithically integrated InP-based coherent photonic transmitter, 2014 Euro, Conf. Opt. Comm. (ECOC) (2014) 5–10.
[20] V. Lal, J. Summers, A. Hosseini, S. Corzine, P. Evans, M. Lauermann, et al., Full C-band tunable coherent transmitter and receiver InP photonic integrated circuits, Euro. Conf. Opt. Comm. (ECOC) (2016) 1–3.
[21] A. Hosseini, M. Lu, R. Going, P. Samra, S. Amiralizadeh, A. Nguyen, et al., Extended C-band tunable multi-channel InP-based coherent receiver PICs, J. Light. Technol. 35 (7) (2017) 18853–18862.
[22] M. Lauermann, R. Going, R. Maher, M. Lu, W. Ko, P. Studenkov, et al., Multi-Channel, widely-tunable coherent transmitter and receiver PICs operating at 88Gbaud/16-QAM, Opt. Fiber Commun. Conf. 1 (2017) Th5C.2.
[23] R. Going, M. Lauermann, R. Maher, H. Tsai, M. Lu, N. Kim, et al., Multi-channel InP-based coherent PICs with hybrid integrated SiGe electronics operating up to 100 GBd, 32QAM, Eur. Conf. Opt. Commun. ECOC 2017 (2018) 1–3.
[24] R. Going, M. Lauermann, R. Maher, H.S. Tsai, A. Hosseini, M. Lu, et al., 1.00 (0.88) Tb/s per wave capable coherent multi-channel transmitter (receiver) InP-based PICs with hybrid integrated SiGe electronics, IEEE J. Quantum Electron. 54 (4) (2018) 1–10.
[25] R.C. Leachman, D.A. Hodges, Benchmarking semiconductor manufacturing, IEEE Trans. Semicond. Manuf. 9 (2) (1996) 158–169.
[26] R. Nagarajan, C. Doerr, F. Kish, Semiconductor photonic integrated circuit transmitters and receivers, in: I.P. Kaminow, T. Li, A.E. Willner (Eds.), Optical Fiber Telecommunications: Components and Subsystems, sixth ed, Academic Press, 2013. Chapter 2.
[27] J. Pleumeekers, E. Strzelecka, K.P. Yap, A. James, P. Studenkov, P. Debackere, et al., Manufacturing progress for InP-based 500 Gb/s photonic integrated circuits,", CS ManTech Conf (2013) 19–22.

[28] S.W. Corzine, P. Evans, M. Fisher, J. Gheorma, M. Kato, V. Dominic, et al., Large-scale InP transmitter PICs for PM-DQPSK fiber transmission systems, IEEE Photonics Technol. Lett. 22 (14) (2010) 1015−1017.

[29] R. Nagarajan, M. Kato, J. Pleumeekers, P. Evans, S. Corzine, A. Dentai, et al., InP photonic integrated circuits, IEEE J. Sel. Top. Quantum Electron. 16 (5) (2010) 1113−1125.

[30] M. Smit, X. Leijtens, H. Ambrosius, E. Bente, J. van der Tol, B. Smalbrugge, et al., An introduction to InP-based generic integration technology, Semiconductor, Sci. Technol. 29 (8) (2014) 1−41.

[31] A. Arriordaz, A. Bakker, R. Cao, C. Cone, J. Ferguson, J. Klein, et al., Improvements in the silicon photonics design flow: collaboration and standardization, 2014 IEEE Photonics Conference, San Diego, CA, 2014, pp. 63−64.

[32] R. Cao, J. Ferguson, Y. Drissi, F. Gays, A. Arriordaz, I. O'Connor, DRC challenges and solutions for non-Manhattan layout designs, 2014 Int. Conf. Opt. MEMS Nanophotonics, 2014, pp. 175−176.

[33] A. Bhardwaj, J. Ferrara, R.B. Ramirez, M. Plascak, G. Hoefler, V. Lal, et al., An integrated racetrack colliding-pulse mode-locked laser with pulse-picking modulator, Conf. Lasers Electro-Optics, CLEO 2017, San Jose, CA, 2017, Paper SM2O.1.

[34] R.B. Ramirez, M.E. Plascak, K. Bagnell, A. Bhardwaj, J. Ferrara, G.E. Hoefler, et al., Repetition rate stabilization and optical axial mode linewidth reduction of a chip-scale MLL using regenerative multi-tone injection locking, J. Lightwave Technol. 36 (14) (2018) 2948−2954.

[35] I. Artundo, Photonic circuit design, coming of age in a fabless ecosystem, EeNews Eur. <http://www.eenewseurope.com/news/photonic-circuit-design-coming-age-fabless-ecosystem>, 2018 (accessed 07.02.18).

[36] J.S. Kilby, Miniaturized electronic circuits, US3138743A, <http://patft.uspto.gov/netacgi/nph-Parser?Sect2 = PTO1&Sect2 = HITOFF&p = 1&u = /netahtml/PTO/search-bool.html&r = 1&f = G&l = 50&d = PALL&RefSrch = yes&Query = PN/3138743>, 1959 (accessed 07.08.18).

[37] K. Greene, Silicon photonics comes to market, MIT Technol. Rev. <https://www.technologyreview.com/s/408520/silicon-photonics-comes-to-market/>, 2007 (accessed 07.08.18).

[38] L. Chrostowski, M. Hochberg, Silicon Photonics Design From Devices to Systems, first ed, Cambridge University Press, Cambridge, 2015.

[39] A. Seeds, S. Chen, J. Wu, M. Tang, H. Liu, Monolithic integration of quantum dot lasers on silicon substrates, Semicon West, <http://www.semiconwest.org/sites/semiconwest.org/files/data15/docs/5_AlwynSeeds_UniversityCollegeLondon.pdf>, 2016 (accessed 07.03.18).

[40] M. Liehr, AIM Photonics: manufacturing challenges for photonic integrated circuits, Semicon West, <http://www.semi.org/eu/sites/semi.org/files/events/presentations/02_MichaelLiehr_SUNY.pdf>, 2017 (accessed 07.03.18).

[41] M. Hochberg, T. Baehr-Jones, Towards fabless silicon photonics, Nat. Photonics. 4 (2010) 492−494.

[42] A.E.J. Lim, J. Song, Q. Fang, C. Li, X. Tu, N. Duan, et al., Review of silicon photonics foundry efforts, IEEE J. Sel. Top. Quantum Electron. 20 (2014) 405−416.

[43] P. Muñoz, G. Micó, L.A. Bru, D. Pastor, D. Pérez, J.D. Doménech, et al., Silicon nitride photonic integration platforms for visible, near-infrared and mid-infrared applications, Sensors (Switzerland) 17 (2017) 1−25.

[44] M. Pantouvaki, H. Yu, M. Rakowski, P. Christie, P. Verheyen, G. Lepage, et al., Comparison of silicon ring modulators with interdigitated and lateral p-n junctions, IEEE J. Sel. Top. Quantum Electron. 19 (2013) 386−393.

[45] Z. Ying, Z. Wang, Z. Zhao, S. Dhar, D.Z. Pan, R. Soref, et al., Comparison of microrings and microdisks for high-speed optical modulation in silicon photonics, Appl. Phys. Lett. 112 (2018).

[46] S. Fathpour, Emerging heterogeneous integrated photonic platforms on silicon, Nanophotonics 4 (2015) 143−164.

[47] D. Thomson, F. Gardes, S. Liu, H. Porte, L. Zimmermann, J.M. Fedeli, et al., High performance Mach Zehnder based silicon optical modulators, IEEE J. Sel. Top. Quantum Electron. 19 (2013) 3400510.

[48] K. Yamada, T. Tsuchizawa, H. Nishi, R. Kou, T. Hiraki, K. Takeda, et al., High-performance silicon photonics technology for telecommunications applications, Sci. Technol. Adv. Mater. 15 (2014).
[49] H. Park, M.N. Sysak, H.W. Chen, A.W. Fang, D. Liang, L. Liao, et al., Device and integration technology for silicon photonic transmitters, IEEE J. Sel. Top. Quantum Electron. 17 (2011) 671–688.
[50] M.S. Rasras, D.M. Gill, M.P. Earnshaw, C.R. Doerr, J.S. Weiner, C.A. Bolle, et al., CMOS silicon receiver integrated with Ge detector and reconfigurable optical filter, IEEE Photonics Technol. Lett. 22 (2010) 112–114.
[51] M. Piels, J.E. Bowers, A. We, S. Ge, 40 GHz Si / Ge uni-traveling carrier 32 (2014) 3502–3508.
[52] L. Vivien, M. Rouvière, J.-M. Fédéli, D. Marris-Morini, J.F. Damlencourt, J. Mangeney, et al., High speed and high responsivity germanium photodetector integrated in a silicon-on-insulator microwaveguide, Opt. Express 15 (2007) 9843.
[53] J. Brouckaert, G. Roelkens, D. Van Thourhout, R. Baets, Thin-film III-V photodetectors integrated on silicon-on-insulator photonic ICs, J. Light. Technol. 25 (2007) 1053–1060.
[54] M.J. Byrd, E. Timurdogan, Z. Su, C.V. Poulton, D. Coleman, N.M. Fahrenkopf, et al., Mode-evolution based coupler for Ge-on-Si photodetectors, in: 2016 IEEE Photonics Conf. IPC 2016, 2017.
[55] D. Thomson, Roadmap on silicon photonics, J. Opt. 18 (2016) 073003.
[56] M. Lipson, B. Guha, Fiber-waveguide evanescent coupler, US9746612B2, 2013.
[57] M. Hochberg, N. Harris, R. Ding, Y. Zhang, A. Novack, Z. Xuan, et al., Silicon photonics: the next fabless semiconductor industry, IEEE Solid-State Circuits Mag. 5 (2013) 48–58.
[58] T. Baehr-Jones, T. Pinguet, P. Lo Guo-Qiang, S. Danziger, D. Prather, M. Hochberg, Myths and rumours of silicon photonics, Nat. Photonics. 6 (2012) 206–208.
[59] Y.H.D. Lee, M. Lipson, Back-end deposited silicon photonics for monolithic integration on CMOS, IEEE J. Sel. Top. Quantum Electron. 19 (2013).
[60] J.M. Fedeli, L. Di Cioccio, D. Marris-Morini, L. Vivien, R. Orobtchouk, P. Rojo-Romeo, et al., Development of silicon photonics devices using microelectronic tools for the integration on top of a CMOS wafer, Adv. Opt. Technol. (2008 () (2008) 1–15.
[61] A. Rickman, The commercialization of silicon photonics, Nat. Photonics. 8 (2014) 579–582.
[62] H. Rong, R. Jones, A. Liu, O. Cohen, D. Hak, A. Fang, et al., A continuous-wave Raman silicon laser, Nature 433 (2005) 725.
[63] R. Soref, The past, present, and future of silicon photonics, IEEE J. Sel. Top. Quantum Electron. 12 (2006) 1678.
[64] J. Liu, X. Sun, R. Camacho-Aguilera, L.C. Kimerling, J. Michel, Ge-on-Si laser operating at room temperature, Opt. Lett. 35 (2010) 679–681.
[65] S.A. Srinivasan, C. Porret, M. Pantouvaki, Y. Shimura, P. Geiregat, R. Loo, et al., Analysis of homogeneous broadening of n-type doped Ge layers for laser applications, IPC (2017).
[66] D. Jung, Z. Zhang, J. Norman, R. Herrick, M.J. Kennedy, P. Patel, et al., Highly reliable low threshold InAs quantum dot lasers on on-axis (001) Si with 87% injection efficiency, ACS Photonics 5 (3) (2017) 1094–1100.
[67] D. Miller, Device requirements for optical interconnects to CMOS silicon chips, OSA Technical Digest (CD), 2010, paper PMB3.
[68] J.C. Norman, D. Jung, Y. Wan, J.E. Bowers, The future of quantum dot photonic integrated circuits, APL Photonics 3 (2018) 030901.
[69] Intel® Silicon Photonics 100G CWDM4 Optical Transceiver Brief, <https://www.intel.com/content/www/us/en/architecture-and-technology/silicon-photonics/optical-transceiver-100g-cwdm4-qsfp28-brief.html?wapkw = cwdm4>, 2017 (accessed 07.08.18).
[70] H. Park, B.R. Koch, E.J. Norberg, J.E. Roth, B. Kim, A. Ramaswamy, et al., Heterogeneous integration of silicon photonic devices and integrated circuits, in: 2015 Conf. Lasers Electro-Optics Pacific Rim, 2015, paper 25J3_2.
[71] G. Kurczveil, D. Liang, M. Fiorentino, R.G. Beausoleil, Robust hybrid quantum dot laser for integrated silicon photonics, Opt. Express 24 (2016) 16167–16174.

[72] M. Tran, T. Komljenovic, D. Huang, L. Liang, M.J. Kennedy, J.E. Bowers, A widely-tunable high-SMSR narrow-linewidth laser heterogeneously integrated on silicon, in Conference on Lasers and Electro-Optics, OSA Technical Digest, 2018, paper AF1Q.2.
[73] A. Beling, J.C. Campbell, Heterogeneously integrated photodiodes on silicon, IEEE J. Quantum Electron. 51 (11) (2015) 1–6.
[74] C. Zhang, S. Zhang, J.D. Peters, J.E. Bowers, 8 × 8 × 40 Gbps fully-integrated silicon photonic network-on-chip, Optica 3 (7) (2016) 785–786.
[75] T. Komljenovic, D. Huang, P. Pintus, M.A. Tran, M.L. Davenport, J.E. Bowers, Heterogeneous III-V silicon photonic integrated circuits, IEEE Proc. 34 (1) (2018).
[76] T. Kageyama, K. Nishi, M. Yamaguchi, R. Mochida, Y. Maeda, K. Takemasa, et al., Extremely high temperature (220°C) continuous-wave operation of 1300-nm-range quantum-dot lasers, in: CLEO/Europe and EQEC 2011 Conference Digest, OSA Technical Digest, 2011, paper PDA_1.
[77] J. Duan, H. Huang, D. Jung, Z. Zhang, J. Norman, J.E. Bowers, et al., Semiconductor quantum dot lasers epitaxially grown on silicon with low linewidth enhancement factor, Appl. Phys. Lett. 112 (2018) 251111.
[78] A.Y. Liu, T. Komljenovic, M.L. Davenport, A.C. Gossard, J.E. Bowers, Reflection sensitivity of 1.3 μm quantum dot lasers epitaxially grown on silicon, Opt. Express 25 (9) (2017) 9535–9543.
[79] Y. Wan, D. Jung, J. Norman, C. Shang, I. Macfarlane, Q. Li, et al., O-band electrically injected quantum dot micro-ring lasers on on-axis (001) GaP/Si and V-groove Si, Opt. Express 25 (22) (2017) 26853–26860.
[80] K. Sun, D. Jung, C. Shang, A. Liu, J. Morgan, J. Zang, et al., Low dark current III–V on silicon photodiodes by heteroepitaxy, Opt. Express 26 (2018) 13605–13613.
[81] P. Bhasker, J. Norman, J.E. Bowers, N. Dagli, Intensity and phase modulators in epitaxial III-V layers directly grown on silicon operating at 1.55μm, Front. Opt., FTh4A.3 (2017).
[82] E.R.H. Fuchs, E.J. Bruce, R.J. Ram, R.E. Kirchain, Process-based cost modeling of photonics manufacture: the cost competitiveness of monolithic integration of a 1550-nm DFB laser and an electroabsorptive modulator on an InP platform, IEEE J. Lightwave Technol 24 (8) (2006) 3175–3186.
[83] J. Song, H. Fernando, B. Roycroft, B. Corbett, F. Peters, Practical design of lensed fibers for semiconductor laser packaging using laser welding technique, IEEE J. Lightwave Technol. 27 (11) (2009).
[84] D. Taillaert, H. Chong, P.I. Borel, L.H. Frandsen, R.M. De La Rue, R. Baets, A compact two-dimensional grating coupler used as a polarization splitter, IEEE Photonics Technol. Lett. 15 (9) (2003).
[85] H. Fukuda, K. Yamada, T. Tsuchizawa, T. Watanabe, H. Shinojima, S. Itabashi, Silicon photonic circuit with polarization diversity, Opt. Express 16 (7) (2008).
[86] M. Pu, L. Liu, H. Ou, K. Yvind, J.M. Hvam, Ultra-low-loss inverted taper coupler for silicon-on-insulator ridge waveguide, Optics Comm. 283 (19) (2010).
[87] T. Barwicz, Y. Taira, T. Lichoulas, N. Boyer, Y. Martin, H. Numata, et al., A novel approach to photonic packaging leveraging existing high-throughput microelectronic facilities, IEEE J. Sel. Top. Quantum Electron 22 (6) (2016).
[88] Nufern, Ultra-High NA Single-Mode Fibers, <http://www.nufern.com/pam/optical_fibers/spec/id/988/>, 2014 (accessed 07.08.18).
[89] S. Preble, UHNA Fiber – Efficient Coupling to Silicon Waveguides, Application Note NUAPP-3, <www.nufern.com/library/getpdf/id/485/>, 2016 (accessed 07.08.18).
[90] B. Snyder, P. O'Brien, Packaging process for grating-coupled silicon photonic waveguides using angle-polished fibers, IEEE Trans. Components, Packaging Manuf. Technol. 3 (6) (2013) 954–959.
[91] C. Scarcella, K. Gradkowski, L. Carroll, J.S. Lee, M. Duperron, D. Fowler, et al., Pluggable single-mode fiber-array-to-PIC coupling using micro-lenses, IEEE J. Lightwave Technol. 29 (22) (2017) 1943–1946.
[92] R. Dangel, J. Hofrichter, F. Horst, D. Jubin, A. La Porta, N. Meier, et al., Polymer waveguides for electro-optical integration in data centers and high-performance computers, Opt. Express 23 (4) (2015) 4736–4750.

[93] S.M. Eaton, M.L. Ng, J. Bonse, A. Mermillod-Blondin, H. Zhang, A. Rosenfeld, et al., Low-loss waveguides fabricated in BK7 glass by high repetition rate femtosecond fiber laser, Appl. Optics 47 (12) (2008) 2098–2102.
[94] M. Rakowski, M. Pantouvaki, P. De Heyn, P. Verheyen, M. Ingels, C. Hongtao, et al., A 4x20Gb/s WDM ring-based hybrid CMOS silicon photonics transceiver, ISSCC Conference (2015) 408–409.
[95] J.S. Lee, L. Carroll, C. Scarcella, N. Pavarelli, S. Menezo, S. Bernabe, et al., Meeting the electrical, optical, and thermal design challenges of photonic-packaging, IEEE J. Sel. Top. Quantum Electron. 22 (6) (2016).
[96] N. Pavarelli, J.S. Lee, M. Rensing, C. Eason, P.A. O'Brien, Optical and electronic packaging processes for silicon photonic systems, 2014 Euro, Conf. Opt. Comm. (ECOC), (2014) 1–4.
[97] E. Timurdogan, Z. Su, C.V. Poulton, M.J. Byrd, S. Xin, R. Shiue, et al., AIM process design kit (AIMPDKv2.0): silicon photonics passive and active component libraries on a 300mm wafer, in: Opt. Fiber Comm. Conf. OSA Technical Digest, 2018, paper M3F.1.
[98] A.I.M. Photonics, Silicon photonics process design kit (APSUNY PDKv2.0a), <http://www.aimphotonics.com/pdk/>, (accessed 07.08.18).
[99] Analog Photonics, Process development kit, <http://www.analogphotonics.com/pdk/>, 2017 (accessed 07.08.18).
[100] T. Barwicz, M.R. Watts, M. Popovic, P.T. Rakich, L. Socci, F.X. Kaertner, et al., Polarization transparent microphotonic devices in the strong confinement limit, Nature Photonics 1 (1) (2006) 57–60.

CHAPTER 5

Software tools for integrated photonics

James Pond[1], Gilles S.C. Lamant[2] and Rich Goldman[1]
[1]Lumerical Inc., Vancouver, BC, Canada
[2]Cadence Design Systems, Inc., San Jose, CA, United States

5.1 The growing need for integration and associated challenges

Integrated optical interconnects are anticipated to provide solutions for the ever-growing demand for low-cost, power-efficient, high-bandwidth interconnects. Initially these needs are most acute in data centers for communication between servers, but optical interconnects are poised to also provide solutions for even shorter-reach communication links such as chip to chip as well as processor to memory. While data communications applications are driving significant investment, there is a wide range of other applications of integrated photonics that take advantage of this investment to enable many future technologies, particularly in sensing. The cost, power, and bandwidth requirements of data communications as well as the breadth of general applications create significant challenges for software tool providers. Here we review some of these challenges in more detail.

5.2 The need to support multiple material systems

A variety of material systems for integrated photonics have been developed over many years [1]. Recently there has been a great deal of interest and investment in silicon photonics [2,3]; however, no single material system has proven optimal for all applications due to the wide range of desired wavelengths and functionalities [4,5]. As a result, there continues to be significant debate about which material system is best in terms of performance and cost for various applications and a general recognition that no single material system will dominate for all applications [6,7]. Furthermore, to achieve all desired functionality, some circuits are designed for hybrid integration, for example, with indium phosphide and silicon on insulator [8].

Since no material system is likely to emerge as the clear winner for all applications, the design and simulation software for integrated photonics must be capable of supporting all of them. Indeed, it is even desirable that a circuit designed at a functional level could be optimized through simulation and layout experimentation for the best material system (and corresponding foundry) to provide the optimal performance, yield, and cost.

Optical Fiber Telecommunications V11
DOI: https://doi.org/10.1016/B978-0-12-816502-7.00005-1

5.3 Applications extend well beyond data communications

A key driver for integrated photonics is the need for low-cost, low-power, high-speed data communications, particularly in the data center where volumes are significant [9]. However, there are a number of other applications of interest [10], including analog RF signal applications, sensing (biological and chemical) [11], LIDAR, and phased array imaging [12], as well as quantum information, particularly quantum communications [13].

The wide variety of applications combined with the wide variety of material systems requires versatile design tools and simulators. Layout and simulation requirements for sensing versus data communications applications can vary widely.

5.4 Challenges specific to photonics

Many challenges of integrated photonics mirror those of integrated electronics; therefore, the many decades of investment in electronic design automation (EDA) tools can be leveraged. However, due to fundamental differences between photonics and electronics, there are other challenges that are specific to photonics and cannot be solved with existing tools and approaches. One important difference is the size of devices compared to the wavelength. In most integrated optical applications, the free space wavelength is 1–2 μm (150–300 THz frequency), which is much smaller than typical devices. By contrast, in most electronic applications, the free space wavelength is typically larger than 3 cm (10 GHz frequency), which is much larger than modern devices. This difference in scale compared to the wavelength has important consequences.

The operating principle behind many photonic devices (particularly modulators, sensors, and filters) is constructive and destructive interference. In other words, these devices rely on precise phase control of the light traveling along different paths. As a result, photonic devices are both large (compared to the wavelength) and yet extremely sensitive to small errors in geometry. They are large because it takes many wavelengths to accumulate a significant phase change between optical paths and typical devices vary from several microns to hundreds of microns in length. Indeed, many Mach-Zehnder interferometers (MZI) are several mms in length. However, the necessary phase sensitivity means that these devices can be sensitive to nm scale errors in geometry, as well as environmental factors such as temperature.

One advantage, of course, is that they make ideal sensors, but this comes with extreme manufacturing challenges as well as stability challenges. For example, most devices must be thermally controlled through active PID controllers, because a 1°C change in temperature causes a change in waveguide effective index on the order of 2e-4. For comparison, a common Mach-Zehnder modulator (MZM) is created by interfering light through two different waveguides and modulating the phase of one of them using the plasma dispersion effect (which results in a final amplitude modulation by constructive and destructive interference when the light is recombined in a single waveguide). In a typical device, a 1 V bias across the PN junction in the waveguide can cause a change in refractive index on the order of 4e-5, sufficient to modulate an optical signal for transceiver applications, but about five times smaller than the change induced by a 1°C temperature variance.

Another challenge posed by photonics is that shapes are nonrectilinear and follow non-Manhattan geometries. Waveguide bends must be carefully controlled arcs with, at the very least, a minimum bend radius to avoid reflections, loss, and polarization conversion. These curvy shapes are challenging for existing EDA software both in terms of layout and Design Rule Checks (DRC). Waveguides that are 500 nm wide but misconnected by only nms can cause unacceptably high reflections and loss, which is not an issue with electrical connections except at very high frequencies. Even discontinuities in the derivative of arcs and bends can cause problems. These devices tend to create many false DRC errors. For example, a typical delay line is created by coiling a waveguide tightly to create a large propagation length in a small area. Even if the original shape respects the design rules for waveguide spacing, after conversion to GDS polygons, the DRC tool may detect many small violations due to the discretization of the original curvy shape.

The curvy waveguide connections and phase sensitive devices mandate back-annotation of the schematic in many design flows. This is necessary because the waveguides that connect different components are typically hundreds of wavelengths in length. The precise phase and group delays through these connectors must be known in order to correctly simulate the circuit, but these are determined only at the layout stage and this information must then be back-annotated to the schematic. A related issue is waveguide crossings which may be introduced during layout and should then be back-annotated to the schematic.

In photonics, waveguides can carry more than one mode of light. Each mode has a particular cross-sectional electromagnetic field profile and propagates down the waveguide accumulating only a phase (as well as some loss). The phase in radians after propagation of length L is $2\pi n_{\mathrm{eff}}L/\lambda_0$ where λ_0 is the free space wavelength. The effective index, n_{eff}, differs for each waveguide mode. In nonabsorbing waveguides, the modes are power orthogonal in the sense that the total optical power in the waveguide is the sum of the power carried by each individual mode. (There is no contribution from cross terms.) Furthermore, in ideal waveguides that have no variation in cross section with length, there is no possible coupling between the modes. However, any deviation from this ideal condition can result in coupling between the modes. This occurs in practice in waveguide bends and tapers, and even due to manufacturing errors such as sidewall roughness or thickness variations.

Single-mode waveguides would generally be ideal; however, this is challenging in practice because it requires very narrow waveguides which tend to have high propagation losses due to sidewall roughness. Wider waveguides have much lower losses but support multiple modes.

Designers typically choose waveguides that support two modes, often called Transverse Electric (TE) and Transverse Magnetic (TM). The terms TE and TM have very different meaning than in microwave waveguides. In photonics, TE means that the electric field is mainly transverse to the chip normal (i.e., the electric field is mainly in the plane of the chip, while the magnetic field is mainly normal to the plane of the chip). The TM modes are the opposite, with the magnetic field mainly in the plane of the chip and the electric field normal to it. These terms originate from the 2D analysis of Maxwell's equations (appropriate for infinitely thick structures) where there is perfect separation of these types of mode. In real 3D devices, this separation is not perfect, and all modes, whether TE or TM, contain all electromagnetic field components. Some authors therefore refer to them as TE-like and TM-like. While it is typical to choose waveguides that support only these two modes, there are exceptions, particularly for longer propagation lengths where low losses are required. In that case, wider waveguides may be used that support more than two modes, and care must be taken to adiabatically taper the waveguide transitions to avoid coupling to the higher-order modes. This multimode nature of waveguides creates an additional level of complexity for photonics.

Finally, for simulation, photonics involves effects that differ on timescales by many orders of magnitude. While the optical frequency is on the order of 200 THz, the electrically modulated signal (and modulated phase or amplitude of the optical carrier) is on the order of tens of GHz, and there are thermal effects on the order of MHz or lower that need to be accounted for. In addition, wavelength division multiplexing (WDM) can be used to encode many different signals in a single optical waveguide, and typical channel spacing is 100 GHz. These vast differences of timescale and frequency make brute force and fully self-consistent simulation of all effects nearly impossible in most practical cases.

5.5 The need for an integrated, standard methodology

Despite recent advances, many photonic circuit designers begin directly with the layout of the circuit and may draw a schematic only for the purpose of simulating the circuit. This is reminiscent of the early days of electronic design. This is possible because many current circuit designs involve only tens to hundreds of components; however, circuit complexity is rapidly increasing, bringing with it an increasing need to eliminate sources of error through automation. Therefore, it is becoming increasingly necessary to design a single "golden" circuit schematic first and automate the

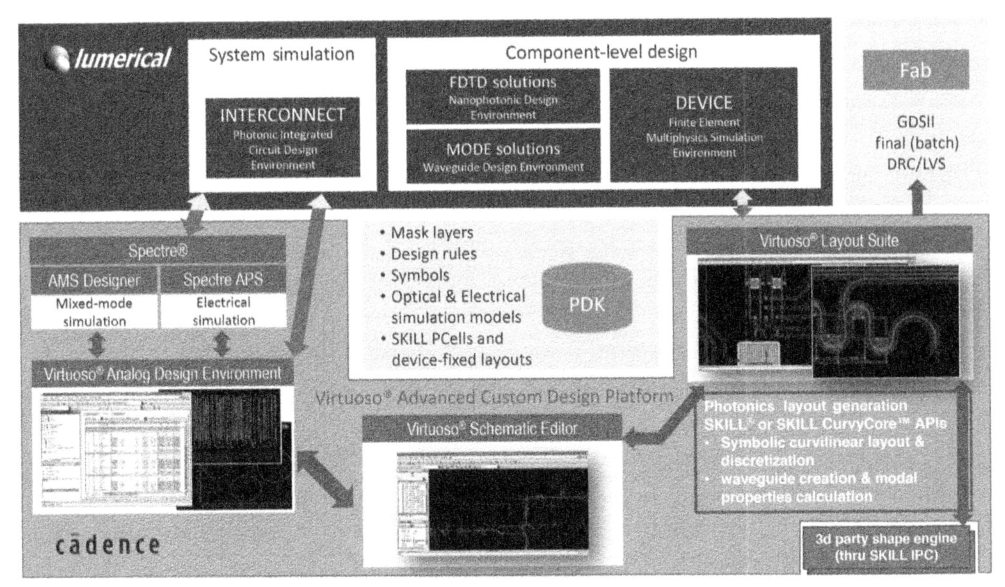

Figure 5.1 Electronic Photonic Design Automation environment commercially available from Cadence and Lumerical Inc.

simulation, layout, layout versus schematic (LVS), design for manufacture, and DRC all the way to fabrication. The golden circuit schematic may be back-annotated with information determined only upon layout, but this process must be automated rather than manual. This evolution is already underway and essentially leverages the decades of development of EDA tools, along with modifications to resolve challenges specific to photonics. This integrated and standard design flow, often called Electronic Photonic Design Automation (EPDA) [14], is absolutely necessary for photonics to continue to scale. In addition to the golden schematic, a key part of the EPDA flow is mature Process Design Kits (PDKs) provided by foundries that provide mask layers, technology files, design rules, symbols, compact models, and parameterized cells (P-Cells).

An example of a commercially available EPDA design flow by Cadence [15] and Lumerical Inc. [16] is shown in Fig. 5.1. PDKs supporting this design flow are becoming increasingly available for both R&D and commercial foundries, such as AIM [17], IMEC [18], and TowerJazz [19].

5.6 Mixed-mode, mixed-domain simulation

5.6.1 Physical simulation

Full 3D physical simulation of photonic components is done for the purposes of designing and optimizing individual photonic components. It is not practical to perform 3D physical simulations to simulate larger circuits and systems. Instead, behavioral or compact models are required to rapidly simulate the response of a component for given electrical, optical, and thermal inputs. These compact models must be calibrated using a combination of parameter extraction from 3D physical simulation, complemented by experimental results when available. Upon calibration, the compact models allow for rapid simulation of circuits and systems containing many components in the frequency and time domains, which we discuss in detail below. Ultimately, a circuit designer with a calibrated compact model library in a PDK should never need to perform 3D simulation, as the PDK developers and component designers already completed this task.

Physical 3D simulation requires a range of solvers: electromagnetic, charge transport, and heat transport. Furthermore, a range of electromagnetic solvers are required for different components. For example, eigenmode solvers are required to solve waveguide modes, while full 3D methods such as the finite-difference time-domain (FDTD) and the eigenmode expansion method are used for devices such as grating couplers, directional waveguide couplers, edge couplers, and multimode interferometers (MMIs). Finally, physical simulation often requires multiple solvers; for example, solving a waveguide phase shifter requires a charge transport solver to determine the carrier density as a function of bias voltage as well as an eigenmode solver to calculate the change in effective index due to the change in carrier density. High-performance physical solvers are commercially available; for example, Lumerical Inc. offers a broad range of solver products (FDTD Solutions, MODE Solutions, and DEVICE) that cover all physical simulation needs for integrated photonics.

5.6.2 S-parameter-based simulation of photonic circuits

Frequency domain simulation of photonic components is typically done using S-parameters. Fig. 5.2 shows an example of a schematic representation of a Mach-Zehnder interferometer, made from two waveguides of different lengths as well as the transmission of the device as a function of wavelength simulated with S-parameters.

In photonics, the S-parameter simulation is slightly more complex than in electronics, as the waveguides can carry multiple modes of light. For example, a device connected to two waveguides such as the MZI shown in Fig. 5.2 will typically have a 4×4 S-matrix (at each wavelength) because each waveguide supports both TE and TM modes. If the connecting waveguides support even more modes, then the S-matrix becomes even larger. The bookkeeping of S-matrix indices to physical waveguides and modes is not well standardized.

The S-parameter analysis of photonic circuits is critically important, but it is insufficient in itself for many applications, as the S-parameter analysis works only for linear circuits where all non-optical ports (electrical voltages and currents) and environmental factors such as temperature are constant. In Fig. 5.3, we consider a MZM made from

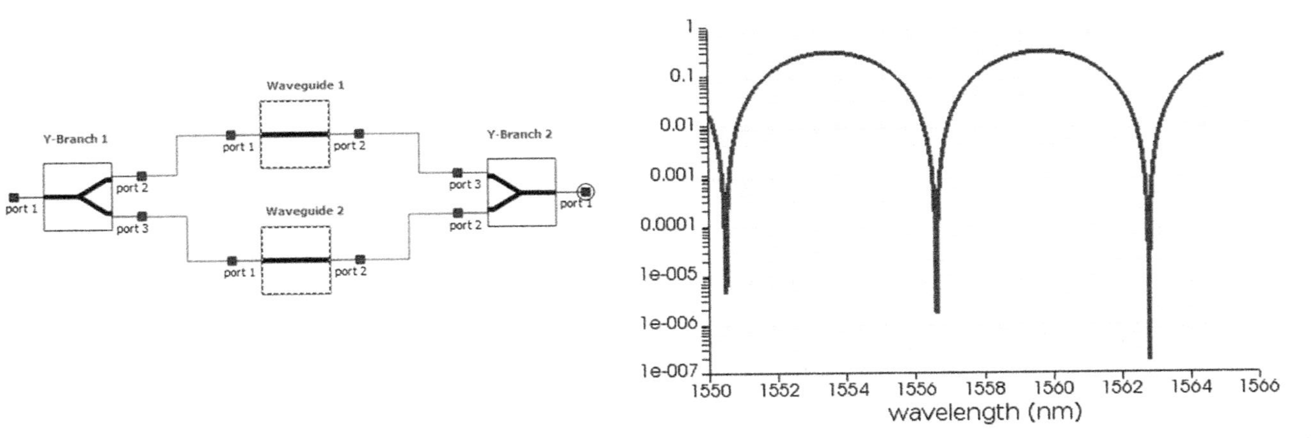

Figure 5.2 A schematic representation of a Mach-Zehnder interferometer made from waveguides with mismatched lengths (*left*) and the optical transmission of the device as a function of wavelength on a log scale simulated with S-parameters (*right*).

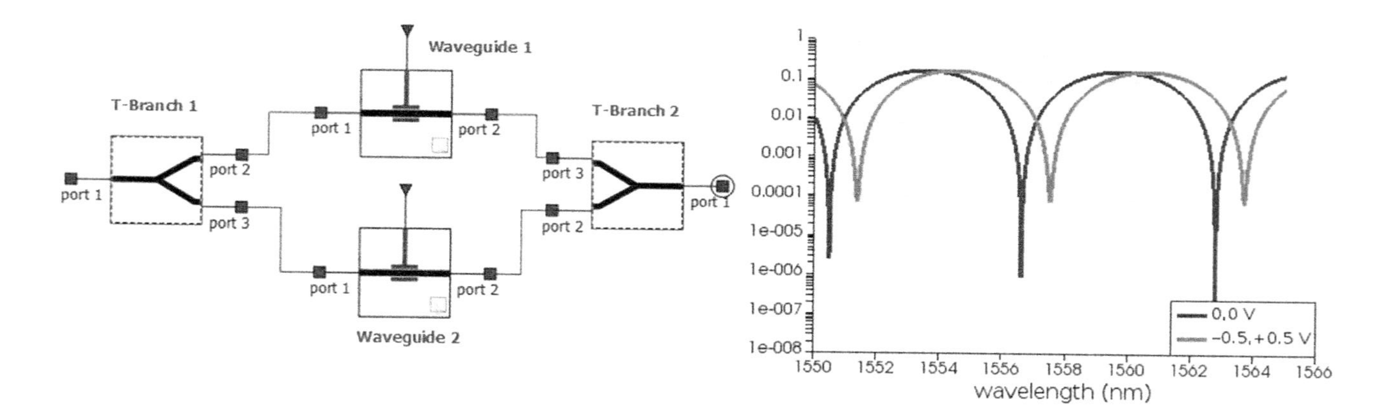

Figure 5.3 A schematic representation of a Mach-Zehnder modulator made from waveguides with mismatched lengths where a portion of the waveguide is also a PN junction with an electrical voltage input (*left*), and the optical transmission of the device as a function of wavelength on a log scale simulated with S-parameters for two different bias voltage conditions of the PN junctions (*right*).

the same MZI shown in Fig. 5.2 but where the two waveguide arms are also PN junctions where the density of charge carriers in the waveguide can be controlled with a bias voltage. We can see the different response under different DC bias conditions. This is useful to determine the ideal operating point and calculate the extinction ratio (ER) and transmitter penalty (TP). But it does not allow us to calculate the transient electrical response of this modulator to a given bit sequence to calculate quantities like eye diagrams and bit error rate (BER). Furthermore, it does not allow us to perform cosimulation with electrical circuit simulators. For that we need transient photonic circuit simulation.

5.6.3 Transient simulation of photonic circuits

To capture the transient response of a circuit, we use time domain simulation. In typical time domain simulations of photonic circuits, laser sources with specified frequencies (typically around 100–300 THz) are modulated either directly or with external modulators, using electrical signals with frequencies in the range of tens of GHz. The modulation may be either amplitude modulation, phase modulation, or both. The electrical signals are typically voltage or current signals representing pseudorandom bit sequences but can be any type of signal such as sinusoidal or delta functions for impulse response calculations. To avoid prohibitively small timesteps for the simulation, we typically simulate only the envelope of the optical carrier, $A(t)$. The true optical carrier signal is then given by

$$S(t) = A(t)\exp\left[-2i\pi f_0 t\right]$$

where f_0 is the frequency of the carrier. For single channel communication, this allows us to use timesteps that are comparable to the timesteps used for simulation of the electrical signal (which contains frequencies of tens of GHz). Typical, single-channel transient results are shown in Fig. 5.4.

However, when WDM is used, such that the waveguide or fiber mode contains more than one carrier frequency, we must reduce the timestep to approximately $1/(N\Delta f)$ where Δf is the channel spacing and N is the number of channels. This can be seen qualitatively in the optical power time signal, for the schematic shown in Fig. 5.5, on the right of Fig. 5.6. Since the typical channel spacing is 100 GHz or more, this can result in a much smaller timestep than would be used for the original electrical signals, shown on the left of Fig. 5.6. Alternatively, the simulator can use a multiband approach by internally handling multiple signals. Each signal has a different carrier frequency which is chosen to match one of the channel frequencies, resulting in the simulation of multiple frequency bands that overlap with the carrier frequencies. This multiband approach allows for larger timesteps, but there is additional overhead because there are multiple time signals for each waveguide

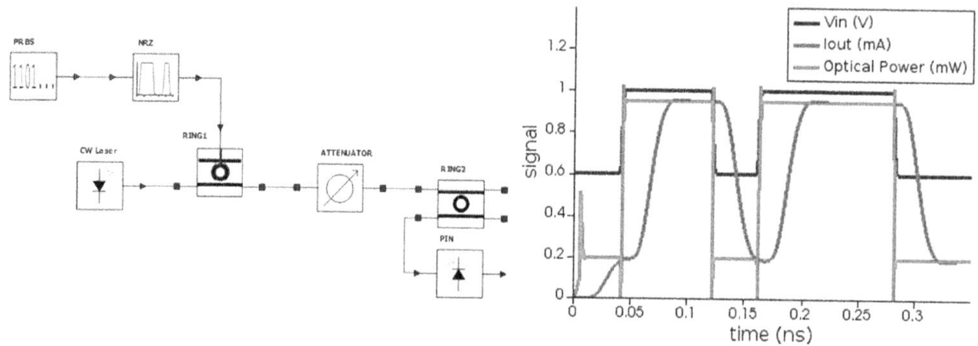

Figure 5.4 A simple transceiver circuit schematic (*left*) showing a pseudorandom bit sequence generator feeding a nonreturn-to-zero pulse generator used to modulate a CW laser with a ring modulator. The signal then passes through an attenuator (representing a fiber link) and is then fed to a ring filter where the transmission channel is dropped and fed to a PIN photodetector. On the right we see typical simulation results: the input voltage to the ring modulator, the optical power at the output of the ring modulator, and the output current of the PIN PD. The optical power contains overshoot effects at the rise and fall, which is a result of the high quality factor of the ring modulator. The final current is offset in time compared to the input voltage due to delays in the circuit. Other than the overshoot effects (which are specific to ring modulators), the optical power tracks the driving voltage signal relatively closely and a relatively large timestep can be used.

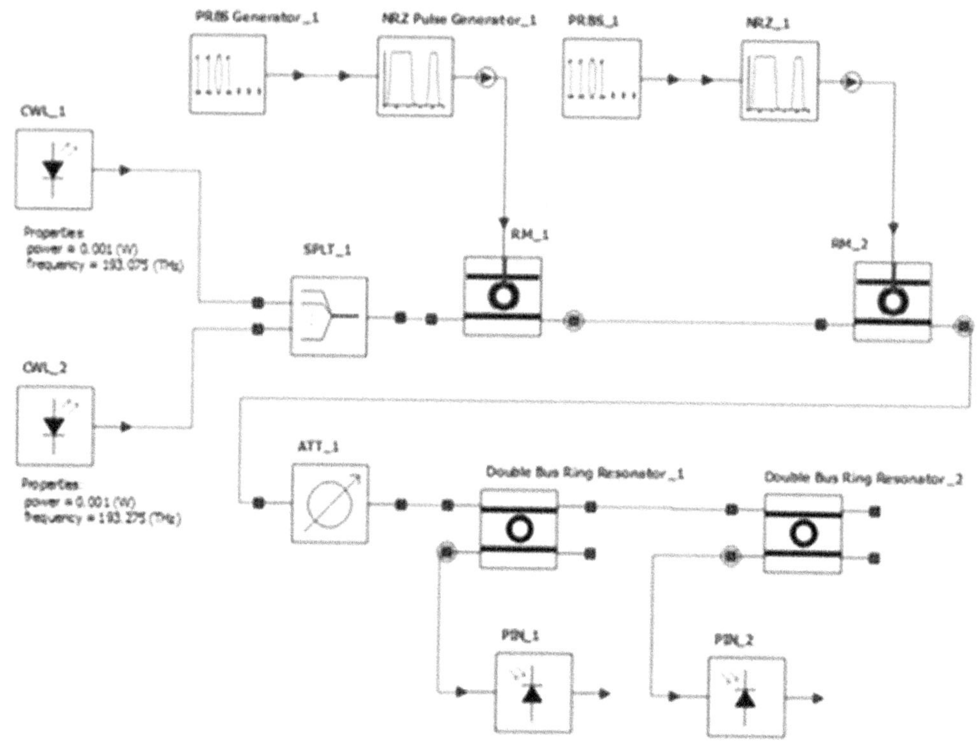

Figure 5.5 A two-channel transceiver circuit similar to the single channel version shown in Fig. 5.4. Here there are two CW lasers offset by 200 GHz in frequency. The lasers are combined into a single waveguide, then modulated by two ring modulators (which modulate only a specific frequency). The ring demultiplexers demultiplex the two signals, which are converted back to a current by two PIN PDs.

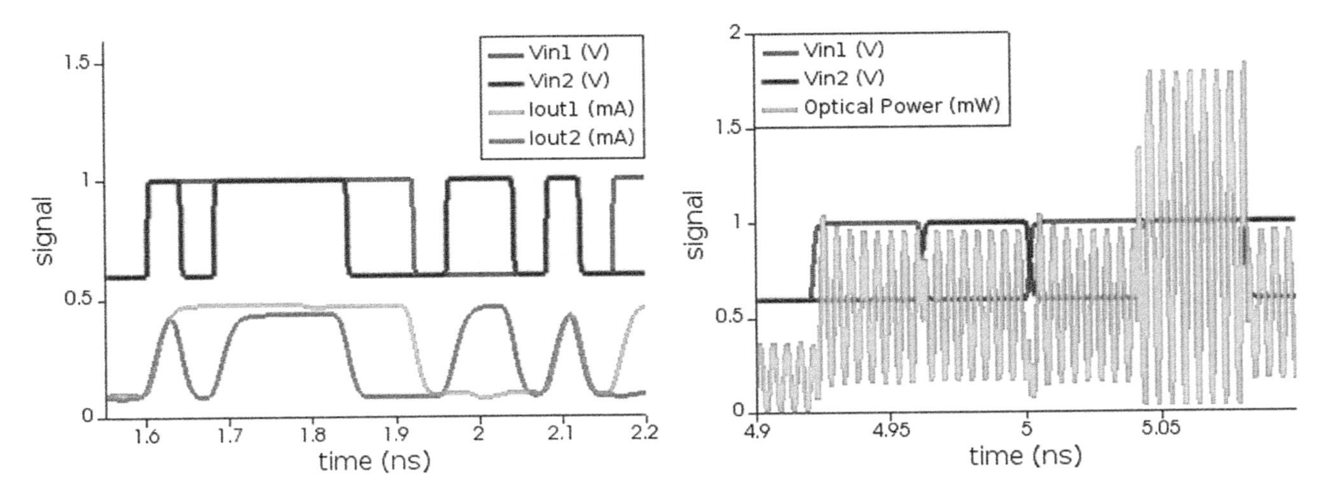

Figure 5.6 Typical simulation results for a two-channel transceiver. The electrical input voltages and output currents are shown on the left, and the optical power in the multiplexed waveguide is shown on the right, along with the two driving voltages. The signals have been time-shifted to align all bit sequences in time for easier viewing. We can see the high-frequency beating of the two laser signals in the optical power, which necessitates special handling for simulation: either a smaller timestep, or each channel of the optical signal must be internally handled as a separate frequency band, as shown in Fig. 5.7.

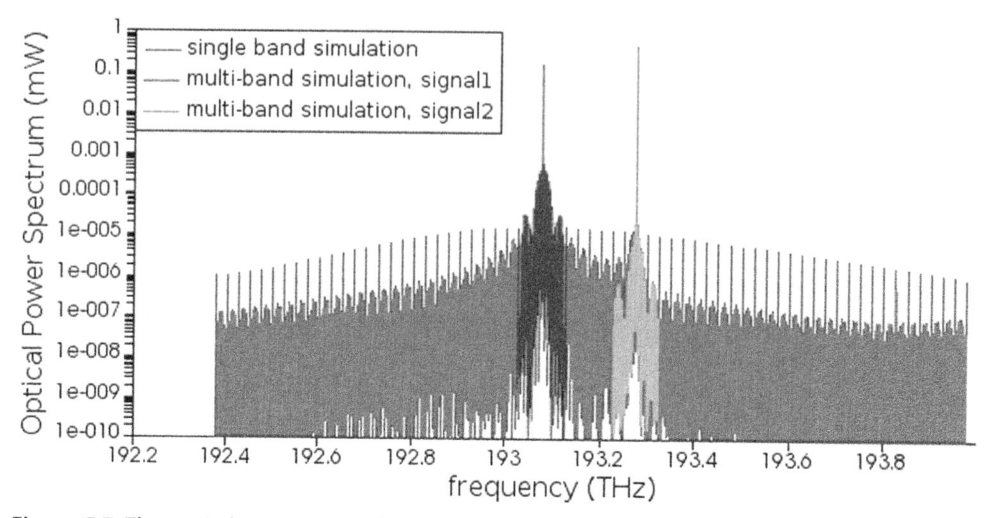

Figure 5.7 The optical spectrum in the waveguide of a two-channel transceiver at the output of the transmitter module. In the single-band simulation, a timestep is chosen that is sufficiently small so that the entire spectrum, including both transmission frequencies, can be handled internally by a single optical signal covering a single frequency band. In the multiband simulation, the optical power inside the waveguide is handled by two distinct signals, covering two distinct frequency bands that are chosen to be centered on the channel frequencies. The multiband simulation allows for much larger timesteps; however, crosstalk effects may not be correctly modeled without special treatment. A photonic circuit simulator must be able to handle both approaches seamlessly.

mode and, furthermore, nonlinear and crosstalk effects require special treatment. Ultimately, it is necessary for a time domain simulator to handle both of these complementary approaches, as both have advantages and disadvantages depending on the circuit, the components used, and the desired simulation results. Fig. 5.7 shows the optical spectrum in the waveguide calculated with both methods.

In photonic transient simulation for data communications, a common result to calculate is the eye diagram or constellation diagram for advanced modulation formats as shown in Fig. 5.8. From these diagrams, key information such as BER estimates can be obtained, for example, by using a Gaussian method to estimate the signal BER from the eye diagram [20].

5.6.4 Sample mode and block mode

Transient photonic circuit simulation can be done in two modes: sample mode and block mode. In sample mode, each element exchanges a single waveform sample at every timestep. In this manner, a transient signal is passed, sample by sample, between the elements. Dispersive elements, with frequency-dependent S-parameters or transfer functions, must be handled using either finite impulse response or infinite impulse response filters. This is ideal for bidirectional simulation and resonant feedback

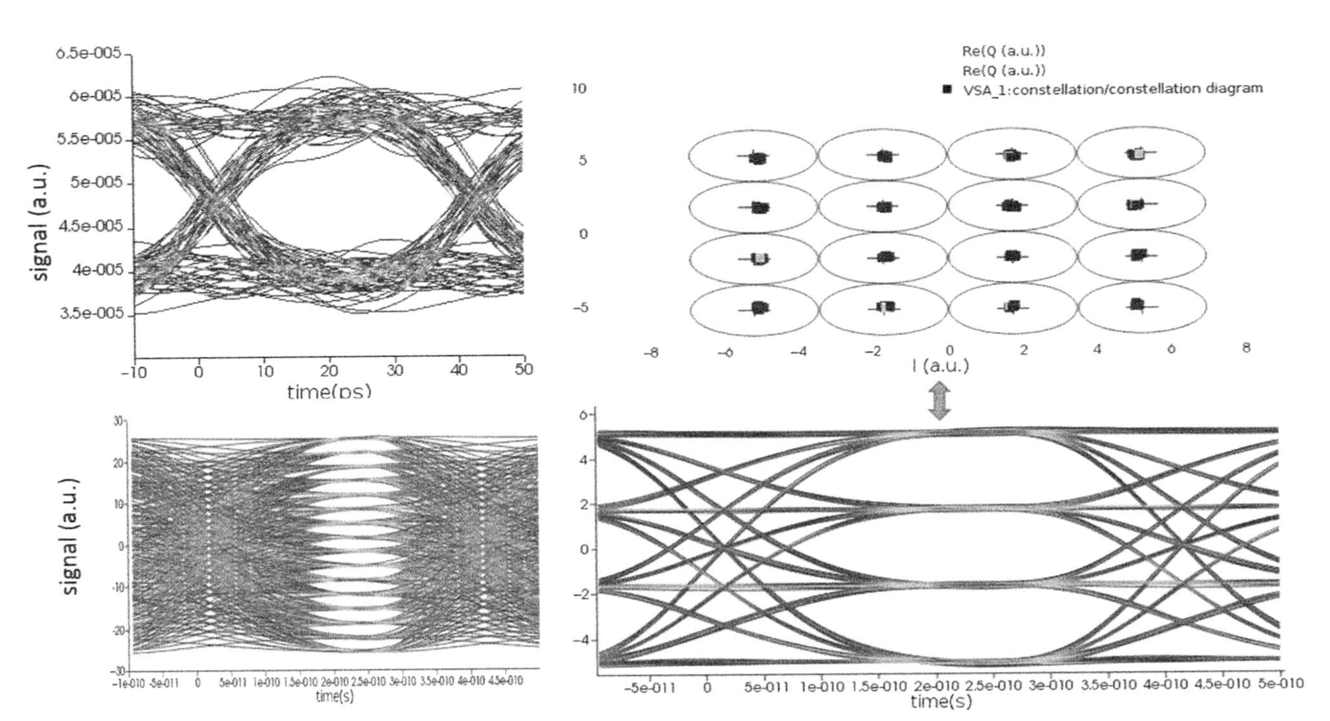

Figure 5.8 Typical results of transient simulation for data communications applications are eye diagrams and constellation diagrams. The upper left eye diagram is typical for a pulse amplitude modulation two-level system (PAM-2). The lower left shows an example of a PAM-16 system. The bottom right shows the eye diagram of a quadrature amplitude modulation (QAM)-16 system that uses both amplitude and phase to encode 16 channels. The QAM-16 system can be studied with a constellation diagram (upper right) to quantify the signal integrity for the 16 channels.

structures such as ring modulators. In block mode simulation, an entire waveform, in the form of *N* samples, is passed from one element to the next. This has the advantage of being faster overall and allows for conversion to and from the frequency domain using fast Fourier transforms. This makes it possible to handle dispersive elements directly in the frequency domain without using digital filter approaches. Furthermore, block mode allows for the treatment of nonlinear, dispersive fiber elements. Typically block mode is used for unidirectional simulation in situations where the overall time delay of propagation is not critical to the functionality of the device. Bidirectional circuits can be handled, but some care must be taken to obtain correct results. Finally, it might be necessary to use both sample mode and block mode in the same system simulation, for example, in a transceiver design with a nonlinear, dispersive fiber link where the transmitter and receiver may be simulated in sample mode but the fiber link is simulated in block mode. Transition elements are therefore required in a photonic integrated circuit simulator.

5.6.5 Electro-optical cosimulation

Thus far, we have dealt with photonic integrated circuit simulation without considering any of the detailed effects of the electrical circuitry. In most photonic circuits and systems there is electrical circuitry that serves two purposes:

- Control: The photonic circuit is typically controlled by an electrical circuit. For data communications, there are transmitting and receiving electronics, while for sensing applications, there is often electrical readout of the results and/or electrical control of the input light.
- Tuning: The sensitivity of photonic devices to temperature and other environmental factors requires active tuning performed typically with PID controllers.

We will first discuss using control circuitry to modulate an optical envelope at GHz frequencies, typical in data communications applications. Fig. 5.9 shows a simple data communications system consisting of an electrical driver (which is itself a subcircuit) driving a MZM fed by a CW laser. The output optical signal is then fed to a PD, creating a photocurrent. The photocurrent would then normally be fed to a TIA and subsequent electronics, but that is not shown in this simple example. The simulation of even such a simple system requires both electrical simulation (for the driver) and photonic simulation.

Various strategies have been employed to simulate these devices in a single simulator. For example, Verilog-A models [21–23] have been used for the optical devices, enabling simulation of the entire system with Verilog-A compatible simulators. However, Verilog-A models are challenging to scale more generally to multimode, multichannel, bidirectional waveguides. Another strategy is to simulate electrical components within commercially available photonic circuit simulators such as Lumerical Inc.'s INTERCONNECT [16]. Both approaches have fundamental limitations

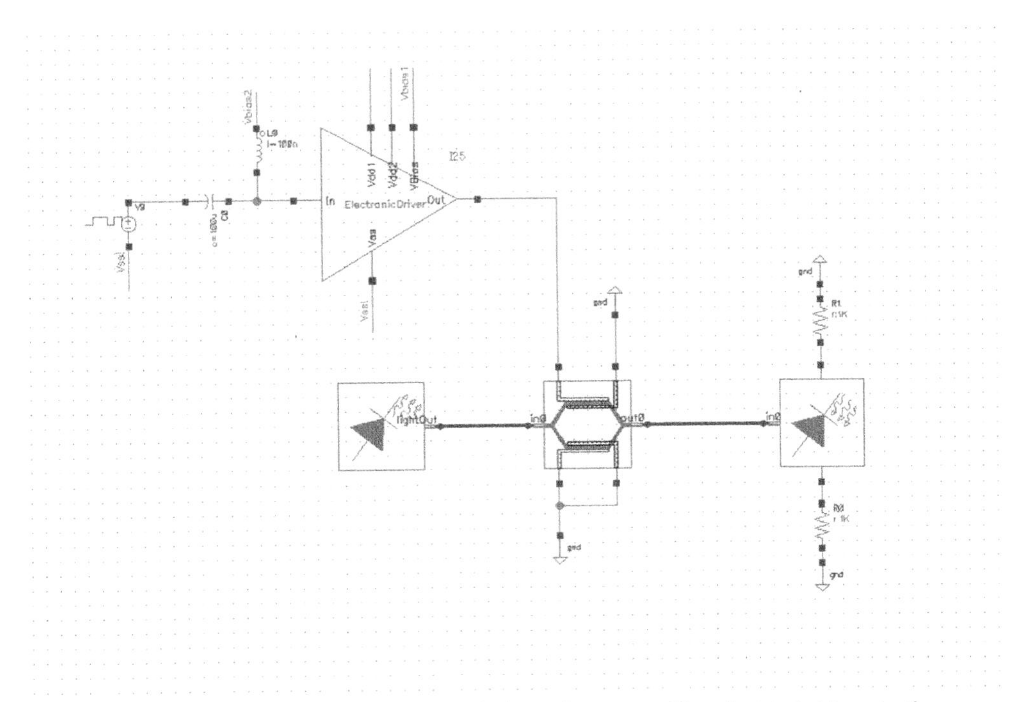

Figure 5.9 A schematic of a simple electrical/photonic system. The electrical driver, in the upper left, is itself a subcircuit consisting of many electrical components. It drives the voltage signal that is used to modulate a MZM, fed by a CW laser. The modulated optical signal is then fed to a PD where it is converted to a photocurrent. Typically, this photocurrent is fed to a TIA, not shown here.

because they require stretching a simulator beyond its intended purpose. Instead, it makes sense to combine the best simulators for each application through codesign or cosimulation.

In codesign, we can separate the circuits into an electrical portion and a photonic portion. The electrical simulation can be run first (with appropriate equivalent circuit loading to represent the electrical load of the optical modulator) to create a waveform that is subsequently passed to the photonic simulation. The waveform output (typically photocurrent) is then passed to a final electrical simulation for the receiving electronics. This approach is cumbersome for the end-designer, who must manage multiple simulations and pass waveforms between them. Furthermore, it does not allow for feedback between the optical and electrical simulators. As a result, full cosimulation is often necessary.

Cosimulation involves running both an electrical and photonic simulation at the same time, exchanging information while the simulation progresses, typically in the form of voltage or current. The timesteps of the simulators do not necessarily have to be identical. This is similar to digital/analog simulation in EDA but has an important difference: there is not always a clean separation of the electrical components from the photonic components. For example, the MZM shown in Fig. 5.9 is clearly a photonic component that translates an applied voltage into a modulation of the optical power. However, it is also an electrical component that loads the electrical driving circuitry with resistors, capacitors, and a diode. In other words, we cannot assume that it behaves as an infinite impedance device, and an electrical equivalent circuit should be included in the electrical simulation. Furthermore, some modulators, such as electro-absorption modulators, also generate a photocurrent that depends on both bias voltage and laser power. These challenges create a need for a symbol to be represented by compact models in both simulators, with the data connection points between them clearly defined.

This type of cosimulation flow is commercially available from Cadence [15] and Lumerical Inc. [16]. In Figs. 5.10 and 5.11, we can see the electrical and optical signals from a full cosimulation of a circuit similar to that shown in Fig. 5.9 but with an additional grating coupler to in-couple the laser light. The actual simulated circuit itself uses the TowerJazz photonic PDK for the photonic components [19] as well as a TowerJazz electrical PDK for the electrical circuitry.

5.6.6 Dealing with varying timescales

Electro-optical cosimulation provides an excellent solution for situations where the electrical signal is driven on comparable timescales (typically ns) to the modulation

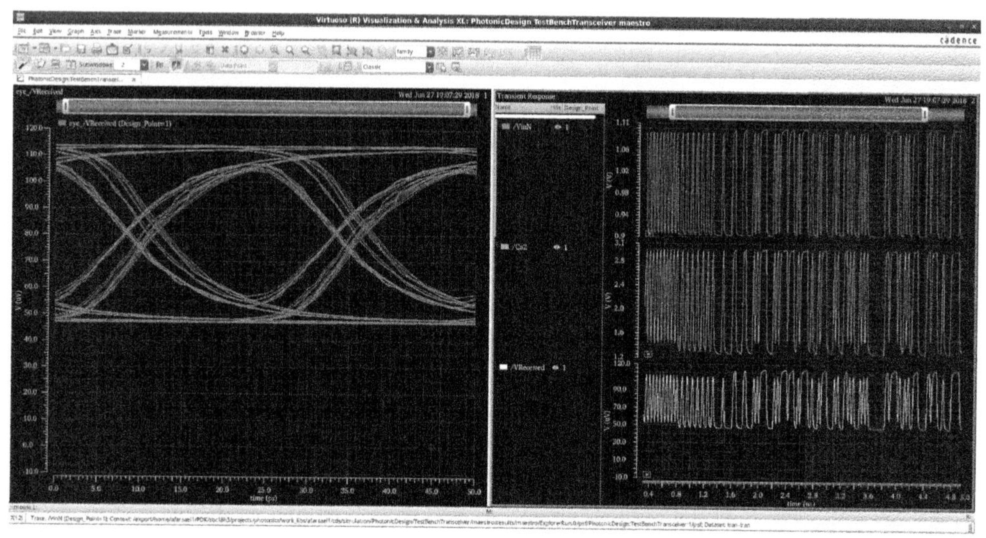

Figure 5.10 Various cosimulation results from with Cadence's Spectre (www.cadence.com) for the electrical simulation and Lumerical Inc.'s INTERCONNECT (www.lumerical.com) for the photonic portion of a circuit similar to that shown in Fig. 5.9. On the right-hand side, top to bottom, we see the input signal to the driver electronics, the driving signal that is fed to the modulator and the received signal, which is also used to create the eye diagram on the left.

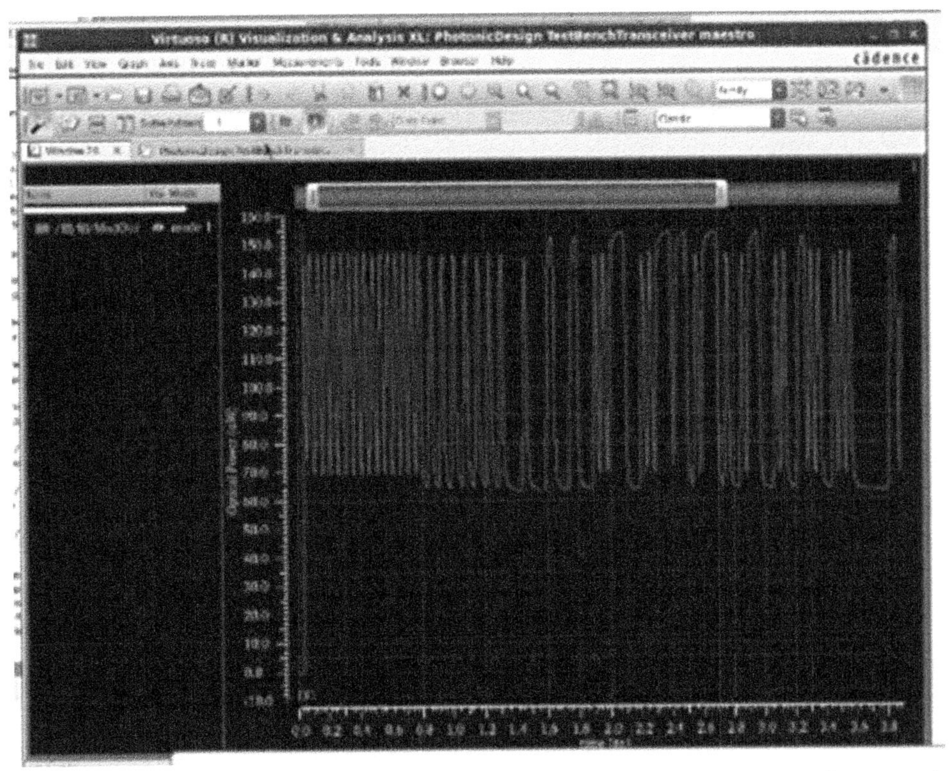

Figure 5.11 The optical power in the waveguide corresponding to the same simulation used to generate Fig. 5.10. Both electrical and photonic signals can be monitored from a single cockpit.

of the optical envelope or phase, typical in data communications, but also in other applications such as opto-electronic oscillators, which have strong feedback between optical and electrical circuits.

The simulation of PID controllers and other effects that occur on much longer timescales of milliseconds to microseconds can be more challenging because the simulation times may be unacceptably long, particularly when multiple simulations need to be completed, such as for optimization. This situation is typical of tuning electronics, although it can also occur for control electronics in sensing applications such as LIDAR. In some cases, it is appropriate to run long simulations, but it is also possible to employ an iterative strategy by having one simulation operating with a large timestep (such as microseconds) internally drive a sequence of secondary simulations that occurs on ns timescales. For example, in a PID controller simulation, each timestep of the PID controller might initiate a new simulation of a photonic circuit where a pseudorandom bit sequence is simulated on ns timescales, while the average optical power in a tap is monitored. The average optical power over the ns scale simulation is then used to update the PID controller simulation, allowing it to progress on the much larger timestep. Similar, iterative simulation strategies can be employed for thermal simulation.

5.6.7 Electrical, optical, thermal, mechanical

The cosimulation strategies discussed above (both direct and iterative) for electro-optical cosimulation can be extended to include other important effects such as thermal and mechanical. Indeed, it is possible for multiple simulators to cosimulate and exchange information such as voltage, current, temperature, thermal power, stress, and strain to simulate complex circuits involving many effects. An approach to this type of modeling has been patented [24], and we anticipate that it will become increasingly important, particularly for system-level simulation.

5.6.8 Circuit and system level

The cosimulation example presented above uses Spectre and INTERCONNECT. However, this can be easily extended to include larger-scale systems, typically using Cadence's Analog Mixed Signal simulator. This allows us to extend our simulation capabilities to include a variety of system-level effects such as thermal self-heating and electromagnetic parasitic effects from the copper pillars connecting the electrical and photonic chips, which we will discuss in more detail in "A System-level Vision" later in the chapter.

5.6.9 Other simulation types

In electrical circuit simulation, it is common to perform other types of simulation rather than transient. For example, in AC analysis an electrical circuit can be linearized around a given operating point such that small signal S-parameters can be rapidly extracted. Similarly, noise analysis is typically used to characterize a circuit. With electro-optical cosimulation, the traditional approaches must be extended in ways that are compatible with photonic circuit simulation. These new simulation types will be the subject of future work.

5.7 Photonics layout in electronic design automation

5.7.1 Photonics layout: curvilinear, non-Manhattan, and extremes of scale

As stated in the introduction, photonics layout brings into the design platform the need for curvilinear layout. We will review a commercially available approach developed by Cadence (supported for simulation by Lumerical Inc.) to provide a solution. Cadence's approach, driven by the need for accuracy and precision, supports a methodology that has a mathematical foundation and a discretization algorithm that minimizes error while maintaining the on-grid characteristic of the design. It is, however, possible to plug in alternative shape generation engines into the Cadence's Virtuoso Layout Editor Tool suite and have, for example, a different strategy, such as manhattanization of the shapes [25]. Both approaches have been proven with working design tapeouts. Indeed, the Virtuoso Design Framework offers the ability to integrate a third-party shape generation engine through interprocess communication in addition to allowing users to write their own in SKILL or use built-in functions.

Discretization of the underlying mathematical model of the design brings two challenges: generating the optimal number of points to represent the curve and ensuring that different element "stitch" perfectly together. The second challenge is accentuated in an EPDA system that targets allowing hierarchical separation of the design elements, as mathematical representations of different elements is done independently. In a system where capacity is not really an issue (like for small design, PC-based systems), it is possible to represent the full design in one mathematical space, resolve the continuity problem in that space, and have to deal with only the discretization of a continuous-by-design mathematical model. In our case, we must ensure that the methodology we follow guarantees continuity "by abutment" across independent mathematical models. The main issue of such discontinuity is snapping of the endpoints of the facets of a waveguide to database points (required for manufacturing) in the independent representations. A single database unit (DBU) (usually 1 μm, or less) difference can introduce a sliver between the facet and as a result significant light reflection, an example of which can be seen in Fig. 5.12.

As a result of this constraint, which is unavoidable to enable hierarchical design handling and the existence of precharacterized PDK elements, Cadence has invested

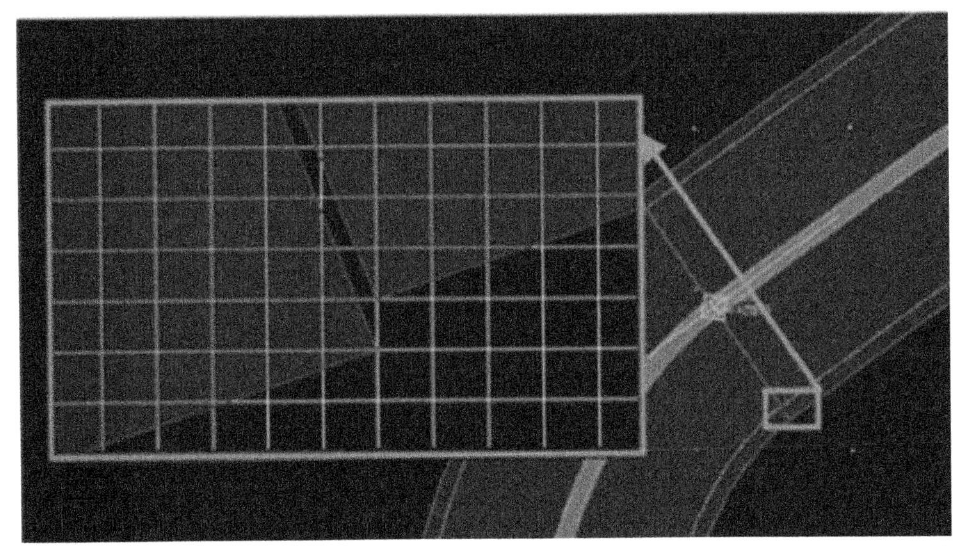

Figure 5.12 A waveguide abutment between two sections of the waveguide. The inset, zoomed-in view shows that there is an empty sliver between the waveguides, which is not acceptable for photonic devices and must be avoided.

significant effort early on to develop algorithms to convert the mathematical model to the physical model without creating such issues, regardless of the width or the angle of the waveguide facet.

At the other end of the size scale, most devices in photonics are several thousands of user units long. Thus, the mathematical geometrical object representations are both very large spatially and yet must be computed with very high accuracy, as discussed previously. In addition, manufacturing (the mask-making process) introduces a limitation on the number of points that a polygon can have.

Cadence's approach to jointly address the very large objects in the geometrical mathematical model (both in term of dimension and in term of vertices/points used to represent them) versus the point limitation in the physical/manufacturing model is to separate the two, and to defer the creation of the physical representation (OpenAccess polygons usually) to the last stage of the photonics layout synthesis process. Again, care needs to be taken to use points that are relevant to the actual layer manufacturing grid, not just the database grid.

Initially driven by the EDA user base to create a design environment for enabling photonics integration, Cadence and Lumerical Inc.'s strategy has been to extend the electronics design environment. There are three advantages to this strategy:

- A designer familiar with doing layout in the current environment can more easily augment his or her capacity to include photonics elements in the design, using already familiar tools, flows, and commands.
- The mixing of electronics and photonics is naturally enabled. Even for non-monolithic processes, there is generally a need to include heating resistors, pads, and wires for connections to modulators. All of this can take advantage of the electronics design tools, including automated routing, assisted wired editing, etc. This becomes even more valuable when the design uses a monolithic process allowing the mixing of active electronic components with photonics on the same substrate.
- The tool providers can significantly accelerate the delivery of the solution and build upon an existing, well-known, fully functional, and high-capacity design cockpit.

This strategy does have a price, forcing upon the photonics designers what some would call a heavy design methodology with which they are unfamiliar. Being aware of this is important while constructing the solution. Still, in the long term, this strategy is a good one, as it brings into the design cockpit the union of the needs on solid foundations.

5.7.2 Schematic driven layout

One of the largest benefits of the shared methodology is the reuse of the schematic driven methodology developed for electronics. This allows the user to create complex electro-photonics designs and identify LVS discrepancies at every step, from a connectivity perspective as well as a device sizing/parameter perspective. This capability is interactive and incremental, allowing in-design LVS as the works progresses. It is also bidirectional. Changes in the golden schematic are highlighted to the layout engineer as differences requiring resolution.

For photonics this can be even more efficient, since waveguides themselves are represented as devices, enabling direct forward and backward annotation of optical interconnect key parameters, rather than a postlayout extraction as for electrical circuit design.

5.7.3 The generation, characterization, and simulation of waveguides and connectors

In the schematic driven flow, all waveguides have a representation in the schematic which is used as a placeholder for the simulation model and its parameters. Based upon end user feedback, two strategies for creating waveguides have been developed that also ensure that the data required for simulation is always up-to-date:

- Composite waveguides, hierarchical elements composed of waveguide primitives (usually precharacterized), enable a very robust design methodology based upon elements with experimentally validated models. It is worth noting that this methodology is called "schematic driven" and, behind the scenes, the composite waveguide is assembled in a layout scratch pad that provides direct geometrical feedback to the designer.
- Generated connectors are much more flexible and allow the generation of any kind of geometries to connect elements. A table-based model parameter generation for such waveguides is provided to enable simulation.

5.7.3.1 Composite waveguides

A composite waveguide is a hierarchical wrapping around a set of waveguide elements. That hierarchical wrapping exists both on the schematic and the layout side. As a result, a set of descriptive parameters is shared between these two representations and that set is subject to in-design LVS and schematic/layout binding, enabling synchronization of the two. The parameters are fed into generators (schematic generators and layout generators using PCells), creating optical subcircuits that can be netlisted for simulation, or for physical layout implementation. The original idea for this comes from a micro-strip solution as described in Ref. [26].

5.7.3.2 Creation/editing

The creation and editing environment for the composite waveguide is a restricted, GUI-driven, interactive editor that has a preview window to see the actual waveguide being created, but that can be used from either the schematic cockpit or the layout cockpit. Because creation is fully controlled through the form, some automatic checks and assistance are built in, for example, to position properly (with respect to angle) a straight after an arc or to verify that widths are compatible.

The same GUI, as seen in Fig. 5.13, provides a visual layout preview in the schematic cockpit, allowing the specification of physically plausible composite waveguide elements, even in a nonlayout cockpit.

Finally, the subcircuit schematic corresponding to the composite waveguide shown in the left of Fig. 5.13 can be seen in Fig. 5.14, and this can be netlisted for simulation with Lumerical's INTERCONNECT.

5.7.3.3 Compose/decompose

There is a class of designer who feels restricted by the GUI-driven nature of the composition step and would prefer a more direct manipulation of the elements. To support them, a capability to decompose (flatten) a composite waveguide in the layout is provided to directly manipulate its constituents. The reverse operation (compose) allows a designer to take a set of connected waveguide components from the layout

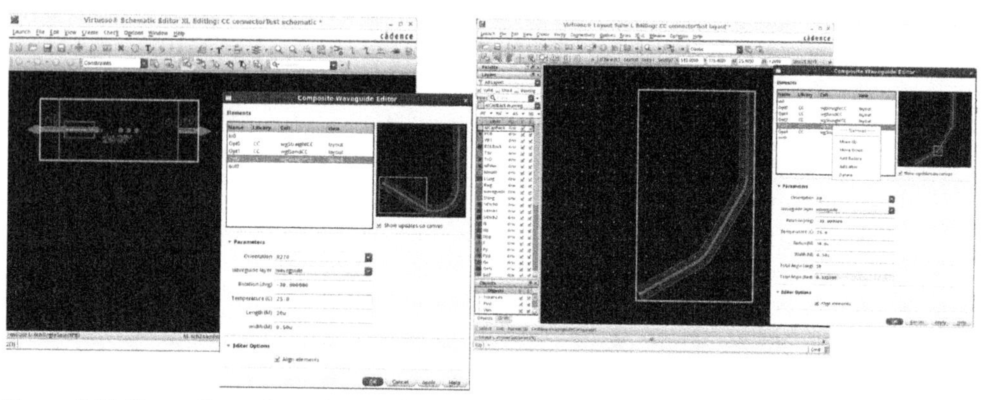

Figure 5.13 These illustrations show, side by side, the composite waveguide assembly GUI from the schematic cockpit and the layout.

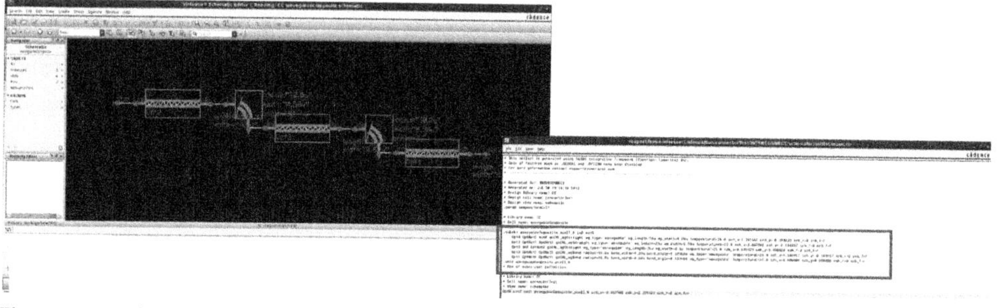

Figure 5.14 The subcircuit schematic for the composite waveguide created in the left of Fig. 5.13.

and "compose" them into a composite waveguide. During all these operations, the software continually monitors the golden reference schematic and the evolving layout to track what is added, changed, or even removed.

5.7.3.4 Generated connectors

Generated connectors enable the creation of a waveguide connection between two points based on mathematical constraints in the 2D plane. These connectors approach the functionality of routing, although it is perhaps aspirational to call this feature routing, as it is unaware of its surroundings (such as blockages or other elements in the way). Multiple types of connectors are required to implement a photonics layout, and we will review a few key and interesting ones here.

5.7.3.5 Curved connector

Curved connectors are composed of lines and circles connected by clothoids, where the curvature changes linearly with the path length. The connectors are constrained by the coordinates, the tangent angles, and the curvatures at the endpoint (end and start facets), and by the minimum allowed curvature radius $R_{\min}$ between the endpoints, which is related to the bending loss. An additional constraint may also be set on the maximum rate of change of the clothoid curvature, which is related to the connection loss.

In general, finding the connector for the given values of constraints is a multidimensional optimization problem. Taking a simplified approach where a suitable connector is found by considering several special cases will require solving only one nonlinear equation for one unknown. Some examples of such special cases are the connection between the straight line and a circle using a single clothoid, between two circles using a pair of clothoids with a line in between, between two straight lines using a pair of apex clothoids with a circle in between, and so on. Most of these special cases and the corresponding equations to be solved have been described previously by D.J. Waltong and D.S. Meek and the references within [27]. In some cases, an additional parameter can be used, such as the relative length of the line between the clothoids, which enables the generation of additional variants of each case by setting the parameter to different discrete values. From all connectors calculated for different cases, the optimal one is finally chosen based on the criteria such as the number of different curve segments, the total length of the connector, the minimum and the average curvature radius, and so on.

Fig. 5.15 shows some curved connectors from a portion of Cadence Virtuoso's regression test suite. These are generated automatically between two facets (width/angle/radius) placed in varying configurations. The color coding is intended to help understand and debug which type of element the connector has "chosen" to assemble itself (straight, arc, clothoids, etc.).

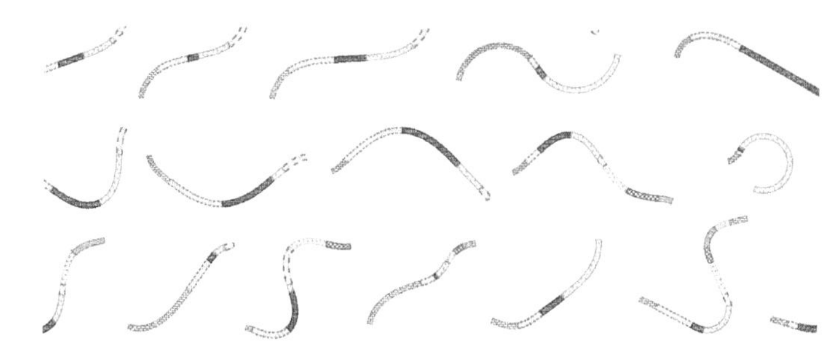

Figure 5.15 Some curved connectors from Cadence Virtuoso regression test suite.

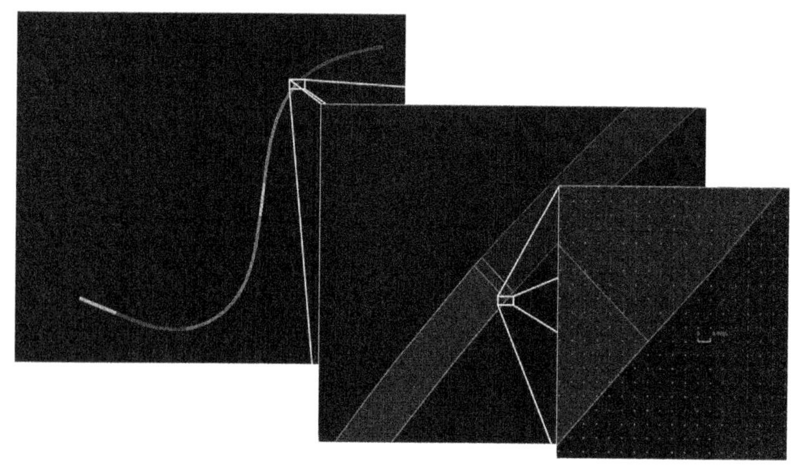

Figure 5.16 Smooth, on manufacturing grid connections between different types of elements in a complex curve connector.

In the case of such connectors, the mathematical model representing the total curve is fully self-contained, making it easier (but no less important) to ensure smoothness of the offset curves around the centerline, across the multiple types and snapping to manufacturing grid (Fig. 5.16).

5.7.3.6 The modal properties of generated waveguides for simulation

Generated connectors have a key distinction compared to precharacterized elements: they do not have a predefined simulation model that captures their exact behavior. This leads to the need for a parameterized model that can be used for the simulator to accurately model the important behavior of the waveguide affecting the design. This model must be driven from physical (geometrically derived) parameters that can be measured on the waveguide generated layout. Lumerical Inc.'s INTERCONNECT makes available such a model.

A typical waveguide model for the simulator consists of the effective index (n_{eff}), the group index (n_g), the dispersion (D), and the loss (α). (Higher-order dispersion terms can be easily added but are typically unnecessary for integrated photonic waveguides.) In addition, each of these properties can be affected by temperature. The transfer function for a waveguide, which represents the output electromagnetic field normalized to the input field, is simply given by

$$H(f) = \exp\left[-\alpha L + i\left(\frac{2\pi f_0}{c} n_{\text{eff}} + \frac{2\pi}{c} n_g (f - f_0) + \frac{\pi c}{f_0^2} D(f - f_0)^2\right) L\right],$$

where f is the frequency, f_0 is the frequency at which n_{eff}, n_g, and D are calculated, and L is the length of the waveguide. If two different waveguide sections are used, and there is no reflection at the interface, the total transfer function is $H(f) = H_1(f)$ $H_2(f)$ where H_1 and H_2 are the transfer functions of each waveguide of length L_1 and L_2 respectively. From this, and the properties of the exponential, we see that we can represent the transfer function of two waveguides as a single waveguide of length $L_1 + L_2$ with average alpha given by

$$\langle\alpha\rangle = \frac{\alpha_1 L_1 + \alpha_2 L_2}{L_1 + L_2},$$

and this is also true for n_{eff}, n_g and D. This can be rigorously generalized to an integral for adiabatically varying waveguides, which allows us to treat long waveguide connectors involving curves and varying widths as a single waveguide model with average properties as long as there are no reflections over the length of the waveguide or coupling between modes.

The task of the EPDA flow is now to ensure that these key waveguide parameters (loss, n_{eff}, n_g, D) are computed and reported on the instance for each relevant specification point (temperature/frequency/mode) required by the simulation. This is where the EPDA flow's direct binding between layout elements and schematic elements comes in handy. Provided the shape generator engine is capable of computing the parameters, transferring them on instances to drive the simulation is no more complicated than pushing a button, and requires no new development. It is the same feature that is used in the electrical flow to back-annotate any type of layout parameter onto the schematic (e.g., the actual source and drain areas of a transistor, after they have been laid out).

As described above, the mathematics to calculate the average modal properties is the same for all parameters. We need to integrate along the centerline path of the waveguide and bring in local values that depend on the local geometrical width and curvature. For example, in the case of n_{eff}, we need

$$\langle n_{\mathrm{eff}} \rangle = \frac{1}{L} \int_0^L n_{\mathrm{eff}}(w(l), \mathrm{curvature}(l), T, h, \lambda) \mathrm{d}l.$$

This expression can be discretized in a form that can be used by the shape engine, namely,

$$\langle n_{\mathrm{eff}} \rangle = \frac{\langle n_{\mathrm{eff}} \rangle_1 L_1 + \langle n_{\mathrm{eff}} \rangle_2 L_2 + \cdots + \langle n_{\mathrm{eff}} \rangle_N L_N}{L_1 + L_2 + \cdots + L_N}.$$

From a PDK creation perspective, the most difficult in all this is to gather the tables capturing the values for n_{eff} based on width, curvature, temperature, mode, and frequency. This can be automated with Lumerical Inc.'s MODE Solutions tool, and, fortunately, the table of values only needs to be calculated once. The loss, which comes mainly from sidewall roughness, is typically determined experimentally.

Fig. 5.17 shows the results of this computation for one of the waveguide connectors in the example design above shown in Fig. 5.18. The specifications for the calculation point (temperature, frequency, and mode) are specified by the user, while the waveguide profile is derived from the actual instance. As noted earlier, these values are available from either the layout or schematic view of the design—although the golden master for running the simulation is always the schematic in this methodology.

We can now run a simulation called Optical Network Analysis (ONA) on this design to calculate the S-parameters and observe the phase and group delays at the rightmost facet of the waveguides. The setup in Virtuoso ADE-Explorer is shown

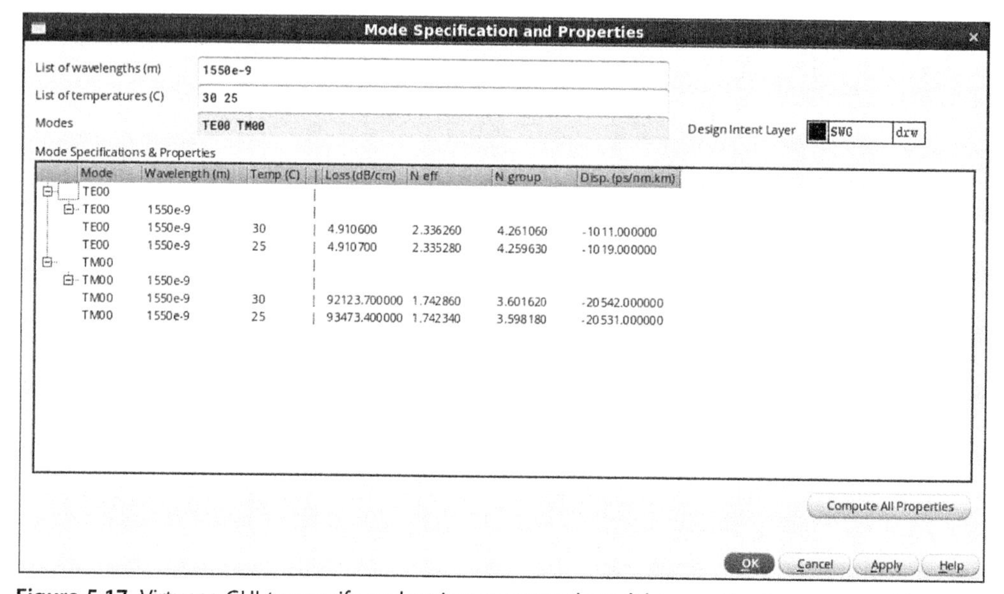

Figure 5.17 Virtuoso GUI to specify and review computed modal properties for a waveguide.

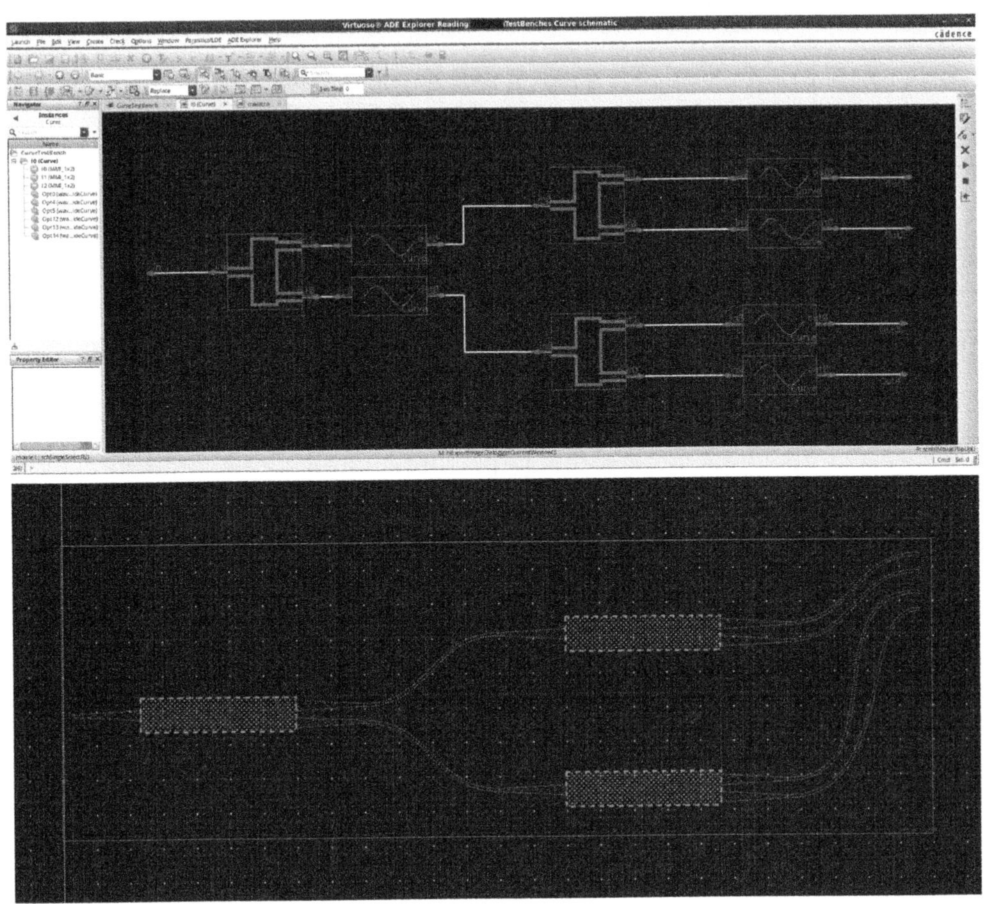

Figure 5.18 A schematic (*above*) and bound layout (*below*) example used to measure phase variation depending on layout for curve connector.

in Fig. 5.19, and the results are shown in Fig. 5.20. This capability is delivered in production PDKs today.

5.7.3.7 Fluid waveguides

Fluid waveguides are a special type of generated connector that takes additional input in the form of a user-defined set of way-points. For that reason, and with user assistance in the form of that extra set of constraints (list of points), it can "avoid" elements and go around, for example, another device or an electrical pad that would be in the way.

Building and editing such a function on top of the Virtuoso Layout Editor is not very complicated, since the editor already has the ability to create and edit devices such as guard-rings based upon the concept of a master centerline. The word "fluid"

Virtuoso® ADE Explorer Editing: PH18MATestBenches CurveTestBench maestro

Launch Session Setup Analyses Variables Outputs Simulation Results Tools Parasitics/LDE Window Help

cādence

27 Replace

Setup

CurveTestBench I0 (Curve) maestro

Name	Value
PH18MATestBenches:CurveTestBench:1	
Simulator INTERCONNECT	
Analyses	
ONA	start_and_stop Start-> 193.29e12 Stop-> 1.
Design Variables	
Parameters	
Corners	
Monte Carlo Sampling	
Checks/Asserts	

Name	Type	Details	Plot	Save	Spec
	signal	/I0/Out1			
	signal	/I0/Out2			
	signal	/I0/Out3			
	signal	/I0/Out4			
	expr	phaseDegUnwrapped(v("/I0:Out1:mode 1" ?result "ona-ona"))			
	expr	phaseDegUnwrapped(v("/I0:Out2:mode 1" ?result "ona-ona"))			
	expr	phaseDegUnwrapped(v("/I0:Out3:mode 1" ?result "ona-ona"))			
	expr	phaseDegUnwrapped(v("/I0:Out4:mode 1" ?result "ona-ona"))			
	expr	(groupDelay(v("/I0:Out1:mode 1" ?result "ona-ona")) * -1e+12)			
	expr	(groupDelay(v("/I0:Out2:mode 1" ?result "ona-ona")) * -1e+12)			
	expr	(groupDelay(v("/I0:Out3:mode 1" ?result "ona-ona")) * -1e+12)			
	expr	(groupDelay(v("/I0:Out4:mode 1" ?result "ona-ona")) * -1e+12)			

mouse L: M: R:

2(3) PH18MATestBenches CurveTestBench schematic Simulator: INTERCONNECT

Figure 5.19 INTERCONNECT ONA simulation setup in Virtuoso ADE-Explorer.

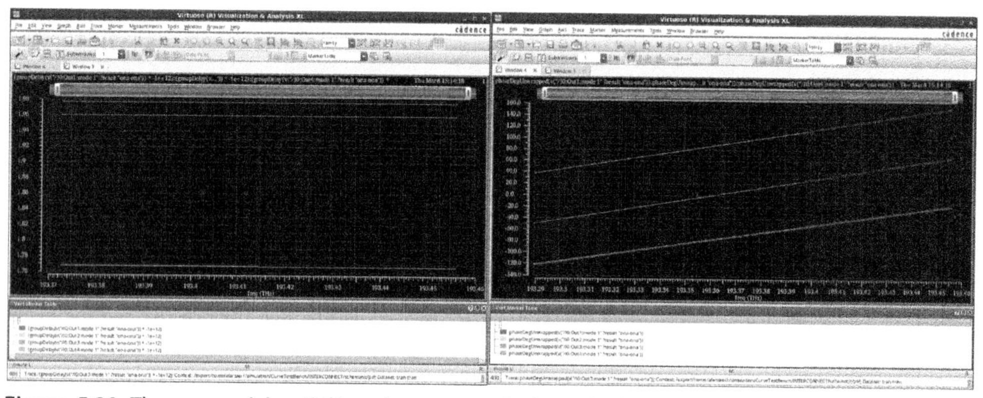

Figure 5.20 The group delay (*left*) and unwrapped phase (*right*) as simulated by INTERCONNECT, based upon the computed waveguide mode property parameters of connector elements.

comes from the fact that the end user can freely edit (stretch, reshape, chop, ...) the centerpath with direct edits in the canvas. After each modification of the centerpath, the generator (PCell) will trigger reevaluation of the waveguide shapes. The challenge is to properly decompose the set of segments into waveguide primitives; however, note that this is not black magic but is based mostly on applying basic trigonometry to compute the arcs in the section between noncollinear segments. Refinements can be added, like enabling S-bends when two consecutive arcs cannot really offer a good (low loss) transition between segments. Similarly, a long segment can be implemented in a way that automatically increases the width, to limit the losses on long paths. In Fig. 5.21 the white line is the master shape. Annotated above is the generated waveguide, as well as some interesting characteristics. It should be noted that the type and number of refinements here are mostly limited by the imagination (and requirements) of the user or PDK provider implementing the mapping between the master central path and the physical (waveguide) realization.

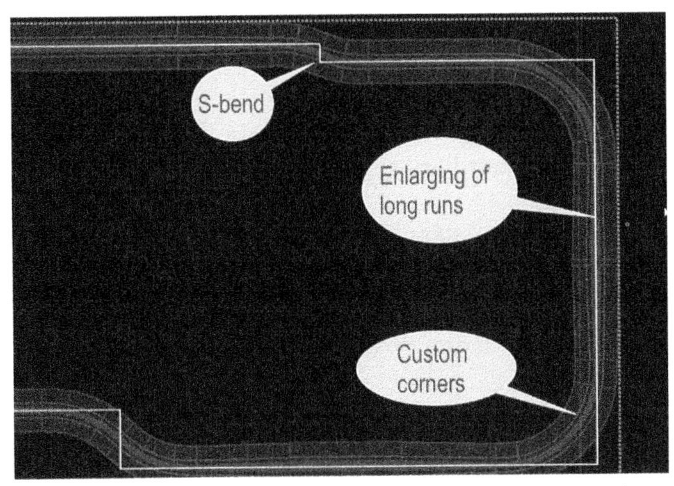

Figure 5.21 Annotated example of fluid waveguide as seen in the interactive layout editor.

The concept of fluid waveguide also brings us closer to automated routing. It is possible to generate the centerline master shape using more traditional EDA routing algorithms and then pass it on to the processing step that decomposes that centerline route and morphs it into a waveguide. This, of course, is a simplified view of the photonics waveguide routing challenge, but it is a good starting point.

Finally, we should note that this type of waveguide can be forward characterized with the method describe above, and therefore a simulation model can be dynamically generated and back-annotated to the schematic for immediate inclusion in simulation.

5.8 Electrical and photonic design in the same platform

The direct benefit of designing photonics in an electronics design platform is the inheritance of all electronics design features. So far, we reviewed how the electronics design features have been adapted or reused for photonics, but they are also usable as originally intended, that is, to say to design electronics parts. This results in a design environment that natively enables codesign of electronics and photonics in the same design space, with the same design paradigms.

5.8.1 A system-level vision

Virtuoso is a standard platform that enables multiple flavors of 2D, 2.5D, and 3D integration, along with the ability to support very complex and advanced mixes of processes. Bringing the photonics design in that environment means that, in addition to the actual PIC design task being performed, the designers can immediately benefit from the rest of the ecosystem tools required to assemble complex systems. We will explore two ways such features can be used and applied to an electro-optical system through an example, developed by Cadence and Lumerical Inc., that was originally presented to users in a workshop at Cadence in 2017. It was developed together with several partners to explore and solve various tool challenges faced by photonics designers.

In the example, an "active" photonics interposer is used that is well aligned with the direction that many companies choose to follow for their designs. The PIC itself acts as the interposer, meaning that the interposer is not limited to interconnect functions. Here it carries the active parts of a photonics quadrature phase shift keying (QPSK) transmitter. The schematic of the design is shown in Fig. 5.22, while a transversal view of the system is shown in Fig. 5.23. This design is similar to those shown in IMEC ITF 2018 presentations [28], confirming the plausibility and relevance of this example.

The system-level analysis explores two important design issues: first, the thermal impact of the CMOS chip on the photonics modulators, and second, the electromagnetic (EM) coupling in the copper pillars connecting the CMOS chip to the PIC and driving the modulators.

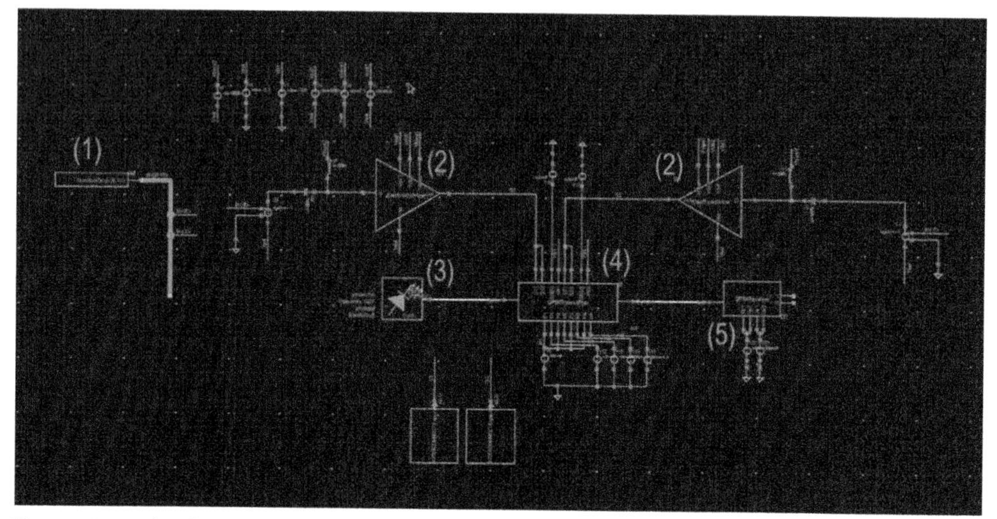

Figure 5.22 The base design for this is a QPSK transmitter/receiver. (1) is the digital source. In our case a random digital (Verilog-D) bit generator. The main point is to demonstrate the ability to take input from the digital domain in the simulation. The amplifiers (2) are used to condition the signal and are simulated in the analog domain. (3) is the light source (laser). In Cadence software, the connections to (4) are in a different color that represents an optical connection in the schematic editor. (4) is the transmit electro-optical circuit. In this example, we used an EAM. Finally, block (5) contains the receiving side of our experiment, diodes, and TIAs.

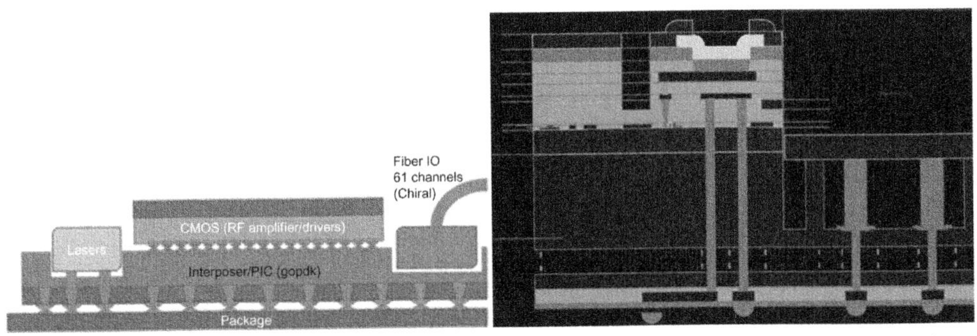

Figure 5.23 Vertical representation of the QPSK setup and transversal (not to scale) cut of our generic process for the active photonic interposer process.

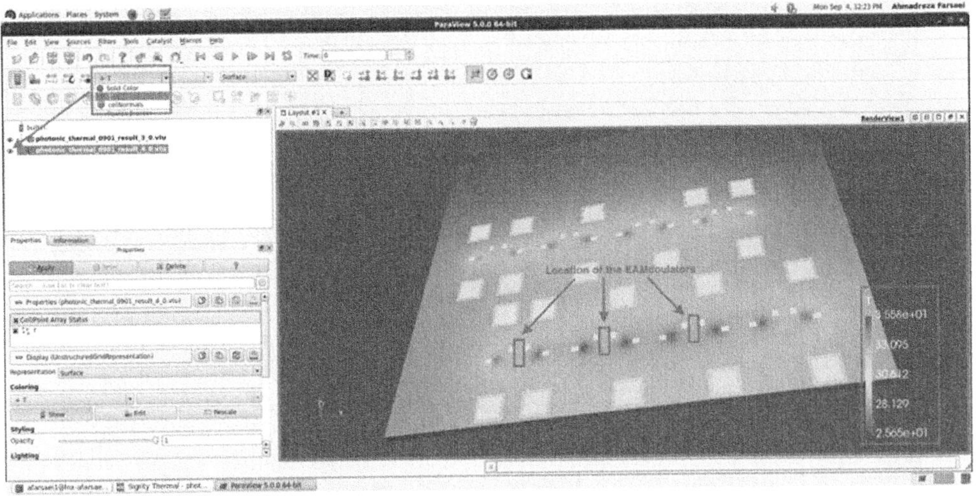

Figure 5.24 The thermal simulation setup and view of the temperature profile. For clarity, the position of the EAM in the PIC has been manually overlaid with rectangles to show their position in the plane of the CMOS chip where the thermal simulation was performed. Although they are not located precisely in the hottest regions, this image shows that they are in a region of the design where there is extreme heat.

5.8.2 Thermal impact analysis

This analysis demonstrates that the environment is capable of showing the thermal impact of switching transistors in the CMOS chip on the overall performance of the system. A regular analog simulation with Spectre-APS is performed, and the transistor activity in the CMOS chip is measured, as they are driving the modulators. The second step consists of extracting the thermal network for the system and performing a thermal simulation with tools like Cadence's Sigrity tool set. An image of the temperature profile of the CMOS chip can be seen in Fig. 5.24. The final outcome of this step is a new ambient temperature for the optical devices.

Finally, the new temperature of the EAMs can be updated and mix-mode mix-domain simulation can be rerun to see the result of the temperature change on the constellation diagram for the output signals. The results are shown in Figs. 5.25–5.27, where we can see in the constellation diagrams that the temperature change due to the heating from the CMOS chip has a detrimental effect on the EAM performance, which will result in a much higher BER for this QPSK transceiver.

This example shows that the tools and the model have are capable of performing this type of thermal system analysis, which is critical to making good decisions in 3D floorplanning. For example, designers should avoid placing the switching logic directly above the modulators.

Figure 5.25 The input signals into the system. The lower two signals are the digital signals, while the upper two signals are the output of the analog drivers.

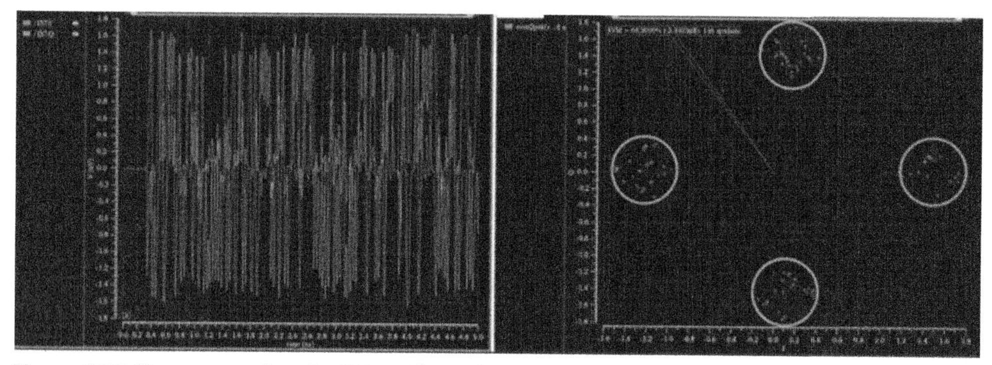

Figure 5.26 The output after the TIAs without thermal back-annotation. The constellation diagram (*right*) shows that the QPSK can transmit with a very low error rate.

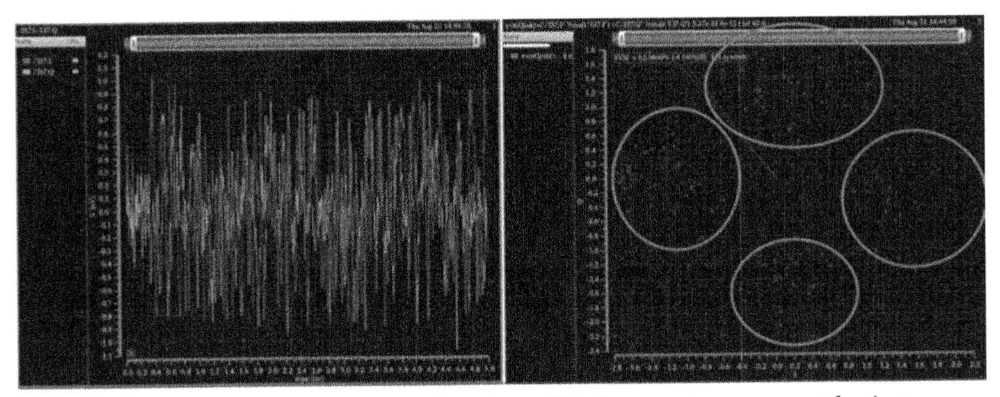

Figure 5.27 The output after the TIAs including thermal back-annotation to account for the temperature increase from the driving electronics (*left*). The constellation diagram (*right*) shows that the QPSK is no longer as focused as the original result of Fig. 5.26 and will have a much higher bit error rate.

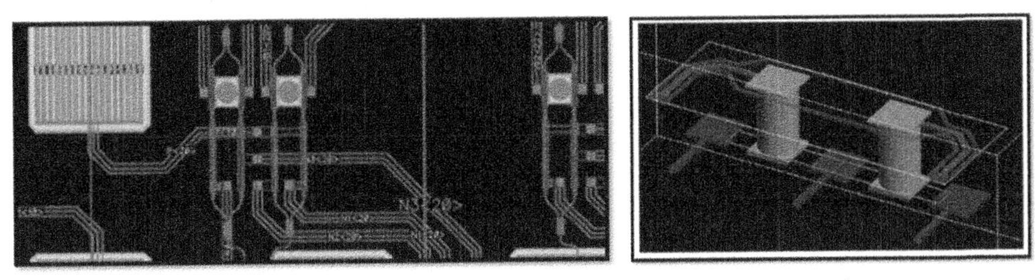

Figure 5.28 Layout and 3D view of the copper pillar connecting the active PIC to the CMOS driver.

5.8.3 Electromagnetic coupling impact analysis

This second example shows the effect of the EM coupling in the copper pillars that are used to get the very high-frequency signals from the CMOS chips to the EAM. The first step is accurate 3D models of the configuration on chip, shown in Fig. 5.28, and of the material stack. Note that these steps are not specific to photonics and are common to other 3D methodologies.

Once the 3D model is constructed, the EM solver extracts the S-parameter matrix representing the coupling of the copper pillars. Note that this is an electrical S-matrix rather than an optical one.

The extracted S-matrix is then back-annotated in the form of an n-port device in our system, as shown in Fig. 5.29.

Finally, the simulation can be rerun but this time with the EM coupling effects considered, and the results, for the materials and configuration chosen, are shown in Fig. 5.30. These can be contrasted with the original simulation results shown in Fig. 5.26. We can see that in this specific case, we have completely lost the constellation diagram and destroyed the performance of the QPSK transmitter!

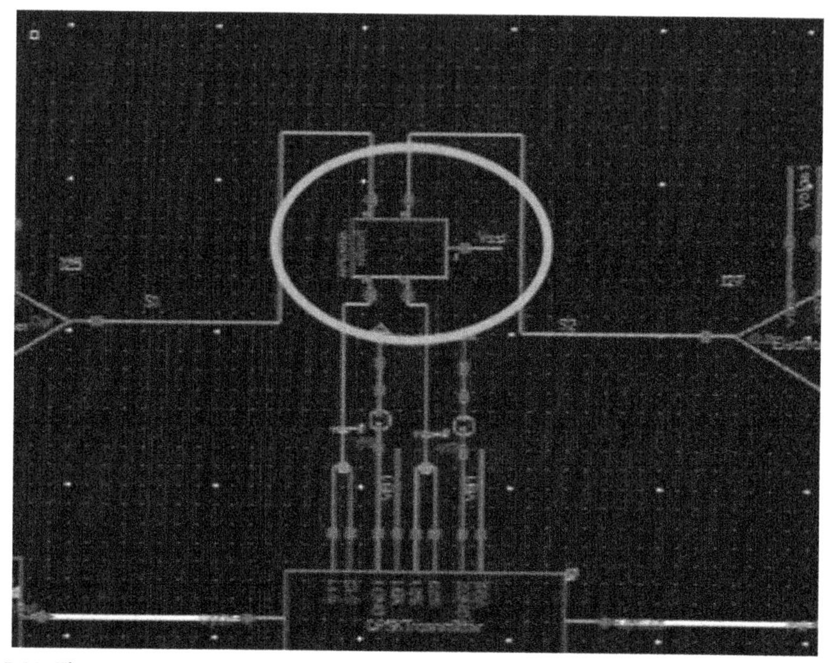

Figure 5.29 The copper pillars couplings are represented by an n-port device (highlighted with ellipse) with an associated S-parameter model.

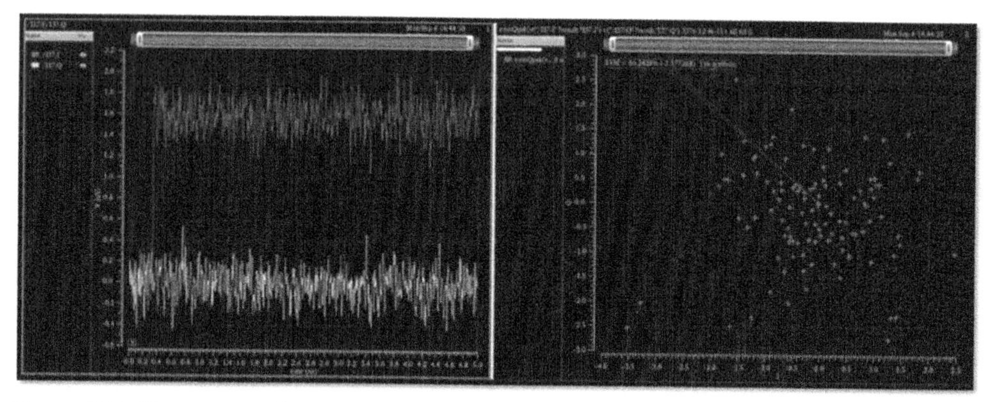

Figure 5.30 The output after the TIAs when the EM coupling effects of the copper pillars are accounted for (*left*). The constellation diagram (*right*) shows that the QPSK can no longer function at all. These results can be contrasted with the original simulation, which does not include the copper pillars, shown in Fig. 5.26.

Both of these examples show that a good QPSK transceiver design can be adversely impacted by thermal effects and EM coupling effects, and therefore it is critical to include these system-level effects in the analysis. Indeed, for this particular QPSK transceiver, the EM coupling effects can destroy the performance entirely.

Furthermore, the development of this type of advanced example allows tool designers to improve the overall design flow and enable a smoother user experience, as well as move toward more dynamic analyses in the future.

5.9 Conclusions

5.9.1 Today versus the future

The first and most important takeaway is that today there is a commercially available platform for doing integrated electro-photonics development. PIC chips and systems have been manufactured using this platform and are currently in use. However, it is clear that the road toward automation and integration is just at its beginning. Indeed, some of the examples we have described are still heavy in scripting and customization. Still, it is clear that as the need for photonics design grows, the tool providers will continue to invest in further development, which will result in ongoing improvements to the design flows and overall productivity.

5.9.2 Layout

Layout automation is one key area that lags behind for photonics design. We should, however, be clear to set proper expectations. Photonics layout is to some extent a close cousin to RF layout, and today there is very little available in term of circuit design automation in that area. But this is not a reason to stop development, and there are clearly many ways to more efficiently layout photonics chip. In particular, automatic routing of non-phase critical waveguides should be enabled. If not for full automation, positive assistance of the user while performing editing tasks will certainly help with productivity. Scripting is very common in the photonics community, and this is an advantage over the analog/RF design communities, which are yet to adopt even Verilog-A as a modeling language. Conversely, that advantage may turn out to also be the biggest factor preventing a migration toward EPDA design flows, as many photonics engineers hold dear their ability to create, in their corner of the lab, in their local version of MATLAB, their secret sauce for making the device magically work, sometimes. And this is a key of what EPDA is trying to bring to the integrated photonics table: efficiency, reproducibility, and predictability.

Finally, a practically efficient DRC and LVS methodology needs to be designed. Today DRC is barely alive for photonics, mostly because the discretized shapes represent a challenge for the highly tuned DRC engines that target advanced Manhattan technologies. The current approach tends to find ways to "wave" false violations based on additional mathematical information. It is entirely possible to think of a new set of photonics layer specific algorithms that would be aware of the underlying mathematical model for the shapes and that could perform true (mathematical) width or spacing checking, for example.

5.9.3 Simulation and design: statistical simulation and design for manufacturing

Photonic devices generally have extreme sensitivity to the environment and manufacturing imperfections. For this reason, it is important to be able to estimate the yield of photonic circuits and attempt to identify design centers where devices have optimal yield while meeting performance specifications. Due to the specific challenges of photonic devices, particularly the large size compared to the wavelength and the phase sensitivity, devices tend to be correlated over large distances, which creates additional challenges. Approaches to statistical modeling have been proposed [29] that can estimate yield. An important future step will be to integrate these approaches into the EPDA flows so that contributions from both electrical and optical circuits can be included. This will be an important area of development for EPDA design flows in the future.

Acknowledgments

The authors would like to thank the many team members working with us who have developed the EPDA design flow and examples shown here. In particular, from the Cadence side, Janez Jacklic and Ahmadreza Farsaei have provided a lot of support to the writing work and illustration-gathering process, and from the Lumerical Inc. side, Jackson Klein has provided support by reviewing drafts and providing key details on photonic circuit simulation.

References

[1] C. Doerr, Silicon photonic integration in telecommunications, Front. Phys. 3 (2015).
[2] W. Bogaerts, L. Chrostowski, Silicon photonics circuit design: methods, tools and challenges, Laser Photon. Rev. 12 (14) (2018) 1700237.
[3] L. Chrostowski, M. Hochberg, Silicon Photonics Design: From Devices to Systems, Cambridge University Press, Cambridge, 2015.
[4] P. Muñoz, Photonic integration in the palm of your hand: generic technology and multi-project wafers, technical roadblocks, challenges and evolution, Proceedings of the 2017 Optical Fiber Communications Conference and Exhibition (OFC), 2017, pp. 1–3.
[5] P. Muñoz, G. Micó, L.A. Bru, D. Pastor, D. Pérez, J.D. Doménech, et al., Silicon nitride photonic integration platforms for visible, near-infrared and mid-infrared applications, Sensors 17 (9) (2017) 2088.
[6] https://www.phiconference.com/market/comparison-between-inp-and-silicon-photonics/.
[7] L. Thylén, L. Wosinski, Integrated photonics in the 21st century, Photon. Res. 2 (2) (2014) 75–81.
[8] M.J.R. Heck, M.L. Davenport, J.E. Bowers, Progress in hybrid-silicon photonic integrated circuit technology, SPIE Newsroom (2013). Available from: https://doi.org/10.1117/2.1201302.004730.
[9] https://www.photonics.com/Articles/Light_Is_the_Ultimate_Medium_for_High-Speed/a61635.
[10] https://aimphotonics.academy/about/what-integrated-photonic.
[11] I. Artundo, Photonic integration: new applications are visible, Optik Photonik 12 (3) (2017) 22–25.
[12] C.V. Poulton, A. Yaacobi, D.B. Cole, M.J. Byrd, M. Raval, D. Vermeulen, et al., Coherent solid-state LIDAR with silicon photonic optical phased arrays, Opt. Lett. 42 (2017) 4091–4094.
[13] P. Sibson, J.E. Kennard, S. Stanisic, C. Erven, J.L. O'Brien, M.G. Thompson, Integrated silicon photonics for high-speed quantum key distribution, Optica 4 (2017) 172–177.

[14] https://www.cadence.com/content/dam/cadence-www/global/en_US/documents/tools/custom-ic-analog-rf-design/photonic-ic-design-wp.pdf.
[15] www.cadence.com.
[16] www.lumerical.com.
[17] http://www.aimphotonics.com/pdk/.
[18] https://www.imec-int.com/en/what-we-offer/iclink/photonic-integrated-circuits.
[19] http://www.towerjazz.com/prs/2018/0313.html.
[20] D. Marcuse, Calculation of bit-error probability for a lightwave system with optical amplifiers and post-detection Gaussian Noise, J. Lightwave Technol. 9 (4) (1991) 505−513.
[21] E. Kononov, Modeling photonic links in Verilog-A, PhD thesis, MIT, 2012.
[22] C. Sorace-Agaskar, J. Leu, M.R. Watts, V. Stojanovic, Electro-optical co-simulation for integrated CMOS photonic circuits with VerilogA, Opt. Express 23 (2015) 27180−27203.
[23] P. Martin, F. Gays, E. Grellier, A. Myko, S. Menezo, Modeling of silicon photonics devices with Verilog-A, in: 29th International Conference on Microelectronics Proceedings - MIEL 2014, 2014, pp. 209−212.
[24] U.S. Provisional Patent No. 62/639,469 filed March 6, 2018.
[25] L. Alloatti, M. Wade, V. Stojanovic, M. Popovic, R.J. Ram, Photonics design tool for advanced CMOS nodes, IET Optoelectron. 9 (2015) 163−167.
[26] Microstrip Design Acceleration, CDNLive 2014 Silicon Valley, Sergey Yevstigneev, Freescale.
[27] D.J. Walton, D.S. Meek, A controlled clothoid spline, Comput. Graph. 29 (2005) 353−363.
[28] IMEC - ITF USA. Dr. Philip Absil, "Scaling bandwidth with photonics," 2018.
[29] J. Pond, J. Klein, J. Flueckiger, X. Wang, Z. Lu, J. Jhoja, et al., Predicting the yield of photonic integrated circuits using statistical compact modeling, Proc. SPIE, 10242, Integr. Opt. Phys. Simul. III, 2017, pp. 102420S.

CHAPTER 6

Optical processing and manipulation of wavelength division multiplexed signals

Leif Katsuo Oxenløwe[1], Frederik Klejs[1], Mads Lillieholm[1], Pengyu Guan[1], Francesco Da Ros[1], Pawel Marcin Kaminski[1], Metodi Plamenov Yankov[1], Edson Porto da Silva[2], Peter David Girouard[1] and Michael Galili[1]
[1]DTU Fotonik, Technical University of Denmark, Lyngby, Denmark
[2]Federal University of Campina Grande, Campina Grande, Paraíba, Brazil

6.1 Introduction

Optical networks today convey hundreds of terabits per second to destinations all over the world, carried by multicolored laser signals [1]. As the need for more and more bandwidth has kept increasing, the utilization of the light signals has become increasingly efficient, to the extent of being only a fraction of a decibel from fundamental limits in terms of spectral efficiency [2]. There is still some capacity gain to obtain from battling nonlinearities in optical transmission [3], but it is becoming increasingly obvious that the energy footprint left by the global communications infrastructure is of such alarming proportions that addressing this must become one of the next big frontiers. Today, 9% of all electricity is estimated to be used for communications worldwide [4,5], and with 20%–30% traffic growth rates [1], reducing the required energy per bit, yet increasing the offered capacity, is of utmost importance. This is already being addressed in several ways, such as making more energy-efficient lasers and switching elements [6,7], making use of single wideband laser sources (frequency combs) flexibly serving multiple channels and formats [8,9] and supporting extremely high data rates [10–12], increasing the rates of spectrally efficient single channels [13,14], spatially combining multiple channels in multicore or multimode fibers [15–18], and addressing the whole network across layers to arrive at optimum architectures [19], to mention a few efforts. It is also becoming increasingly important to utilize the installed network resources optimally. With the Internet of Things, Industry4.0, and cloud computing and storage, there will be heavy demands on supporting many different types of traffic stemming

Optical Fiber Telecommunications VII
DOI: https://doi.org/10.1016/B978-0-12-816502-7.00007-5

from, for example., medical diagnostics, regulating indoor climate, and autonomous cars, which will have different requirements as to latency, bandwidth, reliability, security, and cost or energy consumption. So there is a growing need for using energy efficiently, as well as utilizing network resources efficiently, in a manner that ensures a large degree of network flexibility, enabling heterogeneous flexible bandwidth channels, and that supports changes in demand and quality of service for varying applications [20].

This chapter will discuss several optical processing schemes that aim to provide such flexibility in a network by manipulating wavelength division multiplexed (WDM) signals. In particular, the chapter will focus on concepts that allow for simultaneous processing of multiple parallel wavelength channels in a single optical processing unit, as this allows for multiple channels to share the energy cost and hence collectively drive down the required energy per bit. The focus will be on coherent and phase-sensitive schemes that can handle signals that are modulated both in phase and amplitude.

We will describe the use of optical time lenses for flexible manipulation of WDM signals to, for instance, magnify or compress the WDM grid for better resource utilization in flexible networks. We will describe how a time lens system allows for scalable WDM regeneration by converting the WDM signal into a high-speed time domain multiplexing (TDM) signal, which is straightforward to optically regenerate using a single phase-sensitive amplifier (PSA). We will also describe how a PSA is useful for decomposing a quadrature amplitude modulation (QAM) signal into its quadratures, for potential subsequent processing.

We will describe another way of processing WDM signals to combat nonlinear impairments, namely the use of optical-phase conjugation (OPC). We will also describe a novel way to complement the optical conjugation by digital processing, by the introduction of probabilistic shaping (PS) for the OPC channel.

Finally, we will review the status of research into efficient nonlinear materials that are required to make these optical schemes attractive for real-world applications, and we will introduce a figure of merit (FOM) for practical nonlinear devices used for optical communication systems.

6.2 Time lenses and phase-sensitive processing

Time lenses have proven to be extremely versatile tools. With time lenses, one can convert from frequency to time, and back again, and coherently stretch and compress signals in either dimensions. This can be very useful for optical communication data signals, and in this section, we will describe the basics of time lenses and various interesting applications that utilize the merits of time lenses to arrive at potentially energy-efficient processing units.

6.2.1 Fundamentals: principle and potential benefits

6.2.1.1 Space-time duality

The concept of space-time duality was first reported on in Refs. [21–23], where it was remarked that the equations governing the behavior of thin lenses combined with diffractive propagation are analogous to that of a parabolic phase modulator, called a time lens, combined with chromatic dispersion; this duality is illustrated in Fig. 6.1. Whereas the aperture of the spatial Fourier system at the input focal plane determines the projected image at the output focal plane of a thin lens, a corresponding input pulse shape to a temporal Fourier system dictates the output pulse shape.

Eq. (6.1) shows the phase terms imparted by a thin lens on spatial coordinates (left), as well as that of a time lens (right),

$$\phi(x, y) = \frac{k(x^2 + y^2)}{2f}, \quad \phi(t) = \frac{t^2}{2D} \tag{6.1}$$

where f is the thin lens focal length, k is the wave vector, and D is the accumulated dispersion, which is the temporal analogy to the focal length of a time lens. Just as $1/f$ is the power of a conventional lens, the power of a time lens is given by the imparted linear chirp rate $K = 1/D$. Implementation of a temporal imaging system therefore requires the construction of a system including a time lens which imparts the parabolic phase shift in time mirroring the spatial lens. A special case of such a system is the temporal Fourier system which performs an optical Fourier transformation (OFT) shown in Fig. 6.1, where the input is dispersed by accumulated dispersion D, then receives the parabolic phase shiftaccording to a time lens which imparts a chirp rate K, and then is dispersed by D again to obtain the Fourier transformation of the input pulse at the output temporal focal plane, in complete analogy with a spatial Fourier system. Here the OFT condition $K = 1/D$ must be satisfied. In order to obtain an undistorted image of the input pulse at the output of the time lens system, the temporal imaging condition needs to be fulfilled, analogues to the well-known lens maker's formula. Assuming a thin lens surrounded by air with refractive index ≈ 1, the imaging condition spatial lens system is shown in Eq. (6.2), (left)

$$\textit{Thin lens: } \frac{1}{S_1} - \frac{1}{S_2} = \frac{1}{f'}, \quad \textit{Time lens: } \frac{1}{D_1} + \frac{1}{D_2} = \frac{1}{D} \tag{6.2}$$

where s_1 is distance from the object plane to the lens, and s_2 is the distance from the lens to the image plane, with image magnification factor. Similarly, a pulse subjected to dispersion D_1 will be imaged by a time lens with temporal focal length D after dispersion D_2 with magnification factor if Eq. (2.2) (right) is satisfied. Here the sign of the magnification factors indicate whether or not the image is inverted.

If rather the Fourier transformation of a pulse is required, Eq. (6.3) shows the final form of the temporal and spectral profiles after the so-called "complete" OFT seen in

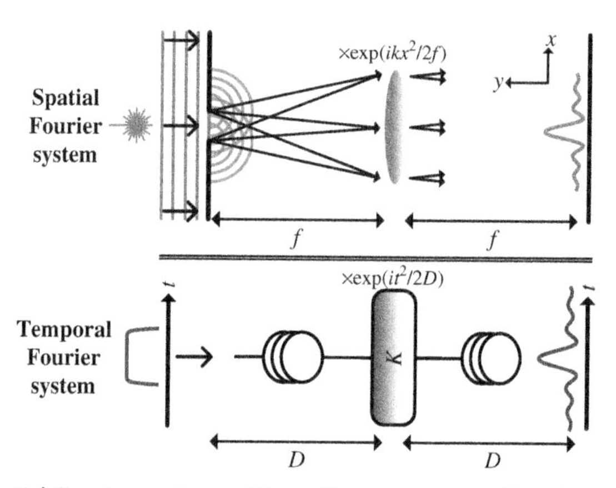

Figure 6.1 Top: A spatial Fourier system with a slit aperture resulting in a sinc-shape in the far field. Bottom: A temporal Fourier system with a square shape in time at the input. The pulse is then transmitted through dispersive fiber and then passed through the time lens that imparts a phase shift, fulfilling the temporal imaging condition on the pulse. Then the pulse is again passed through dispersive fiber. The output of the system is then a sinc-function shaped pulse in time, analogous to the spatial case.

Fig. 6.1 (bottom). In this case, the temporal profile with optical field E_{input}, is transferred to the spectral domain, and vice versa. Equation (6.3) relates the OFT temporal output field, where it can be observed that the output temporal profile is the initial spectral profile with scaling $\omega = t/D$.

$$E_{Output}(t) = \sqrt{\frac{i}{2\pi D}} E_{Input}\left(\omega = \frac{t}{D}\right) \tag{6.3}$$

The corresponding spectral output described in Eq. (6.4) similarly shows that it is a version of the input temporal profile with scaling $t = -\omega D$.

$$E_{Output}(\omega) = \sqrt{i2\pi D E_{Input}}(t = -\omega D) \tag{6.4}$$

Hence, when the complete OFT is carried out, the temporal features are represented in the spectral domain, and vice versa, scaled depending on the parameter $D = 1/K$. If only one domain is of interest after the OFT, simpler "partial" OFT systems can be used, where either the temporal or spectral profile is transferred to the spectral and temporal domains respectively, and not vice versa. For example, analogous to Fraunhofer diffraction, where the Fourier transformation can be obtained after sufficient propagation distance, the spectral profile of a pulse can be transferred to the time-domain, using a sufficient amount of dispersive propagation. However, the minimum dispersion requirement limits the available parameter space of the OFT system design. On the other hand, if either a time-to-frequency conversion or a frequency-to-time conversion is required, these can be accomplished using a single dispersive

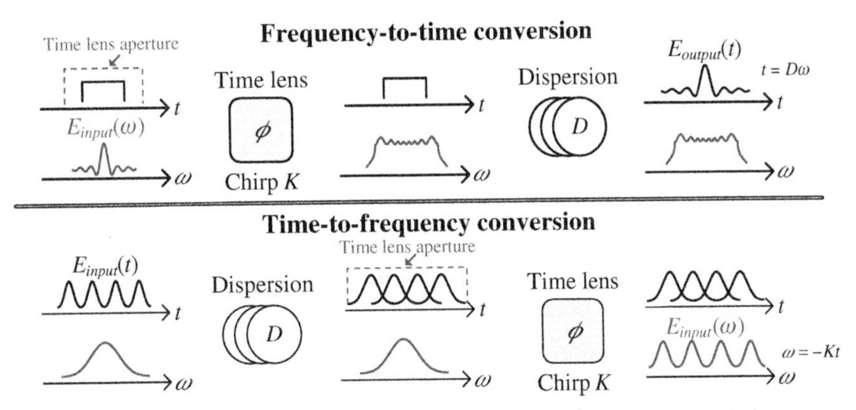

Figure 6.2 Top: Frequency-to-time conversion. Bottom: Time-to-frequency conversion.

element and a time lens. In this way the scaling factors can be designed without the same constraint. Time-to-frequency conversion can be obtained using a time lens with chirp rate K, followed by a dispersive element with dispersion $D = 1/K$, as shown in Fig. 2.2. Also shown, is the corresponding frequency-to-time conversion, which can be obtained by a dispersive element followed by a time lens where D and K again satisfy the OFT condition.

Note that in these cases, residual chirp broadens the output spectrum of the frequency-to-time conversion, whereas residual dispersion broadens the output pulses after time-to-frequency conversion. Hence, these OFTs are not "complete", in the sense that the output is not transform limited. However, in both cases these partial OFTs can be completed simply by adding an element to the output which has identical function to the first element of the system, meaning the time-to-frequency conversion becomes complete by inserting another element with dispersion D at the end (cf. Fig. 6.1 (bottom)). One of the challenges to overcome when operating and constructing these systems is the amount of higher-order dispersion in realistic dispersive elements, which leads to inaccuracies similar to the aberrations of spatial lenses [25]. Another cause of temporal aberrations is edge effects of the chirp implementation, for instance if the signal pulse overlaps imperfectly with the temporal aperture, or if the phase shift is not perfectly parabolic for the full duration. Time lenses have applications in signal processing as they have the potential for energy efficient, all-optical processing of signals. An example of one such application is the temporal equivalent of a refracting telescope, i.e. an arrangement of time lenses in a system for spectral magnification, consisting of two time lenses separated by the sum of their temporal focal lengths [24]. The principle is illustrated spectrographically for spectral magnification of a wavelength-division multiplexing (WDM) signal with subcarrier spacing Dw_{input} in Fig. 2.3. Here the time lenses are based on degenerate four-wave mixing (FWM). In this $c^{(3)}$ nonlinear process a signal field E_s, mixes a pump field E_p, to generate an idler field according to the relation $E_i = hE_p^2E_s^*$, and h is the FWM conversion efficiency. Hence, the idler becomes a phase conjugate copy of the signal, with twice the pump phase imparted on it. For a FWM-based time lens the pump ideally consists of flat-top pulses with a parabolic phase, which are broad enough to encompass the input pulses. Hence, after the FWM process the time lens output becomes the generated idler, which is a linearly chirped copy of the input. This "spectral telescope" is essentially a frequency-to-time conversion followed by a time-to-frequency conversion. By using the scaling relations it is simple to show that the output spectrum is a version of the input spectrum scaled by a factor $M = -K_2/K_1$, where K_i is the chirp rate of the ith pump, with temporal focal length $D_i = 1/K_i$. Thus, the pulses of the first pump have chirp rate $K_1/2$, which imparts chirp rate K_1 to Idler$_1$, the output of the first time lens. Then Idler$_1$ is propagated in a medium with accumulated dispersion D_1 to complete the frequency-to-time conversion, whereby the parallel input signal is converted to a serial signal. The serial signal constitutes the input to the time-to-frequency conversion stage, where it is propagated in a medium with dispersion D_2, before parabolic phase modulation with chirp rate $K_2 = MK_1$ by Pump$_2$ in the second time lens. The resulting Idler$_2$ is then the M times spectrally magnified WDM signal.

The OFT has found applications e. g. for all-optical regenerator of WDM signals [26], as described in detail in section 6.3. Additionally, time lenses have found several applications in sensing where the ability to measure the temporal response of phenomena at very high resolution is beneficial [27]. Finally, the required parabolic phase shift can be realised in several ways but the three most common are FWM, electro-optic modulation or cross-phase modulation (XPM). The implementation of FWM has been described above and a system deploying it shown in Fig. 2.3. An electro-optic modulator driven with a sinusoid, will impart an approximately parabolic phase shift with ~15% duty cycle near the peak or trough of the sine curve. The accuracy of the approximation will be reduced further away from the peak or trough of the sine curve. Hence, if the signal field interacts with the non-parabolic parts of the sinusoidal phase modulation, it will be a source of certain aberrations. Cross-phase modulation (XPM) is another $\chi^{(3)}$ nonlinear process, resulting in the phase of a signal being modulated by the intensity of a pump field. By co-propagating a signal through a nonlinear medium with a pump field which has a parabolic intensity profile, the signal becomes parabolically phase modulated.

6.2.2 Flexible spectral manipulation of wavelength division multiplexed signals

As described above, time lenses can be combined into telescopic arrangements to realize spectral or temporal magnification/compression; the former was shown as an example in Fig. 6.3. The main idea is that one may view it as a frequency-time conversion followed by a time-frequency conversion with a possible scaling between them. Such a tool may be very useful in networks where one may wish to

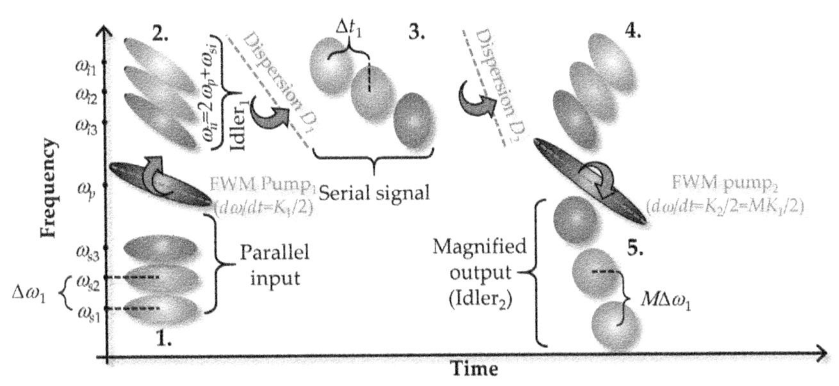

Figure 6.3 Bottom left (1): The input to the system is a WDM signal with three channels which temporally overlap with a FWM pump, shown in black. (2) The generated FWM idler consists of linearly chirped copies of the input channels with frequency overlap. (3) Dispersion is applied to the idler to obtain a serial multiplexed signal, with pulse spacing $\Delta t_{p\text{-}s} = D_1\Delta\omega_{input}$, thus completing a frequency-to-time conversion. (4) The second dispersion stage broadens and linearly chirps the pulses. (5) The second FWM stage compensates for the present chirp converts the dispersed serial signal to a WDM signal without spectral overlap and $M\Delta\omega_{input}$ spacing. Thus, the output consists of the input WDM signal scaled by the magnification of the temporal telescope. It is noted that some residual dispersion is present at the output.

reconfigure the WDM grid following needs for more or less available bandwidth. For example, a number of 25 GBaud WDM channels on a 200 GHz grid may be spectrally compressed to 50 GHz spacing to achieve a higher spectral efficiency and to open up freed bandwidth for other channels to be transmitted on the same line. One may also wish to spread the WDM channels over a larger spectral range to allow for multiplexing new channels in between, or simply to reduce the spectral efficiency to obtain higher optical signal-to-noise ratio (OSNR) tolerance. Such time lens units may be applied directly to installed WDM transmitters and receivers, and thus add a new level of flexibility. In this section, we will discuss the option of spectral compression, noting that spectral magnification is equally simple to implement and has indeed also been demonstrated [28].

6.2.2.1 K-D-K for spectral compression

A time-lens telescope for spectral compression is illustrated in Fig. 6.4. This configuration is also referred to as K-D-K, which indicates the order of the chirping (K) and dispersive (D) elements.

Combining the expressions for the individual K-D and D-K transforms, one can arrive at the expression governing the full imaging system from Fig. 6.4 as

$$U_{KD\text{-}DK}(\omega) = \sqrt{-\frac{C_1}{C_2}}e^{-\frac{i}{2C_2}\left(1+\frac{C_1}{C_2}\right)\omega^2}U\left(-\frac{C_1}{C_2}\omega\right) \tag{6.5}$$

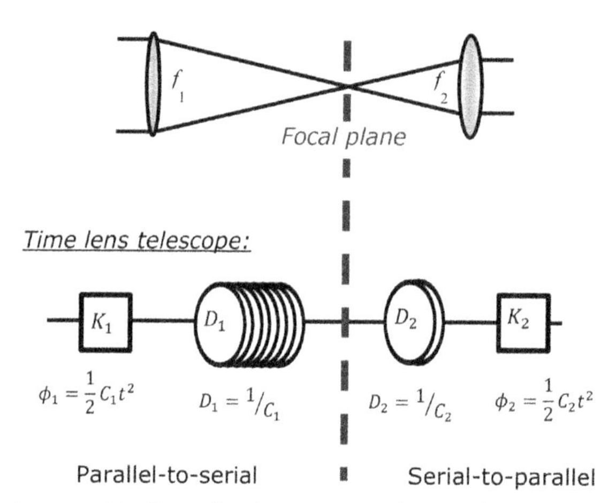

Figure 6.4 Two time lenses with dispersion between can be used to create a "spectral telescope," either magnifying or compressing the outgoing signal based on the chosen chirp rates.

which corresponds to a spectral telescope. Introducing the magnification factor $M=-C_2/C_1$ as a real and positive value, the K-D-K system is expressed as

$$U_{KD-DK}(\omega)=\sqrt{\frac{1}{M}}\,e^{-\frac{i}{2C_2}\left(1-\frac{1}{M}\right)\omega^2}\,U\left(\frac{\omega}{M}\right) \tag{6.6}$$

where $M<1$ corresponds to spectral compression, while $M>1$ corresponds to spectral magnification. The two chirp terms must have opposite signs in order to avoid spectral inversion and to reduce temporal broadening at the second time lens. Total dispersion between the two time lenses is given by

$$D=D_1+D_1=\frac{1}{C_1}\left(1-\frac{1}{M}\right)=-\frac{M}{C_2}\left(1-\frac{1}{M}\right) \tag{6.7}$$

The residual dispersion imposed on the signal in Eq. (6.6) is found to be

$$D_{\mathrm{res}}=-\frac{1}{C_2}\left(1-\frac{1}{M}\right)=\frac{\mathrm{D}}{M} \tag{6.8}$$

The magnification or compression is directly determined by the ratio of the chirp applied in the two time lenses. For spectral compression to occur, the chirp rate of the first time lens must exceed that of the second. Consequently, the dispersion from the D-K stage is larger than in the K-D stage, which may cause the temporal waveform to broaden past the duration of the temporal aperture of the signal chirping. This should be avoided, as it will cause signal distortions. While M only depends on the ratio of the chirp terms, the signal quality may depend on the magnitude. The required dispersion depends on the sum of the reciprocal chirp rates, so using higher chirp rates reduces the required dispersion (allowing for more of the signal to remain within the periodic phase modulation window at the second time lens).

6.2.2.2 Demonstrations of spectral manipulation using time lenses

A number of demonstrations of WDM manipulation using time lenses have been reported. In Ref. [29], spectral compression with a factor two ($M = 0.5$) was shown. More recently, compression of a WDM signal from a 100-GHz grid to a 28-GHz grid was demonstrated in [30]. We will discuss this demonstration in more detail below. In practical implementations it is convenient to implement the chirping of the signal by four wave mixing (FWM) rather than direct electro-optic phase modulation. Using a chirped FWM pump, one can impose chirp on the generated idler that far exceeds what is achievable by electro-optic modulation. Dispersion is quite easily imposed on the signal by transmitting it through optical fiber with the desired dispersion. Alternatively, gratings or other diffractive systems can be used for reduced size or enhanced flexibility. With these elements, a time lens telescope can be constructed for spectral compression, as illustrated in Fig. 6.5. The setup is clearly reminiscent of Fig. 6.3 for spectral magnification, but the chirp rates and dispersions are different here, and the setup corresponds to going backwards through Fig. 6.3.

In Ref. [29], a 100 GHz grid is compressed to 50 GHz grid for eight WDM channels with 10 Gbit/s differential phase shift keying (DPSK) data, and in Ref. [30], the

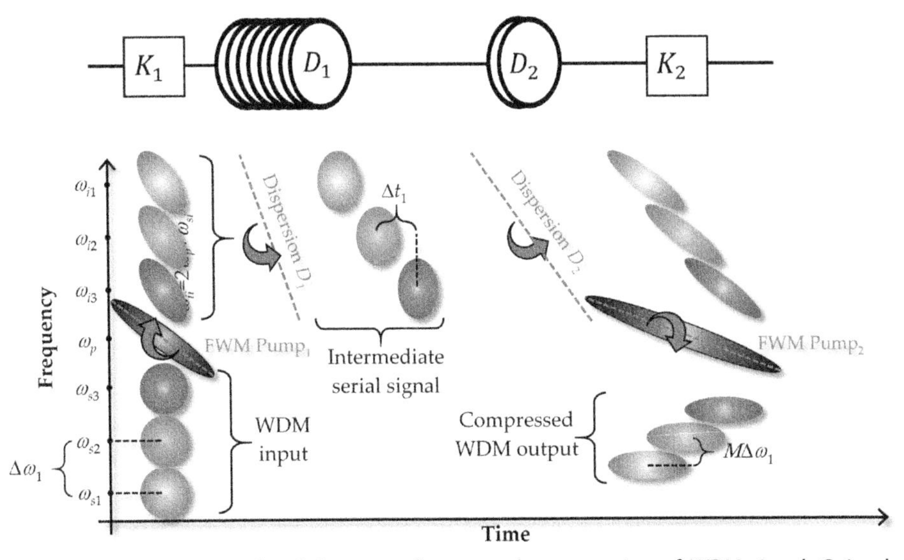

Figure 6.5 Operating principle of the setup for spectral compression of WDM signal. Going backwards through this setup results in spectral magnification, as in Fig. 6.3.

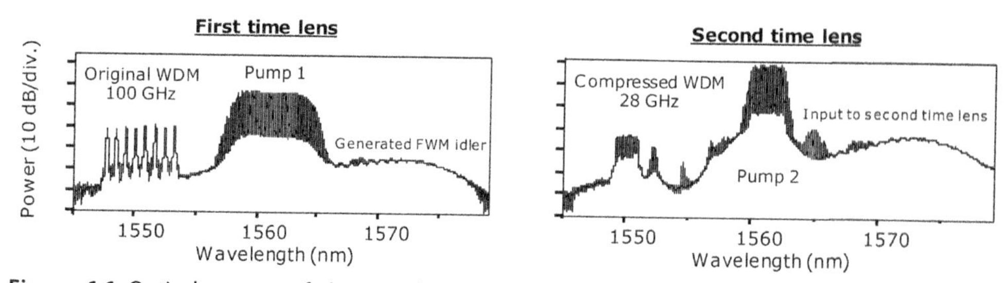

Figure 6.6 Optical spectra of the signals exiting the two time lenses forming the spectral telescope. *Source: From Ref. [30] .M. Røge, P. Guan, H.C.H. Mulvad, N.-K. Kjøller, M. Galili, et al., Flexible DWDM grid manipulation using four wave mixing-based time lenses, in: Proc. 2014 IEEE Photonics Conference, IPC 2014, 2014.*

spacing of the eight channels is further compressed to 28 GHz. In the experimental implementation in Ref. [30], the pumps for FWM are realized by spectrally shaping short pulses from a mode-locked laser and subsequently applying chromatic dispersion to them to achieve pulses with a flat-top temporal envelope and a constant frequency chirp. The optical signals exiting the two time lenses in the telescope are shown in Fig. 6.6. Looking at the WDM signal around 1550 nm, it is clear to see that the grid has been compressed. The grid has been narrowed to 28 GHz, a nonmultiplum of the original grid to show that the compression is not limited in that regard—the (de-)magnification is completely flexible and continuously tunable. All channels after conversion have BER performance below 1E-9, for both compression to 50 GHz and to 28 GHz, with an average penalty of about 1.5 dB for the 50 GHz case and 5 dB for the 28 GHz case. The latter increased penalty owing to increased crosstalk for the smaller frequency spacing, which is, however, not inherent to the scheme and may be reduced with further optimization. As seen in Fig. 6.6, the compressed grid channels have left room for free bandwidth for other purposes. There are some residues from the OFT process, which a passive filter can easily clean out, and additional channels may then be added.

Spectral magnification is obtained by going backwards through the principle sketch of Fig. 6.5, as in Fig. 6.3, and examples are shown in Fig. 6.7. Fig. 6.7A shows the input and output spectra for the case of a 4 × magnification of a five-channel WDM signal consisting of 10 GBaud 16-QAM data signals. The spectral spacing is magnified from 50 GHz (covering 2 nm) to 200 GHz (covering 8 nm) [31]. In Ref. [31], the bandwidth available for this process was characterized, and the available bandwidth for magnified signals reached about 20 nm, meaning that the magnified channels could fall into a range of 20 nm within a 3-dB conversion bandwidth, limited by the FWM bandwidth of the highly nonlinear fibers (HNLFs) used in this setup. As seen, only 8 nm of this available bandwidth was used for this experiment, and one could thus either use more channels or increase the spacing of the used channels further. In Ref. [28], an

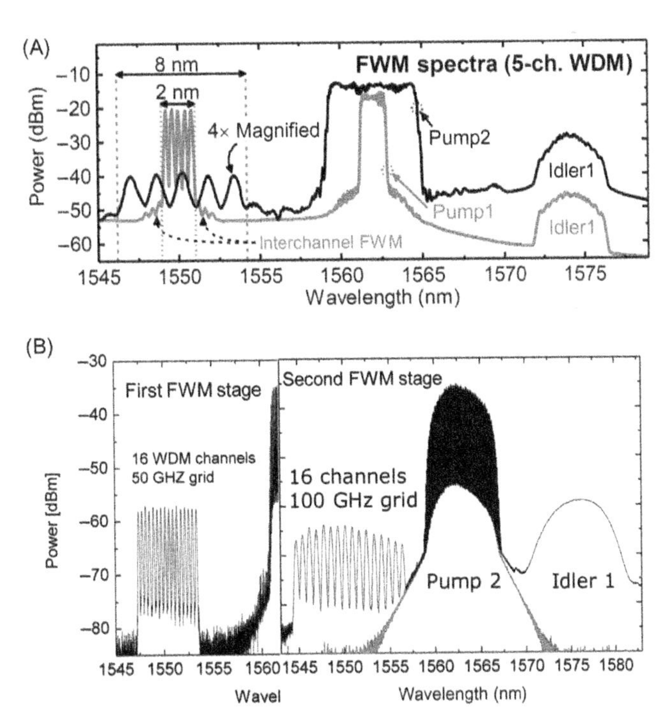

Figure 6.7 Spectral magnification of WDM signals. (A) 5 WDM channels with 50 GHz spacing magnified 4× to 200 GHz spacing. (B) 16 WDM channels magnified 2× from 50 to 100 GHz spacing. Source: *(A) From Ref. M. Lillieholm, B. Corcoran, M. Galili, L. Grüner-Nielsen, L.K. Oxenløwe, A.J. Lowery, Characterization of spectral magnification based on four-wave mixing in nonlinear fibre for advanced modulation formats, in: ECOC2017, Paper P1.SC4.65 [31]; (B) from Ref. L.K. Oxenløwe et al., Energy-efficient optical signal processing using optical time lenses, in: S. Wabnitz, B.J. Eggleton (Eds.), Chapter 9 in All-Optical Signal Processing - Data Communications and Storage Applications, Springer Series in Optical Sciences vol. 194, ISBN 978-3-319-14992-9, 2015 [28]*

example of exploiting the bandwidth for more channels is described, reaching 16 WDM channels. In that case, the 16 channel consisted of 10 GBaud DPSK-modulated channels at 50 GHz spacing, and the grid was magnified to 100 GHz, as shown in Fig. 6.7B. The setup used in Ref. [28] was a folded spectral telescope grid manipulator using only a single nonlinear element in a counter-propagating scheme, where one direction corresponded to lens-1 and the opposite direction to lens-2. All 16 channels were well behaved, with a 1-dB power penalty at the HD-forward error correction (FEC) limit and a 2-dB penalty at a BER 1E-9, showing that this scheme could find practical use.

Spectral magnification may also be useful in situations where it is hard to discern narrow spectral features, both in the communication regime and elsewhere. For communications, one such example could be for highly spectrally efficient data signals, such as orthogonal frequency division multiplexed (OFDM) signals, where the individual channels are spectrally sinc-function shaped and overlapping. For all-optical

OFDM signals, one would usually require both a spectral and temporal *filtering*, such as obtained by a discrete Fourier transformation (DFT) using, for instance, cascaded delay interferometers (DIs) [32] followed by time domain gating. Using optical gates can be very energy consuming, so it was suggested in Ref. [33] to use a time lens system to spectrally magnify the OFDM signal, which separates the spectral channels from each other and simultaneously compresses the channels in time so as not to overlap with each other. One can then apply a spectral filter to each channel and passively filter out each channel with greatly reduced crosstalk compared to trying to filter on a non-magnified OFDM spectrum; see Fig. 6.8. This is possible because the whole OFT process is a coherent process, so the data channels remain transform limited. So squeezing in one dimension broadens in the other. Thus, it is possible to spectrally magnify an OFDM signal and subsequently passively filter out each channel using some standard WDM filter—that is, without using any time domain gates.

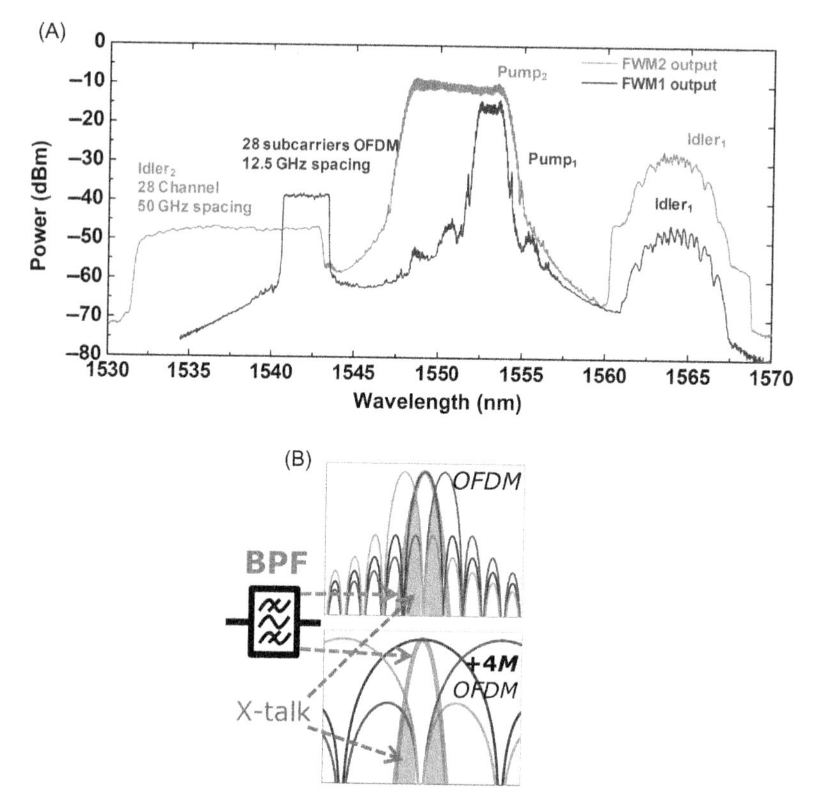

Figure 6.8 Spectral magnification. (A) 28-subcarrier OFDM super-channel magnified 4× from 12.5 to 50 GHz spacing. (B) Basic principle of OFDM magnification for OFDM demultiplexing with reduced crosstalk. Source: *(A) From reference P. Guan, S. Lefrancois, M. Lillieholm, H.C.H. Mulvad, K.M. Røge, H. Hu, et al., All-optical OFDM system using a wavelength selective switch based transmitter and a spectral magnification based receiver, in: ECOC 2014, doi: 10.1109/ECOC.2014.6964089 [34]*

This principle was first demonstrated in Ref. [33] for an OFDM super-channel comprising 10 subcarriers, and later in Ref. [34] the principle was firmly corroborated by spectrally magnifying a 28-subcarrier OFDM super-channel, with each subcarrier being 10 GBaud DPSK; see Fig. 6.8 (left). The spectrum is magnified a factor of 4, allowing for WDM demultiplexing with greatly reduced crosstalk from neighboring subcarriers and error-free performance.

To compare the expected performance of this spectral magnification scheme to the ideal DFT gated receiver, a detailed simulation analysis was presented in Ref. [35]. It was found that for a magnification of 4 × or higher, the OSNR penalties compared to the ideal case would be reduced to 0.5 dB or lower. For DPSK data modulation, $M=4$ gave 0.5 dB and $M=8$ gave 0.2 dB penalties, and about 0.2 dB higher for QPSK. And compared to a *truncated* delay-interferometer (DI) DFT receiver, where the number of DIs is reduced for lower complexity, similar performance is obtained. However, for both the ideal DFT and the reduced-complexity version, one would still need an optical gate per subcarrier, whereas one would only need a single spectral telescope irrespective of the number of subcarriers. So the performance is comparable to the ideal OFDM receiver, and the complexity may be greatly reduced. The final energy consumption evaluation would have to be carried out based on the chosen implementation, considering which nonlinear elements would be used and the power requirements of those.

6.2.3 Wavelength division multiplexed phase-sensitive regeneration

An optical data signal will unavoidably get distorted during fiber transmission due to fiber nonlinearity, amplified spontaneous emission from optical amplifiers, and interference originating from various processes. These distortions may become a major limitation on the transmission reach. Optical regenerators that can restore the quality of the signal have been considered as a solution to inhibit the degradation of the signal (e.g., [36]). All-optical regeneration can reduce signal impairments without optical-electrical-optical conversion and has a processing bandwidth beyond THz [37,38]. In the 1990s, most optical regeneration research focused on amplitude regeneration, as the communication systems were using amplitude modulation to carry data [36,39–41]. However, modern optical communication systems have switched to using both amplitude and phase modulation to achieve higher transmission rates [42–44]. The transmission performance of these systems is mainly limited by the nonlinear phase noise [45,46]. The capability of phase regeneration therefore becomes essential for an optical regenerator.

Recently, regeneration of coherent data formats using PSAs has been experimentally demonstrated for DPSK [37,47], DQPSK [48,49] and even eight QAM signals [50]. Most optical regeneration techniques only operate on a single or few channels level, but to offer energy-efficient solutions, WDM compatible schemes are necessary [51]. Recent reports have shown optical amplitude regeneration of up to 16-channel WDM signals using a group-delay-managed nonlinear medium [38], and up to 6-channel WDM phase regeneration was successfully demonstrated using multiple PSAs [52]. However, both these WDM regeneration proposals are challenging to scale to higher WDM channel counts, due to the requirement of prefabrication of a set of fixed group delays or the nonlinear interactions between many wavelength channels or pumps by FWM.

Very recently, a new WDM-scalable technique for phase regeneration of WDM signals was proposed with a demonstration of 16 phase-regenerated WDM channels in a single regenerator [53]. The proposal relies on a single PSA in between a parallel-to-serial converter and a serial-to-parallel converter based on time lenses [26,54–57]. As the implementation complexity of this scheme does not increase with the channel numbers, this approach is favorably scalable in WDM channel numbers, enabling scaling from 4 × WDM [26] to 8 × and 16 × WDM regeneration [53].

6.2.3.1 Principle of wavelength division multiplexed phase regeneration using a time lens and phase-sensitive amplifying unit

Fig. 6.9 shows the principle of WDM phase regeneration of DPSK signals. The main idea is to convert the WDM signals to a high-speed serial single wavelength channel, which is then straightforward to regenerate in a single PSA-based optical-phase regenerator without unwanted mixing of many pumps and wavelength channels. After regeneration, the serial signal is simply converted back again to a WDM signal.

The conversion between WDM and serial formats is realized by time lens-based OFT. In particular, different OFTs can be realized by changing the combination of a time lens and dispersion media [58]. Two different OFTs were used in the proposed regeneration scheme. The first one (OFT#1) with a chirp-dispersion (K-D)

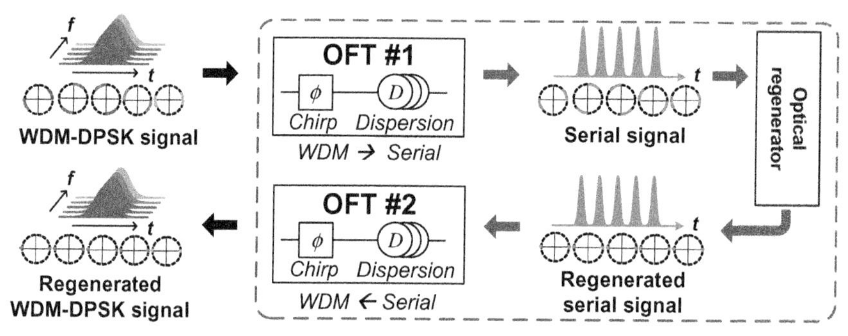

Figure 6.9 Principle of WDM regeneration. The WDM signal is converted to a high-speed serial signal using OFT#1, which is then regenerated in a single optical regenerator. After regeneration, the serial signal is converted back to a WDM signal using OFT#2 [53].

configuration is used for WDM (parallel) to serial conversion. After the optical regeneration, the second OFT (OFT#2) with a D-K configuration is used for converting the regenerated serial signal back to a WDM signal. The signal being serial entails that only one pulse passes through the regenerator at a time, avoiding cross-mixing with other pulses. However, the random carrier phase of the individual WDM channels is carried through to the serial channels, and this poses a challenge to the phase-sensitive optical processing. One solution to this is to convert serial signal onto a new highly coherent carrier; this is achieved by combining phase-to-amplitude modulation conversion (using a 1-bit delay interferometer (DI) with cross-phase modulation (XPM) based optical remodulation prior to the PSA-based regeneration) [59,60]. This principle of using a DI followed by XPM and then PSA is shown in Fig. 6.10 for a single channel.

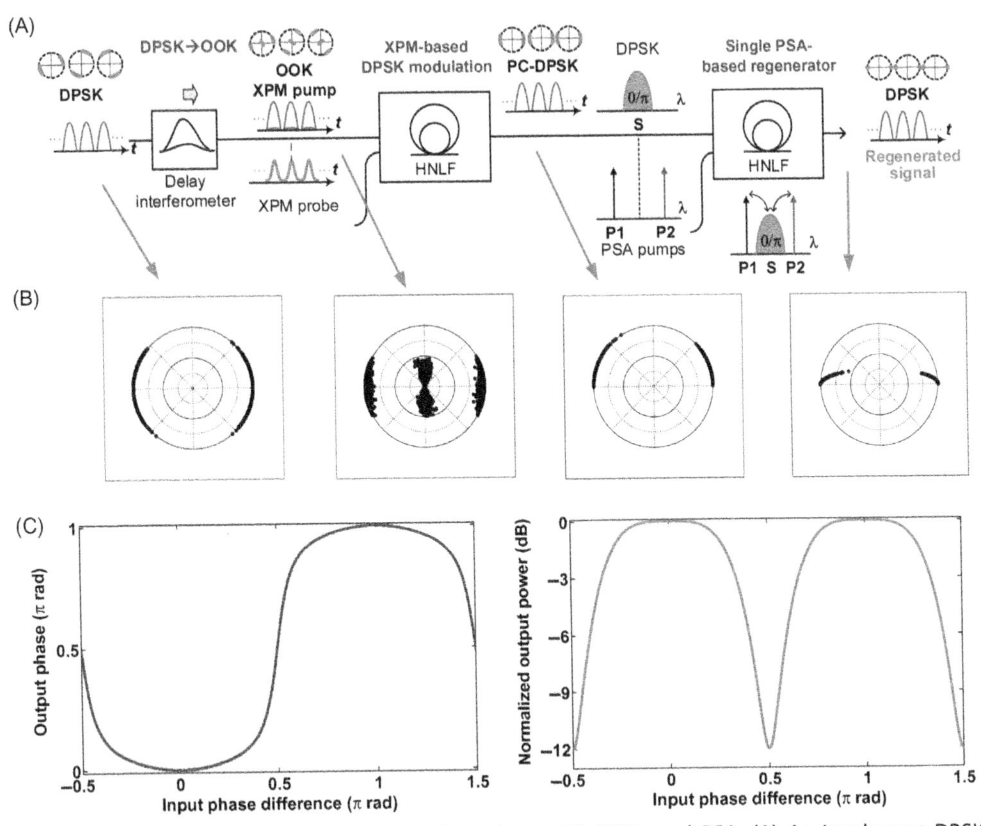

Figure 6.10 All-optical-phase regeneration based on a DI, XPM, and PSA. (A) An incoherent DPSK signal is converted to an on-off keying (OOK) signal using a 1-bit DI. (B) Simulated constellation diagrams at the input of the DI, and the outputs of the DI, the XPM unit, and the PSA. (C) Analytic output phase and power transfer function of the DI-XPM and PSA units with a gain of $g = 2$ [53].

In Fig. 6.10 the DI-XPM-PSA principle is explained for a single WDM channel—that is, without parallel-to-serial conversion. A 1-bit DI at the symbol rate of the WDM signals is used to convert the incoherent WDM DPSK signal to an on-off keying (OOK) signal. XPM in a HNLF is employed to transfer the data modulation from the OOK signal to the phase of a coherent pulsed probe, resulting in a phase-coherent (PC)-DPSK signal. Finally, after combination with two phase-locked pumps, the obtained PC-DPSK signal is sent to a single PSA, where the phase is regenerated. For multiple WDM channels, the DI is placed just before the time lens conversion to serial, such that the DI operates on directly on the WDM signals. See Fig. 6.11 for the detailed experimental implementation of WDM DPSK regeneration.

To analyze the effect of each processing step, DI, XPM, PSA, an analytical model is derived [53]. The results are shown in Fig. 6.10B, where each bit is represented by a point in the constellation diagram. At the input, Gaussian distributed noise is added to the phase of the input DPSK signal, and the theoretical transfer function for each step is applied to the bit sequence in the given order. Fig. 6.10C shows the resulting transfer function of the DI-XPM- and PSA-based optical regenerator with a phase-sensitive gain of $g = 2.0$. The output phase after regeneration is a steplike function alternating between the two values 0 and π. And the output power has flat areas of equal amplification around 0 and π. As seen from Fig. 6.10B, the DI, XPM, and PSA-based optical regenerator results in a squeezing of the constellation points towards 0 and π, confirming the operating principle of phase regeneration. It should be mentioned that when the noise causes a phase shift on a bit that exceeds π, in other words, an error, this will not be corrected by the regenerator, as it will be squeezed to the opposite position on the constellation diagram. Therefore, to extend the transmission reach, the optical regenerator has to be installed within the transmission link.

6.2.3.2 Experimental demonstration of simultaneous regeneration of 8 and 16 wavelength division multiplexed differential phase shift keying channels

The experimental setup of WDM regeneration is shown in Fig. 6.11. To emulate regeneration in a transmission link, the regenerator is placed between two phase noise emulators. In the transmitter, 16 (8) independent continuous wave (CW) carriers are DPSK modulated with a 10 Gbit/s 2^{31}-1 pseudo-random bit sequence (PRBS) in a Mach-Zehnder modulator. The WDM channels are then de-correlated using four different paths and a wavelength-selective switch. Within the regenerator, a 1-bit (100 ps) DI is used to simultaneously convert all DPSK WDM channels to OOK signals. The obtained OOK WDM signal is converted to a 160 (80)-Gbaud serial signal by the first OFT.

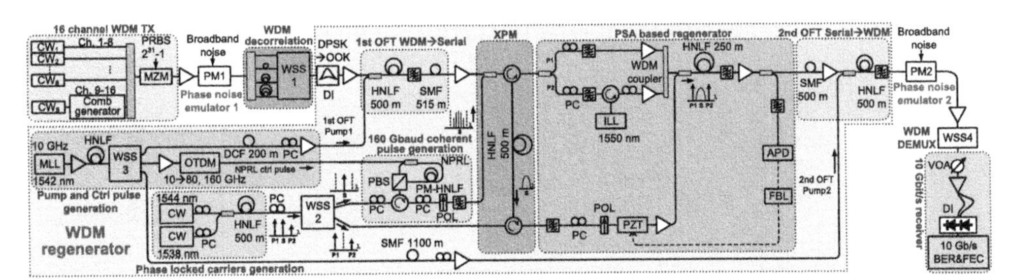

Figure 6.11 Experimental setup for demonstration of simultaneous regeneration of 8 and 16 WDM DPSK channels [53].

The output spectra of the first OFT for both 8- and 16-channel regeneration are shown in Fig. 6.12A, and the waveform of the 80 GBd OOK signal is shown in Fig. 6.12B. The quadratic phase modulation is implemented by a FWM process in a HNLF. For the PSA stage, three CW phase-locked carriers are generated by single-pump FWM in a 500 m HNLF; P1 (1538 nm), S (1544 nm), and P2 (1550 nm). The signal carrier S is optically carved into a 1.2 ps 160 Gbaud coherent pulse train in a fiber-based nonlinear polarization-rotating loop switch. The pulse train S and the PSA pumps are sent into a 500-m HNLF in counter-propagating directions using optical circulators. The obtained 160 Gbaud OOK signal is coupled into the HNLF to act as an XPM pump co-propagating with the coherent pulse train. By carefully adjusting the pump power and time delay, the 160 Gbaud pulse train S is DPSK modulated optically, generating a phase-coherent DPSK signal. Before the PSA, the pumps P1 and P2 are further split for independent amplification and an injection locked laser is used to increase the power, and thus the OSNR of carrier P2. The signal and pumps are then launched into the PSA consisting of a 250 m HNLF with stable phase matching for improved nonlinear efficiency (HNLF-SPINE) [61]. After the PSA, the regenerated 80- or 160-Gbaud serial signals are converted back to WDM signals by the second OFT.

The output spectra of the second OFT are shown in Fig. 6.12C with a zoom-in on the idlers in Fig. 6.12D, where the obtained 8- or 16-channel WDM signals are observed with ~50 GHz spacing. Finally, after WDM demultiplexing, the BER of each channel is measured in a 10-Gbit/s DPSK receiver including a DI and balanced photo-detection. Two broadband phase noise emulators, consisting of a phase modulator (PM) driven by broadband phase noise with approximately Gaussian distribution obtained by detecting the amplified spontaneous emission noise of an EDFA, are inserted before and after the regenerator. The phase noise is quantified by the variance of optical phase, estimated by $D_i = (\pi \times \sigma_e^i / V_\pi^i)^2$, with σ_e^i being the standard deviation of the electrical driving voltage for PM_i and V_π^i being the half-wave voltage.

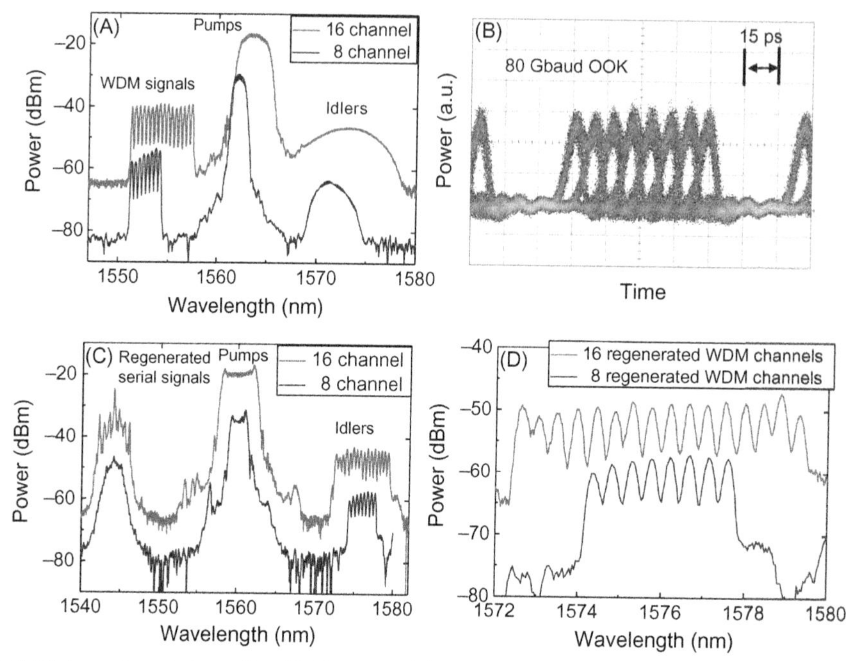

Figure 6.12 Experimental results. (A) Spectra of first OFT output, (B) waveform of the converted 80-Gbaud OOK signal, (C) second OFT output, (D) a zoom-in on the idler.

6.2.3.2.1 Experimental results

The performance of the regenerator is shown in Fig. 6.13A. The WDM regeneration is successfully achieved for both the 8- and the 16-channel cases, as confirmed by BER measurements. Starting from 8-channel WDM regeneration, the BER versus received power of one regenerated 10-Gbit/s channel is measured. The regenerator is benchmarked against the back-to-back (B2B) BER curves with and without phase noise. The regenerator power penalty without noise is 1.2 dB at BER $= 10^{-7}$. For the B2B signal, when the phase noise is added by the two emulators with combination $(D1, D2) = (0.079, 0.051)$, the error floors appear. With regeneration in between the two noise emulators, significant BER improvement is achieved. In particular, without regeneration there is an error floor at BER $= 10^{-6}$. With regeneration, the BER curve is improved by 1.5 orders of magnitude. The 16-channel WDM regeneration performance is also measured with the same phase noise combination. Almost the same BER improvement is achieved for the 16-channel regeneration. The regenerator power penalty without noise is 0.5 dB worse than the 8-channel case at BER $= 10^{-7}$. We measured the regenerated channel performance of all channels for both 8-channel, in Fig. 6.13B, and 16-channel WDM regeneration, in Fig. 6.13C, at a fixed received power of -32 dBm. The BER is improved by 0.8–1.5 orders of magnitude for

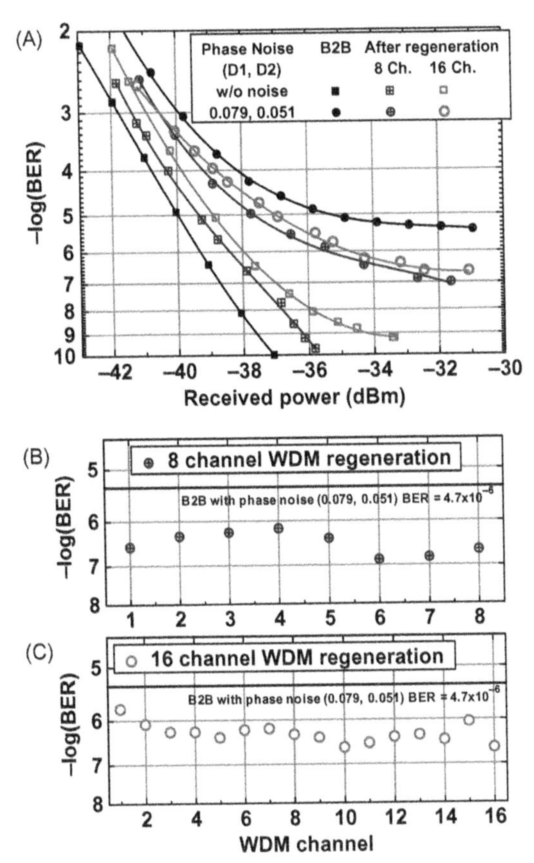

Figure 6.13 Experimental results. (A) BER performance of 8- and 16-channel regeneration with dual noise emulators (*D*1, *D*2), all channel regeneration performance at a received power of −32 dBm with noise (0.079, 0.051) for (B) 8-channel and (C) 16-channel WDM regeneration.

8-channel WDM regeneration, and 0.4–1.3 orders of magnitude for 16-channel WDM regeneration.

To summarize, a scalable WDM phase regeneration scheme has been described. The scheme relies on OFT and a single PSA. The presented results demonstrate the principle for 8 and 16 WDM channels using a realistic WDM-grid spacing of 50 GHz.

6.2.4 Field-quadrature decomposition by polarization-assisted phase-sensitive amplifier

In this section we introduce and discuss field-quadrature decomposition of complex optical modulation formats using a polarization-assisted phase-sensitive amplifier (PA-PSA). We explain how a PA-PSA can simultaneously decompose a signal into its two constituent field-quadrature components. PA-PSA enables coherent addition of a signal with its complex conjugate, accurately matching the power of the two waves. In theory, the result is an ideal binary staircase-like phase-to-phase transfer function suitable for both phase regeneration and field-quadrature decomposition of the signal. The technique of PA-PSA was developed in different places in the same period of time, and presented in Refs [62–64], with slight variations. The scheme was initially demonstrated to perform binary phase-quantization in Refs. [62,63,65] and quadrature decomposition in Refs. [64,66]. Additional details may be found in Refs. [62,67]. Decomposition of a QPSK signal into two BPSK signals at different wavelengths had been proposed [68] and demonstrated [69,70] using FWM with four pump waves and resulting in wavelength conversion of the decomposed outputs. In contrast, in the PA-PSA scheme the number of pumps does not increase with the size of the constellation diagram and the signal wavelength is preserved. Phase squeezing of a QPSK signal into a single BPSK signal, corresponding to either the I or the Q quadrature component, has been demonstrated using scalar PSA with signal, pump, and idler waves in the same polarization [71]. The principal advantages of PA-PSA are the ability to equalize the relative power levels in the signal-idler interference, increasing the phase sensitivity, and the ability to extract both field-quadratures simultaneously. In this way PA-PSA has been demonstrated with high phase sensitivity in several material platforms where low FWM conversion efficiency would otherwise not allow this [72].

6.2.4.1 Principle

The operating principle of polarization-assisted PSA is illustrated in Fig. 6.14. A phase-conjugated idler is generated by degenerate vector FWM between a phase modulated signal (S) and two orthogonally polarized pumps—one aligned parallel to the signal and one orthogonal to the signal. The generated idler (IS) is proportional to the phase conjugate of the signal (S*), it can have the same wavelength as the signal, and it is orthogonally polarized to the signal. More details of this process can be found in Ref. [73]. Next, the signal and idler waves are projected onto the same state

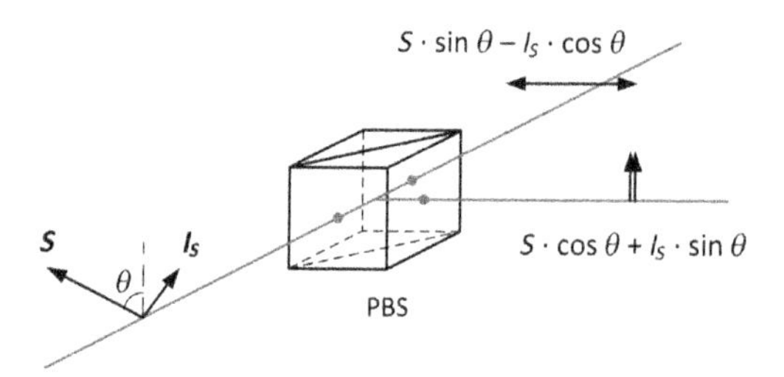

Figure 6.14 Principle of PA-PSA based on vectorial FWM. Source: *Adapted from N.K. Kjøller, M. Galili, K. Dalgaard, H.C.H. Mulvad, K.M. Røge, L.K. Oxenløwe, Quadrature decomposition by phase conjugation and projection in a polarizing beam splitter, in: European Conference on Optical Communication, ECOC, 2014 [64].*

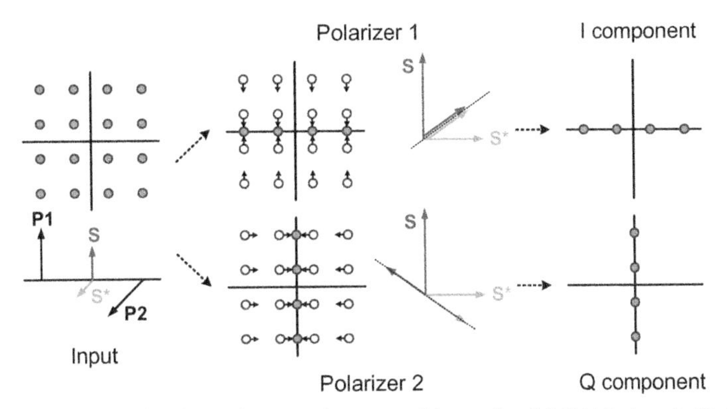

Figure 6.15 Simultaneous field-quadrature decomposition of a 16-QAM signal. "Polarizer 1" and "Polarizer 2" illustrate the two projections of signal and idler required to achieve quadrature decomposition—illustrated in the general case where signal and idler are not equal in amplitude. Adjusting the angles relative to the polarizer transmission axes the signal and idler projections can achieve equal amplitude. Source: *Adapted from N.K. Kjoller, M. Piels, F. Da Ros, K. Dalgaard, M. Galili, L.K. Oxenlowe, 16-QAM field-quadrature decomposition using polarization-assisted phase sensitive amplification, in: 2016 IEEE Photonics Conference, IPC 2016, 2017, pp. 513–514. [74]*

of polarization in a polarizer or polarizing beam splitter (PBS), resulting in the coherent addition or subtraction of the two projections. The I and Q components of the signal is proportional to $S + S^*$ and $- S^*$, respectively. Ideal phase squeezing is obtained when the projections are equal in power. As signal and idler are orthogonally polarized, the FWM process is phase insensitive. The subsequent coherent addition of signal and idler creates a phase-sensitive process, requiring the signal and pumps to be phase-locked.

Fig. 6.15 illustrates how PA-PSA decomposes a 16-QAM signal into its quadrature components. If the signal and idler power is equalized before the polarizer, a PBS can be used to obtain the I and Q components simultaneously at the two PBS outputs. Assuming constant relative phases of the pumps and no pump depletion, the coherent addition of the signal and idler results in an output signal with the complex amplitude:

$$A_{PA-PSA} \propto e^{i\varphi(t)} + \mathcal{M}e^{-i\varphi(t)} \tag{6.9}$$

where $\varphi(t)$ is the input signal phase and $\mathcal{M}$ is the weight of the complex conjugate idler with respect to the signal. We recognize this as the transfer function of a binary phase-quantizer, hence the name "polarization-assisted PSA."

The theoretical phase-to-phase and phase-to-power transfer functions are shown in Fig. 6.16 for different values of $\mathcal{M}$. The phase transfer function converges to ideal binary phase-quantization as $\mathcal{M}$ approaches unity. The power transfer function is sinusoidal with peak levels of amplification and deamplification corresponding to the plateaus and slopes of the phase transfer function. As $\mathcal{M}$ approaches unity, the

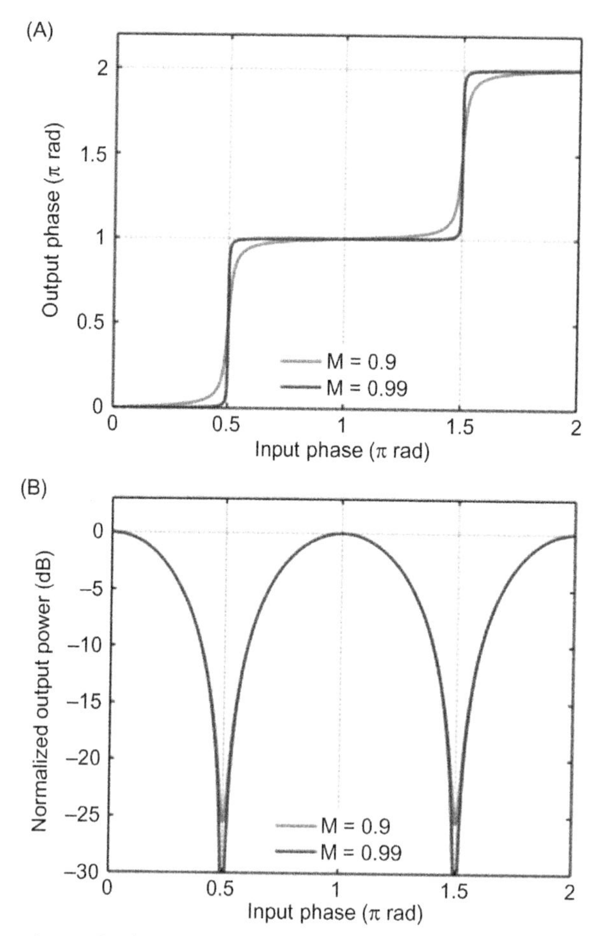

Figure 6.16 Theoretical transfer functions showing the output phase (A) and power (B) of the PA-PSA. Signal-idler power ratios of $\mathcal{M} = \mathbf{0.9}$ (softer steps (A) and smaller extinction ratio (B) and $\mathcal{M} = \mathbf{0.99}$ are shown.

phase-sensitive response increases reaching a phase-sensitive extinctions ratio (PSER) of 25.6 dB for $\mathcal{M} = 0.9$, increasing to 46.0 dB for $\mathcal{M} = 0.99$.

6.2.4.2 16-Quadrature amplitude modulation field-quadrature decomposition

Optical decomposition of a 16-QAM signal in a PA-PSA has been experimentally demonstrated following the scheme outlined in Fig. 6.14 [74]. The resulting outputs are modulated in a format having two amplitude levels ($\frac{1}{2}$ and $\frac{3}{2}$ of the average amplitude) and two phase levels (0, π), which will be referred to as four-level amplitude-phase shift keying (4-APSK). Due to the use of both amplitude and phase modulation, we refrain from using the name four-level pulse-amplitude modulation, which is occasionally used to describe this format.

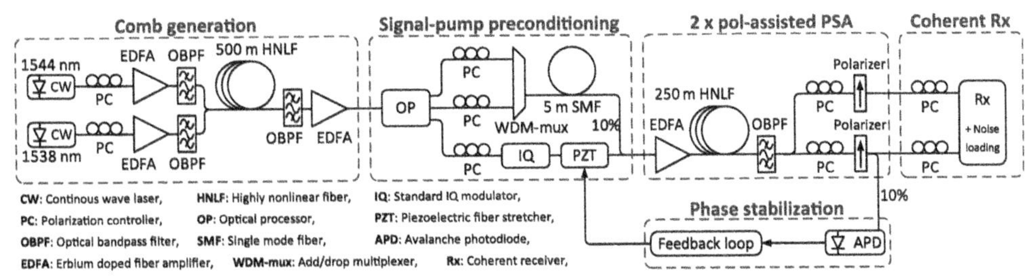

Figure 6.17 Experimental setup for 16-QAM field-quadrature decomposition. Source: *Adapted from N.K. Kjoller, M. Piels, F. Da Ros, K. Dalgaard, M. Galili, L.K. Oxenlowe, 16-QAM field-quadrature decomposition using polarization-assisted phase sensitive amplification, in: 2016 IEEE Photonics Conference, IPC 2016, 2017, pp. 513–514. [74]*

In the experimental demonstration, a comb consisting of three phase-locked waves is generated through FWM between two co-polarized CWs in 500 m HNLF-SPINE. A diagram of the experimental setup is shown in Fig. 6.17. The three waves intended to form pumps and signal are separated to individual outputs of an optical processor, where higher order FWM components are also suppressed. The central of the three lines (labeled S in Fig. 6.3) is modulated into a 10-GBd NRZ-16-QAM data signal composed of PRBSs of lengths $2^{11}-1$ and $2^{15}-1$, on the I and Q components. The three waves are recombined and their polarizations are adjusted so that P1 and S are parallel and orthogonal to P2. The separate path lengths of the three waves are matched to within one bit-slot. The recombined pumps and signal are amplified and launched into 250 m of strained HNLF-SPINE. Here dual-pump vector FWM generates an idler wave at the same wavelength as the signal but with orthogonal polarization, forming the conjugate copy of the signal required for quadrature decomposition.

The output is split in a 3 dB coupler and launched into two polarizers to achieve the projections required for field decomposition as illustrated in Fig. 6.3. Thermal drift of the relative phases between signal and pumps is compensated with a piezoelectric fiber stretcher in an active feedback phase-control loop.

For a total input power of 23.8 dBm into the 250 m HNLF-SPINE and a pump-signal power ratio of 10.0 dB an output signal-idler power ratio of 5.1 dB was achieved. Field-quadrature decomposition of the 16-QAM modulated input signal is performed using these parameters, and a PSER of 17.6 dB is measured. Constellation diagrams and error rates are generated using a coherent receiver with offline DSP. Measured constellation diagrams are shown in Fig. 6.18A for the 16-QAM input and the two outputs with no noise loading in the receiver. The outputs are clearly 4-APSK modulated. In the receiver, the original $2^{11}-1$ and $2^{15}-1$ PRBSs of the input signal are recovered on the two output signals, verifying simultaneous uncorrupted extraction of I and Q components.

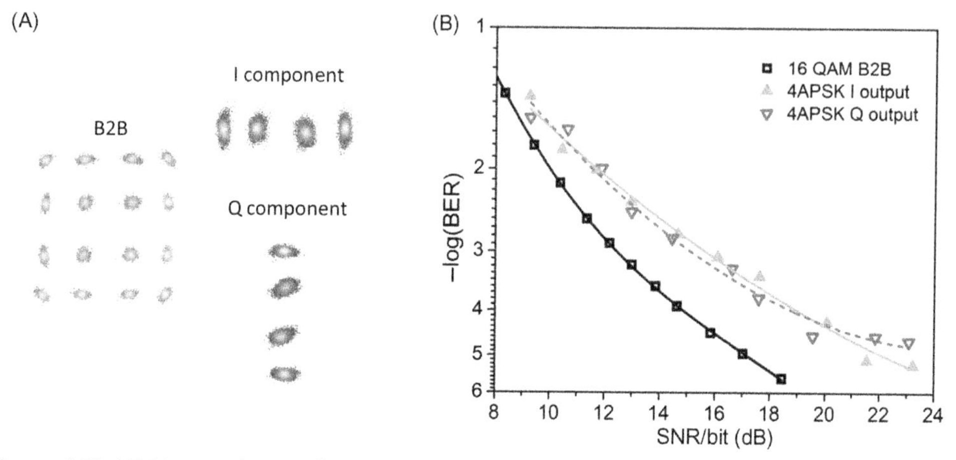

Figure 6.18 (A) Measured constellation diagrams for the 16-QAM signal B2B and the two outputs. The outputs are clearly 4-APSK modulated. A performance penalty is evident by the enlarged distributions around each constellation point of the outputs. (B) Measured BER as a function of SNR/bit for the 16-QAM signal B2B (*black*) and the two 4-APSK outputs corresponding to the I (*solid gray*) and Q (*open gray*) components of the input signal. Source: *Adapted from N.K. Kjoller, M. Piels, F. Da Ros, K. Dalgaard, M. Galili, L.K. Oxenlowe, 16-QAM field-quadrature decomposition using polarization-assisted phase sensitive amplification, in: 2016 IEEE Photonics Conference, IPC 2016, 2017, pp. 513–514 [74].*

In Fig. 6.18B, the BER performance of the two output signals are compared to the original 16-QAM signal. To facilitate this comparison, the curves are plotted as a function of received SNR/bit, where penalty-free conversion would see the curves coincide. We see that the extracted I and Q components show similar performance, and for high error rates, the performance is within 1 dB of the original 16-QAM signal. For lower BERs, however, the curves diverge and for $BER = 10^{-3}$ the quadrature decomposition results in an SNR/bit penalty of around 3 dB, increasing further for decreasing BER.

Optical signal processing can accomplish numerous signal processing tasks. Here we have presented and demonstrated the application of PA-PSA to achieve electric field-quadrature decomposition. As with many other optical signal processing schemes, this approach is transparent to the rate of the signal. This feature holds great promise for energy-efficient signal processing of high-speed optical data signals.

6.2.5 Summary on optical time lenses

In this section we have tried to give an impression of how optical time lenses function and how versatile they are in terms of applications. In most of our work, a theme has been to explore applications where the optical signal processing is performed on multiple data channels simultaneously, taking advantage of this unique capability of optics, as opposed to using multiple parallel processors. Broadband ultrafast optical signal processing does rely on optical nonlinearities, which is typically very energy-hungry. So OSP only seems to make sense when the energy associated with processing multiple channels optically is less than what would otherwise be required using other parallel processors. With the use of time lenses, we have seen that it becomes possible, in an elegant way, to process multiple WDM channels for a plethora of applications; see Fig. 6.19. We have described how time lenses can convert between the temporal and the spectral domains, such as for performing serial-to-parallel conversion (optical TDM to WDM) [57]. This was originally conceived in order to simultaneously demultiplex an ultrahigh-speed optical time division multiplexed signal in a single time lens unit, and indeed Tbit/s single wavelength channels have been demonstrated demultiplexed into several parallel WDM channels, which can then easily be separated using passive filters [80]. This type of system turns out to be more energy-efficient that a standard WDM system, in that only a single laser is needed in the transmitter, and the receiver is just like a WDM receiver with one nonlinear unit added to it. Recently, this serial-to-parallel conversion seems very promising for passive optical networks (PONs), where most often a single serial transmitter is used emitting data at the highest available ETDM rate, and the whole serial optical signal is then transmitted out to potential users via passive power splitters. Each user then needs to have a high-bandwidth receiver, to detect the full signal, and then some intelligent and secure system to only pick out the data relevant for them, without peeking at the neighbors' content. In Ref. [75], it was demonstrated that a single TDM transmitter could have its output converted through a time lens to produce 128 WDM channels running at 2 Gbit/s each. These 2 Gbit/s channels can then be passively managed in WDM-PON scenario, reducing the bandwidth required at each user, enhancing the transmission performance extending the reach of this PON systems to 150 km, and allowing for secure separation between each user, as each is only ascribed a specific wavelength. In [76], it was furthermore demonstrated that these conversions needn't involve a wavelength dislocation, as it was shown that by designing a time lens system relying on orthogonally polarized pumps, the converted output can be made to appear at the same wavelength as the original input, simply following energy and momentum conservation. Going from time to frequency is one possibility, and one may also go from frequency to time [77]. The latter is useful for receiving WDM signals and then converting them to a single wavelength channel, albeit an ultrahigh-bitrate channel, which then in turn lends itself well to other advanced optical signal processing. It's typically easier to perform nonlinear processing on a single channel, rather than many wavelength channels, which may result in multiple FWM crosstalk terms. We described the case of performing phase-sensitive regeneration of such a converted signal, and then convert it back to WDM afterwards. This scheme has the advantage of

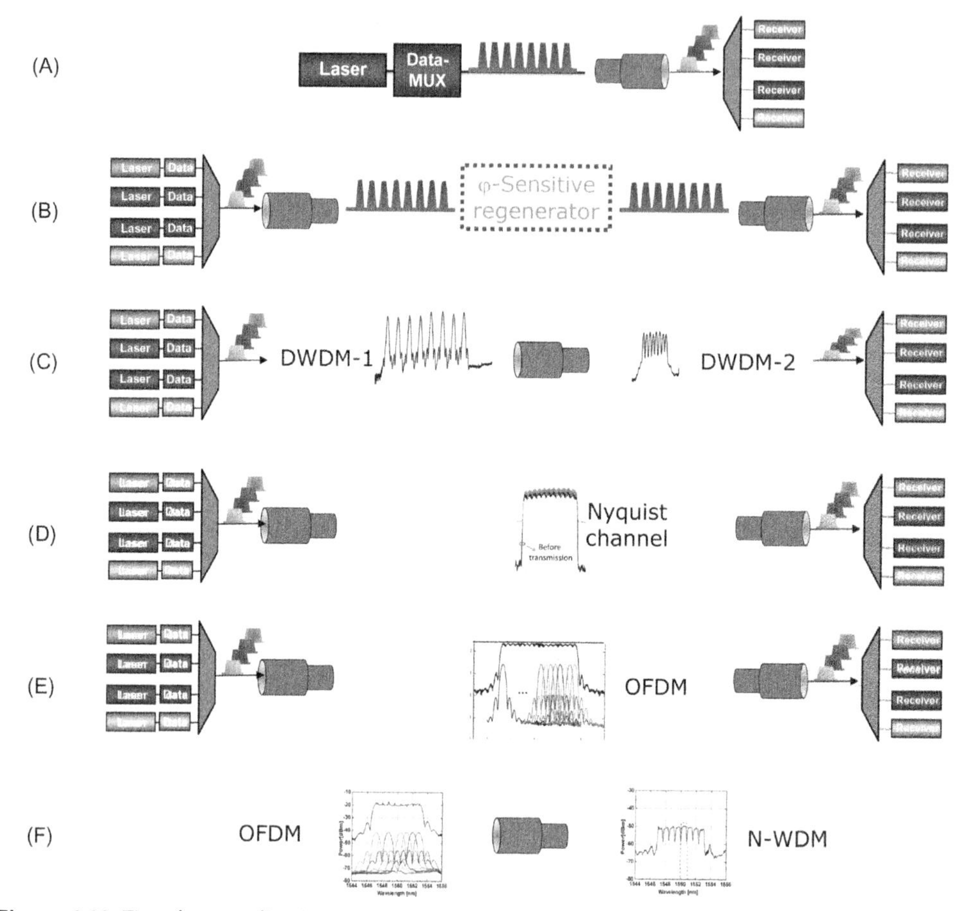

Figure 6.19 Time lens applications overview. (A) Serial-to-parallel conversion [24], for example, for single-laser WDM signal generation, such as for PON applications [75]. Can be wavelength preserving [76]. (B) WDM-to-serial conversion [77] for convenient ultrafast single wavelength channel regeneration followed by serial-to-WDM conversion for inline WDM regeneration [53]. (C) WDM-grid manipulation for spectral magnification or compression, allowing for flexible grids and reconfigurable link throughputs [78]. (D) Format conversion of, for instance, WDM to a single Nyquist channel for increased spectral efficiency, and optionally conversion back to WDM [58]. (E) Format conversion from WDM to OFDM for increased spectral efficiency and optional conversion back again [58]. The latter allows for an energy-efficient means to demultiplex OFDM channels without using multiple time gates [33]. It is also possible to go from a serial OTDM signal to OFDM or Nyquist WDM. (F) Conversion between OFDM and Nyquist WDM for flexible translation between networks potentially supporting only one format [79].

being scalable to the limit of the FWM bandwidth of the employed nonlinear material, and there have been several demonstrations of more than 500 nm FWM bandwidth in integrated nonlinear waveguides. So in principle, one could expect that this scheme could be operational on, for instance, a full C-band or L-band or even both together, if the nonlinear material is efficient enough. One may also simply immediately convert the signal back to WDM using a different time lens, and one may then arrive at spectral manipulation, such as magnifying or compressing the spectrum [31,78]. This adds a whole new flexibility to the network, where link capacity may be reconfigured following the needs at any given time. One may also simply exploit the spectral manipulation to increase the spectral efficiency of a WDM transmitter, for example, a 200 GHz grid legacy WDM transmitter, by converting to highly spectrally efficient formats like Nyquist channels, or OFDM channels [58]. It has been shown that one may convert from most formats to others, such as WDM to OFDM (and Nyquist-WDM) or Nyquist-OTDM and from OFDM to Nyquist-WDM [79]. OFDM to N-WDM conversion opens a possibility of translating between networks that only run either of the formats. We have also seen that we can use the spectral manipulation to efficiently demultiplex all-optical OFDM signals, alleviating the need for multiple time gates [33]. This chapter is mostly focused on FWM-based time lenses for optical communications. But there are several other interesting applications based on electro-optic phase modulation, such as dispersion compensation [55], data packet compression [81], data packet synchronization [82], timing jitter compensation [83], and fast oscilloscopes [56].

6.3 Optical-phase conjugation

OPC has reemerged in recent years as a means to help drive out the last few dB in transmission performance, by compensating the nonlinear distortions picked up by signals transmitted through optical fibers. In this section, we review state-of-the-art in OPC and discuss the benefits and performance limitations, and also arrive at a proposal for adapting digital-signal processing (DSP) to match the OPC channel for improved performance.

6.3.1 Fundamentals—principle and potential benefits

To respond to the ever increasing demand for capacity, current optical networks are migrating toward higher bit rates by the use of more complex modulation formats, such as QAM. Transmitting complex modulation formats requires an increase in the received signal OSNR, which in turn limits the transmission distance. A higher OSNR through the transmission can be maintained by increasing the launched optical power. However, nonlinear distortion caused by the Kerr effect in optical fibers sets an upper bound to the total power that can be transmitted before degrading the signal performance [3,84]. Being able to compensate for such nonlinear distortion would allow increasing the received OSNR without compromising the transmission reach. Several techniques have been proposed, with two main categories of solutions showing great potential, acting either in the digital or in the optical domain. From one side, digital backpropagation (DBP) algorithms, in different flavors, are being investigated [85]. However, this approach scales poorly in complexity with backpropagation bandwidth, and it is therefore mainly employed to single-channel scenarios where only intrachannel nonlinearity is compensated for. As effects such as XPM and FWM cause significant degradation, alternative methods addressing interchannel nonlinearity are of paramount interest. OPC can be effectively used to compensate for both intra- and interchannel nonlinear distortion over a wider bandwidth than the one available with practical receivers. When the OPC operation is performed in the middle of the link, the spectral inversion allows compensating for the nonlinear distortion accumulated in the first half of the link through propagation in the second half. OPC was first introduced by Yariv et al. in 1979, initially as a promising technique to provide dispersion compensation [86], but shortly thereafter its benefits in terms of Kerr nonlinearity compensation were pointed out by Pepper et al. in 1980 [87].

Conjugating an optical signal corresponds, in time domain, to reversing the sign of its chirp and thus temporally inverting the position of its frequency components, as shown in Fig. 6.20A.

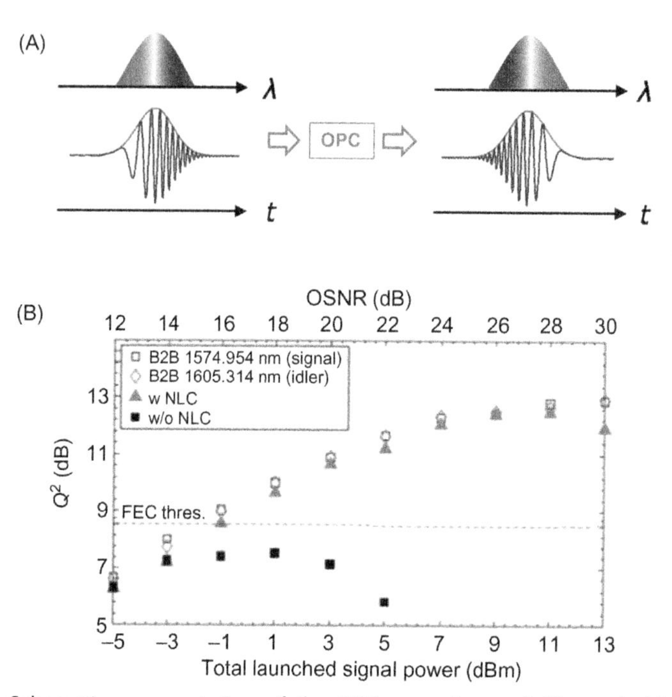

Figure 6.20 (A) Schematic representation of the OPC operation and (B) nearly ideal nonlinearity compensation by optimizing the symmetry of the transmission link. Source: *Reproduced from K. Solis-Trapala, T. Inoue, M. Pelusi, H.N. Tan, S. Namiki, Implementing ideal nonlinear compensation through nonlinearity, in: IEEE Summer Topical Meeting 2015, paper TuF3.1 [88].*

The simplest scenario where OPC enables to provide dispersion and nonlinearity compensation is a homogeneous transmission link with the OPC operation placed within the link. In such a scenario, assuming a lossless link with negligible third-order dispersion, dispersion and signal—signal nonlinear interaction accumulated in the first half of the link can be fully compensated by propagating the conjugated signal through the second half. This is clearly shown by the performance in Fig. 6.20B, where by using an almost perfectly symmetrical transmission link, the B2B performance can be almost fully recovered after nonlinearity compensation, for a 24-km transmission link [88]. In general, for the case of ideal compensation with mid-link (ML) OPC, the expected performance improvement would be of $1.27 \cdot (SNR_0)^{2/3}$, corresponding to a transmission reach enhancement of $1.17 \cdot SNR^{1/3}$, where SNR_0 is the signal-to-noise ratio (SNR) without nonlinearity compensation [89]. For a generic transmission link configuration, however, only partial compensation can be achieved, depending on the degree of link symmetry with respect to the OPC position. In particular, the power and dispersion evolutions throughout the link need to ensure that the same nonlinear distortion is experienced by equally dispersed waveforms before and after conjugation. This, in turn, sets more stringent requirements on the design of the transmission link and highlights the additional benefits provided by employing distributed Raman amplification (DRA). Whereas DRA-based systems already outperform systems with lumped amplification by maintaining a higher power and thus OSNR through the transmission, the lower power variations achievable provide a further advantage in terms of nonlinearity compensation. The use of Raman amplification, especially optimizing it beyond first-order Raman pumping, enables achieving power variations of the order of 6 dB for a 75-km SMF span [90].

As an alternative to DRA, dispersive elements can be employed to enhance the symmetry of the transmission link by offsetting the position of the OPC operation [90]. Whereas this approach enables enhancing the performance improvement with minimal changes to the transmission link, it does not enable to reach the same gains as with DRA.

A comparison between the improvement provided by OPC, in terms of nonlinear compensation (NLC) efficiency (see ref. [10]), can be seen in Fig. 6.21. The compensation is evaluated as a function of the equivalent length of the dispersion added prior to OPC stage and two cases are compared: lumped EDFA-based amplification (Fig. 6.21A) and DRA (Fig. 6.21B). As can be seen, whereas some degree of compensation can be achieved for systems with lumped amplification, the use of DRA significantly reduces the required span length to see a significant compensation.

Remarkably a lack of symmetry in the transmission link can be partially compensated for in the digital domain by applying digital backpropagation. As shown in Ref. [91], OPC and DBP can be combined so that OPC compensates for the bulk of nonlinearity while low-precision DBP compensates for the residual distortion not

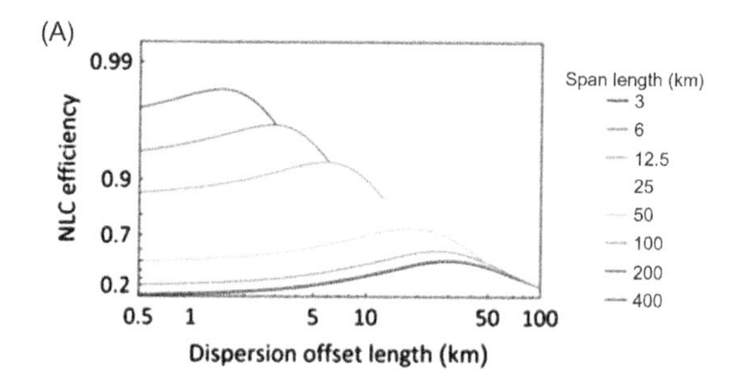

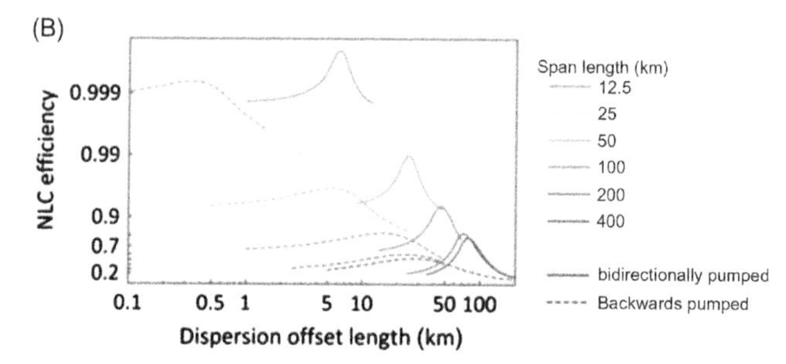

Figure 6.21 Nonlinear compensation efficiency for lumped (A) and distributed (B) amplification. Source: *Reproduced from A.D. Ellis, M.E. McCarthy, M.A.Z. Al Khateeb, M. Sorokina, N.J. Doran, Performance limits in optical communications due to fiber nonlinearity, Adv. Opt. Photon. 9 (2017) 429–503 [89].*

addressed by OPC, due, for example, to a slight offset from ML operation [91]. Therefore, aside from improved performance, OPC might also reduce overall power consumption of novel optical systems [89]. Although high power is required for the conversion itself, it will lead to nonlinearity and chromatic dispersion compensation over the entire telecommunication band. As a result, it will reduce digital-signal processing and simplify receiver designs, leading to net power savings on top of the increased performance [89].

The major advantage of OPC compared to digital techniques is its broad bandwidth of operation, if the power-dispersion symmetry conditions can be fulfilled over the full bandwidth. As the conjugation operation is usually implemented through either FWM or cascaded second-harmonic generation (SHG) and difference frequency generation (DFG), the conjugated signal is generally frequency-shifted compared to the original signal. Nevertheless, a so-called "band-swapping" approach can be used where the signal band is divided into two half-bands that are independently conjugated and thus swapped as shown in Fig. 6.22 [89,92,93]. The input signal is separated into high and low frequencies with a wavelength-selective switch, such that high signal frequencies enter one device to be converted to low idler frequencies, and vice versa. In this manner, the complete telecom band can be occupied with independent

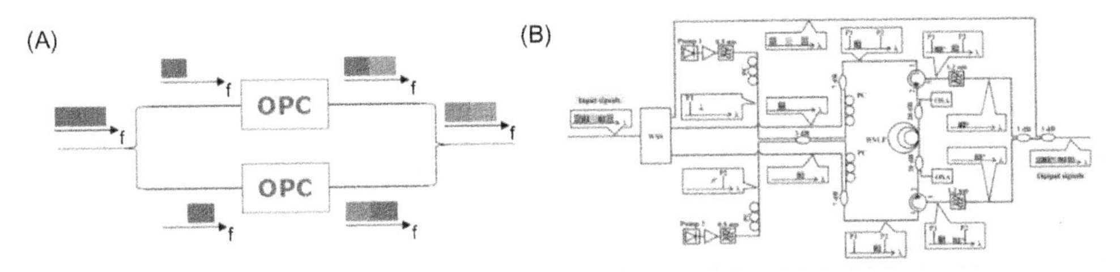

Figure 6.22 Principle of band-swapping OPC (A) and implementation with a single nonlinear device (B). Source: *Reproduced from S. Yoshima, Y. Sun, Z. Liu, K.R.H. Bottrill, F. Parmigiani, D.J. Richardson, et al., Mitigation of nonlinear effects on WDM QAM signals enabled by optical phase conjugation with efficient bandwidth utilization, J. Lightwave Technol. 35 (2017) 971–978 [92].*

data channels and phase-conjugated without losing half of the bandwidth. This "trick" ensures that two OPC devices are sufficient to cover the full telecommunication C-band as highlighted by a number of recent demonstrations [90,94].

The choice of using FWM or SHG/DFG additionally provides the required modulation and bitrate agnostic operation; however, these processes are inherently polarization dependent. Therefore, schemes based on either polarization diversity or orthogonal pumping are commonly used. In the polarization diversity case, the orthogonal signal and pump polarizations are separated in a polarization beam splitter, and enter the nonlinear medium from the opposite sides. OPC is performed between co-propagating, and thus co-polarized, pump and signal components. Assuming identical pump powers in both directions (polarizations), polarization-insensitive operation can be achieved. As an alternative, FWM with two orthogonal pumps can be used such that vectorial FWM is used. In order to minimize the number of nonlinear devices required, the most compact approach is shown in Fig. 6.22B, where orthogonal pumping provides the dual-polarization operation and bidirectional propagation into the nonlinear medium enables the band-swapping configuration [92].

Whereas the inherent frequency shift introduced by FWM or SHG/DFG can be easily dealt with by the band-swapping technique, a broad frequency shift will result in a slightly different dispersion experienced by the original signal and its conjugated copy through the two halves of the link. Along this direction, a technique has been developed to enable waveband-shift free OPC [95]; however, this method scales poorly with overall OPC bandwidth, eventually leading to the same frequency shift if the full band is considered. Nevertheless, recent studies of the impact of the frequency shift on the improvement for practical links with lumped amplification showed a limited degradation [96]. Such an impact is, however, believed to increase with the transmission link optimization, for instance, by using DRA.

Another physical effect that is strongly related to the broadband operation of the OPC is polarization mode dispersion (PMD). The impact of PMD causes a stochastic

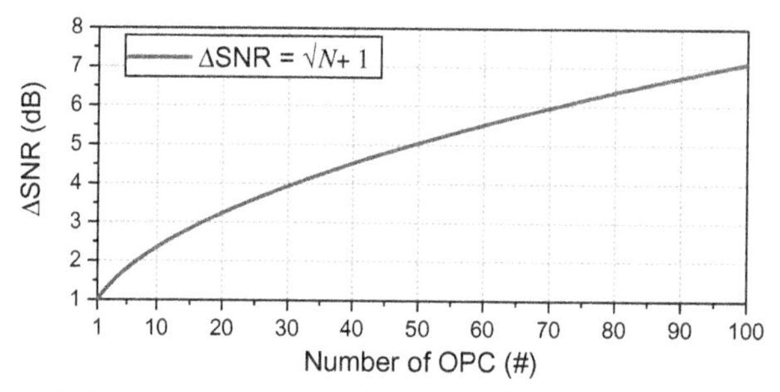

Figure 6.23 Normalized SNR improvement as a function of the number of OPC.

polarization evolution, which is additionally wavelength dependent, leading to different polarization rotations for the different channels. The statistical and wavelength dependence of PMD is an additional source of asymmetry in the transmission link, as NLC through OPC assumes a constant relative state of polarization for the different channels. It has been numerically shown that the presence of PMD can significantly reduce the effectiveness of the nonlinearity compensation [97]. The most effective way to address the PMD limitation relies on performing multiple OPC operations throughout a link—in other words, placing OPC devices spaced less than half of the correlation length [90,97]. Spreading multiple OPC stages within the link not only mitigates the challenges introduced by OPC but overall increases the improvement, as the SNR improvement scales with the square root of the number of OPC ($\Delta SNR = \sqrt{N+1}$), as shown in Fig. 6.23 [97].

The favorable scaling with the number of OPC stages is directly related to another fundamental advantage of OPC compared to digital techniques: its ability to partially address stochastic effects such as parametric noise amplification. Whereas nonlinearity compensation techniques mainly address the deterministic signal–signal interaction, OPC enables to reduce noise-noise interaction. As parametrically amplified noise is conjugated and further transmitted, the parametric amplification is reversed [90]. In the case of multiple OPC within a link, the noise amplification is therefore limited in its growth to the OPC spacing [90].

Experimental demonstration of OPC-based transmission have been demonstrated with OPC stages implemented using different nonlinear platforms, from the more mature HNLFs [90,92,95,96] to periodically poled lithium niobate (PPLN) [94,98] and silicon waveguides [99–101]. Each nonlinear platform presents different benefits and challenges, and to this date, no ideal material has been pointed out, as discussed in Section 4. Regardless of the material platform, it is paramount to ensure minimal penalty in the generation of the conjugate signal as any penalty at the OPC stage would directly reduce the achievable gain. Currently, the implementation penalty from the OPC stage itself is still a challenge. Even though narrow-band low-penalty conversions have been reported, the material platform that enables the most favorable scaling to broadband operation with high signal-to-conjugate conversion is still an open question. This point will be briefly touched upon in the following subsection discussing the most recent advances in the experimental demonstration of OPC.

6.3.2 Examples from literature and recent demonstrations

As discussed earlier, addition of OPC devices into a link has a potential of reversing fiber nonlinearities and chromatic dispersion, possibly increasing total throughput or transmission distance. Under the most basic scenario with OPC in the middle, reach enhancements by over a factor of two have been numerically predicted for idealized systems with high-order modulation formats [89,97,102]. Nonetheless, real-world imperfections such as an imperfect phase conjugator or PMD will limit the achievable gains. In Ref. [103], OPC-aided system in 400 km Raman-amplified link is demonstrated with Q-factor improvements of 1.1 dB at a single 28-GBd polarization-division multiplexed channel, and 0.8 dB improvement at five WDM channels transmitted simultaneously. As shown in Fig. 6.24, the optimum power shifts to higher values, and partial nonlinearity compensation is successfully achieved.

The measured data are in good agreement with the developed numerical model, which predicts that implementation of OPC into the system is capable of extending the maximum transmission distance by up to 80% for five channels before the FEC threshold is reached.

This demonstration is extended to longer distance and higher bit rates in Ref. [90], where OPC-aided system with ten 400 Gb/s super-channels in Raman-amplified link is tested for up to 20% extended reach as compared to the case without OPC. When only two super-channels were transmitted simultaneously, addition of OPC leads to 60% increase instead due to reduction in signal bandwidth and lower PMD impact. The complete results for the analyzed system are given in Fig. 6.25.

Aside from polarization-insensitive operation for PDM, the setup also employs a pair of OPC devices for performing the band-swapping operation introduced in Section 3.1 [89,90]. The demonstration set the record bit rate-distance product for an OPC-based system at 4 Tb/s × 2000 km at the time, which was subsequently overtaken by NTT Japan in 2016 [94]. The group demonstrated transmission, OPC and performance improvement across 92 channels with 22.5 GBd 16-QAM PDM modulation, achieving 13.6 Tb/s total capacity over 3840 km Raman-amplified link. OPC based on the second-order nonlinearity in periodically poled $LiNbO_3$ is inserted ML, and provides polarization-insensitive and band-swapping operation in order to support high data rates and bandwidth. The measured gains for each channel are presented in Fig. 6.26. This astonishing result holds the record bit rate-distance product for OPC-aided systems to date.

The gains from a single OPC device are evident, despite significant implementation limitations. As discussed in Section 6.3.1, the improvement scales with square

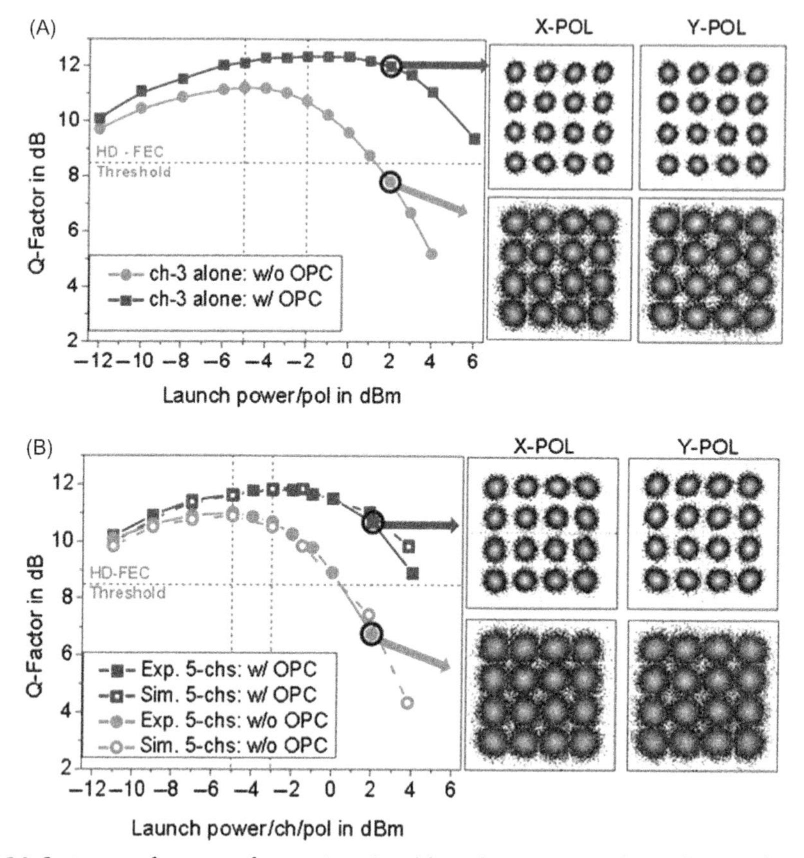

Figure 6.24 System performance for varying signal launch power per channel per polarization for 1 (A) and 5 (B) channels transmitted simultaneously with (*square*) and without (*dot*) ML OPC. Numerical simulations (*dashed*) accompany experimental results (*solid*) for WDM scenario. Source: *Reproduced from I. Sackey, F.D. Ros, M. Jazayerifar, T. Richter, C. Meuer, M. Nölle, et al., Kerr nonlinearity mitigation in 5 × 28-GBd PDM 16-QAM signal transmission over a dispersion-uncompensated link with backward-pumped distributed Raman amplification, Opt. Express 22 27381–27391 (2014) [103].*

root of the number of OPC. Other than having been predicted analytically, it has been confirmed experimentally that increasing the number of OPCs will generally lead to higher gains due to improved symmetry and lower impact of PMD [89,104,105]. In Ref. [104], the nearly ideal OPC scheme discussed in Section 6.3.1 has been extended for a cascaded OPC system. The OPC-based transmission is based on a nonzero dispersion-shifted fiber. The fiber parameters allowed for identical second-order dispersion, and opposite third-order dispersion for signal and idler in the two corresponding fiber spans directly before and after OPC. Both spans used backward Raman pumping for distributed amplification, with greatly enhanced power symmetry due to short transmission fiber lengths (12 km for both spans). The system

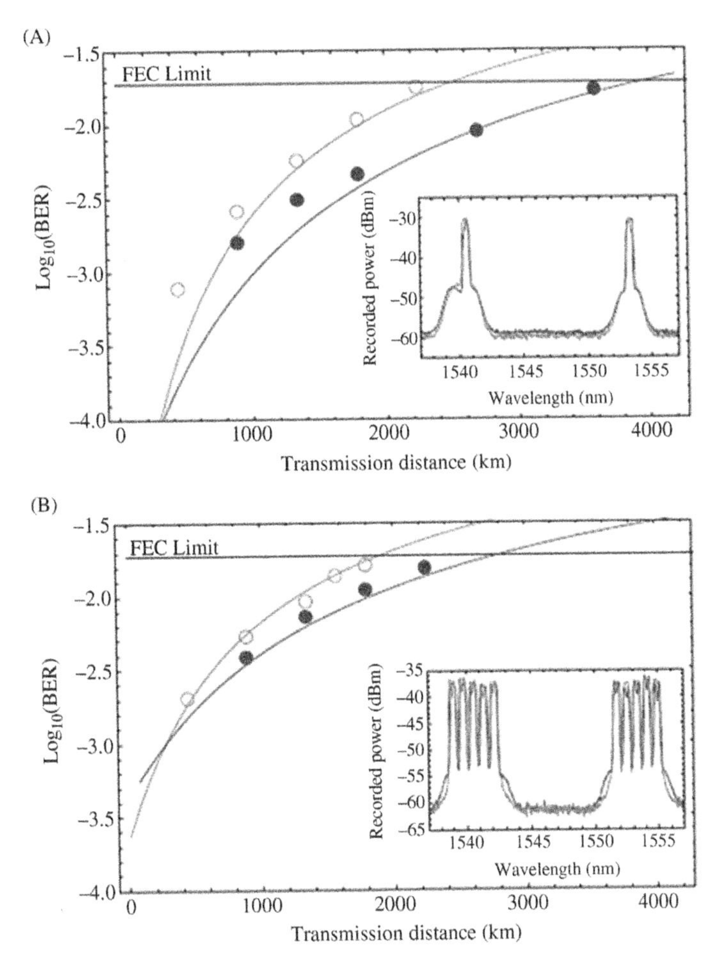

Figure 6.25 Measured (*dots*) and simulated (*lines*) BER of the central channel of the high-frequency band as a function of transmission distance with (*solid*) and without (*red*) ML OPC for 2 (A) and 10 (B) super-channels transmitted. Inset: received spectra with (*blue*) and without (*red*) OPC at maximum distance. Source: *Reproduced from A.D. Ellis, M. Tan, M.A. Iqbal, M.A.Z. Al-Khateeb, V. Gordienko, G.S. Mondaca, et al., 4 Tb/s transmission reach enhancement using 10 × 400 Gb/s super-channels and polarization insensitive dual band optical phase conjugation, J. Lightwave Technol. 34 1717–1723 (2016) [90].*

performance was evaluated for 12 subsequent OPCs at 12 times the transmission distance (144 km in total) at 1.1 Tb/s QPSK PDM signal split among 24 channels. As the conditions for perfect compensation imposed by NLSE are largely fulfilled, near perfect nonlinearity cancellation has been achieved, leading to approximately linear transmission at 10 dBm input power.

A more realistic OPC-cascaded system is demonstrated in Ref. [105] with moderate span lengths and a more practical transmission fiber. The sequential phase

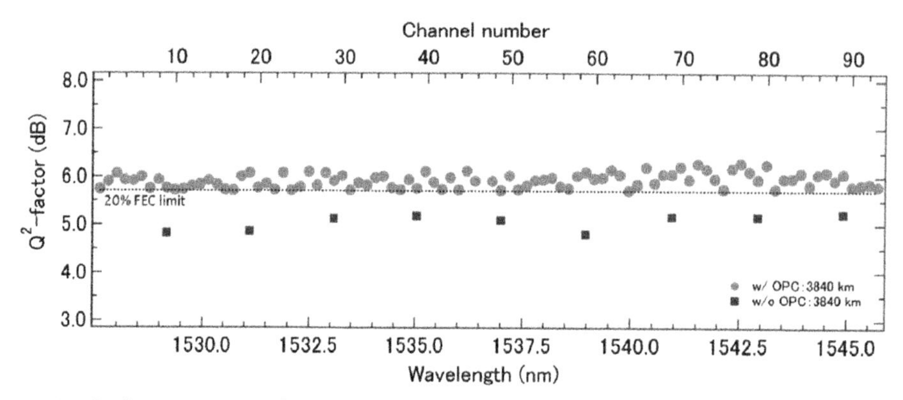

Figure 6.26 Performance gains due to ML OPC across 92 WDM channels after 3840 km transmission. Source: *Reproduced from T. Umeki, T. Kazama, A. Sano, K. Shibahara, K. Suzuki, M. Abe, et al., Simultaneous nonlinearity mitigation in 92 × 180-Gbit/s PDM-16QAM transmission over 3840 km using PPLN-based guard-band-less optical phase conjugation, Opt. Express 24 16945–16951 (2016) [94].*

conjugation is shown to further improve the signal quality as compared to a single ML OPC for 3600 km link at 1 Tb/s data rate. Indeed, performance gains of over 2 dB have been readily demonstrated for practical systems employing OPC cascades [89].

Some of the listed demonstrations reproduce realistic conditions but have not been deployed as such. However, a number of actual field trials of OPC-aided systems have been performed. One such demonstration in Japan with a single ML OPC enabled doubling the reach for a 12-GBd 64-QAM PDM signal from 350 to 700 km, despite only EDFA amplification and legacy standard single-mode fiber (SSMF) used for transmission [106]. Another field trial in England over 843 km EDFA-amplified link evaluated the OPC performance gains for WDM scheme [107], where it led to Q-factor improvements of up to 3 dB at six 10 GBd 16-QAM PDM signals. The corresponding gains per channel are shown in Fig. 6.27. It is, therefore, reasonable to conclude that OPC has a potential to also serve as an upgrade to existing networks, despite the demanding symmetry requirements.

In order to improve the power consumption and footprint of the OPC stage itself, as discussed in Section 4.1, a significant focus has been devoted to the nonlinear platform used to perform the conjugation. As mentioned in Section 4.1, several material platforms can be used. In particular, other than the impressive demonstration reported using HNLFs [90,103–107] and PPLN waveguides [94], a few initial investigations have started focusing on integrated waveguides by using silicon on insulator as nonlinear material. As discussed in Section 4.1, silicon has the advantage of stronger nonlinearity compared to HNLF and PPLN, as well as tighter light confinement that enables using very compact devices.

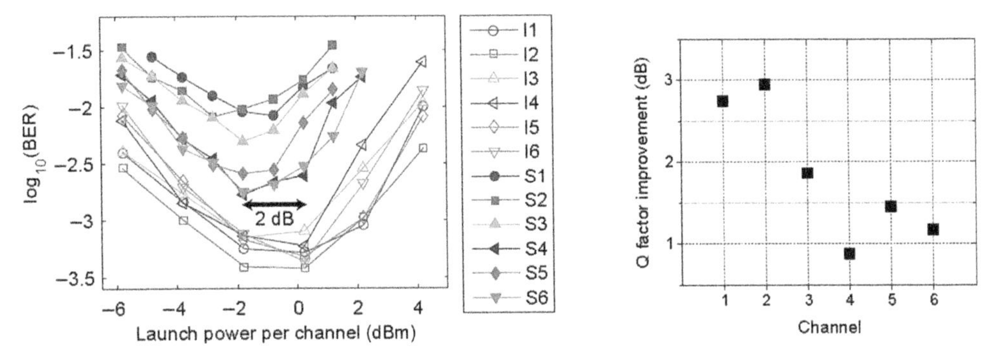

Figure 6.27 Left: BER for each channel as a function of varying launch power without OPC (denoted by S) and with OPC (denoted by I). Right: Performance gains per channel in terms of Q-factor. Source: *Reproduced from Y. Sun, A. Lorences-Riesgo, F. Parmigiani, K.R.H. Bottrill, S. Yoshima, G. D. Hesketh, et al., Optical nonlinearity mitigation of 6 × 10 GBd polarization-division multiplexing 16 QAM signals in a field-installed transmission link, in: Optical Fiber Communication Conference, OSA Technical Digest (online) (Optical Society of America, 2017), paper Th3J.2 [107].*

A first demonstration of OPC-based transmission with the OPC stage using a silicon waveguide was reported in Ref. [99]. Whereas the experiment clearly show the higher nonlinear tolerance of the signals when OPC is employed, the improvement can be seen only for high launched power and at the optimum the signal quality is strongly limited by the quality of the conjugated signal. Silicon is indeed a strongly nonlinear material, but other than Kerr nonlinearity, its bandgap results in two-photon absorption (TPA) at telecom wavelengths (1550 nm). TPA, in turn, leads to the accumulation of free carriers within the waveguide region that causes additional loss through free carrier absorption (FCA). These nonlinear loss mechanisms significantly impair the efficiency of FWM, resulting in the generation of low power conjugated signals that are OSNR limited. In order to mitigate the impact of FCA, a well-known method relies on fabricating a p−i−n diode across the waveguide region. When a reverse bias is applied to such a diode, the electrical field sweeps the free carriers out of the waveguide significantly decreasing FCA and therefore increasing the FWM efficiency. By using this technique, conversion efficiency as high as −0.7 dB has been reported [108]. As the efficiency is improved, so is the OSNR of the conjugated signal; therefore, by employing a silicon waveguide with a p−i−n diode, the first demonstration of signal quality improvement through OPC in an integrated waveguide was recently reported [100]. Such a first demonstration was limited to single-polarization signals due to the polarization sensitivity of the fabricated device. However, by using a polarization-diversity scheme similar to the one described in Fig. 6.2, dual-polarization OPC was demonstrated reporting 1.2 dB of SNR improvement for five-channel transmission [101].

6.3.3 Coding for the optical-phase conjugation channel: complementary digital and optical signal processing—probabilistic shaping for optical phase–conjugated link

6.3.3.1 Motivation for the study

As discussed, OPC is a technique that attempts to cancel the nonlinearities in the optical fiber and thereby provides a higher effective SNR at the receiver. As mentioned previously, ideal OPC on a noiseless fiber provides ideal compensation, rendering the channel additive white Gaussian noise (AWGN)-like (noise only coming from receiver front-end). For AWGN channels, the optimal transmission strategies are well known [109]. The improved channel conditions provided by OPC can be combined with PS as it attempts at optimizing the transmission for given channel conditions. The combination of the gains provided by both techniques and their independence is thus of interest. However, practical OPC on a real fiber channel only partially cancels the nonlinearities, and furthermore it can only partially address parametric noise amplification. Additionally, an analytical channel model representative of the entire link including OPC is, thus, far from straightforward. Therefore, tailoring the PS directly to the implemented channel is of great interest.

6.3.3.2 Basics of information theory

In order to optimize the probability mass function (PMF) of the transmitted constellation (PS), the data transmission through an optical channel, including linear and nonlinear devices, and indeed, any channel, is quantified using achievable information rate (AIR) as metric. If channel input comes from a finite size constellation $\mathcal{X}$, which can be, for instance, the set of QAM symbols with a given PMF P_X, the probability of occurrence of each symbol is defined for short notation as $P_X(X=x)=p(x)$. The AIR can be estimated from the mutual information (MI) between the channel input X and output Y

$$\mathcal{I}(X;Y)=\mathcal{H}(X)-\mathcal{H}(X|Y), \tag{6.10}$$

where $\mathcal{H}(X)$ is the entropy of the input, which describes how much information is present in it, and $\mathcal{H}(X|Y)$ is the conditional entropy, which describes how much information is still present once we observe the output. For ideal noiseless channels $Y=X$, $\mathcal{H}(XY)=0$ and the entire information content of X can be transmitted through the channel. The MI can be estimated as

$$\mathcal{I}(X;Y)=\lim_{K\to\infty}\frac{1}{K}\left[-\log_2 p\left(\boldsymbol{x}_1^K\right)+\log_2 p\left(\boldsymbol{x}_1^K|\boldsymbol{y}_1^K\right)\right], \tag{6.11}$$

where the entropies are estimated from infinitely long sequences $\boldsymbol{x}_1^K$ and $\boldsymbol{y}_1^K$ of length K and $p(\cdot)$ is their PDF [109]. The input sequence typically originates in independent, identically distributed (i.i.d.) data, and the entropy $\mathcal{H}(X) = \lim_{K\to\infty} \frac{1}{K}\left[-\log_2 p\left(\boldsymbol{x}_1^K\right)\right]$ is thus straightforward to calculate. Since the optical fiber channel typically has memory due to the interactions between dispersion and nonlinearities, the calculation of $\mathcal{H}(X|Y)$ is more involved and is rarely available in closed form. Typical approximations assume that the channel is memoryless AWGN, for which the posterior probability function $p(\boldsymbol{x}_1^K|\boldsymbol{y}_1^K)$ is approximated as

$$q\left(\boldsymbol{x}_1^K|\boldsymbol{y}_1^K\right) = \frac{q\left(\boldsymbol{y}_1^K|\boldsymbol{x}_1^K\right)p\left(\boldsymbol{x}_1^K\right)}{q\left(\boldsymbol{y}_1^K\right)} = \prod_k \frac{q(y_k|x_k)p(x_k)}{\sum_i q(y_k|x^i)p(x^i)}, \tag{6.12}$$

where the auxiliary likelihood $q\left(\boldsymbol{y}_1^K|\boldsymbol{x}_1^K\right)$ is factorized into the Gaussian functions $q\left(y_k|x^i\right) = \mathcal{N}\left(y_k; \mu_i, \sigma_i^2\right)$ with mean μ_i and variance σ_i^2. Replacing $p\left(\boldsymbol{x}_1^K|\boldsymbol{y}_1^K\right)$ with $q\left(\boldsymbol{x}_1^K|\boldsymbol{y}_1^K\right)$ results in an upper bound on the entropy $\mathcal{H}_q(X|Y) \geq \mathcal{H}(X|Y)$ and thereby a lower bound $\mathcal{I}_q(X;Y) \leq \mathcal{I}(X;Y)$ [110]. An optimal maximum a-posteriori (MAP) receiver, which maximizes $q\left(\boldsymbol{x}_1^K|\boldsymbol{y}_1^K\right)$ when making decisions [111] can achieve a rate of at most $\mathcal{I}_q(X;Y)$ bits/symbol, and $\mathcal{I}_q(X;Y)$ is therefore referred to as an AIR.

The channel capacity C is defined by Shannon [109] as

$$C = \max_{p(X)} \mathcal{I}(X;Y), \tag{6.13}$$

where the maximization is over the PMF. Since the MI is difficult to evaluate and thus maximize, the AIR can be maximized instead. *PS* refers to the optimization of the input PMF, and the *shaping gain* refers to the gain in AIR, achieved with the optimized PMF w.r.t. standard uniform PMF, for which all symbols from the alphabet occur with the same probability $p\left(x^i\right) = 1/|\mathcal{X}|$.

6.3.3.3 Algorithm for maximizing the achievable information rate

The classical method for finding the optimal memoryless PMF and the capacity of a memoryless channel is the Blahut-Arimoto algorithm (BAA) [112,113]. BAA applies expectation-maximization type update rules iteratively and exploits the fact that $\mathcal{I}(X;Y)$ is concave in $p(X)$ for fixed $p(y|x)$ and $p(x|y)$. In order to apply it to the optical fiber, the likelihoods $q\left(y_k|x^i\right)$ must first be estimated. This can be done by first transmitting a long sequence of training data, and then fitting a Gaussian distribution to the sample distribution around each constellation symbol. Variants of this method exist, in which the Gaussian can be considered correlated across the I and Q dimensions [114], or even across the spatial dimensions (e.g., X and Y polarizations). For simplicity, here we consider the circularly symmetric Gaussian distribution. The BAA is an iterative algorithm that first initializes the PMF and then estimates the likelihoods as above, and based on those likelihoods, attempts to find a new PMF that increases the MI. For several standard channels, for instance, discrete memoryless channel or the memoryless AWGN [115], the algorithm is proven to converge to the optimum.

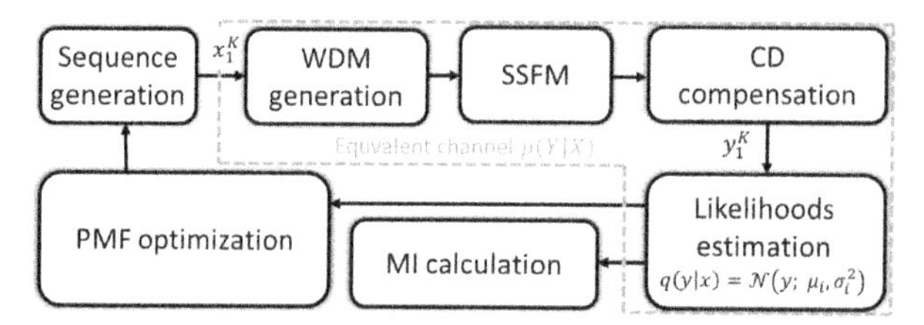

Figure 6.28 The modified BAA for PMF optimization on the optical fiber. Adapted from M.P. Yankov, F. Da Ros, E. P. da Silva, S. Forchhammer, M. Galili, and L. Oxenløwe, "Optimizing the Achievable Rates of Tricky Channels: A Probabilistic Shaping for OPC Channel Example", in IEEE Photonics Conference (IPC), 2018 [117].

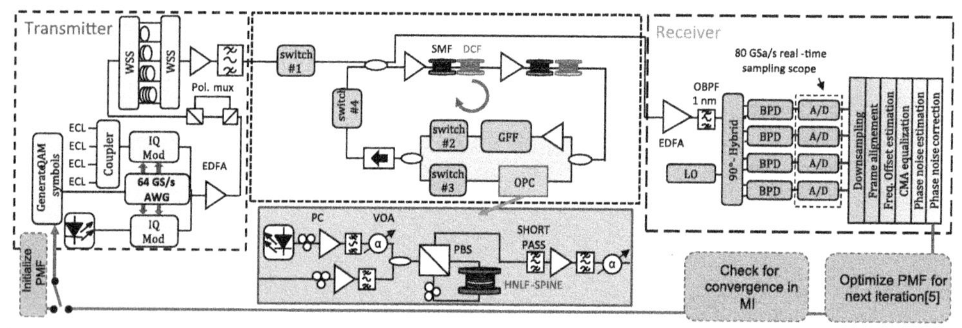

Figure 6.29 2 Experimental setup to optimize the input PMF for straight-based (no OPC) and OPC-based transmission. Source: *Reprinted from M.P. Yankov, F. Da Ros, E. P. da Silva, S. Forchhammer, M. Galili, and L. Oxenløwe, "Optimizing the Achievable Rates of Tricky Channels: A Probabilistic Shaping for OPC Channel Example", in IEEE Photonics Conference (IPC), 2018 [117].*

In the case of the fiber, due to the nonlinearities, the likelihood depends on the input PMF and $p(y|x)$ can no longer be considered fixed. The BAA can thus only be considered approximate. However, it can still be used to find optimized PMFs and improved AIRs [116]. The modified BAA is summarized in Fig. 6.28.

6.3.3.4 The optical-phase conjugation case

A typical laboratory OPC experimental setup includes various devices in order to arrive at fair comparison and OPC gain estimation, as shown in Fig. 6.29 and detailed in [117]. The setup consists mainly of a five-channel WDM PDM transmitter, a recirculating transmission loop with dispersion compensated spans, and a standard preamplified coherent receiver. At each recirculation in the loop, optical switches can select either the original signal or the conjugated signal at the output of the HNLF-based OPC stage. For this analysis, the OPC operation has been performed only in the middle of the link (ML OPC). Given the complexity of such setups, they are increasingly difficult to model accurately and thus generate the sample distribution and its outcome

$\boldsymbol{y}_1^K$. The channel model [e.g., the split-step Fourier method (SSFM) in Fig. 6.27] can thus be replaced by the true experimental setup [117]. Such a process entails performing a measurement of the received samples at each BAA iteration. Each measurement includes training data (used for estimation of the likelihoods) and testing data (for optimization of the PMF to be transmitted on the next iteration).

Similar to the simulation case of the modified BAA, in this case the PMF is initialized to, for instance, uniform. Symbols $\boldsymbol{x}_1^K$ are generated and transmitted through the channel, which may or may not include the OPC stage. The received waveform after analog-to-digital conversion is processed with DSP receiver techniques until the corresponding sequence $\boldsymbol{y}_1^K$ is obtained. The likelihoods are then obtained from the training data and used to find a PMF on the next iteration. The process repeats until convergence either of the PMF or the AIR.

Typical AIRs achieved with this algorithm are given in Fig. 6.30 for straight (*solid lines*) and ML OPC (*dashed lines*) 64-QAM transmission. As mentioned, the two techniques achieve gain through different means: ML-OPC aims at reducing the nonlinear interference noise (NLIN) variance, while PS aims at optimizing the transmission for a given NLIN. The optimal launch power with ML-OPC is thus increased, while with PS, the system is pushed to operate at a lower launch power region, for which the noise is more Gaussian and i.i.d. in time. For the considered setup, ML-OPC provides ≈ 0.19 bits/symbol of gain and PS ≈ 0.3 for both cases of straight and ML-OPC transmission. Furthermore, the gains of both techniques appear to add up. We therefore conclude that the two techniques are independent to the extent of this analysis and that even though the optimal launch power is different, the NLIN statistics of ML-OPC links and standard links at the optimal launch power are qualitatively similar.

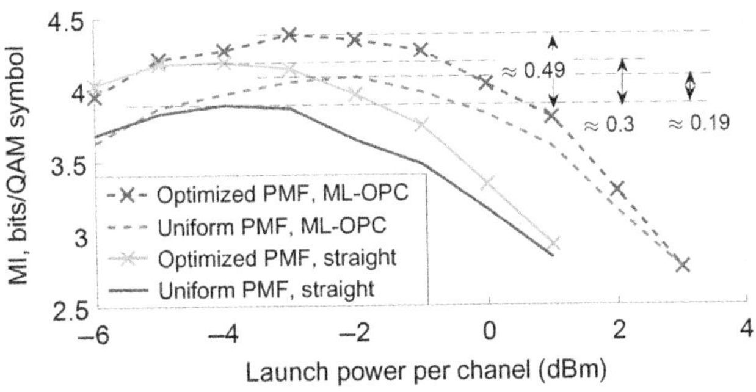

Figure 6.30 AIRs for straight and ML-OPC-based nonlinearities compensation, in the cases of standard uniformly distributed 64-QAM and probabilistically shaped 64-QAM. Source: *Reprinted from M. P. Yankov, F. Da Ros, E. P. da Silva, S. Forchhammer, M. Galili, and L. Oxenløwe, "Optimizing the Achievable Rates of Tricky Channels: A Probabilistic Shaping for OPC Channel Example", in IEEE Photonics Conference (IPC), 2018 [117].*

6.4 Nonlinear material platforms for optical processing

The foundation of optical signal processing is the nonlinear materials used to achieve the desired effect. There has been a lot of research into numerous materials platforms to find the optimum. It is hard, though, to define a single FOM that suits all applications, and one will thus probably always need to consider several FOMs and apply them to the relevant application. This section discusses several nonlinear materials, old and new, and defines a new FOM relevant for OSP. We first start out by discussing one of the most successful materials, the highly nonlinear optical fiber, and then move on to discussing integrated waveguides, and all along we try to discuss pros and cons.

6.4.1 Highly nonlinear fiber: Efficiency and limitations

The silica-based dispersion-shifted germania-doped HNLF [118–120] has been used extensively for optical signal processing based on the Kerr effect, due to the many advantages afforded optical fiber in general, including ease of fiber-optic integration, ultralow propagation losses and availability of mature fabrication techniques which allow for efficient mass fabrication. In SSMF the nonlinear coefficient, γ, is typically in the range of 1–2 $(\mathrm{W\,km})^{-1}$, whereas γ attains values of 5–20 $(\mathrm{W\,km})^{-1}$ in HNLF. The increased nonlinear coefficient is achieved by confining the light in a much smaller core area, with a very high core-cladding index contrast ($>2\%$), Δ. In a typical HNLF the mode field diameter is 4–5 μm compared to ~10 μm diameter for SSMF. The fiber nonlinear coefficient is orders of magnitude smaller than, for instance, silicon-based chips, which have greater confinement and larger intrinsic nonlinearity, and benefit from a very small device footprint. Whereas HNLF are typically much bulkier than chip-based devices, the nonlinear efficiency of HNLF can be higher in practice due to the long effective nonlinear lengths enabled by low propagation losses (<0.8 dB/km). Note that highly compact fiber-based devices have been realized using, for instance, bismuth-oxide glass instead of silica for its high intrinsic nonlinear properties [121], or using HNLF with reduced cladding coiled to approximately coin-sized bobbins [122]. Although HNLF can realize extremely efficient fiber-optic parametric amplifiers (FOPAs) with records of 70 dB gain [123], and up to 270-nm FWM bandwidth reported [124], significantly lower numbers are generally achieved, and it is a major practical challenge to reliably fabricate fibers with uniform transverse geometry over hundreds of meters with typical core dimensions. The resulting unpredictable dispersion fluctuations affect the FWM phase mismatch, resulting in FWM efficiency variations in the direction of propagation [125,126]. This effect contributes to a reduced nonlinear bandwidth compared to compact chip-based devices. However, there exist methods that can mitigate the effects of the dispersion fluctuations, including optimized arrangements of short HNLF pieces with different dispersion to manage the accumulated phase mismatch [127,128], longitudinal temperature control or varying local tensile stress to equalize the dispersion [129,130], or special fiber designs where the dispersion is insensitive to core diameter fluctuations [61]. Methods based on altering the fiber properties after fabrication, generally require customized treatment of individual HNLFs. Finally, the achievable efficiency is often limited by stimulated Brillouin scattering (SBS) [130–132] for applications that require a high-intensity continuous-wave (CW) pump, such as FOPAs or FWM-based wavelength conversion of communication signals. SBS causes the buildup of acoustic phonons which induces propagating grating-like structures in the fiber, thus reflecting power near the pump frequency. Each fiber has a certain SBS threshold of the pump power, at which point most of the injected power is reflected backwards. Since the SBS gain bandwidth in silica is extremely narrow ($\sim$30 MHz), a useful technique to increase the SBS threshold is to distribute the CW power over multiple lines with $>$30 MHz spacing, by phase modulation using multiple RF tones with carefully selected spacing [133]. For applications that are sensitive to phase distortions, techniques that smear out the fiber SBS gain spectrum have been shown to be effective, for instance, by applying a tensile stress gradient [134] or a temperature gradient [135]. Alternatively, doping the core with aluminum instead of germanium has been shown to efficiently increase the SBS threshold without affecting the dispersion properties, by reducing the core confinement of the phonons [136]. In another complementary method, shorter pieces of HNLF are separated by isolators [137]. In this way, the effective SBS gain is reduced by preventing strong SBS amplification of the backscattered light, compared to the aggregated pieces of HNLF without isolators.

6.4.1.1 Design and variations

The index profile of a typical HNLF is shown in Fig. 6.31A. Here the core refractive index (n_c) is increased by heavily doping the silica (SiO_2) core with germania (GeO_2), typically with a $\sim$20 mol.% concentration, whereas a so-called depressed cladding with reduced index (n_{cl}) compared to silica is obtained by fluorine doping. In this way, large index differences of 3% can be obtained [120]. Such fibers are often fabricated using the modified chemical vapor deposition method.

There are a number of variations on the HNLF design, which mainly serve to tailor the dispersion parameters, or to induce a strong fiber birefringence, which gives the HNLF polarization-maintaining (PM-HNLF) properties to maximize the field alignment between interacting waves during propagation. HNLFs are typically fabricated with low tolerance to birefringence; however, fiber straining techniques, for example, tend to induce significant birefringence, which can reduce the FWM bandwidth [138,139]. PM-HNLF can be realized conventionally by inserting, for instance, borosilicate (B_2O_3) rods, which induce a directional material stress due to an increased thermal expansion coefficient compared to fused silica. Other methods induce birefringence via an elliptically shaped core [140]. Transverse profiles for step-index,

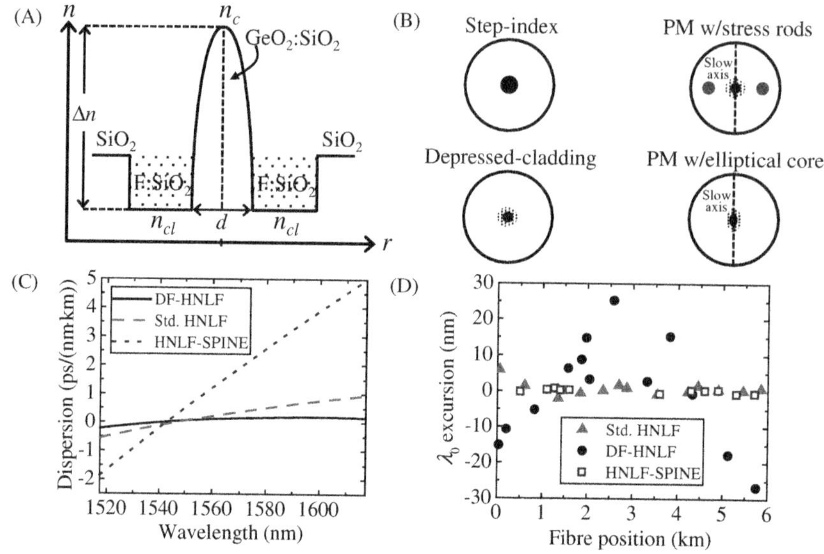

Figure 6.31 (A) Index profile of a typical HNLF with core diameter *d*. (B) Transverse profiles of different HNLF. (C) Dispersion of different HNLF variations. (D) Mean λ_0 excursion measurements over 6 km for different HNLF variations.

depressed-cladding HNLF and PM-HNLFs are illustrated in Fig. 6.31B. The dispersion for three different HNLF variants are shown in Fig. 6.31C, including dispersion-flattened HNLF (DF-HNLF) [141,142], standard HNLF (by fiber manufacturer OFS), and HNLF with stable phase matching for improved nonlinear efficiency (HNLF-SPINE) [61]. The DF-HNLF is designed to have an ultralow dispersion slope (typically ~ 0.005 ps/(nm^2 km)), which is beneficial for applications sensitive to short-pulse walk-off effects. In addition, the dispersion is as low as possible in the nonlinear bandwidth of interest where phase matching over a large bandwidth is required, such as for time lenses based on FWM with a chirped pump. The standard HNLF is versatile and useful for a wide variety of nonlinear applications, whereas the HNLF-SPINE with its relatively large dispersion slope [~ 0.07 ps/(nm^2 km) compared to ~ 0.02 ps/(nm^2 km) for std. HNLF] is less suitable for walk-off sensitive applications. However, with careful pump wavelength selection, a relatively large slope may not affect the efficiency if a CW pump is used. Furthermore, the HNLF-SPINE is a type of fiber where the waveguide geometry is designed to counteract the effects of random core radius fluctuations. For FWM in particular, the zero-dispersion wavelength, λ_0, is an important parameter to take into account when selecting an HNLF for a specific operational bandwidth. To demonstrate the effectiveness of the HNLF-SPINE design, λ_0 was measured for different pieces of fiber originating from the same preform for the three HNLF designs. The results for several pieces spanning 6 km of the drawn preforms are illustrated in Fig. 6.31D. Here the λ_0 excursion relative to the mean is shown at various positions along the fibers. The stability of λ_0 can be seen to be extremely good for the HNLF-SPINE, whereas very large excursions are observed for the DF-HNLF in particular. The reason is partly that λ_0 experiences larger shifts for the same dispersion change with a lower dispersion slope. However, the chromatic dispersion shift turns out to be larger as well, and a stable λ_0 is important for optimum FWM efficiency. Therefore, such dispersion-stable HNLFs are very useful for applications that do not depend on the dispersion slope, whereas DF-HNLF may be required for applications relying on ultrashort or strongly chirped pulses. Hence, the choice of nonlinear application determines which HNLF design is the most suitable.

6.4.2 Photonic chips: broadband and compact

Where optical fibers benefit from accumulating a relatively small nonlinearity over long lengths to arrive at a very high nonlinear efficiency, photonic chips can have a very high nonlinearity and thus achieve a desired effect over a much shorter distance (on chip). This shorter distance also makes phase matching easier over a very wide bandwidth, and a 750 nm FWM bandwidth has been reported [143], enabling an ultrafast optical signal processing of a 1.28 Tbit/s serial data signal. In this section, we review various promising integrated platforms.

6.4.2.1 Aluminum gallium arsenide

In this section we will discuss the AlGaAs-OI platform which, with proper dispersion and bandgap engineering, is not affected by many of the limitations suffered by other integrated platforms for nonlinear signal processing. We will outline the properties of the platform and key fabrication steps. We will then highlight a few demonstrations of applications for nonlinear signal processing: the first demonstration of optical wavelength conversion of 256 QAM data, demonstration of high-performance integrated phase-sensitive FWM and high-quality spectral broadening enabling single-source 661 Tbit/s data transmission.

6.4.2.1.1 The aluminum gallium arsenide on insulator platform

AlGaAs-OI has appeared as an attractive platform for nonlinear optical signal processing. The bandgap of the material can be engineered by adjusting the aluminum concentration avoiding TPA at telecom wavelengths (around 1550 nm). In this way films of high-index material ($n_{AlGaAs} \approx 3.3$) without TPA can be realized on an insulator substrate. These films then allow fabrication of highly compact integrated waveguide devices by lithography.

The AlGaAs-OI wafer is realized by growing the desired AlGaAs film on a GaAs substrate. The wafer is then covered by glass and bonded to another substrate using a layer of BCB. The original GaAs substrate is then removed to expose the AlGaAs film

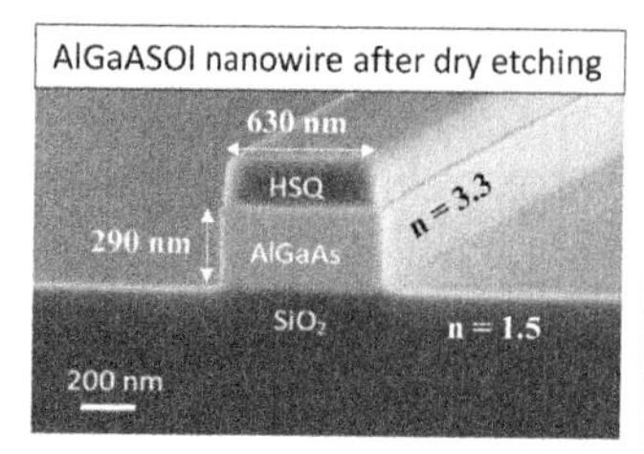

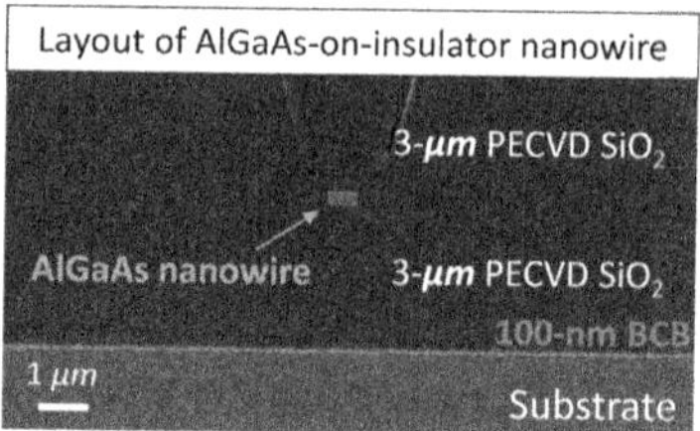

Figure 6.32 Left: SEM picture of a fabricated AlGaAsOI nano-waveguide. Right: End view of a fabricated waveguide after plasma-enhanced chemical vapor deposition (PECVD) glass cladding. M. Galili, F. Da Ros, H. Hu, M. Pu, K. Yvind, L. K. Oxenløwe, Ultra-broadband optical signal processing using AlGaAs-OI devices, *2017 Optical Fiber Communications Conference and Exhibition (OFC)*, Los Angeles, CA, 2017, pp. 1–3 [144].

to subsequent lithography. The nano-waveguides are defined by electron beam lithography and dry etching using a hydrogen silsesquioxane (HSQ) hard mask. Details about the material preparation and device fabrication can be found in Ref. [145].

Fig. 6.32 (left) shows an SEM image of an etched AlGaAs waveguide under the HSQ mask prior to deposition of the SiO2 overcladding. Fig. 6.32 (right) shows an end view of an AlGaAs waveguide after it has been fully surrounded by the SiO_2 cladding material. The BCB layer used for wafer bonding is also visible and is shown here highlighted in purple.

The high index AlGaAs core surrounded by insulator cladding ($n \approx 1.5$) achieved by this fabrication approach creates a strong confinement of the optical field in the waveguide. The resulting high field intensity in combination with a very high nonlinear refractive index ($n_2 \approx 10^{-17}$ W/m^2) [146] allows for a very strong nonlinear Kerr effect to be achieved in the AlGaAs-OI waveguide. Similar to other integrated platforms, the waveguide dispersion can be engineered through careful dimensioning of the waveguide cross section. Propagation losses of ~1.5 dB/cm are typically achieved in these waveguides, and inverse tapers [147] and lensed fibers are used in order to reduce losses when coupling light between optical fiber and the AlGaAs waveguide. Coupling loss down to 1.4 dB/facet is achieved in this way.

6.4.2.1.2 256-Quadrature amplitude modulation wavelength conversion

The first ever demonstration of optical wavelength conversion of a 256-QAM data signal was realized in an AlGaAs waveguide. The results of this demonstration are summarized in Fig. 6.33. A 10 Gbaud 256-QAM data signal was wavelength-converted by degenerate FWM in a 9-mm-long AlGaAs waveguide. As seen in Fig. 6.33A wavelength conversion spanning close to 30 nm can be performed with less than 2 dB penalty in the OSNR required to reach a BER of 2×10^{-2} after FEC [148]. For a given pump power, and thus conversion efficiency, the idler OSNR scales linearly with the input signal power. Fig. 6.33B and C shows the high achievable

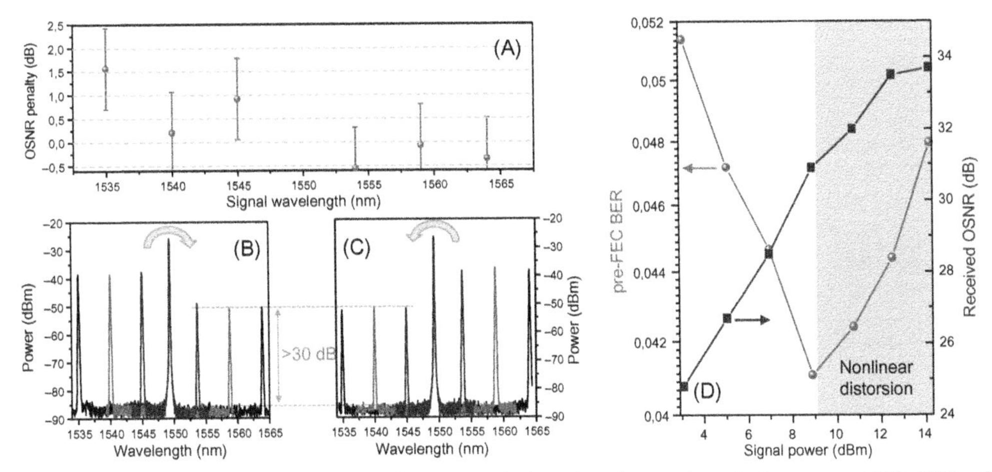

Figure 6.33 (A) OSNR penalty as a function of the signal wavelength for a post-FEC BER of 2×10^{-2}; (B and C) optical spectra at the waveguide output for signals on the short (B) and long (C) wavelength side of the pump. Idler OSNRs above 30 dB are shown for all the cases. (D) Trade-off between increasing OSNR and nonlinear distortion. Source: *Figure is adapted from F. Da Ros, M. P. Yankov, E. Porto da Silva, M. Pu, L. Ottaviano, H. Hu, et al., Characterization and optimization of a high-efficiency AlGaAs-on-insulator-based wavelength converter for 64- and 256-QAM signals., J. Lightwave Technol. 35(17) (2017) 3750–3757 [148].*

OSNR after wavelength conversion. For a given pump power there is, however, an optimum signal power given by the trade-off between increased idler OSNR and the nonlinear distortions caused by excessive signal power in the nonlinear waveguide. This optimum is clearly illustrated in Fig. 6.33D where a detailed characterization of this trade-off is presented. The increase in OSNR results in an improved pre-FEC BER only up to an input signal power of 9 dBm. Beyond this point, self-phase modulation begins to dominate signal performance. A maximum achievable signal quality corresponding to an OSNR after conversion of ~31 dB is found. Achieving higher idler OSNR values by increasing the input signal power will degrade the idler signal quality (but still increase the idler OSNR). This is clearly recognized as the minimum of the red curve showing the pre-FEC BER in Fig. 6.33D. Achieving further improvement of the converted signal stronger pump power or higher accumulated nonlinearity in the waveguide is required.

Thanks to the high achievable nonlinear interaction, low propagation loss, and accurate dispersion engineering, it has been possible to demonstrate wavelength conversion of a 256-QAM data signal over the entire telecom C-band with very low loss of signal quality.

6.4.2.1.3 Phase-sensitive four-wave mixing

The AlGaAs-OI platform has enabled phase-sensitive FWM demonstrating a high PSER of 7.7 dB without the need for any enhancement through, for instance,

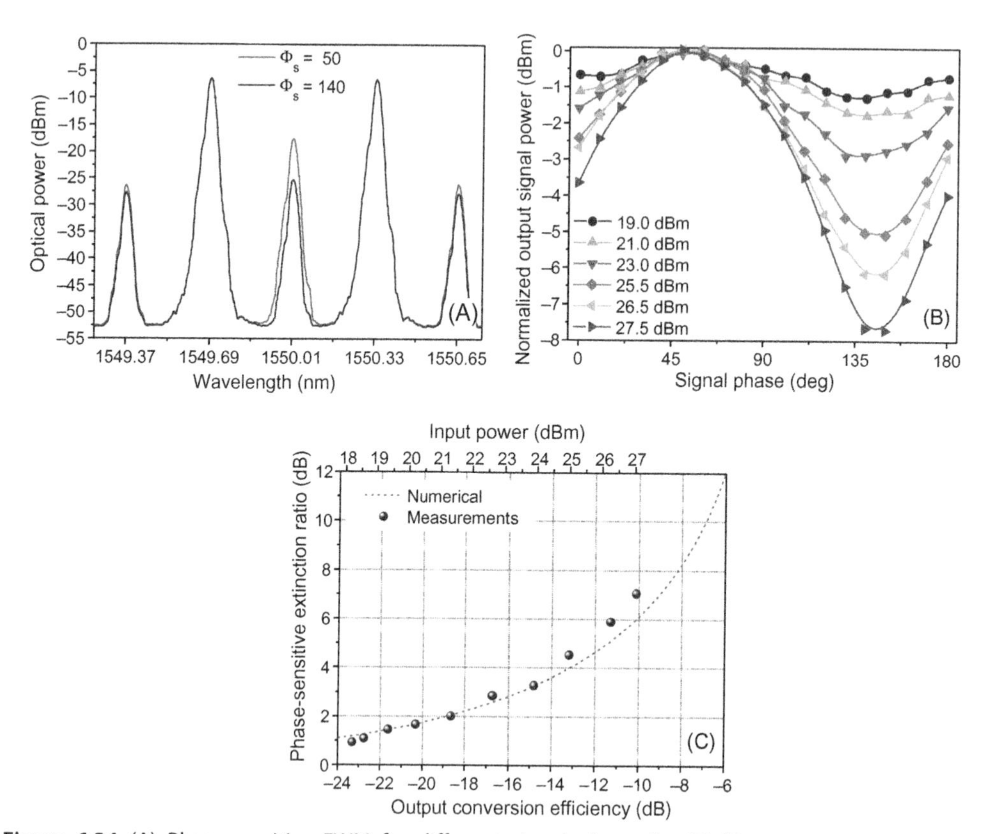

Figure 6.34 (A) Phase-sensitive FWM for different signal phase Φs. (B) Phase-sensitive extinction ratio (PSER) measured at varying total launch powers. (C) Comparison between measured and calculated extinction ratio. Source: *Adapted from F. Da Ros, M. Pu, L. Ottaviano, H. Hu, E. Semenova, et al., Phase-sensitive four-wave mixing in AlGaAs-on-insulator nano-waveguides, in: Proc. 2016 IEEE Photonics Conference, IPC'2016, Paper WB1.1, 2016 [150].*

polarization filtering [149]. Phase-sensitive FWM is achieved when the signal and idler interfere with each other by, for example, having them at the same wavelength and the same polarization. In this way an input signal can be amplified or deamplified depending on the signal-idler interference being constructive or destructive. Fig. 6.34A shows the configuration of the optical waves for achieving phase-sensitive FWM. The wavelength of the signal light is placed halfway between two stronger pumps, and the interference with the idler light generated by FWM causes a net interaction of the waves, which depends on the phase between signal and pumps. The phase-sensitive response can be observed by measuring the signal power at the output of the waveguide as the input signal phase is varied with respect to the pumps.

Fig. 6.34B shows how the phase-sensitive interaction depends on the power of pumps and signal into the waveguide. The normalized output signal power is plotted as a function of the signal phase for different input power levels into the AlGaAs-OI waveguide. The sine-square shape of the curves shows the phase-sensitive nature of the process as the phase is varied between constructive to destructive interference. The phase sensitivity increases with power as the generated idler becomes stronger, causing stronger interference with the original signal. The strength of the interference increases as the difference in power between the two interfering fields (signal and idler) reduces. The phase sensitivity reaches a maximum of 7.7 dB, for a total input power of 27.5 dBm. In Fig. 6.34C, the measured PSER is compared to numerical calculations for changing output conversion efficiencies—meaning different ratios between signal and idler waves. The numerical results are generated by solving the nonlinear Schrödinger equation using the SSFM. Consequently, they include contributions from higher order FWM processes taking place in the waveguide. Good agreement is found between numerical and measured results, indicating that a further increase in conversion efficiency could enable even stronger phase-sensitive interaction and confirming the great potential of the AlGaAs-OI platform for this application.

6.4.2.1.4 661 Tbit/s signal source

Finally, SPM in an AlGaAs-OI waveguide has been demonstrated for spectrally broadening a narrow optical frequency comb seed. The resulting broad frequency comb has been used as a high-quality signal source for the generation of 661 Tbit/s data (80 ch. WDM, 16-QAM, 40 Gbaud, 30 cores), which is simultaneously transmitted over a single multicore fiber [12]. The seed frequency comb is generated as 10-GHz pulses from a mode-locked laser, which are amplified and launched into a waveguide with a peak power of ~5.6 W (19.3 dBm average power). The very strong nonlinearity of the AlGaAs waveguide broadens the spectrum of the light through self-phase modulation, as seen in Fig. 6.35A. The seed comb is seen as the input spectrum, and the

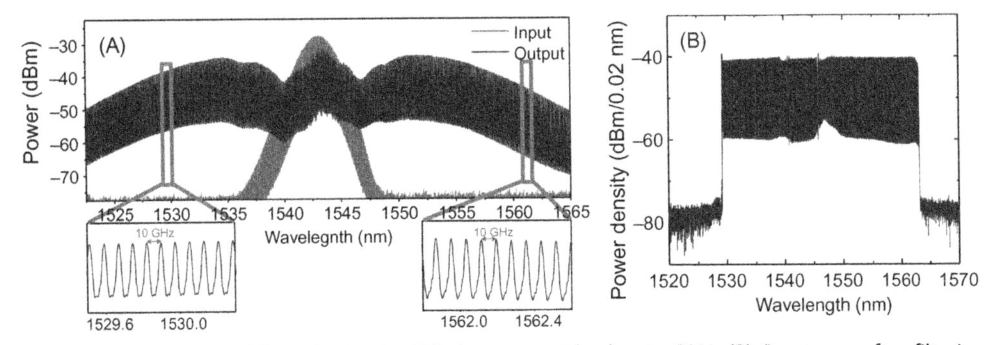

Figure 6.35 (A) Spectral broadening in AlGaAs waveguide due to SPM. (B) Spectrum after filtering and equalization. Source: *Figure is adapted from H. Hu, F. Da Ros, F. Ye, M. Pu, K. Ingerslev, E. Porto da Silva, et al., Single-source AlGaAs frequency comb transmitter for 661 Tbit/s data transmission in a 30-core fiber, Nat. Photon. 12 (2018) 469–473 [12].*

much broadened frequency comb is measured at the waveguide output. The OSNR of the individual lines in the comb spectrum is estimated to be >40 dB at the center of the spectrum (1552 nm) and ~30 dB in the worst part of the equalized spectrum shown in Fig. 6.35B (1563 nm). The equalized spectrum of Fig. 6.35B is achieved by applying a programmable filter to the output spectrum shown in Fig. 6.35A. The noise properties of the individual lines in the equalized spectrum are sufficient to allow all parts of the spectrum to carry polarization multiplexed 16-QAM modulation at 40 Gbaud and be split 30 ways to fully load the 30-core fiber with identically modulated temporally de-correlated data channels.

All the spectral and spatial channels simultaneously propagating in the fiber were characterized with a fixed transmitter configuration—that is, no specific "channel under test" was defined in the system configuration. All channels could be correctly received after transmission through the multicore fiber employing FEC. When the strength (and overhead) of the FEC is adapted to the individual channel performance, a total net data rate of 661 Tbit/s was received. This is by far the highest amount of data transmitted using the light from a single integrated device, and it shows the great potential of the AlGaAs-OI platform for use also in multi-frequency light sources for optical communication.

6.4.2.1.5 Summary on aluminum gallium arsenide

Through a number of record-breaking achievements in optical signal processing, the AlGaAs-OI platform has been confirmed as one of the most promising technologies for optical signal processing. As device fabrication research advances, this platform is expected to significantly impact the role of nonlinear signal processing in communication systems and other high-performance applications.

6.4.2.2 Figure of merit of nonlinear materials for optical signal processing

The ideal material for nonlinear optical signal processing is one having high nonlinearity, low linear and nonlinear losses, and a compact footprint. For practical purposes, it is desirable to realize these properties in a platform that can be inexpensively produced using, for example, standard CMOS fabrication techniques, and integrated flexibly and easily onto other photonic and electronic platforms. Several materials have been investigated over the past decades, and several have attractive features for specific nonlinear applications. It is often desirable to identify a single parameter on which to compare all these different materials, a universal FOM. This is, however, a difficult task, as the desired performance metrics vary greatly from application to application. For instance, a material ideal for parametric amplification may not be as useful for broadband optical signal processing, and some materials suffer from detrimental nonlinear absorption where others do not. In this section, we compare reported nonlinear $\chi^{(3)}$ materials from a performance and systems application perspective by defining a new FOM that indicates whether a certain material is capable of yielding a considerable nonlinear phase shift- a π phase shift or the equivalent phase shift needed to achieve a -10 dB FWM conversion efficiency, which is deemed to be useful for real applications, how much power this requires, and how long the nonlinear waveguide will need to be to achieve this. The FOM is defined as $P_{\pi} \cdot L_{\pi}$. This way one can quickly estimate if the material is low power and short enough for practical components (if that is a priority for a specific application) and is basically intended as a $\chi^{(3)}$ version of the $V_{\pi} \cdot L_{\pi}$ used for $\chi^{(2)}$ electro-optic modulators.

6.4.2.2.1 Nonlinear figure of merit for nonresonant structures

The standard FOM used to compare the efficiencies of nonlinear materials is defined as

$$FOM_{TPA} = \frac{n_2}{\lambda \cdot \beta_{TPA}}, \tag{6.14}$$

where n_2 is the Kerr nonlinearity, λ is the free space wavelength, and β_{TPA} is the TPA coefficient [151]. While suitable to use for comparing nonlinear materials with nonzero β_{TPA}, FOM_{TPA} provides neither a meaningful comparison between weakly nonlinear materials having no TPA nor a fair comparison between strongly nonlinear materials with moderate TPA and weakly nonlinear materials. Furthermore, the FOM_{TPA} does not explicitly reflect two important parameters for telecommunications systems applications: the launch power (P) and the required device length. The launch power is important from a practical perspective where limited on-chip laser or EDFA power is available, and a small device length is important for creating compact devices.

A more meaningful FOM for nonresonant structures (e.g., straight waveguides) used for OSP is one based on the accumulated nonlinear phase shift. Considering the importance of the launch power and device length for OSP systems applications, we propose here the FOM $P_{\pi} \cdot L_{\pi}$, which is the product of the launch power required for a π nonlinear phase shift and the corresponding effective length of the device at that power. A π phase shift represents a benchmark for significant spectral broadening, high FWM conversion efficiency (approximately -1.5 dB output-to-output), and useful all-optical switching. The FOM is introduced as the $\chi^{(3)}$ analogue to $V_{\pi} \cdot L_{\pi}$ used for $\chi^{(2)}$ electro-optic modulators; for both nonlinearities, a lower FOM indicates higher efficiency. We define a similar FOM corresponding to a -10 dB output-to-output conversion efficiency for continuous-wave FWM, $P_{-10dB} \cdot L_{-10dB}$, which requires a phase shift of approximately 0.4π radians. A -10 dB FWM conversion efficiency is a benchmark for decent FWM performance to execute several OSP applications, and most materials are able to reach a -10 dB conversion efficiency, so it possible to compare performances more easily than the $P_{\pi} \cdot L_{\pi}$, which only some materials can reach. The $P_{\pi} \cdot L_{\pi}$ FOM is derived from the nonlinear phase (ϕ_{NL}) experienced by a dual-frequency beat signal propagating in a nonlinear waveguide [152] as

$$P_{\pi} \cdot L_{\pi} = \frac{\pi A_{eff} c}{2 n_2 \omega_0}, \tag{6.15}$$

where A_{eff} is the effective mode area and c is the speed of light in free space. L_{eff} is the effective length, defined as

$$L_{eff} = \frac{1 - e^{-\alpha L}}{\alpha}, \tag{6.16}$$

where L is the device length and $\alpha = \alpha(P)$ is the total loss coefficient, including linear and nonlinear terms, at the power P. The loss coefficient can be written explicitly as

$$\alpha = \alpha_0 + \alpha_1 \left(\frac{P}{A_{eff}} \right) + \alpha_2 \left(\frac{P}{A_{eff}} \right)^2, \tag{6.17}$$

where α_0 is the loss due to scattering and single photon absorption (SPA), α_1 includes contributions from TPA and FCA of single-photon generated carriers, and α_2 includes contributions from FCA of two-photon generated carriers. See Table 6.1.

The dependence of these terms on material properties is described in Table 6.1. SPA mediated by either bulk or surface midgap states is a recent topic of interest for lower bandgap nonlinear optical materials. Its impact on the absorption coefficient is included in Eq. (6.3) as described in Table 6.1; however, it is left out of the calculations since it has not been widely reported yet. Calculating $P_{\pi} \cdot L_{\pi}$ is straightforward for materials where there is no power dependence to the loss coefficient. For materials having power-dependent losses, the FOM is calculated numerically as the minimum power necessary for a π phase shift. For materials where the nonlinear losses are sufficiently large, a π phase shift cannot be obtained, and the material cannot be included in the plot. This is an indication that the material, while likely highly nonlinear, is not suitable for practical telecommunication OSP applications. Finally, we note that the calculations done here are assuming a CW pump since we deem a high FOM obtained under such conditions indicates robust performance for a host of relevant OSP applications.

We calculate $P_{\pi} \cdot L_{\pi}$ and $P_{-10\text{dB}} \cdot L_{-10\text{dB}}$ for various leading candidate integrated nonlinear optical materials in Fig. 6.36 using the data and references provided in Table 6.2

Table 6.1 Linear loss coefficients used in Eq. (6.3).

Symbol	Expression
α_0	$\alpha_{\text{scattering}} + \alpha_{SPA}$
α_1	$\frac{\sigma \tau \alpha_{SPA}}{\hbar \omega} + \beta_{TPA}$
α_2	$\frac{\sigma \tau \beta_{TPA}}{2 \hbar \omega}$

The coefficients $\alpha_{\text{scattering}}$ and α_{SPA} are the scattering and single photon absorption (SPA) coefficients, respectively.

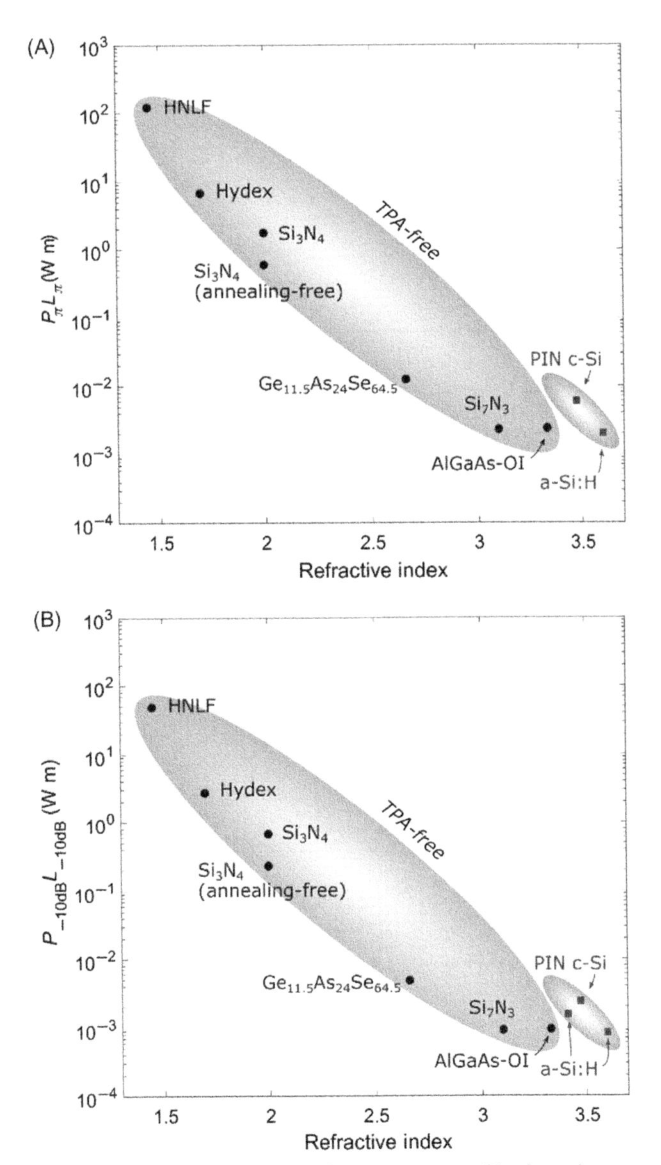

Figure 6.36 Figure of merit (FOM) $P_\pi \cdot L_\pi$ (A) and $P_{-10dB} \cdot L_{-10dB}$ (B) plotted versus the linear refractive index for various leading candidate integrated nonlinear platforms. The *smaller ovals* surround materials with nonlinear loss in the C-band. Calculations are based on information in Table 6.2.

[108,146,158–162]. An HNLF fiber is also included in the plot for reference. A linear trend is observed when plotting $\log(P_\pi \cdot L_\pi)$ versus the linear refractive index when only TPA-free materials with no nonlinear losses are considered. In general, more highly nonlinear materials have a lower FOM. The same trends are observed in the $P_{-10\text{dB}} \cdot L_{-10\text{dB}}$ plot, with all materials having a lower $P_{-10\text{dB}} \cdot L_{-10\text{dB}}$ FOM due to the lower required

Table 6.2 Material properties used for calculating the $\boldsymbol{P_\pi} \cdot \boldsymbol{L_\pi}$ figure of merit (FOM) plotted in Fig. 6.36.

Material	n	n_2 (m^2/W)	α_0 (dB/cm)	A_{eff} (μm^2)	β_{TPA} (m/W)	τ (ns)	σ (m^2)
HNLF [153]	1.45	2.70×10^{-20}	9×10^{-7}	8.5	–	–	–
Hydex [154]	1.7	1.15×10^{-19}	0.06	2.00	–	–	–
Si_3N_4 [155]	2	2.50×10^{-19}	0.07	1.13	–	–	–
Si_3N_4 (annealing free) [156]	2	3.60×10^{-19}	1.7	0.548	–	–	–
$Ge_{11.5}As_{24}Se_{64.5}$ [157]	2.66	8.60×10^{-18}	1.5	0.275	–	–	–
AlGaAs [146]	2.9	3.33×10^{-17}	1.4	0.16	–	–	–
Si_7N_3 [158]	3.1	2.80×10^{-17}	4.5	0.165	–	–	–
a-Si:H [159]	3.41	1.00×10^{-17}	8.6	0.100	1.40×10^{-12}	0.16	1.00×10^{-21}
c-Si [160]	3.47	4.50×10^{-18}	2	0.180	7.00×10^{-12}	0.8 [160]	1.45×10^{-21} [161]
PIN c-Si [108]	3.47	6.50×10^{-18}	2	0.100	9.00×10^{-12}	0.012	1.45×10^{-21} [161]
a-Si:H [162]	3.6	2.37×10^{-17}	3.6	0.125	–	1.87	1.00×10^{-21} [159]

Source: Numbers extracted from Refs. [108,146,158–162].

phase shift. For materials where nonlinear losses are significant, the calculated FOM is either larger than that predicted by the linear trend for TPA-free materials or cannot be calculated due to the inability to acquire a π nonlinear phase. This is the case for amorphous silicon, which either lies significantly above the line or cannot be plotted, and for crystalline silicon, which also cannot obtain a π phase change.

From Fig. 6.36 it is clearly observed that small efficient materials are those with high index, and eventually we face the problem of nonlinear losses. Even so, these materials have the best FOM. On the other hand, HNLFs can easily accumulate a large phase shift over long lengths of fiber, and Hydex and silicon nitride waveguides are sufficiently low-loss that they can be made very long. Chalcogenides are very varied, and the one quoted here, from Ref. [157], has a very low FOM, so if this is stable, it could be very attractive. AlGaAs is very good, as is silicon-rich nitride [158], and amorphous silicon is potentially also very attractive, if it can made reproducible stable. PIN c-Si has one of the highest reported FWM conversion efficiencies and a very good FOM [108].

In this section we introduced a new FOM for nonlinear materials aimed at being used for nonlinear optical signal processing in optical communication systems. We find the $P_{\pi} \cdot L_{\pi}$ and $P_{-10\text{dB}} \cdot L_{-10\text{dB}}$ FOM gives a good estimate of which materials are attractive for nonlinear optical signal processing; however, these FOMs do not evaluate the materials for specific applications. For instance, the bandwidth is not included, nor is the maximum achievable nonlinear phase shift. So, in the end, for a more careful analysis of which material one would like for a specific application, one would have to define that application first and then compare how well the various materials would serve that purpose.

6.4.2.3 Amorphous silicon

AlGaAs, described in previous sections, is a leading candidate material for performing numerous optical signal processing applications, and when placed on an insulator, lends itself to cointegration with other components and materials; however, it is not as easily integrated onto complex photonic integrated circuits as some other CMOS-compliant materials.

To overcome this challenge, we have sought other candidate materials that have similar properties to AlGaAs but possess greater integration flexibility. Amorphous silicon is an attractive alternative comparable material; however, its performance is unreliable due to inconsistent stability under moderate to high optical powers, which are needed for OSP applications. It is widely accepted that the observed photo-induced degradation is caused by the Staebler-Wronski effect [163], a degradation mechanism caused by the interaction of free carriers with weak bonds of the silicon network. The photo-induced degradation causes an increase in the insertion loss and reduction in nonlinear performance over time under high-intensity exposure. Kuyken et al. have reported that the damage is reversible upon annealing in air at 200°C for 30 minutes [164]. Due to the strong presence of free carriers in amorphous silicon, most nonlinear demonstrations to date have used a pulsed pump to prevent carrier buildup and subsequent FCA [164–168]; however, degradation has still been observed. The reported stability of various a-Si:H materials is illustrated in Fig. 6.37 on a plot of average intensity versus peak intensity. While no clear boundary between stable and unstable performance can yet be deduced, reports of instability have occurred at the extreme values of peak or average power.

We have addressed the stability of a-Si:H waveguides by adopting an approach that has been successfully used to stabilize amorphous silicon solar cells, memory, and transistors: passivation with deuterium, creating a-Si:HD films [169]. It has been proposed that the enhanced stability of amorphous silicon films passivated during growth with deuterium is due to the overlapping of Si–D wagging bond modes with the phonon modes of the Si–Si bonds in the network [170]. The films are grown using PECVD with a concurrent flow of silane (SiH_4) and deuterium (D_2) gases at a temperature of 300°C [169]. The deuterium content is controlled by varying the ratio of the gases, allowing for nearly 100% of the hydrogen to be replaced. Larger deuterium fraction films produced using this deposition technique are correlated with a larger surface roughness; hence, we have used films with 49% isotopic fraction of deuterium having a root mean square surface roughness less than 0.5 nm for devices. Waveguides were fabricated from the a-Si:HD films using deep UV lithography and reactive ion etching and were cladded with silica grown by PECVD.

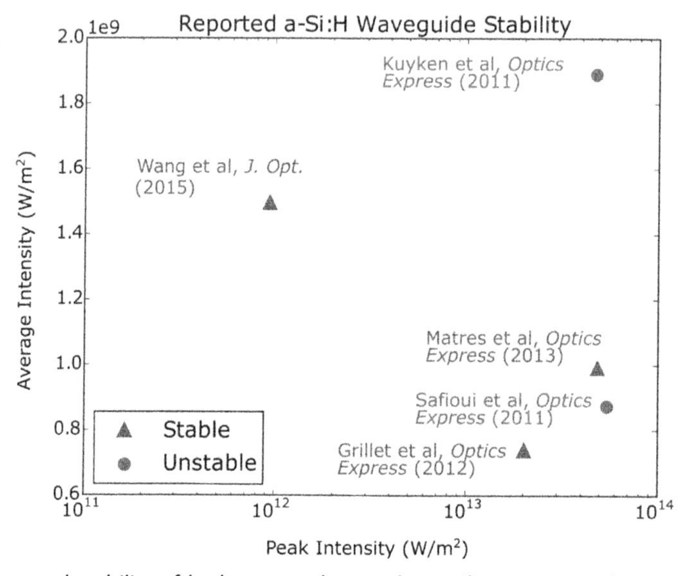

Figure 6.37 Reported stability of hydrogenated amorphous silicon waveguides exposed to pulsed light with a frequency within the C-band [163,165–168].

The stability of the deuterated films was evaluated by measuring the insertion loss of waveguides exposed to consecutive power sweeps. The power sweeps were done using CW light at 1550 nm. The scheme used is illustrated in Fig. 6.38A. The first and third trials were done quickly, while the second trial was done with a longer dwell time at each power level. The coupling to the waveguides was optimized before the start of each trial. After the second trial, the waveguides were exposed just below the burning threshold for a period of 10 hours. Both the hydrogenated and partially deuterated waveguides incur an additional insertion loss increase after the first trial; however, the deuterated waveguide does not suffer from additional degradation after the 10-hour exposure period, while the hydrogenated one continues to degrade, as shown in Fig. 6.38B and C. While the experiment did not rigorously account for the different free carrier concentrations generated, it is expected to be nearly the same for the two materials due to the total amount of additional nonlinear loss incurred over the intensity range of the first trial. Therefore, the increased stability is tentatively attributed to the deuteration; however, a rigorous separation of the degradation due to the Staebler-Wronski effect and enhanced stability due to deuteration is yet to be demonstrated.

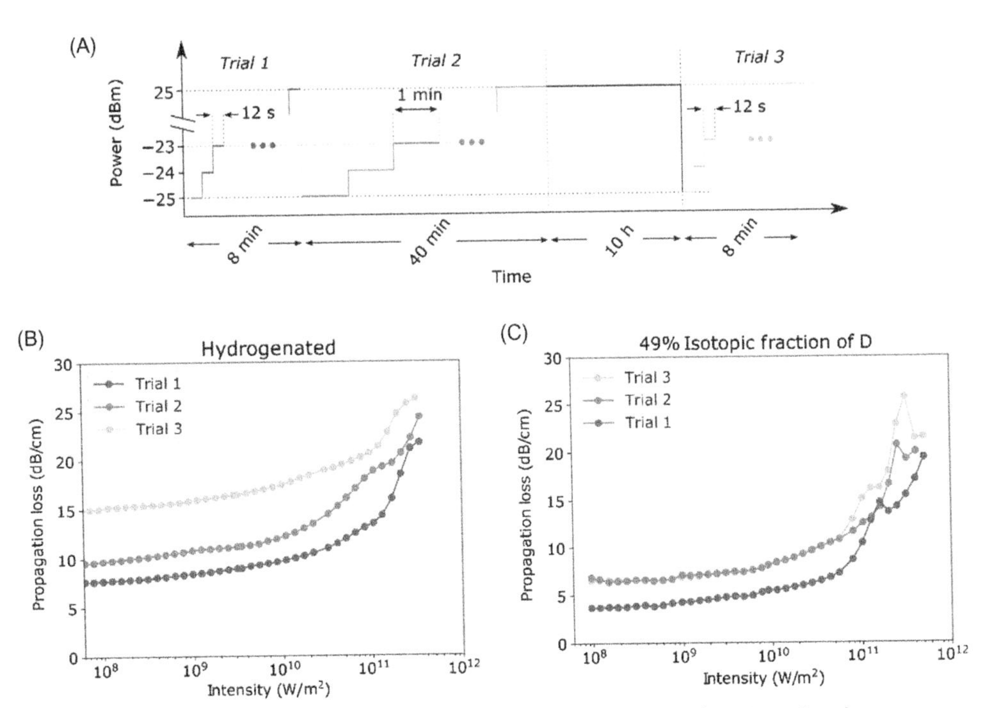

Figure 6.38 Scheme implemented for the power- and time-dependent insertion loss measurements (A). Measured insertion loss for the three trials for hydrogenated amorphous silicon (B) and partially deuterated amorphous silicon (C) where 49% of the hydrogen content has been replaced with deuterium.

In this section, we spent some time describing recent advances in fabricating amorphous silicon, as this is a very promising material with one of the best FOMs. The main challenge is to make highly nonlinear stable materials, and for this we explored the use of deuterium in replacement of hydrogen. At the moment, though, deuterated a-Si has yet to produce a π phase shift or even achieve a -10 dB FWM CW conversion efficiency, to be comparable with the other materials.

6.5 Conclusions

This chapter described some recent trends in nonlinear optical signal processing for optical communication systems. This topic involves efforts in several disciplines. First, there must be use case scenarios exploring and potentially defining where OSP may be useful. Then the device elements needed for these applications must be developed, which may rely on entirely new materials research.

We provided overviews of promising nonlinear materials useful for OSP and of OSP applications we believe are promising. In terms of nonlinear materials, we described several materials platforms and defined a new FOM that we find useful for OSP, namely the power and device length product required to obtain a π phase shift. For many OSP functionalities, it is desirable with a short and highly nonlinear device, and this new FOM helps to identify materials that will enable this. However, not all applications require a short device, so one should always consider the desired application. We have described one of the most promising materials for nonlinear OSP, namely the AlGaAs-OI materials platform, which has already displayed record-breaking performance, and we speculate that given the right progress in fabrication maturity, this could become a large-impact material for OSP.

We also highlighted the emerging deuterated amorphous silicon, which is expected to have a better FOM than hydrogenated amorphous silicon, which is just below AlGaAs, but here further studies are needed. Many other nonlinear materials were mentioned in the chapter, and although they may have worse FOMs than those highlighted, they might still be suitable for specific applications. One such example is Hydex, which is very well suited for ring resonators for frequency comb generation.

HNLFs have also been devoted considerable space in this chapter, as this material gives the highest parametric gain, owing to the low loss enabling long lengths. The only performance drawback is the limited bandwidth obtained with HNLF compared to integrated waveguides. But the final choice of material really depends on the application.

We discussed various OSP applications that could have potential to support optical communication systems, and we focused on schemes where multiple data channels share a single nonlinear unit, since, after all, the strength of optics is that it is broadband and ultrafast. We described recent progress in OPC, where multiple WDM channels are phase conjugated simultaneously in a single nonlinear unit, and we have seen solid evidence of improved transmission performance in terms of extended reach and higher mutual information. A reach of Up to 92 WDM channels has been reported. We have recently seen that tailoring the DSP to the optical signal processing further improves the system performance. In particular, PS adapted to an OPC channel yielded an overall improvement of transmission performance. Both OPC and PS improve transmission performance individually, and their effects seem to be additive, so that PS for an OPC channel further improves the transmission performance. Further studies of DSP matched with OSP will be interesting to follow.

Another case of OSP for multiple channels is the use of optical time lenses. These are found to be very versatile tools allowing for conversion between signal formats like WDM, OFDM, O-TDM, and N-WDM, and allow for flexible WDM spectral manipulation and smart single-source WDM transmitters, as well as scalable WDM regeneration.

Overall, OSP offers potential advantages to installed networks such as potential energy efficiency, increased flexibility, and improved transmission performance. The nonlinear materials required to get there are being developed on many fronts, and the development of these promising materials will be fascinating to follow.

References

[1] Cisco "Cisco Visual Networking Index: Forecast and Methodology, 2016–2021" (2016). Available from: <http://www.cisco.com/c/en/us/solutions/collateral/service-provider/visual-networking-index-vni/complete-white-paper-c11-481360.html>.

[2] S.L.I. Olsson, J. Cho, S. Chandrasekhar, X. Chen, E.C. Burrows, P.J. Winzer, Record-high 17.3-bit/s/Hz spectral efficiency transmission over 50 km using probabilistically shaped PDM 4096-QAM, OFC 2018 postdeadline paper Th4C.5.

[3] A.D. Ellis, J. Zhao, D. Cotter, Approaching the non-linear Shannon limit, J. Lightwave Technol. 28 (2010) 423–433.

[4] G. Cook, Greenpeace, "clicking clean", Green Internet Report 2017. Available from: <http://www.clickclean.org/international/en/>.

[5] S.G. Anders, Andrae (Huawei), the Nordic Digital Business Summit, October 5, 2017. Available from: <https://www.researchgate.net/publication/320225452/download>.

[6] Y. Yu, W. Xue, E. Semenova, K. Yvind, J. Mørk, Demonstration of a self-pulsing photonic crystal Fano laser, Nature Photon. 11 (2) (2017) 81–84.

[7] Y. Yu, W. Xue, H. Hu, L.K. Oxenløwe, K. Yvind, J. Mørk, All-optical switching improvement using photonic-crystal fano structures, IEEE Photon. J. 8 (2) (2016) 0600108.

[8] V. Ataie, E. Temprana, L. Liu, E. Myslivets, B.P.P. Kuo, N. Alic, et al., Ultrahigh count coherent WDM channels transmission using optical parametric comb-based frequency synthesizer, J. Lightwave Technol. 33 (3) (2015) 694–699.

[9] J. Schröoder, L.B. Du, M.M. Morshed, B.J. Eggleton, A.J. Lowery, Colorless flexible signal generator for elastic networks and rapid prototyping. OFC 2013, paper JW2A.44.pdf OFC/NFOEC Technical Digest © 2013 OSA.

[10] B.J. Puttnam, R.S. Luís, W. Klaus, J. Sakaguchi, J.-M. Delgado Mendinueta, Y. Awaji, et al. 2.15 Pb/s transmission using a 22 core homogeneous single-mode multi-core fiber and wideband optical comb, in: European Conference on Optical Communication (ECOC) 2015 postdeadline paper PDP.3.1 (2015). doi: 10.1109/ECOC.2015.7341685.

[11] P. Marin-Palomo, J.N. Kemal, M. Karpov, A. Kordts, J. Pfeifle, et al., Microresonator-based solitons for massively parallel coherent optical communications, Nature 546 (2017) 274–279.

[12] H. Hu, F. Da Ros, M. Pu, F. Pu, K. Ye, E. Ingerslev, Porto da Silva, et al., Single-source chip-based frequency comb enabling extreme parallel data transmission, Nature Photonics 12 (2018) 469–473.

[13] K. Harako, D. Seya, T. Hirooka, M. Nakazawa, 640 Gbaud (1.28 Tbit/s/ch) optical Nyquist pulse transmission over 525 km with substantial PMD tolerance, Opt. Express 21 (18) (2013) 21062–21075. Available from: https://doi.org/10.1364/OE.21.021062.

[14] H.N. Tan, T. Inoue, T. Kurosu, S. Namiki, Transmission and pass-drop operations of mixed baudrate Nyquist OTDM-WDM signals for all-optical elastic network, Opt. Express 21 (17) (2013) 20313–20321. Available from: https://doi.org/10.1364/OE.21.020313.

[15] T. Morioka, Y. Awaji, R. Ryf, P. Winzer, D. Richardson, F. Poletti, Enhancing optical communications with brand new fibers, IEEE Commun. Mag. 50 (2) (2012) S31–S42.

[16] D. Soma, Y. Wakayama, S. Beppu, S. Sumita, T. Tsuritani, T. Hayashi, et al., 10.16 Peta-bit/s dense SDM/WDM transmission over low-DMD 6-mode 19-core fibre across C + L band, ECOC 2017, postdeadline paper Th.PDP.A.1.

[17] N.K. Fontaine, R. Ryf, H. Chen, A.V. Benitez, J.E.A. Lopez, R.A. Correa, et al., OFC, Postdeadline paper Th5C.1. <https://doi.org/10.1364/OFC.2015.Th5C.1>, 2015.

[18] N. Bozinovic, Y. Yue, Y. Ren, M. Tur, P. Kristensen, H. Huang, et al., Terabit-scale orbital angular momentum mode division multiplexing in fibers, Science 28 340 (6140) (2013) 1545–1548. Available from: https://doi.org/10.1126/science.1237861.

[19] K. Ishii, K. Ishii, J. Kurumida, J. Kurumida, S. Namiki, S. Namiki, et al., Energy consumption and traffic scaling of dynamic optical path networks, in: Proc. SPIE 8646, Optical Metro Networks and Short-Haul Systems V, 86460A (5 February 2013). doi: https://doi.org/10.1117/12.2003591.

[20] A.E. Willner, All-optical signal processing techniques for flexible networks, in: Tutorial at Optical Fiber Communication Conference OSA Technical Digest (online) (Optical Society of America, 2018), paper W3E.5. doi: https://doi.org/10.1364/OFC.2018.W3E.

[21] B.H. Kolner, Space-time duality and the theory of temporal imaging, IEEE J. Quantum Electron. 30 (8) (1994) 1951–1963.

[22] P. Tournois, Analogie optique de la compression d'impulsion, Comptes Rendus L'Académie des Sci 258 (1964) 3839–3842.

[23] S.A. Akhmanov, Nonstationary phenomena and space-time analogy in nonlinear optics, Sov. Phys 28 (1969) 748–757.

[24] I.P. Christov, Theory of a 'time telescope, Opt. Quantum Electron. 22 (5) (1990) 473–479.

[25] C.V. Bennett, B.H. Kolner, Aberrations in temporal imaging, IEEE J. Quantum Electron. 37 (1) (2001) 20–32.

[26] P. Guan, K.M. Røge, N.K. Kjøller, et al., WDM regeneration of DPSK signals using optical Fourier transformation and phase sensitive amplification, in: European Conference and Exhibition on Optical Communication (ECOC 2015), Paper We364, 2015.

[27] A. Tikan, S. Bielawski, C. Szwaj, S. Randoux, P. Suret, Single-shot measurement of phase and amplitude by using a heterodyne time-lens system and ultrafast digital time-holography, Nat. Photon. 12 (4) (2018) 228–234.

[28] L.K. Oxenløwe, et al., Energy-efficient optical signal processing using optical time lenses, in: S. Wabnitz, B.J. Eggleton (Eds.), Chapter 9 in All-Optical Signal Processing - Data Communications and Storage Applications, vol. 194, Springer Series in Optical Sciences, 2015.

[29] E. Palushani, H.C.H. Mulvad, M. Galili, F. Da Ros, H. Hu, et al., Spectral compression of a DWDM grid using optical time-lenses, in: 18th OptoElectronics and Communications Conference, OECC 2013, Paper ThO2-1.

[30] K.M. Røge, P. Guan, H.C.H. Mulvad, N.-K. Kjøller, M. Galili, et al., Flexible DWDM grid manipulation using four wave mixing-based time lenses, in: Proc. 2014 IEEE Photonics Conference, IPC 2014, 2014.

[31] M. Lillieholm, B. Corcoran, M. Galili, L. Grüner-Nielsen, L.K. Oxenløwe, A.J. Lowery, Characterization of spectral magnification based on four-wave mixing in nonlinear fibre for advanced modulation formats, in: ECOC2017, Paper P1.SC4.65.
[32] M. Marhic, Discrete Fourier transforms by single-mode star networks, Opt. Lett. 12 (1) (1987).
[33] E. Palushani, H.C.H. Mulvad, D. Kong, P. Guan, M. Galili, et al., All-optical OFDM demultiplexing by spectral magnification and band-pass filtering, Opt. Express 22 (1) (2014) 136–144. Available from: https://doi.org/10.1364/OE.22.000136.
[34] P. Guan, S. Lefrancois, M. Lillieholm, H.C.H. Mulvad, K.M. Røge, H. Hu, et al., All-optical OFDM system using a wavelength selective switch based transmitter and a spectral magnification based receiver, in: ECOC 2014, doi: 10.1109/ECOC.2014.6964089.
[35] M. Lillieholm, H.C.H. Mulvad, M. Galili, L.K. Oxenløwe, Comparison of delay-interferometer and time-lens-based all-optical OFDM demultiplexers, IEEE Photon. Technol. Lett. 27 (11) (2015) 1153–1156.
[36] P.V. Mamyshev, All-optical data regeneration based on self-phase modulation effect,in: European Conference on Optical Communications (ECOC1998), 1998, p. 475.
[37] R. Slavík, F. Parmigiani, J. Kakande, C. Lundström, M. Sjödin, P.A. Andrekson, et al., All-optical phase and amplitude regenerator for next-generation telecommunications systems, Nat. Photon 4 (2010) 690.
[38] L. Li, P.G. Patki, Y.B. Kwon, V. Stelmakh, B.D. Campbell, M. Annamalai, et al., All-optical regenerator of multi-channel signals, Nat. Commun. 8 (2017) 884.
[39] M. Jinno, M. Abe, All optical regenerator based on nonlinear fibre Sagnac interferometer, Electron. Lett. 28 (1992) 1350.
[40] W.A. Pender, P.J. Watkinson, E.J. Greer, A.D. Ellis, 10 Gbit/s all optical regenerator, Electron. Lett. 31 (1995) 1587.
[41] S. Bigo, O. Leclerc, E. Desurvire, All-optical fiber signal processing and regeneration for soliton communications, IEEE J. Select. Topics Quantum Electron. 3 (1997) 1208.
[42] A.H. Gnauck, P.J. Winzer, Optical phase-shift-keyed transmission, J. Lightw. Technol. 23 (2005) 115.
[43] P.J. Winzer, High-spectral-efficiency optical modulation formats, J. Lightw. Technol. 30 (2012) 3824.
[44] A. Sano, et al., 102.3-Tb/s (224 × 548-Gb/s) C- and extended L-band all-Raman transmission over 240 km using PDM-64QAM single carrier FDM with digital pilot tone, in: Optical Fiber Communication Conference (OFC2012), PDP5C.3, 2012.
[45] H. Kim, A.H. Gnauck, Experimental investigation of the performance limitation of DPSK systems due to nonlinear phase noise, IEEE Photon. Technol. Lett. 15 (2003) 320.
[46] X. Yi, W. Shieh, Y. Ma, Phase noise effects on high spectral efficiency coherent optical OFDM transmission, J. Lightwave Technol. 26 (2008) 1309.
[47] M. Gao, T. Kurosu, T. Inoue, S. Namiki, Efficient phase regeneration of DPSK signal by sideband-assisted dual-pump phase sensitive amplifier, Electron. Lett 39 (2013) 140.
[48] J. Kakande, et al., First demonstration of all-optical QPSK signal regeneration in a novel multi-format phase sensitive amplifier, in: European Conference on Optical Communications (ECOC2010), Postdeadline paper PD 3.3, 2010.
[49] Z. Zheng, et al., All-optical regeneration of DQPSK/QPSK signals based on phase-sensitive amplification, Opt. Commun. 281 (2008) 2755.
[50] T. Richter, R. Elschner, C. Schubert, QAM phase regeneration in a phase sensitive fiber-amplifier, in: European Conference on Optical Communications (ECOC2010), We3A2, 2013.
[51] F. Parmigiani, L. Provost, P. Petropoulos, D.J. Richardson, W. Freude, J. Leuthold, et al., Progress in multichannel all-optical regeneration based on fiber technology, IEEE J. Sel. Top. Quantum Electron. 18 (2012) 689.
[52] F. Parmigiani, K.R.H. Bottrill, R. Slavík, D.J. Richardson, P. Petropoulos, Multi-channel phase regenerator based on polarization-assisted phase-sensitive amplification, IEEE Photon. Technol. Lett. 28 (2016) 845.
[53] P. Guan, F.D. Ros, M. Lillieholm, N.-K. Kjøller, H. Hu, K.M. Røge, et al., Scalable WDM phase regeneration in a single phase-sensitive amplifier through optical time lenses, Nat. Commun. 9 (2018) 1049.

[54] B.H. Kolner, M. Nazarathy, Temporal imaging with a time lens, Opt. Lett. 14 (1989) 630–632.
[55] M. Nakazawa, T. Hirooka, F. Futami, S. Watanabe, Ideal distortion-free transmission using optical Fourier transformation and Fourier transform-limited optical pulses, IEEE Photon. Technol. Lett. 16 (2004) 1059–1061.
[56] M.A. Foster, R. Salem, D.F. Geraghty, A.C. Turner-Foster, M. Lipson, A.L. Gaeta, Silicon-chip-based ultrafast optical oscilloscope, Nature 456 (2008) 81–84.
[57] Mulvad, H.C.H. et al., OTDM-WDM conversion based on time-domain optical Fourier transformation with spectral compression, in: Optical Fiber Communication Conference (OFC 2011), OThN2, 2011.
[58] P. Guan, K.M. Røge, M. Lillieholm, M. Galili, H. Hu, T. Morioka, et al., Time lens based optical Fourier transformation for all-optical signal processing of spectrally-efficient data, J. Lightwave Technol. 35 (4) (2017) 799–806. Invited.
[59] N.K. Kjøller, K.M. Røge, P. Guan, H.C.H. Mulvad, M. Galili, L.K. Oxenløwe, A novel phase-locking-free phase sensitive amplifier based regenerator, J. Lightwave Technol. 34 (2016) 643–652.
[60] A. Fragkos, A. Bogris, D. Syvridis, All optical regeneration based on phase sensitive non-degenerate FWM in optical fibers, IEEE Photon. Technol. Lett 22 (2010) 1826–1828.
[61] B.P.-P. Kuo, J.M. Fini, L. Grüner-Nielsen, S. Radic, Dispersion-stabilized highly-nonlinear fiber for wideband parametric mixer synthesis, Opt. Express 20 (17) (2012) 18611–18619.
[62] F. Parmigiani, G. Hesketh, R. Slavik, P. Horak, P. Petropoulos, D.J. Richardson, Polarization-assisted phase-sensitive processor, J. Light. Technol. 33 (6) (2015) 1166–1174.
[63] A. Lorences-Riesgo, C. Lundström, F. Chiarello, M. Karlsson, P.A. Andrekson, Phase-sensitive amplification and regeneration of dual-polarization BPSK without polarization diversity, in: European Conference on Optical Communication, ECOC, 2014.
[64] N.K. Kjøller, M. Galili, K. Dalgaard, H.C.H. Mulvad, K.M. Røge, L.K. Oxenløwe, Quadrature decomposition by phase conjugation and projection in a polarizing beam splitter, in: European Conference on Optical Communication, ECOC, 2014.
[65] F. Parmigiani, R. Slavik, G. Hesketh, P. Horak, P. Petropoulos, D.J. Richardson, Efficient binary phase quantizer based on phase sensitive four wave mixing, in: European Conference on Optical Communication, ECOC, 2014.
[66] F. Parmigiani, R. Slavik, G. Hesketh, P. Petropoulos, D.J. Richardson, Quadrature decomposition of optical fields using two orthogonal phase sensitive amplifiers, in: European Conference on Optical Communication, ECOC, 2014.
[67] A. Lorences-Riesgo, et al., Quadrature demultiplexing using a degenerate vector parametric amplifier, Opt. Express (2014).
[68] R.P. Webb, J.M. Dailey, R.J. Manning, A.D. Ellis, Phase discrimination and simultaneous frequency conversion of the orthogonal components of an optical signal by four-wave mixing in an SOA, Opt. Express 19 (21) (2011) 20015–20022.
[69] F. Da Ros, K. Dalgaard, L. Lei, J. Xu, C. Peucheret, QPSK-to-2 × BPSK wavelength and modulation format conversion through phase-sensitive four-wave mixing in a highly nonlinear optical fiber, Opt. Express 21 (23) (2013).
[70] F. Da Ros, K. Dalgaard, Y. Fukuchi, J. Xu, M. Galili, C. Peucheret, Simultaneous QPSK-to-2 × BPSK wavelength and modulation format conversion in PPLN, IEEE Photon. Technol. Lett. 26 (12) (Jun. 2014) 1207–1210.
[71] T. Inoue, N. Kumano, M. Takahashi, T. Yagi, M. Sakano, Widely wavelength-tunable femtosecond pulse generation based on comb-like profiled fiber comprised of HNLF and zero dispersion-slope NZDSF, in: 2006 Opt. Fiber Commun. Conf. Natl. Fiber Opt. Eng. Conf., 2006, pp. 1–3.
[72] M. Ettabib, K. Bottrill, F. Parmigiani, A. Kapsalis, A. Bogris, M. Brun, et al., All-optical phase regeneration with record PSA extinction ratio in a low-birefringence silicon germanium waveguide, J. Light. Technol. (2016).
[73] C.J. McKinstrie, S. Radic, Phase-sensitive amplification in a fiber, Opt. Express 12 (20) (2004) 4973.

[74] N.K. Kjoller, M. Piels, F. Da Ros, K. Dalgaard, M. Galili, L.K. Oxenlowe, 16-QAM field-quadrature decomposition using polarization-assisted phase sensitive amplification, in: 2016 IEEE Photonics Conference, IPC 2016, 2017, pp. 513–514.
[75] P. Guan, F. Da Ros, M. Pu, M. Lillieholm, Y. Zheng, E. Semenova, et al., 128 × 2 Gb/s WDM PON System with a single TDM time lens source using an AlGaAs-on-insulator waveguide, CLEO 2018, Paper SM2C.3, doi: 10.1364/CLEO_SI.2018.SM2C.3.
[76] M. Galili, E. Palushani, H.C.H. Mulvad, H. Hu, L.K. Oxenløwe, Wavelength preserving optical serial-to-parallel conversion, in: 2013 Optical Fiber Communication Conference and Exposition and the National Fiber Optic Engineers Conference (OFC/NFOEC). IEEE, 2013. paper OM2G.4.
[77] H.C. Hansen, H. Hu, M. Galili, H. Ji, E. Palushani, A. Clausen, et al., DWDM-TO-OTDM conversion by time-domain optical fourier transformation, in: European Conference on Optical Communication (ECOC) 2011. paper Mo.1.A.5.
[78] K.M. Røge, P. Guan, N.-K. Kjøller, M. Lillieholm, M. Galili, T. Morioka, et al., Characterization of spectral compression of OFDM symbols using optical time lenses, in: Proceedings of 2015 IEEE Photonics Conference. IEEE, 2015, pp. 303–304.
[79] P. Guan, K.M. Røge, H.C.H. Mulvad, M. Galili, H. Hu, M. Lillieholm, et al., All-optical ultra-high-speed OFDM to nyquist-WDM conversion based on complete optical Fourier transformation, J. Lightwave Technol. 34 (2) (2016). Available from: https://doi.org/10.1109/JLT.2015.2495188. 15, 15.
[80] H. Hu, D. Kong, E. Palushani, J.D. Andersen, A. Rasmussen, B.M. Sørensen, et al., 1.28 Tbaud Nyquist signal transmission using time-domain optical fourier transformation based receiver, CLEO 2013, postdeadline paper CTh5D.5.
[81] M.A. Foster, R. Salem, Y. Okawachi, A.C. Turner-Foster, M. Lipson, A.L. Gaeta, Generation of 270 Gb/s NRZ data packets from a 10-Gb/s signal using a temporal telescopic system, in: Conference On Optical Fiber Communication 2009, Technical Digest Series. doi: 10.1364/OFC.2009.OWS4.
[82] C.W. Chow, A.D. Ellis, F. Parmigiani, Time-division-multiplexing using pulse position locking for 100 Gb/s applications, Opt. Express 17 (8) (2009) 6562–6567.
[83] L.F. Mollenauer, C. Xu, Time-lens timing-jitter compensator in ultra-long haul DWDM dispersion managed soliton transmissions, in: Conference on Lasers and Electro-Optics OSA Technical Digest (Optical Society of America, 2002), paper CPDB1.
[84] A.R. Chraplyvy, Limitations on lightwave communications imposed by optical-fiber nonlinearities, J. Lightwave Technol. 8 (1990) 1548–1557.
[85] E. Ip, J.M. Kahn, Compensation of dispersion and nonlinear impairments using digital backpropagation, J. Lightwave Technol. 26 (2008) 3416–3425.
[86] A. Yariv, D. Fekete, D.M. Pepper, Compensation for channel dispersion by nonlinear optical phase conjugation, Opt. Lett. 4 (1979) 52–54.
[87] D.M. Pepper, A. Yariv, Compensation for phase distortions in nonlinear media by phase conjugation, Opt. Lett. 5 (1980) 59–60.
[88] K. Solis-Trapala, T. Inoue, M. Pelusi, H.N. Tan, S. Namiki, Implementing ideal nonlinear compensation through nonlinearity, in: IEEE Summer Topical Meeting 2015, paper TuF3.1.
[89] A.D. Ellis, M.E. McCarthy, M.A.Z. Al Khateeb, M. Sorokina, N.J. Doran, Performance limits in optical communications due to fiber nonlinearity, Adv. Opt. Photon. 9 (2017) 429–503.
[90] A.D. Ellis, M. Tan, M.A. Iqbal, M.A.Z. Al-Khateeb, V. Gordienko, G.S. Mondaca, et al., 4 Tb/s transmission reach enhancement using 10 × 400 Gb/s super-channels and polarization insensitive dual band optical phase conjugation, J. Lightwave Technol. 34 (2016) 1717–1723.
[91] J.C. Cartledge, A.D. Ellis, A. Shiner, A.I. Abd El-Rahman, M.E. McCarthy, M. Reimer, et al., Signal processing techniques for reducing the impact of fiber nonlinearities on system performance, OFC 2016, paper Th4F.5.
[92] S. Yoshima, Y. Sun, Z. Liu, K.R.H. Bottrill, F. Parmigiani, D.J. Richardson, et al., Mitigation of nonlinear effects on WDM QAM signals enabled by optical phase conjugation with efficient bandwidth utilization, J. Lightwave Technol. 35 (2017) 971–978.
[93] I. Kim, et al., Analysis of nonlinearity mitigation using spectral inversion for superchannel transmission, IEEE Photon. J. 8 (2016) 7200508.

[94] T. Umeki, T. Kazama, A. Sano, K. Shibahara, K. Suzuki, M. Abe, et al., Simultaneous nonlinearity mitigation in 92 × 180-Gbit/s PDM-16QAM transmission over 3840 km using PPLN-based guard-band-less optical phase conjugation, Opt. Express 24 (2016) 16945–16951.

[95] I. Sackey, C. Schmidt-Langhorst, R. Elschner, T. Kato, T. Tanimura, S. Watanabe, et al., Waveband-shift-free optical phase conjugator for spectrally efficient fiber nonlinearity mitigation, J. Lightwave Technol. 36 (2018) 1309–1317.

[96] F. Da Ros, M. Lillieholm, M.P. Yankov, P. Guan, H. Hu, S. Forchhammer, et al., Impact of signal-conjugate wavelength shift on optical phase conjugation-based transmission of QAM signals, ECOC (2017), paper P1.SC4.66.

[97] A.D. Ellis, M.E. McCarthy, M.A.Z. Al-Khateeb, S. Sygletos, Capacity limits of systems employing multiple optical phase conjugators, Opt. Express 23 (2015) 20381–20393.

[98] P. Minzioni, Nonlinearity compensation in a fiber optic link by optical phase conjugation, Fiber Integr. Opt. 28 (2009) 179–209.

[99] D. Vukovic, J. Schröder, F. Da Ros, L. Bangyuan, Du, C.J. Chae, D.-Y. Choi, et al., Multichannel nonlinear distortion compensation using optical phase conjugation in a silicon nanowire, Opt. Express 23 (2015) 3640–3646.

[100] A. Gajda, F. Da Ros, E.P. da Silva, A. Peczek, E. Liebig, A. Mai, et al., Silicon waveguide with lateral p-i-n diode for nonlinearity compensation by on-chip optical phase conjugation, OFC (2018), paper W3E.4.

[101] F. Da Ros, E.P. da Silva, A. Gajda, P.M. Kaminski, V. Cristofori, A. Peczek, et al., Nonlinearity compensation for dual-polarization signals using optical phase conjugation in a silicon waveguide, CLEO (2018), paper STu4C.1.

[102] A.D. Ellis, M.A.Z. Al Khateeb, M.E. McCarthy, Impact of optical phase conjugation on the nonlinear Shannon limit, J. Lightwave Technol. 35 (2017) 792–798.

[103] I. Sackey, F.D. Ros, M. Jazayerifar, T. Richter, C. Meuer, M. Nölle, et al., Kerr nonlinearity mitigation in 5 × 28-GBd PDM 16-QAM signal transmission over a dispersion-uncompensated link with backward-pumped distributed Raman amplification, Opt. Express 22 (2014) 27381–27391.

[104] K. Solis-Trapala et al., Transmission optimized impairment mitigation by 12 stage phase conjugation of WDM 24 × 48 Gb/s DP-QPSK signals, in: Proc. OFC, Los Angeles, CA, Mar. 2015, paper Th3C.2

[105] H. Hu, R.M. Jopson, A. Gnauck, M. Dinu, S. Chandrasekhar, X. Liu, et al., Fiber nonlinearity compensation of an 8-channel WDM PDM-QPSK signal using multiple phase conjugations, in: Optical Fiber Communication Conference, OSA Technical Digest Series (Optical Society of America, 2014), paper M3C.2.

[106] K. Solis-Trapala et al., Doubled transmission reach for DP-64QAM signal over field-deployed legacy fiber systems enabled by MSSI, in: Proc. ECOC, Valencia, Spain, Sep. 2015, paper Mo.3.6.2.

[107] Y. Sun, A. Lorences-Riesgo, F. Parmigiani, K.R.H. Bottrill, S. Yoshima, G.D. Hesketh, et al., Optical nonlinearity mitigation of 6 × 10 GBd polarization-division multiplexing 16 QAM signals in a field-installed transmission link, in: Optical Fiber Communication Conference, OSA Technical Digest (online) (Optical Society of America, 2017), paper Th3J.2.

[108] A. Gajda, L. Zimmermann, M. Jazayerifar, G. Winzer, H. Tian, R. Elschner, et al., Highly efficient CW parametric conversion at 1550 nm in SOI waveguides by reverse biased p-i-n junction, Opt. Express 20 (2012) 13100–13107.

[109] T.M. Cover, J.A. Thomas, Elements of Information Theory, Wiley Interscience, 2006.

[110] M. Dieter, M. Arnold, H.-A. Loeliger, P.O. Vontobel, W.Z. Aleksandar Kavcic, Simulation-based computation of information rates for channels with memory, IEEE Trans. Inform. Theory 52 (8) (2006) 3498–3508.

[111] J.G. Proakis, Digital Communications, McGraw-Hill, 2001.

[112] S. Arimoto, An algorithm for calculating the capacity of an arbitrary discrete memoryless channel, IEEE Trans. Inform. Theory IT-18 (1972) 14–20.

[113] R. Blahut, Computation of channel capacity and rate distortion functions, IEEE Trans. Inform. Theory IT-18 (1972) 460–473.

[114] T.A. Eriksson, T. Fehenberger, P.A. Andrekson, M. Karlsson, N. Hanik, E. Agrell, Impact of 4D channel distribution on the achievable rates in coherent optical communication experiments, J. Lightwave Technol. 34 (9) (2016) 2256–2266.
[115] N. Varnica, X. Ma, A. Kavcic, Capacity of power constrained memoryless AWGN channels with fixed input constellations, in: Global Telecommunications Conference, 2002. GLOBECOM '02. IEEE, 2002.
[116] M.P. Yankov, F.D. Ros, E.P. da Silva, S. Forchhammer, K.J. Larsen, L.K. Oxenløwe, et al., Constellation shaping for WDM systems using 256QAM/1024QAM with probabilistic optimization, J. Lightwave Technol. 34 (22) (2016) 5146–5156.
[117] M.P. Yankov, F. Da Ros, E. P. da Silva, S. Forchhammer, M. Galili, L. Oxenløwe, Optimizing the achievable rates of tricky channels: a probabilistic shaping for OPC channel example, in IEEE Photonics Conference (IPC), 2018.
[118] S. Sudo, H. Itoh, Efficient non-linear optical fibres and their applications, Opt. Quant. Electron. 22 (3) (1990) 187–212.
[119] M.J. Holmes, D.L. Williams, R.J. Manning, Highly nonlinear optical fiber for all optical processing applications, IEEE Photon. Technol. Lett. 7 (9) (1995) 1045–1047.
[120] M. Onishi, T. Okuno, T. Kashiwada, S. Ishikawa, N. Akasaka, M. Nishimura, Highly nonlinear dispersion-shifted fibers and their application to broadband wavelength converter, Opt. Fiber Technol. 4 (2) (1998) 204–214.
[121] J.H. Lee, T. Nagashima, T. Hasegawa, S. Ohara, N. Sugimoto, K. Kikuchi, Bismuth-oxide-based nonlinear fiber with a high SBS threshold and its application to four-wave-mixing wavelength conversion using a pure continuous-wave pump, J. Light. Technol. 24 (1) (2006) 22–27.
[122] M. Takahashi, Y. Mimura, J. Hiroishi, M. Tadakuma, R. Sugizaki, M. Sakano, et al., Investigation of a downsized silica highly nonlinear fiber, J. Light. Technol. 25 (8) (2007) 2103–2107.
[123] T. Torounidis, P.A. Andrekson, B.E. Olsson, Fiber-optical parametric amplifier with 70-dB gain, IEEE Photon. Technol. Lett. 18 (10) (2006) 1194–1196.
[124] M. Jamshidifar, A. Vedadi, M.E. Marhic, Continuous-wave one-pump fiber optical parametric amplifier with 270 nm gain bandwidth, in: 2009 35th Eur. Conf. Opt. Commun., 2009, pp. 1–2.
[125] M. Karlsson, Four-wave mixing in fibers with randomly varying zero-dispersion wavelength, J. Opt. Soc. Am. B 15 (8) (1998) 2269–2275.
[126] F. Yaman, Q. Lin, S. Radic, G.P. Agrawal, Impact of dispersion fluctuations on dual-pump fiber-optic parametric amplifiers, IEEE Photon. Technol. Lett. 16 (5) (2004) 1292–1294.
[127] K. Inoue, Arrangement of fiber pieces for a wide wavelength conversion range by fiber four-wave mixing., Opt. Lett. 19 (16) (1994) 1189–1191.
[128] M. Lillieholm, P. Guan, M. Galili, M.S. Møller-Kristensen, L. Grüner-Nielsen, L.K. Oxenløwe, Optimization and characterization of highly nonlinear fiber for broadband optical time lens applications, Opt. Express 25 (11) (2017).
[129] M. Takahashi, M. Tadakuma, T. Yagi, Dispersion and brillouin managed HNLFs by strain control techniques, J. Light. Technol. 28 (1) (2010) 59–64.
[130] E. Myslivets, C. Lundström, J.M. Aparicio, S. Moro, A.O.J. Wiberg, C.-S. Brès, et al., Spatial equalization of zero-dispersion wavelength profiles in nonlinear fibers, IEEE Photon. Technol. Lett. 21 (24) (2009) 1807–1809.
[131] E.P. Ippen, R.H. Stolen, Stimulated Brillouin scattering in optical fibers, Appl. Phys. Lett. 21 (11) (1972) 539–541.
[132] A. Kobyakov, M. Sauer, D. Chowdhury, Stimulated Brillouin scattering in optical fibers, Adv. Opt. Photon. 2 (1) (2010) 1.
[133] S.K. Korotky, P.B. Hansen, L. Eskildsen, J.J. Veselka, Efficient phase modulation scheme for suppressing stimulated Brillouin scattering, in: Tech. Dig. Int. Conf. Integrated Optics and Optical Fiber Communications, 1995, pp. 110–111.
[134] J.M. Chavez Boggio, J.D. Marconi, H.L. Fragnito, Experimental and numerical investigation of the SBS-threshold increase in an optical fiber by applying strain distributions, J. Light. Technol. 23 (11) (2005) 3808–3814.

[135] J. Hansryd, F. Dross, M. Westlund, P.A. Andrekson, S.N. Knudsen, Increase of the SBS threshold in a short highly nonlinear fiber by applying a temperature distribution, J. Lightwave Technol. 19 (11) (2001) 1691–1697.
[136] L. Grüner-Nielsen, D. Jakobsen, S. Herstrøm, B. Pálsdóttir, S. Dasgupta, D. Richardson, et al., Brillouin suppressed highly nonlinear fibers, Eur. Conf. Exhib. Opt. Commun. (2) (2012). p. We.1.F.1.
[137] S. Takushima, T. Okoshi, Supression of the simulated brillouin scattering using optical isolator, Electron. Lett. 8 (12) (1992) 1155–1157.
[138] Q. Lin, G.P. Agrawal, Vector theory of four-wave mixing: polarization effects in fiber-optic parametric amplifiers, J. Opt. Soc. Am. B 21 (6) (2004) 1216.
[139] C.J. McKinstrie, H. Kogelnik, R.M. Jopson, S. Radic, A.V. Kanaev, Four-wave mixing in fibers with random birefringence, Opt. Express 12 (10) (2004) 2033–2055.
[140] L.G. Nielsen, T. Veng, J. Bjerregaard, Elliptical core polarization maintaining highly nonlinear fibres, in: Eur. Conf. Opt. Commun. ECOC, vol. 1, no. September, 2008, pp. 83–84.
[141] C.G. Jørgensen, T. Veng, L. Grüner-Nielsen, M. Yan, Dispersion flattened highly non-linear fiber, in: Eur. Conf. Exhib. Opt. Commun., 2003, p. We3.7.6.
[142] T. Okuno, M. Hirano, T. Kato, M. Shigematsu, M. Onishi, Highly nonlinear and perfectly dispersion-flattened fibres for efficient optical signal processing applications, Electron. Lett. 39 (13) (2003) 972–974.
[143] M. Pu, H. Hu, L. Ottaviano, E. Semenova, D. Vukovic, L.K. Oxenløwe, et al., Ultra-efficient and broadband nonlinear AlGaAs-on-insulator chip for low-power optical signal processing, accepted for publication in Laser & Photonics Reviews, 2018.
[144] M. Galili, F. Da Ros, H. Hu, M. Pu, K. Yvind, L.K. Oxenløwe, Ultra-broadband optical signal processing using AlGaAs-OI devices, *2017 Optical Fiber Communications Conference and Exhibition (OFC)*, Los Angeles, CA, 2017, pp. 1–3.
[145] L. Ottaviano, M. Pu, E. Semenova, K. Yvind, Low-loss high-confinement waveguides and microring resonators in AlGaAs-on-insulator, Opt. Lett. 41 (17) (2016) 3396.
[146] M. Pu, L. Ottaviano, E. Semenova, K. Yvind, Efficient frequency comb generation in, Optica 3 (8) (2016) 8–11.
[147] M. Pu, L. Liu, H. Ou, K. Yvind, J.M. Hvam, Ultra-low-loss inverted taper coupler for silicon-oninsulator ridge waveguide, Opt. Commun. 283 (19) (2010) 3678–3682. 2010.
[148] F. Da Ros, M.P. Yankov, E. Porto da Silva, M. Pu, L. Ottaviano, H. Hu, et al., Characterization and optimization of a high-efficiency AlGaAs-on-insulator-based wavelength converter for 64- and 256-QAM signals, J. Lightwave Technol. 35 (17) (2017) 3750–3757. 2017.
[149] F. Parmigiani, G. Hesketh, R. Slavík, P. Horak, P. Petropoulos, D.J. Richardson, Polarization-assisted phase-sensitive processor, J. Lightwave Technol. 33 (6) (2015) 1166.
[150] F. Da Ros, M. Pu, L. Ottaviano, H. Hu, E. Semenova, et al., Phase-sensitive four-wave mixing in AlGaAs-on-insulator nano-waveguides, in: Proc. 2016 IEEE Photonics Conference, IPC'2016, Paper WB1.1, 2016.
[151] T. Vallaitis, S. Bogatscher, L. Alloatti, P. Dumon, R. Baets, M.L. Scimeca, et al., Optical properties of highly nonlinear silicon-organic hybrid (SOH) waveguide geometries, Opt. Express 17 (20) (2009) 17357.
[152] A. Boskovic, S.V. Chernikov, J.R. Taylor, L. Gruner-Nielsen, Oa Levring, Direct continuous-wave measurement of n(2) in various types of telecommunication fiber at 1.55 μm, Opt. Lett. 21 (24) (1996) 1966–1968.
[153] T. Nakanishi, M. Hirano, T. Okuno, M. Onishi, Silica-based highly nonlinear fiber with y = 30 lW / km and its FWM-based conversion efficiency, OFC 2006, OTuH7, 2006.
[154] D.J. Moss, R. Morandotti, A.L. Gaeta, M. Lipson, New CMOS-compatible platforms based on silicon nitride and Hydex for nonlinear optics, Nat. Photon. 7 (8) (2013) 597–607.
[155] M.H.P. Pfeiffer, J. Liu, A.S. Raja, T. Morais, B. Ghadiani, T.J. Kippenberg, Ultra-smooth silicon nitride waveguides based on the Damascene reflow process: fabrication and loss origins, Optica 5 (7) (2018) 884.
[156] H. El Dirani, et al., Annealing-free Si_3N_4 frequency combs for monolithic integration with Si photonics, Appl. Phys. Lett. 113 (8) (2018) 081102.

[157] X. Gai, et al., Progress in optical waveguides fabricated from chalcogenide glasses, Opt. Express 18 (25) (2010) 26635.
[158] K.J.A. Ooi, et al., Pushing the limits of CMOS optical parametric amplifiers with USRN:Si_7N_3 above the two-photon absorption edge, Nat. Commun. 8 (2017) 13878.
[159] J.S. Pelc, K. Rivoire, S. Vo, C. Santori, D.A. Fattal, R.G. Beausoleil, Picosecond all-optical switching in hydrogenated amorphous silicon microring resonators, Opt. Express 22 (4) (2014) 3797–3810.
[160] I. Aldaya, A. Gil-Molina, J.L. Pita, L.H. Gabrielli, H.L. Fragnito, P. Dainese, Nonlinear carrier dynamics in silicon nano-waveguides, Optica, 4 (10) (2017), 1219–1227.
[161] Q. Lin, O.J. Painter, G.P. Agrawal, Nonlinear optical phenomena in silicon waveguides: modeling and applications, Opt. Express 15 (25) (2007) 16604.
[162] B. Kuyken, et al., Nonlinear properties of and nonlinear processing in hydrogenated amorphous silicon waveguides, Opt. Express 19 (26) (2011) B146.
[163] D.L. Staebler, C.R. Wronski, Reversible conductivity changes in discharge-produced amorphous Si, Appl. Phys. Lett. 31 (4) (1977) 292–294.
[164] B. Kuyken, et al., On-chip parametric amplification with 26.5 dB gain at telecommunication wavelengths using CMOS-compatible hydrogenated amorphous silicon waveguides, Opt. Lett. 36 (4) (2011) 552.
[165] C. Grillet, et al., Amorphous silicon nanowires combining high nonlinearity, FOM and optical stability, Opt. Express 20 (20) (2012) 22609–22615.
[166] K.-Y. Wang, A.C. Foster, GHz-rate optical parametric amplifier in hydrogenated amorphous silicon, J. Opt. 17 (9) (2015) 094012.
[167] J. Safioui, et al., Supercontinuum generation in hydrogenated amorphous silicon waveguides at telecommunication wavelengths, Opt. Express 22 (3) (2014) 3089.
[168] J. Matres, G.C. Ballesteros, P. Gautier, J.-M. Fédéli, J. Martí, C.J. Oton, High nonlinear figure-of-merit amorphous silicon waveguides, Opt. Express 21 (4) (2013) 3932–3940.
[169] P. Girouard, L.H. Frandsen, M. Galili, L.K. Oxenløwe, A deuterium-passivated amorphous silicon platform for stable integrated nonlinear optics,CLEO 2018, SW4I.2, 2018.
[170] J.H. Wei, M.S. Sun, S.C. Lee, A possible mechanism for improved light-induced degradation in deuterated amorphous-silicon alloy, Appl. Phys. Lett. 71 (11) (1997) 1498–1500.

Further reading

J.C. Geyer, C. Rasmussen, B. Shah, T. Nielsen, M. Givehchi, Power efficient coherent transceivers, in: Proceedings of the 42nd European Conference on Optical Communication (VDE, 2016), pp. 109–111.
S. Watanabe, F. Futami, R. Okabe, Y. Takita, S. Feber, R. Ludwig, et al., 160 Gbit/s optical 3R re-generator in a fiber transmission experiment, in: Optical Fiber Communication Conference (OFC2003), PD16, 2003.

CHAPTER 7

Multicore and multimode optical amplifiers for space division multiplexing

Yongmin Jung[1], Shaif-ul Alam[1], David J. Richardson[1], Siddharth Ramachandran[2] and Kazi S. Abedin[3]
[1]Optoelectronics Research Centre (ORC), University of Southampton, Southampton, United Kingdom
[2]Electrical and Computer Engineering Department, Boston University, Boston, MA, United States
[3]OFS Laboratories, Somerset, NJ, United States

7.1 Introduction

The explosive growth in global data consumption is placing a tremendous strain on telecommunications networks, including the core, metro, and access sectors. Global IP traffic is currently increasing at a rate of 60% per year (average rate over a 5-year period). Moreover, the traffic from wireless and mobile devices is also growing at a rate of more than 60% per annum according to the Cisco VNI global mobile forecast [1]. This rate of traffic growth is predicted to extend well in to the 2020s due to the underlying growth in social networking and machine-to-machine communications. At the same time, the total power consumption of the installed network equipment is continuously rising as the total data consumption increases. The direct energy use of ICT devices is currently estimated to be $\sim$4% of all electricity consumption and to give rise to $\sim$2% of all CO_2 emission [2].

Conventional single-mode fibers (SMFs) have served as reliable long-haul transmission media of high data rate signals within core optical networks for the past 30 years and network operators have kept up with the increased data traffic through a sequence of technical innovations associated with ever better exploitation of the SMF transmission capacity. However, it is widely recognized that the maximum transmission capacity of a single SMF is rapidly approaching its fundamental information-carrying limit ($\sim$100 Tbit/s) imposed by a combination of the fiber nonlinearity and the bandwidth of the erbium doped fiber amplifier (EDFA) [3,4]. Increasing the number of SMFs is one solution, but it comes at a high cost and has significant energy consumption requirements that may be unsustainable. Recently, space division multiplexing (SDM) technology [5–9] has attracted considerable attention in the fiber optic communication community as a very promising approach to significantly increase the transmission

Optical Fiber Telecommunications V11
DOI: https://doi.org/10.1016/B978-0-12-816502-7.00008-7

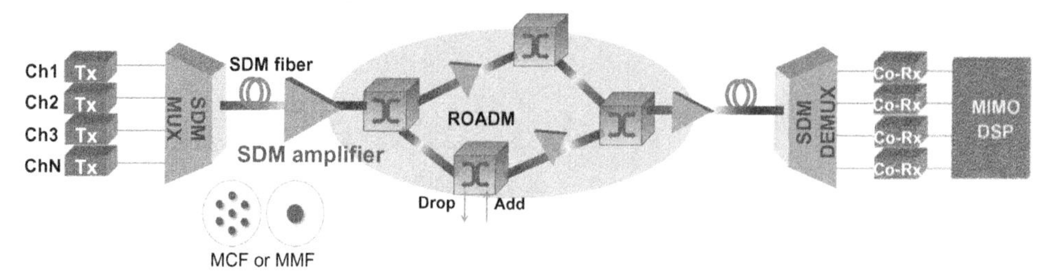

Figure 7.1 Conceptual diagram of an optical network infrastructure based on space division multiplexing (SDM) technology.

capacity of the fiber and to reduce the overall cost per transmitted bit of information. The concept of SDM is to transmit multiple information channels along a single optical fiber, each taking an independent spatial pathway. In principle, N SMFs can be replaced by a single SDM-fiber with *N*-spatial channels.

Fig. 7.1 shows a conceptual diagram of a proposed optical network infrastructure based on SDM technology. A number of optical signals from the *N*-transmitters are combined with a spatial multiplexer and transmitted across a single SDM fiber. These *N* multiplexed channels can be simultaneously amplified and switched through several inline optical SDM amplifiers and SDM-compatible reconfigurable optical add-drop multiplexers (ROADMs). The resultant signals are then demultiplexed at the other end and delivered to the appropriate output line for multiple input multiple output (MIMO) digital signal processing (DSP). Logically, this is just a simple superposition of many SMF systems (i.e., parallel SMF systems) onto a single SDM fiber, but there is great scope for device integration and cost/energy/space saving benefits. For example, parallel SMF systems require lots of duplication of optical components, optical amplifiers, and ROADMs, increasing the overall system cost, energy consumption, and space in direct proportion to the capacity increase. On the other hand, in an SDM system, just one or two optical amplifier/ROADM devices are required to simultaneously amplify/route all the spatial channels in the transmission fiber, and this offers great potential for reducing the cost per transmitted bit. The use of not only the SDM transmission fiber but also many optical components and subsystems make the SDM approach very attractive with regard to cost, space, and power requirements. It should be noted that SDM technology can be used in conjunction with other multiplexing techniques such as wavelength division multiplexing (WDM), and we contend that spatially multiplexed WDM systems might be employed to overcome the current fiber transmission capacity limit.

Up to now, several types of SDM fiber have been introduced, but two mainstream approaches have been extensively investigated: multicore fibers (MCFs) and multimode fibers (MMFs). In MCFs [10–12], multiple optical cores are incorporated in a single fiber cladding and multiple data streams can be simultaneously transmitted through the individual cores. In MMFs [8,13,14], the core (with a slightly enlarged size—typically a few times larger than that of conventional SMFs) is designed to support several transverse spatial modes. These modes can then be used to define multiple independent optical paths through the fiber. Very recently, these two different SDM fiber concepts have been combined to further enhance the data- carrying capacity of a single fiber, specifically, to realize multimode MCFs [15–17]. This combined approach has led to the demonstration of a more than 100-fold data capacity increase (~10 Pbit/s) relative to conventional SMF systems. However, in order to realize any long-haul (>1000 km) SDM transmission system and to add the cost, energy, and space-saving benefits promised by SDM technology, it is necessary to develop high-performance SDM amplifiers.

Very recently, several important inline SDM components (e.g., optical isolators and signal/pump combiners) have been developed with low insertion loss and fully integrated SDM amplifiers have been successfully realized in a compact all-fiberized format. Improved sharing of the optical components and significant device integration was efficiently achieved in these devices, as will be required to ultimately realize the anticipated cost reduction benefits of SDM technology. In this chapter, we review the state of the art in optical amplifiers for the various SDM approaches under investigation with a particular focus on multicore and multimode devices.

7.1.1 Cost, space, and energy benefits of space division multiplexing amplifiers

First of all, it is worth evaluating the potential cost, space, and energy-saving benefits of SDM amplifiers compared to a parallel array of SMF amplifiers. As shown in Fig. 7.2, the parallel SMF amplifier architecture requires a large number of duplicate optical components, electronics, and heat management units. It incorporates N or $2N$

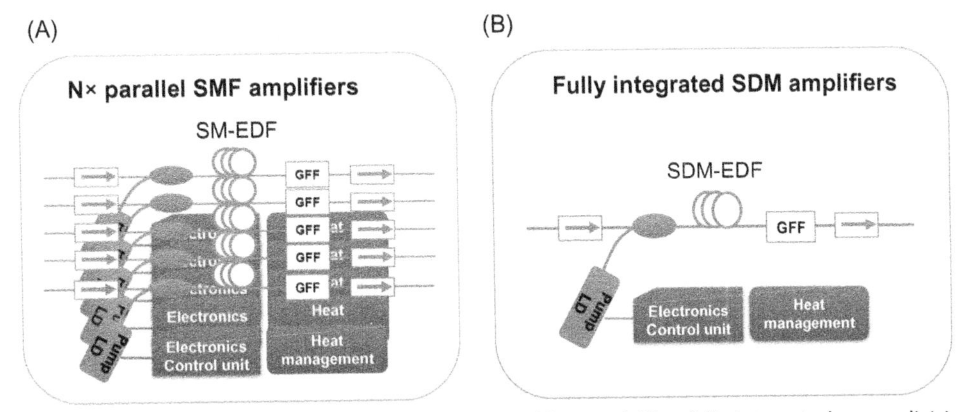

Figure 7.2 Comparison between (A) parallel SMF amplifiers and (B) a fully integrated space division multiplexing (SDM) amplifier.

times each key optical component [e.g., optical isolator, WDM coupler, single-mode erbium-doped fiber (EDF), gain flattening filter (GFF), and pump laser diode], as well as of electronic components (e.g., current driver, thermoelectric cooler), and so on. However, in the case of SDM amplifiers, only one or two optical components, electronic units, and heat control units are necessary to enable multiple spatial channel amplification, where the same device can be shared with multiple spatial channels in the most cost-efficient manner. For example, for the $N = 32$ case, 2 SDM fiber isolators and 1 GFF that exploit essentially the same micro-optic components as SMF devices can replace 64 single-mode isolators and 32 GFFs, respectively. If we assume the passive devices in a SMF amplifier cost \$400 and, just for the sake of argument, we assume SDM devices cost four times more than single-mode devices (a significant markup due to the more critical alignment and necessary optical characterization needed during manufacture), then we would immediately save $(32 - 1 \times 4) \times \$400 = \$11{,}200$ in inline passive component costs by adopting an SDM approach. Likewise, one simple SDM signal/pump combiner can replace 32 WDM couplers (likely saving $(32 - 1 \times 4) \times \$100 = \$2800$) and a single >20 W multimode pump laser diode (costing \$1000) can replace 32 single-mode pump diodes (at a cost of \$500 per device), equating to a total \$15,000 saving. These savings alone sum to a total component saving per amplification stage of \$29,000. Additional savings in electric current drivers/microcontrollers/mechanical packaging and connectors will also add to this optics only saving, say, another saving of \$21,000 to keep the math simple, giving a total amplification stage saving of \$50,000. In a 1000 km transmission link, one would need ~20 such amplifier stages, giving rise to a potential amplifier saving of ~\$1 million by moving to SDM amplifier modules. The preceding calculation is obviously just an example, and one may argue about the various cost assumptions made; however, the undeniable point is that constructing ultrahigh capacity links with SMF technology requires enormous amounts of components, and this component count can be substantially reduced by moving to an SDM technology base. From an electric energy consumption perspective, the SDM amplifier also provides the potential to reduce power consumption. The optical efficiency of SDM amplifiers may be similar to that of parallel SMF amplifiers; however, the main potential for power savings results from simplification of the control electronics. In particular, in a cladding-pumped SDM amplifier configuration, the electrical power required to drive and control a single multimode pump laser is generally much less than that required to drive N single mode pumps, particularly as N gets large. Admittedly, there is an associated reduction in power-control per amplifier as the number of pumps is reduced, and this may place restrictions on end usage for particular applications. Further study is needed; however, the potential for power savings per-se is self-evident. Moreover, the physical space required to package and ultimately house the amplifier in use may also be substantially reduced. The typical dimensions of a standalone SMF amplifier are $19'' \times 1'' \times 11''$

(height × width × length), and the total space occupied for 10 parallel SMF amplifiers will thus be ~19″ × 10″ × 11″. However, a single SDM amplifier can be housed in a 19″ × 2″ × 11″rack mountable unit and even greater savings may be anticipated in further high spatial channel SDM amplifiers. Again, one can argue about the detailed numbers used above and point to the miniaturized SMF amplifiers now on the market after many years of development; however, such miniaturization approaches could equally be applied to SDM amplifiers in due course. Therefore the potential for appreciable capital expenditure cost saving is significant, as is the potential for significant operational expenditure savings given the reduced electric power and space requirement.

7.2 Enabling optical components for space division multiplexing amplifiers

Optical isolators, GFFs, and pump/signal combiners are essential passive fiber-optic components in optical amplifiers used to prevent backward propagating light, to equalize the amplified signal power in WDM systems, and to combine/separate the signals from the pump laser light, respectively. Unlike the widely used SMF components, only a restricted range of fiber-optic components are currently available for SDM fibers, and the prototype modules used in many experiments to date have been implemented only with the aid of numerous free-space optical components mounted on large optical benches. They are not only bulky but also expensive and introduce high optical losses. From this perspective, a fiber-optic platform will inevitably be the preferred route forward, and the fabrication of fully integrated components is a prerequisite for the realization of practical SDM systems. Recently, micro-optic fiber collimator technology has been investigated as a platform to realize compact, integrated SDM components, and compact fiber-optic isolators have been demonstrated without significant core (or mode) dependent losses [18,19]. Due to the simple fabrication process, good beam transfer quality, and low cost, these devices represent an extremely attractive means to reduce the cost of operating and building SDM components and subsystems. In addition, an all-fiber side pump coupler approach has been shown to offer a convenient route toward a fully integrated SDM amplifier in a cladding pumped configuration. Here the pump radiation can easily be coupled into the active fiber using a fully fiberized approach and the SDM amplifier can be directly spliced to the SDM transmission fibers. These two key enabling components for fully integrated SDM amplifiers are discussed in more detail in the following section.

7.2.1 Space division multiplexing components based on micro-optics

Micro-optic technology is an important platform that is already used widely in SMF components such as optical isolators, circulators, GFFs, WDM couplers, switches, and variable optical attenuators. Fig. 7.3 shows a basic schematic of a representative micro-optic

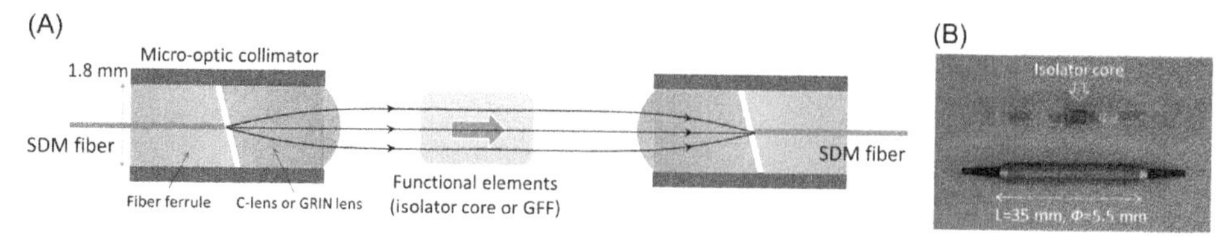

Figure 7.3 (A) Schematic of a micro-optic fiber collimator assembly for compact space division multiplexing (SDM) components and (B) fully integrated device package.

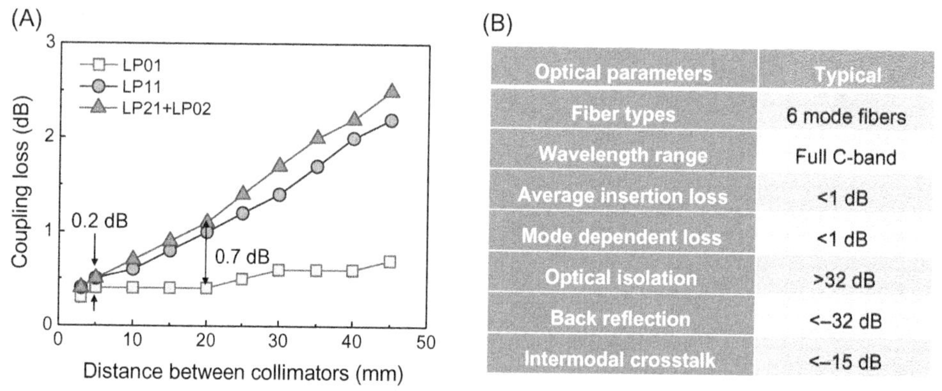

Optical parameters	Typical
Fiber types	6 mode fibers
Wavelength range	Full C-band
Average insertion loss	<1 dB
Mode dependent loss	<1 dB
Optical isolation	>32 dB
Back reflection	<–32 dB
Intermodal crosstalk	<–15 dB

Figure 7.4 Multimode fiber isolators: (A) mode-dependent coupling loss as a function of distance between two multimode fiber (MMF) collimators and (B) typical specification of MMF isolators.

fiber collimator assembly consisting of two fiber-optic collimators to transform the emergent light from an input fiber into a collimated free space beam and then to refocus it into another output fiber using a second collimator assembly operated in reverse. Optical elements can then be inserted into the free space region (e.g., an isolator core or filter chip) to provide in-line functionality. This concept can be extended to SDM fibers (e.g., MMFs and MCFs), and an array of new and practical SDM components can be developed with optical performance comparable to existing equivalent SMF devices in terms of function and insertion loss, whilst at the same time ensuring low levels of intercore (or intermodal) crosstalk.

To prove the principle, a MMF collimator assembly was fabricated by using a graded-index 6-mode fiber supporting LP_{01}, LP_{11}, LP_{21}, and LP_{02} mode groups, and the mode-dependent coupling loss was examined as a function of distance between the two collimators [19]. As shown in Fig. 7.4A, higher order modes experience progressively slightly higher coupling losses relative to lower order modes due to their increased mode field diameter and beam divergence. For relatively short gap distances, however, the mode-dependent loss was not significant—only 0.2 dB (0.7 dB) for 5 mm (20 mm) gap distance, respectively. Note that the working distance range from

5 to 20 mm is long enough to allow the incorporation of most functional optical elements in the free space region between the two collimators. Also, it is expected that the mode-dependent loss can be further improved in the future by optimizing the design/choice of the micro-lens and by adopting micro-optic components with a slightly larger clear aperture. To demonstrate the feasibility of functional devices, a polarization-insensitive optical isolator core was inserted between two MMF collimators. The insertion loss was slightly increased by about 0.1 dB for each spatial mode, but identical optical isolation performance was achieved for all spatial modes. Fig. 7.4B summarizes the typical optical specifications of the fabricated MMF isolators, which is very close to that of their commercial SMF counterparts.

The micro-optic collimator platform can also be extended to multicore fiber devices, and an exemplary 32-core multicore fiber isolator was fabricated using the same procedure [18]. The fiber incorporates 32-cores at high density in a square lattice with a 29 μm core pitch distance and a 243 μm cladding diameter, as shown in the inset of Fig. 7.5A. Importantly, the MCF collimators intrinsically require rotational alignment, and a pair of MCF collimators was mounted on a multiaxis precision micro-stage (offering translation, tilt, and rotation adjustments) for relative alignment. As shown in Fig. 7.5A, all the 32-cores show less than 2.1 dB insertion loss (including two extra splices between the MCFs). Note that the outer cores have a slightly larger insertion loss relative to the inner cores, which may be due to a slight angular orientation misalignment of the MCF splices and/or collimator assembly. As shown in Fig. 7.5B, the initial device specification looks most encouraging with an average insertion loss of ~1.5 dB, core-to-core loss variation of ~1.5 dB, and an intercore crosstalk of less than −40 dB. These results indicate that conventional micro-optic technology can be applied/integrated into MCF based subcomponents such that the same optical function can be achieved on all cores simultaneously with near identical

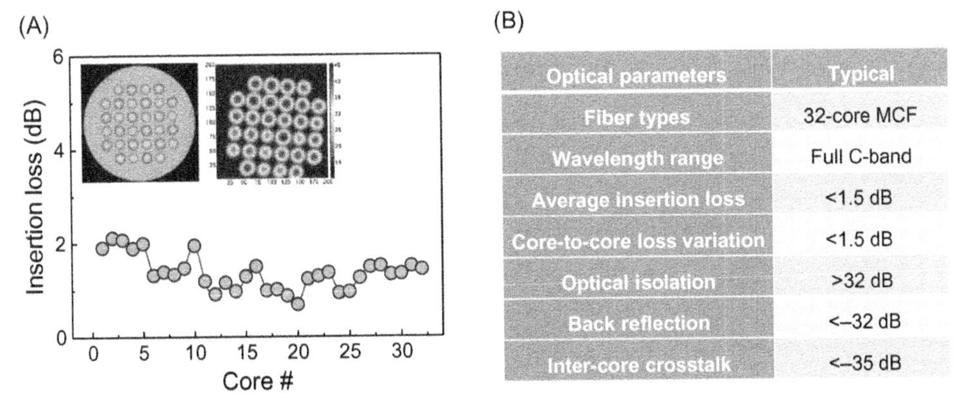

Optical parameters	Typical
Fiber types	32-core MCF
Wavelength range	Full C-band
Average insertion loss	<1.5 dB
Core-to-core loss variation	<1.5 dB
Optical isolation	>32 dB
Back reflection	<−32 dB
Inter-core crosstalk	<−35 dB

Figure 7.5 Multicore fiber isolators: (A) the measured core-to-core insertion loss variation and (B) initial specification of the 32-core multicore fiber (MCF) isolator.

performance for each. The approach can be further extended to other optical components such as circulators, beam splitters, WDM filters, and routing switches.

7.2.2 Pump and signal combiners

This micro-optic collimator platform can be further extended to 3-port SDM fiber components such as a pump/signal combiner (also referred to as a WDM coupler) as shown in Fig. 7.6A. Here a dichroic mirror can be used to reflect the pump light to the reflected fiber port, whilst the signals to be transmitted pass to a collimator that couples the remaining light to the output transmission fiber. One of the collimators (RHS in Fig. 7.6A) has two SDM fibers in a dual-hole glass ferrule (one being the pump input and the other the reflected output) and a GRIN lens with a flat surface for easy placement of the mirror on top. In the case of a MMF device, the center-to-center fiber alignment is critical to minimize the intermodal crosstalk whilst the angular alignment is critical for MCF devices (to ensure the reflected core arrangement is matched to that of the input MCF). Note that this micro-optic collimator approach is very useful for core-pumped amplifiers providing for an efficient pump/signal combiner but is not suitable for cladding-pumped amplifiers, as it requires a high-power handling capability (up to a few tens of watts). For cladding-pumped amplifiers, various forms of pump coupler have already been developed for high-power fiber laser applications, including both end-pumped and side-pumped configurations [20]. All of these approaches are viable, but a tapered fiber side-coupling method [21] has recently received considerable attention for SDM amplifier development due to the ease of fabrication, scalability to multiple combiners, and possibility to provide for a relatively uniform pump distribution

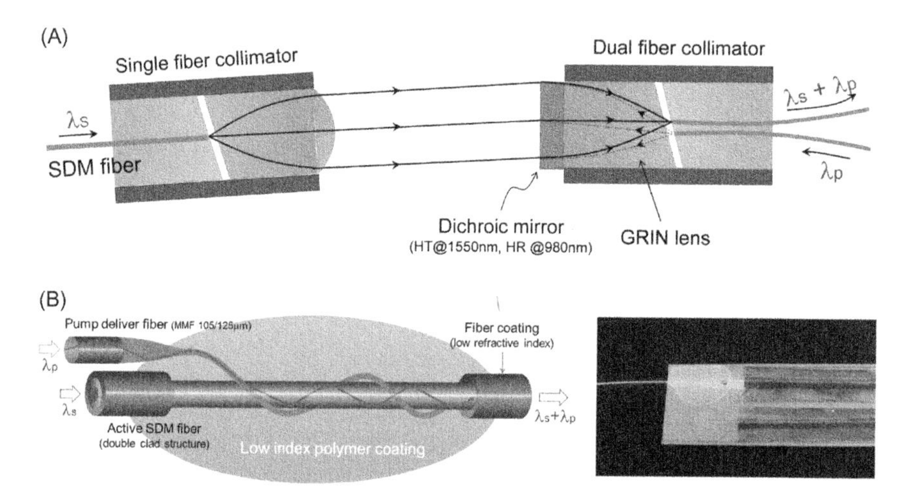

Figure 7.6 Pump and signal combiners: (A) a micro-optic wavelength division multiplexing (WDM) coupler for core-pumped amplifiers and (B) a side pump coupler for cladding-pumped amplifiers.

along the amplifier length without having access to the fiber core. To fabricate this device, as shown in Fig. 7.6B, a multimode pump delivery fiber (typically with a core/cladding diameter of 105/125 μm and NA of 0.15 or 0.22) was first tapered down to a waist diameter of 10–20 μm and then wrapped around a stripped section of the active SDM fiber. A high pump coupling efficiency from the pump fiber to the active fiber can readily be achieved (typically in the range of 70%–90%), and this structure can be subsequently recoated with a low-index polymer coating (e.g., ultraviolet curable acrylate or silicon rubber) to enhance the mechanical/thermal stability. Note that multiple pump combiners can easily be introduced in such an architecture at various locations along the amplifier fiber with minimal impact on the signals, and the overall pump intensity and thermal load can be appropriately distributed.

7.3 Multicore fiber amplifiers

7.3.1 Design considerations for multicore fiber amplifiers

In conventional SMF amplifiers, the key optical properties are small signal gain, noise figure, saturated output power, gain bandwidth, and gain flatness. Each of these optical properties are also important for the development of MCF amplifiers, but a few additional parameters should be considered for simultaneous amplification of multiple spatial channels: for example, intercore crosstalk between neighboring cores, core-to-core gain/noise figure variation, cost, and energy consumption. In this section, we will focus on the most important parameters and essential design aspects of MCF amplifiers.

- *Intercore crosstalk* (*XT*): In most cases, the physical geometry of the active MCF is chosen to match the passive transmission fiber—specifically, the number of cores, geometry, and core pitch should be equal to that of the passive fiber. Today the highest capacity performance passive MCFs incorporate a heterogeneous core arrangement (by incorporating several different types of core material) or trench assisted core designs to minimize the intercore crosstalk per unit length. However, it is too complex to adopt heterogeneous active fiber fabrication because of the delicate nature of the rare earth doping process. Given the fact that active fiber lengths are only a few meters long, intercore crosstalk is likely to be less of an issue and use of a homogenous core design is generally acceptable. For these reasons, simple homogenous step-index core arrangements have been frequently employed to fabricate active MCF preforms with a critical aspect of ensuring that the splice loss between the active homogeneous core design and the various passive heterogeneous cores is low enough (note that the MFDs of the individual cores in an inhomogeneous MCF fiber are usually matched, which helps in ensuring uniform splice losses to a homogeneous active core MCF).
- *Pumping methods*: Both core- and cladding-pumped techniques are viable options to realize MCF amplifiers. Core-pumped optical amplifiers offer the potential to achieve

high-gain, low-noise figures and inherently provide the ability to independently control the gain of the individual cores. On the other hand, the cladding-pumped approach can potentially reduce the cost and complexity of the amplifier by reducing the number of optical components, as well as enabling the use of high-power, low-cost multimode pump laser diodes. Furthermore, as multimode diodes tend to be more electrically efficient than their single-mode counterparts, it follows that high-power multimode pump sources can be operated with forced air cooling instead of Peltier-based temperature control, promising further reduction in energy consumption. The ability to straightforwardly couple the pump radiation into an MCF amplifier is extremely important, particularly if a side-coupling approach is used, since it then becomes possible to splice an MCF amplifier directly onto the MCF transmission fiber—opening a route toward fully fiberized MCF transmission lines.

- *Chemical composition of the core dopants*: For core-pumped MCF amplifiers, an erbium only doped core in conjunction with single-mode pumping has proven to be the preferred option, allowing for high levels of independent gain control per core and full C-band operation. However, under a cladding-pumped configuration, the absorption of erbium only doped fiber is generally very low at the preferred pump wavelength of 978 nm. Consequently the amplifier length either becomes very long (tens of meters), favoring L-band operation and otherwise leading to an unsatisfactory level of pump absorption to achieve full C-band operation. To ensure adequate core absorption per unit length under a cladding-pumped configuration it is generally necessary to adopt an ytterbium-sensitized EDF core [i.e., erbium/ytterbium co-doped fiber (EYDF)]. The use of ytterbium as a co-dopant provides far stronger pump absorption and energy transfer to erbium than can be achieved using pure erbium doping alone. This allows for higher inversion and greatly reduced device lengths. However, to realize a high concentration of erbium/ytterbium ions within the core, it is generally necessary to incorporate additional dopants into the core, notably phosphorus, to assist in the energy transfer process, and this results in a relatively high numerical aperture and therefore a relatively small core size for single-mode operation. In addition, it can result in slightly compromised C-band operation (i.e., reduced gain bandwidth) due to the modified spectroscopy associated with such a glass composition.
- *Optimizing the power conversion efficiency (PCE)*: Due to its relatively small absorption coefficient of the pump radiation in a cladding-pumped EDFA, a long length of gain fiber is required to increase the absorption of the pump light. However, it is not possible to increase the length arbitrarily because the signal intensity in the core becomes large compared to the pump intensity after a certain length of gain fiber, at which point the signal output power begins to decrease even though a large amount of pump light still remains unabsorbed. This results in inefficient utilization of the pump light and poor PCE. However, the presence of multiple cores in a multicore

EDFA can enhance the pump absorption efficiency and a greater fraction of the pump light can be converted to signal radiation, since amplification simultaneously occurs in multiple individual cores. Numerical simulations [22] have shown that a larger PCE can be achieved for a MCF amplifier compared with a single-core EDFA. In the case of a single core, erbium only doped EDFA, the maximum achievable PCE was only 8.6%, corresponding to a length of 90 m, following which the efficiency starts to decline. For a seven-core EDFA, the PCE was 34% for a length of 85 m, about four times higher than that of a single core EDFA. Note that the PCE of the MCF amplifier can be further improved by co-doping with ytterbium ions (owing to the efficient energy transfer mechanism between ytterbium and erbium ions), leading to a much higher pump to signal conversion efficiency in just a few meters' length of gain fiber.

- *Splice loss between active and passive fibers*: In core-pumped amplifiers, the core diameter and the numerical aperture of the fibers are chosen to guarantee robust single-mode operation at both the pump and signal wavelengths. Consequently, optical fiber designs with a smaller core and larger numerical aperture are commonly used to meet these requirements. In cladding-pumped amplifiers, the core diameter can be further enlarged because the single-mode condition for the pump light is no longer needed. In order to increase pump absorption in cladding-pumped amplifiers, however, a higher doping concentration is required, which inevitably increases the core refractive index relative to the cladding refractive index. Therefore a large mode field diameter mismatch between active and passive fibers is to be anticipated for both pumping configurations, and the splice loss should be considered as a fundamental aspect of the MCF amplifier design. It may be possible to reduce the splice loss by using an intermediary fiber [23] and/or by optimizing the splicing process parameters to accommodate the large mode field diameter mismatch between the two fiber types; however, one must keep in mind that it is usually quite challenging to reduce the splice loss between dissimilar MCFs to the levels traditionally associated with SMFs.

7.3.2 Recent progress in multicore fiber amplifiers

MCF amplifiers can broadly be divided into two categories as shown in Fig. 7.7: uncoupled core and coupled core. Both types of fiber have multiple cores inside a single cladding, but they are different types of MCFs with completely different optical properties and transmission principles. In uncoupled-core MCF transmission systems, the light propagates independently in each core, which can be thought of as independent optical signal paths. Therefore low intercore crosstalk is very important in these kinds of MCF (and their associated amplifiers). Various uncoupled-core optical amplifiers have been reported with different core arrangements (e.g., hexagonal, square

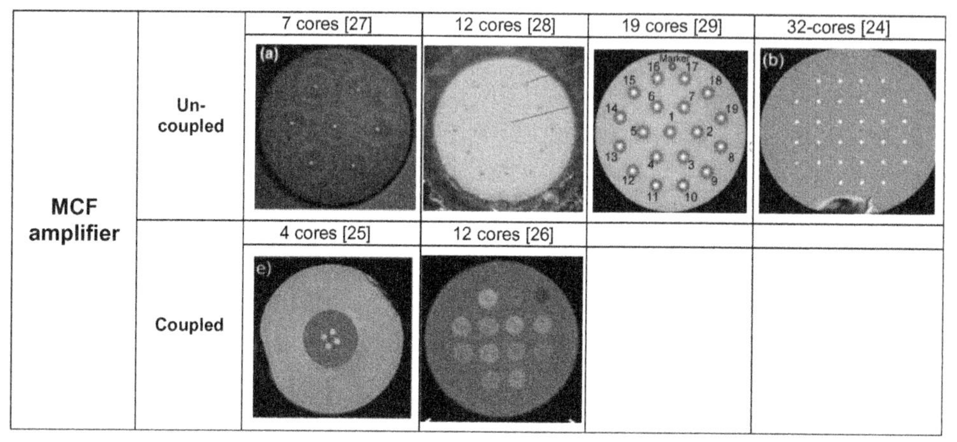

Figure 7.7 Classification of the multimode fiber (MCF) amplifiers using various types of MCFs reported so far.

lattice, or ring structure) and different pumping configurations (e.g., core, cladding, or hybrid pumping). Up until now, a cladding pumped 32-core MCF amplifier [24] represents the highest core density uncoupled MCF amplifier so far reported. On the other hand, in the coupled MCF case, each core is positioned close enough to be strongly coupled to each other. In this regime, the coupled cores can generate coupled waveguide modes which can be considered as supermodes defined by the array of cores. As a consequence, coupled-core MCFs generally require DSP to recover the original input signals from the output signals at the end of the transmission line. Similarly to the case of MMFs, the impact of any mode-dependent loss/gain can be inherently minimized due to the strong mode coupling (essentially an averaging effect). The first coupled (4-core) core MCF [25] was demonstrated using a core-pumped configuration, and a coupled 12-core MCF amplifier [26] has recently been reported with a cladding-pumped configuration.

Table 7.1 summarizes the optical characteristics of recently reported uncoupled MCF amplifiers. The first MCF amplifier for SDM applications was introduced in 2011 [a 7-core erbium-doped fiber amplifier (EDFA) [27] with a hexagonal close-packed core structure]. This EDFA employed a core-pumped configuration with the active erbium-doped cores successfully integrated within a single MCF, and the device was shown to give excellent amplifier performance similar to that of conventional single mode EDFAs. However, a pair of fan-in/fan-out devices were used to connect into the individual SMFs, and the system was not much different in component count from the case of 7-parallel SMF amplifiers. For example, seven single-mode pump laser diodes and 14 isolators/WDM couplers were included for this particular example. Whilst technologically impressive, the result was perhaps not so unexpected given that other than the multicore EDF each channel exploits state-of-the-art SMF components, with each core ordinarily pumped using a dedicated pump laser on a

Table 7.1 Characteristics of recently reported uncoupled multicore fiber amplifiers.

Pumping scheme	# of cores	Gain (dB)	NF (dB)	Variation (dB)	Intercore crosstalk (XT) (dB)	Special approach	References
Core	7	>23	<4	4	< −30	End pumping with fused pump coupler	[27]
	19	>19	<7	4	< −42	Integrated isolator, WDM coupler	[29]
Cladding	7	>20	8	3	< −45	Side pump coupler	[21]
	12	>11	<7.8	3.5	< −39	EYDF	[28]
	32	>17	<6.5	<2	–	EYDF, side coupler, integrated isolator	[24]
Hybrid	7	>20	<10	<2	–	EYDF, hybrid pumping	[31]

per-channel basis. The benefits from a cost-per-bit perspective were thus not entirely self-evident for this approach and can only really be realized by sharing of the expensive constituent components such as the pump-diodes, isolators, filters, electronic, and temperature control units. Work devoted to improving the component sharing really started in a core-pumped 19-core MCF amplifier experiment [29], where only one integrated optical isolator and a single WDM coupler were used for all 19 spatial channels to suppress backward light propagation and to combine the pump laser light with the signal light. This demonstration provided a good example of component sharing amongst the cores in an SDM amplifier. However, note that 10 expensive single-mode pump diodes and current drivers were still required to pump the 19 individual cores in this experiment.

To better optimize the advantages offered by SDM technology, cladding-pumped MCF amplifiers have recently been introduced, which require the use of just one or two high-power multimode pump laser diodes to provide simultaneous amplification for all spatial cores. Although the amplifier performance is inevitably somewhat compromised (in terms of noise figure) by the reduced pump brightness, good optical performance has been reported in both 7-core [30] and 12-core MCF amplifiers [28]. Note that EYDF was employed in the 12-core MCF amplifier to enhance the pump absorption and to reduce the length of active fiber. So far, the highest core count MCF amplifier is a 32-core MCF amplifier operated in a cladding-pumped configuration [24] in which improved component sharing and significant device integration were shown, We will discuss these results in more detail in Section 7.3.3.

Table 7.2 Transmission performance of inline amplified multicore fiber (MCF) experiments with matching amplifiers.

# of cores	Span length (km)	Capacity (Tb/s)	Distance (km)	Data format	Used MCF amplifiers	References
7	45.5	140.7	7326	DP-16QAM	Core-pumped dual-stage 7c-EDFA	[32]
7	60	10.5	2520	QPSK	Two cladding-pumped 7c-EYDFAs	[33]
19	30	–	1200	PDM-QPSK	Core-pumped 19c-EDFA	[29]
32	51.4	–	1850	QPSK	Cladding-pumped 32c-EYDFA	[24]

The cladding pumping scheme provides an excellent opportunity to simplify the MCF amplifier architecture, but the further complication here is that there is less scope to control the pump power delivered to each core (e.g., as needed to tailor the gain in any particular core), and this is potentially problematic in many real-world networking scenarios. For this reason, a hybrid pumping approach has recently been introduced in a 7-core MCF amplifier [31] where the fiber is pumped using cladding pumping from one end to provide constant gain but with core pumping from the other end to achieve fine individual core gain control.

Even though many high-capacity MCF transmission experiments have been reported so far, only a few experiments have been conducted with a matching inline amplifier, as summarized in Table 7.2. The most popular 7-core MCF amplifiers were realized with both core- and cladding-pumped configuration, and long-distance MCF transmission was demonstrated/evaluated utilizing a recirculating loop configuration. For the core-pumped case, 140.7 Tbit/s super-Nyquist-WDM transmission over 7326 km fiber was achieved with a dual-stage EDFA [32], where gain-flattening filters were included between the two stages with the aid of fan-in/fan-out devices. For the cladding-pumped case, 2520 km transmission was reported with a fully integrated 7-core EYDFA with inline MCF isolators [33]. So far the highest core count MCF ever reported is a 32-core MCF amplifier in a cladding pumped configuration [24], and its operation in an amplified multicore loop system with transmission distances over 1850 km was successfully demonstrated.

7.3.3 Fully fiberized 32-core multicore fiber amplifier

Fig. 7.8 shows a schematic of a fully integrated 32-core MCF amplifier in a cladding-pumped configuration [24]. As can be seen, the MCF amplifier looks essentially just like a conventional SMF amplifier, containing input and output optical isolators (which may include optional gain shaping filtering elements, if required, in the same or a separate package), a side pump coupler (as a WDM coupler), and a pump diode. The key difference relative to a conventional EDFA is that all of the inline components are MCF fiber compatible devices and the pump diode and WDM coupler are multimode devices. In this experiment, a 6 m length of erbium/ytterbium co-doped 32-core MCF was used as the gain medium. The fiber was fabricated using the standard stack-and-draw method. For ease of fiber fabrication, a simple step-index erbium/ytterbium-doped preform was used with a large glass volume (i.e., cladding area) of the preform. A significant volume of the undoped cladding was first etched away to obtain the desired core-to-cladding diameter ratio, and the etched preform was then drawn into 32 thin rods and the rods stacked in a square arrangement to obtain a 32-core erbium/ytterbium-doped fiber preform. The resulting preform was then drawn into a fiber with an outer diameter of 243 μm, and a low refractive index acrylate polymer was applied to form a double-clad fiber structure. The core pitch of the fabricated 32-core erbium/ytterbium-doped fiber was 28.8 μm, which was well matched to that of the passive fiber to be used for data transmission. This result demonstrates the accuracy of manufacture of the high core count MCF. Such precise matching of the core pitch and geometry is critical to enable direct splicing of passive and active fibers and to realize fully fiberized multicore transmission lines. The cross sections of both the active fiber and passive fiber are shown in the inset of Fig. 7.8. Note that no cross-sectional cladding shaping is applied prior to fiber drawing (as conventionally used in active fiber fabrication to break the circular symmetry of the cladding in order to increase the pump absorption). This reduces the potential risk of inducing core pitch variation through the drawing of a noncircularly symmetric shaped preform (an important factor in MCF fabrication/splicing). Moreover, the location of the cores is not centrosymmetric, so one does not require shaping of the cladding. A side pump coupler scheme was employed in order to couple multimode pump light

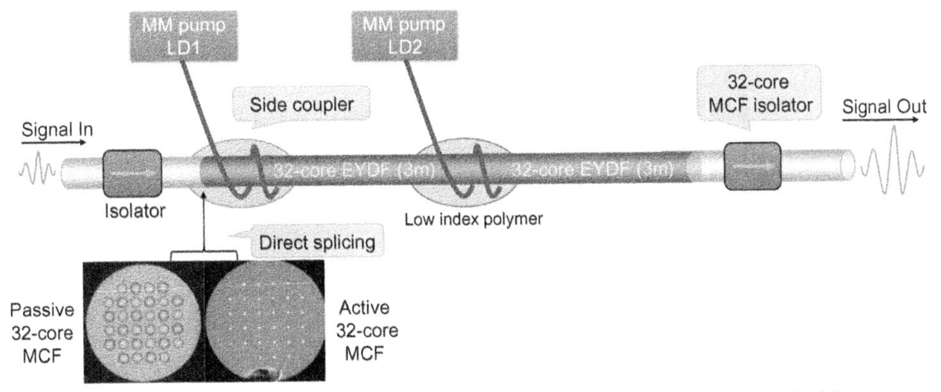

Figure 7.8 Fully integrated 32-core multicore fiber (MCF) amplifier in a cladding-pumped configuration.

into the active fiber as described in Section 7.2.2. However, the pump light was depleted in a very short length of fiber (~6 m) due to the high pump absorption originating from the large number of cores in the fiber cross section, resulting in nonuniform population inversion along the length of the amplifier. Also, the pump power of a single pump diode could not be increased beyond 15–20 W due to the damage threshold of the pump couplers being used. To address these issues, two pump couplers were employed as shown in the schematic setup in Fig. 7.8, one at the beginning of the active 32-core MCF and the other at the center of the active fiber (both in the forward pumping configuration) in order to improve the overall population inversion and also to increase the total pump power coupled into the active fiber. The matching passive transmission fiber was directly spliced to the active fiber, and the average splice loss was estimated to be about 1.3 dB due to the unavoidable mode field diameter mismatch between the two fibers. Note that the fully integrated 32-core MCF isolators [18] are placed at both input and output ends of the amplifier to suppress parasitic lasing and to avoid any unwanted feedback into the amplifier.

To characterize the MCF amplifier, fan-in/fan-out devices were spliced to the input and output of the fabricated 32-core MCF amplifier. Fig. 7.9 shows the gain and noise figure performance of the amplifier at an input signal power of −4 dBm. A minimum gain of >17 dB and an average noise figure of 6.5 dB was measured over all cores in the wavelength range 1534–1561 nm. The core-to-core variation for both amplifier gain and noise figure was measured to be less than 2 dB. This variation is mainly due to the core-to-core insertion loss variation in the passive MCFs, optical isolators and fiber splice imperfections. This amplifier was tested in an MCF recirculating loop experiment, and transmission over distances >1850 km was successfully demonstrated for 100 Gbit/s QPSK signals.

Another important aspect of any optical amplifier is the gain dynamics behavior. The gain dynamics of the previously described cladding pumped 32-core MCF-EDFA

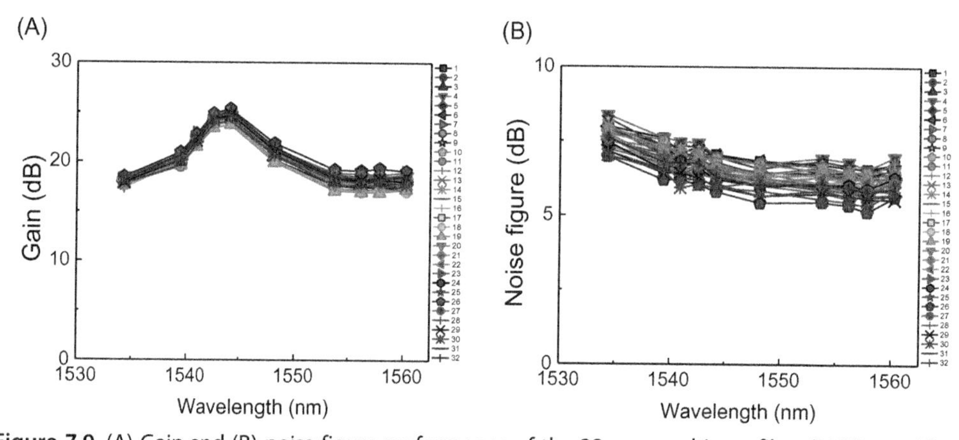

Figure 7.9 (A) Gain and (B) noise figure performance of the 32-core multicore fiber (MCF) amplifier.

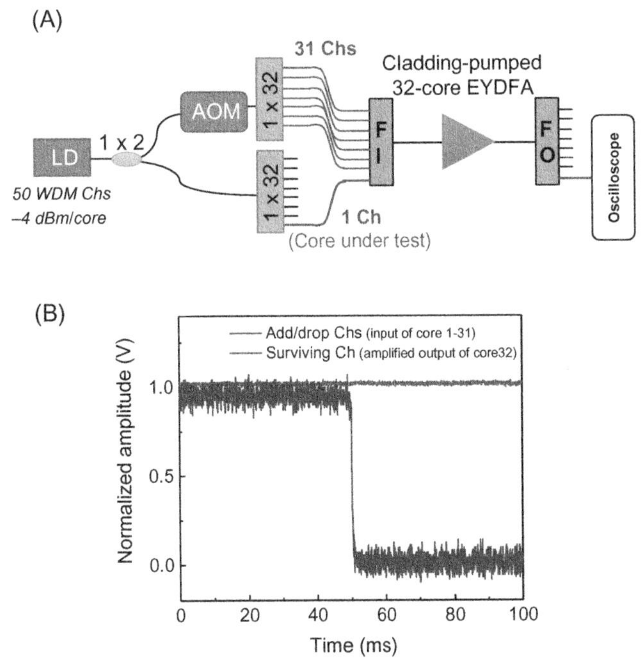

Figure 7.10 (A) Schematic setup and (B) measured transient performance of the cladding-pumped 32-core multicore fiber (MCF) amplifier.

was investigated under different channel loading conditions. To emulate the channel add/drop in the neighboring cores, 50 WDM channels were preamplified and split into two sets. As shown in Fig. 7.10A, one set was for add/drop spatial channels and the other for surviving (or monitoring) channels. An acousto-optic modulator (AOM) was placed before the splitter to simultaneously switch on/off signals in these cores, and the amplified output of the core under test was measured using a photodiode and an oscilloscope. Fig. 7.10B shows the gain dynamics for the worst scenario when the remaining 31 spatial cores are simultaneously turned off. No visible effect was observed on the core under test, and the output power remained constant. This observation indicates that the amplifier's cores are truly independent of each other and that the output power depends only on the injected pump power. Obviously, intracore cross wavelength-dependent gain, just as for any conventional single-mode EDFAs operating without active gain control, was observed as expected.

7.4 Multimode fiber amplifiers

7.4.1 Design concept of multimode fiber amplifiers

Mode-dependent loss or gain is a critical factor in determining the overall performance of an amplified MMF transmission line, and consequently much research effort has been devoted to minimizing the mode-dependent gain (MDG) of MMF amplifiers. In principle, the modal gain of a MMF amplifier is associated with the overlap integral between the signal and pump mode profiles and the rare earth dopant distribution. Consequently, different spatial modes of the input signal will experience different mode overlaps with the launched pump profile, and dopant distribution of the core within the amplifier and consequently MMF amplifiers frequently show rather large MDGs. There are several important considerations/approaches to minimizing MDG, as illustrated in Fig. 7.11. These can largely be divided into three basic strategies: (1) dopant distribution control within the active fiber, (2) pump mode profile control by manipulating the pump field distribution, and (3) signal mode profile control by changing the refractive index profile of the core.

All three approaches have been actively investigated over the years, but the first two schemes are the most widely used and will be briefly introduced in this section. The first approach is to control the dopant distribution of the active fiber. Various interesting fiber doping structures have so far been introduced, and a few examples of demonstrations of MDG reduction in MMF amplifiers are summarized in Fig. 7.12. The simplest fiber design for an MMF amplifier is to use a uniformly doped step-index profile as shown in Fig. 7.12A. In this fiber structure, the higher order modes exhibit a longer extension of the evanescent field (or tail) into the cladding region as compared to lower order modes, resulting in less gain for the higher order modes. In 2011 the first experimental demonstration of the MMF amplifier [34] used a simple uniformly doped step-index fiber, and a very large MDG ($>5-10$ dB) was observed. Therefore the authors proposed and fabricated a fiber with a core exhibiting a "raised-edge core refractive index profile, in which the erbium ion concentration was raised at the edges of the fiber core relative to the center to help mitigate MDG. After that, a ring doped fiber design (depicted in Fig. 7.12B) [35,36] was introduced to further address this issue, and a good MDG was achieved in both 3-moded [37] and 6-moded MMF amplifiers [38]. In this design, the rare earth ions are locally confined in a ring

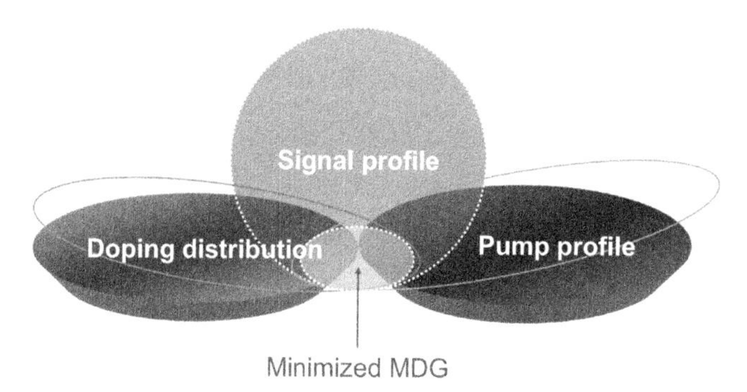

Figure 7.11 The three key physical strategies to minimize the mode-dependent gain in MMF amplifiers.

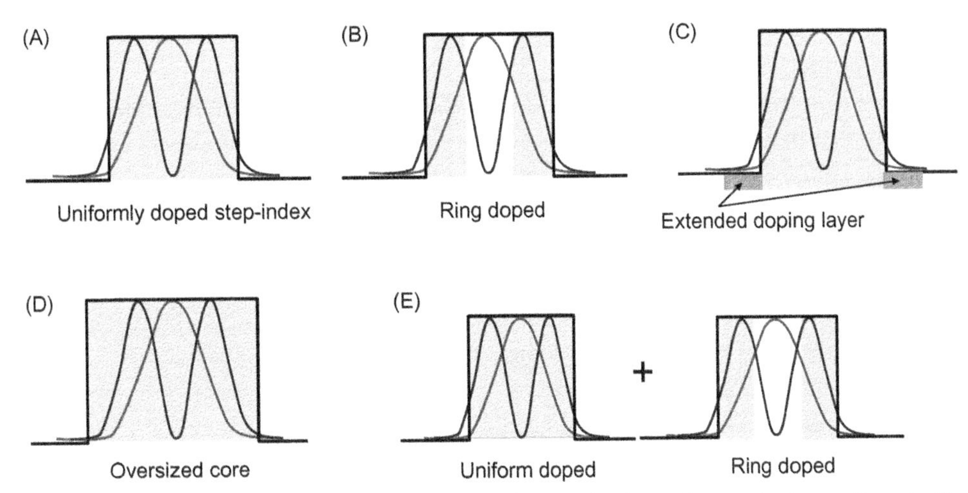

Figure 7.12 Various fiber designs for reducing the mode-dependent gain in multimode fiber amplifiers: (A) uniformly doped step-index fiber, (B) ring-doped fiber, (C) extended doping layer structure, (D) oversized core, and (E) cascaded erbium-doped fibers (EDFs).

inside the fiber core, and accurate modal gain control can be achieved by carefully tuning the thickness of the ring-doped layer. A multiple ring doping architecture may be considered for gain equalization for even larger mode count MMF amplifiers; however, this certainly increases the complexity of the fiber design and fabrication. Current active fiber fabrication processes are typically based on the conventional modified chemical vapor deposition technique coupled with solution doping. However, the solution doping method requires a simple step-index like profile, and only a few layers of silica can be deposited with rare earth ions at a time. A multiple layer solution doping technique was proposed in the literature [39], but it is neither realistic nor economical at this moment. The gas phase deposition technique [40,41] offers the potential to realize complex multilayer profiles but is not well established yet. A relatively simple fiber doping design was recently demonstrated and an extended doping layer structure was introduced into the cladding region [42–44] as illustrated in Fig. 7.12C. With this extended doping layer, the overlap difference between the rare earth ions and the different signal modes can be reduced and the MDG can be improved. Note that this approach can be useful for cladding-pumped amplifier configurations but that the low population inversion may be an issue for a core-pumped system. With this approach, a cladding-pumped 6-mode MMF amplifier was recently demonstrated with less than 3.3 dB MDG [42]. Another interesting fiber approach is the use of an oversized core in an MMF amplifier. When the fiber core dimensions are intentionally oversized to support a greater number of lower-order spatial modes than desired for transmission, the desired subset of supported modes can be better confined inside the core and hence to provide reduced MDG. Here the oversized active fiber is directly spliced to passive MMFs supporting the desired number of spatial modes at both the input and output ends of the amplifier. This has the effect of minimizing the excitation of unwanted higher order modes in the oversized amplifier core at the amplifier input and mode selective or filtering of any unintentional higher order mode excitation at its output. A 10-mode fiber EDFA [45] has recently been demonstrated with an MDG <2 dB exploiting this approach. Here the fiber core was intentionally increased to support 28 spatial modes, but only the first 10 modes were to be used for transmission—all other higher order modes were efficiently suppressed by more than 20 dB. The final fiber design example relates to a cascaded active fiber configuration [46,47], where a combination of multiple EDFs with different doping profiles are spliced to each other to further reduce the total accumulated MDG along the composite fiber length. For example, a 1 m length of uniformly doped step-index EDF was serially cascaded with a 3.5 m length of ring-doped EDF in a 6-mode fiber EDFA and an MDG of ~ 4 dB realized [46].

The second key approach to minimize the MDG of MMF amplifiers is to tailor the shape of the spatial mode(s) of the pump beams. Lower order signal modes generally have a better overlap with lower order pump modes, and vice versa. Consequently, lower order signal modes typically experience higher gain than higher order signal modes when lower order pump modes are applied into the system. As an example, for a uniformly doped step-index EDF supporting 3 spatial modes (i.e., LP_{01}, LP_{11a}, and LP_{11b}), the gain of the LP_{01} signal is much higher than that of LP_{11} when an LP_{01} pump mode is used in the amplifier. However, the gain difference can be significantly reduced by employing an LP_{11} pump mode, and the gain of the LP_{11} signal can be even higher than LP_{01} when a LP_{21} pump mode is used. Detailed simulations and an experimental demonstration are well described in Refs. [48] and [38], respectively. In practice, once the active fiber is fabricated, fiber dopant distribution control is no longer possible and controlling the pump mode distribution is the only option to address the MDG in core-pumped MMF amplifiers. Generally speaking, pump mode profiling can be achieved either by launching the pump laser into the active fiber with a core offset relative to the fiber axis or by using phase plate based mode conversion. However, the latter case results in higher coupling losses for the pump light (e.g., a ~ 3.5 dB coupling loss was observed for LP_{21} pump mode excitation in a recent demonstration [38]). In the future, it is anticipated that inline mode converters such as long period fiber gratings or fiber Bragg gratings could be employed to reduce such losses. Another important issue for MMF amplifiers is the increased pump power requirements as the number of spatial modes is increased. Core-pumped 3-mode or 6-mode EDFAs typically employ one or two single-mode pump laser diodes having a maximum output power of 750 mW (the largest laser output power currently available on the market). Higher mode count MMF amplifiers require the use of even greater numbers of expensive high-power single-mode pump diodes. For this reason, the cladding-pumped architecture has recently attracted significant attention for the development of MMF amplifiers. Cladding pumping allows the use of low-cost, high-power multimode pump laser diodes and can easily satisfy the high pump power requirements. Cladding-pumped MMF amplifiers can significantly reduce the complexity and can amplify many spatial channels simultaneously in a single device. Recent high mode count MMF amplifiers (e.g., 6-mode and 10-mode amplifiers) [45,49,50] have been realized in a cladding-pumped configuration, offering a convenient route toward fully integrated MMF amplifiers by adopting the side-coupling approach described previously. Here we have discussed two main approaches (doping distribution control and pump profiling) to minimize the MDG of the MMF amplifiers, but refractive index profiling can also be applied and new types of SDM fiber structures (e.g., ring-core fibers [51–53] and orbital angular momentum fibers [54–57]) have recently been introduced and exhibit inherent advantages of low MDG in its associated amplifiers due to the similar intensity profiles of the spatial modes within the fiber. OAM modes in fibers have the additional advantage of enabling MIMO-free data transmission [58], hence offering the potential of realizing SDM amplifiers that are compatible with both MIMO-DSP based systems as well as conventional transmission methodologies that do not require MIMO. While OAM amplification has been demonstrated only for two modes to date, with passive fibers, MIMO-free data transmission has been achieved for as many as 12 modes over km-length fibers [59], and mode-coupling-free light transmission has been achieved for 24 modes over 10-m lengths [60]. This provides confidence for the prospects of future developments in MIMO-free SDM amplifiers of high mode counts.

7.4.2 Recent progress in multimode fiber amplifiers

Table 7.3 summarizes the optical characteristics of recently reported multimode fiber amplifiers. The first MMF amplifier simulation results were reported in 2011 using a step-index EDF in conjunction with a reconfigurable pumping configuration incorporating the pump beam in a higher order mode, in an attempt to reduce the MDG [48]. In the same year, the first experimental demonstration of a 3-mode EDFA was reported using a raised-edge refractive index profile to obtain low MDG, high gain, and good broadband performance [34]. Beyond this, improved performance for a 3-mode EDFA was achieved by moving to fibers with a localized ring-doped structure to obtain even lower MDG (<1 dB) for simplified pumping configurations (e.g., using a fundamental mode pump beam only) [35,37]. Similar concepts have been extended to higher mode count amplifiers. For example, a 6-mode EDFA has been demonstrated and a <2 dB MDG has been achieved using a ring-doped concept combined with bidirectional higher order mode pumping [38]. More recently a core-pumped 10-mode EDFA was reported using combined lengths of both a step-index and a ring-doped fiber to obtain <3.5 dB MDG for all 10 spatial modes [47]. However, as the number of spatial modes

Table 7.3 Characteristics of recently reported multimode fiber amplifiers.

Pumping scheme	# of modes	Gain (dB)	MDG (dB)	NF (dB)	Technical approach	References
Core	3	>20	<2	–	Ring-doped EDF, LP_{11} pump mode	[37]
	6	>20	<2.5	–	Ring-doped EDF, LP_{21} pump mode	[38]
	6	20	<4	<7	Cascaded EDFs	[46]
	10	17	<3.5	<7	Cascaded EDFs, LP_{11} pump mode	[47]
Cladding	6	<20	<3	<7	Ring-doped EDF	[49]
	10	20	<2	<6	Oversized core	[45]

is scaled up beyond 10, it becomes increasingly challenging to meet the associated pump power requirements with single-mode pump laser diodes, and a cladding-pumped amplifier configuration needs to be adopted. First, a cladding pumped 6-mode EDFA [49] was introduced in conjunction with a ring-doped core and reasonable amplifier performance was demonstrated (gain >20 dB and MDG <3 dB). Importantly this amplifier was further upgraded into a fully integrated format with the aid of a side pump coupler and inline MMF isolators, as will be discussed in more detail in Section 7.4.3. Recently a cladding-pumped 10-mode EDFA [45] was reported using an oversized core approach and <2 dB MDG was successfully achieved.

Numerous outstanding high-capacity long-haul multimode fiber transmission experiments have been reported incorporating inline MMF amplifiers and are summarized in Table 7.4. In 2013, a record transmission capacity of 73.7 Tbit/s was demonstrated over 115 km of 3-mode fiber with a core-pumped 3-mode, C-band EDFA. After that, several long distance (>1000 km) 3-mode fiber transmission experiments using an MMF amplifier were demonstrated in a recirculating loop [61,62]. These experiments highlighted the importance of achieving low mode-dependent loss/gain per-span across broad bandwidths. Ideally, such amplifiers would include means for active control of MDG and GFFs/isolators to enhance overall system flexibility and performance. Moreover, in order to realize the expected cost benefits of SDM technology, a fully fiberized 6-mode EDFA was realized in a cladding pumped configuration in 2015. Using these two 6-mode EDFAs, all fiberized 6-mode fiber transmission over a fully integrated transmission line was successfully demonstrated over 179 km [50].

7.4.3 Fully integrated 6-mode erbium doped fiber amplifier

As an example of a fully integrated multimode fiber amplifier, a cladding-pumped 6-moded EDFA is introduced in this section. As shown in Fig. 7.13A, the basic

Table 7.4 Transmission performance of inline amplified multimode fiber experiments with multimode fiber amplifiers.

# of modes	Span length (km)	Capacity (Tb/s)	Distance (km)	Data format	FM-EDFA type used	References
3	119	73.7	119	DP-16QAM	Core-pumped 3-mode EDFA	[63]
3	60	—	1000	DP-QPSK	Core-pumped 3-mode EDFA	[62]
3	50	3.04	1000	QPSK	Core-pumped 3-mode EDFA	[61]
6	179	72	179	PDM-QPSK	Two cladding-pumped 6-mode EDFAs	[50]

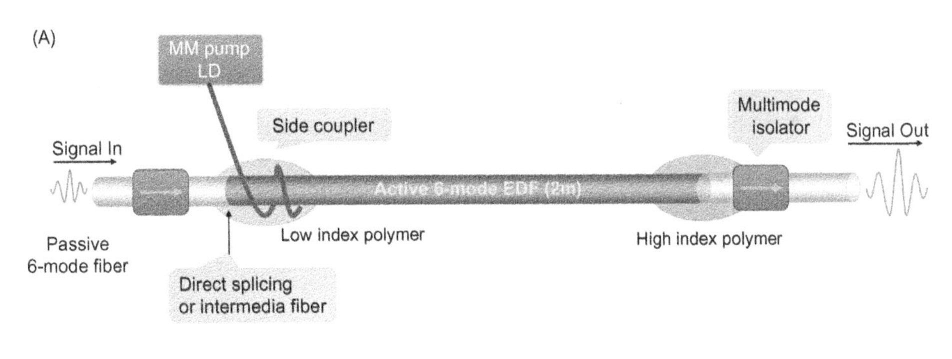

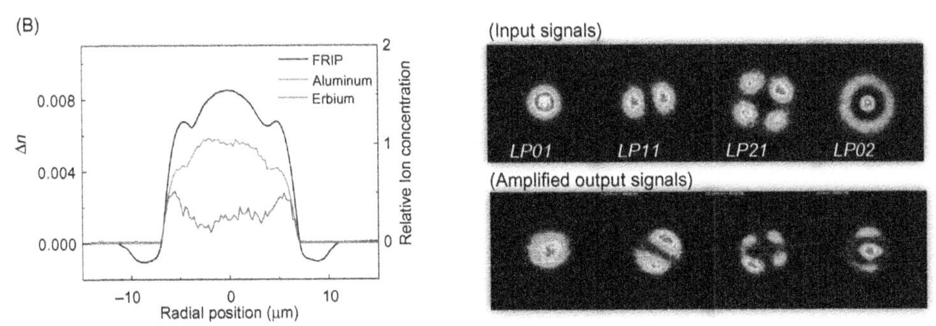

Figure 7.13 (A) Schematic of a fully integrated cladding-pumped 6-mode erbium doped fiber amplifier (EDFA), (B) fiber refractive index profile (FRIP) and doping distribution of the active fiber, and (C) measured input/output mode profiles of the amplifier.

schematic of the amplifier is similar to that the cladding-pumped MCF amplifier illustrated in Fig. 7.8, except that both the passive and active fibers are now MMFs. The multimode pump module could deliver an optical power of up to ~10 W and was wavelength-stabilized at 976 nm with a volume Bragg grating. Using a side pump coupler approach, the pump light was effectively coupled into the inner cladding of the active fiber with a coupling efficiency of more than 70%. To reduce the MDG of the MMF amplifier, a ring-doped EDF design was employed with an appropriate ratio of ring thickness to core radius (~0.48). The designed fiber was fabricated using an in-house modified chemical vapor deposition process coupled with solution doping. Fig. 7.13B shows the measured refractive index profile and aluminum/erbium dopant distribution of the fabricated fiber. Using secondary ion mass spectroscopy, the dopant distribution was measured across the fiber preform core, showing that the refractive index profile closely follows the doping concentration profile of the aluminum. However, note that although a ring-doped EDF was targeted, the fabricated fiber was not a complete ring shape, as some level of erbium diffusion was observed. The degree of the dopant diffusion usually depends on several factors such as the composition of the glass host, doping concentration, and temperature of the fiber fabrication process. Importantly, note that fine radial control of the doping distribution is not readily obtained using the current state-of-the-art solution doping technique and further careful control of the doping process and/or other fabrication approaches (e.g., gas phase deposition) should be explored in the future. For efficient pump light absorption, the fabricated preform was made with a D-shaped cladding and was drawn into fiber incorporating a low-index acrylate polymer coating to guide the pump light. The passive 6-mode transmission fiber was spliced directly to a 2 m length of the ring-doped EDF; however, a significant mode-dependent splice loss was observed (0.6 dB for the LP_{01}, 1.0 dB for the LP_{11}, 1.2 dB for the LP_{21}, and 1.3 dB for the LP_{02} mode, respectively). This variation is mainly due to the mode field diameter mismatch between the passive fiber and active EDF. Recently a graded index fiber-based mode field diameter adaptor [23,64] has been introduced and the mode-dependent splice loss was significantly reduced. Actually this technique was previously introduced to accommodate the mode field diameter mismatch between two dissimilar SMFs [65], but the same technology can be employed for low-loss optical interconnection between dissimilar MMFs. By splicing an appropriate length of graded index fiber between the passive and active MMFs, a compact all-fiber mode field diameter adaptor can easily be realized using only a simple cleaving and fusion splicing procedure. In this experiment, a 250 μm length of graded index fiber was used and the splice loss was reduced considerably to less than 0.4 dB for all spatial modes. Two integrated MMF optical isolators were placed at both ends of the amplifier and mode multiplexer/demultiplexers were connected to test the developed MMF amplifier. To confirm clean amplification of the input signals, mode images of the various input and output signals to and from

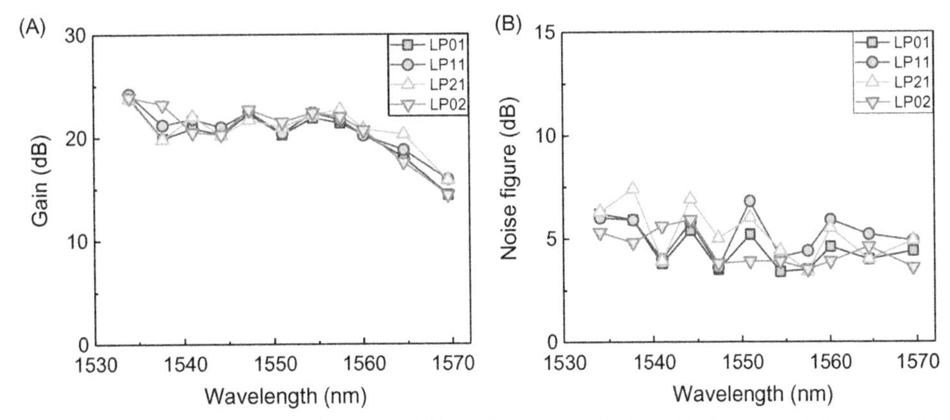

Figure 7.14 (A) Measured modal gain and (B) noise figure of the fully integrated 6-mode erbium doped fiber amplifier (EDFA).

the amplifier were taken using a charge coupled device. As shown in Fig. 7.13C, all spatial mode profiles are well preserved before and after amplification.

Fig. 7.14 shows the measured gain and noise figure of the fully integrated 6-mode EDFA. The amplifier covers the majority of the C-band with an average modal gain of >20 dB and a MDG of ~3 dB. The average noise figure was measured to be between 6 and 7 dB for all guided modes. Note that the developed MMF amplifier is a fully fiberized module having a side pump coupler and inline optical isolators without any bulky free-space components. Also, these integrated amplifiers were tested in a recirculating loop experiment, and inline amplified 6 spatial mode transmission was successfully demonstrated over a record span length of 179 km in 2015 [50].

Similar to the MCF amplifiers, the gain dynamics of an MMF amplifier was also tested in a 3-mode EDFA to investigate both power excursions due to cross-gain modulation and transient times for different spatial modes. The experimental setup for the investigation of the transient effects of the amplifier is shown in Fig. 7.15A. An acousto-optic modulator (AOM) was placed in one of the LP_{01} lines (Ch1) of the mode multiplexer and was used to generate a square wave with 50% duty cycle in order to simulate the impact of adding/dropping spatial channels (6 WDM channels were used to see the difference clearly). Two surviving spatial channels (Ch2 for the LP_{01} mode and Ch3 for the LP_{11} mode) were monitored to investigate the MDG transient effects. Fig. 7.15B shows the normalized input/output transient response when the LP_{01} spatial mode signal is modulated. All spatial channels (both Ch2 and Ch3) experience significant signal power excursions as a result of cross-gain saturation in the MMF amplifier. Note that the spatial modes experience a broadly similar response under a range of add/drop conditions, but some evidence of mode-dependent sensitivity was observed (~0.8 dB). Unlike the case of uncoupled MCF amplifiers, the spatial modes of a MMF amplifier have large field overlaps with each other and the intermodal gain dynamics becomes a critical issue and will need to be carefully addressed in any practical MMF transmission network.

7.5 Multimode multicore fiber amplifiers

To achieve even higher spatial multiplicity in a single SDM fiber, the two major SDM fiber concepts (i.e., MCFs and MMFs) can be merged, and >100 spatial channel transmission [16,17] has already been demonstrated over relatively short distances (~10 km). To explore long-distance transmission, matching amplifiers (i.e., multimode MCF) need to be developed. Recently, there have been two reports of such multimode MCFs, as summarized in Table 7.5. The first demonstration was of a 3-mode 6-core fiber amplifier supporting 18 spatial channels in a cladding-pumped configuration. In this fiber design, an annular fiber cladding was employed to reduce the core-to-cladding area ratio and to enhance the pump absorption. To achieve this, the refractive index of the inner cladding was depressed using fluorine-doped silica. More than 20 dB gain was reported, with an MDG of <3 dB. The second demonstration was of a 3-mode 7-core EDFA supporting 21 spatial channels in a core-pumped configuration. To reduce the MDG within the three spatial mode supporting core, a ring-core EDF (i.e., ring shape in both refractive index and doping distribution) was employed and an LP_{21} pump mode was used. The measured average modal gain was

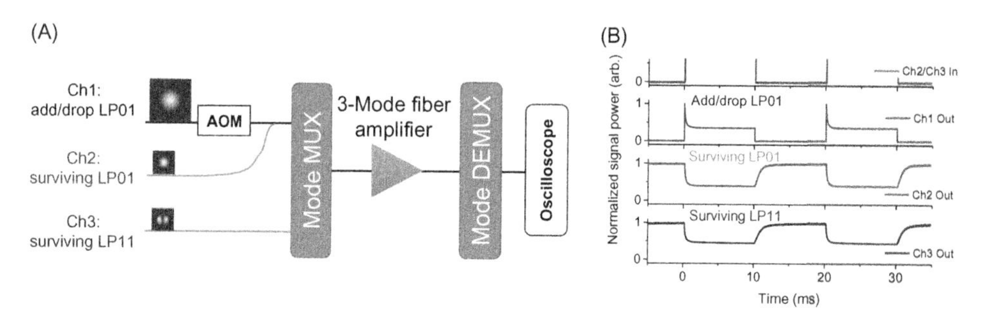

Figure 7.15 (A) Experimental setup and (B) transient gain response of the multimode fiber (MMF) amplifier.

Table 7.5 Characteristics of recently reported multimode multicore fiber amplifiers.

Pumping scheme	# of Chs	Gain (dB)	MDG (dB)	NF (dB)	XT (dB)	Special approach	References
Cladding	18 (3M × 6C)	>20	<3	<7	–	Annular cladding	[66]
Core	21 (3M × 7C)	>20	<3.2	<7	−49	Ring core fiber, LP_{21} pump	[67]

more than 20 dB with an MDG <3.2 dB. Note that the channel multiplicities achieved have so far been relatively modest, and substantial improvements in both multiplicity and performance should be possible in the near future.

7.6 Future prospects

7.6.1 Current key issues and challenges of space division multiplexing amplifiers

- *Further development of a wide range of SDM fiber components*: Using the micro-optic platform, a number of functional SDM fiber components have been recently developed; however, these are still not enough to fulfill the requirements needed to realize a properly engineered SDM fiber network. Also, the earlier device development has been mainly focused on two port devices having a single input and a single output; however, further extension to 3-port or 4-port devices is needed, for example, SDM fiber-compatible circulators, 1 × 2 or 2 × 2 fiber-optic couplers, WDM filters, and optical switches. Importantly, to develop a wide range of SDM components, various fiber-optic component fabrication technologies should be explored/tested to process the SDM fibers and to produce functional devices, such as fiber fusion tapering, fiber gratings, and three-dimensional femtosecond direct laser writing techniques.
- *Independent spatial channel control within the amplifier*: Fully fiberized SDM amplifiers have recently been realized with the aid of the cladding-pumped configuration and inline optical components. This development is a huge technical leap, illustrating the substantial potential benefits of SDM technology in reducing network complexity, cost, space, and energy consumption. However, although the considerable advantages and good performance of cladding-pumped amplifiers have been shown, the practical complications of reducing the scope to control the pump power delivered to each core (or mode) in real-world networking scenarios have not been considered. Independent control per spatial channel is a very important and likely essential functionality for practical SDM amplifiers, and future research efforts should be devoted in this direction. Hybrid pumping schemes may provide one solution; however, the possibility of dynamic control of mode (or core) dependent loss or low loss variable optical attenuator development may be the easiest and most cost-effective option to address this issue.
- *Further increase in the number of spatial channels in SDM amplifiers*: The critical next steps to scale the integrated amplifier approach to higher spatial channel counts are obviously to incorporate more spatial cores/modes in the SDM fiber cross section (whilst maintaining a suitable geometry match to the associated transmission fiber to enable low loss splicing) while at the same time managing distributed crosstalk along the fiber length and ensuring identical gain properties for each spatial

channel. The current highest spatial multiplicity of SDM amplifiers is 32 for MCF amplifiers, 10 for MMF amplifiers, and 21 for multimode MCF amplifiers. The spatial multiplicities achieved in the case of MMF amplifiers and multimode MCF amplifiers are still relatively modest, and substantial improvements in both multiplicity and performance should be possible and are worthy of further study.

7.6.2 Potential applications of space division multiplexing amplifier technology

- *Passive optical networks*: Passive optical networks are based on a point-to-multipoint architecture, in which optical line terminals (OLTs) distribute signals to a large number of end user terminals. As the number of new fiber-to-home subscribers grows across the world, the large number of individual optical network units connected to each OLT has implications for both the cost and volume of amplifiers included in the OLTs. SDM amplifier topologies can offer efficiency savings in terms of space occupied, complexity, and number of components as compared to their conventional single-mode counterparts, and these benefits are directly relevant to passive optical network systems. An SDM amplifier with N cores will support N times more optical network units than its single-mode counterpart, thus offering immediate cost benefits. Cladding-pumped large core count multicore fiber amplifiers could provide substantial economic benefits in such a context.
- *Optical fiber lasers*: SDM amplifiers have multiple independent amplifier channels, and this can also be a very useful feature for developing new types of fiber laser having multiple laser outputs or generating multiwavelengths. For example, a multi-wavelength fiber laser can be produced using a MCF amplifier module, in which each laser wavelength is generated and guided along an individual core within the MCF. An exemplary 7-wavelength fiber laser [68] has recently been constructed using a 7-core erbium-doped fiber, and stable multiwavelength operation was confirmed without any significant transverse crosstalk. Spatially multiplexed laser operation has also been demonstrated in 7-core EDF by inscribing a distributed feedback structure in it [69]. Note that it is difficult to achieve multiwavelength operation within a conventional single-core erbium doped fiber due to the strong wavelength competition caused by the homogenous gain broadening of the erbium-doped fiber. Note MCF amplifiers can also support multiple laser beams emitted from the individual cores that propagate in different angular directions, which may be useful for multiparameter optical sensors or light detection and ranging applications. Current multicore SDM fiber amplifiers can easily be power scaled to produce high-power, ultrashort pulse fiber lasers. Furthermore, MCFs contain multiple cores inside a single optical fiber, and hence they can inherently provide a very similar mechanical and thermal environment for multiple laser beams. The individual cores of the MCF act as independent, parallel gain

media and can provide nearly identical optical path lengths, laser performance, and group/phase velocity for all spatial cores. Therefore the amount of phase change between multiple laser beams within a MCF can be significantly lower than that of the equivalent number of parallel SMFs, and hence MCF amplifiers provide the potential to reduce the complexity of the dynamic phase control required for coherent beam combination. Also, SDM amplifiers can allow common pump and phase modulators such that more compact and cost-effective coherent laser beam combination systems can be implemented.

- *Optical fiber sensors*: SDM fibers have multiple spatial channels (i.e., modes or cores) within a single optical fiber and hence can be used not only as parallel optical waveguides for light transmission/detection but also for independent sensing arms or elements [70,71]. Therefore multiple sensing functionalities can be incorporated within a single fiber, and this can open new possibilities for compact, fully integrated, long-length optical fiber sensors. For example, one core can be used for optical coherence tomography, another core for fluorescence detection. another for temperature sensing or light illumination, and so on. Therefore multiparameter sensing functionality can easily be realized in MCFs with a very small form factor. Also, multiple discrete or distributed fiber Bragg gratings can be inscribed along the length of a MCF, and the local state of strain in each core can be measured and used for curvature or three-dimensional shape sensing [72–74]. Recently there has been increased interest in such fiber-optic shape sensing to address potential applications in medical robotics and aerospace safety. In the case of MMFs, the different temperature and stain coefficients of the individual transverse modes of a MMF have been explored via Brillouin scattering. A long-distance multiparameter distributed sensor was successfully demonstrated with high discrimination accuracy [75–77].

7.7 Conclusions

Over the last few years, remarkable technological advances have been made in the development of SDM components and SDM amplifiers for both multicore and MMF designs, and their combination. Importantly, SDM amplifiers have demonstrated excellent gain/noise figure performance per spatial channel comparable to those of conventional single-mode EDFAs. Moreover, these SDM amplifiers have recently been integrated into a fully fiberized format by employing inline SDM components and cladding-pumped configurations, and portable prototype amplifiers have been constructed to allow a wide range of device, transmission, and system tests. However, although very good performance has been shown, both from a device and system perspective, the impact of the associated loss of independent control per spatial channel has yet to be properly assessed and the envisaged cost and power savings have yet to be properly quantified. This needs to be a major focus of research in the coming years.

References

[1] Cisco Systems, 2017. Cisco Visual Networking Index: Global Mobile Data Traffic Forecast Update, 2016–2021 White Paper, 2017.
[2] G. Fagas, J.P. Gallagher, L. Gammaitoni, D.J. Paul, Energy challenges for ICT, ICT— Energy Concepts Energy Efficiency and Sustainability, InTech, 2017.
[3] R.-J. Essiambre, G. Kramer, P.J. Winzer, G.J. Foschini, B. Goebel, Capacity limits of optical fiber networks, J. Lightwave Technol. 28 (2010) 662–701.
[4] A.D. Ellis, J. Zhao, D. Cotter, Approaching the non-linear Shannon limit, J. Lightwave Technol. 28 (2010) 423–433.
[5] D.J. Richardson, J.M. Fini, L.E. Nelson, Space-division multiplexing in optical fibres, Nat. Photonics 7 (2013) 354–362.
[6] P.J. Winzer, Making spatial multiplexing a reality, Nat. Photonics 8 (2014) 345–348.
[7] G. Li, N. Bai, N. Zhao, C. Xia, Space-division multiplexing: the next frontier in optical communication, Adv. Opt. Photonics 6 (2014) 413–487.
[8] R. Ryf, S. Randel, A.H. Gnauck, C. Bolle, A. Sierra, S. Mumtaz, et al., Mode-division multiplexing over 96 km of few-mode fiber using coherent 6 × 6 MIMO processing, J. Lightwave Technol. 30 (2012) 521–531.
[9] K. Saitoh, S. Matsuo, Multicore fiber technology, J. Lightwave Technol. 34 (2016) 55–66.
[10] T. Hayashi, T. Taru, O. Shimakawa, T. Sasaki, E. Sasaoka, Design and fabrication of ultra-low crosstalk and low-loss multi-core fiber, Opt. Express 19 (2011) 16576.
[11] S. Matsuo, K. Takenaga, Y. Sasaki, Y. Amma, S. Saito, K. Saitoh, et al., High-spatial-multiplicity multicore fibers for future dense space-division-multiplexing systems, J. Lightwave Technol. 34 (2016) 1464–1475.
[12] T. Mizuno, H. Takara, A. Sano, Y. Miyamoto, Dense space-division multiplexed transmission, J. Lightwave Technol. 34 (2016) 582–592.
[13] P. Sillard, Next-generation fibers for space-division-multiplexed transmissions, J. Lightwave Technol. 33 (2015) 1092–1099.
[14] L. Grüner-Nielsen, Y. Sun, J.W. Nicholson, D. Jakobsen, K.G. Jespersen, R. Lingle, et al., Few mode transmission fiber with low DGD, low mode coupling, and low loss, J. Lightwave Technol. 30 (2012) 3693–3698.
[15] Y. Sasaki, K. Takenaga, S. Matsuo, K. Aikawa, K. Saitoh, Few-mode multicore fibers for long-haul transmission line, Opt. Fiber Technol. 35 (2017) 19–27.
[16] K. Igarashi, D. Souma, Y. Wakayama, K. Takeshima, Y. Kawaguchi, T. Tsuritani, et al., 114 Space-division-multiplexed transmission over 9.8-km weakly-coupled-6-mode uncoupled-19-core fibers, in: Optical Fiber Communication Conference Post Deadline Papers OSA, Washington, D. C., 2015, paper. Th5C.4.
[17] J. Sakaguchi, W. Klaus, J.M. Delgado Mendinueta, B.J. Puttnam, R.S. Luis, Y. Awaji, et al., Large spatial channel (36-core × 3 mode) heterogeneous few-mode multicore fiber, J. Lightwave Technol. 34 (2016) 93–103.
[18] Y. Jung, S. Alam, Y. Sasaki, D.J. Richardson, Compact 32-core multicore fibre isolator for high-density spatial division multiplexed transmission, in: ECOC, VDE, Dusseldorf, Germany, 2016, p. W2.B4. <http://ieeexplore.ieee.org/document/7767645/> (accessed 22.09.16).
[19] Jung, Y., Alam, S., Richardson, D.J., Compact few-mode fiber collimator and associated optical components for mode division multiplexed transmission, in: Optical Fiber Communication Conference, 2016, Paper W2A.40.
[20] M.N. Zervas, C.A. Codemard, High power fiber lasers: a review, IEEE J. Sel. Top. Quantum Electron. 20 (2014) 219–241.
[21] K.S. Abedin, J.M. Fini, T.F. Thierry, B. Zhu, M.F. Yan, L. Bansal, et al., Seven-core erbium-doped double-clad fiber amplifier pumped simultaneously by side-coupled multimode fiber, Opt. Lett. 39 (2014) 993.
[22] K.S. Abedin, J.M. Fini, T.F. Thierry, V.R. Supradeepa, B. Zhu, M.F. Yan, et al., Multicore erbium doped fiber amplifiers for space division multiplexing systems, J. Lightwave Technol. 32 (2014) 2800–2808.

[23] Y. Jung, J. Hayes, Y. Sasaki, K. Aikawa, S. Alam, D.J. Richardson, All-fiber optical interconnection for dissimilar multicore fibers with low insertion loss, in: Optical Fiber Communication Conference. OSA, Washington, D.C., 2017, paper W3H.2.

[24] S. Jain, C. Castro, Y. Jung, J. Hayes, R. Sandoghchi, T. Mizuno, et al., 32-core erbium/ytterbium-doped multicore fiber amplifier for next generation space-division multiplexed transmission system, Opt. Express 25 (2017) 32887.

[25] N.K. Fontaine, J.E.A. Lopez, H. Chen, R. Ryf, D. Neilson, A. Schulzgen, et al., Coupled-core optical amplifier, in: Optical Fiber Communication Conference Postdeadline Papers. OSA, Washington, D.C., 2017, paper Th5D.3.

[26] M. Wada, T. Sakamoto, T. Yamamoto, S. Aozasa, S. Nozoe, Y. Sagae, et al., Cladding pumped randomly coupled 12-core erbium-doped fiber amplifier with low mode-dependent gain, J. Lightwave Technol. 36 (2018) 1220–1225.

[27] K.S. Abedin, T.F. Taunay, M. Fishteyn, M.F. Yan, B. Zhu, J.M. Fini, et al., Amplification and noise properties of an erbium-doped multicore fiber amplifier, Opt. Express 19 (2011) 16715.

[28] H. Ono, K. Takenaga, K. Ichii, Matsuo S., Takahashi T., H. Masuda, et al., 12-Core double-clad Er/Yb-doped fiber amplifier employing free-space coupling pump/signal combiner module, in: 39th European Conference and Exhibition on Optical Communication (ECOC 2013). Institution of Engineering and Technology, pp. 588–590.

[29] J. Sakaguchi, W. Klaus, B.J. Puttnam, J.M.D. Mendinueta, Y. Awaji, N. Wada, et al., 19-Core MCF transmission system using EDFA with shared core pumping coupled via free-space optics, Opt. Express 22 (2014) 90–95.

[30] K.S. Abedin, T.F. Taunay, M. Fishteyn, D.J. DiGiovanni, V.R. Supradeepa, J.M. Fini, et al., Cladding-pumped erbium-doped multicore fiber amplifier, Opt. Express 20 (2012) 20191–20200.

[31] M. Yamada, H. Ono, T. Hosokawa, K. Ichii, Gain control in multi-core erbium/ytterbium-doped fiber amplifier with hybrid pumping, OECC/PC2016, 2016, WC1-2.

[32] K. Igarashi, T. Tsuritani, I. Morita, Y. Tsuchida, K. Maeda, M. Tadakuma, et al., Super-Nyquist-WDM transmission over 7,326-km seven-core fiber with capacity-distance product of 103 Exabit/s · km, Opt. Express 22 (2014) 1220.

[33] C. Castro, S. Jain, E. De Man, Y. Jung, J. Hayes, S. Calabro, et al., 100-Gb/s transmission over a 2520-km integrated MCF system using cladding-pumped amplifiers, IEEE Photonics Technol. Lett. 29 (2017) 1187–1190.

[34] Y. Jung, S. Alam, Z. Li, A. Dhar, D. Giles, I.P. Giles, et al., First demonstration and detailed characterization of a multimode amplifier for space division multiplexed transmission systems, Opt. Express 19 (2011) B952–B957.

[35] Q. Kang, E.-L. Lim, Y. Jung, J.K. Sahu, F. Poletti, C. Baskiotis, et al., Accurate modal gain control in a multimode erbium doped fiber amplifier incorporating ring doping and a simple LP_01 pump configuration, Opt. Express 20 (2012) 20835.

[36] E. Ip, M. Li, C. Montero, Experimental characterization of a ring-profile few-mode erbium-doped fiber amplifier enabling gain equalization, in: Optical Fiber Communication Conference. OSA, Washington, D.C., 2013, paper JTh2A.18.

[37] Y. Jung, Q. Kang, V.A.J.M. Sleiffer, B. Inan, M. Kuschnerov, V. Veljanovski, et al., Three mode Er3 + ring-doped fiber amplifier for mode-division multiplexed transmission, Opt. Express 21 (2013) 10383–10392.

[38] Y. Jung, Q. Kang, J.K. Sahu, B. Corbett, J. O'Callagham, F. Poletti, et al., Reconfigurable modal gain control of a few-mode EDFA supporting six spatial modes, IEEE Photonics Technol. Lett. 26 (2014) 1100–1103.

[39] A.S. Webb, A.J. Boyland, R.J. Standish, S. Yoo, J.K. Sahu, D.N. Payne, MCVD in-situ solution doping process for the fabrication of complex design large core rare-earth doped fibers, J. Non Cryst. Solids 356 (2010) 848–851.

[40] S. Unger, F. Lindner, C. Aichele, M. Leich, A. Schwuchow, J. Kobelke, et al., A highly efficient Yb-doped silica laser fiber prepared by gas phase doping technology, Laser Phys. 24 (2014) 035103.

[41] A.J. Boyland, A.S. Webb, M.P. Kalita, S. Yoo, C.A. Codemard, R.J. Standish, et al., Rare earth doped optical fiber fabrication using novel gas phase deposition technique, in: Conference on Lasers and Electro-Optics 2010. OSA, Washington, D.C., paper CThV7.

[42] Y. Wakayama, K. Igarashi, D. Soma, H. Taga, T. Tsuritani, Novel 6-mode fibre amplifier with large erbium-doped area for differential modal gain minimization, in: ECOC, VDE, 2016. <http://ieeexplore.ieee.org/document/7766214/> (accessed 18.07.17).
[43] D. Askarov, J.M. Kahn, Design of transmission fibers and doped fiber amplifiers for mode-division multiplexing, IEEE Photonics Technol. Lett. 24 (2012) 1945–1948.
[44] K.S. Abedin, M.F. Yan, J.M. Fini, T.F. Thierry, L.K. Bansal, B. Zhu, et al., Space division multiplexed multicore erbium-doped fiber amplifiers, J. Opt. 19 (2015) 16665–166671.
[45] N.K. Fontaine, B. Huang, Zahoora Sanjabieznaveh, H. Chen, C. Jin, B. Ercan, A. et al., Multimode optical fiber amplifier supporting over 10 spatial modes, in: Optical Fiber Communication Conference Post Deadline Papers. OSA, Washington, D.C., 2016, paper Th5A.4.
[46] M. Salsi, D. Peyrot, G. Charlet, S. Bigo, R. Ryf, N.K. Fontaine, et al., A six-mode erbium-doped fiber amplifier. In: European Conference and Exhibition on Optical Communication. OSA, Washington, D.C., 2012, paper Th.3.A.6.
[47] M. Wada, T. Sakamoto, S. Aozasa, T. Mori, T. Yamamoto, K. Nakajima, Core-pumped 10-mode EDFA with cascaded EDF configuration, in: ECOC, VDE, Dusseldorf, Germany, 2016, paper M2.A4. <http://ieeexplore.ieee.org/document/7766215/> (accessed 18.07.17).
[48] N. Bai, E. Ip, T. Wang, G. Li, Multimode fiber amplifier with tunable modal gain using a reconfigurable multimode pump, Opt. Express 19 (2011) 16601–16611.
[49] Y. Jung, E.L. Lim, Q. Kang, T.C. May-Smith, N.H.L. Wong, R. Standish, et al., Cladding pumped few-mode EDFA for mode division multiplexed transmission, Opt. Express 22 (2014) 29008.
[50] R. Ryf, N.K. Fontaine, H. Chen, A.H. Gnauck, Y. Jung, Q. Kang, et al., 72-Tb/s transmission over 179-km all-fiber 6-mode span with two cladding pumped in-line amplifiers, in: 2015 European Conference and Exhibition on Optical Communication. IEEE, 2015, pp. 1–3.
[51] Q. Kang, E. Lim, Y. Jun, X. Jin, F.P. Payne, S. Alam, et al., Gain equalization of a six-mode-group ring core multimode EDFA, in: 2014 European Conference and Exhibition on Optical Communication, IEEE, pp. 1–3.
[52] Y. Jung, Q. Kang, H. Zhou, R. Zhang, S. Chen, H. Wang, et al., Low-loss 25.3 km few-mode ring-core fiber for mode-division multiplexed transmission, J. Lightwave Technol. 35 (2017) 1363–1368.
[53] Y. Jung, Q. Kang, L. Shen, S. Chen, H. Wang, Y. Yang, et al., Few mode ring-core fibre amplifier for low differential modal gain, in: 2017 European Conference and Exhibition on Optical Communication, IEEE, 2017, pp. 1–3.
[54] Q. Kang, P. Gregg, Y. Jung, E.L. Lim, S. Alam, S. Ramachandran, et al., Amplification of 12 OAM modes in an air-core erbium doped fiber, Opt. Express 23 (2015) 28341–28348.
[55] Y. Jung, Q. Kang, R. Sidharthan, D. Ho, S. Yoo, P. Gregg, et al., Optical orbital angular momentum amplifier based on an air-hole erbium-doped fiber, J. Lightwave Technol. 35 (3) (2017) 430–436. < https://www.osapublishing.org/jlt/abstract.cfm?uri = jlt-35-3-430> (accessed 19.07.17).
[56] P. Gregg, P. Kristensen, S. Ramachandran, Conservation of orbital angular momentum in air-core optical fibers, Optica 2 (2015) 267–270.
[57] S. Ramachandran, P. Kristensen, Optical vortices in fiber, Nanophotonics 2 (2013) 455–474.
[58] N. Bozinovic, Y. Yue, Y. Ren, M. Tur, P. Kristensen, H. Huang, et al., Terabit-scale orbital angular momentum mode division multiplexing in fibers, Science 340 (2013) 1545–1548.
[59] K. Ingerslev, P. Gregg, M. Galili, F. Da Ros, H. Hu, F. Bao, et al., 12 mode, WDM, MIMO-free orbital angular momentum transmission, Opt. Express 26 (2018) 20225.
[60] P. Gregg, P. Kristensen, A. Rubano, S. Golowich, L. Marrucci, S. Ramachandran, Spin-orbit coupled, non-integer OAM fibers: unlocking a new eigenbasis for transmitting 24 uncoupled modes, in: Conference on Lasers and Electro-Optics. OSA, Washington, D.C., 2016, paper JTh4C.7.
[61] A. Tanaka, A. Korolev, E. Mateo, E. Ip, J. Hu, K. Bennett, et al., 146λ × 6 × 19-Gbaud wavelength- and mode-division multiplexed transmission over 10 × 50-km spans of few-mode fiber with a gain-equalized few-mode EDFA, J. Lightwave Technol. 32 (4) (2014) 790–797.
[62] V.A.J.M. Sleiffer, Y. Jung, M. Kuschnerov, S.U. Alam, D.J. Richardson, L. Grüner-Nielsen, et al., Optical chopper-based re-circulating loop for few-mode fiber transmission, Opt. Lett. 39 (2014) 1181.
[63] V.A.J.M. Sleiffer, Y. Jung, V. Veljanovski, R.G.H. van Uden, M. Kuschnerov, H. Chen, et al., 73.7 Tb/s (96 × 3 × 256-Gb/s) mode division multiplexed DP-16QAM transmission with inline MM-EDFA, Opt. Express 20 (2012) B428–B438.

[64] Y. Jung, S.U. Alam, D.J. Richardson, Enabling component technologies for space division multiplexing, in: Optical Fiber Communication Conference. OSA, Washington, D.C., 2018, paper M4D.3.

[65] A. Mafi, P. Hofmann, C.J. Salvin, A. Schülzgen, Low-loss coupling between two single-mode optical fibers with different mode-field diameters using a graded-index multimode optical fiber, Opt. Lett. 36 (2011) 3596.

[66] H. Chen, C. Jin, B. Huang, N.K. Fontaine, R. Ryf, K. Shang, et al., Integrated cladding-pumped multicore few-mode erbium-doped fibre amplifier for space-division- multiplexed communications, Nat. Photonics 10 (2016) 529–533.

[67] Y. Amma, T. Hosokawa, H. Ono, K. Ichii, K. Takenaga, S. Matsuo, et al., Ring-core multicore few-mode erbium-doped fiber amplifier, IEEE Photonics Technol. Lett. 29 (2017) 2163–2166.

[68] Y. Jung, J.R. Hayes, S.U. Alam, D.J. Richardson, Multi-wavelength fiber laser using a single multicore erbium doped fiber, in: Optical Fiber Communication Conference. OSA, Washington, D.C., 2018, paper M2J.6.

[69] P.S. Westbrook, K.S. Abedin, T.F. Taunay, T. Kremp, J. Porque, E. Monberg, et al., Multicore fiber distributed feedback lasers, Opt. Lett. 37 (2012) 4014.

[70] Y. Weng, E. Ip, Z. Pan, T. Wang, Advanced spatial-division multiplexed measurement systems propositions—from telecommunication to sensing applications: a review, Sensors 16 (2016) 1387.

[71] A. Li, Y. Wang, Q. Hu, W. Shieh, Few-mode fiber based optical sensors, Opt. Express. 23 (2015) 1139.

[72] J.P. Moore, M.D. Rogge, Shape sensing using multi-core fiber optic cable and parametric curve solutions, Opt. Express. 20 (2012) 2967.

[73] X. Sun, J. Li, D.T. Burgess, M. Hines, B. Zhu, A multicore optical fiber for distributed sensing, in: H.H. Du, G. Pickrell, E. Udd, C.S. Baldwin, J.J. Benterou, A. Wang (Eds.), Fiber Optic Sensors and Applications XI, International Society for Optics and Photonics, 2014, p. 90980W.

[74] P.S. Westbrook, T. Kremp, K.S. Feder, W. Ko, E.M. Monberg, H. Wu, et al., Continuous multicore optical fiber grating arrays for distributed sensing applications, J. Lightwave Technol. 35 (2017) 1248–1252.

[75] S. Li, M.-J. Li, R.S. Vodhanel, All-optical Brillouin dynamic grating generation in few-mode optical fiber, Opt. Lett. 37 (2012) 4660.

[76] Y. Weng, E. Ip, Z. Pan, T. Wang, Single-end simultaneous temperature and strain sensing techniques based on Brillouin optical time domain reflectometry in few-mode fibers, Opt. Express 23 (2015) 9024.

[77] A. Li, Q. Hu, W. Shieh, Characterization of stimulated Brillouin scattering in a circular-core two-mode fiber using optical time-domain analysis, Opt. Express 21 (2013) 31894.

PART II

System and Network Technologies

CHAPTER 8

Transmission system capacity scaling through space-division multiplexing: a techno-economic perspective

Peter J. Winzer
Nokia Bell Labs, Holmdel, NJ, United States

8.1 Introduction

This chapter presents a unified view of all possible optical network capacity scalability options known today, with a focus on their techno-economics. The material presented here draws heavily from some of the author's recent accounts on the topic [1–5], where the interested reader will find further details and valuable supplementary information.

8.2 Traffic growth and network capacity scalability options

The fact that optical networks are unavoidably steering toward fundamental capacity scalability limits in the face of continuing exponential traffic growth, owing to nonlinear Shannon limits of silica optical transmission fiber, has been abundantly discussed for more than a decade [1,2,6–9] and is now widely referred to as the "optical networks capacity crunch" [7]. This capacity crunch is rooted in the disparity between two key long-term technology scaling trends [1]: *Moore's Law scaling* and *high-speed interface scaling.*

8.2.1 Moore's Law scaling

Technologies that are used to generate, process, and store, and to some extent also to locally access digital information are typically closely linked to complementary metal-oxide-semiconductor (CMOS) processing capabilities, and as such to Moore's Law scaling [10]. Examples of multiple-decades-long consistent scaling trends between 40% and 90% per year are summarized in Refs. [1,2] for such sectors as microprocessors, supercomputers, and datacenters; various kinds of storage and memory; Internet Protocol (IP) routers and Ethernet switches; and wireless and wireline access technologies. Moore's Law scaling is expected to continue at least for another decade, i.e., the

Optical Fiber Telecommunications V11
DOI: https://doi.org/10.1016/B978-0-12-816502-7.00009-9

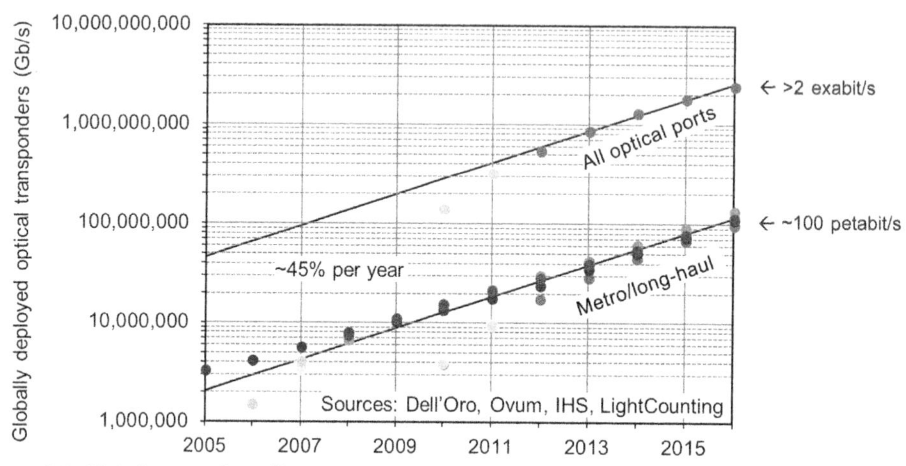

Figure 8.1 Global network traffic growth estimated from deployed optical transponders. Source: *P.J. Winzer, D.T. Neilson, A.R. Chraplyvy, Fiber-optic transmission and networking: the previous 20 and the next 20 years, Opt. Express 26 (18) (2018) 24190–24239.*

functional scaling of technological capabilities is expected to keep scaling (e.g., "storage capacity") rather than any particular *technological implementation* (e.g., "hard disks"). In various long-term industry roadmaps, these expectations are expressed as "More Moore," "More than Moore," and "Beyond CMOS" [10–12].

As the very purpose of information is to always be shared with other entities (humans as well as machines), the need to transport information has been increasing at the same rate as information processing capabilities, reflected in, e.g., Amdahl's rule of thumb [13], which states that a balanced computer architecture requires interface capabilities proportional to its processing power. Network traffic growth rates depend significantly on applications, network segments, operators, and geographies [1–3,14], with 60% per year (a doubling every 1.5 years) being a frequently quoted value [8]. Looked at from a globally and application-averaged network traffic growth rate, estimated from the amount of globally deployed optical networking hardware, results in the scaling shown in Fig. 8.1 [2]. The numbers are based on analyst reports for global annual transponder sales across the entire industry and reveal a remarkably constant 45% per year growth rate (a doubling every 1.9 years), consistent across four major analyst firms, both in the metro/long-haul part of the network and across all installed optical ports, including client and short-reach interfaces. The former reveals a total deployed metro/long-haul capacity of ~100 Petabit/s (Pb/s), and the latter of ~2 Exabit/s (Eb/s) as of 2016, with a consistent 45% per year growth reported since then as well. Note that another frequently referenced source in this context, Cisco's Visual Networking Index (VNI) [14], only reports 371 Terabit/s of traffic for 2017 (122 exabytes/month), with a growth at a mere ~26% per year (a doubling every 3 years).

The discrepancy likely arises because Cisco's VNI only accounts for end-to-end IP traffic, while the reportedly deployed optical transponder capacity captures all traffic types, accounts for over-provisioning of operational networks to accommodate peak-to-average traffic variations and diurnal fluctuations, and comprises the fact that an end-to-end transported information bit typically touches many transponders on its way from source to destination. While the globally averaged transponder capacity growth of ~45% per year at a first glance seems incompatible with some much lower growth numbers of 20%–30% per year reported by major telecom operators [15,16], the fact that large webscale operators see traffic growth well in excess of 45%/year [17] balances the equation.

From a packet processing point of view, this global traffic growth trend is well supported by the ~40% per year growth of IP router blade capacities (cf. Fig. 8.2A [2]) and Ethernet switch chip capacities (cf. Fig. 8.2B [2,18]), which are both rooted in Moore's Law, as packet processors are highly sophisticated special-purpose compute engines. The highest capacity IP router blade available today (e.g., in Nokia's 7950-XRS-20e high-end core router using the FP4 routing chip) supports 2.4 Tb/s, and the highest capacity Ethernet switch application-specific integrated circuit (ASIC) (Barefoot Network's Tofino-2) supports 12.8 Tb/s. As Moore's Law will still support CMOS scaling for at least a decade [10], we expect IP router blades and Ethernet switch chip capacities of 35 and 142 Tb/s by 2025 and 188 and 790 Tb/s by 2030. The big question is how to get such enormous amounts of data into and out of these processors, as we will discuss next.

8.2.2 High-speed interface scaling

In stark contrast to the continuing ~40% per year scaling of CMOS-based packet processing technologies at Moore's Law's pace, the ability to transport traffic has, for

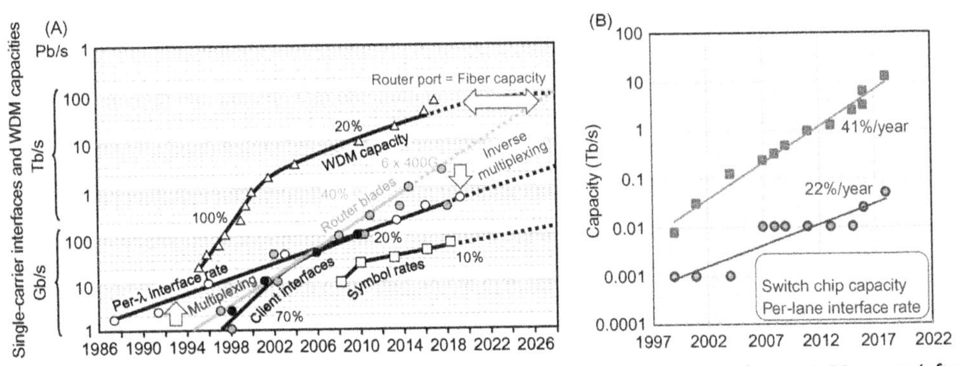

Figure 8.2 (A) Evolution of commercial optical transmission systems over the past 30 years (after [1,2]); (B) Ethernet switch chip capacity growth and associated growth of per-lane line rates [2,18].

decades, only been scaling at 20% per year (a doubling every 3.8 years), leading to an increasingly worrisome 40%/20% growth rate disparity that is starting to affect communications across length scales and applications, from chip-to-chip interconnects to ultralong-haul submarine fiber-optics links.

Chip-to-chip interfaces using ball-grid arrays (BGAs) and electrical signaling have been growing at a rate of ~20% per year, typically operating at 25 Gb/s today for high-capacity ASICs, and pushing toward 50 Gb/s, cf. Fig. 8.2B. However, a 40% per year growth in switching/routing capacities at only a 20% per year growth in per-lane interface rates requires another 17% per year increase in the number of I/O lanes to keep up with switching capacities, which will soon lead to an unmanageable number of lanes at unrealistic densities and will put an end to electrically interfaced ASICs, ultimately resulting in fiber-in-fiber-out (FIFO) hybrid optics/electronic engines [19]. At the same time, the power consumption of off-chip I/Os is starting to dominate a chip's power consumption [20]. A recent DARPA challenge for full-duplex 100-Tb/s and 1-Pb/s chip-to-chip interface capacities, which at a continuing ~40% per year switch chip capacity growth rate will be required by ~2025 and 2032, respectively, and which can only be provided optically, crystalizes this severe industry problem [20]. Note that while today's electrical BGA interfaces limit high-speed chip-to-chip communications to the centimeter range, the low loss and high bandwidth of optical-fiber-based chip-to-chip interconnects will open up entirely new distributed system architectures: once optical and in a fiber, signals can traverse hundreds of meters or even kilometers without appreciable losses and per channel bandwidth limitations.

Client interfaces plugging into router blades to optically connect routers locally to each other and to long-haul networking hardware are falling behind due to the mentioned 40%/20% scaling disparity. From an architectural point of view, IP networks in the 1990s could multiplex several router ports onto a single carrier wavelength for cost-efficient long-haul transport, but today's router blades require inverse multiplexing of a single router port onto multiple carrier wavelengths, cf. Fig. 8.2A [2]. For example, the previously mentioned highest-capacity 2.4-Tb/s IP router blade available today must use 6×400 Gb/s due to the unavailability of higher-speed client interfaces. By ~2024, a router blade will be able to process ~20 Tb/s of packet traffic, which will have to be mapped onto 20×1-Tb/s client interfaces if technology continues on its long-term 40%/20% path, cf. Fig. 8.2A. (And there is presently no reason to believe it shouldn't.)

Line interfaces for long-haul optical transport have scaled at 20% per year for decades, cf. Fig. 8.2A, reaching 40-Gb/s interfaces with binary modulation in the early 2000s. The widespread introduction of coherent detection in ~2010 afforded a four-fold to eight-fold reduction of the underlying per-lane electronics rates relative to the resulting bit rates through the use of (1) multiplexing [in-phase/quadrature (I/Q): 2x; polarization-division multiplexing (PDM): 2x] and (2) higher order modulation (4-level and 8-level electrical signals per quadrature, i.e., 16- to 64-QAM: 2x to 3x), which initially lowered symbol rates of 40-Gb/s coherent transponders to 11.5 GBaud [21]. Coherent symbol rates requiring CMOS-integrated analog-to-digital and digital-to-analog converters (ADCs, DACs) have since been scaling at only 10% per year, cf. Fig. 8.2A, which makes the historic 20% per year scaling of per carrier interface rates questionable going forward, as geographically fixed transmission distances prevent the use of much higher-order modulation. This points to the use of alternate multiplexing dimensions, resulting in spectral or spatial superchannel technologies discussed in Section 8.4.2.

Fiber capacities, enabled by wavelength-division multiplexing (WDM) typically between 50 and 200 individual line interfaces into a single transmission fiber, allowed for a 100% per year scaling in the 1990s, afforded by advances in both erbium-doped fiber amplifier (EDFA) technologies and by dispersion management techniques that were essential to exploit the available amplification bandwidth in a systems context [2]. However, this scalability dimension is starting to be exhausted as well due to systems closely approaching their Shannon limits [9] and bandwidth expansion being a difficult endeavor in practice, as discussed in more detail in Section 8.3.3. The result has been a scaling of WDM capacities at only ~20% per year, akin to single-channel bit rates, cf. Fig. 8.2A. Comparing the scaling of router blade capacities and WDM capacities, we also see that a single router blade will require a full WDM system in the not too distant future, making a *fiber* as oppose to a *wavelength* the logical switching entity within the optical network. This evolution is also addressed from a switching point of view in Chapter 18, Networking and routing in space-division multiplexed systems, in this book.

8.3 Five physical dimensions for capacity scaling

The common basis to address the above four scalability challenges is the optimum use of the (only!) five physical dimensions available for modulation and multiplexing in electromagnetic communication systems [5]:

- *Time*—Modulation symbol rate and pulse shape
- *Quadrature*—Use of real and imaginary parts of an optical wave
- *Polarization*—Two (and only two) orthogonal polarization states
- *Frequency*—Number of WDM channels and spectral bands
- *Space*—Parallel transmission on more than one spatial path

The resulting optimization problem is fully captured by Shannon's famous channel capacity [22], which in its most well-known form applicable to classical[1] memoryless additive white Gaussian noise (AWGN) channels reads

[1] A full quantum-mechanical version of Shannon's capacity formula was derived based on entropy arguments by Gordon in 1962 [23] and formally proven to be the capacity of an arbitrary quantum channel by Holevo in 1973 [24]; see Ref. [3] for a detailed discussion of quantum capacities in the context of this chapter.

$$C = 2MB\log_2(1 + SNR) = 2MB\log_2(1 + P/(2MBN_0)). \quad (8.1)$$

The channel's effective *signal-to-noise ratio* (SNR) is given by the ratio of the average overall signal power P (spread across 2 polarizations, M spatial paths, and a system bandwidth B) to the overall received noise power $2MBN_0$. The noise power spectral density per polarization and spatial path, N_0, is assumed to be signal-independent, additive, complex Gaussian, and white across the system's bandwidth, its two polarizations, and its spatial paths. Colored noise can be accommodated by spectral and spatial waterfilling [25,26]. In practice, N_0 captures a variety of noise sources[2]:

- Optical amplifier noise (amplified spontaneous emission, ASE) [9]
- In-band and out-of-band linear crosstalk (from filters, switches, and double-Rayleigh scattering [9,27–29])
- Nonlinear interference noise (NLIN) in the widely applicable equivalent Gaussian noise model [30–32]
- Deficiencies of high-speed opto-electronic hardware components and of transmit/receive digital signal processing (DSP) algorithms, all of which are typically summarized under the notion of an *implementation penalty* relative to textbook performance [33]

Note that while "memoryless AWGN" is in general a reasonably adequate model for the above noise sources, most of them do not strictly possess this property. This consequently lets Eq. (8.1) only represent the capacity of an *equivalent memoryless AWGN channel*, to be approached by receivers whose decoding is based on this assumption. Receivers that exploit residual channel memory or residual non-AWGN characteristics of the involved noise sources could in principle achieve slightly higher capacities.

With the quantity

$$SE = C/(2MB), \quad (8.2)$$

referred to as the *spectral efficiency* per spatial and polarization mode, i.e., the information capacity of a single-mode signal per unit bandwidth, Eq. (8.1) can be written as

$$SE = C/(2MB) = \log_2(1 + E_b\, SE/N_0), \quad (8.3)$$

with the energy per information bit E_b defined as

$$P = E_b C = 2MBE_b\, SE. \quad (8.4)$$

[2] Impairments that are left uncompensated within a system (even if they are deterministic and *could* therefore in principle be compensated) are generally treated as random noise (if they vary quickly relative to the interleaving depth of the underlying decoding scheme), or they are treated through a margin/outage analysis (if they vary slowly relative to the interleaving depth of the underlying decoding scheme).

Note that E_b is only the received *optical* signal energy per bit used for information transport on the channel, not counting various signal amplification, conditioning, and processing energies, nor any (typically by far dominating) system inefficiencies [34].

The goal of transmission system design is usually to maximize or minimize one of the parameters C, SE, P, E_b, B, or M while imposing constraints on some or all of the others [26,35]. From an operational point of view, the fine granularity in bit rates and *SE*s afforded by recently productized probabilistic constellation shaping (PCS) techniques [36–38] allows the direct application of Eq. (8.1) in many practical systems contexts.

As we will see below, the most energy-efficient capacity scaling method in the absence of any further constraints is multiplexing, i.e., the use of independent (orthogonal) physical dimensions to send multiple signals across parallel physical channels. Except for the quadrature dimension[3], all other multiplexing dimensions are represented by the three "pre-log" factors 2, B, and M: The factor of 2 reflects PDM; multiplexing across the system bandwidth B can be implemented in the form of WDM, through time-division multiplexing (TDM), or through pulse-shaping and bandwidth-expansion techniques [35]; and multiplexing in M refers to space-division multiplexing (SDM) across parallel spatial paths. Identifying the best-suited modulation and multiplexing strategy for a given communication scenario is the subject of communication system design, and the underlying trade-offs take on different forms, depending on additional techno-economic constraints, as we shall examine throughout the remainder of this chapter.

8.3.1 Increasing capacity through SNR —constraints on *M* and *B*

Terrestrial fiber-optic networks are typically constrained

- physically by fiber nonlinearities limiting the power spectral density per spatial path ($P/2MB$) [9],
- techno-economically by the high costs of leased or newly-to-be-deployed fiber limiting the number of spatial paths that may be used, and
- in bandwidth by amplification technologies limiting commercial systems to the combined C + L bands.

As a consequence of these constraints, terrestrial system capacity scaling first and foremost targets an increase in SE, as $C = 2\ B\ SE$. This results in a trade-off between the SE and the unregenerated transmission reach L, as shown in Fig. 8.3A [1,39]. Note that the figure shows the dual-polarization SE (SE_2), as is customary when reporting experimental results, while theoretical results (including our notation

[3] Multiplexing in the quadrature dimension is implicitly captured in the Shannon capacity, Eq. (8.1) by considering *complex* passband signals and *complex* Gaussian noise, and properly accounting for signal bandwidths and the per- quadrature noise power spectral densities on a passband channel.

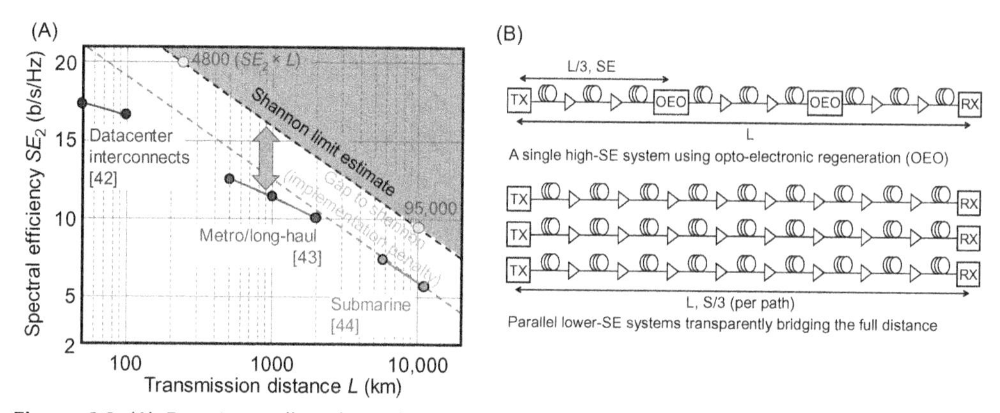

Figure 8.3 (A) Experimentally achieved record *SE*s (in two polarizations) versus transmission distance (*markers*), shown with the Shannon limit estimate on SSMF (*dashed*); figure after [1,2]; (B) a regenerated system versus a parallel system (after [39]).

throughout this chapter unless explicitly denoted as SE_2) are often specified in terms of a *SE* per polarization. The figure also shows an estimate for the (dual-polarization) Shannon limit of standard single-mode fiber (SSMF) [9,40] and summarizes recent experimental research records from <100 km short-reach datacenter interconnect (DCI) systems up to >10,000 km transpacific systems. The trade-off between *SE* and reach is seen to follow a logarithmic relationship, with each doubling in transmission reach reducing the dual-polarization *SE* by ~2 bits/s/Hz, owing to the fact that the accumulated ASE scales linear with the number of spans (or with transmission distance in the case of uniform amplification [9]) and the SNR at and the optimum signal launch power (including NLIN) scales to an excellent approximation inversely proportional to L [41]. The best achieved experimental records across all transmission distances make use of PCS with constellations as large as 4096-QAM [42] for short-reach applications, going down to 256-QAM for terrestrial systems [43], and 64-QAM for submarine distances [44], the latter performed on a deployed transatlantic cable operated by Facebook.

Sometimes, optical transmission records are still being quoted in terms of their "capacity × distance" product ($C \times L$), or, for a fixed system bandwidth B, in terms of their "$SE \times$ distance" product ($SE \times L$). It is important to note, though, that these product metrics date back to early multimode transmission systems that were limited by modal dispersion, whose impact on transmission system performance is to first-order proportional to the signal bandwidth, but they are inadequate for modern single-mode fiber-optic transmission systems, which are limited by the Shannon capacity, Eq. (8.1). In particular, the product metrics $C \times L$ and $SE \times L$ suggest that doubling the *SE* at a fixed transmission reach should in some relevant way be equivalent to doubling the transmission reach at a fixed *SE*, cf. Fig. 8.3B. This is not the case, however, owing to the fact that the SE depends *logarithmically* on L, cf. Eq. (8.1). Typically, doubling capacity is associated with much higher levels of complexity than doubling transmission reach. For example, looking at the Shannon limit estimate of Fig. 8.3A, the highlighted point at 10,000 km corresponds to $SE_2 \times L = 95{,}000$ km, while the highlighted point at 240 km only yields $SE_2 \times L = 4800$ km. The correct system performance metric is given by the product [39]

$$p = 2^{SE} \times L = 2^{SE_2/2} \times L, \tag{8.5}$$

which is derived directly from Eq. (8.1) by invoking the usually applicable high SNR regime (to neglect the "1 + " term within the logarithm) and taking note of the $\sim L^{-1}$ noise scaling at optimum SNR [41]. This metric also shows why it is much harder to double SE than to double L in a modern fiber-optics communication context. An important consequence of this scaling is that systems using multiple opto-electronic regeneration (OEO) points within a link are generally more costly compared to parallel systems that are able to transparently bridge the desired transmission distance [39]; for example, in order to achieve 20 b/s/Hz over 1500 km, one may use ~75 transponders operating at 20 b/s/Hz and spaced ~20 km apart, or one may just use two parallel lines at 10 b/s/Hz each [2]. Since the regenerated system uses 35 times as many transponders (which in addition are significantly more involved than in the parallel case), the parallel approach is much more attractive. Note that "parallelism" can be achieved in any of the pre-log multiplexing dimensions, and in particular in B and M, with trade-offs that we will discuss throughout the remainder of this chapter.

When assessing how close a particular practical system is to the Shannon limit, we may compare systems in terms of their p-metric, Eq. (8.5). The Shannon limit of Fig. 8.3A corresponds to $p = 2.5 \times 10^5$ km, while experimental records currently reach $p = 7.2 \times 10^4$ km (gray dashed line in Fig. 8.3A); significant transponder implementation penalties prevent higher order formats at shorter distances from reaching this limit. However, we note that the experimental records shown in Fig. 8.3A only have fairly modest gaps to the Shannon limit (~7 dB in terms of distance or SNR, or a ~40% headroom in terms of capacity at ~1000 km reach), which lets system engineering approach a point of diminishing returns (an "economic Shannon limit"), as will become even more evident in the following discussion.

Returning to systems constrained by the unavailability of additional spatial paths, scaling SE can be accomplished by implementing communication schemes that more closely approach the underlying assumptions of Shannon's capacity formula, such as (1) Gaussian-shaped modulation, approached by PCS modulation [36–38], (2) close-to-ideal forward error-correction coding (FEC) [45,46]), and (3) increasing the effective SNR of the received and fully electronically processed signal prior to entering the

FEC decoder [1,33]. Measures to improve the effective SNR in the context of fiber-optic communication systems include the following:

- Lower loss optical fiber;
- Lower noise amplification;
- Lower nonlinearity fiber (such that higher signal launch powers can be used to increase the SNR);
- Nonlinearity compensation techniques (digital and analog);
- Improved digital signal processing algorithms;
- Improved high-speed hardware (ADCs, DACs, modulators, drivers, preamplifiers, lasers, etc.).

Importantly, all strategies to improve the SNR work *within* the logarithm of Eq. (8.1), and hence only show logarithmic (sublinear) gains in the fairly high SNR regimes typical of fiber-optic communication systems. Using Eq. (8.1), one may show that a system originally operating at SNR_0 will experience a capacity gain of

$$\frac{C - C_0}{C_0} = \frac{\log_2(1 + SNR_0\sigma_{SNR}) - \log_2(1 + SNR_0)}{\log_2(1 + SNR_0)} \approx \frac{\Delta SNR_{\mathrm{dB}}}{SNR_{0,\mathrm{dB}}}, \tag{8.6}$$

when the original SNR is improved by a factor $\sigma_{SNR} > 1$ (or by a dB *SNR* difference $\Delta SNR_{\mathrm{dB}} = 10 \log_{10} \sigma_{SNR}$). Table 8.1 shows some such capacity gains, revealing that even heroic SNR gains of 10 dB in a long-haul system operating at an SNR of 10 dB ($SE_2 = 7$ b/s/Hz) will at best be able to double system capacity (92% capacity gain), but will not be able to resolve the capacity crunch. Note in this context that all of the above-mentioned techniques to improve the SNR try to eke out between a few tenths of a dB to a few dB, far off any prospect of a 10-dB SNR improvement. Hence, any research on the topics itemized above can only benefit system capacity in very limited ways, and the use of the two linear pre-log factors B and M is the only significant capacity scalability option. As bandwidth scaling is not a long-term option either, cf. Section 8.3.3), there is no way around the deployment of massively parallel spatial paths on the long term. This will introduce a techno-economic step-function into terrestrial system design, as operators will want to drive up *SE* as much as possible in order to delay new fiber deployments, thereby asking for all of the above SNR-improving system sophistications. Once new cables are being deployed (with a massive amount of parallel fiber strands at little extra cost, as cable costs are dominated

Table 8.1 Capacity gains with SNR increase.

		SNR_0		
		20 dB	**15 dB**	**10 dB**
ΔSNR	1 dB	5%	6%	9%
	3 dB	15%	19%	27%
	10 dB	50%	65%	92%

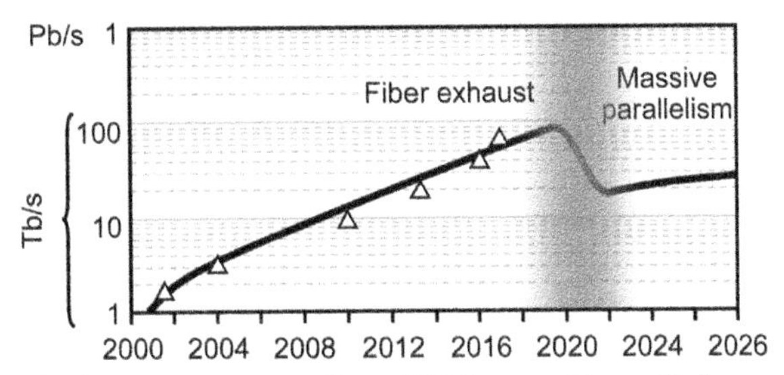

Figure 8.4 Limits to further increase per-fiber WDM capacities will force operators to deploy new cables. As the cost of deployment depends only weakly on the number of fiber strands within a cable, operators are expected to deploy high-fiber-count cables and ask for massively parallel transmission systems instead of systems with ultra-high per-fiber capacities.

deployment costs and not by the cost of bare fiber), operators will ask for massively parallel, low-cost, and in terms of their SNR improvement techniques unsophisticated transponder solutions. This paradigm shift from ultra-high per-fiber capacities in a fiber-exhaust scenario to moderate-capacity massively parallel systems emerging beyond "the cliff" is visualized in Fig. 8.4. We will discuss this interesting techno-economic situation in some more detail in Section 8.5.3.

8.3.2 Power-constrained system scaling—parallelism in *M* and *B*

In some systems, the overall power allocated to communications is limited. Of the systems mentioned in Section 8.2.2, this applies particularly to two classes of systems that lie on the diametrically opposite end of the transmission distance spectrum: chip-to-chip interfaces (constrained by the maximum possible power a chip can afford for communications) and submarine systems (constrained by the maximum possible electrical supply power for all the optical amplifiers within a submarine cable, which must be powered from high-voltage power feeds at each end of the submersed cable [47]). For such systems, Eq. (8.1) can be used to determine how to best distribute the available power across a system bandwidth B and a number of parallel spatial paths M.

Strictly mathematically, Eq. (8.1) is completely symmetric in polarization, bandwidth, and spatial paths, which suggests the substitution $x = 2MB$,

$$C = x\log_2(1 + P/(xN_0)). \tag{8.7}$$

Fig. 8.5A shows the Shannon capacity (8.7) as a function of x for the case of a single-span un-amplified link without NLIN (e.g., a free-space optical link) using heterodyne or intradyne coherent detection (shot-noise limited detection with $N_0 = hf_c$; h is Planck's constant, $f_c = 193$ THz is the system's optical carrier frequency) [3], and three different received signal power levels P are assumed. As the signal energy is being spread across bandwidth and/or spatial paths, the system's capacity monotonically increases with x and approaches its asymptotic value of

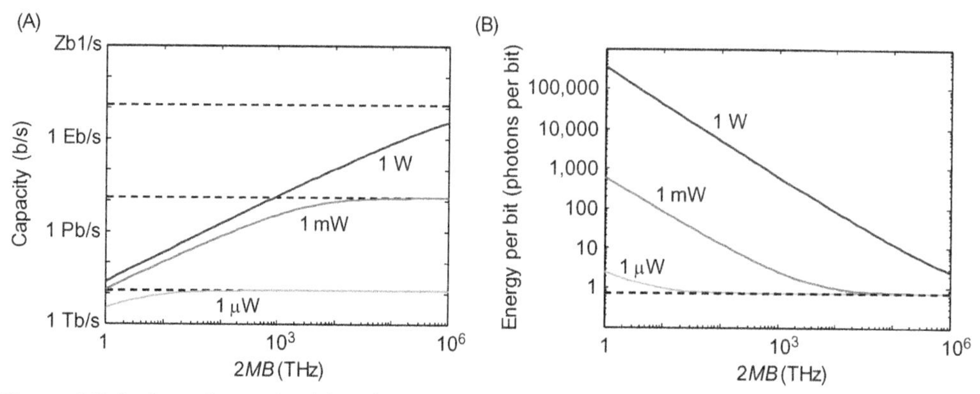

Figure 8.5 Scaling of capacity (A) and energy per bit (B) for different received signal power levels on a single-span, unamplified, shot-noise limited coherent detection channel.

$$\lim_{x\to\infty} C = (P/N_0)\log_2 e = P/(N_0 \ln 2). \tag{8.8}$$

Fig. 8.5B shows that the corresponding energy per bit $E_b = P/C$, cf. Eq. (8.4), normalized by hf_c to photons per bit, monotonically decreases with x and approaches the familiar asymptotically minimum required signal energy per bit of [35]

$$\lim_{x\to\infty} E_b = \lim_{x\to\infty} P/C = N_0/\log_2 e = N_0 \ln 2. \tag{8.9}$$

Importantly, we note that although the above considerations are performed for the *classical* Shannon capacity of Eq. (8.1), the same behavior, yet with $N_0 = kT$ (k being Boltzmann's constant and T the applicable noise temperature), is obtained for the maximum achievable capacity and the minimum achievable energy per bit when *arbitrary* quantum communication techniques are taken into account [3].

While the benefit of bandwidth-spreading modulation is amply known in classical communications engineering [35], the complete analogy to spreading signals within a fixed system bandwidth B across a variable number of parallel spatial paths M may not be as widely appreciated. Which spreading mechanism is best to be used in practice depends on physical and techno-economic considerations, which are unique to each system, as we shall see throughout the remainder of this chapter.

8.3.3 Bandwidth and space are not created equal

Although Section 8.3.2 suggests through the use of the product $x = 2MB$ that B and M are completely equivalent (which they are indeed mathematically), several practical considerations let the two multiplexing dimensions be markedly different in a fiber-optic communications context, revealing that spatial parallelism is clearly to be preferred in as much as the required parallel transport infrastructure can be provided;

spectral parallelism should only be used if a sufficient number of spatial paths is unavailable or techno-economically too costly. We proceed to discuss the trade-off between M and B in general terms in this section and provide some concrete techno-economic examples of this general trade-off in Section 8.5.

Reuse of the available infrastructure

The aspect of existing infrastructure reuse is practically one of the most important differences between the spectral and the spatial multiplexing dimension and strongly favors near-term bandwidth scaling for systems operating over already deployed fiber in order to utilize the installed base as much as possible, cf. Section 8.3.1. New fiber deployments can be prohibitively expensive, dominated not by the cost of the fiber itself, but by the cost of installing a cable. Reported costs vary widely depending on the deployment scenario, with ~\$20,000 per km assuming available duct space being a realistic assumption (see, e.g., [48] and references therein for respective cost analyses). This lets the deployment of a 1000-km cable (~\$20 million) cost more than the WDM system operating over it! Despite the high associated deployment cost, fiber is being amply deployed around the globe, with the installed base growing at an average compound annual rate of ~15%, dominated by fiber-to-the-home deployments (growing at ~25%/year) [2]. While the aspect of fiber reuse is still important today, it is expected to gradually lose its importance over the coming decade(s), as many operators will be forced to deploy new fiber throughout their networks to keep up with traffic demands. Since the cost of extra fiber pairs to a newly installed cable is not very high (legacy fiber being priced at a few dollars per kilometer), operators should opt to deploy massive fiber-count cables, leading to the sudden availability of a large number of parallel spatial paths. Some techno-economic consequences for the fiber and transponder ecosystem are discussed along with Fig. 8.4 and in Section 8.5.3.

Channel power equalization

An important difference between spatial and spectral parallelism lies in the equalization of channel powers across the multiplex. While parallel spatial paths across a fairly small bandwidth can be assumed to experience similar propagation characteristics in terms of fiber and component losses or amplifier gains and noise figures, wavelength channels typically experience strong power variations across a significant WDM bandwidth. These variations are in practice counteracted by periodically inserted gain-flattening filters. However, as these filters are passive components, *loss* is inherently introduced into the system. Especially for heavily power-constrained systems like chip-to-chip interconnects and submarine cables, loss due to gain flattening stands against the imperative to utilize all available power as much as possible. Hence, power-limited systems naturally favor an increase in spatial multiplicity over an increase in bandwidth. As such, system bandwidths should be kept low whenever enough parallel spatial paths can be made available. This is the reason why recent research proposes the use of only half the C-band instead the combined C + L bands for massively parallel submarine systems, cf. Section 8.5.4 and Refs. [4,49,50].

Bandwidth limitations of fiber and system components

Disregarding the above loss considerations affecting power-limited systems, as applicable for terrestrial fiber-optic networks, widening the system bandwidth in theory linearly increases system capacity through the pre-log factor B, cf. Eq. (8.1), provided that such a scaling does not introduce additional impairments and is supported by the bandwidth of *all* the underlying optical system components, including transmission fiber, optical amplifiers, optical switch elements, and laser sources [1].

Regarding the bandwidth of optical fiber, Fig. 8.6 [2] shows typical loss coefficients across the low-loss window of commercial fiber with (*red*) and without (*blue*) the characteristic hydroxyl group (OH) absorption peak around 1380 nm. A factor of ~12 in bandwidth could potentially be gained if operating deployed fiber from the O-band all the way to the L-band (~1260–1625 nm, corresponding to 53.5 THz), as opposed to using the C-band only (~1530–1565 nm, corresponding to 4.4 THz), as is done in the vast majority of today's commercially deployed systems. However, it is very unlikely that this factor of ~12 in bandwidth would actually translate into a similar factor of capacity gain, owing to several fundamental and practical problems associated with ultra-broadband systems that make a factor of up to ~5 in capacity more realistic, as discussed in detail in Ref. [1].

Regarding the deployment of new fiber types, one must distinguish between *green-field* situations (where new fiber is installed as part of new system deployments) and *brown-field* situations (where new systems are installed to augment and upgrade an already existing fiber infrastructure). The former applies to many submarine and DCI situations, while the latter is (and will for a while remain) typical of terrestrial long-haul and metro networks. Brown-field deployments always dictate a smooth upgrade path, letting newly deployed fiber gradually increase the capacity of an already vast installed base [5]. This makes the introduction of radically new fiber very problematic in brown-field scenarios and leaves this option only for green-field deployments. An example for radically new fiber (in a telecommunications context) is photonic crystal hollow-core fiber, which could in principle be designed to operate across a ~37-THz bandwidth in the 2-μm wavelength range, outside the standard 1.55-μm telecom window (cf. *dashed green curve* in Fig. 8.6, showing potentially achievable theoretical loss predictions [51,52]). Note that the wavelength scaling of the x-axis of Fig. 8.6 can be deceiving, cf. second x-axis showing *carrier frequency* (in units of Hz), which is what counts in the context of the system bandwidth B; in fact, the fiber represented by the dashed green curve opens up *less* bandwidth than the wavelength region from the O-band to the L-band in existing standard telecom fiber! Potentially wider-band and

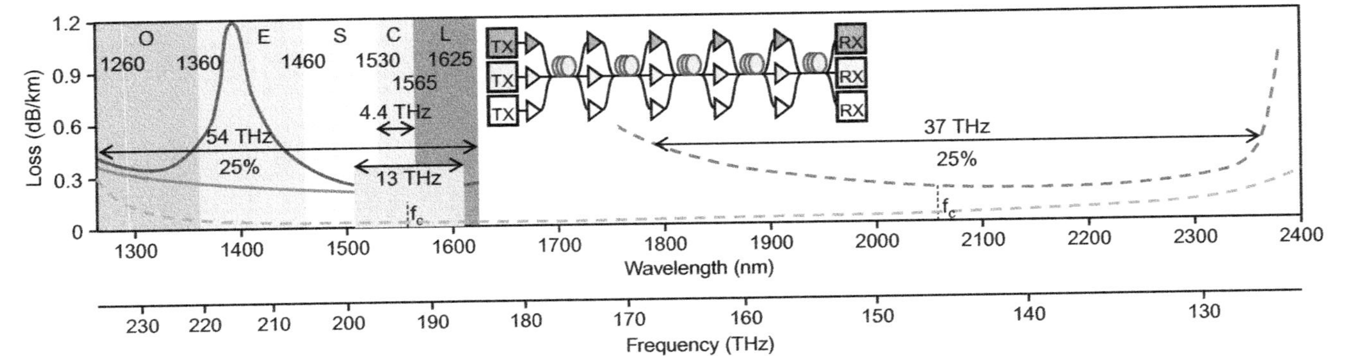

Figure 8.6 Commercially achievable low-loss window of conventional single-mode fiber in the wavelength range from 1260 to 1625 nm (O-band to L-band) using legacy fiber with (*red*) and without the hydroxyl absorption peak (*blue*). The *dashed green* and *dashed orange curves* are redrawn from Ref. [52] and represent model predictions for low-loss photonic crystal hollow-core fiber (*green*) and nested antiresonant nodeless hollow-core fiber (*orange*). Note that the wavelength scaling (*x*-axis) does not represent the frequency bandwidth (*double arrows*). *Figure from P.J. Winzer, D.T. Neilson, A.R. Chraplyvy, Fiber-optic transmission and networking: the previous 20 and the next 20 years. Opt. Express 26 (18) (2018) 24190–24239.*

even lower-loss nested antiresonant nodeless hollow-core fibers (which would also include the traditional telecom bands) have also been studied, with a potentially achievable loss profile shown by the *dashed orange curve* in Fig. 8.6 [52,53]. Note, though, that regardless of practical achievability, even a significantly lower fiber loss does not resolve the capacity scalability problem, as discussed along with logarithmic scaling of capacity with SNR, cf. Table 8.1 in Section 8.3.1, and the need for new deployments of such fiber (once available) begs the question of its advantage compared to massively parallel legacy-fiber cables, which yield linear capacity gains (M).

While the use of wider WDM system bandwidths is frequently posed as a fiber design problem, constraints in system component bandwidths, including optical amplifiers, filters, and lasers, turn out to pose much more of a problem in practice. In this context, it is important to consider the *relative bandwidth* B_{rel}, defined as the absolute system bandwidth B divided by the system's center frequency f_c,

$$B_{\mathrm{rel}} = B/f_c. \tag{8.10}$$

In most fields of engineering, optical communications not excluded, the complexity of components and subsystems grows with their relative bandwidth. As shown in Fig. 8.6, the C-band (corresponding to the gain bandwidth of EDFAs) has a relative bandwidth of 2.3%; a recent 100-nm-wide demonstration of semiconductor optical amplifiers for coherent transmission [54] spanned 13 THz of bandwidth (as indicated by the *blue shaded region* in Fig. 8.6), corresponding to a relative bandwidth of 6.6%; and a Raman-based amplifier in Tellurite fiber was reported with a relative bandwidth of 10% [55]. The frequency region from O-band to L-band occupies a relative bandwidth of 25%, as does the 2-μm region indicated in Fig. 8.6; the relative bandwidth of the nested antiresonant nodeless hollow-core fiber indicated by the *dashed orange curve* in Fig. 8.6 would require octave-spanning components (a relative bandwidth of 67%). Building amplifiers, tunable lasers, or tunable filters that individually span such large relative bandwidths is complicated, which forces a banded approach for the subsystems needed to construct an ultra-broadband system, i.e., a splitting of the system bandwidth into multiple individual sub-bands at amplifier sites or switching nodes, as indicated in the inset to Fig. 8.6. A recent C + L + S-band experiment achieving 150 Tb/s at an aggregate dual-polarization *SE* of 9.2 b/s/Hz (11.05 b/s/Hz when inappropriately counting only the sub-band bandwidths and neglecting guard bands) over a fairly modest distance of 40 km impressively illustrates the problem [56].

Multiband systems are not truly parallel

As discussed, system components are generally increasingly difficult to build the larger the relative bandwidth becomes, which often forces ultra-wideband systems toward a banded approach, i.e., the installation of several logically separated systems that only reuse the fiber as a common wideband transmission medium but otherwise split signals into individual bands at each amplification site and at each switching node, cf. inset to Fig. 8.6. While this approach maximally reuses the expensive installed fiber infrastructure, it has several major drawbacks:

- Multiband systems are harder to design than single-band systems, as (Kerr and Raman) nonlinearities act across the entire system bandwidth, and band-splitting components introduce additional complexity, cost, and loss. This lets multiband systems generally perform worse than single-band systems.
- True parallelism is geared at reducing component costs through volume. However, volume implies a common technology base for all system components. Producing different components for individual spectral sub-bands cannot be considered true parallelism from a techno-economic point of view and will not reduce system costs through volume. On the contrary, a multiband system will typically be more expensive than two single-band systems of the same aggregate capacity.
- Banded systems result in overlays of parallel networks that are physically unable to connect with each other and are therefore undesirable from a higher-layer network architecture point of view.

To summarize, banded systems typically have both lower performance and higher cost-per-bit compared to single-band (e.g., C-band) systems. Banded systems can only be justified as a stop-gap solution allowing the operator to postpone the deployment of new fiber, but scaling systems in the frequency domain cannot solve the long-term capacity scaling problem, which requires cost-effective capacity scaling factors of 100 or even 1000.

Higher carrier frequencies

To address the above-mentioned relative-bandwidth problem on a long-term horizon, higher carrier frequencies could in principle be investigated, allowing for larger absolute system bandwidths at reasonable relative bandwidths; combining Eqs. (8.2) and (8.10), this capacity scaling approach can be cast as

$$C = 2\ B_{\mathrm{rel}}\, f_c\, M\ \ SE. \tag{8.11}$$

In analogy to the 3 to 7 orders of magnitude increase in carrier frequencies when optical fiber replaced microwave transmission technologies in the late 1970s, the possibility of scaling communication capacities through the use of significantly higher carrier frequencies from optical frequencies to the extreme ultraviolet or the soft x-ray spectral region could at least be a (long-term) possibility. However, such a scaling would come at the expense of a roughly capacity-proportional increase in the required energy per bit, as the photon energy (hf_c) is proportional to the system's carrier frequency, and it is the number of photons that determines receiver sensitivity in the quantum regime

$hf_c \gg kT$. Note that the scaling from microwave to optical frequencies in the 1970s was done to a large extent across the classical, i.e., thermal noise dominated frequency range, $hf_c < kT$. In fact, a recent study on the scaling of information capacities on a classical as well as on an arbitrary quantum channel by going to higher carrier frequencies [3] has shown that only fairly modest capacity gains on the order of 10x may be achieved, assuming that carrier frequency scaling is done at a constant energy per bit. Appreciable capacity gains on the order of 1000 would require a 200–600 times higher received energy per bit, which would have to be compensated for by orders of magnitude lower waveguide loss coefficients at the scaled-up carrier frequencies compared to fiber at telecom wavelengths. As such low-loss waveguides and other system components, including the required quantum communication techniques, are out of sight in the extreme ultraviolet and soft X-ray region, spatial multiplexing at telecom wavelengths remains the only viable alternative for system capacity scaling, even in the long term.

Crosstalk

Optical communication systems may exhibit crosstalk, either *within a frequency slot*, e.g.,

- linear crosstalk within add/drop multiplexers and switches [27,28];
- linear crosstalk due to reflections and scattering [29];
- linear or nonlinear crosstalk between parallel spatial paths upon signal generation [57] or transmission [58–60],

or *across frequency slots*

- nonlinear crosstalk due to signal-signal WDM nonlinearities [9,31,32];
- linear crosstalk generated by densely packed modulators operating on different tributaries of a spectral superchannel [61].

The different nature and scaling of linear and nonlinear crosstalk mechanisms, as well as needs and benefits of their compensation further differentiates the spectral and spatial multiplexing dimensions. For example, (1) nonlinear crosstalk in the form of NLIN scales logarithmically with system bandwidth [31,62] and its complex compensation, e.g., split-step Fourier methods or perturbation-based approaches [63,64], yield relatively modest gains [65]; (2) the compensation of reflection-induced linear multipath crosstalk may be prohibitive owing to its potentially long memory; (3) linear crosstalk with short enough memory can often be almost perfectly compensated through multiple-input-multiple-output (MIMO) DSP techniques [66,67].

Irrespective of the nature of crosstalk, a key requirement for its effective compensation is the availability of all crosstalking channels within the same receiver DSP. This requirement for effective crosstalk compensation has two immediate system-level consequences:

- As crosstalk can only be compensated across as many channels as can be interfaced to a single CMOS ASIC, and as CMOS ASICs start to be interface-bandwidth-limited as opposed to processing-power-limited (cf. the 40%/20% per year scaling disparity discussed in Section 8.2), effective crosstalk compensation is evolving toward becoming an ASIC interfacing problem more than a processing problem. Today's coherent DSP ASICs for long-haul transmission typically incorporate two DSP cores, i.e., are able to interface to two separate coherent channels. Interfacing to significantly more channels asks for "fiber-in-fiber-out (FIFO) opto-electronic engines," i.e., the hybrid or monolithic co-integration of photonics and electronics to resolve the interface bottleneck [1,2].

- As all crosstalking signals need to be detected by the same receiver (and for most effective crosstalk compensation should have propagated along the same physical path from transmitter to receiver), crosstalking physical entities *cannot* simultaneously be used as switching dimensions in an optical networking context. For example: (1) The use of PDM entails polarization crosstalk, which is close to perfectly undone by the coherent receiver receiving *both* signal polarizations; a network node cannot use "polarization" as a switching dimension to route the two polarizations to different end points in the network; (2) SDM systems over fibers with spatial crosstalk, e.g., coupled-core fibers or few-mode fibers (FMFs), must ensure that *all* spatial paths can be co-processed in a single MIMO-DSP, and coupled-path SDM networks must therefore not also use the spatial dimension for switching; (3) if cross-channel nonlinearities are to be compensated, the respective frequency channels should have propagated together from a common transmitter to a common receiver, which prohibits their use as wavelength-switching entities.

Switching

From a routing and switching point of view, spectrum and space are not created equal either. This is mostly rooted in the fact that spatial switching is a fairly simple linear process, while spectral switching requires a nonlinear process known as wavelength conversion, implemented either through opto-electronic regeneration or through (as of yet impractical but amply researched) all-optical techniques. The resulting consequences on node blocking probabilities are discussed in [2] and Chapter 18, Networking and routing in space-division multiplexed systems, in this book.

8.4 Architectural aspects of WDM × SDM systems

8.4.1 A Matrix of unit cells and their scaling

As evident from our preceding fundamental Shannon considerations and our discussions on practical limitations of bandwidth scaling, parallelism in space is the only option to significantly scale system capacities by appreciable factors in the long run, complementing wavelength scalability (WDM) using the WDM × SDM matrix shown in Fig. 8.7 [1]. We note in this context that SDM denotes the use of parallel

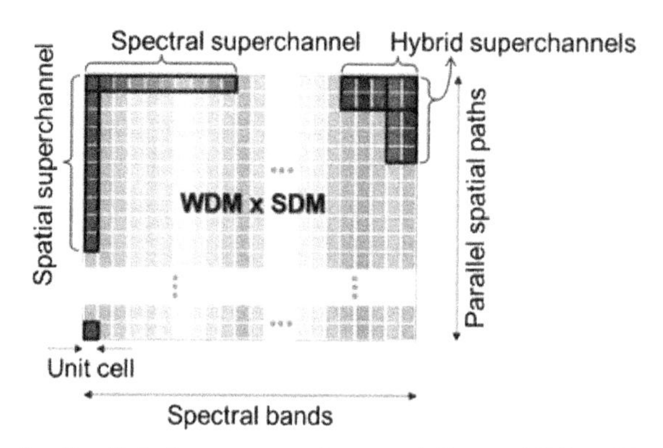

Figure 8.7 A matrix of unit cells in frequency and space defines a WDM × SDM system, whose logical channels are spectral, spatial, or hybrid superchannels. Source: *P.J. Winzer, D.T. Neilson, From scaling disparities to integrated parallelism: a decathlon for a decade, J. Lightwave Technol. 35 (5) (2017) 1099–1115.*

spatial paths, irrespective of whether or not crosstalk compensation is being used. (In the same spirit, we speak of WDM whenever multiple wavelength channels are being used, irrespective of whether or not cross-channel nonlinearities are being compensated.)

Each row of the WDM × SDM matrix represents wavelength multiplexing on a common spatial path, and each column denotes multiple parallel spatial paths using the same carrier frequency. Each unit cell stands for an optical signal modulated onto a single optical carrier using a single optical modulator and detected by a single optical receiver. Assuming that the scaling of opto-electronic modulation and detection hardware continues along its long-term trajectory (cf. Section 8.2.2), commercial (CMOS ASIC integrated) symbol rates of 120 and 300 GBaud should be available by 2027 and 2037, respectively, which implies a bandwidth per unit cell on the order of 120 and 300 GHz when using digital electronic pulse shaping. Depending on the target transmission distance (and hence on the obtainable *SE*, cf. Fig. 8.3), this may then correspond to a bit rate per unit cell of around 1 Tb/s in 2027 and up to 5 Tb/s in 2037. Depending on whether systems will use simple C-band technologies or expand to the entire S + C + L-band region, the requirement for Pb/s systems in 2027 and 100-Pb/s systems in 2037 dictates tens (2027) to hundreds and even thousands (2037) of parallel spatial paths, as summarized in Table. 8.2 [2].

8.4.2 Spatial and spectral superchannels

Whether a single logical interface is constructed out of unit cells on a common optical path across different wavelengths (*spectral superchannel*), at a common wavelength across parallel spatial paths (*spatial superchannel*), or as a mixture of both

Table 8.2 Possible system evolutions over next 10 and 20 years.

	2017	2027	2037
Symbol rate (GBaud)	50	120	300
Bit rate (Gb/s)	200–400	600–1600	2000–6000
System bandwidth (THz)	5	5–12	5–20
Capacity per spatial path (Tb/s)	20–40	25–160	32–400
Unit cells per spatial path	100	40–100	16–66
Target system capacity (Pb/s)	0.02–0.04	1–2	50–100
Required number of spatial paths	1	6–80	125–3125

Source: P.J. Winzer, D.T. Neilson, A.R. Chraplyvy, Fiber-optic transmission and networking: the previous 20 and the next 20 years, Opt. Express 26 (18) (2018) 24190–24239.

(*hybrid superchannel*), cf. Fig. 8.7, depends on the application scenario in which the system is being deployed.

The immediate benefit of *spectral superchannels* [68] is that they can be deployed irrespective of the number of parallel spatial paths available on a link, which greatly benefits terrestrial metro and long-haul networks that are usually confronted with a brown-field situation, as discussed in Section 8.3. Existing infrastructure will be reused until all available parallel spatial paths are filled up and new cables (with a massive number of parallel fiber strands) will have to be added to upgrade the network. We note that as interface rates start approaching the capacity of a full WDM systems (e.g., the 10-Tb/s interface rates expected within a decade), the notion of a *channel* becomes equivalent to that of a *fiber* as opposed to that of a *wavelength*. This evolution will eliminate the need for wavelength switching within the core of the network and will lead to purely spatially-switched core network architectures with wavelength switching used only for edge grooming [2] cf. Fig. 8.2A. Spectral superchannels offer the benefit of closer subcarrier spacings in optically routed networks as well as the possibility to digitally compensate for nonlinear crosstalk among its subcarriers, albeit at only small system performance gains [65]. On the downside, generating spectral superchannels with a significant number of subcarriers requires managing multiple source wavelengths, either by integrating multiple highly stabilized lasers or by demultiplexing an externally supplied frequency comb (with sufficient spectral flatness and with sufficient power per comb line [69]) and subsequently remultiplexing the modulated signals onto a common fiber within the spectral superchannel transponder, which becomes particularly problematic when the superchannel's spectral tunability is to be maintained, as it either requires a WSS per transponder or a (lossy) passive combiner [2,5]. In addition, the wavelength variability of the integrated components and the need for gain-flattened amplification within a superchannel transponder can pose practical problems and add complexity to the solution.

Spatial superchannels by definition operate on a single laser carrier, which makes integrated designs significantly easier [2]. For example, only a single laser is needed per superchannel transponder, all integrated components operate at the same wavelength, and gain-flattening of amplifiers within the superchannel transponder is not required; a single-source wavelength can be passively split and used across all modulators within the superchannel (as well as across all receivers to act as a local oscillator for coherent detection), with lower overall laser power requirements compared to a spectral superchannel [2,5]. In addition, spatial superchannels can more readily share common housekeeping functions such as laser frequency/phase recovery or clock recovery among their unit cells [70]. Linear crosstalk arising from the dense integration of spatial paths can be digitally compensated in a spatial superchannel architecture through MIMO-DSP [66,67].

8.4.3 Array integration and a holistic DSP-electronics-optics co-design

Across technologies, parallelism has always benefited from the integration of simple unit cells, e.g., in integrated circuits and multicore processors. This general fact lets integration aspects become a central theme associated with SDM systems, in addition to purely architectural aspects leveraging the WDM × SDM matrix. In an optical communication system, integration pertains to all subsystems, resulting in transponder arrays, amplifier arrays, optical switch arrays, and the integration of parallel transmission paths into more compact SDM-specific fiber (i.e., arrays of integrated parallel spatial paths), cf. Fig. 8.8 [2]. Regarding SDM-specific fiber, various options have been studied over the past years, including coupled-core and uncoupled-core multicore fiber (MCF) [71,72] and FMF [73–75]. A major benefit of uncoupled-core MCFs is that they don't require any MIMO-DSP at their end points, making the solution compliant with conventional transceivers, even with simple intensity-modulated on/off keying. Similarly, mode group multiplexing over MMFs has been shown to work without MIMO-DSP over distances of several kilometers [76]. In a long-haul coherent transmission context, the additional complexity of MIMO-SDM has been shown to be acceptable [77], and even a benefit in terms of fiber nonlinearities through coupled-mode transmission has been demonstrated [60]. From a techno-economic

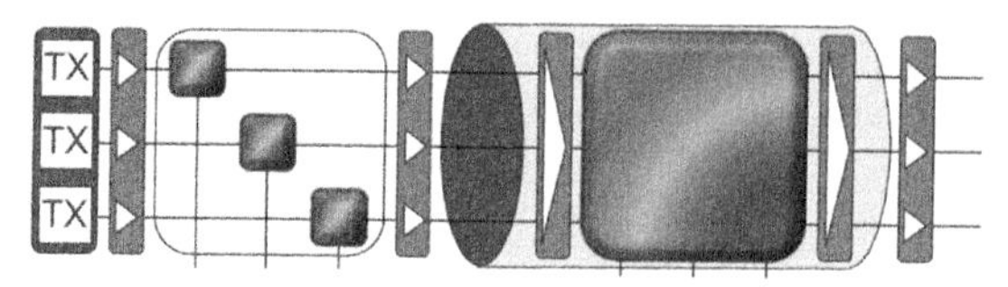

Figure 8.8 Array integration across all system elements. Source: *P.J. Winzer, D.T. Neilson, A.R. Chraplyvy, Fiber-optic transmission and networking: the previous 20 and the next 20 years, Opt. Express 26(18) (2018) 24190–24239.*

point of view, it is not obvious today that SDM-specific fiber, and in particular uncoupled-core MCF, can be produced at a lower cost per spatial path than an SSMF bundle. (The situation is different for coupled-core fiber, which can be cheaper than a fiber bundle, albeit at the cost of MIMO-DSP in the transponders.) In order to reduce the cost of MCF, *standardization* and reasonable manufacturing *volumes* are required. In this context, recent results have revealed that the optimum MCF design is independent of transmission distance, from DCI to submarine, which is expected to facilitate standardization due to the prospect of a larger application space for the same fiber design [78]. Even if MCF remains more expensive per spatial path compared to a bundle of SSMF, the use of MCF may still result in lower-cost overall systems, provided that the higher core density of MCFs enables a cost reduction in other architectural system aspects, such as massive spatial parallelism in submarine cables [4] or highly integrated fiber-to-fiber [79] or fiber-to-chip inter-connections [80,81]. Note in this context that the reason why MMF has prevailed in short-reach links for a long time was not that MMF is cheaper than SSMF (in fact, MMF is several times the cost of SSMF!), but its interconnection to multimode vertical cavity surface emitting lasers (VCSELs) can be done more cost-effectively.

Regarding optical amplifier array integration, significant benefits can be derived from sharing costs and energy overheads across multiple amplifiers in an array. This is illustrated in Fig. 8.9, showing the required energy contributions within a commercial optical amplifier line card typical of a terrestrial long-haul system. The optical signal power at the amplifier output is represented by the thin red slice in (A). The energy consumption within an amplifier module is dominated by various overheads, e.g., for transient control functions (*blue slice*), followed by the power of the thermoelectric cooler (TEC, *cyan slice*) associated with the pump laser. The *green* and *white slices* are

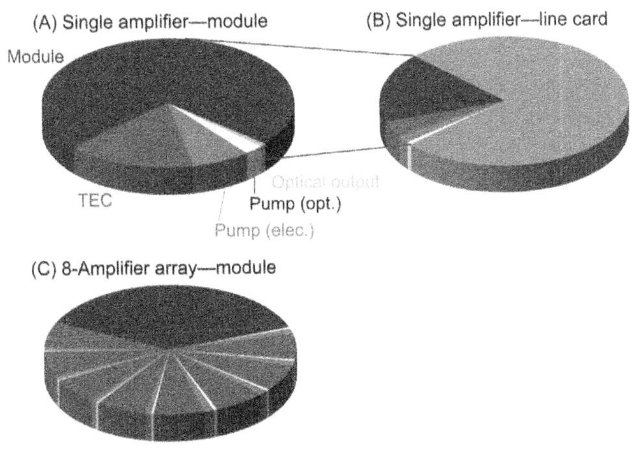

Figure 8.9 Array integration across all system elements for the case of optical amplifier line cards.

the electrical and optical pump powers, respectively. Once the amplification module is placed inside a line card, further overheads dominate the line card energy consumption, such as supply power conditioning as well as control and monitoring functions, cf. *yellow slice* in (B). Once multiple amplifiers are integrated within a common module, sharing of only the module overheads (and not even accounting for potential TEC consolidation in this particular example) results in the much more favorable situation shown in (C). Another example of the benefits of array amplification, pertaining to optimally using all available optical pump power, is given in Ref. [82].

We skip an in-depth discussion on integration aspects of optical switching elements here, as these are discussed in the context of a WDM × SDM matrix in Chapter 18, Networking and routing in space-division multiplexed systems, of this book.

Transponder array integration benefits both spectral and spatial superchannel architectures. For example, in order to construct a 10-Tb/s superchannel, ten 1-Tb/s unit cells must be integrated into a single transponder. If a unit cell cannot achieve 1 Tb/s due to trade-offs resulting from array integration, one has to transition to a correspondingly larger number of lower-rate unit cells, e.g., 100 × 100 Gb/s instead of 10 × 1 Tb/s, as the aggregate superchannel rate must be maintained. As shown in Fig. 8.10, and as further discussed in [1,2], there are three aspects of transponder array integration: (1) Opto-electronic array integration, (2) optics-electronics co-integration, i.e., the hybrid or monolithic integration of the opto-electronic array with the CMOS DSP ASIC, and (3) holistic end-to-end DSP integration, i.e., the co-design of the DSP to compensate for performance shortcomings due to the high integration density [57]. The result is a Fiber-In-Fiber-Out (FIFO) coherent processing engine that optically interfaces via client-side and line-side fibers, as depicted in Fig. 8.10 [1,2].

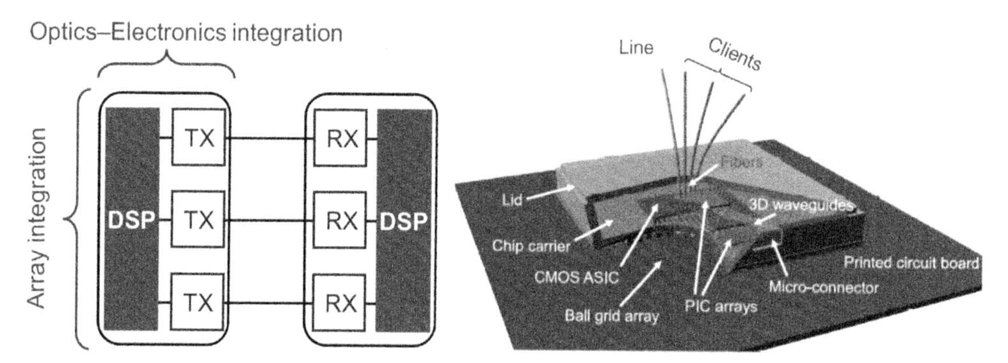

Figure 8.10 Array integration, optics-electronics co-integration, and holistic DSP-opto-electronic co-design will be needed for any superchannel transponder to realize fiber-in/fiber-out (FIFO) engines. Source: *P.J. Winzer, D.T. Neilson, From scaling disparities to integrated parallelism: a decathlon for a decade, J. Lightwave Technol. 35 (5) (2017) 1099–1115 and P.J. Winzer, D.T. Neilson, A.R. Chraplyvy, Fiber-optic transmission and networking: the previous 20 and the next 20 years, Opt. Express 26 (18) (2018) 24190–24239.*

Very similar FIFO architectures are expected to enter switch chips as well, owing to the disparity between switch scaling and interface scaling discussed along with Fig. 8.2.

8.5 Techno-economic trade-offs in WDM × SDM systems

We conclude this chapter with some examples of techno-economic analyses relating to WDM × SDM system architecture trade-offs. All these concrete examples draw from the more general observations made in the previous sections.

8.5.1 Chip-to-chip interconnects

In order to find the optimum architecture for optical chip-to-chip interconnects, we apply the findings of Sections 8.3.2 and 8.3.3. As chip-to-chip communication is inherently a power-limited application, the lowest energy-per-bit solution maximizes multiplexing and minimizes *SE*. As such, simple binary bandwidth-expanding modulation formats such as return-to-zero on/off keying are an attractive solution that provide extra receiver sensitivity per spatial path [83,84], coupled with as large a number of parallel spatial paths as is feasible from a packaging point of view, e.g., using MCFs to achieve the required interface density. As the thermal environment on-chip is challenging for components and is dynamically changing with the chip workload, any temperature-sensitive functions should be removed from within the package. This includes the avoidance of copackaged resonant modulators, wavelength (de)multiplexers, and lasers, leading to simple absorption-based modulation technologies that operate at a single wavelength per spatial path. The removal of lasers from within the package in the spirit of an "optical power supply" [1,2] not only allows for more energy-optimized photon generation solutions off-chip but also removes the dissipation associated with photon generation from within the package.

8.5.2 Datacenter interconnects

DCI systems are often confronted with a green-field situation and hence can deploy the required number of parallel fibers at the time of system installation. Today's DCI systems employ a massive number of parallel spatial paths (in some cases exceeding 10,000 parallel fibers linking regional datacenter buildings), and both 100 and 400 GbE standards include parallel single-mode (PSM) interfaces [85], hence already embracing SDM.

To optimize wavelength versus space allocations in a WDM × SDM transmission system, we first write the cost per b/s of the entire system as

$$c_{\text{System}} = c_{\text{TRX}}(W, M, R) + c_{\text{Path}}(W, M, R)/(WR), \tag{8.12}$$

where c_{TRX} (W, M, R) is the cost per b/s of a transponder, which may depend on the number of wavelengths W, the number of spatial paths M, and the transponder interface rate R; the cost per spatial path c_{Path} (W, M, R) may depend on the same three variables and includes cabling and deployment. The cost of additional line system components such as amplifiers or multiplexers may be either folded into the line cost or apportioned to the transponder cost in the model of Eq. (8.12); we will use the latter approach in our analysis. The cost per spatial path is amortized across the capacity (WR) it is supporting. While transponder costs per bit can be determined fairly easily, the cost of deployed fiber is much harder to asses and depends significantly on the deployment scenario. For example, if ample parallel strands of fiber are already deployed between two datacenter sites, or if the cost of fiber deployment is covered by a different cost center than the one paying for the optical transmission system (both cases being real-world scenarios), the cost of fiber becomes essentially zero to the transmission system designer. On the other end of the spectrum, if the deployment of additional fiber requires the construction of new ducts in metropolitan areas, the cost of fiber can be exceedingly high. As a further complication (not taken into account in our model), the cost of deployed fiber may exhibit a step-function behavior: The cost per additional spatial path may remain prohibitively high (to avoid new deployments), up to the point where the first new fiber strand must be deployed to meet other architectural constraints. At that point, the cost of additional spatial paths may drop to the cost of a bare fiber (a few dollars per kilometer), as deployment costs are dominated by cabling and installation costs. Owing to the wide variations in the cost per spatial path, we treat c_{Path} as a free variable in our assessment.

Fig. 8.11 shows the total cost per b/s c_{System} of a 40-km DCI system as a function of the cost per spatial path c_{Path} for various assumptions on transponder costs c_{TRX} (factoring in line system components) as per the inset table. If fiber costs are negligible, the system cost depends exclusively on the transponder cost per b/s (including the necessary line system components to make the transponders work over the assumed 40 km of SSMF). With the transponder cost assumptions made here, the lowest-cost system uses direct detection (DD) 10-Gb/s black and white interfaces, albeit at massive requirements on the available parallel fiber infrastructure. (For example, a 50-Tb/s DCI system built this way requires 10,000 parallel fiber strands.) Going to a 100-Gb/s DD solution (e.g., using 25-GBaud dual-wavelength PAM-4 [86], and including then needed dispersion-compensating fiber (DCF), *solid red* in Fig. 8.11) increases system costs relative to 10-Gb/s if fiber costs are negligible. If fiber costs are significant, the 100-Gb/s-based system becomes cheaper than the 10-Gb/s solution. The cost range of an additional fiber strand (provided that cabling and installation are already absorbed elsewhere, as discussed above) is shown as a *shaded region* in Fig. 8.10. Using moderate (e.g., 8-channel) WDM, the cost of the

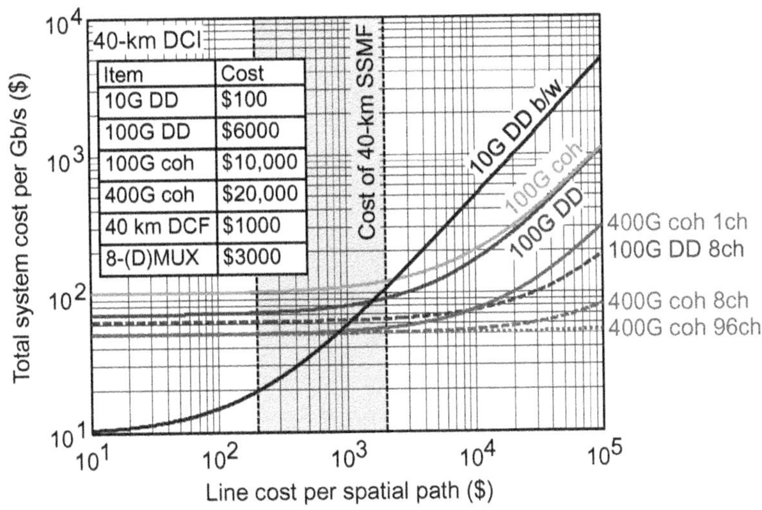

Figure 8.11 Total cost c_{System} of deploying a datacenter interconnect (DCI) system as a function of the cost of a spatial path c_{Path} for various transponder technologies whose assumed costs are summarized in the inset table.

system is further reduced. Coherent transponders at 100 Gb/s are not cost-advantageous under our cost assumptions (*green curve* in Fig. 8.11), but packing more bits onto a laser carrier amortizes transponder costs across more transmitted information bits, making 400-Gb/s coherent transponders an attractive solution (*solid blue*). Depending on the cost per spatial path, the cost reduction of WDM can be very limited, though, and massive WDM systems only prove in at very high costs for deployed fiber (*dashed* and *dotted blue curves*). This restricts the frequently made claim that DCI systems should have high spectral efficiencies to achieve high per-fiber capacities to systems where new fiber deployments are not an option.

8.5.3 Metro and long-haul networking

In complete analogy to the small capacity advantage from increasing the SNR in a terrestrial long-haul system (cf. Table 8.1, Section 8.3.1), systems with reduced SNR performance relative to a reference system only show fairly modest capacity losses, as shown in Table. 8.3. For example, a system operating at 15 dB SNR (corresponding to a dual-polarization *SE* of 10 b/s/Hz) only suffers a 19% capacity loss when reducing the SNR by 3 dB. This finding is particularly important in massively parallel WDM × SDM systems using lower-cost average-quality fiber, line components, and transponders, as the capacity loss due to suboptimum components can be compensated by additional parallel spatial paths. Such a strategy makes sense whenever the capacity penalty due to lower-quality components is outweighed by the resulting cost savings.

Table 8.3 Capacity losses in reduced SNR systems.

		SNR_0		
		20 dB	**15 dB**	**10 dB**
ΔSNR	−1 dB	5%	6%	9%
	−3 dB	15%	19%	25%
	−10 dB	48%	59%	71%

To see this, consider the total cost of all transponders within a fully loaded system, leading to a (transponder only) system cost of

$$c_{\text{System}} = 2MB\ SE\ c_{\text{TRX}} \tag{8.13}$$

A loss in SNR performance, manifested by $SE' < SE$ but compensated by $M' > M$ parallel spatial paths, results in an equal-capacity system if

$$2M'B\ SE' = 2MB\ SE \tag{8.14}$$

or

$$M' = M\ SE/SE' \tag{8.15}$$

Hence, the cost of the new system is given by

$$c'_{\text{system}} = 2M'B\ SE'\ c'_{\text{TRX}} = c_{\text{system}} \cdot c'_{\text{TRX}}/c_{\text{TRX}} \tag{8.16}$$

In other words, any cost savings due to lower-performing transponders translates directly into total system cost savings, provided that enough parallel spatial paths are available to make $M' = M\ SE/SE'$. Using our above example, a 19% capacity loss can be offset by a 23% increase in the number of spatial paths in a massively parallel system. If these additional spatial paths are available, the resulting system with 3-dB worse-performing transponders is cheaper compared to the reference system as long as the worse-performing transponders cost 19% less than the reference transponders, which puts a monetary value on a 3-dB transponder SNR penalty in the presence of a sufficient number of parallel spatial paths. As the incremental cost of installing an additional strand of fiber as part of a cable deployment is to first order given by the cost of an additional bare fiber (~0.1% or less of the cable installation cost per additional fiber strand [40]), and as the cost ratio between fiber types can be as high as a factor of 10 [4], operators should deploy the *largest possible number of legacy-quality fibers per cable* for the lowest-cost long-term evolution of their networks. The paradigm shift in system design from installing high-fiber-count cables "beyond the cliff" is discussed along with Fig. 8.4.

8.5.4 Submarine systems

For submarine systems, which are supply-power constrained due to the fact that the DC power supply for the submersed amplifiers needs to be fed from the two ends of the cable [47], it has recently been shown [4,49,50] that massive SDM with ~50 parallel fibers per direction leads to higher capacity, lower cost system architectures, even if state-of-the-art technology without any further SDM integration is being used, a reflection of the trade-off discussed along with Eq. (8.9). Fig. 8.12 shows a key result of [4], indicating that about 44% of cost savings can be achieved when evolving from a state-of-the-art 8-fiber-pair submarine system to a 50-fiber-pair system. While doing so, the available optical power is diluted across the parallel spatial paths, pushing the system from being limited by fiber nonlinearities (dashed) into the purely linear propagation regime (solid and dotted). The inset in Fig. 8.12 illustrates this fact by comparing the nonlinearity-optimized point of operation of state-of-the-art systems with the linear operating point of cost-optimized submarine SDM systems. This very important finding has wide-reaching and rather counterintuitive consequences, such as the fact that low-nonlinearity (large effective area) fiber will *not* be needed for future submarine systems, and *neither* will digital nonlinearity compensation. Additional findings of [4] include the fact that narrower-band systems than deployed today will be preferred in future submarine systems (such as half-C-band system), as these avoid power loss due to loss-inducing gain flattening filters. Lastly, the fact that the ratio of optical amplifier energy efficiency to amplifier noise figure appears as an amplifier figure of merit in submarine system cost optimizations [4] implies that poorer noise figure amplifiers that are more energy efficient may be

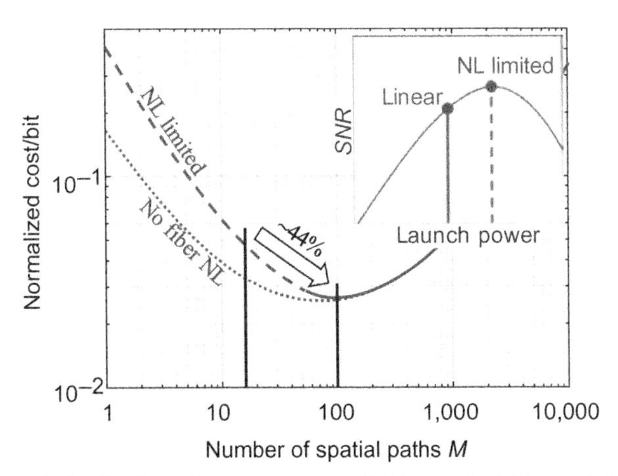

Figure 8.12 A cost-optimized submarine system uses 100 instead of 16 parallel spatial paths, yielding 44% in cost-per-b/s savings. Such massive SDM submarine systems operate in the linear transmission regime.

preferable for future submarine applications, which leaves ample room for impactful device research.

Acknowledgments

I would like to acknowledge many valuable discussions on the material presented in this paper with David Neilson, Andy Chraplyvy, Bob Tkach, and many others.

References

[1] P.J. Winzer, D.T. Neilson, From scaling disparities to integrated parallelism: a decathlon for a decade, J. Lightwave Technol. 35 (5) (2017) 1099–1115.
[2] P.J. Winzer, D.T. Neilson, A.R. Chraplyvy, Fiber-optic transmission and networking: the previous 20 and the next 20 years, Opt. Express 26 (18) (2018) 24190–24239.
[3] P.J. Winzer, Would scaling to extreme ultraviolet or soft X-ray communications resolve the capacity crunch? J. Lightwave Technol. 36 (24) (2018) 5786–5793.
[4] R. Dar, P.J. Winzer, A.R. Chraplyvy, S. Zsigmond, K.-Y. Huang, H. Fevrier, et al., Cost-optimized submarine cables using massive spatial parallelism, J. Lightwave Technol. 36 (18) (2018).
[5] P.J. Winzer, Spatial multiplexing in fiber optics: the 10x scaling of metro/core capacities, Bell Labs Tech. J. 19 (2014) 22–30.
[6] E.B. Desurvire, Capacity demand and technology challenges for lightwave systems in the next two decades, J. Lightwave Technol. 24 (12) (2006) 4697–4710.
[7] A.R. Chraplyvy, The coming capacity crunch, in: European Conference on Optical Communication (ECOC), Vienna, Austria, Plenary Session (2009).
[8] R.W. Tkach, Scaling optical communications for the next decade and beyond, Bell Labs Tech. J. 14 (4) (2010) 3–9.
[9] R.-J. Essiambre, G. Kramer, P.J. Winzer, G.J. Foschini, B. Goebel, Capacity limits of optical fiber networks, J. Lightwave Technol. 28 (2010) 662–701.
[10] More Moore, The international roadmap for devices and systems. IEEE (2017). Available from: <https://irds.ieee.org/images/files/pdf/2017/2017IRDS_MM.pdf>.
[11] More-than-Moore, Available from: <http://www.itrs2.net/uploads/4/9/7/7/49775221/irc-itrs-mtm-v2_3.pdf>.
[12] Beyond CMOS, The international roadmap for devices and systems, IEEE (2016). Available from: <https://irds.ieee.org/images/files/pdf/2016_BC.pdf>.
[13] J. Gray, P. Shenoy, Rules of thumb in data engineering. Tech. Rep. MS-TR-99-100. Microsoft Research, Redmond, WA, United States, 2000.
[14] CISCO, Visual networking index: forecast and methodology, 2013–2018 (2016). Available from: <https://www.cisco.com/c/en/us/solutions/collateral/service-provider/visual-networking-index-vni/white-paper-c11-741490.html>.
[15] G. Wellbrock, T.J. Xia, How will optical transport deal with future network traffic growth? in: Proceedings of European Conference on Optical Communications (ECOC), Paper Th.1.2.1 (2014).
[16] A. Gerber, R. Doverspike, 2011. Traffic types and growth in backbone networks, in: Proceedings of Optical Fiber Communications Conference (OFC), Paper OTuR1.
[17] A. Singh, J. Ong, A. Agarwal, G. Anderson, A. Armistead, R. Bannon, et al., 2015. Jupiter rising: a decade of Clos topologies and centralized control in Google's datacenter network, in: Proceedings of Sigcomm, pp. 183–197.
[18] R.J. Stone, Use of embedded optics to decrease power consumption in IO dense systems. In: Proceedings of Optical Fiber Communications Conference (OFC), Paper Th3G.5 (2017).
[19] M.P. Li, J. Martinez, D. Vaughan, Transferring High-Speed Data over Long Distances with Combined FPGA and Multichannel Optical Modules. White Paper, Avago, Altera (2012).

[20] G. Keeler, Photonics in the Package for Extreme Scalability (PIPES), (2018). Available from: <https://www.darpa.mil/attachments/DARPA_PIPES_Proposers_Day_Slides_Gordon_Keeler.pdf>.
[21] H. Sun, K.-T. Wu, K. Roberts, Real-time measurements of a 40 Gb/s coherent system, Opt. Express 16 (2008) 873–879.
[22] C.E. Shannon, A mathematical theory of communication, Bell Syst. Tech. J. 27 (1948) 379–423. and 623–656.
[23] J.P. Gordon, Quantum effects in communications systems, Proc. IRE 50 (1962) 1898–1908.
[24] S. Holevo, Bounds for the quantity of information transmitted by a quantum communication channel, Probl. Peredachi Inf. 9 (3) (1973) 3–11.
[25] C.E. Shannon, Communication in the presence of noise, Proc. IRE 37 (1) (1949) 10–21.
[26] T.M. Cover, J.A. Thomas, Elements of Information Theory, second ed., Wiley, 2006.
[27] S. Tibuleac, M. Filer, Transmission impairments in DWDM networks with reconfigurable optical add-drop multiplexers, J. Lightwave Technol. 28 (2010) 557–568.
[28] S. Gringeri, B. Basch, V. Shukla, R. Egorov, T.J. Xia, Flexible architectures for optical transport nodes and networks, IEEE Commun. Mag. (2010) 40–50.
[29] J. Bromage, P.J. Winzer, R.-J. Essiambre, "Multiple-path interference and its impact on system design,", in: M.N. Islam (Ed.), Raman Amplifiers and Oscillators in Telecommunications, Springer Verlag, 2003.
[30] A. Splett, C. Kurzke, K. Petermann, Ultimate transmission capacity of amplified optical fiber communication systems taking into account fiber nonlinearities, in: Proceedings of European Conference on Optical Communication (ECOC), Paper MoC2.4 (1993).
[31] P. Poggiolini, G. Bosco, A. Carena, V. Curri, Y. Jiang, F. Forghieri, The GN model of fiber nonlinear propagation and its applications, J. Lightwave Technol. 32 (4) (2014) 694–721.
[32] R. Dar, M. Feder, A. Mecozzi, M. Shtaif, Properties of nonlinear noise in long, dispersion-uncompensated fiber links, Opt. Express 21 (2013) 25685–25699.
[33] P.J. Winzer, High-spectral-efficiency optical modulation formats, J. Lightwave Technol. 30 (2012) 3824–3835.
[34] R.S. Tucker, Green optical communications—Part I: energy limitations in transport, IEEE J. Sel. Top. Quantum Electron 17 (2) (2011) 245–260.
[35] J.G. Proakis, Digital Communications, fourth ed., McGraw-Hill, New York, 2001.
[36] F. Buchali, F. Steiner, G. Böcherer, L. Schmalen, P. Schulte, W. Idler, Rate adaptation and reach increase by probabilistically shaped 64-QAM: an experimental demonstration, J. Lightwave Technol. 34 (7) (2016) 1599–1609.
[37] G. Boecherer, et al., On joint design of probabilistic shaping and forward error correction for optical systems, J. Lightwave Technol. (2019).
[38] J. Cho, P. Winzer, Probabilistic constellation shaping for optical fiber communications, J. Lightwave Technol. 37 (6) (2019) 1590–1607.
[39] P.J. Winzer, Energy-efficient optical transport capacity scaling through spatial multiplexing, Photonics Technol. Lett 23 (13) (2011) 851–853.
[40] R. Dar, M. Shtaif, M. Feder, New bounds on the capacity of the nonlinear fiber-optic channel, Opt. Lett. 39 (2014) 398–401.
[41] R. Dar, M. Feder, A. Mecozzi, M. Shtaif, Accumulation of nonlinear interference noise in fiber-optic systems, Opt. Express 22 (12) (2014) 14199–14211.
[42] S.L.I. Olsson, J. Cho, S. Chandrasekhar, X. Chen, E.C. Burrows, and P.J. Winzer, Record-high 17.3-bit/s/Hz spectral efficiency transmission over 50 km using probabilistically shaped PDM 4096-QAM, in: Proceedings of Optical Fiber Communications Conference (OFC), Paper Th4C.5 (2018).
[43] S. Chandrasekhar, B. Li, J. Cho, X. Chen, E.C. Burrows, G. Raybon, et al., High-spectral-efficiency transmission of PDM 256-QAM with parallel probabilistic shaping at record rate-reach trade-offs, in: Proceedings of European Conference on Optical Communications (ECOC), Paper Th.3.C.1 (2016).
[44] J. Cho, X. Chen, S. Chandrasekhar, G. Raybon, R. Dar, L. Schmalen, et al., Trans-Atlantic field trial using high spectral efficiency probabilistically shaped 64-QAM and single-carrier real-time 250-Gb/s 16-QAM, J. Lightwave Technol. 36 (1) (2018) 103–113.

[45] A. Leven, L. Schmalen, Status and recent advances on forward error correction technologies for lightwave systems, J. Lightwave Technol. 32 (2014) 2735–2750.

[46] A. Alvarado, E. Agrell, D. Lavery, R. Maher, P. Bayvel, Replacing the soft-decision FEC limit paradigm in the design of optical communication systems, J. Lightwave Technol. 33 (2015) 4338–4352.

[47] K. Takehira, Submarine system powering, in: J. Chesnoy (Ed.), Undersea Fiber Communication Systems, second ed., Academic Press, New York, NY, 2016.

[48] S.K. Korotky, Price-points for components of multi-core fiber communication systems in backbone optical networks, J. Opt. Commun. Netw. 4 (2012) 426–435.

[49] A. Pilipetskii, High capacity submarine transmission systems, in: Proceedings of Optical Communication Fiber Conference, 2015, Paper W3G.5.

[50] O.V. Sinkin, A.V. Turukhin, W.W. Patterson, M.A. Bolshtyansky, D.G. Foursa, A.N. Pilipetskii, Maximum optical power efficiency in SDM based optical communication systems, IEEE Photon. Technol. Lett 29 (13) (2017) 1075–1077.

[51] M.N. Petrovich, F. Poletti, J.P. Wooler, A.M. Heidt, N.K. Baddela, Z. Li, et al., Demonstration of amplified data transmission at 2 μm in a low-loss wide bandwidth hollow core photonic bandgap fiber, Opt. Express 21 (23) (2013) 28559–28569.

[52] D.J. Richardson, Hollow core fibres and their applications, in: Proceedings of Optical Fiber Communication Conference (OFC), Paper Tu3H.1 (2017).

[53] F. Poletti, Nested antiresonant nodeless hollow core fiber, Opt. Express 22 (20) (2014) 23807–23828.

[54] J. Renaudier, A. Carbo Meseguer, A. Ghazisaeidi, P. Tran, R. Muller, R. Brenot, et al., First 100-nm continuous-band WDM transmission system with 115 Tb/s transport over 100 km using novel ultra-wideband semiconductor optical amplifiers, in: Proceedings of European Conference on Optical Communication (ECOC), Paper Th.PDP.A.3 (2017).

[55] A. Mori, H. Masuda, K. Shikano, K. Oikawa, K. Kato, M. Shimizu, Ultra-wideband tellurite-based Raman fibre amplifier, Electron. Lett. 37 (24) (2001) 1442–1443.

[56] F. Hamaoka, K. Minoguchi, T. Sasai, A. Matsushita, M. Nakamura, S. Okamoto, et al., 150.3-Tb/s ultra-wideband (S, C, and L bands) single-mode fibre transmission over 40-km using >519 Gb/s/λ PDM-128QAM signals, in: Proceedings of ECOC, Paper Mo4G.1 (2018).

[57] X. Chen, P. Dong, S. Chandrasekhar, K. Kim, B. Li, H. Chen, et al., Characterization and digital pre-compensation of electro-optic crosstalk in silicon photonics I/Q modulators, in: Proceedings of European Conference on Optical Communications (ECOC), Paper Tu3.A.5 (2016).

[58] C. Antonelli, A. Mecozzi, M. Shtaif, P.J. Winzer, Stokes-space analysis of modal dispersion in fibers with multiple mode transmission, Opt. Express 20 (11) (2012) 11718–11733.

[59] A. Mecozzi, C. Antonelli, M. Shtaif, Coupled Manakov equations in multimode fibers with strongly coupled groups of modes, Opt. Express 20 (21) (2012) 23436–23441.

[60] R. Ryf, J.C. Alvarado, B. Huang, J. Antonio-Lopez, S.H. Chang, N.K. Fontaine, et al., Long-distance transmission over coupled-core multicore fiber, in: Proceedings of European Conference on Optical Communications (ECOC), Paper Th.3.C.3 (2016).

[61] T. Zeng, Superchannel transmission system based on multi-channel equalization, Opt. Express 21 (12) (2013) 14799–14807.

[62] C. Xia, et al., Impact of channel count and PMD on polarizationmultiplexed QPSK transmission, J. Lightwave Technol. 29 (21) (Nov. 2011) 3223–3229.

[63] A. Mecozzi, R.-J. Essiambre, Nonlinear Shannon limit in pseudolinear coherent systems, J. Lightwave Technol. 30 (2012) 2011–2024.

[64] J.C. Cartledge, F.P. Guiomar, F.R. Kschischang, G. Liga, M.P. Yankov, Digital signal processing for fiber nonlinearities [Invited], Opt. Express 25 (2017) 1916–1936.

[65] R. Dar, P.J. Winzer, Nonlinear interference mitigation: methods and potential gain, J. Lightwave Technol. 35 (2017) 903–930.

[66] P.J. Winzer, G.J. Foschini, MIMO capacities and outage probabilities in spatially multiplexed optical transport systems, Opt. Express 19 (17) (2011) 16680–16696.

[67] S. Randel, R. Ryf, A. Sierra, P.J. Winzer, A.H. Gnauck, C. Bolle, et al., 6 × 56-Gb/s mode-division multiplexed transmission over 33-km few-mode fiber enabled by 6 × 6 MIMO equalization, Opt. Express 19 (17) (2011) 16697–16707.

[68] S. Chandrasekhar, X. Liu, OFDM based superchannel transmission technology, J. Lightwave Technol. 30 (24) (2012) 3816–3823.
[69] B.J. Puttnam, R.S. Luís, G. Rademacher, J. Sakaguchi, W. Klaus, Y. Awaji, et al., High-capacity MCF transmission with wideband comb, in: Optical Fiber Communication Conference, OSA Technical Digest (online), Optical Society of America, (2017), Paper M2J.4.
[70] M.D. Feuer, L.E. Nelson, X. Zhou, S.L. Woodward, R. Isaac, B. Zhu, et al., Joint digital signal processing receivers for spatial superchannels, IEEE Photon. Technol. Lett 24 (21) (2012) 1957–1960.
[71] B. Zhu, T.F. Taunay, M.F. Yan, J.M. Fini, M. Fishteyn, E.M. Monberg, et al., Seven-core multicore fiber transmissions for passive optical network, Opt. Express 18 (11) (2010) 11117–11122.
[72] K. Saitoh, S. Matsuo, Multicore fiber technology, J. Lightwave Technol. 34 (1) (2016) 55–66.
[73] R. Ryf, S. Randel, A.H. Gnauck, C. Bolle, A. Sierra, S. Mumtaz, et al., Mode-division multiplexing over 96 km of few-mode fiber using coherent 6x6 MIMO processing, J. Lightwave Technol. 30 (4) (2012) 521–531.
[74] R. Ryf, N. Fontaine, Space-division multiplexing and MIMO processingChapter 16 Enabling Technologies for High Spectral-efficiency Coherent Optical Communication Networks, Wiley, 2016pp. 547–607.
[75] M. Bigot, D. Molin, K. de Jongh, D. Van Ras, F. Achten, P. Sillard, Next-generation multimode fibers for space division multiplexing, in: Proceedings of Advanced Photonics Congress (IPR), Paper NeM3B.4 (2017).
[76] K. Benyahya, C. Simonneau, A. Ghazisaeidi, N. Barré, Pu Jian, J.-F. Morizur, et al., 14.5 Tb/s mode-group and wavelength multiplexed direct detection transmission over 2.2 km OM2 fiber, in: Proceedings of ECOC (2017).
[77] S. Randel, P.J. Winzer, M. Monoliu, R. Ryf, Complexity analysis of adaptive frequency-domain equalization for MIMO-SDM transmission, in: Proceedings of European Conference on Optical Comm. (ECOC), Paper Th.2.C.4 (2013).
[78] J.M. Gene, P.J. Winzer, A universal specification for multicore fiber crosstalk, IEEE Photonics Tech. L. 31 (9) (2019), 673–676.
[79] Y. Geng et al., High-speed, bi-directional dual-core fiber transmission system for high-density, short-reach optical interconnects, Proc. SPIE9390, 939009 (2015).
[80] B.G. Lee, et al., 120-Gb/s 100-m transmission in a single multicore multimode fiber containing six cores interfaced with a matching VCSEL array, in: Proceedings of IEEE Summer Topical Meeting, 2010, pp. 223–224.
[81] T. Hayashi, A. Mekis, T. Nakanishi, M. Peterson, S. Sahni, P. Sun, et al., End-to-end multi-core fibre transmission link enabled by silicon photonics transceiver with grating coupler array, in: Proceedings of ECOC (2017).
[82] A. Gnauck, P.J. Winzer, R. Jopson, E.C. Burrows, Efficient pumping scheme for amplifier arrays with shared pump laser, in: Proceedings of European Conference on Optical Communication (ECOC'16), Paper M2.A.1 (2016).
[83] P.J. Winzer, A. Kalmár, Sensitivity enhancement of optical receivers by impulsive coding, J. Lightwave Technol. 17 (2) (1999) 171–177.
[84] G.A. Keeler, et al., The benefits of ultrashort optical pulses in optically interconnected systems, J. Sel. Top. Quant. Electron 9 (2) (2003) 477–484.
[85] Parallel single mode four lane per direction 100 Gbit/s optical interface (Online). Available from: <http://psm4.org/>.
[86] N. Eiselt, J. Wei, H. Griesser, A. Dochhan, M. Eiselt, J.-P. Elbers, et al., First real-time 400G PAM-4 demonstration for inter-datacenter transmission over 100 km of SSMF at 1550 nm, in: Proceedings of Optical Fiber Communications Conference, Paper W1K.5D (2016).

CHAPTER 9

High-order modulation formats, constellation design, and digital signal processing for high-speed transmission systems

Tianhua Xu[1,2], Gabriele Liga[3] and Polina Bayvel[2]
[1]School of Engineering, University of Warwick, Coventry, United Kingdom
[2]Optical Networks Group, University College London, London, United Kingdom
[3]Signal Processing Systems Group, Department of Electrical Engineering, Eindhoven University of Technology, Eindhoven, The Netherlands

9.1 Fiber nonlinearity in optical communication systems with higher order modulation formats

9.1.1 Introduction to optical fiber nonlinearity

Over the past four decades, data rates in optical communications systems have seen a dramatic increase. Currently, over 95% of the digital data traffic is carried over optical fiber networks, which form a substantial part of the national and international communication infrastructure. The development of a series of new technologies contributed to this increase. These include techniques such as wavelength-division multiplexing (WDM), advanced modulation formats, optical amplifiers and novel fibers, and coherent detection [1−3]. These developments promoted the revolution of optical communication systems and the growth of the Internet toward the direction of high-capacity and long-distance transmissions. The performance of long-haul high-capacity optical fiber communication systems is significantly degraded by transmission impairments, such as chromatic dispersion (CD), polarization mode dispersion (PMD), laser phase noise, and Kerr fiber nonlinearities [4,5]. With the ability to capture of the amplitude and the phase of the signals using coherent optical detection, the powerful compensation and effective mitigation of the transmission impairments can be implemented using the digital signal processing (DSP) in the electrical domain. This has become one of the most promising techniques for next-generation optical communication networks to achieve a performance close to the Shannon capacity limit. DSP combined with coherent detection is a very promising solution for long-haul, high-capacity optical fiber communication systems, offering great flexibility in the design, deployment, and operation of optical communication networks.

Optical Fiber Telecommunications V11
DOI: https://doi.org/10.1016/B978-0-12-816502-7.00010-5

CD can be compensated using the digital filters in both the time domain and the frequency domain [6,7]. PMD can be equalized adaptively using the least-mean-square method [8], the constant modulus algorithm (CMA), and other approaches [9]. Phase noise from laser sources can also be estimated and compensated using the feed-forward and the feedback carrier phase estimation (CPE) approaches [10,11]. However, the achievable capacity of optical communication systems is limited by the nonlinear distortions inherent to transmission using optical fibers, such as self-phase modulation (SPM), cross-phase modulation (XPM) and four-wave mixing (FWM). These signal degradations are more significant in systems using larger transmission bandwidths (BWs), closer channel spacing, and higher order modulation formats. Optical fiber nonlinearities are the major bottleneck to optical fiber transmission performance [3,12,13]. This chapter focuses on the introduction and investigation of DSP employed for fiber nonlinear impairments compensation based on the coherent detection of optical signals, to provide a roadmap for the design and implementation of real-time optical fiber communication systems.

The optical fiber channel is fundamentally nonlinear, which means that the refractive index of the propagation medium n_{eff} changes in response to the strength of the square of the electric field $E(t, z)$ [4],

$$n_{\mathrm{eff}} = n_0 + n_2|E(t,z)|^2 \tag{9.1}$$

where n_0 and n_2 are the linear and nonlinear refractive indices of the fiber, respectively. For silica, $n_0 \approx 1.5$ and $n_2 \approx 3 \times 10^{-20}\ \mathrm{m}^2/\mathrm{W}$. For Eq. (9.1), note that the value of $|E(t, z)|$ has been normalized, such that $|E(t, z)|^2$ is equal to the optical power I. After a transmission distance of L, the nonlinear optical Kerr effect will result in a phase shift, which, when added to the shift from linear propagation, leads to a total phase shift, given by

$$\phi = k_0 n_0 L + \phi_{\mathrm{NL}} \tag{9.2}$$

where $k_0 = 2\pi/\lambda_0$ is the free-space wave number, λ_0 is the optical wavelength, and

$$\phi_{\mathrm{NL}}(t,z) = j\gamma \int_0^z |E(t,z')|^2 dz' = j\gamma |E(t,0)|^2 L_{\mathrm{eff}} \tag{9.3}$$

and $L_{\mathrm{eff}} \triangleq \frac{1-e^{-\alpha z}}{\alpha}$ is the so-called *effective length*. The fiber nonlinear coefficient γ, a measure of nonlinearity, is described by

$$\gamma = \frac{k_0 n_2}{A_{\mathrm{eff}}} \tag{9.4}$$

where A_{eff} is the effective area of the fiber.

Transmission of more data within a finite BW requires greater optical power levels $|E(t, z)|^2$, which in turn leads to the power-dependent nonlinear distortion of the transmitted data. This imposes a Kerr nonlinearity limit, sometimes referred to as the nonlinear Shannon limit. Strictly speaking, this limit is only a lower bound to the channel capacity, as it assumes that nonlinear interference (NLI) is noise (whereas, as shown above, it is deterministic).

The propagation of signals through an optical fiber is described by the well-known nonlinear Schrödinger equation [4] given, in the single polarization case, by

$$\frac{\partial E(t,z)}{\partial z} = -j\frac{\beta_2}{2}\frac{\partial^2 E(t,z)}{\partial^2 t} - \frac{\alpha}{2}E(t,z) + j\gamma|E(t,z)|^2\, E(t,z) \tag{9.5}$$

where $E(t, z)$ is the optical field as a function of time t and distance z, and β_2, α, and γ are the dispersion, the attenuation, and the nonlinearity coefficient of the fiber, respectively. In the general dual-polarization case, Eq. (9.5) can be generalized to the Manakov equation [14,15]:

$$\frac{\partial \mathbf{E}(t,z)}{\partial z} = -j\frac{\beta_2}{2}\frac{\partial^2 \mathbf{E}(t,z)}{\partial^2 t} - \frac{\alpha}{2}\mathbf{E}(t,z) + j\frac{8}{9}\gamma||\mathbf{E}(t,z)||^2\, \mathbf{E}(t,z) \tag{9.6}$$

where $\mathbf{E}(t, z)$ denotes the dual-polarization optical field vector. The effects of propagation through an optical fiber on an optical signal can be understood by breaking Eq. (9.5) into two different equations obtained by setting β_2 and γ to zero, in turn. For $\gamma = 0$ Eq. (9.6) becomes, for each polarization of the optical field,

$$\frac{\partial E(t,z)}{\partial z} = -j\frac{\beta_2}{2}\frac{\partial^2 E(t,z)}{\partial^2 t} - \frac{\alpha}{2}E(t,z) \tag{9.7}$$

which can be easily solved in closed form in the frequency domain and whose solution is given by

$$E(\omega,z) = E(\omega,0) \times \exp\left(-\frac{\alpha}{2}z\right) \tag{9.8}$$

Eq. (9.8) physically corresponds to a spreading of the optical pulse in the time domain which is quadratically dependent on its optical BW. On the other hand, when $\beta_2 = 0$ but $\gamma \neq 0$ we have

$$\frac{\partial E(t,z)}{\partial z} = -j\frac{8}{9}\gamma|E(t,z)|^2 E(t,z) - \frac{\alpha}{2}E(t,z) \tag{9.9}$$

whose solution in the time domain is given by

$$E(t,z) = E(t,0)\ \exp\left(j\phi_{\mathrm{NL}}(t,z)\right) \tag{9.10}$$

The combination of the above-described effects produces nonlinear distortion or interference [4], which become more significant as the intensity of the optical field increases. Known as the nonlinear Kerr effect, it is the main theoretical bottleneck to reliable transmission of data in optical fiber systems, especially in the presence of amplifier noise where mixing of nonlinear distortion and noise occurs.

The nonlinear term in Eq. (9.5) is responsible for the so-called Kerr effect [4]. High data rate optical transmission uses WDM channels; that is, parallel data streams are carried by different wavelengths, spaced apart by a given frequency spacing to avoid channel crosstalk [16,17]. Thus, traditionally, it has been of great interest to classify the effects of the Kerr term in Eq. (9.5) based on which channels are involved in the generation of such effects. A first broad categorization can be made by separating nonlinear effects based on whether or not channels other than the one of interest are involved in their generation. We therefore refer to either interchannel nonlinearity and intrachannel nonlinearity. Furthermore, different types of nonlinear interactions (still within the scope of the Kerr effect) are possible when two or more channels copropagate in a nonlinear optical fiber.

In order to identify these different effects, let us assume that the transmitted electrical field is given by

$$E(t,z) = E_0(t,z) + E_1(t,z) + E_2(t,z) \tag{9.11}$$

where E_0, E_1, and E_2 are the (scalar) complex envelopes of the transmitted channels at three different wavelengths and E_0 is assumed to be the channel of interest. Replacing Eq. (9.11) into Eq. (9.5) we find

$$\frac{\partial(E_0 + E_1 + E_2)}{\partial z} = -j\frac{\beta_2}{2}\frac{\partial^2(E_0 + E_1 + E_2)}{\partial t^2} + j\gamma|E_0 + E_1 + E_2|^2(E_0 + E_1 + E_2) \tag{9.12}$$

where the attenuation term has been dropped for simplicity of notation. A standard approach consists of assuming a small nonlinearity (perturbative approach) [4], hence Eq. (9.12) can be rewritten as a set of three differential equations. The equation relative to the channel of interest reads as

$$\frac{\partial E_0}{\partial z} = -j\frac{\beta_2}{2}\frac{\partial^2 E_0}{\partial t^2} + j\gamma|E_0|^2 E_0 + 2j\gamma(|E_1|^2 + |E_2|^2)E_0 + j\gamma E_1^2 E_2^* \tag{9.13}$$

As shown in Eq. (9.13), the original nonlinear phase shift acting on the channel of interest is split into three components: (1) a term dependent only on the power of the channel of interest itself, hence called SPM; (2) a term dependent only on the power of the interfering channels called XPM; and (3) a term dependent on a cross-product between the two interfering channels called FWM. The derivation above can be repeated for three or more channels. However, the arising terms are similar to the ones in Eq. (9.13). The only difference consists of the XPM term adding up multiple channels, and additional pairwise FWM cross-products arising with more channels. While both SPM and XPM depend on the power of the channels, the FWM effect is dependent on the actual optical field. Due to this feature, the phase relationship between the copropagating channels is crucial for the accumulation of the FWM.

The derivation above can be repeated on a frequency component basis, that is, assuming the propagating field is composed of four continuous-wave (CW) components at frequencies f, f_1, f_2, and f_3. The CW optical fields at frequency f_1, f_2, and f_3 interact with each other to form a new wave at frequency f. The phenomenon is possible whenever the relationship between the four frequencies is [4]

$$f = f_1 + f_2 - f_3 \tag{9.14}$$

Defining $E(z) = E(f, z)$ and $E_j(z) = E_j(f, z)$ and solving Eq. (9.13) in the Fourier domain for the field component at frequency f, it can be shown that [5]

$$E(z) = jd\frac{(2\pi)^2 f}{nc} E_1^* E_2 E_3 \exp\left(\frac{-\alpha}{2} z\right) \frac{1 - \exp(j\Delta\beta z - \alpha z)}{\alpha - j\Delta\beta} \tag{9.15}$$

where c is the speed of light, d is a degeneracy coefficient either equal to 3 for $f_1 = f_2 \neq f_3$ or to 6 for $f_1 \neq f_2 \neq f_3$, n is the refractive index (as a function of the frequency f), and

$$\Delta\beta = \beta(f_1) + \beta(f_2) - \beta(f_3) - \beta(f) \tag{9.16}$$

From Eq. (9.15) it can be observed that the amplitude of the FWM strongly depends on $\Delta\beta$. For $\Delta\beta << 1$, the rightmost fraction in Eq. (9.15) significantly increases in magnitude and, as a result, the FWM product is amplified. The condition where $\Delta\beta$ is close to zero is commonly referred to as phase matching condition, and it is typically met in systems where the dispersion parameter is low. As a result, in scenarios where transmission is performed far from the zero-dispersion point and standard single-mode fiber (SSMF) is used, FWM can be considered as a minor nonlinear effect compared to SPM and XPM. In Section 9.1.2, the performance of receivers compensating for intrachannel nonlinearity (SPM) is analyzed, as well as compensation of interchannel nonlinearity, including both XPM and FWM.

9.1.2 Nonlinear distortions and modulation dependency

Considering the contributions from amplified spontaneous emission (ASE) noise and fiber nonlinearities (optical Kerr effect), the performance of a dispersion-unmanaged optical communication system can be described using a so-called effective signal-to-noise ratio (SNR) [14,15], which after fiber propagation can be described as

$$\mathrm{SNR} = \frac{P}{\sigma_{\mathrm{eff}}^2} = \frac{P}{\sigma_{\mathrm{ASE}}^2 + \sigma_{\mathrm{S-S}}^2 + \sigma_{\mathrm{S-N}}^2} \tag{9.17}$$

where P is the average optical power per channel, σ_{ASE}^2 represents the total power of ASE noise within the examined channel due to the optical amplification process, $\sigma_{\mathrm{S-S}}^2$ represents the distortions due to signal–signal nonlinear interactions, and $\sigma_{\mathrm{S-N}}^2$ represents the distortions from signal–ASE noise interactions.

For dual-polarization multispan Erbium-doped optical fiber amplifier (EDFA) amplified Nyquist-spaced WDM transmission systems, the Gaussian ASE noise from EDFAs at the end of each transmission span can be expressed as [1]

$$\sigma_{\mathrm{ASE}}^2 = N_{\mathrm{spans}}(G-1)\cdot F_n \cdot h\nu_0 R_S \tag{9.18}$$

where N_s is the total number of spans in the link, G is the EDFA gain, F_n is the EDFA noise figure, $h\nu_0$ is the average photon energy at the optical carrier frequency ν_0, and R_S denotes the symbol rate of the transmitted signal. The contribution of the nonlinear distortions due to nonlinear signal–signal interaction can be described as follows [18–20]:

$$\sigma_{\mathrm{S-S}}^2 = N_s^{\epsilon+1}\eta P^3 \tag{9.19}$$

where η is the nonlinear distortion coefficient and ϵ is the coherence factor, which is responsible for the increasing signal–signal interaction with transmission distance. This factor lies between 0 and 1 depending on the decorrelation of the nonlinear distortions between each fiber span [18]. In contrast to the $\sigma_{\mathrm{S-S}}^2$ term, the corresponding noise contribution due to signal–ASE interaction grows quadratically with launched power [21,22] and can be modeled as

$$\sigma_{\mathrm{S-N}}^2 = 3\zeta\eta\sigma_{\mathrm{ASE}}^2 P^2 \tag{9.20}$$

with the distance dependent prefactor of ζ, which accounts for the accumulation of signal–ASE distortions with the transmission distance, and can be effectively truncated as [23]

$$\zeta = \sum_{k=1}^{N_s} k^{1+\epsilon_{\mathrm{S-N}}} \approx \frac{N_s^{\epsilon+1}}{2} + \frac{N_s^{\epsilon+2}}{\epsilon+2} \tag{9.21}$$

In the framework of the first-order perturbation theory, the coefficient η can be defined depending on the model assumptions. In conventional approaches, the impact of optical fiber nonlinearities on signal propagation is typically treated as additive circularly symmetric Gaussian noise (GN) [18,19,24]. In particular, the effective variance of nonlinear distortions is clearly supposed to be entirely independent of the signal modulation format. However, it has been theoretically shown that the nonlinear distortions depend on the channel input symbols, and the Gaussian assumption of optical nonlinear distortions cannot therefore be sufficiently accurate [25–31]. Furthermore, significant discrepancies have been recently observed, especially for low-order signal modulation formats [27–30]. Assuming that all WDM channels are equally spaced, the modulation format dependent nonlinear distortion coefficient η over a single fiber span can be expressed in closed form using the approximation [32]

$$\eta(N_{\mathrm{ch}}) \approx \eta_0(N_{\mathrm{ch}}) - \frac{80}{81}\frac{\kappa\gamma^2 L_{\mathrm{eff}}^2}{\pi|\beta_2|\, L_s R_S^2}\left[\Psi\left(\frac{N_{\mathrm{ch}}+1}{2}\right) + C + 1\right] \tag{9.22}$$

where β_2 is the group velocity dispersion coefficient, γ is the fiber nonlinear coefficient, L_{eff} is the effective fiber span length, L_s is the fiber span length, N_{ch} is the total number of WDM channels, $\Psi(x)$ is the digamma function, $C \approx 0.577$ is the Euler–Mascheroni constant. The constant κ is related to the fourth standardized moment (kurtosis) of the input signal constellation. For Gaussian, quadrature phase shift keyed (QPSK) (4QAM), 16QAM, 64QAM, and 256QAM, its values are 0, 1, 17/25, 13/21, 121/200, respectively [30,32]. Finally, the coefficient η_0 quantifies the influence of optical fiber nonlinearity per fiber span under the Gaussian assumption, and can be analytically approximated as [18–20], where α is the fiber attenuation parameter.

$$\eta(N_{\mathrm{ch}}) \approx \left(\frac{2}{3}\right)^3 \frac{\alpha\gamma^2 L_{\mathrm{eff}}^2}{\pi|\beta_2|R_S^2}\,\mathrm{arsinh}\left(\frac{\pi^2}{2}|\beta_2|L_s N_{\mathrm{ch}}^2 R_S^2\right) \tag{9.23}$$

Parameters used in the analytical model are specified in Table 9.1. The SNR performance of transmission systems considering signal-signal interactions (using dispersion compensation only) for different modulation formats is shown in Fig. 9.1. The effectiveness of the modulation-dependent analytical model for predicting the performance of electronic dispersion compensation (EDC) and multichannel backpropagation (MC-DBP) has been verified in previous work [30,31,33–35], where detailed comparisons between the analytical and simulation results in terms of SNR have been investigated under various transmission schemes.

9.2 Digital schemes for fiber nonlinearity compensation

9.2.1 Principle of digital backpropagation

One way to improve the transmission quality is carry out mitigation of the above effects at the receiver, that is, after the optical field $\mathbf{E}(t, z)$ is detected. If, to a first approximation, we assume that the received signal is noise-free, the distortion

Table 9.1 Parameters in model of fiber nonlinearities considering modulation formats.

Parameter	Value
Symbol rate	32 GBd
Channel spacing	32 GHz
Central wavelength	1550 nm
Number of channels	9
Roll-off	0
Attenuation coefficient (α)	0.2 dB/km
Chromatic dispersion (CD) coefficient (D)	17 ps/nm/km
Nonlinear coefficient (γ)	1.2/(W.km)
Span length	80 km
Erbium-doped optical fiber amplifier (EDFA) noise figure	4.5 dB
Planck constant (h)	6.626×10^{-34} J s
Euler–Mascheroni constant (C)	0.577

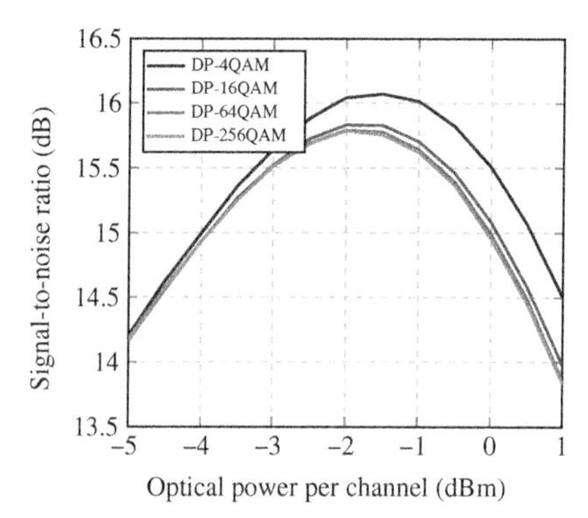

Figure 9.1 Performance of optical communication systems with different modulation format considering fiber nonlinearities.

generated by Eq. (9.6) is deterministic and can be fully predicted. As a result, it can also be, in principle, fully compensated. The effect of Eq. (9.6) can be reversed by integrating it with a reversed sign of z, which is equivalent to solve such equation with reversed sign parameters:

$$\alpha \rightarrow -\alpha \tag{9.24}$$

$$\beta_2 \rightarrow -\beta_2 \tag{9.25}$$

$$\gamma \rightarrow -\gamma \tag{9.26}$$

Therefore, when coherent detection is used, the transmitted signal at the beginning of the fiber can be reconstructed by using the detected signal $\mathbf{E}(t, L)$ as initial value of

Eq. (9.6) and using DSP at the receiver to integrate it numerically, hence obtaining the optical field at the initial section of the fiber $E(t, 0)$. Such integration effectively digitally reverses the forward propagation of the optical fiber and for this reason it is commonly referred to as digital backpropagation (DBP) [36–40].

The numerical implementation of DBP is based on the split-step Fourier (SSF) algorithm [4]—the most widespread approach to solving Eq. (9.6). Eq. (9.6) can be rewritten as

$$\frac{\partial \mathbf{E}(t,z)}{\partial z} = \left(\hat{D} + \hat{N}\right)\mathbf{E}\,(t,z) \tag{9.27}$$

where $\hat{D}$ and $\hat{N}$ are respectively a linear and nonlinear operator defined as

$$\hat{D} = -j\frac{\beta_2}{2}\frac{\partial}{\partial^2 t} - \frac{\alpha}{2} \tag{9.28}$$

$$\hat{N} = j\frac{8}{9}\gamma||\,\mathbf{E}(t,z)||^2 \tag{9.29}$$

The linear operator $\hat{D}$ contains the dispersion and attenuation effects, whereas the nonlinear operator $\hat{N}$ causes the nonlinear effects. The SSF method is based on breaking the integral solution into dz pieces small enough such that the solution can be approximated by

$$E(t, z + dz) = \exp\left[(\hat{D} + \hat{N})dz\right]E(t,z) \approx \exp\,(\hat{D}dz)\,\exp\,(\hat{N}dz)E(t,z) \tag{9.30}$$

that is, the effect of the sum of the two operators can be thought of as the composition of each of the two acting alone. In the limit of dz tending to zero, Eq. (9.30) converges to the solution of Eq. (9.6) [41]. Therefore, in order to obtain the wanted solution, one can iteratively find the solution for one operator and then for the other, covering the required integration length by small steps dz. The solution given by the operator $\hat{D}$ can be conveniently represented in the frequency domain as

$$\exp\,(\hat{D}dz) = \exp\left(j\frac{\beta_2}{2}\omega^2 dz\right) \tag{9.31}$$

where it can be efficiently implemented using a fast Fourier transform (FFT) and multiplication blocks. Instead, the solution given by the operator $\hat{N}$ is easily implemented as (complex) multiplication in the time domain. As a result, as shown in Fig. 9.2, the SSF fundamental step consists of an FFT to pass the signal in the frequency domain, a multiplication block, an inverse FFT to switch back into the time domain, and a multiplication block to implement $\exp(\hat{N}dz)$.

However, in a real DBP implementation, due to complexity constraints, the number of steps for given link length L needs to be reduced as much as possible,

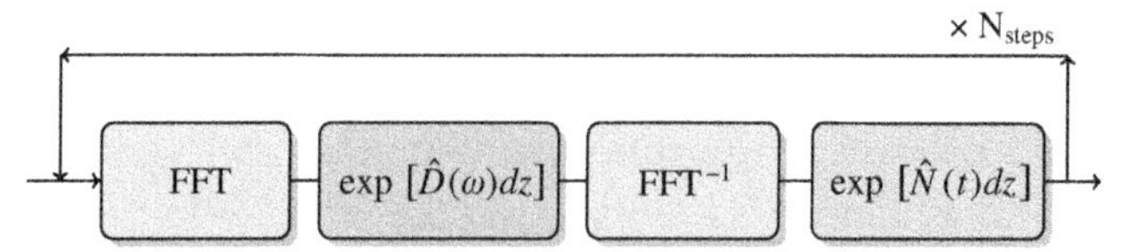

Figure 9.2 Schematic diagram of the split-step Fourier (SSF) algorithm.

resulting in a certain degree of inaccuracy in the calculation of the SSF step dz. This has been shown to significantly reduce the performance of DBP especially as the optical field BW increases [38,39,42]. Large DBP BWs are to be considered when two or more WDM channels are backpropagated together in order to compensate jointly for intrachannel (SPM) and interchannel (XPM and FWM). Modified versions of the basic algorithm discussed above have been shown in Refs. [43–46] with the aim of decreasing the number of steps required to maintain a given backpropagation accuracy.

As mentioned where precisely, the ideal linearization of the fiber channel through DBP is fundamentally limited by the nonlinear interaction or mixing between signal and ASE noise as well as PMD. The detected optical field is then a combination of signal and ASE which arises from the optical amplification along the fiber link. In an EDFA-amplified fiber system an amplifier is typically placed at the link of each fiber span. As a result, when DBP is applied, both signal and noise are backpropagated. While the signal–signal nonlinear effects within the DBP BW are fully compensated, the signal–ASE cannot be due to the way noise is added across the link. This residual term affects and ultimately limits the performance of system employing DBP at high transmitted powers. PMD has been shown to be an additional major bottleneck to the performance of systems employing DBP [42,47]. The optical field experiences random rotations of the polarization state as it propagates through the optical fiber due to its birefringence [14,15]. The average effect of the interaction between these rotations and nonlinear fiber effects is already accounted for in Eq. (9.6). However, PMD results in random differential group delays (DGDs) accumulating over the fiber length between the two polarization states of the propagating optical signal. This effectively results in different frequency components with a polarization state drifting apart as the signal propagates. Due to the stochastic nature of PMD, DBP can only reverse the signal propagation in the fiber assuming that the polarization states are not moving relatively to each other. This effectively results in a residual uncompensated NLI term which is more significant as the expected value of the DGD of the fiber link increases [42,47].

9.2.2 Achievable digital backpropagation gain

Nonlinear distortions can be modeled as additive noise. Therefore, the effective SNR at the sampled output of a matched filter (MF) can be defined as

$$\mathrm{SNR} = \frac{P}{\sigma^2_{\mathrm{ASE}} + \sigma^2_{\mathrm{S-S}} + \sigma^2_{\mathrm{S-N}}} \tag{9.32}$$

where P is the transmitted signal power, σ^2_{ASE} is the overall ASE noise power accumulated along the link, $\sigma^2_{\mathrm{S-S}}$ is the signal–signal NLI power, and $\sigma^2_{\mathrm{S-N}}$ is the signal-noise NLI power. All these noise components are to be considered integrated over the MF BW. The overall ASE noise is the contribution of multiple additive and uncorrelated components coming from each EDFA at the end of the fiber span; then

$$\sigma^2_{\mathrm{ASE}} = N_{\mathrm{spans}}\, P^2_{\mathrm{ASE}} \tag{9.33}$$

and [1]

$$P_{\mathrm{ASE}} = G \cdot \mathrm{NF} \cdot h\nu B \tag{9.34}$$

where G is the amplifier gain, NF its noise figure, $h\nu$ the photon energy, and B the signal optical BW. This quantity remains fixed once the span losses, the amplifiers, and the link length are chosen.

In addition to the system parameters, the NLI terms are also dependent on the transmitted power. According to any first-order perturbative approximation, the two terms $\sigma^2_{\mathrm{S-S}}$ and $\sigma^2_{\mathrm{S-N}}$ can be expressed as [20,21,25,27,48,49]

$$\sigma^2_{\mathrm{S-S}} = \eta_{\mathrm{S-S}} P^3 N_{\mathrm{spans}}^{(1+\epsilon_{\mathrm{S-S}})} \tag{9.35}$$

$$\sigma^2_{\mathrm{S-N}} = \eta_{\mathrm{S-N}} \zeta P_{\mathrm{ASE}} P^2 \tag{9.36}$$

where

$$\zeta = \sum_{k=1}^{N_{\mathrm{spans}}} k^{1+\epsilon_{\mathrm{S-N}}} \tag{9.37}$$

is a coefficient asymptotically growing as $N_{\mathrm{spans}}^{(2+\epsilon_{\mathrm{SN}})}$ Depending on the specific model within the first-order perturbative framework, the values of $\eta_{\mathrm{S-S}}$, $\eta_{\mathrm{S-N}}$, $\epsilon_{\mathrm{S-S}}$, and $\epsilon_{\mathrm{S-N}}$ can change significantly. The GN model [20,48], for instance, considers nonlinear coefficients η that are modulation format independent (due to the initial assumption of Gaussianity of the propagating signal). This approach has been shown not to be strictly correct ([25,27,50]), and significant inaccuracies have been demonstrated for low-order modulation formats [27–29].

However, as shown in Ref. [49], the values of the signal–signal parameters (η and ϵ) are typically close to their signal-noise counterparts. As a result, to a first (but accurate!) approximation, it can be assumed that

$$\eta_{\text{S-S}} \approx \eta_{\text{S-N}} = \eta \tag{9.38}$$

$$\epsilon_{\text{S-S}} \approx \epsilon_{\text{S-N}} = \epsilon \tag{9.39}$$

As a result, Eq. (9.32) becomes

$$\text{SNR}_{\text{EDC}} \triangleq \frac{P}{N_{\text{spans}}P_{\text{ASE}} + \eta P^3 N_{\text{spans}}^{(1+\epsilon)} + \eta\zeta P_{\text{ASE}}P^2} \approx \frac{P}{N_{\text{spans}}P_{\text{ASE}} + \eta P^3 N_{\text{spans}}^{(1+\epsilon)}} \tag{9.40}$$

where the approximation is due to the fact that for all power values and distances of interest the signal–signal NLI term, growing cubically with P, dominates the signal–ASE NLI, which grows quadratically. Eq. (9.40) represents the SNR when EDC only is used at the receiver to compensate for fiber CD. On the other hand DBP, by mitigating nonlinear distortions, can effectively reduce the cubic term. When it is possible to backpropagate the entire forward-propagated field $\mathbf{E}(t, L)$ at a distance L (commonly referred to as FF-DBP), the effect of DBP can be thought as generating a first-order term with opposite sign to the nonlinear distortion term [25, Sect. 2], hence effectively canceling the signal–signal NLI power $\sigma^2_{\text{S-S}}$. The signal–ASE NLI term can be thought of as made up of multiple components, each one generated from the ASE noise of each amplifier along the link. As shown in Refs. [49,51,52], when the nonlinear compensation is performed entirely at the receiver, DBP undoes the signal–ASE NLI generated by the ASE noise term at the first span of the link. It also partially cancels the signal–ASE NLI power for the ASE noise component arising later in the link, up to the point where it was added. An artificial signal–ASE NLI is generated for the ASE noise terms backpropagated beyond the point where they were added. The balance is equal to the same $\sigma^2_{\text{S-N}}$ occurring before DBP is applied [49,51].

The SNR after FF-DBP is applied is thus given by

$$\text{SNR}_{\text{FF-DBP}} = \frac{P}{3\eta\zeta P_{\text{ASE}}P^2} \tag{9.41}$$

Comparing Eqs. (9.40) and (9.41) at the transmitted powers where each of the expressions reaches its maximum, we find that the theoretical gains achievable by FF-DBP can be well approximated (for $N_{\text{spans}} \geq 2$):

$$\Delta\text{SNR} \approx \frac{K}{\eta^{1/6}P_{\text{ASE}}^{1/3}\sqrt{2N_{\text{spans}}^{\left(1+\frac{2}{3}\epsilon\right)} + (2+\epsilon)N_{\text{spans}}^{\frac{2}{3}\epsilon}}} \tag{9.42}$$

where K is a constant given by

$$K = \frac{3}{4}\sqrt[3]{2}\sqrt{2\cdot\frac{2+\epsilon}{3}} \tag{9.43}$$

Eq. (9.42) shows that the SNR gain introduced by the full compensation of the signal–signal nonlinear term is weakly dependent on the fiber parameters (η) and on the amount of the amplifiers noise. On the contrary, it has a stronger dependence on the link distance (N_{spans}): since typically $\epsilon << 1$, the SNR gain decreases approximately as the square root of the distance (as long as the span length is kept constant). Based on Eq. (9.42), one can observe that the SNR gains experienced by forcing the deterministic signal–signal nonlinearities to zero are very substantial. For instance, replacing the numbers for a 11×10 GBd superchannel system spaced at 10.2 GHz, when FF-DBP is applied, the SNR gain at 40×80 km of SSMF fiber is, according to the GN model, approximately equal to 8 dB. Additional gain can be obtained by splitting the backpropagation process between the transmitter (predistortion) and the receiver, as shown in Ref. [49]. This is due to the reduction of the $\sigma^2_{\text{S-N}}$ term, which in the case of a 50%–50% split (optimal case), effectively accumulates only over half of the number of spans of the fiber link. This results in a further asymptotic (long distances) gain of 1.5 dB [49].

However, in most cases backpropagating the entire transmitted WDM BW might be unrealistic, both for complexity and networking reasons. Thus, it is interesting to calculate the SNR performance of DBP when a given portion of the transmitted BW is backpropagated. It can be shown that [25,52] to a first-order approximation, applying DBP over a given BW coherently subtracts the NLI term that the forward propagation would generate over the same BW. As a result, the residual signal–signal NLI term when DBP is operated over a certain BW is given by

$$\sigma^2_{\text{S-S}} = [\eta(B) - \eta(B_{\text{DBP}})]P^3 \, N_{\text{ASE}}^{1+\epsilon} \tag{9.44}$$

where $\eta(\cdot)$ denotes the dependency of the NLI coefficient either on the forward-propagated BW (B) or backpropagated BW (B_{DBP}). The relationship between η and the BW depends on the adopted model (see Refs. [20,25,27,48,50]) and remains an open research topic. However, for dispersion-unmanaged SSMF fibers using EDFA amplifiers, the scaling of η as a function of the propagated BW is given by

$$\eta \propto (B^2) \tag{9.45}$$

The resulting SNR applying DBP over only a portion of the transmitted BW B_{DBP} is therefore given by

$$\text{SNR}_{\text{DBP}} = \frac{P}{N_{spans}P_{\text{ASE}} + [\eta(B) - \eta(B_{\text{DBP}})]P^3 N_{spans}^{(1+\epsilon)} + \eta\zeta P_{\text{ASE}}P^2} \tag{9.46}$$

where we assume that using DBP over the BW B_{DBP} does not affect the η coefficient multiplying the residual signal–ASE term.

Differently from Eq. (9.40) we notice that for any nonzero $\sigma^2_{\text{S-S}}$, the cubic term is always present at the denominator along with the signal–ASE NLI quadratic term.

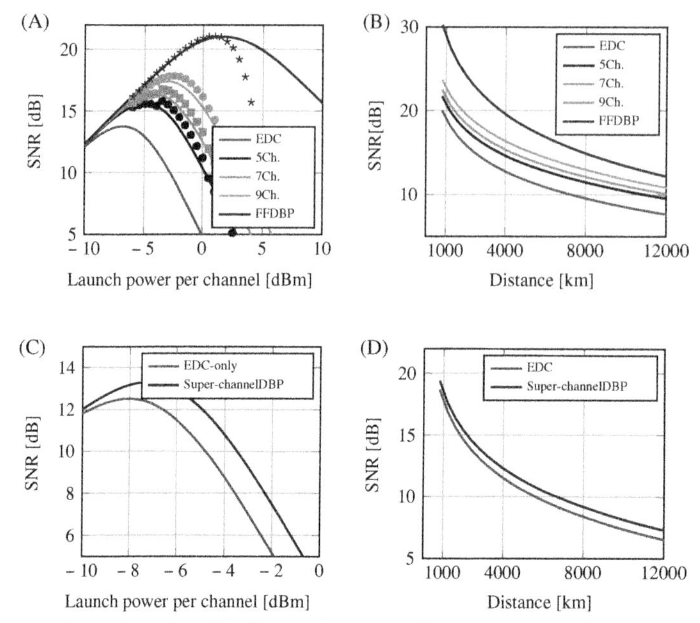

Figure 9.3 (A) Signal-to-noise ratio (SNR) as a function of transmitted power for the central subcarrier, for electronic dispersion compensation (EDC) only, 5-, 7-, and 9-channel digital backpropagation (DBP), and finally for full-field (FF) DBP. (B) Received SNR as a function of transmission distance for various DBP BWs. (C) SNR of the central subcarrier for 5.5 THz transmission system, both for EDC only and when the entire superchannel in backpropagated. (D) Corresponding SNR performance as a function of transmission distance.

At optimum launch power, the cubic term dominates the quadratic one (for all distances of interest) and as a result the signal–ASE term can be dropped from Eq. (9.46). We then obtain a DBP gain as a function of B_{DBP} and B, given by

$$\Delta\mathrm{SNR}(B_{\mathrm{DBP}}, B) \propto \sqrt[3]{\frac{\eta(B)}{\eta(B) - \eta(B_{\mathrm{DBP}})}} \tag{9.47}$$

The significance of Eqs. (9.42) and (9.47) is in explicitly quantifying the SNR gains achievable as a function of backpropagated BW. In Fig. 9.3 we illustrate the result of the above equations for the transmission of the 11×10 GBaud superchannel (parameters in Table 9.2) as used experimentally in this submission and is described in the next section. Fig. 9.3A and B show the DBP performance of a single superchannel transmission, in terms of (Fig. 9.3A) SNR versus transmitted power P_{TX} at 3200 km (40×80 km), and then (Fig. 9.3B) in terms of optimum SNR at a given distance, for different values of B_{DBP}. In order to explore the validity of Eq. (9.42) for the achievable SNR gain, we additionally show the results of numerical simulations based on SSF algorithm using the same parameters as before. It can be seen that there is excellent agreement for all curves.

Table 9.2 Analyzed system parameters.

Parameter description	Value
Subchannel symbol rate	10 GBd
Superchannel subcarriers	11
Roll-off	0%
Channel spacing	10.2 GHz
EDFA noise figure	4.5 dB
Span length	80 km
Fiber loss (α)	0.2 dB/km
Fiber dispersion coefficient (β_2)	-21 ps^2/km
Fiber nonlinear coefficient (γ)	1.2/W/km

However, a discrepancy is seen for powers >2 dBm per channel when the transmission moves into the highly nonlinear regime due to second-order nonlinearities, and the first-order perturbative analysis breaks down.

Fig. 9.3A and B indicates that the DBP gain does not saturate with the DBP BW, but instead its rate of increase is larger as the FF BW B is approached. For example, in Fig. 9.3A 4 dB of additional gain can be obtained by backpropagating the outermost subchannels of the superchannel (9−11 subchannels). This additional "jump" can be explained because the residual signal−signal NLI term when backpropagating nine channels is still significantly larger than the signal−ASE term at optimum launch power, although being just a small fraction of the overall signal−signal NLI. Fig. 9.3C and D show instead the benefit of backpropagating the entire superchannel when transmitted along with 50 other superchannels in the C-band with no guard-band between the superchannels. Still, a gain of approximately 1 dB can be observed at 3200 km (see Fig. 9.3C), while a gain in distance of approximately 1000 km can be obtained at 10 dB SNR (see Fig. 9.3D). MC-DBP has been implemented as part of DSP functionality shown in Fig. 9.14, and its effectiveness was investigated for a range of modulation formats, as described in Refs. [16,53].

Overall, the use of DBP can offer significant improvements in performance. SNR gains of up to 8 dB over 3200 km are possible (decreasing to 5 dB at 10,000 km) for a single superchannel system, resulting in an additional 1.6 bits/sym in terms of spectral efficiency (SE) (assuming $\log_2(1 + \text{SNR})$) and 176 Gbit/s higher capacity. For fully populated EDFA BW (50 Nyquist-spaced subchannels), this results in additional 6.25 Tbit/s of capacity at 3200 km. However, in practice, gains are reduced by the transceiver noise interactions.

9.3 Digital nonlinearity compensation in presence of laser phase noise

In dispersion-unmanaged transmission systems, laser phase noise can be converted into amplitude noise, due to the phase modulation to amplitude modulation conversion induced by the group velocity dispersion in the fiber [54,55]. In DSP-based coherent system, this phenomenon leads to an interaction between laser phase noise and EDC, inducing an effect of equalization enhanced phase noise (EEPN) [56,57]. The EEPN has already been identified as a source of significant degradation in the performance of single-channel transmission systems, increasing with fiber dispersion, local oscillator (LO) laser linewidth, modulation format, and symbol rate [56–65].

In multichannel transmission, the effects of fiber nonlinearities are more significant as the channel spacing is decreased to Nyquist spacing. Compared to conventional single-channel DBP for SPM compensation, considerable benefits can be obtained by applying MC-DBP over the entire superchannel BW, for example, see Section 9.2.1. This allows the compensation for interchannel nonlinear effects such as XPM and FWM across the entire superchannel. MC-DBP can also be operated at different digital BWs, involving different numbers of subchannels to achieve a compromise between the performance improvement and the computational complexity [42,66]. Several factors influencing superchannel transmission, MC-DBP, and nonlinear pre-compensation have been studied, such as the impact of nonlinear signal-noise interaction, accumulated PMD, comb carrier frequency uncertainty, and LO phase imperfect synchronization [21,47,67,68]. However, to date, no investigation of the effects of EEPN on the performance of multichannel transmission and MC-DBP has been reported. In fact, EEPN may significantly distort the performance of such schemes, especially when the CD must be simultaneously compensated over the entire superchannel BW, such as in the case of the use of FF-DBP. The impact of EEPN may differ for different subchannels within the same superchannel, which must be taken into consideration for the optimization of optical fiber networks. Therefore, it is of both great practical importance and interest to study the impact of EEPN on the performance of superchannel transmission systems for both EDC and MC-DBP.

In this section, the origin and influence of EEPN on the performance of the long-haul Nyquist-spaced superchannel transmission system are described, with and without the use of MC-DBP [69]. Analysis has been carried in a Nyquist-spaced 9-channel 32-Gbaud dual-polarizations 64-ary quadrature amplitude modulation (DP-64QAM) WDM superchannel system, with a total raw capacity of 3.456-Tbit/s. The achievable transmission distance of this superchannel system was evaluated both numerically, using the SSF algorithm, and analytically, using the perturbation GN model. The performance of each subchannel in the 9-channel DP-64QAM superchannel transmission system was investigated in detail to assess the origin and the influence of EEPN. Our results indicate that with both EDC and MC-DBP, EEPN causes a significant deterioration in the performance of all subchannels, with penalties more severe for the outer subchannels. Meanwhile, it is also shown that the source of EEPN, from the transmitter laser or the LO laser, depends on the relative position between the CPE and the MC-DBP (or the EDC) modules.

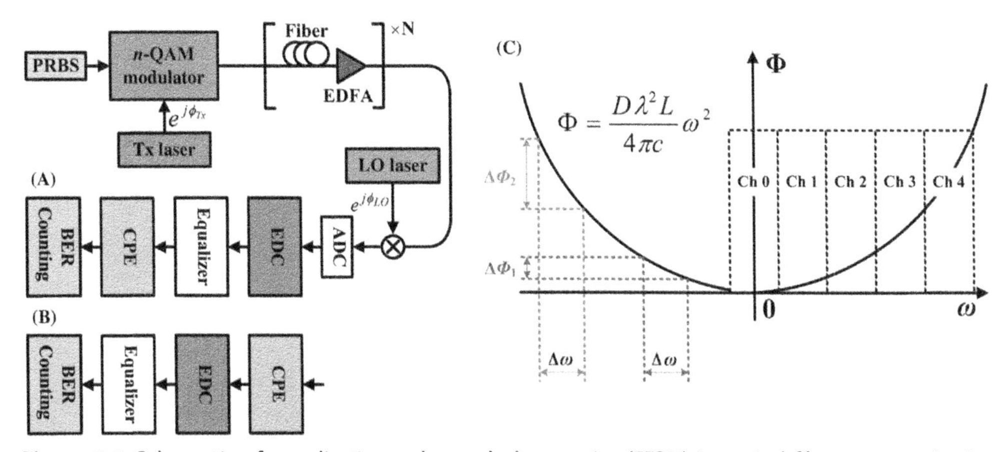

Figure 9.4 Schematic of equalization enhanced phase noise (EEPN) in optical fiber communication system using electronic dispersion compensation (EDC) (or digital backpropagation, DBP): (A) the DSP scenario for carrier phase estimation (CPE) applied after EDC; (B) the digital signal processing (DSP) scenario for CPE applied prior to EDC; and (C) the phase variation due to frequency perturbation in dispersion compensation in EDC (or multichannel backpropagation, MC-DBP). *PRBS*, pseudorandom bit sequence; *Ch*, channel.

Fig. 9.4 schematically shows the origin of EEPN in long-haul coherent optical communication systems, in which electronic CD postcompensation and CPECPE are employed. In Fig. 9.4, two scenarios are considered within the DSP modules: (A) CD compensation applied prior to CPE and (B) CD compensation applied after CPE. Generally, scenario (A) is more common for DSP operation in digital coherent receivers, since the required minimum oversampling rate (1 sample/symbol) in the CPE operation is less than that in the EDC and the MC-DBP operations and some adaptive CPE algorithms can also affect the performance of the dispersion compensation [10,62,70]. It has been indicated that laser phase noise can be converted into amplitude noise by CD [54,55]. Taking scenario (A) in Fig. 9.4 as an example, the phase noise from the transmitter laser passes through both the transmission fiber and the EDC module, so the net experienced dispersion is close to zero. The phase noise from the LO laser only goes through the EDC module in the receiver, which is heavily dispersed in systems without any optical dispersion compensation. Thus, the LO phase noise will interplay with the EDC module leading to the effect of phase noise to amplitude noise conversion, and the induced EEPN will affect the performance of long-haul high-speed optical transmission systems [56–65]. Previous work has also demonstrated that the EEPN arises from the nonzero net dispersion experienced by the laser phase noise, either from the transmitter laser or the LO laser [71,72]. Therefore, the EEPN originates from the interaction between the LO laser phase noise and the EDC (or the DBP) module in case (A), or from the interplay between the transmitter laser phase and the fiber dispersion in case (B). According to the different origins of EEPN, the effect can be categorized into equalization enhanced LO phase noise (EELOPN) for case (A) and equalization enhanced transmitter phase noise (EETXPN) for case (B).

The representations of the EELOPN (EEELOPN(t)) and the EETXPN (EETXPN (t)) in the time domain can be described by the following equations:

$$E_{\mathrm{EELOPN}}(t) = [A_{\mathrm{TX}} A_{\mathrm{LO}} \cdot e^{j\phi_{\mathrm{LO}}(t)}] \otimes g_{\mathrm{EDC}}\,(L, t) \tag{9.48}$$

$$E_{\mathrm{EETXPN}}(t) = [A_{\mathrm{TX}} A_{\mathrm{LO}} \cdot e^{j\phi_{\mathrm{TX}}(t)}] \otimes g_{\mathrm{Fibre}}\,(L, t) \tag{9.49}$$

where A_{TX} and A_{LO} are the amplitudes of the transmitter laser carrier and the LO laser optical wave, respectively, $\phi_{\mathrm{TX}}(t)$ and $\phi_{\mathrm{LO}}(t)$ are the phase fluctuation in the transmitter laser and the LO laser, respectively, $g_{\mathrm{EDC}}(L, t)$ is the time-domain transfer function of the EDC filter, $_{\mathrm{Fiber}}(L, t)$ is the time-domain transfer function of the fiber, L is the fiber length, t represents the temporal variable, and ⊗ indicates the convolution operation.

Taking scenario (A) as an example. The transfer function $G_{\mathrm{EDC}}(L, \omega)$ of the EDC filter in the frequency domain, which is the Fourier transform of the time-domain transfer function $g_{\mathrm{EDC}}(L, t)$, can be expressed as follows: [6,7]

$$G_{\mathrm{EDC}}(L, \omega) = \exp\left(j \frac{D\lambda^2 L}{4\pi c} \omega^2\right) \tag{9.50}$$

As shown Fig. 9.4C, the phase of the EDC filter in Eq. (9.50) varies quadratically with frequency, which increases the gradient at a higher relative frequency (further from the central channel carrier frequency). For the same perturbation $\Delta\omega$ in the frequency (e.g., the frequency shift due to the laser linewidth), the phase variation $\Delta\phi_2$ caused by the perturbation at a higher relative frequency is larger than the phase variation $\Delta\phi_1$ induced by the frequency perturbation at a lower relative frequency. Therefore, the interplay between the laser phase noise and the EDC module will be greater at a higher relative frequency and will generate a more significant EEPN. Correspondingly, for the superchannel transmission systems, the outer subchannels (e.g., channel 4 in Fig. 9.4C) will be impacted by more serious EEPN than the central subchannel (e.g., channel 0 in Fig. 9.4C), when the dispersion compensation is applied over the entire superchannel simultaneously. It is clear then that, when only linear EDC is applied, it is beneficial to compensate only the intrachannel dispersion. However, it is a prerequisite of MC-DBP that the EDC is applied over all subchannels simultaneously. The implications of this are investigated in more detail in the following sections.

The setup of the 9-channel 32-Gbaud DP-64QAM superchannel transmission system is schematically illustrated in Fig. 9.5, and all numerical simulations were carried

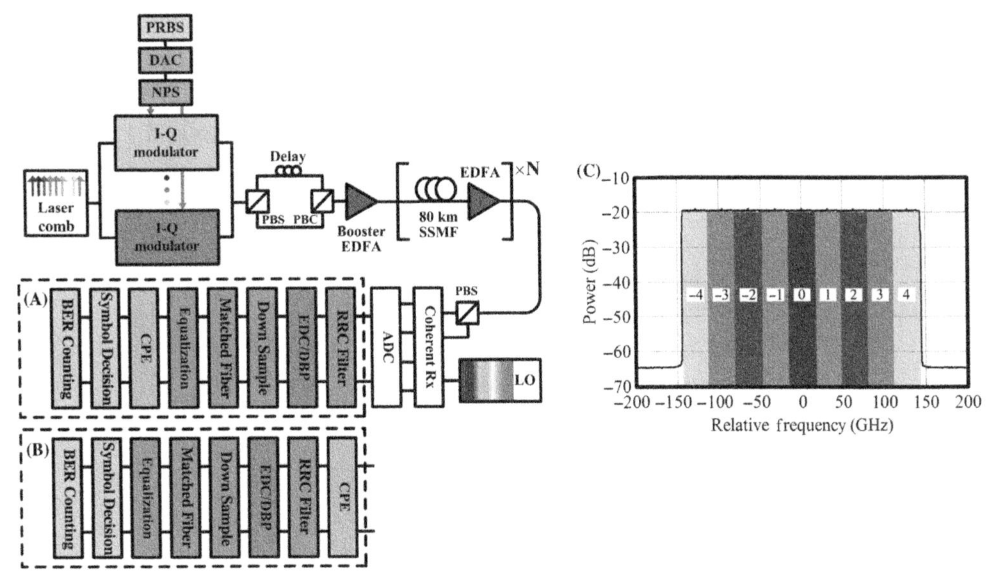

Figure 9.5 Schematic of 9-channel dual-polarizations 64-ary quadrature amplitude modulation (DP-64QAM) superchannel transmission system using the electronic dispersion compensation (EDC) or the multichannel backpropagation (MC-DBP): (A) the digital signal processing (DSP) scenario for carrier phase estimation (CPE) applied after EDC (or MC-DBP); (B) the DSP scenario for CPE applied prior to EDC (or MC-DBP); (and C) the simulated transmission spectrum and the schematic of back-propagated BW in the 9-channel DP-64QAM superchannel transmission system, where the frequency 0 Hz refers to the superchannel central frequency (wavelength of 1550 nm). *PBS*, polarization beam splitter; *PBC* polarization beam combiner.

out using the SSF algorithm to solve the Manakov equation with a digital resolution of 32 sample/symbol. In the transmitter, a 9-line 32-GHz spaced laser comb (centered at 1550 nm) is used as the phase-locked optical carrier for each subchannel. Digital-to-analog converter (DAC) with a resolution of 16-bit (to ensure no B2B implementation penalty) and root raised cosine (RRC) filter with a roll-off of 0.1% were used for the Nyquist pulse shaping. The transmitted symbol sequences were decorrelated with a delay of 256 symbols using a cyclical time shift to emulate the independent data transmission in each subchannel, and the sequences in each polarization were also decorrelated with a delay of half the sequence length. The SSMF is simulated based on the SSF method with a step size of 0.1 km, and the detailed parameters are span length of 80 km, attenuation coefficient of 0.2 dB/km, CD coefficient of 17 ps/nm/km, nonlinear coefficient of 1.2/W/km. The dispersion slope and PMD effect were neglected. The noise figure of the EDFA was set to 4.5 dB. At the receiver, the received signal was mixed with a free running LO laser and sampled at 32 samples/symbol, without any BW limitation, which allows an ideal and synchronous detection of all the in-phase and quadrature signal components over the whole superchannel

BW. The DSP modules include a RRC filter for selecting the MC-DBP BW, followed by MC-DBP (or linear EDC), down-sampling (to 2 samples/symbol), MF, multiple modulus algorithm equalization, ideal CPE, symbol de-mapping, and bit-error-rate (BER) measurement. The ideal CPE was realized by using the conjugate multiplication between the received signal and laser carrier phase, to isolate the influence of EEPN from the intrinsic laser phase noise. Corresponding to Fig. 9.4, the same two scenarios are considered in the simulations: (A) ideal CPE applied after the EDC/MC-DBP and (B) ideal CPE applied prior to the EDC/MC-DBP, as shown in Fig. 9.5A and B, respectively. The spectrum of the 9-channel DP-64QAM coherent transmission system is shown in Fig. 9.5C, where the number represents the subchannel within the superchannel. In all simulations, the MC-DBP algorithm was operated with the 800 steps per span and the nonlinear coefficient of 1.2 W/km to ensure optimum operation of nonlinear compensation.

Influence of EEPN in superchannel transmission using EDC and MC-DBP. The performance of all the subchannels in the 9-channel DP-64QAM superchannel transmission system was investigated in terms of Q^2 factors (converted directly from BERs), evaluated both using the EDC and the 9-channel (full-BW) DBP. Corresponding to previous descriptions, two DSP scenarios were considered to assess the origin and the impact of EEPN: (A) CPE implemented after the DBP/EDC and (B) CPE implemented prior to the DBP/EDC.

In the practical long-haul superchannel transmission system where a nonzero linewidth of the TX and the LO lasers exists, the impact of EEPN should be considered. As described above, due to the larger phase slope for the outer subchannels in the dispersion compensation filter (or the fiber dispersion) transfer function, the outer subchannels will have a more significant EEPN than the central subchannel, which will generate a more serious distortion. Therefore, the side subchannels will exhibit a worse performance than the central subchannel due to the more significant EEPN. The performance of all the subchannels in the Nyquist-spaced 9-channel DP-64QAM superchannel system with a transmission distance of 880 km (11 fiber spans) is illustrated in Fig. 9.6, in which a significant EEPN has been applied. Again, two DSP scenarios are considered in the numerical simulations: Fig. 9.6A shows results for the ideal CPE realized after the DBP/EDC, and Fig. 9.6B shows the results for the ideal CPE implemented prior to the DBP/EDC. To investigate the origin and the impact of EEPN, different distributions of the linewidths from the TX laser and the LO laser have been applied: (1) the TX laser linewidth is 100 kHz, and the LO laser linewidth is 0 Hz, (2) both the TX laser and the LO laser linewidths are 50 kHz, (3) the Tx laser linewidth is 0 Hz, and the LO linewidth is 100 kHz. It can be found in Fig. 9.6A that the EEPN originates from the interaction between the LO laser phase noise and the EDC/DBP module, corresponding to the case of equalization enhanced LO phase noise. With the increment of the LO laser linewidth (from 0 Hz to 100 kHz) in

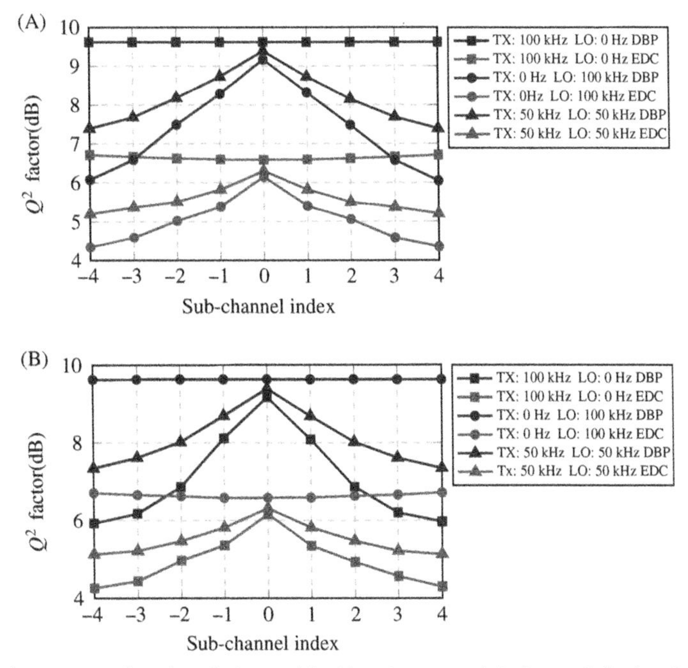

Figure 9.6 Performance of each subchannel in Nyquist-spaced 9-channel dual-polarizations 64-ary quadrature amplitude modulation (DP-64QAM) transmission system influenced by equalization enhanced phase noise (EEPN). The simulation is carried out under different distributions of the transmitter (TX) and the local oscillator (LO) laser linewidths, and both the electronic dispersion compensation (EDC) and the digital backpropagation (DBP) have been applied over the whole superchannel: (A) the scenario for ideal carrier phase estimation (CPE) implemented after EDC/DBP and (B) the scenario for ideal CPE implemented before EDC/DBP.

Fig. 9.6A, the outer subchannels exhibit a significantly worse performance than the central subchannel due to the EEPN induced additional noise, in both cases of the EDC and the FF-DBP. When the TX laser linewidth is 100 kHz and the LO laser linewidth is 0 Hz, the performance of all the subchannels shows a similar behavior, and is almost the same as for the ideal case (both Tx and LO laser linewidths are 0 Hz). Thus, there is no EEPN influence in this scenario, since the interplay between the TX laser phase noise and fiber dispersion can be fully compensated either by EDC or DBP. The performance of all the subchannels in the superchannel transmission system in Fig. 9.6B is exactly reversed compared to Fig. 9.6A, where the EEPN arises from the interaction between the TX laser phase noise and fiber dispersion, corresponding to the case of equalization-enhanced TX phase noise.

Fig. 9.7 shows the degradation in the performance of all the subchannels in the 9-channel DP-64QAM transmission system with the increment of the EEPN, where an increase in the TX or the LO laser linewidths is applied. The transmission distance is again 880 km (11 fiber spans). In Fig. 9.7A, the TX laser linewidth is kept at 0 Hz and

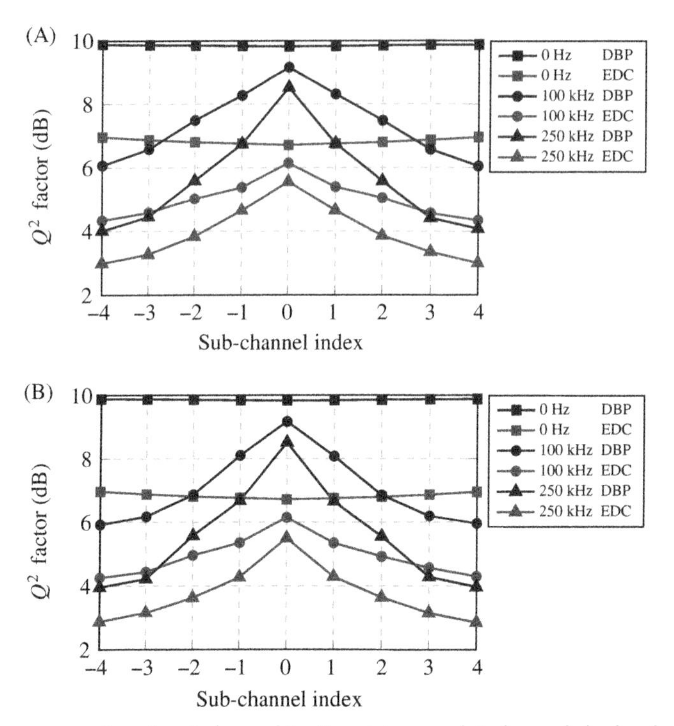

Figure 9.7 Performance of each subchannel in Nyquist-spaced 9-channel dual-polarizations 64-ary quadrature amplitude modulation (DP-64QAM) transmission system influenced by equalization enhanced phase noise (EEPN), where both the electronic dispersion compensation (EDC) and the digital backpropagation (DBP) have been applied over the whole superchannel. (A) The carrier phase estimation (CPE) implemented after the electronic dispersion compensation/digital backpropagation (EDC/DBP); the transmitter (TX) laser linewidth is 0 Hz and the indicated linewidth is for the local oscillator (LO) laser. (B) The carrier phase estimation (CPE) implemented before the EDC/DBP; the LO laser linewidth is 0 Hz and the indicated linewidth is for the TX laser.

the LO laser linewidth is varied from 0 to 250 kHz. In Fig. 9.7B, the LO laser linewidth is kept at 0 Hz and the TX laser linewidth is varied over the range from 0 to 250 kHz. It can be found that, in both cases using the EDC and the FF-DBP, the performance of the transmission system is significantly degraded by EEPN with the increment of the laser linewidth, and the performance of the outer subchannels behaves worse than the central subchannel with the increment of the laser linewidth, due to the more severe EEPN induced from the quadratic phase distribution of the EDC (or the fiber dispersion) transfer function. It is also found that the EEPN rises with the increment of the LO laser linewidth in Fig. 9.7A since the EEPN originates from LO laser phase noise for the case of CPE implemented after the EDC/DBP, and the EEPN rises with the increment of the TX laser linewidth in Fig. 9.7B since the EEPN originates from TX laser phase noise for the case of CPE implemented prior to the EDC/DBP. In addition, it has also been verified in our simulation that, to ensure that the Q^2 factor in all the subchannels is above the 20% overhead hard-decision pre-forward-error correction (pre-FEC) BER threshold (1.5×10^{-2}) in the FF-DBP scheme, the EEPN puts a limitation on the maximum tolerable linewidth (around 60 kHz for our system) on the LO laser in the case of CPE implemented after DBP, and the TX laser in the case of CPE implemented before DBP.

It is worth noting that the dispersion compensation in both the EDC and the FF-DBP were applied over the entire superchannel BW in all the above analysis. In the forward propagation along the fiber, the phase delay in each subchannel is different, since the fiber dispersion profile (quadratic shape) is different in different frequency bands. Thus, the carrier phase will have different delays in the fiber for different subchannels. In the receiver, if the dispersion is compensated over the entire superchannel simultaneously, the EDC filter will have the same dispersion profile as the fiber over the superchannel BW, and thus the carrier phase in all subchannels will be synchronized, and the phase noise can be compensated perfectly by applying the conjugate multiplication with the original carrier phase. According to the above investigations, the outer subchannels will suffer more pronounced EEPN than the central subchannel in such case. On the other hand, if the dispersion is compensated on an individual subchannel basis (using a frequency shift and a MF applied before EDC), the EDC filter will be designed separately by only considering the frequency range within each subchannel. Thus the overall phase response will be different from the fiber response over the whole superchannel BW. The carrier phase after the EDC will have different delays in different subchannels, and the phase noise in the outer subchannels cannot be perfectly removed by applying the conjugate multiplication with the original carrier phase, which is only synchronized with the central subchannel, and additional phase delays need to be considered in the ideal CPE. In this case, the impact of EEPN on the specific outer subchannel can be reduced to some extent, and will be the same as that in the central subchannel. Note that practical CPE algorithms work in both cases for compensating the intrinsic carrier phase noise, since the CPE algorithms only consider the current carrier phase noise in the measured subchannel. However, as previously noted, the dispersion compensation in the MC-DBP (or FF-DBP) has to be applied simultaneously over the whole back-propagated subchannels, since all the information of the NLI in the involved subchannels is required for the MC-DBP to cancel both the intrachannel and the interchannel nonlinearities. In this case, the EEPN in the outer subchannels in Nyquist superchannel transmission systems cannot be reduced.

EEPN originates from the interplay between the laser phase noise and the dispersion in the long-haul optical communication system using the EDC or the DBP, and significantly impacts the performance of the transmission system. The digital coherence enhancement based approaches can achieve an effective mitigation for the EEPN, while an independent measurement (or a complicated phase decorrelation) to estimate the LO laser phase fluctuation is necessary [65,73,74]. The EEPN can also be mitigated efficiently by applying a certain cutoff frequency to the LO laser, whereas a high-pass filtering based on electrical feedback or digital coherence enhancement is required to suppress the frequency noise [75].

9.4 Signal design for spectrally efficient optical transmission

9.4.1 Coded modulation

Coded modulation (CM) is the combination of multilevel modulation formats and forward error correction (FEC) codes, and is nowadays one of key techniques to maximize the SE of communication systems. This technique is typically used together with digital spectral shaping, which is aimed at reducing, as much as possible, the BW occupancy of information channels operating at a given symbol rate, or equivalently maximize the information throughput for a given transmission BW. As the optical BW becomes an increasingly precious commodity, this is a crucial aspect in the design of modern optical communications systems.

Research on CM dates back to the 1980s, originating from seminal works in Refs. [76,77] that highlighted the importance of combining coding and higher order modulation formats in order to achieve capacity in the so-called *BW-limited transmission regime*, that is, where (theoretically) unlimited transmission power can be used but only limited BW is available. Due to the increasing data rate demand and the consequent introduction of high-speed optical coherent detection systems, CM has also become a key technology in optical communications [78–83]. CM has enabled all of the most recent optical transmission *hero* experiments achieving SEs in excess of 7 b/s/Hz over transoceanic distances [84–86].

Fig. 9.8A illustrates an example of a generic CM system in the context of optical fiber transmission. The CM transmitter (CM TX) can be seen as a nonbinary encoder that maps a sequence of uniformly distributed information bits[1] ($\underline{i}$) of length N_b to sequences of vectors ($\underline{x}$) of length N_s. Each vector of complex symbols belongs to the codebook, $B = \{\underline{x}_1, x_2, \ldots, \underline{x}_{2^{N_b}}\}$, where 2^{N_b} is the number of possible transmitted messages. The sequence of symbols is subsequently traxnsmitted over the optical channel, which includes a single-mode optical fiber, optical amplification devices, and the transceiver subsystems with their corresponding DSP implementations. The CM receiver (CM RX) attempts to estimate the transmitted vector, $\underline{x}$, from the received noisy symbols $\underline{y}$.

The CM TX and RX in Fig. 9.8A can be in practice implemented in different ways, each showing a different trade-off between performance and complexity. The most popular approaches to CM can be classified within three main categories: trellis-

[1] The information bits are assumed to have passed through a source encoder for data compression.

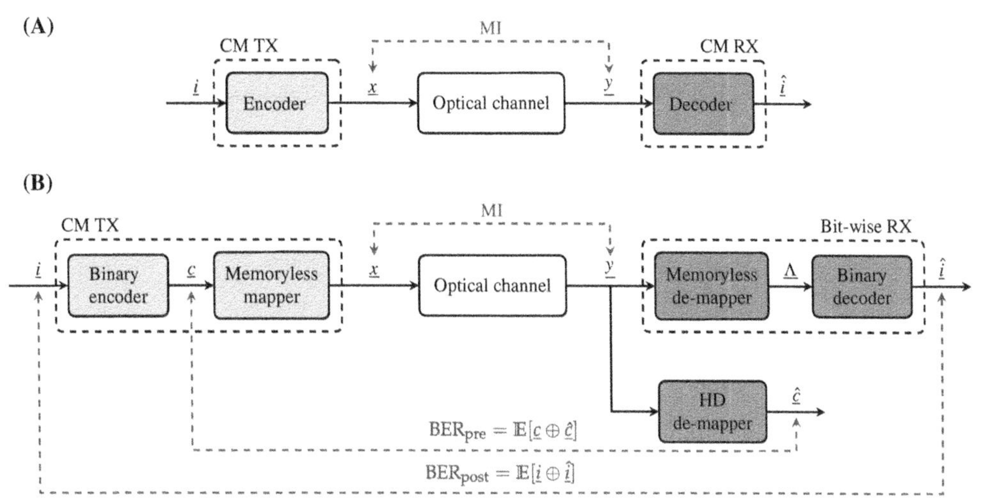

Figure 9.8 Coded modulation (CM). (A) CM system using the optimal maximum likelihood (ML) decoder. The mutual information (introduced in Sec. 9.4.2) is estimated using the transmitted and received symbols. (B) Bit-interleaved coded modulation (BICM) system illustrating the variables used (discussed in Sec. 9.4.2), pre-SD-FEC BER (BER_{pre}) and post-SD-FEC bit-error-rate (BER_{post}).

coded modulation (TCM) [76], multilevel coding (MLC) [87], and BICM [88]. In TCM, convolutional codes are used and jointly designed with the mapper in order to maximize the Euclidean distance between different codewords. Due to the structure of the transmitted codewords, the maximum likelihood (ML) decoder for TCM can be implemented using the Viterbi algorithm. In the case of MLC, the sequence of information bits is separated into a number of parallel streams. A set of parallel encoders (using in general different codes) are used to encode the different streams. The decoder is usually implemented using the so-called multistage decoding approach, where the decoding of each bit stream is used recursively as an input for the decoder of the subsequent bit stream. In both TCM and MLC the encoding and mapping stages need to be jointly designed, thus resulting in a monolithic and inflexible CM structure which also requires a relatively complex decoding stage. Finally, in the BICM case, the CM TX is implemented as a concatenation of a binary encoder and a memoryless mapper. This approach significantly simplifies the decoder which can be also split in a memoryless demapper and a binary decoder. More importantly, BICM decouples completely modulation and FEC, giving the designer increased flexibility and allowing the optimization of the FEC and modulation structures in a disjoint manner. This is the reason why BICM is the most popularly adopted CM technique, including the optical communications world.

The BICM architecture is shown in greater detail in Fig. 9.8B. At the CM TX, information bits are passed into the binary encoder, which adds redundancy (for error correction at the receiver) and operates at a code rate, $R = N_b/N_c$, where N_c is the number of coded bits (c). The coded bits are mapped to a set of discrete constellation points using a memoryless mapper and are then transmitted over the optical channel. Although the ML sequence decoder can be used also for this transmitter architecture, the high complexity associated with this decoder renders its implementation in real-world systems infeasible. Therefore, a more feasible (albeit suboptimal) receiver would use a simple two-step decoding process that decouples the decoder and demapper into two distinct processes. This bit-wise receiver (Bit-Wise RX) structure is the key functional block in a BICM-based system [89], as shown in Fig. 9.8B. The Bit-Wise RX consists of a memoryless demapper and a soft decision (SD) FEC decoder. The demapper calculates a ratio of probabilities [known as log likelihood ratios (LLRs)], Λ, that are passed into the binary decoder and are used to estimate the transmitted bits. The decoded bits are subsequently used to calculate the post-SD-FEC bit-error-rate (BER_{post}), while the pre-SD-FEC BER (BER_{pre}) is calculated by passing the received symbols through a hard-decision demapper. These two BER quantities are schematically shown in Fig. 9.8B and will be used throughout this work, along with the mutual information (MI) and generalized mutual information (GMI) introduced in the following section, to demonstrate the performance of our complete system.

9.4.2 Mutual information and generalized mutual information

Information theory addresses the fundamental question as to the maximum amount of information that can be reliably transmitted through any channel [90,91]. The mutual information (MI) between the transmitted and received symbols is a key quantity, as it represents the maximum achievable information rate (IR)[2] for a CM system based on the ML sequence decoder. The MI has been analyzed in the context of the optical channel and has been previously employed as a performance metric for digital optical communications systems [92–96]. Experimental demonstrations have verified the use of MI to predict the BER_{post} in optical systems based on the DP-QPSK format [97] and recently for DP-8QAM [98]. The MI has also been used as a figure of merit for WDM transmission systems based on the m-ary QAM format [99,100]. In this chapter, an experimental measurement of information theoretic quantities such as MI and GMI is performed, showing that modulation format and code rate of a practical CM optical system must be simultaneously optimized to match the specific SNR achieved after digital coherent detection. In addition to this, we depart from uniformly shaped constellations to consider the use of probabilistic shaping (Section 9.4.2.1) to maximize the IR of our system.

Let X be a random variable representing the transmitted symbol taking values in $\mathcal{X} = \{x_1, x_2, \ldots, x_m\}$ and $p_X(x_i)$ for $i = 1, 2, \ldots M$ its probability mass function (PMF).

[2] A rate is said to be achievable if there exists an encoder operating at that rate and a decoder giving a vanishing error probability as the block length tends to infinity.

Let then Y be a random variable taking values in the complex field C and modeling the received symbol, where $p_{Y|X}(y|x_i)$ for $i=1, 2,\ldots,m$ is a family of conditional probability density functions statistically defining the channel. The MI for discrete constellations and memoryless channels is defined as

$$\mathrm{MI} = \sum_{i=1}^{m} p_X(x_i) \int_{\mathbb{C}} p_{Y|X}(y|x_i) \log \frac{p_{Y|X}(y|x_i)}{\sum_{j=1}^{m} p_X(x_j) p_{Y|X}(y|x_j)} dy. \quad (9.51)$$

The MI defines the maximum rate (typically in bit/symbol) at which reliable communication is possible over a given channel for a given modulation format and an optimum CM scheme [91, Theorem 7.7.1]. Such an optimal CM scheme, in turn, entails an optimum bit-to-symbol mapping, optimal FEC (including potentially infinitely long codewords), and an ML receiver.

As discussed in Section 9.4.1, practical CM structures typically use the BICM scheme. Under this constraint, it was shown that the MI is not an accurate indicator, either in terms of rate achievability or in terms of post-FEC BER performance at a given CM rate [83]. An alternative information theoretic quantity was proposed in Ref. [88] to characterize the performance of BICM systems. This quantity is referred to as *bit-wise* MI or more notably as GMI. The GMI is defined as [89]

$$\mathrm{GMI} = \sum_{k=1}^{\log_2 m} I(B_k; Y) \quad (9.52)$$

where $I(B_k; Y)$ is the MI between the kth bit B_k in a transmitted symbol X and the corresponding received symbol Y. The GMI was shown to be an achievable IR (although not necessarily the largest) for the BICM system in Fig. 9.8B [88], [89, Chapter 4]. For equally likely transmitted symbols Eq. (9.52) can be rewritten as

$$\mathrm{GMI} = \log_2 m + \frac{1}{m} \sum_{k=1}^{\log_2(m)} \sum_{b\in\{0,1\}} \sum_{i\in\mathcal{I}_k^b} \int_{\mathbb{C}} f_{Y|X}(y|x_i) \log_2 \frac{\sum_{j\in\mathcal{I}_k^b} f_{Y|X}(y|x_j)}{\sum_{p=1}^{m} f_{Y|X}(y|x_k)} dy \quad (9.53)$$

where $\mathcal{I}_k^b$ is the subset of points in X having kth bit equal to b. From Eq. (9.53) it can be deduced that, unlike the MI, the GMI is not only a function of the channel law and the constellation shape (coordinates and symbol probability of occurrence), but also of the bit-to-symbol mapping. As shown in [83], the GMI is a reliable indicator to estimate the post-FEC BER of good BICM schemes operating at a given IR.

The computation of the MI and GMI for the additive white Gaussian noise (AWGN) channel is typically performed using two different numerical integration strategies [101]: (1) Monte Carlo integration and (2) Gauss–Hermite quadrature. For the AWGN channel, the channel law is given by

$$p_{Y|X}(y|x) = \frac{1}{\pi\sigma_z^2} \exp\left(-\frac{|y-x|^2}{\sigma_z^2}\right) \tag{9.54}$$

where σ_z^2 is the noise (complex) noise variance. Using Eq. (9.54) in Eq. (9.51), the MI becomes

$$\begin{aligned} \mathrm{MI} &= \sum_{i=1}^{m} p_X(x_i) \log_2 \frac{1}{p_X(x_i)} \\ &- \frac{1}{m}\sum_{i=1}^{m}\int_{\mathbb{C}} f_Z(z) \log_2 \sum_{j=1}^{m} \exp\left(-\frac{|x_i-x_j|^2 + 2\Re\{(x_i-x_j)z^*\}}{\sigma_z^2}\right) dz \end{aligned} \tag{9.55}$$

where $Z \sim \mathcal{CN}(0, \sigma_z^2)$. In a similar way the GMI for the AWGN channel and equally likely symbols is given by

$$\begin{aligned} \mathrm{GMI} &= \log_2 m \\ &- \frac{1}{m}\sum_{k=1}^{\log_2 m}\sum_{b\in\{0,1\}}\sum_{i\in\mathcal{I}_k^b}\int_{\mathbb{C}} f_Z(z) \log_2 \frac{\sum_{p=1}^{m} \exp\left(-\frac{|x_i-x_p|^2 + 2\Re\{(x_i-x_p)z^*\}}{\sigma_z^2}\right)}{\sum_{j\in\mathcal{I}_k^b} \exp\left(-\frac{|x_i-x_j|^2 + 2\Re\{(x_i-x_j)z^*\}}{\sigma_z^2}\right)} dz \end{aligned} \tag{9.56}$$

Eqs. (9.55) and (9.56) can be computed using the Monte Carlo approximation

$$\int_{\mathbb{C}} f(x) g(x)\, dx \approx \frac{1}{N_s}\sum_{n=1}^{N_s} g(x_n) \tag{9.57}$$

where x_n are samples from a random variable $X \sim f(x)$. Using Eq. (9.57) in both Eqs. (9.55) and (9.56) we obtain for the MI and GMI, respectively,

$$\begin{aligned} \mathrm{MI} &\approx \sum_{i=1}^{m} p_X(x_i) \log_2 \frac{1}{p_X(x_i)} \\ &- \frac{1}{mN_s}\sum_{i=1}^{m}\sum_{n=1}^{N_s} \log_2 \sum_{j=1}^{m} \exp\left(-\frac{|x_i-x_j|^2 + 2\Re\{(x_i-x_j)z^*[n]\}}{\sigma_z^2}\right) \end{aligned} \tag{9.58}$$

$$\text{GMI} \approx \log_2 m - \frac{1}{mN_s}\sum_{k=1}^{\log_2 m}\sum_{b\in\{0,1\}}\sum_{i\in\mathcal{I}_k^b}\sum_{n=1}^{N_s}\log_2\frac{\sum_{p=1}^{m}\exp\left(-\frac{|x_i-x_p|^2+2\Re\{(x_i-x_p)z^*[n]\}}{\sigma_z^2}\right)}{\sum_{j\in\mathcal{I}_k^b}\exp\left(-\frac{|x_i-x_j|^2+2\Re\{(x_i-x_j)z^*[n]\}}{\sigma_z^2}\right)} \tag{9.59}$$

where $z[n]$ are samples from the random variable Z, $\Re\{\cdot\}$ denotes real part and N_s is the number of Monte Carlo samples.

In Fig. 9.9 the MI and GMI are computed for AWGN and mQAM modulation formats for $m\in\{4, 16, 64, 256\}$. The common binary reflected Gray code is here assumed to be the labeling choice. For 4QAM and 16QAM modulation formats, MI and GMI assume very similar values across the whole range of rates (0 to $\log_2 m$). For higher order QAM (64 and 256 QAM), the gap between MI and GMI is more significant at low to medium code rates ($R\leq 0.8$). This indicates a larger SNR penalty when BICM as opposed to other (more complex) CM schemes is used at low FEC rates and higher order modulation formats. This is the case for instance when 256QAM is used in the SNR range of 5–12 dB (see Fig. 9.9). At high rates ($R\geq 0.8$), the gap between MI and GMI is negligible for all modulation formats.

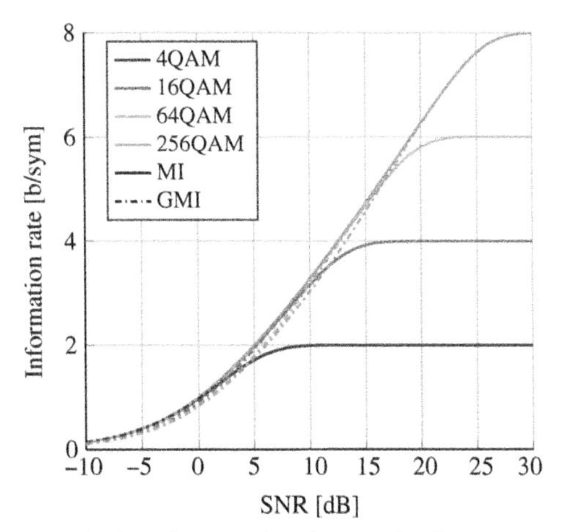

Figure 9.9 Mutual information (MI) and generalized mutual information (GMI) for mQAM modulation formats in the additive white Gaussian noise (AWGN) channel.

9.4.3 Constellation shaping

A technique that can be used in combination with CM to further increase SE is *constellation shaping* [102]. Constellation shaping consists in varying the transmitted symbols distribution, that is, their location and probability of occurrence, in order to maximize the IR through a given communication channel. One of the main results in information theory shows that, for asymptotically large constellation cardinalities, there is a 1.53 dB SNR gap between the highest achievable IR for a FEC scheme over a regular mQAM constellation and the Shannon capacity in the AWGN channel [103]. This gap can be reduced (and asymptotically closed) using constellation shaping. Constellation shaping techniques fall into two main categories: *probabilistic shaping* and *geometrical shaping*. In the following sections both approaches are discussed.

9.4.3.1 Probabilistic shaping

Using CM schemes where symbols can be transmitted using nonuniform probabilities while retaining a regular constellation geometry (e.g., mQAM modulation) is known as *probabilistic shaping* (see, e.g., [77,102,104] and a detailed review in Ref. [105, Sect. II]). For the AWGN channel with an average power constraint, this shaping techniques yield a sensitivity gain of up to 1.53 dB relative to uniform QAM as the constellation cardinality tends to infinity [87, Sect. VIII-A]. This SNR margin results in increased achievable IR for SNR-limited digital coherent transceivers [106]. Therefore, probabilistic shaping is a viable candidate to transmit high SE over longer transmission distances in future optical fiber systems and has already been demonstrated in several works for mQAM systems (see, e.g., [82,107,108]).

Geometrical shaping has also been demonstrated in fiber experiments (e.g., [109,110]); however, this shaping alternative imposes a constraint on the required effective number of bits (ENOB) of the DAC due to the unequally spaced constellation points. Instead, probabilistic shaping does not require any modification of the constellation geometry and its complexity is moved to the encoder and decoder, making it more compatible with existing QAM systems. Additionally, probabilistic shaping is also known to give larger gains than geometrical shaping for square QAM with a fixed number of constellation points, as shown in Ref. [89, Fig. 4.8].

For fiber transmission in the presence of nonlinearities, probabilistic shaping has been shown to improve transmission performance. Different input PMFs have been studied in simulations. These include dyadic PMFs and PMFs obtained from the Blahut–Arimoto algorithm [111, Sect. II], ring-like constellations [82, Sect. III], the Maxwell–Boltzmann PMF [112, Sect. V-B], [113, Sect. 2.3], [95, Sect. 3.5], and other nonlinearity-tailored distributions [114]. Recently, reach increases were experimentally demonstrated in Refs. [107,115] using a finite number of Maxwell–Boltzmann distributions. These distributions were chosen to target different net data rates, and shaping gains over a transmission range of more than 4500 km were reported.

Finding a good nonuniform input distribution for QAM constellations is a two-dimensional (2D) optimization problem. However, because of the symmetry of the 2D constellation, of the binary reflected Gray labeling that is typically used, and the AWGN channel for which the optimization is carried out, it is sufficient to consider a one-dimensional (1D) constellation [116].

Under these assumptions, mQAM constellations can be decomposed into a product of two constituent 1D (pulse amplitude modulation) constellations, each with $M = \sqrt{m}$ constellation points. Without loss of generality, we therefore consider only one of the quadratures for the shaping optimization; however, we emphasize that the analysis in Section 9.4.3.3 is performed for 2D (mQAM) constellations on each polarization.

Let $\boldsymbol{x} = [x_1, x_2, \ldots, x_M]$ denote the real-valued constellation symbols, represented by the random variable X. We assume that the symbols are distributed according to a PMF $\boldsymbol{P}_X = [P_X(x_1), P_X(x_2), \ldots, P_X(x_M)]$ and that they are sorted in ascending order (i.e., $x_i < x_{i+1}$, $i = 1, 2, \ldots, M-1$). To obtain the results presented later in Section 9.4.3.3, the input symbols are shaped using a PMF from the family of Maxwell–Boltzmann distributions, which are well known to be optimal for the AWGN channel (see, e.g., [104, Sect. IV], [87, Sect. VIII-A], [105, Sect. III-C]). Following the approach of [105, Sect. III-C], the shaped input is distributed as

$$P_X(x_i) = \frac{1}{\sum_{k=1}^{M} e^{-\nu x_k^2}} e^{-\nu x_i^2} \tag{9.60}$$

where ν is a scaling factor. To find the optimal PMF among all distributions given by Eq. (9.60), let the positive scalar Δ denote a constellation scaling of X. Fixing Δ and the SNR for which the optimization is carried out (denoted *shaping SNR* in the following), the scaling factor is chosen such that $E[|\Delta X|^2] = \text{SNR} = E_s/N_0$ where E_s represents the signal power and N_0 represents the noise power. The MI between the scaled channel input ΔX and the AWGN channel output is unimodal in Δ, and thus, the scaling factor is chosen to maximize the MI at a given rate. Although this optimization is carried out for symbol-wise MI, the loss from considering a bit-wise decoder [89] gives a negligible achievable IR penalty [105, Table III].

In order to implement probabilistic shaping, the data sequence to be transmitted must be transformed into a sequence with unequal per-symbol probabilities. A constant composition distribution matcher [117] imposes a desired distribution on a block of uniform bits, and this shaped sequence is then encoded with a systematic FEC encoder and mapped onto QAM symbols [105]. At the receiver, the received shaped bits after FEC decoding are fed into a distribution dematcher that undoes the

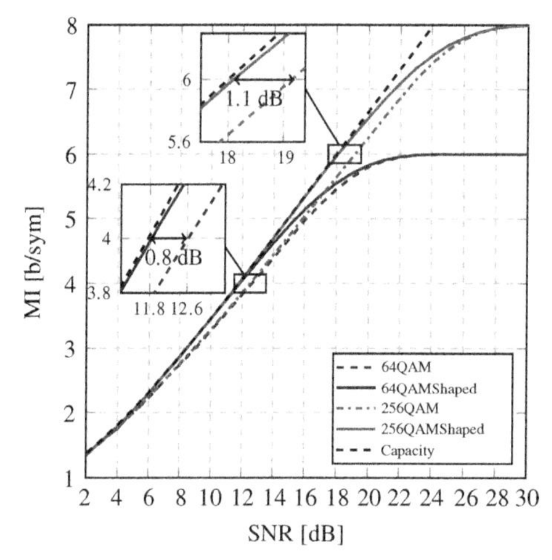

Figure 9.10 MI as a function of SNR for probabilistically shaped 64- and 256-QAM formats in the additive white Gaussian noise (AWGN) channel. The black dashed line shows the Shannon capacity of the AWGN channel.

matching operation of the transmitter. In order to see the full shaping gain after decoding, the LLR calculation must take into account the input distribution, and the bit mapping must be optimized for each FEC scheme used.

Fig. 9.10 shows numerical simulations for the AWGN channel for 64QAM and 256QAM input with uniform and probabilistically shaped PMFs. The shaped constellations use the Maxwell–Boltzmann distribution in (60) with optimized ν for the SNR values corresponding to $R = 0.75$ (6 b/sym) for both 64QAM and 256QAM. The probabilistically shaped constellations gain up 0.8 dB for 64QAM at 4 b/sym and 1.1 dB for 256QAM at 6 b/sym. Significant SNR gains compared to uniformly distributed QAM constellations can also be observed throughout the entire medium/high rate region. In can also be noticed that using probabilistic shaping closes almost entirely the gap to capacity up to $R \approx 0.75$. Probabilistic shaping is thus a near optimum approach for the design of CM schemes. Unfortunately, the IR illustrated in Fig. 9.10 only represents an upper bound on the performance of CM systems employing probabilistic shaping. Low complexity CM implementations closely approaching the performance gains predicted in Fig. 9.10 are still an open research topic [105,118,119].

9.4.3.2 Geometrical shaping

In the previous section, it was shown how modifying the probability of occurrence of each symbol in a given transmitted constellation can lead simultaneously to SNR savings and increase in MI for a fixed SNR. This section instead focuses on the design of the geometry of a constellation where all symbols are assumed to be equally likely. This approach is referred to as *geometrical shaping*. Similarly to probabilistic shaping, geometrical shaping can improve the transmission performance and it can closely approach Shannon channel capacity when coupled with strong FEC schemes.

For the AWGN channel, probabilistic shaping can reach an SNR gain that is achievable through geometrical shaping only when the latter is applied over a sufficiently large number of signal dimensions [77,89,103]. This maximum shaping gain is given by 1.53 dB and it corresponds to the shaping gain achieved by carving constellation points out of an N -dimensional lattice via an N -dimensional hypersphere. Such an N -dimensional constellation induces a constituent 2D constellation which is Gaussian distributed [103]. Thus, using probabilistic shaping in 2D is equivalent in terms of shaping gain to geometrical shaping for an asymptotically high number of dimensions. However, although probabilistic shaping might appear as a *2D shortcut* to achieving optimal shaping gains, the geometrical shaping approach leads in general to simpler implementations of the CM scheme. In particular, the advantage of using geometrical shaping as opposed to probabilistic shaping lies in the fact that geometrical shaping is typically easier to interface to powerful off-the-shelf FEC schemes. This can be done, for instance, via simple memoryless bit-to-symbol mapping, thus allowing the FEC and the signal modulation blocks to remain decoupled, which enables a much deeper design flexibility.

Another reason for exploring geometrical shaping in the optical fiber channel is given by the strong impact of the constellation geometry on the fiber nonlinear effects [27,31]. Different constellation geometries result in different higher order statistical moments of the constellation (such as the constellation kurtosis and 6th-order moment), which in turn play a significant role on the NLI generated through the fiber. The achievable IR of a shaped constellation in the optical fiber channel is then not only the result of the *linear shaping gain* previously discussed, but also of the increased nonlinearity tolerance of the shaped constellation [120,121].

In Fig. 9.11 the MI of three different constellations with cardinality $M = 64$ are compared in terms of MI as function of the SNR in an AWGN channel: 64QAM (blue); a randomly shaped constellation (red); and a ring constellation (green). The results show that for low rates (below 3 bit/symbol (b/sym)) the three different formats have very similar performance. However, for rates in between 3 and 5 b/sym, the ring constellation achieves a gain of up to 0.57 dB (at 4.1 b/sym) and up to 0.17 b/sym over 64QAM. Conversely, at very high rates (≥5.5 b/sym), 64QAM still outperforms the other two modulation formats. Surprisingly, it can be observed that the MI performance of 64QAM are comparable to a very irregular (randomly generated) modulation format for rates up to 4 b/sym. However, at higher rates, the gap becomes quite significant as the performance of the randomly generated constellation deteriorates.

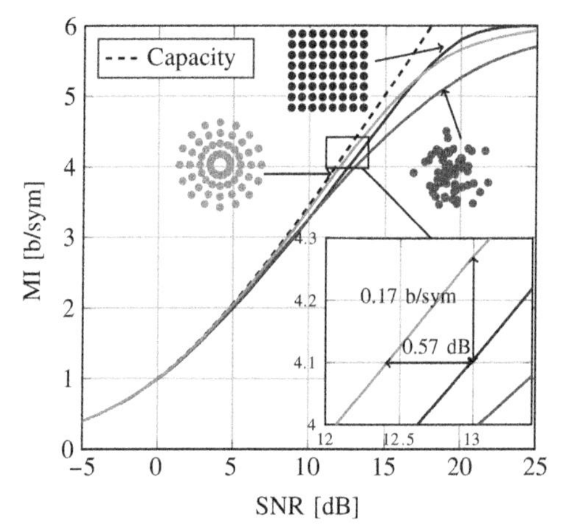

Figure 9.11 Mutual information (MI) versus signal-to-noise ratio (SNR) for different geometrically shaped constellations in the additive white Gaussian noise (AWGN) channel.

9.4.4 Experimental investigation of high spectral efficiency coded modulation systems for optical communications

In this section, experimental demonstrations of high SE CM schemes for fiber systems are presented, as reported in our previous works in Refs. [16,122].

9.4.4.1 Nyquist wavelength-division multiplexing

As the frequency spacing between optical carriers in a dense WDM system is reduced, interchannel interference begins to cause significant performance penalties due to linear crosstalk. Although tight filtering can be employed to constrain the BW of each WDM carrier, the filtering process itself results in significant intersymbol interference (ISI) within each channel. However, if an appropriate filter shape is used, for example, a sinc-shaped pulse with a corresponding rectangular spectrum, then the Nyquist criterion for ISI is met. In particular, the Nyquist criterion for zero ISI implies that the matched filtering output has a raised cosine spectral shape. This ensures that at the correct sampling instant, that is, where the pulse of interest reaches a maximum, the adjacent pulses reach a zero, thus avoiding ISI. A root-raised cosine (RRC) filter and a corresponding matched RRC filter at the receiver satisfy the Nyquist criterion and are selected to maximize BW use. Additionally, by decreasing the roll-off factor of the RRC filters, the WDM channel spacing can be reduced closer to the Nyquist frequency, without incurring significant penalties due to linear crosstalk (due to the signal power below the 3 dB BW). To mitigate the linear crosstalk induced penalties in a Nyquist-spaced system and maximize the SE requires the choice of optimum roll-off factors in the transmitter and receiver RRC filters, as a function of the WDM channel spacing.

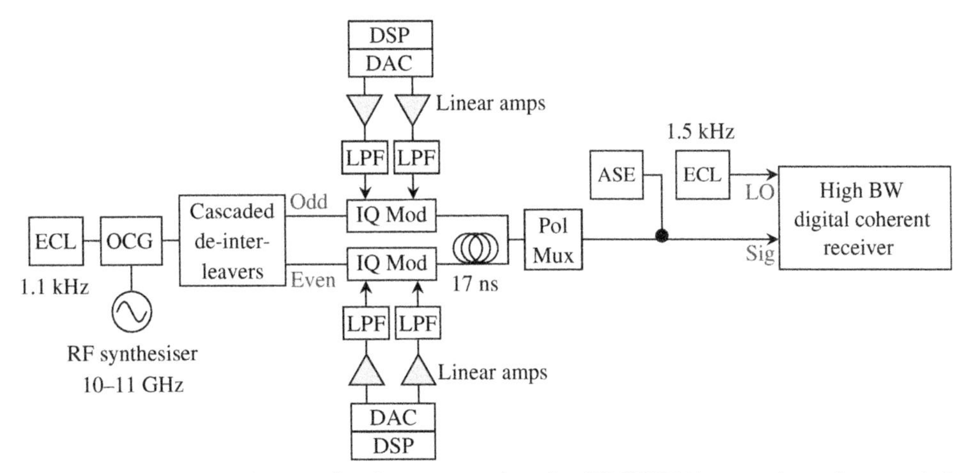

Figure 9.12 Experimental setup for the seven subcarrier DP-256QAM superchannel transmission system. *LPF*, Low pass filter.

As an example, in Ref. [16], a 7 subcarrier superchannel was used to demonstrate the transmission performance of a high throughput optical communications system. However, as free running lasers typically vary in frequency relative to each another, it is difficult to experimentally investigate the performance penalty caused by the RC filter roll-off factor in isolation. To overcome this practical discrepancy, the performance of the digital RRC filter as a function of subcarrier spacing was verified using an optical comb source. As the comb lines are inherently frequency- and phase-matched, it was possible to isolate the performance degradation caused by the digital filtering process itself. Fig. 9.12 illustrates the schematic of a 7 × 10 GBd dual polarization (DP)-16QAM system used to investigate the optimum filter characteristics and subcarrier spacing for our superchannel transmission system. The output of external cavity laser (ECL) with a linewidth of 1.1 kHz was passed through an optical comb generator that consisted of two cascaded Mach–Zehnder modulators, both overdriven with an amplified sinusoidal wave. This generated seven evenly spaced, frequency locked comb lines, with the channel spacing equal to the frequency of the applied sine wave. The frequency comb was separated into odd and even subcarriers using three cascaded micro-interferometer interleavers. Each set of comb lines were independently modulated using two in-phase/quadrature (IQ) modulators. Four decorrelated random binary sequences of length 2^{16} were digitally generated offline and combined to provide two 4-level driving signals (required for the 16QAM format), which were spectrally shaped using a truncated RRC filter. The filter design was minimum order with a specified stop-band attenuation and roll-off factor. The roll-off factor ranged from 10% down to 0.1% and the number of filter taps increased from 320 for the 10% filter to ~1000 for the lowest roll-off factor of 0.1%. There was also a corresponding decrease in stop-band attenuation from 40 dB down to 20 dB as the filter roll-off factor was reduced. The RRC filtered in-phase and quadrature signals were loaded onto a pair of field-programmable gate arrays (FPGAs) and output using two DACs, identically to that described in Section 9.4.3.2. The modulated odd and even channels were decorrelated by 170 symbols before being combined and polarization multiplexed to form a 7 × 10 GBd DP-16QAM signal.

The back-to-back performance of the digital transmitter was initially verified using a single channel and is illustrated in Fig. 9.13A. The roll-off factor of the RRC filter was set to 0.1%, and an implementation penalty of 1.3 dB relative to the theoretical optical signal-to-noise ratio (OSNR) limit was achieved at a BER of 1.5×10^{-2}. The

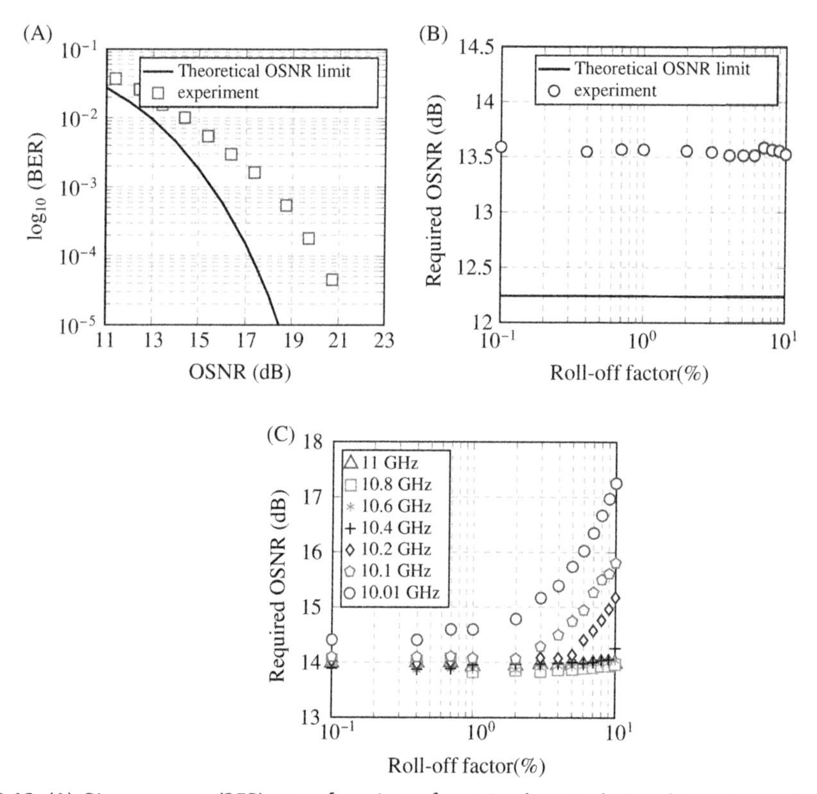

Figure 9.13 (A) Bit-error-rate (BER) as a function of received optical signal-to-noise ratio (OSNR) [0.1 nm resolution bandwidth (BW)] for a single-channel with a roll-off factor of 0.1%. (B) Required OSNR to achieve a BER of 1.5×10^{-2} as a function of the root raised cosine (RRC) filter roll-off factor for a single-channel system. (C) Required OSNR to achieve a BER of 1.5×10^{-2} as a function of both the RRC filter roll-off factor and the channel spacing, for the central channel, in the 7-channel wavelength-division multiplexing (WDM) system.

required OSNR to achieve a BER of 1.5×10^{-2}, as a function of the filter roll-off factor, was experimentally measured and is shown in Fig. 9.13B. The average OSNR was 13.5 dB as the roll-off factor increased from 0.1% to 10% and varied by $\pm$ 0.05 dB, which was within the measurement accuracy of our system. This demonstrates that there was no intrinsic penalty associated with the RRC filter characteristics and provides a baseline level of performance for the 10 GBd DP-16QAM digital transmitter. Fig. 9.13C shows the back-to-back performance for the DP-16QAM superchannel transmitter as a function of both the WDM channel spacing and RRC filter roll-off factor. An additional implementation penalty of 0.4 dB was measured at a channel spacing of 11 GHz and for a roll-off factor of 0.1%, which was due to the finite stop-band attenuation of the RRC digital filter in the transmitter. At this WDM channel spacing (11 GHz), the required OSNR (13.9 dB) to achieve a BER below 1.5×10^{-2} was constant as a function of the RRC filter roll-off factor. This can be intuitively understood by considering the highest roll-off factor of 10%, which provides an optical BW that did not exceed 11 GHz; therefore, no linear crosstalk induced OSNR penalty was experienced at this spacing.

However, as the channel spacing was reduced, there was a corresponding increase in the required OSNR to achieve a BER below 1.5×10^{-2} for a RRC filter roll-off factor of 10%. This penalty was due to linear crosstalk between neighboring superchannel subcarriers and caused the required OSNR to gradually increase to 14.3 dB as the channel spacing approached 10.4 GHz. Below this spacing there was a sharp degradation in performance, thus requiring a reduction in the channel BW or decrease in the RRC filter roll-off factor. The OSNR penalty was found to reduce linearly as the RRC filter roll-off factor was decreased and an acceptable level of performance was achieved for a roll-off factor of 1% at any channel spacing greater than 10.1 GHz. However, in order to incur the minimum performance penalty due to linear crosstalk in a Nyquist-spaced (10 GHz) 10 GBd DP-16QAM WDM transmitter, while simultaneously achieving the highest information spectral density, the roll-off factor must be reduced to 0.1%. In the following experimental verifications, a subcarrier spacing of 10.2 GHz was chosen to allow for the low frequency drift of the free running lasers to be used in the MC-DBP demonstration. A RRC roll-off factor of 1% was therefore chosen to complement the subcarrier spacing of 10.2 GHz.

9.4.4.2 Higher order modulation formats for the optical fiber channel: experimental and numerical demonstration

State-of-the-art commercial transmission links typically employ DP-QPSK and DP-16QAM modulations format, with some more advanced shorter links utilizing DP-64QAM. In this section, DP-mQAM formats are considered, where $m \in \{4, 16, 64, 256\}$ (can be higher order) and demonstrate that, for any received SNR, the highest throughput or capacity will always be achieved by using the highest order

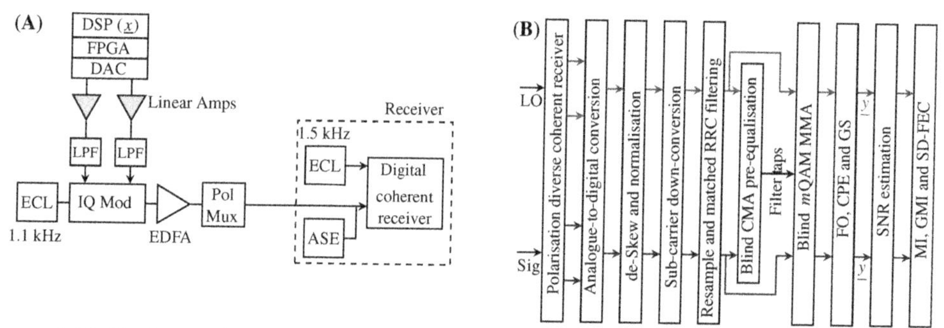

Figure 9.14 Single wavelength DP-mQAM transceiver: (A) experimental setup and (B) digital signal processing (DSP) functions within digital coherent receiver.

modulation format. This demonstration was initially performed in a single-channel, back-to-back experimental test-bed. The single optical transceiver experimental testbed used in this work is shown in Fig. 9.14A. The multilevel drive signals required for each mQAM format were generated by combining $\log_2 m$ decorrelated random binary sequences of length 2^{16}. The coded bits were mapped to uniformly shaped QAM constellations using a memoryless mapper based on the binary reflected Gray code, which generated the transmitted symbols, $\underline{x}$, as shown at the output of the mapper in Fig. 9.8B. The complex symbols were subsequently upsampled to 2 Sa/sym, which provided a symbol rate of 10 GBd as the DAC operated at 20 GSa/s. The complex signals were filtered using an RRC filter with a roll-off factor of 1% and a stop-band attenuation in excess of 30 dB. The filtered mQAM signals were preemphasized to overcome the BW response of the electrical and optoelectronic components in the transmitter, before being quantized to 2^6 discrete levels, which corresponding to the 6 physical bits of resolution available within the DAC. Finally, the in-phase (I) and quadrature (Q) components were loaded onto an FPGA and output using the DAC. Each electrical signal was amplified using a linear electrical amplifier and passed through an 8th-order Bessel low pass filter (LPF) with a rejection ratio in excess of 20 dB/GHz and a 3 dB BW of 6.2 GHz, before being applied to an IQ modulator. The output of an external cavity ECL with a 1.1 kHz linewidth was passed directly into the modulator before being optically amplified and polarizations multiplexed to form an 10 GBd Nyquist shaped DP-mQAM optical carrier. The DP-mQAM signal was passed directly into the signal port of the digital coherent receiver, which had a sampling rate of 160 GSa/s and an analog electrical BW of 62.5 GHz. The ASE noise was added to the signal to vary the received SNR and a second ECL (1.5 kHz linewidth) was used as a LO.

The key blocks of the blind DSP implementation, including the SNR and MI estimation, are illustrated in Fig. 9.14B. The received signals (sampled at 160 GS/s) were initially corrected for receiver skew imbalance and normalized to overcome the varying responsivities of the 70 GHz balanced photodiodes within the coherent receiver. Each polarization was resampled to 2 Sa/sym before matched RRC filtering. A blind 51-tap multimodulus algorithm (MMA) equalizer [123] was used to equalize the signal and to undo polarizations rotations, with the CMA equalizer [124] used for tap-weight preconvergence. The symbols at the output of the equalizer were decimated to 1 Sa/sym and the intermediate frequency was estimated and removed using a 4th-order nonlinearity algorithm [125]. The carrier phase was estimated per polarizations using a decision directed phase estimation algorithm and the complex field was averaged over a 64 T-spaced sliding window to improve the estimate [126]. Gram–Schmidt orthogonalization [127] was performed in order to correct for suboptimal phase bias in the transmitter IQ modulators, which occurred over time due to temperature variations. The symbols at the output of the Gram–Schmidt orthogonalization stage represented the received symbols, $\underline{y}$, as shown in Fig. 9.8B, and were used to calculate the SNR and MI. The SNR (γ) was ideally estimated as: $\gamma = E[|\underline{x}|^2]/\sigma_z^2$, where $E[|\underline{x}|^2]$ is the average received symbol energy and σ_z^2 is the noise variance.

The MI was calculated over both polarizations as a function of the estimated SNR for 4QAM, 16QAM, 64QAM, and 256QAM, respectively and is shown in Fig. 9.15A. The experimental measurement was recorded by adding ASE noise to the signal and measuring the MI for 20 discrete values of SNR ranging from 0 to 26 dB (depending on the order of the format). The Shannon capacity of the AWGN channel is also shown to provide a performance reference for the experimental results. For DP-4QAM (QPSK), the MI increased from 1.95 b/sym at a received SNR of 0 dB, to 3.17 b/sym at an SNR of 4 dB, and finally to a maximum MI of 4 b/sym (2 b/sym

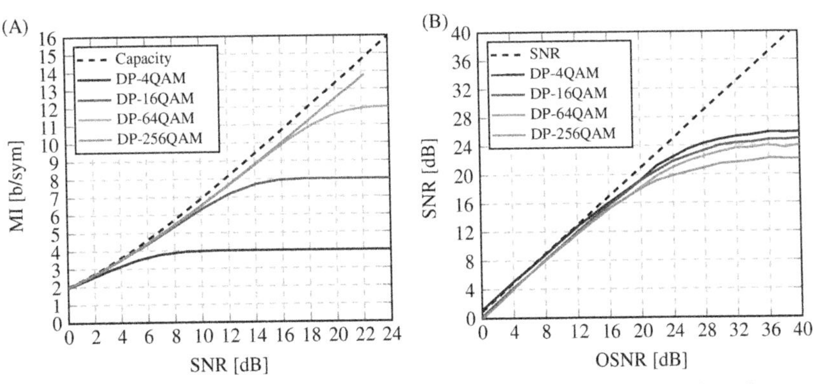

Figure 9.15 (A) Experimentally measured MI as a function of received SNR. (A) MI (measured over two polarizations) for DP-4QAM (DP-QPSK), DP-16QAM, DP-64QAM and DP-256QAM. The dotted line illustrates the Shannon capacity of an AWGN channel. (B) Corresponding received OSNR as a function, where Fb is the symbol rate and N0 is the reference noise BW corresponding to 0.1 nm.

for 4QAM over two polarizations) at an SNR of 12 dB. The maximum SNR obtained for the DP-4QAM format was 26 dB and this remained constant even as the power of the power of the ASE noise loading stage was reduced toward zero. This is clear from Fig. 9.15B, which shows the received OSNR as a function of the measured SNR. The main contributions to the limited SNR within the digital coherent transceiver were the DAC and linear electrical amplifiers in the transmitter and the analog-to-digital converters (ADCs) in the real-time sampling oscilloscope at the receiver. The DAC used in this work exhibited a physical resolution of 6 bits. However, the measured ENOB of the DAC was ~5 bits over the frequency range from 0 to 5 GHz (BW of the spectrally shaped IQ drive signals), which corresponds to an SNR of ~32 dB. The linear amplifiers had a noise figure of 6 dB at a frequency of 5 GHz; therefore, the maximum attainable SNR from the electrical components within the transmitter was ~26 dB. The ADCs in the real-time sampling oscilloscope also exhibited a frequency-dependent ENOB, which was 4.8 bits at a frequency of 4 GHz (reduced to 4.3 bits at a frequency of 60 GHz).

Similar MI performance was achieved for the DP-16QAM format, which exhibited a MI of 1.98 b/sym at an SNR of 0 dB and reached a maximum of 8 b/sym at an SNR of 19.5 dB. The maximum SNR for The DP-64QAM format realized the maximum MI of 12 b/sym at an SNR of 24 dB. The upper limit on the SNR that could be practically achieved using this transceiver setup was lower for the DP-64QAM format than that of the DP-4QAM format. The reduction in SNR of ~2 dB was caused by the blind DSP implementation. As the order of QAM is increased, blind decisions made in the DSP are more likely to be incorrect as the order of the format is increased. This is exasperated for the DP-256QAM format, which never achieved the maximum possible MI of 16 b/sym. The MI for the DP-256QAM format was limited to 13.7 b/sym due to both the saturation in the SNR within the coherent optical transceiver and an additional penalty arising from the blind DSP implementation. Therefore, as the power of the ASE noise loading stage was reduced towards zero, the SNR remained constant at 22 dB. The MI curves shown in Fig. 9.15A never intersect. Therefore, in a transceiver-limited CM optical communication system, the largest constellation size will always provide the greatest achievable IR and thus the highest throughput. It is important to note that the finite SNR of a coherent optical transceiver is an inherent property that cannot be mitigated or compensated and currently represents a significant obstacle to increasing the throughput of light wave communications systems. However, if practical improvements are made to the transceiver subsystems, greater IRs could be achieved. Currently, high resolution ADCs and DACs that are capable of achieving an SNR in excess of 50 dB are commercially available, albeit at lower sampling rates of approximately 3 GSample/s (GSa/s). However, the availability of these components will become commonplace at higher sampling rates in the near future.

9.4.4.3 Numerical investigation of shaped DP-64QAM and DP-256QAM in the optical fiber channel

To demonstrate the performance of the probabilistic shaping, numerical simulations using DP-64QAM and DP-256QAM were performed. Single-channel simulations were carried out in a back-to-back (B2B) system configuration using similar physical configurations as described in Section 9.4.3.2 and compared to uniform QAM. For a given SNR, QAM symbols ($\underline{x}$), as shown at the output of the mapper in Fig. 9.8B were generated based on Eq. (9.60). The signal was then upsampled to 2 sample/symbol (Sa/sym) and filtered with an RRC filter with a roll-off factor of 1%, generating a signal at 12.5 GBd. Additionally the signal was filtered by a 5th-order Bessel filter to emulate limited electronic BW conditions. The signal was then noise loaded to desired SNR and detected with a coherent receiver introducing a 200 MHz frequency offset and 5 kHz linewidth. The DSP blocks used to process the signal are the same used in Section 9.4.3.2, illustrated in in Fig. 9.14B. A matched RRC was applied before 11-taps MMA equalizer [123], preconverged with CMA equalizer [124]. Frequency offset is then compensated using a 4th-order nonlinearity and carrier phase is estimated using a decision directed algorithm with an averaging window of 64 T-spaced samples. The blind algorithms for equalization and CPE rely on regular structure of the uniform QAM constellation to recover the signal; however, for low SNRs, due to the markedly uneven probabilities of the shaped constellation do not perform as reliably (Fig. 9.16 *insets*). Therefore, for the results presented in this

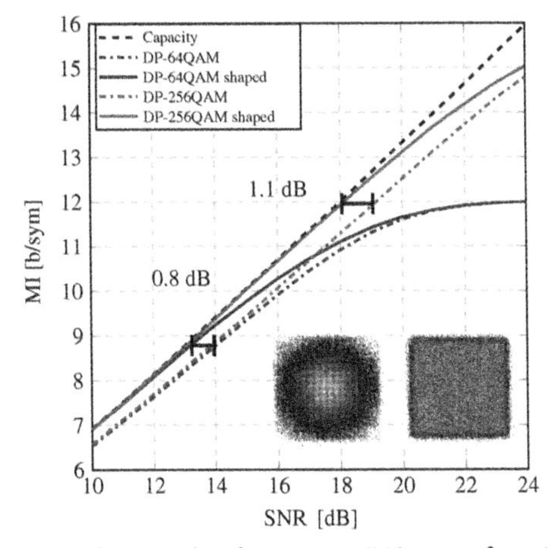

Figure 9.16 Simulation measured mutual information (MI) as a function of signal-to-noise ratio (SNR) for dual-polarizations 64-ary quadrature amplitude modulation (DP-64QAM) and dual-polarizations 256-ary quadrature amplitude modulation (DP-256QAM) with and without shaping, with data-aided (DA) digital signal processing (DSP). The dotted line illustrates the Shannon capacity of an additive white GN channel.

section, both the equalizer and the phase estimator are data-aided, similar to previous investigation of probabilistic shaped signal [107]. Although for the uniform QAM data-aided DSP was not required, it was used for fair comparison. Additionally, training aided and blind algorithms that handle shaped constellations at no additional penalty are currently under investigation. At the output of the phase estimator the symbols $\underline{y}$ are used to calculate the MI.

Fig. 9.16 shows the performance when filtering, frequency offset, and phase noise are taken into consideration, with the data-aided DSP discussed above. The maximum SNR gains due to probabilistic shaping are 0.8 dB for DP-64QAM and 1.1 dB for DP-256QAM. A comparison with the theoretical results showed in Section 9.4.3.1 for the AWGN channel demonstrate that no SNR penalty was incurred due to the optical transceiver impairments.

9.5 Conclusions

Optical fiber nonlinearities have been seen as the major bottleneck to the achievable IR of optical transmission networks. The signal degradations due to the nonlinear distortions limit the achievable transmission distances and become more significant in systems with larger transmission BWs, closer channel spacing, and higher order modulation formats. This chapter discussed a series of techniques developed to overcome the capacity barriers in optical fiber communication systems. The key goal was to describe the DSP-based nonlinearity compensation and quantify possible increases in the achievable capacity and transmission distances, depending on the modulation format used, that come from the combination of nonlinearity compensation and signal constellation shaping. This rapidly changing field has seen tremendous progress, and it remains to be seen whether the significant point-to-point system gains, successfully achieved with the described techniques, can be extended to network-wide regime. Much hard work remains in applying them to transform the optical networks for the cloud to become more intelligent and adaptive, to be able to tailor network topology, capacity and delay to the rapidly changing, and ever increasing, array of applications.

Acknowledgments

The authors are grateful to many colleagues and support in the UK EPSRC Program Grants TRANSNET (EP/R035342/1) and UNLOC (EP/J017582/1) as well as EU Horizon 2020 RISE Grant DAWN4IoE (778305) and the EU ERC Grant (757791) for numerous contributions to the work described. The authors would like to thank Dr Domaniç Lavery and Professor Robert I. Killey (UCL) for suggestions to improve this chapter.

References

[1] G.P. Agrawal, Fibre-Optic Communication Systems, Wiley, 2010.
[2] I. Kaminow, T. Li, A.E. Willner, Optical Fiber Telecommunications VIA: Component and Subsystems, Academic Press, 2013.
[3] R.-J. Essiambre, G. Kramer, P.J. Winzer, G.J. Foschini, B. Goebel, Capacity limits of optical fiber networks, J. Lightwave Technol. 28 (2010) 662–701.
[4] G.P. Agrawal, Nonlinear Fiber Optics, Wiley, 2001.
[5] G.P. Agrawal, Nonlinear Fiber Optics, Academic Press, 2007.
[6] S.J. Savory, Digital filters for coherent optical receivers, Opt. Express 16 (2008) 804–817.
[7] T. Xu, G. Jacobsen, S. Popov, J. Li, E. Vanin, K. Wang, et al., Chromatic dispersion compensation in coherent transmission system using digital filters, Opt. Express 18 (2010) 16243–16257.
[8] O. Zia-Chahabi, R. Le Bidan, M. Morvan, C. Laot, Efficient frequency-domain implementation of block-LMS/CMA fractionally spaced equalization for coherent optical communications, IEEE Photon. Technol. Lett. 23 (2011) 1697–1699.
[9] E. Ip, J.M. Kahn, Digital equalization of chromatic dispersion and polarization mode dispersion, J. Lightwave Technol. 25 (2007) 2033–2043.
[10] Y. Mori, C. Zhang, K. Igarashi, K. Katoh, K. Kikuchi, Unrepeated 200-km transmission of 40-Gbit/s 16-QAM signals using digital coherent receiver, Opt. Express 17 (2009) 1435–1441.
[11] E. Ip, J.M. Kahn, Feedforward carrier recovery for coherent optical communications, J. Lightwave Technol. 25 (2007) 2675–2692.
[12] P. Bayvel, R. Maher, T. Xu, G. Liga, N.A. Shevchenko, D. Lavery, et al., Maximizing the optical network capacity, Phil. Trans. R. Soc. A 374 (2016) 20140440.
[13] D. Semrau, T. Xu, N.A. Shevchenko, M. Paskov, A. Alvarado, R.I. Killey, et al., Achievable information rates estimates in optically amplified transmission systems using nonlinearity compensation and probabilistic shaping, Opt. Lett. 42 (2017) 121–124.
[14] P.K.A. Wai, C.R. Menyuk, Polarization mode dispersion, decorrelation, and diffusion in optical fibers with randomly varying birefrin-gence, J. Lightwave Technol. 14 (1996) 148–157.
[15] D. Marcuse, C.R. Menyuk, P.K.A. Wai, Application of the Manakov–PMD equation to studies of signal propagation in optical fibers with randomly varying birefringence, J. Lightwave Technol. 15 (1997) 1735–1745.
[16] R. Maher, T. Xu, L. Galdino, M. Sato, A. Alvarado, K. Shi, et al., Spectrally shaped DP-16QAM super-channel transmission with multi-channel digital back-propagation, Sci. Rep. 5 (2015) 8214.
[17] T. Xu, T. Xu, P. Bayvel, I. Darwazeh, Non-orthogonal signal transmission over nonlinear optical channels, IEEE Photon. J. 11 (2019) 7203313.
[18] P. Poggiolini, The GN model of non-linear propagation in uncompensated coherent optical systems, J. Lightwave Technol. 30 (2012) 3857–3879.
[19] P. Johannisson, M. Karlsson, Perturbation analysis of nonlinear propagation in a strongly dispersive optical communication system, J. Lightwave Technol 31 (2013) 1273–1282.
[20] P. Poggiolini, G. Bosco, A. Carena, V. Curri, Y. Jiang, F. Forghieri, The GN-model of fiber nonlinear propagation and its applications, J. Lightwave Technol. 32 (2014) 694–721.
[21] D. Rafique, A.D. Ellis, Impact of signal-ASE four-wave mixing on the effectiveness of digital back-propagation in 112 Gb/s PM-QPSK systems, Opt. Express 19 (2011) 3449–3454.
[22] L. Beygi, N.V. Irukulapati, E. Agrell, P. Johannisson, M. Karlsson, H. Wymeersch, et al., On nonlinearly-induced noise in single-channel optical links with digital backpropagation, Opt. Express 21 (2013) 26376–26386.
[23] N.A. Shevchenko, T. Xu, D. Semrau, G. Saavedra, G. Liga, M. Paskov, et al., Achievable information rates estimation for 100-nm Raman-amplified optical transmission system, in: Proc. Eur. Conf. Opt. Commun. (ECOC), 2016, pp. 878–880.
[24] P.P. Mitra, J.B. Stark, Nonlinear limits to the information capacity of optical fibre communications, Nature 411 (2001) 1027–1030.

[25] M. Secondini, E. Forestieri, G. Prati, Achievable information rate in nonlinear WDM fiber-optic systems with arbitrary modulation formats and dispersion maps, J. Lightwave Technol. 31 (2013) 3839–3852.
[26] A. Mecozzi, R.-J. Essiambre, Nonlinear Shannon limit in pseudolinear coherent systems, J. Lightwave Technol. 30 (2012) 2011–2024.
[27] R. Dar, F. Meir, A. Mecozzi, M. Shtaif, Properties of nonlinear noise in long, dispersion-uncompensated fiber links, Opt. Express 21 (2013) 25685–25699.
[28] P. Serena, A. Bononi, On the accuracy of the Gaussian nonlinear model for dispersion-unmanaged coherent links, in: Eur. Conf. Opt. Commun. (ECOC) 2013, p. Th.1.D.3.
[29] P. Serena, A. Bononi, N. Rossi, The impact of the modulation dependent nonlinear interference missed by the Gaussian noise model, in: Eur. Conf. Opt. Commun. (ECOC) 2014, p. Mo.4.3.1.
[30] R. Dar, M. Feder, A. Mecozzi, M. Shtaif, Accumulation of nonlinear interference noise in fiber-optic systems, Opt. Express 22 (2014) 14199–14211.
[31] A. Carena, G. Bosco, V. Curri, Y. Jiang, P. Poggiolini, F. Forghieri, EGN model of non-linear fiber propagation, Opt. Express 22 (2014) 16335–16362.
[32] P. Poggiolini, G. Bosco, A. Carena, V. Curri, Y. Jiang, F. Forghieri, A simple and effective closed-form GN model correction formula accounting for signal non-Gaussian distribution, J. Lightwave Technol. 33 (2015) 459–473.
[33] J.C. Cartledge, F.P. Guiomar, F.R. Kschischang, G. Liga, M.P. Yankov, Digital signal processing for fiber nonlinearities, Opt. Express 25 (2017) 1916–1936.
[34] P. Poggiolini, Y. Jiang, A. Carena, F. Forghieri, Analytical modeling of the impact of fiber non-linear propagation on coherent systems and networks, Enabling Technologies for High Spectral-Efficiency Coherent Optical Communication Networks, Wiley, 2016, pp. 247–310.
[35] R. Dar, M. Feder, A. Mecozzi, M. Shtaif, Inter-channel nonlinear interference noise in WDM systems: modeling and mitigation, J. Lightwave Technol. 33 (2015) 1044–1053.
[36] X. Li, et al., Electronic post-compensation of WDM transmission impairments using coherent detection and digital signal processing, Opt. Express 16 (2008) 880–888.
[37] E. Mateo, L. Zhu, G. Li, Impact of XPM and FWM on the digital implementation of impairment compensation for WDM transmission using backward propagation, Opt. Express 16 (2008) 16124–16137.
[38] E.F. Mateo, F. Yaman, G. Li, Efficient compensation of inter-channel nonlinear effects via digital backward propagation in WDM optical transmission, Opt. Express 18 (2010) 15144–15154.
[39] E. Ip, J.M. Kahn, Compensation of dispersion and nonlinear impairments using digital backpropagation, J. Lightwave Technol. 26 (2008) 3416–3425.
[40] E. Ip, Nonlinear compensation using backpropagation for polarization-multiplexed transmission, J. Lightwave Technol. 28 (2010) 939–951.
[41] O.V. Sinkin, R. Holzlöhner, J. Zweck, C.R. Menyuk, Optimization of the split-step Fourier method in modelling optical fiber communications systems, J. Lightwave Technol. 21 (2003) 61–68.
[42] G. Liga, T. Xu, A. Alvarado, R.I. Killey, P. Bayvel, On the performance of multichannel digital backpropagation in high-capacity long-haul optical transmission, Opt. Express 16 (2014) 1217–1226.
[43] L.B. Du, A.J. Lowery, Improved single channel backpropagation for intra-channel fiber nonlinearity compensation in long-haul optical communication systems, Opt. Express 18 (2010) 17075–17088.
[44] Y. Gao, J.H. Ke, K.P. Zhong, J.C. Cartledge, S.S.H. Yam, Intra-channel nonlinear compensation for 112 Gb/s dual polarization 16QAM systems, J. Lightwave Technol. 30 (2012) 390–391.
[45] Y. Gao, J.H. Ke, J.C. Cartledge, K.P. Zhong, S.S.H. Yam, Implication of parameter values on low-pass filter assisted digital back propagation for DP 16-QAM, IEEE Photon. Technol. Lett. 25 (2013) 917–920.
[46] M. Secondini, S. Rommel, F. Fresi, E. Forestieri, G. Meloni, L. Poti, Coherent 100G nonlinear compensation with single-step digital back-propagation, in: Proc. Opt. Netw. Design Modelin (ONDM), 2015, pp. 1–5.
[47] G. Gao, X. Chen, W. Shieh, Influence of PMD on fiber nonlinearity compensation using digital back propagation, Opt. Express 20 (2012) 14406–14418.

[48] A. Carena, V. Curri, G. Bosco, P. Poggiolini, F. Forghieri, Modeling of the impact of nonlinear propagation effects in uncompensated optical coherent transmission links, J. Lightwave Technol. 30 (2012) 1524–1539.
[49] D. Lavery, D. Ives, G. Liga, A. Alvarado, S.J. Savory, P. Bayvel, The benefit of split nonlinearity compensation for optical fiber communications, IEEE Photon. Technol. Lett. 28 (2016) 1803–1806.
[50] A. Mecozzi, R.J. Essiambre, Nonlinear Shannon limit in pseudolinear coherent systems, J. Lightwave Technol. 30 (2012) 2011–2024.
[51] A.D. Ellis, M.E. McCarthy, M.A.Z. Al-Khateeb, S. Sygletos, Capacity limits of systems employing multiple optical phase conjugators, Opt. Express 23 (2015) 20381.
[52] P. Serena, Nonlinear signal-noise interaction in optical links with non-linear equalization, J. Lightwave Technol. 34 (2016) 1476–1483.
[53] T. Xu, N.A. Shevchenko, D. Lavery, D. Semrau, G. Liga, A. Alvarado, et al., Modulation format dependence of digital nonlinearity compensation performance in optical fibre communication systems, Opt. Express 25 (2017) 3311–3326.
[54] W. Marshall, B. Crosignani, A. Yariv, Laser phase noise to intensity noise conversion by lowest-order group-velocity dispersion in optical fiber: exact theory, Opt. Lett. 25 (2000) 165–167.
[55] S. Yamamoto, N. Edagawa, H. Taga, Y. Yoshida, H. Wakabayashi, Analysis of laser phase noise to intensity noise conversion by chromatic dispersion in intensity modulation and direct detection optical-fiber transmission, J. Lightwave Technol. 8 (1990) 1716–1722.
[56] W. Shieh, K.-P. Ho, Equalization-enhanced phase noise for coherent-detection systems using electronic digital signal processing, Opt. Express 16 (2008) 15718–15727.
[57] C. Xie, Local oscillator phase noise induced penalties in optical coherent detection systems using electronic chromatic dispersion compensation, in: Proc. Opt. Fiber Comm. Conf. (OFC), 2009, p. OMT4.
[58] A.P.T. Lau, T.S.R. Shen, W. Shieh, K.-P. Ho, Equalization-enhanced phase noise for 100Gb/s transmission and beyond with coherent detection, Opt. Express 18 (2010) 17239–17251.
[59] K.-P. Ho, A.P.T. Lau, W. Shieh, Equalization-enhanced phase noise induced timing jitter, Opt. Lett. 36 (2011) 585–587.
[60] K.-P. Ho, W. Shieh, Equalization-enhanced phase noise in mode-division multiplexed systems, J. Lightwave Technol. 31 (2013) 2237–2243.
[61] C. Xie, WDM coherent PDM-QPSK systems with and without inline optical dispersion compensation, Opt. Express 17 (2009) 4815–4823.
[62] T. Xu, G. Jacobsen, S. Popov, J. Li, A.T. Friberg, Y. Zhang, Analytical estimation of phase noise influence in coherent transmission system with digital dispersion equalization, Opt. Express 19 (2011) 7756–7768.
[63] I. Fatadin, S.J. Savory, Impact of phase to amplitude noise conversion in coherent optical systems with digital dispersion compensation, Opt. Express 18 (2010) 16273–16278.
[64] R. Farhoudi, A. Ghazisaeidi, L.A. Rusch, Performance of carrier phase recovery for electronically dispersion compensated coherent systems, Opt. Express 20 (2012) 26568–26582.
[65] G. Colavolpe, T. Foggi, E. Forestieri, M. Secondini, Impact of phase noise and compensation techniques in coherent optical systems, J. Lightwave Technol. 29 (2011) 2790–2800.
[66] L.B. Du, D. Rafique, A. Napoli, B. Spinnler, A.D. Ellis, M. Kuschnerov, et al., Digital fiber nonlinearity compensation: toward 1-Tb/s transport, IEEE Signal Process. Mag. 31 (2014) 46–56.
[67] S.-G. Park, Effect of the phase-error of local oscillators in digital back-propagation, IEEE Photon. Technol. Lett. (2015) 363–366.
[68] N. Alic, E. Myslivets, E. Temprana, B.P.-P. Kuo, S. Radic, Nonlinearity cancellation in fiber optic links based on frequency referenced carriers, J. Lightwave Technol. 32 (2014) 2690–2698.
[69] T. Xu, G. Liga, D. Lavery, B.C. Thomsen, S.J. Savory, R.I. Killey, et al., Equalization enhanced phase noise in Nyquist-spaced superchannel transmission systems using multi-channel digital back-propagation, Sci. Rep. 5 (2015) 13990.
[70] M.G. Taylor, Phase estimation methods for optical coherent detection using digital signal processing, J. Lightwave Technol. 27 (2009) 901–914.

[71] G. Jacobsen, T. Xu, S. Popov, J. Li, A.T. Friberg, Y. Zhang, EEPN and CD study for coherent optical nPSK and nQAM systems with RF pilot based phase noise compensation, Opt. Express 20 (2012) 8862–8870.

[72] G. Jacobsen, M. Lidn, T. Xu, S. Popov, A.T. Friberg, Y. Zhang, Influence of pre-and post-compensation of chromatic dispersion on equalization enhanced phase noise in coherent multilevel systems, J. Opt. Commun 32 (2011) 257–261.

[73] M. Secondini, G. Meloni, T. Foggi, G. Colavolpe, L. Poti, E. Forestieri, Phase noise cancellation in coherent optical receivers by digital coherence enhancement, in: Eur. Conf. Opt. Commun. (ECOC), 2010, p. P4.17.

[74] T. Yoshida, T. Sugihara, K. Uto, DSP-based optical modulation technique for long-haul transmission, SPIE Next-Generation Optical Communication: Components, Vol. 9389, Sub-Systems, and Systems IV, 2015, pp. 1–3.

[75] A. Kakkar, R. Schatz, X. Pang, J.R. Navarro, H. Louchet, O. Ozolins, et al., Impact of local oscillator frequency noise on coherent optical systems with electronic dispersion compensation, Opt. Express 23 (2015) 11221–11226.

[76] G. Ungerboeck, Channel coding with multilevel/phase signals, IEEE Trans. Inf. Theory 28 (1982) 55–67.

[77] G. Forney, R. Gallager, G. Lang, F. Longstaff, S. Qureshi, Efficient modulation for band-limited channels, IEEE J. Sel. Areas Commun. 2 (1984) 632–647.

[78] M. Magarini, R. Essiambre, B.E. Basch, A. Ashikhmin, G. Kramer, A.J. de Lind van Wijngaarden, Concatenated coded modulation for optical communications systems, IEEE Photon. Technol. Lett. 22 (2010) 1244–1246.

[79] I.B. Djordjevic, M. Cvijetic, L. Xu, T. Wang, Using LDPC-coded modulation and coherent detection for ultra highspeed optical transmission, J. Lightwave Technol. 25 (2007) 3619–3625.

[80] H.G. Batshon, I.B. Djordjevic, L. Xu, T. Wang, Multidimensional LDPC-coded modulation for beyond 400 Gb/s per wavelength transmission, IEEE Photon. Technol. Lett. 21 (2009) 1139–1141.

[81] P.J. Winzer, High-spectral-efficiency optical modulation formats, J. Lightwave Technol. 30 (2012) 3824–3835.

[82] B.P. Smith, F.R. Kschischang, A pragmatic coded modulation scheme for high-spectral-efficiency fiber-optic communications, J. Lightwave Technol. 30 (2012) 2047–2053.

[83] A. Alvarado, E. Agrell, Four-dimensional coded modulation with bit-wise cecoders for future optical communications, J. Lightwave Technol. 33 (2015) 1993–2003.

[84] J. Cai, H.G. Batshon, M.V. Mazurczyk, O.V. Sinkin, D. Wang, M. Paskov, et al., 70.4 Tb/s capacity over 7600 km in C + L band using coded modulation with hybrid constellation shaping and nonlinearity compensation, in: Opt. Fiber Commun. Conf. (OFC), 2019, p. Th5B.2.

[85] J.X. Cai, H.G. Batshon, M.V. Mazurczyk, C.R. Davidson, O.V. Sinkin, D. Wang, et al., 94.9 Tb/s single mode capacity demonstration over 1,900 km with C + L EDFAs and coded modulation, in: Eur. Conf. Opt. Commun. (ECOC), 2018, pp. 1–3.

[86] M. Ionescu, D. Lavery, A. Edwards, E. Sillekens, L. Galdino, D. Semrau, et al., 74.38 Tb/s transmission over 6300 km single mode fiber with hybrid EDFA/Raman amplifiers, in: Opt. Fiber Commun. Conf. (OFC), 2019, pp. 1–3.

[87] U. Wachsmann, R.F.H. Fischer, J.B. Huber, Multilevel codes: theoretical concepts and practical design rules, IEEE Trans. Inf. Theory 45 (1999) 1361–1391.

[88] G. Caire, G. Taricco, E. Biglieri, Bit-interleaved coded modulation, IEEE Trans. Inf. Theory 44 (1998) 927–946.

[89] L. Szczecinski, A. Alvarado, Bit-Interleaved Coded Modulation: Fundamentals, Analysis and Design, John Wiley & Sons, 2015.

[90] C.E. Shannon, A mathematical theory of communication, Bell Syst. Tech. J. 14 (1948) 306–423.

[91] T.M. Cover, J.A. Thomas, Elements of Information Theory, Wiley, 2006.

[92] R.-J. Essiambre, G. Kramer, P.J. Winzer, G.J. Foschini, B. Goebel, Capacity limits of optical fiber networks, J. Lightwave Technol. 28 (2010) 662–701.

[93] E. Agrell, A. Alvarado, F.R. Kschischang, Implications of information theory in optical fibre communications, Philos. Trans. R. Soc. A 374 (2016) 20140438.

[94] I.B. Djordjevic, B. Vasic, M. Ivkovic, I. Gabitov, Achievable information rates for high-speed long-haul optical transmission, J. Lightwave Technol. 23 (2005) 3755–3763.
[95] T. Fehenberger, A. Alvarado, P. Bayvel, N. Hanik, On achievable rates for long-haul fiber-optic communications, Opt. Express 23 (2015) 9183–9191.
[96] G. Liga, A. Alvarado, E. Agrell, P. Bayvel. Information Rates of Next-Generation Long-Haul Optical Fiber Systems Using Coded Modulation. Journal of Lightwave Technology 35(1) (2017) 113–123. https://doi.org/10.1109/JLT.2016.2603419.
[97] A. Leven, F. Vacondio, L. Schmalen, S.T. Brink, W. Idler, Estimation of soft FEC performance in optical transmission experiments, IEEE Photon. Technol. Lett. 23 (2011) 1547–1549.
[98] L. Schmalen, A. Alvarado, R. Rios-Muller, Predicting the performance of nonbinary forward error correction in optical transmission experiments, in: Opt. Fiber Commun. Conf. (OFC), 2016, pp. 1–3.
[99] R. Maher, D. Lavery, D. Millar, A. Alvarado, K. Parsons, R. Killey, et al., Reach enhancement of 100% for a DP-64QAM super-channel using MC-DBP, in: Opt. Fiber Commun. Conf. (OFC) 2015, pp. Th4D–5.
[100] D.S. Millar, R. Maher, T. Koike-Akino, M. Pajovic, A. Alvarado, M. Paskov, et al., Detection of a 1 Tb/s superchannel with a single coherent receiver, in: Eur. Conf. on Opt. Commun. (ECOC), 2015, pp. 1–3.
[101] A. Alvarado, T. Fehenberger, B. Chen, F.M. Willems, Achievable information rates for fiber optics: applications and computations, J. Lightwave Technol. 36 (2018) 424–439.
[102] R.F. Fischer, Precoding and Signal Shaping for Digital Transmission, Wiley, 2005.
[103] G.D. Forney, L.-F. Wei, Multidimensional constellations—Part I: Introduction, IEEE J. Sel. Areas Commun. 7 (1989) 877–892.
[104] F.R. Kschischang, S. Pasupathy, Optimal nonuniform signaling for Gaussian channels, IEEE Trans. Inf. Theory 39 (1993) 913–929.
[105] G. Böcherer, F. Steiner, P. Schulte, Bandwidth efficient and rate-matched low-density parity-check coded modulation, IEEE Trans. Commun. 63 (2015) 4651–4665.
[106] R. Maher, A. Alvarado, D. Lavery, P. Bayvel, Modulation order and code rate optimisation for digital coherent transceivers using generalised mutual information, in: Eur. Conf. Opt. Commun. (ECOC), 2015, pp. 1867–1875.
[107] F. Buchali, F. Steiner, G. Bcherer, L. Schmalen, P. Schulte, W. Idler, Rate adaptation and reach increase by probabilistically shaped 64-QAM: an experimental demonstration, J. Lightwave Technol. 34 (2016) 1599–1609.
[108] T. Fehenberger, A. Alvarado, G. Bocherer, N. Hanik, On probabilistic shaping of quadrature amplitude modulation for the nonlinear fiber channel, J. Lightwave Technol. 34 (2016) 5063–5073.
[109] J. Estaran, D. Zibar, A. Caballero, C. Peucheret, I.T. Monroy, Experimental demonstration of capacity-achieving phase-shifted superposition modulation, in: Eur. Conf. Opt. Commun. (ECOC), 2013, pp. 1–3.
[110] T.H. Lotz, et al., Coded PDM-OFDM transmission with shaped 256-iterative-polar-modulation achieving 11.15-b/s/Hz intrachannel spectral efficiency and 800-km reach, J. Lightwave Technol. 31 (2013) 538–545.
[111] M.P. Yankov, D. Zibar, K.J. Larsen, L.P.B. Christensen, S. Forch-hammer, Constellation shaping for fiber-optic channels with QAM and high spectral efficiency, IEEE Photon. Technol. Lett. 26 (2014) 2407–2410.
[112] L. Beygi, E. Agrell, J.M. Kahn, M. Karlsson, Rate-adaptive coded modulation for fiber-optic communications, J. Lightwave Technol. 32 (2014) 333–343.
[113] T. Fehenberger, G. Böcherer, A. Alvarado, N. Hanik, LDPC coded modulation with probabilistic shaping for optical fiber systems, in: Opt. Fiber Commun. Conf. (OFC), 2015, pp. 1–3.
[114] E. Sillekens, D. Semrau, G. Liga, N.A. Shevchenko, Z. Li, A. Alvarado, et al., A simple nonlinearity-tailored probabilistic shaping distribution for square QAM, in: Optical Fiber Communications Conference and Exposition (OFC) 2018, pp. 1–3.
[115] F. Buchali, G. Böcherer, W. Idler, L. Schmalen, P. Schulte, F. Steiner, Experimental demonstration of capacity increase and rate-adaptation by probabilistically shaped 64-QAM, in: Eur. Conf. Opt. Commun. (ECOC), 2015, pp. 1–3.

[116] M.F. Barsoum, C. Jones, M. Fitz, Constellation design via capacity maximization, in: IEEE Int. Symp. Inf. Theory, 2007, pp. 1821–1825.
[117] P. Schulte, G. Bocherer, Constant composition distribution matching, IEEE Trans. Inf. Theory 62 (2016) 430–434.
[118] T. Fehenberger, D.S. Millar, T. Koike-Akino, K. Kojima, K. Parsons, Multiset-partition distribution matching, IEEE Trans. Commun. 67 (2019) 1885–1893.
[119] T. Yoshida, M. Karlsson, E. Agrell, Hierarchical distribution matching for probabilistically shaped coded modulation, J. Lightwave Technol. 37 (2019) 1579–1589.
[120] K. Kojima, T. Yoshida, T. Koike-Akino, D.S. Millar, K. Parsons, M. Pajovic, et al., Nonlinearity-tolerant four-dimensional 2a8psk family for 57 bits/symbol spectral efficiency, J. Lightwave Technol. 35 (2017) 1383–1391.
[121] B. Chen, O. Chigo, H. Hafermann, A. Alvarado, Polarization ring-switching for nonlinearity-tolerant geometrically-shaped four-dimensional formats maximizing generalized mutual information, J. Lightwave Technol. 37 (2019) 3579–3591.
[122] R. Maher, A. Alvarado, D. Lavery, P. Bayvel, Increasing the information rates of optical communications via coded modulation: a study of transceiver performance, Sci. Rep. 6 (2016) 21278–21288.
[123] M.J. Ready, R.P. Gooch, Blind equalization based on radius directed adaptation, in: Int. Conf. Acoustics, Speech, Signal Process (ICASSP), 1990, pp. 1699–1702.
[124] D. Godard, Self-recovering equalization and carrier tracking in two-dimensional data communication systems, IEEE Trans. Commun. 28 (1980) 1867–1875.
[125] S.J. Savory, G. Gavioli, R.I. Killey, P. Bayvel, Electronic compensation of chromatic dispersion using a digital coherent receiver, Opt. Express 15 (2007) 2120–2126.
[126] T. Pfau, S. Hoffmann, R. Noé, Hardware-efficient coherent digital receiver concept with feedforward carrier recovery for M-QAM constellations, J. Lightwave Technol. 27 (2009) 989–999.
[127] I. Fatadin, S.J. Savory, D. Ives, Compensation of quadrature imbalance in an optical QPSK coherent receiver, IEEE Photon. Technol. Lett. 20 (2008) 1733–1735.

CHAPTER 10

High-capacity direct-detection systems

Xi Chen[1], Cristian Antonelli[2], Antonio Mecozzi[2], Di Che[1,3] and William Shieh[3]
[1]Nokia Bell Labs, Holmdel, NJ, United States
[2]Department of Physical and Chemical Sciences, University of L'Aquila, L'Aquila, Italy
[3]Department of Electrical and Electronics Engineering, The University of Melbourne, Parkville, VIC, Australia

10.1 Direct-detection systems and their applications

A direct-detection (DD) system is a communication system based on detecting modulated optical power (also referred to as the optical field intensity or simply the optical intensity). In conventional DD systems the receiver consists of a single photodiode (PD) and, correspondingly, the transmitter modulates the optical power. Therefore such systems are often referred to as intensity-modulation and direct-detection (IM-DD) systems.

The architecture of a typical IM-DD transceiver is illustrated in Fig. 10.1. The transmitter has a laser as a light source. Light can be directly modulated inside the laser or can be externally modulated using a separate modulator. The receiver consists of a PD, a clock recovery module, and symbol decision module. This electronic circuit is usually referred to as a clock and data recovery unit. Compared with coherent systems that use digital-to-analog converters (DACs) and a dual-polarization I/Q modulator at the transmitter, an optical hybrid plus four pairs of balanced photodetectors and four analog-to-digital converters (ADCs) at the receiver, DD systems have simpler electronics and optics.

At the time when the required rates per transceiver interface were 10 and 40 Gb/s, DD systems were widely deployed for various kinds of fiber communication systems ranging from short-reach metro connections to long-reach links in core networks. Later, core networks transitioned to coherent systems, as they offer higher receiver sensitivity and better spectral efficiency, and therefore are more suitable for long-distance transmission at interface rates of 100 Gb/s and beyond. The applications of DD systems were then limited to cost-sensitive short-reach applications such as metro transports, intra- and interdatacenter interconnects, and passive optical networks. Nowadays, typical transmission distances for DD systems range from few meters to about 100 km.

In this chapter, we review the principles of conventional DD systems and discuss their limitations. We then introduce advanced DD systems where novel transmitter and receiver designs are employed, so as to enable ultra-high interface rates for short-

Optical Fiber Telecommunications V11
DOI: https://doi.org/10.1016/B978-0-12-816502-7.00012-9

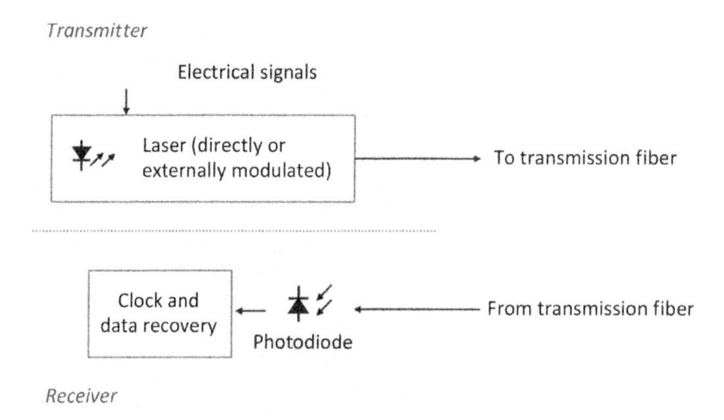

Figure 10.1 Schematic structure of a typical transceiver used in a direct-detection system.

reach applications. We conclude the chapter by presenting our view on the future of short-reach transmission systems.

10.2 Principle of conventional direct-detection systems

To modulate optical power, a directly modulated laser (DML) can be used. In a DML, the laser output power can be changed by varying the pump current in the laser gain medium. As shown in Fig. 10.2, the pump current is controlled by the electrical driving signal. The modulation format for such DD systems is typically on-off keying. In other words, binary signals are applied to change the pump current of the DML.

After fiber transmission and photo detection, the receiver conducts clock recovery and then sets a proper threshold to decide whether the received symbol is a logical 0 or 1. In a typical case, no adaptive digital filtering is used and intersymbol interference, if any, is not mitigated before the symbol decision. Additionally, for short-reach purposes, no forward-error correction (FEC) coding is used.

The advantage of DMLs is that modulation is implemented inside the laser source, and there is no need to couple light between a laser and a modulator. This way, fabrication and packaging costs are minimized. DML-based DD systems are used for 10 Gb/s systems. Research and lab demonstrations have shown that DMLs can have bandwidths beyond 50 GHz [1,2] and are capable of modulating data at >100 Gb/s [2]. However, applications for DML-based DD systems are limited primarily due to the DML limited extinction ratio and other constraints such as laser chirp where the modulation changes the laser frequency directly.

Higher signal integrity can be obtained by using external modulators, as illustrated in Fig. 10.3. A popular type of external modulators for DD systems is referred to as an

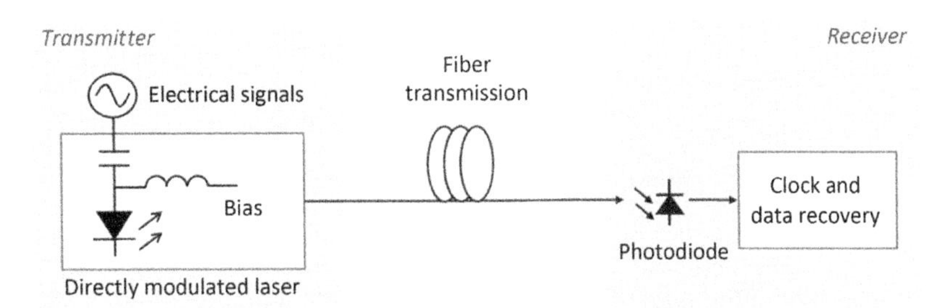

Figure 10.2 Schematic structure of a directly modulated laser (DML)-based intensity-modulation and direct-detection (IM-DD) system.

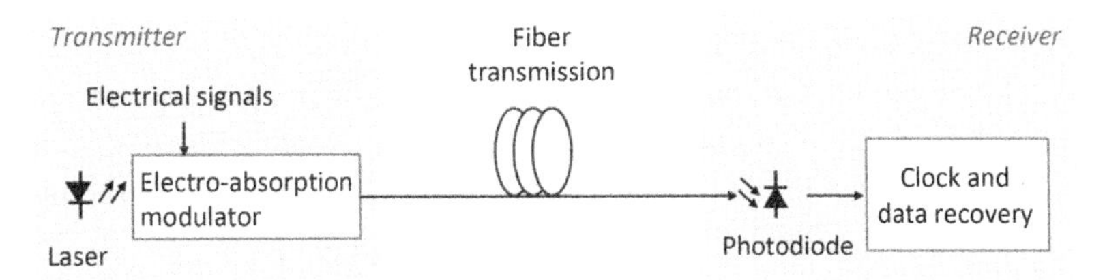

Figure 10.3 Schematic structure of an electro-absorption modulator (EAM)-based intensity-modulation and direct-detection (IM-DD) system.

electro-absorption modulator (EAM). An EAM is often made in the form of a semiconductor waveguide with electrodes for applying an electric field in a direction perpendicular to the propagation direction of the light beam [3]. The semiconductor's optical absorption spectrum (thus the modulator output power) is changed by the applied electric field, a property known as the Franz–Keldysh effect [4]. EAMs can operate with low drive voltages (in the order of 2 V) and can provide more than 100 GHz bandwidth [5]. They can be integrated with distributed feedback laser on the same chip, and the combined laser and EAM is referred to as an electro-absorption modulator laser (EML).

Though EMLs have better extinction ratio than DMLs, they still produce modulation chirp. To avoid this, Mach-Zehnder modulators (MZMs) can be used. A MZM is a different type of external modulators that achieves intensity modulation by combining two phase modulators in a Mach-Zehnder interferometer scheme [6]. Utilizing differential driving or a push-pull configuration [6], the MZM can be chirp-free. A schematic of an MZM-based IM-DD system is shown in Fig. 10.4.

10.3 Limitations of conventional direct-detection systems

As mentioned in the previous section, an optical power/intensity modulator can operate at data rates as high as 100 Gb/s. Unfortunately, such high-speed signals have a

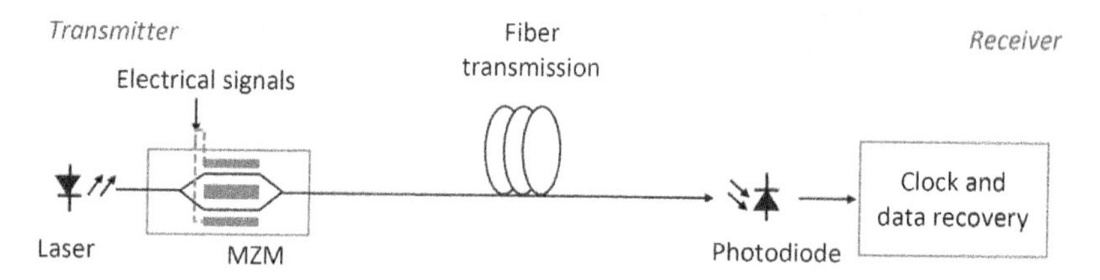

Figure 10.4 Schematic structure of a Mach–Zehnder modulator-based intensity-modulation and direct-detection (IM-DD) system.

very limited reachable transmission distance because fiber chromatic dispersion introduces severe nonlinear distortion to the signal due to the square-law photo detection. In other words, the fiber chromatic dispersion broadens the optical pulse and eventually leads to electrical signal spectral fading when direct detection is used. Linear propagation in a single-mode fiber is characterized by the propagation constant β of the fundamental mode. β can be expanded in a Taylor series as

$$\beta(\omega) = \beta_0 + \beta_1(\omega - \omega_{cw}) + \frac{1}{2}\beta_2(\omega - \omega_{cw})^2 + \cdots \tag{10.1}$$

where ω_{cw} is the center angular frequency of the optical carrier. In Eq. (10.1), β_1 is the inverse group velocity and β_2 describes group velocity dispersion, which is responsible for pulse broadening. The coefficient β_2 is related to the fiber dispersion parameter D via the familiar formula [7]

$$D = -\frac{2\pi c}{\lambda^2}\beta_2 \tag{10.2}$$

Assuming large carrier-to-signal power ratio (CSPR, defined as the power ratio between the CW and the modulated signal excluding the CW), the detected power P at a given baseband frequency f can be expressed as [8]

$$P \propto \cos^2\left[\pi c D L\left(\frac{f}{f_{cw}}\right)^2\right] \tag{10.3}$$

where c is the speed of light in vacuum, L is the fiber length, and $f_{cw} = \omega_{cw}/2\pi$ is the optical carrier frequency.

Typical received electrical power spectra of an IM-DD system for different transmission distances (10, 50, and 100 km) are plotted in Fig. 10.5A. The optical carrier wavelength is assumed to be 1550 nm, and a dispersion coefficient of 17 ps/km/nm is used. It can be observed that at any given distance, the wider bandwidth (the higher

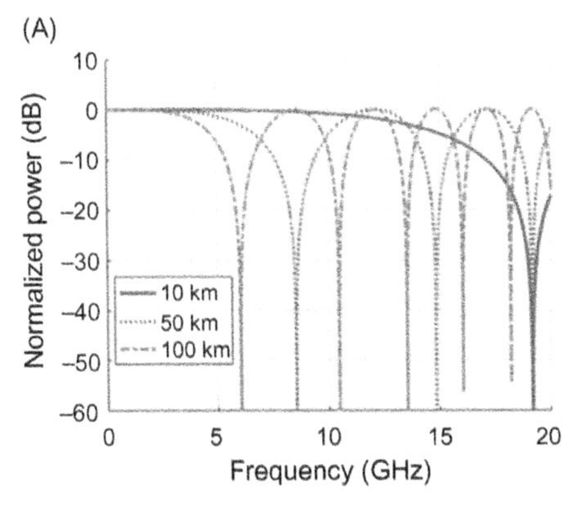

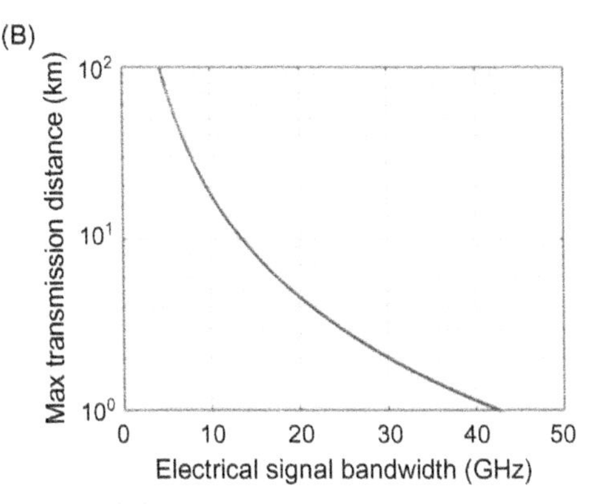

Figure 10.5 (A) Spectral fading on intensity-modulated and direct detected signals with standard single-mode fiber (SSMF) transmission at 1550 nm; (B) maximum SSMF transmission distances for intensity-modulation and direct-detection (IM-DD) signals modulated at 1550 nm.

symbol rate) a signal has, the more fading it undergoes. At a given signal bandwidth, fading is more severe for longer distance.

We can also study the dispersion-induced fading in a different perspective, for example, by finding the maximum allowed transmission distance for various signal bandwidths f_{BW}. We define the "maximum allowed distance," $L_{\max}$, as the transmission distance that gives at most 3-dB power attenuation on any of the frequency components of the baseband signal. The distance $L_{\max}$ can be expressed as

$$L_{\max} = \frac{\arccos\left(\sqrt{1/2}\right)}{\pi c D}\left(\frac{f_{cw}}{f_{BW}}\right)^2 \tag{10.4}$$

where the factor of 1/2 accounts for the 3-dB power attenuation. Fig. 10.5B shows $L_{\max}$ as a function of f_{BW}. It can be seen that the maximum reachable distance of IM-DD systems reduces dramatically as the symbol rate increases. For instance, at 1-km transmission distance, the electrical signal can have a bandwidth of 42.5 GHz, corresponding to a symbol rate of up to 85 Gbaud. However, at 100-km transmission distance, the signal bandwidth has to be reduced to about 4 GHz, which limits the symbol rate to 8 Gbaud or less.

One way of reducing the impact of chromatic dispersion is to transmit signals around 1310 nm, which has almost zero dispersion for standard single-mode fiber (SSMF) [9,10]. However, this limits IM-DD systems to carrying one or only several channels due to the absence of low-cost amplification at that wavelength and due to the increase in dispersion for edge channels. Staying at 1550 nm, where erbium-doped fiber amplifiers are available, dispersion-compensation module (DCM) can be used to reduce the impact of dispersion. A DCM is an optical module with a specialty fiber or Bragg gratings that provide negative dispersion around 1550 nm, making the overall optical channel (transmission fiber plus DCM) have zero dispersion. A DCM can usually compensate dispersion very well for a single channel, but it is challenging to provide accurate full-band compensation and allow wavelength-division multiplexing (WDM). For example, a 56-Gbaud 4-ary pulse-amplitude modulated (PAM-4) IM-DD system carrying 100 Gb/s per wavelength is expected to have a dispersion tolerance of about ± 30 ps/nm [11], which is less than the ± 35 ps/nm of dispersion-slope induced dispersion mismatch across the C-band for a nominally perfectly compensated 100-km SSMF [12]. Moreover, adding optical compensation modules increases system complexity and costs.

Without shifting the operating wavelength or using a DCM, spectral fading can be avoided by receiving the optical field (power and phase) rather than detecting its power only. To receive the field, a coherent receiver or a so-called self-coherent DD receiver needs to be used. In the remainder of this chapter we focus on novel self-coherent DD receivers that enable high-speed DD systems with transmission distances up to several hundreds of kilometers.

10.4 Advanced direct-detection systems

10.4.1 Self-coherent systems: detecting the optical field with a single photodiode

Under certain circumstances, a complex-valued field impinging upon the receiver can be reconstructed after simple intensity detection with a single PD. This requires that the information-carrying signal is single-sideband (SSB) with respect to a sufficiently strong CW (called carrier) component. The signal obtained by adding a CW component at the spectral edge of a modulated signal is often referred to as a "self-coherent" signal, as the receiver is detecting an optical field like a coherent receiver, but the CW is provided by the transmitter side.

Field detection can be described as follows. We denote the optical carrier amplitude by A_0, and consider a field-modulated signal S with a bandwidth B. The photocurrent $I(t)$ resulting from square-law detection can be expressed as

$$I(t) = |A_0 + S|^2 = |A_0|^2 + |S|^2 + 2Re(S \cdot A_0^*) \tag{10.5}$$

The third term at the right-hand side of the equality is the desired linear copy of the complex signal S. The first term in the right side of Eq. (10.5) is a DC component. The potential contamination comes from the second term, which introduces single-signal beating interference (SSBI). Exemplary spectra of an SSB signal before and after power detection are shown in Fig. 10.6. As can be seen, the SSBI is a broadband nonlinear product and interferes with linear copy of the signal.

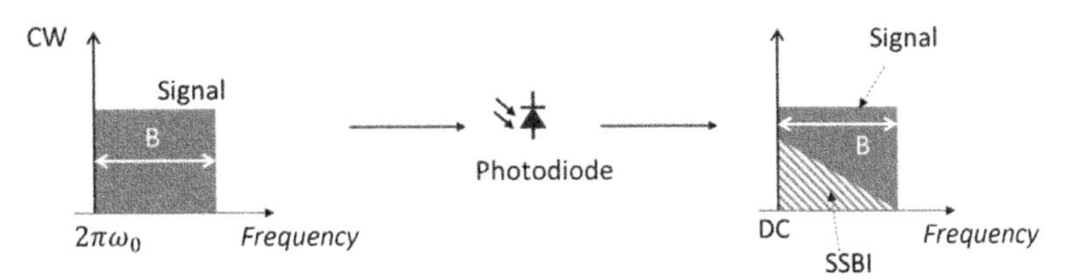

Figure 10.6 Typical optical and received electrical spectra for self-coherent systems.

Early work [13] leaves a frequency gap between the carrier and the signal. If the frequency gap is larger than the signal bandwidth, the SSBI falls entirely into the gap and does not affect signal reconstruction. Clearly, the utilization of the frequency gap reduces spectral efficiency. An approach to avoiding this gap is to increase the CSPR, to the extent that the SSBI term becomes negligible. However, high CSPR values reduce the system OSNR sensitivity and therefore limits the transmission reach. In order to keep the CSPR relatively low, specially designed nonlinear equalization schemes for SSBI cancellation can be applied. For instance, an early attempt used a decision-feedback algorithm [14] to cancel the SSBI. This algorithm first performs symbol decision with the interference of SSBI and then uses the decision results to reconstruct SSBI, which is finally subtracted from the original received photocurrent signal. The cancellation can be accurate after several iterations. A variety of such iterative nonlinear equalization algorithms have been proposed and demonstrated [15,16] in recent years.

Self-coherent systems require the transmitter to generate an SSB signal and then add a CW component. This can be done with various methods, which can be categorized into the following main types: (1) adding the RF tone digitally in the electrical domain [14] or by means of RF up-conversion [13]; (2) in the optical domain, by combining a frequency-shifted replica of the optical carrier with the modulated-signal field using an optical coupler [17]; (3) suppressing one sideband of a double-sideband (DSB) signal using an optical I/Q modulator [18,19] or an optical filter. Fig. 10.7 illustrates some examples for the above approaches.

10.4.2 Kramers–Kronig receivers: rigorous field reconstruction

As discussed in the previous section, the key challenge in the reconstruction of a complex-valued SSB signal after intensity detection with a single PD is to minimize the effect of SSBI while maintaining the CSPR to a reasonable value. The Kramers–Kronig (KK) receiver [20] is one of the best candidates to achieve this. The KK receiver has the same minimal hardware architecture as other self-coherent receiver schemes, but makes use of a different signal reconstruction algorithm. The KK receiver yields a rigorous reconstruction of the signal impinging upon the receiver,

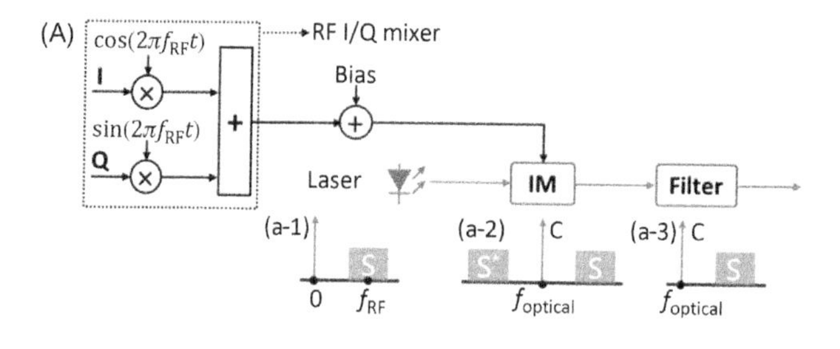

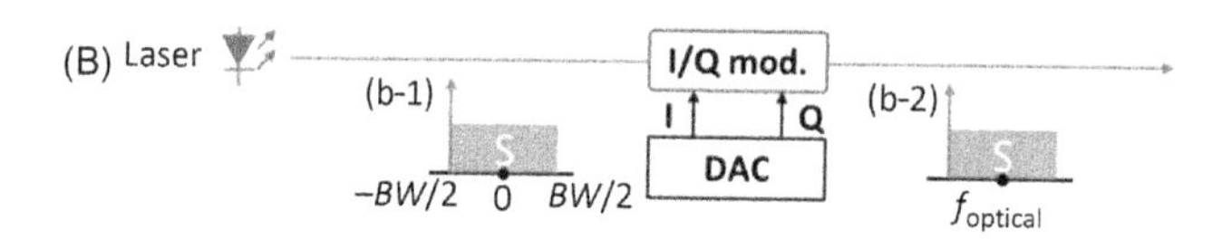

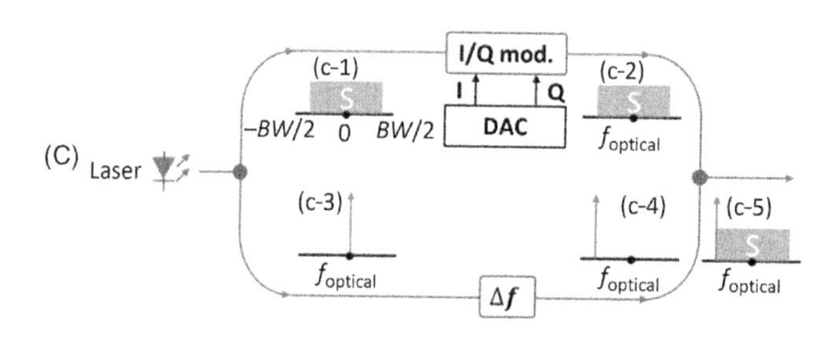

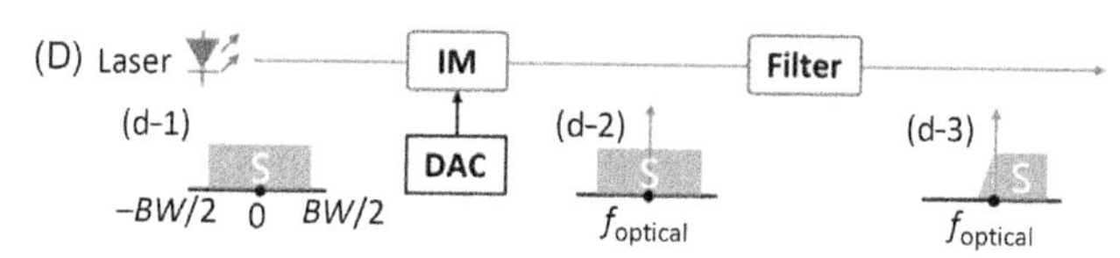

Figure 10.7 Examples of single-sideband (SSB) signal generation for self-coherent transmission schemes. (A) Insertion of a carrier in the analog RF domain [13]; spectrum (A-1) after RF I/Q mixer, (A-2) after optical intensity modulator, (A-3) after optical filter. (B) Insertion of a carrier in the digital RF domain; spectrum (B-1) in the RF domain and (B-2) in optical domain. (C) Combination of signal and carrier in the optical domain; spectrum (C-1): RF-domain signal; (C-2) optical-domain signal; (C-3) optical carrier; (C-4) frequency-shifted optical carrier; (C-5) spectrum after optical coupler. (D) Filtering one sideband of a double-sideband (DSB) signal: spectrum (D-1) in the RF domain; (D-2) optical-domain DSB spectrum; (D-3) optical-domain SSB spectrum. S, signal; C, carrier; IM, intensity modulator; I/Q mod., I/Q modulator; DAC, digital-to-analog converter; Δf, optical frequency shifter; BW, signal bandwidth. Source: *(A to B) From D. Che, Coherent Optical Short-Reach Communications, https://minerva-access.unimelb.edu.au/handle/11343/129708. (C to D) From D. Che, A. Li, X. Chen, Q. Hu, W. Shieh, Rejuvenating direct modulation and direct detection for modern optical communications, Opt. Commun. 409 (2018) 86–93.*

if this fulfills the minimum-phase condition described in the next section. The signal reconstruction algorithm that it uses can therefore be considered as an effective SSBI cancellation algorithm operating in a noniterative fashion. In the absence of implementation-related nonidealities, the KK algorithm yields perfect SSBI cancellation.

10.4.2.1 Principle of operation

The KK relations [21,22] are known to relate the real and imaginary parts of the transfer function of a linear system under only one requirement, which is causality. We explain the causality as follows. If we denote by $h(t)$ the system impulse response and by $\tilde{h}(\omega)$ its Fourier transform, causality implies that $h(t) = 0$ for $t < 0$, and the KK relations yield $\tilde{h}_{R/I}(\omega) = \mathrm{H}\{\tilde{h}_{I/R}(\omega)\}$, where with the subscripts R and I we label the real and imaginary part, respectively, and by H we denote a Hilbert transform [23]. The use of the KK relations in the context of optical communications is readily understood by noting that any band-limited signal can be considered to meet a "causality requirement" in the frequency domain for any center frequency chosen at the left of its spectrum. In this case its real and imaginary parts are related to each other through the relation

$$\tilde{s}_{R/I}(t) = \mathrm{H}\{\tilde{s}_{I/R}(t)\} \tag{10.6}$$

Although this relation cannot be used to extract the phase of an optical signal from its intensity, it can be used for this purpose if applied to the logarithm of the same signal. Indeed, if we express the signal impinging upon the receiver as $E(t) = \sqrt{I(t)}\exp[i\varphi(t)]$, where $\varphi(t)$ and $I(t)$ are the signal phase and intensity, respectively, the signal logarithm is

$$\log[E(t)] = \frac{1}{2}\log[I(t)] + i\varphi(t), \tag{10.7}$$

and its real and imaginary parts obey the KK relations if $\log[E(t)]$ is SSB. The property of a signal that fulfills this requirement is called of *minimum-phase*, and for a minimum-phase signal

$$\varphi(t) = \frac{1}{2}\mathrm{H}\{\log[I(t)]\}. \tag{10.8}$$

A simple way of constructing a minimum-phase signal from a SSB signal $s(t)$ of bandwidth B is to add a sufficiently intense tone at the low-frequency side of its spectrum. To see why this is the case, we can expand the logarithm of the overall signal $E(t) = E_0 + s(s)$, as follows, $\log[E_0 + s(t)] = \log(E_0) + s(t)/E_0 - s^2(t)/2E_0^2 +$

$s^3(t)/3E_0^3 \cdots$, where by E_0 we denote the complex amplitude of the CW tone, and perform a Fourier transform, which yields

$$\mathcal{F}\{\log[E_0 + s(t)]\} = \log(E_0)\delta(\omega) + \frac{\tilde{s}(\omega)}{E_0} - \frac{\tilde{s}(\omega) * \tilde{s}(\omega)}{2E_0^2} + \frac{\tilde{s}(\omega) * \tilde{s}(\omega) * \tilde{s}(\omega)}{3E_0^3} \cdots \tag{10.9}$$

Since $\tilde{s}(\omega)$ is band-limited, its self-convolution products are individually and collectively SSB as well and so is $\log[E_0 + s(t)]$. The condition $|E_0| > |s(t)|$ ensures that the expansion of $\log[1 + s(t)/E_0]$ converges, thereby setting the magnitude of the CW tone. The addition of a sufficiently intense CW at the high-frequency side of the spectrum produces a maximum-phase signal, which is the conjugate of a minimum-phase signal. Being the intensity of such signal equal to the intensity of its conjugate, for a maximum-phase signal the reconstruction procedure with the logarithmic Hilbert transform would return the phase of the signal with a change of sign.

A rigorous characterization of minimum-phase signals in the context of fiber-optic transmission can be found in Ref. [24], where it is shown that a necessary and sufficient condition for a SSB signal to be minimum-phase is that its time trajectory never encircles the origin of the complex plane. In this case too one can readily see that a minimum-phase signal can be constructed by adding a CW tone at one edge of the signal spectrum. This is illustrated in Fig. 10.8, where we consider a signal of bandwidth B, whose spectral occupancy extends from $f = 0$ to $f = B$, and a real-valued positive CW tone is added at $f = 0$. As seen in the figure, the addition of the CW moves the time trajectory of the optical signal to the right, and if it is sufficiently intense the origin of the plane is not encircled, thereby fulfilling the minimum-phase condition.

To explain the implementation of the KK-based phase-retrieval digital signal processing (DSP), denote by $E_s(t)$ the information-carrying (complex-valued) signal of

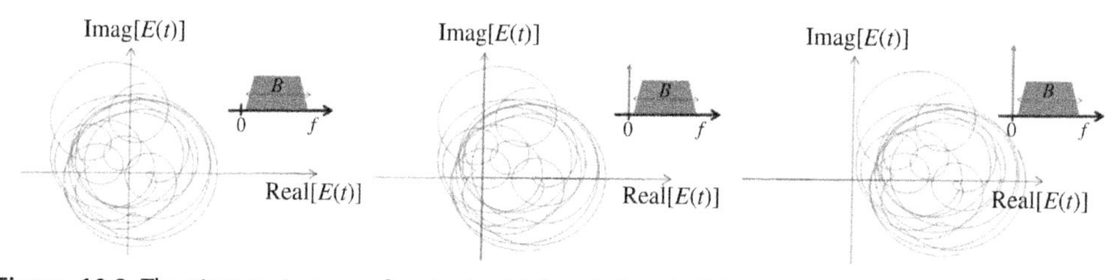

Figure 10.8 The time trajectory of a single-sideband signal of bandwidth B in the complex plane. The addition of a CW tone (center and right panels) prevents the time trajectory from encircling the origin of the plane, thereby fulfilling the minimum-phase condition that is necessary for the implementation of the KK algorithm.

bandwidth B, whose spectrum extends between $-B/2$ and $B/2$. The transmitted signal, which includes the CW term, can then be expressed as

$$E(t) = E_0 + E_s(t)e^{i\pi Bt} \tag{10.10}$$

where E_0 is the amplitude of the CW tone, which we assume with no loss of generality to be real-valued. The reconstructed information-carrying signal $E_r(t)$, as obtained after photodetection and KK processing (which includes phase reconstruction, CW removal, and frequency shift), is then given by

$$\begin{aligned} E_r(t) &= \left[\sqrt{I(t)}e^{i\varphi(t)} - E_0\right]e^{-i\pi Bt} \\ \varphi &= \frac{1}{2}\mathrm{H}\{\log[I(t)]\} \end{aligned} \tag{10.11}$$

where $I(t) = |E(t)|^2$. The term $\sqrt{I(t)}\exp[i\varphi(t)]$ constitutes the reconstructed optical field impinging upon the receiver and digital chromatic dispersion compensation, as well as the mitigation of other propagation effects, should be applied to this term, prior to CW removal and frequency shift. Note that $E_s(t)$ is an arbitrary complex-valued signal of bandwidth B and its generation requires in general an I/Q modulator, unless special solutions are adopted [25,26].

It is worth pointing out that the perfect fulfillment of the minimum-phase condition may require impractical CW power values, mainly in the case of signals characterized by high peak-to-average power ratios. However, in practice the CW power level is not so high, as it is set so as to make sure that field reconstruction errors are negligible at bit error ratio (BER) values of interest. For example, at a pre-FEC BER of 10^{-2}, a CSPR of the order of 6 dB is adequate to fulfill this requirement [20,27]. Such CSPR is several dB lower than what is required by other schemes [16,28].

Fig. 10.9 shows some simulation results for a DD receiver that uses the KK algorithm. The OSNR is measured in the absence of the CW component.[1] The simulated signals are 24-Gbaud 16-quadrature amplitude modulation (QAM) transmitted over 100 km of single-mode fiber (chromatic dispersion of 21 ps^2/km, nonlinearity coefficient of 1.3 $\text{W}^{-1}\,\text{km}^{-1}$, loss coefficient of 0.22 dB/km), assuming an overall span loss budget of 26 dB. The OSNR was varied by changing the signal launch power. Chromatic dispersion was compensated digitally at the receiver, after signal reconstruction. The simulated signal consisted of a pseudorandom sequence of 2^{15} symbols. Prior to reception, amplification noise was added to the signal and then a 12th order

[1] The use of the equivalent OSNR facilitates the comparison with the performance of a coherent homodyne receiver, which would be less transparent by accounting also for the CW power in the OSNR.

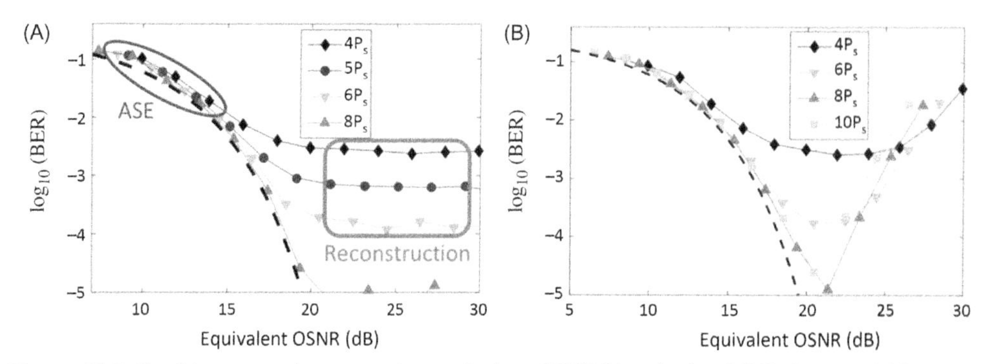

Figure 10.9 The bit error ratio versus the equivalent OSNR (that is the OSNR that would be measured if the CW tone were not transmitted) in the linear (A) and nonlinear (B) regime of operation. The various symbols refer to different values of the CW signal power, which is expressed in terms of the information-carrying signal power P_S. The dashed curve is the BER of a coherent homodyne receiver.

super-Gaussian optical filter with a 3-dB bandwidth of 36 GHz was applied. The filter center frequency was set to 16.6 GHz above the CW tone's frequency, higher than the center frequency of the data carrying signal. This was done to suppress ASE noise at negative frequencies while minimizing the attenuation of the CW tone. The left and right panels refer to the cases of linear and nonlinear transmission, respectively. The simulations in the nonlinear case were based on the Manakov equation assuming five WDM channels spaced by 40 GHz. The dashed curve is the reference, which is the BER of the same 24-GBd 16-QAM signal detected by a coherent receiver [29]. Fig. 10.9 allows gaining insight on the performance of the KK receiver:

1. In the low-OSNR region, where the BER is dominated by ASE noise-induced errors, the KK results approach those of a coherent system.
2. In the high-OSNR region, reconstruction errors become dominant and the BER becomes practically independent of the OSNR (appearing as a floor in the BER curve). This effect is less pronounced at higher CSPR values. Note that at a CSPR value of 7 dB ($P_{CW} = 5P_s$), the BER is already well below the hard-decision FEC threshold of 10^{-3}.
3. In the nonlinear transmission region, the CSPR cannot be increased indefinitely, since large CW power levels may cause nonlinear distortions whose effect prevails that of reconstruction errors. Indeed, Fig. 10.9B shows that the BER increases when the CSPR goes from 9 to 10 dB.

It is important to mention that the KK receiver aims to achieve rigorous signal reconstruction. For this reason, the KK scheme is intrinsically immune to the SSBI, whose mitigation is the main goal of a number of approximate signal reconstruction techniques.

10.4.2.2 Kramers–Kronig receiver-based experimental demonstrations

The first experimental implementation of the KK receiver has been presented in Ref. [27], where Z. Li and co-authors showed the superior performance of the KK receiver in the comparison with various other SSBI mitigation techniques. With a KK receiver, they achieved 112 Gb/s transmission per single wavelength, polarization, and PD over 80 and 240 km, at a pre-FEC BER of 3.8×10^{-3}.

Later, a line rate of 218 Gb/s over 100 km of SSMF at a pre-FEC BER of 6.2×10^{-3} was achieved in Ref. [17]. The experimental setup used in the experiment is sketched in Fig. 10.10A. An external cavity laser at 1550 nm was used as the optical input of an I/Q modulator to generate an optical DMT signal. A CW tone is added at the long-wavelength side of the information-carrying signal spectrum. The transmitted and received spectra of the so-generated minimum-phase signal is shown in Fig. 10.10B.

In order to demonstrate the ability of the KK receiver to suppress the SSBI, a test-signal with only odd-indexed DMT subcarriers was loaded. This way, the SSBI and its mitigation could be monitored by reading the power level of the even-indexed discrete Fourier transform (DFT) bins. The comparison of the SSBI suppression with and without KK processing is shown in Fig. 10.10C. Subsequently, all subcarriers were loaded with QAM-modulated signals. The measured BER as a function of CSPR and the measured SNRs on all subcarriers are shown in Fig. 10.10D and E, respectively.

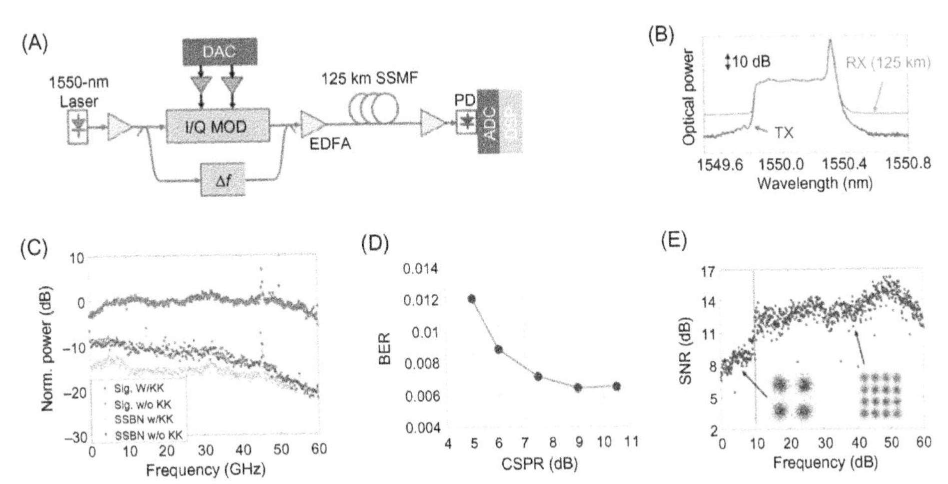

Figure 10.10 (A) Experimental setup used in Ref. [11] to demonstrate a line rate of 218 Gb/s per single wavelength, polarization, and PD, over 100 km of SSMF. (B) Transmitted and received signal spectra. (C) Measured SSBN, (D) average BER, and (E) SNR. Source: *After N. Eiselt, J. Wei, H. Griesser, A. Dochhan, M. Eiselt, J. Elbers, et al., First real-time 400G PAM-4 demonstration for inter-data center transmission over 100 km of SSMF at 1550 nm, in: Optical Fiber Communications Conference and Exhibition (OFC), 2016, paper W1K.5; L. Zhang, T. Zuo, Y. Mao, Q. Zhang, E. Zhou, G.N. Liu, et al., Beyond 100-Gb/s transmission over 80-km SMF using direct-detection SSB-DMT at C-band, J. Lightw. Technol. 34 (2) (2016) 723–729.*

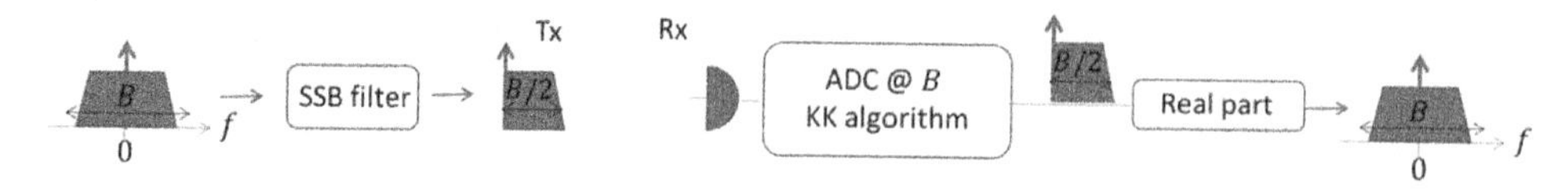

Figure 10.11 Operation principle of the KK-PAM transceiver. Source: *After C. Antonelli, M. Shtaif, A. Mecozzi, Kramers-Kronig PAM transceiver, in: Optical Fiber Communication Conference, 2017.*

A similar setup was used in Ref. [30] to demonstrate a line rate of 440 Gb/s per single wavelength, polarization, and PD, over 100 km of SSMF, using probabilistically shaped subcarriers with entropy- and power-loading.

It is worth nothing that the KK receiver can also be used to transmit real-valued signals with the goal of compressing their spectral occupancy. This is the case in the so-called KK-PAM transceiver scheme [31–33], whose operation principle is illustrated in Fig. 10.11. At the transmitter end, the real-valued PAM-modulated signal passes through a SSB filter so as to suppress half of the signal spectrum. At the receiver end, the photocurrent signal is sampled at a sampling rate of B and the use of the KK algorithm returns the reconstructed SSB signal, whose real part coincides with the original PAM-modulated signal.

10.4.2.3 Discussion

Below we discuss a number of requirements entailed by the KK algorithm.

The logarithm operation in Eq. (10.11) is necessary to the extraction of the signal's phase from the optical power. This implies a bandwidth expansion that must be accommodated by means of digitally up-sampling the digitized photocurrent. A digital up-sampling factor in the order of two to three has been shown to be sufficient in most relevant cases [20,34]. This up-sampling operation is applied to the samples of the signal taken at rate $2B$ by the ADC and is entirely performed in the digital domain. Schemes with reduced digital up-sampling requirements at the expenses of some power penalty have been recently proposed [35,36].

The SSB nature of the received signal that is necessary to fulfill the minimum-phase condition implies relatively strict requirements on the optical filtering. Indeed, insufficient suppression of the spectral content at the opposite side of the information-carrying signal may cause reconstruction errors.

The KK algorithm relies on negligible electrical distortions from the receiver. In other words, distortions from bandwidth limitations of the PD, ADC, and other RF components may introduce distortions and therefore result in field reconstruction errors [37,38]. For this reason, receiver electrical distortions need to be equalized digitally prior to performing the KK processing.

We conclude this discussion by pointing out that the KK receiver should not be confused with compatible SSB transmission [28]. In fact, the latter approach aims to transmit spectrally compressed intensity-modulated signals that are directly detected with no signal processing. The technique that is used to compress the signal spectrum is such that the transmitted signal turns out to be minimum-phase; however, this property is not exploited at the reception stage. For this reason, compatible SSB transmission is only suitable for short links where chromatic dispersion is negligible.

10.4.3 Stokes vector receivers: polarization recovery without a local oscillator

Typical self-coherent receivers including the KK receiver only handle single-polarization signals, as they cannot recover the transmitted signals' original state of polarization. Polarization recovery is essential if a polarization multiplexed (POL-MUX) signal is transmitted. Performing polarization recovery in Jones space without a local oscillator at the receiver end and without active optical polarization control is generally not possible. To achieve polarization-division multiplexing (PDM) in a DD system, the receiver should detect the signal in Stokes space, which is known to be related to Jones space by means of an isomorphism [39]. The Stokes vector receiver (SVR) is a receiver that works in Stokes space and can achieve polarization recovery without using a local oscillator. In this section, we introduce its structure and the necessary signal processing in an SVR. The next section focuses on how to combine the SVR with the KK technique to demultiplexing a POL-MUX optical field and optimize the use of a DD receiver's electrical bandwidth.

10.4.3.1 System architecture

Stokes space modulation (SSM) was proposed in the 1990s for optical communication compatible with direct detection [40,41]. The Jones space polarizations X and Y can be characterized by the intensity of three orthogonal components in Stokes space, expressed as:

$$|s\rangle = \begin{bmatrix} X \\ Y \end{bmatrix}, \quad \boldsymbol{S} = \begin{bmatrix} S_0 \\ S_1 \\ S_2 \\ S_3 \end{bmatrix} = \begin{bmatrix} XX^* + YY^* \\ XX^* - YY^* \\ XY^* + X^*Y \\ i(XY^* - X^*Y) \end{bmatrix} \tag{10.12}$$

Fig. 10.12A illustrates a receiver structure that detects a Stokes vector defined by Eq. (10.12), which is first used in Ref. [42]. In fact, an SVR can be realized by three or four detections of polarization states as long as they are nonsingular superposition of the Stokes components (S_0, S_1, S_2, S_3). Fig. 10.12B–D presents three other possible configurations [43–45].

The SSM is a general concept containing all polarization modulations that can be recovered by an SVR. Starting with a simple example, we keep Y as a constant carrier and modulate only X with a complex-valued DSB signal. A transmitter that can generate such signal is shown in Fig. 10.13. An exemplary spectrum is shown as an inset to Fig. 10.13. At the receiver, with an SVR of any kind shown in Fig. 10.12, Stokes vector $S_1(t)$, $S_2(t)$, and $S_3(t)$ can be received. After digitally rotating the Stokes vector

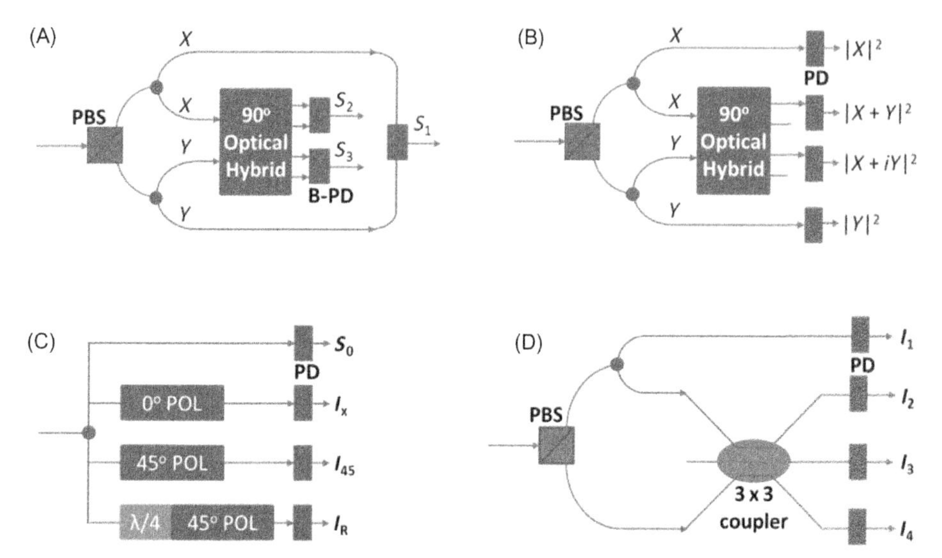

Figure 10.12 Various architectures for Stokes vector receiver. (A) A Stokes vector receiver realized by a 90-degree optical hybrid and three balanced PDs; (B) A Stokes vector receiver realized by a 90-degree optical hybrid, two balanced PDs, and two single-ended PDs; (C) A Stokes vector receiver realized by polarization controllers and four single-ended PDs; (D) A Stokes vector receiver realized by a 3×3 optical coupler and four single-ended PDs.

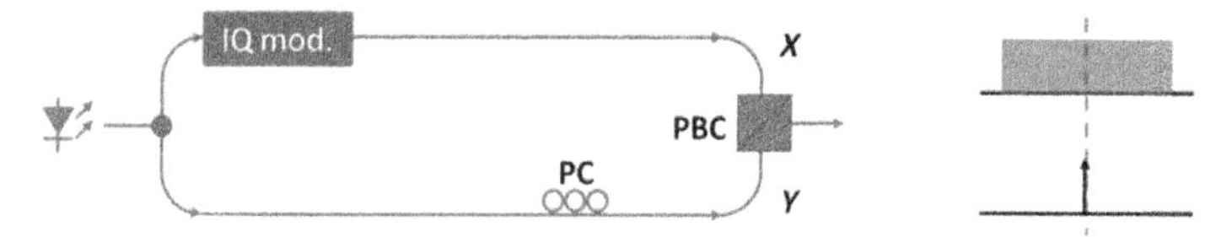

Figure 10.13 Transmitter structure for SSM using self-coherent single polarization modulation. PC, polarization controller; PBC, polarization beam combiner. The *right column* shows the spectrum utilization in Jones space.

$\vec{S}(t) = S_1(t)\hat{s}_1 + S_2(t)\hat{s}_2 + S_3(t)\hat{s}_3$ (where the unit vectors $\hat{s}_1$, $\hat{s}_2$, and $\hat{s}_3$ are a right-handed basis of Stokes space), so that its time-averaged value is aligned to $\hat{s}_1$, the Stokes vector can be converted back to Jones space with a simple operation $X = S_2(t) \pm iS_3(t)$.

10.4.3.2 Receiver-side digital signal processing for stokes vector receivers

The receiver DSP procedures for the SVR is shown in Fig. 10.14. Detailed illustration of each step can be found in Fig. 10.15. Overall, there are three steps. The first step is to map the received signal to the standard definition of Stokes vector. Depending on the actual implementation of an SVR, the received photocurrent may not directly represent the Stokes vector. For instance, among the four exemplary structures shown in Fig. 10.12, only Fig. 10.12A directly detect the Stokes vector. The other three require a transformation matrix to convert the received photocurrent to Stokes vector.

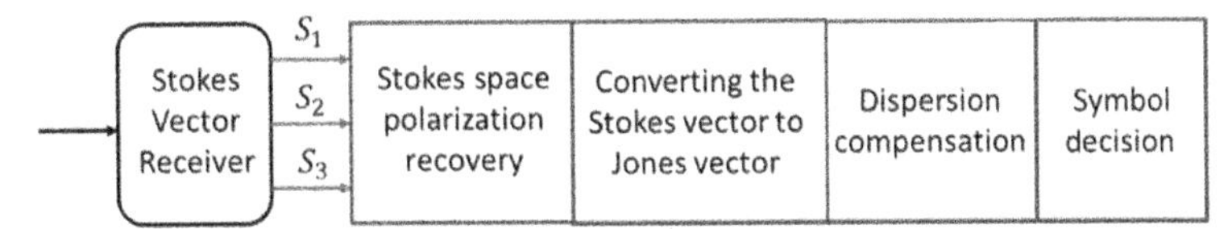

Figure 10.14 Digital signal processing procedures for Stokes vector receiver (SVR).

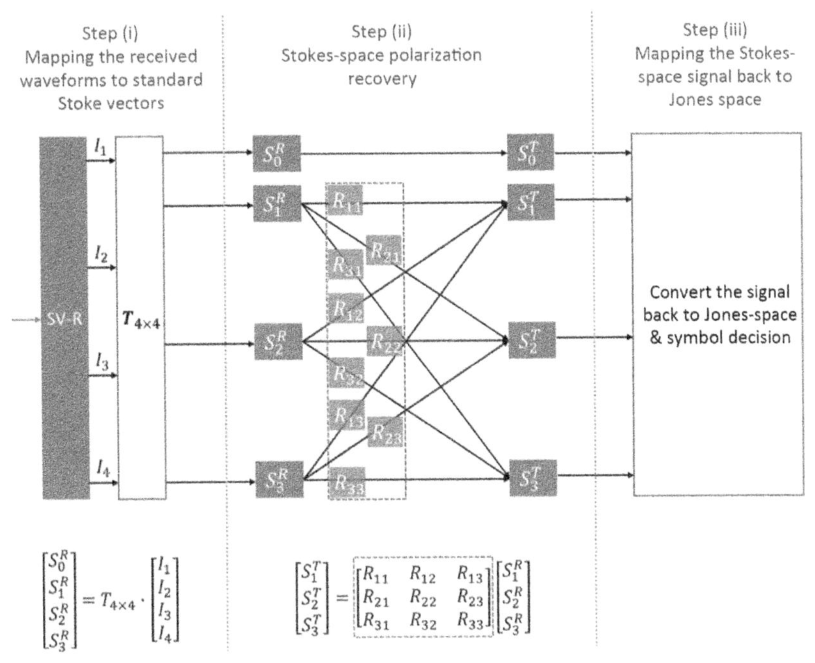

Figure 10.15 Detailed digital signal processing procedures for Stokes vector receiver (SVR).

Note that such transformation matrix is fixed and defined by the realization of the SVR and is independent of the fiber polarization rotation.

The second step of the DSP is Stokes space polarization recovery. This purpose of this step is to derotate polarization variation from the fiber channel. This step is equivalent to the Jones-space multi-in-multi-out (MIMO) processing. In Stokes space, the fiber channel with random polarization variation can be characterized by a real-value 3×3 rotation matrix (RM). Such matrix rotation can be summarized by a general MIMO equalizer concept. The MIMO can be performed either in time domain or frequency domain [46] due to the linearity of DFT operation shown as Eq. (10.13)

$$\mathrm{DFT}\left(\boldsymbol{RM}_{3\times3}\cdot\begin{bmatrix}S_1\\S_2\\S_3\end{bmatrix}\right)=\boldsymbol{RM}_{3\times3}\cdot\begin{bmatrix}DFT(S_1)\\DFT(S_2)\\DFT(S_3)\end{bmatrix} \tag{10.13}$$

The Stokes space MIMO can be realized by various established MIMO algorithms, such as the least-mean-square method, steep-decent adaptive algorithm, or even blind search. Training sequences can be used for adaptive MIMO filter convergence. One of the popular training sequences can be found in Ref. [42] where the training sequence represents three orthogonal bases in Stokes space. Generating these training sequences, however, imposes special requirements for the transmitter. In particular, this generation requires the transmitter producing symmetric power spectrum density on X- and Y-polarization in Jones space, and can also control the interpolarization differential phase (e.g., adjusting it to 0 or 90 degrees). Many SSM schemes including the example shown in Fig. 10.13 have asymmetric power on X- and Y-polarization. Such asymmetry is also true for the many other SSM schemes that will be introduced in the next section. In this case, instead of sending training sequences, the asymmetric characteristic on the power distribution itself can be used to identify the transmitter polarization axis as demonstrated in Ref. [47], and such polarization recovery can be done without knowledge of the digital signal.

After the polarization recovery, the third step in the DSP is to convert the Stokes vector back to Jones vectors. The conversion is related to transmitter structure. For instance, for the signal shown in Fig. 10.13, the Jones space complex signal X can be obtained by $X = S_2(t) \pm iS_3(t)$. After the Jones vectors are recovered, digital dispersion compensation can be done, followed by symbol decision.

The exemplary modulated signal (Fig. 10.13) in this section shows only a single polarization signal. However, as explained, the SVR can recover and derotate the fiber polarization variations and recover the original transmitted signal polarization states. This essentially enables detecting PDM signals. We will focus on a special PDM signal that will be introduced in 13.4.4 where an SVR is combined with KK algorithm for recovery.

It is worth mentioning that, compared with single-PD based self-coherent receivers, the SVR supports polarization multiplexing at the price of increased number of hardware. Comparison of various self-coherent receivers, KK receivers, and SVRs can be found in publications, for example, Refs. [48,49]. For instance, an examination on efficiency of electrical bandwidth utilization assuming a fixed OSNR budget and a fixed target data rate can be found in Ref. [50]. A comprehensive OSNR sensitivity evaluation for different self-coherent systems carrying 100 Gb/s 16-QAM signal can be found in Ref. [51].

10.4.4 Kramers-Kronig Stokes receivers

The SVR can be combined with the KK field reconstruction technique to achieve efficient receiver electrical bandwidth utilization [50–52]. One way to do so is by transmitting a minimum-phase signal in one polarization of the optical field and a SSB signal (without a CW component) in the orthogonal polarization [52], as shown in

Fig. 10.16A. This scheme is still not optimal in terms of receiver bandwidth utilization. The receiver electrical bandwidth is fully exploited when a minimum-phase signal is transmitted in one polarization and a DSB signal is transmitted in the orthogonal polarization [52] as depicted in Fig. 10.16B.

The receiver architecture of the KK-Stokes receiver is illustrated in Fig. 10.17. As shown in the figure, the receiver is mostly the same as a typical Stokes receiver of the kind described in Section 13.4.3. The difference is that the X-polarization field component $E_x(t)$ is obtained by applying the KK algorithm to $I_x(t)$, whereas the y-polarization field component follows from the relation between the Jones and Stokes representation of field, namely

$$E_y(t) = \frac{S_2(t) - iS_3(t)}{2E_x^*(t)}. \tag{10.14}$$

It is worth noting that the modulation scheme in Fig. 10.16A allows optical noise rejection on the other side of the SSB signal, which is not possible for the scheme of Fig. 10.16B.

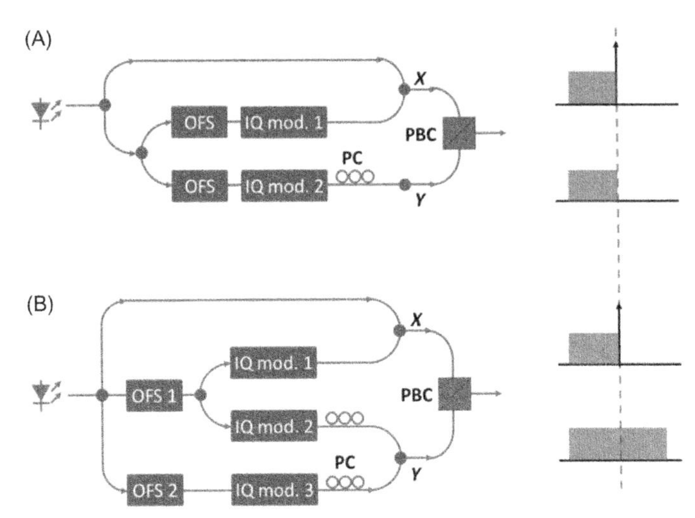

Figure 10.16 Transmitter structure for generating signals that can be detected by KK-Stokes receivers. (A) SSB band signals on both polarizations; (B) one SSB signal on X polarization, and one DSB signal on Y polarization; PC, polarization controller; PBC, polarization beam combiner; OFS, optical frequency shifting. The *right column* illustrates the spectrum of the X and Y polarization signals in Jones space.

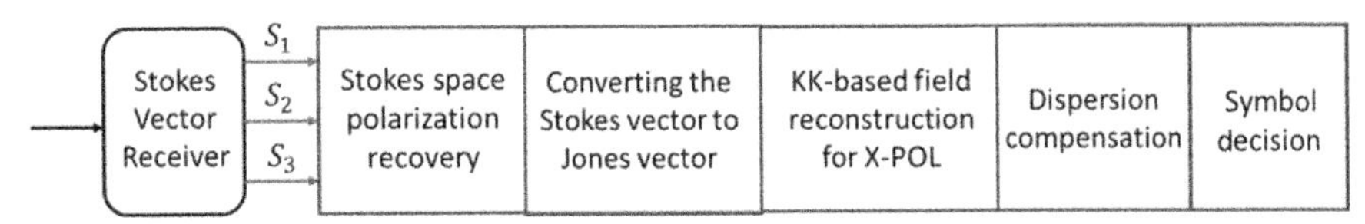

Figure 10.17 Principle of operation of the KK-Stokes receiver.

10.5 The future of short-reach transmission systems

High-capacity DD systems are now facing unprecedented competition from coherent detection (CD) technologies. Since the first deployment of 40 Gb/s CD transponders [53], CD systems have replaced DD as workhorse from ultra-long haul to metropolitan networks, or for any networks beyond the reach of hundreds of kilometers and at rates beyond 100 Gb/s. Today, 400ZR is being standardized to carry coherent 400 Gb/s 16 QAM modulation targeting intra-datacenter applications with reach up to 100 km. Consequently, CD systems have been gradually encroaching into the traditional application sphere of DD at a fast pace. The pertinent question to ask is what the future is holding for high-performance DD systems.

In our view, there are two intrinsic advantages of DD over CD systems:

1. Cost. Among all the constituent optoelectronic components, we identify the laser as a major differentiator of the "game." Lasers for CD applications need to be narrow-linewidth and temperature-stabilized, and therefore tend to be bulky and expensive. In contrast, lasers for DD can be uncooled and therefore inexpensive. Barring any breakthrough in photonic integrated circuits, high-performance lasers form the fundamental cost barrier for CD systems.
2. Energy efficiency. As both symbol rate and capacity continue to grow, energy consumption becomes a critical issue. CD systems rely on energy-intensive DSP for signal generation and reception, which is undesirable for large-capacity high-density applications.

Based on the above two observations, we foresee the two applications for high-capacity DD systems:

1. Intradatacenter transport. These applications are mostly characterized by short reach of below one-kilometer, high-density, high-volume, and cost-sensitiveness fiber links. As cloud computing has become ever predominant, such high-speed transport has become an active research field for both industry and academia.
2. On-chip or off-chip interconnects. The electronic IC industry is driven by Moore's law, with the number of transistors per chip roughly doubling every 24 months. The interconnect bandwidth including I/O and clock distribution becomes a major bottleneck. Optical interconnects have been proposed and demonstrated as an alternative to copper-based interconnects for both on-chip and off-chip applications [54].

Therefore we believe that high-capacity DD systems will continue to play a critical role in intra-datacenter and on/off-chip communications. Those systems will march on the path to higher symbol rates augmented by WDM [55]. The next interesting development for those systems would be exploring more modulation dimensions such as phase, polarization, and space. This includes the techniques of KK receiver, Stokes vector detection, and space-division-multiplexing using multimode or multicore fibers. Lastly, in order to improve the space density and reduce the cost, high-performance DD systems will leverage the advance in photonic integration [56,57].

References

[1] W. Kobayashi, T. Ito, T. Yamanaka, T. Fujisawa, Y. Shibata, T. Kurosaki, et al., 50-Gb/s direct modulation of 1.3-μm InGaAlAs based DFB laser with ridge waveguide structure, IEEE J. Select. Topics Quant. Electr. 19 (2013).

[2] Y. Matsui, T. Pham, W.A. Ling, R. Schatz, G. Carey, H. Daghighian, et al., 55-GHz bandwidth short-cavity distributed reflector laser and its application to 112-Gb/s PAM-4, in: Optical Fiber Communication Conference, 2016, paper PDP Th5B.4.

[3] C. Palais Joseph, Fiber Optic Communications, Prentice Hall, Englewood Cliffs, 1988.

[4] J.I. Pankove, Optical processes in semiconductors. Chapter 2-C-2, Courier Corporation, 1971.

[5] M. Chacinski, U. Westergren, B. Stoltz, L. Thylen, R. Schatz, S. Hammerfeldt, Monolithically integrated 100 GHz DFB-TWEAM, IEEE J. Lightw. Technol. 27 (2009) 3410−3415.

[6] J.C. Cartledge, Performance of 10 Gb/s lightwave systems based on lithium niobate Mach-Zehnder modulators with asymmetric Y-branch waveguides., IEEE Photon. Technol. Lett. 7 (9) (1995) 1090−1092.

[7] G.P. Agrawal, Nonlinear fiber optics., Nonlinear Science at the Dawn of the 21st Century, Springer, Berlin, Heidelberg, 2000, pp. 195−211.

[8] G. Meslener, Chromatic dispersion induced distortion of modulated monochromatic light employing direct detection., IEEE J. Quant. Electr. 20.10 (1984) 1208−1216.

[9] W. Yan, T. Tanaka, B. Liu, M. Nishihara, L. Li, T. Takahara, et al., 100 Gb/s Optical IM-DD transmission with 10G-class devices enabled by 65 GSamples/s CMOS DAC Core, OFC 2013, Paper OM3H.1.

[10] M. Chagnon, M. Osman, M. Poulin, C. Latrasse, J.-F. Gagné, Y. Painchaud, et al., Experimental study of 112 Gb/s short reach transmission employing PAM formats and SiP intensity modulator at 1.3 μm., Opt. Express 22 (2014) 21018−21036.

[11] N. Eiselt, J. Wei, H. Griesser, A. Dochhan, M. Eiselt, J. Elbers, et al., First real-time 400G PAM-4 demonstration for inter-data center transmission over 100 km of SSMF at 1550 nm, in: Optical Fiber Communications Conference and Exhibition (OFC), 2016, paper W1K.5.

[12] Available from: http://fiber-optic-catalog.ofsoptics.com/Asset/OFS_SMFDK_S.pdf.

[13] A.J. Lowery, J. Armstrong, Orthogonal-frequency-division multiplexing for dispersion compensation of long-haul optical systems, Opt. Express 14 (6) (2006) 2079−2084.

[14] W.R. Peng, X. Wu, K.M. Feng, V.R. Arbab, B. Shamee, J.Y. Yang, et al., Spectrally efficient direct-detected OFDM transmission employing an iterative estimation and cancellation technique, Opt. Express 17 (11) (2009) 9099−9111.

[15] Z. Li, M.S. Erkilinc, R. Maher, L. Galdino, K. Shi, B.C. Thomsen, et al., Two-stage linearization filter for direct-detection subcarrier modulation, IEEE Photon. Technol. Lett. 28 (24) (2016) 2838−2841.

[16] S. Randel, D. Pilori, S. Chandrasekhar, G. Raybon, P. Winzer, 100-Gb/s discrete-multitone transmission over 80-km SSMF using single-sideband modulation with novel interference-cancellation scheme, in: Proc. ECOC' 2015, paper Mo.4.5.2.

[17] X. Chen, C. Antonelli, S. Chandrasekhar, G. Raybon, J. Sinsky, A. Mecozzi, et al., 218-Gb/s Single-wavelength, single-polarization, single-photodiode transmission over 125-km of standard singlemode fiber using Kramers-Kronig detection, in: Optical Fiber Communication Conference Postdeadline Papers, 2017.

[18] G.H. Smith, D. Novak, Z. Ahmed, Technique for optical SSB generation to overcome dispersion penalties in fibre-radio systems, Electr. Lett. 33 (1) (1997) 74−75.

[19] L. Zhang, T. Zuo, Y. Mao, Q. Zhang, E. Zhou, G.N. Liu, et al., Beyond 100-Gb/s transmission over 80-km SMF using direct-detection SSB-DMT at C-band, J. Lightw. Technol. 34 (2) (2016) 723–729.
[20] A. Mecozzi, C. Antonelli, M. Shtaif, Kramers–Kronig coherent receiver, Optica 3 (11) (2016) 1220–1227.
[21] H.A. Kramers, La diffusion de la lumiere par les atomes, in: Atti Cong. Intern. Fisica (Transactions of Volta Centenary Congress) Como, vol. 2, pp. 545–557, 1927.
[22] R.L. Kronig, On the theory of dispersion of X-rays, J. Opt. Soc. Am. 12 (6) (1926) 547–557.
[23] M. Cini, The response characteristics of linear systems, J. Appl. Phys. 21 (1950) 8–10.
[24] A. Mecozzi, C. Antonelli, M. Shtaif, Kramers–Kronig receivers, Adv. Opt. Photon. 11 (2019) 480–517.
[25] L. Shu, J. Li, Z. Wan, F. Gao, S. Fu, X. Li, et al., Single-lane 112-Gbit/s SSB-PAM4 transmission with dual-drive MZM and Kramers-Kronig detection over 80-km SSMF, IEEE Photon. J. 9 (2017) 1–9.
[26] S.T. Le, K. Schuh, M. Chagnon, F. Buchali, R. Dischler, V. Aref, et al., 1.72-Tb/s virtual-carrier-assisted direct-detection transmission over 200 km, J. Lightw. Technol. 36 (2018) 1347–1353.
[27] Z. Li, M.S. Erkilinç, K. Shi, E. Sillekens, L. Galdino, B.C. Thomsen, et al., SSBI mitigation and the Kramers–Kronig scheme in single-sideband direct-detection transmission with receiver-based electronic dispersion compensation, J. Lightw. Technol. 35 (2017) 1887–1893.
[28] M. Schuster, S. Randel, C.A. Bunge, S.C.J. Lee, F. Breyer, B. Spinnler, et al., Spectrally efficient compatible single-sideband modulation for OFDM transmission with direct detection, IEEE Photon. Technol. Lett. 20 (2008) 670–672.
[29] J.G. Proakis, Digital Communications, third ed., McGraw-Hill, 1995.
[30] X. Chen, J. Cho, S. Chandrasekhar, P. Winzer, C. Antonelli, A. Mecozzi, et al., Single-wavelength, single-polarization, single- photodiode Kramers-Kronig detection of 440-Gb/s entropy-loaded discrete multitone modulation transmitted over 100-km SSMF, in: 2017 IEEE Photonics Conference (IPC) Part II, 2017.
[31] C. Antonelli, M. Shtaif, A. Mecozzi, Kramers-Kronig PAM transceiver, in: Optical Fiber Communication Conference, 2017.
[32] C. Antonelli, A. Mecozzi, M. Shtaif, Kramers–Kronig PAM transceiver and two-sided polarization-multiplexed Kramers–Kronig transceiver, J. Lightw. Technol. 36 (2018) 468–475.
[33] S. Ohlendorf, R. Joy, S. Pachnicke, W. Rosenkranz, Flexible PAM in DWDM transmission with Kramers-Kronig DSP, in: 2018 European Conference on Optical Communication (ECOC), 2018.
[34] Z. Li, M.S. Erkilin, K. Shi, E. Sillekens, L. Galdino, B.C. Thomsen, et al., Joint optimisation of resampling rate and carrier-to-signal power ratio in direct-detection Kramers-Kronig receivers, in: 2017 European Conference on Optical Communication (ECOC), 2017.
[35] T. Bo, H. Kim, Kramers-Kronig receiver without digital upsampling, in: Optical Fiber Communication Conference, 2018.
[36] T. Bo, H. Kim, Kramers-Kronig receiver operable without digital upsampling, Opt. Express 26 (2018) 13810–13818.
[37] X. Chen, C. Antonelli, S. Chandrasekhar, G. Raybon, A. Mecozzi, M. Shtaif, et al., Kramers–Kronig receivers for 100-km datacenter interconnects, J. Lightw. Technol. 36 (2018) 79–89.
[38] X. Chen, S. Chandrasekhar, S. Olsson, A. Adamiecki, P. Winzer, Impact of O/E front-end frequency response on Kramers-Kronig receivers and its compensation, in: 2018 European Conference on Optical Communication (ECOC), 2018.
[39] J.P. Gordon, H. Kogelnik, PMD fundamentals: polarization mode dispersion in optical fibers, Proc. Natl Acad. Sci. 97 (2000) 4541.
[40] S. Betti, G. De Marchis, E. Iannone, Polarization modulated direct detection optical transmission systems, J. Lightw. Technol. 10 (12) (1992) 1985–1997.
[41] S. Benedetto, R. Gaudino, P. Poggiolini, Direct detection of optical digital transmission based on polarization shift keying modulation, IEEE J. Select. Areas Commun. 13 (3) (1995) 531–542.

[42] D. Che, A. Li, X. Chen, Q. Hu, Y. Wang, W. Shieh, Stokes vector direct detection for linear complex optical channels, J. Lightw. Technol. 33 (3) (2015) 678–684.
[43] W. Shieh, H. Khodakarami, D. Che, Polarization diversity and modulation for high-speed optical communications: architectures and capacity, APL Photon. 1 (2016) 040801.
[44] D. Che, A. Li, Q. Hu, X. Chen, W. Shieh, Implementing simplified Stokes vector receiver for phase diverse direct detection, in: Optical Fiber Communication Conference, Los Angeles, CA, paper Th1E.4, 2015.
[45] K. Kikuchi, S. Kawakami, Multi-level signaling in the Stokes space and its application to large-capacity optical communications, Opt. Express 22 (7) (2014) 7374–7387.
[46] D. Che, F. Yuan, W. Shieh, 200-Gb/s Polarization-multiplexed DMT using Stokes vector receiver with frequency-domain MIMO, in: Optical Fiber Communication Conference, Los Angeles, CA, paper Tu3D.4, 2017.
[47] D. Che, C. Sun, W. Shieh, Analog polarization identification for asymmetric polarization modulations in Stokes space, in: Proc. European Conference on Optical Communication, Rome, Italy, paper We2.38, 2018.
[48] D. Che, C. Sun, W. Shieh, Optical field recovery in Stokes space, J. Lightw. Technol. 37 (2019) 451–460.
[49] X. Chen, S. Chandrasekhar, P. Winzer, Self-coherent systems for short reach transmission, in: 2018 European Conference on Optical Communication (ECOC), Rome, 2018.
[50] D. Che, C. Sun, W. Shieh, Maximizing the spectral efficiency of Stokes vector receiver with optical field recovery, Opt. Express 26 (22) (2018) 28976–28981.
[51] C. Antonelli, A. Mecozzi, M. Shtaif, X. Chen, S. Chandrasekhar, P.J. Winzer, Polarization multiplexing with the Kramers-Kronig receiver, J. Lightw. Technol. 35 (24) (2017) 5418–5424.
[52] D. Che, C. Sun, W. Shieh, Direct detection of the optical field beyond single polarization mode, Opt. Express 26 (3) (2018) 3368–3380.
[53] H. Sun, K.-T. Wu, K. Roberts, Real-time measurements of a 40 Gb/s coherent system, Opt. Express 16 (2) (2008) 873–879.
[54] G. Roelkens, L. Liu, D. Liang, R. Jones, A. Fang, B. Koch, et al., III-V/silicon photonics for on-chip and intra-chip optical interconnects, Laser Photon. Rev. 4 (6) (2010) 751–779.
[55] C. Cole, Beyond 100G client optics, IEEE Commun. Maga.e 50 (2) (2012).
[56] P. Dong, X. Chen, K. Kim, S. Chandrasekhar, Y.-K. Chen, J.H. Sinsky, 128-Gb/s 100-km transmission with direct detection using silicon photonic Stokes vector receiver and I/Q modulator, Opt. Express 24 (13) (2016) 14208–14214.
[57] M.-J. Li, B. Hoover, V.N. Nazarov, D.L. Butler, Multicore fiber for optical interconnect applications, In: Opto-Electronics and Communications Conference (OECC), 2012, pp. 564–565.

CHAPTER 11

Visible-light communications and light fidelity

Harald Haas, Elham Sarbazi, Hanaa Marshoud and John Fakidis
Li-Fi Research and Development Center and Institute for Digital Communications, School of Engineering, The University of Edinburgh, Edinburgh, United Kingdom

11.1 Introduction

The available radio-frequency (RF) spectrum below 10 GHz has become very limited because of the exponential increase in wireless data traffic during the last 15 years [1]. The wireless communication industry has responded to this challenge by using frequencies above 10 GHz in the millimeter wave region. However, the use of higher frequencies results in increased path losses according to the Friis free-space equation. Also, the effects of blockage and shadowing at higher frequencies become more significant in terrestrial wireless communications. Therefore wireless data systems need to exploit line-of-sight (LoS) paths by using beamforming techniques and to cover even smaller areas. To this end, the use of small cells has been widely acknowledged as the next step for an increased system capacity in wireless cellular networks [2]. However, the main challenge is the provision of a supporting infrastructure to these small cells. Motivated by the recent advancements in light-emitting diodes (LEDs) and the unique possibility to encode data at speeds of tens of Gb/s to their intensity [3–5], visible-light communications (VLC) and light fidelity (LiFi) emerged as integral components of fifth-generation (5G) networks and beyond, as well as the internet-of-things (IoT) era [6]. VLC and LiFi can be considered as a special form of temporal structured lighting [7] whereby the challenge is the optimal encoding of binary information into the structure of light (of one, or in the case of LiFi, of multiple light sources) while considering the transformation of light by the optical wireless channel. While VLC refers to the data transfer between two single points using visible light, LiFi stands for a complete bidirectional multiuser networking technology that supports user mobility and random device orientations. LiFi uses ultra-small cells with radii in the order of meters. Therefore these ultra-dense optical wireless networks have been referred to as optical attocells [8] which provide ubiquitous high-speed and secure coverage to mobile users [9]. Considering the wide deployment of LEDs in indoor environments such as offices, hospitals, conference venues, and museums, as well as for outdoor

Optical Fiber Telecommunications V11
DOI: https://doi.org/10.1016/B978-0-12-816502-7.00013-0

street lighting, there is a unique potential to harness this existing infrastructure for LiFi. Therefore LiFi attocells are envisioned to form an additional layer to the current heterogeneous wireless networks without interfering with their RF counterparts. Off-the-shelf passive components are available for beamforming in LiFi, while sophisticated and computationally involved signal processing techniques are required to shape RF beams in traditional wireless systems.

In this chapter a review of the various channel modeling techniques in VLC is provided. Channel models will form the basis for the design of the analog optical front ends. In this context, available optical sources such as LEDs and laser diodes (LDs), and data receivers including single-photon avalanche diodes (SPADs) and photovoltaic (PV) cells are discussed. The latest advancements in the communication bandwidth of LEDs and LDs are presented with reported values of several GHz [4,10]. Moreover, traditional digital modulation techniques for point-to-point communication in intensity modulation/direct detection (IM/DD) systems are reviewed. In addition, more recently developed modulation schemes are discussed. These include optical orthogonal frequency division multiplexing (O-OFDM) with adaptive bit and power loading that use the communication channel optimally and boost the achievable data rates. Furthermore, state-of-the-art VLC studies are given with data rates in the order of several Gb/s including the use of micrometer-sized LEDs (μLEDs) and multiple LDs. Novel results on the simultaneous optical wireless energy and data transfer are given with achievable data rates of hundreds of Mb/s using monochromatic PV cells. Various multichannel transmission techniques in the spatial and wavelength domain are discussed for an increased system capacity. Since multiple users can be located within a LiFi attocell, a review of multiple access techniques is provided with an emphasis on nonorthogonal multiple access (NOMA). Power-domain NOMA is particularly shown to be suitable for LiFi, as it provides the highest data rate out of all the multiple access schemes in scenarios with a high signal-to-noise ratio (SNR), small number of users, and perfect channel state information (CSI) [11]. The use of multiple optical attocells introduces cochannel interference in a LiFi network. Thus, various interference mitigation methods are introduced to ensure high data densities to mobile and moving users. In addition, different strategies for the provision of a seamless coverage when users move from one attocell to another are discussed. Finally, concluding remarks are provided.

11.2 An optical wireless communications taxonomy

In optical wireless communications (OWC), lasers and LEDs are used to transmit data wirelessly. A number of different components can be used as receivers, such as positive-intrinsic-negative (PIN) photodiodes (PDs), avalanche photodiodes (APDs), SPADs, complementary metal-oxide-semiconductor (CMOS) sensors, and PV solar cells [12]. With the advancements in optical devices and component technologies during the last two decades, there has been a strong diversification of use cases in OWC. Therefore a general taxonomy for OWC technologies is needed. Such a taxonomy exists in RF communications where, for example, Bluetooth serves different use cases compared with cellular long-term evolution (LTE) systems. A taxonomy for OWC is introduced in Fig. 11.1. There are four principal technologies: (1) free-space optics (FSO), (2) VLC, (3) optical camera communications (OCC), and (4) LiFi. Typical applications for each of these technologies can be classified according to four basic parameters: (1) link data rates, (2) range, (3) duplex mode, and (4) communication mode.

FSO is an outdoor, long-range bidirectional point-to-point and high-speed communications technology that primarily is a fiber replacement. Transmission speeds greater than 1 Gb/s are required in FSO systems. VLC serves more general high-speed point-to-point communication use cases predominantly focusing on LoS indoor applications. It has applications in data centers, aircraft cabins, and industry 4.0 manufacturing environments providing point-to-point gigabit per second (Gbps) connectivity between static or quasistatic system units. VLC is also used for more general wireless indoor backhaul use cases, for example, providing gigabit data links to lightbulbs [13]. All these applications benefit from advanced security features as a result of the spatial containment of light signals and blockage by opaque objects such as

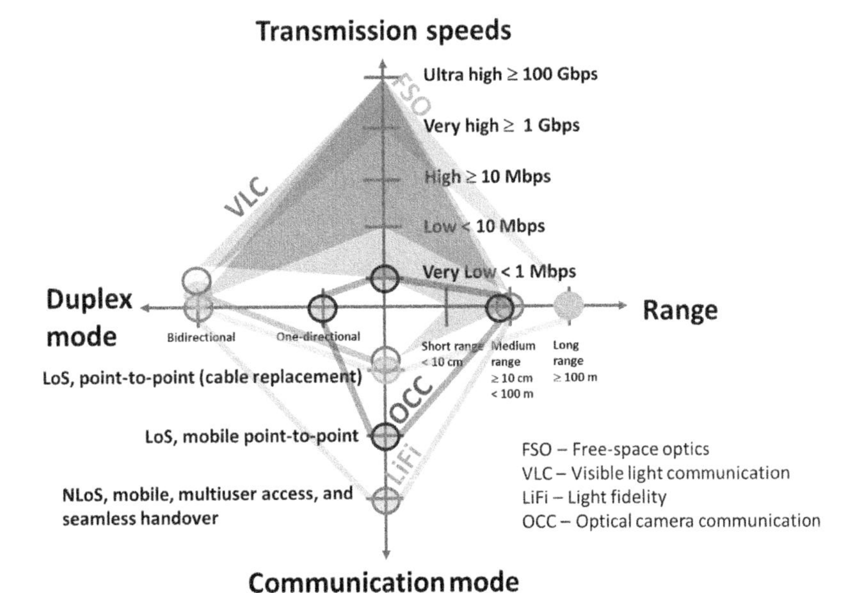

Figure 11.1 A taxonomy of optical wireless communications (OWC); the feature that clearly differentiates the various OWC technologies is the *communication mode*.

walls. OCC uses CMOS camera sensors to receive data encoded in intensity variations of light. The rolling shutter effect [14] has been harnessed to achieve data rates higher than the frame rate of CMOS sensors. OCC is used in conjunction with embedded camera sensors in mobile devices [15]. The achievable data rates are low and the communication link is one-directional (simplex). However, since existing CMOS sensors of mobile devices are used, OCC enables applications on existing smartphones. Typical applications are indoor positioning, navigation, asset tracking, and the broadcast of barcodes [16]. LiFi extends the concept of VLC to achieve fully networked OWC [17]. LiFi supports mobility of users, which includes scenarios where communication links are blocked and the LoS path is obstructed. Consequently, LiFi will continue to provide a connection in nonline-of-sight (NLoS) scenarios. A key feature of LiFi is that it provides simultaneous independent data links to multiple users, which is referred to as multiuser access. A LiFi system is composed of multiple communication cells. The access point (AP) at the center is a light source that can simultaneously provide room illumination. This means LiFi piggybacks on existing infrastructures to form LiFi attocell networks. The cell sizes can be ultra-small with radii in the meter region. Mobility support in this context means that mobile communication sessions such as a Skype call must not be interrupted when a user moves from one corner of a room to another. LiFi attocellular systems therefore, require handover mechanisms that involve multiple layers of the communication protocol stack. As a result, an optical front end that is optimized for VLC may only be suboptimum in a LiFi network. As in RF networks, there are issues surrounding interference management and mitigation in LiFi networks. However, since there is no multipath fading because the detector sizes are much larger than the wavelength, techniques developed for RF systems may only be suboptimum. There are also fundamental differences due to the use of IM/DD in LiFi systems in that signals can only be positive and real-valued. Consequently, new LiFi-bespoke wireless networking methods must be developed. Moreover, so long as light can be confined spatially by using very simple and inexpensive optical components, interference can be controlled much easier. This feature also allows step-change improvements of the small-cell concept, as single cells might cover subm^2 areas. Therefore LiFi not only benefits from an additional free spectrum that is 2600 times larger than the entire RF spectrum, but it also takes the small-cell concept to new levels that are not easily achievable in RF communications. Furthermore, due to the extremely small wavelength, the active detector sizes are very small, and massive multiple-input multiple-output (MIMO) structures can be implemented at chip-level. This property can be used to develop unique LiFi-bespoke MIMO systems, networked MIMO methods, and new angular-diversity techniques in conjunction with coordinated multipoint systems of low computational complexity. Diversity techniques in LiFi systems are especially powerful to combat random blockages, which naturally occur in a mobile scenario.

VLC is currently being standardized in the institute of electrical and electronics engineers (IEEE) 802.15.13, while OCC is being standardized in IEEE 802.15.7r1. There is a new task group in IEEE 802.11 that is set out to standardize LiFi. The reference for this new standard, which is planned to be released by 2021, is IEEE 802.11bb.

11.3 Channel models

Optical wireless systems are typically categorized as either LoS or NLoS based on the degree of directionality and the orientation of the transmitter and the receiver [18]. In a LoS system, an unobstructed optical path exists between the transmitter and the receiver. A NLoS system does not rely upon the LoS path, but rather on specular and diffuse reflections from surrounding surfaces such as the walls, the ceiling, and objects. In the path from the transmitter to the receiver, the optical signal experiences temporal dispersion due to these reflections. This effect is known as multipath dispersion. For any specified transmitter and receiver locations, the multipath dispersion is characterized by a channel impulse response (CIR).

In RF systems, the NLoS paths are formed by specular reflections and diffraction from surrounding objects, whereas in VLC systems, reflections contain both specular and diffusive components. In diffuse reflections, any ray incident on the reflective surface is scattered at many angles rather than just one, as in the case of specular reflections [19]. With diffuse reflections, and hence, the countless number of NLoS paths, the CIR of VLC systems is a continuous waveform. The characteristics of this continuous CIR depend on various parameters, such as the transmitter and receiver locations as well as orientations and the specifications of the reflective surfaces (e.g., spectral reflectance).

As wireless infrared (IR) and VLC systems operate in adjacent parts of the optical spectrum, the signal propagation characteristics in these systems are similar. Therefore a number of methods developed for simulating IR channels can be applied for VLC channels [20]. In the following, the models for the transmitter, reflective surfaces, and the receiver are defined.

11.3.1 Transmitter model

A transmitter is mathematically described by a position vector $\boldsymbol{r}_S$, a normalized orientation vector $\hat{\boldsymbol{n}}_S$, a power value P_S, and an axially symmetric radiation pattern $R(\phi)$. The radiation pattern is defined as the optical power per unit solid angle emitted from the

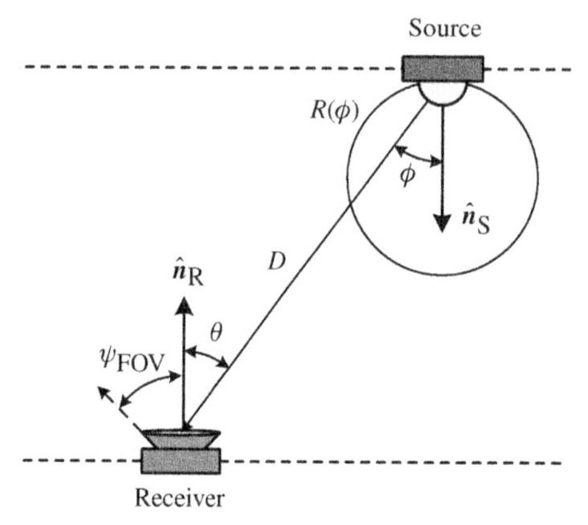

Figure 11.2 Geometry of the source and the receiver, without reflectors.

transmitter at an angle ϕ with respect to $\hat{\boldsymbol{n}}_S$ (see Fig. 11.2). It is commonly assumed that a LED transmitter has a generalized Lambertian radiation pattern [21]:

$$R(\phi) = \frac{n+1}{2\pi} P_S \cos^n(\phi), \tag{11.1}$$

where n is the Lambertian order, determining the directionality of the emitted beam. Parameter n is related to the semiangle of half power of the light source $\phi_{1/2}$ by $n = -1/\log_2[\cos(\phi_{1/2})]$. A light source $\mathcal{S}$ that emits a unit impulse of optical intensity at time zero is denoted by [21]:

$$\mathcal{S} = \{\boldsymbol{r}_S, \hat{\boldsymbol{n}}_S, n\}. \tag{11.2}$$

11.3.2 Receiver model

Similarly, a receiver with position $\boldsymbol{r}_R$, orientation vector $\hat{\boldsymbol{n}}_R$, physical area A_R, and field of view (FOV) ψ_{FOV} is denoted by [21]:

$$\mathcal{R} = \{\boldsymbol{r}_R, \hat{\boldsymbol{n}}_R, A_R, \psi_{FOV}\}. \tag{11.3}$$

11.3.3 Reflector model

In practice, the reflections contain both specular and diffusive components; however, experimental measurements (e.g., [21,22]) show that a purely diffusive Lambertian pattern can well approximate the reflective characteristics of many typical materials such as plaster walls, carpets, and unvarnished wood. This simplifying assumption has been made in the majority of IR and VLC channel modeling studies [19,21–23].

11.3.4 Channel impulse response

Given an optical source $\mathcal{S}$ and a receiver $\mathcal{R}$, light from the transmitter can reach the receiver after any number of reflections. Therefore the CIR at time t can be written as an infinite sum [19,21]:

$$h(t; \mathcal{S}, \mathcal{R}) = \sum_{k=0}^{\infty} h^{(k)}(t; \mathcal{S}, \mathcal{R}), \tag{11.4}$$

where $h^{(k)}(t; \mathcal{S}, \mathcal{R})$ is the response of the light undergoing exactly k reflections. The LoS impulse response, $h^{(0)}(t; \mathcal{S}, \mathcal{R})$, is given by [21]:

$$h^{(0)}(t; \mathcal{S}, \mathcal{R}) = \frac{(n+1)\cos^n(\phi)\cos(\theta)A_{\mathrm{R}}}{2\pi D^2}\mathrm{rect}\left(\frac{\theta}{\psi_{\mathrm{FOV}}}\right), \tag{11.5}$$

where $D = \| \boldsymbol{r}_{\mathrm{S}} - \boldsymbol{r}_{\mathrm{R}} \|$, $\cos(\phi) = \langle \hat{\boldsymbol{n}}_{\mathrm{S}}, \boldsymbol{r}_{\mathrm{R}} - \boldsymbol{r}_{\mathrm{S}} \rangle / D$, $\cos(\theta) = \langle \hat{\boldsymbol{n}}_{\mathrm{R}}, \boldsymbol{r}_{\mathrm{S}} - \boldsymbol{r}_{\mathrm{R}} \rangle / D$, and $\langle \cdot, \cdot \rangle$ denotes the scalar product of two vectors. The symbol rect(x) is the rectangular function and is equal to 1 if $|x| \leq 1$, or 0, otherwise.

11.3.5 Existing methods for visible-light communications channel modeling

To date, several methods have been proposed for obtaining $h(t; \mathcal{S}, \mathcal{R})$ for IR and VLC systems. In this section, the most prevalent methods for VLC channel characterization are summarized.

11.3.5.1 Deterministic algorithms

A widely used recursive algorithm for CIR calculation was first proposed in Ref. [21] where a cuboid room with reflective walls is considered. It is assumed that reflections are purely diffuse and the transmitter has an ideal Lambertian radiation pattern. In the proposed numerical algorithm, the surfaces are discretized into a large number of small reflective pixels, where each pixel is assumed to have constant reflectance. With the given locations and orientations of the transmitter and receiver, the interaction between each pair of pixels is calculated. The CIR corresponding to the kth order reflections is given by

$$h^{(k)}\left(t; \mathcal{S}, \mathcal{R}\right) = \int_A h^{(0)}\left(t; \mathcal{S}, \left\{\boldsymbol{r}, \hat{\boldsymbol{n}}, dr^2\frac{\pi}{2}\right\}\right) \otimes h^{(k-1)}(t; \{\boldsymbol{r}, \hat{\boldsymbol{n}}, 1\}, \mathcal{R}\Big), \tag{11.6}$$

where symbol $\otimes$ denotes the mathematical convolution. The integration in (11.6) is performed with respect to $\boldsymbol{r}$ on the area A of all reflective surfaces. Here, $\hat{\boldsymbol{n}}$ is the normal to the surface A at position $\boldsymbol{r}$. By dividing the surfaces into numerous small pixels, each with area ΔA, $h^{(k)}(t; \mathcal{S}, \mathcal{R})$ in (11.6) is approximated as

$$\begin{aligned} h^{(k)}(t;\mathcal{S},\mathcal{R}) &\approx \sum_{i=1}^{N} h^{(\theta)}(t;\mathcal{S},\mathcal{E}_i)\otimes h^{(k-1)}(t;\mathcal{E}_i,\mathcal{R}) \\ &= \sum_{i=1}^{N} \frac{\rho_i(n+1)\cos^n(\phi)\cos(\theta)\Delta A}{2\pi d_i^2}\operatorname{rect}\left(\frac{20}{\pi}\right) h^{(k-1)}\left(t-\frac{d_i}{c};\{\boldsymbol{r},\hat{\boldsymbol{n}},1\}\mathcal{R}\right), \end{aligned} \tag{11.7}$$

where $\mathcal{E}_i$ denotes the ith pixel and N is the total number of pixels. Constant c is the speed of light, ρ_i is the reflectivity of the ith pixel, $d_i = ||\boldsymbol{r} - \boldsymbol{r}_S||$, $\cos(\phi) = \langle \hat{\boldsymbol{n}}_S, \boldsymbol{r} - \boldsymbol{r}_S \rangle / d_i$, and $\cos(\theta) = \langle \hat{\boldsymbol{n}}_S, \boldsymbol{r}_S - \boldsymbol{r} \rangle / d_i$.

According to (11.7), to compute $h^{(k)}(t;\mathcal{S},\mathcal{R})$, first $h^{(k-1)}(t;\mathcal{S},\mathcal{R})$ should be obtained, and so on. However, direct implementation of (11.7) is not efficient for reflection orders k greater than 1, because identical computations would then be performed multiple times. Various algorithms have optimized the number of necessary computations to reduce the time required to compute $h^{(k)}(t;\mathcal{S},\mathcal{R})$ [24].

In practice, the optical sources used in LiFi systems (e.g., white LEDs) radiate wideband visible light—that is, their radiant power is widely distributed over the visible spectrum (from 380 to 780 nm)—while an IR LED has a narrow emission spectrum and emits invisible monochromatic light with a typical operation wavelength of 850 or 950 nm. Moreover, the reflectance of some materials depends on the wavelength, and the assumption of constant reflectance may not be accurate for some scenarios. In Ref. [22], some modifications are made in the aforementioned algorithm to account for the power spectral distribution (PSD) of optical sources as well as the spectral reflectance of surfaces.

With various optimizations and modifications, the main drawback of the deterministic algorithms is still their extremely high computational complexity, which is a direct function of the desired resolution. It takes several days to calculate a CIR for a small cuboid room considering up to third-order reflections with a moderate resolution [21]. It is also noticeable that the CIR depends not only on the positions and the orientations of the transmitter and the receiver but also on many other parameters. With the deterministic algorithms, only simplified scenarios can be addressed, as any modifications to the geometry of the room, characteristics of reflective surfaces, and so on result in an even higher computational complexity and longer simulation time. Therefore the capability of these methods is limited.

11.3.5.2 Monte Carlo ray-tracing

More efficient methods using Monte Carlo ray-tracing have been proposed to significantly reduce the time it takes to calculate a CIR [25]. Specifically, these methods provide more accurate CIR results within several minutes and offer the possibility of simulating more complicated scenarios compared to classical deterministic algorithms. In ray-tracing methods, although the computational time depends on the desired accuracy, it is still much shorter than that of deterministic recursive methods.

In ray-tracing methods, many rays are generated randomly at the transmitter position according to the emission profile of the transmitter [e.g., the Lambertian distribution as in (11.1)]. Each generated ray is then traced—the ray follows a random path, impinging on various surfaces such as obstacles, walls, and the ceiling. The point on a reflective surface where the ray hits is treated as a new optical source and a new ray is generated. At these points, the power of the ray is reduced by the reflection coefficient of the surface. This process continues for each ray until it arrives at the receiver or a specified maximum time is met (t_{max}). Two variables are associated with every ray: the time of flight (i.e., the time from its generation) and the power it carries.

In this Monte Carlo simulation method, for each ray the probability of reaching the receiver in the specified time (t_{max}) is very small (in the order of 10^{-7}). Therefore millions of rays have to be generated to get reliable results. The CIR is then calculated using the power carried by the detected rays and their times of flight:

$$h(t;\mathcal{S},\mathcal{R}) \approx \sum_{i=1}^{K_r} p_i\delta(t-\tau_i), \tag{11.8}$$

where p_i is the power of the ith ray, τ_i is the propagation time of the ith ray, $\delta(t)$ is the Dirac delta function, and K_r is the total number of rays received at the detector.

Software tools of optical design such as Zemax OpticStudio can be used to facilitate the simulation of more complicated scenarios [26]. The nonsequential ray-tracing features of this software provide an accurate description of the interaction of rays inside a specified confined space. The power and path length of each ray are accurately provided, so that the calculation of the CIR is straightforward. Various attempts demonstrate that this software yields more accurate results with less computational time [27].

OpticStudio allows us to specify the complicated geometry of the environment with the objects inside. Another added value of this software is the possibility of addressing practical specifications, such as realistic reflection characteristics of the surface materials (i.e., mixed specular-diffuse reflections) and a non-Lambertian emission pattern for transmitters. Therefore with this software, the effect of the PSD of optical sources and the spectral reflectance of surfaces can be easily addressed.

11.3.5.3 Analytical methods

In some cases the performance evaluation of VLC systems may require many channel samples, such as in MIMO VLC systems or networked VLC systems. In such cases the use of deterministic or Monte Carlo methods requires a long simulation time, whilst a rough estimation of the NLoS channel can be sufficient [19].

Some analytical models can provide an estimation of the CIR of the VLC channel. The channel models considering only the LoS component are widely used, as the VLC channel is generally dominated by the LoS paths and the calculation has a very low computational complexity. In Ref. [28], the cubic indoor environment has been approximated by a spherical space with equivalent size and reflectivity and a simple closed-form expression is obtained for the NLoS CIR. An analytical closed-form expression for a NLoS CIR with a ceiling-bounce model in a diffused link has been derived in Ref. [29]. Motivated by Ref. [29], an efficient analytical method is proposed to calculate the NLoS CIR for VLC links in Ref. [19]. It has been shown in Ref. [19] that this method provides higher accuracy compared to the sphere and the ceiling-bounce models.

In Ref. [19], first, the complicated NLoS channel is decomposed into multiple components with respect to the number of reflections as in (11.4). It is assumed that the reflections are purely diffuse and the reflective surfaces have constant reflectance. The light source is assumed to be incoherent, so that the power of the optical signals can be added in the CIR calculation, as the phases of optical signals are uncorrelated. In order to further decrease the analytical complexity, each CIR component $h^{(k)}(t;\ \mathcal{S}, \mathcal{R})$ is further decomposed according to the light propagation category. A number of major light propagation categories introduced in Ref. [19] are as follows:

- $h^{(1)}(t)$: In this case, the signal undergoes a single reflection. The mathematical derivations are tractable and there is only one propagation category:
  - Transmitter-to-wall-to-receiver (TWR): Light travels from the transmitter to one of the four walls, and then toward the receiver. For convenience, this category is referred to as the TWR and is shown in Fig. 11.3A.
- $h^{(2)}(t)$: In this case, two reflections happen. The analytical complexity is higher than that of TWR. However, via appropriate approximations and simplifications, analytical CIR expressions are obtained. Depending on the order of reflective surfaces, there are four light propagation categories:
  - Transmitter-to-floor-to-ceiling-to-receiver (TFCR): This category is shown in Fig. 11.3B.

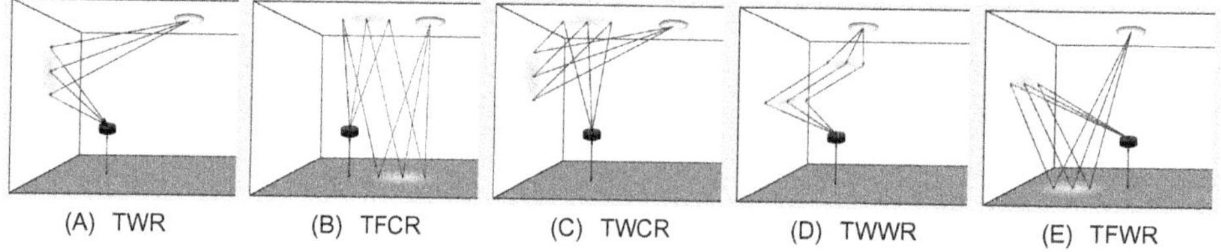

Figure 11.3 Light propagation categories. (A) Transmitter-to-wall-to-receiver (TWR) (B) Transmitter-to-floor-to-ceiling-to-receiver (TFCR) (C) Transmitter-to-wall-to-ceiling-to-receiver (TWCR) (D) Transmitter-to-wall-to-wall-to-receiver (TWWR) (E) Transmitter-to-floor-to-wall-to-receiver (TFWR) [19].

- Transmitter-to-wall-to-ceiling-to-receiver (TWCR): This category is shown in Fig. 11.3C.
- Transmitter-to-wall-to-wall-to-receiver (TWWR): This category is shown in Fig. 11.3D.
- Transmitter-to-floor-to-wall-to-receiver (TFWR): This category is shown in Fig. 11.3E.

The total CIR is approximately the superposition of the CIRs of the above categories. The expressions for the CIR with TWR, TFCR, and TWCR categories are derived in Ref. [19] and used to estimate the NLoS CIR. For the case of TFWR, the rays interact with the wall and the floor. Therefore the CIR is only perceivable when the link is next to one of the walls. In addition, the reflectivity of the floor is typically low (0.2–0.4), which further decreases the significance of the channel of the TFWR category. For the case of TWWR, the CIR is only perceivable when the link is next to one of the room corners. As long as the room is not small, this can be treated as a minor case.

The channel categories of higher order reflections can also be defined in a similar manner but are omitted due to the mathematically involved derivations. It is shown in Ref. [19] that the omission of these higher order reflection components causes an acceptable loss in the accuracy of the CIR results. The proposed CIR model provides a good approximation for NLoS VLC links considering up to second-order reflections in a typical indoor environment such as a cuboid room.

The required computational time for this method is significantly shorter than that of deterministic and Monte Carlo methods. However, this method offers lower accuracy.

11.3.6 Results of the visible-light communications channel models

In Fig. 11.4 the NLoS CIR results based on the analytical models of [19] are compared with the results of the Monte Carlo method. Four different scenarios are

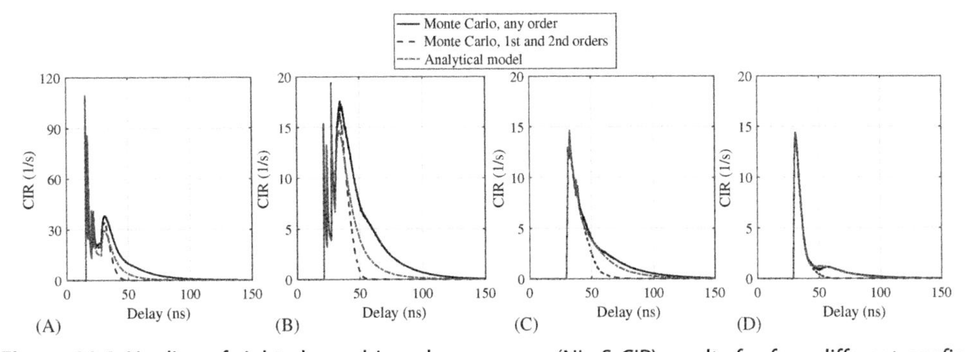

Figure 11.4 Nonline-of-sight channel impulse response (NLoS CIR) results for four different configurations, as in Table 11.1: (A) setup 1, (B) setup 2, (C) setup 3, and (D) setup 4 [19].

Table 11.1 Link configurations corresponding to the NLoS CIR results in Fig. 11.4 [19].

Setup	Room size (m^3)	Transmitter Coordinates (m)	Receiver Coordinates (m)	$\phi_{1/2}$ (degrees)	ψ_{FOV} (degrees)
1	5 × 5 × 3	(2.5, 2.5, 3)	(2, 1.5, 0.75)	60	90
2	5 × 8 × 3	(2, 2, 3)	(2.5, 6, 0.75)	60	90
3	10 × 10 × 3	(4, 4, 3)	(5, 7, 0.75)	60	90
4	10 × 10 × 3	(4, 4, 3)	(5, 7, 0.75)	40	45

considered, and the specifications are listed in Table 11.1. The following results are for empty rooms where the effect of objects and furniture is not taken into consideration. The following observations are made for these four setups:

- In setup 1, the size of the room is relatively small such that the transmitter and the receiver are close to each other, and the link is sufficiently close to the walls. Consequently, the overall magnitude of the NLoS CIR is relatively high, and the responses due to first-order reflections dominate the NLoS CIR, especially for the responses with short delays.
- In setup 2, a longer rectangular room is considered such that the increased transmitter-receiver separation leads to a NLoS CIR with a decreased magnitude.
- In setups 3 and 4, the considered transmitter/receiver positions are further away from the walls. Therefore the magnitude of the CIRs due to first-order reflections decreases significantly.
- Compared to setup 3, smaller $\phi_{1/2}$ and ψ_{FOV} are used in setup 4. Hence, the magnitude of responses with longer delays is decreased. This is because the majority of the detected signal with a long delay is launched from the transmitter in a very large radiant angle, and very low optical power is radiated in the side directions of the transmitter with a small $\phi_{1/2}$.

11.4 Analog optical front-end designs

In this section, the most important analog optical transmitter and receiver designs for LiFi systems are introduced. Novel receiver designs including SPADs and PV cells are discussed.

11.4.1 Transmitter front end

In this section, state-of-the-art technologies of LEDs, LDs, and configurations of optics are given. Also, the signal distortion caused by the nonlinear characteristic of LEDs is discussed along with new methods to overcome it.

Table 11.2 Comparison of the main LED technologies; updated version of Table III in Ref. [30].

Technology	Phosphor coated	Multichip	Micrometer sized	Resonant cavity
Application	Illumination		Displays, biosensors	Polymer optical fibers
Complexity	Low	Moderate	High	High
Cost	Low	High	High	High
Electrical bandwidth	2–5 MHz	10–20 MHz	From tens of MHz to 1.5 GHz	From tens to hundreds of MHz
Luminous efficiency	150 lm/W	65 lm/W	0.7 lm/W	0.3 lm/W

11.4.1.1 Light-emitting diodes

The tremendous increase in the penetration rate of LEDs to the global lighting market makes them the strongest candidate for VLC [30]. The use of high-brightness LEDs has gained momentum for indoor illumination, street lighting, and outdoor displays. White light has been widely adopted in current illumination systems because of its natural characteristics perceived by the human eye [31]. The most promising LED technologies for high-speed OWC are summarized in Table 11.2. The electrical bandwidth of an optical device is a very important performance metric, since it indicates the speed at which data can be modulated and transmitted. The 3-dB point determines the frequency where the electrical power becomes half of its highest value. The luminous efficiency, also known as efficacy in the lighting community, is an important figure of merit for visible LEDs [31]; it is defined by the ratio of output luminous flux over the consumed electrical power and is measured in lumen per Watt (lm/W). The luminous efficiency is calculated using [31]:

$$\eta_{\mathrm{LED}} = \frac{683}{V_{\mathrm{f}} I_{\mathrm{f}}} \int_{\lambda} p(\lambda) V_{\mathrm{es}}(\lambda) d\lambda, \tag{11.9}$$

where V_{f} is the forward bias voltage, I_{f} is the forward bias current, λ is the wavelength in the operation spectrum, $p(\lambda)$ is the PSD of the LED, and $V_{\mathrm{es}}(\lambda)$ denotes the eye-sensitivity function.

A traditional method to generate white light is the combined use of a blue-color-emitting chip with a yellow-phosphor coating. Part of the blue light is transformed to red, green, and yellow, while the rest passes through the phosphor. The thickness of the phosphor layer determines the correlated color temperature (CCT).[1] The light

[1] A metric of the appearance of light in respect with a heated source [31]. The typical CCT of cool, neutral, and warm white LEDs is around 6000, 4000, and 3000 K, respectively [30].

quality of phosphor-coated (PC) LEDs is not affected by the rapid fluctuations of their intensity[2] for data communication, so long as a direct-current (DC) offset is applied [32]. The electrical 3-dB bandwidth of PC LEDs is of the order of 2–3 MHz because of the slow photon absorption and reemission of the phosphor [33]. A typical way to enhance the communication bandwidth of a PC-LED-based system is to use a blue filter at the receiver at the cost of reduced power efficiency [33,34]; thus, increased values of up to 20 MHz have been reported for the modulation bandwidth [35]. A record luminous efficiency of 303 lm/W has been reported for a PC white LED lamp by Cree in 2014, while typical efficiency values are of the order of 150 lm/W [31].

An alternative method to create white light is the use of separate monochromatic such as red, green, and blue (RGB) LED sources. The color of a RGB LED can be controlled by modifying the irradiance of the multiple emitting chips. Thus, the color-rendering index (CRI)[3] is shown to vary from 60 and 70 to 95 for trichromatic and tetrachromatic white LEDs, respectively, while the CRI of PC LEDs scales from 55 to 95 [31]. RGB LEDs exhibit a higher 3-dB modulation bandwidth than PC LEDs with values between 10 and 20 MHz [34]. However, multichip LEDs are considered to be more expensive and complex in their structure compared with PC LEDs [30]. The highest luminous efficiency for a single visible-light-emitting chip is reported to be 180 lm/W for a current density of 35 A/cm^2 at the wavelength of 565 nm [36]. In terms of high-speed data communication, the multichip structure of white LEDs is most desirable because of the great potential of using wavelength-division multiplexing (WDM) [37].

The main limitation of the frequency response of LEDs is imposed by the minority carrier lifetime and the junction capacitance of the diode [31]. In order to boost the modulation bandwidth of LEDs, μm-sized light-emitting chips have been introduced [38]. Micro-LEDs based on gallium nitride (GaN) have been proposed for high-speed VLC and communications using polymer optical fibers (POFs) [38,39]. These devices mostly emit in the wavelength window between 370 and 520 nm and can be combined with wavelength converters for the creation of white light. Because of their very small size, μLEDs can be used in an array form for displays; each emitting chip is effectively an individual pixel with a typical diameter ranging from 14 to 84 μm [39,40]. The optical modulation bandwidth of μLEDs scales from 40 MHz [39] to record values of more than even 800 MHz [41]; the frequency response of μLEDs varies with the injected current density and chip diameter. The feasibility of 3-Gb/s VLC for a blue μLED with a 60-μm diameter is shown in Ref. [40]. The luminous

[2] The intensity or irradiance of an optical source is defined by the ratio of the optical power per unit area.

[3] A metric of the accuracy of the white-light source to show—render—the true colors of physical objects. A CRI varying from 90 to 100 is desirable for all of the illumination applications [31].

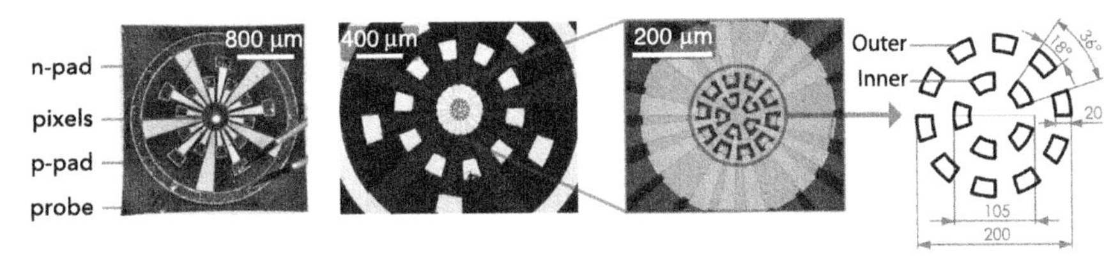

Figure 11.5 Plan view micrographs of the segmented arrays of violet micrometer-sized LEDs (μLEDs) [4]; the magnified micrographs show the setup of the two arrays and the individual pixels, and the units are in micrometer (μm). *Provided through the courtesy of Prof. M. Dawson who is with the Institute of Photonics, University of Strathclyde, Glasgow, United Kingdom.*

efficiency of this optical source is calculated to be $\eta_{\mathrm{LED}} = 0.7$ lm/W by applying $V_f = 5.2$ V, $I_f = 40$ mA, $\lambda = 450$ nm, $p(\lambda) = 4.5$ mW, and $V_{es} = 0.05$ [31,40] to (11.9). A state-of-the-art violet μLED array is demonstrated to be capable of transmitting information with a rate of 12 Gb/s in Ref. [4]; the structure of the array is given in Fig. 11.5. Two circular arrays are used to shape the LED transmitter: the inner and outer rings consist of five and 10 pixels, respectively. The optical bandwidth of 655 MHz for the inner pixel of the μLED array is considered to be the highest in the violet wavelength region [4]. A record 3-dB modulation bandwidth of 1.5 GHz is reported for a 450-nm nonpolar *m*-plane indium GaN (InGaN)/GaN μLED at a current density of 1 kA/cm^2 [10]. While the minority carrier lifetime is reduced by injecting higher current densities to μLEDs, the "droop", that is a gradual decrease of the power efficiency, is enhanced significantly [31]. For optical devices made of III−V[4] nitride, the efficiency droop is mostly attributed to the unique polarization charges and electric fields rather than the delocalization of carriers and Auger recombination.[5]

Resonant cavity LEDs (RCLEDs) are a promising technology for LiFi because of the highly narrow spectral width, directive light transmission, and increased output power [31]. They consist of an optical cavity with a typical thickness of a fraction of 1 μm. The emission wavelength of the active layer is the same with the resonance wavelength of the cavity. Although RCLEDs with a central wavelength of 650 nm were mainly used due to their increased coupling efficiency with POFs, these devices have also become available at IR wavelengths, for example, 850 and 930 nm [31]; therefore, they are promising candidates for uplink communication in LiFi systems. A record data rate of 4.2 Gb/s for RCLEDs is reported in Ref. [42] using an off-the-shelf

[4] III−V compounds contain at least one chemical element of group III (Boron group) and at least one element of group IV (nitrogen group) of the periodic table.

[5] This is a nonradiative mechanism, where the energy generated by the recombination of an electron-hole pair is consumed by the excitation of a free electron higher to the conduction band or by a hole excited deeper to the valence band [31].

red-colored chip. However, the luminous efficiency of this device is determined to be $\eta_{\mathrm{LED}} = 0.3$ lm/W by applying $V_f = 2$ V, $I_f = 20$ mA, $\lambda = 650$ nm, $p(\lambda) = 0.2$ mW, and $V_{es} = 0.1$ [31,42] to (11.9). The main disadvantage of RCLEDs based on the aluminum InGaN (AlInGaN) alloy is the complexity in manufacturing the microcavities [38]. Organic LEDs have also been proposed as a potential technology at the transmitter of VLC systems [43]. However, their low modulation bandwidth, that is in the order of hundreds of kHz, is an important drawback for high-speed data transfer along with the degradation time of their organic layer.

The light-output—current characteristic of a LED is typically nonlinear, and thus the communication signal is often clipped at particular minimum and maximum levels [44—46]. The clipping noise of an O-OFDM signal is determined by the use of the Bussgang and the central-limit theorem in Ref. [44]. The clipping noise is modeled in the frequency domain as an attenuation of the data subcarriers and an addition of zero-mean complex-valued Gaussian noise. Also, the nonlinear transfer characteristic of the VLC transmitter is modeled by the use of a generalized piecewise polynomial function in Ref. [45]. A closed-form analytical solution for the bit-error rate (BER) of a wireless O-OFDM system is derived for the first time in Ref. [46] considering an arbitrary memoryless nonlinear distortion. An interesting solution to overcome the highly nonlinear LED characteristic is given in Ref. [47] by the use of a discrete power-level stepping VLC transmitter or data-to-light modulator. In particular, the emitter is comprised of several on-off switchable groups of LEDs. These groups are controlled separately and transmit specific stepped optical power levels in parallel that constructively add up at the receiver. Hence, the transmission of intensity-modulated signals using for example pulse-amplitude modulation (PAM) or orthogonal frequency division multiplexing (OFDM) is enabled without the need for digital-to-analog converters (DACs) or amplifiers. The capability of the generation of 16 intensity levels at switching speeds of up to 500 MHz is shown in Ref. [48] for a novel digital-to-light converter.

11.4.1.2 Laser diodes

The inevitable trade-off between the modulation bandwidth and the power efficiency of LEDs restricts the design of both energy-efficient and high-speed transmitters for VLC [3,49]. Thus, LDs have recently been acknowledged as the most prominent candidate for increasing the data rate by one order of magnitude in LiFi systems [3]. In addition, the quality of white light using four-color laser mixing has been experimentally shown to be comparable to that of state-of-the-art white LEDs [50]. Laser sources offer significant advantages over LEDs: they do not experience any power efficiency droop but increase at higher injection currents; they provide a larger modulation bandwidth, that is in the order of hundreds of MHz or even of a few GHz [51]; they are more directive and therefore, their beam can be steered more efficiently by using

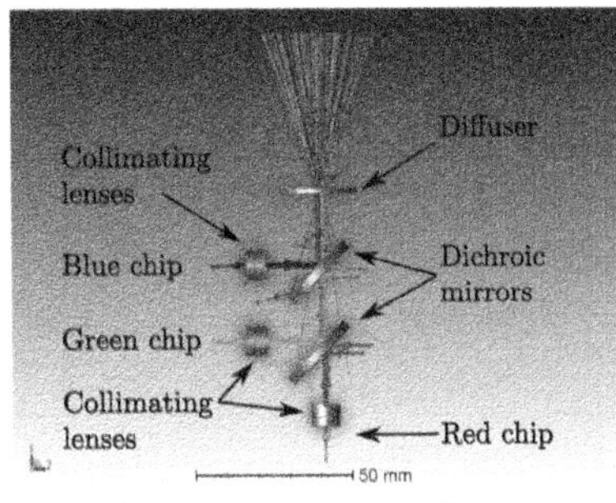

(A) Simulation design in Zemax.

(B) Experimental setup.

Figure 11.6 Optical setup of a red, green, and blue (RGB) laser transmitter used for visible-light communications (VLC) of Gbps data rates [3].

optics; and they have a linewidth[6] of the order of 2 nm that is 25 times smaller than that of LEDs, [52] and thus they allow for an optimal use of the optical spectrum with the application of dense WDM [3]. However, the main drawbacks of LDs compared with LEDs are the higher cost, the more stringent requirements for eye safety, and the higher complexity of color mixing and uniform illumination [49].

In order to generate white light at high-current densities with high power efficiency, the blue-photon source can be a LD instead of a LED. The record power efficiency of a blue LD is 38%, while the respective highest value for a state-of-the-art blue LED is more than 70% [49]. However, the peak efficiency of the blue LD is achieved for an electrical power density of about 25 kW/cm^2 that is at least three orders of magnitude higher than that of the blue LED. This means that the irradiance on the epitaxial area of the chip is significantly increased by the use of a laser device. The second common method to create a white laser beam is to mix the RGB colors of three separate emitting chips by using a diffuser. In Fig. 11.6, the optical setup of a laser transmitter that provides white illumination and data speeds in the order of Gb/s is given [3]. In particular, three pairs of lenses are used to collimate the output light from a triplet of elliptically divergent RGB LDs. The laser beams pass through two dichroic mirrors and are scattered by an engineered diffuser.[7] Ray tracing with reflections, refraction, and scattering using the optical-design software Zemax is shown in Fig. 11.6A. The coherent laser from the RGB-colored diodes is significantly distorted by the scattering diffuser. However, the remaining coherence of the white laser causes the appearance of speckles in the illumination pattern [49]. Different

[6] The linewidth or spectral width of an optical source is defined by the wavelength window over which the irradiance is reduced by a fraction of its peak value; it is typically expressed using the full width at half the maximum intensity [52].

[7] The engineered diffuser consists of an array of microlenses that are randomly positioned according to probability density functions in order to ensure the desired shape of the output beam.

methods have been introduced to reduce the speckle contrast including the use of a homogenizer and beam splitter, piezoelectric actuators and a rotating diffuser [49]. The electrical 3-dB bandwidth of three bespoke RGB LDs used for VLC in Ref. [3] is measured to be 230 MHz, 780 MHz and approximately 1 GHz, respectively; these bandwidth values are at least one order of magnitude higher than those of μLEDs [40]. A novel semipolar GaN ridge 410-nm LD is shown to achieve a record modulation bandwidth of 6.8 GHz in the blue-violet part of the optical spectrum [51].

In mainstream communication technologies using RF waves, complicated beamforming techniques are applied in the electrical domain for a directional transmission of the radiation. On the other hand, off-the-shelf passive elements are widely available in the optical spectrum, allowing for an uncomplicated steering of the light beam. Various lenses such as aspheric and Fresnel, reflectors, and parabolic mirrors are some of the optical devices used to collimate the wide radiation pattern of LEDs. The minimum semiangle of half power that light can be collimated is determined by the etendue law [52]. According to the etendue law, the area of the source multiplied by the solid angle subtended by the entrance pupil of the system as seen by the source either remains constant or increases in a passive optical system. Since the area of the laser-emitting chips is typically smaller than that of LEDs, their light can be collimated efficiently by using optics of smaller dimensions. However, the typical radiation profile of visible LDs is elliptical because of the rectangular shape of the active layer [52]. Thus, additional optical components might be required for the transformation of the elliptical beam to a circular or uniform beam. Aspheric lenses are typically used to collimate the output beam of an optical source or increase the optical power that is incident to the PD. Hemispherical lenses and compound parabolic concentrators are used to increase the FOV of a VLC receiver [18].

11.4.2 Receiver front end

In RF receivers, one or more antennas are used for heterodyne or homodyne down-conversion of the signal using a local oscillator and a mixer. The most common detection technique in VLC systems is direct detection (DD) [18]; a PD is used to produce a current that is proportional to the instantaneous received optical power. While the interference patterns of light incident on the receiver have dimensions of μm scale [30], the typical size of PDs ranges from hundreds to thousands of μm. Thus, the PD averages the collected interference patterns and no Doppler shift is experienced.

PDs are the most prominent technology of a VLC receiver because of their fast response that allows for modulation bandwidths up to the order of GHz. The typical spectral response[8] of silicon (Si) PDs takes the highest values in the wavelength

[8] The spectral response or responsivity of a PD is defined by the ratio of the generated DC photocurrent over the collected constant-wave optical power.

window of 0.8 − 0.9 μm and has a cutoff wavelength of 1.1 μm. Either germanium or III−V alloys such as InGa arsenide (InGaAs) are used for PDs with spectral response at longer wavelengths. Two main categories of PDs have been widely considered in VLC: the PIN and the APD. While PIN PDs offer some advantages over APDs such as the higher temperature tolerance and lower cost, APDs can detect much lower levels of optical power. This is because of the internal gain mechanism, as APDs operate at a higher reverse voltage than PIN PDs. Therefore APDs are mostly used in scenarios of received radiation of low intensity, while PIN PDs are suitable in cases of high received irradiance. However, one of the main drawbacks of APDs is the excessive shot noise because of the high photocurrent induced by either ambient light or the information signal [18]. The operation of an APD at the Geiger mode[9] results in such a high internal gain that the collection of a single photon is sufficient to trigger the avalanche effect and generate a large current; this is the principle that rules the function of SPADs. The spectral efficiency[10] and receiver sensitivity of a wireless O-OFDM system in the absence of background light is investigated in Ref. [53]. The SPAD sensitivity is shown to be superior to that of a conventional PD with values approaching those of standard RF antennas. Significantly low optical power of 0.4 nW is experimentally detected by a SPAD receiver with a 100-kb/s signal for a BER of 10^{-5} in Ref. [54]. Thus, SPADs are able to approach the quantum limit the most out of all the detectors. A quenching circuit is used to revert the SPAD to the operation mode below the breakdown threshold; this process introduces a recovery period, the dead time, during which the receiver does not respond to the collected photons. The counting statistics of active and passing quenching in the presence of dead time are modeled for the first time in Ref. [55]. Sunlight is shown not to interfere with the OFDM communication signal in a VLC system [56]; however, it results in an increased shot noise at the APD. The use of optical band-pass filters at the detector is shown to eliminate the shot noise induced by ambient light.

The use of PDs in a typical VLC receiver requires a reverse bias, and this process leads to an additional power consumption. An alternative approach to the design of a VLC receiver considers the use of PV cells[11] [58,59]; these devices are passive and generate electrical energy only from an optical bias. A novel design of a solar panel receiver for simultaneous power harvesting of mW levels and VLC with data rates of the order of Mb/s is shown for the first time in Ref. [60]. The main disadvantage of most Si solar cells is their limited bandwidth that is in the kHz region; this is due to

[9] The Geiger mode occurs when the applied reverse bias is far above the breakdown voltage.

[10] The spectral efficiency is one of the most important performance metrics in communication systems. It determines the data rate that can be achieved within a particular frequency bandwidth and is, therefore, measured in bits per second per Hertz (b/s/Hz).

[11] PV cells are designed to harvest energy either from sunlight, in which case they are referred to as solar cells, or from laser sources, in which case they are known as laser power converters [57].

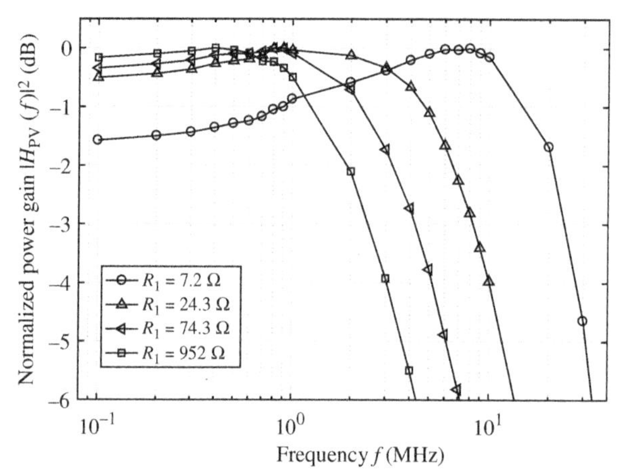

Figure 11.7 Frequency response of a gallium arsenide photovoltaic (GaAs PV) cell with a 1-mm diameter for various load resistors [61].

their large area and, therefore, large capacitance. The measured frequency response of a GaAs laser power converter (LPC) of a 0.8-mm^2 area is given in Fig. 11.7 for different values of load resistance R_l. An inductive behavior of the LPC can be observed at small load resistances because of the impedance mismatch between the receiver and the oscilloscope. The upper cutoff frequency of 3 dB for $R_l = 7.2\ \Omega$ is determined to be 24.5 MHz using linear interpolation between the consecutive points (20 MHz, −1.7 dB) and (30 MHz, −4.6 dB). This is the largest reported value for PV cells used for optical wireless transfer of data and power along with the achievable data rate of 522 Mb/s using OFDM [61]. A new reliable and low-cost technology with FSO and solar cell receivers for 5G wireless backhaul communication is introduced for the first time in Ref. [62]. A real-world pilot case within the '5G RuralFirst' project in the U.K. is introduced. The use of PV cells in OWC receivers offers the great potential to exploit the natural source of sunlight that is absent in RF energy harvesting and data communication systems.

11.5 Digital modulation techniques

In VLC, the amplitude of the optical signals is modulated by the transmitted data. This process is known as intensity modulation (IM). The fluctuations of light intensity occur at sufficiently high frequencies that cannot be perceived by the human eye. Thus, the illumination function of the LEDs is not affected. The incoherent radiation of LEDs, that is, optical waves of different phase, does not allow for any phase modulation but only IM. On the other hand, both intensity and phase modulation schemes are applicable to LDs because of their ability to produce coherent light output.

11.5.1 Single-carrier modulation schemes

Real-valued single-carrier modulation schemes such as on-off keying (OOK), PAM, and pulse-position modulation have been widely used in VLC; the analog modulated signals fit the dynamic range of the LED by the addition of a DC bias [63]. Also, to avoid significant clipping because of the DAC and the LED characteristic, the alternating-current power of the signals is scaled accordingly. Since complex-valued signals cannot be used to directly modulate the driving current, well-known complex modulation techniques such as quadrature-amplitude modulation (QAM) cannot be directly applied to VLC systems. Carrier-less amplitude and phase (CAP) modulation, which is closely related to QAM, has been recently considered for VLC. This modulation scheme employs two orthogonal finite-impulse-response digital filters to form a Hilbert transform pair in order to generate a low-frequency DC-biased QAM signal [64]. Fig. 11.8 shows the basic principle of CAP signaling. The complex QAM signal is mapped into two uncorrelated in-phase and quadrature signals. These components are then independently passed through two shaping filters whose impulse responses form a Hilbert transform pair. After the DAC, the CAP signal at the modulator output can be expressed as:

$$s(t) = \sum_{m=-\infty}^{\infty} [I(m)f(t - mT) - Q(m)\hat{f}(t - mT)], \tag{11.10}$$

where $I(m)$ and $Q(m)$ are the corresponding in-phase and quadrature symbols and T indicates the symbol duration. Also, $f(t)$ and $\hat{f}(t)$ are the analog impulse responses of the shaping filters. At the receiver side, the in-phase and quadrature signals are recovered after passing through two shaping filters with impulse responses that are

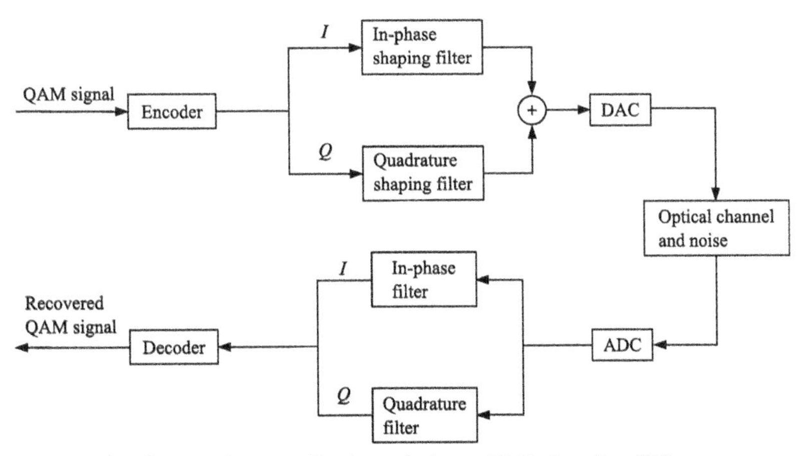

Figure 11.8 Principle of carrier-less amplitude and phase (CAP) signaling [65].

equal to the complex conjugates of the respective transmitter filters. A detailed description of the CAP transceiver operation can be found in Ref. [65].

Few VLC-specific modulation techniques have also emerged to convey the information signal by altering the color of the LEDs. Color intensity modulation (CIM), for example, encodes the transmitted bits into the instantaneous color intensity, while maintaining a constant perceived color [66]. Color-shift keying (CSK) uses different combinations of the RGB color bands to convey information, keeping the transmitted signal power at a fixed level. As a result, CSK signals have low intensity fluctuations compared to CIM, which reduces the flicker in the optical signal [67].

11.5.2 Multicarrier modulation

Regarding the multicarrier modulation techniques, O-OFDM has been the most widely used scheme due to its ability to boost the achievable data rates of VLC systems [68]. In order to adapt O-OFDM to the requirements of IM, Hermitian symmetry is used to generate real OFDM signals in the time domain. The block diagram of a Hermitian-symmetric OFDM (HS-OFDM) system is given in Fig. 11.9. First, the information bits are framed and mapped to complex symbols based on the selected modulation scheme that is typically QAM. Then the complex bipolar signal at the input of the inverse fast Fourier transform (IFFT), $\mathbf{X} = [X(0), X(1), \ldots, X(N_{\mathrm{FFT}} - 1)]$, is forced to have Hermitian symmetry by applying the following condition [69]:

$$X(l) = X^{*}(N_{\mathrm{FFT}} - l) \text{ for } 0 < l < \frac{N_{\mathrm{FFT}}}{2}, \tag{11.11}$$

and setting the subcarriers $X(0)$ and $X(N_{\mathrm{FFT}}/2)$ to zero, where l is the index of the digital samples in the frequency domain, and N_{FFT} is the size of the IFFT. Thus, a total number of $N_{\mathrm{FFT}}/2 - 1$ out of N_{FFT} subcarriers is actually used to carry

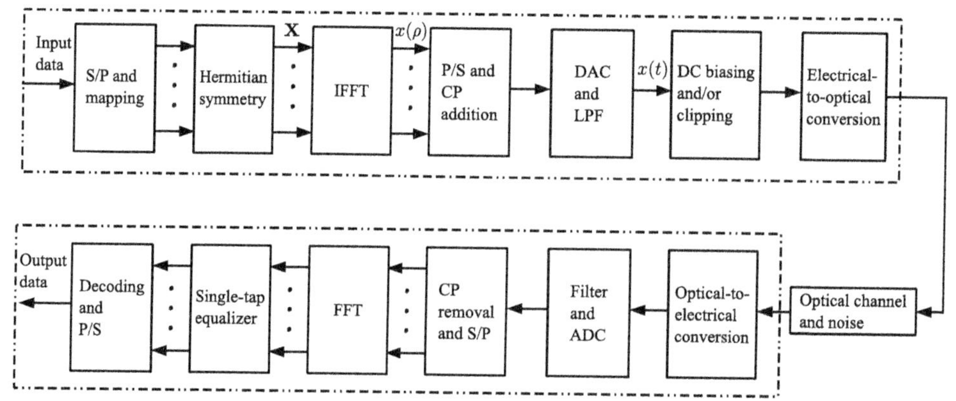

Figure 11.9 Block diagram of a Hermitian-symmetric orthogonal frequency division multiplexing (HS-OFDM) system [69].

information, yielding to a spectral loss of around half of the available bandwidth. After performing IFFT, the resulting time-domain samples given by [69]

$$x(\rho) = \frac{1}{\sqrt{N_{\mathrm{FFT}}}} \sum_{l=0}^{N_{\mathrm{FFT}}-1} X(l) \exp\left(\frac{j2\pi l\rho}{N_{\mathrm{FFT}}}\right) \quad \text{for} \quad 0 \le \rho \le N_{\mathrm{FFT}} - 1, \tag{11.12}$$

are real. In (11.12), ρ is the index of the digital samples in the time domain, j is the imaginary unit and exp(.) denotes the exponential factor. In general, a cyclic prefix (CP) is added in order to combat inter-symbol interference (ISI) and transform the dispersive wireless channel into a flat fading channel. After the parallel-to-serial conversion (P/S), $x(\rho)$ is transformed to an analog waveform using a DAC and is passed through a low-pass filter (LPF) to remove high-frequency noise components.

To satisfy the positivity constraint needed for modulating the intensity of the LED current, different variations of HS-OFDM have been proposed. Direct-current-biased O-OFDM (DCO-OFDM) [70], for example, has been widely adopted in several Gbps VLC implementations as it offers a straightforward solution by adding a DC bias, resulting in the signal [69]:

$$x_{\mathrm{DCO}}(t) = x(t) + B_{\mathrm{DC}}, \tag{11.13}$$

where

$$B_{\mathrm{DC}} = \mu\sqrt{E\{x(t)^2\}}, \tag{11.14}$$

is the DC bias level, μ is a proportionality constant and $E\{.\}$ denotes the expectation operator. It is noted here that the DC bias needed to guarantee a positive signal may result in high signal amplitudes, leading to signal distortion and/or overheating. It is, however, possible to mitigate this effect by clipping the signal at a minimum and maximum value, which introduces the so-called clipping noise. At the receiver terminal, the received signal goes through an optical-to-electrical conversion using a PD before being processed in a way similar to a conventional OFDM receiver.

Asymmetrically clipped O-OFDM (ACO-OFDM) [71] is a power-efficient alternative to DCO-OFDM. In ACO-OFDM, only the odd subcarriers are used to carry information signals, leading to a further degradation of the spectral efficiency of HS-OFDM by 50%. To this end, the Hermitian-symmetric signal in ACO-OFDM comprises only odd components, that is, $\mathbf{X} = [0, X(1), 0, X(3), \ldots, 0, X(N_{\mathrm{FFT}} - 1)]$, which, after IFFT, generates an antisymmetric real-valued digital signal $x(\rho)$ such that [69]

$$x(\rho) = -x\left(\rho + \frac{N_{\mathrm{FFT}}}{2}\right) \quad \text{for} \quad 0 < \rho < \frac{N_{\mathrm{FFT}}}{2}. \tag{11.15}$$

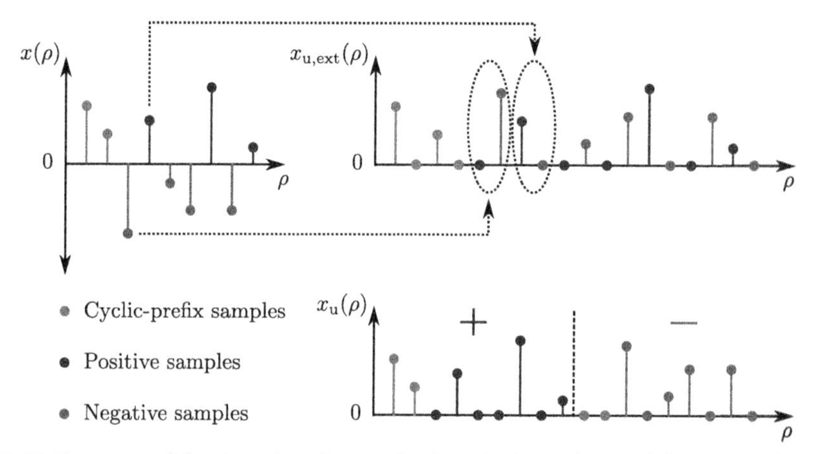

Figure 11.10 Structure of the time-domain samples in unipolar orthogonal frequency division multiplexing (U-OFDM) [73].

The antisymmetry of $x(\rho)$ ensures that no data is lost as a result of signal clipping at the zero level. The transmitter front end of ACO-OFDM is similar to the DCO-OFDM transmitter with the DC bias removed, and the receiver processing is similar to the DCO-OFDM receiver except that only the odd subcarriers are demodulated in ACO-OFDM. An alternative energy-efficient modulation technique to ACO-OFDM is PAM discrete multitone (PAM-DMT) [72]. In PAM-DMT, the real part of the complex OFDM signal is set to zero and only the imaginary part is pulse-amplitude modulated.

Asymmetrically clipped DC biased O-OFDM (ADO-OFDM) combines the advantages of both ACO-OFDM and DCO-OFDM [69]. In ADO-OFDM, the input Hermitian-symmetric signal $\mathbf{X}$ is divided into an even component $\mathbf{X}_{\text{even}} = [X(0), 0, X(2), \ldots, X(N_{\text{FFT}} - 2), 0]$ and an odd component $\mathbf{X}_{\text{odd}} = [0, X(1), 0, \ldots, 0, X(N_{\text{FFT}-1})]$. DCO-OFDM is performed on $\mathbf{X}_{\text{even}}$ while $\mathbf{X}_{\text{odd}}$ undergoes an ACO-OFDM operation. As a result, ADO-OFDM has higher power efficiency compared to DCO-OFDM, and higher bandwidth efficiency compared to ACO-OFDM. Another HS-OFDM technique is unipolar OFDM (U-OFDM) [73], in which the OFDM samples are rearranged and sent in separate positive and negative blocks as shown in Fig. 11.10. In particular, the real-valued OFDM signal $x(\rho)$ is transformed to a unipolar extended OFDM frame $x_{\text{u,ext}}(\rho)$ by mapping each bipolar sample to a pair of nonnegative samples. If the original OFDM sample is positive, the first sample of the new pair takes the original value and the second is set to zero. If, however, the original OFDM sample is negative, the first sample of the new pair is set to zero and the second takes the absolute value of the original sample. The actual U-OFDM frame $x_{\text{u}}(\rho)$ is formed by grouping the samples of each pair in two separate blocks, one "positive" and one "negative."[12] The positive group is sent first and the negative group is transferred

[12] The terms "positive" and "negative" are used for the nonnegative samples of $x_{\text{u}}(\rho)$ with respect to the sign of the original samples in the bipolar OFDM signal $x(\rho)$.

second. U-OFDM offers a significant energy advantage over DCO-OFDM, but it sacrifices half of the spectral efficiency. Enhanced U-OFDM (eU-OFDM) compensates for the spectral efficiency loss in U-OFDM while maintaining high power efficiency at the expense of increased computational complexity [74]. An experimental proof of concept is given in Ref. [75] for U-OFDM and eU-OFDM. Results show that both schemes offer a high energy efficiency but their performance can be highly affected by nonlinear distortion. In an eU-OFDM system, multiple U-OFDM streams are superimposed at the transmitter using the same constellation order. It can be noted that eU-OFDM approaches the spectral efficiency of DCO-OFDM, but this would require the superposition of an infinite number of streams in the modulated signal. However, it can be shown that 99.2% of the spectrum efficiency of DCO-OFDM can be achieved with seven superimposed data streams. The concept of eU-OFDM is generalized in Ref. [76] using data streams of arbitrary QAM constellation orders and arbitrary levels of allocated energy. As a result, the gap in spectral efficiency between eU-OFDM and DCO-OFDM is shown to be completely closed with three information streams.

To avoid the spectral loss involved in HS-OFDM schemes, non-Hermitian symmetry O-OFDM (NHS-OFDM) has been introduced [77]. The idea is based on utilizing the spatial dimension to convey complex-valued OFDM signals by means of positive and real-valued signals. A pair of LEDs could be used to transmit the real and imaginary parts of a complex symbol. At the receiver, the real and imaginary parts are received by two separate PDs and are recombined to recover the complex OFDM symbol. By eliminating the need for Hermitian symmetry in the frequency-domain OFDM signal, NHS-OFDM can utilize all the available subcarriers for data transmission. However, the spectral efficiency gain of NHS-OFDM comes at the expense of additional hardware complexity involved in using a MIMO configuration. Also, the associated system performance might be highly dependent on the locations of the transmitting and receiving terminals.

Although the majority of the theoretical work on O-OFDM has been conducted under the assumption of flat AWGN channels, many experimental implementations report low-pass characteristics in the optical wireless channels [78]. This implies that high SNR statistics may exist at low-frequency subcarriers. The work in Ref. [79] shows that DCO-OFDM and ACO-OFDM exhibit performance degradation when tested under a frequency-selective channel compared to a flat channel. This is expected due to the higher attenuation affecting the high-frequency subcarriers. With such a frequency-selective behavior, the transmission rate of O-OFDM can be enhanced by using bit and/or power loading techniques. Bit loading involves the use of larger constellations on subcarriers with high SNR characteristics and smaller constellations on subcarriers with low SNR characteristics. Based on the widely used approximated BER formula for M-order QAM given by [80]:

$$P_e \leq 4Q\left[\sqrt{\left(\frac{3}{M-1}\gamma\right)}\right], \tag{11.16}$$

where $Q(x)$ is the Q-function, $M=2^b$ is the modulation order of QAM, and γ is the SNR, the spectral efficiency can be expressed as

$$b \cong \log_2\left(1+\frac{\gamma}{\Gamma}\right), \tag{11.17}$$

where $\Gamma=[Q^{-1}(P_e/4)]^2/3$. Hence, the transmission rate of O-OFDM can be enhanced by optimizing the number of bits carried by each subcarrier based on (11.17). This enhancement can be further increased by optimizing the transmit power allocated to each subcarrier. Several algorithms have been used to perform bit and power loading in O-OFDM such as the water-filling-based Chow algorithm, exhaustive-search-based Krongold algorithm, and the optimal greedy Hughes-Hartogs algorithm [81]. Detailed description and analysis of adaptive bit and power loading in DCO-OFDM and ACO-OFDM can be found in Ref. [82]. The work in Ref. [40] demonstrated that DCO-OFDM with adaptive bit and power loading outperforms pre- and postequalization techniques. PAM-DMT is shown to be optimally adapted to a typical channel with a Gaussian LPF response outperforming ACO-OFDM with bit loading [72].

11.6 Multichannel transmission techniques

In this section, various methods of multichannel transmission in a VLC system are discussed including MIMO, angular diversity, and WDM.

11.6.1 Multiple-input multiple-output

The concept of MIMO communication has been originally explored for increasing the capacity of RF systems [83] and has also been applied to OWC systems [84]. Multiple light emitters and a single detector form a multiple-input single-output (MISO) system, while a single emitter and many PDs create a single-input multiple-output (SIMO) system. The use of MISO or SIMO to transfer the same data offers an improved performance in the achievable SNR compared with a single transmitter and receiver, known as a single-input single-output (SISO) system. This performance improvement is referred to as diversity gain and is based on the use of the optical channel with the lowest attenuation in an indoor environment. When independent bit sequences are transmitted in parallel from multiple light sources to many detectors, the spectral efficiency increases linearly with the number of channels; this improvement is referred to as multiplexing gain.

For an optical wireless MIMO system that comprises N_t light transmitters and N_r detectors, the received signal can be written in the form of a vector as follows [85]:

$$\mathbf{y}=\mathbf{H}\mathbf{x}+\mathbf{n}. \tag{11.18}$$

The vector **n** denotes the summation of the shot and thermal noise. Shot noise stems from the ambient light and the desired signal itself, while thermal noise is produced by the thermal agitation of charge carriers inside the semiconductor. Thus, **n** is modeled as real-valued additive white Gaussian noise (AWGN) with zero mean and variance $\sigma^2 = \sigma^2_{\text{shot}} + \sigma^2_{\text{thermal}}$, where σ^2_{shot} and $\sigma^2_{\text{thermal}}$ are the shot and thermal noise variances, respectively [18]. The noise power can be calculated by using $\sigma^2 = N_0 B$, where N_0 represents the power spectral density of noise and B is the bandwidth of the receiver. The vector $\mathbf{x} = [x_1, \ldots, x_{N_t}]^T$ is the transmitted signal with $[.]^T$ denoting the transpose operator. The subscripts of the elements of **x** determine the light emitter, that is, x_{n_t} is the transmitted signal of emitter n_t. The channel matrix **H** with a size of $N_t \times N_r$ is given by

$$H = \begin{bmatrix} h_{11} & h_{12} & \cdots & h_{1N_t} \\ h_{21} & h_{22} & \cdots & h_{2N_t} \\ \vdots & \vdots & \ddots & \vdots \\ h_{N_r1} & h_{N_r2} & \cdots & h_{N_rN_t} \end{bmatrix}, \tag{11.19}$$

where $h_{n_r n_t}$ is the DC attenuation factor of the channel between emitter n_t and receiver n_r [18]. Typical LoS conditions occur in an indoor VLC scenario, and therefore, the separate attenuation factors of each channel are calculated using (11.5). In practice, the channel matrix is determined using some training sequences. A number of receiver estimates are multiplied by the inverse of **H** to obtain the transmitted data [86]. Typically, optical MIMO channels are more correlated than a RF channel. As a consequence, sufficiently large receiver arrays are required to achieve a channel matrix of full rank [30]. A full-rank matrix is desired because it is invertible and offers the maximum capacity gain in a MIMO communication system [86].

The simplest MIMO transmission method is the repetition coding (RC), where the same information sequence is transmitted from each source, that is, $x_1 = x_2 = \cdots = x_{N_t}$. Lambertian channels can be highly correlated because of the wide radiation pattern of LEDs; therefore, the application of RC is preferred in many VLC scenarios to achieve a diversity gain. If we consider the use of M-PAM with RC, the spectral efficiency is calculated to be $\log_2(M)$, where M denotes the constellation size. Spatial multiplexing (SMP) is used to transfer independent data streams from the various light emitters of the same wavelength at the same time [87]. This technique can be used primarily with LDs because of their high directivity resulting in an enhanced spectral efficiency, that is, $N_t \log_2(M)$.

Spatial modulation (SM) is a combined MIMO and modulation format originally introduced for RF communication that enhances the system spectral efficiency and reduces the system complexity at the same time [88–90]. The operation principle of SM is based on the expansion of the typical signal constellation to an additional spatial

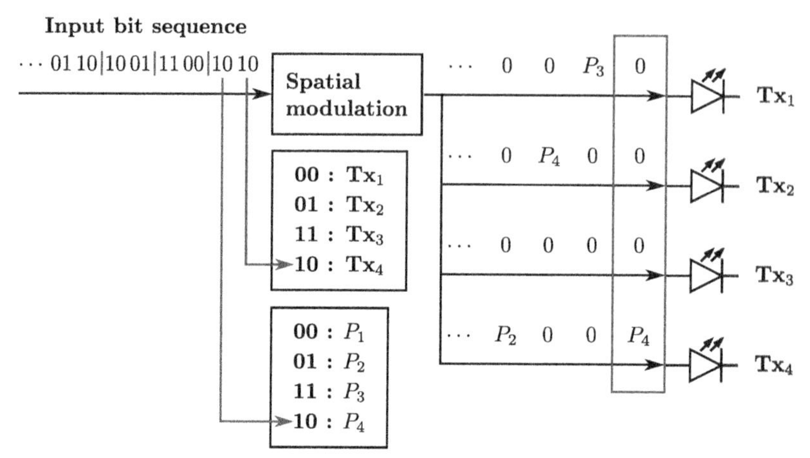

Figure 11.11 An example of optical spatial modulation for transmission with $N_t = 4$ and 4-PAM [85].

dimension. This means that each transmitter is assigned a unique binary sequence, namely a spatial symbol or index, to carry additional information. Only one transmitter is activated during the symbol period to convey the assigned constellation symbol—M-PAM in this example—along with the spatial symbol. Therefore the spectral efficiency of SM is calculated to be $\log_2(N_t) + \log_2(M)$ exceeding that of a SISO system, which is $\log_2(M)$ in this case. In addition to the increased spectral efficiency, interchannel interference (ICI) is effectively mitigated; this is because at any given time instance, only one transmitter is active. Energy efficiency is another important advantage of SM since the power consumption is independent of the number of emitters, while information is still transferred from them. An example of optical SM using four LEDs and 4-PAM at the transmitter is given in Fig. 11.11 [85]. The first two bits activate the respective LED in accordance with Gray mapping, while the last two determine the power of the 4-PAM symbol P_i, where $i \in \{1, 2, 3, 4\}$ that modulates the intensity of the LED. A comparison of RC, SMP and SM for indoor VLC under LoS conditions shows that SM is more robust than SMP under high channel correlation offering a higher spectral efficiency than RC [85].

The scheme of generalized SM (GeSM), first introduced for RF communications [91], allows the activation of a number N_a of LEDs during the symbol period, where $0 < N_a < N_t$ [87]. Since N_a out of a total number of N_t LEDs are active during the symbol period, the number of different possible combinations is $N_c = \binom{N_t}{N_a}$; the symbol $\binom{\cdot}{\cdot}$ denotes the binomial coefficient. However, only $2^{\lfloor \log_2(N_c) \rfloor}$ LEDs can be used to encode the binary data, where $\lfloor . \rfloor$ denotes the floor function [91]. In optical GeSM, each LED transmits the same constellation symbol. Therefore the spectral efficiency is determined to be $\lfloor \log_2(N_c) \rfloor + \log_2(M)$. The use of optical multistream SM allows the activation of more than one LED during the symbol period and each LED transmits a different data symbol [87]. Therefore the spectral efficiency of multistream SM is $\lfloor \log_2(N_c) \rfloor + N_a \log_2(M)$ and is shown to be further enhanced than that of optical GeSM. Optical SM and multistream SM are shown to provide the best BER performance compared with that of SMP in the region of low and high SNR, respectively [87]. A novel generalized LED index modulation (GLIM) is proposed in Ref. [92] to avoid the application of Hermitian symmetry and DC biasing of MIMO-OFDM VLC systems. In particular, separate LEDs are used to encode the real-imaginary and positive-negative counterparts of the complex values of the OFDM signals. The use of a maximum-a-posteriori-probability estimator at the receiver is shown to offer superior BER performance of GLIM compared to benchmark systems such as DCO-SM, non-DCO-OFDM and ACO-SM.

The configuration of a lens with an array of PDs used to distinguish the images arriving from the various optical sources is referred to as imaging-diversity reception. Nonimaging receivers typically consist of an array of spatially separated pairs of an optical element and a PD. The orientation of the elements of a nonimaging receiver is typically vertical. The use of a novel mirror-diversity receiver (MDR) in a 2×2 VLC MIMO system results in a superior performance in terms of the achievable BER compared to other nonimaging receivers such as the vertically oriented and link-blocked receivers [93]. Link blockage is a suitable method to decrease the channel correlation and is achievable by the use of an opaque object in front of the PD such as an iris to decrease the FOV [85]. This can be assumed to be a benchmark system for static receivers used for MIMO communication. Imaging-diversified reception is shown to offer a higher resilience in user mobility than nonimaging systems in Ref. [84]. Also, the use of image-diversity receivers relaxes significantly any alignment requirements between the transmitter and the receiver. A data rate of 1 Gb/s is experimentally achieved in Ref. [94] using a 4×9 OFDM-MIMO system for indoor VLC coverage.

11.6.2 Angular diversity

The separation and increase in the directionality of the transmitted RF beams require the modification of the phase and amplitude of the electrical signals. In a LiFi transmitter, the creation of confined light beams is achieved using simple optical components such as reflectors and lenses. The technique of space division multiple access (SDMA) using angular diversity for VLC transmission is considered for the first time in Ref. [95]. An angular-diversity transmitter (ADT) consisting of 37 LED–based elements is shown in Fig. 11.12 [95]. The LED elements are placed on a hemispherical substrate; the first is mounted at the center of the hemisphere, while the rest are located in the form of ring arrays. Each LED element is pointed toward a particular direction with a semiangle of half power of $\phi_{1/2}$; thus, the optical cell is split to 37 overlapping sub-cells. The parameter a denotes the separation angle between the LED ring arrays.

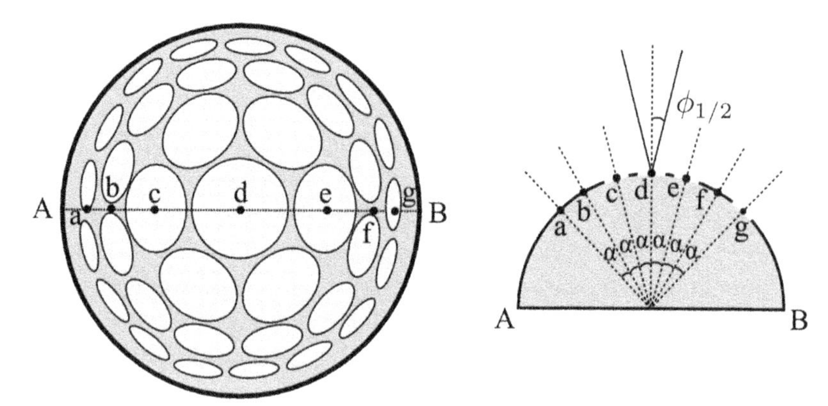

Figure 11.12 Geometry of the 37 light-emitting diode (LED)-based transmitter used for angle diversity in Ref. [95].

The angular-diversity SDMA transmitter is shown to offer gains in the average spectral efficiency by at least one order of magnitude in comparison to time division multiple access (TDMA) [95].

An angular-diversity receiver (ADR) is a nonimaging receiver with an angular separation of the pairs of the optical element and the PD. Recent studies considering ADRs for indoor VLC MIMO communication demonstrate the mobility benefits of users achieving a significantly larger capacity compared with vertically oriented receivers [93,96,97]. In particular, the configurations of a pyramid and hemispheric receiver are shown to achieve a similar channel capacity and BER performance to those of a link-blocked receiver [96]. The throughput benefits of using ADRs compared with vertically oriented receivers in an indoor diffuse VLC environment are shown in Ref. [97]. A combined angle-and-mirror-diversified receiver is experimentally shown to offer SNR gains of 1.5 and 5 dB compared to the use of an ADR and MDR, respectively [93].

While most of the studies on ADRs for VLC MIMO communication are realized on a link level, the networking and interference mitigation mechanisms are explored for the first time in Ref. [98]. In particular, each user equipped with an ADR selects the optical cell that provisions the strongest data signal. Four signal combining techniques are investigated for reception in Ref. [98]: select-best combining (SBC), equal-gain combining (EGC), maximum-ratio combining (MRC), and optimum combining (OC). The PD with the highest SNR is selected for data communication in SBC; the information of the channel state of the desired cell needs to be known at the receiver for mitigation of the interference. The electrical signals received by all of the PDs are equally combined when EGC is applied at the ADR; no information of the channel state is required at the receiver and, thus, EGC has the lowest implementation complexity of all the schemes. The weight of each PD is proportional to its achievable SNR in MRC, and the CSI of the desired cell needs to be known at the ADR. OC

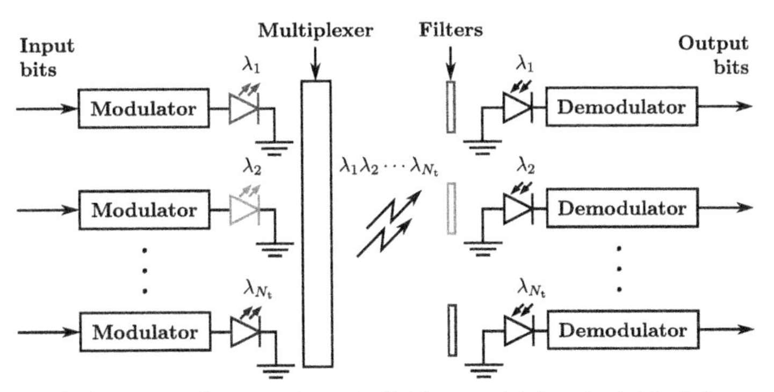

Figure 11.13 Block diagram of a wavelength-division multiplexed visible-light communications (VLC) system.

is shown to offer the best performance in terms of signal-to-interference-plus-noise ratio (SINR) close to that of interference-free systems; however, OC requires the highest computational complexity at the receiver because the CSI from all the attocells of the network needs to be estimated [98]. The techniques of MRC and SBC are shown to be superior to the use of a single-PD receiver at the cost of additional complexity.

11.6.3 Wavelength-division multiplexing

The use of the wavelength as an additional physical dimension to space and time was originally considered for an increased data transfer in optical-fiber networks [99]. This concept, referred to as WDM, is applicable to VLC systems and the respective block diagram is given in Fig. 11.13. The various input bit streams are separately modulated using a number of N_t electrical chains. The analog signals are modulated on the intensity of each optical source with a unique operation wavelength λ_i, where $i = \{1, 2,..., N_t\}$. The optical channels are multiplexed for the generation of white light typically by using dichroic mirrors and a diffuser[13] [3,100]. Demultiplexing of the visible-light channels is performed by the use of filters tailored to the relevant color in front of each PD.

The first demonstration of a high-speed wireless WDM system based on discrete multitones with a 803-Mb/s data rate using a RGB LED luminary and an APD is given in Ref. [37]. A data rate of 11.3 Gb/s is reported in Ref. [100] for a LED-based WDM system using OFDM for each channel. In particular, a commercially available RCLED and a green and blue μLED are used at the VLC transmitter, and the receiver includes a PIN PD. In Ref. [36], a commercially available RGB-yellow-cyan LED is used as a VLC transmitter and a lens with a PIN PD is used for reception. The WDM system achieves a data rate of 10.7 Gb/s using 64-QAM discrete multitones.

[13] Either dichroic mirrors or a single lens are commonly used with multicolored LED chips, while visible LDs are combined with dichroic mirrors and a diffuser.

In Ref. [3], a triplet of bespoke RGB LDs is used for VLC covering a circular area of 1 m^2 with a vertical distance of 3 m. The achievable data rate is 3.4 Gb/s for an illuminance[14] of white light of only 21.3 lx; the data rate is boosted to 14 Gb/s in well-illuminated areas with an illuminance of 843.7 lx and a measured CCT of 8000 K; and it is inferred that the use of 36 parallel WDM channels can provide data rates of over 100 Gb/s under standard illumination conditions. The use of RGB LDs with dichroic mirrors and frosted glass instead of a diffuser generates white light with a CCT of 8328 K, a CRI of 54.4 and an illuminance of 7540 lx in Ref. [101]. Moreover, the white LD-based luminary is used as a WDM-based VLC transmitter using OFDM for each optical channel and a convex lens with an APD is used at the receiver; thus, the data rate reaches a maximum value of 8.8 Gb/s and also the blue-light hazard is shown to be avoided.

11.7 Multiuser access techniques

As a technology enabler for 5G networks and beyond, LiFi systems are expected to support seamless multiuser connectivity where multiple users could access the LiFi attocell by sharing the available system resources. Multiple access techniques can generally be divided into two categories: orthogonal multiple access (OMA) and NOMA. OMA allocates orthogonal, that is, nonoverlapping, time and/or frequency slots to different users, whereas NOMA allows all multiplexed users to utilize the entire time and frequency resources by exploiting the power, code, or space domain. Although the term NOMA has been lately associated with power-domain multiple access, NOMA includes a group of multiple access schemes that allow each user to exploit the full time and frequency resources.

11.7.1 Optical time division multiple access

Optical TDMA (O-TDMA) allows the orthogonal multiuser access to the same frequency channel by allocating different time slots to each user. The application of *M*-PAM to O-TDMA provides a spectral efficiency of $\log_2(M)$ measured in b/s/Hz per user. It was reported in Ref. [17] that the performance of O-TDMA could be severely degraded in multicell LiFi scenarios due to the ICI. Hence, sophisticated interference mitigation techniques are needed to eliminate the ICI that affects the cell-edge users.

11.7.2 Optical orthogonal frequency division multiple access

Optical orthogonal frequency division multiple access (O-OFDMA) [102] has been recently considered as a straightforward extension of O-OFDM to realize multiple

[14] The illuminance is defined by the ratio of luminous flux per unit area and is measured in lux (lx) [52]. The illuminance under typical office lighting conditions scales from 400 to 1000 lx [30].

access in LiFi systems. To this end, O-OFDM can be tailored to the requirements of IM by using the Hermitian symmetry and biasing/clipping techniques based on the different modulation techniques discussed in Section 11.5.2. This implies an inevitable loss of more than half of the available subcarriers. For example, M-QAM-based DCO-OFDMA exhibits a spectral efficiency of $(N_{\mathrm{FFT}}-2)\log_2(M)/(2N_{\mathrm{FFT}})$ measured in b/s/Hz per user [103]. Proper resource allocation mechanisms can lead to full utilization of the benefits of O-OFDMA schemes in the context of LiFi [104]. Such mechanisms involve joint optimization of the subcarriers allocation, power allocation, and DC bias under illumination constraints as presented in Ref. [105]. A multicell downlink LiFi system based on DCO-OFDMA was studied in Ref. [106], where $(N_{\mathrm{FFT}}/2)-1$ available data subcarriers were shared among the network users. Let the time-domain DCO-OFDM signal be:

$$x(t)=\sum_{\rho=0}^{N_{\mathrm{FFT}}-1} x_\rho(t)+B_{\mathrm{DC}},\quad t=0,1,\ \ldots,N_{\mathrm{FFT}}-1. \tag{11.20}$$

The received IM signal of each user on subcarrier ρ can be expressed as

$$y_\rho(t)=h_0x_{0,\rho}(t)+\sum_{i\in\Pi} h_ix_{i,\rho}(t)+n_\rho(t), \tag{11.21}$$

where $x_{i,\rho}(t)$ denotes the transmitted signal from attocell i over subcarrier ρ at time t; parameter h_i represents the DC channel gain from attocell i; and $n_\rho(t)$ is the noise signal received by each user on subcarrier ρ. Also, the symbol Π is the set of all the interfering attocells. Here $i=0$ represents the desired attocell transmission for a given user. Thus, the first term indicates the desired signal while the second term accounts for the interference from other attocells. Based on this, a statistical model for the SINR at a given user was developed in Ref. [106] given the randomness of users' locations. Fig. 11.14 shows the analytical and simulation results for the derived cumulative density function (CDF) of the SINR per subcarrier. It is clear that the SINR behavior is strongly affected by the radius of the optical attocell, R, and the half-power transmitting angle of the light source, $\phi_{1/2}$. Hence, for a given R, the value of $\phi_{1/2}$ needs to be carefully adjusted so as to maximize the SINR at the attocell user terminals without introducing excessive interference to the neighboring cells.

11.7.3 Optical code division multiple access

Optical code division multiple access (O-CDMA) is a NOMA scheme that allows simultaneous transmissions to different users by the assignment of code sequences [107]. Bipolar codes commonly used in RF-CDMA systems cannot be directly applied to O-CDMA because of the constraints of IM/DD. One solution to this limitation is

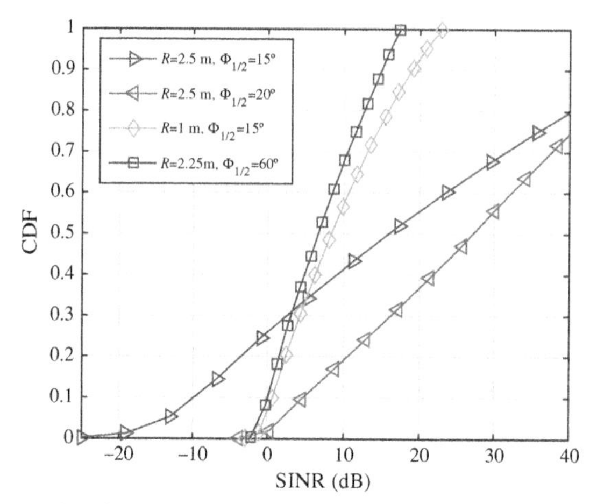

Figure 11.14 Cumulative distribution function (CDF) of the signal-to-interference-plus-noise ratio (SINR) per subcarrier [106].

to modify the bipolar sequences by changing −1 to 0. This, however, might demolish the orthogonality between the code sequences, leading to significant multiple access interference (MAI). Another solution is to use unipolar codes that are specifically designed for O-CDMA systems. Examples of these include random optical codes (ROCs), optical orthogonal codes (OOCs), unipolar *m*-sequences and Walsh-Hadamard codes [103,107,108]. An experimental prototype has been developed in Ref. [108] for an O-CDMA VLC system modulated using OOK. ROCs were assigned to different users with a chip duration $T_c = T/L$, where T is the bit period and L is the code length. During a single bit period, T, only w pulses could be transmitted, where w is the Hamming-weight, and the positions of the w pulses are randomly located in L. Based on this, the code sequence is transmitted only if the data bit is equal to 1, while no transmission is carried out when the data bit is equal to 0. It was shown that the poor correlation characteristics in ROCs can severely affect the achievable system performance, and that ROCs are not optimal for VLC-based O-CDMA systems. Also, the use of OOK results in very limited data rates of the order of 20 kb/s per channel [108].

In Ref. [109], OOCs were utilized to provide multiple access for an indoor VLC system. An OOC, $(L, w, \lambda_a, \lambda_c)$, represents a family of binary sequences with auto- and cross-correlation values λ_a and λ_c, respectively. For any two distinct codewords $b = (b_0, b_1, \ldots, b_{L-1})$ and $d = (d_0, d_1, \ldots, d_{L-1})$, the periodic cross-correlation function can be expressed as

$$\theta_{bd}(\tau) = \sum_{\kappa=0}^{L-1} b_\kappa d_{\kappa \oplus \tau} \leq \lambda_c, \quad 0 \leq \tau \leq L-1, \tag{11.22}$$

where ⊗ denotes the modulo-L addition. OOCs have better correlation characteristics compared to ROCs, but the number of multiplexed users has to be restricted to the number of the available optical orthogonal codewords; this is determined by the Johnson's bound [110]. In order to boost the data rate of O-CDMA-based VLC systems, multicarrier CDMA (MC-CDMA) combines O-CDMA with O-OFDM. As a result, MC-CDMA harnesses the advantage of dense O-OFDM subcarriers along with optical codes so as to boost the system data rate and suppress MAI [107].

11.7.4 Optical space division multiple access

Optical space division multiple access (O-SDMA) is another NOMA scheme that allows each user to fully utilize the available time and frequency blocks. This is achieved by means of an ADT that radiates multiple narrow optical beams to spatially separated users. Fig. 11.15 shows a single-element transmitter versus an ADT. Since the different LED elements are driven by independent electrical signals, each LED conveys an independent data stream to one of the users. The use of ADTs can effectively eliminate ICI as separate beams convey information signals to different users. In Ref. [111], O-SDMA was shown to achieve more than a tenfold throughput enhancement compared to TDMA. Furthermore, the results in Ref. [95] demonstrated that O-SDMA can reach up to 26 times higher average spectral efficiency compared to O-TDMA when the transmitter is equipped with 37 LED elements. This enhancement, however, comes with the added complexity associated with user grouping and resource allocation. Also, O-SDMA requires a more complicated transmitter design in order to supply independent driving signals to the multiple LED elements. Moreover, since the coverage areas of the angle-separated LEDs are smaller compared to the area of a single-LED transmitter, O-SDMA leads to severe handover as the users move within the cells of the ADT.

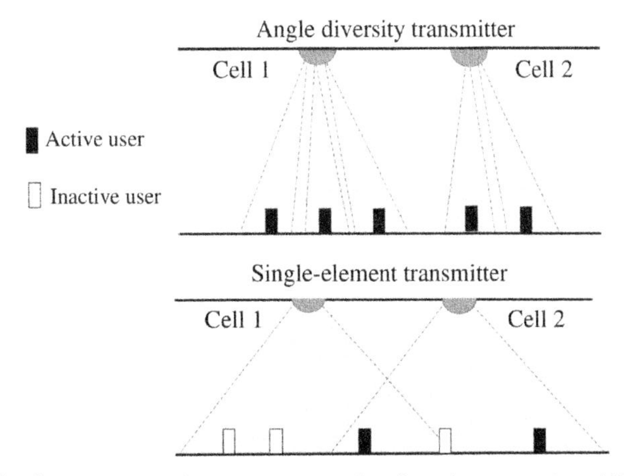

Figure 11.15 Single-element transmitter versus angular-diversity transmitter (ADT) [95].

11.7.5 Power-domain nonorthogonal multiple access

Power-domain NOMA has recently emerged as a spectrum-efficient solution to boost the spectral efficiency of LiFi systems. In power-domain NOMA, the signals of the different users are allocated distinct power values before being superimposed and sent as a single beam. This process is referred to as superposition coding (SC). Fig. 11.16 shows the block diagram of a power-domain NOMA transmitter and receiver. Without loss of generality, assuming that the users $U_1, \ldots, U_N$ are sorted in ascending order according to their channels, that is, $h_1 \leq h_2 \leq \cdots \leq h_N$, power-domain NOMA allocates higher power levels to users with less favorable channel conditions, whereas lower power levels are allocated to users with better channel conditions. Thus, the data signals $s_1,\ldots,s_N$ are allocated power values $P_1,\ldots,P_N$, where s_i conveys information intended for user U_i, where $i = 1,2,\ldots,N$. To this effect, the N transmitted signals are superimposed in the power domain as follows:

$$x = \sum_{i=1}^{N} P_i s_i \tag{11.23}$$

and the sum of the allocated power levels is constrained by the LED total transmit power, that is, $P_{\text{LED}} = \sum_{i=1}^{N} P_i$. At the receiving terminals, multiuser interference at user U_k can be eliminated by means of successive interference cancellation (SIC). Based on this, in order to decode its own signal, U_k needs to successfully decode and subtract the signals of all other users with a lower decoding order, that is, $s_1, \ldots, s_{k-1}$. As a result, the residual interference from $s_{k+1}, \ldots, s_N$ becomes insignificant and can be treated as noise.

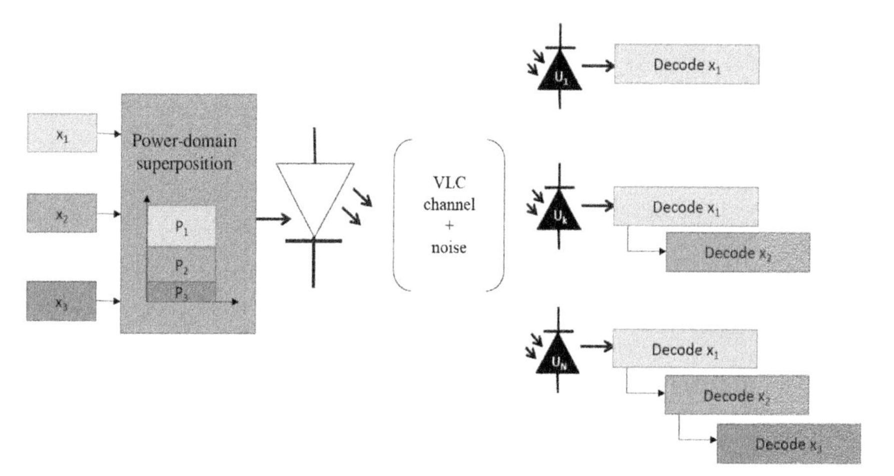

Figure 11.16 Power-domain nonorthogonal multiple access light fidelity (NOMA LiFi) downlink.

It is noted here that power allocation is a key issue in NOMA since appropriate power levels are critical to facilitate SIC and to achieve better trade-off between throughput and fairness. The simplest power allocation strategy is the fixed power allocation (FPA) [112], where users are sorted in an ascending order according to the channel gains. The power assigned to the ith sorted user is then calculated as $P_i = \alpha P_{i-1}$, where $0 < \alpha < 1$. Hence, FPA does not require the exact channel gain values of the users, but rather their respective ordering which mainly depends on the users' distances from the transmitting LED. More sophisticated power allocation techniques have been developed aiming to optimize power allocation according to the users' specific channel conditions [113,114]. It is evident that power-domain NOMA can offer significant capacity gain compared to OMA, which is particularly important for bandwidth-limited LiFi transmissions. For example, it was demonstrated in Ref. [11] that power-domain NOMA can provide a data rate of 2.2 bits per channel use (bpcu) for the case of two multiplexed users, while OFDMA can provide a data rate of only 0.7 bpcu under the same coverage probability. Nonetheless, this gain is constrained by the level of interference affecting the multiplexed users and the errors in the SIC process. As a result, power-domain NOMA is best suited to multiplex a small number of users in order to maintain reliable communication links. This is, in fact, consistent to the nature of LiFi systems where the LiFi attocell is regarded to be a small cell that is used to multiplex only a few users.

11.8 Networking techniques for light fidelity

The ever-increasing demand for high-speed data communication in wireless cellular networks necessitates a further reduction in the cell sizes. The existing highly dense lighting infrastructure and intrinsically limited coverage area of LEDs has propounded the idea of LiFi cellular networks. A LiFi cellular network is composed of many LiFi attocells, which are smaller than RF femtocells. LiFi attocells constitute an additional wireless networking layer within existing heterogeneous wireless networks and they do not interfere with their RF counterparts. LiFi networks are expected to form a substantial part of the heterogeneous future networking landscape beyond 5G and 6G and to carry a large share of wireless Internet connectivity [12]. According to Ref. [115], up to three orders of magnitude throughput improvement can be obtained by LiFi attocell networks compared with RF femtocells. Fig. 11.17 illustrates the concept of a LiFi attocell network. Lights on the ceiling, besides illuminating, act as optical APs forming a cellular network. A single AP can serve multiple user equipments (UEs) simultaneously in its corresponding optical attocell.

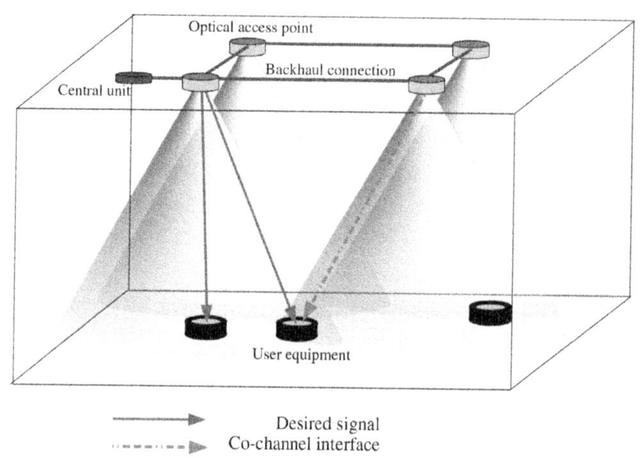

Figure 11.17 Concept of a light fidelity (LiFi) attocell network.

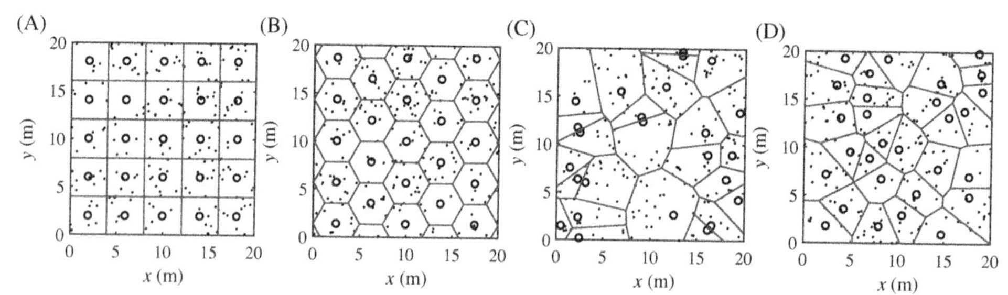

Figure 11.18 Different deployment scenarios: (A) square network, (B) hexagonal network, (C) Poisson point process (PPP) network, (D) hard-core point process (HCPP) network. A room with a size of 20 m × 20 m is considered. The *black circles* represent the positions of the access points (APs), and the *black dots* represent the positions of the user equipments (UEs) [17].

11.8.1 Network deployment

In LiFi attocell networks, the placement of APs affects the system performance. Since lighting and wireless data communication are combined, the location of the APs is mainly determined by the illumination design. Lighting in home and office environments is designed such that the entire space is illuminated in a uniform manner. In Fig. 11.18, four different deployment scenarios are shown. These models are adopted from RF cellular networks with a similar principal optimization objective of complete and uniform signal coverage [17,116,117].

The first potential network deployment in practice is the square lattice model, in which APs are placed on a square lattice as shown in Fig. 11.18A. This arrangement is common for indoor lighting, especially for large offices and public areas. This has several advantages, including design simplicity, illumination uniformity and its compliance with the shape of a room [17,116,117].

Fig. 11.18B shows the conventional hexagonal network deployment that is an idealized model for RF networks. In a hexagonal network, the locations of APs are fully correlated due to spatial symmetries. This guarantees a minimum distance between the desired user and the interfering APs [116]. Therefore cochannel interference (CCI), the interference caused by other APs using the same frequency resources, is minimized in a hexagonal network [116]. However, because of wiring complexity, uneven lighting requirements and aesthetic quality, the use of a hexagonal deployment is not very common for optical attocell networks, especially when lighting and wireless communications are jointly used [17,116,117].

In a typical indoor environment, there are a number of APs that are randomly located (e.g., ceiling luminaries, desktop lamps, and LED screens). Therefore the use of deterministic models for the network performance assessment is not realistic. In such cases, spatial point processes can be used for modeling the topology of the cells in the network. Fig. 11.18C shows the Poisson point process (PPP) cell deployment in which the locations of APs are completely uncorrelated. In this model, the number of APs is assumed to follow a Poisson distribution and the two-dimensional positions of the APs are arranged according to a stationary homogeneous Poisson point process with a fixed density. Each user is then associated with the closest AP. In this topology, APs may be extremely close to each other, causing significant CCI [17,116,117].

The hard-core point process (HCPP) network topology is illustrated in Fig. 11.18D. This model is used when uniform illumination is not required, such as in scenarios of enhanced illumination in particular task areas and reduced lighting in the remaining areas. The HCPP is based on a PPP with the condition that the shortest distance between any two APs is greater than a specified threshold, d_{th}. Therefore the position of APs may be unregulated, but it is unlikely that two APs are placed extremely close to each other [17,116,117].

Fig. 11.19 shows the CDF of the SINR for the above network deployments. The density of each system is assumed to be 0.0353 APs per m^2. The optical power of each LiFi AP is such that the average illuminance in the room is at least 500 lx for reading purposes. Other simulation parameters are provided in Table 11.3.

From this figure, it can be seen that the SINR of the hexagonal network gives the best performance, followed by that of the square network. The random PPP network has the worst SINR performance. By enforcing a minimum distance between APs as in the HCPP network, the SINR performance can be improved. Overall, the results show that the SINR of each user in a LiFi network varies significantly depending on the topology of the optical attocells [17,117].

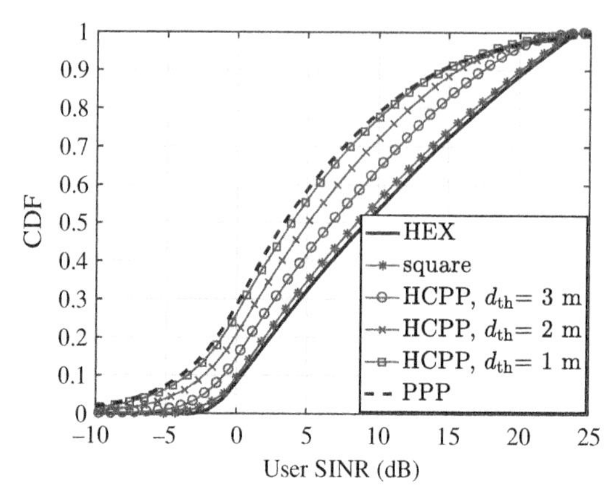

Figure 11.19 Cumulative density function (CDF) comparison of the signal-to-interference-plus-noise ratio (SINR) for different network deployments [17].

Table 11.3 Simulation parameters [17].

Parameters	Values
Vertical separation	2.25 m
LED 3-dB bandwidth	60 MHz
PD responsivity	0.6 A/W
PD physical area	1 cm^2
Receiver FOV	90 degrees
Receiver noise PSD	10^{-19} A^2/Hz

11.8.2 Interference mitigation

In optical attocell networks, the installation of APs very close to each other and the use of the same frequency resources for adjacent APs causes CCI. Since the optical power of a WLED is distributed in accordance with the Lambertian radiation pattern within the LiFi attocell, CCI is primarily expected at the cell edges. However, since the white-light cones of the APs usually overlap, the effect of CCI can be severe. Fig. 11.17 illustrates CCI in an optical attocell network. CCI poses major challenges to the downlink in LiFi attocell networks with dense spatial reuse[15] and significantly degrades the system performance by limiting the SINR of cell-edge users, thereby resulting in higher outage probability and lower data rates. It is therefore of great importance to apply interference mitigation techniques, such as joint transmission (JT), spatial frequency reuse, and busy-burst signaling.

[15] Spatial reuse is the simultaneous transmission of data over two or more spatially nonoverlapping regions in a cellular network. Spatial reuse schemes can improve the area spectral efficiency of multiuser systems.

11.8.2.1 Joint transmission

JT is the concurrent data transmission from multiple coordinated APs to a UE [118]. In this approach, first the primary and secondary APs for the UE are determined based on the received signal strength. Then the received SINR is estimated by assuming the signal from the primary AP as the desired signal and the other signals as the interference. If the SINR is above a predetermined threshold, the UE requests for a single-point transmission from its primary AP. Otherwise, a JT is required. In JT, the secondary APs serve the UE, that is, the interference signals from the secondary APs are substituted with desired signals. Therefore the received SINR for the cell-edge users and the system throughput is significantly improved. A drawback of the JT systems is that they require extra signaling overhead.

11.8.2.2 Spatial frequency reuse

Another interference mitigation technique is spatial frequency reuse [119]. In this method, the entire network is divided into multiple clusters of cells. Each cluster consists of Δ cells. Parameter Δ is also termed as the spatial reuse factor and significantly affects the performance of the corresponding LiFi attocell network. Fig. 11.20A shows an example of the spatial frequency reuse plan with $\Delta = 3$, where different colors of the cells in one cluster represent distinct frequency sub-bands. Since in this plan different frequency sub-bands are used for adjacent cells, a lower CCI level is expected at the cost of a decreased spectral efficiency. This method is also known as static resource partitioning (SRP).

To achieve a higher spectral efficiency, the fractional frequency reuse (FFR) plan can be applied [119]. There are two typical FFR schemes: (1) strict fractional frequency reuse (sFFR) and (2) soft frequency reuse (SFR). In sFFR, the whole frequency band is divided into multiple protected sub-bands and one common

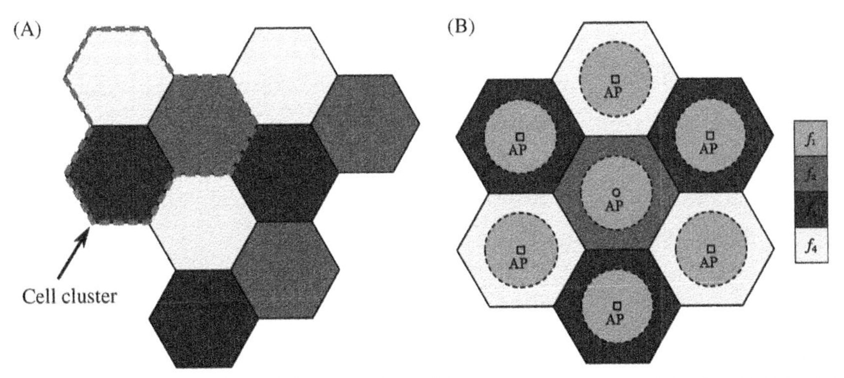

Figure 11.20 An example of spatial reuse plan: (A) static resource partitioning (SRP) with $\Delta = 3$ [12], (B) fractional frequency reuse (FFR) with $\Delta = 1$ for cell-center regions and $\Delta = 3$ for cell-edge regions [102].

sub-band. The common sub-band is assigned to the cell-center users, while the protected sub-bands are assigned to the cell-edge users. The reason is that the cell-center users experience lower levels of interference from nearby APs. The protected sub-bands are arranged such that the geometric reuse distance of each sub-band is maximized. Similarly, in SFR, different sub-bands are used for cell-edge users in adjacent cells. However, the sub-bands that are assigned to the cell-edge users can also be allocated to the cell-center users of an adjacent cell. Also, the transmission power is higher for the cell-edge users because of a higher path loss. Note that in the FFR scheme, the serving AP decides whether the user is at the center or at the edge of the cell by determining the received average power of a downlink pilot signal. If the received power is higher than a threshold value, this particular user is categorized as a cell-center user. Otherwise, it is categorized as a cell-edge user. With the FFR plan, CCI is effectively mitigated and therefore, the downlink SINR is improved. The FFR scheme also offers a significant improvement in the cell-edge spectral efficiency. In addition, it does not require precise instantaneous CSI and has low computational complexity [119].

11.8.2.3 Busy-burst signaling

The busy-burst signaling is proposed in Ref. [120]. In this method, when a UE intends to receive data from the desired AP, a busy-burst signal is broadcast in order that neighboring APs avoid data transmission. Hence, each AP in the system can acquire knowledge of the interference without any central supervision. This method offers improvements in the SINR of cell-edge users and in the average spectral efficiency, but requires additional overhead for the busy-burst slots.

There are also some advanced CCI mitigation techniques in which the LiFi APs are operated by a central controller within the server of a software-defined network (SDN) [121]. In such methods, the central controller adaptively allocates signal power, frequency, time and wavelength resources to achieve multiuser access.

11.8.3 Handover

In cellular networks, in order to provide a seamless connectivity for a mobile UE, the wireless connection may need to be transferred from one AP to another. This process is known as handover. A handover is generally required when: (1) the UE moves between the neighboring cells, that is, exits from the coverage area of one AP and enters that of an adjacent AP; (2) the transmission channel is severely degraded due to a potential rotation of the UE, interference or blockage; and (3) the serving AP is fully loaded and cannot provide any service to more users.

The above handover scenarios are referred to as "horizontal handover," as they occur between APs in the same network. In a heterogeneous network with multiple access technologies such as wireless fidelity (WiFi), LTE, and LiFi, the so-called 'vertical handover' occurs between APs of different network layers. For example, when a UE quits the LiFi coverage in a room, a seamless handover from LiFi to WiFi may occur [12,122,123].

To decide when a handover should take place and to update the users' AP allocation, handover management schemes are used. Generally, handover schemes are categorized as hard and soft. Hard handover schemes follow a break-before-make rule, that is, the UE's connection to the current AP is broken before it connects to the next AP. Soft handover schemes are based on a make-before-break rule. The UE is disconnected from the current AP only after a connection to the next AP is successfully established. Soft schemes provide better user experience at the expense of higher hardware complexity. They also require more wireless transmission resources for preventing service interruptions [12,124].

The handover schemes in the context of LiFi networks rely on estimating the UE position [125] or the received signal strength (RSS) from an AP [126]. The latter is more widely used due to its simplicity. In Ref. [127], the handover probability due to the random movement and rotation of a UE in a LiFi cellular network is studied. In Ref. [128], the handover probability in a hybrid LiFi/RF network is investigated. Some controlling parameters such as a threshold and a hysteresis level are used to avoid unnecessary handovers. This helps to reduce the ping-pong effect due to the rapid fluctuations of the link quality, especially for UEs located near the cell boundaries. Both Refs. [127] and [128] use a RSS-based hard handover scheme. In the RSS-based approach, a handover decision is made when there is a new AP in the receiver FOV whose RSS is greater than that of the primary AP, assuming that all APs transmit the same power. The RSS is directly proportional to the DC gain of the corresponding channel between the AP and the UE receiver. In the following, the RSS-based handover decision method is briefly discussed.

Let $\mathbf{H} = [H_i]$, for $i = 1, 2, \ldots, N_{\mathrm{AP}}$, be the DC gain vector for the channels from APs to the UE receiver at one time sample, where N_{AP} is the total number of APs. The UE receiver is connected to AP_j where $j = \arg_i \max(\mathbf{H})$. The DC gain vector is updated at each time sample while the UE rotates or moves within the network. For the next time sample, denoting j' as the updated index of the AP with the maximum DC gain, a handover occurs if $j' \neq j$. The probability of handover is then given by [127]:

$$P_{\mathrm{handover}} = \Pr\left\{ \bigcup_{j' \neq j} H_{j'} > H_j \right\} \tag{11.24}$$

With specific mathematical models commonly used for describing the UE movements and rotations, P_{handover} can be calculated accordingly.

A handover scenario including one UE and four APs is illustrated in Fig. 11.21. Let us denote the angle of rotation about the x, y and z axes by ω_x, ω_y, and ω_z,

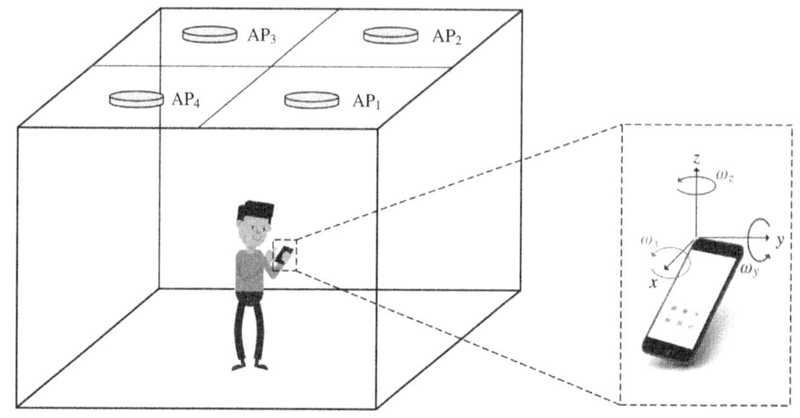

Figure 11.21 A handover scenario: one user equipment (UE) and four access points (APs) in a room. When the orientation of the UE changes, a handover may happen.

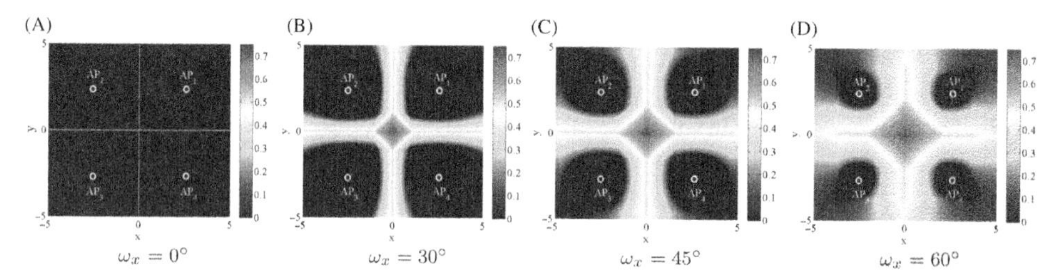

Figure 11.22 Handover probability at different positions in the network due to the random rotation of user equipment (UE) about the *z*-axis, for $\omega_x = 0$, 30, 45, 60, and $\omega_y = 0$ degrees. Other parameters are $\phi_{1/2} = 60$ degrees and $\psi_{\mathrm{FOV}} = 90$ degrees [127].

respectively. Fig. 11.22 illustrates the handover probability due to the UE rotation about the z-axis, at different positions in a room of size $10 \times 10 \times 2.15\ \mathrm{m}^3$. The network area is divided into four quadrants with one AP at the center of each quadrant. It is assumed that in each quadrant, the UE is initially connected to its corresponding AP. The angle ω_z changes according to a uniform distribution in $[0, 2\pi]$. Depending on the position of the UE, ω_x and ω_y, the downlink connection is handed over to one of the three neighboring APs, as ω_z changes. Note that there is a blue zone at the center of each quadrant where the handover probability is equal to zero, and the UE is connected to the corresponding AP for that quadrant regardless of the value of ω_z. This blue zone is smaller for larger values of ω_x, since the DC gain of the channel between the UE and the serving AP is smaller. Therefore a handover happens with a higher probability.

In networks with small number of users, a handover generally happens due to the UE's movements (i.e., a change in the location and/or the orientation of the UE) [127,128]. When the network load is high, users may experience handovers more frequently. For example, if a UE is handed over from AP_1 to AP_2, it will increase the load on AP_2. If AP_2 is highly loaded, then other UEs served by AP_2 may have to be transferred to neighboring APs to decrease the load of the corresponding attocell and to enhance the data rates for the remaining users. In Ref. [122], a dynamic load-balancing scheme in a hybrid LiFi/RF system is presented which considers a handover overhead. However, the handover overhead degrades the system throughput. In Ref. [123], a fuzzy logic (FL) based dynamic handover scheme is proposed to reduce the handover overhead in a hybrid LiFi/RF network.

Although this topic has been well studied for RF cellular networks, for LiFi networks very little research has been reported so far. LiFi attocell networks use ultra-small cells potentially resulting in very frequent handovers, and the orientation of the UE may also result in a handover due to the directional propagation characteristics of the light and also the limited FOV of the receiver. Therefore the handover rate is expected to be higher. This degrades the system throughput and the quality of service. It is then of major importance to properly model the handover processes and develop fast and efficient handover algorithms for LiFi attocell networks. There is much work yet to be done on this topic, and significant contributions are expected in the future.

11.9 Conclusions

The exponential increase in the demand for wireless data traffic has led to new research and the development of novel wireless communication systems that use the IR and visible-light parts of the electromagnetic spectrum. VLC has been widely acknowledged as a potential complementary technology to traditional RF systems for point-to-point data transfer. However, future applications such as autonomous vehicles, industry 4.0, virtual, mixed, and augmented reality will require ubiquitous high-speed and wide-coverage wireless networks able to serve many mobile devices. LiFi has been conceived to achieve this by extending VLC systems to provide bidirectional networking.

Extensive research on the physical layer of VLC and LiFi systems during the last decade has shown the feasibility of delivering data rates of tens and hundreds of Gb/s. In this chapter, channel modeling techniques for VLC were reviewed. Channel models were the basis for the design of analog optical front ends. In this context, high-speed μLEDs and LDs were discussed with a communication bandwidth up to the order of GHz. Furthermore, SPADs were introduced as potential receivers for long-distance VLC with ambient light of low power levels. PV cells were shown to be suitable components for simultaneous energy harvesting and data reception. Sophisticated digital modulation schemes such as O-OFDM have been developed to offer enhanced spectral and power efficiency meeting the illumination requirements for VLC. It has been shown that MIMO techniques are able to boost the achievable data rates at the cost of increased hardware and computational complexity. New approaches for channel modeling and multiuser access have also been highlighted for LiFi networks which enable seamless coverage by supporting user mobility. The method of NOMA introduced for RF systems has been considered for VLC systems because of the high spectral efficiency compared to TDMA, OFDMA, and SDMA. A LiFi network consists of multiple optical access points, referred to as attocells, causing interference. Therefore interference mitigation methods have been introduced to ensure high data densities for the entire coverage area.

References

[1] P.J. Winzer, D.T. Neilson, From scaling disparities to integrated parallelism: a decathlon for a decade, J. Lightwave Technol. 35 (5) (2017) 1099–1115.
[2] H. Claussen, L.T.W. Ho, L.G. Samuel, An overview of the femtocell concept, Bell Labs Tech. J. 13 (1) (2008) 221–245.
[3] D. Tsonev, S. Videv, H. Haas, Towards a 100 Gb/s visible light wireless access network, Opt. Exp. 23 (2) (2015) 1627–1637.
[4] M.S. Islim, R.X. Ferreira, X. He, E. Xie, S. Videv, S. Viola, et al., Towards 10 Gb/s orthogonal frequency division multiplexing-based visible light communication using a GaN violet micro-LED, Photonics Res. 5 (2) (2017) A35–A43.
[5] T. Koonen, Indoor optical wireless systems: technology, trends, and applications, J. Lightwave Technol. 36 (8) (2018) 1459–1467.
[6] N. Serafimovski, R. Lacroix, M. Perrufel, S. Leroux, S. Clement, N. Kundu, et al., Light communications for wireless local area networking, IEEE 5G Tech. Focus 2 (2) (2018).
[7] H. Rubinsztein-Dunlop, A. Forbes, M.V. Berry, M.R. Dennis, D.L. Andrews, M. Mansuripur, et al., Roadmap on structured light, J. Opt. 19 (1) (2017) 013001. URL http://stacks.iop.org/2040-8986/19/i = 1/a = 013001.
[8] H. Haas, High-speed wireless networking using visible light, SPIE Newsroom (2013). Available from: https://doi.org/10.1117/2.1201304.004773.
[9] L. Yin, H. Haas, Physical-layer security in multiuser visible light communication networks, IEEE J. Sel. Areas Commun. 36 (1) (2018) 162–174.
[10] A. Rashidi, M. Monavarian, A. Aragon, A. Rishinaramangalam, D. Feezell, Nonpolar m-plane InGaN/GaN micro-scale light-emitting diode with 1.5 GHz modulation bandwidth, IEEE Electron Device Lett. 39 (4) (2018) 520–523.
[11] L. Yin, W.O. Popoola, X. Wu, H. Haas, Performance evaluation of non-orthogonal multiple access in visible light communication, IEEE Trans. Commun. 64 (12) (2016) 5162–5175.
[12] H. Haas, C. Chen, D. O'Brien, A guide to wireless networking by light, Prog. Quantum Electron. 55 (2017) 88–111.
[13] H. Kazemi, M. Safari, H. Haas, A wireless backhaul solution using visible light communication for indoor Li-Fi attocell networks, in: Proc. IEEE Int. Conf. Commun., Paris, France, 2017, pp. 1–7.
[14] C. Danakis, M. Afgani, G. Povey, I. Underwood, H. Haas, Using a CMOS camera sensor for visible light communication, in: Proc. IEEE Global Commun. Conf. Workshops, Anaheim, CA, USA, 2012, pp. 1244–1248.
[15] P. Luo, M. Zhang, Z. Ghassemlooy, S. Zvanovec, S. Feng, P. Zhang, Undersampled-based modulation schemes for optical camera communications, IEEE Commun. Mag. 56 (2) (2018) 204–212.
[16] T. Fath, F. Schubert, H. Haas, Wireless data transmission using visual codes, Photonics Res. 2 (5) (2014) 150–160.
[17] H. Haas, L. Yin, Y. Wang, C. Chen, What is LiFi? J. Lightwave Technol. 34 (6) (2016) 1533–1544.
[18] J.M. Kahn, J.R. Barry, Wireless infrared communications, Proc. IEEE 85 (2) (1997) 265–298.

[19] C. Chen, D. Basnayaka, X. Wu, H. Haas, Efficient analytical calculation of non-line-of-sight channel impulse response in visible light communications, J. Lightwave Technol. 36 (9) (2018) 1666–1682.
[20] T. Komine, M. Nakagawa, Fundamental analysis for visible-light communication system using LED lights, IEEE Trans. Consum. Electron. 50 (1) (2004) 100–107.
[21] J.R. Barry, J.M. Kahn, W.J. Krause, E.A. Lee, D.G. Messerschmitt, Simulation of multipath impulse response for indoor wireless optical channels, IEEE J. Sel. Areas Commun. 11 (3) (1993) 367–379.
[22] K. Lee, H. Park, J.R. Barry, Indoor channel characteristics for visible light communications, IEEE Commun. Lett. 15 (2) (2011) 217–219.
[23] C. Chen, D. Basnayaka, H. Haas, Non-line-of-sight channel impulse response characterisation in visible light communications, in: Proc. IEEE Int. Conf. Commun., Kuala Lumpur, Malaysia, 2016, pp. 1–6.
[24] F.J. Lopez-Hernandez, M.J. Betancor, DUSTIN: algorithm for calculation of impulse response on IR wireless indoor channels, Electron. Lett. 33 (21) (1997) 1804–1806.
[25] F.J. Lopez-Hernandez, R. Perez-Jimenez, A. Santamaria, Ray-tracing algorithms for fast calculation of the channel impulse response on diffuse IR wireless indoor channels, Opt. Eng. 39 (10) (2000) 2775–2781.
[26] Zemax L.L.C., OpticStudio, 2018. Available from <https://www.zemax.com/products/opticstudio>.
[27] F. Miramirkhani, M. Uysal, Channel modeling and characterization for visible light communications, IEEE Photonics J. 7 (6) (2015) 1–16.
[28] V. Jungnickel, V. Pohl, S. Nonnig, C. von Helmolt, A physical model of the wireless infrared communication channel, IEEE J. Sel. Areas Commun. 20 (3) (2002) 631–640.
[29] J.B. Carruthers, J.M. Kahn, Modeling of nondirected wireless infrared channels, IEEE Trans. Commun. 45 (10) (1997) 1260–1268.
[30] D. Karunatilaka, F. Zafar, V. Kalavally, R. Parthiban, LED based indoor visible light communications: state of the art, IEEE Commun. Surv. Tuts. 17 (3) (2015) 1649–1678.
[31] E.F. Schubert, Light-Emitting Diodes, third ed., Cambridge University Press, Cambridge, 2018.
[32] W.O. Popoola, Impact of VLC on light emission quality of white LEDs, J. Lightwave Technol. 34 (10) (2016) 2526–2532.
[33] A.M. Khalid, G. Cossu, R. Corsini, P. Choudhury, E. Ciaramella, 1-Gb/s transmission over a phosphorescent white LED by using rate-adaptive discrete multitone modulation, IEEE Photonics J. 4 (5) (2012) 1465–1473.
[34] G. Cossu, A.M. Khalid, P. Choudhury, R. Corsini, E. Ciaramella, 3.4 Gbit/s visible optical wireless transmission based on RGB LED, Opt. Express 20 (26) (2012) B501–B506.
[35] J. Grubor, S. Randel, K. Langer, J.W. Walewski, Broadband information broadcasting using LED-based interior lighting, J. Lightwave Technol. 26 (24) (2008) 3883–3892.
[36] X. Zhu, F. Wang, M. Shi, N. Chi, J. Liu, F. Jiang, 10.72 Gb/s visible light communication system based on single packaged RGBYC LED utilizing QAM-DMT modulation with hardware pre-equalization, in: Proc. Opt. Fiber Commun. Conf., San Diego, CA, USA, 2018, pp. 1–3.
[37] J. Vucic, C. Kottke, K. Habel, K. Langer, 803 Mbit/s visible light WDM link based on DMT modulation of a single RGB LED luminary, in: Proc. Opt. Fiber Commun. Conf. Expo., Los Angeles, CA, USA, 2011, pp. 1–3.
[38] J.J.D. McKendry, R.P. Green, A.E. Kelly, Z. Gong, B. Guilhabert, D. Massoubre, et al., High-speed visible light communications using individual pixels in a micro light-emitting diode array, IEEE Photonics Technol. Lett. 22 (18) (2010) 1346–1348.
[39] J.J.D. McKendry, D. Massoubre, S. Zhang, B.R. Rae, R.P. Green, E. Gu, et al., Visible-light communications using a CMOS-controlled micro-light-emitting-diode array, J. Lightwave Technol. 30 (1) (2012) 61–67.
[40] D. Tsonev, H. Chun, S. Rajbhandari, J.J.D. McKendry, S. Videv, E. Gu, et al., A 3-Gb/s single-LED OFDM-based wireless VLC link using a gallium nitride μLED, IEEE Photonics Technol. Lett. 26 (7) (2014) 637–640.

[41] R.X.G. Ferreira, E. Xie, J.J.D. McKendry, S. Rajbhandari, H. Chun, G. Faulkner, et al., High bandwidth GaN-based micro-LEDs for multi-Gb/s visible light communications, IEEE Photonics Technol. Lett. 28 (19) (2016) 2023–2026.
[42] B. Schrenk, M. Hofer, F. Laudenbach, H. Hbel, T. Zemen, Visible-light multi-Gb/s transmission based on resonant cavity LED with optical energy feed, IEEE J. Sel. Areas Commun. 36 (1) (2018) 175–184.
[43] P.A. Haigh, Z. Ghassemlooy, S. Rajbhandari, I. Papakonstantinou, Visible light communications using organic light emitting diodes, IEEE Commun. Mag. 51 (8) (2013) 148–154.
[44] S. Dimitrov, S. Sinanovic, H. Haas, Clipping noise in OFDM-based optical wireless communication systems, IEEE Trans. Commun. 60 (4) (2012) 1072–1081.
[45] S. Dimitrov, H. Haas, Information rate of OFDM-based optical wireless communication systems with nonlinear distortion, J. Lightwave Technol. 31 (6) (2013) 918–929.
[46] D. Tsonev, S. Sinanovic, H. Haas, Complete modeling of nonlinear distortion in OFDM-based optical wireless communication, J. Lightwave Technol. 31 (18) (2013) 3064–3076.
[47] T. Fath, C. Heller, H. Haas, Optical wireless transmitter employing discrete power level stepping, J. Lightwave Technol. 31 (11) (2013) 1734–1743.
[48] A.V.N. Jalajakumari, E. Xie, J. McKendry, E. Gu, M.D. Dawson, H. Haas, et al., High-speed integrated digital to light converter for short range visible light communication, IEEE Photonics Technol. Lett. 29 (1) (2017) 118–121.
[49] F. Zafar, M. Bakaul, R. Parthiban, Laser-diode-based visible light communication: toward gigabit class communication, IEEE Commun. Mag. 55 (2) (2017) 144–151.
[50] A. Neumann, J.J. Wierer, W. Davis, Y. Ohno, S.R.J. Brueck, J. Tsao, Four-color laser white illuminant demonstrating high color-rendering quality, Opt. Express 19 (S4) (2011) A982–A990.
[51] C. Lee, C. Shen, C. Cozzan, R.M. Farrell, S. Nakamura, A.Y. Alyamani, et al., Semipolar GaN-based laser diodes for Gbit/s white lighting communication: devices to systems, in: Proc. SPIE, Vol. 10532, San Francisco, CA, USA, 2018, pp. N1–N11.
[52] R.S. Quimby, Photonics and Lasers: An Introduction, Wiley, New York, NY, 2006.
[53] Y. Li, M. Safari, R. Henderson, H. Haas, Optical OFDM with single-photon avalanche diode, IEEE Photonics Technol. Lett. 27 (9) (2015) 943–946.
[54] O. Almer, D. Tsonev, N.A.W. Dutton, T.A. Abbas, S. Videv, S. Gnecchi, et al., A SPAD-based visible light communications receiver employing higher order modulation, in: Proc. IEEE Global Commun. Conf., San Diego, CA, USA, 2015, pp. 1–6.
[55] E. Sarbazi, M. Safari, H. Haas, Statistical modeling of single-photon avalanche diode receivers for optical wireless communications, IEEE Trans. Commun. 66 (9) (2018) 4043–4058.
[56] M.S. Islim, S. Videv, M. Safari, E. Xie, J.J.D. McKendry, E. Gu, et al., The impact of solar irradiance on visible light communications, J. Lightwave Technol. 36 (12) (2018) 2376–2386.
[57] O. Höhn, A.W. Walker, A.W. Bett, H. Helmers, Optimal laser wavelength for efficient laser power converter operation over temperature, Appl. Phys. Lett. 108 (24) (2016) 241104.
[58] S.-M. Kim, J.-S. Won, Simultaneous reception of visible light communication and optical energy using a solar cell receiver, in: Proc. IEEE Int. Conf. ICT Convergence, Jeju Island, South Korea, 2013, pp. 896–897.
[59] Z. Wang, D. Tsonev, S. Videv, H. Haas, Towards self-powered solar panel receiver for optical wireless communication, in: Proc. IEEE Int. Conf. Commun., Sydney, NSW, Australia, 2014, pp. 3348–3353.
[60] Z. Wang, D. Tsonev, S. Videv, H. Haas, On the design of a solar-panel receiver for optical wireless communications with simultaneous energy harvesting, IEEE J. Sel. Areas Commun. 33 (8) (2015) 1612–1623.
[61] J. Fakidis, S. Videv, H. Helmers, H. Haas, 0.5-Gb/s OFDM-based laser data and power transfer using a GaAs photovoltaic cell, IEEE Photonics Technol. Lett. 30 (9) (2018) 841–844.
[62] H. Haas, S. Videv, S. Das, J. Fakidis, H. Stewart, Solar cell receiver free-space optical for 5G backhaul, in: Proc. Opt. Fiber Commun. Conf., San Diego, CA, USA, 2019, M3G.2.
[63] S. Dimitrov, S. Sinanovic, H. Haas, Signal shaping and modulation for optical wireless communication, J. Lightwave Technol. 30 (9) (2012) 1319–1328.

[64] P.A. Haigh, A. Burton, K. Werfli, H.L. Minh, E. Bentley, P. Chvojka, et al., A multi-CAP visible-light communications system with 4.85-b/s/Hz spectral efficiency, IEEE J. Sel. Areas Commun. 33 (9) (2015) 1771–1779.
[65] F.M. Wu, C.T. Lin, C.C. Wei, C.W. Chen, H.T. Huang, C.H. Ho, 1.1-Gb/s white-LED-based visible light communication employing carrier-less amplitude and phase modulation, IEEE Photonics Technol. Lett 24 (19) (2012) 1730–1732.
[66] K.I. Ahn, J.K. Kwon, Color intensity modulation for multicolored visible light communications, IEEE Photonics Technol. Lett. 24 (24) (2012) 2254–2257.
[67] W. Xu, J. Wang, H. Shen, H. Zhang, Multi-colour LED specified bipolar colour shift keying scheme for visible light communications, Electron. Lett. 52 (2) (2016) 133–135.
[68] M. Afgani, H. Haas, H. Elgala, D. Knipp, Visible light communication using OFDM, in: Proc. IEEE 2nd Int. Conf. Testbeds Res. Infrastructures Develop. Netw. Communities, Barcelona, Spain, 2006, pp. 129–134.
[69] S.D. Dissanayake, J. Armstrong, Comparison of ACO-OFDM, DCO-OFDM and ADO-OFDM in IM/DD systems, J. Lightwave Technol. 31 (7) (2013) 1063–1072.
[70] Y. Jiang, Y. Wang, P. Cao, M. Safari, J. Thompson, H. Haas, Robust and low-complexity timing synchronization for DCO-OFDM LiFi systems, IEEE J. Sel. Areas Commun. 36 (1) (2018) 53–65.
[71] X. Li, J. Vucic, V. Jungnickel, J. Armstrong, On the capacity of intensity-modulated direct-detection systems and the information rate of ACO-OFDM for indoor optical wireless applications, IEEE Trans. Commun. 60 (3) (2012) 799–809.
[72] S.C.J. Lee, S. Randel, F. Breyer, A.M.J. Koonen, PAM-DMT for intensity-modulated and direct-detection optical communication systems, IEEE Photonics Technol. Lett. 21 (23) (2009) 1749–1751.
[73] D. Tsonev, S. Sinanovic, H. Haas, Novel unipolar orthogonal frequency division multiplexing (U-OFDM) for optical wireless, in: Proc. IEEE 75th Veh. Technol. Conf., Yokohama, Japan, 2012, pp. 1–5.
[74] D. Tsonev, H. Haas, Avoiding spectral efficiency loss in unipolar OFDM for optical wireless communication, in: Proc. IEEE Int. Conf. Commun., Sydney, Australia, 2014, pp. 3336–3341.
[75] D. Tsonev, S. Videv, H. Haas, Unlocking spectral efficiency in intensity modulation and direct detection systems, IEEE J. Sel. Areas Commun. 33 (9) (2015) 1758–1770.
[76] M. Islim, D. Tsonev, H. Haas, A generalized solution to the spectral efficiency loss in unipolar optical OFDM-based systems, in: Proc. IEEE Int. Conf. Commun., London, U.K., 2015, pp. 5126–5131.
[77] Y. Li, D. Tsonev, H. Haas, Non-DC-biased OFDM with optical spatial modulation, in: Proc. IEEE 24th Annu. Int. Symp. Personal Indoor Mobile Radio Commun., London, U.K., 2013, pp. 486–490.
[78] A. Sewaiwar, P.P. Han, Y.H. Chung, 3-Gbit/s indoor visible light communications using optical diversity schemes, IEEE Photonics J. 7 (6) (2015) 1–9.
[79] M.M.A. Mohammed, C. He, J. Armstrong, Performance analysis of ACO-OFDM and DCO-OFDM using bit and power loading in frequency selective optical wireless channels, in: Proc. IEEE 85th Veh. Technol. Conf., Sydney, NSW, Australia, 2017, pp. 1–5.
[80] J. Proakis, Digital Communications, fifth ed., McGraw Hill, 2007.
[81] D. Bykhovsky, S. Arnon, An experimental comparison of different bit-and-power-allocation algorithms for DCO-OFDM, J. Lightwave Technol. 32 (8) (2014) 1559–1564.
[82] L. Wu, Z. Zhang, J. Dang, H. Liu, Adaptive modulation schemes for visible light communications, J. Lightwave Technol. 33 (1) (2015) 117–125.
[83] A. Goldsmith, Wireless Communications, Cambridge University Press, 2005.
[84] L. Zeng, D.C. O'Brien, H.L. Minh, G.E. Faulkner, K. Lee, D. Jung, et al., High data rate multiple input multiple output (MIMO) optical wireless communications using white LED lighting, IEEE J. Sel. Areas Commun. 27 (9) (2009) 1654–1662.
[85] T. Fath, H. Haas, Performance comparison of MIMO techniques for optical wireless communications in indoor environments, IEEE Trans. Commun. 61 (2) (2013) 733–742.

[86] D. O'Brien, Optical multi-input multi-output systems for short-range free-space data transmission, in: Proc. 7th Int. Symp. Commun. Syst. Netw. Digit. Signal Process., Newcastle upon Tyne, U. K., 2010, pp. 517–521.
[87] A. Stavridis, H. Haas, Performance evaluation of space modulation techniques in VLC systems, in: Proc. IEEE Int. Conf. Commun. Workshops, London, UK, 2015, pp. 1356–1361.
[88] R. Mesleh, H. Haas, C.W. Ahn, S. Yun, Spatial modulation – a new low complexity spectral efficiency enhancing technique, in: Proc. 1st Int. Conf. Commun. Netw., Beijing, China, 2006, pp. 1–5.
[89] M. Di Renzo, H. Haas, A. Ghrayeb, S. Sugiura, L. Hanzo, Spatial modulation for generalized MIMO: challenges, opportunities, and implementation, Proc. IEEE 102 (1) (2014) 56–103.
[90] R. Mesleh, H. Elgala, H. Haas, Optical spatial modulation, IEEE J. Opt. Commun. Netw. 3 (3) (2011) 234–244.
[91] A. Younis, N. Serafimovski, R. Mesleh, H. Haas, Generalised spatial modulation, in: Conf. Rec. 44th Asilomar Conf. Signals, Syst. Comput., Pacific Grove, CA, USA, 2010, pp. 1498–1502.
[92] A. Yesilkaya, E. Basar, F. Miramirkhani, E. Panayirci, M. Uysal, H. Haas, Optical MIMO-OFDM with generalized LED index modulation, IEEE Trans. Commun. 65 (8) (2017) 3429–3441.
[93] K.H. Park, H.M. Oubei, W.G. Alheadary, B.S. Ooi, M.S. Alouini, A novel mirror-aided non-imaging receiver for indoor 2 × 2 MIMO-visible light communication systems, IEEE Trans. Wirel. Commun. 16 (9) (2017) 5630–5643.
[94] A. Azhar, T. Tran, D. O'Brien, A gigabit/s indoor wireless transmission using MIMO-OFDM visible-light communications, IEEE Photonics Technol. Lett 25 (2) (2013) 171–174.
[95] Z. Chen, D.A. Basnayaka, H. Haas, Space division multiple access for optical attocell network using angle diversity transmitters, J. Lightwave Technol. 35 (11) (2017) 2118–2131.
[96] A. Nuwanpriya, S.W. Ho, C.S. Chen, Indoor MIMO visible light communications: novel angle diversity receivers for mobile users, IEEE J. Sel. Areas Commun. 33 (9) (2015) 1780–1792.
[97] P.F. Mmbaga, J. Thompson, H. Haas, Performance analysis of indoor diffuse VLC MIMO channels using angular diversity detectors, J. Lightwave Technol. 34 (4) (2016) 1254–1266.
[98] Z. Chen, D. Basnayaka, X. Wu, H. Haas, Interference mitigation for indoor optical attocell networks using an angle diversity receiver, J. Lightwave Technol. 36 (18) (2018) 3866–3881.
[99] C.A. Brackett, Dense wavelength division multiplexing networks: principles and applications, IEEE J. Sel. Areas Commun. 8 (6) (1990) 948–964.
[100] H. Chun, S. Rajbhandari, G. Faulkner, D. Tsonev, E. Xie, J.J.D. McKendry, et al., LED based wavelength division multiplexed 10 Gb/s visible light communications, IEEE, J. Lightwave Technol. 34 (13) (2016) 3047–3052.
[101] T.-C. Wu, Y.-C. Chi, H.-Y. Wang, C.-T. Tsai, Y.-F. Huang, G.-R. Lin, Tricolor R/G/B laser diode based eye-safe white lighting communication beyond 8 Gbit/s, Sci. Rep. 7 (11) (2017) 2045–2322.
[102] H. Kazemi, H. Haas, Downlink cooperation with fractional frequency reuse in DCO-OFDMA optical attocell networks, in: Proc. IEEE Int. Conf. Commun., Kuala Lumpur, Malaysia, 2016, pp. 1–6.
[103] J. Fakidis, D. Tsonev, H. Haas, A comparison between DCO-OFDMA and synchronous one-dimensional OCDMA for optical wireless communications, in: Proc. IEEE 24th Annu. Int. Symp. Personal Indoor Mobile Radio Commun., London, UK, 2013, pp. 3605–3609.
[104] Y. Wang, X. Wu, H. Haas, Resource allocation in LiFi OFDMA systems, in: Proc. IEEE Global Commun. Conf., Singapore, 2017, pp. 1–6.
[105] X. Ling, J. Wang, Z. Ding, C. Zhao, X. Gao, Efficient OFDMA for LiFi downlink, J. Lightwave Technol. 36 (10) (2018) 1928–1943.
[106] C. Chen, M. Ijaz, D. Tsonev, H. Haas, Analysis of downlink transmission in DCO-OFDM-based optical attocell networks, in: Proc. IEEE Global Commun. Conf., Austin, TX, USA, 2014, pp. 2072–2077.
[107] M. Shoreh, A. Fallahpour, J. Salehi, Design concepts and performance analysis of multicarrier CDMA for indoor visible light communications, IEEE J. Opt. Commun. Netw. 7 (6) (2015) 554–562.

[108] M.F. Guerra-Medina, O. Gonzalez, B. Rojas-Guillama, J.A. Martin-Gonzalez, F. Delgado, J. Rabadan, Ethernet-OCDMA system for multi-user visible light communications, Electron. Lett. 48 (4) (2012) 227–228.
[109] M. Noshad, M. Brandt-Pearce, High-speed visible light indoor networks based on optical orthogonal codes and combinatorial designs, in: Proc. IEEE Global Commun. Conf., Atlanta, GA, USA, 2013, pp. 2436–2441.
[110] S. Johnson, A new upper bound for error-correcting codes, IRE Trans. Inf. Theory 8 (3) (1962) 203–207.
[111] Z. Chen, H. Haas, Space division multiple access in visible light communications, in: Proc. IEEE Int. Conf. Commun., London, UK, 2015, pp. 5115–5119.
[112] R.C. Kizilirmak, C.R. Rowell, M. Uysal, Non-orthogonal multiple access (NOMA) for indoor visible light communications, in: Proc. 4th Int. Workshop Opt. Wireless Commun., Istanbul, Turkey, 2015, pp. 98–101.
[113] H. Marshoud, V.M. Kapinas, G.K. Karagiannidis, S. Muhaidat, Non-orthogonal multiple access for visible light communications, IEEE Photonics Technol. Lett. 28 (1) (2016) 51–54.
[114] Z. Yang, W. Xu, Y. Li, Fair non-orthogonal multiple access for visible light communication downlinks, IEEE Wirel. Commun. Lett. 6 (1) (2017) 66–69.
[115] I. Stefan, H. Burchardt, H. Haas, Area spectral efficiency performance comparison between VLC and RF femtocell networks, in: Proc. IEEE Int. Conf. Commun., Budapest, Hungary, 2013, pp. 3825–3829.
[116] C. Chen, D. Basnayaka, H. Haas, Downlink SINR statistics in OFDM-based optical attocell networks with a Poisson point process network model, in: Proc. IEEE Global Commun. Conf., San Diego, CA, USA, 2015, pp. 1–6.
[117] C. Chen, D.A. Basnayaka, H. Haas, Downlink performance of optical attocell networks, J. Lightwave Technol. 34 (1) (2016) 137–156.
[118] C. Chen, D. Tsonev, H. Haas, Joint transmission in indoor visible light communication downlink cellular networks, in: Proc. IEEE Global Commun. Workshops, Atlanta, GA, USA, 2013, pp. 1127–1132.
[119] C. Chen, S. Videv, D. Tsonev, H. Haas, Fractional frequency reuse in DCO-OFDM-based optical attocell networks, J. Lightwave Technol. 33 (19) (2015) 3986–4000.
[120] P. Omiyi, H. Haas, G. Auer, Analysis of TDD cellular interference mitigation using busy-bursts, IEEE Trans. Wirel. Commun 6 (7) (2007) 2721–2731.
[121] D. Simeonidou, R. Nejabati, M. Channegowda, Software defined optical networks technology and infrastructure: enabling software-defined optical network operations, in: Opt. Fiber Commun. Conf. Expo. and Nat. Fiber Opt. Eng. Conf., Anaheim, CA, USA, 2013, pp. 1–3.
[122] Y. Wang, H. Haas, Dynamic load balancing with handover in hybrid Li-Fi and Wi-Fi networks, J. Lightwave Technol. 33 (22) (2015) 4671–4682.
[123] Y. Wang, X. Wu, H. Haas, Fuzzy logic based dynamic handover scheme for indoor Li-Fi and RF hybrid network, in: Proc. IEEE Int. Conf. Commun., Kuala Lumpur, Malaysia, 2016, pp. 1–6.
[124] M.S. Demir, F. Miramirkhani, M. Uysal, Handover in VLC networks with coordinated multipoint transmission, in: Proc. IEEE Int. Black Sea Conf. Commun. Netw., Istanbul, Turkey, 2017, pp. 1–5.
[125] J. Xiong, Z. Huang, K. Zhuang, Y. Ji, A cooperative positioning with Kalman filters and handover mechanism for indoor microcellular visible light communication network, Opt. Rev. 23 (4) (2016) 683–688.
[126] A.M. Vegni, T.D. Little, Handover in VLC systems with cooperating mobile devices, in: Proc. IEEE Int. Conf. Comput. Netw. Commun., Maui, Hawaii, USA, 2012, pp. 126–130.
[127] M.D. Soltani, H. Kazemi, M. Safari, H. Haas, Handover modeling for indoor Li-Fi cellular networks: the effects of receiver mobility and rotation, in: Proc. IEEE Wireless Commun. Netw. Conf., San Francisco, CA, USA, 2017, pp. 1–6.
[128] A.A. Purwita, M.D. Soltani, H. Haas, Handover probability of hybrid LiFi/RF-based networks with randomly-oriented devices, in: Proc. IEEE Veh. Technol. Conf., Porto, Portugal, 2018, pp. 1–6.

CHAPTER 12

R&D advances for quantum communication systems

Gerd Leuchs[1], Christoph Marquardt[1], Luis L. Sánchez-Soto[1,2] and Dmitry V. Strekalov[1,3]
[1]Max Planck Institute for the Science of Light, Erlangen, Germany
[2]Departamento de Óptica, Facultad de Física, Universidad Complutense, Madrid, Spain
[3]Jet Propulsion Laboratory, Pasadena, CA, United States

12.1 Communication as transfer of information

12.1.1 Introduction to this chapter

Understanding the nature of light leads to the question of how the principles of quantum physics can be harnessed in practical optical communication. A deeper understanding of fundamental physics has always advanced technology. However, the quantum principles certainly have a distinctly limiting character when looked upon from the engineering point of view. A particle cannot have well-defined momentum and position at the same time. An informative measurement will unpredictably alter the state of a quantum object. One cannot reliably clone an arbitrary quantum state. These and a number of other similar principles give rise to what is commonly known as the quantum "no-go theorems"—a disconcerting term when it comes to building something practical. And yet a search for novel principles of communication enabled by quantum physics began already in its early days and has only intensified since. On this path physicists are faced with a remarkable challenge: to turn a series of negative statements into new technological recipes.

This chapter presents the progress of this search achieved over the past several years. The first section starts with a brief review on communication as dispatching of information. We revisit the general concept of information as a physical parameter and the means of transferring it from a sender to a recipient: the communication channels. Principal characteristics of communication channels, such as their information throughput capacity and noise, are discussed in this section.

In Section 12.2 we revisit those principles of quantum physics that have the most important implications for information theory and technology. We start with the notions of uncertainty, entanglement, and the quantum description of amplification and measurement. The aforementioned no-go theorems are shown to arise as a natural

Optical Fiber Telecommunications V11
DOI: https://doi.org/10.1016/B978-0-12-816502-7.00014-2

consequence of these principles. This section helps us to identify the main aspects of optical communication that are most fundamentally affected by the quantum theory: communication security, information transmission capacity, and novel paradigms of information processing. The next three sections report the R&D advances in these areas.

Section 12.3 on quantum communication security is dedicated to the quantum key distribution (QKD). The QKD protocols are claimed to provide unconditional cryptographic security, wherein the information is protected by the laws of nature rather than by the computation complexity. This is an important example of how a quantum no-go theorem—the no-cloning theorem in this case—can be put to a practical use. One might emphasize that the 3 dB noise limit for a high-gain phase-insensitive amplifier used by the telecom industry is ultimately a result of this no-cloning theorem. We start this section by reminding the readers of the QKD basic principles involving discrete and continuous quantum variables. Next, we proceed to more advanced consideration including noise, losses, and various types of eavesdroppers' attacks in QKD channels.

Discussion of the QKD principles started in Section 12.3 leads us into a review of historic and modern QKD R&D, presented in Section 12.4. We emphasize that QKD is perhaps the most technologically mature area of quantum communication, which in some aspects has already reached the commercialization stage. Therefore Section 12.4 may be viewed as central to our chapter. Here we discuss the state-of-the-art QKD techniques and systems in fiber and free-space. Special emphasis is placed on the satellite communication QKD programs that have shown an explosive worldwide growth in the last couple of years.

Finally, Section 12.5 reviews the information technologies enabled by quantum physics that are proven or expected to surpass their classical alternatives. Here we focus on the technologies that go beyond transferring information and tap into information processing—that is, into applying logic. The most famous, but not the only, example of such technology concepts is the quantum computer. The authors are well aware that this topic alone is worthwhile of a sizable book volume. Therefore we only provide a most fundamental outline of the principles underlying quantum computation and review the modern approaches to its realization. Other examples of specialized quantum-logic systems, such as those allowing for creating counterfeit-proof money or conducting truly anonymous decision making, are also discussed.

12.1.2 Information measures

There is a general consensus that information theory was established in 1948 with Shannon's famous paper [1], which answered two simple questions: what is information, and what are the limits on the transmission of information? According to the

Shannon paradigm, the process of communication, that is, of carrying information between two points, includes the following essential elements:

- A sender, which generates the message and a receiver, which is the recipient of that message.
- An encoder, which converts the message into a physical signal, and a decoder, which interprets the signal to convert it into a message.
- A transmission channel, which is the physical medium used to connect the sender to the receiver.

Furthermore, one has to take into account random ambient noise and imperfections of the signaling process. The channel can be modeled as either a continuous channel, which can be characterized by the noise probability density function $w(t)$, or as a discrete channel that accepts symbols xi of an input alphabet X provided by the channel encoder and produces symbols y_j from an output alphabet Y.

Consider a finite alphabet X formed by M symbols $\{x_i\}$ $(i = 1, \ldots, M)$. If each symbol has a probability of occurrence p_i, it carries the amount of information

$$I(x_i) = \log_2 \frac{1}{p_i}. \tag{12.1}$$

If the alphabet consists of two equiprobable symbols x_1 and x_2, then $I(x_1) = I(x_2) = 1$ and this amount of information is one bit. In this case it is customary to represent x_1 by a 0 and x_2 by a 1.

This definition of information satisfies a number of useful properties; viz [2],

- It is intuitive: the occurrence of a highly probable event carries little information $[I(x_i) = 0$ for $p_i = 1]$.
- It is positive: information cannot decrease upon receiving a message $[I(x_i) \geq 0$ for $0 \leq p_i \leq 1]$.
- More information is gained when a less probable message is received $[I(x_i) > I(x_j)$ for $p_i < p_j]$.
- Information is additive when probabilities are multiplicative, that is, if the messages are independent.

In order to characterize the alphabet X, we introduce its Shannon entropy $H(X)$ as the average content of information expressed in bits/symbol:

$$H(X) = \sum_{i=1}^{M} p_i \, I(x_i) = \sum_{i=1}^{M} p_i \log_2 \frac{1}{p_i}. \tag{12.2}$$

Consider a binary emitter, producing two statistically independent symbols with probabilities p and $1 - p$. The average information per message is

$$H = p \log_2 \frac{1}{p} + (1 - p)\, log_2 \frac{1}{1 - p}. \tag{12.3}$$

The Shannon entropy (12.3) is maximized for $p=½$, with a corresponding entropy of 1 bit/symbol. This result can be generalized for M-symbols alphabets, whose entropy also becomes maximum when all of the symbols are equally likely. Then each symbol has probability $1/M$ and the entropy is $H=\log_2 M$.

When encoding each symbol into a binary code word (using binary symbols 0 and 1) a problem appears. In general, the way of encoding each symbol is not unique, which opens up a question how to minimize the average number of bits (binary symbols) used to transmit the message. A classical source encoding example is the Morse code where the letters A–Z, the numbers 0–9, and some punctuation marks are encoded in binary words constituted by dashes and dots.

To bypass this drawback, we define $\overline{L}$ as the average length of the code words,

$$\overline{L}=\sum_{i=1}^{M} p_i \ell_i, \tag{12.4}$$

where ℓ_i is the length in bits of the code word associated with the symbol x_i. It can be demonstrated that in order for the source alphabet X to be encoded and decoded without ambiguity, the average length has to exceed the alphabet's entropy:

$$\overline{L}\geq H(X). \tag{12.5}$$

There are several procedures that can be used to build uniquely decodable codes. The Huffman algorithm [3], proposed in 1952, requires a probabilistic source. However, the Lempel–Ziv algorithm [4], which is adaptive and does not require knowledge of the source distribution, is currently the most popular technique.

12.1.3 Channel capacity

A system is optimal when it minimizes the bit error probability under certain constraints imposed on the transmitted energy and channel bandwidth. The channel capacity of a given channel is the highest information rate that can be achieved with arbitrarily small error probability.

Let us start this discussion with the case of a continuous channel. A basic noise model used in information theory to mimic the effect of many random processes that occur in nature is the so called additive white Gaussian noise (AWGN). AWGN is often used as a model in which the only impairment to communication is a linear addition of wideband or white noise with a constant spectral density and a Gaussian distribution of amplitude. The model does not account for fading, frequency selectivity, interference, nonlinearity, or dispersion. However, it produces simple and tractable mathematical results. In fact, the AWGN-limited channel capacity (in bits/s) is [5]

$$C_{\mathrm{AWGN}}=B\log_2(1+\Sigma), \tag{12.6}$$

where B is the bandwidth, $\Sigma = \overline{P}/N_0 B$ is the received signal-to-noise ratio (SNR), $\overline{P}$ is the average received power, and N_0 is the power spectral density. Since $\overline{P} = E_b R_b$, where E_b is the average bit energy and $R_b \leq C_{\mathrm{AWGN}}$ is the transmission rate in bits/second, we immediately find the minimum value of the bit SNR for transmissions with arbitrarily small probabilities:

$$\frac{E_b}{N_0} \geq \frac{2^{R_b/B} - 1}{R_b/B}. \tag{12.7}$$

For the infinite bandwidth, the asymptotic value of the capacity is

$$C_\infty = \frac{\overline{P}/N_0}{\ln 2}, \tag{12.8}$$

which can be reformulated as

$$\frac{E_b}{N_0} > \ln 2 \approx -1.6\mathrm{dB} \approx 0.69. \tag{12.9}$$

This SNR value, in photons/bit, is the absolute minimum required for communications over a continuous AWGN-limited channel, known as the *Shannon limit*.

Next, we consider the case of a discrete channel. It is characterized by an input alphabet $X = \{x_i\}$ $(i = 1, \ldots, M)$, an output alphabet $Y = \{y_j\}$ $(j = 1, \ldots, N)$, and a set of conditional probabilities $p_{i\,j} = P(y_j \mid x_i)$ of receiving the symbol y_j when x_i was transmitted. The channel is said to be memoryless when

$$P(y(1), \ \ldots, y(n))|x(1), \ldots, n(n)) = \prod_{i=1}^{n} P(y(i)|x(i)). \tag{12.10}$$

Obviously, we have $\sum_{j=1}^{N} p_{ij} = 1 (i = 1, \ldots, M)$. It is customary to organize the transition probabilities in a channel matrix $\mathbf{P}$ with elements $p_{i\,j}$.

A memoryless channel is symmetric when each row of $\mathbf{P}$ contains the same set of values $\{r_j\}$ $(j = 1, \ldots, N)$ and each column the same set of values $\{q_i\}$ $(i = 1, \ldots, M)$. For each symmetric channel it is usual to define the input entropy

$$H(X) = \sum_{i=1}^{M} P(x_i) \log_2 \frac{1}{P(x_i)}, \tag{12.11}$$

with similar expression for the output alphabet. The joint entropy

$$H(X, Y) = \sum_{i=1}^{M} \sum_{j=1}^{N} P(x_i, y_j) \log_2 \frac{1}{P(x_i, y_j)} \tag{12.12}$$

measures the average information contents of a pair of output and input symbols. Finally, the conditional entropy

$$H(Y|X) = \sum_{i=1}^{M}\sum_{j=1}^{N} P(x_i, y_j)\log_2 \frac{1}{P(x_i, y_j)} \tag{12.13}$$

measures the average amount of information available from the output symbol y when the input (transmitted) symbol x is known. An analogous expression can be defined for $H(X|Y)$.

Using the previous definitions and the fact that $H(X|Y) \leq H(X)$ and $H(Y|X) \leq H(Y)$, we obtain

$$H(X, Y) = H(Y, X) = H(X) + H(Y|X) = H(Y) + H(X|Y). \tag{12.14}$$

To quantifies the reduction of uncertainty relative to a X given the knowledge of Y, the flow of information through the channel, or the mutual information between X and Y, is introduced as

$$I(X; Y) = H(X) - H(X|Y) = H(Y) - H(Y|X) = H(X) + H(Y) - H(X|Y), \tag{12.15}$$

where we have used elementary properties of the (12.14). One can check that

$$I(X; Y) = \sum_{i=1}^{M}\sum_{j=1}^{N} P(x_i, y_j)\log_2 \frac{P(y_j|x_i)}{P(y_j)}, \tag{12.16}$$

and $I(X; Y) = I(Y; X)$.

The capacity of a discrete memoryless channel can be defined as the maximum of mutual information that can be transmitted; that is

$$C = \max_{P(x)} I(X; Y), \tag{12.17}$$

where the maximization has to be carried with respect to the probabilities of the input symbols.

12.2 Quantum physics for communication

12.2.1 Quantum uncertainty

Uncertainty relations are in many respects central to quantum physics. They arise directly from the property of certain physical parameters, such as atomic energy levels, to only take on discrete ("quantum") values, in contrast with the classical physics that allows them to be continuous. This realization is not dissimilar to Democritus insight on the atomic theory of matter around 490–460 BCE, except that in quantum theory the discreteness pertains to the physical observables[1] rather than to the structure of matter. One such discrete observable is the action. The minimum action that can be performed with a physical system is defined as the Planck constant h. The quantized action implies that a physical system state cannot be represented as a point in a phase space spanned by the canonical coordinate x and momentum p, but has to occupy the area equal to at least h.

This says nothing about the marginal distributions of the variables x and p and their uncertainties. For illustration we can assume that the quantum state of our system is described in position representation by a Gaussian wave function $\psi(x)$. This is a very important case, describing a harmonic oscillator or a hydrogen atom electron in their ground states, or a coherent state of an optical mode. Let us assume that the distribution has a width Δx defined as the standard deviation: $(\Delta x)^2 = \langle\psi|x^2|\psi\rangle - \langle\psi|x|\psi\rangle^2$. Calculating the standard deviation $(\Delta p)^2 = \langle\psi|\hat{p}^2|\psi\rangle - \langle\psi|\hat{p}|\psi\rangle^2$ of the momentum operator $\hat{p} = -i\hbar\partial/\partial x$ for this state we quickly arrive at

$$\Delta x \Delta p = \hbar/2, \tag{12.18}$$

where $\hbar = h/(2\pi)$.

Relation (12.18) is known as the *Heisenberg uncertainty* for position and momentum. Similar uncertainty relations can be derived for other quantum-mechanical operators $\hat{P}$ and $\hat{Q}$ directly from the commutation relations between these operators [6]:

$$\Delta P \Delta Q \geq \frac{1}{2}\left|\left\langle[\hat{P}, \hat{Q}]\right\rangle\right|. \tag{12.19}$$

Our result (12.18) for a Gaussian state is just a special case of (12.19). It illustrates another important principle: that the lower limit of the uncertainties product is reached for pure quantum states. Mixed states have larger uncertainties products and consequently occupy larger volumes in the phase space.

Uncertainty relation for energy and time has a form similar to (12.18) [7]

$$\Delta E \Delta t \geq \hbar/2, \tag{12.20}$$

but cannot be derived from the general relation (12.19) because time is not a quantum-mechanical operator.[2] Relation (12.20) is often interpreted as a consequence

[1] It is sometimes said that what's discretized in quantum theory is our knowledge of physical parameters, not the parameters per se. This position affects the philosophy behind quantum mechanics but not its practical aspects. It also leads to the *hidden variables* theories that can be experimentally tested in the so-called Bell measurements.

[2] But one can conceive a "quantum clock" device [8], whose time-indicating pointer will be described by a dynamic variable and will have an associated quantum operator.

of the relation $\Delta\nu\Delta t \geq 1/(4\pi)$ between the time and frequency measurements uncertainties, which arises in classical time-frequency analysis from the Benedicks's theorem [9], combined with $E = h\nu$.

However the time-energy uncertainty relation (12.20) can also be derived from the quantum principles. One way to accomplish this is to define Δt as $\Delta E \times (d\langle E\rangle/dt)^{-1}$ and then proceed with evaluation of ΔE according to the rules of quantum mechanics [10]. Alternatively, Δt can be defined as a quantum state lifetime, that is, such interval of time for which a state survival probability $|\langle\Psi(0)|\Psi(\Delta t)\rangle|$ equals ½. Analysis of quantum state evolution equations in this case also leads to Eq. (12.20) [14,11].

Another important and highly nontrivial uncertainly relation exists between the quantum state number of an oscillator n and its phase ϕ. Commonly the number-phase uncertainty relation is written as

$$\Delta n \Delta \phi \geq 1/2. \tag{12.21}$$

However, an attempt to introduce a quantum-mechanical phase operator will lead to eigenstates that have infinite energy and hence are physically impossible [12]. So (12.21) cannot be directly derived from a commutation relation such as (12.19). Moreover, a physically meaningful phase uncertainty cannot exceed π. Several approaches can be taken to circumvent these difficulties [13]. One can, for example, consider operators for sine and cosine of a phase ($\hat{S}$ and $\hat{C}$, respectively) rather than the phase operator [14]. These operators have the following commutation relations with the number operator $\hat{n}$:

$$[\hat{n}, \hat{S}] = i\hat{C}, \quad [\hat{n}, \hat{C}] = -i\hat{S}, \tag{12.22}$$

which leads to the following uncertainty relations (cf. (12.19)):

$$\Delta n \Delta C \geq \frac{1}{2}|\langle\hat{S}\rangle|, \quad \Delta n \Delta S \geq \frac{1}{2}|\langle\hat{C}\rangle|. \tag{12.23}$$

There is still a long way from the state-dependent relations (12.23) to the general form (12.21). In fact, it is shown [13] that this connection can be made only as an approximation. In a more strict sense, the number-phase uncertainty relation remains state-dependent, and its lower bound (12.21) is reached only for a specific set of parameters within each state. It is common to call the states reaching the minimal uncertainty the *intelligent states*. Importantly for our discussion, coherent and squeezed states of electromagnetic field are intelligent states (12.21).

12.2.2 Measurement and detectors

The process of measurement is important both in classical and quantum communications systems; however, its roles in these two cases are different. In quantum physics, a measurement defines the boundary between the quantum and the classical words. It may be said not just to record but actually to *create* an element of the observer's reality, in the sense of J.A. Wheeler's remark that "No phenomenon is a real phenomenon until it is an observed phenomenon." In particular, the question "what is a photon?" is inseparable from the method we choose to detect one. Consequently the properties of a photon, including its information content, also depend on the type of measurement.

In quantum physics, measurement is described as projection of a quantum system's state onto a state specified by the measuring device. Energy basis is most commonly used for detecting photons, in which case detection of a photon is accomplished by absorbing its energy, that is, by its annihilation. A photon defined by this type of detection is indivisible: it can be absorbed only as a whole and only at one point of space and time.[3] If a single-photon state $|1\rangle$ is incident on a beam splitter, the photon can be detected in one its output or the other—never in both.

The energy-basis, or photon-counting, measurements have certain practical advantages for communications: an optical photon energy is large compared to room temperature, so the thermal noise can be low. Furthermore, the physical transitions caused by a photon absorption can be very fast, so there is a potential for high-bandwidth communications. But photon absorption is not the only possible measurement strategy.

Instead of the energy basis, one may choose the basis of optical field quadratures. This type of measurement is known as homodyne measurement and is widely used in classical communications. In a homodyne measurement a weak signal is mixed with a strong reference beam, conventionally called the *local oscillator* (LO), and directed to a photo detector. The phase between the signal and LO determines the observed quadrature. Note that in this measurement scenario the main portion of the absorbed optical power is carried by the LO.

Homodyne measurement technique redefines the concept of a photon in a way which allows one to detect the photon's presence in more than one location. This was first noticed in 1991 by Tan et al. [15], who proposed to repeat the experiment with a single-photon incident on a beam splitter, but now with two homodyne detectors in the two output ports. According to quite straightforward quantum optics calculations, *both* local measurements should reveal a presence of the same single photon at the first beam splitter. Even though it is commonly accepted that a photon occupies the entire mode volume, the possibility to actually *probe* it in an experiment appears truly remarkable. A modified version [16] of Tan et al.'s proposal was experimentally

[3] The implication that electromagnetic energy initially spread over a finite—possibly large—volume has to be instantly absorbed by a small—possibly atom-size—object is one of conceptual difficulties of quantum theory.

realized in 2003 by Hessmo et al. [17]. In complete agreement with the theory, the experiment showed that with a single-photon input $|1\rangle$, the detection events in the two channels can be correlated, anticorrelated, or uncorrelated depending on the LO phase. If the input photon was not supplied, that is, the input state was that of the electromagnetic vacuum $|0\rangle$, the events were always uncorrelated. This experiment demonstrated presence of the same photon in two different channels, and thereby validated the concept of *entanglement with the vacuum* referring to the superposition state $|\Psi\rangle = (|1\rangle_a|0\rangle_b + |0\rangle_a|1\rangle_b)/\sqrt{2}$.

Detecting a single photon is possible not only in two distant location but also at the same location at two different moments of time. This was experimentally demonstrated in 2014 by Gulati et al. [18] by observing the exponential shape of a single-photon wave packet emitted by a single Rubidium atom.

The direct photon-counting and homodyne measurements are different types of *destructive* measurements wherein a detected photon is eventually absorbed. There is also a family of more delicate *nondestructive* measurement strategies that reveal a photon's presence without absorbing it. These strategies are collectively known as the quantum nondemolition (QND) measurements.[4] A QND measurement is often explained using the following example. Consider a Mach-Zehnder interferometer whose one arm contains Kerr media. Let this interferometer phase be adjusted so that one of its output ports is perfectly dark. Now if a probe photon passes through the Kerr media, it will alter the interferometer phase due to the cross-phase modulation, the dark port will not be perfectly dark any more, and any light emitted from it will indicate the presence of the probe photon—without destroying it.

A practical realization of this QND scheme would require extremely strong Kerr response which cannot be presently achieved. On a more fundamental level, this scheme requires the reference field to be in a photon-number eigenstate $|n\rangle$. Indeed, if the photon number in the reference beam is allowed to fluctuate, the self-phase modulation will mask the signature of the probe photon one attempts to detect.

The problem of self-action of the reference field is overcome in a family of QND schemes based on the so-called quantum zeno blockade (QZB). To understand how the QZB works, consider a linearly polarized optical beam passing through a stack of N half-wave plates aligned so that each plate rotates the beam polarization by a small angle α. The cumulative rotation angle by the entire stack is $N\alpha$. Now let each plate be followed by a polarizer aligned to transmit only the original polarization. Then the output polarization will remain unchanged, as enforced by the last polarizer. But most importantly, the total optical power loss due to all N polarizers will scale as $1-(\cos\,\alpha)^{2N} \approx 2\,N\alpha^2$ and can be made arbitrary

[4] Another commonly encountered term is "interaction-free measurements." Some authors use these two terms interchangeably, while others assign them subtly different meanings.

small for sufficiently small α and large N, even though the cumulative rotation $N\alpha$ may be large.

Each polarizer in this example performs a polarization measurement whose outcome is almost certain and which consequently provides only little information about the measured state, according to (12.1). Such a measurement imparts only an infinitesimal alteration to the measured state. In quantum theory this type of measurement is known as a *weak measurement.* Our classical optics illustration shows how a series of frequently repeated weak measurements can inhibit continuous evolution of a quantum system while causing arbitrary small decoherence. This constitutes the quantum Zeno effect. To use this effect for a QND detection of a single probe photon, one may implement the required weak measurement via a two-photon conversion process involving the probe photon and a photon from a reference optical mode. A suitable conversion process may be incoherent, for example, two-photon absorption [19] or electromagnetically induced transparency in atomic vapor [20]. It also may be coherent, for example, generation of the optical frequency sum or difference [21–23].[5] If the nonlinear interaction takes place in a high-finesse optical cavity, and if it is sufficiently strong at single-photon level, a probe photon's presence will shift the cavity resonance frequency for the reference field *without the actual conversion taking place.* The reference light will become decoupled from the cavity (hence the term "quantum zeno blockade"), which will indicate the presence of the probe photon without destroying it. A QZB QND photon detector can be viewed as a quantum version of an all-optical switch. In quantum information theory, this kind of switch is known as Fredkin logic gate. A practical implementation of a reliable Fredkin gate has not yet been demonstrated.

A different QND technique relies on two-photon entanglement. It was first demonstrated back in 1990s using spontaneous parametric down conversion (SPDC) in two optically nonlinear crystals arranged in such a way that an idler photon from the first crystal "seeds" the idler mode of the second crystal [24]. Emergence of the first idler photon manifests itself through induced coherence between the signal photons that occupy a different spatial mode, and furthermore may have different from the idler polarization and wavelength. Notably, the detected idler photon is neither absorbed nor in any way affected in the process.[6] If, however, the first idler photon does not arrive, the signal photons coming from different crystals remain incoherent. This demonstration gave rise to an interesting version of *ghost imaging* [25], when the idler photons that successfully passed through a mask are never detected. Instead, the image of the mask is reconstructed from interference of their

[5] The coherent QZB can be explained in terms of Autler-Townes splitting [21].

[6] It is not parametrically amplified, as the stimulated parametric emission plays no role in this effect and can be made arbitrarily weak without diminishing the induced coherence [24].

"twin" signal photons [26]. This approach, sometimes called "imaging with undetected light," is distinct from the original ghost imaging technique [25] and its numerous variations that all depend on detecting the photons that have passed through the mask.

The concept of QND measurement can be turned around and used for optical detection of macroscopical objects without any light scattering or absorbed by them, rather than for detection of photons. Let us return to our earlier example of a balanced interferometer which has a dark port. If we place an opaque object in one of its arms, we destroy interference, and a single photon injected into its input port will have a 25% chance of exiting from the formerly dark port. Realization of this possibility indicates the presence of the test object in the interferometer, discovered without actually illuminating the object. This is another example of how the notion of a photon may be introduced via such a measurement that reveals the photon's presence simultaneously at two distant locations. Indeed, if a photon did not somehow "know" of the object's presence, it could never exit through the dark port. Remarkably, with a clever design of this experiment, it is possible to increase the rate of its successful outcomes from 25% to almost certainty [27].

Once reliably implemented, the nondestructive strategies of photons detection may become beneficial for long-range communication scenarios when the signal photons are few and precious, as they potentially may allow for reusing these photons for multiple measurements. However, the information content of such measurements needs to be carefully analyzed and weighted against the information available from a direct measurement for each measurement scenario.

12.2.3 True random numbers generation

Random number sequences are important in mathematics, physics, and communication technology. In mathematics, random numbers are used in Monte-Carlo computations. In this application it is important to have a fair (not biased by inadvertent correlations) random sampling across the entire parameter space. If the parameter space is large and samples are many, pseudo-random sequences may start showing the built-in correlations and lead to biased sampling and computation errors.

Importance of true random numbers in fundamental physics can be exemplified by Bell tests. One of the known loopholes in these tests is associated with the choice of Bell-basis for measurements. The necessity to choose a new random basis for every detection arises from closing another loophole based on the possible causal relation between the preset measurement basis and the "true" (hidden) values of Bell observables. If this choice is based on a pseudo-random process correlated with some other part of the universe, it is conceivable that this correlation will find its way to the hidden values of Bell observables and void the test.

In communication technology, specifically in cryptography, the presence of an inadvertent correlation in the key-generating sequence may become discovered by the eavesdropper and can compromise the communication protocol's security. A classic example is breaking codes by analyzing the characters occurrence frequency.

The standard pseudo-random number generators are inadequate for these tasks. The significance of this problem is indicated by a recent Bell-test experiment involving human conciseness as a source of randomness [28]. In this truly epic effort, some 100,000 volunteers around the world generated 97,347,490 binary choices. These choices were streamed in real time to 12 laboratories on five continents, where they were used to determine the local settings in various types of Bell measurements. All the measurements strongly contradicted the local realism in favor of quantum mechanics.

Besides the proven tendency to impart chaos to every aspect of their lives, there is a deeper reason for using human subjects as sources of true random numbers. It is possible, at least fundamentally, to include any measuring device into the quantum system that is being observed, thereby shifting the boundary between the quantum evolution and the irreversible measurement—the border line between the quantum and classical worlds—closer and closer to the observer's mind. The mind may be the ultimate place where this line can be drawn and where the final projection operation on the quantum state of the outside world is performed. It is therefore plausible that only past this boundary can the true random number sequence originate.

Ruling out the local realism, a much simpler method of true random numbers generation is offered by quantum mechanics. Let us return to a single-photon incident on a balanced beam splitter. According to the quantum theory, it is fundamentally impossible to predict out of which port of the beam splitter the incident photon will emerge. Repeating this measurement again and again, one can obtain a sequence of truly random binaries. This can be implemented by injecting a weak coherent state on the beam splitter. The same experiment can be construed as a quadrature measurement of the vacuum field at the other port of the beam splitter against the coherent LO [29], which similarly leads to a fundamentally random result.

However, the output of any realistic device also contains classical noise that may be partly deterministic. To recover the true randomness of the underlying quantum process, this noise needs to be eliminated. In complexity-based cryptography this can be achieved by using hash functions such as SHA512, Whirlpool, or RipeMD [30,31]. Taking into account composable security for QKD, apropriate 2-universal hash functions should be used. Today the quantum random number generation is a mature technology, with the instruments commercially available in a chip and standard PCI card formats with 1 Gbit/s true random numbers output.

12.2.4 Entanglement and communication

The concept that nonclassical light may be useful for communication that emerged in the early days of quantum optics. Squeezed light was proposed to enhance communication rates [32,33], while two-photon parametric light was proposed to enable noise-resilient dual communication channels [34]. The latter proposal hinges on the idea that the parametric photons are tightly correlated in time, and therefore detecting the intensity modulation encoded in two channels via photon coincidence counting is more robust against background noise than single-channel detection. This technique does not actually rely on entanglement: two-channel communication with synchronized short pulses would achieve practically the same.[7]

An even more appealing advanced communication possibility seems to emerge from the concept of a nonlocal projection that occurs to one part of a bipartite entangled state when a local measurement is performed on the other part. Consider, for example, a two-photon polarization-entangle state $|\Psi\rangle = (|V\rangle_a|H\rangle_b + |H\rangle_a|V\rangle_b)/\sqrt{2}$ where V and H stand for vertical and horizontal polarization, respectively, and a and b label two channels coupled to two distant polarization-resolving detectors. Let the detectors be in possession of two observers whom we will call Alice and Bob. If detector a detects (let us say) horizontal polarization, then according to the quantum theory, the *entire* state $|\Psi\rangle$ instantaneously collapses, and the photon in channel b is instantaneously projected into state $|\Psi\rangle_b = |V\rangle_b$. Can this be used to exchange information between Alice and Bob?

12.2.4.1 No-signaling theorem

Quantum mechanics answers negatively to this question. Not only instantaneous or superluminal, but no communication at all is possible between Alice and Bob, who are allowed only local measurements on their respective parts of the entangled state. This constitutes the essence of the no-signaling or no-communication theorem, which is one of those fundamental no-go theorems that we mentioned in the introduction. The theorem is proven by considering the system's Hilbert space consisting of subspaces a and b. If Alice does nothing at all, Bob's part of the state is described by the density operator $\hat{\rho}_b = \mathrm{tr}_a(\hat{\rho}_{ab})$, where $\hat{\rho}_{ab}$ is the density operator for the initial state Alice and Bob share, and the trace is taken over Alice's subspace. Note that $\hat{\rho}_{ab}$ may represent any quantum state at all: pure or mixed, entangled or not. If Alice performs a local measurement of any kind described by a projector operator $\hat{P}_a$, then Bob's part of the state will be $\hat{\rho}'_b = \mathrm{tr}_a(\hat{P}_a\hat{\rho}_{ab})$. It can be mathematically proven that $\hat{\rho}'_b = \hat{\rho}_b$ [37]. Since $\hat{\rho}_b$ fully describes the probabilities of all possible Bob's measurements outcomes,

[7] Except that under special conditions the narrow correlation function of two-photon entangled light may be immune or resilient to chromatic dispersion, see Refs. [35,36].

this result means that Alice's local manipulations with the state are imperceptible to him, and hence no communications or signaling is possible.

12.2.4.2 Quantum teleportation

Despite the no-signaling theorem, the state projection that occurs in Bob's channel due to a measurement performed in Alice's channel is a real physical phenomenon: it can be observed, if classical communications between Alice and Bob are allowed. A Bell measurement is one type of the observation. In this case Alice and Bob perform their local measurements and compare their notes (via a classical communication channel) to establish a correlation between the results.

Alternatively, Alice may use her part of the entangled state to perform a two-particle measurement involving another unknown state and communicate the result to Bob. For simplicity, let us assume that each local state situates in a two-dimensional Hilbert space $|0\rangle \otimes |1\rangle$. For a photon, the basis states $|0\rangle$ and $|1\rangle$ may correspond, for example, to vertical and horizontal (or left- and right-handed) polarizations. An orthonormal basis for bipartite states may be chosen based on the following four Bell states:

$$|\Psi^{\pm}\rangle = \frac{1}{\sqrt{2}}(|0\rangle_a|1\rangle_b \pm |1\rangle_a|0\rangle_b), \quad |\Phi^{\pm}\rangle = \frac{1}{\sqrt{2}}(|0\rangle_a|0\rangle_b \pm |1\rangle_a|1\rangle_b). \tag{12.24}$$

Any bipartite state of a qubit can be expanded in the Bell basis (15.24). This is true for the entangled state Alice and Bob initially share, as well as for the collective state of Alice's shared photon and a third photon in an unknown state $|\psi\rangle = \alpha|0\rangle + \beta|1\rangle$ (where $|\alpha|^2 + |\beta|^2 = 1$) that she may have at her disposal. Performing a two-photon measurement on her two photons in the Bell basis (15.24), Alice obtains one of four possible outcomes. Note that by doing this she neither collapses the shared entangled state nor measures the unknown state $|\psi\rangle$. Instead, she communicates her measurement result to Bob via a classical channel as 2 bits of information. This information turns out to be sufficient (and also necessary) for Bob to deduce what transformation he needs to perform on his photon to transfer it to the state $|\psi\rangle$. Bob still does not know what this state is but can reliably reconstruct it *once*, based on his knowledge of the initial shared state and using the two bits of classical information he received from Alice. This protocol is known as the quantum teleportation [38]. Quantum teleportation requires classical information to be sent from Alice to Bob and hence makes no claim to being superluminal or achieving the signaling.

The teleported state does not have to be a single-photon state in the $|0\rangle\otimes|1\rangle$ basis, that is, a discrete-variable (DV) state. It can also be a continuous-variable (CV) state, where the observables, for example, optical field quadratures or frequencies, may take on a continuous range of values. In this case the shared resource state for

Alice and Bob is a CV entangled state or two-mode squeezed state, and the measurement Alice needs to perform is a homodyne measurement. The measurements involved in the CV quantum teleportation are usually simpler but more noisy than for DV. As a result, the CV teleportation typically has higher efficiency but lower fidelity than DV teleportation. Polarization-based teleportation of DV qubits over 143 km free-space optical link in 2012 may be considered as a modern benchmark in the field [39].

If the quantum teleportation fidelity is sufficiently high, the teleported state preserves its nonclassical properties, for example, squeezing [40]. Furthermore, the local teleported state itself may be entangled with another optical mode. This entanglement is also observed in high-fidelity quantum teleportation [41]: the quantum state that Bob recovers becomes entangled with the mode Alice's measured state was entangled with. This type of quantum communications is known as *entanglement swapping*.

Quantum teleportation has been shown not only in bipartite but also in tripartite systems, in which case Bob needs to receive classical communications not only from Alice but also from the third party [42]. The number of involved parties can be in principle even larger, which opens up the opportunities for quantum communication among many parties—a quantum network—and for various multipartite quantum protocols.

12.2.4.3 No-cloning theorem

Discussing the quantum teleportation, we emphasized that by performing her local 2-photon measurement, Alice does not measure the unknown state $|\psi\rangle$ that she is going to teleport. Likewise, performing his local transformation, Bob does not know which state he is reconstructing. This is a very important point. If the teleported state $|\psi\rangle$ became known to any party at any point (which in our example amounts to knowing the coefficients α and β), it could be duplicated. And this is prohibited by yet another no-go theorem: the no-cloning theorem.

The no-cloning theorem prohibits reliable copying of an unknown quantum states under the unitary evolution assumption.[8] A most general cloning scenario assumes that an unknown state of a particle a is transferred to a particle b, which is initially in a state $|\phi\rangle_b$. Let the cloning process be described by a unitary evolution operator $\hat{U}(t) = \exp\{-i\int \hat{H}(t)dt/\hbar\}$. Cloning of an arbitrary state $|\psi_j\rangle_a$ is complete at $t = t_0$ if

$$\hat{U}(t_0)|\psi_j\rangle_a|\phi\rangle_b = \alpha_j|\psi_j\rangle_a|\psi_j\rangle_b, \tag{12.25}$$

[8] It does not, however, preclude such cloning if we have even partial knowledge of the initial state.

where unitarity of $\hat{U}(t)$ requires that $|\alpha_j|^{2=}1$. To prove the no-cloning theorem, we consider two arbitrary states we wish to clone: $|\psi_1\rangle_a$ and $|\psi_2\rangle_a$. Multiplying Eq. (12.25) for $j = 1$ by its conjugate for $j = 2$ we obtain

$$\langle\phi|_b\langle\psi_2|_a\hat{U}^\dagger(t_0)\,\hat{U}(t_0)|\psi_1\rangle_a|\phi\rangle_b = \alpha_1\alpha_2^*\,\langle\psi_2|_b\langle\psi_2|_a\psi_1\rangle_a|\psi_1\rangle_b. \tag{12.26}$$

Taking into account unitarity of $\hat{U}(t)$, normalization of $|\phi\rangle_b$, and independence of the states' inner products on the particle designation a and b, (12.26) transforms to

$$\langle\psi_2|\psi_1\rangle = \alpha_1\alpha_2^*\langle\psi_2|\psi_1\rangle^2. \tag{12.27}$$

This indicates that the states $|\psi_1\rangle$ and $|\psi_2\rangle$ can be either orthogonal: $\langle\psi_2\,|\psi_1\rangle = 0$, or different by only a phase: $|\langle\psi_2|\psi_1\rangle| = 1$, in contradiction to our initial assumptions that these states are arbitrary. Therefore we proved that an arbitrary state cannot be cloned by a unitary and state-independent evolution operator $\hat{U}(t)$.

The no-cloning theorem can be also proven as a consequence the no-signaling theorem that we discussed earlier. Suppose Bob and Alice share a singlet entangled state $|\Psi^-\rangle$ (12.24) and Alice may choose to perform or not perform a measurement of her particle in the $\{|0\rangle, |1\rangle\}$ basis. If she performs the measurement, Bob's particle state becomes $|0\rangle$ or $|1\rangle$. If she does not, Bob's particle remains in a mixed state $\rho_b = \mathrm{tr}_a(|\Psi^-\rangle\langle\Psi^-|) = 1/2\,|0\rangle\langle 0| + 1/2\,|1\rangle\langle 1|$. If Bob can clone his state after Alice's decision, he could perform multiple measurements in the $\{|0\rangle, |1\rangle\}$ basis and determine which choice Alice made based on statistics of his results. Indeed, if the projection took place, his particle would be in one of the eigenstates and every measurement would yield the same result. Otherwise, Bob would be performing his measurements with the clones of the mixed state ρ_b, leading to equal probability of either outcome. This would mean that Alice communicated her decision to Bob via a shared entanglement with arbitrarily high fidelity, which contradicts the no-signaling theorem.

12.2.4.4 Quantum clock synchronization

The reality of a nonlocal state projection inspired a proposal to use a shared entanglement as a resource for distant clock synchronization, originally put forth by Jozsa et al. [43]. Just like the quantum teleportation, this protocol requires classical communication between the parties and does not imply signaling. It requires Alice and Bob to share an ensemble of the singlet Bell states $|\Psi^-\rangle$ introduced in (12.24). Let the basis states $|0\rangle$ and $|1\rangle$ be energy eigenstates with the eigenvalues $E_0 \neq E_1$, conveniently mapped on two poles of a Bloch sphere. The superposition states

$$|+\rangle = \frac{1}{\sqrt{2}}(|0\rangle + |1\rangle), \quad |-\rangle = \frac{1}{\sqrt{2}}(|0\rangle - |1\rangle) \tag{12.28}$$

may be called clock states as they precess around the Bloch sphere's equator in two opposite directions with frequency $\Omega = (E_1 - E_0)/\hbar$. Note that the singlet state

$$|\Psi^-\rangle = \frac{1}{\sqrt{2}}(|0\rangle_a|1\rangle_b - |1\rangle_a|0\rangle_b) = \frac{1}{\sqrt{2}}(|-\rangle_a|+\rangle_b - |+\rangle_a|-\rangle_b) \tag{12.29}$$

does not evolve in time. It may be seen as composed from clock states going forward and backward. Now let Alice perform a measurement on her clock ensemble in the $\{|-\rangle, |+\rangle\}$ basis. Such a measurement may consist of a $\pi/2$ rotation on the Bloch sphere with, for example, a $\pi/2$-pulse, followed by the energy-basis measurement. Note that a measurement in the $\{|-\rangle, |+\rangle\}$ basis can be defined only with respect to Alice's LO, which provides a phase reference and defines the instant when the $\pi/2$-pulse is to be applied.

After the measurement, statistically half of Alice's clocks are found in the $|-\rangle$ state, and half in the $|+\rangle$ state. At this instant, the same happens to Bob's clocks, and for him the time evolution begins! Of course, by virtue of the no-signaling theorem, this fact is of no use to Bob, who does not know if any particular clocks of his ensemble are supposed to go forward or backward. This information needs to be communicated to him by Alice. After receiving this communication, Bob can measure the phases of his clocks relative to his LO. From this procedure Alice and Bob should be able to find the phase difference of their LOs and correct it if necessary.

The flaw of this protocol, recognized by Jozsa et al. [43], is that Alice's and Bob's LOs phase discrepancy enters the problem twice: in the definition of their local $\{|-\rangle, |+\rangle\}$ basis measurements and in the measurement result. To eliminate the ensuing ambiguity, they proposed a workaround based on using multiple frequencies. However, it was pointed out [44] that this workaround itself requires the classical clock synchronization as a prerequisite for the quantum clock synchronization to work, which denies the latter its merit. The question whether the quantum clock synchronization can improve the accuracy of the underlying classical synchronization, and at what resource cost, remains open.

In spite of the critique from the clock community [44] as well as from the quantum optics community [45], the quantum protocols for clock synchronization continue to be an active field of research. Its modern emphasis is on synchronization of multiple clocks on a network using multipartite entangled states, pursued both in theory [46] and in experiment [47,48].

12.2.4.5 The pursuit of superluminal heresy

Numerous experimental results demonstrate that local variables models cannot explain the observed correlations between entangled particles under the assumption that the speed of light in vacuum c sets the limit for communications of any kind. Most commonly this leads to rejection of the concept of local realism and accepting the paradigm that an element of physical reality is created by a measurement (cf. the earlier quotation from J.A. Wheeler). However, an alternative interpretation is possible. The local realism can be rescued if the speed-of-light limitation is rejected. Considering a tremendous success of the relativity theory, this sounds almost heretic, but the search for loopholes that would allow for superluminal communication is a valid field of science nonetheless. Here we can provide only a very brief overview of this field. In spite of the strong underlying assumption $v > c$, the superluminal communication theories strive to avoid causal paradoxes. This is possible in the framework of the so-called v-causal models [49] postulating existence of a preferred frame of reference where superluminal signals propagate isotropically. Such a frame of reference would not be nonphysical even in the context of a Lorentz-invariant theory [50]. Moreover, the existence of such a frame is suggested by the observed properties of the cosmic microwave background radiation [51].

The v-causal models allow for superluminal communications between local measurements performed over entangled particles, that is, for the signaling. There is no global consensus among physicists whether this can be reconciled with the relativity or requires its revision. Meanwhile, the experimental research aimed to establish the lower limit on the superluminal communication speed v that could explain quantum correlations between remote measurements is continued. Most recently, this limit was reported [52] to reach an impressive benchmark $v_{min} \approx 5 \times 10^6\ c$. Slower signaling velocities cannot accommodate the modern Bell tests in the local-realistic paradigm.

12.2.5 Linear quantum amplifier basics

An amplifier is a device that takes an input signal and produces an output signal with increased amplitude. The input and output signals are carried by sets of bosonic modes (usually modes of the electromagnetic field). When the amplifier works only in a narrow band, the input and output signals can be approximated as being carried by single modes, which we will assume here. These modes should be interpreted as having duration $1/\delta\omega$, the maximum sampling time consistent with the bandwidth.

In a quantum description, each mode is specified by its creation and annihilation operators. If a and b are the annihilation operators for the input and output modes, the amplifier is specified by a relation $b = \mathcal{D}(a, a^{\dagger})$, where b and $b^{\dagger}$ must preserve the commutation relations.

A particular instance of this situation is the so-called linear amplifier. This is a drastic simplification yielding a simple formalism that still is applicable to a wide class of systems. The output field operator now takes on the form

$$b = Ma + La^\dagger + \mathcal{F}. \tag{12.30}$$

The operator $\mathcal{F}$ is assumed to be independent of the input signal. It represents the noise the amplifier adds to the output signal regardless of the input signal. Only the fluctuations in $\mathcal{F}$ matter, so nothing is lost by assuming $\langle \mathcal{F} \rangle = 0$. The unitarity imposes $1 = |M|^2 - |L|^2 + [\mathcal{F}, \mathcal{F}^\dagger]$, wherefrom we derive

$$\Delta^2 \mathcal{F} \geq 1 - |M|^2 + |L|^2. \tag{12.31}$$

Sometimes it is convenient to introduce quadrature components for a and b:

$$a = X_1 + iX_2, \quad b = Y_1 + iY_2, \tag{12.32}$$

so that (12.30) can be split as

$$Y_1 = (M + L)X_1 + \mathcal{F}_1, \quad Y_2 = (M - L)X_2 + \mathcal{F}_2, \tag{12.33}$$

where $\mathcal{F}_1$ and $\mathcal{F}_2$ are the quadrature components of $\mathcal{F}$ and M and L are assumed to be real. One can define the quadrature and mean gains as $G_1 = (M + L)^2$, $G_2 = (M - L)^2$, and $G = (G_1 + G_2)/2 = |M|^2 + |L|^2$.

The uncertainties in the output quadrature phases have the simple form

$$\Delta^2 Y_i = G_i \Delta^2 X_i + \Delta^2 \mathcal{F}_i, \quad i = 1, 2. \tag{12.34}$$

The amplifier is called *phase insensitive* when its output signal and the noise show no phase preference. The gain of a phase-insensitive amplifier is independent of phase, so that $G = G_1 = G_2$. The total fluctuation at its output is

$$\Delta^2 b = G\Delta^2 a + \Delta^2 \mathcal{F}. \tag{12.35}$$

The added noise is conveniently characterized by an added noise number $A = \Delta^2 \mathcal{F}/G$.

It has been shown that for phase-insensitive linear amplifiers this number satisfies

$$A \geq \frac{1}{2}|1 \mp G^{-1}|, \tag{12.36}$$

where the upper sign holds for phase-preserving amplifiers ($L = 0$) and the lower one for phase conjugating ones ($M = 0$).

This result implies that a high-gain phase-insensitive amplifier must add noise to any signal at least in the amount equivalent to the half-quantum of noise at the input. In contrast, a phase-preserving unity-gain amplifier need not add any noise.

Maser amplifiers and direct-current (DC) SQUID amplifiers are typical examples of phase-preserving amplifiers that can be made to operate near the quantum limit. The parametric amplifier is a typical example of a phase-sensitive amplifier.

12.2.6 Quantum state discrimination

The mysterious quantum world is accessible to us only via measurements, whose outcomes are classical. In consequence, the state of the quantum system must be inferred, which leads to a fundamental problem. When the set of the possible output states is known and these states are mutually orthogonal, quantum interpretation of a classical measurement is a relatively simple task. One can arrange detectors to uniquely respond to each of these states, and then a click in one of the detectors will unambiguously identify the state of the system. However, when the possible output states are not orthogonal. they cannot be discriminated perfectly, and optimum discrimination is far from trivial even if the set of the possible nonorthogonal states is known.

It is not difficult to show that quantum mechanics does not permit perfect discrimination between nonorthogonal states. There are now several possibilities. One possibility is to require conclusive discrimination all the time but to permit errors to occur. When we minimize the error probability, we follow the so-called minimum error (ME) strategy. Another possibility is to insist on error-free discrimination but to allow for inconclusive measurements to occur. When we minimize the probability of inconclusive outcomes, we follow the so-called optimal unambiguous discrimination (UD) strategy. Let 's review these two strategies following [53–55].

12.2.6.1 Minimum error discrimination

We want to distinguish, with minimum probability of error, among $N \geq 2$ given states of a quantum system. The states are specified by the density operators $\hat{\rho}$ and the ith state occurs with an a priori probability η_i, such that $\sum_{i=1}^{N} \eta_i = 1$. We describe the measurement as a positive operator-valued measurement (POVM), which is a set of Hermitian operators $\{\Pi_j\}$ such that $(\mathrm{tr}(\hat{\rho}\Pi_j)$ is the probability to infer the state of the system to be $\hat{\rho}_j$ if it has been prepared in $\hat{\rho}$.

In the ME discrimination strategy the measurement is required to be exhaustive and conclusive; that is, in each single case one of the N possible states is identified with certainty and inconclusive results do not occur. This leads to the requirement

$$\sum_{j=1} \Pi_j = I_D, \tag{12.37}$$

where I_D is the identity operator in the D-dimensional Hilbert space of the system.

The probability P_{err} of making an erroneous guess for any of the incoming states is

$$P_{\mathrm{err}} = 1 - P_{\mathrm{corr}} = 1 - \sum_{j=1}^{N} \eta_j \mathrm{tr}(\hat{\rho}_j \Pi_j), \tag{12.38}$$

where P_{corr} is the probability that the guess is correct. To find the ME strategy, one has to determine the POVM that minimizes P_{err} under the constraint (12.37). Once these optimum detection operators are known, (12.38) gives the ME probability $P_{err}^{min} = P_E$. The explicit solution to this ME problem is, however, not trivial and analytical expressions have been derived only for a few special cases.

When only two states are given, either pure or mixed, the ME probability was derived by Helstrom [56]. The result is $P_E = (1 - \mathrm{tr}|\eta_2\hat{\rho}_2 - \eta_1\hat{\rho}_1|)/2$. For the special case that the states to be distinguished are pure states $|\psi_1\rangle$ and $|\psi_2\rangle$, this expression reduces to

$$P_E = \frac{1}{2}\left[1 - \left(1 - 4\eta_1\eta_2|\langle\psi_1|\psi_2\rangle|^2\right)^{1/2}\right].$$

This is usually recast in the equivalent form

$$P_E = \eta_{min}\left[1 - \frac{2\eta_{max}(1 - |\langle\psi_1|\psi_2\rangle|^2)}{\eta_{max} - \eta_{min} + (1 - 4\eta_{min}\eta_{max}|\langle\psi_1|\psi_2\rangle|^2)^{1/2}}\right], \tag{12.40}$$

where $\eta_{min} = \min\{\eta_1, \eta_2\}$ and $\eta_{max} = \max\{\eta_1, \eta_2\}$. The first factor η_{min} on the right-hand side of (12.40) is what we would get if we always guessed the state that is prepared more often, without performing any measurement at all. The second factor is the result of the optimized measurement.

The optimal measurement is known in only a few special cases, such as, for example, pure states with equal a priori probabilities (two orthogonal detectors placed symmetrically around the two pure states). However, necessary and sufficient conditions that must be satisfied by the optimal POVM are known:

$$\sum_i \eta_i\hat{\rho}_i\Pi_i - \eta_j\hat{\rho}_j \leq 0, \quad \forall j, \quad \text{and} \quad \Pi_i(\eta_i\hat{\rho}_i - \eta_j\hat{\rho}_j)\Pi_j = 0, \quad \forall i,j. \tag{12.41}$$

12.2.6.2 Unambiguous discrimination

In the UD strategy the observer is not allowed to make an error. However, we know that this cannot be done with 100% probability of success. If we consider the case of two states, we have to introduce three POVM elements such that

$$\Pi_1 + \Pi_2 + \Pi_0 = I, \tag{12.42}$$

where Π_1 and Π_2 are detection operators that unambiguously identify the first and second states, respectively, whereas Π_0 can identify both states and thus corresponds to an inconclusive detection result.

Let us denote by $\langle\psi_1|\Pi_1|\psi_1\rangle = p_1$ the probability of successfully identifying $|\psi_1\rangle$ and $\langle\psi_1|\Pi_0|\psi_1\rangle = q_1$ the probability of failing to confidently identify $|\psi_1\rangle$. UD imposes that $p_1 + q_1 = p_2 + q_2 = 1$. The optimum UD strategy imposes that the average failure probability $Q = \eta_1 q_1 + \eta_2 q_2$ is a minimum (or, equivalently, the average success probability is a maximum) with the constraint $q_1 q_2 \geq |\langle\psi_1|\psi_2\rangle|^2$. The minimization procedure is lengthy, but the final result can be condensed into

$$Q^{opt}\begin{cases} Q_{POVM} & \dfrac{\cos^2\Theta}{1+\cos^2\Theta} \leq \eta_1 \dfrac{1}{1+\cos^2\Theta}, \\ Q_1 & \eta_1 < \dfrac{\cos^2\Theta}{1+\cos^2\Theta}, \\ Q_2 & \dfrac{1}{1+\cos^2\Theta} < \eta_1. \end{cases} \tag{12.43}$$

Here, $Q_{POVM} = 2(\eta_1\eta_2)^{1/2} \cos\Theta$ with $\cos\Theta = |\langle\psi_1|\psi_2\rangle|$, whereas $Q_1 = \eta_1 + \eta_2 |\langle\psi_1|\psi_2\rangle|^2$ and $Q_2 = \eta_1 |\langle\psi_1|\psi_2\rangle|^2 + \eta_2$. The optimum detection operators are given by

$$\Pi_1 = \frac{1 - q_1^{opt}}{\sin^2\Theta} |\psi_2^{\perp}\rangle\langle\psi_2^{\perp}|, \Pi_2 = \frac{1 - q_2^{opt}}{\sin^2\Theta} |\psi_1^{\perp}\rangle\langle\psi_1^{\perp}|, \tag{12.44}$$

where $\left|\psi_j^{\perp}\right\rangle$ is the vector orthogonal to $|\psi\rangle$.

A general receipt for optimum UD between two *mixed* states is not available. In contrast to ME discrimination, a compact formula for the minimum probability of inconclusive results, that is, the minimum failure probability, for two arbitrary mixed states does not exist. However, analytical solutions can be obtained for certain special classes of density operators.

12.2.7 Quantum tomography

While introducing the quantum uncertainty in Section 12.2.1, we mentioned that the uncertainty relation does not fix the marginal phase-space distributions of the canonical coordinate x and momentum p for a quantum state. In this section we would like to elaborate on this question considering the example of a harmonic oscillator.

In classical mechanics, a state of a harmonic oscillator is described by a point in the phase space. When we have a large number of identical classical oscillators, we can introduce a classical phase-space probability density function $W(x, p)$. This function is nonnegative and normalized, and has another important property. If we perform a large number of measurements of the position x (but ignoring the momentum p), we obtain a probability distribution associated with the position. This distribution is a marginal of $W(x, p)$; that is, it is just a column density projection of $W(x, p)$ onto a plane associated with the chosen quadrature x.

In the quantum domain, the notion of a localized point in phase space does not make sense anymore. Neither does the phase-space probability density. However, as we discussed in Section 12.2.2, even in the quantum domain one can perform measurements of a single quadrature (x, p or their linear combination). A multiple measurement of a quadrature on a set of identical quantum states will yield a probability density associated with this quadrature, that is, a marginal distribution. Interestingly, even in the quantum domain there exists a phase-space quasiprobability density—called the Wigner function—with exactly the same property as in the classical case.

Just as the classical phase-space probability density, the Wigner function is real and normalized. However, it has one very important difference from its classical counterpart: because by itself it does not have a meaning of a probability density, it does not have to be positive definite. Indeed, the phase space for many states has regions where the Wigner function takes on negative values.

The Wigner function cannot be directly measured, but all its marginal distributions can. Once we know all the marginal distributions associated with different quadratures (i.e., the Wigner function projections upon various planes), we can reconstruct the Wigner function and therefore the quantum state. This reconstruction procedure is associated to a Radon transform, much in the same way as it is performed in computer tomography. For this reason, the procedure has been dubbed as *quantum tomography*.

The trick to do phase-sensitive measurements of the electric field using regular "slow" detectors is called balanced homodyne detection. We overlap the mode whose quantum properties we want to measure (the signal) with a relatively strong laser beam (the LO) on a beam splitter. If the electric fields in the two modes are E_{signal} and E_{LO}, the fields emerging in the two beam splitter output ports are given by

$$E_{\text{out1}} = \frac{1}{\sqrt{2}}(E_{\text{LO}} + E_{\text{signal}}), \quad E_{\text{out2}} = \frac{1}{\sqrt{2}}(E_{\text{LO}} + E_{\text{signal}}). \tag{12.45}$$

Let us now have both beam coming out of the beam splitter hit high-efficiency photodiodes. Our two photodiodes produce photocurrents that we subtract from each other. The photocurrent difference $\mathcal{I}$ is given by

$$\mathcal{I} \propto |E_{\text{out2}}|^2 - |E_{\text{out1}}|^2 = 2E_{\text{LO}}E_{\text{signal}}. \tag{12.46}$$

Since the LO is a very bright classical state, we neglect its quantum noise and therefore the photocurrent difference is proportional to the amplitude of the signal field. By changing the relative optical phase of the LO and signal waves, we measure the electric field at different phases. At each phase, we perform a multitude of electric field measurements (each time preparing an identical quantum state in the signal channel), thus obtaining the marginal distribution. A set of marginal distributions for various phases will provide us with full information about the quantum state and allows us to reconstruct its Wigner function.

12.3 Quantum mechanics for securing communication channels

An important aspect of practical communications is protecting the messages from tampering and eavesdropping—that is, cryptography. The field of cryptography has a long and rich history, which leads to a commonly accepted conclusion that a truly unbreakable cipher requires a random string of key characters privately shared by the two communicating parties. A sender (Alice) performs an XOR operation on her message and the key (in a binary alphabet), and sends the result to the recipient (Bob). It is easy to see that if Bob now performs an XOR operation on the encoded message and the key, he will recover the original message. This encryption technique is called the one-time pad[9] (OTP) encryption. Originally proposed by F. Miller in 1882, it was reinvented by G. Vernam and J. Mauborgne in 1917, and is also known as the Vernam cipher. The "one-time" aspect of the OTP cipher points at the fact that it is secure only if the random key string is used only once and then discarded. Thus the main problem of the OTP-based cryptography is replenishment of the secure key bits. The no-cloning theorem allows for accomplishing this task with fundamentally unconditional security, providing a rare example of a quantum no-go theorem serving as a technology vehicle rather than a hindrance.

12.3.1 Basic principles of quantum key distribution

Quantum key distribution (QKD) is a family of protocols used to generate a secure key sequence shared by two parties: Alice and Bob. All QKD protocols require that besides a quantum communication channel, Alice and Bob share a publicly accessible classical channel. It is assumed that a potential eavesdropper, traditionally called Eve, can listen to the public channel and furthermore has a full control over the quantum channel. This means, in particular, that she can spoof any transmissions in the quantum channel. Moreover, Eve is assumed to have unlimited computation power and access to any possible technology that abides by laws of physics. But she is not allowed to have access to Alice's and Bob's setups or hijack the classical communication channel. The last limitation is important, because if Eve is able to impersonate one legitimate user for the other and prevent their secure authentication, she can successfully intercept the key by the so-called man-in-the-middle attack.

The unconditional security of QKD hinges on the fact that a symbol-encoding quantum state cannot be cloned, whereas simply intercepting it automatically removes

[9] Named after the paper sheets (pads) on which the key string was usually printed.

this symbol from the legitimate user's key string. This can be achieved by employing the *prepare-and-measure*, the *entanglement-sharing*, or the *bidirectional* strategy. We note that the term "unconditional security" should not be misread as not needing any condition at all. Of course, certain fundamental conditions of all security systems have to apply, most importantly matching the security proof model to the experimental implementation.

In the prepare-and-measure strategy Alice encodes her symbol into a quantum state, which she sends to Bob. Bob performs a measurement and extracts the symbol as explained below. Alice needs to make sure that her sequence of transmitted characters is truly random and free from spurious correlations that could potentially supply Eve with information. A true random number generator becomes necessary if Alice and Bob plan to generate an arbitrary long key with unconditional security. In fact, one of the attacks that Eve can launch in order to gain a partial access to the secure key consists of tampering with an unprotected random number generator. This can be done by obtaining the random number seed or influencing the process of random number generation. The attack may be directed at the software by hacking or intentionally planted backdoors, or at hardware through the shared environment such as, for example, power lines. Furthermore, the necessity to encode *known* symbols opens up a number of technical loopholes that Eve can use to her advantage. Perhaps the most classic example of such a loophole is the radio-frequency pulses broadcast by Pockels cells used for polarization encoding or by electro-optical modulators used for phase encoding. Matching the security proof model to practical implementations is an important task of certification processes. This includes using properly working true random number generators, transmitters, detection systems, and shielding against external coupling. Some of that work is known from classical high-security products (TEMPEST requirements), some work directly related to the quality or unintended readout of quantum states and related to the field of "quantum hacking."

In the entanglement-sharing scheme, Alice and Bob derive their identical keys from a shared bipartite entangled state. This approach eliminates the need for encoding known symbols and the consequent security problems. Now the strings of symbols are created by Alice's and Bob's measurements and do not exist until these measurements are complete. Symbols in these strings have truly random order, as we discussed in Section 12.2.3, whereas the quantum correlations guarantees the identity of the generated key sets. The cost of these benefits is the need for the entangled light source. In practice this amounts to using significantly more powerful lasers and specially designed optically nonlinear crystals or waveguides and quantum state generation rates that are still orders of magnitude lower than those of weak laser sources. In addition, especially in satellite-based QKD the requirement of two simultaneous free space channels with potentially large loss renders an efficient implementation challenging. Notably, the entanglement-sharing QKD is also not free from the technical loopholes open for Eve. One of these loopholes is the transient emission of light from an avalanche photo diode (APD, the most common type of photon-counting optical detectors), which occurs upon a photon detection. Again, as with every security product, proper certification measures have to be taken into account that ensure to avoid unwanted loopholes and side-channels.

In the bidirectional strategy Alice and Bob assume highly asymmetric roles. Bob sends light pulses to Alice, who encodes them with the key by performing certain unitary operations on the phase or polarization, and returns the encoded pulses to Bob. Both communication directions require an eavesdropping check to ensure security. The bidirectional approach has a significant practical advantage, as it does not require the exchange of measurement basis information between Alice and Bob, but it also has a major disadvantage of effectively doubling the communication distance. Because of this disadvantage, the bidirectional approach is less common than the prepare-and-measure and entanglement-sharing approaches. However, it is still a field of active research. Recently a new version of bidirectional protocol has emerged that suppresses Eve's information gain by utilizing nonorthogonal unitary transformations for encoding a binary alphabet [57].

QKD protocols may rely on quantum measurements yielding either discrete-value (DV) or continuous-value (CV) results. Both types of protocols rely on the key symbols of a chosen alphabet encoded into quantum states that are transmitted from one party to the other or distributed among them by a third party. Binary alphabets are most common; however, larger alphabets based on three or more mutually unbiased orthogonal quantum states have been also considered [58–60]. Remarkably, time-bin encoding naturally allows for efficient realization of larger alphabets [61,62].

Using discrete time bins also allows for QKD protocols where a macroscopic intensity measurement, which in its nature is continuous, directly generates a discrete-value quantum key. Such are the distributed-phase-reference protocols, for example, the continuous one-way (COW) and differential phase shift (DPS) protocols. In these protocols Alice sends to Bob a train of phase-coherent pulses. In the DPS protocol, each pulse is generated exactly in or out of phase relative to the previous one, which suppresses the slow phase drift. The binary key is encoded in the phase difference of two adjacent pulses; see Ref. [63] for the basic and Refs. [64,65] for more advanced and robust "round-robin" DPS protocols. The key is decoded by using an unbalanced interferometer with the path lengths difference matching the pulse period. In the COW protocol, amplitude modulation is used instead of phase modulation. The binary key is encoded in the sequence of bright (the pulse arrived) and dark (the pulse did not arrive) frames, and Eve's presence is detected by monitoring coherence between the pulses [66].

12.3.1.1 Discrete variables quantum key distribution

Historically, the DV QKD protocols were invented first, and they still remain most widely used in commercial systems. These protocols are based on the energy-basis measurement with a single-photon (or approximately single-photon, e.g., low-power coherent) state. The measurement outcome is discrete: in a binary alphabet a photon is detected either by one detector or by the other. It may be not detected at all, in which case the particular measurement does not contribute to the key. This, as we mentioned earlier, also includes such events when the photon was intercepted by Eve. Information is encoded in the photon's state in a certain basis. For example, using polarization encoding, Alice and Bob may agree to interpret a vertically polarized photon as Boolean "1," and horizontally polarized photon as Boolean "0." Alternatively they may agree to assign the "1" and "0" to left-hand and right-hand polarizations, respectively, or choose a linear basis rotated by 45°, but in DV QKD the quantum states encoding "1" and "0" must be always orthogonal.

A classic example of a prepare-and-measure DV QKD protocol is the BB84 invented by Charles Bennett and Gilles Brassard in 1984 [67]. In the basic version of this protocol Alice randomly chooses one of two polarization basis sets, for example, vertical/horizontal or right-handed/left-handed, and one of two symbols to be sent, for example, "1" or "0". She encodes the given symbol in the given basis according to a predefined alphabet and sends the photon to Bob. Bob randomly chooses one of the same two basis sets and performs a measurement. He can trust a random number generator to make the basis choice, or alternatively he may build two equivalent setups performing measurements each in its own basis and couple them to the input beam by a balanced polarization-insensitive beam splitter. The latter approach solves the problem of true random numbers for Bob by incorporating a quantum random number generator in his setup. It also makes basis switching unnecessary for Bob. But it is obviously more bulky and costly.

Next, Alice and Bob perform the key sifting. They publish their lists of basis choices on the public channel. Statistically in half of the cases their lists are consistent. These are the cases they will use for the secret keys, and the keys themselves will be the symbols encoded by Alice and decoded by Bob. The keys are identical because they were encoded and decoded in the same basis. They are secure because if Eve steals a photon sent to Bob, then Bob does not receive it, and this photon is excluded from the key string. Eve may smartly substitute a stolen photon with another one, thereby performing the so-called intercept-and-resend attack, but to avoid detection, she needs to encode it with respect to the correct basis, which is not yet published. Without the knowledge of the right basis Eve's manipulations will produce errors visible to Alice and Bob. As a countermeasure, Alice and Bob may sacrifice some part of their key to measure the error rate and try to detect Eve's presence. They can also improve their key security by implementing privacy amplification protocols. These steps reduce the initial key length. It is therefore important to distinguish between the raw (or sifted) and the final (secure, or secret) key rates. Both are important characteristics of practical QKD systems typically quoted in literature, but they may differ by orders of magnitude for any given measurement.

A typical entanglement-sharing DV QKD protocol follows the general steps of BB84 or its variations—except for the step when Alice makes a decision as to which character to transmit. This choice is now made for her by nature, and she only needs to perform the same measurement on her photon as Bob performs on his, in a randomly chosen basis. The cases when Alice and Bob accidentally choose the same basis for their measurements will provide them with strings of perfectly correlated for the $|\Phi^{\pm}\rangle$-states, or anticorrelated for the $|\Psi^{\pm}\rangle$-states, symbols, that can be used as a secure key. Using entanglement provides the access to different techniques for an eavesdropper detection and privacy amplification then than in prepare-and-measure protocols. This was pointed out by Artur Ekert, who proposed that Alice and Bob should choose their basis so that their unused subset constitutes a Bell measurement [68]. Deviation of the Bell observable from its theoretical value will indicate Eve's activity, and Alice and Bob will not have to sacrifice their useful key to detect it.[10]

Note that in neither of the two above cases Alice and Bob employ a true single-photon or two-photon state. In the prepare-and-send protocol Alice most commonly uses an attenuated laser, which means that the number of photons detected in a given time interval or a frame has Poisson distribution. If Alice and Bob use SPDC as the entangled photons source in the entanglement-sharing protocol, the photon pairs statistics ranges from Poissonian for the multimode case to Gaussian for the single-mode case. For Poisson distribution, the probabilities to have no photons, exactly one photon, and more than one photon per frame are uniquely determined by the mean number of photons in the frame, as shown in Fig. 12.1.

The frames where more than one photons are present are labeled "bad" because they lead to a security vulnerability to the so-called photon-number-splitting (PNS) attack. Since photons in a "bad" frame are identically encoded, Eve may be able to spirit split away the extra photons and come to possess a fraction of the key string. To reduce this fraction, Alice and Bob must reduce the mean number of photons per frame and predominantly send and receive empty frames. This will also reduce the sifted key rate, which is highly detrimental for practical QKD. One possible solution would be to design a source with sub-Poisson statistics, that is, with a strongly suppressed probability of generating more than one photon (or one photon pair) at a time. The BB84 protocol with such a source based on InAs quantum dot embedded in a micropillar cavity was demonstrated across a 35 km optical fiber link [69].

[10] Since Ekert protocol requires using three basis sets instead of two, the fraction of consistent basis sets is smaller, so Alice and Bob still have to pay some key rate penalty.

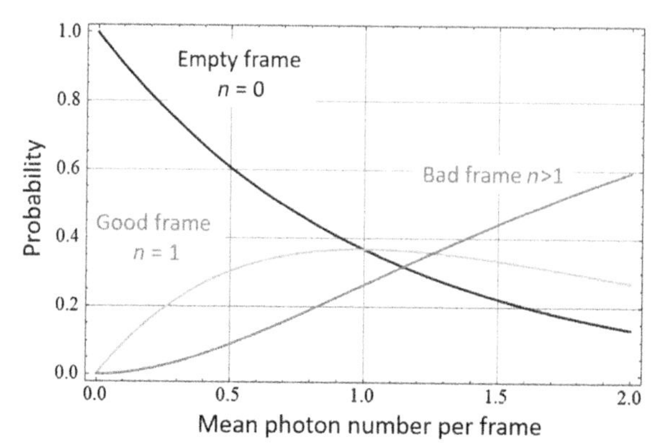

Figure 12.1 Probability to find zero, exactly one, and more than one photons per detection interval (frame) using an ideal detector versus the mean number of photons in the frame, assuming Poisson distribution.

The PNS attack can also be negated if the secret key is encoded in the basis rather than in the measurement result, and the measurement results are disclosed in the key-sifting phase instead of the basis sets. Such an "inverted" protocol is known as SARG04 [70].

However, the most effective countermeasure to the PNS attack is generally agreed to be the *decoy state protocol*. In this protocol, Alice sends weak coherent pulses at different power levels and, hence, with different photon number statistic (see Fig. 12.1). After the transmission, she announces which power level is to be used for the key, whereas the other levels are used for monitoring the bit error rate. It is shown to be fundamentally impossible for Eve to minimize the error rate due to the PNS attack at multiple signal levels [71]. In this way the decoy state protocol allows Alice and Bob to detect Eve's PNS attack and discard the compromised part of the key.

The information leakage from the hardware preparing or detecting the encoded quantum states is conventionally called *side channels*. The side channels can be exploited by Eve employing the *quantum hacking* techniques, which will be further discussed in Section 12.3.2. We already gave some simple examples of side channels and have shown how the entanglement-sharing protocols may help to eliminate those associated with the state-preparing hardware. A similar trick, known as the Measurement-Device-Independent (MDI)[11] QKD [72], can be used to eliminate the side channels associated with the state-detecting hardware. Just like in the entanglement-sharing protocols, the state preparation can be assigned to a third party (say, Charlie) who does not have to be trusted, performing the MDI protocol Alice

[11] Not to be confused with the earlier proposed fully device-independent approach, which is notoriously difficult to implement in practice.

and Bob designate Charlie to perform a measurement for them. Security of the entanglement-sharing protocols is based on the fact that while having a full knowledge of the entangled state that he prepares, Charlie cannot predict its behavior with respect to a local measurement Alice and Bob perform. Similarly, in the MDI protocol, Charlie performs a Bell measurement whose result in itself is insufficient for him or Eve to infer the key. However, being broadcast over a public channel, this result allows Alice and Bob to retrospectively perform a basis rotation operation on the states they have supplied to Charlie and to come up with private and secure copies of a binary key. From a practical perspective, MDI QKD is well suited for multiuser networks, because the most complicated and expensive measurement hardware does not have to be in the exclusive possession of any one party but may be shared among many users.

12.3.1.2 Continuous variables quantum key distribution

In CV QKD, information can be encoded in the optical field quadratures and retrieved via a homodyne measurement requiring an LO. It can be also encoded in polarization Stokes variables and retrieved in a polarization measurement, in which case the LO is transmitted along with the signal [73]. In contrast with the energy-basis (photon-counting) measurement employed in DV QKD, this measurement may return a continuous range of values, which explains the name. Note also that the states representing different symbols are no longer orthogonal.

Originally CV QKD was designed to utilize squeezed optical states in the prepare-and-measure protocols similar to BB84, where instead of the encoding basis, Alice had to choose which quadrature is to be squeezed, and applied a phase-space displacement to encode a symbol [74]. Bob attempts to decode the symbol by measuring the displacement of a randomly chosen quadrature, following which he and Alice perform the key-sifting step closely following that of BB84. Later it was proven that using coherent states is equally secure and efficient as using squeezed states [75]. In both cases Bob works with a *local* state represented by the same density matrix, which leads to a concept of *effective entanglement* and allows one to treat both approaches on equal footing and use the same metric to quantify correlations.

Using coherent states allows for eliminating of the key-sifting step. Instead, Alice may encode symbols in *both* quadratures at the same time and only keep those that Bob announced. Furthermore, the necessity for Bob to shift quadratures can be eliminated as well, if he performs a measurement on both quadratures at once in a heterodyne measurement. This measurement is noisier than the single-quadrature homodyne measurement, so the key rate is not exactly doubled, but it is still improved [76]. As a later development, the concept of a heterodyne measurement performed by Bob has been combined with the original idea of Alice using squeezed states and shown to surpass other Gaussian CV QKD protocols in terms of noise resistance [77].

A prepare-and-measure CV QKD protocol using coherent state allows for using larger than binary alphabets but requires a more involved key-sifting procedure. In its basic form it works as follows. Alice encodes a pair of random symbols into quadratures of a weak coherent light and transmits this signal to Bob over a noisy and unsecured optical channel. The channel SNR determines the optimal bit length of the transmitted symbols. As an illustration, let us assume that Alice and Bob agreed on using a 3-bit alphabet. Alice may encode her symbols into $2^3 = 8$ slices of the X or Y quadrature distribution, as shown in Fig. 12.2. Note that this step is very similar to the binning algorithm used in the quantum random number generator [29].

Bob measures a randomly chosen quadrature of the received signal by homodyne detection and announces his quadrature to Alice over the public channel. He may also measure both quadratures at once, as mentioned earlier. As a result, Alice and Bob obtain the raw key material: two strings of random numbers with Gaussian distribution that are correlated but not identical as in the DV QKD protocol. They need to reconcile the raw key material and to sift the key from it. Suppose in our example that Alice encodes a symbol "010." Most likely, Bob identifies the encoded symbol correctly, but because of the channel noise, he can also identify it mistakenly as "001" or "011" that correspond to the adjacent slices. To eliminate this error, Alice publicly discloses the first (low) bit of her symbol "0." This rules out the "001" and "011" for Bob. There is still a small chance that Bob misinterprets the encoded symbol as "000" or "100." This residual error probability can be estimated from the properties of Gaussian functions. If Alice and Bob find this probability acceptably low, they may finish their reconciliation process, having sifted two key bits out of three bits of raw key material. Otherwise, Alice may disclose the second bit, "1," to rule out the "000" and "100" and to further reduce the error probability. In this case she and Bob are down to one sifted key bit.

When Alice and Bob finish the reconciliation step, they implement a privacy amplification step and obtain the final secret key. In addition to the reconciliation, they can use a subset of their raw key material to evaluate their actual mutual information $I(A;B)$, and to set the upper bound for the information $I(A; E)$, which may be

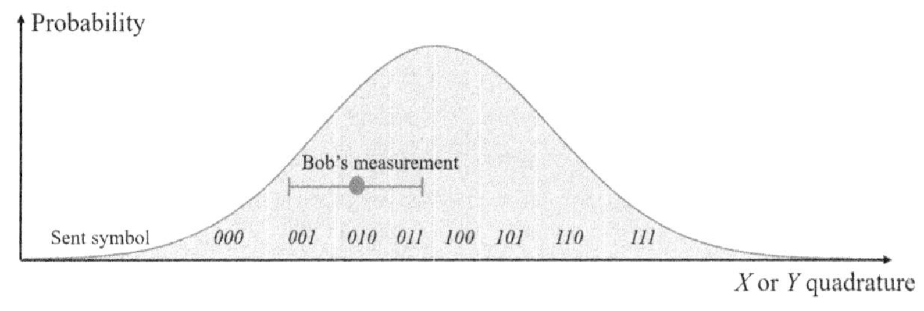

Figure 12.2 Simple reconciliation of raw key material in continuous-variable (CV) quantum key distribution (QKD) using slicing.

shared with Eve. The key material used for these purposes has to be discarded, but its function is important for optimization of the key reconciliation process. For the secret key generation to be successful, Bob should have information advantage over Eve, and an obvious condition $I(A;B) > I(A;E)$ must be fulfilled.

The maximum channel capacity between Alice and Bob and between Alice and Eve is given by Eq. (12.6) where the channel SNR can be introduced as a ratio of the quadrature variances due to the encoding and due to the channel noise: $\Sigma = \sigma_{mod}^2/\sigma^2$.

The condition for Bob's information advantage over Eve is $\Sigma_{AB} > \Sigma_{AE}$. For AWGN-limited channels this is equivalent to $\sigma_{AB} < \sigma_{AE}$. Note that the modulation width σ_{mod} which is common to Bob and Eve cancels out, and only the noise added in a channel due to its imperfect transmission remains. Let us denote this transmission η for Bob's channel, and let us make the most generous assumption to Eve that she collects all the photons that did not make it to Bob. Then her channel transmission is $1-\eta$. The added noises for Bob and Eve are $\sigma_{AB} = \sigma 0\ (1-\eta)/\eta$ and $\sigma_{AE} = \sigma_0 \eta/(1-\eta)$, respectively, where σ_0 is the vacuum noise variance. Then from $\sigma_{AB} < \sigma_{AE}$ we arrive at the information advantage condition in the simple form $\eta > ½$. This is a very important result, setting a well-known 3 dB loss limit on CV QKD quantum channel efficiency, that only could be lifted later.

Practical encoding and reconciliation algorithms are considerably more involved than our simple example. They are discussed in great detail in a number of publications; see for example, Refs. [78,79] and references therein. Two particularly important tools are the reverse reconciliation, when Alice corrects her keys to have the same values as Bob's [75,80], and the postselection, when Alice and Bob only keep the data points providing information advantage over Eve [81]. These tools, especially effective in combination [82], allow for increasing the effective key rate, minimizing the information accessible to Eve, and surpassing the 3 dB loss limit.

CV QKD has an important advantage over DV QKD in that it typically offers a much faster raw key material generation rate. While DV QKD depends on individual photon counting entailing inevitable postdetection dead time, CV QKD is performed with analog photo detectors that are free from this problem. They also have higher quantum efficiency and lower noise. The latter includes both the dark noise and the ambient noise due to background light, which is strongly suppressed in a homodyne measurement. On the other hand, DV QKD is more resilient to losses. For DV protocols losses only reduce the key rate, whereas in CV protocol they contribute to errors and may become a security-limiting factor.

12.3.2 Eavesdropping challenge

Increasingly more sophisticated techniques that might allow Eve to breach the QKD security have been proposed along with the advanced QKD protocols R&D. There is a natural competition between these two fields, and any breakthrough in one stimulates research in the other. Information-theoretical classification distinguishes individual, collective, and coherent attacks. All these attacks consist of three common steps:

1. Allowing an ancillary quantum state to interact with a state transmitted via the quantum channel and storing this state in a quantum memory.
2. Listening to the key sifting or reconciliation on the public channel.
3. Deducing and performing the optimal measurement on the stored quantum state.

To launch an individual attack, Eve uses a single quantum state, for example, a qubit, as the ancillary state. For a more powerful collective attack, she uses an ensemble of identically prepared independent ancillary states. For the most powerful coherent attack, Eve employs a multipartite entangled state. The security against these attacks under various scenarios has been analyzed in a large number of publications; see for example, Ref. [83] for review. Note that it is very difficult to prove security against the coherent attack in a general case for many QKD protocols.

The attacks discussed above are designed to break the QKD protocols at the conceptual level. From a practical standpoint we should also consider the attacks exploiting flaws in the protocol implementations or in the hardware, which is known as quantum hacking. One example of a flawed implementation is the BB84 or a similar DV QKD protocols using weak coherent pulses instead of single photons. Such a "sloppy" implementation is justified by considerable practical benefits of using coherent light sources, but it opens the doors to the PNS attack, which is not possible in the single-photon implementation. Examples of faulty hardware are the already mentioned APDs emitting their own light after a detection and electro-optical devices broadcasting electromagnetic pulses when operated.

In a Trojan horse attack Eve injects bright optical pulses into the quantum channel and monitors the light reflected from Alice's or Bob's setup. Analyzing this light, she may gain knowledge of, for example, polarizers settings and hence of the basis choice for polarization-based DV QKD protocols. Such attacks have to be considered even with the currently available equipment. This has been recently demonstrated [84] on the commercial quantum cryptosystem Clavis2 from ID Quantique, which employs the SARG04 protocol. The attack was directed against Bob, because in this protocol his choice of the basis defines the raw key. By doing so Eve was able to obtain over 90% of the secret key, clearly defeating the security.

Trojan horse attacks can be adapted for breaching entanglement-sharing, MDI, and CV QKD protocols. An efficient defense against the Trojan horse attack requires monitoring the optical power in the quantum channel for anomalies. This could be easily implemented for Alice, who does not expect any incoming signals, but it is more difficult for Bob. In his case such monitoring would be associated with additional attenuation of the quantum signals, which is undesirable. Incidentally, the

Clavis2 system used in the experiment [84] could be easily defended against this attack. The APDs used in this system have strongly anomalous response to bright pulses, which can be easily detected if Bob knows what to look for.

A variation of the Trojan horse attack when instead of bright probing pulses Eve sends to Bob such pulses that he mistakes for legitimate quantum communications from Alice has also been studied [85]. This type of attack is known as the faked states attack. Its goal is to introduce into Bob's list of the basis and bit values those that are known to Eve. This type of attack can be further strengthened if it leverages a mismatch in the detectors' quantum efficiencies [86]. Such a mismatch may arise from gating of the detectors, which can be seen as a temporal modulation of their quantum efficiency. If the gating pulses do not perfectly overlap, and Eve has the knowledge of the relative gating timing, she can launch a time-shift attack. To do so she will delay or advance Alice's pulses in time at random, but in such a way that her delay/advance choice will strongly correlate with one of Bob's detectors being blinded by the timing. For example, it may happen that the gating pulse arrives to the detector set to measure "0" slightly earlier than to the detector set to measure "1." Then Eve knows that the pulses timed to arrive near opening of the gate will likely result in "0" or a nondetection. Conversely, pulses timed to arrive near closing of the gate will likely result in "1" or a nondetection. In this way Eve can obtain a significant fraction of the key without increasing the error rate and hence remain undetected. The time-shift attack will significantly reduce the signal, but in realistic QKD schemes with large channel loss, Eve could easily mask this by making her pulses brighter.

An efficient time-shift attack is feasible with the currently available technology. Remarkably, it is especially efficient against a perfect single-photon source (which is immune to the PNS attack) and detectors with low dark counting rates. Note that not only temporal but also the chromatic or directional mismatch of the quantum efficiencies can be exploited by Eve in a similar way. There also exists a variation of the time-shift attack aimed specifically against the bidirectional QKD systems using phase encoding. This type of attack is called the phase remapping attack [87].

So far we considered only "opportunistic" quantum hacking, when Eve exploits the hardware vulnerabilities overlooked by Alice and Bob. However, Eve may go beyond it and actively modify their hardware parameters. She may, for example, partially but permanently blind one of the detectors with a powerful laser. This will lead to the quantum efficiency imbalance and open the system to the fake states attack. Similarly, destroying a collimating pinhole in a free-space QKD receiver makes it vulnerable to the Trojan horse attack using different incidence angles. Both types of attacks have been successful against commercial QKD systems from ID Quantique and MagiQ with only standard equipment available to Eve; see Refs. [88,89] and references therein.

The detector-blinding attack described above was invented in 2009. Five years later ID Quantique developed and implemented (via a firmware update for the Clavis2 systems) a countermeasure based on randomly reducing a detector efficiency to practically zero during certain time slots. Such a detector will nonetheless remain sensitive to a blinding laser pulse and will indicate the attack. However, in less than a year this countermeasure was defeated by a modified blinding attack. Since then an accelerating series of increasingly advanced attacks and counterattacks has been unfolding, This example highlights the importance of certification procedures and best practices that already have to be taken into account during the development phase of QKD devices. These procedures are standard in classical security devices and currently are being developed for quantum cryptography.

12.3.3 Channel loss, quantum repeaters, and quantum memory

Early in Section 12.3.1 we mentioned that the quantum states transmitted in the QKD channels cannot be amplified for the same reason it cannot be cloned. This makes the quantum channel loss a very important factor in practical QKD implementations. This loss not only limits the key generation rate but also affects the key security, because we have to assume that Eve is able to collect all of the lost photons.

Performance of various CV and DV QKD protocols with respect to the loss is reviewed in Ref. [80]. Since different QKD protocols depend on a large number of different parameters, a direct comparison is difficult. However, a general pattern can be seen from the two examples in Fig. 12.3. Here the channel efficiency, characterized by the ratio of the key generation rate to the pulse rate, is plotted as a function of the channel loss for various protocols: continuous variables with Gaussian modulation

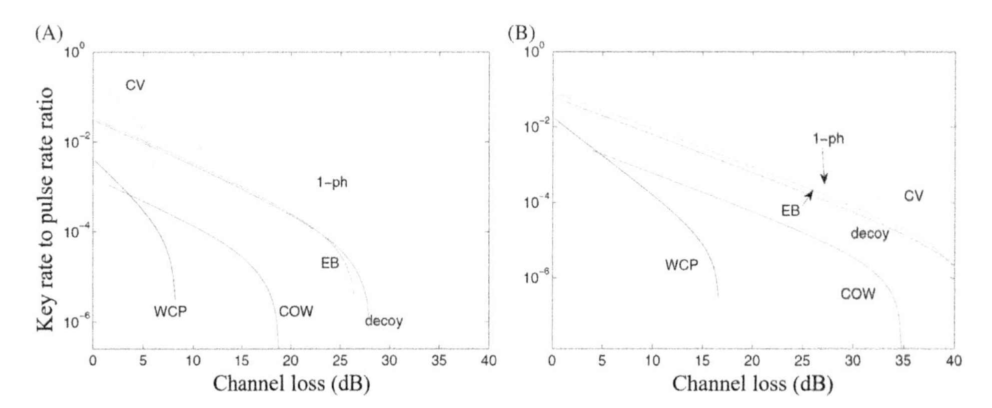

Figure 12.3 The crypto-key rate efficiency of various quantum key distribution (QKD) protocols are evaluated for two sets of parameters given in Table 12.1. Source: *Fig. 4A and B from V. Scarani, H. Bechmann-Pasquinucci, N.J. Cerf, M. Dušek, N. Lütkenhaus, M. Peev, The security of practical quantum key distribution, Rev. Mod. Phys., 81 (3) (2009) 1301–1350 [80].*

(CV), perfect single-photon source (1-ph), weak coherent pulses with and without decoy states (decoy and WCP, respectively), entanglement-based (EB), and coherent one way (COW). In both Fig. 12.3(A) and (B) the mean intensity for DV protocols and variances for CV protocols are assumed to be optimized, and Bob's receiver is assumed to have a unity transmission. The error-correction codes are implemented as described in Ref. [80] (we will talk more about it in Section 12.3.4). Other relevant parameters are listed in Table 12.1.

The most important message of Fig. 12.3 is that the losses degrade performance of all QKD techniques. Up to a certain point, the key rate reduction is approximately a polynomial function of loss, but when the critical loss value is reached, the rate drops catastrophically and the QKD becomes impractical. This happens when the raw key fraction required for the error correction and privacy amplification approaches the entire raw key length. The critical loss value depends on the system's parameters. Notice that as we improved these parameters from the set (a) to the set (b) in Table 12.1, the critical value of the loss increased significantly. For an ideal system the catastrophic key rate drop does not occur at all, even though the key-to-pulse rate ratio still declines as a function of loss. Upper limits of this ratio derived for such an ideal system under different sets of assumptions and known as the TGW [90] and PLOB [91] bounds:

$$R_{TGW} = \log_2\left(\frac{1+\eta}{1-n}\right) \approx 2.89\eta, \quad R_{\mathrm{PLOB}} = \log_2\left(\frac{1}{1-\eta}\right) \approx 1.44\eta. \qquad (12.47)$$

Here η is the channel transmission, and the approximation is made for $\eta < 1$.

Unfortunately, the hard limit imposed by the hardware imperfections on the QKD range is prohibitive for many important applications requiring long-range communications. For example, a 40 dB loss corresponds to some 200 km of a telecom fiber. Such an experiment using a DPS QKD protocol was performed in 2007 [92],

Table 12.1 Parameters for Fig. 12.3A and B.

Platform	Parameter	Fig. 12.3A	Fig. 12.3B
BB84	Visibility (P&M)	0.99	0.99
COW	Visibility (ES)	0.96	0.99
	Detector efficiency	0.1	0.2
Dark counts fraction	10^{-5}	10^{-6}	
	Bit error (COW)	0.03	0.01
CV	Optical noise	0.005	0.001
	Detector efficiency	0.6	0.85
	Electronic noise variance	0.01	0

P&M, Prepare-and-measure; *ES* entanglement-sharing protocols. The electronic noise variance is relative to the shot noise.

demonstrating 12.1 bps secure key rate produced out of 10 GHz raw key pulse rate, that is, approximately 1.2×10^{-9} channel efficiency.

To get around the loss problem, one may introduce a number of communication nodes along the communication channel. With one such extra node (Charlie) the channel topology will be: A–C–B. If Alice shares a private key with Charlie by using a QKD protocol, and Charlie likewise shares another private key with Bob, then Alice and Bob also can share a key. To do so, Charlie can use Alice's key to encrypt Bob's and send the result to Alice via a public channel. Alice now is able to recover Bob's key; this key is also known to Charlie, who therefore has to be trusted. We will review some practical realizations of this approach later, but note that for a large number of intermediate nodes, A–C1–C2–...–B, the requirement that all C nodes (or relays) must be trusted becomes practically equivalent to building a fully secure communication channel, which is assumed impossible under the QKD paradigm.

Quantum mechanics allows Charlie to help Alice and Bob to establish a secure key without learning it in the process. This approach to the long-range QKD is known as the *quantum repeater*. To implement a basic quantum repeater, Charlie, instead of establishing keys with Alice and with Bob, performs a Bell measurement on the two photons sent to him by these two parties, just like he does it in the MDI protocol. Communicating its result to Alice or Bob, Charlie facilitates the entanglement swapping as described in Section 12.2.4.2. Now Alice and Bob share a two-photon entangled state unknown to Charlie and can generate a private and secure key.

Entanglement swapping between three nodes is easily generalized to an arbitrary long chain of nodes and can be made noise resilient by implementing error correction and entanglement distillation or purification steps. In theory, this should not only eliminate the hard limit on the channel loss but also allow for exceeding the TGW bound (12.47). However, a practical realization of such a protocol requires a synchronization of the states' measurement and preparation, and this requires a capability to store quantum states for sufficient intervals of time while preserving their coherence, that is, a *quantum memory*.

A quantum memory is based on strong coupling between photons and matter qubits. The latter may be implemented as individual ions or atoms, atomic ensembles, quantum dots, color centers, and other quantum systems. Here we discuss three examples of different approaches to quantum memory based on atomic ensembles and rare-earth ions. The same approaches can be adapted to other matter qubit implementations.

In our first example [93], quantum state of a weak light pulse was recorded in two cesium vapor cells with the opposite spin polarizations, and preserved for up to 4 ms. The state was then recovered with a better fidelity than could be achieved by a classical memory. The underlying principle of this quantum memory realization was a QND measurement (see Section 12.2.2) on the light pulse and the atomic ensemble, followed by an electro-optical feedback system. To utilize this system for a practical quantum repeater in an entanglement-sharing QKD protocol, one needs to implement the same type of measurement with an entangled photon in place of a weak laser pulse. However, most of the available entangled photons sources have optical bandwidth far exceeding that of a typical atomic transition, and cannot be efficiently used for this purpose. A progress in this direction was achieved recently with building an SPDC source based on a very high-finesse optical resonator, which can match the important cesium and rubidium transitions in optical frequency and bandwidth [94,95].

The second example of a quantum memory realization is based on electromagnetically induced transparency. In this process, the signal pulse group velocity is controlled by a classical optical pump field. When the pump field is gradually turned off, the signal pulse slows down and eventually stops, being transformed into a dark-state polariton wave. By turning the pump back on, the process is coherently reversed and the initial pulse is retrieved. This approach has been used in a number of experiments involving cold atomic ensembles and demonstration of storing DV and CV quantum states. For CV states, a squeezed vacuum is most typically used. The typical storage efficiency in this case is about 10%–15%, but the storage and retrieval can be accomplished practically without adding noise [96].

Our third example is based on a photon echo technique. The advantage of this approach is the increased optical bandwidth, which is allowed by the inhomogeneously broadened absorption in the active medium. The broadening causes the signal dephasing, which is, however, compensated due to the phase-reversing nature of the photon echo. Furthermore, the photon echo allows for control over the release of the stored quantum state. This approach has been experimentally realized using solid host crystal, such as lithium niobate or yttrium orthosilicate, doped with rare-earth ions, such as Pr^{3+} or Er^{3+}; see for example, Refs. [97,98].

12.3.4 Quantum error correction and privacy amplification

Noise in communication channels leads to errors, which need to be corrected. Classical error-correction techniques are typically associated with one or another form of redundancy. A sender may prepare and transmit multiple copies of each symbol and rely on most of them arriving without an error. He may also restrict the use of code words to those that have a sufficiently large minimal Hamming distance (number of different bits), which is also a form of coding redundancy. However, a direct application of classical error-correction techniques to quantum communication channels is disallowed by the no-cloning theorem: one cannot generate copies of arbitrary quantum states to implement the desired redundancy. A more subtle approach is required.

Such an approach was developed in 1995 by Shor [99] and Steane [100] independently, and experimentally validated in 1998 [101]. At the heart of this quantum error-correction technique is entanglement between multiple qubits. If one qubit is affected by noise, it changes the quantum states of all qubits it is entangled with. It is possible to entangle the signal qubit with auxiliary qubits in such a way and then perform such a measurement in the end of a noisy channel that its result will reveal if a signal qubit was altered *without actually measuring this qubit*. This amounts to introducing and exploiting of quantum redundancy; for example, original Shor's protocol [99] requires nine entangled qubits to validate the error-free transmission of one.

Closely related to the quantum error correction is the concept of *entanglement distillation* (or *purification*). Entanglement distillation is a process of extracting of a small ensemble of strongly entangled states from a larger ensemble of less entangled states. Such necessity may arise when an entangled state is shared via a channel affected by loss and noise. In DV, for example, polarization entanglement, this leads to mixing between different Bell states $|\Psi^{\pm}\rangle$ and $|\Phi^{\pm}\rangle$ defined in (12.24). However, the undesired component of the resulting density matrix can be postselectively removed based on an additional Bell-type measurement whose result is distributed between the parties via a classical communication channel [102]. As a result, the initial pure state is recovered, albeit with a less-than-unity multiplier. This factor represents the ensemble reduction associated with entanglement purification.

CV entanglement purification may be desired, for example, when Gaussian quadrature-entangled states are transmitted through a lossy channel. Distilling the initial entanglement from the resulting states requires non-Gaussian transformations that are difficult to deterministically implement. This makes experimental CV entanglement distillation difficult, although a number of approaches have been devised to achieve this goal. Perhaps the most convincing demonstration was based on coherent single-photon subtraction technique [103]. Other approaches include cross-mode Kerr interaction, off-resonant coupling to a Rydberg atom in a cavity, and linear-optics transformations conditioned on single-photon detection. The CV entanglement distillation task is somewhat simpler when the communication channel introduces non-Gaussian noise, for example, by a time-dependent transmission or by phase modulation. In this case distillation can be carried out using linear optics, homodyne detection, and feedforward [104,105].

Privacy amplification is a method for producing an unconditionally secure crypto-key from a shared string of raw key material that may have been partially compromised. Unlike the quantum error correction and entanglement purification, the privacy amplification protocols are completely classical, as they deal only with classical information. Suppose Alice and Bob are aware that they have shared some amount of information with Eve during their secure communication, which is now complete. They do not know how much of information was shared but can place a reasonable upper limit on it. Their goal is to process their information in such a way that the result is completely decoupled from Eve's information. In the QKD scenario, Eve's active presence leads to errors, whose rate increases with increasing of the information that may be available to Eve. Therefore the first step for Alice and Bob is to estimate the error fraction, for example, by estimating the Hamming distance between the transmitted string x and a received string $y \neq x$. This can be done by using various cryptographic hash functions [30,31] computed on the strings x and y and publishing the results. Having found the error fraction, Alice and Bob reduce the length of their keys accordingly and agree on a new, shorter key string. There are several strategies for this setup they may follow in order to minimize or completely eliminate Eve's knowledge of the new key; see for example, Ref. [106] and its references. The choice of the optimal strategy depends on the error fraction and other factors. Note that the privacy amplification task is conceptually similar to the removal of classical entropy from the quantum random number generator output discussed in Section 12.2.3.

12.4 Modern quantum key distribution

As mentioned, QKD is the most mature and commercialized field of quantum communication technology. Presently there are a few companies specializing in providing commercial QKD services. Commercial QKD has also been actively developed by other companies, such as HP, IBM, Toshiba, Mitsubishi, NTT, and NEC alongside their other products and services. In this section we review the status of the field, dividing it into fiber-based, free-space, and satellite-based QKD.

12.4.1 Fiber-based quantum key distribution

The world's first bank transfer using fiber-based entanglement-sharing DV QKD took place in Vienna in 2004. The same year, DARPA launched the world's first fiber-based DV QKD 10 km long network named Qnet. Situated in Cambridge, Massachusetts, the Qnet initially consisted of six nodes, but later was upgraded to ten nodes. The subsequent rapid growth of the QKD industry is illustrated by the following milestones.

2007: ID Quantique used QKD to transmit the national election results for the Swiss canton of Geneva across a nearly 100 km long optical fiber.

2008: The EU-funded Secure Communication Based on Quantum Cryptography (SEC-OQC) network connected six computers in Vienna and in St. Poelten (69 km from Vienna) using a total of 200 km of optical fiber. This network supported a variety of QKD demonstration utilizing different CV and DV protocols; see review [107].

2009: A hierarchical metropolitan QKD network was demonstrated in China [108]. The network consisted of four nodes, one of which was a subnet of three more nodes. The subnet communicated with its nodes through a trusted relay. The sifted key rate on this network ranged from 2.56 to 11 kbps, allowing for the secure key rates from 0.08 to 2.53 kbps for different routes ranging from 0.5 to 10 km. Note that the largest sifted key rate does not necessarily yield the larges secure key rate; in fact, in this demonstration the 11 kbps sifted key rate corresponds to 0.08 kbps secure key rate. The same year ID Quantique launched a three-node Swiss Quantum network project in the Geneva metropolitan area for a long-term testing of the robustness and reliability of QKD. The rectified key rate in this test consistently remained close to 2.5 kbps for one route and 1 kbps to the other two routes for the two-year test duration.

2010: The Tokyo QKD Network was inaugurated. The network was built on the Japan's Gigabit Network JGN2plus and configured as a star connecting the central node in Otemachi with three other nodes via a 12, 13 and a 45 km routes. This project involved an international collaboration of seven partners, each responsible for a certain part of the network: Mitsubishi Electric, NEC, NTT, and NICT from Japan; the European research division of Toshiba from the United Kingdom; ID Quantique from Switzerland; and the "All Vienna" team from Austria. The latter team consisted of researchers from the University of Vienna, the Austrian Institute of Technology, and the Institute for Quantum Optics and Quantum Information. Among other technology demonstration goals, the project has shown a QKD encryption of real-time video communications between the nodes with the maximum distance of 45 km. The average secure bit rate of 293 kbps was achieved and maintained for tens of hours of continuous operation. With 1 GHz clock rate this corresponds to the rate efficiency of approximately 2.9×10^{-4}, consistently with Fig. 12.3.

2011: Los Alamos National Laboratory launched a QKD network that had a hub-and-spoke architecture. With this architecture, all messages from the nodes are routed through the central hub, which establishes a secure key between them by serving as a trusted relay. This approach allows for all bulky and expensive single-photon detectors to be placed at the hub, while the nodes are equipped only with the transmitters. It resembles the MDI approach discussed earlier, although the MDI protocol was not performed in this particular demonstration. However, the benefit of placing all detectors at the hub was utilized: the nodes were made very compact, as shown in Fig. 12.4.

2013: Battelle Memorial Institute in collaboration with ID Quantique installed a 32 km long fiber QKD link between Columbus and Dublin, Ohio.

2016: The first CV QKD network was demonstrated in Shanghai, China [109]. This network used polarization encoding and consisted of four nodes connected by six fiber links. The links lengths ranged from 2.08 to 19.92 km with the link loss from 3.21 to 5.7 dB. The reported point-to-point key rates ranged from 0.25 to 10 kbps.

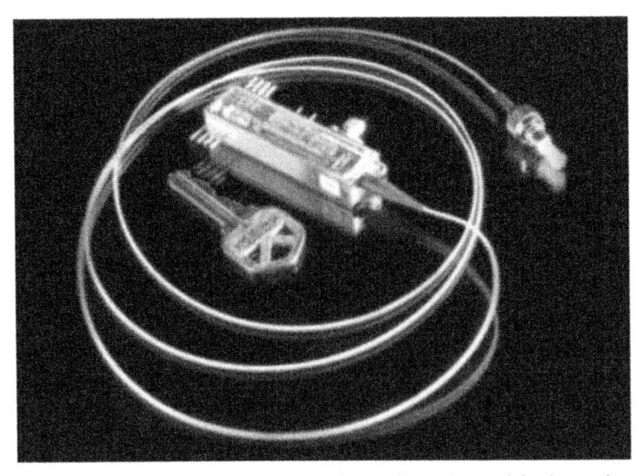

Figure 12.4 A quantum key distribution (QKD) node in fiber-based hub-and-spoke architecture can be made very compact. Source: *From http://arxiv.org/abs/1305.0305.*

2017: The first long-distance QKD fiber link Beijing–Jinan–Hefei–Shanghai came online with the total length of 2000 km. This link required 32 trusted relay nodes to work. Several major Chinese banks are presently using it for sensitive data transfer.

Several trends are evident from these milestones. Perhaps the strongest one is the shift of the emphasis in the QKD architecture from connecting two private parties to a network of users. This trend is apparently driven by broadening the potential QKD clientele and the opportunity to leverage the economy of scale. Today QKD Metropolitan Area Networks (Q-MANs) are well established in Europe, the United States, China, Japan, and South Africa. The next expected step is integrating the Q-MANs into QKD Wide Area Networks (Q-WANs) for long-range and eventually global interconnect. This, as we will see in the following, requires deployment of a satellite QKD relay network.

It is also clear that the main focus of the R&D efforts in contemporary fiber-based DV and CV QKD is no longer made on the fundamental research and algorithms development, but on the field demonstrations and overcoming the known technical and engineering limitations. One such limitation arises from the current silicon photonics technology used in fabrication of commercial telecom-band on-chip devices, in particular modulators. These devices do not always demonstrate adequate performance to manipulate fragile quantum states. A significant progress in this field was recently reported. In one demonstration [110] the secure key rate of up to 916 kbps was achieved in a 20 km long fiber link. With 1.72 GHz clock rate this corresponds to the rate efficiency of approximately 5.3×10^{-4}. In a later demonstration [111] the secure key rate of up to 157 kbps was achieved in a 43 km long fiber link with 16.4 dB channel loss. With 625 MHz clock rate this corresponds to the rate efficiency of approximately 2.5×10^{-4}. Both results are compatible with Fig. 12.3.

Another limitation specific to the fiber-optic QKD is the phase and polarization drift in fibers. These effects and the possible ways of their mitigation have been studied in the framework of the earlier mentioned Tokyo QKD Network demonstration [112]. This research also pointed out the importance of synchronization of Alice's and Bob's detectors' gate pulses, which may be done by distributing a clock pulse through the same fiber bundle that supports the quantum channel [112], or even through the quantum channel itself, using wavelength-division multiplexing (WDM). The WDM approach allowed for two QKD channels and a clock channel in a single fiber [113].

Addressing the long-term stability in fiber-optic quantum links led to a number of demonstration of uninterrupted operation of such links, summarized in Table 12.2. Here the key rate refers to the final secure key, and QBER is the quantum bit error rate in DV protocols. Reference [107] also reports shorter term performance of various QKD protocols over various fiber link lengths.

Both CV and DV approaches are used in fiber-based QKD. CV QKD is much closer to standard optical telecommunication techniques, as it relies on homodyne or heterodyne detection using PIN diodes, rather than on photon-counting detectors. PIN photo diodes typically have higher quantum efficiencies and do not require cooling. They are easily integrated on an opto-electronic silicon chip. Such chips can also incorporate all other essential components of a CV QKD system, including modulators and attenuators. Silicon photonics allows for CMOS-compatible technology and cost-effective production of receivers and transmitters. However, just as with the modulators, specific requirements for CV QKD operation must be taken into account. Most importantly, homodyne detectors must be optimized to reach the shot noise limited performance.

In CV QKD, the raw key rate is limited by the smaller of the sender and receiver bandwidths. In many practical scenarios, the receiver bandwidth is smaller and

Table 12.2 A performance summary of several fiber-optic QKD long-term demonstrations.

References	Year	Length (km)	Loss (dB)	Duration	Key rate	Protocol
[107]	2009	33	7.5	24 h	3.1	BB84
[114]	2009	N/A	3	57 h	8	CV, 2-quad
[115]	2011	3.7	2.5	237 days	2.5	BB84, SARG
[112]	2012	45	14.5	60 h	293	BB84 + decoy
[116]	2012	17.7	5.6	85 days	0.6	CV, 2-quad
[113]	2013	22	12.6	30 days	112.4	BB84
[117]	2014	90	29	25 days	1	DPS
[118]	2014	N/A	18.4	212 days	800	BB84
[119]	2015	45	14.5	34 days	301	BB84

Key rate is in kbps.

Table 12.3 A summary of several recent fiber-link CV QKD demonstrations.

Ref.	Year	Length (km)	Pulse rate	Key rate	Protocol
[122]	2005	55	<10 MHz	1000*	bi-Gaussian, quadrature
[123]	2007	5	100 kHz	30	Gaussian, quadrature
[114]	2009	3 dB loss	500 kHz	8	bi-Gaussian reverse, quadrature
[124]	2009	24.2	100 MHz	3.45	Four-state, quadrature (QPSK)
[125]	2013	80.5	1 MHz	0.7	Gaussian, quadrature
[126]	2015	75	1 MHz	0.49	Gaussian, quadrature, and WDM
[127]	2015	25	50 MHz	1000	Gaussian, quadrature, and WDM
[128]	2016	20	1 MHz	80	bi-Gaussian, polarization
[129]	2016	100	2 MHz	0.5	Gaussian, quadrature
[130]	2017	40	40 MHz	240*	Eight-state, quadrature (8-PSK)
[131]	2018	25	50 MHz	3140	Gaussian, quadrature

Key rate is the secure key rate in kbps, except where marked by * to denote the raw key rate. WDM indicates that the same channel is used for classical communications with bright signals through wavelength-division multiplexing. Some of these references also report results for shorter-length links.

therefore is the limiting factor. However, it is still typically very large compared to photon-counting detectors rates, which allows CV QKD to generate raw key much faster than the DV QKD. With sender and receiver placed back-to-back, and both transmitter and receiver operating at 20 GHz, the secure key rate has been shown to exceed 12 Mbps [120]. Finite channel length leads to loss, which necessitates the privacy amplification and leads to a significant key rate reduction. Some examples of recent CV QKD demonstrations over finite length fiber links are given in Table 12.3, while more theoretical and experimental details of this technology are available from a review [121].

In contrast to CV protocols, the raw key rate in DV QKD is limited by the dead time and jitter of the photon-counting detectors. This typically leads to lower secure key rates, but enables protocols to be more tolerant to transmission loss, thereby allowing for longer fiber links. Some examples of recent DV QKD demonstrations are given in Table 12.4. In this table we included COW and DPS protocols. As discussed earlier, such protocols fall in between of CV and DV QKD concepts because they combine binary encoding of the raw key bits into phase or amplitude, with continuous-value measurement that does not generally require single-photon detection. In some demonstrations, however, such detection was implemented. For example, in Ref. [92] this was done with a superconducting single-photon detectors based on niobium-nitride nanowires. This type of detector offers the dark count rate as low as just a few Hz, and very low (some 60 ps rms) timing jitter. Note that this jitter is still comparable to the raw key pulse period (the inverse of the 10 GHz rate equals 100 ps) and could have presented a problem if not for the strong fiber loss, which allowed on the average only one photon out of 10,000 to arrive at the detector.

Table 12.4 A summary of several recent fiber-link DV QKD demonstrations.

References	Year	Length	Pulse rate	QBER (%)	Key rates	Protocol
[132]	2004	11.07	1000	2.8	8.5/4.07	B92 [133], pol., WDM
[132]	2004	11.07	1000	6.8	11.68/3.05	B92, pol., WDM
[134]	2006	1	625	3.08	2100/?	B92, polarization
[135]	2006	100	1000	3.4	2/0.166	DPS
[92]	2007	200	10,000	4	?/0.0121	DPS
[136]	2009	200	625	4.5	?/0.0025	COW
[137]	2011	45	1000	2.7	304/100	BB84, time-bin
[113]	2013	22	1000	1.7	483/112	BB84, time-bin, and WDM
[119]	2015	45	1000	4.4	1500/301	BB84, time-bin
[138]	2016	404	75	26	3.2×10^{-7}/?	BB84, time-bin, MDI, 4-decoy, and WDM
[110]	2017	20	1720	1.01	3000/916	COW
[110]	2017	20	1000	1.1	4000/329	BB84, polarization
[111]	2018	43	625	2.8	?/157	BB84, polarization

Fiber-link length is in kilometers. Pulse rate is in MHz. Key rates are given in kbps as (sifted key rate)/(secure key rate). Some of these references also report results for shorter-length links.

12.4.2 Free-space quantum key distribution

Tables 12.3 and 12.4 show a dramatic drop of the secure key rate with increasing fiber link length. In Section 12.3.3 we related this expected behavior to the effects of accumulating loss and error rate. Loss in fibers arises due to various absorption and scattering mechanisms. It is described by Beer's law, which means that the loss is exponential function of the fiber length. Considering the scale of the optical fiber industry and the amount of efforts already invested in perfecting the fiber performance over the past decades, it is highly unlikely that the fiber absorption can be significantly reduced. In free space, on the other hand, loss and scattering can be much smaller, and consequently there is a potential for implementing longer QKD links. This warrants developing the free-space QKD, which we discuss in this section.

Optical transmission in free-space channels, especially through a dense and turbulent atmosphere, depends on a number of factors and requires a thorough study in order to understand and optimize its performance. A large volume of such studies has been already performed in the context of classical optical communications, optical sensing, and astronomy. Quantum links, however, set their own specific requirements, which demands a more focused investigation. The most important factor is the type of encoding. One of its types especially resilient to atmospheric turbulence is the polarization-based CV prepare-and-send approach. In this protocol Alice prepares continuous-value polarization states by modulating polarization of the reference beam which she sends to Bob. Because the signal and LO are transmitted in the same spatial mode, they experience the same atmospheric fluctuations that cancel out in Bob's measurement. This type of link has been tested for realization of a two-state and a four-state alphabets under unfavorable atmospheric conditions of a daytime urban environment [139]. No detectable added polarization noise was observed in spite of considerable beam profile distortions. Furthermore, a polarization-squeezed quantum optical state was transmitted through the same link [140]. Such nonclassical light is known to be fragile to decoherence and quickly lose its squeezing properties in the presence of noise. However, in this case a quantum state tomography showed that the squeezing was preserved in transmission. This implies that the atmospheric polarization-based CV QKD channels can be used not only for prepare-and-send but also for entanglement-sharing QKD protocols. This is important because the entanglement-sharing protocols are expected to be more loss-tolerant than the prepare-and-send protocols. It is shown [141] that by using the Gottesman–Lo two-way postprocessing, the entanglement-sharing QKD can tolerate up to 70 dB channel loss, assuming that the entanglement source is half the way between Alice and Bob.

Demonstrations [139,140] were carried out with a stationary sender and receiver located in two buildings 1.6 km apart. However, many practical applications may require QKD between moving platforms, for example, between an aircraft and a ground station. This adds an extra layer of complexity associated with dynamic acquisition and tracking of optical links, as well as mitigating the Doppler effect, which affects the phase and ppm encoding. In 2012 a Dornier 228 utility aircraft was used to distribute the crypto key between two ground stations over a distance of 144 km using a BB84 protocol [142]. The aircraft was moving at 4 mrad/s with respect to the ground station, mimicking a realistic optical communication link to a satellite. A bidirectional optical link was established, and the mean sifted key rate of 145 bps with QBER of 4.8% was maintained for approximately 10 minutes over a 20 km range. Similar demonstrations were performed in 2013 [143] with a hot-air balloon and ground stations implementing a polarization-based decoy-state protocol described in [71]. In this case the key rates exceeding 150 bps with QBER below 2.8% was reported for the 20 and 40 km links, and the key rate of 48 bps with QBER of 4% for the 96 km link.

In all the above demonstration the prepare-and-send downlink approach was adopted, with the moving platform being the transmitter and the stationary platform being the receiver. A team from the Institute for Quantum Computing (IQC, Canada) explored the potential of the *uplink* configuration. In 2015 this team used a pickup truck as a moving receiver. The truck was moving with 33 km/h linear speed, or 13 mrad/s angular speed with respect to the stationary transmitter some 650 m

Figure 12.5 A quantum receiver ("Bob") onboard of an airplane used in the Institute for Quantum Computing (IQC) free-space quantum key distribution (QKD) demonstration. Source: *Quantum Encryption and Science Satellite (QEYSSat) research group, Institute for Quantum Computing (IQC), University of Waterloo.*

away. The secure key rate of 40 bps was reported, with QBER of 6.5%–8% [144]. In 2017 this team performed a similar experiment using a variation of BB84 protocol and an airplane instead of a truck [145]; see Fig. 12.5. In these measurements the optical links of 310 km were maintained for relatively short periods of time, ranging from 30 seconds to almost 6 minutes. The angular velocities were similar to those expected for the LEO satellites. The best average sifted key rate was 32.3 kbps, the best secure key rate was 3.5 kbps.

Not all R&D in free-space QKD is concerned with large distances or continuous motion. A short-range handheld optical QKD transmitter and receiver were recently developed [146] and proposed for using in ATM transactions, secure building access, and similar scenarios. These devices can quickly acquire optical link at a half-meter range and sustain a secure key rate over 30 kbps using a polarization-based BB84 protocol with decoy states. While the practical utility of an unconditionally secure optical link that is less than a meter long may be questioned, it certainly provides an impressive technology demonstration.

12.4.3 Quantum key distribution in satellite communication

The ultimate implementation of long-range free-space QKD is the satellite QKD. Compared to "horizontal" atmospheric links, satellite QKD has to deal only with a relatively thin layer of dusty and turbulent dense atmosphere. A different loss mechanism, mainly associated with collecting a diverging beam with a limited-size aperture, becomes dominant. Note that this type of loss scales quadratically with the link length, which is a far more favorable scaling than exponential. Therefore much longer QKD links may be afforded.

The satellite QKD can be realized in two distinct scenarios. In one scenario, the key is shared with the spacecraft itself, that is, Alice or Bob rides it. This scenario may be relevant when one wishes to provide security for critical data exchanges between a ground station and the spacecraft, or possibly between two spacecraft. In the second scenario, a satellite may be used as an entanglement distribution node or as a relay node, enabling a secure link between Alice and Bob, who both are ground-based. This scenario covers a broader range of possible applications, including a very economic realization of a Q-WAN. Indeed, assuming realistic optics aperture sizes and atmosphere quality, an optical link provided by a satellite at a 1000 km altitude to two ground stations 2000 km apart will have a total loss roughly equivalent to 200 km of standard optical fiber, that is, approximately 40 dB. If we furthermore make a rather acceptable assumption that the satellite is a *trusted* node allowed to share Alice's and Bob's key, then it will be able to support a global Q-WAN by establishing the key with Alice and with Bob in turns as it passes over their locations.

Realization of the satellite QKD required extensive support from the classical optical communication technology, in particular concerned with optical link acquisition and tracking, as well as with dynamical correction of the phase and polarization drifts. Early polarization measurements using satellite-based retroreflectors date back to the end of the 20th century, when a 48 kg Starlette/Stella was launched in 1993. This was followed by a larger 500 kg CHAMP launched in 2000, a 21 kg Larets launched in 2003, and a 510 kg Jason-2 launched in 2008. All these platforms were used to perform polarization measurements in ground-LEO-ground optical links. The goal of these measurements was to demonstrate the optical link quality compatible with polarization-based QKD protocols. A typical QBER was shown to be in the range 4%–7%, or SNR of 16:1 for CHAMP, which fits the requirement.

A more advanced polarization measurement with the satellite assuming the role of active transmitter was performed with a 570 kg OICETS launched in December 2005 by JAXA. This mission pursued multiples goals related to the QKD and a broader field of optical communications: to establish optical links with ESA's *Artemis* GEO, with a NICT stationary optical ground station in Koganei, Japan, and with DLR's mobile ground station. All these goals were achieved. A polarization measurement of the downlink signal performed at the Koganei station demonstrated the polarization direction rms error of 28 mrad and the degree of polarization 99.4% ± 4.4%.

These encouraging results warranted further tests aimed at verifying that the noise in optical satellites links can be quantum-limited. One such a test was performed in August of 2016 with a LEO microsatellite SOCRATES (Space Optical Communications Research Advanced Technology Satellite), a 48 kg cube with a

50 cm side, hosting a Small Optical Transponder terminal [147]. A polarization-based link at 10 MHz rate was established between SOCRATES and a ground station receiver. The received quantum states had an average of 0.146 photons per pulse. The QBER was measured to be under 5%. The link was validated as suitable for the practical QKD.

A phase encoding has been also proven to be quantum noise limited in a test involving a laser communication terminal of Tesat Spacecom on GEO satellite *Alphasat* [148]. In this test the phase-encoded signals transmitted from *Alphasat* to a ground station shown the mean quadrature variance of 1.01 ± 0.03 in shot noise units, thereby validating the concept of the phase or quadrature based QKD between ground stations and satellites.

A more QKD-specific example of satellite technology demonstration is the development of flight-compatible source of polarization-entangled photon pairs carried out by the National University of Singapore in the SPEQS and GALASSIA projects. A 2U SPEQS Cubesat hosting a photon-pair source based on nondegenerate (405 nm $\rightarrow$ 760 nm + 867 nm) collinear type-I SPDC in a BBO crystal was launched in October 2014 in the payload bay of the Cygnus CRS Orb-3 ISS resupply mission. This source was subjected to an inadvertent extreme environmental testing by the carrier rocket explosion. Remarkably, it continued to function after being recovered from the postexplosion debris, albeit with a marginally degraded performance [149]. GALASSIA is a 1.65 kg, 2U Cubesat that carried a very similar source that successfully launched in December 2015. The in-orbit polarization correlation with this source reached a contrast of 97% ± 2% [150].

The success of free-space QKD demonstrations with moving platforms led to a rampant growth of satellite QKD programs worldwide. A recent Optics and Photonics news article [151] graphically captured this phenomenon, as shown in Fig. 12.6. Below we give a brief summary of the existing and upcoming space QKD programs.

QUESS Quantum Experiments at Space Scale (QUESS) is a Chinese-Austrian LEO satellite mission operated by the Chinese Academy of Sciences. In the framework of this mission, a 631 kg satellite *Micius* was launched on August 16, 2016. The mission goal was to demonstrate an LEO-to-ground key distribution with a trusted-node satellite, uplink quantum teleportation, and double downlink entanglement distribution. The QKD part of the mission was performed using the BB84 protocol with decoy states across up to 1200 km downlink. The sifted key rate of 14 kbps was reached, with the QBER of approximately 1% [152,153]. Serving as a trusted node, Micius created a shared secure key with ground stations in Austria and China as it was flying over their locations. Then by performing exclusive OR operations with two keys and distributing the result through a public channel, a secret key was created between the Austria and China stations 7600 km apart, as illustrated in

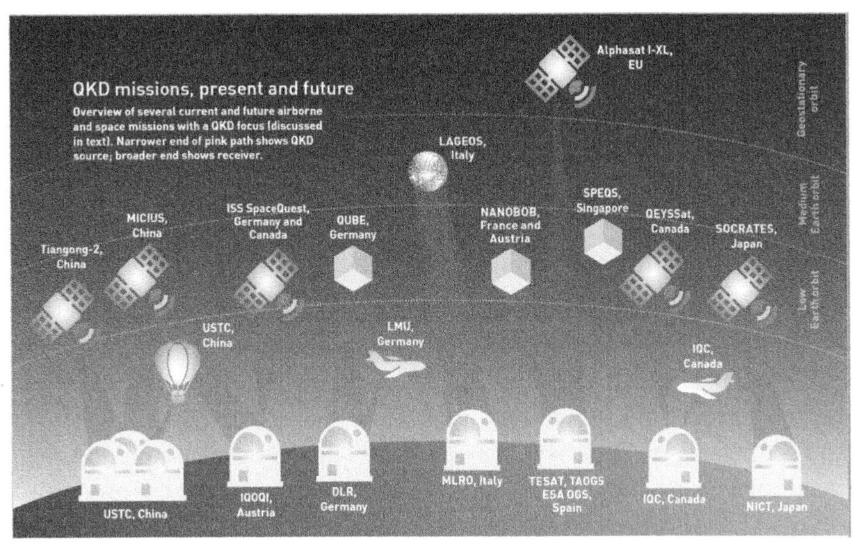

Figure 12.6 Launched and upcoming quantum key distribution (QKD) space missions. Source: *I. Khan, B. Heim, A. Neuzner, C. Marquardt, Satellite-based QKD. Opt. Photonics News 29 (2) (2018) 28–33.*

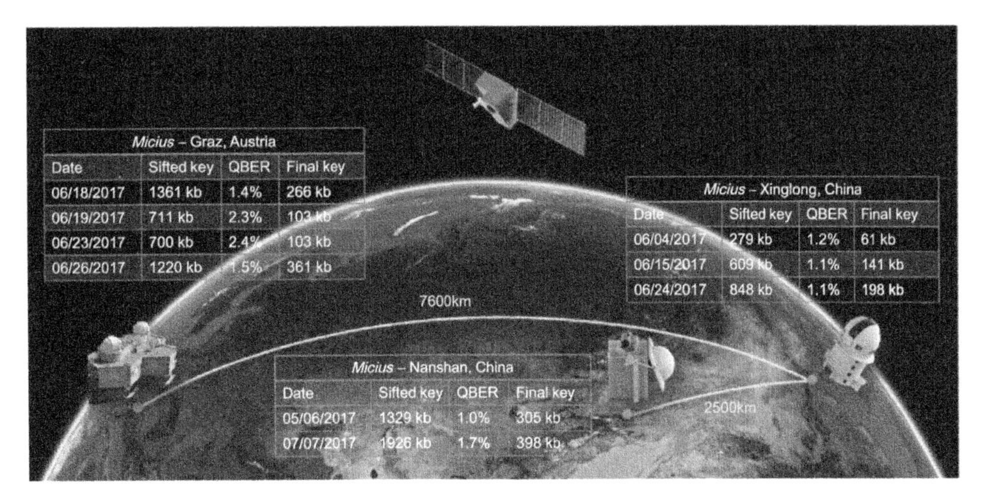

Figure 12.7 The first intercontinental quantum key distribution (QKD) link with Micius as a trusted node, and a summary of various links' performance. Source: *[153] S.-K. Liao, W.-Q. Cai, J. Handsteiner, B. Liu, J. Yin, L. Zhang, et al., Satellite-relayed intercontinental quantum network, Phys. Rev. Lett., 120 (3) (2018) 030501.*

Fig. 12.7. This key was utilized to facilitate the first intercontinental quantum-secure video conference.

Micius also demonstrated polarization-entanglement distribution to two ground locations 1203 km apart through satellite-to-ground downlinks adding up to

1600–2400 km [154]. The entanglement of received photons was verified by a Bell measurement violating the local realism by four standard deviations. The importance of this measurement comes from the fact that it was done under the strict Einstein locality conditions, such that each local measurement was well outside of the other measurement's light cone. This eliminated one of important loopholes in Bell measurements.

The QUESS mission is anticipated to expand, counting up to 10 satellites and enabling a EuropeanAsian Q-WAN by 2020. By 2030 it is expected to support a global QKD network, with the projected secure key cost of 10 cents per kbit.

Tiangong-2 is a large multipurpose space laboratory operated by the Chinese Academy of Sciences launched on September 15, 2016. Its small-sized (57.9 kg) QKD payload has established a quantum downlink with a ground station in Nanshan using a bidirectional tracking system. Key rate over 100 bps was demonstrated across the 719 km distance in daylight [155]. This demonstration also relied on polarization-based BB84 protocol with decoy states.

QUBE. A cubesat QKD program combining efforts of MPL, DLR, OHB Systems, ZFT (Zentrum für Telematic E.V.), and LMU (Ludwig-Maximilians-Univerität München) announced in the Space News press release from September 25, 2017 (page 22). QUBE will launch a cubesat in 2020, that will test different highly integrated QKD transmitters and an integrated photonic chip as a quantum random number generator, see Fig. 12.8.

QUARTZ The Quantum Cryptography Telecommunication System (QUARTZ) is a SES-led consortium consisting of 10 partners: MPL, DLR-IKN, AIT, Tesat-Spacecom, ID Quantique, Itrust consulting, LMU, LuxTrust, Palacky University, and TNO. Inaugurated in May 2018, the QUARTZ project will develop technology for a world-wide commercial satellite QKD service.

SpooQySats A follow-on to the earlier SPEQS and GALASSIA projects designed to demonstrate polarization-entangled two-photon sources in space, packaged in a 3U Cubesats. SpooQy-1 was planned to launch in 2018, SpooQy-2 in 2020.

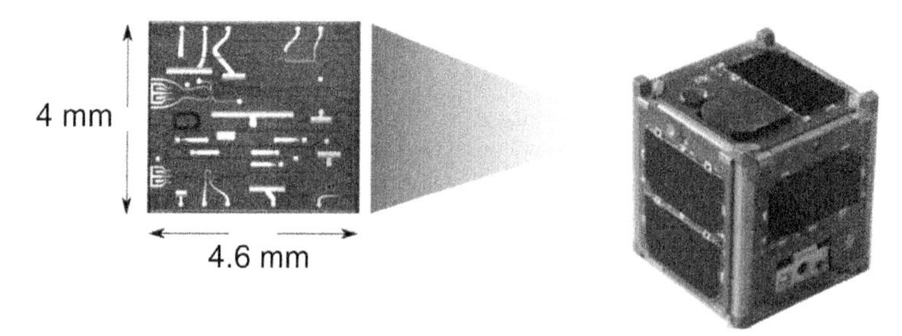

Figure 12.8 The QUBE quantum key distribution (QKD) cubesat featuring a chip-based optical quantum random number generator. Source: *MPL and Wikipedia (adapted picture of cubesat from Cbrandovnt under CC BY 3.0).*

QEYSSat Quantum Encryption and Science Satellite (QEYSSat) is a Canadian Quantum Satellite project [156]. A 23 kg satellite with less than 30 W power budget and under 60 cm^3 volume is planned to be launched in 2021–22. In keeping with the traditional IQC uplink approach, the satellite will host a quantum receiver to detect the optical signals sent from the ground.

CAPSat A NASA-funded mission developed by undergraduate students of University of Illinois. A 3U Cubesat will be launched to test laser annealing of radiation-damaged APDs. APDs, the most common class of photon-counting detectors and critical QKD elements, are known to suffer from radiation damage. The investigation of possible methods to minimize or reverse this damage is important for space QKD.

NanoBob A 12 U CubeSat mission developed by of University of Grenoble. This mission is designed to demonstrate a polarization-based QKD uplink to the CubeSat, which will serve as a trusted node and distribute the key to ground-based users. It will furthermore focus on supplemental tasks such as accurate clock synchronization between the ground station and the CubeSat, accurate pointing of source and receiver, using a standard telecom laser wavelength (1550 nm) for a laser tracking beacons, and for fast classical optical communication using ppm encoding at up to 1 Gbit/s.

NanoQEY A mission proposal from the IQC/SFL team based on the SFLs 16 kg NEMO nanosatellite bus. Its goal is to perform ground-to-LEO Bell tests and QKD demonstration with a trusted-node satellite. The objective is to reach 10 kbps secure key rate.

CQuCoM A mission proposal from University of Strathclyde, Glasgow, to launch a 6U CubeSat from the ISS. The mission goal is to demonstrate LEO-to-ground QKD downlink. The objective is to reach 10 kbps secure key rate.

QUARC Quantum Research CubeSat (QUARC). A University of Strathclyde-led collaboration proposal with the goal to demonstrate a free-space quantum optical link using a CubeSat, and to solve the acquisition, tracking, pointing, and turbulence mitigation challenges.

Further details on these and other CubeSat and microsatellite QKD and supplementary projects can be found in a recent review [157].

12.5 Quantum supremacy in information processing

Can the quantum theory improve communication channels and protocols beyond the classical bounds, besides making them secure against eavesdropping? This question was studied from various perspectives since the early days of quantum optics, occasionally leading to rather exotic proposals. For example, a quantum repeater could be used in observational astronomy [158]. As a long-baseline or synthetic aperture telescope detects a faint signal from a distant object, it needs to send the precious collected photons to a detector across a large distance. If a quantum repeaters could make this process lossless, it would remove the baseline length limitation.

However, the two most popular approaches to reaching the quantum supremacy in communications are to leverage either the reduced noise or the increased information content of quantum channels. The idea of reducing a communication channel noise below classical limits comes from the quantum metrology, where such advantageous possibility is well established; see for example, review [159]. In communication theory, this idea is exploited by designing alphabets and measurements in such a way that the Shannon limit (0.9) can be surpassed, which means that the quantum communication protocol outperforms the classical one. This has been theoretically demonstrated for the phase shift keying (PSK) and the coherent frequency shift keying communication protocols [160]. The proposed quantum measurement strategies allow for these protocols to be simultaneously optimized with respect to the channel SNR and its bandwidth.

The idea of increasing the information content of optical signals beyond the classical limits hinges on the concept of dense or superdense quantum encoding.

12.5.1 Dense and superdense encoding of information

The dense coding in DV quantum communications was proposed in 1992 [161] and experimentally demonstrated in 1996, when a polarization Bell-state measurement of an SPDC photon pair was shown to produce more than one bit of information [162]. In fact, three equally probable different messages were thus encoded, which corresponds to $\log_2 3 \approx 1.585$ bits of information. Importantly, this information was encoded by manipulating only one photon of the pair, even though the second photon was still required to complete the measurement. Soon after the first demonstration of dense coding in a DV channel, it was pointed out that a very similar technique should achieve a dense encoding in a CV channel [163].

As the use of quantum dense encoding for communications requires distribution of entangled photon pairs prior and additionally to sending an actual message, its practical advantage is sometimes disputed in spite of a higher information content of the message (in bits per photon). However, this technique may still have merit in communication scenarios when the entanglement resource can be shared under more favorable conditions than when the communication has to take place. Of course, the capability of storing the entangled photons (a quantum memory) would be required in such scenarios.

Dense coding uses only a three-letter alphabet (approximately 1.585 bits) instead of its full capacity of a four-letter alphabet (2 bits) because a full Bell-state analysis is complicated and cannot be accomplished by means of linear optics. This problem can be circumvented by using *hyperentangled* states: such states that are entangled with respect to more than one physical parameter [164]. Consider, for example, Bell states (0.24) where "0" and "1" are represented not only by the polarization but also by linear or angular momentum, frequency, or time-bin. This does not imply a larger alphabet, as the extra parameters are correlated with polarization, but it provides a platform for a simple measurement to efficiently discriminate the four basis states.

Dense coding based on the hyperentangled states is traditionally called superdense coding. This technique attempts to extract the entire two bits of information inherent in the Bell-state measurement. In practice, however, this limit cannot be reached because of the limited measurements efficiency and fidelity. Instead, the bound (0.17) with empirically determined conditional probabilities applies. The best-known quantum information capacity of 1.665 ± 0.018 bits per photon was recently reported in a fiber-based communication channel [165], improving on the previous-best free-space result of 1.629 ± 0.006 bits per photon [166].

The success of fiber-based and free-space superdense optical encoding inspired an ambitious technology demonstration wherein a source of polarization–time-bin hyperentangled photon pairs would be placed onboard of the International Space Station (ISS) and establish a communication link with a ground station [167]. The goal of this project is to perform a quantum teleportation using a hyperentangled state from the ISS to the ground station, along the lines of a lab-based demonstration [168].

Entanglement in higher dimensions is another way to increase the information capacity of a photon [59,60,169,170]. This can be achieved by using a larger alphabet. We already discussed the larger alphabets in the context of QKD in Section 12.3.1, where we gave references to some important implementation examples. Several types of sources of photon pairs entangled in many dimensions have been demonstrated. One such source produced photon pairs entangled in 100 orbital angular momentum (OAM) states, therefore spanning a 100×100 Hilbert space [171]. OAM entanglement has limited utility for practical communications, especially over a long range, as such states cannot be transmitted in telecom fibers and are subject to strong diffraction in free space. However, they have been recently used in a table-top demonstration of a universal QKD setup capable of supporting several different protocols without reconfiguring [172].

A chip-based source of photon pairs suitable for fiber communications was built in 2017 using a micro-ring resonator [173]. This source produces frequency-entangled photon pairs in 10×10 Hilbert space. In this demonstration the photon pairs are generated in a micro-ring resonator through degenerate four-wave mixing, or spontaneous hyperparametric downconversion, of a pump laser light. A discrete, quasiequidistant spectrum of the resonator gives rise to a frequency comb, whose lines are pair-wise entangled. A much large frequency comb, counting more than 10,000 frequency lines, was generated in a fiber resonator through *non*degenerate four-wave mixing, a process requiring two pump lasers, a year earlier [174]. However, in this demonstration the possible frequency entanglement was not discussed. It was designed to simulate a large Ising spin network for adiabatic quantum computing—an exciting topic we will briefly touch upon in the end of this section.

Let us point out that encoding more than one bit of information in a single photon is not necessarily a quantum phenomenon. It is possible to get greater information capacity in classical communications using large alphabets. The pulse-position modulation encoding is one such example. Using this technique in the Mars Laser Communication Demonstration project, the Jet Propulsion Lab (JPL) team reported the average information capacity of 1.12 bits per photon in 2007 [175]. A year later the group demonstrated 2.8 bits per photon capacity in a 30-Mbps free-space channel used to transmit HDTV video signal between JPL buildings.

12.5.2 Quantum algorithms

Transfer of information is closely related to *processing* of information. In fact, some previously discussed protocols such as error correction already fall into this category. While it would not be possible to provide a comprehensive review of the field of quantum information processing within this chapter, we would like to list several quantum information algorithms that we see as most relevant to the topic.

Quantum bit commitment. In this protocol, Alice needs to commit information to Bob. This information may represent, for example, Alice's intended response to Bob's hypothetical proposal. She may or may not reveal it to Bob later, but she is not allowed to change it after the commitment is made. This is equivalent to Alice sending Bob a safe with a letter inside but holding back the key. Remarkably, quantum mechanics prohibits such bit commitment that is absolutely secure against both Alice's and Bob's possible cheating (alteration and eavesdropping, respectively). This is yet another no-go theorem that can be rigorously proven. However, weaker versions of secure bit commitment can be realized, for example, when the cheating is limited to Gaussian operations [176]. Alternatively, a secure quantum bit commitment can be achieved in a protocol that depends on special relativity principles, specifically, on the limitation of communication speed [177].

Quantum secret sharing. A classical task of secret sharing is to distribute information among a group in such a way that only a specified subset of its members (which may or may not be equal to the entire group) can reconstruct the entire secret message; however, no smaller subset or an individual member may possesses this information. This can be envisioned as a safe with multiple locks, whose keys are distributed to different group members. *Quantum* secret sharing addresses the situations when the secret message is represented by a quantum state, or when quantum principles are used for distributing classical information. Unlike quantum bit commitment, quantum secret sharing is allowed by nonrelativistic quantum theory but requires a meager resource such as multipartite entangled states—typically *cluster* [178] or *graph* [179] states. Quantum secret sharing is fundamentally different from QKD in that it is not geared to facilitate private communication between any two parties. But it can be modified to support advanced multiuser QKD network geometries such as branching, star and loop structures, and others; see for example Ref. [180] and references therein.

Quantum money. This rather evident application of the no-cloning theorem was first published by Stephen Wiesner in 1983, just before the pioneering QKD publications. It was, however, deemed impractical, because it required a long-term quantum memory and because verification of a quantum coin authenticity requires multiple quantum communications with the trusted authority (the bank). Because of the latter requirement, the quantum money algorithm could not be proven secure until 2012, when its version was proposed that only used classical communications between the bank and the quantum coin owner. Further investigations of this protocol established that it remains secure up to 23% error rate [181], which is a practically acceptable level. An experimental attempt to forge a "quantum coin," implemented as an image encoded in the polarization states of single photons, was made [182]. The conclusion of this test was that the coin cannot be ideally copied without alerting the owner. Modern research in the field includes developing of the semidevice-independent protocol (compare to MDI QKD approach discussed in Section 12.3.1.1) to cover the situations when the communication terminal is not trusted [183].

Quantum signature. Similarly to the quantum money, the quantum signature protocol relies on the no-cloning theorem to protect a digital signature from forging. The protocol has two stages. In the distribution stage, the sender provides recipients with a set of nonorthogonal quantum states. There are as many states in the set as there are possible messages. Then in the messaging stage, one of the messages is sent along with a classical "key" revealing which quantum state from the set is paired with this message. The recipients therefore can verify the identity of the sender as the same person who initially distributed the states by performing a state analysis. Its results are conclusive and trustworthy because the states cannot be duplicated by an impostor. The original version of this protocol required the recipients to possess a long-term quantum memory, which made the protocol impractical. However, later this requirement was replaced by a type of measurement known as the *unambiguous quantum state elimination*. Therefore storing of only classical information is required [184]. This gave the quantum signature protocol a considerable practical appeal and stimulated its further research. Its progress is covered in review [185].

Quantum coin tossing. This appears to be a straightforward application for a quantum random number generator discussed in Section 12.2.3. However, the problem becomes nontrivial if the two communicating parties do not trust each other and suspect that whoever has possession of the generator may statistically bias its outcomes.

This difficulty can be avoided by using an algorithm resembling BB84, when Alice and Bob encode and decode randomly chosen quantum states in randomly chosen bases. They abort the protocol if the identical preparation and measurement bases lead to a discrepancy between the prepared and measured states. This protocol is theoretically described in Ref. [186] and experimentally verified in Ref. [187].

Quantum fingerprinting. Fingerprinting is representing a long string of characters by a shorter string in such a way that the original string may be identified from the short string (the fingerprint) with minimal error probability. In contrast with the classical information theory, quantum fingerprints can be made exponentially small compared with the originals, even without a requirement for correlations or entanglement between the parties. This can be achieved using weak coherent states with the number of modes proportional to the initial string length, but with a small mean-number of photons in each mode. These are the photons that constitute the fingerprint, and it is shown theoretically [188] and verified experimentally [189] that this protocol offers an exponential advantage over classical fingerprinting in terms of the fingerprint length [190].

12.5.3 Quantum computing

Quantum computing may be viewed as the apex of quantum information processing algorithms. There are two mainstream paradigms this term may refer to. One is the "digital," or gate-based, quantum computer, and the other is the "analog" quantum computer, which also may be called a quantum simulator.

The first paradigm is more common, and usually a reference to a "quantum computer" defaults to the gate-based concept. This concept was suggested in 1982 by Feynman [191]. It is based on the quantum superposition principle that is harnessed to simultaneously carry out a large number of computations by the same physical device. To achieve this quantum parallelism, the logic operations need to be performed by a quantum system, and the computation result "lives" in Hilbert space. To read the result in the end of such a quantum computation, one has to perform a projection measurement and collapse a quantum state to a classical outcome. Because of this, many quantum algorithms are probabilistic: they give the correct answer with a less-than-unity probability. This does not present a serious problem if the correct answer probability is sufficiently large, a simple verification protocol is available, and the number of attempts is unlimited.

Quantum computers require systems that are well isolated from the decohering effects of their environment, while at the same time allowing precise manipulation during computation. Meeting these demands has proven difficult, which has motivated the search for schemes that can demonstrate quantum-enhanced computation under more favorable experimental conditions. One such a scheme is Boson sampling. This simple task is native to linear optics. It has captured the imagination of quantum scientists because it seems possible that the anticipated supremacy of quantum machines could be demonstrated by a near-term experiment.

The distribution of bosons that have undergone a unitary transformation U is thought to be exponentially hard to sample from a classical standpoint. The probability amplitude of obtaining a certain output is directly proportional to the permanent of a corresponding submatrix of U. The permanent expresses the wave function of identical bosons, which are symmetric under exchange; in contrast, the Slater determinant expresses the wave function of identical fermions, which are antisymmetric under exchange. Whereas determinants can be evaluated efficiently, permanents have long been believed to be hard to compute. The best-known algorithm scales exponentially with the size of the matrix.

An ideal quantum boson-sampling machine (QBSM) creates indistinguishable bosons, physically implements U, and records the outputs. Although the QBSM is not believed to efficiently estimate any individual matrix permanent, for a sufficiently large system it is expected to beat the classical computer in sampling over the entire distribution. This has been demonstrated by a number groups using a variety of platforms; see for example, Ref. [192].

Several other practical problems can be efficiently solved using quantum computing algorithms. One such problem is a database search. Classically, a search of a database of length N requires $O(N)$ operations. Quantum *Grover's algorithm*, proposed by Grover [193], allows this problem to be solved with $O(\sqrt{N})$ operations—that is, it provides a quadratic improvement over the classic search algorithms.

An even more dramatic quantum advantage is achieved for the factorization problem, when a large number needs to be presented as a product of its prime factors. A classical solution to this problem requires exponential resources with respect to the number length. The quantum *Shor algorithm* [194], initially formulated by Peter Shor in 1994, solves this problem with polynomial resources making use of the quantum interference phenomenon. Note that the *classical* optical interference can be used in a similar way [195].

Besides the database search and number factorization, quantum computers could be efficient for computational modeling of other quantum systems. Unlike a practically useful realization of the Grover's or Shor algorithms, this application does not necessarily require a large number of quantum-logic gates, or qubits. For example, the Rigetti and IBM quantum computers used to perform the first nuclear calculation had 19 and 16 qubits, respectively.

Practical realization of the Shor algorithm for large numbers would entail very serious implications for communication security. The commonly used Diffie–Hellman public-key security protocol relies on the computational hardness of the factorization problem, that is, on the exponentially growing demand for the computation resources.

With that demand becoming polynomial, Diffie–Hellman encryption will become easily defeated, and the secure communications as we know it today will become vulnerable. This affects not only the newly communicated encoded messages but also those that were communicated in the past but may be presently stored in a computer memory waiting for a quantum computer to crack them. It is not a coincidence that the surge of interest in practical QKD, which we illustrated in Sections 12.4.1–12.4.3, coincided with a decisive progress in quantum computing.

To the best of our knowledge, the longest number that was factorized by a quantum computer running Shor algorithm to date does not exceed 16 bits [196]. However, the rate of technological progress in the field, and especially its acceleration in the past several years, is very impressive. Several major companies presently have their quantum computing programs. At the end of 2017, Rigetti made its 19-qubit "19Q" quantum processor accessible via cloud services through the company website. At the same time, IBM unveiled its 20 and 50 qubit quantum computers, closely followed by Intel's announcement of a 49-qubit test chip in January 2018. Google's Quantum AI Lab announced a 72-qubit quantum processor at the APS meeting in March 2018. This processor is believed to be capable of achieving the *quantum supremacy*, the point when it could solve problems intractable with the state-of-the-art classical computers.

The second paradigm of quantum computing is based on mapping of a problem at hand onto a quantum system's Hamiltonian, such that, for example, a ground-state energy of the Hamiltonian corresponds to the optimal solution of the problem. For example, many combinatorial optimization problems can be mapped on the Ising Hamiltonian

$$H = \frac{1}{2}\sum_{i,j=1}^{N} J_{ij}\sigma_i\sigma_j + \sum_{i=1}^{N} h_i\sigma_i \tag{12.48}$$

describing N coupled spins in external magnetic field, where $J_{i\ j}$ are the coupling coefficients, h_i are Zeeman shifts, and σ are the spin operators with eigenvalues ± 1. Programming such a quantum computer amounts to setting up a desired array of $J_{i\ j}$ final values. The computation process consists of first initializing the system with all $J_{i\ j} = 0$ and then adiabatically turning the coupling on until it reaches the desired "landscape" of potentials. According to the principles of quantum mechanics, this process (called the *quantum annealing*) should transfer the system to the fundamental state of the final Hamiltonian, hence delivering the optimal solution of the problem mapped onto the Ising Hamiltonian. This type of quantum computing is therefore also known as the *adiabatic computing*.

The main challenge of adiabatic quantum computing is to freely and independently control all coupling coefficients $J_{i\ j}$ for a large system. One approach to this problem is to emulate the spins as superconducting qubits, shaped as long and narrow loops. The loops can be put into an array with their long axes aligned, and two such arrays can be placed on top of one another in perpendicular direction. In this geometry every qubit of the first array can be coupled to every qubit of the second array, with the coupling coefficient controlled with the DC bias.

This type of geometry was implemented by a Canadian company D-Wave Systems, Inc. The approach was successful, and D-Wave became the first (and to best of our knowledge, the only) commercial company to actually sell the adiabatic quantum computers, as well as to offer the web services in quantum problem solving using this technology.

In spite of its success, the two-array geometry of artificial spins is not the best possible implementation of the Ising Hamiltonian. Coupling between the spins within each array is difficult, and therefore the matrix $J_{i\ j}$ is sparse. It is possible that a better connectivity may be achieved in an all-optical system. We already discussed optical frequency combs in the context of multidimensional photon entanglement. This type of comb was also shown to emulate a large-scale Ising spin network [174,197]. A recent report [198] indicates that such an all-optical Ising network may have all-to-all qubits connectivity, achieving an exponential advantage over the state-of-the-art D-Wave 2000Q machine. Moreover, unlike the superconducting cubits, optical cubits do not require liquid helium cooling to operate and are nearly immune to decoherence. A practical realization of the all-optical quantum annealing machine may have implication beyond what we may currently expect.

Concluding this chapter, we would like to mention that quantum communication research is not limited to the problems of transfer and processing of information. Performing quantum optics experiments over large distances at high speeds and in significantly gravitational potential offers a unique opportunity for fundamental research, for example, unifying quantum theory and relativity. On the other hand, understanding the relativistic effects impacts of quantum communications is important for their successful long-term deployment. In Section 12.2.4.5 we already mentioned the experimental attempts to establish the lower limit for the speed of superluminal communications and the wavefunction collapse. The review [199] discusses various other direct tests of quantum theory at large = length scales, approaching that of the radius of curvature of spacetime. At such scales one begins to probe the interaction between gravity and quantum phenomena and gains the access to a wide variety of potential tests of fundamental physics.

Acknowledgments

This research was partly carried out at the Jet Propulsion Laboratory, California Institute of Technology, under a contract with the National Aeronautics and Space Administration.

References

[1] C.E. Shannon, A mathematical theory of communication., Bell Syst. Tech. J. 27 (3) (1948) 379–423.
[2] C. Arndt, Information Measures, Information and its Description in Science and Engineering, Springer, New York, 2004.
[3] D.A. Huffman, A method for the construction of minimum-redundancy codes, Proc. IRE 40 (9) (1952) 1098–1101.
[4] J. Ziv, A. Lempel, Compression of individual sequences via variable-rate coding, IEEE Trans. Inf. Theory 24 (5) (1978) 530–536.
[5] T. Cover, J. Thomas, Elements of Information Theory., Wiley, New York, 1991.
[6] H.P. Robertson, The uncertainty principle, Phys. Rev. 34 (1929) 163–164.
[7] L.I. Mandelshtam, I.E. Tamm, The uncertainty relation between energy and time in nonrelativistic quantum mechanics, J. Phys. IX (4) (1945) 249–254.
[8] J. Hilgevoord, The uncertainty principle for energy and time. II, Am. J. Phys. 66 (May 1998) 396–402.
[9] M. Benedicks, On fourier transforms of functions supported on sets of finite lebesgue measure, J. Math. An. Apps. 106 (1985) 180–183.
[10] L.I. Mandelstam, I.E. Tamm, The uncertainty relation between energy and time in non-relativistic quantum mechanics, J. Phys. 9 (1945) 249–254.
[11] J. Uffink, The rate of evolution of a quantum state, Am. J. Phys. 61 (1993) 935–936.
[12] R. Loudon, The quantum Theory of Light., Oxford university press, 2000.
[13] T. Opatrny, Number-phase uncertainty relations, J. Phys. A Math. Gen. 28 (1995). 6961–6915.
[14] L. Susskind, J. Glogower, Quantum mechanics phase and time operator, Physics 1 (1964) 49–61.
[15] S.M. Tan, D.F. Walls, M.J. Collett, Nonlocality of a single photon, Phys. Rev. Lett. 66 (3) (1991) 252–255.
[16] G. Björk, P. Jonsson, L.L. Sánchez-Soto, Single-particle nonlocality and entanglement with the vacuum, Phys. Rev. A 64 (4) (2001) 042106.
[17] B. Hessmo, P. Usachev, H. Heydari, G. Björk, Experimental demonstration of single photon nonlocality, Phys. Rev. Lett. 92 (18) (2004) 180401.
[18] G.K. Gulati, B. Srivathsan, B. Chng, A. Cerè, D. Matsukevich, C. Kurtsiefer, Generation of an exponentially rising single-photon field from parametric conversion in atoms, Phys. Rev. A 90 (3) (2014) 033819.
[19] S.M. Hendrickson, C.N. Weiler, R.M. Camacho, P.T. Rakich, A.I. Young, M.J. Shaw, et al., All-optical-switching demonstration using two-photon absorption and the zeno effect, Phys. Rev. A 87 (2013) 23808.
[20] B.D. Clader, S.M. Hendrickson, R.M. Camacho, B.C. Jacobs, All-optical microdisk switch using eit, Opt. Express 21 (2013) 6169–6179.
[21] Y.-P. Huang, P. Kumar, Interaction-free all-optical switching in $\chi^{(2)}$ microdisks for quantum applications, Opt. Lett. 35 (2010) 2376–2378.
[22] Y.-Z. Sun, Y.-P. Huang, P. Kumar, Photonic nonlinearities via quantum zeno blockade, Phys. Rev. Lett. 110 (2013) 223901.
[23] D.V. Strekalov, A.S. Kowligy, Y.-P. Huang, P. Kumar, Progress towards interaction-free all-optical devices, Phys. Rev. A 89 (2014) 063820.
[24] X.Y. Zou, L.J. Wang, L. Mandel, Induced coherence and indistinguishability in optical interference, Phys. Rev. Lett. 67 (1991) 318–321.
[25] T.B. Pittman, Y.H. Shih, D.V. Strekalov, A.V. Sergienko, Optical imaging by means of two-photon quantum entanglement, Phys. Rev. A 52 (1995) R3429–R3432.
[26] G.B. Lemos, V. Borish, G.D. Cole, S. Ramelow, R. Lapkiewicz, A. Zeilinger, Quantum imaging with undetected photons, Nature 512 (2014) 409–412.
[27] P.G. Kwiat, A.G. White, J.R. Mitchell, O. Nairz, G. Weihs, H. Weinfurter, et al., High-efficiency quantum interrogation measurements via the quantum zeno effect, Phys. Rev. Lett. 83 (23) (1999) 4725–4728.

[28] The BIG Bell Test Collaboration, Challenging local realism with human choices, Nature 557 (7704) (2018) 212–216.
[29] C. Gabriel, C. Wittmann, D. Sych, R. Dong, W. Mauerer, U.L. Andersen, et al., A generator for unique quantum random numbers based on vacuum states, Nat. Photonics 4 (2010) 711–715.
[30] R. Impagliazzo, M. Luby, One-way functions are essential for complexity based cryptography. In *30th FOCS*, IEEE Computer Soc. (1989) 230–235.
[31] C. Cachin, Hashing a source with an unknown probability distribution, in: Proceedings of the IEEE International Symposium on Information Theory, 1998.
[32] H.P. Yuen, J.H. Shapiro, Optical communication with two-photon coherent states-part I: quantum-State propagation and quantum-noise reduction., IEEE Trans. Inf. Teor., IT 24 (6) (1978) 657–668.
[33] J.H. Shapiro, H.P. Yuen, J.A. Machado, Optical communication with two-photon coherent states - part II: photoemissive detection and structured receiver performance, IEEE Trans. Inf. Teor., IT 25 (2) (1979) 179–192.
[34] C.K. Hong, S.R. Friberg, L. Mandel, Optical communication channel based on coincident photon pairs, Appl. Opt. 24 (22) (1985) 3877–3882.
[35] J.D. Franson, Nonlocal cancellation of dispersion, Phys. Rev. A 45 (5) (1992) 3126–3132.
[36] S.-Y. Baek, Y.-W. Cho, Y.-H. Kim, Nonlocal dispersion cancellation using entangled photons, Opt. Expr. 17 (21) (2009) 19241–19252.
[37] A. Peres, D.R. Terno, Quantum information and relativity theory, Rev. Mod. Phys. 76 (1) (2004) 93–123.
[38] C.H. Bennett, G. Brassard, C. Crépeau, R. Jozsa, A. Peres, W.K. Wootters, Teleporting an unknown quantum state via dual classical and Einstein-Podolsky-Rosen channels, Phys. Rev. Lett. 70 (13) (1993) 1895–1899.
[39] X.-S. Ma, T. Herbst, T. Scheidl, D. Wang, S. Kropatschek, W. Naylor, et al., Quantum teleportation over 143 kilometres using active feed-forward, Nature 489 (7415) (2012) 269–273.
[40] H. Yonezawa, S.L. Braunstein, A. Furusawa, Experimental demonstration of quantum teleportation of broadband squeezing, Phys. Rev. Lett. 99 (11) (2007) 110503.
[41] N. Takei, H. Yonezawa, T. Aoki, A. Furusawa, High-fidelity teleportation beyond the no-cloning limit and entanglement swapping for continuous variables, Phys. Rev. Lett. 94 (22) (2005) 220502.
[42] H. Yonezawa, T. Aoki, A. Furusawa, Demonstration of a quantum teleportation network for continuous variables, Nature 431 (2004) 430–434.
[43] R. Jozsa, D.S. Abrams, J.P. Dowling, C.P. Williams, Quantum clock synchronization based on shared prior entanglement, Phys. Rev. Lett. 85 (2000) 2010–2013.
[44] E.A. Burt, C.R. Ekstrom, T.B. Swanson, Comment on "quantum clock synchronization based on shared prior entanglement", Phys. Rev. Lett. 87 (2001) 129801.
[45] U. Yurtsever, J.P. Dowling, Lorentz-invariant look at quantum clock-synchronization protocols based on distributed entanglement, Phys. Rev. A 65 (5) (2002) 052317.
[46] M. Krco, P. Paul, Quantum clock synchronization: multiparty protocol, Phys. Rev. A 66 (2002) 024305.
[47] A. Tavakoli, A. Cabello, M. Zukowski, M. Bourennane, Quantum clock synchronization with a single qudit, Sci. Rep. 5 (2015) 7982.
[48] X. Kong, T. Xin, S. Wei, B. Wang, Y. Wang, K. Li, et al., Implementation of Multiparty quantum clock synchronization. ArXiv170806050 Quant-Ph, 2017.
[49] J.-D. Bancal, S. Pironio, A. Acín, Y.-C. Liang, V. Scarani, N. Gisin, Quantum non-locality based on finite-speed causal influences leads to superluminal signalling, Nat. Phys. 8 (12) (2012) 867–870.
[50] B. Cocciaro, The principle of relativity, superluminality and EPR experiments. "Riserratevi sotto coverta ...", J. Phys.: Conf. Ser 626 (1) (2015) 012054.
[51] C.H. Lineweaver, L. Tenorio, G.F. Smoot, P. Keegstra, A.J. Banday, P. Lubin, The dipole observed in the COBE DMR 4 year data, Astrophys. J. 470 (1996) 38.
[52] B. Cocciaro, S. Faetti, L. Fronzoni, Improved lower bound on superluminal quantum communication, Phys. Rev. A 97 (5) (2018) 052124.
[53] A. Chefles, Quantum state discrimination, Contemp. Phys. 41 (2000) 401–424.

[54] J. Bergou, Discrimination of quantum states, J. Mod. Opt. 57 (2010) 160–180.
[55] J. Bae, L. Kwek, Quantum state discrimination and its applications, J. Phys. A: Math. Theor. 48 (2015) 083001.
[56] C.W. Helstrom, Quantum Detection and Estimation Theory., Academic Press, New York, 1976.
[57] J.S. Shaari, S. Soekardjo, Indistinguishable encoding for bidirectional quantum key distribution: theory to experiment, EPL Europhys. Lett. 120 (6) (2017) 60001.
[58] H. Bechmann-Pasquinucci, A. Peres, Quantum cryptography with 3-state systems, Phys. Rev. Lett. 85 (15) (2000) 3313–3316.
[59] H. Bechmann-Pasquinucci, W. Tittel, Quantum cryptography using larger alphabets, Phys. Rev. A 61 (2000) 062308.
[60] S.P. Walborn, D.S. Lemelle, M.P. Almeida, P.H.S. Ribeiro, Quantum key distribution with higher-order alphabets using spatially encoded qudits, Phys. Rev. Lett. 96 (2006) 090501.
[61] H. de Riedmatten, I. Marcikic, V. Scarani, W. Tittel, H. Zbinden, N. Gisin, Tailoring photonic entanglement in high-dimensional Hilbert spaces, Phys. Rev. A 69 (5) (2004) 050304.
[62] I. Ali-Khan, C.J. Broadbent, J.C. Howell, Large-alphabet quantum key distribution using energy-time entangled bipartite states, Phys. Rev. Lett. 98 (2007) 060503.
[63] K. Inoue, E. Waks, Y. Yamamoto, Differential-phase-shift quantum key distribution using coherent light, Phys. Rev. A 68 (2003) 022317.
[64] T. Sasaki, Y. Yamamoto, M. Koashi, Practical quantum key distribution protocol without monitoring signal disturbance, Nature 509 (7501) (2014) 475–478.
[65] H. Takesue, T. Sasaki, K. Tamaki, M. Koashi, Experimental quantum key distribution without monitoring signal disturbance, Nat. Phot. 9 (2015) 827–831.
[66] D. Stucki, N. Brunner, N. Gisin, V. Scarani, H. Zbinden, Fast and simple one-way quantum key distribution, Appl. Phys. Lett. 87 (19) (2005) 194108.
[67] C.H. Bennett, G. Brassard, Quantum cryptography: public key distribution and coin tossing, in: Proceedings of IEEE International Conference on Computers, Systems and Signal Processing, vol. 175, p. 8, 1984.
[68] A.K. Ekert, Quantum cryptography based on Bell's theorem, Phys. Rev. Lett. 67 (6) (1991) 661–663.
[69] P.M. Intallura, M.B. Ward, O.Z. Karimov, Z.L. Yuan, P. See, A.J. Shields, et al., Quantum key distribution using a triggered quantum dot source emitting near 1.3 μm, Appl. Phys. Lett. 91 (16) (2007) 161103.
[70] A. Acín, N. Gisin, V. Scarani, Coherent-pulse implementations of quantum cryptography protocols resistant to photon-number-splitting attacks, Phys. Rev. A 69 (1) (2004) 012309.
[71] H.-K. Lo, X. Ma, K. Chen, Decoy state quantum key distribution, Phys. Rev. Lett. 94 (23) (2005) 230504.
[72] H.-K. Lo, M. Curty, B. Qi, Measurement-device-independent quantum key distribution, Phys. Rev. Lett. 108 (13) (2012) 130503.
[73] D. Elser, T. Bartley, B. Heim, C. Wittmann, D. Sych, G. Leuchs, Feasibility of free space quantum key distribution with coherent polarization states, New J. Phys. 11 (4) (2009) 045014.
[74] M. Hillery, Quantum cryptography with squeezed states, Phys. Rev. A 61 (2) (2000) 022309.
[75] F. Grosshans, P. Grangier, Continuous variable quantum cryptography using coherent states, Phys. Rev. Lett. 88 (5) (2002) 057902.
[76] C. Weedbrook, A.M. Lance, W.P. Bowen, T. Symul, T.C. Ralph, P.K. Lam, Quantum cryptography without switching, Phys. Rev. Lett. 93 (17) (2004).
[77] R. García-Patrón, N.J. Cerf, Continuous-variable quantum key distribution protocols over noisy channels, Phys. Rev. Lett. 102 (13) (2009).
[78] N.J. Cerf, S. Iblisdir, G.V. Assche, Cloning and cryptography with quantum continuous variables, Eur. Phys. J. D 18 (2) (2002) 211–218.
[79] G.V. Assche, J. Cardinal, N.J. Cerf, Reconciliation of a quantum-distributed Gaussian key., IEEE Trans. Inf. Theory 50 (2) (2004) 394–400.
[80] V. Scarani, H. Bechmann-Pasquinucci, N.J. Cerf, M. Dušek, N. Lütkenhaus, M. Peev, The security of practical quantum key distribution, Rev. Mod. Phys. 81 (3) (2009) 1301–1350.

[81] C. Silberhorn, T.C. Ralph, N. Lütkenhaus, G. Leuchs, Continuous variable quantum cryptography: beating the 3 dB loss limit, Phys. Rev. Lett. 89 (16) (2002).
[82] M. Heid, N. Lütkenhaus, Security of coherent-state quantum cryptography in the presence of Gaussian noise, Phys. Rev. A 76 (2) (2007) 022313.
[83] U.L. Andersen, G. Leuchs, C. Silberhorn, Continuous-variable quantum information processing, Las. Phot. Rev. 4 (2010) 337–354.
[84] N. Jain, E. Anisimova, I. Khan, V. Makarov, C. Marquardt, G. Leuchs, Trojan-horse attacks threaten the security of practical quantum cryptography, New J. Phys. 16 (12) (2014) 123030.
[85] V. Makarov, D.R. Hjelme, Faked states attack on quantum cryptosystems, J. Mod. Opt. 52 (5) (2005) 691–705.
[86] V. Makarov, A. Anisimov, J. Skaar, Effects of detector efficiency mismatch on security of quantum cryptosystems, Phys. Rev. A 74 (2) (2006).
[87] C.-H.F. Fung, B. Qi, K. Tamaki, H.-K. Lo, Phase-remapping attack in practical quantum-key-distribution systems, Phys. Rev. A 75 (3) (2007).
[88] V. Makarov, J.-P. Bourgoin, P. Chaiwongkhot, M. Gagné, T. Jennewein, S. Kaiser, et al., Creation of backdoors in quantum communications via laser damage, Phys. Rev. A 94 (3) (2016).
[89] A. Huang, S. Sajeed, P. Chaiwongkhot, M. Soucarros, M. Legre, V. Makarov, Testing random-detector-efficiency countermeasure in a commercial system reveals a breakable unrealistic assumption, IEEE J. Quantum Electron. 52 (11) (2016) 1–11.
[90] M. Takeoka, S. Guha, M.M. Wilde, Fundamental rate-loss tradeoff for optical quantum key distribution, Nat. Commun. 5 (2014) 5235.
[91] S. Pirandola, R. Laurenza, C. Ottaviani, L. Banchi, Fundamental limits of repeaterless quantum communications, Nat. Commun. 8 (2017) 15043.
[92] H. Takesue, S.W. Nam, Q. Zhang, R.H. Hadfield, T. Honjo, K. Tamaki, et al., Quantum key distribution over a 40-db channel loss using superconducting single-photon detectors, Nat. Photonics 1 (2007) 343–348.
[93] B. Julsgaard, J. Sherson, J.I. Cirac, J. Fiurášek, E.S. Polzik, Experimental demonstration of quantum memory for light, Nature 432 (7016) (2004) 482–486.
[94] G. Schunk, U. Vogl, D.V. Strekalov, M. Förtsch, F. Sedlmeir, H.G.L. Schwefel, et al., Interfacing transitions of different alkali atoms and telecom bands using one narrowband photon pair source, Optica 2 (2015) 773–778.
[95] G. Schunk, U. Vogl, F. Sedlmeir, D.V. Strekalov, A. Otterpohl, V. Averchenko, et al., Frequency tuning of single photons from a whispering-gallery mode resonator to MHz-wide transitions, J. Mod. Opt. 63 (2016) 2058–2073.
[96] J. Cviklinski, J. Ortalo, J. Laurat, A. Bramati, M. Pinard, E. Giacobino, Reversible quantum interface for tunable single-sideband modulation, Phys. Rev. Lett. 101 (13) (2008) 133601.
[97] M.U. Staudt, S.R. Hastings-Simon, M. Nilsson, M. Afzelius, V. Scarani, R. Ricken, et al., Fidelity of an optical memory based on stimulated photon echoes, Phys. Rev. Lett. 98 (11) (2007) 113601.
[98] G. Hétet, J.J. Longdell, A.L. Alexander, P.K. Lam, M.J. Sellars, Electro-optic quantum memory for light using two-level atoms, Phys. Rev. Lett. 100 (2) (2008) 023601.
[99] P.W. Shor, Scheme for reducing decoherence in quantum computer memory, Phys. Rev. A 52 (4) (1995) R2493–R2496.
[100] A.M. Steane, Error correcting codes in quantum theory, Phys. Rev. Lett. 77 (5) (1996) 793–797.
[101] D.G. Cory, M.D. Price, W. Maas, E. Knill, R. Laflamme, W.H. Zurek, et al., Experimental quantum error correction, Phys. Rev. Lett. 81 (10) (1998) 2152–2155.
[102] J.-W. Pan, S. Gasparoni, R. Ursin, G. Weihs, A. Zeilinger, Experimental entanglement purification of arbitrary unknown states, Nature 423 (6938) (2003) 417–422.
[103] A. Ourjoumtsev, A. Dantan, R. Tualle-Brouri, P. Grangier, Increasing entanglement between Gaussian states by coherent photon subtraction, Phys. Rev. Lett. 98 (2007) 030502.
[104] R. Dong, M. Lassen, J. Heersink, C. Marquardt, R. Filip, G. Leuchs, et al., Experimental entanglement distillation of mesoscopic quantum states, Nat. Phys. 4 (12) (2008) 919–923.
[105] B. Hage, A. Samblowski, J. DiGuglielmo, A. Franzen, J. Fiurášek, R. Schnabel, Preparation of distilled and purified continuous-variable entangled states, Nat. Phys. 4 (12) (2008) 915–918.

[106] C.H. Bennett, G. Brassard, C. Crepeau, U.M. Maurer, Generalized privacy amplification, IEEE Trans. Inf. Theory 41 (6) (1995) 1915–1923.
[107] M. Peev, C. Pacher, R. Alléaume, C. Barreiro, J. Bouda, W. Boxleitner, et al., The SECOQC quantum key distribution network in Vienna, New J. Phys. 11 (7) (2009) 075001.
[108] F. Xu, W. Chen, S. Wang, Z. Yin, Y. Zhang, Y. Liu, et al., Field experiment on a robust hierarchical metropolitan quantum cryptography network, Chin. Sci. Bull. 54 (17) (2009) 2991–2997.
[109] D. Huang, P. Huang, H. Li, T. Wang, Y. Zhou, G. Zeng, Field demonstration of a continuous-variable quantum key distribution network, Opt. Lett. 41 (15) (August 2016) 3511.
[110] P. Sibson, J.E. Kennard, S. Stanisic, C. Erven, J.L. O'Brien, M.G. Thompson, Integrated silicon photonics for high-speed quantum key distribution, Optica 4 (2) (2017) 172–177.
[111] D. Bunandar, A. Lentine, C. Lee, H. Cai, C.M. Long, N. Boynton, et al., Metropolitan quantum key distribution with silicon photonics, Phys. Rev. X 8 (2) (2018) 021009.
[112] J.F. Dynes, I. Choi, A.W. Sharpe, A.R. Dixon, Z.L. Yuan, M. Fujiwara, et al., Stability of high bit rate quantum key distribution on installed fiber, Opt. Express, 20 (15) (2012) 16339–16347.
[113] K.-i Yoshino, T. Ochi, M. Fujiwara, M. Sasaki, A. Tajima, Maintenance-free operation of WDM quantum key distribution system through a field fiber over 30 days, Opt. Express 21 (25) (2013) 31395–31401.
[114] S. Fossier, E. Diamanti, T. Debuisschert, A. Villing, R. Tualle-Brouri, P. Grangier, Field test of a continuous-variable quantum key distribution prototype, N. J. Phys. 11 (4) (2009) 045023.
[115] D. Stucki, M. Legré, F. Buntschu, B. Clausen, N. Felber, N. Gisin, et al., Zbinden. long-term performance of the swissquantum quantum key distribution network in a field environment, N. J. Phys. 13 (12) (2011) 123001.
[116] P. Jouguet, S. Kunz-Jacques, T. Debuisschert, S. Fossier, E. Diamanti, R. Alléaume, et al., Field test of classical symmetric encryption with continuous variables quantum key distribution, Opt. Express 20 (13) (2012) 14030–14041.
[117] K. Shimizu, T. Honjo, M. Fujiwara, T. Ito, K. Tamaki, S. Miki, et al., Performance of long-distance quantum key distribution over 90-km optical lkinks installed in a field environment of Tokyo metropolitan area, J. Lightwave Technol. 32 (1) (2014) 141–151.
[118] S. Wang, W. Chen, Z.-Q. Yin, H.-W. Li, D.-Y. He, Y.-H. Li, et al., Field and long-term demonstration of a wide area quantum key distribution network, Opt. Express 22 (18) (2014) 21739–21756.
[119] A.R. Dixon, J.F. Dynes, M. Lucamarini, B. Fröhlich, A.W. Sharpe, A. Plews, et al., High speed prototype quantum key distribution system and long term field trial, Opt. Express 23 (6) (2015) 7583–7592.
[120] Z. Qu, I.B. Djordjevic, M.A. Neifeld, RF-subcarrier-assisted four-state continuous-variable QKD based on coherent detection, Opt. Lett. 41 (23) (2016) 5507.
[121] E. Diamanti, A. Leverrier, Distributing secret keys with quantum continuous variables: principle, security and implementations, Entropy 17 (12) (2015) 6072–6092.
[122] J. Lodewyck, T. Debuisschert, R. Tualle-Brouri, P. Grangier, Controlling excess noise in fiber-optics continuous-variable quantum key distribution, Phys. Rev. A 72 (5) (2005). 050303®.
[123] B. Qi, L.-L. Huang, L. Qian, H.-K. Lo, Experimental study on the Gaussian-modulated coherent-state quantum key distribution over standard telecommunication fibers, Phys. Rev. A 76 (5) (2007) 052323.
[124] Q.D. Xuan, Z. Zhang, P.L. Voss, A 24 km fiber-based discretely signaled continuous variable quantum key distribution system, Opt. Express 17 (26) (2009) 24244–24249.
[125] P. Jouguet, S. Kunz-Jacques, A. Leverrier, P. Grangier, E. Diamanti, Experimental demonstration of long-distance continuous-variable quantum key distribution, Nat. Photonics 7 (5) (2013) 378–381.
[126] R. Kumar, H. Qin, R. Alléaume, Coexistence of continuous variable QKD with intense DWDM classical channels, N. J. Phys. 17 (4) (2015) 043027.
[127] D. Huang, D. Lin, C. Wang, W. Liu, S. Fang, J. Peng, et al., Continuous-variable quantum key distribution with 1 Mbps secure key rate, Opt. Express 23 (13) (2015) 17511–17519.

[128] D. Huang, P. Huang, T. Wang, H. Li, Y. Zhou, G. Zeng, Continuous-variable quantum key distribution based on a plug-and-play dual-phase-modulated coherent-states protocol, Phys. Rev. A 94 (3) (2016) 032305.

[129] D. Huang, P. Huang, D. Lin, G. Zeng, Long-distance continuous-variable quantum key distribution by controlling excess noise, Sci. Rep. 6 (2016) 19201.

[130] S. Kleis, M. Rueckmann, C.G. Schaeffer, Continuous variable quantum key distribution with a real local oscillator using simultaneous pilot signals, Opt. Lett. 42 (8) (2017) 1588−1591.

[131] T. Wang, P. Huang, Y. Zhou, W. Liu, H. Ma, S. Wang, et al., High key rate continuous-variable quantum key distribution with a real local oscillator, Opt. Express 26 (3) (2018) 2794−2806.

[132] K.J. Gordon, V. Fernandez, P.D. Townsend, G.S. Buller, A short wavelength GigaHertz clocked fiber-optic quantum key distribution system, IEEE J. Quantum Electron. 40 (7) (2004) 900−908.

[133] C.H. Bennett, Quantum cryptography using any two nonorthogonal states, Phys. Rev. Lett. 68 (21) (1992) 3121−3124.

[134] X. Tang, L. Ma, A. Mink, A. Nakassis, H. Xu, B. Hershman, et al., Experimental study of high speed polarization-coding quantum key distribution with sifted-key rates over Mbit/s, Opt. Express 14 (6) (2006) 2062−2070.

[135] E. Diamanti, H. Takesue, C. Langrock, M.M. Fejer, Y. Yamamoto, 100 km differential phase shift quantum key distribution experiment with low jitter up-conversion detectors, Opt. Express 14 (26) (2006) 13073−13082.

[136] D. Stucki, C. Barreiro, S. Fasel, J.-D. Gautier, O. Gay, N. Gisin, et al., Continuous high speed coherent one-way quantum key distribution, Opt. Express 17 (16) (2009) 13326.

[137] M. Sasaki, M. Fujiwara, H. Ishizuka, W. Klaus, K. Wakui, M. Takeoka, et al., Field test of quantum key distribution in the Tokyo QKD Network, Opt. Express 19 (11) (2011) 10387−10409.

[138] H.-L. Yin, T.-Y. Chen, Z.-W. Yu, H. Liu, L.-X. You, Y.-H. Zhou, et al., Measurement-device-independent quantum key distribution over a 404 km optical fiber, Phys. Rev. Lett. 117 (19) (2016) 190501.

[139] B. Heim, C. Peuntinger, N. Killoran, I. Khan, C. Wittmann, C. Marquardt, et al., Atmospheric continuous-variable quantum communication, New J. Phys. 16 (11) (2014) 113018.

[140] C. Peuntinger, B. Heim, C.R. Müller, C. Gabriel, C. Marquardt, G. Leuchs, Distribution of squeezed states through an atmospheric channel, Phys. Rev. Lett. 113 (2014) 060502.

[141] X. Ma, C.-H.F. Fung, H.-K. Lo, Quantum key distribution with entangled photon sources, Phys. Rev. A 76 (2007) 012307.

[142] S. Nauerth, F. Moll, M. Rau, C. Fuchs, J. Horwath, S. Frick, et al., Air-to-ground quantum communication, Nat. Photonics 7 (5) (2013) 382−386.

[143] J.-Y. Wang, B. Yang, S.-K. Liao, L. Zhang, Q. Shen, X.-F. Hu, et al., Direct and full-scale experimental verifications towards ground−satellite quantum key distribution, Nat. Photonics 7 (5) (2013) 387−393.

[144] J.-P. Bourgoin, B.L. Higgins, N. Gigov, C. Holloway, C.J. Pugh, S. Kaiser, et al., Free-space quantum key distribution to a moving receiver, Opt. Express 23 (26) (2015) 33437−33447.

[145] C.J. Pugh, S. Kaiser, J.-P. Bourgoin, J. Jin, N. Sultana, S. Agne, et al., Airborne demonstration of a quantum key distribution receiver payload, Quantum Sci. Technol. 2 (2017) 024009.

[146] H. Chun, I. Choi, G. Faulkner, L. Clarke, B. Barber, G. George, et al., Handheld free space quantum key distribution with dynamic motion compensation, Opt. Express 25 (6) (2017) 6784−6795.

[147] H. Takenaka, A. Carrasco-Casado, M. Fujiwara, M. Kitamura, M. Sasaki, M. Toyoshima, Satellite-to-ground quantum-limited communication using a 50-kg-class microsatellite, Nat. Photonics 11 (8) (2017) 502−508.

[148] K. Günthner, I. Khan, D. Elser, B. Stiller, O. Bayraktar, C.R. Müller, et al., Quantum-limited measurements of optical signals from a geostationary satellite, Optica 4 (6) (2017) 611−615.

[149] Z. Tang, R. Chandrasekara, Y.C. Tan, C. Cheng, K. Durak, A. Ling, The photon pair source that survived a rocket explosion, Sci. Rep. 6 (1) (2016) 25603.

[150] Z. Tang, R. Chandrasekara, Y.C. Tan, C. Cheng, L. Sha, G.C. Hiang, et al., Generation and analysis of correlated pairs of photons aboard a nanosatellite, Phys. Rev. Appl. 5 (5) (2016) 054022.

[151] I. Khan, B. Heim, A. Neuzner, C. Marquardt, Satellite-based qkd, OPN 29 (2) (2018) 28–33.
[152] S.-K. Liao, W.-Q. Cai, W.-Y. Liu, L. Zhang, Y. Li, J.-G. Ren, et al., Satellite-to-ground quantum key distribution, Nature 549 (7670) (2017) 43–47.
[153] S.-K. Liao, W.-Q. Cai, J. Handsteiner, B. Liu, J. Yin, L. Zhang, et al., Satellite-relayed intercontinental quantum network, Phys. Rev. Lett. 120 (3) (2018) 030501.
[154] J. Yin, Y. Cao, Y.-H. Li, S.-K. Liao, L. Zhang, J.-G. Ren, et al., Satellite-based entanglement distribution over 1200 kilometers, Science 356 (2017) 1140–1144.
[155] S.-K. Liao, J. Lin, J.-G. Ren, W.-Y. Liu, J. Qiang, J. Yin, et al., Space-to-ground quantum key distribution using a small-sized payload on tiangong-2 space lab, Chin. Phys. Lett. 34 (2017) 090302.
[156] T. Jennewein, J.P. Bourgoin, B. Higgins, C. Holloway, E. Meyer-Scott, C. Erven, et al., QEYSSAT: a mission proposal for a quantum receiver in space, in: Proceedings of SPIE, vol. 8997, p. 89970A, San Francisco, CA, February 2014.
[157] D.K.L. Oi, A. Ling, J.A. Grieve, T. Jennewein, A.N. Dinkelaker, M. Krutzik, Nanosatellites for quantum science and technology, Contemp. Phys. 58 (2017) 25–52.
[158] D. Gottesman, T. Jennewein, S. Croke, Longer-baseline telescopes using quantum repeaters, Phys. Rev. Lett. 109 (2012) 070503.
[159] V. Giovannetti, S. Lloyd, L. Maccone, Advances in quantum metrology, Nat. Photonics 5 (2011) 222–229.
[160] I.A. Burenkov, O.V. Tikhonova, S.V. Polyakov, Quantum receiver for large alphabet communication, Optica 5 (3) (2018) 227–232.
[161] C.H. Bennett, S.J. Wiesner, Communication via one-and two-particle operators on Einstein–Podolsky–Rosen states, Phys. Rev. Lett. 69 (20) (1992) 2881–2884.
[162] K. Mattle, H. Weinfurter, P.G. Kwiat, A. Zeilinger, Dense coding in experimental quantum communication, Phys. Rev. Lett. 76 (25) (1996) 4656–4659.
[163] S.L. Braunstein, H.J. Kimble, Dense coding for continuous variables, Phys. Rev. A 61 (4) (2000) 042302.
[164] J.T. Barreiro, N.K. Langford, N.A. Peters, P.G. Kwiat, Generation of hyperentangled photon pairs, Phys. Rev. Lett. 95 (2005) 260501.
[165] B.P. Williams, R.J. Sadlier, T.S. Humble, Superdense coding over optical fiber links with complete bell-state measurements, Phys. Rev. Lett. 118 (2017) 050501.
[166] J.T. Barreiro, T.-C. Wei, P.G. Kwiat, Beating the channel capacity limit for linear photonic superdense coding, Nat. Phys. 4 (4) (2008) 282–286.
[167] C. Zeitler, T.M. Graham, J. Chapman, H. Bernstein, P.G. Kwiat, Super-dense teleportation for space applications, in: Free-Space Laser Communication and Atmospheric Propagation XXVIII, vol. 9739, p. 973912, International Society for Optics and Photonics, March 2016.
[168] T.M. Graham, H.J. Bernstein, T.-C. Wei, M. Junge, P.G. Kwiat, Superdense teleportation using hyperentangled photons, Nat. Commun. 6 (2015) 7185.
[169] P.B. Dixon, G.A. Howland, J. Schneeloch, J.C. Howell, Quantum mutual information capacity for high-dimensional entangled states, Phys. Rev. Lett. 108 (2012) 143603.
[170] W. Wasilewski, A.I. Lvovsky, K. Banaszek, C. Radzewicz, Pulsed squeezed light: simultaneous squeezing of multiple modes, Phys. Rev. A 73 (2006) 063819.
[171] M. Krenn, M. Huber, R. Fickler, R. Lapkiewicz, S. Ramelow, A. Zeilinger, Generation and confirmation of a (100 x 100)-dimensional entangled quantum system, PNAS 111 (2014) 6243–6247.
[172] F. Bouchard, K. Heshami, D. England, R. Fickler, R.W. Boyd, B.-G. Englert, et al., Experimental investigation of high-dimensional quantum key distribution protocols with twisted photons, Quantum 2 (2018) 111.
[173] M. Kues, C. Reimer, P. Roztocki, L.R. Cortés, S. Sciara, B. Wetzel, et al., On-chip generation of high-dimensional entangled quantum states and their coherent control, Nature 546 (2017) 622–626.
[174] T. Inagaki, K. Inaba, R. Hamerly, K. Inoue, Y. Yamamoto, H. Takesue, Large-scale ising spin network based on degenerate optical parametric oscillators, Nat. Photonics 10 (2016) 415–419.
[175] A. Biswas, B. Moision, W.T. Roberts, W.H. Farr, A. Gray, K. Quirk, et al., Palomar receive terminal (PRT) for the mars laser communication demonstration (MLCD) project, Proc. IEEE 95 (10) (2007) 2045–2058.

[176] A. Mandilara, N.J. Cerf, Quantum bit commitment under Gaussian constraints, Phys. Rev. A 85 (2012) 062310.

[177] T. Lunghi, J. Kaniewski, F. Bussières, R. Houlmann, M. Tomamichel, A. Kent, et al., Experimental bit commitment based on quantum communication and special relativity, Phys. Rev. Lett. 111 (2013) 180504.

[178] H.-K. Lau, C. Weedbrook, Quantum secret sharing with continuous-variable cluster states, Phys. Rev. A 88 (2013) 042313.

[179] A. Keet, B. Fortescue, D. Markham, B.C. Sanders, Quantum secret sharing with qudit graph states, Phys. Rev. A 82 (2010) 062315.

[180] W.P. Grice, P.G. Evans, B. Lawrie, M. Legré, P. Lougovski, W. Ray, et al., Two-party secret key distribution via a modified quantum secret sharing protocol, Opt. Express 23 (6) (2015) 7300−7311.

[181] R. Amiri, J.M. Arrazola, Quantum money with nearly optimal error tolerance, Phys. Rev. A 95 (2017) 062334.

[182] K. Bartkiewicz, A. Černoch, G. Chimczak, K. Lemr, A. Miranowicz, F. Nori, Experimental quantum forgery of quantum optical money, NPJ Quantum Inf. 3 (2017) 7.

[183] M. Bozzio, E. Diamanti, F. Grosshans, Semi-device-independent quantum money with coherent states. Available from: <arXiv:quant-ph/181209256>, 2018.

[184] R.J. Collins, R.J. Donaldson, V. Dunjko, P. Wallden, P.J. Clarke, E. Andersson, et al., Realization of quantum digital signatures without the requirement of quantum memory, Phys. Rev. Lett. 113 (2014) 040502.

[185] R. Amiri, E. Andersson, Unconditionally secure quantum signatures, Entropy 17 (8) (2015) 5635−5659.

[186] A. Pappa, A. Chailloux, E. Diamanti, I. Kerenidis, Practical quantum coin flipping, Phys. Rev. A 84 (2011) 052305.

[187] A. Pappa, P. Jouguet, T. Lawson, A. Chailloux, M. Legré, P. Trinkler, et al., Experimental plug and play quantum coin flipping., Nat. Commun. 5 (2014) 3717.

[188] J.M. Arrazola, N. Lütkenhaus, Quantum fingerprinting with coherent states and a constant mean number of photons, Phys. Rev. A 89 (2014) 062305.

[189] F. Xu, J.M. Arrazola, K. Wei, W. Wang, P. Palacios-Avila, C. Feng, et al., Experimental quantum fingerprinting with weak coherent pulses, Nat. Commun. 6 (2015) 8735.

[190] J.-Y. Guan, F. Xu, H.-L. Yin, Y. Li, W.-J. Zhang, S.-J. Chen, et al., Observation of Quantum Fingerprinting Beating the Classical Limit, Phys. Rev. Lett. 116 (2016) 240502.

[191] R.P. Feynman, Simulating physics with computers, Int. J. Theor. Phys. 21 (1982) 467−488.

[192] S. Aaronson, A. Arkhipov, The computational complexity of linear optics, in: Proc. ACM Symp. Theory Computing, p. 333, San Jose, CA, 2011.

[193] L.K. Grover. A fast quantum mechanical algorithm for database search. Available from: <arXiv: quant-ph/9605043>, 1996.

[194] P.W. Shor, Polynomial-time algorithms for prime factorization and discrete logarithms on a quantum computer, SIAM J. Comput. 26 (1997) 1484−1509.

[195] J.F. Clauser, J.P. Dowling, Factoring integers with young's n-slit interferometer, Phys. Rev. A 53 (1996) 4587−4590.

[196] N.S. Dattani, N. Bryans, Quantum factorization of 56153 with only 4 qubits. Available from: <arXiv:quant-ph/14116758>, 2014.

[197] Y. Yamamoto, K. Aihara, T. Leleu, K.-i Kawarabayashi, S. Kako, M. Fejer, et al., Coherent Ising machines—optical neural networks operating at the quantum limit, NPJ Quantum Inf. 3 (2017) 49.

[198] R. Hamerly, T. Inagaki, P.L. McMahon, D. Venturelli, A. Marandi, T. Onodera, et al., Scaling advantages of all-to-all connectivity in physical annealers: the coherent ising machine vs. D-Wave 2000Q. Available from: <arXiv:phys.quant-ph/180505217>, 2018.

[199] D. Rideout, T. Jennewein, G. Amelino-Camelia, T.F. Demarie, B.L. Higgins, A. Kempf, et al., Fundamental quantum optics experiments conceivable with satellites—reaching relativistic distances and velocities, Class. Quantum Gravity 29 (2012) 224011.

CHAPTER 13

Ultralong-distance undersea transmission systems

Jin-Xing Cai, Georg Mohs and Neal S. Bergano
SubCom, Eatontown, NJ, United States

For 30 years international telecommunications have been made possible by fiber-optic undersea cable networks, all based on single-mode fibers (SMFs). For many of these years the concept of approaching the Shannon limit was just that: "a concept" that was far in the future. The biggest change the industry has experienced over the past few years is that this concept is now becoming a reality. Modern undersea cable systems are being installed at spectral efficiencies within 30%–50% of the fundamental limits for transmission systems. This is made possible with high-performance coherent transponders and modern optical fibers. Even though growing single-mode capacity is challenging, the need for capacity continues its unrelenting march. While it would appear that these two observations are at odds, they really are not. The challenge for future systems will be to work within these fundamental limits to design, manufacture, and install systems with larger total capacity that are economically viable and provide the flexibility needed by customers.

Over the past few years SMF capacity as demonstrated in the laboratory has increased from 30 to 70 Tb/s [1] (Fig. 13.1). Significant improvements have been achieved using broadband amplification employing both C- and L-bands (>70 nm optical bandwidth), advanced modulation formats, improved nonlinearity compensation (NLC), variable spectral efficiency (SE), ultralow loss and large effective area fiber, forward error correction (FEC) codes, etc. At the same time, space division multiplexing (SDM) concepts have increased single fiber capacity to 520 Tb/s over 8830 km using 12-core multicore fiber (MCF) [2].

This chapter provides an overview of the progress in undersea transmission technology since the last edition of Optical Fiber Telecommunications in 2012 [3]. We start with a brief introduction to the Gaussian noise (GN) model and its application to coherent subsea fiber-optic communication systems. This is followed by discussions on how to improve system performance, including symbol rate optimization, NLC, and nonlinear transmission optimization. Afterward, we review recent progress in driving system capacity in undersea fiber-optic communication systems to higher values; experimental results on improving SE (advanced modulation formats, coded

Optical Fiber Telecommunications V11
DOI: https://doi.org/10.1016/B978-0-12-816502-7.00015-4

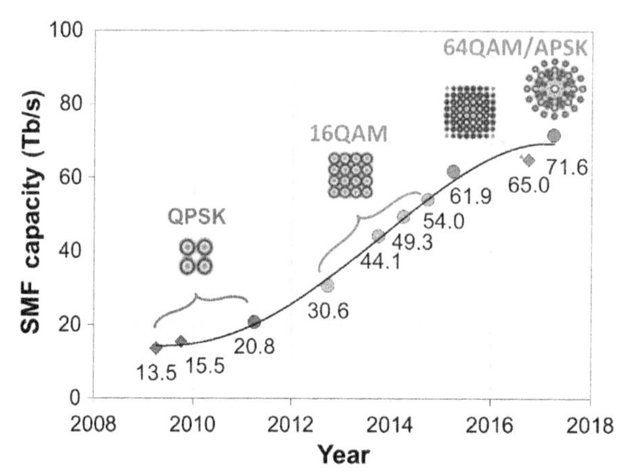

Figure 13.1 Laboratory demonstrations of single-mode fiber (SMF) capacity over transoceanic distances, © 2018 IEEE. Reprinted, with permission, from J.-X. Cai, et al., 70.46 Tb/s over 7,600 km and 71.65 Tb/s over 6,970 km transmission in C + L band using coded modulation with hybrid constellation shaping and nonlinearity compensation, J. Lightw. Technol. 36 (2), 2018, 114–121 [1].

modulation (CM), and constellation shaping), system bandwidth, and overall cable capacity are reviewed. Special emphasis is given to SDM for power-efficient transmission and improving system capacity with limited electrical power. We then move onto system design fundamentals focusing on open cables and modem-independent wet plant–only performance metrics such as generalized optical signal-to-noise ratio (GOSNR). Basic system design trade-offs are discussed, including system length, repeater spacing, repeater output power, and fiber performance in relation to the Shannon Limit. We finish the chapter with a review of new technology now available in subsea applications such as wavelength selective switch (WSS) based reconfigurable optical add-drop multiplexer (ROADM) and new materials in cable design such as aluminum conductors enabling overall system optimization.

13.1 Undersea transmission over dispersion uncompensated fibers

In 1988 the first transatlantic undersea cable using fiber-optic technology (TAT-8) was put into service with 560 Mb/s capacity (2 fiber pairs × 280 Mb/s), which was equivalent to ~40,000 telephone circuits. [4]. Thanks to the intrinsic low loss of optical fiber, the undersea optical 3R (reamplification, reshaping, retiming) repeater spacing had been extended to 40 km (much larger than was needed for a similar microwave system). The first completely cross-the-Pacific optical cable, TPC-3 (Transpac 3), connecting the United States and Japan was built in 1989 using optical 3R regenerators with a total capacity of 560 Mb/s (2 fiber pairs × 280 Mb/s).

Since then, optical cable networks have quickly become the backbone of the global communication network—both in terrestrial and undersea telecommunications. Fig. 13.2 shows the 2018 optical undersea cable map. With more than 450 undersea

Figure 13.2 Undersea cable map in 2018. Source: *Courtesy: TeleGeography. http://www.telegeography.com/.*

cables in services covering ~1.2 million km, undersea communication cables are nearly ubiquitous in all international waters and carry nearly 100% intercontinental communication traffic that links the countries and continents of the world. Moreover, annually laid undersea optical cables are projected to reach 53,000 km by 2022 [5]. For these reasons it has been said that undersea fiber-optic cable systems make the web worldwide.

As we are celebrating the 30th anniversary of the first fiber-optic transoceanic cable, the state-of-the-art transatlantic cable—MAREA connecting the United States to Spain —is in operation with eight fiber pairs and an initial estimated design capacity of 160 Tb/s, an increase of more than 285,000 times compared with TAT-8 [6]. The state-of-the-art transpacific optical cable system, Pacific Light Cable Network (PLCN, connecting Hong Kong, China and Los Angeles, United States), can carry 144 Tb/s (6 fiber pairs × 24 Tb/s) over ~13,000 km [7], an increase of more than 250,000 times compared with TPC-3. Many technologies developed over these 30 years contributed to the >5 orders of magnitude capacity increase, and in this work we will review the changes that took place over the past few years. Key among these were the use of coherent transponders with dispersion uncompensated fiber spans and all the techniques used to optimize these systems. The next section starts the chapter with a brief introduction of the underlying transmission fundamentals.

13.1.1 Linear and nonlinear degradations in optical fiber

In fiber-optic transmission systems the attenuation of the optical fiber is periodically compensated by the gain in an optical amplifier. Both erbium-doped fiber amplifiers (EDFA) and Raman amplifiers can provide gain in the low-loss fiber transmission window around 1550 nm; however, the vast majority of transoceanic length systems use only EDFAs. Typically the spacing between amplifiers ranges from 50 to 100 km, depending on the transmission performance target and distance. EDFAs generate noise that is added to the signal along the transmission path; thus, optical signal-to-noise ratio (OSNR) gradually decreases along the transmission line. The received OSNR (in dB/0.1 nm) of an optical channel after a chain of N_{EDFA} can be estimated based on the channel fiber launch power (in dBm), span loss (in dB), and EDFA noise figure (NF) (in dB):

$$\mathrm{OSNR} \approx 58 - 10\log_{10}(N_{\mathrm{EDFA}}) + P_{\mathrm{Launch}} - \mathrm{Loss} - NF \tag{13.1}$$

The dependence of OSNR on P_{Launch} is straightforward (OSNR increases linearly with launch power). However, fiber launch power is limited by optical signal distortion due to fiber nonlinearity, mainly Kerr effects in coherent transmission system [8]. Kerr effects arise from the weak intensity dependent fiber refractive index n, which leads to modulation of the fiber refractive index, resulting

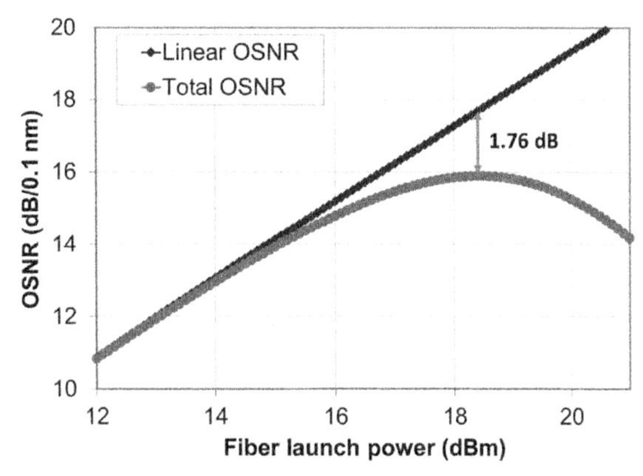

Figure 13.3 Channel performance (OSNR) versus fiber launch power.

in time-varying optical phase change. The optical phase change depends on fiber nonlinear coefficient n_2, time-varying signal power $P_s(t)$, transmission distance, and fiber effective area (A_{eff}).

$$\Delta\phi \propto \frac{n_2 P_S(t)\, L_{eff}}{A_{eff}} \tag{13.2}$$

The induced optical phase modulation makes it impossible to exactly recover the transmitted optical field. Although n_2 is very small ($2.7 \times 10^{-20}\ \mathrm{m^2/W}$), the ultralong transmission distance (10,000 km), large total power from EDFAs, and very small effective area of SMF ($80-150\ \mu\mathrm{m}^2$) makes this a dominant degradation effect for transoceanic distance transmission. As shown in Fig. 13.3, channel performance starts to grow linearly as OSNR increases, but then reaches a maximum and starts to degrade due to nonlinear effects. The understanding of this nonlinear behavior and how to optimize performance of undersea fiber-optic transmission systems occupies the majority of this chapter. The next section reviews the important GN model that quantitatively describes the nonlinear effects.

13.1.2 Gaussian noise model

The best optical fibers for transmission systems have the lowest loss (giving improved OSNR per launch power) and the largest effective area (thus lower nonlinear distortions), and tend to have high positive chromatic dispersion. In prior technology paradigms, it was not possible to use this type of fiber alone because of the large accumulated dispersion. For example, in systems with direct detection transponders, it was necessary to manage dispersion optically by combining positive and negative dispersion fibers on the transmission line, creating complex dispersion maps. Often the negative dispersion fibers had larger attenuation and smaller effective area.

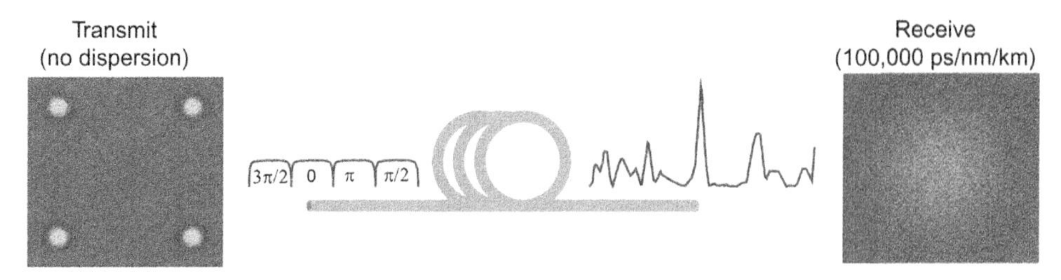

Figure 13.4 Signal randomization by chromatic dispersion.

With the advent of coherent detection and digital signal processing (DSP), transmission links without optical dispersion compensation became possible. Without using the higher loss and more nonlinear negative dispersion fiber, dispersion uncompensated links provide improved OSNR and lower nonlinearity. The large accumulated dispersion encountered by the signal leads to time-domain broadening resulting in the overlap of many symbols (Fig. 13.4) at the receiving terminal. For example, a 6500 km link operating at a symbol rate of 30 GBd can have an overlap of ~1000 symbols. This overlap tends to randomize the time-domain optical field, resulting in a statistically independent zero-mean Gaussian distribution. The amplitude randomized optical field also randomizes the intra- and interchannel nonlinear interactions among symbols, and the result of nonlinear interference can be modeled at the receiver as random noise. Ref. [9] shows that the statistical distribution of the nonlinear interference noise (NLI) also follows a statistically independent zero-mean Gaussian distribution. Therefore, the effect of NLI can be modeled as additive Gaussian noise (GN), at least for low to moderate NLI. There are many published research works related to the GN model [10–15].

From the GN model [which is based on four-wave mixing (FWM), perturbative nonlinearity model], the total noise accumulated during propagation is the sum of amplified spontaneous emission (ASE) noise and nonlinear interference noise P_{NLI}:

$$P_{\mathrm{Noise}} = P_{\mathrm{ASE}} + P_{\mathrm{NLI}} \tag{13.3}$$

$$P_{\mathrm{NLI}} = \int_{-B_n/2}^{B_n/2} G_{\mathrm{NLI}}(f)df \tag{13.4}$$

$G_{\mathrm{NLI}}(f)$ is the power spectral density (PSD) of P_{NLI} in the optical bandwidth B_n. For Nyquist WDM channels over a single EDFA amplified span, $G_{\mathrm{NLI}}(f)$ can be simplified as [14]:

$$G_{\mathrm{NLI}}(0) \approx \frac{8}{27}\Upsilon^2 G_{\mathrm{WDM}}^3 L_{\mathrm{eff}}^2 \frac{\log_e(\pi^2\beta_2 B_{\mathrm{WDM}}^2/\alpha)}{\pi\beta_2/\alpha} \tag{13.5}$$

where G_{WDM} is the PSD of the signal P_{ch}/B_n; α, fiber loss coefficient in km^{-1}; β_2, absolute value (always positive) of dispersion in ps^2/km; γ, fiber nonlinearity coefficient [1/(W km)]; L_{eff}, effective span length (km); B_{WDM}, optical bandwidth of the system.

From Eq. (13.5), NLI is proportional to $P_{ch}{}^3$, and $\log(B_{WDM})$.

The accumulation of NLI after N_s span for perfect Nyquist WDM channels (zero roll-off factor, no WDM channel gap, and transmitted optical amplitude that follows a zero-mean Gaussian distribution) grows linearly with distance (L):

$$P_{NLI} = P_{NLI}{}^{(1)} N_s \tag{13.6}$$

where $P_{NLI}{}^{(1)}$ is the nonlinear noise power from a single span. For real transmitters (non-Nyquist WDM), the accumulated NLI increases faster than linearly [14,16]:

$$P_{NLI} = P_{NLI}{}^{(1)} N_s^{(1+\varepsilon)}; \quad \varepsilon\text{: } 0.035\text{for} + D \text{ fiber} \tag{13.7}$$

Ref. [17] directly measured NLI induced by broadband FWM (equivalent to perfect Nyquist WDM channels) for different power levels, system lengths, and transmission BWs in dispersion uncompensated systems and confirmed that broadband FWM induced NLI is proportional to $P_{ch}{}^3$, L^1, and $\log(B_{WDM})$.

Combining above results, we have

$$P_{NLI} = \eta\, P_{ch}{}^3 \tag{13.8}$$

$$\eta = \frac{8}{27}\left(\frac{\gamma\, L_{eff}}{B_n}\right)^2 \frac{\log_e(\pi^2 \beta_2 L_{eff,a} B_{WDM}^2)}{\pi\, \beta_2 L_{eff,a}} \tag{13.9}$$

and the received OSNR or nonlinear OSNR becomes

$$OSNR_{NL} = \frac{P_{ch}}{P_{noise}} = \frac{P_{ch}}{P_{ASE} + \eta\, P_{ch}{}^3} \tag{13.10}$$

The optimum channel power to achieve the maximum OSNR is

$$P_{ch}{}^{opt} = \left(\frac{P_{ASE}}{2\eta}\right)^{1/3} \tag{13.11}$$

The NLI power to optimize the channel performance is equal to half of the ASE noise:

$$P_{NLI}{}^{opt} = \frac{P_{ASE}}{2} \tag{13.12}$$

The maximum achievable received OSNR is

$$\mathrm{OSNR}_{\mathrm{NL}}^{\mathrm{opt}} = \frac{P_{\mathrm{ch}}}{1.5\,P_{\mathrm{ASE}}} = \frac{2}{3}\,\mathrm{OSNR}_{\mathrm{LN}} \tag{13.13}$$

Since the nonlinear OSNR is 2/3 of the linear OSNR ($\mathrm{OSNR}_{\mathrm{LN}}$), we expect 1.76 dB OSNR degradation due to nonlinearity at the optimum operation point (Fig. 13.3).

One interesting conclusion from the GN model is that the optimum channel power should be independent of system length and modulation format. This is interesting since it does not follow the intuition from systems that use direct detection and dispersion maps. This was experimentally studied in Ref. [18] by measuring the optimal launch conditions as a function of transmission distance. The first observation shown in Fig. 13.5 is that the optimum channel power drops rapidly between very short distances and ~2000 km. Second, for distances greater than ~2000 km the result predicted by the GN model is found, since the optimum power is nearly constant out to a transmission distance of 10,000 km. The deviation for short distances comes from the fact the signal amplitude is constant for quadrature phase shift keying (QPSK) signal at the beginning of the transmission link, which is far away from the Gaussian distribution assumption. Therefore, NLI noise is overstated for the first few hundred kilometers. In general, GN model overestimates the NLI noise, especially in the first few spans of a link. A more rigorous model, enhanced GN (EGN), is developed and can account for all these effect very accurately in Ref. [19].

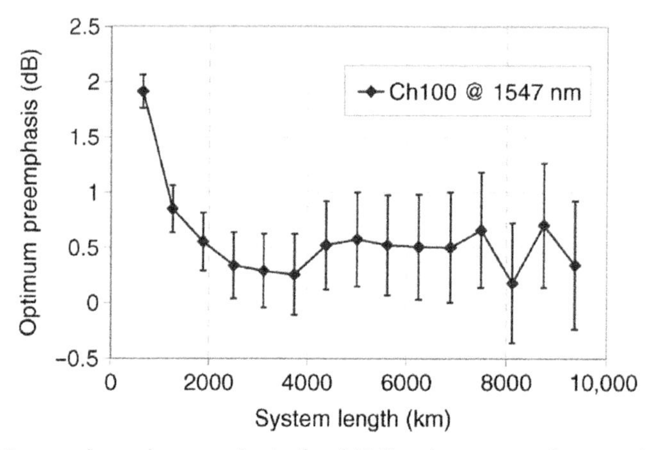

Figure 13.5 Optimum channel preemphasis for 100 G coherent quadrature phase shift keying (QPSK) on a D+ transmission, © 2012 IEEE. Reprinted, with permission, from J.-X. Cai, C.R. Davidson, A. Lucero, H. Zhang, D.G. Foursa, O.V. Sinkin, et al., 20 Tbit/s transmission over 6,860 km with sub-nyquist channel spacing, J. Lightw. Technol. 30 (4), 2012, 651–657 [18].

13.1.3 Symbol rate optimization

Can symbol rate be used as a design parameter to improve system performance? This question was first investigated for orthogonal frequency division multiplexing transmission [20]. The benefit of this optimization for single-carrier modulation was confirmed recently where DSP allowed the use of a single optical transmitter and receiver for generation and detection of multiple subcarriers. However, in practice this benefit is found to be modest.

EGN model predicts there is an optimum symbol rate for Nyquist channel, and the optimum symbol rate depends on total accumulated dispersion [21]:

$$R_{\mathrm{opt}} = \sqrt{\frac{2}{\pi \beta_2 L_s N_s}} \tag{13.14}$$

For pure silica core fiber (PSCF) ($D = 20$ ps/nm) over 10,000 km, the optimum symbol rate is 1.6 GBd.

Of course, higher symbol rate is attractive from a terminal equipment perspective to reduce cost. Future transponders might first electronically multiplex many subcarrier channels at optimum symbol rate and then modulate onto a single optical carrier at 64 GBd or higher. This subcarrier multiplexed (SCM) single carrier could operate at higher symbol rate and larger nonlinear tolerance, and is possibly even advantageous in terms of reducing the receiver DSP load thanks to the relatively lower subcarrier rate.

Fig. 13.6 compares performance for QPSK and 16QAM with different number of subcarriers. The results were measured after ~8600 km with 74 nm optical bandwidth. SCM yields 0.7 dB improvement for QPSK, and ~0.2 dB improvement for 16QAM. The best performance is achieved for 6–10 subcarriers, which corresponds

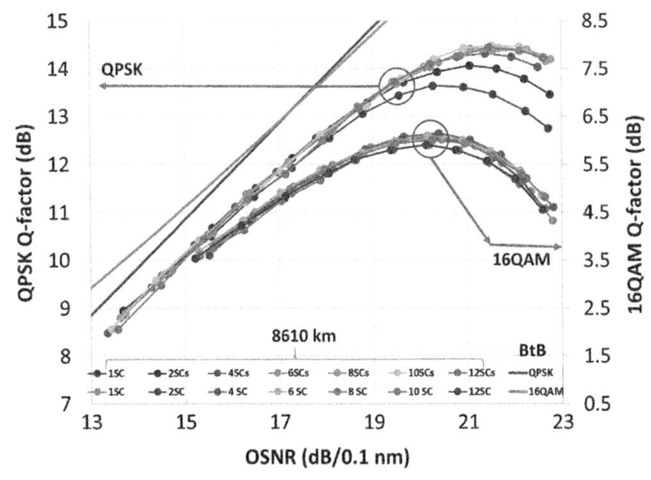

Figure 13.6 Transmitter preemphasis of 16QAM and quadrature phase shift keying (QPSK) with different number of subcarriers [22].

with subcarrier symbol rates between 5.4 and 3.24 GBd, respectively. Thus, some system specific performance improvements are possible especially for lower constellations; however, the benefit reduces for higher constellations.

Also of note: The optimum symbol rate is 3 × that predicted by EGN. From this measurement, the following relationship holds at the optimum baud rate: $D_{\text{total}}\ \Delta\lambda\ \approx 10\Delta T$, where D_{total} is the total accumulated dispersion of the channel, $\Delta\lambda$ is the optimum subcarrier spacing in wavelength domain, and ΔT is the optimum subcarrier period.

13.1.4 Nonlinearity compensation

Even though nonlinear interaction in dispersion uncompensated links can be modeled as noise, much of the interaction is still deterministic. Many works over the past 20 years have been dedicated to the study of nonlinear transmission penalties in fiber-optic systems. The following sections review several techniques for performing NLC within the DSP of the coherent transponder.

As we have discussed in previous sections, transmission performance improves with increasing launch power—until an optimum is reached. From the GN model we know that the total noise power that enters into the SNR calculation is not only given by the linear ASE noise generated by the amplifiers along the transmission path but also the "nonlinear noise" generated by nonlinear interaction among the signals through the Kerr effect (see Section 13.1.2). The leading contributions to this NLI are generated by signal–signal interaction which is not random but deterministic. This provides the opportunity for compensation as long as sufficient knowledge of the signal and physical properties of the transmission system exist. Only the contributions to the NLI that are truly "noiselike" (e.g., ASE–ASE interactions) cannot be predicted in a deterministic way and will therefore ultimately limit the transmission performance, SE, and capacity.

13.1.4.1 Digital back propagation

A NLC method used in many "off-line" transmission experiments is digital back propagation (DBP) [23–25]. Here the received field is digitally propagated backwards through the transmission path by solving the inverse nonlinear Schrödinger equation (NLSE) to arrive at an approximation of the transmitted signal before fiber impairment (Fig. 13.7). In practice, the NLSE is solved numerically by the split-step Fourier method where dispersion and nonlinear phase shift are treated separately and a small step size is required for accurate results.

Fig. 13.8 compares performance with and without DBP in a lab experiment employing all +D fibers [16]. The solid lines are predictions based on the GN model, and the discrete symbols are from lab measurements. Two steps per span were used, and the nonlinear coefficient in the DBP algorithm was optimized for best performance. The optimal signal bandwidth in Fig. 13.8 was 35 GHz for 32 GBd symbol.

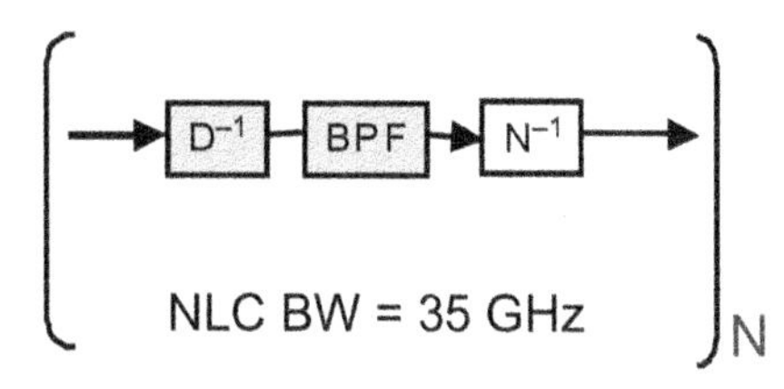

Figure 13.7 Block diagram of digital back propagation (DBP) for 32 GBd signal.

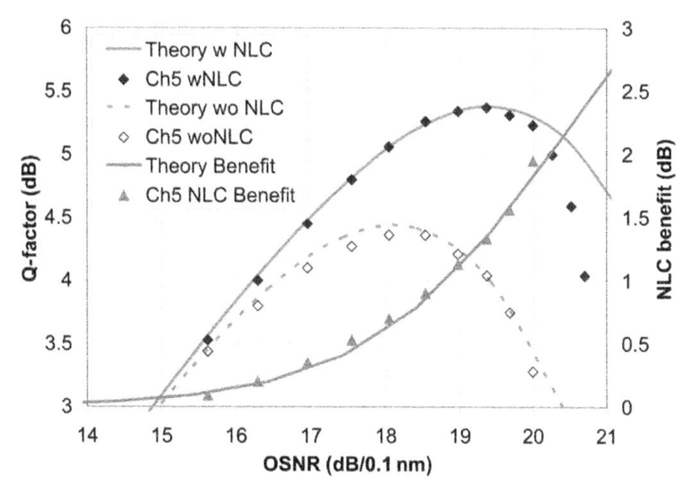

Figure 13.8 Performance and nonlinearity compensation (NLC) benefit versus optical signal-to-noise ratio (OSNR) for a 200 Gb/s channel after 10,290 km, © 2014 IEEE. Reprinted, with permission, from J.-X. Cai, H. Zhang, H. Batshon, M. Mazurczyk, O. Sinkin, D. Foursa, et al., 200 Gb/s and dual wavelength 400 Gb/s transmission over transpacific distance at 6.0 b/s/Hz spectral efficiency, J. Lightw. Technol. 32 (4), 2014, 832–839 [16].

Since the DBP bandwidth is close to the channel bandwidth, NLC mainly compensates intrachannel nonlinear effects in the experiment. Using more DBP bandwidth does not result in further performance improvement. This limitation is likely due to depolarization (i.e., the frequency dependent polarization change across the signal bandwidth) in transmission through polarization mode dispersion which is not accounted for in the DBP algorithm.

To find a theoretical prediction for the NLC benefit, the NLI is calculated according to the GN model using 35 GHz bandwidth (DBP bandwidth) with the assumption that this portion of the NLI can be fully compensated. A good match is achieved with experimental results up to the optimum OSNR after NLC. At higher OSNR values the experimental results decrease more rapidly than predicted by the model, possibly indicating additional nonlinear contributions such as nonlinear interaction of the signal with ASE. The DBP algorithm is computationally intense, since the NLSE must be solved numerically, and even by today's standards exceeds the capability of leading edge transponder circuits. Much effort has been devoted to finding good

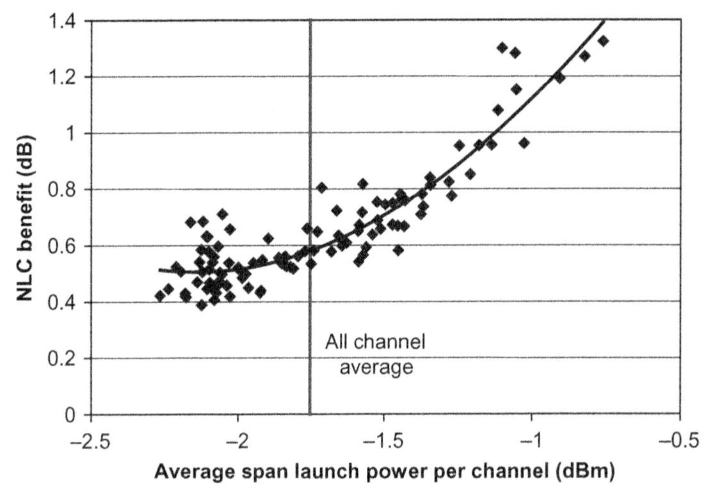

Figure 13.9 Nonlinearity compensation (NLC) benefit vs average channel launch power after 10,290 km, © 2014 IEEE. Reprinted, with permission, from J.-X. Cai, H. Zhang, H. Batshon, M. Mazurczyk, O. Sinkin, D. Foursa, et al., 200 Gb/s and dual wavelength 400 Gb/s transmission over transpacific distance at 6.0 b/s/Hz spectral efficiency, J. Lightw. Technol. 32 (4), 2014, 832–839 [16].

approximations that allow for faster data processing. In terms of computational complexity perturbation NLC is the simplest, followed by adaptive filters and then DBP, which is the most resource-hungry especially for longer transmission distance.

Another interesting result from the same experiment is the correlation between NLC benefit and average channel launch power across the different WDM channels. The average channel launch power is measured experimentally as detailed in Ref. [16]. The solid line in Fig. 13.9 corresponds to a second-order polynomial fit. We find 90% correlation between the NLC benefit and average channel launch power. As confirmed in Fig. 13.9, DBP compensates mostly intrachannel effects, hence channels with higher launch power suffer more intrachannel nonlinearity, while channels with lower launch power suffer less.

13.1.4.2 Perturbation nonlinearity compensation

Perturbation nonlinearity compensation (PNLC) is based on an approximation where the NLSE is expanded in the time-domain and only the first-order term is considered. If X_k and Y_k are the received kth symbols on x and y polarizations after carrier phase recovery, then the estimated or compensated symbols $\hat{X}_k$ and $\hat{Y}_k$ can be expressed as [26]

$$\begin{aligned} \hat{X}_k &= X_k - \rho \sum_{\substack{m=-B \\ m\neq 0}}^{B} \sum_{\substack{n=-B \\ n\neq 0}}^{B} (X_{m+k}X_{n+k}X^*_{m+n+k} + X_{m+k}Y_{n+k}Y^*_{m+n+k})C_{m,n} \\ \hat{Y}_k &= Y_k - \rho \sum_{\substack{m=-B \\ m\neq 0}}^{B} \sum_{\substack{n=-B \\ n\neq 0}}^{B} (Y_{m+k}Y_{n+k}Y^*_{m+n+k} + Y_{m+k}X_{n+k}X^*_{m+n+k})C_{m,n} \end{aligned} \tag{13.15}$$

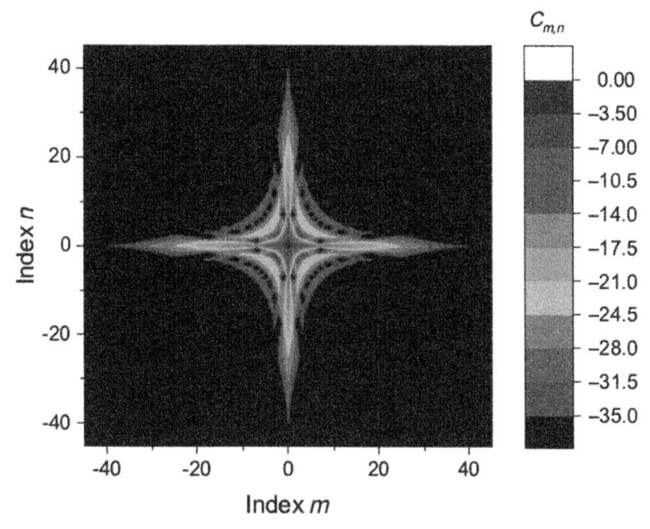

Figure 13.10 Normalized *Cm,n* coefficients (in dB) for 3600 km of standard SMF, © 2017 OSA. Reprinted, with permission, from J.C. Cartledge, F.P. Guiomar, F.R. Kschischang, G. Liga and M.P. Yankov, Digital signal processing for fiber nonlinearities, Opt. Express 25, 2017, 1916–1936, and references therein [23].

The set $\{Cm,n|\ m,\ n=-B,\ldots,\ B\}$ is a collection of precalculated perturbative coefficients stored in a look-up table (LUT). The LUT size is controlled by the parameter B, which can be more than 100 after transoceanic distance transmission. Hence, the number of double multiplications can be very large in the 2D summation term. On the other hand, many of the terms in $Cm,\ n$ can be neglected, as shown in Fig. 13.10 for a 3600 km link using standard SMF with root-raised-cosine pulses. Therefore, the complexity can be reduced significantly.

PNLC can be implemented using one sample per symbol and only needs a single computation step in a link. Ref. [26] shows that PNLC is only slightly inferior to DBP (~0.1 dB), but its implementation complexity is much lower than DBP. Hence, PNLC seems a promising candidate for future generations' transponders.

13.1.4.3 Nonlinearity compensation using fast adaptive filters

The fast decision directed least mean square (LMS) algorithm [1,27,28] is schematically shown in Fig. 13.11. Typically, a fast LMS adaptive filter used by itself has very limited capability to compensate nonlinearity. However, it can be used together with CM to significantly lower the symbol error ratio (SER) of the hard decisions used within the LMS algorithm (see Fig. 13.12). This improves the quality of the LMS error signal, allowing the gain and tracking speed of the LMS filter to be increased which provides compensation of fast varying nonlinear effects.

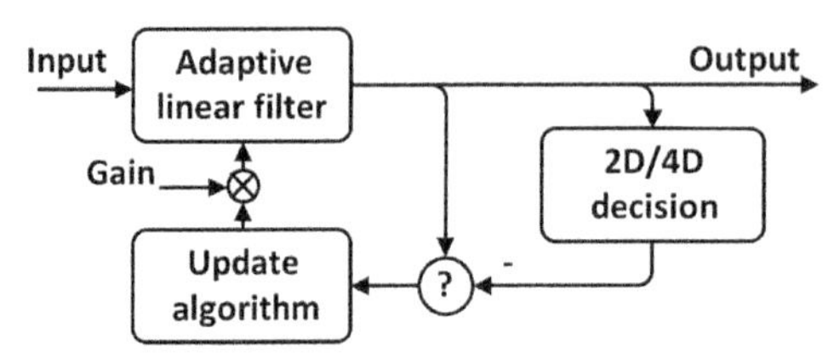

Figure 13.11 Fast least mean square (LMS) with 4D decision, © 2018 SPIE. Reprinted, with permission, from H.G. Batshon, M.V. Mazurczyk, J.-X. Cai, O.V. Sinkin, M. Paskov, C.R. Davidson, et al., Coded modulation based on 56APSK with hybrid shaping for high spectral efficiency transmission, in: Proc. Eur. Conf. Opt. Commun., Gothenburg, Sweden, 2017, Paper Tu.1.D.2. [28].

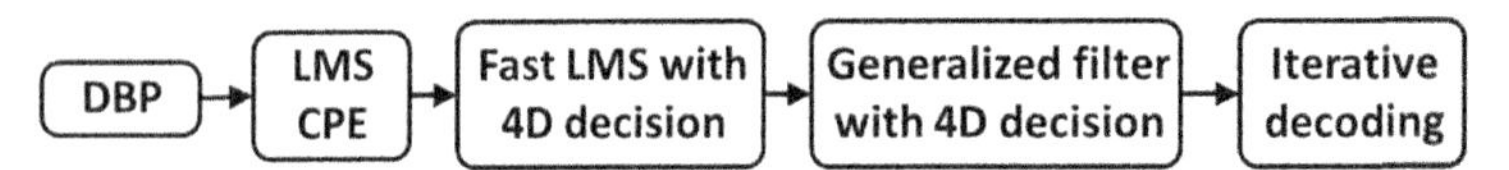

Figure 13.12 DSP chain of a cascading nonlinearity compensation schemes, © 2018 SPIE. Reprinted, with permission, from H.G. Batshon, M.V. Mazurczyk, J.-X. Cai, O.V. Sinkin, M. Paskov, C.R. Davidson, et al., Coded modulation based on 56APSK with hybrid shaping for high spectral efficiency transmission, in: Proc. Eur. Conf. Opt. Commun., Gothenburg, Sweden, 2017, Paper Tu.1.D.2. [28].

Another adaptive filtering technique is a generalized filter that estimates correlated nonlinear noise [1,27,28] and compensates for it using a time-varying linear filter as described in Eq. (13.16). The assumption is that nonlinear noise is at least partially correlated, and the correlated part can be compensated. The time-varying filter coefficients H_t^l are calculated using Eq. (13.17) at each symbol period. The noise estimation ϵ_t at time t is defined in Eq. (13.18), where $D(\cdot)$ is the decision operator, and x_t is the input signal at time t. The decision error rate, ϵ_t, and overall performance can be improved by using CM-based 4D decisions. Averaging with the exponential weight function ${}^N W_i$ is applied in Eq. (13.17) to reduce the influence of uncorrelated noise including ASE. The function ${}^N W_i$ is centered at $i = 0$ (the symbol of interest) and exponentially decays with width N. The compensation is applied using a traditional dual polarization butterfly structure. N and α^k are experimentally optimized for the through and cross-polarization filter paths. The above procedure is applied iteratively, and $D(\cdot)$ improves after correcting nonlinear noise. For the data shown in Fig. 13.13, we used an averaging window N of approximately 350 and a filter length of five taps.

$$x_t^{\text{output}} = x_t - \sum_k H_t^k x_{t+k} \tag{13.16}$$

$$H_t^k = \alpha^k \frac{\left\langle {}^N W_i \, \epsilon_{t+i} \, x_{t+i+k}^* \right\rangle_{\text{avg over } i}}{\left\langle {}^N W_i |x_{t+i+k}|^2 \right\rangle_{\text{avg over } i}} \tag{13.17}$$

$$\epsilon_t = D(x_t) - x_t \tag{13.18}$$

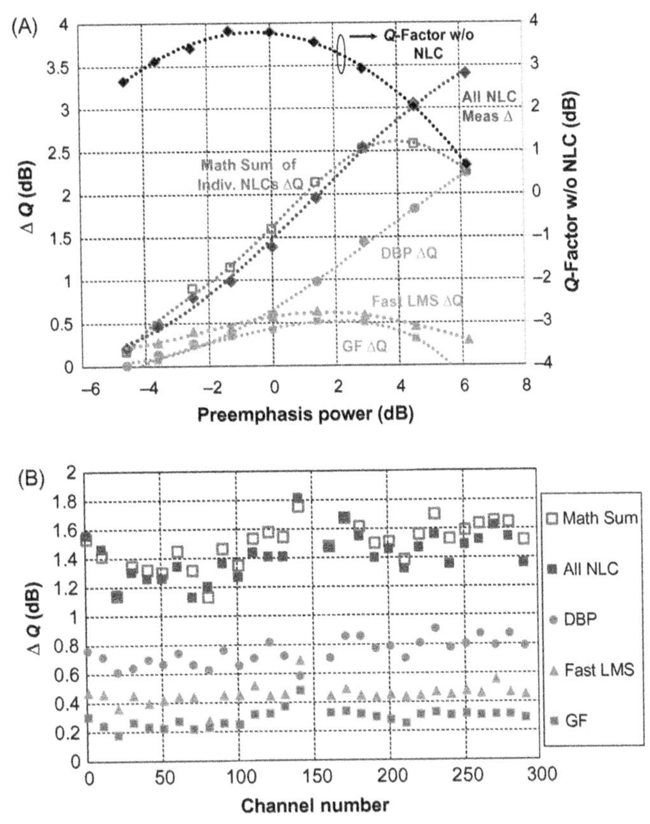

Figure 13.13 (A)Benefit of individual NLC algorithms compared to performance without NLC after 6970 km transmission. (B) Benefit of individual NLC algorithm compared to performance without NLC across transmission bandwidth. © 2017 IEEE. Reprinted, with permission, from J.-X. Cai, et al., 70.46 Tb/s over 7,600 km and 71.65 Tb/s over 6,970 km transmission in C + L band using coded modulation with hybrid constellation shaping and nonlinearity compensation, J. Lightw. Technol. 36 (2) (2018) 114.

13.1.4.4 Other nonlinearity mitigation techniques

Of course, good optical engineering is always important for the design of transoceanic length systems, where limiting nonlinear penalty is paramount. The nonlinear interaction strength is proportional to the ratio of nonlinear refractive index and effective area. To decrease the nonlinear interaction strength fiber, manufacturers have been able to decrease the nonlinear refractive index slightly by removing dopants from the core glass and moving to PSCF types. Modern transmission fibers have effective area up to $\sim 150\ \mu m^2$, a more than twofold improvement compared with legacy dispersion managed undersea transmission fiber. Perhaps in the future hollow-core fibers can improve performance by lowering the nonlinear refractive index by about two orders of magnitude. However, at this point fiber loss of these photonic crystal fibers is still too high to be attractive.

Many other nonlinearity mitigation or compensation schemes are being investigated, including Volterra series transfer function, Kalman filter, multisymbol detection using maximum likelihood sequence detection or maximum a priori receiver, optical phase conjugation, digital phase conjugated twin waves, nonlinear Fourier transform, machine learning nonlinear equalizer, nonlinear tolerant modulation format (subcarrier multiplexing, CM, pulse shaping), and so forth. These techniques are beyond the scope of this chapter; interested readers are referred to Refs. [23,30] for more on these NLC techniques.

13.1.4.5 Combination of nonlinearity compensation techniques

Some of the NLC techniques can be combined to achieve an even higher benefit. Ref. [1] has shown an average NLC benefit of 1.4 dBQ or 1.6 dB SNReff after 6970 km transmission using a combination of three NLC algorithms: DBP, fast adapting LMS equalizer, and an adaptive generalized filter, as shown in the schematic below (Fig. 13.12). Single-channel DBP is performed first, followed by fast LMS equalizer with 4D decisions, and then a generalized filter with 4D decisions. How SNReff is measured is outlined in Section 13.6.1.

To get a better insight into the performance of the different NLC schemes, we analyze the performance improvement of each individual algorithm versus transmitter preemphasis. Fig. 13.13A shows the benefit using one of the NLC methods ("DBP ΔQ," "Fast LMS ΔQ," and "GF ΔQ"), as well as the benefit from simultaneously using all three ("All NLC Meas Δ"). The benefit provided by DBP increases approximately proportional to P^3_{sig} with the increase of the power during the preemphasis. The benefit from adaptive filters (fast LMS and generalized filter) follows that of DBP up to the nominal power (0 dB in Fig. 13.13A); then the adaptive filter benefit starts going down due to the decrease in performance of decision-directed algorithms as the Q-factor drops beyond the optimal preemphasis ("Q-Factor w/o NLC").

Fig. 13.13B shows the benefit of individual NLC algorithms across ~10 THz optical bandwidth (with 33 GHz channel spacing) after 6970 km for every 10th transmission channel. The total NLC benefit ("All NLC") ranges from 1.1 to 1.8 dBQ with an average of 1.4 dBQ or 1.6 dB SNReff. The average NLC benefit across all channels are 0.75, 0.45, and 0.29 dB for DBP, fast LMS, and generalized filter, respectively. On average, the NLC benefit using all three algorithms ("All NLC") is ~0.1 dB less than the "Math Sum" curve of the three individual benefits.

13.1.5 Nonlinear transmission optimization

In a broadband optical transmission systems, the linear noise floor can vary across the usable optical bandwidth by 2 dB, and even up to 3 dB for systems that employ Raman amplification. There are several reasons for this large linear noise floor difference in wide optical bandwidth systems. First, the EDFA NF is wavelength

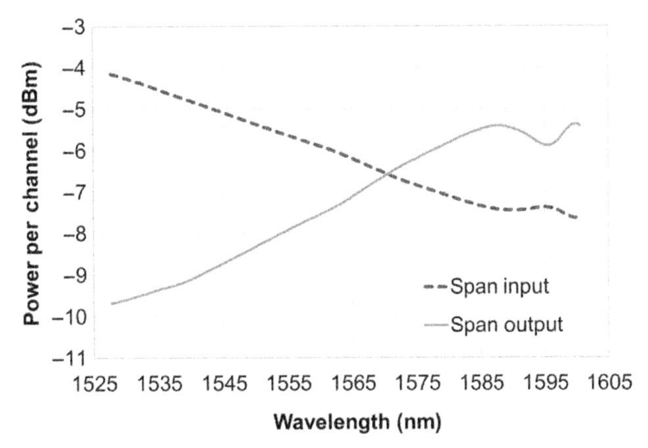

Figure 13.14 Designed optimum power spectral densities at span input and span output, © 2018 IEEE. Reprinted, with permission, from J.-X. Cai, Y. Sun, H. Zhang, H.G. Batshon, M.V. Mazurczyk, O.V. Sinkin, et al., 49.3 Tb/s transmission over 9100 km using C + L EDFA and 54 Tb/s transmission over 9150 km using hybrid-Raman EDFA, J. Lightw. Technol. 33 (13), 2015, 2724–2734 [31].

dependent and higher in the short-wavelength region for single-stage amplifiers, and the NF in the L-band is typically higher than that in the C-band for C + L EDFAs. Second, the fiber loss exhibits a parabolic shape with minimum loss near 1575 nm. Third, Raman effects become significant with wide optical bandwidth. Thus, it stands to reason that the system's design needs to be optimized across the useable bandwidth. Maximum system capacity is achieved when all channels operate at their corresponding optimum power, implying that the nonlinear OSNR differential among WDM channels will also be minimized compared to the nonoptimized case.

Fig. 13.14 shows the designed PSD of a Raman-assisted EDFA transmission system [31] using the GN model. The optimum input PSD function is achieved by considering both noise performance of distributed Raman amplification and the power evolution of channels across the bandwidth. There is ~4 dB negative power tilt at the fiber span input and ~5 dB positive power tilt at fiber span output. The shape of the span input spectrum is determined by the linear and nonlinear interactions inside the span, as well as the gain of the EDFA and Raman amplifiers.

Fig. 13.15 confirms the optimum nonlinear design by transmitter power preemphasis curves for selected channels covering the full transmission bandwidth after 5380 km. Location of zero preemphasis point for each curve indicates that all channels accumulate similar nonlinear penalties after the transmission and operate near the peak of the system performance [32].

These preemphasis curves also show that channel performance varies across the optical bandwidth as expected with an optimal design. System capacity is maximized through the use of variable SE techniques (Section 13.2.2).

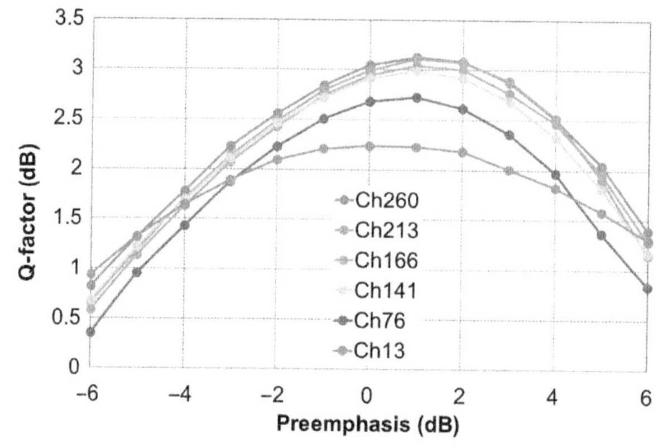

Figure 13.15 64QAM power preemphasis after 5380 km distance [32].

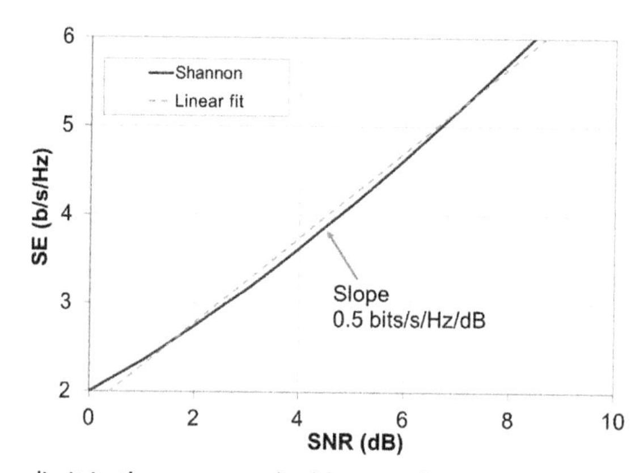

Figure 13.16 Shannon limit in the range applicable to undersea transmission systems.

13.2 Increasing spectral efficiency

Information theory teaches that the maximum channel capacity for an additive white Gaussian noise (AWGN) channel is [33]:

$$C = 2B \log_2\left(1 + \frac{S}{N}\right) \tag{13.19}$$

where S is the channel signal power, N is the noise in the channel bandwidth B. To increase the channel capacity, we need to increase the ratio S/N, which is typically referred to as SNR. The factor 2 comes from the two degenerate polarization modes of single-mode optical fiber used in polarization multiplexing. In the range of interest for undersea communication systems, the Shannon limit in Eq. (13.19) can be approximated simply with a slope of 0.5 b/s/Hz/dB (see Fig. 13.16).

In a WDM system, SE can be increased by transmitting more information using higher order modulation formats. The received OSNR of undersea systems can span a wide range of more than 10 dB. Different techniques are used to design modulation formats and coding at the extremes of the OSNR ranges. At lower OSNR values, for example, low-order signal constellations with strong binary codes are nearly optimal. As the OSNR (and SE) increases, strong coding alone becomes insufficient to achieve optimal capacity and higher order signal constellations are required.

This section examines the state of the art in coherent transmission to maximize SE and to get as close as possible to the Shannon limit in Eq. (13.19).

13.2.1 Advanced modulation formats—increasing channel data rate

Digital coherent detection opens the door for optical communication to use higher order constellations to approach the Shannon capacity. The widely studied modulation formats for undersea systems are quadrature amplitude modulation (QAM) formats and amplitude phase shift keying (APSK) formats with 16, 32, and 64 constellation points. However, increasing the constellation size to create higher order modulation formats results in lower receiver sensitivity such that the achievable distance for the same optical SNR and FEC algorithms rapidly decreases with increasing SE. Until recently, most of the improvements in practical optical coherent systems focused on reducing the implementation penalty of QAM formats using advances in high-speed DSP and hardware including digital-to-analog converters and analog-to-digital converters.

Researchers have studied many modulation and coding techniques to increase receiver sensitivity. Such techniques included trellis-CM, multilevel coding, constellation shaping, capacity achieving binary and nonbinary FEC codes, bit-interleaved coded modulation (BICM) and multidimensional CM. Further reductions in the gap to the Shannon limit now depend on improving the modulation format and coding technique.

13.2.2 Geometric constellation shaping

The use of conventional high-order QAM with uniform signaling on an AWGN channel results in a gap of up to 1.53 dB from Shannon's capacity curve. To approach the Shannon limit, the power distribution of the constellation in Euclidean space needs to follow multidimensional Gaussian distribution. The Gaussian-like power distribution of the constellation can be achieved using probabilistic shaping (PS) or geometric shaping.

Geometric shaping uses equiprobable constellation points at locations that are nonuniformly distributed in Euclidean space. Geometric shaping has been implemented using circular based constellations such as APSK, iterative polar modulation, and rate targeted optimization. These designs achieve Gaussian-like constellation shaping without introducing any redundancy. It also has been shown that this approach achieves the capacity of the AWGN channel if the number of constellation

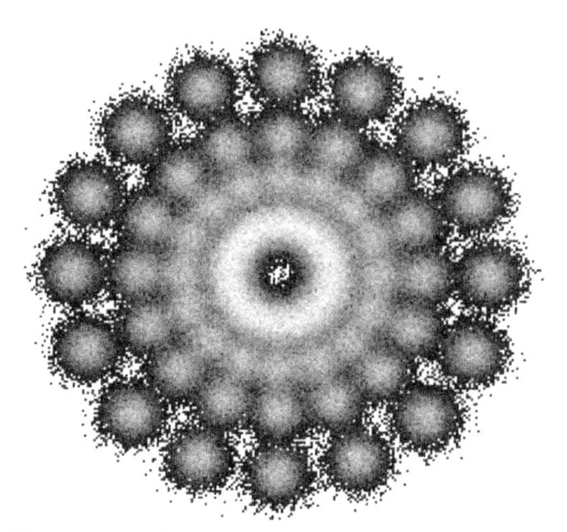

Figure 13.17 64APSK with geometric shaping.

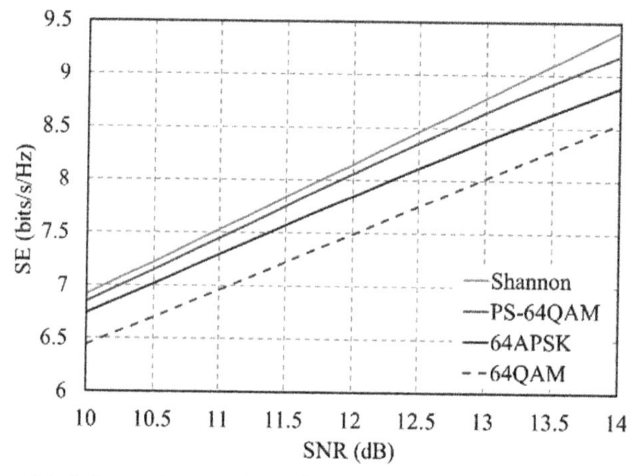

Figure 13.18 Achievable information rate (AIR) for PS-64QAM, 64APSK, and 64QAM, © [2018] SPIE. Reprinted, with permission, from [34].

points goes to infinity. Fig. 13.17 shows the constellation diagram of the radii optimized 64APSK modulation format [34]. The radii are optimized based on Gaussian distribution. The ratios of the radii from inner to the outer rings are 1, 1.88, 2.71, and 3.95, respectively. Fig. 13.18 shows the achievable information rate (AIR) comparing 64APSK with 64QAM. At 8 b/s/Hz, 64APSK is >0.5 dB better than 64QAM.

Ref. [35] demonstrated 64APSK and transmitted 34.5 Tb/s capacity over 6375 km using hybrid quasi-SMF fiber spans, resulting in 8.3 b/s/Hz SE, which is still the highest SE over transoceanic distance reported to date.

13.2.2.1 Probabilistic constellation shaping

Probabilistic shaping (PS) accomplishes Gaussian-like distribution by using a set of equidistant constellation points with a nonuniform probability distribution [36], as shown in the constellation diagram for PS-64QAM in Fig. 13.19. PS requires a transformation of equally distributed input bits into constellation symbols with a nonuniform distribution. Such transformations have been implemented using prefix codes, many-to-one mappings combined with a turbo code, distribution matching, or cut-and-paste method. PS does require redundancy but has attracted more attention due to its capability to approach Shannon faster than geometric shaping with an equivalent number of constellation points. Fig. 13.18 shows the AIR comparing PS-64QAM with 64QAM. At 8 b/s/Hz, PS-64QAM is >1 dB better than 64QAM and is only 0.1 dB away from the Shannon limit for AWGN channel. PS also provides a straightforward ability to achieve variable SE (Section 13.2.2.3).

Experimentally, PS-64QAM has been used to demonstrate a SE of 7.3 bits/s/Hz over 6600 km over C + L bandwidth. In this demonstration, the SE is controlled to maximize capacity over the 8.95 THz optical bandwidth [37]. Moreover, PS-64QAM has been demonstrated in field trials offering 6 b/s/Hz SE over 11,000 km [38].

In general, for limited number of constellation points, PS can outperform circular based geometric shaping by up to few tenths of a dB in shaping gain for the same number of constellation points. From Fig. 13.18, PS-64QAM has ~0.5 dB higher shaping gain than 64APSK. Yet both geometric and PS significantly outperform conventional QAM formats thanks to their superior power efficiency (PE) (i.e., maximum achievable rate per unit power).

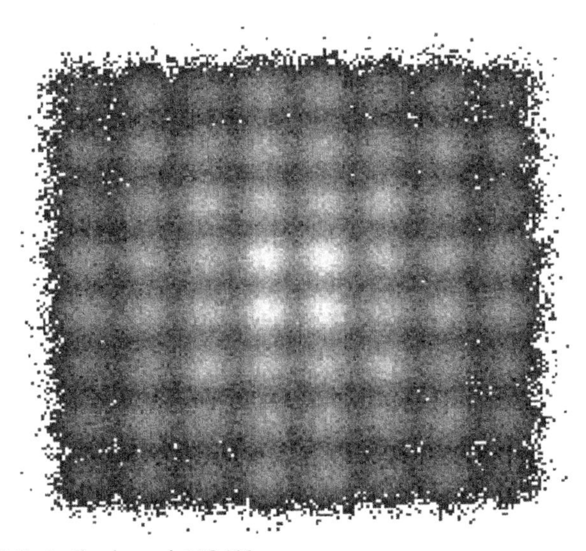

Figure 13.19 Probabilistically shaped-64QAM.

13.2.2.2 Multidimensional coded modulation

With binary transmission formats, error-correcting algorithms are limited to codes that are independent of the modulation format and the maximum SE is determined by the code rate. Coherent transmission techniques make higher order modulation formats practical and allow us to introduce CM, which is the joint design of an error-correcting code and a signal constellation. Here, groups of coded bits are mapped to points of the signal constellation in a way that enhances the distance properties of the code [39]. Typically, CM expands the signal constellation size and then applies error-correcting coding to increase the minimum Euclidean distance between modulated coded sequences [40]. A pragmatic approach to CM design is bit-interleaved coded modulation (BICM) where binary encoding is followed by a pseudorandom bit interleaver [41]. A wide variety of binary codes can be used in this case, increasing the flexibility with which this technique can be implemented, and single parity check (SPC) is one of the powerful such class code [69].

Fig. 13.20A shows a schematic for one such implementation for 9/12 SPC 64QAM. A binary low-density parity check (LDPC) code is used for error correction followed by the bit interleaver. In addition, the resulting bit stream is further encoded with a 9/12 rate SPC encoder before the bits are mapped in groups of 12 onto two consecutive 64QAM symbols using Gray mapping. The SPC encoding results in set partitioning where the available points of the 64QAM constellation are restricted to only 16 possible states in the second symbol depending on the constellation point of the first symbol in the pair. On the receive side, a coherent receiver is used to recover the 64QAM constellation. After equalization, a soft-input/soft-output (SISO) maximum a posteriori probability (MAP) decoder takes advantage of the SPC bit(s) and calculates the symbol log likelihood ratios (LLRs) of the two consecutive symbols based on the SPC codeword book. The symbol LLRs are then converted to bit LLRs and passed onto the LDPC decoders, which calculate the extrinsic information. The next iteration the MAP decoder uses this extrinsic information as a priori information. A schematic of the receiver is shown in Fig. 13.20B.

Fig. 13.21 shows the waterfall curve of for 9/12 SPC 64QAM for up to five iterations. The combination of 9/12 SPC and LDPC FEC generates very sharp waterfall

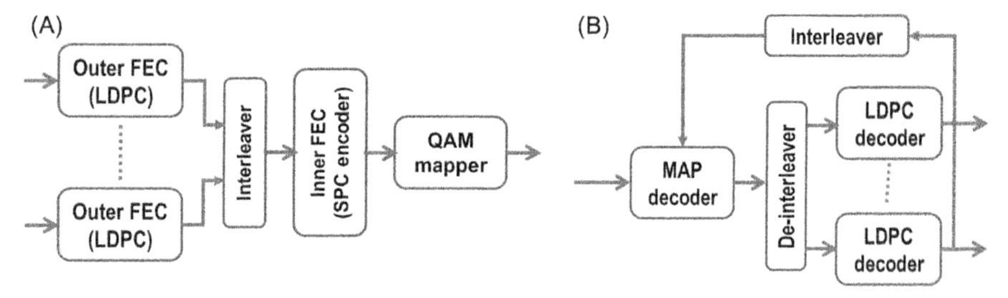

Figure 13.20 (A) Implementation schematic for single parity check bit-interleaved coded modulation (SPC BICM). (B) Receiver schematic.

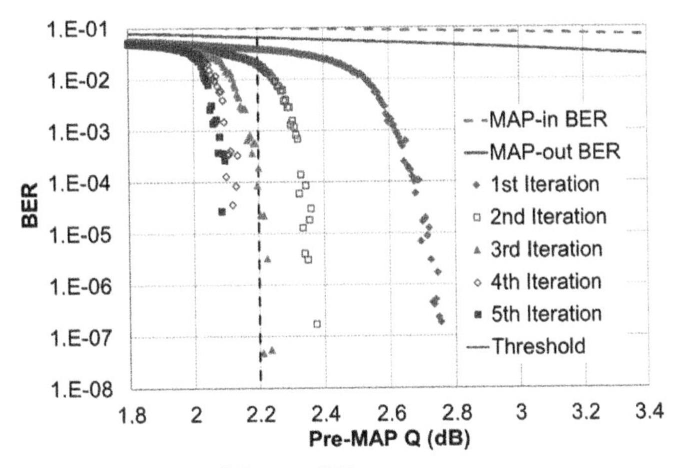

Figure 13.21 SPC-9/12 64QAM waterfall curve [32].

and increases total coding gain significantly. The error correction thresholds ($BER = 10^{-15}$), obtained from Monte Carlo simulations over an AWGN channel, are estimated to be 2.2 dBQ. The error correction waterfall performance is approximately 0.15 dB away from the theoretical threshold. With 9/12 SPC 64QAM CM, 63.5 Tb/s transmission was achieved over 5380 km with 7.1 b/s/Hz SE. CM also provides a straightforward ability to achieve variable SE (Section 13.2.2.4).

13.2.2.3 Coded modulation with both geometric and probabilistic shaping

While coding and constellation shaping can be optimized separately, joint optimization can improve the overall transmission performance. Moreover, geometric shaping and PS can be applied together to further improve channel performance. In Ref. [1], multidimensional CM was optimized to maximize the Euclidean distance under the existence of both geometric and PS. In this scheme, the redundancy is shared by LDPC, PS, and CM. The redundancy used for CM introduces structure in the transmitted signal; and this structure can be used to improve Euclidian distance and reduce SER, hence improve the theoretical limit. Moreover, lower SER can improve the implemented system receiver capability by improving equalization and the performance of adaptive nonlinear compensation algorithms, like the schemes introduced in Section 13.1.4.3. This is in contrast to conventional PS where the redundancy is used to improve format PE. Instead, the multidimensional CM formats achieve PE through geometric shaping only, and no redundancy is needed for geometric shaping.

Fig. 13.22 shows two examples of the new modulation scheme: 4D-PS-7/12–56APSK and 4D-PS-7/12–40APSK. In the projected 2D constellations, different sizes and shapes represent different probability of occurrence of the constellation points in the projection. Both formats are based on geometric shaping of 64APSK using four

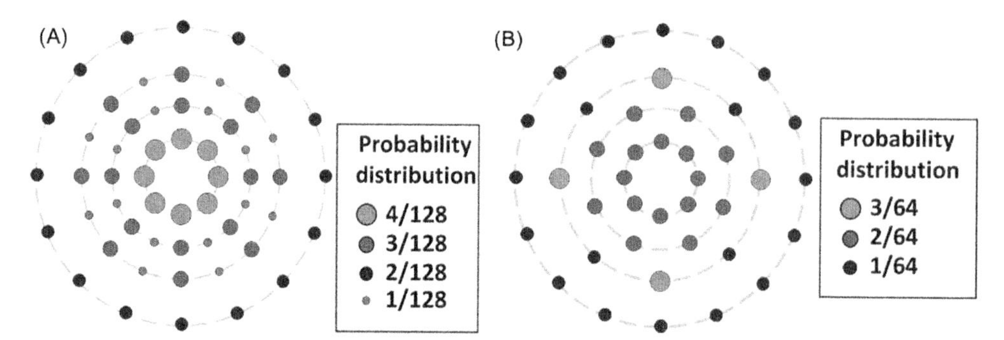

Figure 13.22 (A) 4D-PS-9/12–56APSK, (B) 4D-PS-7/12–40APSK constellation diagram, © [2018] SPIE. Reprinted, with permission, from [34].

rings with equal probability of occurrence. The radii are optimized based on Gaussian distribution. The ratios of the radii from inner to the outer rings are 1, 1.88, 2.71, and 3.95, respectively. As mentioned, PS is mainly used to maximize Euclidean distance between 4D constellation points and reduce SER.

The 4D-PS-9/12–56APSK format is designed for higher SE ~7.2 b/s/Hz (hence higher OSNR application) assuming 25% LDPC. This format uses a nonlinear 9/12 code to gradually vary the probability distribution of the constellation points from 16 equiprobable points on the outer ring to 8 equiprobable points on the inner ring, resulting in 56 constellation points of the 2D symbol. Hence, it is referred to as 4D-PS-9/12–56APSK. The encoding process is as follows: the information bits first goes through a LDPC encoder, then they are encoded by the reversible 9/12 nonlinear code, and the output of the nonlinear encoder is Gray mapped 12 bits at a time and modulated into 4D-56APSK symbols.

The 4D-PS-7/12–56APSK format is designed for lower SE ~5.6 b/s/Hz (hence lower OSNR application) assuming 25% LDPC. This format uses the same geometric shaping as 4D-PS-9/12–56APSK, but PS is redesigned to approach Shannon at the lower SNR value. This format uses a nonlinear code of code rate 7/12, and there are 40 constellation points with varying probability of occurrence in four rings. The two inner rings are composed of 8 equiprobable points each, the third ring is composed of 8 points with two different probability distributions, and the outer ring is composed of 16 equiprobable points.

The transmission performance of 4D-PS-9/12–56APSK and 4D-PS-7/12–40APSK has been demonstrated experimentally showing record capacities of 70.4 Tb/s over 7600 km and 51.5 Tb/s over 17,107 km, respectively. Some of the key enablers of these demonstration are NLC techniques that take advantage of features built into the modulation format as described earlier.

To compensate for time-varying nonlinear effects, LMS requires good-quality hard 2D or 4D decisions. Fig. 13.23 shows a power preemphasis transmission measurement with either 2D or 4D decision in the fast LMS block in comparison with the case

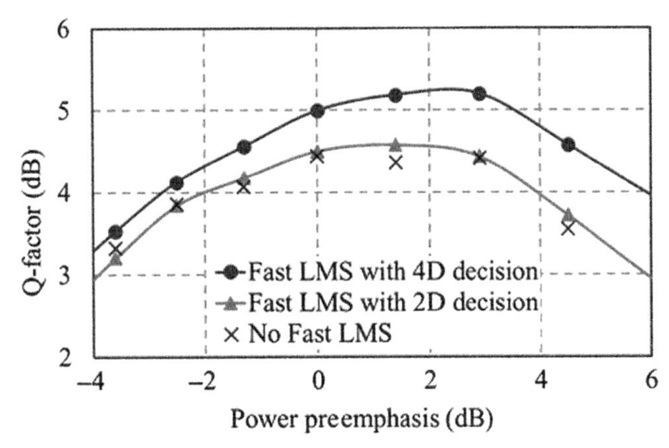

Figure 13.23 Preemphasis performance with fast least mean square (LMS) of 2D and 4D decisions, © [2017] IEEE. Reprinted, with permission, from [28].

where no fast LMS adaptive filter. DBP of the individual channel is applied in all cases and due to the limited bandwidth of back propagation primarily compensates for intrachannel nonlinearity. It is clear from the figure that Q-factor has improved when 4D decisions are used in the fast LMS (blue circle), while no significant improvement is observed in the case with 2D decisions (red triangle). The Q-factor improvement is higher at higher power, which confirms the NLC benefit of the techniques with 4D modulation format.

13.2.3 Variable spectral efficiency

As shown in Section 13.1.5, channel performance varies across the optical bandwidth with optimum nonlinear transmission design. The maximum transmission distance will be limited by the worst performance channel if only one SE is supported. To fully take the advantage of the available OSNR of different WDM channels, multiple spectral efficiencies are needed. Variable SE can be achieved by varying adapting FEC rate, mixing of modulation formats with different SE (time-hybrid QAM), PS, and CM.

13.2.3.1 Adaptive rate forward error correction

Using multiple FECs with different FEC thresholds in a WDM system can squeeze more capacity than using a single FEC. Ref. [43] designed a family of 52 Spatially-Coupled LDPC codes and studied the gain of using different number of FECs. Capacity increase due to using 8 FECs (with respect to single FEC, both without NLC) is between 15.5% and 21% for transmission distances from 10,200 to 6000 km. Further increase of the number of FECs used does not provide much more gain in capacity, as shown in Fig. 13.24. The shortcoming of this scheme is that the FEC implementation penalty increases for stronger FEC code. For example, the implementation penalty increases from 0.5 to >1 dB when the FEC code rate drops from 0.87 down to 0.52.

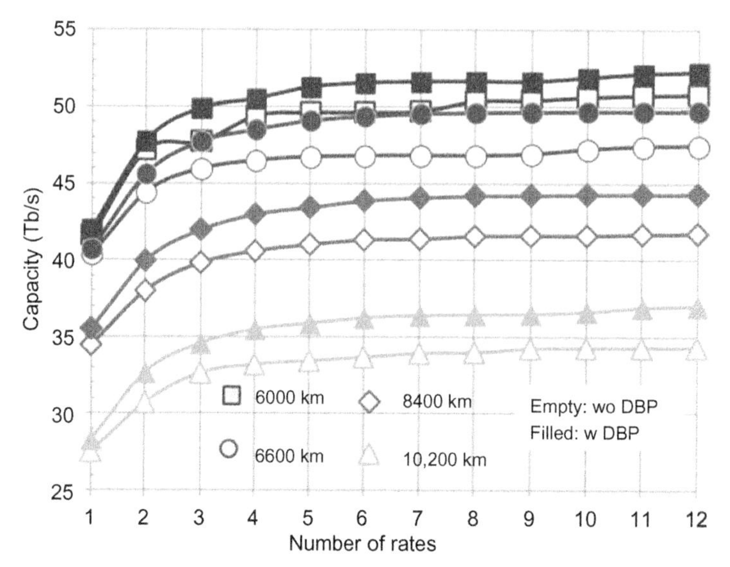

Figure 13.24 Net transmitted capacity versus number of forward error corrections (FECs), © [2015] IEEE. Reprinted, with permission, from [43].

13.2.3.2 Time-domain hybrid quadrature amplitude modulation

Another possible solution to provide a finer granularity to SE is the use of the time-domain hybrid QAM (TDHQ) [44,45]. TDHQ is based on the assignment of two or more regular QAM with different SE in alternating symbol times. Feasibility of TDHQ has been demonstrated in several long-haul transmission experiments at different net spectral efficiencies. For example, successful transmission of 4 and 8 bits/s/Hz were realized by a hybrid polarization division multiplexing (PDM) QPSK/8QAM and PDM 32/64QAM using 20% FEC overhead. Theoretically, a large QAM in conjunction with variable FEC codes can approach Shannon limit more than TDHQ as shown in Fig. 13.25.

13.2.3.3 Probabilistic constellation shaping

Varying the redundancy assigned to PS results in different SE. This allows for minimal changes at the transmitter and receiver to integrate a wider range of spectral efficiencies with fine granularity in SE using one FEC overhead and one base QAM format. Ref. [46] demonstrated widely flexible transmission using probabilistically shaped (PS) PDM 256-QAM. The paper reported 12.6, 11.4, 10.1, and 8.9 b/s/Hz at 500, 1000, 2000, and 4000 km transmission (in a narrow optical window).

13.2.3.4 Coded Modulation

Multidimensional CM with constellation shaping offers a straightforward approach to achieve variable SE [32,34]. For a given FEC code overhead and a constellation size,

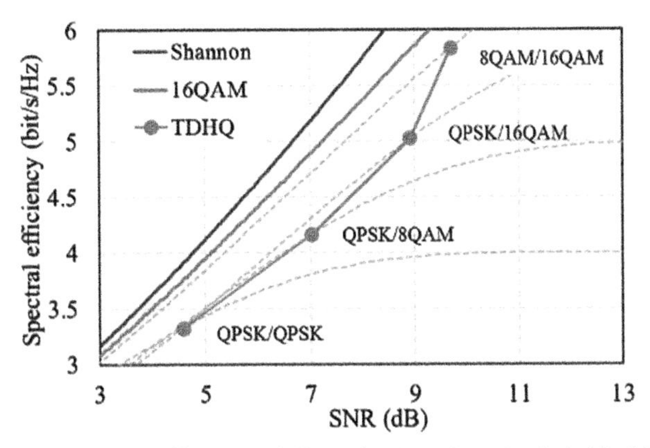

Figure 13.25 Variable spectral efficiency (SE) with time-domain hybrid QAM (TDHQ) over a Gaussian channel, © [2018] SPIE. Reprinted, with permission, from [34].

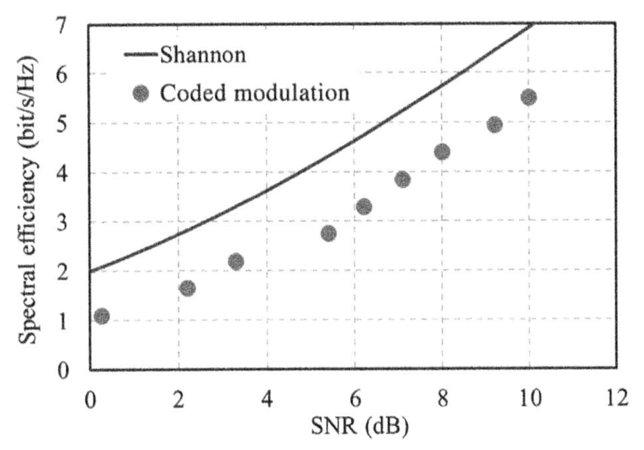

Figure 13.26 Variable spectral efficiency (SE) example using coded modulation (CM) with a single forward error correction (FEC) overhead and 6D SPC based 16QAM CM, © [2018] SPIE. Reprinted, with permission, from [34].

the rate of the inner code (k_s/n_s) is varied. This leaves the outer FEC code and the constellation size intact. The inner code is usually done with simple rules if the code is linear or a short MAP table if the code is nonlinear.

A key benefit of using CM for variable SE is that the structure can be optimized to help equalization at the receiver as shown in Section 13.2.1.4. Fig. 13.26 shows an example of different spectral efficiencies using 6D multistage SPC based 16QAM CM with a single FEC overhead. Ref. [49] experimentally demonstrated SE range from 2.4 to 8.0 b/s/Hz using a subset of the CM schemes shown in Fig. 13.26.

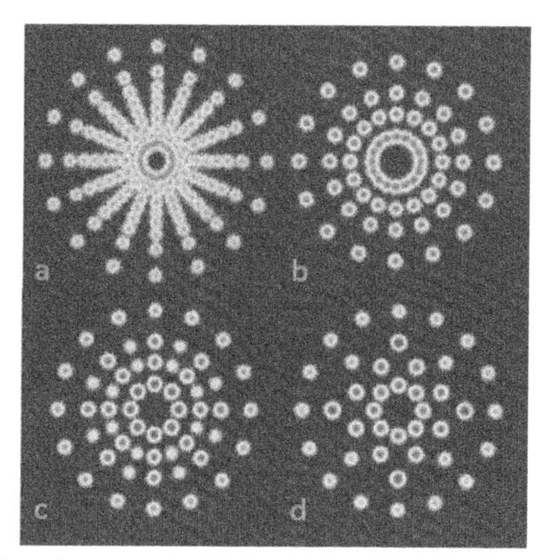

Figure 13.27 Constellation diagrams of APSK-based CM formats, © [2018] IEEE. Reprinted, with permission, from [47].

Table 13.1 APSK-based coded modulation formats (25% FEC), © [2018] IEEE. Reprinted, with permission, from [47].

Coded modulation format	Nyquist SE (b/s/Hz)	Constellation	References
4D-PS-7/12−40APSK	5.6	Fig. d	[29,34]
4D-PS-8/12−56APSK	6.4	Fig. c	[47]
4D-PS-9/12-56APSK	7.2	Fig. c	[1,28,34]
4D-GS-11/12-64APSK	8.8	Fig. b	[47]
4D-GS-12/14-128APSK	9.6	Fig. a	[47]

PS, probablistic shaping; *GS*, geometric shaping.

13.2.3.5 Fine spectral efficiency granularity with coded modulation and adaptive rate forward error correction

Adaptive FEC has higher penalty with stronger FEC code [43], while CM has coarse SE granularity (Fig. 13.26). By combining the two techniques, we can achieve a very wide range of SE with very fine granularity, and most importantly, almost constant gap from the Shannon limit. Fig. 13.27 shows several modulation formats using APSK-based CM using an LDPC code with 25% FEC overhead and different (k_s/n_s) inner codes. The base SE can be varied from 5.6 to 9.6 b/s/Hz. Together with variable SE, we can cover a SE range from 4.5 to 10.2 b/s/Hz (Table 13.1).

Fig. 13.28 shows that the simulated SNR gap with the implemented FEC for the CM family over an AWGN channel is 1.85−2.8 dB away from Shannon limit (diamond symbols). The gap of the mutual information from Shannon limit for these modulation formats is shown by *open square symbols*. We also plot performance of the

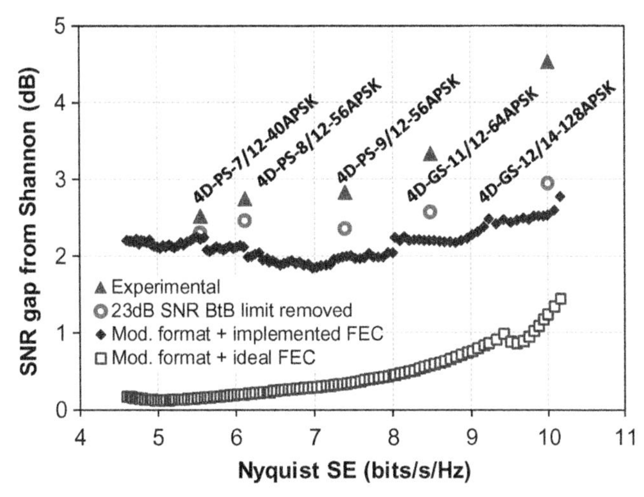

Figure 13.28 Variable spectral efficiency (SE) example using CM with hybrid constellation shaping, © [2018] IEEE. Reprinted, with permission, from [47].

modulation formats with laboratory implementation penalty at the nominal SE for the five CM formats (*triangle symbols*). The difference in performance shown by *triangle* and *solid diamond* symbols reflects the implementation penalty. It ranges from <0.3 dB for 40APSK with 5.6 b/s/Hz SE to >2.0 dB for 128APSK with 10 b/s/Hz SE (20% FEC). This penalty is primarily related to the limited transmitter and receiver SNR ≈ 23 dB imposed by the lab hardware. *Open circles* in Fig. 13.28 show performance of the modulation formats for ideal hardware when the back-to-back SNR limit is removed.

Fig. 13.29 shows lab measured capacity and net SE using the above APSK-CM format and C + L EDFAs [47]. The achieved SE is ~5.4 bits/s/Hz at 17,107 km using 4D-PS-7/12−40APSK modulation format and increases to ~9.75 bits/s/Hz at 1900 km using 4D-GS-12/14−128APSK modulation format. The capacity points are generated from three full capacity measurements of all WDM channels over 9.74 THz optical bandwidth. The maximum capacity is 51.5 Tb/s at 17,107 km using 4D-PS-7/12−40APSK modulation format and increases to 95 Tb/s at 1900 km using the 4D-GS-12/14−128APSK modulation format.

13.3 Increasing optical bandwidth

From the Shannon limit in Eq. (13.19) it is easy to see how to increase capacity on a fiber pair. There are fundamentally two ways: (1) increasing SE by increasing the available SNR and (2) increasing the available transmission bandwidth. This section considers the latter. Note that adding bandwidth is a more effective way to increase system

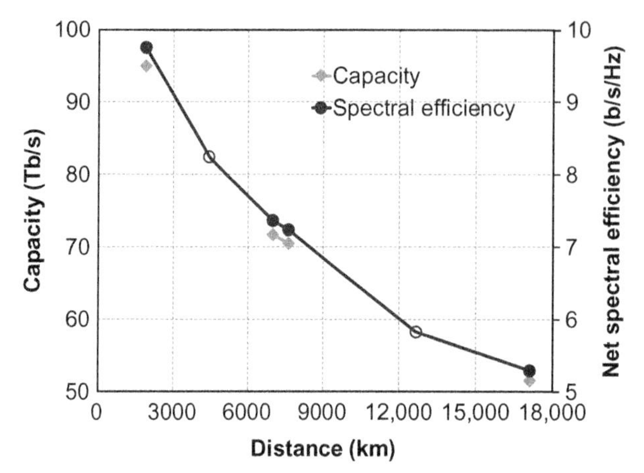

Figure 13.29 Total capacity and average net spectral efficiency (SE) versus distance for the coded modulation family, © [2018] IEEE. Reprinted, with permission, from [47].

capacity than increasing SNR since SE increases linearly with bandwidth and only logarithmically with OSNR (Eq. 13.20).

$$C = \sum_{\substack{i=1 \\ WDM}}^{i=n} 2B_i \log_2(1 + \mathrm{OSNR}_i) \tag{13.20}$$

13.3.1 Maximizing C-band capacity

Much progress has been made over the last years in undersea transmission demonstrations using the ~40 nm and/or ~5 THz of bandwidth in the C-band (for "conventional band"), of EDFAs. Table 13.2 lists in chronological order the achieved record capacity, record SE, record capacity distance product, and record SE distance product over transoceanic distance, with the corresponding records highlighted. The record SE achieved is 8.3 b/s/Hz, and the largest capacity (per core) demonstrated is ~35 Tb/s. From 2009 to 2012, the capacity in C-band has increased from 7 Tb/s up to 30 Tb/s, and it took another 4 years to grow the C-band capacity to 35 Tb/s. Based on current amplification and modulation technology, the maximum capacity in the C-band will be ~40 Tb/s over transoceanic distance.

To maximize C-band capacity, gain equalization is indispensable to remove the intrinsic gain shape of the EDFA . As seen in Fig. 13.30, excess loss rapidly increases for EDFA equalization bandwidth beyond 38 nm. The widest bandwidth of EDFAs deployed in an undersea system is 41 nm [50]. The challenge for wide bandwidth undersea amplifier is to limit the power variation among WDM channels to about 5–10 dB for adequate performance. To achieve <6 dB gain variation after a 9000 km

Table 13.2 Recent demonstrations with large capacity distance and/or large SE distance product over the C-band.

Capacity (Tb/s)	SE (b/s/Hz)	Distance (km)	Capacity distance product (Pb/s · km)	SE distance product (b/s/Hz · Mm)	Modulation format	Year	References
7.2	*2*	7040		*14.08*	QPSK	2009	OFC, PDPB6
9.60	*3*	10,608	*101.84*	*31.82*	QPSK	2010	OFC, PDPB10
11.20	*3.6*	9360	*104.83*	*33.70*	QPSK	2010	ECOC, PD2.1
20.00	*4*	6860	*137.20*	27.44	QPSK	2011	OFC, PDPB4
4.704	*4.7*	10,181	47.89	*47.85*	16QAM	2012	OFC, PDP5C4
16.67	*5.21*	6860	114.35	35.74	16QAM	2012	OE, p. 11688
26.05	5.21	5530	*144.04*	28.81	16QAM	2012	ECOC, Mo.1.C.1
30.58	*6.12*	6627.5	*202.64*	40.54	16QAM	2012	ECOC, Th.3.C.2
21.20	6	10,290	*218.15*	*61.74*	16QAM	2013	OFC, PDP5A.6
30.40	6	9748	*296.34*	58.49	16QAM	2013	OE, p. 9116
30.60	*6.6*	6817.5	208.62	45.00	16QAM	2014	OFC, PDP5B.5
34.9	*8.3*	6375	222.49	52.91	64APSK	2016	OFC, Th5C.2
25.48	6.06	10,285	262.06	*62.33*	PS-64QAM	2018	JLT, p. 1354

Data given in italics show the new record at the time of the publication. *SE*, spectral efficiency.

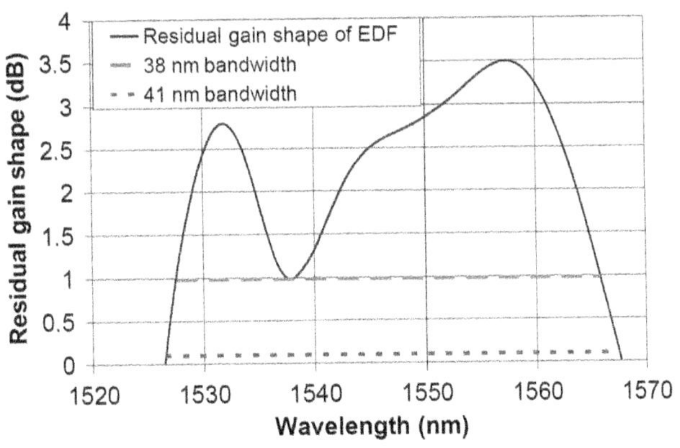

Figure 13.30 EDF gain shape to be corrected by the gain flattening filter (GFF), for 38 nm (*dashed line*) and 41 nm (*dotted line*) [50].

system with 120 EDFAs (75 km repeater spacing), the systematic gain error of individual EDFA gain equalization needs to be within ~50 mdB.

The largest capacity among all in-service undersea systems, MAREA, is deployed with C-band EDFA. It connects Virginia Beach, United States to Bilbao, Spain, via eight fiber pairs, with an initial estimated design capacity of 160 Tb/s, an increase of more than 285,000 times compared with TAT-8 [6].

The widest bandwidth C-band EDFAs employed in lab transmission experiments goes up to 43 nm [51]. In lab experiments, typically a tunable gain equalizer is employed in the loop to remove the residual gain error. Hence the residual gain error is typically <1 dB even after 10,000 km transmission. The maximum capacity will be limited to ~40 Tb/s over transoceanic distance using C-band EDFAs, further capacity increases require extension of the optical transmission bandwidth beyond the C-band into L-band wavelengths [52].

13.3.2 Moving beyond the erbium-doped fiber amplifiers C-band

EDFA can also be designed to amplify the longer wavelength band (the L-Band) from 1570 to 1605 nm covering ~4.5 THz optical bandwidth. Thus, taken together the useable bandwidth of a C+L band, EDFA-based system doubles to >9 THz (Fig. 13.31).

Significant capacity improvement has been demonstrated in recent transoceanic transmission experiments using C+L band EDFA, where 71.6 Tb/s capacity over ~6960 km [1] and 51.5 Tb/s capacity over 17,107 km [29] have been reported. Extension of the bandwidth beyond the C-band can also be achieved with Hybrid Raman-EDFA [53]. The higher OSNR of Hybrid Raman-EDFA due to distributed amplification has been shown in earlier experiments. The largest capacity over transoceanic distance using Raman- assisted EDFA achieved 54 Tb/s with ~9 THz continuous optical bandwidth over ~9150 km [31].

Table 13.3 lists in chronological order the transmission record using C+L band, including record capacity, record SE, record capacity distance product, and record SE distance product for transoceanic distances, with the corresponding records highlighted. The maximum achievable SE goes down to 7.4 b/s/Hz for C+L

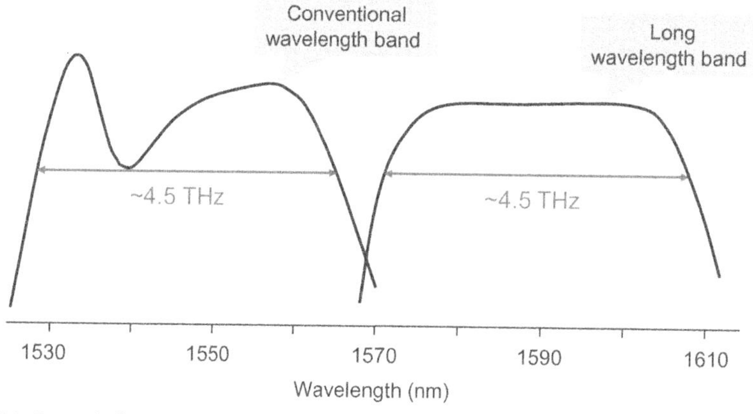

Figure 13.31 Extended optical bandwidth with C+L EDFA.

Table 13.3 Recent demonstrations with large capacity-distance and/or large SE distance product over the C + L bandwidth.

Capacity (Tb/s)	SE (b/s/Hz)	Distance (km)	Capacity distance product (Pb/s · km)	SE distance product (b/s/Hz · Mm)	Modulation format	Year	Reference
13.5		6248	*84.35*		QPSK	2009	OFC, PDPB5
15.5		7200	*111.60*		QPSK	2009	ECOC PD2.5
38.75	*5*	6600	*255.75*		16QAM	2013	ECOC, PD3-E-2
44.10	4.93	9037.5	*398.55*	*44.52*	16QAM	2013	ECOC, PD3-E-1
49.302	*5.245*	9037.5	*445.57*	*47.40*	16QAM	2014	OFC, Th5B.4
52.90	5.94382	6000	317.40	35.66	16QAM	2014	ECOC PD3.4
54	*6.016*	9149.4	*494.07*	*55.04*	16QAM	2014	ECOC, PD3.3
52.2	5.816	10,225.8	*533.79*	*59.47*	16QAM	2014	ECOC, PD3.3
61.9	5.816	5920	366.45	34.43	64QAM	2015	OFC, Th5C.8
63.5	*7.1*	5380	341.63	38.20	64QAM	2015	OFC, Th5C.8
65	*7.3*	6600	429.00	48.18	PS-64QAM	2016	ECOC, Th.3.C.4
71.64	*7.36*	6960	498.61	51.23	PS-64APSK	2017	OFC, Th5B.2
51.5	5.6	17,108	*881.06*	*95.80*	PS-64APSK	2017	ECOC, PDP.A.2

Data in italics show the new record at the time of the publication. *SE*, spectral efficiency.

bandwidth compared with 8.3 b/s/Hz achieved in the C-band. This is partially due to the larger NLI (~0.5 dB) in the C + L bandwidth as indicated by the GN model.

13.3.3 Comparison of C + L erbium-doped fiber amplifiers and Raman amplification

In this section we compare the performance and merits of C + L EDFA and hybrid Raman-EDFA. Fig. 13.32A shows the schematic of a C + L EDFA used in Ref. [31]. It consists of a band splitter, parallel C-band and L-band EDFAs, and a band combiner. Both C- and L-band EDFAs are standard single-stage 980 nm forward-pumped configurations with a gain flattening filter at the output. The C + L EDFAs are equalized to ~77 nm (or ~9.4 THz) bandwidth with a ~4 nm band gap and operate at 22.2 dBm combined output power in C- and L-bands which corresponds to an average PSD of 17.7 mW/THz launched into the transmission fiber. Fig. 13.32B shows the schematics of the hybrid Raman-EDFA used in Ref. [31]. It consists of a backward Raman pumping launched prior to each single-stage EDFA. A pump wavelength of 1484 nm allows the gain spectra of the Raman and highly inverted EDFA sections to complement each other and provide wide continuous bandwidth with minimum gain ripple (no C/L WDM coupler is needed or used). The pump power from two polarization multiplexed grating stabilized lasers is coupled into the span using a circulator to achieve simultaneous low loss and high isolation for both pump and signals.

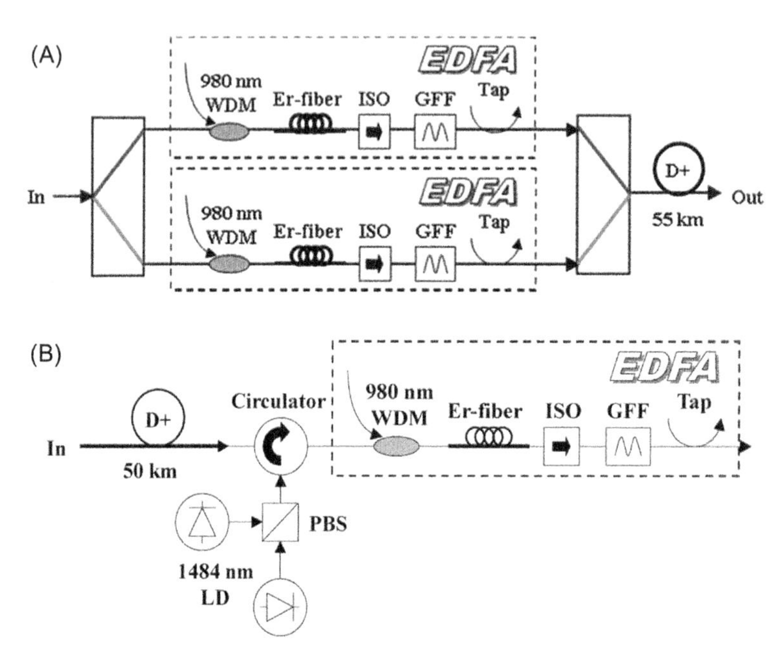

Figure 13.32 Schematic of (A) C + L EDFA, (B) Hybrid Raman-EDFA [31]. *ISO*, Isolator; *GFF*, gain flattering filter.

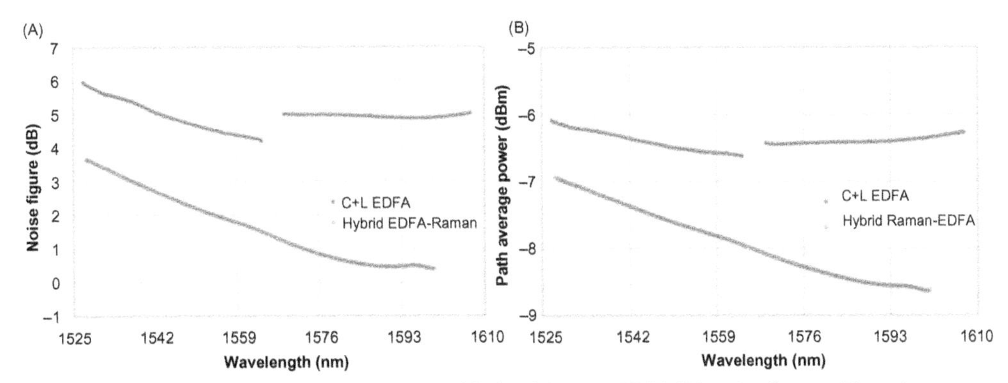

Figure 13.33 Comparison of C + L EDFA and hybrid Raman-EDFA (A) noise figure, (B) path average power, © [2015] IEEE. Reprinted, with permission, from [31].

The EDFA section of the hybrid amplifier is a standard single-stage 980 nm forward-pumped configuration with a gain flattening filter at the output designed to equalize the combined Raman-EDFA gain shape. Amplifiers are equalized over a continuous ~73 nm (~9 THz) bandwidth and operate at 18.4 dBm output power which corresponds to an average PSD of 7.7 mW/GHz launched into the transmission fiber.

Compared to an EDFA, the distributed Raman amplification can achieve a "black-box" NF with up to 5 dB lower value. However, this does not translate into a similar performance improvement. Fig. 13.33A compares designed amplifier NF of hybrid Raman-EDFA and C + L EDFA. The hybrid amplifier shows an average NF advantage of 3.2 dB compared to C + L EDFA.

In systems with fiber nonlinearity, this NF benefit does not entirely translate into performance benefit. Recall from the GN model, channels with higher linear noise can also tolerate more nonlinear noise and allow channels with low OSNR to operate at high power.

Fig. 13.33B compares signal path average powers across the bandwidth for both amplifier schemes, on average, C + L EDFA system can operate 1.5 dB higher power than hybrid Raman-EDFA system. In this case the 1.5 dB higher power tolerance of C + L EDFA comes from two factors: 1.1 dB from 3.2 dB higher NF and 0.3 dB from 13% larger effective area of the transmission fiber (hybrid Raman-EDFA amplifier uses 130 μm^2 fiber to achieve higher Raman gain).

The smaller fiber launch power leads to a reduction of distributed amplification advantage in the presence of nonlinearity. Ref. [31] showed that an optimized hybrid Raman-EDFA provides ~1.8 dB better nonlinear OSNR compared to C + L EDFA (see Table 13.4). However, for similar transmission bandwidth and distance, the experimental capacity increase is only 9.5% because of the limitations caused by nonlinear effects. Analysis of optical pump currents revealed that this capacity increase is achieved at a cost of 2 × greater electrical power compared to C + L EDFA.

Table 13.4 Comparison of C + L EDFA (CL) and hybrid Raman-EDFA (CR) testbeds, © [2018] IEEE. Reprinted, with permission, from [31].

Amp. type	Noise figure (dB)	Optimum path ave. power (dBm)	Linear OSNR (dB)	$OSNR_{eff}$ (dB)	Capacity (Tb/s)	SE (bits/s/Hz)	Electrical power per span (W)
C + L	4.9	− 6.4	17.5	16	49.3	5.24	3.1
CR	1.7	− 7.9	19 .0	17.8	54	6.02	5.5
Delta	− 3.2	− 1.5	1.5	1.8	9.5%	14.7%	2x*

SE, spectral efficiency. All parameters except for capacity are average numbers, *2 × after factor in the span length difference (50 km vs 55 km).

Reliability is also a major concern for 14XXnm pump lasers used in Raman amplification applications. Typically, to maintain a constant output power, the required electrical current gradually increases over time. Furthermore, the optical gain in the Raman system operates in the small signal gain regime. The end-to-end gain changes linearly as the 14xx pump power degrades.

EDFAs designed for undersea applications operate in deep saturation with typically 5−10 dB gain compression. Even if the pump power of a particular EDFA changes by several dBs, the system spectrum and total gain will be recovered automatically due to the strong gain compression in EDFA systems.

For these reasons, C + L EDFA technology becomes the first broadband amplification technology deployed over transoceanic distance. The first undersea system built with C + L EDFA technology, PLCN (connecting Hong Kong, China and Los Angeles, California, United States), can carry at least 144 Tb/s (6 fiber pairs × 24 Tb/s) over ~13,000 km [7], an increase of more than 250,000 times comparing with the first transpacific optical system, TPC-3.

13.3.4 Comparison of C + L erbium-doped fiber amplifier and C + C erbium-doped fiber amplifier

The capacity of one C + L EDFA fiber pair can be achieved using two C-band EDFA fiber pairs (i.e., C + C). As shown in Fig. 13.34, the main difference in the two architectures is that a fiber span is replaced by a pair of C/L band splitters and combiners when moving from C + C EDFA to C + L EDFA. This can lead to cost savings increasing linearly with optical fiber grade, span length, system length, and number of fiber pairs. Therefore, the wet plant cost using C + L EDFAs can be significantly lower than that using C + C EDFAs depending on system parameters.

Note that a C + L EDFA is slightly less power efficient than two C-band EDFAs (Table 13.5). In laboratory experiments for two fiber pairs using C EDFAs, the required 980 nm pump power is 4 × 500 mW = 2.0 W; while the required 980 nm power is 3 × 800 mW = 2.4 W for a single fiber pair using C + L EDFAs. At the same

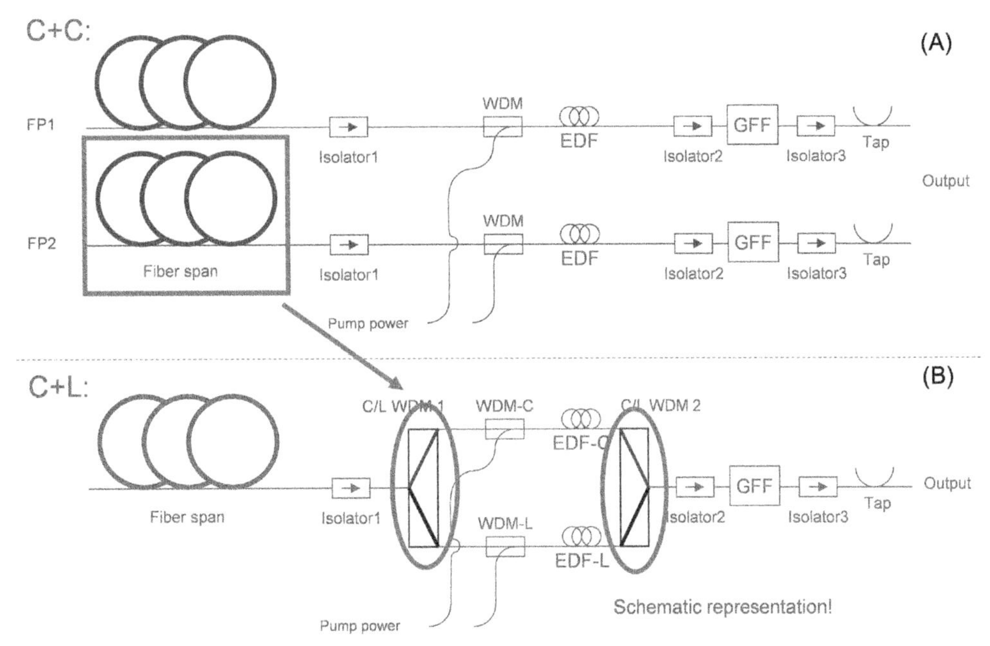

Figure 13.34 Schematic of (A) C + C EDFA, (B) C + L EDFA.

Table 13.5 Performance C + L EDFA relative to C + C EDFA.

	C + L	
	C (dB)	**L (dB)**
Noise figure	0	− 0.2
Raman	− 0.3	0.3
Fiber loss	0	− 0.3
Nonlinearity	− 0.1	− 0.1
Total	− 0.4	− 0.3

repeater spacing the performance with C + L EDFAs can be slightly lower than that of C + C EDFAs due to a higher NF in the L-band and the impact of the C/L splitters. The details depend on the actual system design, including amplifier total output power, span length, etc. Nevertheless, the cost reduction due to the reduced number of fiber pairs which can also enable a more compact cable design can be significant leading to an overall cost advantage for undersea system deployed with C + L EDFAs.

13.4 Increasing cable capacity

The previous sections discussed techniques to increase capacity in single-mode and single-core transmission, by increasing SE or optical bandwidth. However, using SDM techniques, the cable capacity will scale with the number of fibers in the cable, the number of cores in a fiber, and the number of modes in a core. Combining WDM and SDM, the maximum capacity for AWGN channel grows linearly with the number of WDM channels and also grows linearly with the number of spatial modes (Eq. 13.21).

$$C = \underbrace{\sum_{j=1}^{j=m}}_{SDM} \underbrace{\sum_{i=1}^{i=n}}_{WDM} 2B_{ij} \log_2(1 + \mathrm{OSNR}_{ij}) \tag{13.21}$$

13.4.1 Space division multiplexing using multicore fiber

Transmission systems based on MCFs have a host of performance and economic issues. For example, intercore crosstalk-induced penalty is a concern but may be able to be overcome. Calculated Q-factor transmission penalty associated with in-band crosstalk under the assumption of crosstalk being noiselike and additive is shown in Fig. 13.35 along with measurement data. To keep the accumulated crosstalk < -25 dB after 15,000 km, the crosstalk needs to be < -50 dB per 50 km span.

The availability of low crosstalk MCF opens the door for significant capacity increase per fiber. For one particular lab experiment, the worst in-band crosstalk per span including the fan-in and fan-out was -54 dB. Using this 12-core fiber, we demonstrated 520 Tb/s transmitted over 8832 km, and the achieved SE is boosted to 60 bits/s/Hz per fiber [2]. Table 13.6 lists transmission records based on MCF.

The number of cores that can fit into these low crosstalk MCF designs without increasing its cladding diameter is limited to about 7. Beyond seven cores the outside diameter of the fiber has to increase such that the core density can only be increased

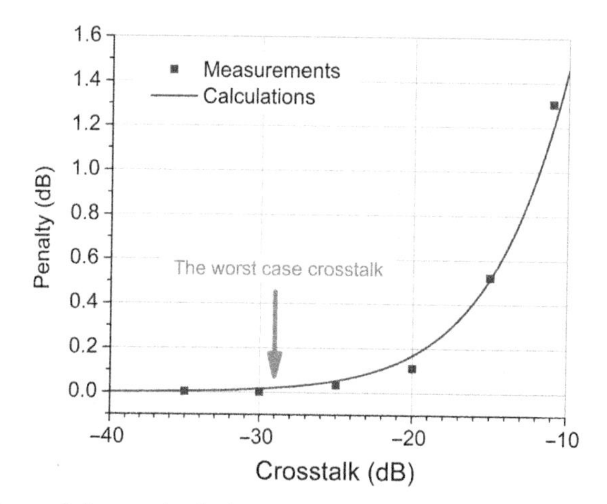

Figure 13.35 Experimental data and calculation results of intercore crosstalk transmission penalty, © [2017] IEEE. Reprinted, with permission, from [54].

Table 13.6 Recent demonstrations with large capacity distance and/or large SE distance product using SDM technology.

Capacity (Tb/s)	SE (b/s/Hz)	Distance (km)	SE distance product (b/s/Hz · km)	Capacity distance product (Pb/s · km)	Modulation format	Year	References
28.84	*14.42*	6160	*177*	*88.83*	QPSK	2012	ECOC, Th.3.C.3
111	*22.42*	6370	*707.07*	*142.84*	QPSK	2013	OE 2013, p. 18053
140.7	*28*	7326	*1030.77*	*205.13*	QPSK	2013	ECOC, PD3-E-3
520	*60*	8832	*4592.64*	*529.92*	16QAM	2016	ECOC, Tu.1.D.3

Data in italics show the new record at the time of the publication. *SE*, spectral efficiency.

by about one order of magnitude compared with conventional fiber bundles. For applications where space is limited such as in an undersea cable, this may be a compelling reason for using MCFs to increase capacity. On the other hand, all the cores in the cable have to be fitted with EDFAs when the cable is first installed. This is in contrast to terrestrial applications where EDFAs are installed along a new path only when the path is being lit for the first time. The number of unlit paths in a undersea cable therefore has to balance the initial system cost and final target capacity. This scenario assumes that separate amplifiers are required for each core, which is the current situation. In the future multicore EDFAs may become available such that MCF can be used in a similar fashion as SMF today. The economics of multicore systems still need to play out; however, it might be challenging for an "N-core" fiber system to be more economic than an N-fiber system.

13.4.2 Space division multiplexing using multimode fiber/few-mode fiber/ coupled core multicore fiber

Another option to increase cable capacity is mode division multiplexing (MDM). Conventional multimode fiber has a very large number of modes that could (in principle) be used for transmission. SMF already supports two orthogonal polarization modes and is used to double SE using PDM techniques. The two polarization modes couple strongly in transmission fiber, but a coherent receiver is able to unscramble the two modes as long as the amount of polarization dependent loss is small and the differential mode delay from polarization mode dispersion does not exceed the length of the linear equalizer. The linear equalizer in this case is a 2×2 multiple-input multiple-output (MIMO) equalizer (two polarization streams in and two demultiplexed polarization streams out).

The same principles apply to transmission on multimode fiber, only now the number of modes can be very large. This is very attractive to increase the fiber capacity; however, this also increases the complexity of the required MIMO equalizer considerably. For an MMF with M modes, the MIMO size goes to 2 M × 2 M, which grows quadratically with M. Moreover, there can be an enormous differential modal group delay (DMGD) among modes, typically on the order of 1000 ps/km. Therefore, the numbers of taps needed in CMA becomes unmanageable over long-haul transmission distance.

This issue can be alleviated by using few-mode fibers (FMF), which has a limited number of modes and reduces mode mixing to a manageable level. Furthermore, the DMGD can be reduced in a managed link, where DMGD is canceled by concatenating multiple fiber segments with DMGDs of opposite sign [55]. Ref. [56] demonstrated 16QAM transmission over 2400 km using low DMGD (27.1 ps/km) and low mode dependent loss (MDL, 1.75 dB) 3-mode FMFs, and showed the FMF outperformed SMF for distances up to 4500 km due to its large effective area.

Another approach in MDM is the newly developed coupled core multicore fiber (CC-MCF). In this new scheme, designers use MCF with strong coupling among the cores to support a few spatial modes. Ref. [57] demonstrated QPSK transmission over 10,000 km and 16QAM transmission over 5000 km using CC-MCF. The group also compared the performance to using similar SMF where the performance of the CC-MCFs were better than SMFs for distance <2000 km.

Another issue in MDM transmission is the MDL, which can be several dBs over a 50 km span of MMF. MDL can only be partially recovered by DSP, and MDL must be mitigated in order to not limit the transmission distance. Also, the stability of the differential loss between modes is important, especially in transoceanic length systems where gain equalization is critical. Recently, multimode/few-mode amplifiers have been demonstrated in transmission experiments including gain equalization among modes.

13.5 Increasing capacity under the constraint of electrical power

As shown by the results in the previous section, cable capacity has much room to grow using SDM techniques. However, what cannot be ignored is that as total capacity is increased, so too is the need for greater optical powers from the optical amplifiers, meaning more electrical power. There are many practical aspects of providing high electrical power to the remote undersea repeaters located along the cable network. Attaining these high optical power levels in long-haul undersea systems is problematic, since the ability to deliver power to the optical amplifiers is limited due to the maximum voltage that can be applied to a cable from the shore ends (Fig. 13.36). Modern undersea cable systems use high-voltage power feed equipment (PFE) at the terminals to supply a constant current and a maximum voltage of ~15 kV to the cable's shore ends. Ref. [58] shows ~1% overall efficiency from PFE to optical power launched to the undersea fibers. Thus finding more efficient ways of using the optical

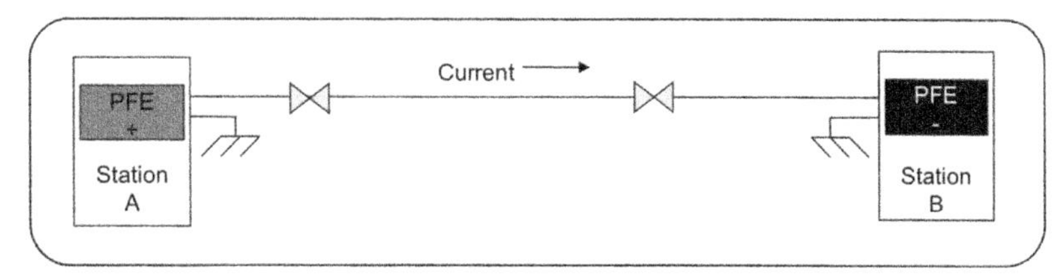

Figure 13.36 Schematic of power feeding in undersea system. *PFE*, Power feeding equipment.

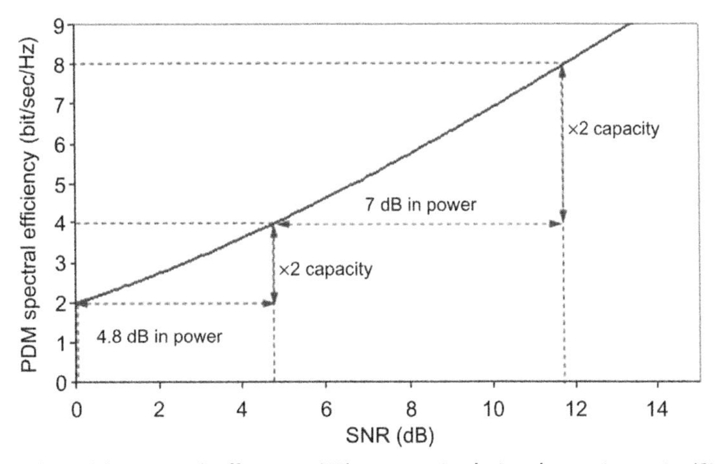

Figure 13.37 Achievable spectral efficiency (SE) vs. required signal-to-noise ratio (SNR), © [2018] IEEE. Reprinted, with permission, from [60].

power becomes imperative [70,59]. Improvement of overall PE of a transmission system depends on electrical power delivery and optical power utilization. We focus on the ways to use optical power in a more efficient manner which means how to transmit more information having overall optical power limitations.

SDM techniques can be used to improve network PE. This is illustrated in Fig. 13.37 using the Shannon capacity curve already discussed in Eq. (13.19), and in Fig. 13.16, more detail has been added to explain how SDM can help. In this example doubling SE using conventional (non-SDM means) from 2 to 4 b/s/Hz requires ~4.8 dB increase in both the SNR and the required signal power, if the noise power stays the same. Another factor of 2 (from SE = 4 to SE = 8) requires ~7 dB further increases in SNR and signal power. Thus, a total signal power increase of 11.8 dB is required for a 4× increase in SE from 2 to 8 b/s/Hz. Alternatively, with an SDM approach, this 11.8 dB total power increase could be divided, for example, between 15 parallel space dimensions (cores, fibers), each carrying the initial SE of 2 b/s/Hz (noting that 11.8 dB is ~15×). The total capacity in this case would be 15× larger as opposed to the 4× larger for the single dimension, or a 375% increase in capacity for the same power. In reality, the SDM efficiency gain is probably understated since the increased optical signal power in single fiber could also incur added nonlinear penalty in the single-dimension case. In the SDM case, this problem is mitigated since the power is distributed over multiple dimensions.

13.5.1 Optimum spectral efficiency

PE of an SDM transmission link is defined here as the ratio of aggregate capacity C to the total optical power P_{tot} summed at the output of all amplifiers of the link:

$$PE = \frac{C}{P_{tot}} = \frac{C}{P_1 N_{dim} N_{amp}} \tag{13.22}$$

where $P_1 = P_{tot}/(N_{dim}N_{amp})$ is the output optical power per amplifier per dimension, N_{dim} is the number of spatial dimensions, and $N_{amp} = L/l$ is the number of SDM amplifiers, where L is the system length and l is the repeater spacing.

Recall the capacity of an SDM-PDM system with N_{dim} and bandwidth B_{amp} is

$$C = B_{amp} \cdot N_{dim} \cdot 2\log_2(1 + \mathrm{SNR}) \tag{13.23}$$

The SNR per dimension defined by P_1 and P_{ase} is defined as

$$\mathrm{SNR} = \frac{P_1 - P_{ase}}{P_{ase}} \tag{13.24}$$

P_{ase} is the forward-propagating ASE noise in any single spatial dimension at the end of the link. The relationship between amplifier (or system) bandwidth B_{amp}, amplifier noise figure NF, amplifier gain G, number of SDM amplifiers in the link N_{amp}, and ASE power P_{ase} is given by

$$P_{ase} \approx \left(e^{\alpha l} NF - 1\right) h\nu B_{amp} N_{amp} \tag{13.25}$$

where α is the fiber loss, h is the Plank's constant, and ν is the center frequency of the ASE spectrum.

Note that all EDFAs operate in deep saturation, where the total output power is shared between signal and spontaneous noise. Hence, the signal power will reduce as the accumulated noise power goes up. Eq. (13.24) indicates that amplifier saturation results in a growing deterioration of SNR at low EDFA power known as "signal droop," as shown in Fig. 13.38. This signal droop effect makes $P_1 \approx P_{ase}$ and forces capacity to go to 0 when either N_{dim} or B_{amp} tends to infinity. Therefore, there exists an optimum system capacity supported by an optimum SNR to achieve the maximum power efficiency.

Combining Eqs. (13.19)–(13.21) we obtain the following expression for PE:

$$PE = \frac{B_{amp}}{N_{amp} P_{ase}} \cdot \frac{2\log_2(1 + \mathrm{SNR})}{1 + \mathrm{SNR}} \tag{13.26}$$

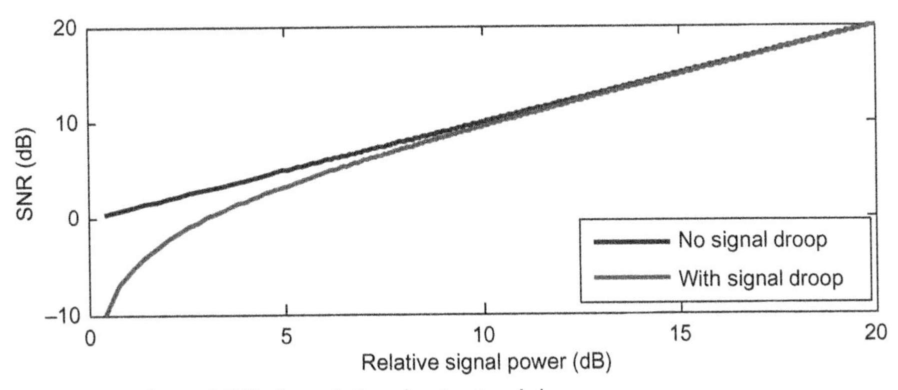

Figure 13.38 Accelerated SNR degradation due to signal droop.

For the next step, we treat P_{ase} as a constant parameter independent from N_{dim} or signal power. One can vary SNR by adjusting the signal power for example by changing N_{dim} while keeping P_{tot} constant. Differentiating Eq. (13.26) with respect to SNR, one can find that PE has a maximum at the values of SNR and corresponding polarization PDM SE of:

$$\begin{aligned} SNR_{opt} &= e - 1 \approx 1.72 \\ SE_{opt} &= \frac{2}{\ln 2} \approx 2.89 \end{aligned} \tag{13.27}$$

SNR_{opt} is 1.72 or 2.36 dB and the optimum SE is 2.89 b/s/Hz for a single SDM dimension and two polarizations. Note that while the value of maximum PE depends on system parameters such as P_{ase}, B and N_{amp} in Eq. (13.26), the values of SNR_{opt} and SE_{opt} are system independent. This implies that Eq. (13.27) defines the optimum PE condition for any system topology.

The maximum capacity of an SDM system at a given power P_{tot} is a capacity with highest PE and can be obtained by combining Eqs. (13.22), (13.26), and (13.27):

$$C_{max} = \frac{P_{tot} B_{amp}}{P_{ase} N_{amp}} \frac{2 \log_2 e}{e} \tag{13.28}$$

The above derivations assume perfect transponders and no system margin. With a practical transponder operating at some distance from the Shannon limit, the optimum SE will be lower while the optimum SNR will be higher. The details on this subject can be found in Ref. [60]. For a transponder that operates 3 dB away from Shannon (transponder penalty + system margin), the optimum SNR is 2.31 or 3.62 dB and the optimum SE is 2.22 b/s/Hz.

Fig. 13.39 shows experimentally obtained PE as a function of SNR, where PE is normalized to its maximum value. The value $M = 1$ (Quantity M defines the

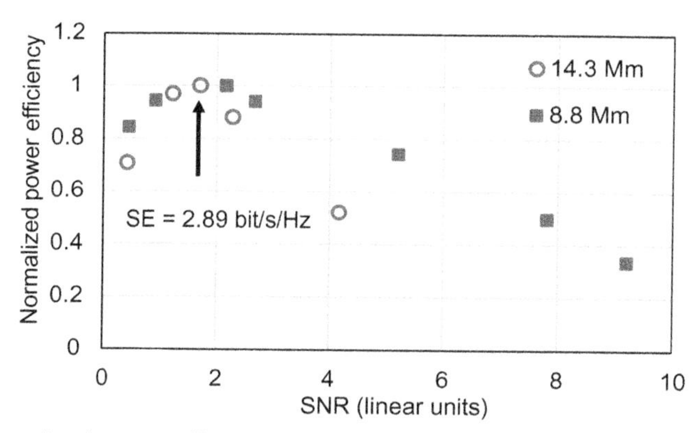

Figure 13.39 Normalized power efficiency (PE) derived from measurement and theoretical dependence. Generalized margin $M = 1$ is assumed. The *arrow* points to the analytically predicted optimum, © [2018] IEEE. Reprinted, with permission, from [60].

generalized margin in linear units) is used for Fig. 13.39 and it corresponds to a theoretically achievable SE given the $\mathrm{SNR}_{\mathrm{eff}}$ of the test bed. The optimum SNR is the same for two different transmission distances, which confirms that the theoretically predicted optimum does not depend on system parameters. Ref. [60] shows the value of optimum SE decreases monotonically with increasing generalized margin M. The optimum PE is reached at a signal power roughly 6 dB below the system's nonlinear limit for 14,350 km distance (and roughly 8 dB for 8830 km) per the GN model.

The PE maximum occurs due to the interplay among amplifier saturation, signal droop, and Shannon's capacity curve. While this deviation is based on EDFA systems, similar dependencies can be observed in any other system with amplifiers operating in saturation.

13.5.2 Optimizing power-efficient undersea systems

If we assume that the total output power is constant, then maximizing PE is equivalent to maximizing capacity as well. In this part, we determine the conditions for maximum PE such as span length, number of space dimensions, and amplifier bandwidth.

The optimum span length is found with $\partial C/\partial l = 0$ as

$$l_{\mathrm{opt}} = \frac{2}{\alpha}. \tag{13.29}$$

It is interesting that the optimum span length depends only on fiber loss α and independent of any other system parameters, including bandwidth, number of dimensions, or noise performance of the amplifier.

The optimum number of dimensions and bandwidth can be found by differentiating Eq. (13.23) and finding solutions to $\partial C/\partial N_{\text{dim}} = 0$ and $\partial C/\partial B = 0$. For convenience, we treat N_{dim} as a continuous variable. The optimum number of dimensions is given by

$$N_{\text{dim,opt}} = \frac{P_{\text{tot}}}{eN_{\text{amp}}^2 B_{\text{amp}}(e^{\alpha l} NF - 1)h\nu} \tag{13.30}$$

In an ideal case with no bandwidth-dependent loss and other effects, capacity depends on the product of N_{dim} and B_{amp} but not their individual values. Therefore, the optimum bandwidth could be obtained from Eq. (13.30) by swapping N_{dim} and B_{amp}. This implies that the optimum capacity could be obtained at any given bandwidth by optimizing the number of spatial dimensions or at any fixed number of dimensions by optimizing the amplifier bandwidth.

Using the optimum values of span length and number of dimensions defined by Eqs. (13.27) and (13.28) and substituting them into Eq. (13.23) or Eq. (13.28), we obtain the following expression for maximum capacity:

$$C_{\max} = \frac{8\alpha^2 P_{\text{tot}}}{\ln 2eL^2(e^2 NF - 1)h\nu} \tag{13.31}$$

Eq. (13.31) represents the maximum theoretical capacity achieved in an EDFA-based SDM system ignoring nonlinearities. Similar to Eq. (13.28), the capacity scales in Eq. (13.31) linearly with power, as opposed to the logarithmic dependence of single-mode system.

13.5.3 Techniques for power-efficient transmission

In practice, we are also interested in the PE of the amplifiers themselves, or how well they convert electrical power into total optical output power P_{tot}, often referred to as amplifier "wall-plug" efficiency. The PE of single-stage C-band EDFAs can be improved by optimizing the width and location of the operating bandwidth. By reducing the EDFA bandwidth down to ~20 nm the gain excursion can be reduced down to only ~0.4 dB. Hence no individual gain equalization filter is needed, and the PE of the EDFA can be increased significantly.

Improvement to the optical PE can also come from the use of modulation formats with higher receiver sensitivity [70,48]. A good example is the 8D-APSK-CM format with SE equivalent of QPSK and ~0.8 dB better receiver sensitivity is shown in Fig. 13.40. The demonstration reviewed in Ref. [54] used many of these techniques to demonstrate 105.1 Tb/s transmission in a 12-core fiber over ~14,350 km using power-efficient modulation format and 20 nm optical amplifier design. This optimized

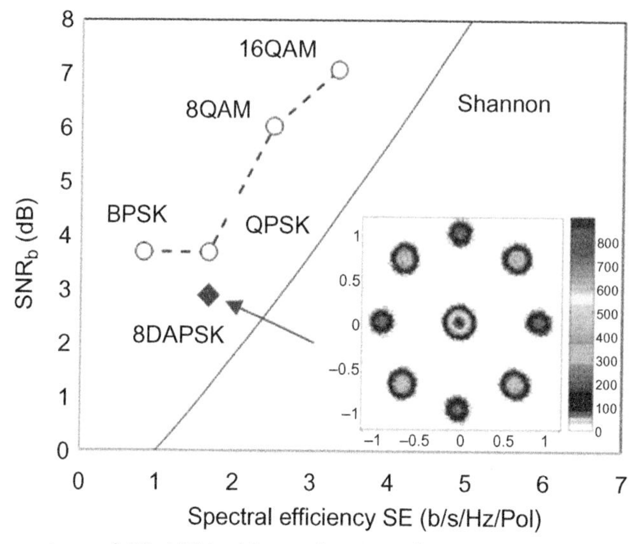

Figure 13.40 Comparison of 8D-APSK with regular QAM formats. Constellation shows nonequal probability of points, © [2016] IEEE. Reprinted, with permission, from [70].

amplifier design allows for the use of a single commercially available 800 mW laser diode for all 12 EDFAs connected to the MCF.

For comparison, we estimate the capacity that can be achieved in a SMF system with equivalent span loss and nonlinear properties to that of the 12-core fiber used in Ref. [54]. With 40 nm bandwidth, similar implementation penalty of the modulation format, and the same pump power restriction, we estimate a capacity of ~22 Tb/s primarily limited by nonlinearity. Using a 12-core fiber, 105.1 Tb/s capacity is achieved, a factor of ~5 times more capacity than SMF system.

13.5.4 Space division multiplexing technologies in undersea

The advantages of SDM technology in undersea systems require several challenges to be overcome before successful field deployment. Recent improvements in fiber parameters that include developments in low crosstalk MCF, low-loss core-coupled MCF [61], and multicore EDF [62] is yet to be matched by progress in integrated SDM components such as amplifiers, couplers, isolators, connectors, and splicing technology. Photonics integration promises improved cost-effectiveness and reduced complexity. Several demonstrations of current state-of-the-art SDM components [63] indicate that practical integrated SDM components may become available in the near future.

While development and undersea qualification of integrated SDM components will take time, the PE advantages of the SDM approach could be realized by adding more SMFs to the cable. The use of bundled low-effective-area low-cost SMFs can be a feasible SDM design option. Recent cost studies indicate that SDM architectures based on a group of multiple SMFs could be a good starting point for a gradual transition of undersea systems to SDM technology [64]. The existence of optimal SE also sets the upper limit on the number of SMFs for PE improvement. Low power transmission and the use of independent fibers may help reduce NLC complexity in DSP algorithms and avoid complex MIMO decoding. As the SDM component base becomes more mature, the industry can potentially evolve to more integrated SDM solutions such as MCF fibers and amplifiers, and possibly to MIMO based approaches.

13.6 Open cables

Today, owners/operators of undersea cable networks would like the flexibility to select their own vendor(s) of choice for the data transponders. The timeline for the implementation of an undersea cable network can sometimes be years between the first concept and the time the system is placed into service. Also, today's systems are implemented to be flexible enough such that only a small fraction of the ultimate capacity might be lit from day 1. Thus, it is desirable for the system design to be transponder agnostic so that selections can be made when the capacity is needed. Fortunately in today's dispersion uncompensated paradigm there is a weak dependence between path design optimization and transponder type. As an example, the optimum optical power depends only weakly on the modulation and transmission distance for a given channel spectrum. Also, for a given dispersion uncompensated path the maximum potential system capacity can be evaluated independently of the particular transponder assumptions. All this means that the optical transmission path can be designed and optimized separately from a particular transponder choice. A system can then allow flexible choice of the transponders, provided minimum performance requirements are met. Moreover, optical fiber transmission lines can be designed for optimum capacity taking advantage of future advances in transponder technology.

MAREA is the first transatlantic and FASTER is the first transpacific undersea system designed and deployed using the open cable concept. MAREA is designed to support 160 Tb/s capacity on eight fiber pairs with 4 b/s/Hz SE [6], and FASTER is designed to support 60 Tb/s on six fiber pairs with 2 b/s/Hz SE. Most recently, Ref. [38] demonstrated 6 b/s/Hz using PS-64QAM over FASTER, which is 3 × the initial designed SE of 2 b/s/Hz using 100 G technology.

13.6.1 $OSNR_{NL}$, $OSNR_{eff}$, and GOSNR

The concept of deploying an optical fiber cable independently from a particular choice of transponder requires a performance metric that is decoupled from the transponder Q-factor metric. This performance metric should satisfy three requirements: (1) it must uniquely characterize the transmission path; (2) it should be derived from

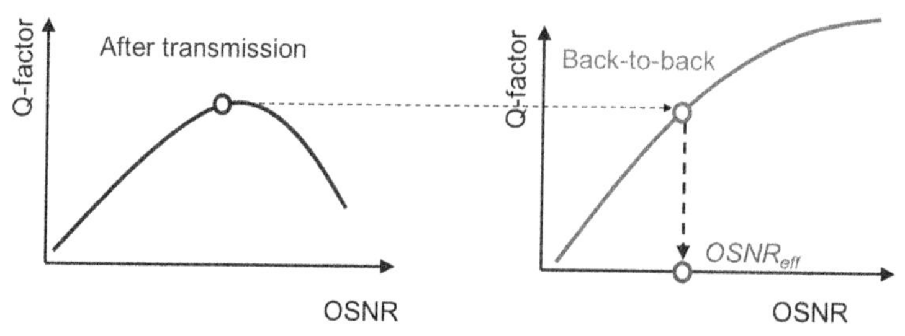

Figure 13.41 Deriving $OSNR_{eff}$ from preemphasis and back-to-back measurements.

measurements, possibly using available transponders; and (3) it must support prediction of performance through the path using other transponders.

The measured channel performance depends on linear noise (P_{ASE}) and nonlinear noise (P_{NLI}), as well as the equivalent noise of transmitter and receiver (N_{TX} and N_{RX}). In general, the total noise at the receiver can be expressed as

$$N_{total} = (N_{TX} + N_{RX}) + (P_{ASE} + P_{NLI}) \quad \text{or} \quad N_{total} = N_{Dry} + (P_{ASE} + P_{NLI}) \qquad (13.32)$$

For design or simulation verification, the channel performance can be estimated from $OSNR_{Dry}$ and $OSNR_{Wet}$ or $OSNR_{NL}$:

$$\frac{1}{OSNR_{total}} = \frac{1}{OSNR_{Dry}} + \frac{1}{OSNR_{NL}} \qquad (13.33)$$

For perfectly Nyquist shaped optical signal, both ASE noise and nonlinear noise are spectrally flat within the channel bandwidth. We can easily derive the $OSNR_{NL}$ or effective OSNR ($OSNR_{eff}$) by removing the penalties from transmitter and receiver hardware implementations back-to-back measurement as shown in Fig. 13.41. $OSNR_{eff}$ was used in Section 13.3.3 to compare the performance of a system using Raman amplification and a system using C + L EDFA amplification. $OSNR_{eff}$ can be used to compare the performance of different wet plant using the same optical transponder.

Fig. 13.42 [49] compares the transmission performance as measured by Q-factor and then by $OSNR_{eff}$ for experiments performed at very different spectral efficiencies. This was done by transmitting channels with the same optical bandwidth and shape but with different CM formats, resulting in different performances and spectral efficiencies. Performance measurement results obtained at a fixed signal power for four different CM formats at four different spectral efficiencies as a function of transmission distance for the same system configuration are shown in Fig. 13.42A. Once again we convert Q-factors to Nyquist $OSNR_{eff}$ simply using $OSNR_{eff} = f_{BtB}^{-1}(Q_{Tx})$ noting that the channel spectrum and spacing in these experiments were very close to Nyquist. The results are shown in Fig. 13.42B [65]. This figure demonstrates that for four different

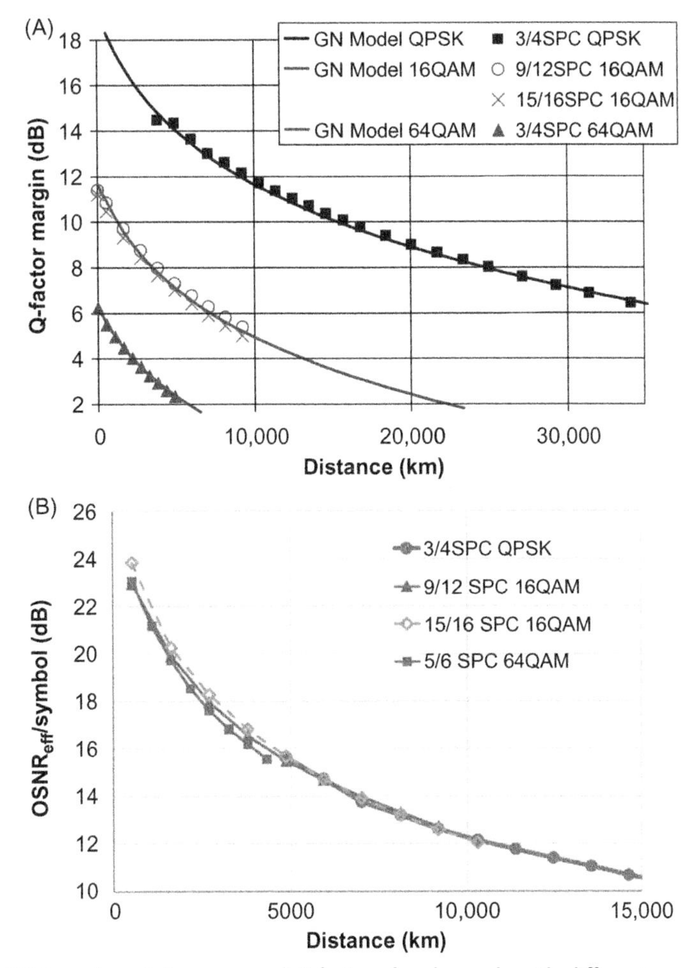

Figure 13.42 (A) Experimentally measured *Q*-factors for channels with different spectral efficiencies but the same Nyquist spectrum [49]; (B) calculated $OSNR_{eff}$, © [2015] IEEE. Reprinted, with permission, from [65].

SE transmitter settings we obtain essentially the same Nyquist $OSNR_{eff}$ within the accuracy of the experiment (accuracy of *Q*-factor measurement), although the performance (*Q*-factor) differs significantly for different modulation formats.

However, $OSNR_{eff}$ cannot be used directly to compare the performance of a wet plant using different transponders since $OSNR_{eff}$ also depends on transmitter pulse shape. Unlike ASE noise, nonlinear noise is not spectrally flat within the channel bandwidth, but weakly depends on the optical pulse shape, as predicted by the more accurate EGN model [19]. Consequently, the receiver filtering function and transmitter pulse shape should be removed when only estimating the wet plant performance. Ref. [65] proposed a GOSNR or $OSNR_{eff}$ for channels with perfectly Nyquist shaped spectra as the transponder independent performance metric for dispersion uncompensated optical paths. The use of the Nyquist channel spectrum creates no ambiguity since the spectrum is flat and allows decoupling from a particular transponder. It should be noted that $OSNR_{eff}$ approaches GOSNR in a dense WDM case with nearly rectangular spectral shapes, such as a signal with a raised cosine spectrum with a very small roll-off factor. Furthermore, GOSNR can be used to calculate the capacity of a transmission path.

The relation between GOSNR and $OSNR_{eff}$ measured with a real transponder relies on linear and nonlinear noise contributions. The ideal Nyquist channel will have equal ASE noise power density as the physical channel but different nonlinear noise for the same channel power and channel spacing. To remove the interaction of the spectral shape of the nonlinear noise and the receiver filter function from $OSNR_{eff}$, we define the ratio k

$$k = \frac{P_{NL,Nyq}}{P_{NL}} \tag{13.34}$$

This ratio k can be estimated using the EGN model which includes the spectral shape of the channels and system parameters. One can then relate GOSNR and $OSNR_{eff}$ as follows:

$$\frac{1}{GOSNR} = \frac{1}{OSNR} + k\left(\frac{1}{OSNR_{eff}} - \frac{1}{OSNR}\right) \tag{13.35}$$

With GOSNR, we can predict the performance using a second transponder with k_2 using the following equation:

$$\frac{1}{OSNR_{eff}} = \frac{1}{OSNR} + \left(\frac{1}{GOSNR} - \frac{1}{OSNR}\right)/k_2 \tag{13.36}$$

The GOSNR concept has been verified by many groups. Ref. [65] varied the channel spacing and the spectrum of the channels by changing the roll-off factor of the raised-cosine shape. Three spectral conditions were considered for 34.2 GBd QPSK and 16QAM modulated channels: (1) channel spacing $\Delta f = 35$ GHz and raise-cosine roll-off factor $\beta = 0.01$; (2) $\Delta f = 50$ GHz and $\beta = 0.01$; and (3) $\Delta f = 50$ GHz with $\beta = 0.5$. The transmission distance was set to 11,527 km for QPSK, and 4247 km for 16QAM. The results for GOSNR versus preemphasis are shown in Fig. 13.43. The agreement between values of GOSNR derived from measurements with different spectral shapes and channel spacing is excellent across the whole range of signal powers.

The key in GOSNR measurement is to find the correct k factor for a particular transponder, and there are still debates on how to characterize the k factor.

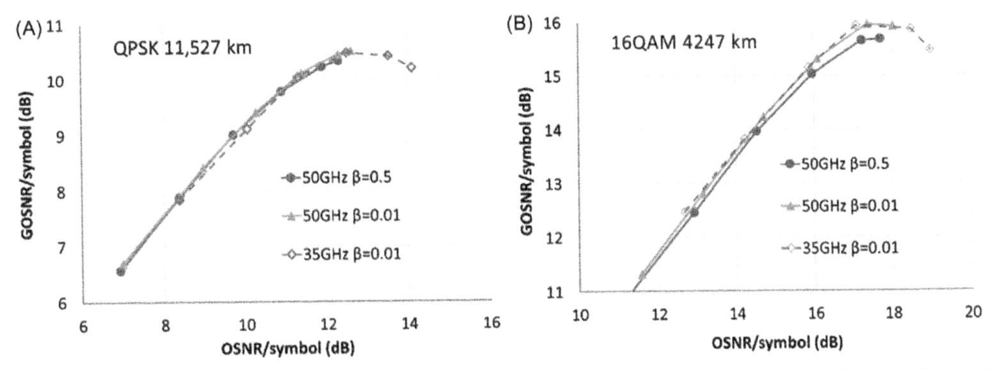

Figure 13.43 GOSNR calculated for (A) QPSK and (B) 16 QAM transponders for different channel spacing and β-factors, © [2015] IEEE. Reprinted, with permission, from [65].

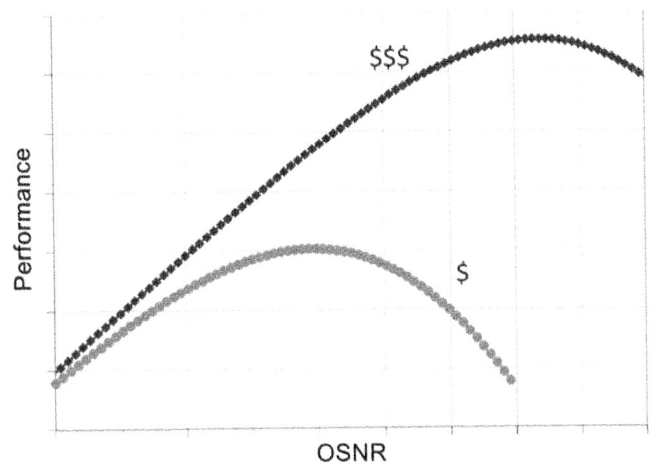

Figure 13.44 Two wet plant designs for the same path with different transmission performance.

13.6.2 System design trade-offs

Optical path design balances transmission impairments and cost to meet customer requirements for capacity, reach, and performance margin. Low-noise single-stage C-band amplifiers are well established, but repeater spacing and power as well as fiber type are still key design parameters that must be optimized for every design. Fibers with low loss and/or large effective area are good for performance, but fibers with best-in-class values for these parameters have a price premium. To optimize design for a given system, sophisticated and qualified design, manufacturing, and installation processes are needed to deliver a compliant, cost-effective system.

Fig. 13.44 illustrates a typical output from design study. It shows transmission performance (Q-factor) versus received OSNR for two path designs. The designs have the same length, but use different fiber types at significantly different performance

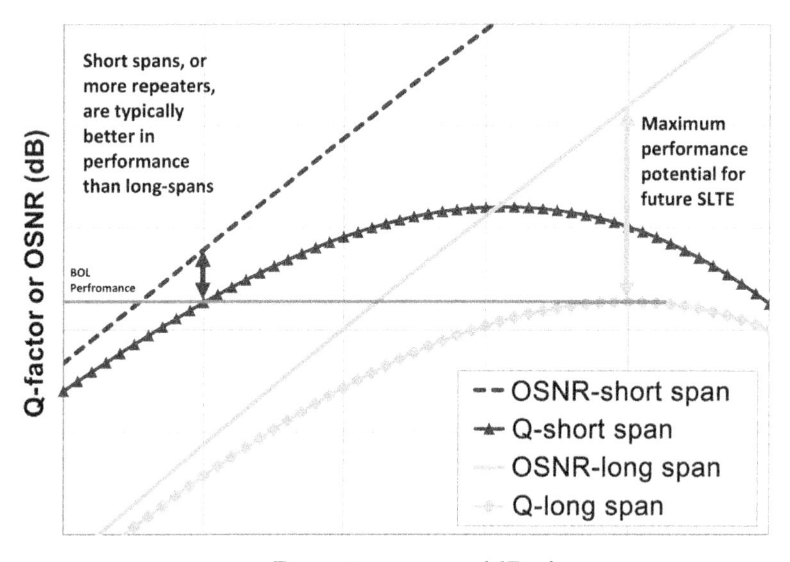

Figure 13.45 System design for optimum future potential.

values and therefore price points. Only one of the two designs may actually be able to meet the desired transmission performance.

The design trade-offs become even more complex when considering different operating points. What is the difference between two systems with equal GOSNR but different amounts of nonlinear impairments. This situation is schematically captured in Fig. 13.45 which shows two different designs: the *green diamond curve* operates at the optimum point with longer fiber spans, while the *black triangle curve* operates in linear region with shorter fiber spans. Even though the two designs can have the same performance at the beginning of life, the system at the optimum point (*green diamond*) has significantly more capacity upside potential than the linear system (*black triangle*) due to its high OSNR and higher nonlinear impairment. With future NLC technology, the additional OSNR could be accessed and transmission performance and capacity improved. It is therefore not true that shorter repeater spacing is always better. Longer repeater spacing may provide equivalent optical performance on day one and better upgrade potential with future improvements for NLC.

The two designs shown in Fig. 13.45 also have different PE. The balance of system aggregate capacity, overall system electrical PE, and system cost is very complex and is beyond the scope of this chapter.

13.7 System value improvements

Besides capacity improvements, there have also been other significant enhancements to undersea transmission systems over the past few years. For example, systems with full ROADM capability are now becoming available for submarine systems. While in use already for several years in terrestrial networks, the technology is new for undersea systems mostly due to the stringent reliability requirements for undersea applications.

The second is more design flexibility in cable conductor materials. Traditionally, copper is used in undersea cables for its excellent electrical and mechanical properties. However, copper is expensive and significant savings can be achieved using Aluminum. Furthermore, system economics can be optimized by engineering of the average cable conductivity to match the power requirement of the system and available power of the power feed equipment.

13.7.1 Wet wavelength selective switch–based reconfigurable optical add-drop multiplexer

Undersea OADM technology allows multiple landing points to efficiently share capacity on a single fiber (Fig. 13.46). There are two major advantages from introducing reconfigurable OADM technology based on WSSs. First, due to the steep edges of the WSS filter, wasted bandwidth (or "guard bands") between OADM bands are much smaller than OADM nodes implemented with previous generations of filter technology. For older types of filters (such as thin film filters or fiber Bragg gratings) about 200 GHz of bandwidth was needed between OADM bands; reducing available bandwidth for data transmission. For C-band amplifiers that corresponds to ~4% of the available bandwidth not available due to guard bands, or 4% of the value of the fiber pair. Additional OADM bands increase the unavailable bandwidth correspondingly.

Second, the wavelength flexibility given by the WSS ROADM reduces the probability of "stranded capacity" on the fiber pair. In previous generation undersea ROADMs, the reconfigurability was limited to a few preselected OADM ratios.

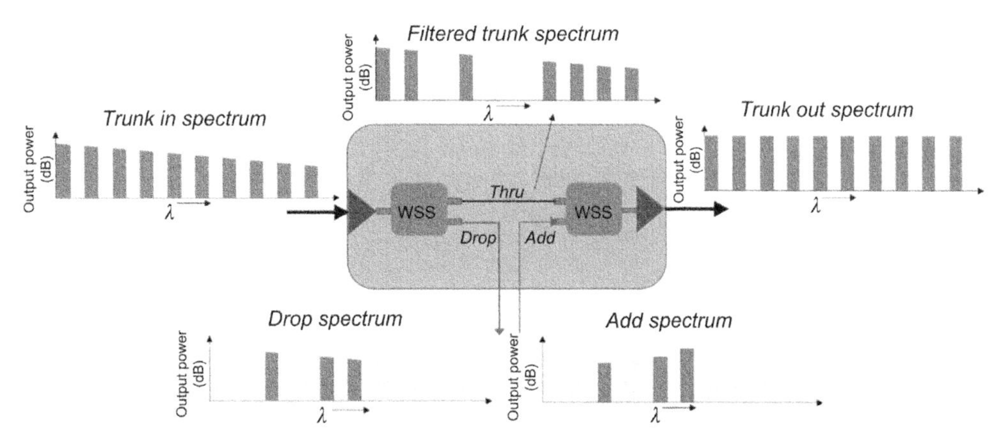

Figure 13.46 Schematic wavelength selective switch–based reconfigurable optical add-drop multiplexer (WSS ROADM node).

Now with the WSS, the choice of OADM ratios is practically unlimited. This has significant economic impact, since the OADM ratio can be adjusted to optimally reflect the traffic requirements today and can be reconfigured for future needs.

Some of the new WSS based OADM nodes also have the capability to measure and adjust undersea spectra. Traditionally the operator was restricted to the transmit and receive spectra only, with very limited information about the evolution of the signal spectrum along the transmission line. Now available undersea spectra will have advantages and benefits in the optimization and fine-tuning of a transmission path to achieve optimum transmission performance and capacity over the life of the system.

13.7.2 New cable types: lower cost, higher direct current resistance trade-offs

In Section 13.5 we discussed optical design techniques for making the transmission system more power efficient. Here we consider trade-offs in the electro-mechanical design of the cable to optimize cost and electrical PE. Specifically, we look at the material used to make the electrical conductor in the cable. An aluminum conductor can reduce the cost per length of submarine cable due to its lower price compared to the standard copper conductor. Fig. 13.47 shows photographs of light weight submarine cables (with the conductors exposed) made with aluminum and copper.

The significance of the submarine cable with aluminum conductor becomes clear when looking at Fig. 13.48. The cost of the cable is related to the direct current resistance (DCR) of the cable, meaning that it is more expensive to make the cable's electrical resistance lower. Due to the large amount of cable in a transoceanic project, even small changes in cable cost have the potential to lead to significant overall cost improvements.

13.8 Future trends

Today's best transmission fibers have an attenuation of ~0.14 dB/km, which is very close to the intrinsic Rayleigh scattering loss limit, with an effective area of ~150 μm^2 [42]. From the GN model, 1 dB of fiber loss change can bring us 1/3 of dB in

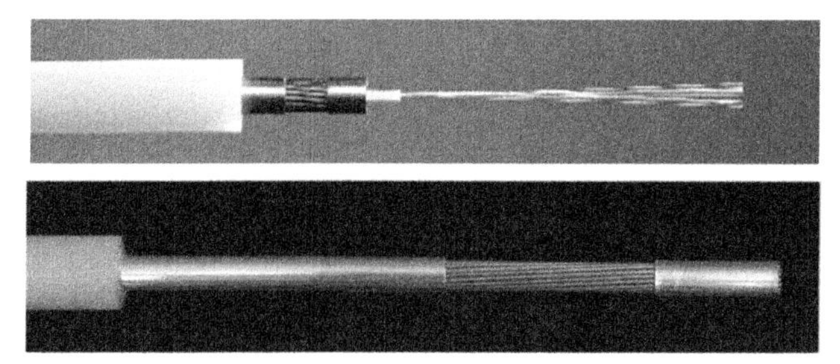

Figure 13.47 Light weight submarine cable with copper conductor on top and aluminum conductor on the bottom.

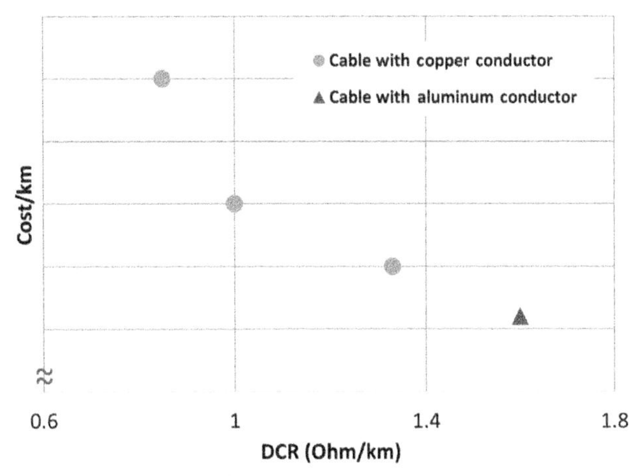

Figure 13.48 Cable cost as a function of direct current resistance (DCR).

performance, and 1 dB of A_{eff} change can bring us 2/3 dB in performance. So, there could still be ~1 dB improvement with the traditional PSCF. Hollow-core photonic crystal fiber (HC-PCF) has been considered as a potential candidate to achieve ultra-low loss, even below the fundamental Rayleigh scattering loss limit of PSCF. However, the current state-of-the-art HC-PCF has a loss of 1.7 dB/km and usable bandwidth of <20 nm.

As stated earlier, system capacity can be increased by using more optical bandwidth. C + L EDFA provides ~80 nm optical bandwidth, but it is still only a fraction of silica fiber bandwidth (1260–1625 nm). Many optical amplification technologies are being studied to cover the whole range, including the original (O-), extended (E-), short-wavelength (S-), conventional (C-), and long wavelength (L-) bands. Semiconductor optical amplifier (SOA) could be an alternative for cost- and power-efficient solutions in optical amplifications. With proper design, SOA can provide more than 100 nm optical bandwidth with high gain, high output power, and low NF [66]. There is also a lot of research focusing on transmission using the 2 μm optical window—employing thulium-doped fiber amplifier (TDFA). TDFA provides a potential amplification window from ~1700 to ~2100 nm covering an optical bandwidth of 30 THz.

NLC techniques have been focused on narrow bandwidth signal–signal interaction induced transmission penalty. Broadband signal–signal interactions, and even signal–ASE interaction, could be mitigated when more powerful DSP is available. The complexity of the current NLC techniques need to be further reduced to be able to be implemented with current ASIC technology.

Although PS QAM can approach the Shannon limit with only a gap of ~0.1 dB assuming an ideal distribution matcher, the gap to Shannon is typically ~0.5 dB with realistic distribution matchers. This gap increases to >2 dB with current available FEC. Nonbinary FEC and irregular FEC has the potential to close this gap significantly—irregular LDPC has been shown to come as close as 4.5 mdB to the Shannon limit.

Finally, WDM, SDM, and MDM are techniques that can grow system capacity [67,68]. To deploy systems using MCF, MMF, FMF, or CC-MCF, multicore and multimode optical amplifiers need to improve in NF, PE, and differential modal gain. With many parallel optical amplifiers in a single repeater, the integration of optical components and amplifiers is indispensable to keep the physical volume reasonable. The trade-off among aggregate capacity, electrical power, and total cost will provide the optimum design to suit for different customer's need.

13.9 Conclusions

Significant progress has been made over the last few years in undersea optical fiber telecommunication systems. Most importantly, all new transoceanic distance undersea systems are deployed with high-dispersion fiber and digital coherent transponders. New technologies and inventions have allowed designs to approach the Shannon limit, and greatly increase capacity to new records. In the laboratory, NLC techniques improve the NLC benefit to 2 dB over ~10 THz optical bandwidth. Combinations of probabilistically shaping, geometric shaping, CM, adaptive rate FEC, advanced NLC provides record capacity 95, 71.6, 51.5 Tb/s over 1900, 6970, and 17,107 km in SMF. WDM, PDM, and SDM pushed fiber capacity to an unprecedented 520 Tb/s over 8832 km. We note that the electrical power needed for this capacity is way beyond the power that can be delivered by the current power feeding equipment. With smart power-efficient transmission concepts, 105 Tb/s capacity has been transmitted in a single fiber using electrical power that can be provided with current PFE. Going forward, combinations of PDM, WDM, SDM, and MDM and repeaters with integrated optical components will offer more flexibility to trade off system cost and system capacity.

Acknowledgments

The authors wish to thank all of their colleagues in optical fiber telecommunications around the world for their significant achievements in undersea transmission. We also express our heartful thanks to the SubCom team for their valuable contributions to this work and for kindly providing the material and fruitful discussions that made this chapter possible.

List of acronyms

3R reamplification, reshaping, retiming
AIR achievable information rate
APSK amplitude phase shift keying
ASE amplified spontaneous emission

AWGN	additive white Gaussian noise
BER	bit error ratio
BICM	bit-interleaved coded modulation
BOL	beginning of life
CC-MCF	coupled-core multicore fibers
CM	coded modulation
DBP	digital back propagation
DCR	direct current resistance
DSP	digital signal processing
EDFA	erbium-doped fiber amplifier
EGN	enhanced Gaussian noise
FEC	forward error correction
fLMS	fast least mean square
FWM	four-wave mixing
GN	Gaussian noise
GOSNR	generalized optical signal-to-noise ratio
GS	geometrical shaping
LDPC	low-density parity check
LLR	log likelihood ratio
LMS	least mean square
LUT	look-up table
MAP	maximum a posteriori probability
MCF	multicore fiber
MDL	mode dependent loss
MDM	mode division multiplexing
MI	mutual information
MIMO	multiinput multiple-output
NF	noise figure
NLC	nonlinearity compensation
NLI	nonlinear noise
NLSE	nonlinear Schrödinger equation
OADM	optical add-drop multiplexer
OSNR	optical signal-to-noise ratio
PDM	polarization division multiplexing
PE	power efficiency
PFE	power feeding equipment
PNLC	perturbation nonlinearity compensation
PS	probabilistic shaping
PSCF	pure silica core fiber
PSD	power spectral density
QAM	quadrature amplitude modulation
QPSK	quadrature phase shift keying
ROADM	reconfigurable optical add-drop multiplexer
SCM	subcarrier multiplexing
SDM	space division multiplexing
SER	symbol error ratio
SISO	soft-input/soft-output
SLTE	submarine line terminating equipment

SMF single-mode fiber
SNR signal-to-noise ratio
SPC single parity check
TDHQ time-domain hybrid QAM
WDM wavelength division multiplexing
WSS wavelength selective switch

References

[1] J.-X. Cai, et al., 70.46 Tb/s over 7,600 km and 71.65 Tb/s over 6,970 km transmission in C + L band using coded modulation with hybrid constellation shaping and nonlinearity compensation, J. Lightw. Technol. 36 (2) (2018) 114–121.

[2] A. Turukhin, H.G. Batshon, M. Mazurczyk, Y. Sun, C.R. Davidson, J.-X. Cai, et al., Demonstration of 0.52 Pb/s potential transmission capacity over 8,830 km using multicore fiber, in: European Conference on Optical Communication, Tu.1.D.3, Dusseldorf, Germany, 2016.

[3] J.-X. Cai, K. Golovchenco, G. Mohs, Modern undersea transmission technology, in: I. Kaminow, T. Li, A. Willner (Eds.), Chapter 25 in Optical Fiber Telecommunications VIB, Academic Press, 2013. and references therein.

[4] S. Abbott, Review of 20 years of undersea optical fiber transmission system development and deployment since TAT-8, in: Proc. ECOC 2008, Mo.4.E.1.

[5] Submarine Optical Fiber Cables - Global Strategic Business Report, <https://www.researchandmarkets.com/research/k23w3x/2018_submarine?w = 5>.

[6] Marea, <https://www.submarinenetworks.com/en/systems/trans-atlantic/marea>.

[7] Submarine Cable PLCN - PLCN - Submarine Cable Networks, <https://www.submarinenetworks.com/systems/trans-pacific/plcs>.

[8] G.P. Agrawal, Fiber-Optic Communication Systems, John Wiley and Sons, Inc, 2002.

[9] F. Vacondio, C. Simonneau, L. Lorcy, J.C. Antona, A. Bononi, S. Bigo, Experimental characterization of Gaussian-distributed nonlinear distortions, in: Proc. ECOC 2011, We.7.B.1.

[10] G. Bosco, V. Curri, A. Carena, P. Poggiolini, F. Forghieri, On the performance of nyquist-WDM terabit superchannels based on PM-BPSK, PM-QPSK, PM-8QAM or PM-16QAM subcarriers, J. Lightw. Technol. 29 (1) (2011) 53–61.

[11] A. Carena, V. Curri, G. Bosco, P. Poggiolini, F. Forghieri, Modeling of the impact of nonlinear propagation effects in uncompensated optical coherent transmission links, J. Lightw. Technol. 30 (10) (2012) 1524–1539.

[12] R. Dar, M. Feder, A. Mecozzi, M. Shtaif, Inter-channel nonlinear interference noise in WDM systems: modeling and mitigation, J. Lightw. Technol. 33 (5) (2015) 1044–1053.

[13] E. Grellier, A. Bononi, Quality parameter for coherent transmissions with gaussian-distributed nonlinear noise, Opt. Exp. 19 (13) (2011) 12781–12788.

[14] P. Poggiolini, The GN model of non-linear propag.ation in uncompensated coherent optical systems, J. Lightw. Technol. 30 (24) (2012). 3857–575.

[15] P. Serena, A. Bononi, An alternative approach to the gaussian noise model and its system implications, J. Lightw. Technol. 31 (22) (2013) 3489–3499.

[16] J.-X. Cai, H. Zhang, H. Batshon, M. Mazurczyk, O. Sinkin, D. Foursa, et al., 200 Gb/s and dual wavelength 400 Gb/s transmission over transpacific distance at 6.0 b/s/Hz spectral efficiency, J. Lightw. Technol. 32 (4) (2014) 832–839.

[17] J.-X. Cai, O. Sinkin, D. Foursa, G. Mohs, A. Pilipetskii, Direct measurement of broadband FWM induced noise in dispersion uncompensated systems, in: Conference on Optical Fiber Communication (OFC), Paper OM3B.5, Anaheim, CA, Mar. 2013.

[18] J.-X. Cai, C.R. Davidson, A. Lucero, H. Zhang, D.G. Foursa, O.V. Sinkin, et al., 20 Tbit/s transmission over 6,860 km with sub-nyquist channel spacing, J. Lightw. Technol. 30 (4) (2012) 651–657.

[19] A. Carena, et al., EGN model of non-linear fiber propagation, Opt. Express 22 (13) (2014) 16335.

[20] W. Shieh, Y. Tang, Ultrahigh-speed signal transmission over nonlinear and dispersive fiber optic channel: the multicarrier advantage, IEEE Photon. J. 2 (3) (2010) 276–283. Available from: https://doi.org/10.1109/JPHOT.2010.2043426.

[21] P. Poggiolini, et al., Analytical results on system maximum reach increase through symbol rate optimization, in: Proc. OFC, Th3D.6, Los Angeles, 2015.
[22] J.-X. Cai et al., Experimental study of subcarrier multiplexing benefit in 74 nm bandwidth transmission up to 20,450 km, in: ECOC 2016; W.3.D.4, Dusseldorf, Germany, 2016.
[23] J.C. Cartledge, F.P. Guiomar, F.R. Kschischang, G. Liga, M.P. Yankov, Digital signal processing for fiber nonlinearities, Opt. Express 25 (2017) 1916–1936. and references therein.
[24] E. Ip, J.M. Kahn, Compensation of dispersion and nonlinear impairments using digital back propagation, J. Lightw. Technol. 26 (20) (2008) 3416–3425.
[25] X. Li, X. Chen, G. Goldfarb, E. Mateo, I. Kim, F. Yaman, et al., Electronic post-compensation of WDM transmission impairments using coherent detection and digital signal processing, Opt. Exp. 16 (2) (2008) 880.
[26] A. Ghazisaeidi, et al., Submarine transmission systems using digital nonlinear compensation and adaptive rate forward error correction, J. Lightw. Technol. 34 (8) (2016) 1886–1895.
[27] M.V. Mazurczyk, J.-X. Cai, H.G. Batshon, M. Paskov, O.V. Sinkin, D. Wang, et al., Performance of nonlinear compensation techniques in a 71.64 Tb/s capacity demonstration over 6,970 km, in: Proc. Eur. Conf. Opt. Commun., Gothenburg, Sweden, 2017, Paper Th.2.E.1.
[28] H.G. Batshon, M.V. Mazurczyk, J.-X. Cai, O.V. Sinkin, M. Paskov, C.R. Davidson, et al., Coded modulation based on 56APSK with hybrid shaping for high spectral efficiency transmission, in: Proc. Eur. Conf. Opt. Commun., Gothenburg, Sweden, 2017, Paper Tu.1.D.2.
[29] J.-X. Cai, et al., 51.5 Tb/s capacity over 17,107 km in C + L BW using single mode fibers and nonlinearity compensation, J. Lightw. Technol. 36 (11) (2018) 2135.
[30] P. Bayvel, C. Behrens, D. Millar, DSP and its application optical communication systems, in: I. Kaminow, T. Li, A. Willner (Eds.), Chapter 5 in Optical Fiber Telecommunications VIB, Academic Press, 2013. and references therein.
[31] J.-X. Cai, Y. Sun, H. Zhang, H.G. Batshon, M.V. Mazurczyk, O.V. Sinkin, et al., 49.3 Tb/s transmission over 9100 km using C + L EDFA and 54 Tb/s transmission over 9150 km using hybrid-Raman EDFA, J. Lightw. Technol. 33 (13) (2015) 2724–2734.
[32] J.-X. Cai, H.G. Batshon, M. Mazurczyk, H. Zhang, Y. Sun, O.V. Sinkin, et al., 64QAM based coded modulation transmission over transoceanic distance with > 60 Tb/s capacity, in: Optical Fiber Communication Conference, Th5C.8, Los Angeles, CA, USA, 2015.
[33] C.E. Shannon, A mathematical theory of communication, Bell Syst. Tech. J. 27 (3) (1948) 379–423.
[34] H.G. Batshon, Approaching Shannon capacity limit using coded modulation, in: Proc. SPIE 10561, SPIE OPTO, 2018.
[35] S. Zhang et al., Capacity-approaching transmission over 6375 km at spectral efficiency of 8.3 bit/s/Hz, in: Proc. Opt. Fiber Commun. Conf., Anaheim, CA, USA, 2016, Paper Th5C.2.
[36] F. Buchali, F. Steiner, G. Böcherer, L. Schmalen, P. Schulte, W. Idler, Rate adaptation and reach increase by probabilistically shaped 64-QAM: an experimental demonstration, J. Lightw. Technol. 34 (7) (2016) 1599–1609.
[37] A. Ghazisaeidi, I. Fernandez de Jauregui Ruiz, R. Rios-Muller, L. Schmalen, P. Tran, P. Brindel, et al., Advanced C + L-band transoceanic transmission systems based on probabilistically shaped PDM-64QAM, J. Lightw. Technol. 35 (7) (2017) 1291–1299.
[38] V. Kamalov, et al., Evolution from 8QAM live traffic to PS 64-QAM with neural-network based nonlinearity compensation on 11000 km open subsea cable, in: 2018 Optical Fiber Communications Conference and Exposition (OFC), Paper Th4C.5, San Diego, CA, 2018, pp. 1–3.
[39] R.H. Morelos-Zaragoza, The Art of Error Correcting Coding, second ed., 2006, John Wiley & Sons, Ltd. ISBN: 0-470-01558-6.
[40] G. Ungerboeck, Trellis-coded modulation with redundant signal sets, IEEE Commun. Mag. 25 (2) (1987) 5–21.
[41] I.B. Djordjevic, M. Cvijetic, L. Xu, T. Wang, Proposal for beyond 100 Gb/s optical transmission based on bit-interleaved LDPC-coded modulation, IEEE Photon. Technol. Lett. 19 (12) (2007) 874–876.

[42] Y. Yamamoto, Y. Kawaguchi, M. Hirano, Low-loss and low-nonlinearity pure-silica-core fiber for C- and L-band broadband transmission, J. Lightw. Technol. 34 (2) (2016) 321−326.
[43] A. Ghazisaeidi, et al., Transoceanic transmission systems using adaptive multirate FECs, J. Lightw. Technol. 33 (7) (2015) 1479−1487.
[44] W.R. Peng, et al. OECC2011, 8D2-4.
[45] X. Zhou, et al., High spectral efficiency 400 Gb/s transmission using PDM time-domain hybrid 32−64 QAM and training-assisted carrier recovery, IEEE J. Lightw. Technol. 31 (7) (2013) 999−1005.
[46] S. Chandrasekhar, B. Li, J. Cho, X. Chen, E. Burrows, G. Raybon, et al., High-spectral-efficiency transmission of PDM 256-QAM with Parallel Probabilistic Shaping at Record Rate-Reach Trade-offs, in: Proc. Eur. Conf. Opt. Commun., Düsseldorf, Germany, 2016, Paper Th.3.C.1.
[47] J.-X. Cai et al., 94.9 Tb/s single mode capacity demonstration over 1,900 km with C + L EDFAs and coded modulation, in: Proceedings of ECOC 2018, Mo4G.3.
[48] E. Agrell, M. Karlsson, Power-efficient modulation formats in coherent tranmission systems, J. Ligthw. Technol. 27 (22) (2009) 5115−5126.
[49] J.-X. Cai, H. Batshon, H. Zhang, M. Mazurczyk, O. Sinkin, D. Foursa, et al., Transmission performance of coded modulation formats in a wide range of spectral efficiencies, Presented at the Opt. Fiber Commun. Conf., Los Angeles, CA, 2014, paper M2C.3.
[50] S. Abbott, D. Kovsh, G. Harvey, Key features of the undersea plant for high capacity and flexibility, in: Proc. SubOptic, Tu1C-1, 2013.
[51] C.R. Davidson, C. Chen, M. Nissov, A. Pilipetskii, N. Ramanujam, H. Kidorf, et al., 1800 Gb/s transmission of one hundred and eighty 10 Gb/s WDM channels over 7,000 km using the full EDFA C-band, in: Proc. OFC 2000, PD25.
[52] D.G. Foursa, H.G. Batshon, H. Zhang, M. Mazurczyk, J.-X. Cai, O. Sinkin, et al., 44.1 Tb/s transmission over 9,100 km using coded modulation based on 16QAM signals at 4.9 bits/s/Hz spectral efficiency, in: Proc. Eur. Conf. Opt. Commun., London, UK, 2013, Paper Th.3.E.1.
[53] D.G. Foursa, C.R. Davidson, M. Nissov, M.A. Mills, L. Xu, J.-X. Cai, et al., 2.56 Tb/s (256x10 Gb/s) transmission over 11,000 km using hybrid Raman/EDFAs with 80 nm of continuous bandwidth, in: OFC'2002, post-deadline paper FC3, Anaheim, California, March 2002.
[54] A.V. Turukhin, et al., High capacity ultralong-haul power efficient transmission using 12-core fiber, J. Lightw. Technol. 35 (4) (2017) 1028−1032.
[55] S. Randel, et al., Mode-multiplexed 6 × 20-GBd QPSK transmission over 1200-km DGD-compensated few-mode fiber, in: OFC/NFOEC, Los Angeles, CA, 2012, pp. 1−3.
[56] J. van Weerdenburg *et al.*, "Mode-Multiplexed 16-QAM transmission over 2400-km large-effective-area depressed-cladding 3-mode fiber, in: OFC 2018, San Diego, CA, 2018, pp. 1−3.
[57] R. Ryf, et al., Long-distance transmission over coupled-core multicore fiber, ECOC Th.3.C3, 2016.
[58] T. Frisch, S. Desbruslais, Electrical power, a potential limit to cable capacity, in: SubOptic TU1C-04, Paris, France, 2013.
[59] A. Pilipetskii, et. al., Optical designs for greater power efficiency, Paper TH1A.5, SubOptic 2016.
[60] O.V. Sinkin, et al., SDM for power-efficient undersea transmission, J. Lightw. Technol. 36 (2) (2018) 361−371.
[61] T. Hayashi, et al., Record-low spatial mode dispersion and ultra-low loss coupled multi-core fiber for ultra-long-haul transmission, J. Lightw. Technol. 35 (3) (2017) 450−457.
[62] D. Richardson et al., Optical amplifiers for space-division-multiplexed systems, in: Proc. ECOC, Th.2.D.1, 2017.
[63] N. Fontaine, Components for space-division multiplexing, in: Proc. ECOC, W.3.F.1, 2017.
[64] M. Bolshtyansky, M. Cantono, L. Jovanovski, O. Sinkin, A. Pilipetskii, G. Mohs, V. Kamalov and V. Vusirikala, "Cost-optimized Single Mode SDM Submarine Systems," in SubOptic, OP18-1, 2019.

[65] O. Sinkin, C.R. Davidson, H. Wang, L. Richardson, J.-X. Cai, D. Kovsh, et al., Effective signal to noise ratio performance metric for dispersion-uncompensated links, in: Proc. Eur. Conf. Opt. Commun., 2015, Valencia, Spain, Paper p. 5.3.

[66] J. Renaudier, et al., First 100-nm continuous-band WDM transmission system with 115Tb/s transport over 100 km using novel ultra-wideband semiconductor optical amplifiers, in: ECOC 2017, Gothenburg, 2017, pp. 1–3.

[67] P.J. Winzer, D.T. Neilson, From scaling disparities to integrated parallelism: a decathlon for a decade, J. Lightw. Technol. 35 (5) (2017) 1099–1115.

[68] A. Pilipetskii, High capacity submarine transmission systems, in: Optical Fiber Communication Conference, W3G.5, Los Angeles, CA, USA, 2015.

[69] H. Zhang, H. Batshon, BICM coded modulation and 4D modulation for long-haul optical transmission system, in: IEEE Photonics Conference (IPC) 2013, paper MG2.4, 08 Sep–12 Sep 2013, Bellevue, WA, USA.

[70] H. Zhang, A. Turukhin, O.V. Sinkin, W. Patterson, H.G. Batshon, Y. Sun, et al., Power-efficient 100 Gb/s transmission over transoceanic system, J. Lightw. Technol. 34 (8) (2016) 1859–1863.

CHAPTER 14

Intra-data center interconnects, networking, and architectures

Saurabh Kumar[1], George Papen[2], Katharine Schmidtke[3] and Chongjin Xie[4]
[1]Amazon, Seattle, WA, United States
[2]University of California, San Diego, CA, United States
[3]Facebook, Menlo Park, CA, United States
[4]Alibaba Group, Sunnyvale, CA, United States

14.1 Introduction to intra-data center interconnects, networking, and architectures

Internet services have become part of our daily life. Almost all of the internet services are run in data centers, where a massive number of servers are connected through intra-data center networks (DCNs). The majority of internet traffic is inside data centers and the traffic in hyperscale data centers doubles every one to two years, which presents a scalability challenge for intra-data center interconnects and networking. The rapid rise of web-based applications and cloud computing over the last decade has led to a growing interest in data center technologies. A tremendous amount of research and development effort, both from academia and industry, has enabled deployment of hyperscale data centers all over the world. A critical piece of this infrastructure is the network connectivity between storage and compute elements within the data center.

In this chapter we first give an overview of data centers, their history and current status including the key features of the DCN. The initial sections focus on network architecture and various topology choices available to designers. We also highlight the challenges faced in deploying and operating these networks. We discuss scale-out network architecture and describe how it was introduced into DCNs to solve the scalability issues. The specific interconnect technologies utilized in today's networks are discussed in Section 14.3. We show that cabling is not only affected by network architecture but by retrofitting and the building refresh cycle as well. Section 14.4 focuses on optical interconnect technology development. Many technologies are used in intra-data center interconnects, such as direct attach cables (DAC), active optical cables (AOC), and optical transceivers. The characteristics of each technology are described and their application areas are discussed. Pluggable optics have been widely used in intra-data center interconnects, which have evolved from 40 to 100 G in the last two years and will move to 400 G in 1 or 2 years. We describe technologies of different

Optical Fiber Telecommunications V11
DOI: https://doi.org/10.1016/B978-0-12-816502-7.00016-6

generations of pluggable optics, including 40, 100, and 400 G, their physical media dependent (PMD), and form factors. Finally, in Section 14.5 we conclude by presenting our perspectives on future development of intra-data center networking and interconnects, including coherent technology, mid-board optics, electronic and optical copackaging, and optical switching technology.

14.2 Intra-data center networks

The DCN directly impacts the speed, reliability, and latency of communication essential to running applications on the compute and storage infrastructure housed in the data center. The capital cost of deploying the network and the cost associated with operating it in a reliable manner are a significant component of overall data center expense. Hence, a network architecture that can minimize capital expenditure (capex) while not compromising on reliability and scalability of the network is a key goal. Before discussing various topology options, we examine the dynamics of the traffic that these networks intend to serve.

14.2.1 Data center network growth drivers

As new types of services that run on data center infrastructure have emerged, the traffic patterns observed in these networks have also evolved. The term north-south traffic is used to describe traffic between servers in a data center to entities outside the DCN, for example, users connected via the internet, peering with other networks, etc. East-west traffic has been used as a moniker for traffic that stays within the DCN, that is, from one server to another. It is worth noting that the phrase "within the DCN" need not mean within a building. Depending on the specific physical architecture, the inter-server switching fabric that comprises the DCN could be spread out over multiple buildings. Also, requirements for replication of data and interconnecting different cloud networks generate east-west traffic that is not confined to a single data center. Due to the rapid rise of data-intensive services like web search, high-performance computing, social networking, and video streaming, the growth in the east-west traffic has dwarfed the growth in the north-south traffic. For example, Ciscos Global Cloud Index [1] that tracks internet traffic growth predicts (Fig. 14.1) that east-west traffic will account for 85% of all data center traffic by 2021. This does not include traffic within a data center that stays within a rack of servers, which by itself is twice the amount of traffic that is transported across racks within a data center.

According to the Cisco report, the traffic between data centers is growing faster than the traffic within a data center or the traffic between data centers and users. However, in this chapter we will focus on the network between servers within a data center. The data center's connectivity to the internet, other data centers across the

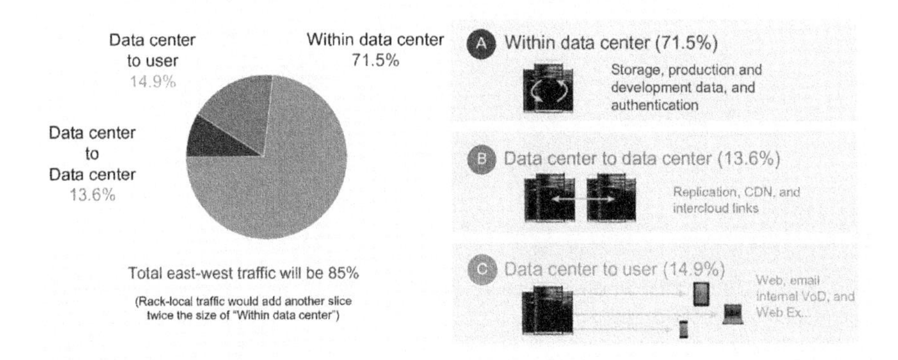

Figure 14.1 Ciscos global cloud index prediction for data center traffic distribution. Source: *Cisco Global Cloud Index, 2016–2021.*

world, and peering with other networks are equally important topics, which will not be discussed in this chapter.

14.2.2 Characteristics and classification of data center networks

Traditional enterprise data centers were intended to serve specific use cases, for example, data storage, email, etc. The scaling requirements were also well understood over a finite period of time. With the shift toward data center virtualization and cloud computing, more and more companies are choosing to transition from their own private data centers and into the public cloud. The companies that run these public clouds, Amazon, Microsoft, Google, and Alibaba, to name a few, are building massive (called hyperscale) data centers with the intent to serve diverse workloads from millions of customers. The underlying infrastructure needs to be "generic" in order to realize benefits of scale, while allowing for enhanced capabilities to be incorporated for specialized workloads when needed. These companies have made a push toward simplified hardware and networking software that caters specifically to the DCN instead of complex "jack of all trades" networking technology that can handle multiple network protocols. As noted in Ciscos Global Cloud Index [1], by 2021 the number of hyperscale data centers will surpass the number of traditional data centers worldwide. Further, more than 90% of workloads will run in the cloud instead of on-premises data centers, while >70% of the cloud workloads will be run in the public cloud instead of private cloud networks (Fig. 14.2). These trends motivate in-depth understanding of DCN architecture in the context of hyperscale cloud service providers.

DCNs can be evaluated based on a host of metrics. Several works [2–8] have documented such metrics as well as compared different DCN architectures based on these metrics [7]. From the perspective of a hyperscale cloud service provider, the

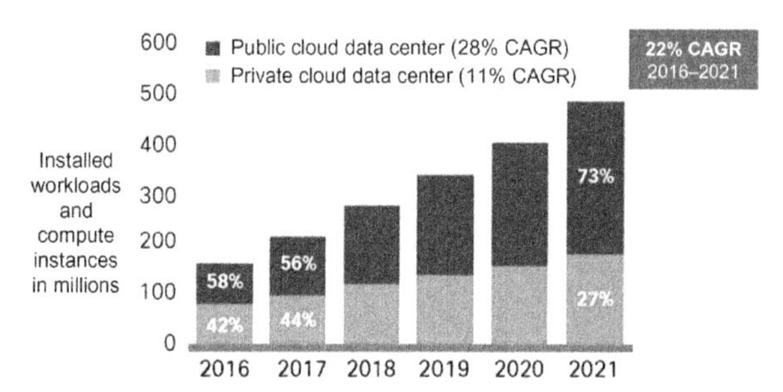

Figure 14.2 Public cloud data centers will serve a growing percentage of cloud workloads. Source: *Cisco Global Cloud Index, 2016–2021.*

network should exhibit the following characteristics enabling efficient data transfer from one server to another:

- High bandwidth;
- Low latency;
- Low latency variability;
- Resilience against device or link failures;
- Low power consumption and related cooling requirements;
- High degree of scalability without disruption to the existing network;
- Backward compatibility to enable parts of the network to adopt new technology;
- Low cost.

To achieve these goals, network architecture has transitioned from a hierarchical switching fabric to a flatter topology that enables interconnections between a large number of endpoints without relying on expensive, high-performance switches in the higher tiers. The details of the switching fabric form the basis for classification of DCNs. Several published works have attempted to create a taxonomy for this classification [7,8].

A broad classification can be used to separate DCNs into two groups: ones that use electronic switching only and ones that also include optical switching. The latter is sometimes referred to as a dynamic architecture since the topology can be reconfigured using the optical switches while the purely electronic topology is "fixed." This section focuses on electronic switching based DCNs only, and optical switching technologies that are starting to emerge will be covered in Section 14.5.3.

Within the domain of electronically switched DCNs, another broad classification is based on the type of network element that is responsible for packet forwarding [9]. The conventional approach utilizes switches for packet forwarding while servers running applications are connected to these switches. The other extreme is a fabric that connects servers directly to each other and the packet forwarding function also runs on the servers. Hybrid topologies that use a mix of these two approaches have also been reported. We will first review topologies that rely on conventional packet switching and then touch upon the server-driven and hybrid approaches.

14.2.2.1 Switch-centric topologies

Data centers are comprised of a large number of compute and storage endpoints that need to be interconnected. Traditional enterprise data centers achieved this via a hierarchical topology where servers were connected to the bottom tier of switches and multiple tiers provided connectivity between the switches of the first tier. Fig. 14.3 shows this type of topology, called a "basic tree." In this three-tier interpretation, the edge tier of switches connects to the servers, aggregation switches in the second tier connect to the edge switches and are in turn connected by the core tier at the top of the hierarchy (root of the tree). Note that there is no direct connectivity between switches in the same tier. As one moves up the tree, the reliability and performance requirements grow since the switches higher up in the network are involved in traffic flow for a larger number of servers. Redundancy is often built in for the core and aggregation tiers which also serves as a means to increase capacity. The scalability of the basic tree topology is limited by the port count of the switches. For this topology the same type of switches cannot be used in all the tiers, leading to an expensive network that requires very large, high-performance switches. While this approach may have worked for traditional enterprise data centers, hyperscale data center operators have moved away from this topology, instead choosing to build switching fabrics using common switches across all tiers.

A variant of the basic tree is the "fat tree" [10] wherein the number of links connecting a switch to the lower tier is equal to the number connecting it to the upper tier (Fig. 14.4). When such a topology is multirooted, a very large endpoint count can be accommodated and multiple paths exist between these endpoints. This kind of switching fabric is highly reliable due to the redundancy gained from overprovisioning of resources.

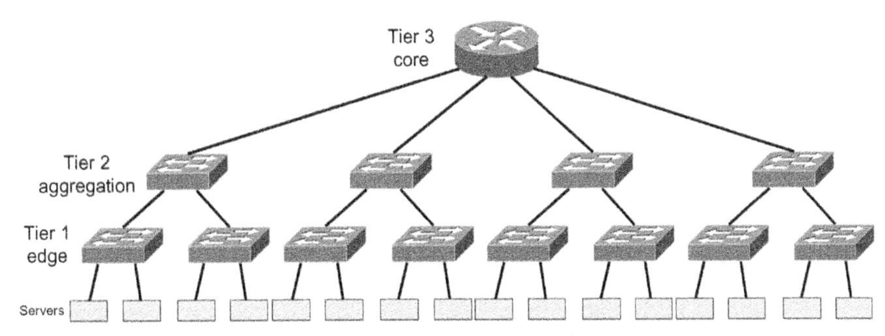

Figure 14.3 Traditional multitier single-rooted basic tree topology.

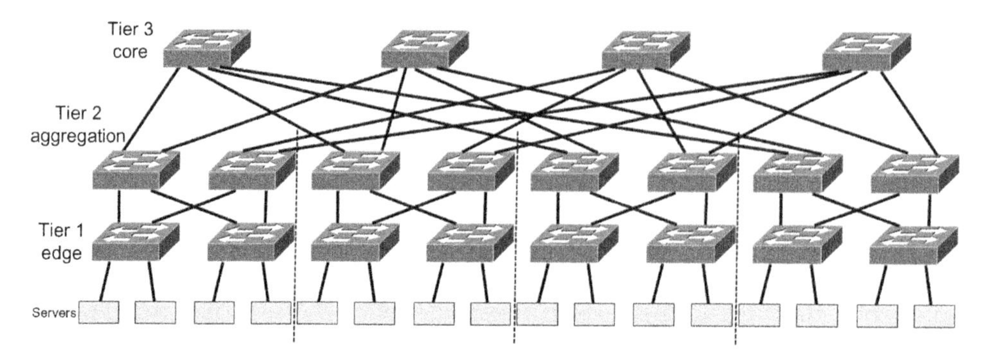

Figure 14.4 Multirooted fat-tree topology.

A variant of the multirooted tree topology that leverages a nonblocking architecture originally proposed by Charles Clos in 1953 for telephony applications [11] has also become a popular [12] choice for high throughput, high reliability DCNs. The core switches are connected to every switch in the aggregation layer, enabling multipath traffic flow optimization. An example of the Clos topology is shown in Fig. 14.5.

To enable rapid deployment and scalability, commercial data center designs have leaned toward units of server racks with a top of rack (ToR) switch that connects to these servers. The switching fabric that is used to connect the ToRs to each other across a data center is also deployed in units that enable incremental scaling. The tree topologies discussed here lend themselves to such an incrementally scalable architecture. The scaling is achieved by organizing the fabric in groups of switches called leaves and spines, leading to the term leaf-spine architecture. While the specific details of how the hardware scaling units are designed and what role is played by the top of rack (ToR) within those scaling units varies between the data center operators, most commercial deployments are currently based on minor variants of this topology.

Another variant called the elastic tree that focuses on minimizing power consumption by dynamically turning off switches based on traffic patterns has also been proposed [13]. Other switch-centric topologies that do not involve a tree structure have also been proposed as a way to minimize network cost by reducing the number of switches needed to connect a given number of servers. The flattened butterfly is one such topology.

14.2.2.2 Server-centric and server-switch hybrid topologies

The extreme case of server-to-server direct connect topology has also been explored by researchers. One such proposal is called CamCube [14], and it completely eliminates switches from the data center fabric. A fully connected mesh between all the servers would be the ideal choice to implement any logical topology. However, such a fabric would not be very scalable, as the number of degrees for each server would

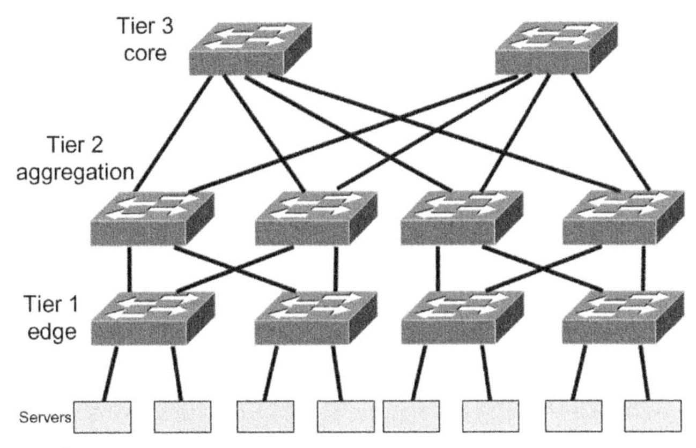

Figure 14.5 Clos topology.

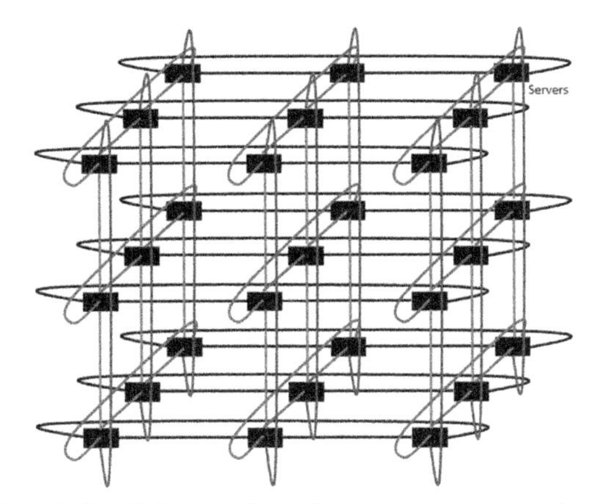

Figure 14.6 3D Torus-based CamCube topology. Servers are connected to two neighbors in each direction of 3D space.

need to grow as the network size increased. Instead, CamCube proposes a 3D Torus-based design shown in Fig. 14.6 where each server is connected via short links to two neighbors in each of the three dimensions (six in total).

The servers need multiple network interfaces to realize such a fabric, and the packet forwarding function resides on the servers in the absence of switches. For small networks, CamCube offers lower initial cost by eliminating switches while providing link redundancy and incremental scalability. However, latency suffers due to long routing paths with multiple hop counts as the number of servers grow.

Compared to the ordered CamCube topology, other architectures within this group do not follow a specific network pattern. Instead, these networks rely on arbitrary connections between switches and servers. Examples include Jellyfish, which involves random graph connectivity between ToRs utilizing a subset of their ports while the remaining ports are connected to servers. The benefits compared to organized tree topologies include the ability to grow the network in more granular increments instead of large step sizes dictated by switch port radix and lower average latency. Server counts that can be supported also exceed those for an equal cost fat-tree topology. Another random architecture generation algorithm that breaks free from symmetric design and homogeneous switching equipment is Scafida [15]. This architecture derives from scale-free networks that are inspired by asymmetry found in biological instances. Scafida maintains high failure tolerance while respecting nodal degree limitations of practical switches.

Another approach to scalability where switches and servers share the switching functions is to build recursive characteristics in the network. There are several such topologies where low-level cells are used to build higher level cells of a similar structure. The DCell architecture utilizes multiple network interfaces per server to create recursion by connecting a set of servers to switches and each server to another server on the same level [16]. As the network scales, new levels are added by building out the higher levels using the previous level as a unit cell connected via the same topology. An example of a layer of DCell with three servers per switch is shown in Fig. 14.7. A modified version, called FiConn [17] that limits the number of interfaces on each server to two and trades off cost and link complexity with reduced capacity and redundancy has also been proposed.

A similar recursive approach called BCube [18] that grows layers by adding more switches instead of servers has also been proposed. The topology is shown in Fig. 14.8

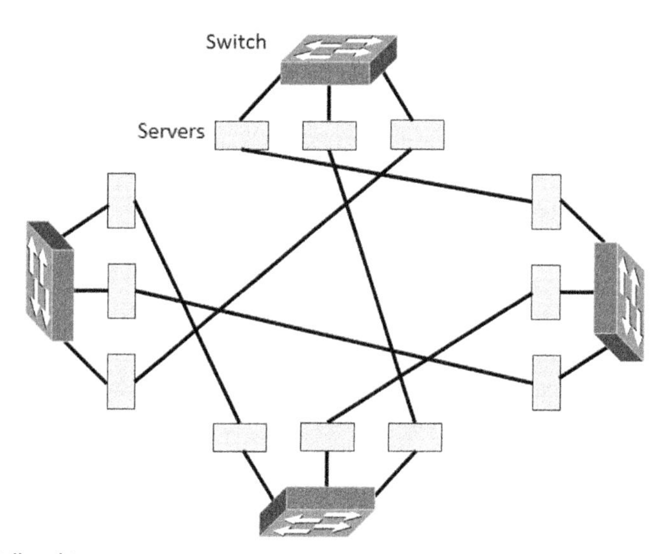

Figure 14.7 DCell architecture.

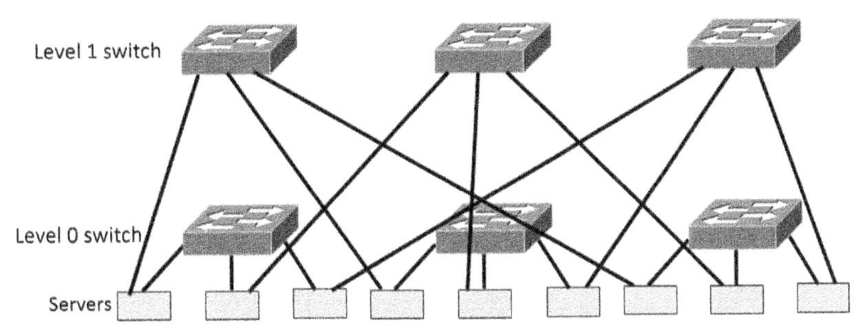

Figure 14.8 BCube architecture.

where the base unit cell is identical to Dcell with a set of servers connected to a switch. However, unlike Dcell, to grow the network, another layer of switches is added with the switch connecting to one server in each of the lower unit cells.

To build a hyperscale data center using BCube-based modular elements, a high-performance interconnection structure called MDCube [19] has also been proposed.

14.2.2.3 Metrics to compare topologies

While a direct comparison of the diverse set of topologies described in the earlier sections is challenging, one can identify specific metrics that should be included in such an exercise. There are several studies [7,9] that have compared and contrasted the cost and performance of these topologies. These surveys include the following metrics:

- *Scalability*: Allowing addition of individual components to expand the network instead of an upgrade of all existing infrastructure.
- *Modularity*: Ability to start with small units and incrementally expand with the same units.
- *Energy consumption*: Optimizing usage across servers and switches.
- *Hardware redundancy*: Resilience across switch and interconnect failures.
- *Number of elements*: Switches, servers, and interconnects that drive cost/capacity.
- *Number of network interfaces per server.*
- *Number of hops in multiple paths*: Drives latency.
- *Bandwidth*: One-to-one and all-to-all.
- *Oversubscription*: Trade-off between cost and server-to-server bandwidth across racks.
- *Load balancing*: To ensure high throughput.
- *Cabling complexity*: Determines deployment velocity and ability to reuse.
- *Cost*: Capex as well as scaling/managing opex.

Data center topologies are continuously evolving due to the dynamic nature of the traffic, which is driven by new applications. Some architectures emphasize performance-related metrics, while others focus more on cost reduction. The challenge for data center architects is in finding the right balance that can provide an optimum return on investment.

14.2.3 Traffic routing in data center networks

In order to use the physical network topology most efficiently, data center operators also need to focus on the process of getting the data from one node to another. This involves three important aspects of traffic management:

- *Addressing*: Intelligent address assignment for all the nodes.
- *Routing*: Protocols/algorithms to determine the next hop for the packets.
- *Forwarding*: Transferring packets from one switch port to another.

There is a tremendous amount of ongoing research in all three domains. Hyperscale operators tend to use commodity application-specific integrated circuits (ASICs) for the switching function but layer their own traffic management intelligence to improve the efficiency of their networks.

14.2.3.1 Addressing

The "regular" topology of most data center architectures lends itself to algorithmic generation of addresses for servers and switches in the network. Once the topology has been chosen, it is easy to determine the number of switches and servers that will be used. For example, in an approach using edge and aggregation switches organized into pods [20], the address allocation can follow the dotted decimal notation *10.pod.switch.1* where *pod* is the number of the pod within the network and *switch* corresponds to the location of the switch within its pod.

However, it is not necessary to mimic internet addressing, as long as the alternative approach is scalable. For example, multirooted trees have been approached with layer-2 addressing utilizing pseudo-MAC [21]. Modified address resolution protocols are needed to implement MAC-to-PMAC and PMAC-to-IP address translations. This flexibility is essential for virtual machine migration within the physical network infrastructure. Alternative proposals to achieve the same goal include usage of location-specific and application-specific IP addresses with a directory system to store the mapping [12].

Recursive architectures like BCube and Dcell can also be dealt with using simple algorithms that can generate the addresses based on arrays of identifiers linked to the levels in the network topology.

14.2.3.2 Routing and forwarding

Traditional routing protocols do not apply directly to DCNs due to special features and traffic patterns which are better handled through modifications. For scalability, multiple paths between nodes are required. To maximize throughput and avoid congestion, single flows could pass through multiple paths between source and destination. One of the major decisions a routing architect has to make is the choice between centralized or distributed routing.

Distributed routing does not rely on knowledge of link states across the network. Communicating nodes route traffic on their own by identifying congested paths and routing around them based on local decisions. This also reduces overhead and provides for a robust and scalable routing process. Equal cost multi-path routing (ECMP) is an example wherein multiple paths are identified and traffic is spread across these paths [22]. An extension [23] that can be deployed on server-centric topologies unlike the switch-centric ECMP is called distributed adaptive routing for data centers.

Centralized routing is based on the complete view of the network. This allows optimal routing decisions splitting flows across multiple paths with high efficiency. Hedera [24] is one such scheme that relies on a central controller to distribute traffic and update routing tables across the network. Scalability is challenging, however, since it increases the congestion on the links due to controller communication. Since the central controller has complete visibility of the network, it can accurately assess the impact of failures and perform appropriate routing updates.

For tree topologies the regularity in the structure can again be leveraged to design an efficient routing scheme. A two-level lookup involving a longest prefix match to the destination IP followed by a potential second table lookup has been proposed. This allows pods to identify servers that don't belong to the same pod and achieve balanced flow distribution through multiple core switches.

For recursive structures like DCell the routing algorithms need to determine the link that allows two cells in the lower layer of a given cell to communicate. This procedure needs to be repeated to identify all subpaths creating a direct link. Routing in BCube is a little simpler due to the well-defined digit offset between addresses of servers connected to the same switch.

Traditional transport control protocols (TCP) present several challenges when applied to data center topologies [25]. High bandwidth many-to-one communication, low round-trip times, the need for fast convergence, shallow buffers in commodity switches, and the simultaneous presence of long and short flows lead to TCP performance degradation. Some of the limitations that have been observed include the following:

- Throughput degradation when a large number of servers send data simultaneously to one receiver, leading to buffer overflows and catastrophic packet loss
- Timeouts caused due to TCP features like delayed ACK intended to limit the amount of ACK traffic in the reverse direction of data flow

- Port blackouts and high latency for short flows due to queuing delays introduced by long flows
- Unfairness introduced due to single-path TCP limiting bandwidth utilization
- TCP retransmission forced due to volatile nature of cloud workloads

Alternatives to conventional TCP have been proposed in literature targeting specific weaknesses identified above [26–28]. A comprehensive discussion of the traffic management aspects of the topologies discussed in this chapter can be found in Ref. [7].

14.2.4 Network cabling

As described in previous sections, hyperscale data centers consist of complex networks full of compute, storage, and switching equipment. For many hyperscale data centers with switch networks operating above 10 Gb/s, the physical connections are made by optical signals carried over optical fiber cables. As an example, a photograph of network cabling inside a hyperscale data center is shown in Fig. 14.9. The photograph shows a view down one of the aisles in Facebook's Altoona data center in 2014. The different types of optical fiber used in data centers will be discussed in Section 14.4.1. The *aqua* and *purple* cable jackets in the photograph indicate the types of multimode fiber (MMF) that is predominantly used in this data center.

The cabling structure and arrangement of the optical fiber in a data center depends on the network architecture and the building layout of each particular data center. In general, data center layouts seek to reduce the length of cable and minimize the number of connections between endpoints while achieving the desired network topology. An example of a network cabling layout is shown in Fig. 14.10. The schematic shows the arrangement of different types of switch hardware in Facebook's Fabric network topology [29]. It also shows the layout of the network cabling inside the data center. The network cabling consists of structured cabling with longer lengths of trunk fibers

Figure 14.9 The optical fiber cabling inside Facebook's Altoona data center in 2014.

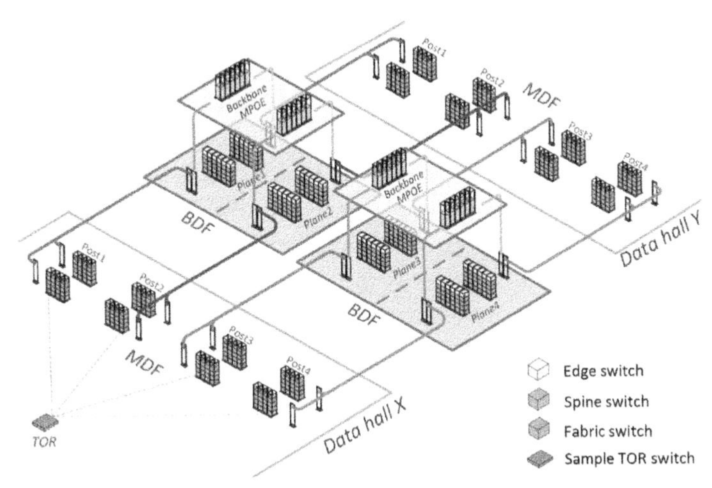

Figure 14.10 Schematic of fiber cabling physical layout for a data center built using Facebook Fabric architecture.

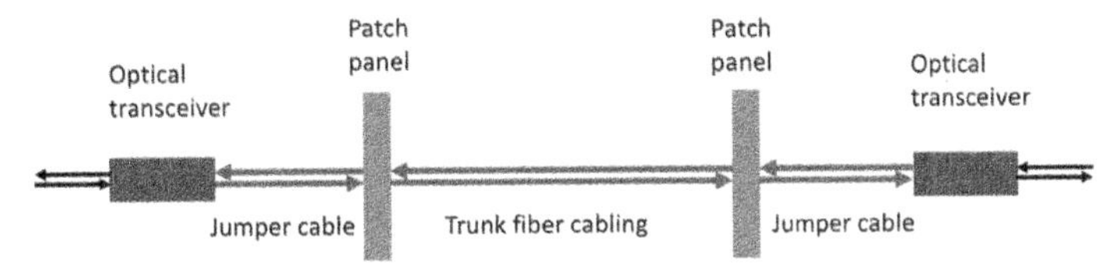

Figure 14.11 The key elements in a typical data center optical link.

(shown as colored pipes) that terminate at patch panels. The patch panels act as rack connection points, from which shorter jumper cables (not shown) connect the patch panels to the endpoints in the network equipment.

The network cabling in a hyperscale data center can become very complex. There are typically hundreds of fibers in each of the trunk cables and the patch panels consist of hundreds of connection points. Despite the overall complexity, the logical connection for a single fiber is actually fairly simple. The simplicity of the optical link is illustrated in Fig. 14.11. The signal starts and ends in the networking equipment where it is connected through an electrical connector to an optical transceiver plugged into the front faceplate of the equipment chassis. At each of these endpoints, the optical transceiver converts electrical signals into optical signals to be transmitted over optical fiber. The optical transceiver also receives optical signals from the other end of the link and converts them back into electrical signals to complete the link. In between the two endpoints there are different trunk fibers and jumper cables depending on the route to be taken by the signals. The routing is determined by the connections in the patch panels. The design goal is to minimize the number of connections in each link so in most cases the signals pass through just two patch panels, as represented in Fig. 14.11.

The network connections between switching equipment and other network elements such as compute or storage equipment usually occur at the rack level. Connections within a rack usually involve shorter distances on the order of a few meters and can be made using cables such as DAC or AOC. These different types of cables will be discussed in Sections 14.3.1 and 14.3.2 of this chapter.

The network cabling becomes complex because of the number of fibers and connections that must be made. In the example of Facebook data center shown in Fig. 14.10, each of the optical links is less than 500 m. With thousands of switches in a data center, and parallel fiber configurations, the total amount of optical fiber deployed in a data center to connect all these switches might reach tens of thousands of kilometers. Given this scale of the cabling infrastructure, it is not surprising that the choice of fiber type for the intra-data center interconnects must be carefully considered based on economic as well as technical factors [30].

The economic factors are driven by changing, retrofitting, or upgrading the large amount of fiber, which becomes a large operational and financial challenge. It takes a significant amount of installation time to pull cables and patch them together between network switches and then test interconnects for insertion loss. This is a manual process that requires a large labor force trained in handling optical fiber to maintain cleanliness of the connection. For this reason, changes in the fiber cabling or retrofits to older data centers incur significant expense. Ideally, the fiber plant should remain installed for the whole life of the data center and support many technology innovation cycles. When determining the best solution for 100 G optical connections, one must consider the total link cost, which is made up of the cabled strands of fiber, patch panels to connect fiber throughout the data center, the optical transceivers at each end, and the operational expense of installing or retrofitting the network cabling.

The technical factors are driven by signal-to-noise ratio and distance. As the data rate increases beyond 10 Gb/s, dispersion (modal and chromatic) starts to limit the length over which optical signals can be transported [31]. For example, at 40 Gb/s the data center is usually cabled with OM3 MMF, which provides sufficient bit error rate transmission over 100 m. To reach 100 m at 100 Gb/s using standard optical transceivers usually requires re-cabling with OM4 MMF. This is a workable solution inside some data centers, but it does not allow any flexibility for longer link lengths or for a data rate evolution beyond 100 Gb/s.

For these reasons, there has been strong motivation to move to deploying single-mode fiber in data centers that would allow the same fiber to support many data rate evolution cycles. Single-mode fiber eliminates modal dispersion problems and has been used for decades in the telecommunications industry to support longer reaches at higher data rates. This has led many data center operators to invest in single-mode fiber which will support multiple technology generations.

Figure 14.12 Example of switch chassis showing the pluggable modules on the front faceplate. Source: *Facebook's BackPack switch submitted to the Open Compute Project.*

14.3 Interconnect technologies

The majority of intra-data center interconnects can be separated into three broad categories; DAC, AOC, and optical transceivers. Each of these solutions is designed to fit into a package with an electrically pluggable cage located on the front of the equipment that can be the same for each type of interconnect. This provides design flexibility to optimize the type of connection used to interconnect different pieces of network hardware in the data center such as switch chassis, compute racks, and the storage racks without having to design a separate piece of equipment for each type of connection. An example of a switch chassis design showing the pluggable cage ports at the front is shown in Fig. 14.12. In this case, the pluggable optics are QSFP28 operating at 100 Gb/s.

The flexibility provided by the pluggable devices has been a key factor in the design and construction of modern data centers. The hardware network device, whether a switch, compute server, or storage unit, is fitted with a common electrical receptacle in the form of a connector cage assembly that the interconnect technology plugs into. Several examples of different applications using pluggable form factors are shown in Fig. 14.13. In the figure the equipment box at the bottom is a rack switch fitted with pluggable interconnects, and the devices on the top are storage adapters and network interface cards which are deployed inside compute and storage equipment [32].

14.3.1 Pluggable form factors

The type of pluggable connection is chosen depending on the application. There are several types of electrical connectors in use in data centers that have been standardized by forums such as standards bodies like the small-form-factor (SFF) or multisource agreements (MSA). These pluggable optical transceivers gained market traction after the development of the small form-factor pluggable (SFP) specification in the year 2000 at data rates of 1 Gb/s and below [33]. One of the most used form factors for interconnects operating up to 28 Gb/s is the SFP28 form factor. For higher data rate

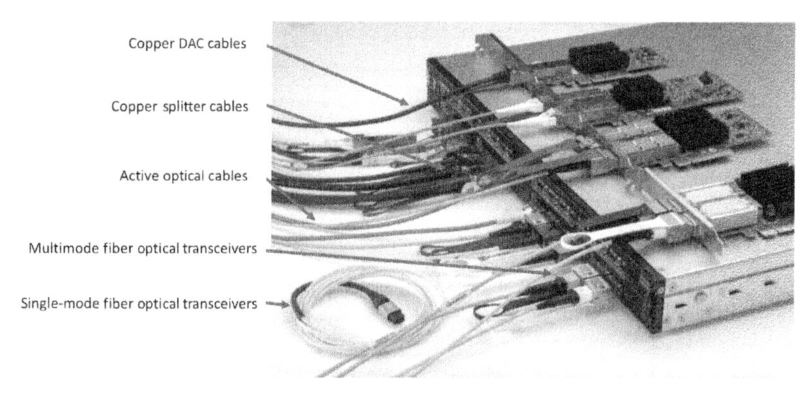

Figure 14.13 Examples of pluggable modules for different networking applications. Source: *Mellanox.*

links, four lanes are combined together into one package in a module interface called quad-SFP (QSFP) [34]. In 100 Gb/s networks the predominant form factor is QSFP28. For interconnects above 100 Gb/s there are several pluggable formats that are being developed including a double density version of the QSFP called QSFP-DD [35] and the octal SFP (OSFP) [36]. The QSFP-DD is gaining traction because it allows backward compatibility and interoperability with the QSFP form factor, whereas the OSFP is attractive because it is slightly larger and supports more cooling options and therefore supports applications with higher power consumption. In the following sections we describe each of these different interconnects and explain where each of them is typically used in large data centers.

14.3.2 Direct attach cables (DAC)

DAC consist of a passive copper cables between two electrical connectors that plug into the ports on the front of a switch or server chassis. The DAC is usually used for the shorter reach applications in the data center from the server to the top-of-rack switch. The applications where a DAC is used are determined by the bandwidth of the signal and the length of the connection. The bandwidth-length product has been identified as a useful metric to indicate suitable applications [37]. DAC cables have typically been used in applications where the product of the link length and the signal frequency is less than 100 Gb/s.m. For example, at 10 Gb/s per lane, a DAC would support applications <10 m. A serial 100 Gb/s connection would only support lengths up to 1 m length, so 100 Gb/s is typically made up of four parallel lanes each of 25 Gb/s which supports connections <4 m. The connector package used for 4×25Gb/s links is the QSFP28.

There are several configurations of DACs that are used depending on the application. A DAC can be used in a point-to-point connection to interconnect two devices operating at 100 Gb/s. Or because in practice the cable is made up of four 25 Gb/s lanes, there is the option to bifurcate a single 100 Gb/s connection into two 50 Gb/s endpoints, for example, in the case of connecting two 50 Gb/s servers to a 100 Gb/s ToR switch port. Servers operating at 25 Gb/s can similarly be connected in groups of four into a single 100 Gb/s switch port.

Even though the DAC is a simple passive connection, different DAC cables have different properties and signal responses. For this reason, the equipment may need to be tuned to optimize the signal on a port-by-port basis. This need for tuning may also cause problems with the interoperability of different pieces of network equipment. Testing to determine which cables and equipment will work together has been carried out by a number of vendors and independent organization such as the University of New Hampshire's Interoperability Laboratory [38].

14.3.3 Active optical cables

The distances that a DAC can cover decrease with bit rates. For a link with a distance beyond the reach of DAC, one simple interconnect technique is AOC. An AOC uses optical technology to improve the speed and distance performance of a cable without sacrificing compatibility with standard electrical interfaces. An AOC consists of two optical transceivers with an optical cable permanently attached to them. The basic idea of an AOC is to embed active optical transceiver components, which perform electrical-to-optical conversion, into an electronic connector, and use an optical cable to extend the distance and speed of the connector.

AOC can be treated as a simple "plug and play" copper cable, as it has a standard electrical interface and does not expose any optical characteristics, but it has many advantages over a DAC, including (1) a lower weight and a smaller bend radius, which enables simpler cable management in high-density deployments; (2) a thinner cable size, which frees up space for increased air flow for better thermal management in dense systems; (3) a longer reach with distances up to 200 m depending on the AOC technology; (4) better EMI resistance.

Fig. 14.14 can help understand the cost advantage of an AOC as compared to two optical transceivers with a connectorized fiber. The figure is a block diagram of optical transmit/receive path defined in IEEE 802.3 [39]. For an AOC, there is no patch cord, and the medium dependent interface (MDI) and all the optics are contained inside the AOC and not exposed outside. AOC only needs to comply to IEEE, the InfiniBand Trade Association and SFF industry standards for the electrical, mechanical, and thermal requirements but not for optical requirements, which is the hardest part. Therefore AOC designers do not have to comply with any industry standards for

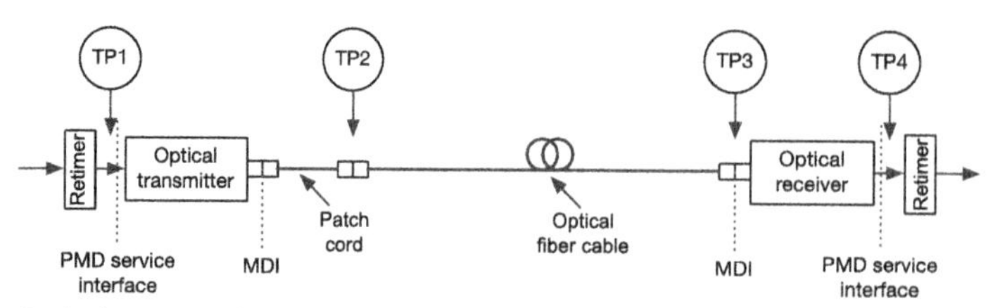

Figure 14.14 Block diagram of optical transmit/receive path defined in IEEE 802.3. TP1–4 are four test points along the path. *PMD*, Physical media dependent; *MDI*, media dependent interface.

transceiver interoperability with other vendors. This provides design freedom to pick and choose the technology with the lowest cost and best performance. In addition, the two test points TP2 and TP3 in Fig. 14.14 are not accessible for an AOC and almost no optical testing is needed. Note that optical testing accounts for a significant part of a transceiver's cost. Because of the above benefits, AOC is considered as the most cost- effective optical interconnect technology.

Although AOC can achieve the same distances as optical transceivers, installing a long AOC in a crowded data center is difficult, as an AOC has two bulky ends. In data centers, AOC are typically deployed in open access areas such as from servers to TOR switches or from TOR switches to leaf switches, with reaches of 3–30 m. As the reach is short, most AOCs are made from cost-effective multimode vertical cavity surface emitting lasers (VCSELs) and MMF.

14.3.4 Optical transceivers

Most interconnects in high-speed data centers with link distances longer than 10 m are provided by optical transceivers and fibers. These links are either beyond the distance limit of a DAC or not in open access areas where it is difficult to deploy and route an AOC. Unlike DACs and AOCs, optical transceivers have to be combined with optical fiber cables to provide a link, as indicated in Fig. 14.14. An optical transceiver contains an optical transmitter and receiver and has both electrical and optical interfaces that have to comply with industry standards for interoperability with other vendors.

There are many types of optical transceivers, and they can be separated into different categories according to distances [such as split ratio (SR) and LR], optical wavelengths (such as short and long wavelengths), form factors (such as SFP and QSFP), speeds (such as 40 and 100 G), etc. For example, optical transceivers can be divided into six categories according to the maximum distances they can reach, as shown in Table 14.1. SR transceivers use multimode technologies, that is, VCSELs and MMF, and they can only reach 100–300 m for 40 G and above, depending on the speeds of optical transceivers and types of MMF. As the wavelength of most multimode technologies are around 850 nm, which are also called short-wavelength transceivers. All the other reach optical transceivers use single-mode technologies, with wavelengths

Table 14.1 Optical transceiver categories in reaches.

	SR	DR	FR	LR	ER	ZR
Distances	100–300 m	500 m	2 km	10 km	40 km	80 km

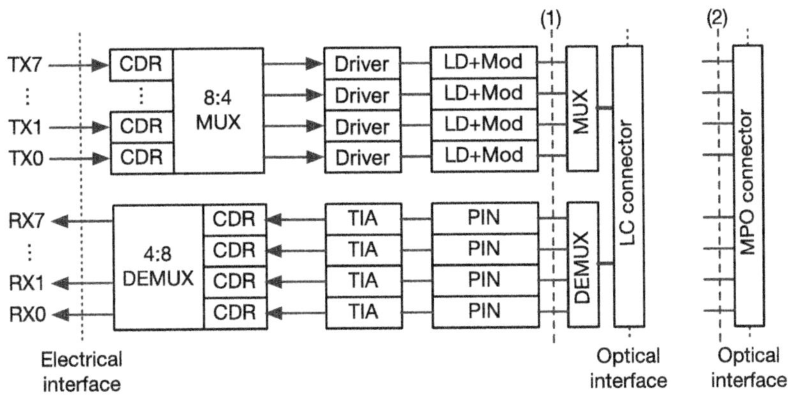

Figure 14.15 Block diagram of an optical transceiver with LC and MPO connectors. *LD*, Laser diode; *Mod*, modulator; *CDR*, clock and data recovery; *MUX*, multiplexer; *DEMUX*, demultiplexer.

either at O-band (1310 nm) or C-band (1550 nm), which are also called long-wavelength transceivers. Single-mode transceivers with different reaches use different technologies to optimize the cost and performance. For example, DR and FR typically use uncooled lasers, whereas lasers with temperature control are required for LR, extinction ratio (ER), and ZR to control laser wavelengths and modulation performance.

The block diagram of a typical optical transceiver is shown in Fig. 14.15. In the figure, the speed of an optical lane is two times that of an electrical lane, so there is an electrical multiplexer and demultiplexer between the electrical interface and optical lanes, which multiplex two input electrical lanes to one optical lane at the input of the electrical interface and demultiplex one optical lane to two electrical lanes at the output. Depending on the speed of the transceiver, clock and data recovery (CDR) may be needed and some equalization and preemphasis functions such as continuous time linear equalizer, decision feedback equalizer, and feed-forward equalizer may be included in the CDR module. There are two types of optical lanes. One uses wavelength lanes, which means that each optical lane corresponds to a wavelength, and a wavelength multiplexer combines different wavelength lanes into one fiber and a wavelength demultiplexer splits the input wavelength-division-multiplexed (WDM) signal to different wavelength lanes at the optical output and input, respectively. The optical connector in this case is a duplex fiber connector such as a duplex LC connector. The other type uses parallel fiber lanes, which means that each optical lane is carried by a fiber, and no wavelength multiplexer and demultiplexer are needed. The fibers are directly connected to a multi-fiber Push On connectors (MPO) connector, which is the optical interface of the transceiver. The advantage of a transceiver using multiple wavelength lanes is that the optical connector is simple and it only needs one pair of fibers. The drawback is that the wavelength multiplexer and demultiplexer introduce additional loss and some wavelength control may be needed, which increases the requirements on lasers and the overall cost of the transceiver. A transceiver using multiple fiber lanes needs multiple pairs of fibers, which increases fiber cost, especially for a long link, but it has a lower transceiver cost due the reduced requirements on the lasers.

Another advantage of an optical transceiver with multiple fiber lanes is that it easily breaks out to multiple fiber lanes, which provides flexibility for the network architecture, as illustrated in Figs. 14.16 and 14.17. The figures show the architecture of a network with spine and leaf switches, where each spine and leaf switch has sixteen and four 100 G downlink and uplink ports, respectively. If the 100 G optical transceiver cannot provide breakout function. The network can only have 4 spine switches and 16 leaf switches, as show in Fig. 14.16. Whereas if a 100 G transceiver can be split into four 25 G ports with breakout fibers, one transceiver can effectively provide four 25 G ports, which is equivalent to four 25 G transceivers. Therefore each spine and leaf switch has sixty-four and sixteen 25 G downlink and uplink ports, respectively, and the network can have 16 spine switches and 64 leaf switches, increasing the number of addressable network interfaces by a factor of four using the same switches, as is shown in Fig. 14.17. Compared with four separate 25 G transceivers, one 100 G transceiver that can breakout has a better cost, a higher bandwidth density and lower power consumption.

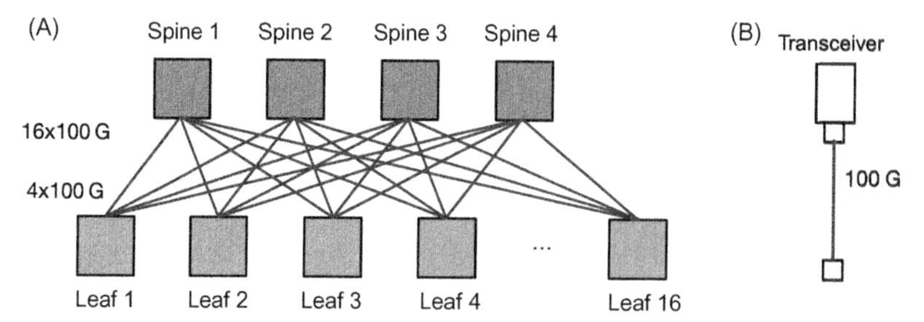

Figure 14.16 (A) The network architecture of spine and leaf switch with 100 G transceivers. Each spine switch has 16 × 100 G downlink ports and leaf switch 4 × 100 G uplink ports. (B) 100 G transceiver without breakout.

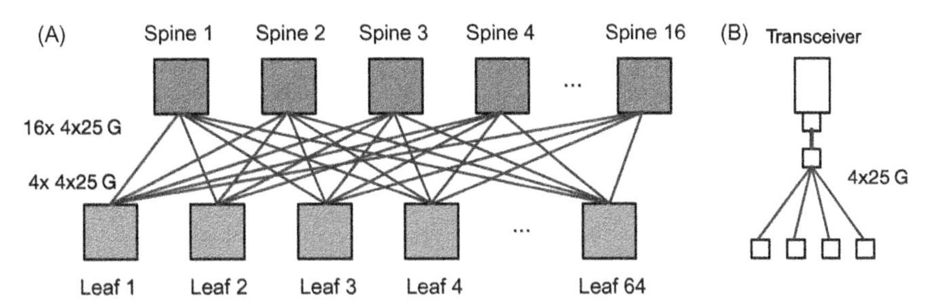

Figure 14.17 (A) The network architecture of spine and leaf switch with 100 G breakout transceivers. Each spine switch has 16 × 100 G downlink ports and leaf switch 4x100G uplink ports. (B) 100 G transceiver with 4 × 25 G breakout cable.

14.4 Development of optical transceiver technologies

14.4.1 40 G technologies

Up to 40 G, most intra-data center optical interconnects use multimode technology, that is, VCSELs combined with MMF. Not only can VCSELs be processed and tested on wafer scale due to its vertical structure, but they can have low power consumption as well. Compared with single-mode fiber (SMF), MMF has a larger core size and numerical aperture so that it is much easier to couple light in and out of MMF compared to SMF, thus making light coupling and alignment for multimode transceivers much easier compared to single-mode transceivers. Together with wafer processing and testing, VCSEL-based multimode transceivers are much more cost-effective than single-mode transceivers. Accordingly, multimode technology is widely used in short-reach applications. However, light launched into MMF excites different modes, which travel at different group velocities and therefore limit the bandwidth of MMF, as shown in Fig. 14.18. To increase the bandwidth of MMF, four generations of MMF have been developed, from OM1 to OM4. Table 14.2 gives the characteristics of different MMF. The bandwidths and link distances have been significantly improved from OM1 to OM4 [40]. For example, the link distance of a 10 G system has been increased from 33 m for OM1 to 550 m for OM4. Note that the link distances listed in Table 14.2 are defined by IEEE standard [40,41], and the effective mode bandwidths (EMB) of MMF in Table 14.2 are defined at 850-nm wavelength. The link distances can be further increased using distance-enhancement methods such as low-linewidth VCSELs. To accommodate VCSELs that operate at wavelengths that are different than 850 nm, a wide-band MMF has been developed [42].

There are two main types of 40 G optical transceivers, one uses multimode technology called SR4, and the other single-mode technology called LR4. Both are standardized by IEEE [41]. The schematic of a 40 G SR4 optical transceiver is depicted in Fig. 14.19. It consists of four electrical lanes and optical lanes in each direction, with each lane signaling at 10 G. For each lane, an input 10 G nonreturn-to-zero (NRZ)

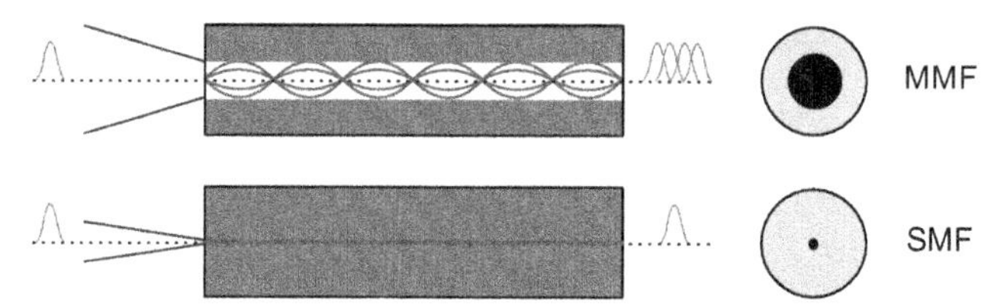

Figure 14.18 Schematic of pulse propagation in multimode and single-mode fibers.

Table 14.2 Characteristics of different multimode fibers.

Fiber	Core diameter (μm)	EMB (MHz.km)	Link distance (m)			
		850 nm	1 G	10 G	40 G	100 G
OM1	62.5	N/A	275	33	N/A	N/A
OM2	50	N/A	550	82	N/A	N/A
OM3	50	2000	N/A	300	100	70
OM4	50	4700	N/A	550	150	100

EMB, effective mode bandwidth.

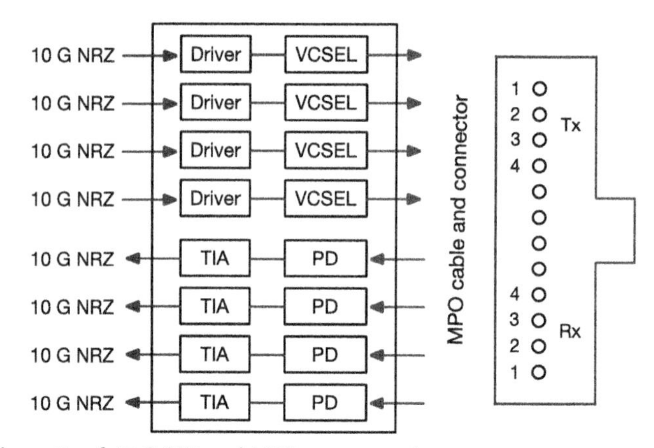

Figure 14.19 Schematic of 40 G SR4 and MPO connector layout.

electrical signal is amplified by a driver amplifier, which drives a VCSEL to convert the electrical signal into 10 G NRZ optical signal. The wavelength of the VCSELs is in the range of 840–860 nm, and the four output optical signals from the VCSEL are carried by four MMFs. The optical transceiver uses a 12-fiber MPO connector, with four fiber ports not connected or empty, as shown in Fig. 14.19. According to IEEE standard, 40 G SR4 can only reach 100 and 150 m over OM3 and OM4 fibers, respectively. The distance of 40 G can be significantly improved using distance-enhancement techniques such as electrical equalization and/or low-linewidth VCSELs. Up to 300 and 550 m over OM3 and OM4 fibers can be achieved, respectively, using distance-enhanced 40 G SR4 transceivers, which cover most of the links in data centers.

For links of distances exceeding the range of multimode technology, one can use 40 G LR4 transceivers and SMF. The schematic of a 40 G LR4 transceiver is illustrated in Fig. 14.20. It also has four electrical and four optical lanes, with each lane working at 10 G. Unlike SR4, it uses either directly modulated lasers (DMLs) or electro-absorption-modulated lasers (EMLs) and the four optical lanes have different wavelengths according to a course wavelength division multiplexing (CWDM) specification. The wavelengths are multiplexed into a single fiber with a wavelength multiplexer. At the receiver, the wavelengths are separated with a wavelength demultiplexer. Instead of an MPO cable, a duplex fiber cable is used for a LR4 optical transceiver with a link distance up to 10 km. The wavelengths of the four lanes are given in Table 14.2. The wavelength spacing of the four lanes is 20 nm and wavelength range is ± 6.5 nm, which permits the use of uncooled lasers (Table 14.3).

Optical transceivers are characterized by the optical modulation amplitude (OMA) because it is modulation amplitude not the total optical power that conveys information. The definition of OMA for a NRZ signal is illustrated in Fig. 14.21 [43] and is given by

$$\mathrm{OMA} = \mathrm{P1} - \mathrm{P0}, \tag{16.1}$$

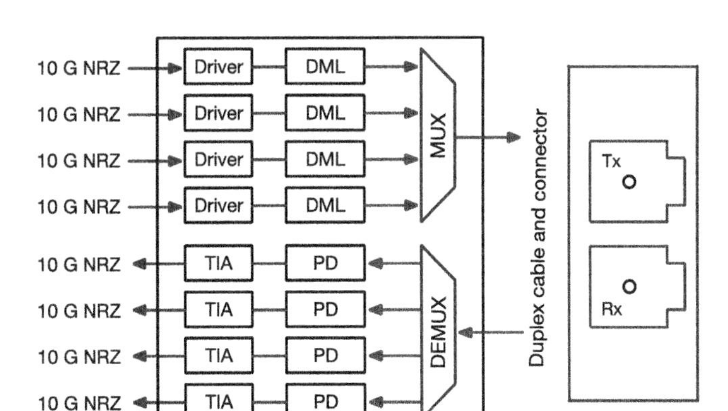

Figure 14.20 Schematic of 40 G LR4 and LC connector layout.

Table 14.3 Wavelength assignment of 40 G LR4 course wavelength division multiplexing (CWDM).

Lane	Center wavelength (nm)	Wavelength range (nm)
0	1271	1264.5–1277.5
1	1291	1284.5–1297.5
2	1311	1304.5–1317.5
3	1331	1324.5–1327.5

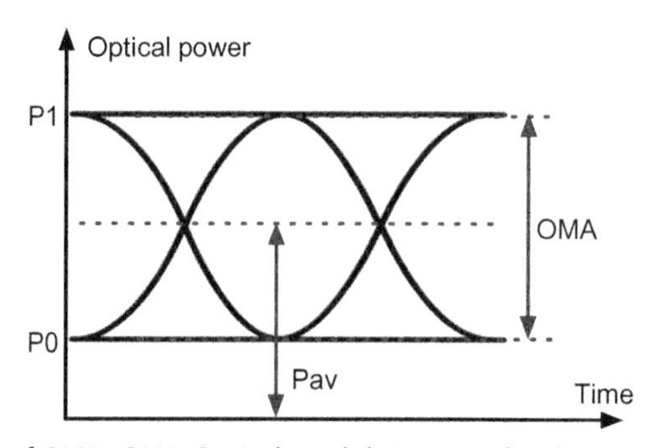

Figure 14.21 Definition of OMA. *OMA*, Optical modulation amplitude.

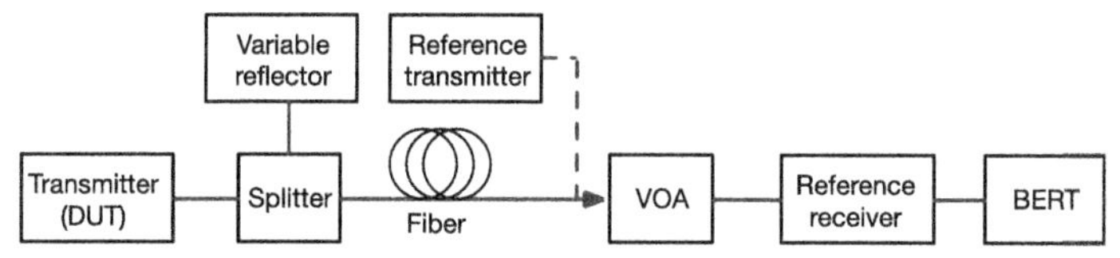

Figure 14.22 Test setup for dispersion penalty (TDP) measurement. *DUT*, Device under test; *VOA*, variable optical attenuator; *BERT*, BER tester.

where P1 and P0 are the powers of level "1" and "0," respectively. The average optical power is

$$\mathrm{Pav} = (\mathrm{P1} + \mathrm{P0})/2. \tag{16.2}$$

OMA can be calculated from the average optical power and ER, which is defined as

$$\mathrm{ER} = \mathrm{P1}/\mathrm{P0}. \tag{16.3}$$

At high ER, OMA is approximately two times the average power.

Another important parameter for an optical transceiver is the transmitter and dispersion penalty (TDP), which is used to quantify transmitter impairments along with dispersion. For SMF, the dispersion is chromatic dispersion. For MMF the dispersion is mostly modal dispersion. Fig. 14.22 shows the test setup defined in [43] to measure TDP. In the measurement, first the nominal sensitivity of a reference receiver, S, is measured using the OMA of a reference transmitter without a test fiber. Both the reference receiver and transmitter are specified in IEEE standard [43]. Then the reference transmitter is replaced with a transmitter under test and a test fiber. The sensitivity using this configuration is measured, denoted as P_DUT. The TDP is defined as

$$\mathrm{TDP} = \mathrm{P\_DUT} - \mathrm{S}. \tag{16.4}$$

Table 14.4 Link power budget of 40 G SR4 and LR4 (in dB).

	SR4 (for OM3)	LR4
Power budget	8.3	9.3
Channel insertion loss	1.9	6.7
Allocation for TDP	6.4	2.6

Because no optical amplifier and dispersion compensation are used for a short link, the link power budget has to be considered in the design of the optical transceiver. Table 14.4 gives the link power budget specified in 40 G SR4 and LR4 [41]. The total power budget for SR4 and LR4 are 8.3 and 9.3 dB, respectively. Most of the power budget is allocated to the TDP for SR4 because of the large modal dispersion of MMF. For LR4, most of the power budget is allocated to insertion loss because of the long transmission length. Note that the channel insertion loss is calculated using the maximum distances specified in IEEE standard and cabled optical fiber attenuation, which is 3.5 dB/km for MMF at 850 nm and 0.47 dB/km for SMF at 1264.5 nm, plus an allocation for connection and splice loss, which is 1.5 dB and 2 dB for MMF and SMF, respectively.

14.4.2 100 G technology

IEEE defines two types of multimode technologies for 100 G, 100 G SR10 and 100 G SR4. 100 G SR10 uses the same technology as 40 G SR4, but uses ten optical 10 G lanes instead of four [44]. This technology can reach the same distance as 40 G SR4. While SR10 was used in early 100 G deployments, 100 G SR4 is now more widely deployed. This standard uses four electrical lanes and four optical lanes each carrying a 25 G signal [41]. With the increase of bit rates from 10 to 25 G, two techniques have been added to increase the performance of a 100 G SR4 interconnect. One is the addition of CDR at each input and output electrical lane, which is used to improve the signal quality at the electrical interface. The second technique is the use of forward error correction (FEC). To use a 100 G SR4 transceiver, a Reed-Solomon FEC RS(528, 514) is required at the host, which can decrease an uncoded bit error rate (BER) of $5.0E-5$ down to $1.0E-12$. Although some distance-enhancement techniques can increase reach of 100 G SR4 beyond 100 m over OM4 fiber specified by IEEE, most deployed 100 G SR4 transceivers are limited to 100 m because it is much more challenging to reduce the modal dispersion penalties of 100 G SR4 than 40 G SR4.

Since 100 G SR4 is limited to 100 m, a large number of links have to use single-mode technologies for 100 G. Currently, the IEEE has only one single-mode 100 G technology standard, that is, 100 G LR4, which covers up to 10 km over SMF. To reduce the chromatic dispersion penalties, 100 G LR4 uses the local area network

Table 14.5 Wavelength assignment of 100 G LR4 (local area network wavelength-division-multiplexed, LAN WDM).

Lane	Center frequency (THz)	Center wavelength (nm)	Wavelength range (nm)
0	231.4	1295.56	1294.53–1296.59
1	230.6	1300.05	1299.02–1301.09
2	229.8	1304.58	1303.54–1305.63
3	229.0	1309.14	1308.09–1310.19

(LAN) WDM, whose wavelength assignment is given in Table 14.5 [44]. The channel spacing of LAN WDM is about 4.5 nm (800 GHz) and the wavelength range about ± 1 nm, so uncooled lasers cannot be used and temperature control is required. Considering these facts, 100 G LR4 does not meet the bandwidth cost and power consumption requirements for data center applications. Therefore many different 100 G single-mode technologies for data center optical interconnects are under development by MSAs.

One single-mode 100 G transceiver MSA is CWDM4 [45]. The required reach of CWDM4 is up to 2 km on SMF. Compared with LR4, two changes are made in 100 G CWDM4 to reduce cost. First, it uses course WDM wavelengths instead of LAN WDM. The wavelengths are the same as those of 40 G LR4, that is, from 1291 to 1331 nm with a 20-nm channel spacing and a wavelength range of ± 6.5 nm. Therefore uncooled lasers can be used. Second, the use of FEC that allows the system to operate at a pre-FEC BER of 5.0E − 5, instead of 1.0E − 12 as specified by LR4, which relaxes the requirements on the transmitter power and receiver sensitivity, but RS(528,514) FEC is required to be implemented on the host in order to ensure reliable operation of the host system. Another similar 100 G transceiver technique is CLR4 [46]. The only difference between CLR4 and CWDM4 is that CLR4 supports two operation modes, a FEC enabled mode for an increased link margin and a FEC disabled mode for lower power consumption and a lower latency.

To further reduce the cost of 100 G CWDM4 transceivers, open compute project (OCP) defines a relaxed version of CWDM4, called CWDM4-OCP [47]. The specification differences between CWDM4 MSA and CWDM4-OCP are given in Table 14.6. Considering the data center environment, the specifications in three main areas are relaxed: reach, loss budget, and temperature. As most data center links are less than a few hundred meters, CWDM4-OCP reduces the reach down from 2 km to 500 m. With this distance, the link loss budget is reduced from 5 to 3.5 dB, which relaxes both transmitter OMA and receiver sensitivity. In addition, most data centers operate under a very predictable thermal environment and the air-inlet temperature to the switching equipment is well controlled; so is the thermal environment inside the

Table 14.6 Specification comparison between 100 G course wavelength division multiplexing multisource agreement (CWDM4 MSA) and course wavelength division multiplexing open compute project (CWDM4-OCP).

	CWDM4-OCP	CWDM4 MSA
Reach	500 m	2000 m
TX OMA	− 5.0 dBm	− 4.0 dBm
TX OMA TDP	− 6.0 dBm	− 5.0 dBm
RX sensitivity	− 9.5 dBm	− 10.0 dBm
Link loss budget	3.5 dB	5.0 dB
Operating case temperature	15°C−55°C	0°C−70°C

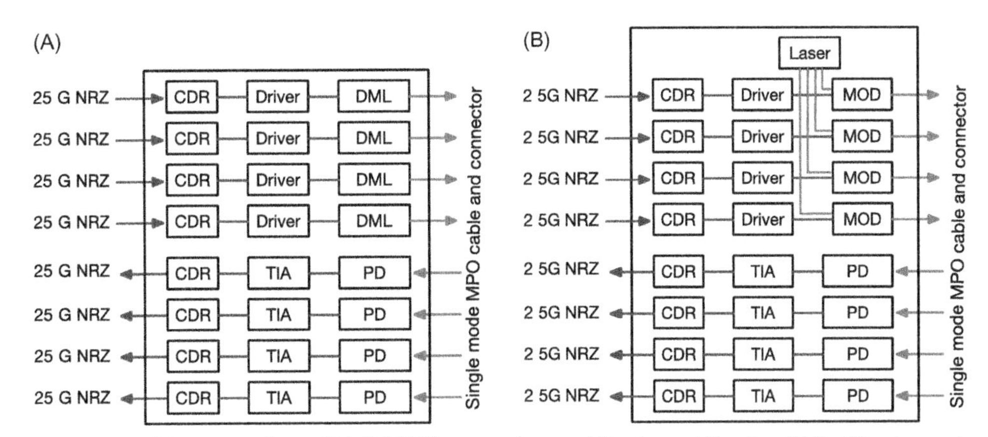

Figure 14.23 Diagrams of two 100 G PSM4 transceiver architectures. (A) using DML, (B) using external modulators and one laser.

switch. In the consideration of this fact, the operating case temperature of optical transceivers is relaxed down from 0°C−70°C to 15°C−55°C.

Another single-mode 100 G transceiver MSA is PSM4 (PSM stands for parallel single mode), targeting a low-cost solution capable of reach up to 500 m over a parallel SMF infrastructure [48]. For PSM4, each optical lane is in a separate fiber instead of a separate wavelength. The laser wavelength for each fiber is in range is from 1295 to 1325 nm, which further reduces the requirements on lasers compared with CWDM4. In addition, when external modulators are used, which is the case for silicon photonics technologies, lasers can be shared by two or four optical lanes. Fig. 14.23 shows the diagrams of two different PSM4 transceivers, one uses four DMLs and the other one has one laser shared by four external modulators. Note that in the case of four lasers, external modulators can also be used. Like CWDM4, PSM4 also specifies the optical link to operate at a pre-FEC BER of 5.0E − 5.

Table 14.7 Main characteristics of different 100 G technologies.

	SR4	CWDM4	PSM4	LR4
Reach (m)	2–100	2-2,000	2–500	2–10,000
Mode	Multimode	Single mode	Single mode	Single mode
Fiber	MMF MPO	SMF, Duplex	SMF, MPO	SMF, Duplex
BER	5.0E − 5	5.0E − 5	5.0E − 5	1.0E − 12
Power budget (dB)	8.2	8.0	6.2	8.5

Table 14.7 lists the main characteristics of four 100 G technologies, including reaches, single-mode or multimode, fiber cables type, BER requirements, and power budget. Note that the reaches listed here are defined by IEEE and MSAs, and actual products may have a longer reach.

When choosing technologies, one of the main factors needs to consider is the total cost of ownership (TCO). Optical interconnect cost includes the cost of optical transceivers and link fibers. SR4 has the least expensive transceivers but most expensive fibers, as MMF is much more expensive than SMF, whereas for CWDM4, the cost of fibers is the lowest but the cost of transceivers is the highest. TCO of optical interconnects depends on many factors, including data center sizes, lifetime of data centers, and the cost of transceivers and fibers. It must be calculated based on the parameters of each data center to get the most cost-effective solution. One needs to choose the suitable technologies to optimize their TCO considering their network architecture and data center environment.

14.4.3 400 G technology

There are three techniques to increase the speed of a transceiver: increasing the symbol rate, increasing bits/per symbol using a higher order modulation format, and increasing the number of lanes, which can be fiber lanes, wavelength lanes, or a combination of both, as shown in Fig. 14.24. In practice, a high-speed transceiver is typically realized with a combination of different techniques. For example, combining 100 G CWDM4 and SR4 transceiver using four wavelength lanes and four fiber lanes with each lane running at 25 Gbaud can lead to a link that supports 400 Gb/s.

Intuitively, a simple way to realize a 400 G transceiver is to use four 100 G lanes, that is, to scale up the speeds of both electrical and optical lanes from 25 G, currently used by a 100 G transceiver, to 100 G. However, it is still technically challenging to realize a 100 G lane speed. A 50 G electrical lane speed is ready for commercialization, and standard bodies are working on a 100 G electrical lane speed [49]. Although an optical lane with 100 G was demonstrated several years ago using various modulation formats and is mature for commercialization, they typically need an external modulator such as Mach-Zehnder modulators or an electro-absorption modulator [50–52].

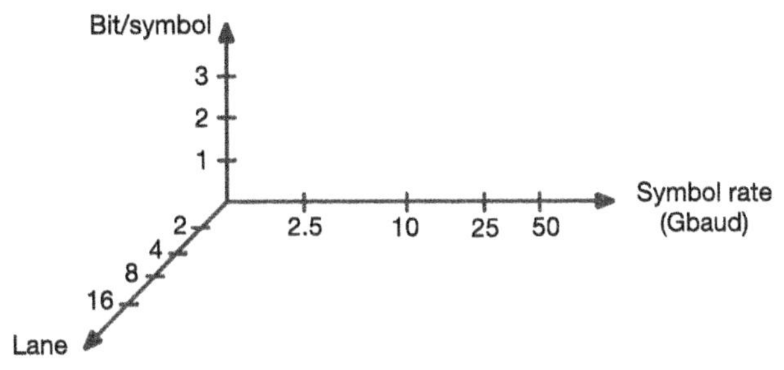

Figure 14.24 Methods to increase transceiver speeds.

Table 14.8 Generations of 400 G transceiver technologies.

Generation	Elec. I/O	Multimode technology		Single-mode technology	
		Opt. I/O	Name	Opt. I/O	Name
1st	16 × 25 G NRZ	16 × 25 G NRZ	SR16	8 × 50 G PAM4	FR8, LR8
2nd	8 × 50 G PAM4	8 × 50 G PAM4	SR8/ SR4.2	8 × 50 G PAM4	FR8, LR8
3rd	8 × 50 G PAM4	4 × 100 G PAM4	SR4	4 × 100 G PAM4	DR4, FR4, and LR4
4th	4 × 100 G PAM4	4 × 100 G PAM4	SR4	4 × 100 G PAM4	DR4, FR4, and LR4

Directly modulated VCSEL-based interconnects operating at 100 G per lane are still in a research phase [53,54].

Until directly modulated 100 G lane speeds are commercially available, one can achieve 400 G by increasing the number of lanes with lower lane speeds, such as 16 × 25 G or 8 × 50 G. According to the speeds of electrical lanes (SERDES speeds) and optical lanes, 400 G optical transceiver technologies can be divided into four generations, as illustrated in Table 14.8, with lane speeds evolving from 25 G for the first generation to 100 G for the fourth generation. One should note that the optical lane speed may be different from the electrical lane speed and in this case, a gear box that matches the signaling rate of the electrical lanes to the signaling rate of the optical lanes is needed.

There are several ways to achieve 400 G using multimode technology before 100 G per lane is available, such SR16, SR8, and SR4.2, as shown in Table 14.8. A technically easy method to realize 400 G is SR16, which uses the same technology as 100 G SR4 but increases the number of lanes from 4 to 16. SR16 is standardized in IEEE 802.3bs and is the first available 400 G technology [55]. To reduce the footprint and cost of an optical transceiver, one needs to reduce the number of optical lanes.

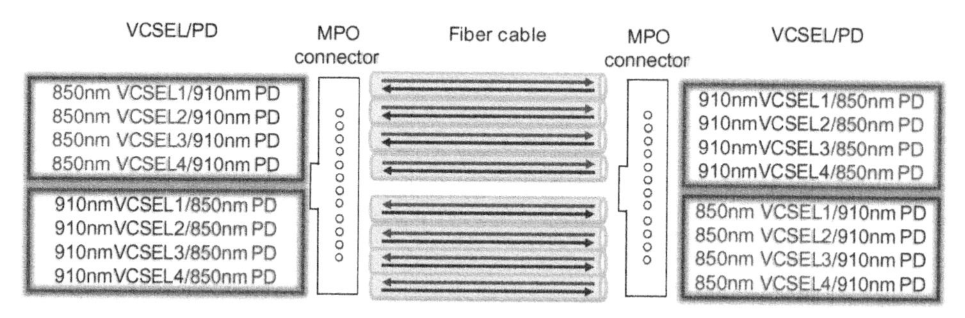

Figure 14.25 The vertical cavity surface emitting laser (VCSEL), photodiode (PD), and wavelength arrangement for a 400 G BiDirectional (BiDi) link.

SR8 and SR4.2 use eight electrical and optical 50 G lanes [56]. SR8 has eight optical lanes with each lane carried on a separate fiber. Each lane uses a 50 G 4-level pulse-amplitude-modulation (PAM4) signal and has a wavelength around 850 nm. SR4.2 also has eight optical lanes, but it uses the combination of fiber lanes and wavelength lanes. There are two wavelengths for SR4.2 with each fiber carrying two wavelength lanes—one around 850 nm and the other around 910 nm—so that only four pairs of fibers are required for SR4.2. The two wavelengths can either copropagate (Co-Directional) in one fiber or counterpropagate (BiDi–BiDirectional) in one fiber. The schematic of a link using 400 G BiDi transceiver is shown in Fig. 14.25. It shows that two wavelengths counterpropagate in each fiber. The MPO12 connectors and fiber cables used for 100 G SR4 can be reused for 400 G BiDi and 400 G SR4.2.

IEEE specifies three single-mode 400 G transceiver technologies, that is, DR4, FR8, and LR8, with reaches up to 500 m, 2 km, and 10 km over SMF, respectively. DR4 is equivalent to PSM4, which uses four parallel SMF lanes with a lane speed of 100 G. FR8 and LR8 uses eight wavelength lanes with a lane speed of 50 G. Additional four wavelengths shorter than LAN WDM are assigned to FR8 and LR8, and the wavelength assignments are the same for FR8 and LR8. as shown in Table 14.9. The wavelength spacing in the upper and lower four lanes is 800 GHz, same as LAN WDM, but the spacing between the upper four wavelengths and lower four wavelengths is 1.6 THz. Temperature control may be needed due to the narrow wavelength range requirement. FR8 and LR8 may not meet data center requirements on cost and footprint, and a transceiver with four wavelength lanes and a large channel spacing is desired for data center operators. To meet this demand, 400 G FR4 is specified by 100 G lambda MSA, which uses the CWDM wavelengths with a 20-nm channel spacing so that uncooled lasers can be used [57].

To reduce the requirements on the bandwidths of electronic and optical components, IEEE suggests using PAM4 for both electrical and optical signals when lane speeds are larger than 25 G [55]. Except for 400 G SR16, all the 400 G transceivers

Table 14.9 Wavelength assignment of 400 G FR8 and LR8.

Lane	Center frequency (THz)	Center wavelength (nm)	Wavelength range (nm)
0	235.4	1273.54	1272.55–1274.54
1	234.6	1277.89	1276.89–1278.89
2	233.8	1282.26	1281.25–1283.27
3	233.0	1286.66	1285.65–1287.68
4	231.4	1295.56	1294.53–1296.59
5	230.6	1300.05	1299.02–1301.09
6	229.8	1304.58	1303.54–1305.63
7	229.0	1309.14	1308.09–1310.19

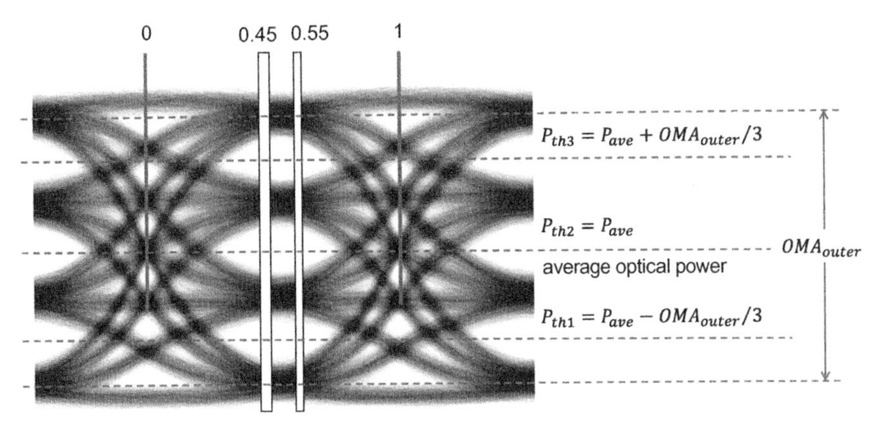

Figure 14.26 Outer optical modulation amplitude (OMA_{outer}), thresholds, and average power of a pulse-amplitude-modulation (PAM4) signal.

use PAM4 signals. A PAM4 signal has a higher requirement on the power and on the relative intensity noise of the transmitters compared to an NRZ signal [58]. A higher gain FEC is suggested by IEEE for 400 G, which has a FEC threshold at BER of 2.0E − 4 [55].

A different parameter, transmitter, and dispersion eye closure for PAM4 (TDECQ) is used to quantify 400 G transceivers. TDECQ is a measure of each optical transmitter's vertical eye closure when transmitted through a worst-case optical channel, as measured through an optical to electrical converter and oscilloscope with the combined frequency response and equalized with the reference equalizer given in IEEE 802.3bs [55]. The reference receiver and equalizer may be implemented in software or may be part of an oscilloscope. Fig. 14.26 illustrates the eye diagram of a PAM4 signal and corresponding parameters including the average optical power, thresholds, and outer optical modulation amplitude (OMA_{outer}). The details on the definition and measurement of TDECQ can be found in IEEE standard 802.3bs [55].

Table 14.10 Characteristics different 400 G transceiver form factors.

	CFP8	OSFP	QSFP-DD	QSFP
Elec. I/O	16 × 25 G/ 8 × 50 G	8 × 50 G	8 × 50 G	4 × 100 G
Size (mm)	102 × 40 × 9.5	107.7 × 22.6 × 9	78.3 × 18.4 × 8.5	72.4 × 18.4 × 8.5
Capacity/ RU	6.4 T	12.8 T	14.4 T	14.4 T

Existing 40 and 100 G optical transceivers for data center application predominantly use QSFP form factor, which has four electrical lanes. For 400 G transceivers, at least eight electrical lanes are needed before 100 G SERDES is available, and there are a few form factors, including CFP8 [59], OSFP [36], and QSFP-DD [35]. The main difference between these form factors are the size. A transceiver of a larger size form factor has a higher power dissipation capability and a lower bandwidth density. The characteristics of these form factors are listed in Table 14.10. A form factor of a bigger size is available earlier; it may also be able to accommodate more complex modules such as coherent modules.

14.5 Future development

14.5.1 Coherent detection inside data centers

Coherent detection can not only achieve higher sensitivity than direct detection but can significantly increase the spectral efficiency (encoding more bits on each symbol) as well because it uses phase, amplitude, and polarization of an optical carrier to carry information. Although coherent detection has enjoyed great success in optical transport networks, today virtually all the optical interconnects inside data centers use intensity modulation and direct detection (IM-DD) techniques because of the cost, power consumption, and footprint.

There has been tremendous reduction in cost, power, and density on the coherent technology over the last decade due to the advances in photonics and electronics [60]. Coherent modulation detection techniques are beginning to be deployed in metro networks as well as shorter links [61]. For example, 400 G ZR, which is defined by Optical Internetworking Forum (OIF), is targeted for ~100-km interdata center interconnect applications [62]. On the other hand, with the increase of bit rates, it becomes challenging to use simple IM-DD techniques and some advanced techniques such as external modulation, more complex signals, and detection have to be used, which increases the complexity and cost of IM-DD systems. For example, PAM4 modulation is suggested for 400 G transceivers and some digital signal processing (DSP) is required for 400 G DR4 and FR4 [63]. Note that the DSP complexity of a

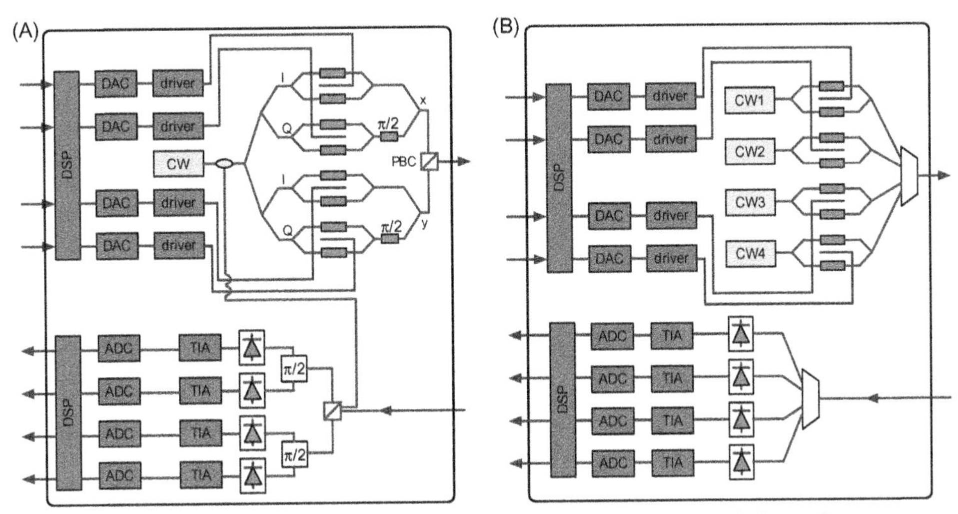

Figure 14.27 Block diagram of a coherent-detection transceiver (A) and direct-detection transceiver (B). *ADC*, Analog-to-digital converter; *DAC*, digital-to-analog converter.

coherent transceiver can be significantly reduced for short-reach applications, for example, the digital chromatic dispersion compensation could be removed and the adaptive filter tap length could be shortened, etc. Fig. 14.27 depicts block diagrams of a high-speed coherent detection transceiver and an IM-DD transceiver that requires DSP. In the figure, the IM-DD transceiver has four optical lanes, similar to the four lanes in the coherent transceiver (in-phase, quadrature, x polarization, and y polarization). The figure shows that the complexity of the two transceivers are similar except that the coherent transceiver has some phase requirements between different lanes (modulators in the transmitter and 90-degree hybrid in the receiver) and the IM-DD transceiver needs a wavelength multiplexer and demultiplexer. We also note that the IM-DD transceiver has four lasers in order to use the same number of fibers as the coherent one. Fig. 14.28 gives the estimated ASIC power consumption for a 400 G IM-DD PAM4 and a coherent 16-ary quadrature-amplitude modulation (16QAM) transceiver [64]. The details of the parameters used in the figure can be found in the reference. This reference shows the power consumption of the IM-DD and coherent detection transceiver are comparable. One challenging part for a coherent system is that it requires a low linewidth and wavelength stable laser, which increases the cost and power consumption of the transceiver. However, a coherent transceiver needs one laser instead of four lasers, which can reduce the packaging cost in a silicon-photonic-based transceiver.

Compared with direct detection, coherent detection offers many advantages: (1) better sensitivity, which can be used to reduce the laser power requirement, to trade for a higher level modulation format to reduce the bandwidth requirement of

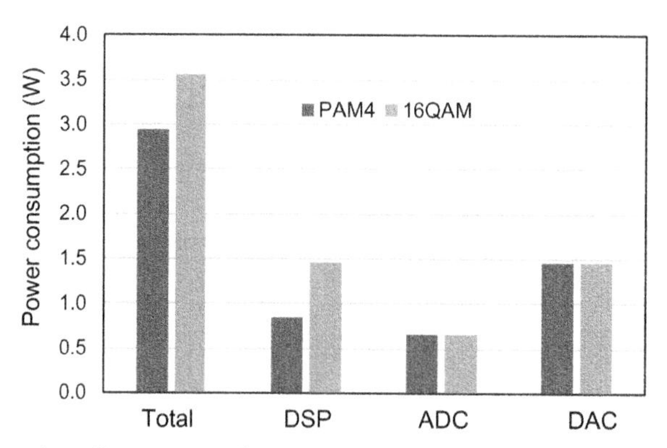

Figure 14.28 Estimated application-specific integrated circuit (ASIC) power consumption for 400 G IM-DD pulse-amplitude-modulation (PAM4) and coherent 16QAM transceiver based on 7-nm CMOS technology. *ADC*, Analog-to-digital converter; *DAC*, digital-to-analog converter.

component, or a high-power margin for a longer link; (2) more tolerance toward optical impairments such as chromatic dispersion. For an IM-DD transceiver beyond 100-Gbaud, chromatic dispersion can induce serious impairment for a couple of kilometer link; (3) higher spectral efficiency, which may become an issue in the future. With continuing CMOS technology improvements, there is a potential for coherent detection technology to be used for intra-data center interconnects.

14.5.2 Mid-board and copackaged optics

Continual traffic growth driven by data center applications puts pressure on the network to provide higher data rates to connect the network equipment. However, as the data rate increases, the interconnect become more challenging as we have discussed in Section 14.3 on the choice of interconnect technology. The difficulties are not limited to the optical domain because at higher data rates, the electrical signals inside the equipment chassis also encounter more severe signal integrity impairments. These impairments can be compensated by adding electrical functions at each end of the link to boost or recover the signals but has disadvantages because of increased complexity and power consumption. Signal integrity is a complex topic, but the dominant terms are transmission-loss, return-loss, and crosstalk. In order to increase the data rate while mitigating these impairments, the signal path should be kept as short as possible and discontinuities such as at connections should be avoided. Mitigating electrical signal integrity impairments is one of the main drivers to integrate optics closer with the host chip inside the equipment chassis. There are many applications for this technology. For example, the host chip could be either the switch chip die or the network interface controller chip at the compute node or the media controller at the storage device.

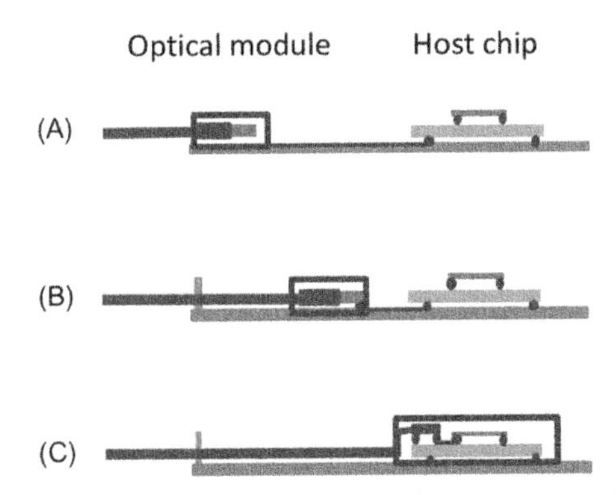

Figure 14.29 Illustration of the evolution of optics towards integration with host chip.

The evolution of optics deeper into network devices is illustrated in Fig. 14.29. The picture shows the evolution of optical transceivers from pluggable devices at the edge of the chassis (a), to on-board optics connected directly onto the printed circuit board in the middle of the chassis (b), to optics copackaged together with the host chip (c). As the optics move deeper into the chassis, optical fiber rather than electrical traces carry the high-speed signal to the faceplate, improving the signal integrity and freeing up faceplate area.

Since the early 2000s, the majority of short-reach interconnect optics used in telecommunication, datacom and high-performance computing applications have been pluggable optics as described in Section 14.3.4 of this chapter. These pluggable modules were in turn adopted by data centers for network applications. On-board optical modules [65] were developed starting around 2010 to meet the needs of the supercomputer industry. This need was driven by the doubling of the compute capability every two or three years in step with Moore's Law. This increase required more networking bandwidth. Highly integrated, power and cost efficient solutions that could meet these requirements could be realized by moving the optical connections deeper into the system [66]. Other modules were also developed [67,68] for these applications, but the designs were all proprietary and lacked a common interface to provide interchangeability. This lack of interoperability stalled the broader adoption of on-board modules in these markets.

Hyperscale data centers have grown to the point where they face similar problems to those of super computers in terms of scaling to higher data rates. Power, size and cost are again the three key performance metrics to optimize because these resources are limited and are prioritized for the revenue generating functions of the data center. On-board optics has the potential to improve power and cost efficiency and also increase the density of network equipment. On-board optics would also overcome system challenges by simplifying the routing of electrical signals to the front panel of the motherboard. Moving the optics module deeper into the equipment would also free up space on the front panel and allow for more cooling air to pass through the equipment.

Despite the potential advantages, there are several disadvantages associated with the practicalities of deploying and servicing on-board optics. Changes would need to occur in the way the network equipment is manufactured and installed. Equipment manufacturers would need to have expertise in handling optical fiber and testing optics. The data center operators would also need to change their operational models to handle different failure patterns. When pluggable modules fail, their location at the front panel allows them to be easily replaced, whereas on-board optics would be inaccessible inside the networking equipment. The network would need to accommodate loss of that link and it is not clear if the reliability of optical modules and in particular the laser sources will be sufficient to support this operational model.

To explore these use cases and serviceability models, the consortium for on-board optics (COBO) [69] was founded in 2015 with the aim of creating a common interface specification for on-board optics. The consortium led by Microsoft grew rapidly and attracted over 60 participants, including system equipment manufacturers, optical modules companies and electrical interconnect companies. COBO has specified three classes of on-board module to support applications ranging from client-side multimode optics to line-side coherent optics. Each form-factor class has a different size which allows different thermal management options to support varying power consumption regimes.

In addition to the three form-factor classes, the COBO specification has detailed two types of electrical interfaces. The first is an eight-lane interface based on the IEEE's 400GAUI-8 developed for 400 Gb/s applications. The second interface is 16-lane version for 800 Gb/s modules where one single 800 Gb/s module rather than two separate 400 Gb/s modules would reduce packaging costs and simplify thermal management. Modules supporting 400 Gb/s would use eight 50 Gb/s PAM4 signals and those supporting 800 Gb/s modules would use sixteen 50 Gb/s PAM4 signals. When IEEE 100 Gb/s signals are defined, it is expected that the two electrical interfaces would support 800 Gb/s and 1.6 Tb/s modules respectively.

The success of the COBO modules for intra-data center applications will be determined based on the key performance metrics: power, cost and size, and the adoption of suitable operational models. Even if there is no broad-scale adoption of COBO, the exploration of on-board optics is an important stepping stone that provides critical learning for future generations of embedded optics.

Beyond on-board optics, copackaged optics would take a step closer to the host chip by packaging the optics together with the host chip as shown in Fig. 14.30. Copackaged optics would allow the electrical signal integrity between the host chip and the optical module to be further improved. Copackaged optics has been identified as a core area of focus for next generation products and is in early investigation and

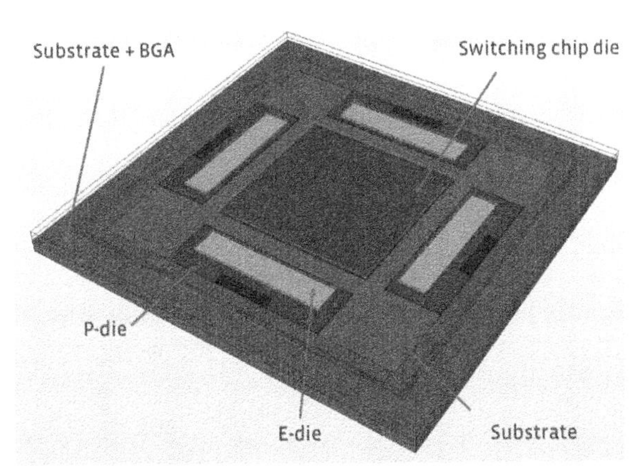

Figure 14.30 Schematic of a copackaged optical module including switching chip die as well as photonic (p −) and electronic (e −) dice.

development phase. A number of different scenarios are being explored ranging from embedded modules which use an interposer material to mount the host chip and optics chips with separate packages, or copackaged where the optics are located on the same package substrate and housed together in one package.

Copackaged optics might also be able to leverage the high-volume manufacturing processes developed for the silicon chip industry. Highly automated manufacturing techniques and process control have been shown to reduce defects and have several benefits including: avoiding human error, minimizing contamination, and preventing improper handling. These problems have been found to be the root cause of a large proportion of failures in the manufacture of pluggable modules using traditional manufacturing techniques [70].

Higher levels of integration would also provide higher density and save board space for other components. However, increasing the physical density without significantly reducing the power would increase power density and, in turn, lead to thermal challenges. Rather than traditional air cooling, new methods of cooling might be required such as liquid-assisted cooling techniques. High levels of integration also compound the serviceability challenge because a failure of a single component affects the whole package and repairing it brings down the whole assembly. If the reliability can be improved sufficiently and depending on the resiliency of the network, it might be possible to accommodate failures while leaving the device in place.

Perhaps the biggest hurdle to the widescale adoption of copackaged optics is that it requires a significant change in the way the supply chain is organized. The evolution from on-board to copackaged optics presents an additional challenge because the traditional demarcation between photonics module supplier and integrated circuit designer would need to change. The current supply-chain ecosystem fosters specialization in a particular area for example photonics or ASIC design. New business models will need to be developed that blur the lines between photonics, electronic integrated circuits and semiconductor packaging. There are few companies today with the capability to combine these.

Interest in the area of copackaged optics is growing and it is discussed in many industry technology forums such as the Optical Fiber Conference, the European Conference on Optical Communication and the IEEE Optical Interconnects conference. Agreement on supporting common standard interfaces which enable interoperability and interchangeability will be important in supporting a diverse supply-based ecosystem and enabling wide-scale adoption.

14.5.3 Optical switching inside data centers

The Cisco Global Cloud Index [1] mentioned in Section 14.2 estimated that the internal network bandwidth within hyperscale data centers is over 70% of the total worldwide bandwidth. This internal DCN demand has led to a divergence between the traffic growth rate within a single hyperscale data center, which is doubling approximately every year [71], and the single switch chip switching capacity growth rate, which is doubling approximately every two years. This trend, shown in Fig. 14.31, implies that future hyperscale data centers based on existing network topologies will become increasingly difficult to scale based on current technology projections.

The divergence between the required bandwidth within a data center and the single switch chip switching capacity is likely to increase because of the limitations of horizontally scaling merchant silicon switch hardware. This trend has led to efforts to design networks based on optical switching that can mitigate some of the limitations

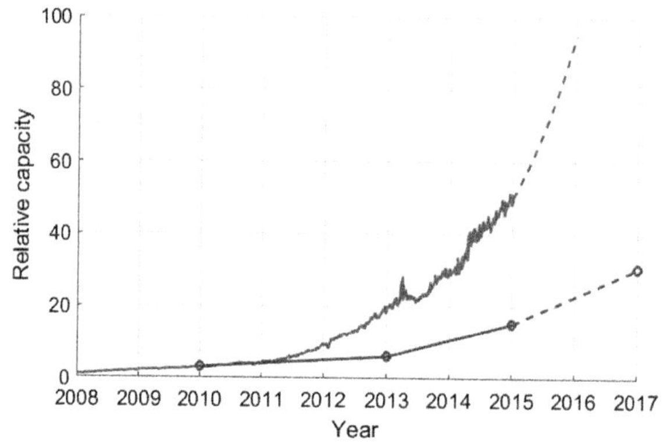

Figure 14.31 *Blue curve*: Increase in data center traffic as a function of time (adapted from Ref. [71]). *Red curve*: Several generations of electrical switch chips.

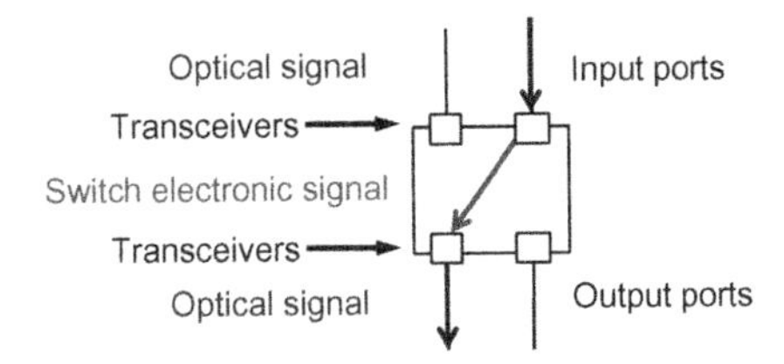

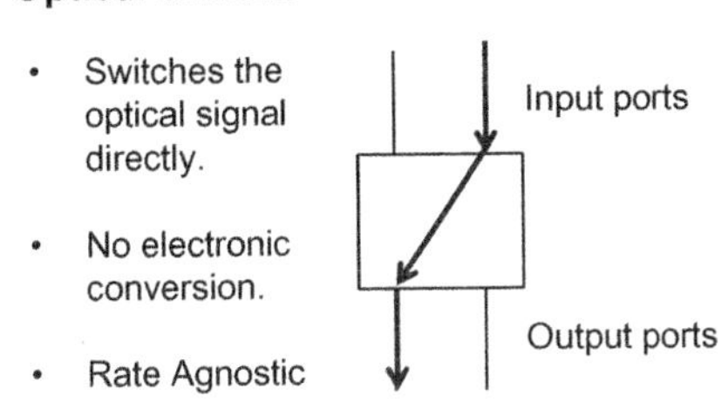

Figure 14.32 Comparison of electrical and optical switching.

of existing electronic packet-switch-based networks. This section discusses the potential and pitfalls of optical switching in future data centers.

The basic attributes of an optical switch are compared with an electronic packet switch in Fig. 14.32. In a standard DCN, the input and output ports of a core packet switch are populated by optical transceivers that operate at a fixed data rate. The packets to be routed are converted from the optical domain to the electrical domain in the transceivers and queued in an electronic buffer at an input port to the switch. The header of each packet is read and based on a set of rules such as a routing table stored in the switch, a local switch configuration is calculated that routes each input packet to an appropriate output port.

All practical optical switches work on a different principle because it is currently infeasible to store photons for any meaningful period of time. This means that all practical optical switches are circuit based with the optical switch simply creating light paths or circuits between input ports and output ports based on a schedule. This type of circuit switching does not require transceivers and is rate agonistic in the sense that the data rate is limited only by the bandwidth of the optical components used for the switch and not by the electronics in the transceivers and the switch chips as is the case for a packet switch.

The simplicity of this kind of data plane and the fact that the switch is rate agonistic are attractive features of optical circuit switching (OCS). However, the simplicity of the data plane is offset by a potentially more complicated control plane because the control of the switch is no longer based on local decisions derived from the information in the packet headers. Instead, the OCS must be centrally controlled, or the switch must operate in a deterministic fashion, which may lead to inefficient network performance. Practical optical circuit switching for a hyperscale data center must balance the simplicity of an optically-switched data plane with the potential complexity of a control plane that cannot use local decisions. Each attribute is considered separately in this section.

There are several complementary technologies that can be used for the data plane of an optical switch depending on the intended use of the switch. These technologies can be compared using the metrics of port count, switching speed, crosstalk, and loss. Several such switch technologies are compared in Table 14.11. Electro-optic Mach-Zehnder switches based on silicon photonics have the fastest switching speed among the switches considered but are currently limited in port count because of crosstalk and loss. MEMs-actuated-waveguide switches balance switching speed, port count, and loss. In these switches, a small MEMS-actuated adiabatic coupler is placed at each waveguide crossing. An N by N switch is an N^2 matrix of such waveguide crossings. In the off-state, the MEMs-actuated coupler is raised, and the waveguide crossing is in a through state. In the on-state, the coupler is lowered, and the waveguide crossing is in a cross state. This switch can scale to hundreds of ports and has low on-chip loss and excellent crosstalk [75,76].

A switch with a larger port count and lower overall fiber-to-fiber loss can be built using nonplaner free-space technologies based on piezoelectronic actuation or 3D beam steering MEMs [77]. These technologies have switching speeds measured in tens of milliseconds. Currently these switches are used to reconfigure the overall topology of some hyperscale DCNs to match anticipated workloads, but these switches cannot dynamically reconfigure the network on the time scale associated with an individual flow or an isolated packet.

Table 14.11 Comparison of several optical switch technologies.

Technology	Ports	Speed (μs)	Crosstalk (dB)	On-chip loss (dB)	Fiber-to-fiber loss (dB)	References
Electro-optic Mach-Zehnder	32	$\sim < 0.001$ (estimated)	-15	6.4	11 (average)	[72]
MEMS-actuated waveguide	64	1	-60	4	10 (estimated)	[73]
Piezoelectric	384	2.5×10^4	< -50	N/A	2.7 (maximum)	[74]
3D beam steering MEMS	1100	1×10^5	-60	N/A	4	[75]

The data plane is controlled by the control plane. Ideally the control-plane technology is matched to the data-plane technology. Large port-count, slow-speed switches such as piezoelectronic switches or 3D beam steering MEMs switches can be used as part of a software-defined network that reconfigures the network on long-duration time scales to match the time-averaged or anticipated data center workloads. This kind of reconfiguration is more aptly described as dynamic network provisioning instead of switching because the network does not dynamically respond to individual packets or flows. This network can be constructed using standard transceivers that can take milliseconds to lock onto the clock phase of the incident optical signal.

The control-plane technology required for a switch to dynamically respond to an individual flow or packet is more involved. These networks require burst-mode receivers [78] that can quickly lock onto the optical signal. A link consisting of both an OCS and a burst-mode receiver requires precise synchronization of the data path to the control plane to ensure that the data is sent when the switch is in the correct state. For this kind of network, an appropriate metric is the system-level switching time, which includes both the lock-time of the burst-mode receiver and the physical transition time of the switch. Switched links using electro-optic Mach-Zehnder switches at 12.5 and 20 Gb/s with measured system-level switching times of 90 and 60 ns, respectively, has been reported [79]. The resulting duty cycle of the circuit switch, which quantifies the throughput of the network, is the ratio $T_{\text{dwell}}/(T_{\text{dwell}} + T_{\text{switch}})$ where T_{dwell} is the dwell time the switch remains in one configuration and T_{switch} is the system-level switching time. Duty cycles in excess of 90% are desired for high-throughput networks.

The system-level switching time is just one aspect of the design of the control plane. Another important aspect is determining the sequence of switch states. To this point, the difference between a control plane that determines the sequence of switch states for a packet switch and a control plane for an OCS designed to route flows or packets is shown in Fig. 14.33. For the electronic packet switch shown in Fig. 14.33A,

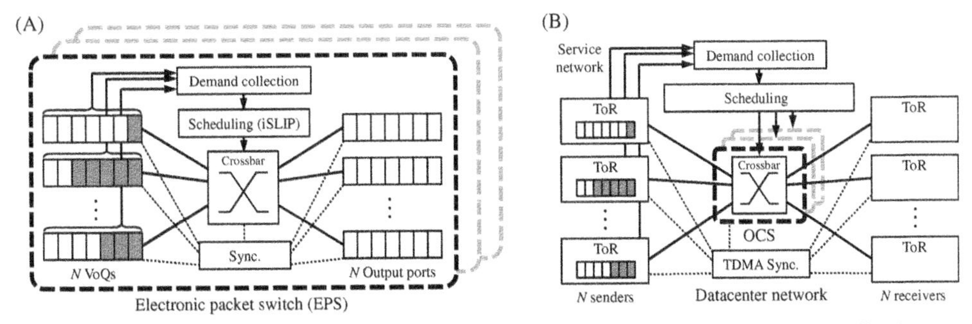

Figure 14.33 (A) The control plane of an electronic packet switch is within the switch itself. (B) A conventional control plane for an optical circuit switch (OCS) requires some form of centralized control.

the packets arrive at N input ports (left), where they are buffered in virtual output queues. A scheduling algorithm examines the packets in these queues and chooses the appropriate sequence of input-output port matchings needed to route packets to the N output ports (right).

OCSs that dynamically reconfigure the network topology in response to observed or predicted traffic must carry out the same tasks as an electronic packet switch. For an electronic packet switch, the routing decisions are made locally based on the information stored in the packet header and on locally stored rules such as a routing table. In contrast, a practical OCS-based topology cannot buffer and inspect packets.

Fig. 14.33B illustrates the distinction between these two types of control plane, depicting the networkwide demand collection to a central point required to schedule each OCS in the network [80]. Because the centralized scheduling of this kind of network does not scale, alternative network architectures have been investigated. As an example, RotorNet [81] combines a sequence of static network configurations with packet forwarding during the dwell time for each network configuration. This type of network has the potential to dynamically support both latency-sensitive and latency-insensitive network traffic.

In summary, optical switches are currently being used for dynamic bandwidth provisioning in DCNs on time scales of minutes or longer. Based on this trend, it is a reasonable conjecture that as the data plane and the control plane technology matures, a DCN based on OCS will be able to support both latency-sensitive and latency-insensitive network traffic, thereby alleviating future networking bottlenecks in hyperscale data centers.

References

[1] Cisco Global Cloud Index, <https://www.cisco.com/c/en/us/solutions/collateral/service-provider/global-cloud-index-gci/white-paper-c11-738085.html>, 2018 (accessed 16.10.18).

[2] K. Wu, J. Xiao, L. Ni, Rethinking the architecture design of data center networks, Front. Comput. Sci. 6 (2012) 596–603.

[3] K. Kant, Data center evolution: a tutorial on state of the art, issues, and challenges, Comput. Netw. 53 (17) (2009) 2939–2965.

[4] K. Chen, C. Hu, X. Zhang, K. Zheng, Y. Chen, A. Vasilakos, Survey on routing in data centers: insights and future directions, IEEE Netw. 25 (4) (2011) 6–10.

[5] M. Bari, R. Boutaba, R. Esteves, L. Granville, M. Podlesny, M. Rabbani, et al., Data center network virtualization: a survey, Commun. Surv. Tutor., IEEE 15 (2) (2013) 909–928.

[6] C. Kachris, I. Tomkos, A survey on optical interconnects for data centers, IEEE Commun. Surv. Tutor. 14 (4) (2012) 1021–1036.

[7] Y. Liu, J.K. Muppala, M. Veeraraghavan, D. Lin, M. Hamdi, Data center networks: topologies, architectures and fault-tolerance characteristics, Springer Briefs in Computer Science, Springer, New York, 2013.

[8] W. Xia, P. Zhao, Y. Wen, H. Xie, A survey on data center networking (DCN): infrastructure and operations, IEEE Commun. Surv. Tutor. 19 (1) (2017) 640–656.

[9] L. Popa, S. Ratnasamy, G. Iannaccone, A. Krishnamurthy, I. Stoica, A cost comparison of datacenter network architectures, in: Proceeding ACM CoNEXT, Philadelphia, PA, USA, 30 November–3 December, 2010.
[10] C.E. Leiserson, Fat-trees: universal networks for hardware-efficient supercomputing, IEEE Trans. Comput. 34 (10) (1985) 892–901.
[11] C. Clos, A study of non-blocking switching networks, Bell Syst. Tech. J. 32 (2) (1953) 406–424.
[12] A. Greenberg, J.R. Hamilton, N. Jain, S. Kandula, C. Kim, P. Lahiri, et al., VL2: a scalable and flexible data center network, in: Proceedings of the ACM SIGCOMM 2009, Barcelona, Spain, 16–21 August 2009, pp. 51–62.
[13] B. Heller, S. Seetharaman, P. Mahadevan, Y. Yiakoumis, P. Sharma, S. Banerjee, et al. Elastictree: saving energy in data center networks, in: USENIX NSDI, April 2010.
[14] H. Abu-Libdeh, P. Costa, A. Rowstron, G. O'Shea, A. Donnelly, Symbiotic routing in future data centers, ACM SIGCOMM Comput. Commun. Rev. 40 (4) (2010) 51–62.
[15] L. Gyarmati, T.A. Trinh, Scafida: a scale-free network inspired data center architecture, SIGCOMM Comput. Commun. Rev. 40 (5) (2010) 4–12.
[16] C. Guo, H. Wu, K. Tan, L. Shi, Y. Zhang, S. Lu, DCell: a scalable and fault-tolerant network structure for data centers, ACM SIGCOMM Comput. Commun. Rev. 38 (4) (2008) 75–86.
[17] D. Li, C. Guo, H. Wu, Ficonn: using backup port for server interconnection in data centers, in: IEEE INFOCOM, 2009, pp. 2276–2285.
[18] C. Guo, G. Lu, D. Li, H. Wu, X. Zhang, Y. Shi, et al., Bcube: a high performance, server-centric network architecture for modular data centers, ACM SIGCOMM Comput. Commun. Rev. 39 (4) (2009) 63–74.
[19] H. Wu, G. Lu, D. Li, C. Guo, Y. Zhang, MDCube: a high performance network structure for modular data center interconnection, in: Proceeding ACM CoNEXT, Rome, Italy, 01–04 December 2009, pp. 25–36.
[20] M. Al-Fares, A. Loukissas, A. Vahdat, A scalable, commodity data center network architecture, in: Proceedings of the ACM SIGCOMM 2008, 2008, pp. 63–74.
[21] R. Niranjan Mysore, A. Pamboris, N. Farrington, N. Huang, P. Miri, S. Radhakrishnan, et al., Portland: a scalable fault-tolerant layer 2 data center network fabric, ACM SIGCOMM Comput. Commun. Rev. 39 (4) (2009) 39–50.
[22] D. Thaler, C. Hopps, Multipath issues in unicast and multicast next-Hop selection, in: RFC 2991 (Informational), Internet Engineering Task Force, November 2000.
[23] X. Wu, X. Yang, Dard: distributed adaptive routing for datacenter networks, in: 2012 IEEE International Conference on Distributed Computing Systems, Macau, China, 18–21 June 2012.
[24] M. Al-Fares, S. Radhakrishnan, B. Raghavan, N. Huang, A. Vahdat, Hedera: dynamic flow scheduling for data center networks, in: Proceedings of the 7th USENIX conference on Networked systems design and implementation, 2010.
[25] P. Sreekumari, J. Jung, Transport protocols for data center networks: a survey of issues, solutions and challenges, Photonic Netw. Commun. 31 (1) (2016) 112–128.
[26] M. Alizadeh, A. Greenberg, D.A. Maltz, J. Padhye, P. Patel, B. Prabhakar, et al., Data center TCP (DCTCP), SIGCOMM Comput. Commun. Rev. 40 (4) (2010) 63–74.
[27] W. Chen, P. Cheng, F. Ren, R. Shu, C. Lin, Ease the queue oscillation: analysis and enhancement of DCTCP, in: 2013 IEEE International Conference on Distributed Computing Systems, Philadelphia, PA, USA, 8–11 July 2013, pp. 450–459.
[28] T. Das, K.M. Sivalingam, TCP improvements for data center networks, in: 2013 Fifth International Conference on Communication Systems and Networks, 7–10 January 2013, pp. 1–10.
[29] A. Andreyev, Introducing data center fabric, the next generation Facebook data center network, in: Facebook blogpost posted. <https://code.fb.com/production-engineering/introducing-data-center-fabric-the-next-generation-facebook-data-center-network/>, 2014 (accessed 10.16.18).
[30] K. Schmidtke, Designing 100G optical connections, in: Facebook blogpost. <https://code.facebook.com/posts/1633153936991442/designing-100g-optical-connections/>, 2017 (accessed 16.10.18).
[31] G. Keiser, Optical Fiber Communications, McGraw-Hill, Boston, MA, 2008, p. 578.

[32] A. Martin, SiP optical engines for 400G/800G COBO on-board optics and next generation plug-gables, in: ECOC'2016 Market Focus. <http://www.mellanox.com/related-docs/solutions/ECOC_Market_Focus-Arlon_Martin.pdf>, 2016 (accessed 16.10.18).

[33] SFP Multi-source agreement. <http://schelto.com/SFP/SFP%20MSA.pdf>, 2000 (accessed 16.10.18).

[34] SNIA SFF Specifications, <www.snia.org/technology-communities/sff/specifications>.

[35] The QSFP-DD Multi-Source Agreement, <www.qsfp-dd.com> (accessed 16.10.18).

[36] The OSFP Multi-Source Agreement, <www.osfpmsa.org> (accessed 16.10.18).

[37] A. Krishnamoorthy, K. Goossen, W. Jan, X. Zheng, R. Ho, G. Li, et al., Progress in low-power switched optical interconnects, IEEE J. Sel. Top. Quant. Electron. 17 (2) (2011) 357–376.

[38] Open Networking Integrators List, <www.iol.unh.edu/registry/opennetworking> (accessed 16.10.18).

[39] IEEE 802.3by, IEEE Standard for Ethernet, Amendment 2: Media Access Control Parameters, Physical Layers, and Management Parameters for 25 Gb/s Operation, Section 112, 2016.

[40] M.-J. Li, D.A. Nolan, Optical transmission fiber design evolution, J. Lightwave Technol. 26 (9) (2008) 1079–1092.

[41] IEEE 802.3bm, IEEE standard for Ethernet, Amendment 3: Physical Layer Specifications and Management Parameters for 40 Gb/s and 100 Gb/s Operation over Fiber Optic Cables, 2015.

[42] Y. Sun, R. Lingle, F. Chang, A.H. McCurdy, K. Balemarthy, R. Shubochkin, et al., SWDM PAM4 transmission from 850 to 1066 nm over NG-WBMMF using 100G PAM4 IC chipset with real-time DSP, J. Lightwave Technol. 35 (15) (2017) 3149–3158.

[43] IEEE 802.3, IEEE standard for Ethernet, section 52.9, 2012.

[44] IEEE 802.3ba, IEEE standard for Ethernet, Amendment 4:Media Access Control Parameters, Physical Layers, and Management Parameters for 40 Gb/s and 100 Gb/s Operation, 2010.

[45] The 100G CWDM Multi-Source Agreement, <www.cwdm4-msa.org> (accessed 16.10.18).

[46] The 100G CLR4 Alliance, <www.clr4-alliance.org>.

[47] CWDM4-OCP standard, <http://files.opencompute.org/oc/public.php?service = files&t = 4a3b7fe672003c1a2fe2f4b624bcc749> (accessed 16.10.18).

[48] The 100G PSM4 Multi-Source Agreement, <www.PSM4.org> (accessed 10.16.2018).

[49] OIF workshop – 100G Serial Electrical Interconnect needs, March 2017, Los Angeles, CA, USA, <http://www.oiforum.com/meetings-and-events/march-2017-workshop-100g-serial-electrical-interconnect-needs/>, 2017 (accessed 16.10.18).

[50] J. Lee, S. Shahramian, N. Kaneda, Y. Baeyens, J. Sinsky, L. Buhl, et al., Demonstration of 112-Gbit/s optical transmission using 56G Baud PAM-4 driver and clock-and-data recovery iCs, in: Proceeding 2015 European Conference on Optical Communications (ECOC'2015), Valencia, Spain, 27 September–1 October 2015, Paper Mo.4.5.4.

[51] M. Verplaetse, R. Lin, J. Van Kerrebrouck, O. Ozolins, T. De Keulenaer, X. Pang, et al., Real-time 100 Gb/s transmission using three-level electrical duobinary modulation for short-reach optical interconnects, J. Lightwave Technol. 35 (7) (2017) 1313–1319.

[52] J. Verbist, M. Verplaetse, S.A. Srivinasan, P. De Heyn, T. De Keulenaer, R. Pierco, et al., First real-time 100-Gb/s NRZ-OOK transmission over 2 km with a silicon photonic electro-absorption modulator, in: Proceeding 2017 Optical Networking and Communication Conference (OFC'2017), Los Angeles, CA, USA, 19–23 March 2017, paper PDP Th5C.4.

[53] J. Lavrencik, S. Varughese, J.S. Gustavsson, E. Haglund, A. Larsson, S.E. Ralph, Error-free 100 Gbps PAM-4 transmission over 100 m wideband fiber using 850 nm VCSELs, in: Proceeding 2017 European Conference on Optical Communications (ECOC'2017), Gothenburg, Sweden, 17–21 September 2017, paper M.1.A.

[54] P. Li, L. Yi, L. Xue, W. Hu, 100Gbps IM/DD transmission over 25 km SSMF using 20G-class DML and PIN enabled by machine learning, in: Proceeding 2018 Optical Networking and Communication Conference (OFC'2018), San Diego, CA, USA, 11–15 March 2018, paper W2A.46.

[55] IEEE 802.3bs, IEEE Standard for Ethernet, Amendment 10: Media Access Control Parameters, Physical Layers, and Management Parameters for 200 Gb/s and 400 Gb/s Operation, 2017.

[56] IEEE 802.3bm, 400 Gb/s over Multimode Fiber Task Force, <http://www.ieee802.org/3/cm/> (accessed 16.10.18).

[57] 100G lambda MSA, <http://100glambda.com/> (accessed 16.10.18).

[58] D. Sadot, G. Dorman, A. Gorshtein, E. Sonkin, O. Vidal, Single channel 112 Gbit/sec PAM4 at 56 Gbaud with digital signal processing for data centers applications, Opt. Express 23 (2) (2015) 991–997.

[59] The CFP Multi-Source Agreement, <www.cfp-msa.org> (accessed 16.10.2018).

[60] X. Zhou, H. Liu, R. Urata, Datacenter optics: requirements, technologies, and trends, Chin. Opt. Lett. 15 (5) (2017) 120008–120011.

[61] T.N. Nielsen, C. Doerr, L. Chen, D. Vermeulen, S. Azemati, G. McBrien, et al., Engineering silicon photonics solutions for metro DWDM, in: Proceeding 2014 Optical Networking and Communication Conference (OFC'2014), San Francisco, CA, USA, 9–13 March 2014, paper Th3J.1.

[62] 400G White Paper, Technology Options for 400G Implementation, in: OIF-Tech-Options-400G-01.0. <www.oiforum.com>, 2015 (accessed 16.10.18).

[63] F. Chang, S. Bhoja, New Paradigm Shift to PAM4 Signaling at 100/400G for Cloud Data Centers: A Performance Review, in: Proceeding 2017 European Conference on Optical Communications (ECOC'2017), Gothenburg, Sweden, 17–21 Septermber 2017, paper W.1.A.5.

[64] J. Cheng, C. Xie, M. Tang, S. Fu, Power consumption evaluation of ASIC for short-reach optical interconnects, in: Proceeding 2018 Opto-Electronics and Communications Conference (OECC'2018), Jeju, Korea, 2–6 July 2018, paper 6B2-3.

[65] M. Fields, J. Foley, R. Kaneshiro, L. McColloch, D. Meadowcroft, F. Miller, et al., Transceivers and optical engines for computer and datacenter interconnects, in: Proceeding 2010 Optical Fiber Communications Conference (OFC'2010), San Diego, US, 21–25 March 2010, paper OTuP1.

[66] M. Taubenblatt, Optical interconnects for high-performance computing., J. Lightwave Technol. 30 (4) (2012) 448–457.

[67] K. Schmidtke, F. Flens, Trends and future directions in optical interconnects for datacenter and computer applications, in: Proceeding 2010 Optical Fiber Communications Conference (OFC'2010), San Diego, US, 21–25 March 2010, paper OTuP4.

[68] End-to-End Communications with Advanced Fiber Optic Technologies, Whitepaper, <www.te.com/data-communications-coolbit-te-25g-active-optics-white-paper.pdf>, 2014 (accessed on 16.10.18).

[69] The Consortium for On-Board Optics, <onboardoptics.org> (accessed 16.10.18).

[70] A. Chakravarty, K. Schmidtke, V. Zeng, S. Giridharan, C. Deal, R. Niazmand, 100Gb/s CWDM4 optical interconnect at facebook data centers for bandwidth enhancement, in: Proceeding Frontiers in Optics 2017, Washington, D.C., US, 18–21 September 2017, paper JW4A.65.

[71] A. Singh, J. Ong, A. Agarwal, G. Anderson et al., Jupiter rising: a decade of Clos topologies and centralized control in Google's Datacenter Network, in: Proceedings of the ACM SIGCOMM Conference 2015, London, UK, 17–21 August 2015.

[72] K. Suzuki, R. Konoike, J. Hasegawa, S. Suda, H. Matsuura, K. Ikeda, et al., Low insertion loss and power efficient 32×32 silicon photonics switch with extremely-high-Δ PLC connector, in: Proceeding 2018 Optical Fiber Communication Conference (OFC'2018), San Diego, CA, USA, 11–15 March 2018, paper PDP Th4B.5.

[73] T.J. Seok, N. Quack, S. Han, R.S. Muller, M.C. Wu, Large-scale broadband digital silicon photonic switches with vertical adiabatic couplers, Optica 3 (1) (2016) 64–70.

[74] Polatis Series 7000n datasheet, <http://www.polatis.com/series-7000-384 × 384-port-software-controlled-optical-circuit-switch-sdn-enabled.asp> (accessed 16.10.18).

[75] J. Kim, C.J. Nuzman, B. Kumar, D.F. Lieuwen, et al., 1100×1100 Port MEMS-based optical crossconnect with 4-dB maximum loss, Photon. Technol. Lett. 15 (11) (2003).

[76] M.C. Wu, T.J. Seok, High-radix silicon photonic switches, in: Proceeding 2017 Conference on Lasers and Electro-Optics (CLEO'2017), San Jose, CA, US, 14–19 May 2017, paper SW1N.1.

[77] W. Mellette, J.E. Ford, Scaling limits of MEMS beam-steering switches for data center networks, J. Lightwave Technol. 33 (15) (2015) 3308–3318.

[78] A. Rylyakov, J.E. Proesel, S. Rylov, B.G. Lee, J.F. Bulzacchelli, A. Ardey, et al., A 25 Gb/s burst-mode receiver for low latency photonic switch networks, IEEE J. Solid-State Circuits 50 (12) (2015) 3120–3132.

[79] A. Forencich, V. Kamchevska, N. Dupuis, B.G. Lee, C. Baks, G. Papen, et al., System-level demonstration of a dynamically reconfigured burst-mode link using a nanosecond Si-photonic switch, in: Proceeding 2018 Optical Fiber Communication Conference (OFC'2018), San Diego, CA, USA, 11–15 March 2018, paper Th1G.4.

[80] H. Liu, F. Lu, A. Forencich, R. Kapoor, M. Tewari, G.M. Voelker, et al., Circuit switching under the radar with REACToR, in: USENIX Symposium on Networked Systems (NSDI) '14.

[81] W.M. Mellette, R. McGuinness, A. Roy, A. Forencich, G. Papen, A.C. Snoeren, et al., RotorNet: a scalable, low-complexity, optical datacenter network, in: SIGCOMM '17 Proceedings of the Conference of the ACM Special Interest Group on Data Communication, Los Angeles, CA, USA, 21–25 August, 2017, pp. 267–280.

CHAPTER 15

Innovations in DCI transport networks

Loukas Paraschis[1] and Kannan Raj[2]
[1]Systems Engineering, Cloud Transport, Infinera, Sunnyvale, CA, United States
[2]Infrastructure and Region Build, Oracle Cloud Infrastructure, San Diego, CA, United States

15.1 Introduction

The capacity of the networks interconnecting data centers, typically referred to as DCI for inter-Data-Center Interconnects, has grown more than any other transport network traffic type over the last 5–10 years [1], has by now come to dominate the global fiber infrastructure (Fig. 15.1) [2], and has been projected to grow by at least two more orders of magnitude [3].

The main reason behind the growth of DCI is the proliferation of data centers (DCs) of massive compute and storage resources [4], and the subsequent, intimately related, proliferation of the "cloud"-based service delivery models [5]. In many ways, the explosive growth of cloud services since 2007[1] has been the biggest evolution in networking since the proliferation of the internet in the 1990s. In this respect, the growth of DCI is correspondingly the most significant evolution in transport networks since the transitions from time-domain multiplexing (TDM) to Internet Protocol and Multi-Protocol Label Switching (IP/MPLS (Multiprotocol Label Switching)) and wavelength-division multiplexing (WDM) which was reviewed in Optical Fiber Telecommunications VI [6].

Although it's not the scope of this OFT chapter to elaborate on the (many interesting) details motivating the proliferation of DCs and cloud services, it is useful to note here that the DCI definition typically encompasses both "public cloud" and "private cloud" networks, since both can be served by the same transport architecture.[2] In this sense, DCI includes interconnections between DCs belonging to the same operator (typically referred to as private cloud), or between different operators (hybrid cloud), or between a "cloud" DC and an Internet points-of-presence (PoP) that

[1] August 3, 2006, when Amazon first announced its EC2 service offering, can serve as a good "birthday" of "cloud" services.

[2] This is true for optical transport, but not always true for packet transport in which the routing table size of a private network can be orders of magnitude smaller than the routing table size of most public networks.

Optical Fiber Telecommunications V11
DOI: https://doi.org/10.1016/B978-0-12-816502-7.00017-8

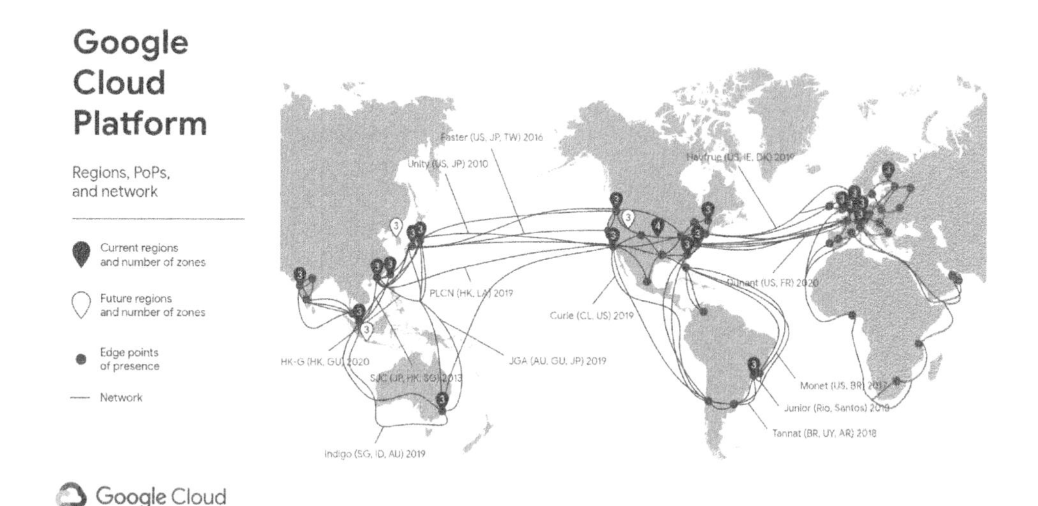

Figure 15.1 The global transport infrastructure of a major cloud and internet content provider (ICP) network operator that includes hundreds of regions[3] and points-of-presences (PoPs), and thousands of miles of fiber. *With permission from U. Holzle, A ubiquitous cloud requires a transparent network, in: OFC 2017 (Plenary), and V. Vusirikala, SDN Enabled Programmable, Dynamic Optical Layer, ECOC 2017, Plenary.*

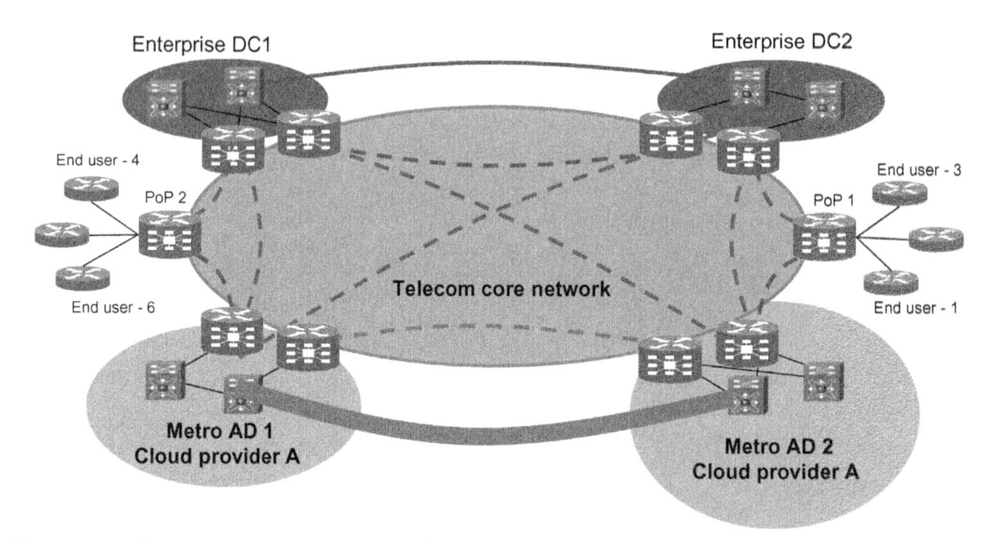

Figure 15.2 Data-center interconnect (DCI) main use cases: private, public, and hybrid cloud connectivity [7].

[3] https://cloud.google.com/compute/docs/regions-zones/.

connects this DC to end users (public cloud). Fig. 15.2 illustrates these three main DCI use cases, which are analyzed further in Section 15.2. It is important to highlight here that the capacity requirements of intra-cloud applications (e.g., MapReduce Hadoop clusters) constitute the overwhelming majority of DCI traffic; for example, Ref. [8] reported that each bit of Facebook end user traffic generates 900 bits of DCI traffic among its DCs. Serving this intra-cloud, also known as "east-west," traffic was arguably the main motivation for the proliferation of the new intra-DC[4] "leaf-spine" architectures [9], and the proliferation of inter-DCI networks that is the focus of this chapter. This DCI private wide area network (WAN) infrastructure not only has become crucial to the operation of cloud and Internet services, but for the largest cloud and internet operators, it has also grown to massive global scale, as shown in Fig. 15.1. Therefore economics have necessitated the building of dedicated DCI networks and of a new class of purpose-built systems that are optimized for the unique DCI requirements [2,3,9–12].

While DCI has a few commonalities with traditional telecommunication transport, for example most of the current WDM technologies employed in DCI have been practically the same as the technologies employed in telecom networks, the DCI transport networks have a substantial amount of unique characteristics (both architectural and operational) to have motivated the development of a new class of DCI-optimized packet and optical transport systems [2,3,9–12]. More specifically, DCI-optimized transport systems have been developed to address the DC operational environment, the simpler DCI routing requirements that focus on maximizing throughput rather than routing scale, and the desire for high-capacity WDM systems employing state-of-the-art coherent transmission, with the added requirements for lower power per Gb/s. Moreover, DCI transport has also pioneered the extensive adoption of significant software innovations in network programmability, automation, management abstraction, and control-plane disaggregation, typically referred collectively as software-defined networking (SDN). This new DCI-optimized SDN-enabled infrastructure is being increasingly deployed globally and has proliferated important network innovations in "open" multi-terabit transport architectures with spectrally efficient WDM systems that can exceed 6 b/s/Hz [13].

This chapter reviews the main innovations in technology, system, and network architectures that have facilitated the explosive DCI evolution, from its early "humble" days of low-speed connectivity (initially Gb/s or less, often TDM-based) to the current multi-Tb/s systems that account for some of the most spectrally efficient fiber networks. We focus our discussion on optical DCI, which practically always leverage

[4] There is a separate chapter in this book that focuses on intra-DC networking.

WDM. Packet DCI transport, which is also based mostly on dedicated DCI routers, is considered only to the extent that it has significance to optical DCI. Section 15.2 reviews the main attributes that make DCI networks and transport systems unique, and Section 15.3 summarizes the main innovations of the current, first-generation DCI-optimized transport systems. Then, Section 15.4 highlights the most important emerging DCI transport innovations, including the emphasis in Metro DCI, as well as innovations from advancements in SDN, and from coherent WDM and photonic integrated circuit (PIC). Finally, Section 15.5 attempts an outlook of the most important DCI topics currently under investigation in R&D, and the related trade-offs. For example, we review the potential of 400GE coherent WDM pluggables in DCI routers that scale to 12.8 Tb/s and beyond. We also discuss the system and control-plane disaggregation in open (multiple-vendor) network architectures that at the same time can minimize the operational overhead, and operate as close as possible to the fiber Shannon limit, and we summarize the debate on the potential value from more traffic engineering, as for example from optical restoration in addition to IP/MPLS protection, particularly for DCI transport.

15.2 Data-center interconnect transport networks

The growth of DCI transport, as mentioned in the introduction, has been a direct consequence of the proliferation of massively scalable DCs [4] and "cloud" services [5]. The main driver behind this DC proliferation has been the substantial economic benefit arising from centralizing compute and storage, and employing "virtual machines" to maximize the utilization of the available resources for most software applications.[5] The economic benefit of such centralization is directly proportional to scale, at least until failure domain considerations start diminishing returns [3,4]. Therefore building, running, and maintaining such massively scalable "cloud" DCs is a large investment in capital expenditure (CAPEX) that typically accounts for $50 M to $80 M, and can even exceed $200 M per DC in some of the largest scale deployments, as well as an equally high yearly operational expenditure (OPEX) that can reach tens of millions of dollars, especially when it includes leasing facilities and "24 × 7" staffing. It is difficult to generalize on the scale and characteristics of such DCs, given the still new, quite diverse, and very dynamic nature of the underlying businesses; for example, internet content providers (ICPs), like Netflix or Facebook, have occasionally different

[5] Typical "cloud" services include virtualized elastic compute and object store, with workloads for relational database (structured query language), distributed database (NoSQL), map-reduce (e.g., Hadoop), and big-data analytics including more recently artificial intelligence and machine learning.

needs from cloud providers, like Amazon, Oracle, or Microsoft, and often the same operator is both an ICP and a cloud provider (e.g., YouTube and Google Cloud). Most cloud providers develop DCs with power budgets in the range of a few MWs to tens of MW. Some of the large cloud and ICP providers, however, have started developing even larger scale DCs with power budgets of the order of 100 MW, which allows for more than 2 million servers to be deployed in a single DC. It is important to note here that the DC power budget has increasingly become the hardest constraint of the cloud and ICP DCs and in this sense a very important performance characteristic of the systems employed by such DCs, including the DCI transport systems. We discuss more the implications of power-constrained DC performance in next sections. First we focus next on the networking aspects of DCI transport.

There are several considerations in building data centers that support online transactions and handle and store "cloud" data. The most important requirements are the following:

- Scalable networking infrastructure that aggregates effectively high-performance compute and storage [4,8–10]. Particularly high throughput intra-DC networking is paramount in effectively interconnecting the compute and storage resources.
- High availability, typically to ensure five nines (99.999%) availability [14–16]. Moreover, in cloud and ICP DCs the goal is typically to build a highly available infrastructure using unreliable parts, employing automatic failover [15]. In this sense, high availability leads to the requirement of failure resiliency—with no disruption to service even in the presence of hardware failures (such as power outage, disk failures, memory corruption, network switch/router or optical equipment failures) or software failure (such as software programming errors), misconfiguration, change management upgrades/patches, or most importantly operator errors.
- Automation is also very important in order to enable scale, modular functionality for services, run time operations, and monitoring. Note here that automation has been a key focus area of "cloud" operators because their technical innovations in automation have catalyzed a lot of the economic benefits of centralizing compute and storage. Such automation frameworks initially employed by compute were then applied to networking inside the DCs, and have recently been extended to DCI transport, starting with routers, and more recently becoming a requirement even for optical systems, with important implications in optical transport (well beyond DCI, as we elaborate later).
- Security, and particularly encryption, is also paramount because cloud and ICP services are inherently less "safe," hence the need to protect data, whether it is in motion or at rest, and the need for end-to-end encryption, including all links in the network. Per the NIST Zero-Trust Model for Cybersecurity, all network traffic is untrusted, whether it comes from within the (enterprise) network or from outside the network. At the heart of secure networks are three main requirements

of encryption: (1) confidentiality of information; that is, the need to prevent anyone from accessing and reading any content of a message; (2) integrity of information; that is, messages when modified are detected by checksum; (3) authenticity of information; that is, verify that the sender is connected to the intended receiver.

Building a data center that can optimize simultaneously for scalability, availability, and modularity is a nontrivial network engineering challenge. Having a single large data center facility offers cost and efficiency gains. On the other hand, it is very difficult to avoid outages as the size of a single fault-domain grows; that is, the group of infrastructure with correlated failures. Outages of services happen for multiple reasons, including hardware/power/cooling failures, software bugs, operator errors, application overload degradation when handling large-scale events (such as with a database migration or with the addition of compute/storage resources), and failure to limit "blast radius" of any of these problems. To ensure high availability and failure resiliency, cloud operators have thus employed at least one other redundant data center, with the user unaware of where the data originates from (Refs. [3,14,16]). For the large cloud providers, having 2+1 redundancy has actually been even more efficient than 1+1 redundancy [3,14], and often redundancy extends among more than three DCs. These (three or more) DCs constitute a region with multiple availability domains[6] (ADs), as shown in Fig. 15.3. AD is the largest group of infrastructure that may fail as a whole unit (e.g., failures due to power/cooling; floods; fire, etc.). Therefore ADs are located far enough apart within each region to minimize correlated failures (e.g., an

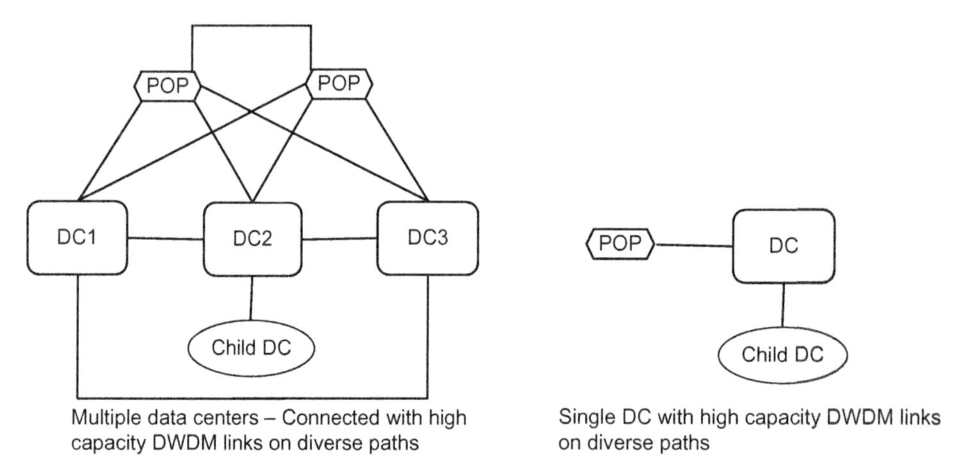

Figure 15.3 A typical high-availability network interconnection architecture of a "cloud" region is shown, comprised of three data centers, each constituting an availability domain.

[6] Also referred to as availability zones or availability regions by Amazon and others.

earthquake) and are interconnected with a high-capacity WDM network, with additional path-level diversity for each WDM link. Basic hardware redundancy for fault tolerance is also introduced, ensuring there is no single point of failure. During a failure in one AD, all application services in the region remain uninterrupted by switching over to one of the other ADs. Within a region, data is replicated across each AD, and those replicas need to be synchronized and consistent, to allow for uninterrupted service against failures.

In view of the above, synchronous replication across location-diverse DCs within the same geographic region is one of the most important functions of designing a DCI network of ADs. Synchronous replication requires that total round-trip latency at the application layer typically shall not exceed few milliseconds. Latency arises from several functions, most notably including switching delay, serialization delay, smartNIC delay, buffering delay, time of flight in optical fiber delay, and application read/write times. From a Hardware (HW) perspective, the networking application-specific integrated circuits (ASICs) in the path (dealing with switching, buffering) contribute a substantial amount of latency. After accounting for all such contributions to the total application latency, the allowed latency budget specifically for optical fiber transmission shall not exceed few hundreds of microseconds, which essentially requires that the data centers within an AD region be physically less than 100 km apart. This becomes a very important requirement for DCI transport architectures. In this sense, the AD synchronous replication latency constraint requirement has defined and proliferated the "Metro" DCI use-case with typical distances of up to 100 km. These Metro DCI networks, as depicted in Fig. 15.3, are simple point-to-point network graphs, with only a limited number of nodes (usually three to five). Even when such Metro DCI networks are deployed over a physical ring fiber infrastructure (as is often the case for fiber leased from traditional non-DCI operators), Metro DCIs operate as logical point-to-point Ethernet links and are therefore operationally very different from the typical multiservice metro networks of traditional telecom providers [17].

In addition to the Metro DCI, dedicated long-haul (LH) and subsea DCI transport networks have also been increasingly deployed globally (Fig. 15.1) [2,10–12,18], aiming to maximize the connectivity and optimize the cost-performance of the cloud and ICP compute and storage infrastructure. While the LH and subsea transport in DCI networks has a lot of commonalities with those in traditional telecom, for example, both employ practically the same WDM technologies, DCI still has a few unique attributes, particularly in terms of the DC-based operational environment [2,10,14,18]. Specifically, in LH DCI networks, the scarcity of fiber has typically constrainted the operations of most cloud and ICP networks (unlike traditional telecom operators) and has motivated deployment of WDM systems with maximized as much as cost-effectively possible fiber spectral efficiency [11,19]. For example, LH DCI transport networks have been the earliest adopters of coherent WDM technologies

[9–11,19], accounting for the majority of initial deployments for every new generation of coherent WDM.

The requirement for high spectral efficiency WDM networks has also motivated the DCI architectures, in the majority of cases, to maintain a distinct separation between the packet routing and the WDM layer, in order to enable each of these transport network layers to maximize its respective performance. Note that in the early days of DCI transport, there has been a substantial amount of analysis about the potential benefits of a converged DCI transport infrastructure [19]. The conclusions of this analysis has led most cloud and ICP operators to build distinct IP/MPLS packet transport networks,[7] separate from their underlying high-capacity WDM networks.

The explosive capacity growth and eventually dominant role of DCI in the global networking infrastructure has also motivated an evolution of the overall internet transport toward flatter and more densely connected wireline architectures [6]. As a result, the global IP traffic patterns have evolved from "any-to-any" to "hub-and-spoke" between a few "cloud" DCs and end users. Also, hyperscale has resulted in the majority of traffic to be between data centers than between end user and the "cloud." The implications of this evolution have so far been more profound for the IP/MPLS packet transport. For example, advanced traffic-engineering techniques have been developed to help scale the IP/MPLS network, and were often combined with innovations in centralized SDN control plane [20,21]. The recent advancements in new packet traffic engineering technologies, like SDN controllers, programmability (e.g., path computation element protocol (PCEP)), and the renewed interest in source routing (i.e., segment routing) have been primarily motivated by the DCI evolution [7]. Extending a few of these innovations also to the WDM layer is now being considered by the major ICP and cloud network operators (as we discuss further in Sections 15.4 and 15.5).

In addition to these innovations which are common in most DCI networks, when we take a closer look each DCI network would still have few differentiations arising from the specific application and customization needs of each operator. The most notable distinction arises from the nature of the services offered. For example, cloud operators (like Amazon, Microsoft, Oracle, or IBM) care most for scale of their data center fabric across their ADs, while ICPs (like Facebook or Netflix) are primarily interested in high-performance delivery of stored content, and if we wish to also include here DC hosting operators (like Equinix or Rackspace) they care a lot to maintain service flexibility and network granularity. These distinctions, however, become

[7] There are typically two separate packet transport networks in cloud and ICP operators [2,11,19,20]. The private WAN packet transport network with simpler routing characteristics aims for the highest available packet throughput to serve the very high-capacity inter-DC traffic, while the public packet transport WAN, with higher scale RIBs, enables the connectivity of data centers to the internet.

increasingly less important, as many large "cloud" operators employ both private and public DCI networks, for example, Google Search and Google Cloud. In this sense, notwithstanding distinct service and operational characteristics, most DCI transport networks have enough important common requirements to motivate the development of a new class of DCI-optimized infrastructure. We identify the following main common attributes of DCI networks:

- Scale with high availability has been the most important goal of every cloud networking architecture, including DCI transport networks [2–4,9–11,14–16]. This is based on (A) the desire to leverage the significant economic benefit of massively scalable centralized compute/storage [3,4] and (B) the fact that at every large-scale deployment, failure will always occur inevitably, often more than once a day [15].
- High capacity, which when combined with the scarcity of fiber (which has typically characterized the operations of most cloud and ICP operators), has motivated the deployment of commercial WDM networks that can scale to very high fiber spectral efficiency [10–13,19].
- DC operational environment. For example, unlike traditional telecom transport networks, network equipment-building system (NEBS) environmental requirements and direct-current power provisioning have not been hard requirements for many DCI networks [12]. Another even more important example is the power-constrained nature of most cloud and ICP DC operations, which makes power efficiency, measured in bits/sec/Watt (i.e., bits/Joule), an increasingly important performance metric, by now as important as packet throughput (in routing), or fiber spectral efficiency (in WDM transport) [2,12].
- Software automation. The extensive use of software as part of the DC operational environment, and particularly the use of "DevOps" and "automation," has been applied also to DCI transport networks, motivating DCI to pioneer the extensive adoption of significant software innovations in network programmability, automation, management abstraction, and control plane disaggregation [7,22–24]. These software and SDN transport innovations were often less appreciated initially, but they are a very important characteristic of DCI transport. For example, unlike the initial use of SDN inside DCs for service overlays, SDN in DCI transport has mostly aimed to advanced system automation and abstraction, and occasionally traffic engineering optimization [7,20–24]. To this end, new innovative wireline transport automation and abstraction frameworks have been developed by DCI network operators [24], with Openconfig [23] being the most notable such example. Most of these innovations were first introduced in packet transport. Soon, however, they became important also in optical DCI [23], defining a radical new operational and management paradigm for WDM systems and networks. We discuss further the current adoption and future promise of these transport SDN innovations in the next sections.

- Reduced transport system complexity has been another important DCI requirement. Because DCI networks are typically much simpler, point-to-point links among few nodes, with Ethernet only service requirements, there is hardly any benefit from multiservice bandwidth aggregation, and optical transport network (OTN) switching. Therefore DCI transport has sought to minimize as much as possible transport system complexity [11,12,19]. Moreover, system simplicity also becomes an important enabler for network disaggregation toward more "open" transport architectures, which has also become an important innovation focus for DCI networks, discussed in detail in the next section.

15.3 Data-center interconnect optimized system

The early deployments of DCI networks had employed transport systems that were originally developed for traditional telecom networks and internet transport. This was natural because, initially, DCI networks did not distinguish between internet and inter-DC traffic, as shown in Fig. 15.4 (left). The explosive growth of DCI traffic, and the increasingly unique requirements of the DCI transport networks (summarized in Section 15.2) eventually led to the development of dedicated DCI networks and a new class of purpose-built DCI-optimized transport systems, as shown in Fig. 15.4 (right), which recently have increasingly been employed in the DCI global infrastructure (Fig. 15.1). This section reviews the unique characteristics of this first-generation DCI-optimized systems and their primary differences from the traditional telecom transport systems [10–12].

15.3.1 Requirements and innovations in data-center interconnect systems

The most distinguishing initial characteristic of DCI-optimized transport systems has been their small footprint, typically 1–2 RU, that scale to high capacity through "stacking" of multiple such units [10–12], as shown in Fig. 15.4. Small DCI transport systems have become possible because of the simpler DCI system requirements. As we summarized in the previous section, DCI applications employ mainly Ethernet (with the occasional addition of Fiber Channel for storage applications), without the need for multiservice granular aggregation and capacity management based on an OTN switching layer used in telecom networks. This allows DCI systems to require a minimal switching fabric, just enough to map the "client" Ethernet ports to the "line-facing" WDM ports, which currently operate at 100–200 Gb/s in the majority of deployed DCI networks. Given that practically all deployed DCI systems employ currently 10, 40, and 100 GE client ports, their mapping to the WDM "line-side" ports have been quite straightforward. We discuss more details of the client and line-side characteristics, along with the optical system building blocks, and

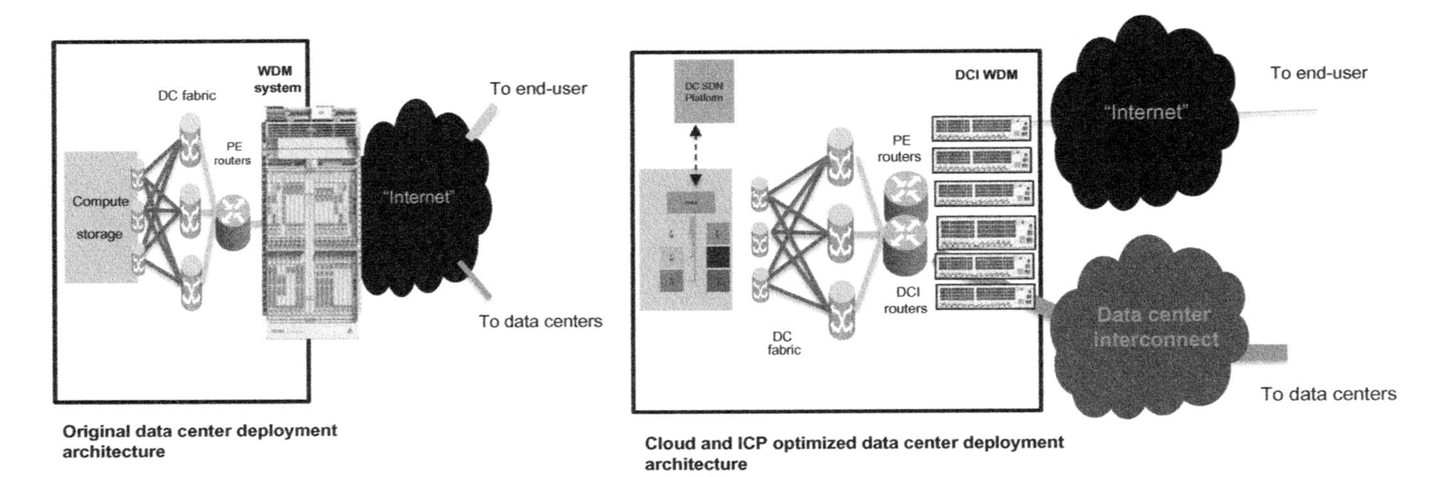

Figure 15.4 Evolution to dedicated data-center interconnect (DCI) networks and DCI-optimized transport systems [12].

the related fiber transmission innovations in Section 15.3.2, after we first complete the review of the main distinguishing characteristic of DCI-optimized transport systems. Another noteworthy characteristic of DCI systems has been the relaxed environmental operational requirements (known in the industry jargon as NEBS compliance). Because DCI systems are deployed in the controlled DC environment, with significant airflow (facilitated by chillers and forced-air cooling), operating case temperatures typically do not exceed 50°C, which also help simplify the DCI system design [11,12].

The unique operational characteristics of the hyperscale "cloud" DCs have also given rise to novel software requirements and innovation opportunities for DCI. Cloud and ICP operators pioneered the pervasive use of "dev-ops" and software "automation" techniques, initially to serve their hyperscale compute infrastructure needs [4]. These techniques were introduced soon also to networking infrastructure inside the DCs, primarily leveraging programmability and separation of the control plane from the data plane (often referred to as SDN) to create logical application overlays and abstractions that were decoupled from the underlying physical infrastructure and network topology. Incorporating equivalent software innovations to advance the functionality of DCI networks has motivated the DCI transport systems to increasingly introduce advancements in network programmability, automation, management abstraction, and eventually also control plane disaggregation, typically referred collectively as SDN transport [7,22–24]. Notable initial such example has been the use in DCI network elements of (A) open operating systems, typically Linux based, and (B) extensive application program interface (API) programmability frameworks, based primarily on YANG data models and the related Netconf, Restconf, or gRPC protocols [23,24]. Moreover, new wireline transport automation and abstraction frameworks have been also developed by DCI network operators [7,22–24]. Openconfig has been the most widely adopted such framework [23], being currently supported by all major DCI transport system vendors (both routing and optical). The underlying motivation of these open software frameworks has been primarily the evolution toward a new operational paradigm that leverages extensive programmability of the DCI network elements using nonproprietary data models and APIs [7,22–24]. The focus on such network element innovations has been a very important distinguishing characteristic of DCI systems. While many of these "SDN" innovations were initially introduced in DCI packet transport, their more recent adoption in DCI optical systems [11,14,22–24] has been an even more radical innovation because traditional optical network management [25] has previously been based on proprietary (vendor-specific) network management system (NMS).[8] From a

[8] Refer also to Chapter 21, Software-defined networking and network management, that focuses entirely on optical transport SDN and NMS.

DCI system perspective, the initial goal of these software innovations has been to improve operational efficiency, and thus reduce OPEX. At the same time, by allowing open and vendor-agnostic network management these innovations can also improve CAPEX by becoming the first important step toward open line-system (OLS) architectures, which are discussed in detail in Section 15.3.3.

As important as modularity and SDN innovations are, the most important consideration for the first-generation DCI systems has been the ability to cost-effectively scale the network capacity [9–11,19]. The modular DCI system architecture is, of course, advantageous for allowing DCI networks to start with small building blocks, but in addition it requires achieving significant scaling. The graceful per-channel scaling of WDM capacity enables a modular optical transport system unit to be easily defined by incorporating a small number of WDM ports. The required line system WDM technology (Fig. 15.6), particularly optical amplification, but even multiplexing, can be easily designed into separate system building blocks (separate 1–2 RU network element) and leverage control plane among all separate network elements to coordinate functionality needed for the end-to-end system provisioning and optimization. This separation of a DCI WDM system into physically distinct network elements, combined with some of the SDN innovations discussed in the previous paragraph, has also been the first practical step toward the optical transport disaggregation, which has now become an increasingly popular architectural requirement in transport networks worldwide, not just in DCI. We review DCI disaggregation along with the OLS innovations in Section 15.3.3, after summarizing the main building blocks and current innovations of WDM system performance.

The advancements in coherent WDM transmission, whose successful commercialization [26] serendipitously preceded by a couple of years the rise of DCI networks, has enabled significantly increased WDM channel capacity [26–28]. Coherent WDM deployments currently can scale from 100 Gb/s channels employing quadrature phase-shift keying (QPSK) to 400 Gb/s channels employing 32QAM, and will soon extend to 600 Gb/s channels employing 64QAM. Leveraging such coherent WDM implementations, commercial DCI transport systems can currently achieve more than a Tb/s per RU (Fig. 15.5), in modular systems architectures, that moreover has continuously improved power consumption that is already well below a watt per Gb/s [12]. Initial DCI system capacity level of the order of Tb/s has been good enough for most DCI applications. These systems can, of course, scale linearly by adding WDM channels to installed capacities and, for most distances, can exceed 26 Tb/s per fiber [13]. More specifically, combining innovations in coherent digital signal processing (DSP) modulation, including sharp rolloff (1% root-raised cosine filters) of subcarrier modulated channels, with digitally generated control tones, and photonic-integration technology, the resulting state-of-the-art coherent WDM transmission systems, leveraging Nyquist filtering superchannels, can with 16QAM scale to fiber capacities exceeding 27 Tb/s

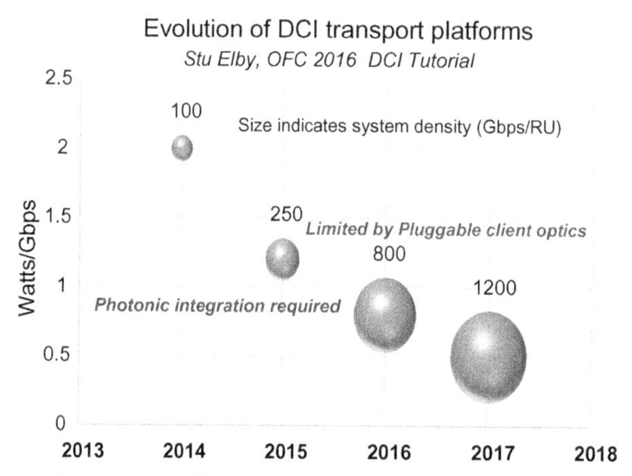

Figure 15.5 The increasing power efficiency and density of data-center interconnect (DCI)-optimized transport systems [12].

in the extended C-band (4.82 THz bandwidth), and even in subsea networks exceed 6 b/s/Hz [13]. The same modular 1–2 RU WDM systems have also been employed in the LH, and even subsea transport. Here, as mentioned earlier, the WDM technologies employed are practically identical to the WDM technologies employed in other (non-DCI) LH and subsea networks, including state-of-the-art coherent and subcarrier technology, Raman amplification, channel equalization, and noise-loading in the longer reach subsea links. The most distinguishing WDM transmission characteristic of this first-generation DCI systems in LH and subsea DCI networks is arguably their achievement of some of the highest spectral efficiency fiber network deployments globally; for example, [29]. Recently, within the last 1–2 years, the significant proliferation of Metro DCI cloud transport requiring no more than 80–100 km link distances (due to the synchronous replication latency requirements discussed in Section 15.2) has increased the desire for additional optimization in Metro DCI transport. We review more of this metro-optimized DCI transport development in Section 15.4, along with other important emerging DCI innovations. But first we consider it valuable to summarize next the most important optical technology building blocks that enable a typical DCI WDM system. (Note that readers who are already familiar with the WDM technology employed in DCI may wish to skip parts of Section 15.3.2 and proceed directly to the discussion of OLS and open transport in Section 15.3.3).

15.3.2 Wavelength-division multiplexing technology building blocks

In this subsection we review the building blocks that make up the DCI WDM system and recent related developments. Readers who are already familiar with the WDM technology employed in DCI may wish to skip parts of Section 15.3.2. We particularly discuss the technology details associated with the WDM transponder and the progression to WDM system disaggregation. Fig. 15.6 summarizes a typical WDM link, wherein all building blocks function as a single system.

A transponder[9] aggregates the input from the client ports, such as 100G QSFP28 (or in the future 400G QSFP-DD/OSFP ports), to line-side WDM ports connected by fiber spans. Fig. 15.7 shows the main elements that make up such a high-speed WDM digital link. The transponder employs a tunable laser that can cover the extended C-band and feeds into a WDM multiplexer. In a typical WDM system, the WDM multiplexer will most often include a built-in erbium-doped fiber amplifiers (EDFA) to enable span reaches of about 80 km. Using the most basic example of an 80 km fiber link, we illustrate the concept of the typical link budget engineering required to close the link between the transmitter and the receiver: In this very simple model, assuming fiber loss of 0.25 db/km, the link loss is 20 dB. Adding a mux/demux loss on each side of 6 dB, an aging/repair margin of 1–2 dB, and a patch-panel loss of 1–2 dB leads to a total of 35 dB link budget between transmit and

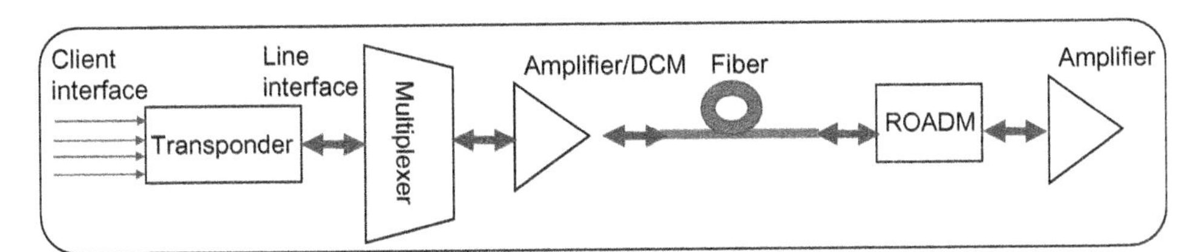

Figure 15.6 The basic system building blocks of a data-center interconnect (DCI) optical link.

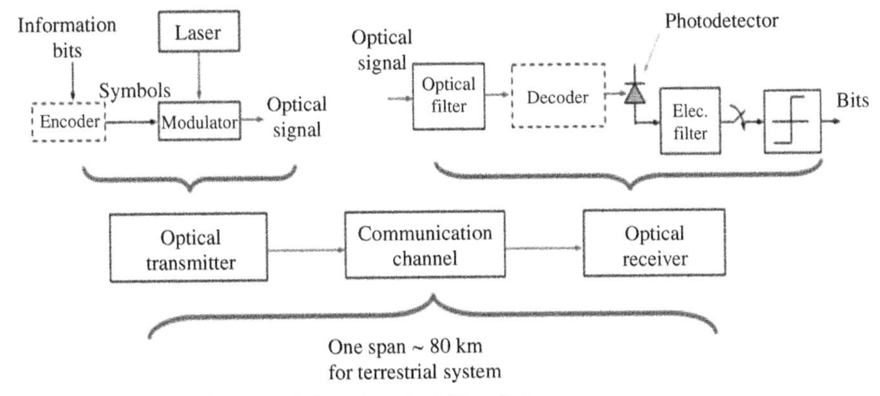

Figure 15.7 Schematic of a typical digital optical fiber link.

[9] Our analysis does not distinguish between a transponder and a muxponder and uses the term transponder for both of these functions.

receive ends. The range of transmit powers required depends on whether amplification (with EDFA or an semiconductor optical amplifiers) is used, the number of channels getting multiplexed, eye safety limit requirements (typically Class 1 M). On the receiver side, the receiver sensitivity required is dependent on the baud rate and modulation format. Some of these link budget considerations are captured in Ref. [30]. This is, of course, an overly simplified analysis, as many additional factors may often need to be accounted for. Particularly in longer reach WDM systems, the complete analysis becomes much more complex, as we also need to account for amplified spontaneous emission (ASE) noise, nonlinear interactions and dispersion penalties, as well as the additional transmitter and receiver technology details, such as the forward-error correction (FEC) or DSP.

The optical technologies used for client-side and line-side interfaces can vary considerably (Fig. 15.8), as optical transmission systems can manipulate independently (or in combination) the amplitude, phase, polarization, or frequency of each channel. The initial optical fiber communication systems used intensity modulation-direct detection (IM-DD) technology (Fig. 15.9 left), with NRZ format, also known as binary amplitude modulation. IM-DD technology deployment has progressed well from a few Mb/s to 100 Gb/s and is currently scaling to 400 Gb/s with PAM4 modulation. A laser source (such as in VCSELs) is directly modulated, or intensity modulated such as with Mach-Zehnder modulators (MZM), and is directly detected by a photodetector. IM-DD technology today remains the basis of most client-side optical interfaces (pluggable transceivers or active optical cables, as described further in Chapter 16, Intra-data center interconnects, networking and architectures, which focuses on intra-DC optical interconnects). However, IM-DD technology cannot overcome cost-effectively the fiber impairments limitations of high-capacity WDM transmission, more specifically signal-to-noise degradation, chromatic and polarization mode dispersion, self-phase modulation, cross-phase modulation, and associated requirements for launched optical power, eye safety, eye mask, and extinction ratio. Therefore IM-DD technology is not employed anymore in optical systems requiring multi-Tb/s capacity. Instead, coherent dense wavelength-division multiplexing (DWDM) technology with higher

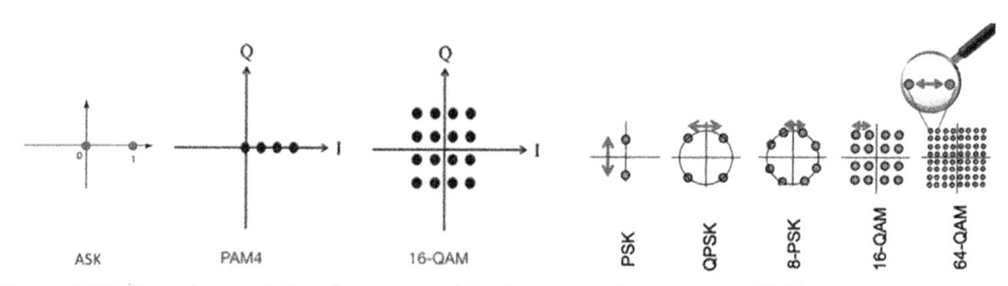

Figure 15.8 Typical modulation formats used in data-center interconnect (DCI) transport.

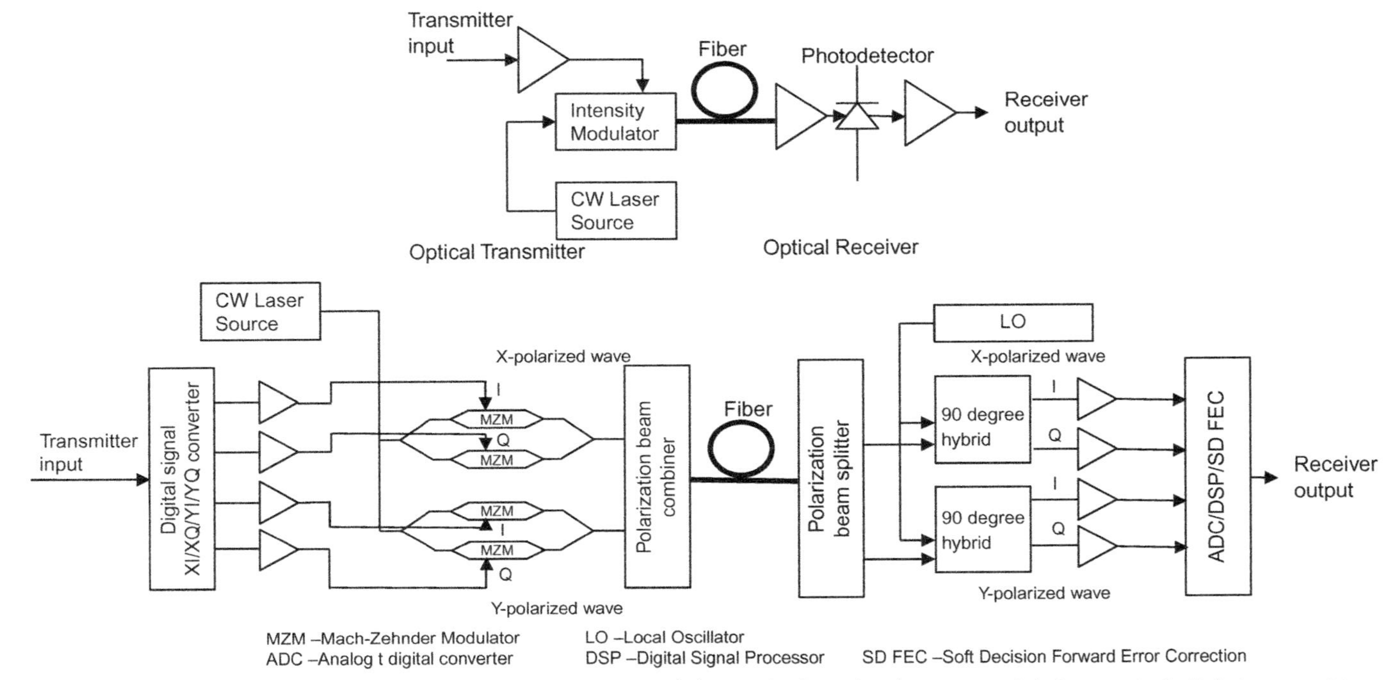

Figure 15.9 Schematic block diagram of a typical intensity modulation (*top*) and coherent modulation optical digital transmitter and receiver.

order modulation has been used in high-capacity WDM system implementations the last several years [26–28].

Coherent optical digital transmission achieves error-free operation at lower OSNR (longer distance) links in high-capacity WDM systems by efficiently utilizing the phase and polarization of the optical signal. Fig. 15.9 shows the block diagram of a coherent optical transponder, in which a digital signal is converted into the in-phase (I) and quadrature (Q) components of the X- and Y-polarized waves, respectively. This results in four signals. The electrical signals XI and XQ drive the MZMs for X-polarized wave, and the electrical signals YI and YQ drive the MZMs for Y-polarized wave. Then, the X- and Y-polarized waves are combined to generate the optical signal with phase-shift keying and dual polarization, which creates a dual polarization quadrature phase-shift keying (DP-QPSK) optical signal. In the optical receiver, the DP-QPSK optical signal is separated into each of the two polarizations and then interferes with a laser light source employed as a local oscillator (LO) in order to detect the in-phase I-components and the quadrature phase Q-components of the X- and Y-polarized waves. This process is called coherent detection because it detects signals by causing interference between the signal and the LO. Coherent detection preserves completely the information of the received optical signal; that is, amplitude, phase, frequency, and polarization, thus enabling higher spectral efficiency. The detected I and Q-components of the X- and Y-polarized waves are converted into electrical signals by the photodetectors and then converted into digital data by an analog-to-digital converter (ADC). The sampled digital data is fed to a DSP to compensate for many of the signal transmission impairments, such as chromatic and polarization mode dispersion.

The use of DSP also allows higher order modulation to be supported, such as quadrature amplitude modulation with 8-ary QAM, 16-ary QAM, or 64-ary QAM implementations, as shown in Fig. 15.8. For example, a 16-ary QAM system represents 16 bit values and can thus transmit twice the amount of information compared to the QPSK system. Note, however, that the higher the modulation order, the fewer photons per bit, and the more advanced the signal processing requirements. The DSP can also perform pulse shaping, predispersion, FEC, nonlinear compensation, and high-speed ADC in the receiver, enhancing operation at the higher order amplitude/phase modulation with the corresponding higher spectral efficiencies (bits/sec/Hz). The symbol rate takes into account the framing overhead, the FEC overhead, the data rate, and number of bits per symbol. For example, with QPSK modulation of 2 bits/symbol, 2 polarization states, with a 100 Gbit/s throughput requirement, a framing overhead of 9%, FEC overhead of 27%, we arrive at 33 Gbaud. WDM systems increasingly employ modulation formats that can be reconfigured to maximize the channel spectral efficiency for each channel based on the specific fiber link OSNR characteristics, as shown, for example, in Fig. 15.10. However, higher SNR is required for higher symbol resolution, because the constellation points between symbols are

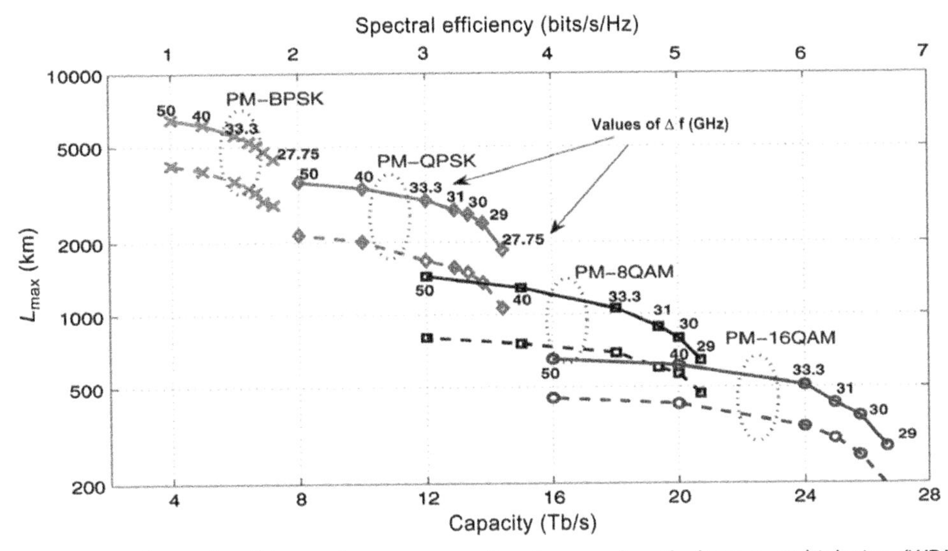

Figure 15.10 The trade-off in reach versus capacity of a wavelength-division multiplexing (WDM) system for different coherent modulation formats [28].

closer than in the QPSK implementation making it more susceptible to noise, and the trade-off between channel capacity and reach is nonlinear. As Fig. 15.10 indicates, 16QAM is limited to 500% shorter reach versus QPSK, for only 200% increase in spectral efficiency. Also as Fig. 15.9 shows, a design of coherent WDM is significantly more complex, and hence costly, than IM-DD technology.

At the same time, the advancements in CMOS technology, commonly referred to as Moore's law, have been benefiting significantly the DSP ASICs, and thus coherent WDM transmission. Progressively, since the initial generation of coherent WDM DSP at 40 nm CMOS technology just under a decade ago [26], each new generation of CMOS process has provided more DSP capabilities, and lower power dissipation [27]. The latest generation coherent WDM ASICs would support powerful DSP for up to 64QAM modulation formats, FEC, dynamic compensation of channel impairments, and active digital monitoring of the health of the links [12]. All these factors combined have significantly advanced the coherent WDM system performance ever closer to the fiber Shannon limit, and also enhanced the network operational flexibility. In addition to OSNR optimization, advanced DSP technology can also decrease the spectral width of each WDM channel, allowing superchannels employing Nyquist filtering to improve further the WDM system spectral efficiency [13]. Next- generation coherent DSP would advance channel performance even further based on different implementations of constellation shaping (CS), as Section 15.4 reviews in details.

In addition to the benefits from Moore's law, optical industry efforts have aimed to improve the cost and complexity of commercial implementations of coherent WDM by defining related multisource agreements (MSAs). For example, OIF defined a coherent WDM MSA module and also specified some of the key optical components inside the module [31]. The specific OIF module was defined to be 5 × 7 in., with power consumption to be less than 80 W, and use polarization-multiplexed (PM)-QPSK with reach up to 2000 km [31]. At the same time, the IEEE had been standardizing the C form-factor pluggable (CFP) transceiver modules of different sizes, whose power dissipation varied progressively from 80, 40, 32 and 12 W [32]. The CFP2 form factor has been adopted also for WDM coherent modules in a few DCI-optimized systems. However, the coherent WDM CFP2 implementations have been challenged by the high power consumption of the coherent DSP and optics [12]. To reduce the coherent WDM module power budget, the initial coherent CFP2 implementations moved the DSP out of the module and placed it on the host card. These CFP2−analog coherent optics (ACO) modules were challenged though by the need for a high-speed analog interface between the CFP2 and host card. So, attention has focused on the CFP−DCO (digital coherent optics) implementations that can accommodate the coherent DSP (and its power budget) inside the module. To this end, a higher degree of optical integration and better power-optimized DSP has been important. Employing lower power DSP, tailored for DCI metro reach, and a silicon photonics integrated Mach-Zehnder modulator, the most recent implementations of CFP2−DCO have been able to productize 16QAM 200 Gb/s with power dissipation of 20 W, which was the lowest reported power consumption per bit for coherent WDM module [33]. This module has also integrated a tunable narrow linewidth laser, a single-chip silicon PIC for quad-parallel and polarization diversity 90 degrees optical hybrid, trans-impedance amplifiers, RF drivers, and a DSP based on 16-nm CMOS [33]. When 7 nm DSP ASICs become cost-effective, it is expected that the related advanced pluggable modules could scale to 400 Gb/s. We discuss these advancements and the related emerging technology and system innovations, notably the 400 Gb/s ZR coherent WDM proposal, in Section 15.4.

Aside the specific implementation details, the progress in the CFP2 pluggable coherent WDM modules has enabled additional modularity at the system level. This modularity has been an important motivation for the CFP2 adoption in DCI-optimized WDM systems. The availability of coherent CFP2 pluggables also made attractive the opportunistic adoption of WDM in DCI routers [34]. The integration of WDM in routers, also known as converged IP + WDM, is not new [6,35]. It has been pursued for more than 10 years (well before the rise of DCI) due to its benefits, particularly the elimination of the short-reach optical interconnections between the router and WDM system, but it has enjoyed limited success due to some well-established technical and business challenges [6]. For the high growth DCI infrastructure, the WDM integration in routers has been evaluated to constrain the cost-effective network evolution [19]. Most notably, one of the biggest technical drawbacks of employing coherent WDM CFP2 (let alone bigger modules) in routers has been their size, which is substantially bigger than the QSPF28 typically employed in today's DCI routers [34]. The port density of routers employing CFP2 has been more than 50% lower than the typically router current density of 36 QSPF28 100 Gb/s ports per RU [36]. Nevertheless, one major DCI cloud operator has reported that the benefit of integrating the currently available coherent WDM in DCI routers still outweighs the related limitations [34]. Even so, because compromising system density is particularly important in the Metro DCI, the same cloud operator has also implemented a 100 Gb/s noncoherent WDM QSFP28 modules optimized for less than 80 km [37].

This 100 Gb/s noncoherent WDM transceiver optimized for less than 80 km employs PAM4 direct-detect technology [37]. The modules are compatible with the QSFP28 MSA (as in SFF-8665), and have the standard CAUI4 electrical interface, supporting 100GE payloads. The electrical input to the module is 4×25.78125 Gb/s NRZ modulation, and the WDM output is 2×28.125 Gbaud with PAM4 modulation. The module has an integrated FEC and a PAM4 DSP engine. To accommodate the DSP ASIC power in the QSFP28 power budget, a nontunable WDM laser is used. The direct-detect nontunable WDM implementation lowers cost/bit, and most importantly the power/bit, at the expense, however, of significantly reduced WDM system capacity (typically 4 Tb/s in C-band) and increased operational complexity. For example, the lack of wavelength tuning necessitates fixed port multiplexing, complicating the WDM system planning and provisioning, and increases the inventory for sparing as each wavelength corresponds to a distinct module. Also, the less than 200 ps/nm chromatic dispersion tolerance of DD PAM4 is very narrow (particularly when compared to the hundreds of thousands ps/nm in coherent WDM), and thus such DCI network implementations require per channel tunable dispersion compensation technology for the most common fiber types and link distances. Mostly because of these operational limitations, the use of PAM4 WDM QSFP28 modules has not been widely adopted. Instead, the consensus among DCI network operators, including the ones that have employed PAM4 WDM QSFP28, has been that it is important to identify a next-generation metro-optimized coherent WDM interface that does not comprise system capacity and network operational simplicity (e.g., employing full wavelength tunability). The OIF definition of the 400GE ZR coherent WDM has been the most notable such effort currently under development [30,38], and is discussed further in Section 15.4.

More important than the specific details of the PAM4 QSPF28 WDM system limitations, or the coherent DSP power consumption evolution, is the general conclusion that the first generation of DCI transport systems employed mostly WDM technology originally developed for non-DCI networks. As our analysis of the optical building blocks makes clear, such technology did not introduce major limitation for longer reach DCI links. For Metro DCI transport, however, initial adoption of such technology has often been suboptimal, particularly as Metro DCI started becoming increasingly important for the fabric extension of cloud networks with relatively shorter links. For example, Fig. 15.11 illustrates this WDM technology "gap" in DCI metro links that are typically between 40 and 100 km (due the AD synchronous replication latency requirement within each cloud region discussed in Section 15.2). This gap has existed mainly because (1) the power of most coherent DSP ASIC implementations so far has been too high for metro-optimized DCI systems, and (2) the initial WDM transceiver MSAs, employing discrete optical components, required larger form factors. The two limitations combined have compromised the implementation of more dense, lower power metro-optimized DCI systems.

In fact, the most widely adopted initial metro-optimized DCI system (introduced in late 2014) benefited from photonic integration [12]. Its PIC specifically combined 10 WDM QPSK channels with their multiplexing and optical amplification, to implement a 2 RU WDM transponder of 500 Gb/s (over 250 GHz), which corresponded to density more than twice better than the closest alternative [12]. However, as 16QAM and higher order coherent WDM started becoming increasingly applicable for the shorter metro DCI links, the coherent DSP started to dominate the WDM system design power and cost, and photonic integration became less important than DSP optimization. This led, in the last 1–2 years, to a concerted effort to define lower power WDM DCI designs, with a particular focus on reducing the DSP power. DSP design optimization for the shorter reach metro links allows for lower equalization with fewer digital taps, and lower sampling rate of ADCs/DACs [33]. Benefits would also arise from designing in the newer CMOS 7 nm nodes. At the same time, to improve the density of metro designs, integrating many optical building blocks into PICs has been increasingly explored, leveraging advancements in InP or silicon photonics. For example, a single laser array has been used as a source for the modulator and the local oscillator. Such integration and copackaging of lasers, modulators, polarization combiners, and drivers has enabled substantial reduction in design size and cost, while simultaneously improving channel capacity, reach, and spectral efficiencies. Section 15.4 discusses in more details how the progress in DSP designs in 7 nm CMOS technology, and in photonic integration, are promising to improve the Metro DCI WDM transport over the next 1–2 years. Before reviewing the emerging DCI transport innovations, however, it is important to summarize the significant progress in OLS that the current generation of DCI WDM has pioneered.

15.3.3 Data-center interconnect open line system

The rapid DCI capacity growth has made very important the deployment of systems that scale to tens of Tb/s per fiber, and leverage the most advanced and cost-effective

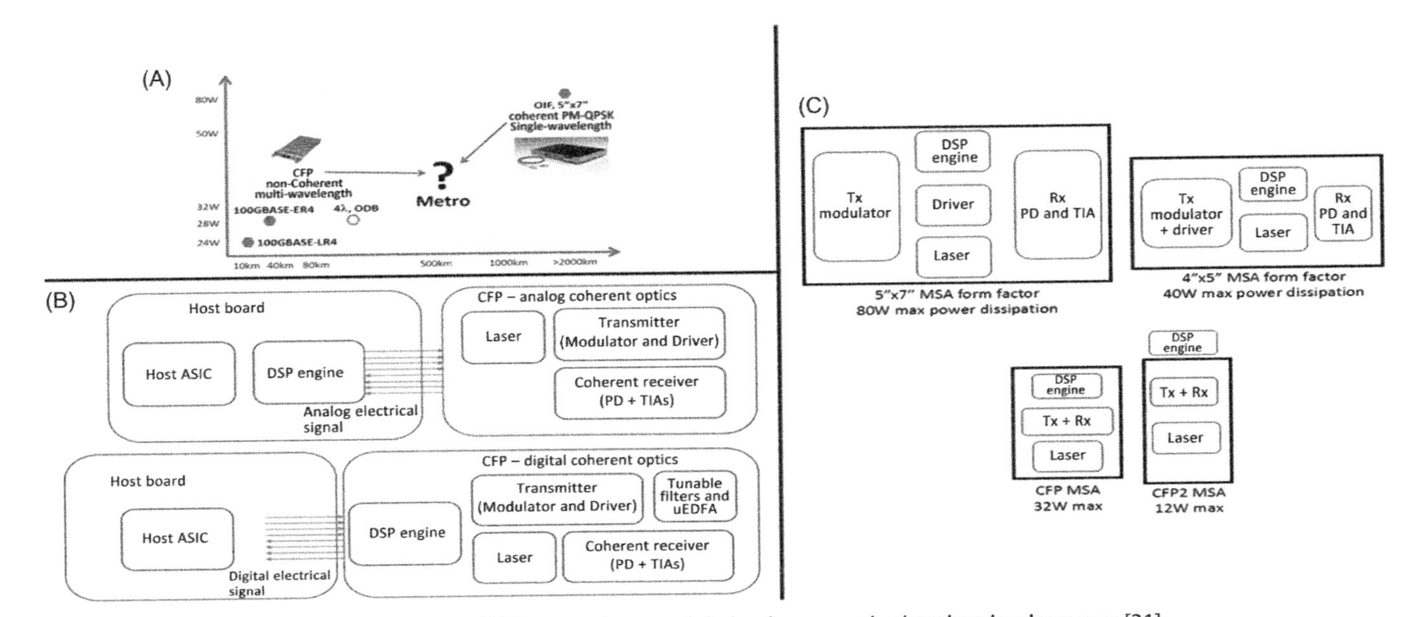

Figure 15.11 Metro data-center interconnect (DCI) transceiver module landscape and related technology gap [31].

technology as soon as it's available. To meet this key business need, DCI network operators have pioneered Open Line System (OLS) networks, that is, transport architectures that allow the operation of different WDM transponders over the same line system. An OLS enables the network operator to control cost, not just at the system level, as for example by using a newer lower cost or lower power transponder, but also at the network level, as for example by employing the highest spectral-efficiency transponder available to minimize the number of lit fibers.

A successful OLS architecture has a few significant WDM system requirements, some of them more nuanced than other. For example, while the concept of WDM alien waves was demonstrated more than 10 years ago [35] and such alien waves have since been opportunistically deployed in networks, the definition of a complete OLS architecture was first pioneered by DCI transport networks [11,39]. First of all, OLS requires to successfully disaggregate the WDM line system from the WDM transponders used over this line system. As Fig. 15.12 illustrates, an OLS network has two distinct blocks, with the transponder(s) being completely separate from the OLS. By disaggregating these two main functional blocks of the WDM system, an OLS implementation can employ not just transponders from different vendors but ideally even different generations of transponders as each vendors' transponder technology may advance independently. A true OLS also requires little, ideally no, control plane interactions between the transponders and line system. The modular transponder-only (1−2 RU) implementations pioneered by the DCI WDM systems were a very helpful first step toward this control plane disaggregation requirement, because they enabled a well-defined demarcation, with usually minimal control plane interaction, within the DCI WDM system. Also, the simpler point-to-point topologies of most DCI networks help the OLS control plane disaggregation to be much less impactful to the overall network performance.

At the same time, achieving "future-proof" physical layer interoperability of different types of transponders requires that an OLS system is able to successfully control all the important physical system parameters. Most notably, a robust OLS implementation is required to independently monitor and control: (1) the optical power of each WDM carrier in order to accommodate any differences in the launch powers of the

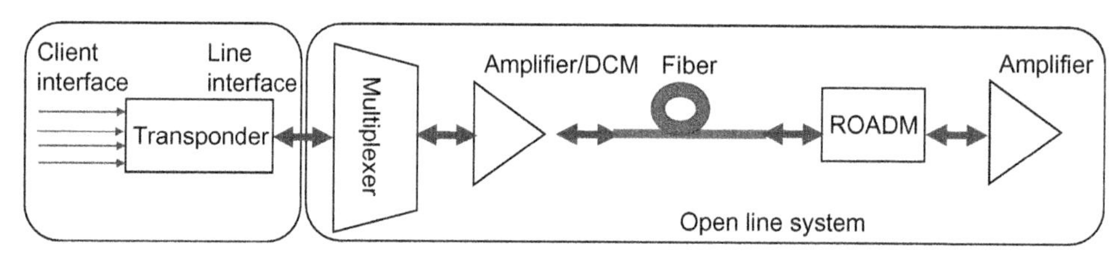

Figure 15.12 Schematic of the basic building blocks of a disaggregated transponder and open line-system (OLS).

different transponders, or to recover after loss of some channels, (2) the spectral width of each WDM port (at the OLS ROADM nodes) in order to be agnostic to baud rates and modulation of each channel, or superchannel, and in the most advanced system (3) the WDM carriers spacing, as this may increase or decrease in order to respectively improve system capacity or channel reach. Thankfully the innovations in flex-grid WDM line systems, particularly in per carrier control of optical power, gain, tilt, and performance monitoring, when combined with the advancements in coherent WDM technology, particularly the chromatic and polarization dispersion compensation, have made the latest generation of WDM line systems capable of accommodating most of these physical layer OLS requirements.

For successful OLS deployment, however, accommodating the OLS physical layer requirements is not sufficient. As important, and in many cases even more critical, is the requirement to manage the OLS network effectively. However, managing a multivendor infrastructure is usually very challenging with the traditional, mostly proprietary, optical NMS [23,25]. To this end, the SDN innovations pioneered by DCI transport toward network element programmability, automation, and vendor-agnostic data model (e.g., YANG) abstractions, are instrumental for successful OLS deployments. Particularly some of the most sophisticated DCI operators have pioneered the development of several of these new transport network management abstraction frameworks required by OLS [23,39], which arguably constitutes a very important innovation to the advancement of deployable OLS architectures.

The ability to manage effectively OLS networks have enabled DCI operators to also explore transponder interoperability between different vendors. Such flexibility would be particularly useful in simplifying network planning and provisioning, especially when there is uncertainty in traffic patterns [40]. Although the different commercial DCI WDM system implementations are comparable in form factors, power, and performance and share many common technology building blocks, there are still enough differences in key specifications (such as spectral slicing, DSP, SD or HD FEC, encryption algorithms or PIC implementation) to prevent interoperability. One approach to achieve transponder interoperability would be to use the same DSP in multiple WDM transponders. In one such recent demonstration [41], successful interoperability among WDM CFP2–ACO interfaces of three different router vendors employing the same coherent DSP was established over a WDM OLS test bed with links up to few thousand km, for 100 GE client services. Given the DSP commonality, the main physical layer challenge was to tune appropriately the lasers in the CFP2 WDM ports.

More generally, OLS deployments are considered the first step toward more disaggregated transport systems and networks. For example, in principle, the different line system building blocks, such as amplifiers, gain equalizers, reconfigurable add/drop multiplexers (ROADMs), any dispersion compensation modules, or optical channel monitors, could also be disaggregated among each other. However, such a complete disaggregation of the OLS would be much more complicated, and would require significant control plane simplifications that may compromise the overall system performance, particularly in longer reach networks. Therefore in practice, the OLS network deployments so far have mostly focused on architectures based on a single supplier providing all the building blocks of the line system (both hardware and control plane). The main focus in these implementations has been to define cost-effective OLS architectures that are optimized for each specific network characteristics (e.g., a metro OLS may be different from a LH OLS), and can accommodate as many types of (current and future) WDM transponders. In this sense, the most successful definition of an OLS architecture would also account carefully for next-generation technologies and the related upgrade cycles. In the next section, we review some of these important emerging DCI technologies, including DSPs, PICs, and SDN, and their potential impact in the evolution of the DCI transport OLS network architectures.

15.4 Emerging data-center interconnect transport innovations

The very successful adoption of the initial generation of DCI transport systems by the global cloud and ICP network infrastructure [2,42], and at times even by traditional telecom networks [43], has motivated additional R&D toward advanced next-generation DCI transport. This section reviews some of these important emerging innovations in network architectures, systems, and enabling technologies.

The most important emerging network architecture innovation may arguably be the proliferation in cloud providers of Metro DCI [44]. As summarized in Section 15.2, the interconnection among the 3–4 distinct ADs in each cloud metro (or city) region, which are predominantly less than 100 km apart, has been a very important DCI transport use case, which has been characterized by the sub 100 ms latency requirement for synchronous replication of the applications running in such a region. In this sense, Metro DCI has also been very different from traditional metro telecom networks [17]. This Metro DCI use case is becoming even more important in scope, as cloud operators extend their metro networks to be more "in-region distributed," henceforth referred to as IRD, motivated by their desire to operate such a region as a continuous DC that can allow more than five 9's (i.e., 99.999%) availability [14]. This proliferation of metro IRD networks will result in a 100–1000 growth in the number of DC nodes and corresponding DCI links [3,44]. To best accommodate this growth, DCI IRD networks require (1) simplicity in design and (2) scale with operational efficiency. Hence, a few system innovations are being actively pursued to meet these requirements. For example, the short distance of IRD WDM links has motivated for a new generation of metro-optimized line-system and transponder technologies [39,44]. Also, the interest in OLS architectures (discussed in Section 15.3.3) has increased because the OLS CAPEX, and OPEX benefits are even more attractive in IRD DCI deployments [39]. We review the IRD related technology innovations, like the 400GE ZR coherent DSP development effort, later in this chapter, after first summarizing other DCI network and system level innovations.

15.4.1 Software-defined network advancements

The benefits of SDN automation and abstraction has been a very important focus of next- generation DCI transport [14,23,24]. Network element provisioning is being automated by leveraging frameworks initially developed for compute, then introduced to switching and routing systems, and now being extended to optical transport, pioneered by DCI WDM systems [24]. Also, network element programmability is advanced beyond APIs (discussed already in Section 15.3) by enhancing the DCI network element operating systems to accommodate "third-party" software "agents" that would allow applications[10] developed by a network operator to interact with the DCI transport system [45,46]. Along with the increased openness and programmability comes also the need to advance the system control plane and network management abstraction beyond vendor-specific, usually proprietary, implementations. This effort is part of a much wider networking effort towards intent-based configuration frameworks [14,46], which is not specific to transport. It requires that network layers are controlled by an SDN controller or NMS that maintains global network state and monitors the entire network (including optical transport) for changes. Based on this global state machine (and the operator's intent), the controller decides when the network needs to transition from one state to another; for example, from A to B, where A is a steady-state optical network operating at capacity X, and B is the optical network with increased capacity $X' > X$. As part of the state transition, the controller sends the entire configuration (including ones that don't need change) to each network element (NE), which is expected to identify and apply only the required changes, that is, turn-up extra wavelengths to increase capacity to X'. Much like other SDN innovations, such "declarative" frameworks have been initially introduced in network switching and routing systems but are now being extended to optical transport, pioneered by next-generation DCI WDM systems [14,45,46]. Note that these SDN advancements can benefit all transport use cases, not just DCI, but they become particularly valuable in improving the OPEX of large-scale deployments, as for example in the emerging IRD networks.

[10] One can think of such "software agent applications" as being similar to the applications available on top of the operating systems in our smartphones.

15.4.1.1 Network monitoring

Complementary to SDN automation and abstraction, the ability to effectively monitor the state of the network is becoming increasingly important [24]. Particularly as cloud and ICP networks scale in size and complexity, network failures are becoming inevitable [15]. Therefore the ability to monitor effectively the aspects of the network that can affect an application, from the Network Interface Card (NICs) and switches inside the DC to the DCI transport network elements and links, is becoming critical. Network failures can be grouped under two main categories: (1) catastrophic failures, which can manifest without warning, and can take down portions of a network, or even all of it, causing significant service disruption and downtime, and (2) degradation failures, which would typically result in slow impairment of system performance without necessarily impacting network traffic (often over several days or weeks) and may not trigger alarms even when they relate to imminent failures. In traditional telecom networks, simple network management protocol (SNMP) polling has been the main mechanism to monitor the state of the network. However, as cloud and ICP networks have increasingly utilized new more sophisticated software frameworks to improve operational efficiency at scale, they have also introduced new monitoring frameworks to account more dynamically for all network failures, including DCI transport degradation. For example, Openconfig has identified streaming telemetry as its top priority and has explicitly aimed to replace SNMP "data-pull" by using an open source framework based on YANG data modeling and the gNMI [23]. It has already enjoyed significant adoption in the major DCI vendors, initially in routing systems, and more recently also in most DCI WDM systems. Beyond the specific Openconfig implementation details, these telemetry frameworks, pioneered by DCI transport, have enabled two important innovations more generally in WDM systems: First, these new monitoring frameworks can report a much more extensive amount of network and system parameters, such as optical power on transmit and receive, pre-FEC and post-FEC, amplifier parameters, dispersion, client and line-side temperatures, device/port up/down status, etc. Second, and even more important, such data model based nonproprietary frameworks allow the end user extensive flexibility in defining the desired content (more or less info), the method (e.g., data encoding mechanism), and the granularity (from milliseconds to hours) of the network monitoring mechanisms; for example, [47].

This new advanced monitoring flexibility also allows transport to be more effectively integrated in the network management and control planes, and more easily combined with network analytics [46]. Network analytics aim to identify, based on operator defined trigger points, potential drifts in parameters and notify network operations for actionable remedial steps that would allow for sufficient time to anticipate and plan repair maintenance and recovery, minimizing potential down times [46,47]. For cloud and ICP operators such new network analytics frameworks, extensively based on innovations in streaming telemetry methodologies, have been increasingly considered an important evolution of network management and mediation for wireline transport and lately even combined with the potential of cognitive systems [24]. We discuss more the promise of predictive network analytics and cognitive system for DCI in Section 15.5.3.

15.4.2 Optical protection switching

Complementary to the new SDN monitoring frameworks, to ensure maximum availability of the cloud services, and achieve service level agreement (SLA)'s of 99.995% availability, DCI networks have employed protection typically based on (1) hardware-level redundancy to protect from node failures and (2) diversity of routing paths based on layer 3 routing protocols (such as MPLS Fast Re-Routing) that would reroute traffic in the event of a link failure. Nevertheless, even after such node-level redundancy and routing path protection, service availability can remain below the cloud service SLA goals when the occurrence of simultaneous failures is too high, as for example cloud services expand globally in emerging market regions where mean time to repair may be high (or unpredictable). In metro regions with such limited network availability, optical protection switching (OPS) is being increasingly combined with restoration to improve network survivability.

The OPS methods employed in such DCI transport networks are the same as the previously well-established connection-oriented network protection mechanisms [48]. These methods include 1:1 (one-to-one) wavelength protection and 1+1 (one plus one) wavelength protection switching. In 1:1 wavelength protection switching, a dedicated protection channel on a diverse path is available, on standby, for each working WDM channel. (Note that to be truly 1:1, the same wavelength must be reserved across both paths, although only one path is lit at any given time). An interesting lower CAPEX architecture of this protection mechanism, most appropriate for networks limited by link failures, employs a single transponder whose line side is connected to a directionless ROADM. So when a failure on the primary path is detected, the system will trigger the directionless ROADM to automatically switch to the protect path. This scheme further allows for 1:N protection where multiple working paths can be protected by a single protection path. In 1+1 wavelength protection switching, each working channel has a dedicated protection channel, and the same payload is sent across both channels simultaneously. In the event of a failure, the receiver switches to the signal from the protection path. This protection scheme is commonly implemented, and referred to, as Y cable protection. With 1+1, there are two active wavelength channels, not necessarily on the same wavelength, between endpoints. Note that 1+1 provides faster recovery than 1:1, as both ROADM endpoints continuously receive data and failover by switching the transponders feeding the Y cable. However, 1+1 protection requires twice the number of transponders. Note finally that both of these two mechanisms can be combined with restoration, leading respectively to 1:1+R or 1+1+R, which would add an extra layer of redundancy during a failure. Restoration would seek to dynamically rebuild a new path around a failed link, to reestablish 1+1 or 1:1 redundancy, while the protection channel has become the working channel. Given the extensive amount of prior art in such protection schemes, including most recently extensive work on advanced 1:1+R and 1+1+R in the context of WSON (wavelength switched optical network) [49], the adoption of OPS in DCI transport systems has been quite straightforward. Also, in most DCI networks, and particularly in IRD networks, given their relatively small size and few nodes, optical restoration within few tens of ms has been possible by most system implementations. In this sense, the main focus of OPS in the DCI implementation has been to maintain the overall goal of a highly scalable OLS architecture, through simplicity and control plane disaggregation.

15.4.3 Data encryption

Much like increased network availability, leveraging OPS, and advanced network monitoring frameworks, have all been motivated by the proliferation of cloud services for business-critical application, encryption has also become important in cloud networks. In the majority of DCI implementations, encryption has been using layer 2/3 standard protocols, mainly Media Access Control security (MACsec) per IEEE 802.1ae in hardware implementation that allow for line-rate operations, or IPsec (typically when an IP/MPLS VPN is used instead of a dedicated DCI network). Occasionally, layer-1 encryption has also been considered appealing because it could offer (1) lower bandwidth overhead, especially for small-size packets in highly utilized links (i.e., when there are very few idlers, in which case MACsec overhead may exceed 20%), and (2) line-rate throughput (compared to any nonline-rate MACsec implementation). Nevertheless, WDM encryption has been deployed less often so far because it cannot account for any non-WDM DCI (e.g., traditional IP/MPLS VPN) and could add complexity to the WDM operations due to the associated requirements for key authentication, rotation, and management. Moreover, WDM encryption implementations, whether based on an FPGA or incorporated in the coherent DSP ASIC, had been associated with increased cost, power, and latency—all critical parameters in IRD DCI transport. Recently, however, as WDM has become the pervasive underlay transport layer in DCI networks, and the latest coherent DSP technology (implemented at 16 nm or soon 7 nm CMOS nodes) can accommodate encryption at minimal extra cost and power, WDM encryption is being considered an attractive option for high-capacity DCI transport, for offering line-rate operations with low overhead.

15.4.4 Advancements in wavelength-division multiplexing digital signal processing and photonic integration

Moving on from network to system advancements, next-generation DCI focus has been primarily toward improving WDM capacity and power efficiency. Power efficiency (W/b/s) has become a particularly important requirement because many cloud and ICP deployments of next-generation DCI systems will be limited more by the power available per rack for hosting equipment, rather than the actual physical size of the equipment. In this sense, new coherent DSPs aim to improve power efficiency by employing power optimizations in the DSP designs and also the latest CMOS technology at 16 nm and soon 7 nm nodes. The adoption of such CMOS technology, however, increases substantially the R&D cost of developing these ASIC designs, as well as the overall system complexity. For example, high-end next-generation coherent DSP designs will be dual carrier implementations, with aggregate bandwidth exceeding 1 Tb/s and power consumption exceeding 100 watts, raising important system design trade-offs. To improve power efficiency, system complexity, and ultimate cost, PIC technology is becoming increasingly important. For example, for the next-generation power-limited system designs, advancements in photonic integration can be as important as gains from coherent DSP performance. Advancements in PIC can serve ideally the dual goal of multichannel integration, as well as operating at higher Gbaud rates without significant system implementation penalty. For example, a recent demonstration reported a coherent 2 channel InP-based PIC design with hybrid integrated SiGe drivers, capable of achieving 1 Tb/s per-channel capacity by operating at 100 GBd and 32QAM [50]. The same PIC also demonstrated 200 km transmission operating at 66 GBd and 64QAM, with performance exceeding that of previously reported discrete InP modulators and approaching the performance of LiNbO3 modulators.

The benefits of PIC advancements would be complementary to the spectral efficiency benefits of coherent. More specifically, next-generation coherent DSP technology is incorporating higher order modulation formats up to 64QAM and is also proliferating the use of subcarrier modulation whose early adoption (as we discussed in Section 15.3) has been instrumental in enabling record spectral efficiencies of more than 6 b/s/Hz [13]. Subcarrier modulation allows for smaller amount of nonlinear interference penalty and can benefit system reach by more than 10% versus single-carrier [51]. Moreover, combining photonic integration with coherent DSP enabled sharp rolloff (1% root-raised cosine filters) of subcarrier modulated channels modulation has provided state-of-the-art WDM superchannel spectral efficiency [13], which is particularly appropriate for point-to-point DCI networks.

15.4.5 Constellation shaping

More recently, and typically combined with subcarriers in coherent modulation, advancements in CS is also starting to be employed to increase the spectral efficiency of WDM systems. Although a few different CS techniques have been proposed [52], in general an adjustable distribution matcher at the transmitter, combined with an inverse distribution matcher at the receiver, optimizes the data rate of each coherent (sub)carrier to minimize optical transmission impairment for each channel independently, primarily by reducing the "peak-to-average" optical power. The addition of CS modulation in a coherent DSP enables the WDM system to optimize per wavelength capacity much more granularly (than fixed-QAM modulation). In principle, CS has been reported to deliver more than 1.5 dB of improvement in a linear transmission channel [52]. In practice, emerging WDM systems with 60+channels in the C-band operating at 67 Gbauds per channel have reported benefits from CS to be slightly less than 1 dB [53]. Most of these CS coherent DSP implementations are capable of modulation formats with adjustable shaping combined with a fixed-QAM modulation, and a fixed FEC code, to realize WDM channels with adjustable net data rate that can each operate in 50 Gb/s increments, up to 64QAM and 600 Gb/s [52]. Moreover, CS offers an additional complementary "degree of freedom" in system optimization. For example, in a recently proposed alternative CS design [54], a constellation-shaped 1024QAM transceiver operating at 66 Gbauds, allowing for a coded bit rate of 1.32 Tb/s, was demonstrated to achieve error-free 680 Gb/s net data rate transmission over 400 km, corresponding to a spectral efficiency of 9.35 b/s/Hz, which is 12% higher than any previously reported spectral efficiency at this SNR. It is exactly this ability to achieve high spectral efficiency and approach even closer the fiber Shannon limit that has made CS modulation techniques increasingly attractive in WDM transmission in general (see also Chapter 10: High-order modulation formats, constellation design, and DSP for high-speed transmission systems) and in DCI transport in particular (due to the fiber scarcity explained in Section 15.2). From a practical perspective, however, CS DSP implementations also increase substantially the ASIC complexity and power consumption, which may raise an important DCI system trade-off for lower distances and IRD networks.

15.4.6 L-band, and open line-system disaggregation

Motivated by the drive to maximize fiber capacity, some of the DCI LH and subsea networks have also been among the first to employ L-band WDM transmission [11,18,19]. L-band is, of course, complementary to the coherent modulation advancements discussed in the last few paragraphs, allowing the total fiber capacity to essentially double, exceeding 50 Tb/s for most networks. Much like most other LH WDM innovations, the optical technology advancements for L-band systems, for example, [55], are not specific to DCI transport. In the context of next-generation line systems, of which L-band is a key advancement, the most important innovation being pioneered specifically by DCI transport networks, has been the OLS extensions to a disaggregated line-system control plane [39,56].

Disaggregating the line-system control plane is the natural next important step in the OLS architecture evolution. As discussed in Section 15.3.3, the initial OLS efforts have focused on the most immediate practical priorities, namely: (1) decoupling of the transponders from the line-system control plane and (2) creating SDN-enabled management abstraction frameworks (like Openconfig) that will facilitate the operation of multivendor transport networks. With the advent of C+L-bands, as well as the increasing interest in new types of fiber that offer lower loss (0.17–0.19db/km, as for example, in SMF-28 ULL and Terawave), or in the future larger core fiber or multiple core fibers, as a way to advance the available fiber capacity, the further disaggregation of the OLS control plane could allow useful network flexibility. For example, the L-band system vendor could be separate from the vendor employed in the C-band. In another example, as the original related proposal astutely observed [39,56], disaggregation of the line-system control plane would enable different control plane instances to optimize each distinct fiber path per node "degree" (i.e., disaggregation per different fiber direction). For many of these use cases, achieving effective line-system operation with a disaggregated control plane calls for new WDM system features, for example, employing much more extensively noise-loading and optical monitoring, as well as a substantial amount of control plane functionality being now centralized (rather than distributed) [39,56]. Disaggregation is of course much simpler as network size gets smaller, because physical impairments are less pronounced. OLS control plane disaggregation actually becomes quite straightforward in the DCI of the emerging IRD networks, and other such small metro networks. Hence, metro OLS architectures with disaggregated control plane are now being adopted more widely, including telecom operators, for example, [57].

15.4.7 400GE wavelength-division multiplexing ZR

The fact that most advancements in coherent WDM technologies (e.g., CS discussed in Section 15.4.1.5) have been most applicable to longer distance transmission (Fig. 15.11) has recently motivated alternative designs optimized for smaller metro and IRD DCI networks. For the IRD DCI and short-reach metro networks, innovations in transponder designs optimized for lower noise and dispersion requirements aim to reduce power, size, and cost, without the PAM4 operational limitation (discussed in Section 15.3.2). The coherent WDM 400GBASE-KP4 ZR [30,38] is the most important such example. It provides a single-carrier DP-16QAM modulation PMD definition for a DCO module with unamplified/passive single channel reach up to 40 km, or an amplified WDM reach up to 120 km without external dispersion compensation and aggregate fiber (C+L) capacity of 52 Tb/s. The specification includes the digital format and processing required for a 400 Gb/s ZR interface, along with some basic analog parameters with the goal to maximize the reuse of the O/E components from the DWDM ecosystem in support of symbol rates between 40 and 70 Gbaud/s. The client side supports IEEE P802.3bs 400GBASE-R, 400GAUI-8 C2M clock and data recovery, alignment lock, lane ordering, FEC coding, TX RS-FEC SD/SF signaling, and RX SC-FEC SD/SF signaling. The proposed FEC is based on concatenated staircase FEC + Hamming with post-FEC error floor $<1E-15$. The intended form factors are mainly QSFP-DD, OFSP, and CFP8, targeting port densities equivalent to gray client optics, which currently are 14.4 Tb/s per 1U line card. The proposed designs plan to leverage photonic integration, and simpler coherent DSP ASIC implementation at 7 nm CMOS that support for absolute (nondifferential) phase encoding/decoding. Most importantly, these

ZR designs aim for low total power dissipation that would not exceed 15–20 W (optics + DSP), in which case this DCO can leverage one of the emerging 400GE small pluggable module form factors of OSFP or even DD-QSFP. Also, note that at ECOC 2018, an additional yet-to-be-specified use case, referred to as ZR + , was proposed that claimed WDM reach extending beyond 1000 km by operating the ZR DCO module at the higher end of this power range (around 20 W). Regardless of the specific details of each of these ZR use cases, achieving 400GE WDM in an OSFP, let alone the DD-QSFP form factor, would certainly be a significant advancement in terms of system (faceplate) density and could become the catalyst for the adoption of such a WDM interface directly on 400GE DCI routers expected to start being deployed by 2020. Even at 20 W power consumption, such a 400GE DCO transceiver would correspond to 5 W per 100 Gb/s, which is a significant (approximately 2–4 ×) advancement in coherent WDM power efficiency, and which is particularly important given that power efficiency has become a very critical constraint in the scale of DC operations. Also, note here, that motivated by such potential benefits of 400GE ZR, and of course the explosive needs of DCI growth, a new similar OIF initiative, aiming to define an 800 Gb/s coherent ZR DCI interface, was proposed in May 15, 2019 (oif2019.184) by Google, Microsoft and other network operators.

15.4.8 Implications of intra-data center networking and Moore's law

As summarized throughout Section 15.4, DCI-optimized transport has motivated many recent WDM innovations and raised new system trade-offs. To better appreciate these trade-offs, we need to also consider that the main driver behind the overall cloud and ICP infrastructure has been the advancements in compute, storage, and networking inside the data center, all of which have been scaling very effectively, albeit at different timescales, leveraging Moore's law. To a great extent, DCI aims to best accommodate the Moore's law scaling of the DC switching. Therefore in concluding the analysis of the emerging DCI innovations, it can be insightful to summarize also the evolution and implications of intra-data center technologies and Moore's law.

The achievement of the semiconductor industry to deliver the projection of Moore's law by doubling the transistor density every 2 years has resulted in a 100 million- fold density increase over the last 50 years, and it is projected to achieve one billion-fold density increase over the 60-year period from 1971 to 2031. Specifically, central processing unit (CPU) performance grew 2 × every 2 years. During similar periods, Ethernet networking speeds grew from 1 to 10G over a 12-year period and 10 to 100G over another 10-year period (Fig. 15.13). The question is why networking did not keep pace with Moore's law. Moore's law applies to transistor scaling and not speed. Transistor speed is only increasing slowly, while CPU processor speed growth eventually plateaued for power reasons. In addition, the number of I/O pins per processor ASIC die package is basically fixed, constrained by the silicon die area (limited by max reticle size) and packaging technology, thus limiting overall throughput; where throughput-per-chip = (number

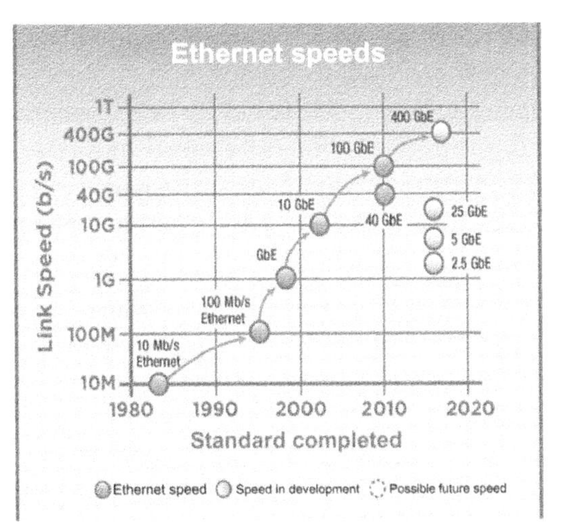

Figure 15.13 Ethernet speed evolution. Source: *Ethernet Alliance https://www.ethernetalliance.org/.*

of IO pins) × (speed per IO) bandwidth is limited by I/O capability irrespective of how many transistors are on the processor ASIC. Only improvement option available is to increase the I/O speed, which is achieved by improvements in SERDES (Serializer-Deserializer). Typically, SERDES speeds have progressed at a rate of 200% every 4 years [58]. I/O speed therefore scales much slower than Moore's law. The industry focus over multiple generations of SERDES has been to employ higher order modulation to improve aggregate throughput, energy/bit, and cost.

DC networking infrastructure has correspondingly evolved to accommodate the semiconductor industry progress and challenges. In cloud networking, the explosive growth in "east-west" traffic [8] led to the proliferation of the "spine-and-leaf" intra-DC architecture, a folded-Clos (often multitiered) switching fabric that is a highly scalable nonblocking network design that lends itself well to minimal IP routing protocols with low, and more importantly deterministic, latency. In such scalable DC designs, all servers within a rack are typically connected, through the NICs,[11] to top-of-rack switches, which are then connected to the "leaf" switches (typically with an oversubscription of about 150%). Each switch has 32–36 ports in a single RU form factor.

[11] Note that this boundary between network and computations is becoming increasingly blurred by the introduction of "SmartNICs," which is a network interface card that includes a programmable network processor that can be used to handle virtual switching and operate as a coprocessor to the host general-purpose CPU. Essentially the SmartNIC serves as programmable network adapter that in cloud environments can help accelerate a wide range of networking and security functions (encryption of data in motion), allowing CPU cores to be used exclusively on compute and applications. The broad scale deployment of such NICs, soon at 50 Gb/s (and then 100 Gb/s), is closely related with the impending transition to 200G and 400G network ports and has thus implications for East-West and North-South traffic traversing the DCs.

Each port is connected with direct attach copper (DAC) cables or active optical cables (AOCs) using industry MSA or standards-based QSFP, QSFP28, QSFP-DD, or OSFP form factors. The switches in such multitier fabric have progressively (over multiple generations) increased the port speeds from 10 to 40G to 100G today. The next wave of buildout is going to migrate to 200 and 400G data center fabrics in order to support emerging workloads, and also to avoid costly recabling changes of the fiber plant topology.[12] In the context of DCI transport, the most important aspect is that cloud DC switching systems already scale to tens of Tb/s aggregate throughput, in compact (1RU−9 RU) designs that provide full line-rate layer 2 and layer 3 forwarding logic, and also "tunneling" technology (GRE, MPLS, VXLAN, etc.), which allows for SDN techniques to build optimized logical (overlay) networks over the existing physical (underlay) intra-DC fabric. Integral to these switch platforms are switch ASICs which perform the functions of packet forwarding, routing, tunnel termination, and access control list (ACL) at progressively higher port speeds (i.e., 10GbE, 40GbE, 100GbE, 200GbE, and 400GbE). These switch ASICs are characterized by high bandwidth, high scalability, and low latency (Fig. 15.14). The figure shows clearly the interdependence of Server I/O with networking, with the bottom line representing the per-lane SERDES throughput and the top line representing the aggregate switch ASIC throughput. It also provides insights about the amount of bandwidth per port that DCI transport needs to support. Next-generation switching ASICs (by multiple vendors) will support 12.8 Tb/s aggregate throughput, with 50G SERDES, and 400G/

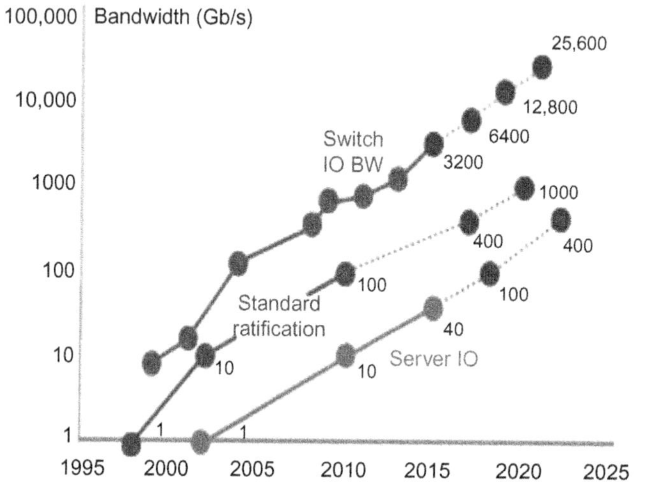

Figure 15.14 Switch application-specific integrated circuits (ASIC) bandwidth evolution.

[12] See also Chapter 16, which focuses on intra-data center optics and networking.

ports, while the designs after that (4−5 years later) are expected to scale to 25.6 Tb/s (with 100G SERDES, and 800G/ports) and then to >50 Tb/s some years later.

Moreover, power dissipation of these DC switches ranges typically from 125 to 250 W, which corresponds to a power efficiency range of 60−30 pJ/bit and can be supported with conventional forced-air cooling. It is anticipated, however, that the next-generation switch ASICs that scale to 25.6 Tb/s (and beyond) will be challenged by thermal dissipation, requiring advanced cooling techniques. To appreciate better the related challenge, it is insightful to perform a high-level power budget analysis of such high-performance switching ASIC. The main building blocks of a switch ASIC include the crossbar, buffer memory, L2 (MAC) hash table, processor and forwarding logic, TCAM, and SERDES. Specifically, high throughput switching ASICs are characterized by large allocation for memory (buffer and TCAM), which currently consumes about 50% of their power (30−15 pJ/bit). About 25% of power (15−7 pJ/bit) is consumed by logic and processor. The remaining power dissipation, about 25% (15−7 pJ/bit) accounts for the SERDES I/O. Until recently the switch ASIC I/O has typically supported NRZ modulation, but the higher throughput would require the use of the much more power-consuming 56G-PAM4 modulation, leading to future 100 Gb/s SERDES expected power to increase to as high as 40-50% of the total ASIC power budget [58,61,62].

The high throughput and related power constraints mentioned are important also for the evolution of DCI-optimized transport and the associated system trade-offs. For example, would the optimal DCI evolution aim for the integration of WDM technology in the DCI router system, or instead aim to optimize WDM and packet transport technology building blocks in separate systems that follow distinct evolution paths? Also, the advent of power-constrained switch-ASICs with capacities above 25 Tbps would call for next-generation DCI transport systems with fiber capacities that approach ever closer to the fiber Shannon limit even for the shorter interconnection distances. We explore these implications for next-generation DCI-optimized transport in Section 15.5.

15.5 Outlook

Motivated by the explosive growth of cloud services, DCI has pioneered a few of the most important recent innovations in optical transport systems and networks. As summarized in Section 15.3, DCI-optimized WDM systems have led the advancements in system modularity, state-of-the-art coherent transmission, achieving some of the highest fiber spectral efficiency deployments, as well as open line systems and open transport network architectures, which also pioneered the adoption of SDN automation, abstraction, and disaggregation in optical transport. Also, the DCI need for further scale and operational efficiency is driving active research and development in multiple areas summarized in Section 15.4. This section reviews four additional important longer term innovation topics currently under investigation, and the associated debate regarding their adoption and trade-offs.

15.5.1 Power efficient photonics-electronics integration

Power efficiency has become an increasingly prominent topic for DCI transport for multiple reasons. First, as discussed in Section 15.2, DCI deployments are becoming limited most often by the available power budget in a rack, rather than the physical size of the systems. Moreover, the overall DC power budget is getting further constrained by deployment of new compute servers specifically optimized for machine learning and artificial intelligence applications, which are more power consuming (currently > 30%) than traditional general-purpose servers. Third, and most important for networking, the power efficiency of high-capacity switching ASICs has not improved substantially over the last 5 years, as analyzed in Section 15.4.2. It has practically plateaued at about 3 Tb/s per 100 W [58], which corresponds to about 33 pJ/bit. There is some quite interesting physics [59] behind the limited improvements in power efficiency from successive ASICs generations, mostly related to the inherent SERDES speed and power limitations.

At the system level, which is our main focus here, the key implication is that in future switches, scaling to 25.6 Tb/s and beyond, the optical I/O (with SERDES blocks being a dominant contributor to power) may be exceeding the 1 KW, becoming well over 50% of the system power budget. Therefore the current commercial system implementations, which focus on maximizing the faceplate density and throughput (in Gb/s per mm) of pluggable optical modules, may not be the most appropriate future system design optimization goal. The SERDES speed and power limitations have motivated the industry to explore mid-board optical modules (MBOM) [60], on-board optics (COBO) [61], and more recently even copackaged optics designs, that is, optics integrated with switch silicon on an interposer package in the form of a multichip module (MCM) [62]. By reducing the physical distance between ASIC and optics, copackaging could offer significant (25–30%) reduction in SERDES power. Successful implementations of MCM copackaging, however, call for multiple new standardization agreements from a large ecosystem of players regarding optical and electrical connectors, tunable laser integration, physical form factor, and designs that efficiently address challenges in manufacturability, reliability, and serviceability (including field repairs) [62]. These co-packaging efforts have naturally focused on intra-DC optical technology integration with the DC switches.

For DCI WDM transport, copacking and integration could offer some initial benefit to the coherent WDM modules, depending on the specific implementation details. Longer term, however, more interesting may be the debate around system optimization trade-off from integrating coherent WDM DSP technology into switching ASIC, instead of DCO implementations that copackage DSP closer with WDM PIC. From a technical perspective, the integration may offer different system advantages for transport networks with shorter versus longer distances, or ones with high versus lower spectral efficiency needs. Whether such integration will also be commercially viable is another important aspect of this debate (probably one that is more complicated, but conveniently outside the scope of this chapter). For high-end transport systems that scale as closely as possible to the fiber Shannon limit for a wide range of network distances, that is, employ the most sophisticated available coherent DSPs to allow for near optimal system performance at wide range of "space-bandwidth product" networks (e.g., by including subcarriers and CS), the integration of DSP with the switching ASIC would be technically much more challenging. For such systems, instead, a DCO module, with closer copackaging of coherent DSP with advanced WDM PIC, for example, using interposer and MCM technology along the lines of discussion in Ref. [50], would likely prove the optimal evolution path.

15.5.2 Open transport model-driven networking

Complementary to the optimal system integration debate and the related hardware integration efforts, a few other, potentially more immediate, network optimization trade-offs around the optimal "SDN" transport evolution are also being increasingly debated in the context of DCI. Notably, for example, the drive for open transport, which DCI has pioneered, raises the challenge of maintaining operational simplicity in the presence of more diverse multivendor OLS architectures, for example [39]. Because SDN automation and abstraction frameworks based on model-driven networking have offered a few enabling technologies, the debate concerns how best to evolve the current model-driven networking paradigm to a higher level of abstraction. The related proposals approach network programmability "top-down" (akin to object-oriented programming); that is, the network control logic (e.g., automation or path computation) employs only abstract data model entities (links and nodes) without needing to know domain specifics (DWDM link or IP link), allowing for a separation between specifying network operators intent (what) and the system-level actuation (how). The domain specific information is known only by the systems at the nodes of the domain involved in a specific operation, based on deterministic schema translation between generic and specific data models. Such an SDN evolution (Fig. 15.15) offers two very attractive benefits: (1) Network automation is even more (ideally completely) abstracted, which improves substantially the service provisioning and availability because 70% of cloud failures are reported to happen when a network element (thus vendor specific) management operation is in progress [14], and (2) such a unified operational framework would further enable a common networkwide operating system for

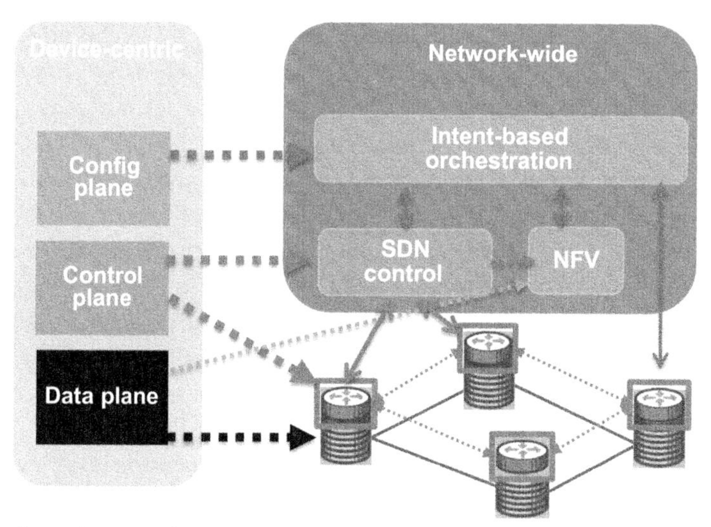

Figure 15.15 The software-defined network (SDN) evolution in automation, disaggregation and abstraction of device-centric to networkwide functions [7].

both WDM and packet [11], which again would substantially improve network provisioning and availability by minimizing multilayer inefficiencies (e.g., SRLG).

While this SDN evolution vision is increasingly being adopted by most cloud and ICP network operators, however, the specific phases and implementation details are still being debated. For example, proposals regarding the most appropriate data model definitions has recently proliferated [24] beyond YANG and Openconfig [23] to a few different data modeling "languages," such as Thrift [45] or YAML [34], practically calling for future DCI transport systems to ideally be able to provide "multilingual" vendor-neutral abstraction instantiations of their underlying domain specific data model based on programmatic representations[13] of the device manageable entities [46].

15.5.3 Network analytics, optimization, and traffic engineering

Network analytics is another important new frontier being pioneered by cloud and ICP operators, again motivated by their desire to use novel software technologies to enhance operational efficiency. New network analytics frameworks, extensively leveraging telemetry methodologies, and lately even combined with the potential of machine learning, for example, [63], or even cognitive systems [64], have been increasingly considered an important evolution of network management and

[13] This specific outlook is introducing an important new era for optical transport, one in which computer science expertise would be needed complementary to the optical physics and electrical engineering expertise.

mediation for cloud transport [24]. The underlying motivation for this evolution in cloud and ICP network management, as we mentioned in Section 15.2, is to enable an autonomic (e.g., Ref. [65]), policy-driven (e.g., Ref. [46]) operational paradigm with little, ideally minimal, human intervention. However, employing such autonomic operations in WAN transport raises a few important questions. Notably, WANs are characterized by significant heterogeneity in technology (both hardware and software), in failure modes (with typically more stringent availability requirements, e.g., up to 5 9s), and in performance metrics (e.g., latency variation in the WAN can be three orders of magnitude more than in compute). Also, many related implementation details are being debated. For example, the divergence in viewpoints on the most appropriate data modeling implementations (discussed in Section 15.5.2) applies also to network analytics, as cloud network analytics frameworks too seek to leverage the benefit from model-driven network abstraction. Moreover, a divergence among leading cloud and ICP operators on the most appropriate telemetry architecture has been recently identified [23 vs 45]. Despite these debates in the implementation details, network analytics are certainly a very exciting prospect for enhancing the DCI operational efficiency and management.

Complementary to the operational benefits, combining insights from network analytics with traffic engineering could enable DCI transport SDN optimization [7,20–22]. Such optimization could be particularly powerful for the emerging coherent WDM transport systems with very granular rate-adaptive channel capability to trade-off capacity for reach and operate close to the fiber Shannon limit. Note that while traffic engineering has been well established in IP/MPLS networks [20,21], and the synergies of packet transport traffic engineering with SDN control and analytics innovations are being actively explored [66,67], such traffic engineering in optical transport has been rarely employed in actual networks. Note also that while the concept of adaptive rate WDM transmission has been well established, for example, [68,69], combining it with network analytics to explore network optimization through traffic engineering is just starting. For example, a very recent analysis [70] that attempted to maximize spectral efficiency over a 3-year period in the actual DCI network of a leading cloud operator has reported that using dynamic rate-adaptive WDM transmission, the capacity of 64% of 100 Gb/s IP links can be augmented leading to an overall capacity gain of over 134 Tb/s. Moreover, adapting link capacity to a lower rate can prevent up to 25% of link failures. The same analysis then used data-driven simulations to show that the overall network throughput can improve by 40% while also improving the average link availability.

Network analytics raises also the even more exciting prospect for predicative network optimization, for example, based on machine learning or even artificial intelligence. In the most obvious use case of predictive optimization of transport networks, analytics could enable proactive protection [71] or enhance network restoration. This is particularly interesting for DCI WDM transport, because combined with advancements in SDN could catalyze the use of optical restoration, which despite its considerable value in networks limited only by link failures [72] has so far found limited applicability in real-world deployments.

Of course, the adoption of such networkwide optimizations would require new technologies and would also raise important implementation trade-offs. For example, rate-adaptive WDM calls for sufficient flexibility in the corresponding IP/MPLS layer, such as proposed by FlexE technology [11,73], or alternatively by inverse multiplexing as described in Ref. [74] when Ethernet clients at 400GE and beyond may exceed WDM single-channel capacity. Even more importantly, successful network optimization is generally a nontrivial multilayer endeavor [69,72] and raises important trade-offs. Specifically for DCI transport, a notable such trade-off would be how to best serve the desire for an open multi-vendor network with maximum scale and operational simplicity and be able to optimize the network availability and capacity, often with very high spectral efficiency, close to the fiber Shannon limit, let alone employing also new power-efficient systems, as discussed in Section 15.5.1.

15.5.4 Edge cloud evolution

In addition to the specific system and networking technology efforts discussed so far, to complete an outlook for cloud transport, it is important to highlight the emerging proliferation of cloud compute to the edge of the network. This edge cloud evolution, beyond the current hyperscale DCs, is motivated by multiple distinct use cases, which often refer to edge cloud evolution with different names. For example, the impending transition to 5G increasingly refers to mobile edge compute as part of the new 5G infrastructure. Similarly, the advent of Internet of Things (IoT) networks has often introduced the need for "Fog" compute. Well before the 5G or IoT services start driving actual network deployments, most telecom network operators have already started to transform their central offices to small DC, usually combining this transformation with the adoption of SDN and network functions virtualization frameworks, as for example discussed in CORD (which stands for central office rearchitected as a data center). While the details behind such a networkwide evolution are well beyond the scope of this chapter, note that the edge cloud evolution has many similarities, as Fig. 15.16 tries to summarize, to the proliferation of IRD (discussed in Section 15.4.1). Therefore much of our analysis in Sections 15.4 and 15.5 is quite applicable also to the edge cloud networks. Of course, scale would be dramatically different. For example, note that by 2021, 94% of compute instances are projected to be in cloud data centers [1]. Fig. 15.16 below provides a reference for all the implementations in DCI transport network application domains identifying the reach, physical hardware and specific innovations.

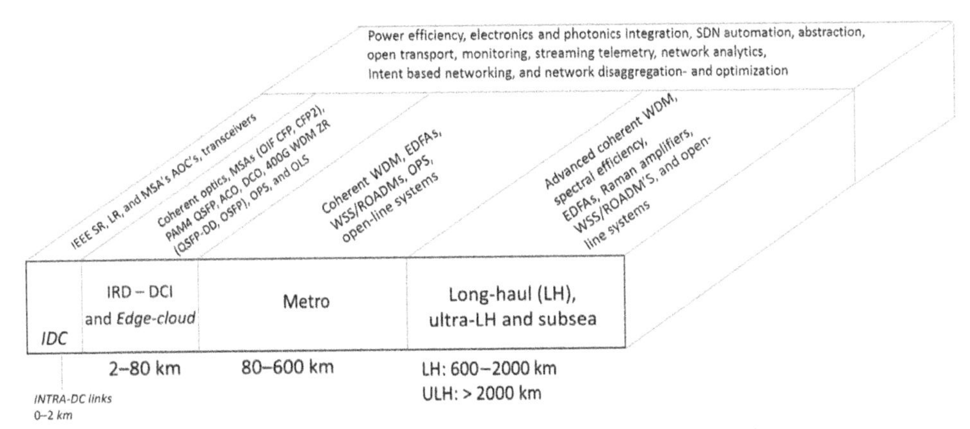

Figure 15.16 Data-center interconnect (DCI) transport network application domains.

Acknowledgments

We would like to acknowledge insightful conversations related to this work with many colleagues in industry and academia, including J. Rahn, A. Sadasivarao, Matt Mitchell, P. Kandappan, Harshad Sardesai, P. Dale, Steve Pegg, J. Gill, S. Elby, D. Welch, V. Vusirikala, Bikash Koley, Anees Shaikh, Tad Hofmeister, K. Tse, R. Doverpsike, Axel Clauberg, G. Nagarajan, J. Gaudette, Brad Booth, M. Filer, G. Rizzelli, Katharine Schmidtke, Chongjin Xie, J. Larikova, B. Zeljko, A.V. Krishnamoorthy, X. Zheng, J. Brar, Ed Crabbe, A. Arcuino, Kyle Edwards, F. Yusuf, Gary Seagraves, T. Wallace, Tony Grayson, and C. Newcombe.

References

[1] Global Cloud Index. <www.cisco.com/c/en/us/solutions/collateral/service-provider/global-cloud-index-gci/white-paper-c11-738085.html>.

[2] U. Holzle, A ubiquitous cloud requires a transparent network, in: OFC 2017 (Plenary), and V. Vusirikala, SDN Enabled Programmable, Dynamic Optical Layer, ECOC 2017, Plenary.

[3] J. Hamilton, How Many Data Centers Needed World-Wide. <https://perspectives.mvdirona.com/2017/04/how-many-data-centers-needed-world-wide>.

[4] L.A. Barroso, et al., The Datacenter as a Computer - An Introduction to the Design of Warehouse-Scale Machines, Morgan & Claypool, 2009.

[5] National Institute of Standards and Technology. <www.csrc.nist.gov/groups/SNS/cloud-computing/index.html>.

[6] L. Paraschis, Advancements in metro regional and core transport network architectures for the next-generation Internet, Chapter 18, Optical Fiber Telecommunications VI Volume B, Systems and Networks, Elsevier, 2013, pp. 793–817. 978-0123969606.

[7] L. Paraschis, SDN innovations in WAN, in: Optical Internetworking Forum (Plenary presentation oif2015.083), January 20, 2015.

[8] N. Farrington, et al., Facebook's data center network architecture, in: 2013 Optical Interconnects Conference, TuB5 (Invited), Santa Fe, NM, 2013, pp. 49–50.

[9] C.F. Lam, et al., Fiber optic communication technologies: what's needed for datacenter network operations, IEEE Commun. Mag. 48 (7) (2010) 32–39.

[10] B. Koley, et al., 100GbE and beyond for warehouse scale computing interconnects, Opt. Fiber Technol. 17 (2011) 363–367.

[11] B. Koley, et al., Future needs of WDM transport for inter-datacenter interconnections, in: M2E1 (invited) OFC, 2014.
[12] S. Elby, Requirements of inter data center networks to meet the explosive growth of cloud services, in: OFC2016 (Tutorial), W3J.1.
[13] J. Rahn, et al., DSP-enabled frequency locking for near-Nyquist spectral efficiency superchannels utilizing integrated photonics, in: IEEE/OSA Conference on Optical Fiber Communications (OFC) 2018 W1B.3 and S. Grubb et al., Real-time 16QAM transatlantic record spectral efficiency of 6.21 b/s/Hz enabling 26.2 Tbps Capacity, in: IEEE/OSA Conference on Optical Fiber Communications (OFC), 2019, M2E.6.
[14] B. Koley, The zero touch network, in: IEEE 12th International Conference on Network and Service Management, Keynote, 2016.
[15] V. Gill, et al., Worse is better, in: NANOG 49, San Francisco, June 14, 2010.
[16] <https://cloud.oracle.com/iaas/architecture>.
[17] L. Paraschis, et al., Metro networks: services and technologies, Chapter 12, Optical Fiber Telecommunications V, vol. B, Academic Press, 2008, 978-0-12-374172-1pp. 477–509.
[18] T. Stuch, et al., Open Undersea cable systems for cloud scale operation, in: Proc. OFC, M2E.1, Los Angeles, 2017.
[19] V. Vusirikala, et al., Drivers and applications of optical technologies for Internet data center networks, in: NThD2, OFC, 2011.
[20] S. Jain, et al., B4: experience with a globally-deployed software defined WAN, in: SIGCOMM'13, 2013.
[21] C. Hong, et al., Achieving high utilization with software-driven WAN, in: SIGCOMM'13, 2013.
[22] IEEE Future Directions Activities and IEEE Cloud Computing Panels at IEEE/OSA, Optical Fiber Communication Conferences on "SDN beyond hype" (24.03.15), "Disaggregation" (23.03.16), and from "Network Analytics" (22.03.17).
[23] A. Shaikh, et al., Vendor-neutral network representations for transport SDN, in: OFC2016, Th4G.3, 2016.
[24] Symposium on "Transport Network Management and Analytics innovations", IEEE/OSA Conference on Optical Fiber Communications (OFC) 2018 Tu3H.
[25] R. Doverspike, et al., Optical network management and control, Proc. IEEE 100 (5) (2012).
[26] K. Roberts, et al., 100G and beyond with digital coherent signal processing, IEEE Commun. Mag. 48 (2010) 62–69.
[27] K. Kikuchi, Fundamentals of coherent optical fiber communications, J. Lightwave Technol. 34 (1) (2016) 1–23. and also summarized in M. Junichiro et al., Development of the Digital Coherent Optical Transmission Technology. NEC Tech. J. 10 (3), Special Issue on Telecom Carrier Solutions for New Value Creation, pp. 67–69.
[28] G. Bosco, et al., On the performance of Nyquist-WDM terabit superchannels based on ... or PM-16QAM subcarriers, J.Lightwave Technol. 29 (1) (2011).
[29] V. Kamalov, et al., "Faster Open Submarine Cable", Th.2.E.5, European Conference on Optical Communications (ECOC), 2017.
[30] <http://www.ieee802.org/3/B10K/public/17_11/williams_b10k_01b_1117.pdf>.
[31] OIF-CEI-3.0, <http://www.oiforum.com/public/documents/OIF_CEI_03.0.pdf>.
[32] CFP MSA Hardware Specification, Revision 1.4, June 7, 2010. IEEE P802.3bm, 40Gbit/s and 100Gbit/s Operation Over Fiber Optic Cables Task Force. <http://www.ieee802.org/3/bm/index.html>.
[33] H. Zhang, et al., Real-time transmission of 16 Tb/s over 1020 km using 200 Gb/s CFP2-DCO, 26, 6 | 19 March 2018 | OPTICS EXPRESS 6943.
[34] M. Filer, et al., IEEE/OSA Conference on Optical Fiber Communications (OFC), 2017, W4H.1.
[35] D. Ventori, et al., Demonstration and evaluation of IP-over-DWDM networking as "Alien-Wavelength" over existing carrier DWDM infrastructure, in: IEEE/OSA Conference on Optical Fiber Communications (OFC) '08, paper NME3.
[36] <https://www.cisco.com/c/en/us/products/collateral/routers/network-convergence-system-5500-series/datasheet-c78-736270.html>.

[37] M.M. Filer, et al., Demonstration and performance analysis of 4 Tb/s DWDM Metro-DCI system with 100G PAM4 QSFP28 modules, in: IEEE/OSA Conference on Optical Fiber Communications (OFC) 2017 W4D.4. <https://www.inphi.com/portal/products/product-listing.php?filt = ColorZ>.
[38] OIF-400ZR Draft IA document; e.g. OIF2017.245.10 at: <http://www.oiforum.com/technical-work/current-oif-work/>.
[39] V. Kamalov, et al., Lessons learned from open line system deployments, in: IEEE/OSA Conference on Optical Fiber Communications (OFC) 2017, M2E.2.
[40] L. Paraschis, Advancements in data-center networking, and the importance of optical interconnections, in: Invited Tutorial, European Conference on Optical Communications (ECOC), London, England, September 2013.
[41] M. Filer, et al., Toward transport ecosystem interoperability enabled by vendor-diverse coherent optical sources over an open line system, J. Opt. Commun. Netw. 10 (2) (2018) A216–A224.
[42] G. Rizzelli, et al., Pizzabox transponders deployment in the field and related issues, in: IEEE/OSA Conference on Optical Fiber Communications (OFC) 2018 Th3A.2 (Invited).
[43] G. Grammel, et al., IEEE/OSA Conference on Optical Fiber Communications (OFC) 2018, Tu2H4.
[44] M. Filer, Panel: Direct vs. Coherent Detection for Metro-DCI, IEEE/OSA Conference on Optical Fiber Communications (OFC) 2017 Tu3A.
[45] V. Dangui, Proc. IEEE/OSA Conference on Optical Fiber Communications (OFC), 2018 Tu3H.3.
[46] A. Sadasivarao, et al., Demonstration of advanced open WDM operations and analytics, based on an application-extensible, declarative, data model abstracted instrumentation platform, in: IEEE/OSA Conference on Optical Fiber Communications (OFC), 2019, M3Z.1.
[47] A. Sadasivarao, et al., High performance streaming telemetry in optical transport networks, in: IEEE/OSA Conference on Optical Fiber Communications (OFC), 2018, Tu3D.3.
[48] W. Grover, Mesh-Based Survivable Networks: Options and Strategies for Optical, MPLS, SONET and ATM Networking, Prentice Hall PTR, Upper Saddle River, NJ, 2003.
[49] Draft-ietf-ccamp-wson-impairments-10.txt. Available at: <http://www.rfc-editor.org/rfc/rfc6566.txt>.
[50] R. Going, et al., Multi-channel InP-based PICs... up to 100GBd, 32QAM, in: European Conference on Optical Communication (ECOC) PDP, 2017.
[51] A. Nespola, et al., Effectiveness of digital back-propagation and symbol-rate optimization in coherent WDM optical systems, in: Th3D.2.pdf OFC 2016.
[52] G. Böcherer, et al., Probabilistic shaping for fiber-optic communication, IEEE Photonics Soc. Newslett. (2017) 4–9.
[53] F. Buchali, et al., Rate adaptation ... probabilistically shaped 64-QAM: an experimental demonstration, J. Lightwave Technol. 34 (7) (2016) 1599–1609.
[54] R. Maher, et al., Constellation shaped 66 GBd DP-1024QAM transceiver with 400 km transmission over standard SMF, in: Proc. ECOC, PDPB2, 2017.
[55] L. Chuang, et al., Demonstration of fully integrated 6 lambda x 200 Gbps PICs and transceivers in L-band, in: ECOC, 2018.
[56] R. Schmogrow, Flexible grid deployments, in: IEEE/OSA Conference on Optical Fiber Communications (OFC) 2018, Tu3I.
[57] M. Birk, Open roadm, in: OFC IEEE/OSA Conference on Optical Fiber Communications (OFC) 2018 Tu2H2 (Invited), and <http://www.openroadm.org/home.html>.
[58] R. Velaga, Enabling future network system scaling, in: OIDA Executive Forum OFC 2018, IEEE/OSA Conference on Optical Fiber Communications (OFC) 2018 and <http://www.ieee802.org/3/ad_hoc/ngrates/public/17_11/cfi_kochuparambil_1117.pdf>.
[59] D. Miller, Attojoule optoelectronics for low-energy information processing and communications, J. Lightwave Technol. 35 (3) (2017) 346–396.
[60] A.V. Krishnamoorthy, et al., From chip to cloud: optical interconnects in engineered systems, J. Lightwave Technol. 35 (15) (2017) 3103–3115.
[61] <http://onboardoptics.org/>.
[62] A. Bechtolscheim, <https://www.opencompute.org/files/OCP2018-AndyBechtolsheim-Final-1.pdf>, or the related Co-Packaged Optics initiative announced in March 2019 by Facebook and Microsoft as part of the Joint Development Foundation.

[63] X. Wu, et al., Applications of artificial neural networks in optical performance monitoring, J. Lightwave Technol. 27 (16) (2009) 3580–3589.
[64] V. Chan, Cognitive optical networks, in: 2018 IEEE International Conference on Communications (ONS.05).
[65] A. Sadasivarao, et al., Optonomic: architecture for autonomic optical transport networks, in: 2019 IFIP/IEEE International Symposium on Integrated Network Management.
[66] C. Hong, et al., B4 and After: managing hierarchy, partitioning, and asymmetry for availability and scale in google's software-definedWAN, in: SIGCOMM, 2018.
[67] G. Landi, et al., Inter-domain optimization and orchestration for optical datacenters networks, IEEE/OSA J. Opt. Commun. Netw. (JOCN) 10 (7) (2018) B140–B151.
[68] D. Geisler, et al., The first testbed demonstration of a flexible bandwidth network with a real-time adaptive control plane, in: European Conference on Optical Communication (ECOC), 2011.
[69] S. Balasubramanian, et al., Demo demonstration of routing and spectrum assignment automation in a transport SDN framework, in: IEEE/OSA Optical Fiber Communications Conference (OFC), 2018, Tu3D.2.
[70] R. Singh, et al., RADWAN: rate adaptive wide area network, in: SIGCOMM 2018.
[71] O. Gerstel, et al., Proactive protection mechanism based on advanced failure warning. US Patent 7,724,676.
[72] M. Khaddam, et al., SDN multi-layer transport benefits, deployment opportunities, and requirements, in: IEEE/OSA Optical Fiber Communications Conference (OFC), 2015.
[73] T. Hofmeister, et al., How can flexibility on the line side best be exploited on the client side? in: IEEE/OSA Conference on Optical Fiber Communications (OFC), 2016, W4G.4.
[74] C. Xie, Alibaba network overview: current, future, challenges, in: Optical Internetworking Forum (oif2017.573.00), October 31, 2017.

CHAPTER 16

Networking and routing in space-division multiplexed systems

Dan M. Marom[1], Roland Ryf[2] and David T. Neilson[2]
[1]The Hebrew University of Jerusalem, Jerusalem, Israel
[2]Nokia Bell Labs, Holmdel, NJ, United States

16.1 Introduction

Optical networks serve as the cornerstone of our connected society, enabling the information superhighway that delivers the internet all across the globe. As the number of users and data services increase, the network technology and architecture must adapt to continue to efficiently and economically support the larger traffic loads. Currently these optical networks consist of optical transceivers of different wavelengths whose signals are wavelength-division multiplexed (WDM) together and transported using fiber-optic cables with optical amplifiers. Each fiber can potentially support more than 100 wavelengths, each carrying hundreds of gigabits per second for a per fiber capacity in the tens of terabits per second. The paths these optical signals traverse through the network can be selected by using reconfigurable optical add-drop multiplexers (ROADM) at network nodes. In this chapter we address current architecture of WDM networks and how it may evolve in the future to support even greater capacities through the use of additional spatial paths, an approach referred to as space-division multiplexing (SDM) as an analogy to WDM.

While the primary focus of this chapter is on the challenges of switching SDM channels, we also address some of the fundamental aspects of SDM optical communications where they have consequences for switching, particularly the choice between coupled and uncoupled spatial modes.

16.1.1 Network growth

Network traffic has been consistently experiencing rapid growth over many decades. This growth has been driven by a seemingly endless stream of digital applications and services, each generation consuming more data resources than the former. This traffic growth trend has continued despite many failures in predicting the next "killer application" or placing limits on future traffic requirements based on extrapolating from existing applications and services into the future [1]. Today's growth is dominated by

Optical Fiber Telecommunications V11
DOI: https://doi.org/10.1016/B978-0-12-816502-7.00018-X

video streaming and machine-to-machine traffic; the emerging artificial intelligence and Internet of Things are expected to dominate data traffic in the near future. It seems reasonable to assume that we will continue to see similar growth rates, 40% per annum (pa), in network traffic capacity over the next decade and likely the one beyond. Indeed, per Ref. [1], "the success and demand for existing applications continuously drives scale and capacity of the underlying network infrastructure to points where further applications are enabled, renewing the cycle."

Optical communication systems as well have demonstrated tremendous capacity scaling over the last decades (see Fig. 16.1), growing in lockstep with the carried traffic load. This orders-of-magnitude capacity enhancement was achieved by the introduction of higher channel baud rates, more efficient modulation formats, denser packing of wavelength channels, and extended spectral support. With said technologies, today's commercial long-haul C + L-band lightwave systems, carrying up to 192 channels at up to 250 Gb/s/ch on a 50-GHz grid, support an aggregate longhaul capacity of ~48 Tb/s at a spectral efficiency (SE) of 5 b/s/Hz. For short-reach applications, up to 400 Gb/s/ch at 8 b/s/Hz are multiplexed for a total capacity of up to 76 Tb/s [2].

While some modest improvement in capacity per fiber core is expected over the next few years, it is anticipated that continuing to achieve a 40% pa growth rate in capacity will only be possible by using multiple spatial channels: SDM. This may take

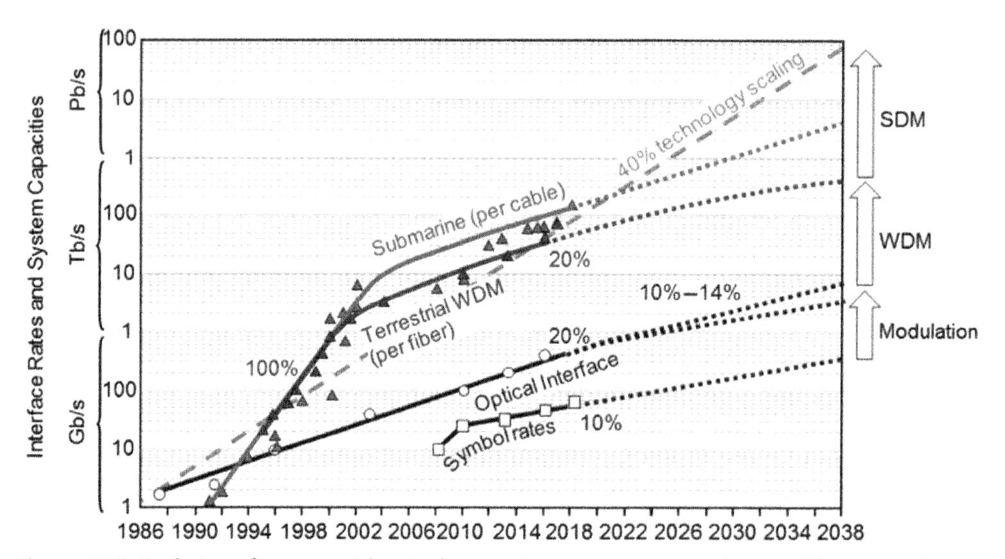

Figure 16.1 Evolution of commercial optical transmission systems over the past 30 years and extrapolations for the coming 20 years. Historical capacity growth has resulted from the use of increased interface rates and the use of wavelength-division multiplexing wavelength-division multiplexed (WDM). Future capacity growth will require the use of spatial channels space-division multiplexing (SDM). Source: *After P.J. Winzer, D.T. Neilson, From scaling disparities to integrated parallelism: a decathlon for a decade, J. Lightwave Technol. 35 (5) (2017) 1099–1115.*

the form of discrete single-mode fibers (SMF) or by using fibers supporting multiple spatial modes, which may further be classified as coupled or uncoupled. Under any of these scenarios, the physical fiber foundation of the network is altered, a scenario that has not been contemplated since the adoption and standardization of the SMF. Complementing the fiber infrastructure change and the larger transported capacity, the network elements and functionality must evolve as well. As of this writing, there are longhaul networks in which multiple SMF are required on a given link to provide the capacity, illustrating that the dawn of the SDM era is already upon us.

16.1.2 Current optical networking

Optical networks follow a mesh topology having nodes located at points of interest (cities, data centers, etc.) and fiber links interconnecting these nodes. Data connections across the network are managed by network operators at the wavelength-channel level. Connection requests between any two nodes are provisioned through an available resource, a wavelength channel that can be routed along a network path in which the utilized fiber links have not yet been assigned this wavelength. Once an available network path for the wavelength channel is identified by the routing and wavelength assignment (RWA) algorithms, the network is reconfigured by setting the wavelength switching gear in all intermediate nodes to create a transparent light path between the communicating endpoints. The functionality allowing network nodes to route wavelength channels all-optically is termed ROADM. The ROADM supports both routing at intermediate mesh network nodes and cross-connect functionality from ingress fiber links to egress fiber links, as well as the local extraction (drop function) and new channel insertion (add function) in support of the communicating endpoints (at wavelength-channel receivers and transmitters).

The key element comprising the ROADM node is the wavelength-selective switch (WSS), which performs the tasks of demultiplexing, switching on a wavelength basis, and remultiplexing in a single compact module utilizing free-space optics [3–7]; see Fig. 16.2. The light from an SMF array is spatially dispersed with a diffraction grating and superimposed at the focal plane. Wavelength-channel selection is achieved by a phase spatial light modulator (SLM), typically incorporating liquid crystal on silicon (LCoS) technology [8], for beam steering to the desired fiber port. The WSS further incorporates dynamic equalization of signal power, by imperfectly switching to the output fiber, balancing uneven optical broadband amplification to improve system optical signal-to-noise ratio (OSNR) in order to maximize reach. The operation of conventional WSS is restricted to a single input, multiple output case, denoted $1 \times K$ WSS, allowing the input wavelengths to be independently distributed to the K output fibers; alternatively, the WSS can operate with multiple (K) inputs and a single output, denoted $K \times 1$ WSS, in which case the connection from the desired input to the

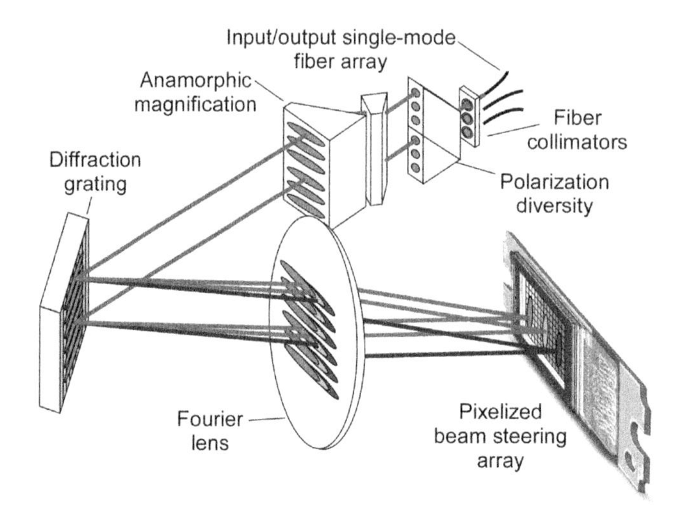

Figure 16.2 Layout of a wavelength-selective switch, consisting of dispersive imaging of all input/output fibers onto a phase spatial light modulator (SLM) for beam steering wavelength channels to desired fiber ports.

output fiber is established independently for each wavelength channel. While initial WSSs only supported fixed channel allocation plans of the standard International Telecommunication Union grids used in the early 2000s (100 and 50 GHz channel plans), it was recognized [9] that there would be a need for optical networks with flexible wavelength allocation by subdividing this grid to finer granularity. Currently [10], the grid is defined on a 6.25-GHz granularity with a minimum 12.5-GHz slot width, to match flexible data rates and modulation formats for a fine-grained rate/reach optimization on any transmission link, but in principle the technology allows for channel widths to be adjusted with sub-GHz granularity. This is possible due to the multitude of fine pixels available on today's LCoS panels, a derivative of the display industry. A flexible channel grid can allow for, for example, 30% of additional capacity on smaller (metro) networks. A flexible grid also supports the transmission of optical superchannels to improve SE, by routing closer spaced subcarriers as one entity with reduced interchannel guard bands.

Using WSS modules at a network node's ingress and egress fiber ports, wavelengths can be routed following the "route-and-select" (R&S) arrangement of Fig. 16.3, left [11]. This introduces cross-connection capability at the ROADM node between the fiber links, and offers direct channel add/drop attachment at the fiber ports. Here, a $1 \times K$ WSS at each ingress fiber port distributes the incoming wavelength channels toward egress fiber ports or to colorless drop ports. At the egress fiber ports, a $K \times 1$ WSS combines the routed and added channels to the output WDM fiber. The R&S topology with multiple modular WSS elements at each node's fiber

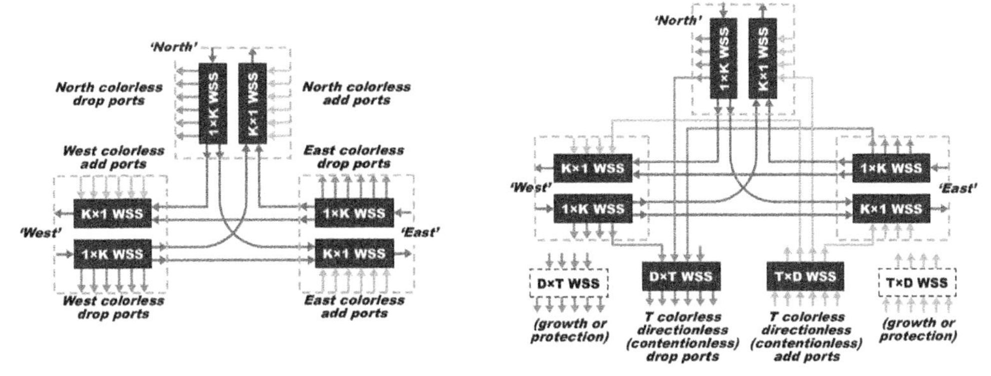

Figure 16.3 Conceptual topologies of multidegree reconfigurable optical add-drop multiplexers (ROADM) (here $D = 3$) in optical mesh networks, offering wavelength routing through the node, as well as channel add and drop. *Left*: colorless add/drop transceivers bound to attached direction. *Right*: colorless and directionless add/drop transceivers. Contentionless add/drop features depend on hardware capability.

ports provides robustness against possible component failure. This topology fulfills a requirement that traffic in one node direction is independent of traffic in other topological directions. In bidirectional line systems (the precursor to mesh networks), this was referred to as East-West separability. As a consequence, optical elements handling traffic coming from, and going toward, a given topological direction (e.g., East) must be capable of continuing to function if any component supporting any other direction (e.g., West) fails, is being replaced, or is undergoing maintenance. The practical implication is that a single switching device cannot simultaneously provide both the add and the drop function for a given through-path in the network; rather, the switching function must be split between switching elements attached to each fiber direction. However, a single module can be used per direction (e.g., West), allowing the ingress $1 \times K$ WSS and egress $K \times 1$ WSS to be combined onto a single optical unit or placed on a common line card (as shown by dashed line surrounding directional hardware in Fig. 16.3).

The add/drop transceivers, being directly attached to the WSS at the ingress and egress fibers as shown in Fig. 16.3, left, are physically associated with that direction. This can be considered as a limitation, as unutilized transceivers disposed at other ports are inaccessible. To best utilize deployed node transceivers, it is desired that they be "directionless," hence able to serve drop (add) channels arriving from (directed to) any network link connected to the node. This can be achieved by adding additional switching hardware to the node design that serves to aggregate the directions and distribute the add/drop channels to the available transceivers; see Fig. 16.3, right. Other desired attributes are that they be colorless and contentionless, the latter attribute allowing any add/drop assignment regardless of other simultaneous node switching assignments, thereby achieving colorless, directionless, and contentionless (CDC)-ROADM functionality. Having a shared pool of local transponders at standby enables protection/restoration functionalities. As little as 15%–20% of overhead capacity can provide sufficient restoration and protection within a large network framework. Moreover, available add/drop network transceivers can serve as effective wavelength converters, potentially resolving blocking situations in RWA algorithms, thereby increasing network throughput.

There are several ways to introduce directionless access to add/drop transceivers [12]. One possibility is to introduce optical cross-connect functionality using a fiber space switch for the add/drop ports [13,14]. Another possibility is to use a multi-port input, multi-port output WSS, with D input fiber ports, matching the node direction count, and multiple (T) output fiber ports for transceiver access (i.e., $D \times T$ WSS), as shown in Fig. 16.3, right. However, conventional WSS cannot handle multiple inputs and outputs simultaneously for independent switching since all ports are imaged to the same position on the LCoS beam steering element. Hence, the caveat in this implementation of an $D \times T$ WSS is that any particular optical wavelength can only appear on one of its input ports, at most. Thus, the node architecture is restricted to drop only a single occurrence of a particular wavelength among all the node's ingress links, which represents contention for the drop wavelength choice in network routing. Such a solution is thus termed colorless, directionless (CD)-ROADM.

A possible route to a CDC-ROADM solution is based on a multicast switch (MCS) solution [15] for the add and drop wavelength channels. An MCS splits its D input dropped signals and delivers a copy of all the drop channels to each attached transceiver out of T. A multiple(D)-input, single-output fiber switch (nonwavelength-selective) at each of T transceivers determines which direction will be connected to corresponding transceiver. A tunable wavelength filter may be required to isolate a single drop channel from the selected direction. However, for coherent detectors this filter may be eliminated. The tunable filter at the drop ports may limit the channel bandwidth, hence impacting the support for channel flexibility and elastic optical networking. Since most elastic networking solutions utilize coherent detection, this should not be a detriment. While the MCS-based solution provides the full CDC functionality using rudimentary elements of splitters and selector switches, it is hampered by the high splitting losses ($1/T$) inherent to the MCS solution. Therefore MCS with high-count transceiver connectivity must be accompanied by an optical amplifier array to compensate for the inherent losses, which adds to the cost of this solution, increases the power consumption due to the optical amplifiers, and degrades the signal's OSNR. Alternatively, MCS with low-count transceiver connectivity (e.g., 4) may offer enough power budget to forego amplification but offer poor add/drop port scaling. These shortcomings have hampered the widespread acceptance of the MCS-based CDC-ROADM.

A WSS alternative that is instrumental for achieving the full CDC-Flex-ROADM requirements is a contentionless $D \times T$ WSS. A contentionless $D \times T$ WSS can support multiple input ports with the same wavelength channel appearing across more than one of its input ports. To achieve this, the WSS design is modified to support multiple inputs, each dispersed onto disjoint positions on the LCoS phase SLM with additional means to route each output (drop) port to a selected dispersed spectrum. Contentionless WSSs conforming to this mode of operation have been demonstrated using hybrid guided-wave and free-space optics arrangements [16], as well purely free-space optics solutions [17,18]. Contentionless WSS are particularly more implementation efficient for CDC-ROADMs when serving many add/drop ports (e.g., 24 or higher), in comparison to MCS-based solutions.

The different CDC access alternatives can be compared for a particular implementation. Consider a degree 8 node, a connectivity scale representing the highest supported neighboring node count today. While degree 8 nodes are rare, due to the way network nodes are geographically distributed in almost any region, the additional available ports can be used to support link overlay. Network traffic is not uniformly distributed, and "hot links" (links reaching full utilization) may arise, especially between two major data sources (adjacent cities or neighboring datacenters). In such circumstances, an additional fiber pair may be utilized for the link, doubling the capacity and still benefiting from the full network connectivity and shared add/drop transceivers without blocking. Furthermore, consider the case where fiber links are terminated with 1×20 WSS, a popular WSS scale, and that the node supports 24 add/drop nodes; see Fig. 16.4. Seven of the WSS ports are required for R&S cross-connect functionality, leaving 13 ports for CDC add/drop connectivity (it is customary to leave one port for future expansion, hence 12 ports remain). If a large port

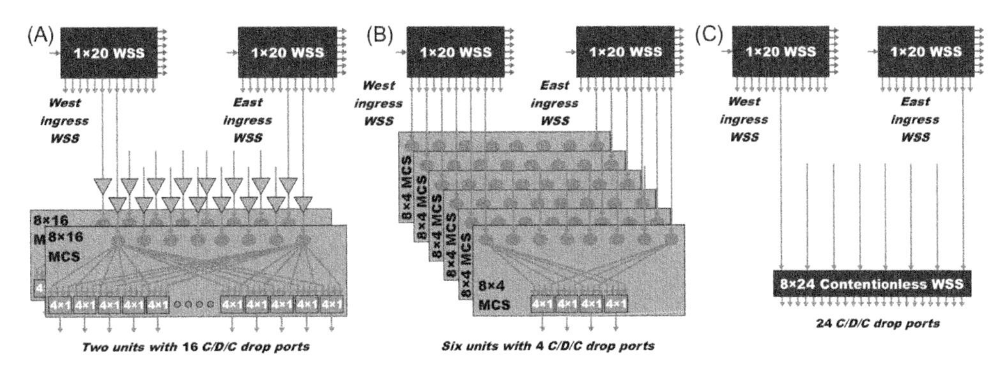

Figure 16.4 Colorless, directionless, and contentionless (CDC) access alternatives to 24 drop transceivers in a degree 8 node. Only two ingress ports are shown, for simplicity, and the add path is omitted. (A) Two 8×16 multicast switch (MCS) with booster amplifiers, (B) six 8×4 MCS, and (C) single 8×24 contentionless wavelength-selective switch (WSS).

count MCS, with an 8×16 MCS being the norm due to the amplification required to overcome the large splitting loss, then two such MCS units would be required to support the 24 transceivers. Low-loss and low port count MCS, such as an 8×4 MCS that can operate without amplification, would require the deployment of six such MCS units. And a single contentionless 8×24 WSS can provide all the required CDC functionality. With 12 WSS drop fiber ports available (see above), it is easy to show that with 8×16 MCS the node can support up to $12 \times 16 = 192$ add/drop transceivers, with 8×4 MCS it can support only $12 \times 4 = 48$ add/drop transceivers, and with 8×24 WSS it can support up to $12 \times 24 = 288$ CDC transceivers.

While a CDC-ROADM allows available transceivers to be connected to any nodal degree on any wavelength with no blocking, network blocking on the path routing may still occur. The network RWA algorithm may face a wavelength continuity impasse at a network node, when the set of available wavelengths on an inbound link does not overlap with the set of available wavelengths on the outbound link. Hence, while there's available capacity on the desired links (e.g., shorted path routing), there is no possible wavelength assignment to fulfill the routing request. Generally, a network reoptimization may resolve this impasse, but this would incur a service disruption, which is typically not allowed. One of the ways to visualize this is to consider each wavelength to be a separate plane of mesh connected switches at the network level, with the CDC-ROADM allowing the add/drop traffic to select from these available wavelength networks. As such, the add/drop transceiver can serve to resolve wavelength contention, being accessible to drop a blocked request, receive, and retransmit on a new wavelength that can then be added back to the network, since the CDC accessibility can bridge the different wavelength planes.

16.1.3 Wavelength-selective switch optical system

A wavelength-selective switch, shown conceptually in Fig. 16.2, is an independent $1 \times K$ (or $K \times 1$) selector switch for each wavelength in the system. This requires the optics to separate the wavelengths in space (spatially dispersed) and image all the input and output fiber ports to the same spatial position (on a wavelength basis) and differing by their incidence angle. Switching is performed by an array of beam steering elements, one per wavelength channel, which can be implemented using many LCoS phase pixels or tilting microelectromechanical system (MEMS) micromirrors. Switching is accomplished by steering an input beam in a direction commensurate with the desired output port position. The main advantage of the WSS over a discrete element implementation (demultiplexer, multiple fiber switches, and multiplexers) is the broader and flatter bandwidth of each wavelength channel being determined by the signal's spectral components contained within the spatial channel extent defined by the SLM. The channel characteristics (sharpness of the edge roll-off) can be shown to depend on the spectral resolution of the resolving optics.

A more detailed schematic of the WSS optical system is shown in Fig. 16.5, where in one dimension (top view) it acts a spectrometer for each of the ports and in the orthogonal (side view) it is a $1 \times K$ switch. The input/output ports are formed by using fibers and microlenses to modify their beam divergence. In the top view a 4f relay arrangement with a diffraction grating at the telecentric stop can be used to create an imaging spectrometer. The focal length of the two lenses in the relay system may be adjusted to determine the illumination size on the grating, which together with the grating frequency determines the system's spectral resolution and illumination extent. In the side view a lens one focal length (f) away from ports and one focal length (f) away from steering element is used to form the switch. This is a 2f or Fourier transform system, which transforms angle steering of the MEMS or LCoS to position and therefore selects the output port. By using cylindrical optics, these two systems can be merged.

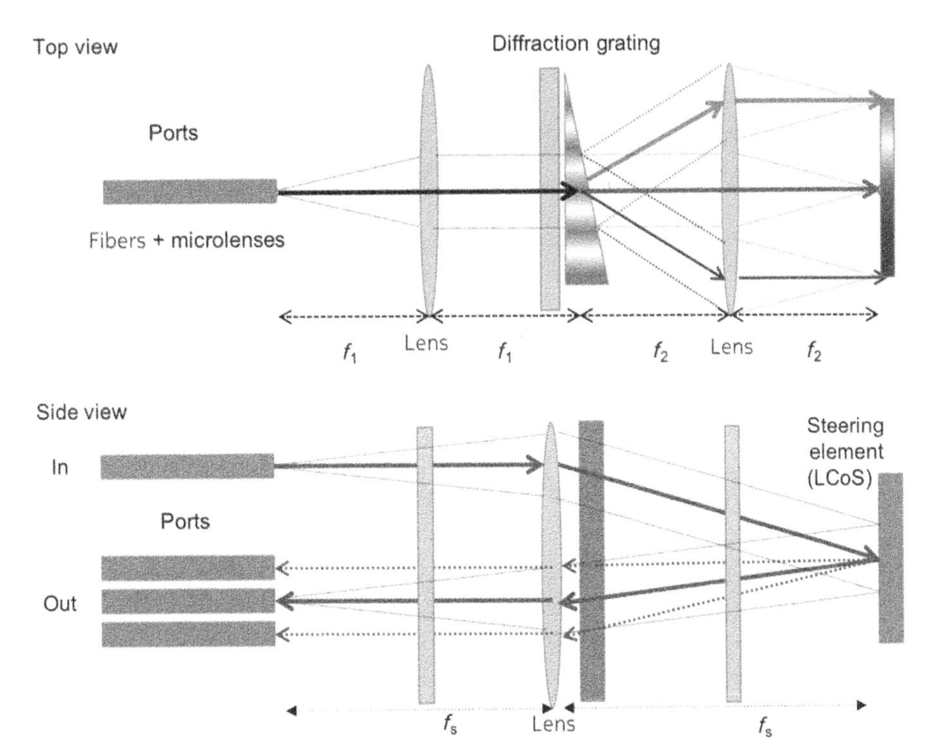

Figure 16.5 Schematic of wavelength-selective switch optical system. The *top view* shows 4-f imaging spectrometer which separates the wavelengths using a diffraction grating. The *side view* shows a $1 \times K$ switch function which is implemented for each wavelength using a 2-f Fourier transform system. The lenses shown in blue are cylindrical.

In actual implementations the diffraction grating may be used in reflection or transmission and the lenses may be replaced by curved mirrors. It is also possible to add more 4f relays to control beam sizes, and most systems will also incorporate some form of polarization diversity (not shown), since the performance of some elements such as the LCoS and gratings are polarization dependent.

It should be noted that a WSS efficiently utilizes the spatial degrees of freedom offered by free-space optics. In one dimension (top view) WDM channels are converted to spatially dispersed channels to allow separate switching. In the other dimension (side view) the output ports form a spatial array. This means that introducing additional support for SDM operation, as will be described in subsequent sections, poses a challenge for switch realizations.

16.2 Spatial and spectral superchannels

16.2.1 Spatial parallelism

As noted, WDM capacities on SMF are quickly approaching their fundamental nonlinear Shannon limits [19], and ways to resolve the looming capacity crunch can only rely on the four physical properties available for modulation and multiplexing of electromagnetic waves: amplitude, time (often also partitioned into frequency, phase, and quadrature), polarization, and space. With these four degrees at hand, and neglecting nonlinear effects, the overall fiber channel capacity, that is, the maximum capacity that can be reliably communicated over an equivalent additive white Gaussian noise channel, can be written as

$$C = 2M \cdot B \cdot \log_2(1 + \mathrm{SNR}),$$

where SNR denotes the signal-to-noise ratio and the logarithmic term captures the maximum possible SE of a single-polarization complex optical signal (in both in-phase and quadrature components). The preceding factor of 2 accounts for polarization multiplexing, and the final capacity is obtained by multiplying the dual-polarization SE with the system bandwidth, B, and the number of parallel spatial paths, M. As reflected in the long-term scaling slowdown clearly visible in Fig. 16.1, the gap to Shannon in terms of SE has become small, and efforts to increase the capacity through improving the SNR with lower noise optical amplification, lower loss or lower nonlinearity fiber, and digital nonlinearity compensation offer only logarithmic (rapidly diminishing) returns. Big linear capacity gains can only be achieved using the "prelog" factors of bandwidth and spatial channels. Parallelism in space is the only option to significantly scale system capacities by appreciable factors in the long run [2].

16.2.2 Partitioning spatial and wavelength space

SDM denotes the use of parallel spatial paths, to complement wavelength scalability (WDM) using the WDM × SDM channel matrix shown in Fig. 16.6 [38]. Each row of the WDM × SDM channel matrix represents wavelength multiplexing within one spatial path, and each column represents multiple parallel spatial paths at the same carrier frequency. Each unit square ("unit cell") represents an optical signal modulated onto a single optical carrier using a single optical modulator and detected using a single optical receiver.

Whether a logical interface is constructed out of unit cells on a common optical path but at different contiguous wavelengths (spectral superchannel), at a common wavelength but across several parallel paths (spatial superchannel [20]), or at a mixture of both (hybrid superchannel), as shown in Fig. 16.6, depends first and foremost on the availability of spatial paths throughout the network at the time of transponder deployment. Terrestrial longhaul and metro networks, which rely on the reuse of already deployed fiber, possibly with very different numbers of fibers available on different links of the network, are likely to prefer spectral superchannels to build out the network by gradually adding more spatial paths once the available spectrum on the existing paths is filled. On the other hand, submarine cables and data-center interconnect (DCI) systems are typically confronted with a greenfield situation, hence can deploy the required number of parallel fibers at the time of system installation, usually at little extra cost in the context of the overall project. In fact, DCI systems are already employing a massive number of parallel spatial paths today (in some cases with more

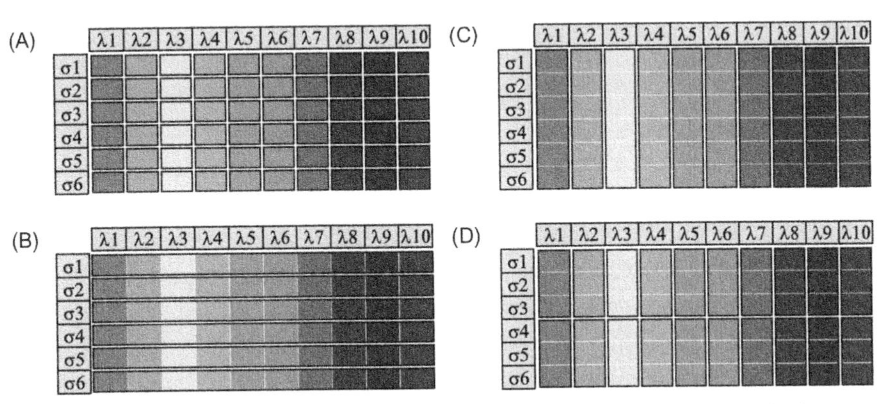

Figure 16.6 Parsing the fiber's space-division multiplexing (SDM) and wavelength-division multiplexed (WDM) channels for switching, where space modes ($\sigma_1, \ldots, \sigma_M$) and wavelength channels ($\lambda_1, \ldots, \lambda_N$) are fully utilized. (A) Space-wavelength granularity (each mode/wavelength channel can be independently switched), (B) space granularity switching performed on mode basis across all wavelengths), (C) wavelength granularity (switching performed on wavelength basis across all modes), and (D) fractional space-full wavelength granularity (switching performed on wavelength basis and spatial mode subgroups).

than 10,000 parallel fibers linking regional datacenter buildings), and both 100 and 400 GbE standards include parallel single-mode interfaces. In the longhaul space, some operators are close to deploying fully loaded WDM systems on parallel fibers, and ROADMs are offered with more degrees than required to accommodate physically diverse paths, thereby opening the possibility to support multiple parallel fibers per physical nodal direction.

16.2.3 Coupled and uncoupled modes

SDM systems are typically divided into two major categories: those with uncoupled modes (low crosstalk) and those with coupled modes (strong coupling). In all cases, bringing parallel light paths in a fiber with a common cladding will introduce crosstalk between the light paths, and the fiber can be operated in two possible regimes:

- Low crosstalk regime, where the crosstalk is kept low enough such that the quality factor degradation of the transmitted signal due to crosstalk leakage is within the margin of the transmission budged
- Strong coupling, where the spatial channels are strongly coupled and multiple-input, multiple-output (MIMO) digital signal processing (DSP) is applied after transmission to undo the coupling which occurred during transmission

Note that the level of crosstalk is in most practical cases linearly dependent on the fiber length, such that the same fiber type could be used in either regime depending on the required transmission range.

A clear example of a coupled mode system would be a multimode core fiber in which each of the different modes of the fiber mix with each other over a distance much less than the transmission distance. Coupled systems are not restricted to single multimode cores but can be composed of many subcores or other index structures which result in supermodes with beat lengths much less than the transmission distance.

For uncoupled mode system an example would be a cable composed of several discrete SMF in which there is negligible crosstalk between modes over the transmission distance. Again, uncoupled systems are not restricted to discrete fibers but also include fibers with multiple cores where the beat length of the supermodes of those cores is longer than the distance over which they maintain phase coherence and coupling between fibers takes the form of crosstalk.

The boundary between coupled and uncoupled mode systems is ultimately a function of the overall system [21], and a coupled system can be defined by the need to perform joint MIMO processing of the spatial channels.

Which regime is used has, however, important implication on the optical switching architecture. In the case of strong coupling, all spatial channels must be routed together from transmitter to receiver for MIMO DSP to function. Therefore also the used optical switches have to switch the strongly coupled spatial channels as an ensemble and routing them along the same path, as depicted in Fig. 16.6C. If no spatial channel mixing occurs, then routing and switching assignments can be applied to each spatial channel and wavelength channel, as shown in Fig. 16.6A, representing the finest switching granularity. As such, it requires the most switching hardware. Also for the no mixing case, optical cross-connects may be applied, switching the entire fiber content comprising all WDM channels, Fig. 16.6B. Finally, if spatial channel mixing is contained within subgroups, then only the subgroups have to be jointly switched throughout the network (Fig. 16.6D). Such can occur for isolated multimode cores in a multicore fiber (MCF).

16.2.4 Switching and blocking considerations

From a switching point of view, it is important to note that the two dimensions making up the WDM × SDM channel matrix are conceptually not equivalent: a connection cannot independently choose any available wavelength on a link but must use the same wavelength end-to-end (unless wavelength conversion is performed, which is a costly undertaking). In contrast, in the absence of coupling between spatial modes, any available spatial path may be chosen independently on each link, provided the ROADM switching equipment allows spatial path switching between ingress and egress fiber ports. If spatial paths are coupled together then routing is restricted to always occur in spatially coupled groups. As a simple example, consider a polarization division multiplexed system on a SMF, where there is strong coupling between polarization modes in a fiber then both polarizations must be routed together as a group within a network.

There are many possible schemes for allocating wavelength and spatial channels in the network links, which lead to the occurrence of various blocking circumstances. This is quantified by a blocking probability metric that depends on the exact details of the system. We can illustrate the general blocking probabilities of wavelength and spatial superchannels [2], as shown in Fig. 16.7, using the following example:

Consider a network node having SDM−WDM links. The number of parallel spatial paths is M and the number of wavelengths is N. We consider spatial superchannels consisting of $M_{SC} \leq M$ contiguous spatial channels, where M/M_{SC} is an integer. A spectral superchannel consists of $N_{SC} \leq N$ contiguous spatial channels, where N/N_{SC} is an integer as well. While M_{SC} and N_{SC} are in general not linked, we consider the case where they are equal, $M_{SC} = N_{SC}$, representing allocation of equal amounts of bandwidth or capacity in either case. A service request requires the allocation of a superchannel from an ingress to an egress link. Let's assume a fraction $\gamma \leq 1$ of the aggregate link capacity in the desired egress direction of the ROADM node is utilized by other traffic.

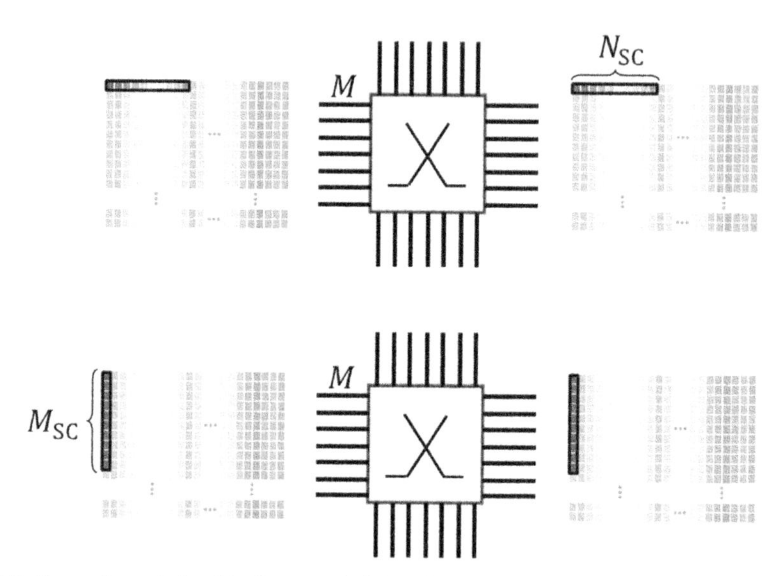

Figure 16.7 Illustration of the blocking probability for spectral (*top*) and spatial (*bottom*) superchannels. The node has *M* spatial paths on its attached fiber links. Each spatial superchannel occupies M_{SC} contiguous paths. Each spectral superchannel occupies N_{SC} contiguous spectral blocks. The number of equivalent paths for a spectral superchannel is *M*, while for spatial superchannels the number of equivalent paths is M/M_{SC}. Larger numbers of equivalent paths in the spectral superchannel system results in much lower blocking probabilities.

For spectral superchannels, since we are allocating blocks of N_{SC} wavelengths, the probability that the particular, identical, wavelength block of a spectral superchannel is already occupied on any one of the M parallel egress links is γ. Hence, the probability that this wavelength range is already occupied on all M parallel egress paths is γ^M, which is the blocking probability for the spectral superchannel.

In contrast, a spatial superchannel occupying M_{SC} spatial paths at a given wavelength will encounter blocking at a probability $\gamma^{M/M_{SC}}$, as the number of available egress options (blocks of M_{SC} spatial lanes) is now restricted M/M_{SC}. Thus the blocking probability for spatial superchannels is larger than that for spectral superchannels, as $\gamma^{M/M_{SC}} \geq \gamma^M$.

The example is somewhat simplified by using fixed size contiguous blocks, but it illustrates the differences in blocking probabilities between spatial and spectral superchannels. However, for all cases where you treat them equivalently, the spectral superchannel allocation has equal or lower blocking probability compared to spatial superchannels. A lower blocking probability not only means more capacity, but also simplifies provisioning algorithms for the spectral superchannel. The presence of a larger set of equivalent paths in the spectral superchannel system results in much lower blocking probabilities than the spatial superchannel allocation system.

16.3 Coupled mode space-division multiplexing

While capacity increase can be achieved by simply using more SMFs, new fibers have been proposed, as an alternative, that support multiple parallel optical paths either by using fibers with multiple cores, by using cores that support multiple spatial modes, or a combination of the two.

16.3.1 Multimode switches

The conventional WSS, based on SMFs (see Fig. 16.2), can be generalized to support multimode fibers (MMFs), radiating into free-space propagating modes. In the simplest case, where only 3 spatial modes are supported, this can be achieved by replacing the input SMF array with a few-mode fiber array and a readjustment of the collimation microlenses [22], whereas for a larger number of modes, a redesign of the optics might be required to make sure that all working planes of the device (diffractive grating and switching element) are at either a direct image plane or a spatial Fourier transform plane of the MMFs' end facet [23] (see also Fig. 16.5). A disadvantage of multimode WSSs built this way is the appearance of mode-dependent passbands caused by the different lateral dimensions of the modes along the dispersed direction of the LCoS switching element.

It is therefore instructive to analyze in more detail the mode profile of MMFs. The mode profiles for the two most common MMFs, graded-index (GI) and step-index (SI) MMF, are shown in Fig. 16.8 in terms of the linear polarized (LP) modes LP_{nm}, where n is the radial mode number and m is the azimuthal mode number.

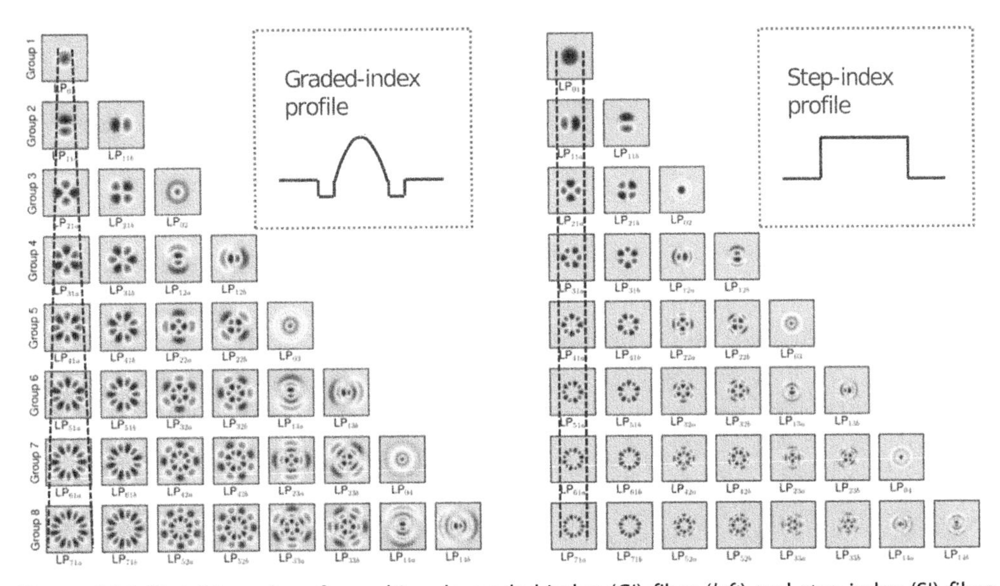

Figure 16.8 First 36 modes of a multimode graded-index (GI) fiber (*left*) and step-index (SI) fiber (*right*).

For the GI−MMF the modes are divided in groups, where each added mode group consists of one added degenerate mode. The modal-field diameter grows proportionally to the group number, where the modal-field diameter of the last supported mode group is comparable to the core diameter. In contrast, the modes of the SI−MMF are organized differently and consists in nondegenerate modes for the LP_{0n} modes and twofold degenerate mode groups for the LP_{nm} modes with for $n\neq0$. Note that in Fig. 16.8 we arrange the SI-fiber modes in the same triangular fashion as the GI-fiber modes even if the degeneracy groups are different, just to allow for a simple direct graphical comparison between the two fiber types. The mode-field diameters of the SI−MMF are found to be mostly constant and defined by the core diameter.

The design of a multimode WSS (see, e.g., Fig. 16.5) will generally depend on both the mode profile as present at the end of the fiber but also on the so-called far field (spatial Fourier transform of the modes), as several elements in the switch are arranged in Fourier transform configuration. Therefore in Fig. 16.9 we also report the far field of the modes. The far-field modal profiles look identical to the near field in the case of the GI−MMFs, whereas for SI fibers the far-field mode diameter grow as function of the mode number, but faster than in the case of the GI fiber.

To get a better qualitative understanding of how the mode-field profile impacts the switch performance in terms of channel passbands and dimension and deflection angle requirements of the switching element, we analyze mode clipping by a finite, one-dimensional aperture, as set by the SLM for WDM channel definition.

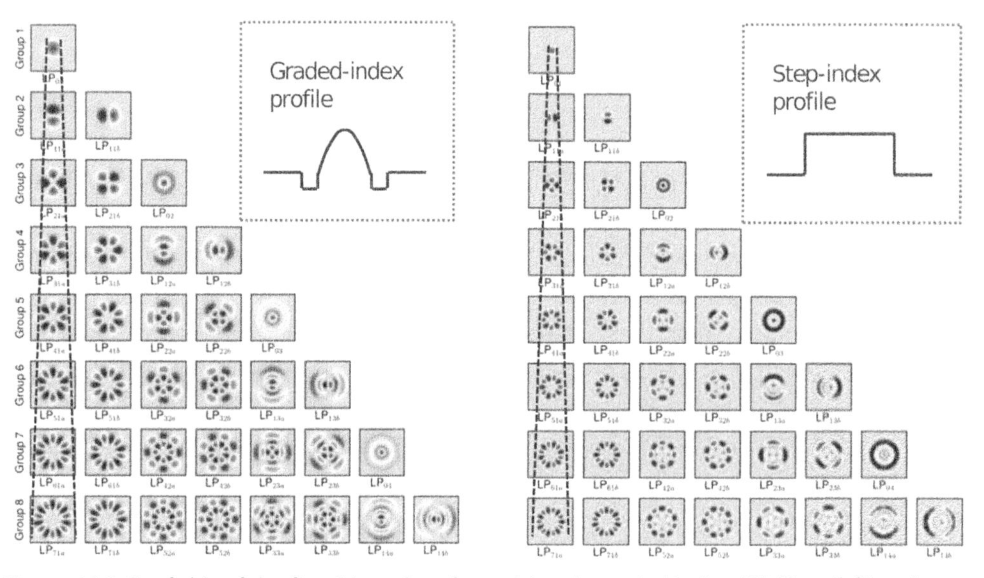

Figure 16.9 Far fields of the first 36 modes of a multimode graded-index (GI) fiber (*left*) and step-index (SI) fiber (*right*). Far field represents the angular space of the mode.

In Fig. 16.10, we show the relative width *d* of the slot required to transmit 98% of the light as function of the mode number, where the width *d* is set to 1 for the fundamental mode (LP01). As described and indicated by the dotted lines in Figs. 16.8 and 16.9, the lateral dimensions of the modes grow as function of the mode number for the GI fiber, and is almost constant for the SI fiber in the near field, whereas growth is faster than in GI fiber when the far-field is considered.

Note that the modal extension in both the near- and the far field are of relevance in determining the spectral resolution and the switch capability in terms of number of ports of the WSS [24]. It is therefore instructive to analyze the relative angle-aperture product obtained by multiplying the relative aperture widths of the near- and far-field case (see Fig. 16.11). As can be seen, the angle-aperture product is similar for both GI

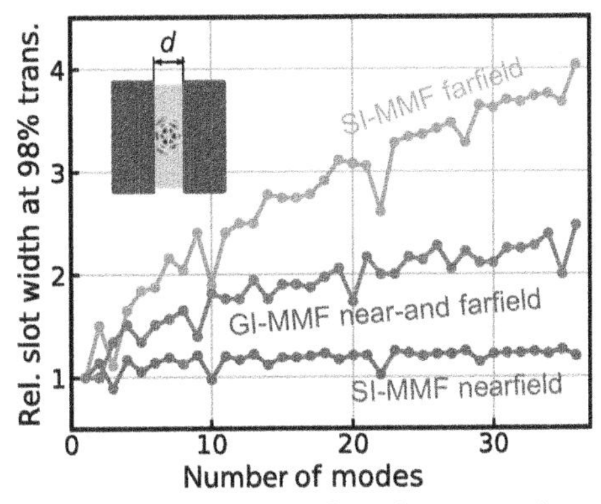

Figure 16.10 Slots width for 98% transmission after clipping with a rectangular aperture (slot) relative to the fundamental mode (LP01) for the near and far field of both graded-index (GI) and step-index (SI) fibers.

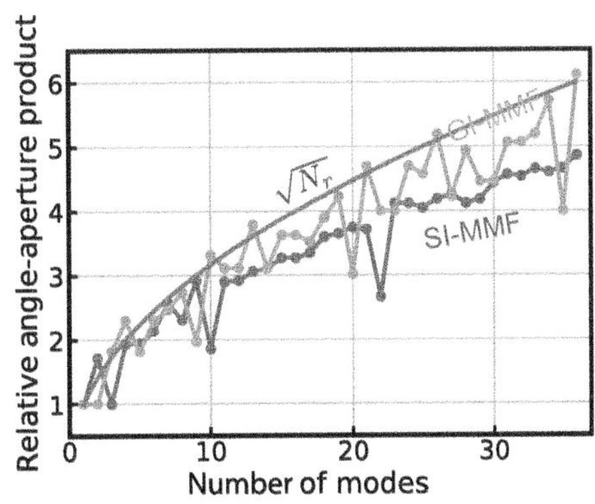

Figure 16.11 Relative angle-aperture products calculated for step-index (SI) and graded-index (GI) multimode fibers (MMFs) as a function of the number of modes. Angle-aperture product obtained by multiplying the relative aperture widths of the near- and far-field case.

and SI fibers, and both grow approximately proportionally to the $\sqrt{N_r}$, where N_r is the mode number.

As a first fundamental result, we learn that multimode wavelength-selective switches will require $\sqrt{N_r}$ times the spectral resolution of a single-mode WSS to achieve comparable passband width. Note that for the case of multimode WSSs, a reduction in clear passband could be considered as a viable trade-off, as the device now supports N_r times the switching capacity and the bandwidth occupied by transceivers is expected to grow to 200 GHz or wider allowing for wider guard bands then currently required in high-SE SMF systems.

The second fundamental result concerns the number of ports that can be addressed. The multimode WSS will require $\sqrt{N_r}$ times the number of pixels to steer to an equivalent number of ports. Note that this is more favorable then utilizing N_r single-mode WSS in parallel, which would require N_r times the number of pixels.

As the channel passband performance of a multimode WSS is modal dependent, it is instrumental to further study the impact of a WSS cascade, as would be experienced in a networking environment, when the signal traverses multiple network nodes having WSS elements. In a conventional WSS (single-mode based), the impact of successive WSS passages is equivalent to multiple filtering events with the same filter function (more realistic statistical modeling takes into account random channel misalignments). In a multimode WSS cascade, the impact of mode mixing between successive WSS passages plays a very important role [25]. Without mode mixing, the higher order modes (that experience the strongest filtering due to their larger mode-field diameters) accumulate stronger channel filtering, with the divergence between the fundamental and higher order modes growing [resulting in mode-dependent loss (MDL)-defined passband narrowing]. With strong mixing between multimode WSS passages, signals carried on higher order modes get transferred to lower order modes (and vice versa) in random fashion, resulting in slower accumulation of bandwidth reduction from mode-dependent filtering. In this sense, the reported advantages of strong mode mixing in the propagation of SDM signals (lower MDL, differential group delay, and nonlinear distortion accumulation) extend also to passband narrowing in multimode WSS passages.

While the discussion here focused on multimode WSS design, the same methodology can be applied to strongly coupled cores (hence, rather densely packed). The fiber cores support supermodes, that can be propagated (imaged and Fourier transformed) within a WSS, to spectrally resolve and beam steer the fiber supermodes. However, since MCFs do not possess rotational symmetry (as opposed to MMFs), the channel passband characteristics depend on the MCF angular orientation [26]. Flatter, wider passbands are achieved by minimizing the mode-field diameter in the dispersion direction. For a four-core fiber, placing it in a square arrangement (aligned with the dispersion axis) versus a 45 degree orientation makes an appreciable difference as the mode-field diameter varies by $\sqrt{2}$.

16.3.2 Joint-switching architecture

An approach for reducing the SDM−ROADM hardware complexity is to employ WSSs that operate in a joint-switching mode. A joint-switching WSS switches all the spatial channels from an SDM input (MMF, MCF, or SMF array) as a group, on a wavelength-channel basis, to desired output SDM destinations. As described, the design of a conventional WSS has all the fiber ports dispersed and imaged to the same positions at the switching plane, where the input beam is switched to a desired output port or fiber on a wavelength basis (see Section 16.1.2). Since all the ports are imaged to the same location on the steering element, the beam steering can simultaneously redirect multiple beams with the same angular shift. If the WSS input/output ports are arranged in a regularly spaced array, then the steering of a set of inputs is reimaged onto different sets of outputs. Associating the groups of ports with SDM fiber interfaces allows parallel switching of all the spatial channels to their destinations. This allows an increase in the SDM capacity without increasing the WSS count compared to today's SMF implementations. More importantly, the channel passband characteristic of the joint switching is identical to that of conventional WSS, and no mode-dependent passbands are introduced.

However, there is an increase in the number of ports, which results in an increase in the aperture of the lenses and overall system size in the steering direction by the SDM multiplier. Additionally there may be increased crosstalk into undesired ROADM ports from the joint switching. Since LCoS beam steering is diffraction based, it produces spurious diffraction orders in addition to the primary beam steering one that can couple to other ports and cause crosstalk [27]. With the multiple parallel input and output ports for joint switching, it will become more challenging to ensure the other orders do not couple to other ports. It should be noted that this joint-switching approach can be thought of as an extension of an approach present within conventional SMF-based WSS in the form of the polarization diversity where the polarization modes are converted to discrete spatial modes and jointly switched. The mapping of the SDM into the SMF array also impacts the WSS performance. The SDM group can be mapped to a contiguous fiber port group, with multiple groups spanning the fiber array; see Fig. 16.12, left. This arrangement is less favorable as the beam steering range is increased, and the performance of LCoS-based beam steering decreases for larger angles. Placing the SDM groups in an interleaved fashion, as shown in Fig. 16.12, right, requires the same beam steering range as conventional WSS. Hence while the number of switched channels has grown by the SDM group size, the same number of LCoS pixels is performing the switching.

One variation of the WSS design to allow an increase of the port count involves arranging the fiber ports in a two dimensional array [28]. Switching remains in the vertical axis (orthogonal to dispersion). If SDM groups are associated with rows of the

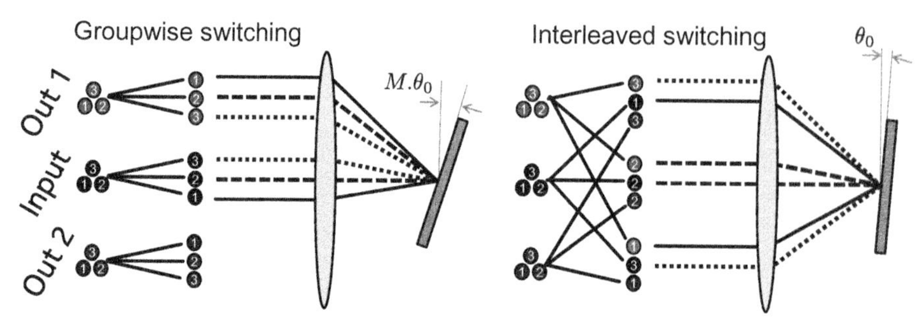

Figure 16.12 Joint-switching architecture for wavelength-selective switches. Two approaches with groupwise and interleaced port positions are shown. Figure shows function in switching direction.

2D array, and an input SDM fiber is spatially demultiplexed and fed to one row, then the entire SDM group can be jointly switched to output SDM fibers associated with other rows.

In order to use the joint-switching architecture with MMFs, the modes have to be multiplexed and demultiplexed from multimode space into N_r single-mode channels, which can introduce a significant amount of insertion loss depending on the multiplexing technology used [29,30]. Mode multiplexer supporting three spatial modes can have an insertion loss ≤ 1 dB, whereas number of modes as large as 36 the minimum demonstrated insertion loss is around 5 dB, which would result in an added loss of 10 dB, and multimode WSS using a multimode input with much smaller loss have been demonstrated [23].

16.4 Uncoupled mode space-division multiplexing

In this section we will describe the approaches and challenges in switching systems composed of many spatially multiplexed channels that are independent of each other. This means that the spatial channels do not have to be jointly MIMO processed and so do not have to be routed together or to the same final destination. As described in Section 16.2, this allows us to consider several switching scenarios: independent switching of spatial/spectral units, assigned per demand, and the options of spatial or spectral superchannels. Generally, the former will require more complicated switching equipment to access the finest switching granularity. Routing spatial superchannels in uncoupled SDM would use essentially the same approaches as those used for coupled mode SDM, and covered in preceding sections, except that the input to the optical systems starts with set of discrete single modes. Finally, we elaborate on the case of spectral superchannels, which have lower blocking probabilities (see Section 16.2), and the prospect to reduce switch complexity by reduction in spectral resolution with increasing superchannel extent.

16.4.1 Uncoupled space-division multiplexing fibers

For uncoupled mode SDM the most obvious path is to use large numbers of SMF fibers. This has the advantage of backward compatibility and takes advantage of the existing connector technology and standards. There are now ultrahigh count cables with up to 3456 fibers that can fit within standard ducts and be spliced in groups of 12 using ribbon splicers [31]. Higher spatial densities may be achieved using uncoupled MCFs [32]. These could be joint switched in the same manner as for coupled mode or spatially remapped and treated as discrete single modes. One of the more potentially attractive uncoupled MCF is that of a 7- core hexagonally packed fiber in a 125 μm cladding, since the center core can interoperate with existing SMF infrastructure while allowing future 7 × capacity upgrade. It has been recently observed that the crosstalk requirements for such fibers [33] may allow a universal design suitable for all system reaches.

16.4.2 Transitioning to space-division multiplexing-wavelength-division multiplexed reconfigurable optical add-drop multiplexers nodes

As networks evolve to support higher capacities by introducing SDM links, so too must the node architecture and its role in routing communication channels toward their ultimate destination. With uncoupled transmission fiber, channels assigned to spatial paths and wavelengths are to be independently switched within modified ROADM solutions expanded to support SDM and WDM channel plans. At small--scale SDM deployments, as contemplated today, the R&S architecture for cross-connecting D ingress and D egress fiber ports continues to be effective(with D^2 inter-connecting fibers). Access to add/drop transceivers, currently the challenge in single-mode ROADM nodes, is even more taxing with the introduction of SDM. We propose a few SDM−WDM ROADM designs that can be network implemented in the near term, as shown in Fig. 16.13.

In the general case, the R&S topology requires individual interconnecting fiber for each spatial channel between all ingress and egress ports that are each served with conventional WSS. For a degree D node, with M spatial channels, $(D-1)\times M$ of the WSS ports are required to support cross-connect functionality including spatial lane change (i.e., routing of a WDM signals from one spatial channel to another). This number can grow very quickly and surpass the number of WSS output ports (consider $D=M=6$). To curtail the WSS port consumption for cross-connect functionality, we can restrict the switching to remain at the same spatial channel, hence requiring only $(D-1)$ WSS ports, with the remaining WSS ports available to support add/drop functionality. The theme of placing limitations on the SDM-ROADM capabilities will repeat itself in all following node architectures, in effort to match desired network functions to switching gear capabilities. Fig. 16.13A depicts a directional and colorless

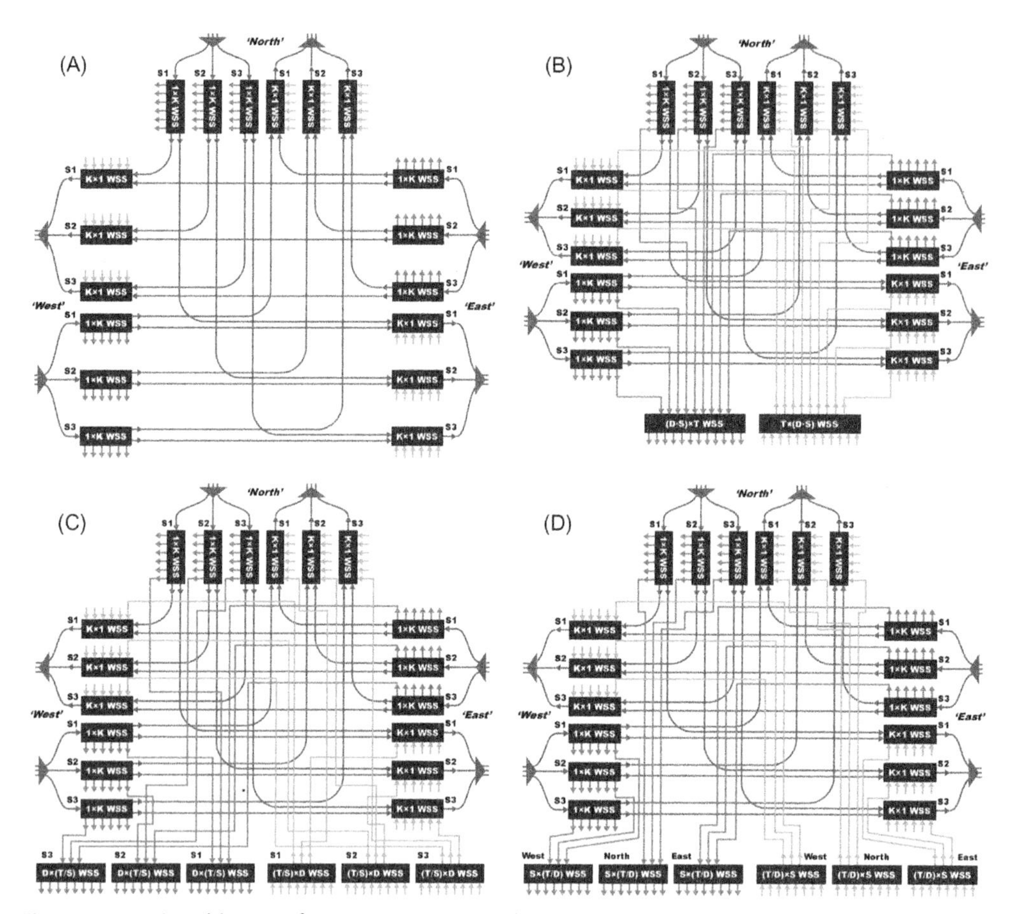

Figure 16.13 Possible ways for supporting space-division multiplexing (SDM)-expanded reconfigurable optical add-drop multiplexers (ROADM) nodes. (A) Route and select node topology with directional transceiver access. (B) Colorless, directionless, and contentionless (CDC)-ROADM operating over space-division multiplexing-wavelength-division multiplexed (SDM-WDM). (C) Layered SDM-ROADM, with colorless/contentionless access per spatial degree. (D) ROADM with colorless/directionless access per nodal direction.

SDM-WDM ROADM, supporting R&S cross-connect functionality within the spatial channel level (or degree), that is, without spatial degree change. As in the WDM-ROADM case, having the add/drop transceivers bound to a particular direction—here direction has a double meaning, both nodal direction and spatial degree—is considered inefficient as the transceivers are not a shared resource, which would ease network routing and failure recovery.

CDC-ROADM access to add/drop transceivers can be obtained with a multi-port intput, multi-port output WSS serving as a port aggregator. Such a drop port aggregator WSS would accept wavelength multiplexed channels on interconnecting fibers originating from each ingress spatial degree and nodal direction, and distribute dropped channels to any of the large number of attached receivers in a contentionless manner. A $(D \times S) \times T$ WSS allows desired CDC access to T transceivers. This scenario was previously mentioned with respect to today's state-of-the-art multi-port WSS supporting 8×24 ports, and since most network node degrees are smaller than 8, the additional inputs can be assigned to spatial degrees; see Fig. 16.13B. However, even for the small node case shown in the figure, with $D = 3$ and $M = 3$ (on all node links), the WSS requires nine input ports, already exceeding the capabilities available today.

To allow some level of directionless access to shared transceivers, we again restrict some of the permissible connections to transceivers at the network node. Just as the R&S was limited to within the same spatial degree, we can restrict the CDC access to shared transceivers to a single spatial degree (Fig. 16.13C). This requires S parallel port aggregators, one per spatial degree, with each supporting $D \times (T/S)$ WSS, commensurate with existing technology. The resulting architecture is that of a layered SDM-ROADM node, with each spatial degree being completely decoupled from the others, forming parallel planes with each devoted a spatial degree. Such SDM-ROADM implementations can be deployed as overlays on existing optical networks, enhancing the network capacity by the introduction of additional spatial degrees. A second form of restricted access to shared transceivers is possible by assigning the transceivers to a nodal direction (Fig. 16.13D). This requires D parallel port aggregators, one per nodal direction degree, with each supporting $S \times (T/D)$ WSS (enabling access for all spatial degrees). Such SDM-ROADM implementations are better matched to new SDM network installations, where each link fully supports SDM parallelism and new links are turned on as network develops (node degree increases by one, and additional transceivers assigned to serve new nodal direction are introduced).

Since SDM is encroaching upon us and certain network links already utilize more than one fiber for the link, SDM-enabled nodes need to be installed now. SDM-ROADM nodes supporting a relatively small number of spatial degree parallelism can be implemented in one of the ways described, using switching elements available today. The offered capacity increases in these first SDM networks, projected to provide $\times 4-8$ capacity increase, will support network growth for the next 4–6 years (under the assumption of 40% pa growth). It is interesting to note that first generation WDM lightwave systems, deployed in the mid-1990s, provided eight wavelength channels. Twenty years later, channel count in WDM networks has increases more than tenfold. In the next section we investigate how SDM can scale to higher spatial degrees.

16.4.3 Scaling space-division multiplexing switches

As noted, there are networks in which more than one SMF is required to provide the required capacity, on a network link. Currently this can be managed by treating the additional fibers as an increase in nodal degree. It is therefore reasonable to consider whether this approach is scalable.

In order to understand the scaling challenge, we must look again at the CDC ROADM described above and illustrated in Fig. 16.4. The 1×20 WSS can be thought of as a $1 \times (D+P)$ WSS, where D is the nodal degree (typically $D=8$) and P is the number of drop ports (typically $P=12$). The number of drop ports P is actually set by the desired drop fraction, which is typically 25%, and the size of the MCS is $D \times T$. For the MCS, D is the number of splitters which must equal the nodal degree and T is the split ratio which is limited by the ability of the amplifier to compensate the loss, $T \approx 16$. The number of drop port is then $P \times T = 12 \times 16 = 192$ and the drop fraction is $F_D = (P \times T)/(D \times L)$, where there are L spectral channels (today $L=96$), so $F_D = (P \cdot T)/(D \cdot L) = (12 \times 16)/(8 \times 96) = 25\%$.

For a constant number of spectral channels L, split ratio T and drop fraction F_D, the number of drop ports P scales proportionately with D. For example, if we scaled a degree 8 ROADM node to four fibers per direction, we would require 1×80 WSS at the ingress/egress ports, since $D=32$ and $P=48$, which may still be possible with today's WSS platforms. However, at 16 fibers per direction, we would require 1×320 WSS, which is beyond current WSS technology approaches. Therefore we will consider if there are methods to alleviate this scaling challenge by reducing the number of spectral channels or by modifying the switching architecture.

In terms of its internal architecture, a ROADM requires the node to first perform a wavelength separation of the ingress signal before switching. As architecturally explained along with Fig. 16.5, wavelengths are first transformed to W individual spatially resolved channels inside a WSS by means of a diffractive element and are then spatially switched before recombining back into a common egress fiber. The number of spatially resolved channels W is a significant factor in determining the size and complexity of a wavelength-selective switch. Importantly, the number of wavelength-equivalent spatial paths W, is greater than the number of spectral superchannels L, as it is dictated by the spectral resolution ("steepness") of the ROADM's optical filtering function, commonly defined using a 0.5- to 6-dB roll-off; the value of 6 dB is chosen because the filter characteristics of adjacent channels typically cross at their 6-dB points. With today's typical values of 40-GHz 0.5-dB passbands on a 50-GHz grid, implying a 5-GHz spectral resolution, a ROADM supporting a 5-THz system bandwidth (C-band) thus needs $W \approx 1000 \approx 10\,L$ wavelength-equivalent spatial paths. This number depends only on the ratio of system bandwidth to spectral resolution and not on the width of the (super)channels to be switched. Therefore W can be scaled with L while still maintaining overall system SE.

While spatial superchannels, as shown in Fig. 16.6, require high-resolution spectral switching, with the resolution of a single unit slice, spectral superchannels can reduce the spectral resolution of the underlying system components through wider absolute guard bands at the same relative guard bandwidth. (In order to detect individual subcarriers within a superchannel, digital coherent detection can be used, but true subcarrier add/drop requires either steep optical filtering [39] or interferometric opto-electronic processing [40].) The implications of reducing the number of superchannels L across a given system bandwidth are visualized in Fig. 16.14 with spectral resolution and spectral utilization as parameters, the latter defined as the available spectrum within a 0.5-dB filter transmission bandwidth divided by the total system bandwidth. Fig. 16.14A shows that the required spectral resolution of the switch W, at a constant spectral utilization scales inversely with $(L-1)$. As a consequence, W can be reduced, which reduces the spatial

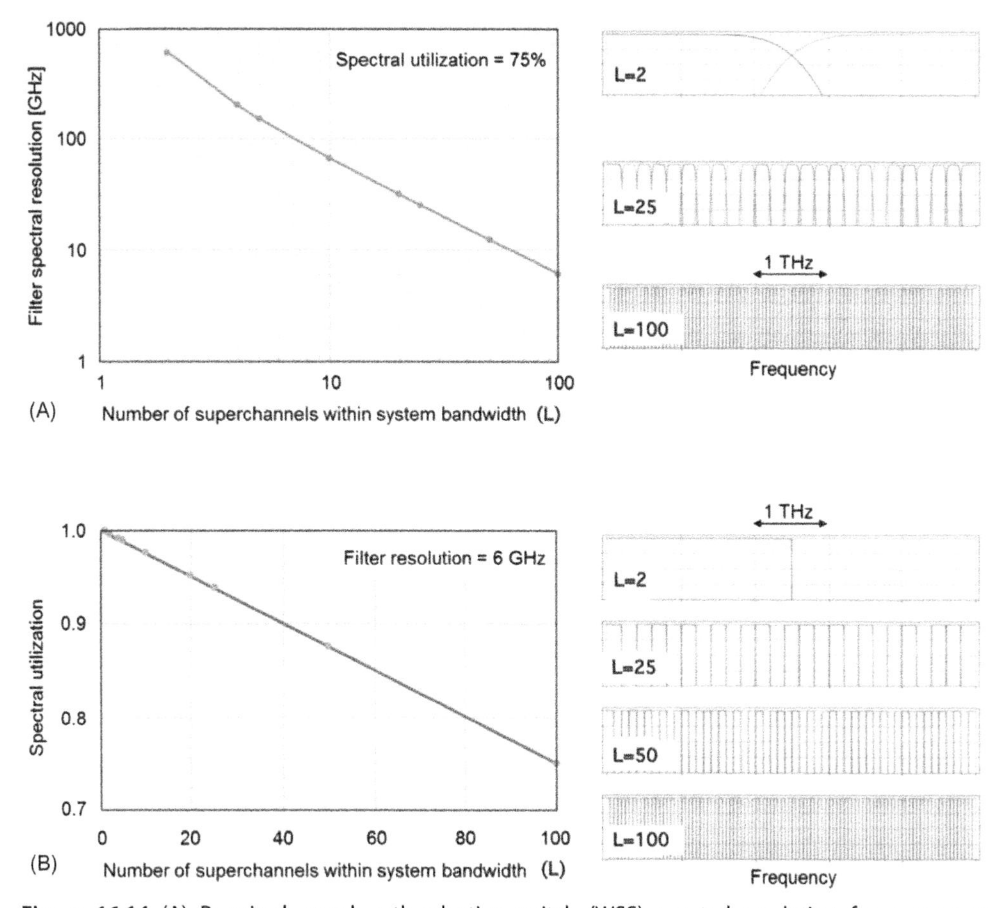

Figure 16.14 (A) Required wavelength-selective switch (WSS) spectral resolution for a constant spectral utilization of 75% as a function of the number of spectral superchannels L within the system bandwidth. Example spectra are given for $L = 2$, 25, and 100 superchannels. (B) Spectral utilization as a function of the number of spectral superchannels with int system bandwidth for a constant WSS spectral resolution of 6 GHz. Example spectra are given for $L = 2$, 25, 50, and 100. Source: *After P.J. Winzer, D.T. Neilson, A.R. Chraplyvy, Fiber-optic transmission and networking: the previous 20 and the next 20 years (invited), Opt. Express 26 (18) (2018) 24190–24239.*

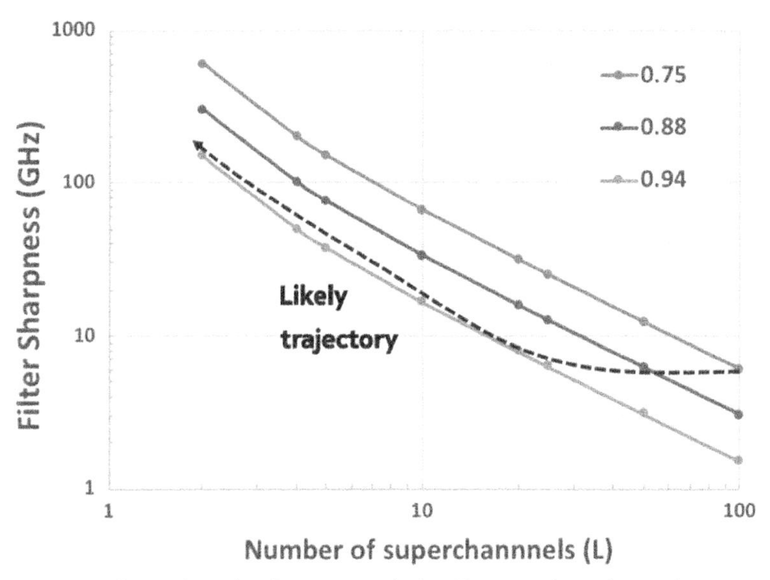

Figure 16.15 Required wavelength-selective switch (WSS) spectral resolution for a constant spectral utilization of 75%, 88%, and 94% as a function of the number of spectral superchannels within the system bandwidth. Also likely evolution path shown, from left to right over time, where filter sharpness is initially maintained resulting in higher spectral utilization and then a transition wider filtering functions as number of superchannels drops below 20, corresponding to 250 GHz wide superchannels.

complexity of the wavelength switch in the wavelength direction. Alternatively, as illustrated in Fig. 16.14B, if the spectral resolution of the ROADM's filtering components is kept constant, higher spectral utilization is obtained for a reduced number of superchannels. It is likely that the evolution will be a mixture of the two approaches, shown in Fig. 16.15, with an initial evolution at constant spectral resolution until system utilization is over 90% for $20 \approx 30$ superchannels, followed by a reduction in required SE to manage the increased complexity of the greater number of fibers.

A reduction in the number of spectral superchannels L not only offers the opportunity to reduce W, but since L contributes to the number of required drop ports, it reduces the complexity in the switching direction. If it were possible to reduce the number of wavelength channels L, at the same rate as the number of fibers increases (hence degree D increases), then $D \times L$ product would maintain at a constant, leading to a fixed number of drop ports on a WSS. For the case of four fibers per direction, $L = 96/4 = 24$, we would require 1×48 WSS since $D = 32$ and $P = 16$, and at 16 fibers per direction $L = 96/16 = 6$, we would require 1×144 WSS since $D = 128$ and $P = 16$. While this approach partially mitigates the scaling problem, the switch scaling rapidly becomes dominated by degree switching D, which is independent of the number of superchannels L.

For the future scenarios of [2], scaling to 22×8 spatial paths (by 2027) requires 176 units of 1×440 WSSs, and scaling to 625×8 spatial paths (by 2037) requires 5000 units of 1×125000 WSSs! While these numbers are large, they are a consequence of permitting nonblocking switching between all the fibers that is possible with spectral superchannels. While it may seem that the spectral superchannels requires higher port count switches than the spatial superchannels, making it seem a less attractive option, we note it is simply a consequence of implementing the possible lower blocking probability architecture. If blocking comparable to the spatial superchannels were assumed, the port count of the switches would also be comparable. The large port counts of the individual WSS is also a consequence of the inefficiency of building an $N \times N$ cross-connect out of two stages of discrete $1 \times N$ switches.

In the limit where a spectral superchannel occupies the entire system bandwidth ($L = 1$), no spectral switching is required ($W = 1$) and the node-internal number of spatial paths reduces to just $(D + P)2\ M$. Once spectral superchannel systems evolve to this limit of pure fiber switching, several additional advantages can be expected on a system level: (1) switches do not need to split wavelengths at all (except for subcarrier add/drop functions performed at the edge of the network), leading to lower-loss and broader-band operation (e.g., photonic cross-connects (PXCs) can cover 1270–1650 nm of bandwidth); (2) the SE of superchannels can then be increased to its maximum potential, as filtering guard bands are no longer required; (3) fiber cuts affect entire amplifiers as opposed to fractions of an amplifier's bandwidth, making amplifier transients due to fiber cuts no longer an issue; (4) the characteristics of a path are largely fixed once that path is configured, making the physical-layer performance independent of optical path reconfiguration and consequently allowing for operation at a reduced (almost zero) margin as well as easier system automation.

To gain an understanding of the scale and devices required for a future ROADM, we consider the case of spectral superchannels in the year 2037 according to Ref. [2], assuming 625 fibers per nodal direction, and interfaces filling the entire system bandwidth (i.e., pure fiber switching). The switching architecture would then be entirely based on PXCs, potentially with flexible wavelength multiplexing on the add/drop side, should optical subcarrier grooming be required (see Fig. 16.16). The switching architecture for the enormous space switch lends itself to a strictly nonblocking three-stage Clos network, where the first and third stages are switches associated with one of the ROADM directions, and the mid-stage contains multiple parallel cross-connects. (While there are other possible switch architectures, the Clos architecture has the benefit of keeping the number of spatial paths (i.e., fibers) between stages at only twice the number of input fibers). A degree 8 node in the year 2037 with 25% add/drop functionality will therefore need 6250 fibers. Implementing this directly in the Clos network of Fig. 16.16 with $M = 625$, $D = 8$ and $P = 2$ leads to 625×1250

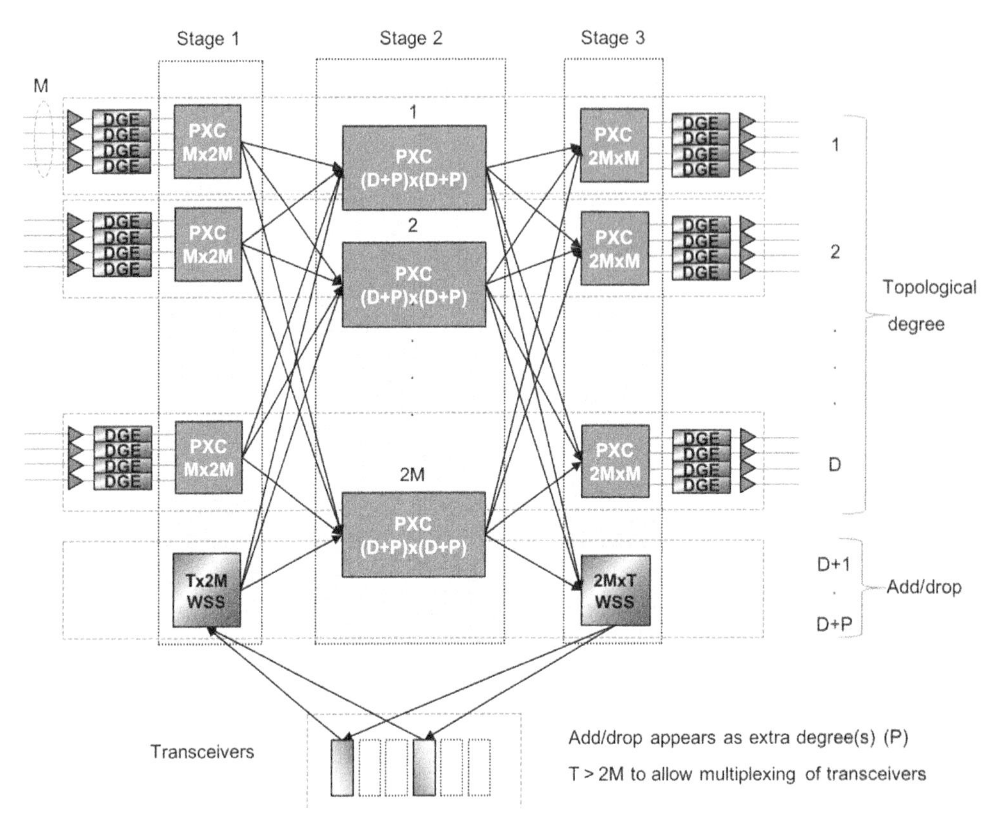

Figure 16.16 Possible future spatial switching node showing amplifier and dynamic gain equalization on a per-spatial-link basis and a spatial photonic cross-connect (PXC) composed of a Clos network with the first and third stages associated with switching among the fibers (M) on each nodal degree (D) or add drop (P) and central stages providing degree connectivity $(D+P)\times(D+P)$. Add/drop functionality uses extra spatial degrees and may retain wavelength switching to provide flexible subcarrier multiplexing. Source: *After P.J. Winzer, D.T. Neilson, A.R. Chraplyvy, Fiber-optic transmission and networking: the previous 20 and the next 20 years (invited), Opt. Express 26 (18) (2018), 24190–24239.*

and 10×10 cross-connects for edge and center stage, respectively. While these are potentially practical devices, it is worth noting that the Clos network [34] allows for an internal resizing; for example, we can maintain 625 fibers per direction but increase the logical degree $(D+P)$ by a factor of 5 (i.e., we partition the number of fibers per degree into five groups), so the effective M becomes 125 and the effective D and P become 40 and 10, respectively. This leads to 125×250 and 50×50 cross-connects for edge and center stages, which are moderate-sized cross-connects that are available even today [35–37]. This also reduces the failure group size for the PXC associated with the degrees.

16.5 Future networks

Future growth in optical network capacity will require the use of multiple spatial paths. For terrestrial systems we expect the most likely path that will be taken is the use of large numbers of uncoupled spatial modes. This will probably be in the form of large numbers of SMF deployed in cables, though uncoupled MCFs may also be deployed in situations where duct space is limited. There are several historical factors that make this the most likely approach: the embedded base of SMF which has large fiber counts on certain routes, the relative low cost of SMF compared to MCF, the mature connector and splicing infrastructure for SMF, the ability to incrementally deploy capacity by lighting additional fibers, and the ability to reuse today's hardware; it also builds on the approach that is already being used for routes with multiple fibers. We expect the bundling of channels to be in the form of spectral superchannels, which is only possible for uncoupled spatial mode approaches. This leads to lower blocking probabilities and simpler network control and management. Additionally the reduction of filtering allows for greater per fiber capacity. Ultimately this path leads to the elimination of wavelength switching and it being replaced with purely spatial or fiber switching. One of the benefits of a purely spatially switched scenario is that it eliminates the blocking that is present today because wavelengths are not interchangeable. While this results in a more fully connected network, it also requires switches with higher degrees. Implementing these higher degree switches requires a more efficient approach than the R&S approach of today's WSS and will lead to the use of Clos switching architectures. The elimination of spectral switching would further simplify optical networks since the optical links would be independent of each other, with no static or variable impairments from the introduction or removal of wavelengths along their path. Additionally, requirements to suppress transient in amplifiers and provide per channel equalization along the path would be reduced or eliminated. Furthermore, having all WDM channels traverse the same fiber, end to end, without wavelength add/drop allows additional nonlinear distortion compensation techniques that consider cross phase modulation and four-wave mixing between carriers, such as digital back propagation [], to be deployed [41].

16.6 Conclusions

The use of SDM in terrestrial networks is an inevitable consequence of the demand for increased bandwidth. While the addition of SDM to optical transport simply adds linearly to the number of channels in an unused degree of freedom, the challenges in switching are more significant. The spatial degrees of freedom are already being used in the optical switches to enable wavelength demultiplexing and switching to multiple ports. This means that adding SDM compromises some aspect of the switching performance. To manage increased numbers of spatial channels, it may be necessary to compromise on the spectral resolution and therefore the channel count.

It seems that the most likely approach to SDM in terrestrial networks, at least initially, will be to use increasing numbers of uncoupled spatial channels, again most likely in the form of multiple SMFs. It also seems likely that as the number of these spatial channels increases, there will be a corresponding reduction in the number of wavelength channels to be switched, ultimately leading us to a purely spatial switching network scenario. A purely spatial switched network will lead to significant changes not only in switch architecture and network control, but also due to the elimination of combining diverse wavelength routed signals, in optical line system design.

References

[1] P.J. Winzer, D.T. Neilson, From scaling disparities to integrated parallelism: a decathlon for a decade, J. Lightwave Technol. 35 (5) (2017) 1099–1115.

[2] P.J. Winzer, D.T. Neilson, A.R. Chraplyvy, Fiber-optic transmission and networking: the previous 20 and the next 20 years (invited), Opt. Express 26 (18) (2018) 24190–24239. Available from: https://doi.org/10.1364/0E.26.024190.

[3] J.E. Ford, V.A. Aksyuk, D.J. Bishop, J.A. Walker, Wavelength add-drop switching using tilting micromirrors, J. Lightwave Technol. 17 (5) (1999) 904–911. Available from: https://doi.org/10.1109/50.762910. ISSN: 0733-8724.

[4] D.M. Marom, D.T. Neilson, D.S. Greywall, N.R. Basavanhally, P.R. Kolodner, Y.L. Low, et al., Wavelength-selective 1 x 4 switch for 128 WDM channels at 50 Ghz spacing, in: Optical Fiber Communication Conference and Exhibit, March 2002, FB7–FB7. https://doi.org/10.1109/0FC.2002.1036767.

[5] T. Ducellier, J. Bismuth, S.F. Roux, A. Gillet, C. Merchant, M. Miller, et al., The MWS 1 × 4: a high performance wavelength switching building block, in: 2002 28th European Conference on Optical Communication, vol. 1, September 2002, pp. 1–2.

[6] B.P. Keyworth, ROADM subsystems and technologies, in: OFC/NFOEC Technical Digest. Optical Fiber Communication Conference, 2005, vol. 3, March 2005. doi:https://doi.org/10.1109/0FC.2005.192706.

[7] D.T. Neilson, C.R. Doerr, D.M. Marom, R. Ryf, M.P. Earnshaw, Wavelength selective switching for optical bandwidth managementBell Labs Tech. J. 11 (2) (2006) 105–128ISSN: 1538-7305 . Available from: https://doi.org/10.1002/bltj.20164.

[8] G. Baxter, S. Frisken, D. Abakoumov, I. Clarke, A. Bartos, S. Poole, Highly programmable wavelength selective switch based on liquid crystal on silicon switching elements, in: 2006 Optical Fiber Communication Conference and the National Fiber Optic Engineers Conference, March 2006. doi: https://doi.org/10.1109/0FC.2006.215365.

[9] Y. Su, S. Chandrasekhar, R. Ryf, C.R. Doerr, L. Moller, I. Widjaja, et al., A multirate upgradable 1.6-Tb/s hierarchical OADM network, IEEE Photonics Technol. Lett. 16 (1) (2004) 317–319. Available from: https://doi.org/10.1109/LPT.2003.820495. ISSN: 1041-1135.

[10] ITU-T G.694.1, Spectral grids for WDM applications: DWDM frequency grid, International Telecommunications Union, Geneva, CH, Standard, February 2012.

[11] B. Collings, New devices enabling software-defined optical networks, IEEE Commun. Mag. 51 (3) (2013) 66–71. Available from: https://doi.org/10.1109/MC0M.2013.6476867. ISSN: 0163-6804.

[12] B.C. Collings, Wavelength selectable switches and future photonic network applications, in: 2009 International Conference on Photonics in Switching, September 2009, pp. 1–4. doi:https://doi.org/10.1109/PS.2009.5307841.

[13] J.M. Simmons, A.A.M. Saleh, Wavelength-selective cdc roadm designs using reduced-sized optical cross-connects, IEEE Photonics Technol. Lett. 27 (20) (2015) 2174–2177. Available from: https://doi.org/10.1109/LPT.2015.2455931. ISSN: 1041-1135.

[14] R. Jensen, A. Lord, N. Parsons, Colourless, directionless, contentionless ROADM architecture using low-loss optical matrix switches, in: 36th European Conference and Exhibition on Optical Communication, September 2010, pp. 1–3. doi:https://doi.org/10.1109/ECOC.2010.5621248.

[15] S. Tibuleac, ROADM network design issues, in: 2009 Conference on Optical Fiber Communication, March 2009, pp. 1–48. doi:https://doi.org/10.1364/NFOEC.2009.NMD1.

[16] Y. Ikuma, K. Suzuki, N. Nemoto, E. Hashimoto, O. Moriwaki, T. Takahashi, 8 x 24 wavelength selective switch for low-loss transponder aggregator, in: 2015 Optical Fiber Communications Conference and Exhibition (OFC), March 2015, pp. 1–3.

[17] L. Pascar, R. Karubi, B. Frenkel, D.M. Marom, Port-reconfigurable, wavelength-selective switch array for colorless/directionless/contentionless optical add/drop multiplexing, in: 2015 International Conference on Photonics in Switching (PS), September 2015, pp. 16–18. https://doi.org/10.1109/PS.2015.7328938.

[18] P.D. Colbourne, S. McLaughlin, C. Murley, S. Gaudet, D. Burke, Contentionless twin 8 × 24 WSS with low insertion loss, in: Optical Fiber Communication Conference Postdeadline Papers, Optical Society of America, 2018, Th4A.1. https://doi.org/10.1364/OFC.2018.Th4A.1. [Online]. Available from: <http://www.osapublishing.org/abstract.cfm?URI = OFC-2018-Th4A.1>.

[19] R.-J. Essiambre, G. Kramer, P.J. Winzer, G.J. Foschini, B. Goebel, Capacity limits of optical fiber networks, IEEE/OSA J. Lightwave Technol. 28 (4) (2010) 662–701.

[20] L. Nelson, M. Feuer, K. Abedin, X. Zhou, T. Taunay, J. Fini, et al., Spatial superchannel routing in a two-span ROADM system for space division multiplexing, IEEE/OSA J. Lightwave Technol. 32 (4) (2014) 783–789. Available from: https://doi.org/10.1109/JLT.2013.2283912 [Online]. Available from: <http://ieeexplore.ieee.org/stamp/stamp.jsp?arnumber = 6623116>.

[21] P.J. Winzer, From first fibers to mode-division multiplexing (invited paper), Chin. Opt. Lett. 14 (12) (2016) 120002 [Online]. Available: http://col.osa.org/abstract.cfm?URI = col-14-12-120002.

[22] D.M. Marom, J. Dunayevsky, D. Sinefeld, M. Blau, R. Ryf, N.K. Fontaine, et al., Wavelength-selective switch with direct few mode fiber integration, Opt. Express 23 (5) (2015) 5723–5737. Available from: https://doi.org/10.1364/OE.23.005723.

[23] H. Chen, N.K. Fontaine, R. Ryf, B. Huang, A. Velazquez-Benítes, C. Jin, et al., Wavelength selective switch for commercial multimode fiber supporting 576 spatial channels, in: ECOC 2016; 42nd European Conference on Optical Communication, September 2016, pp. 1–3.

[24] K.-P. Ho, J.M. Kahn, J.P. Wilde, Wavelength-selective switches for mode-division multiplexing: scaling and performance analysis, J. Lightwave Technol. 32 (22) (2014) 3724–3735.

[25] M. Blau, D.M. Marom, Channel passband broadening via strong mixing in cascaded few-mode fiber wavelength-selective switches, in: Optical Fiber Communication Conference, Optical Society of America, 2017, Tu2C.2. doi:https://doi.org/10.1364/OFC.2017.Tu2C.2 (Online). Available from: <http://www.osapublishing.org/abstract.cfm?URI = OFC-2017-Tu2C.2>.

[26] M. Blau, D.M. Marom, Benefits of a coupled-core wavelength-selective switch, in: Optical Fiber Communication Conference (OFC) 2019, Optical Society of America, 2019, W2A.5. doi:https://doi.org/10.1364/OFC.2019.W2A.5 (Online). Available from: <http://www.osapublishing.org/abstract.cfm?URI = OFC-2019-W2A.5>.

[27] H. Yang, B. Robertson, D. Yu, Z. Zhang, D.P. Chu, Origin of transient crosstalk and its reduction in phase-only LCOS wavelength selective switchesJ. Lightwave Technol. 31 (23) (2013) 3822–3829ISSN: 0733-8724 . Available from: https://doi.org/10.1109/JLT.2013.2288153.

[28] N.K. Fontaine, T. Haramaty, R. Ryf, H. Chen, L. Miron, L. Pascar, et al., Heterogeneous space-division multiplexing and joint wavelength switching demonstration, in: Optical Fiber Communication Conference Post Deadline Papers, Optical Society of America, 2015, Th5C.5. doi:https://doi.org/10.1364/OFC.2015.Th5C.5 (Online). Available from: <http://www.osapublishing.org/abstract.cfm?URI = OFC-2015-Th5C.5>.

[29] N.K. Fontaine, R. Ryf, C. Liu, B. Ercan, J.R.S. Gil, S.G. Leon-Saval, et al., Few-mode fiber wavelength selective switch with spatial-diversity and reduced-steering angle, in: Optical Fiber Communication

Conference, Optical Society of America, 2014, Th4A.7. https://doi.org/10.1364/0FC.2014.Th4A.7 (Online). Available from: <http://www.osapublishing.org/abstract.cfm?URI = 0FC-2014-Th4A.7>.

[30] J. Carpenter, S.G. Leon-Saval, J.R. Salazar-Gil, J. Bland-Hawthorn, G. Baxter, L. Stewart, et al., 1×11 few -mode fiber wavelength selective switch using photonic lanterns, Opt. Express 22 (3) (2014) 2216–2221. Available from: https://doi.org/10.1364/0E.22.002216 [Online]. Available from: <http://www.opticsexpress.org/abstract.cfm?URI = oe-22-3-2216>.

[31] F. Sato, K. Tsuchiya, Y. Nagao, T. Hirama, R. Oka, K. Takahashi, Ultra-high-fiber-count optical cable for data center applications, SEI Tech. Rev. 86 (2018) 45–50.

[32] T. Hayashi, T. Taru, O. Shimakawa, T. Sasaki, E. Sasaoka, Uncoupled multi-core fiber enhancing signal-to-noise ratio, Opt. Express 20 (26) (2012) B94–B103. Available from: https://doi.org/10.1364/0E.20.000B94.

[33] J.M. Gené, P.J. Winzer, A universal specification for multicore fiber crosstalk, IEEE Photonics Technol. Lett. 31 (9) (2019) 673–676. Available from: https://doi.org/10.1109/LPT.2019.2903717. ISSN: 1041-1135.

[34] C. Clos, A study of non-blocking switching networks, Bell Syst. Tech. J. 32 (2) (1953) 406–424. Available from: https://doi.org/10.1002/j.1538-7305.1953.tb01433.x. ISSN: 0005-8580.

[35] CALIENT S series optical circuit switch 320×320 port. Available from: <www.calient.net/products/s-series-photonic-switch/>, 2019 (archived 08.03.19).

[36] X. Zheng, V. Kaman, O. Jerphagnon, R.C. Anderson, H.N. Poulsen, J.R. Sechrist, et al., Three-dimensional MEMS photonic cross-connect switch design and performance, IEEE J. Sel. Top. Quantum Electron. 9 (2) (2003) 571–578. Available from: https://doi.org/10.1109/JSTQE.2003.813321. ISSN: 1077-260X.

[37] Polatis series 7000, 384×384 port software-defined optical circuit switch. Available from: <www.polatis.com/series-7000-384x384-port-software-controlled-optical-circuit-switch-sdn-enabled.asp>, 2018 (archived 23.08.18).

[38] D.M. Marom, M. Blau, Switching Solutions for WDM-SDM Optical Networks, IEEE Commun. Mag. 53 (2) (2015) 60–68.

[39] R. Rudnick, A. Tolmachev, D. Sinefeld, O. Golani, S. Ben-Ezra, M. Nazarathy, et al., Sub-GHz Resolution Photonic Spectral Processor and its System Applications, J. Lightwave Technol. 35 (11) (2017) 2218–2226.

[40] P.J. Winzer, An Opto-Electronic Interferometer and Its Use in Subcarrier Add/Drop Multiplexing, J. Lightwave Technol. 31 (2013) 1775–1782.

[41] R. Maher, T. Xu, L. Galdino, M. Sato, A. Alvarado, K. Shi, et al., Spectrally Shaped DP-16QAM Super-Channel Transmission with Multi-Channel Digital Back-Propagation, Scientific Reports 5 (2015). Article number: 8214.

CHAPTER 17

Emerging optical communication technologies for 5G

Xiang Liu[1] and Ning Deng[2]
[1]New Jersey Research Center, Futurewei Technologies, Bridgewater, NJ, United States
[2]Huawei Technologies, Shenzhen, P.R. China

17.1 Introduction on 5G requirements and 5G-oriented optical networks

In this section we will introduce the key features of 5G and its requirements on optical networking. An overview of 5G-oriented optical networking will also be provided.

17.1.1 Introduction to 5G requirements

5G is expected to offer unprecedented improvements over the previous generations of mobile networks in multiple aspects, such as a 1000-fold increase in system capacity, 100-fold improvement in energy efficiency, millisecond-level end-to-end network latency, and massive connectivity for numerous devices [1−5]. For instance, downlink peak data throughput could reach 20 Gb/s, while uplink peak data rates could be as high as 10 Gb/s. 5G will also reduce latency and improve overall network efficiency. Streamlining network architectures will deliver end-to-end latency requirements of less than 5 ms. This will allow 5G to offer ultra-reliable low-latency communication for machine-to-machine type and public safety type applications. As it delivers new infrastructure solutions, 5G will depend on an end-to-end digital service transformation, which will minimize operating cost, deliver efficiencies, and drive revenue growth. A widely used diagram to describe the main application scenarios to be supported by 5G is shown in Fig. 17.1. These 5G application scenarios can be divided into three categories:

1. Enhanced mobile broadband (eMBB), for applications such as smart phones, augmented reality (AR), virtual reality (VR), and wireless-to-the-X to provide broadband access for residential and business users, requiring a throughput in the order of 10 Gb/s.

Optical Fiber Telecommunications V11
DOI: https://doi.org/10.1016/B978-0-12-816502-7.00019-1

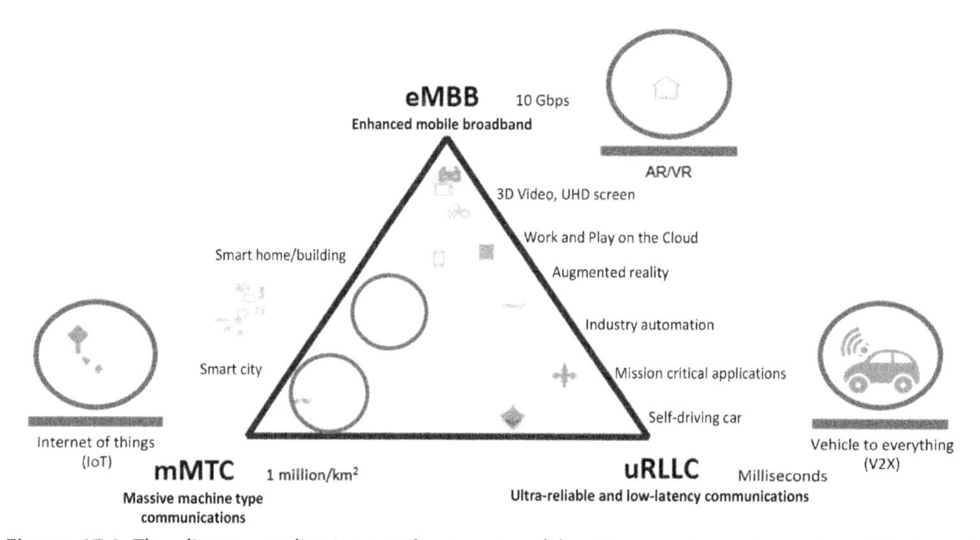

Figure 17.1 The diverse applications to be supported by 5G, covering enhanced mobile broadband (eMBB), ultra-reliable and low-latency communications (uRLLC), and massive machine type communications (mMTC). *AR*, Augmented reality; *IoT*, internet of things; *UHD*, ultra-high definition; *VR*, virtual reality.

2. Ultra-reliable and low-latency communication (uRLLC), for applications such as industrial applications, autonomous vehicles, stock trading communications etc., requiring an end-to-end network latency in the order of 1 ms.
3. Massive machine type communication (mMTC), for applications such as intelligent sensing and monitoring, intelligent lighting, intelligent meter reading, etc., requiring millions of connections per square kilometer.

From the preceding description, we can see that the 5G business use cases are mainly driven by the need for better user communication experiences and more connectivity for the internet of things (IoT).

Currently, 5G wireless network and 5G transport network are being extensively studied and standardized by organizations such as the Third Generation Partnership Project (3GPP) [1] and International Telecommunication Union (ITU) [2,3], with an anticipated global launch in 2020, as shown in Fig. 17.2. This important goal has attracted extensive research efforts from telecommunication industries and academia. The infrastructure of the current wireless communications will not be able to meet the requirements of 5G, so a set of novel radio access technologies (RATs) will be required. In addition, the Common Public Radio Interface (CPRI) industry cooperation defines publicly available specification for the key internal interface of radio base stations between the radio equipment control (REC) and the radio equipment (RE) to facilitate worldwide efforts toward 5G [6].

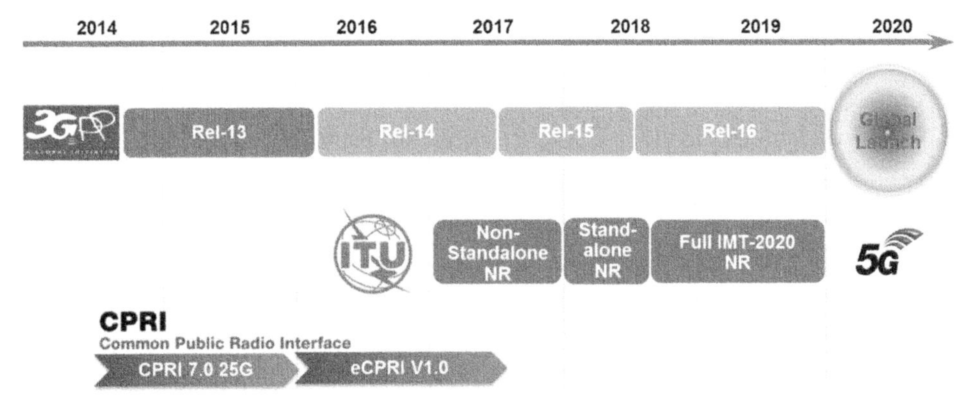

Figure 17.2 The evolution of wireless networks toward 5G with global efforts coordinated by organizations such as Third Generation Partnership Project (3GPP), International Telecommunication Union (ITU) and common public radio interface (CPRI). *IMT*, International mobile telecommunications; *NR*, new radio; *Rel*, release.

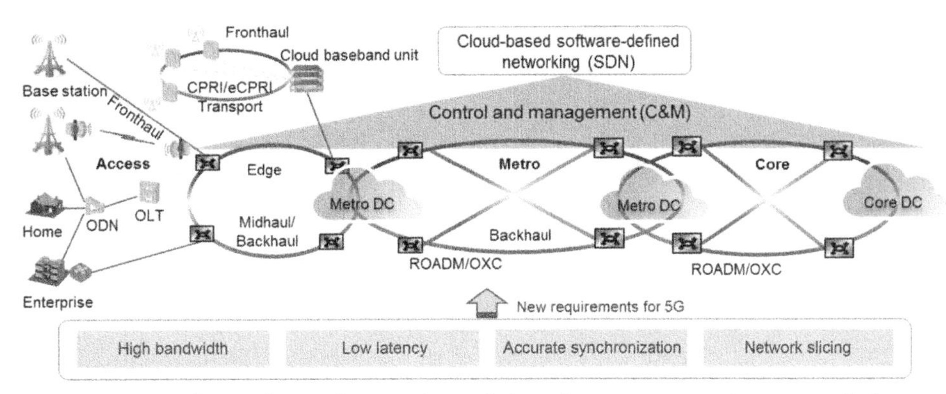

Figure 17.3 An end-to-end optical network consisting of core, metro, and access optical networks to support diverse applications including radio access and fixed access applications. *DC*, Data center; *ODN*, optical distribution network; *OLT*, optical line terminal.

17.1.2 Introduction on 5G-oriented optical networking

A typical end-to-end optical communication network consists of core, metro, and access optical networks, as shown in Fig. 17.3. Most of today's optical networks are built to carry diverse services including mobile services, residential services, and enterprise services, among others. 5G brings to optical networking new requirements such as high bandwidth, low latency, accurate synchronization, and the ability to perform network slicing to optimize 5G-oriented the resource utilization for any given application. To better address these requirements for 5G, optical networks need to be evolved and improved, leading to the so-called optical networking [7–9].

A new trend in optical networking is the widespread buildout of data centers, effectively supporting both core and metro networks via cloud computing and centralized resource sharing. As mobile network migrates from the fourth-generation (4G) to 5G, cloud radio access network (C-RAN) becomes even more important, providing desirable features such as centralized baseband processing for CoMP and cross-cell interference cancellation, resource pooling at central offices, and simplification at remote radio units (RRUs) [10]. Thus, a key enabler of 5G-oriented optical networking is the widespread distribution of cloud-based data centers. In fact, not only have internet content providers based their infrastructure on data centers in recent years, but also telecommunication service providers have started to evolve their traditional optical networks towards data center-centric networks, as illustrated in Fig. 17.4. For data center interconnection (DCI), it demands high traffic volume, nearly full mesh connectivity, and dynamic bandwidth reconfiguration capability.

In the wireless network evolution from the second generation (2G), the third generation (3G), the fourth-generation, (4G) to the upcoming 5G, there have also been important architecture changes, as illustrated in Fig. 17.5. Following are the key features of these generations of networks.

- In 2G, BSC: base transceiver station (BTS) and base station controller (BSC) are separated. The supporting optical network is primarily based on synchronous optical networking (SONET) and synchronous digital hierarchy (SDH).
- In 3G, BTS and BSC evolved to Node B and radio network controller (RNC), respectively, which remained to be separated. The supporting optical network is primarily based on the multiservice transport platform.

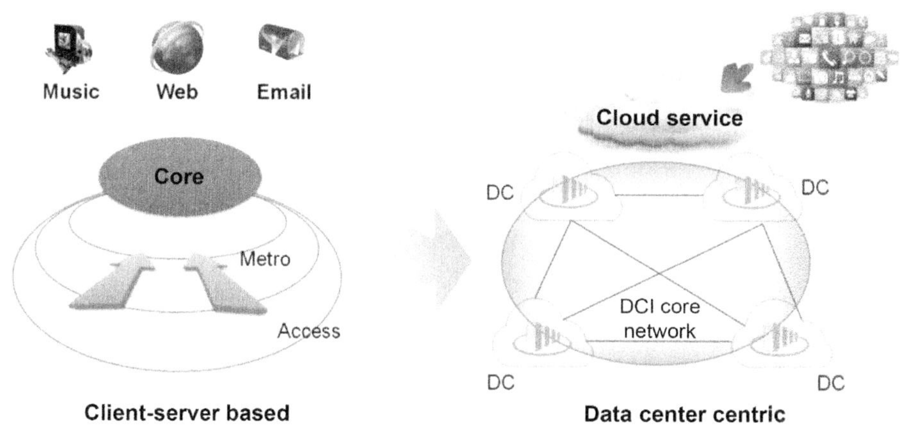

Figure 17.4 Data- enter supported modern optical networking to better provide cloud services. *DC*, Data center; *DCI*, data center interconnection.

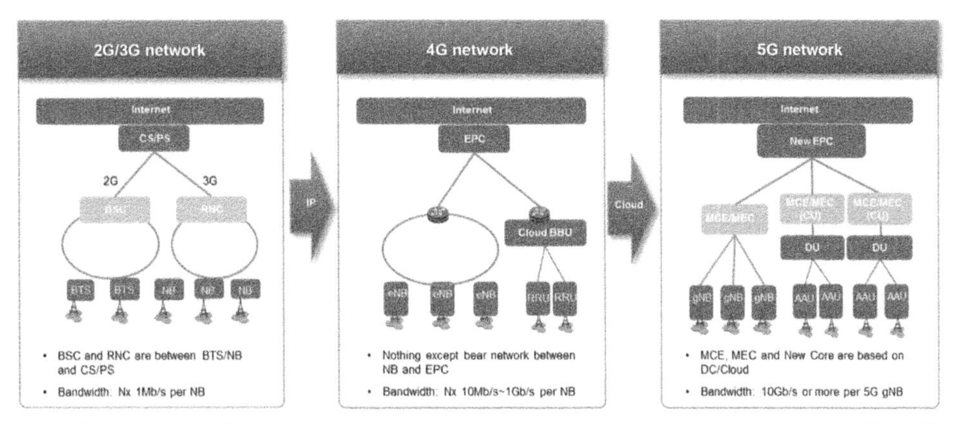

Figure 17.5 Major wireless network architecture changes during the evolution from 2G/3G, 4G to 5G. *AAU*, Active antenna unit; *BBU*, baseband unit; *BSC*, base station controller; *BTS*, base transceiver station; *CS/PS*, circuit switched core/packet switched core; *CU*, centralized unit; *DU*, distributed unit; *eNB*, evolved node B; *EPC*, evolved packet core; *gNB*, next-generation node B for 5G; *IP*, internet protocol; *MCE*, mobile cloud engine; *MEC*, mobile edge computing; *NB*, node B; *RNC*, radio network controller; *RRU*, remote radio unit.

- In 4G, BSC disappeared, due to the emergence of C-RAN (https://en.wikipedia.org/wiki/C-RAN). C-RAN is a centralized cloud computing-based architecture for radio access networks (RANs) that supports 2G, 3G, 4G, and future wireless communication standards. It supports centralized processing, collaborative radio, and simplified (clean) RRUs. C-RAN is widely regarded to bring not only network performance improvements but also substantial reduction in the total cost of ownership (TCO) of mobile networks. The supporting optical network is primarily internet protocol based radio access network (IP-RAN).
- In the upcoming 5G, the wireless architecture will be based on the cloud architecture, with the addition of mobile cloud engine (MCE) and mobile edge computing (MEC) in distributed data centers. With MCE and MEC, multiple RATs can be readily supported. Connection between multiple RATs can be realized for resource integration and superior user experience. Moreover, elastic network architecture with both distributed and centralized controls can be formed, allowing for on-demand deployment of new network functions. The supporting optical network would be the aforementioned 5G-oriented optical network.

From the description above, it is evident that both 5G wireless networks and 5G-oriented optical networks share important common elements such as distributed data centers and cloud computing. In the following sections, we will present optical transmission technologies for fronthaul, midhaul, and backhaul (referred to here as X-haul) that are required in C-RAN, optical access networks for 5G and fixed-mobile convergence, and 5G-oriented optical transport networks (OTNs).

17.2 Optical interfaces for fronthaul, midhaul, and backhaul

17.2.1 The partition of fronthaul, midhaul, and backhaul

C-RAN is playing an important role in modern mobile networks [10] by improving network performance via CoMP and increasing network energy efficiency via capacity sharing and optimization. In 4G, optical transmission has been used to support mobile fronthaul and mobile backhaul [11,12], as shown in Fig. 17.6. Mobile fronthaul is a key network element in the C-RAN architecture, as it connects centralized baseband units (BBUs) with RRUs. Mobile fronthaul could also be used to support massive M-MIMO, which is considered as a key technology for 5G networks [4,5]. On the other hand, mobile backhaul connects BBUs with the core networks to transport the baseband data streams to their respective destinations.

In 5G, BBU and RRU may evolve into centralized unit (CU), distributed units (DU), and active antenna unit (AAU), and the transmission link is expected to be partitioned to three sections as follows:

1. The fronthaul section connecting AAUs and DUs
2. The midhaul section connecting DUs and CUs
3. The backhaul section connecting CUs and the core network

These three sections are generally referred to as X-haul sections. CPRI is the commonly used interface for mobile fronthaul, connecting the radio frequency (RF) equipment with the physical layer (PHY) processing in REC. To obtain a balanced trade-off between transmission throughput and processing complexity for a given application scenario, at RE site, there are more functional split options available, as also shown in Fig. 17.6.

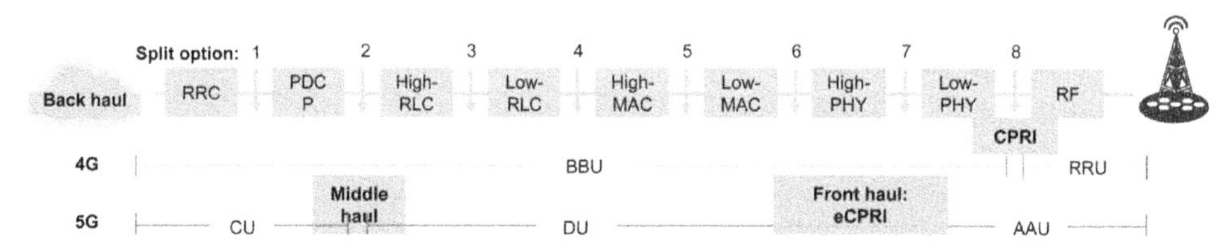

Figure 17.6 Various functional slit options for 4G and 5G. *AAU*, Active antenna unit; *BBU*, baseband unit; *CPRI*, common public radio interface; *CU*, centralized unit; *DU*, distributed unit; *eCPRI*, evolved CPRI; *MAC*, media access control; *PDCP*, packet data convergence protocol; *PHY*, physical layer; *RF*, radio frequency; *RLC*, radio link control; *RRC*, radio resource control; *RRU*, remote radio unit. Source: *After The common public radio interface. Available from: <http://www.cpri.info/>.*

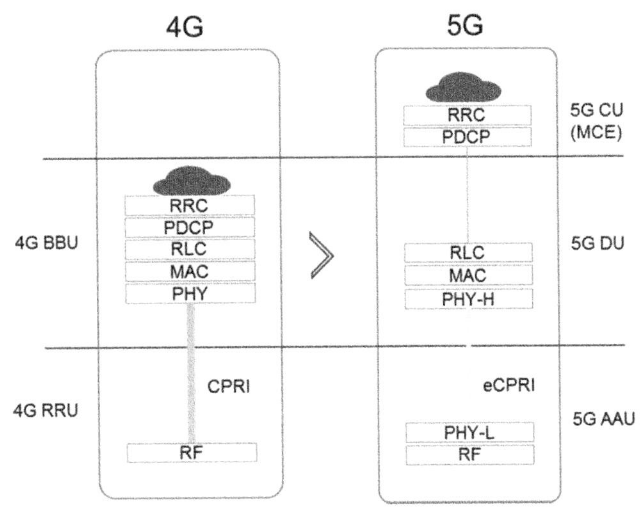

Figure 17.7 Commonly used common public radio interface (CPRI) and eCPRI inferences for 4G and 5G. *MCE*, Mobile cloud engine; *PHY-H*, higher level in the PHY; *PHY-L*, lower level in the PHY.

A popular functional split for the fronthaul section is defined by the CPRI industry association as the evolved (or Ethernet-based) CPRI, referred to as eCPRI, which connects the lower level portion of the PHY layer with the higher level portion of the PHY layer [6]. Fig. 17.7 illustrates the placement of the eCPRI in comparison to that of the traditional CPRI. We will describe CPRI and eCPRI in more depth in the following sections.

17.2.2 The common public radio interface

CPRI is based on time-domain multiplexing (TDM) of the digitized I/Q bits of multiple antennas and carriers, together with the control words (CWs). Each time-domain sample of a wireless signal is typically digitized to 15 bits for each of the I and Q components of the complex waveform of the signal. The ratio between the bits used for CWs and the I/Q bits is 1:15. Each CRPI basic frame lasts 0.26 μs (1/3.84 MHz), and 256 basic frames form a superframe. Then 150 superframes form a 10-ms frame conventionally used in universal mobile telecommunications system (UMTS). For optical transmission, binary on-off keying (OOK) is used. 8b/10b line coding is often used to facilitate OOK clock recovery and error detection. Table 17.1 shows the typical options defined by CPRI. For a commonly used option 7, the CPRI data rate needed for aggregating 8 20-MHz long-term evolution (LTE) signals (with 30.72-MHz sampling rate) in the 8×8 MIMO case is as high as 9.8304 Gb/s ($= 8 \times 2 \times 16 \times 30.72 \times 10/8$ Mbit/s), although the wireless payload data rate is only 0.6 Gb/s. This indicates that CPRI is not bandwidth efficient. One way to increase the bandwidth efficiency of CPRI is to

Table 17.1 Typical options defined by common public radio interface (CPRI).

Option	Bit rate (Gb/s)	No. of CPRI containers	Typical application example	Wireless rate (Gb/s)
1	0.6144	1	2G/3G RF channel	0.0375
2	1.2288	2	LTE 20-MHz channel	0.075
3	2.4576	4	20-MHz, 2 × 2 MIMO	0.15
4	3.0720	5		
5	4.9152	8	20-MHz, 4 × 4 MIMO	0.3
6	6.144	10	5 × 20-MHz, 2 × 2 MIMO	
7	9.8304	16	20-MHz, 8 × 8 MIMO	0.6
7 A	8.11008	16		
8	10.1376	20	5 × 20-MHz, 4 × 4 MIMO	0.75
9	12.16512	24	3 × 20-MHz, 8 × 8 MIMO	0.9
10	24.33024	48	6 × 20-MHz, 8 × 8 MIMO	1.8
A 5G use case	786.432	1280	200-MHz, 64 × 64 MIMO	48

Options 1 ~ 7 use 8b/10b line coding, while Options 7 A ~ 10 use 64b/66b line coding; the bit rate of each CPRI container is 491.52 Mb/s, for transmitting 16 bytes (128 bits) within each 3.84-MHz UMTS chip period.
Source: http://www.cpri.info/downloads/CPRI_v_7_0_2015-10-09.pdf.

performance CPRI compression [13]. It was found that a compression ratio of 3:1 can be realized with small degradation of the overall CPRI performance.

For a typical 5G use case with 64 × 64 massive MIMO and 200-MHz carrier bandwidth, the needed CPRI data rate is as high as 786.432 Gb/s, which is too high to be realized cost-effectively. Even with 3:1 CPRI compression, the required interface data rate is over 200 Gb/s, still too large to be realized cost-effectively. This leads the industry to consider new functional split options, such as NGFI and eCPRI, to substantially reduce the interface data rate, albeit at the expense of somewhat compromised network performance and increased complexity at RRUs. We will discuss the eCPRI option and compare its bandwidth efficiency with CPRI in the following section.

17.2.3 The evolved common public radio interface

To address the aforementioned bandwidth inefficiency issue of CPRI, the CPRI industry association, consisting of Ericsson, Huawei Technologies, NEC, and Nokia, had released the first version of eCPRI specification on August 31, 2017. The new specification is designed to support the 5G fronthaul use cases and provide enhancements to meet the increased requirements of 5G. The eCPRI specification offers several advantages to the base station design as follows [6]:

1. The new interface may enable about 10 times reduction of the required bandwidth.
2. Required bandwidth can scale flexibly according to the user plane traffic.

3. Use of packet based transport technologies will be enabled. Mainstream technologies like Ethernet open the possibility to carry eCPRI traffic and other traffic simultaneously, in the same switched network; for example, one Ethernet network can simultaneously carry eCPRI traffic from several system vendors. In addition, the use of well-established protocols, such as Ethernet- operations, administration, and maintenance (OAM), is possible for operation, administration, maintenance, provisioning, and troubleshooting of the network.
4. The new interface is a real-time traffic interface enabling use of sophisticated coordination algorithms guaranteeing best possible radio performance.
5. The interface is future proof, allowing new feature introductions by software (SW) updates in the radio network.

The eCPRI specification is based on new functional partitioning of the cellular base station functions, positioning the split point inside the PHY Layer, as shown in Fig. 17.8 [6]. Process stages marked with gray text are optional. The traditional CPRI is at split position E, which is between the RF equipment and the PHY processing layer and requires the highest interface data rate. At the other end, the split position D is between the MAC and the PHY processing layer, requiring the lowest interface data rate. The eCPRI specification focuses on three different functional splits, two

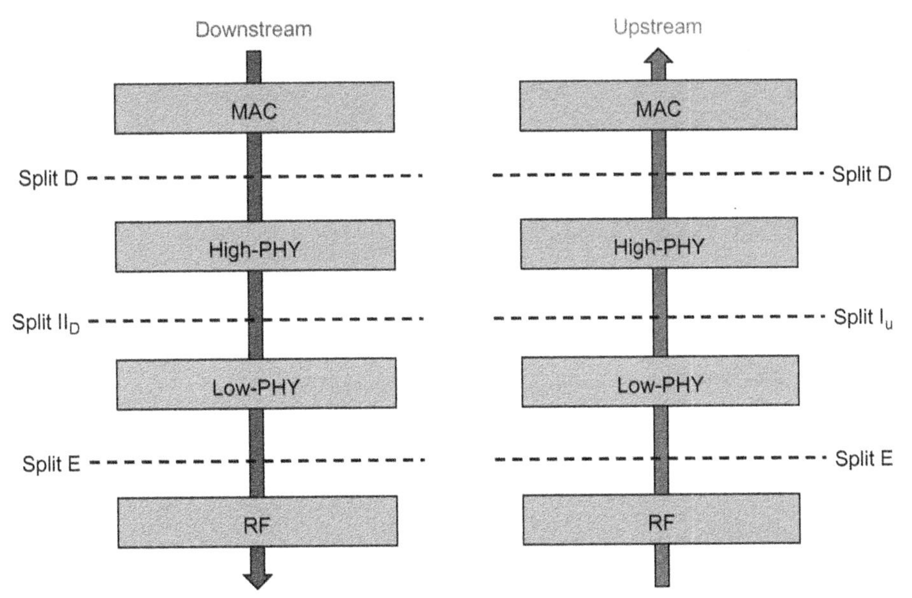

Figure 17.8 Functional split options for Ethernet-based CPRI (eCPRI). The E interface is the traditional common public radio interface (CPRI) interface, while II_D and I_U are the preferred eCPRI interface options for downstream (DS) and upstream (US), respectively. *RF*, Radio frequency equipment; *TX*, transmitter. Source: *After http://www.cpri.info/press.html.*

splits in downlink (I_D, II_D) and one split in uplink (I_U). The major difference between I_D and II_D is that the data in Split ID is bit oriented and the data in split II_D and I_U is IQ oriented. Moving the split position upward in Fig. 17.8 leads to lowered fronthaul bit rate, and thus increased fronthaul bandwidth efficiency. On the other hand, the gain from CoMP will be less when the split position is moved closer to the MAC. So, there is a trade-off that one needs to make to balance between the interface data rate and the network performance for a given application scenario.

When going to a split that is above the split E, there are many factors that will have an impact on the interface data rate, such as:

- Throughput (closely related to the wireless signal bandwidth);
- Number of MIMO layers;
- Multi-user (MU) MIMO support;
- Code rate;
- Modulation scheme;
- Selection of beam-forming algorithm; and
- Number of antennas.

In the specification of eCPRI 1.0, the data rate needed by eCPRI is compared with that needed by CPRI for a typical 5G use case. The following values are used for an exemplary interface data rate calculation:

- *Throughput**: 3 Gb/s for downlink (DL) and 1.5 Gb/s for uplink (UL)
  (*: the throughput here is the end user data rate at the interface connected to the MAC);
- *Wireless signal bandwidth*: 100 MHz;
- *Number of downlink MIMO-layers*: 8;
- *Number of uplink MIMO-layers*: 4 (with two diversity streams per uplink MIMO layer);
- *MU-MIMO*: No;
- *TTI length*: 1 ms;
- Digital beam-forming where BF-coefficients calculation is performed in eREC;
- *Code rate*: ~0.80;
- *Modulation scheme*: 256-QAM;
- *Number of antennas*: 64;
- *Subcarrier spacing*: 15 kHz;
- *IQ sampling frequency*: 122.88 Msps (3.84×32);
- *IQ-format*: 30 bits per IQ-sample; and
- No IQ compression.

The user data rates and control signal data rates needed are summarized in Table 17.2. For split E, a high interface data rate of 236 Gb/s is needed for transmitting the user data in either the DL or the UL direction. When the overheads for control signal and line coding (e.g., the 8b/10b line coding) are considered, the total

Table 17.2 Comparison of interface data rates at key split positions for a given 5G use case [after (eCPRI 1.0)].

Downstream direction	Split E (CPRI)	Split II_D (An eCPRI option)		Split D (Another eCPRI option)	
	User data	User data	Control data	User data	Control data
	236 Gb/s	~20 Gb/s	< 10 Gb/s	3 Gb/s	≪1 Gb/s
Upstream direction	Split E (CPRI)	Split I_u (An eCPRI option)		Split D (Another eCPRI option)	
	User data	User data	Control data	User data	Control data
	236 Gb/s	~20 Gb/s	< 10 Gb/s	1.5 Gb/s	≪1 Gb/s

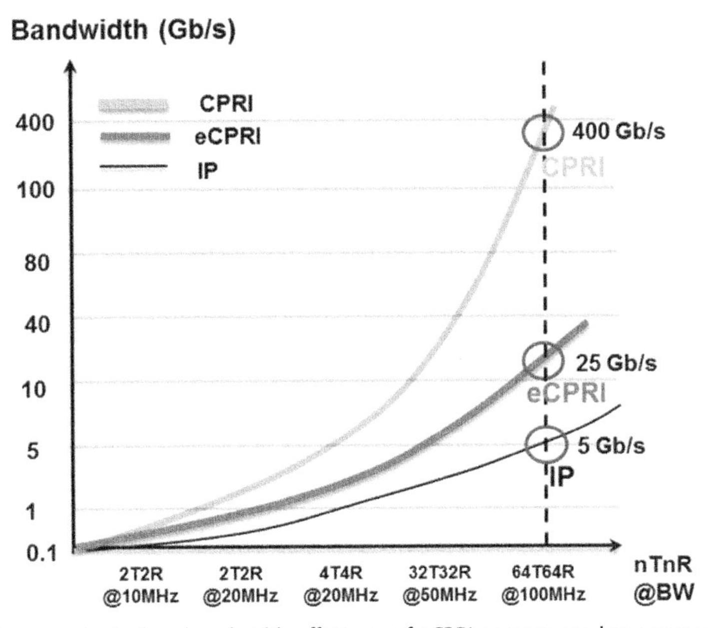

Figure 17.9 Improved interface bandwidth efficiency of eCPRI as compared to common public radio interface (CPRI). *BW*, RF signal bandwidth; *nTnR*, n-by-n MIMO with n transmitters and n receivers.

interface data rate at split E (or the traditional CPRI) would be ~400 Gb/s. On the other hand, the total interface data rate at the eCPRI option (II_D, I_U) is only ~25 Gb/s, representing an eightfold data rate reduction as compared to CPRI.

The improved interface bandwidth efficiency of the eCPRI option as compared to CPRI is illustrated in Fig. 17.9 for a series of use cases with the wireless signal bandwidth ranging from 10 to 100 MHz and the number of MIMO antennas ranging from 2 to 64. In all these cases, eCPRI shows an eightfold reduction in interface data rate, or an eightfold increase in bandwidth efficiency.

It is worth noting that although eCPRI is more bandwidth-efficient than CPRI, CPRI has the advantage of providing better CoMP gain and more simplified remote equipment. The CPRI industry association also noted that "in addition to the new eCPRI specification, the work continues to further develop the existing CPRI specifications to keep it as a competitive option for all deployments with dedicated fiber connections in Fronthaul including 5G." On June 25, 2018, the CPRI association had announced its plan to work for an updated specification eCPRI (2.0). The new specification will enhance the support for the 5G fronthaul by providing functionality to support CPRI (7.0) over Ethernet, allowing for CPRI and eCPRI interworking [6]. This means that both CPRI and eCPRI may coexist in future 5G networks to provide the optimal interface for any given application.

17.3 Optical transmission technologies for X-haul

17.3.1 X-haul via direct fiber connection

To support the X-haul, the most straightforward optical connection approach is based on direct fiber connection, as shown in Fig. 17.10 [14]. As the name implies, each X-haul port at the remote site is directly connected to the corresponding port at the DU or CU. There are three main drawbacks of this approach. First, it requires one fiber per port in case of bidirectional (BiDi) transmission in fiber, or one fiber pair per port in case of unidirectional transmission in fiber. Thus it requires significant fiber resource when the number of ports is large. Second, the number of optical transceivers needed is as large as the number of the ports, so low-cost optical transceivers are desired. Third, it does not offer protection and network functionalities such as OAM. So, this approach is only viable in deployment scenarios where the number of ports to be connected is small and sufficient fibers are available to make the connections.

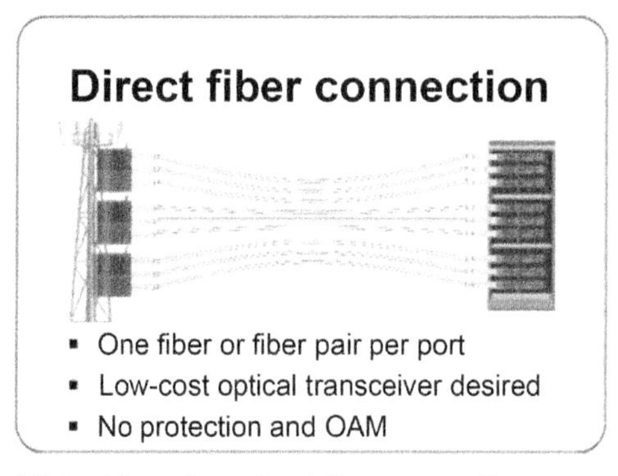

Figure 17.10 Schematic of X-haul based on direct fiber connection.

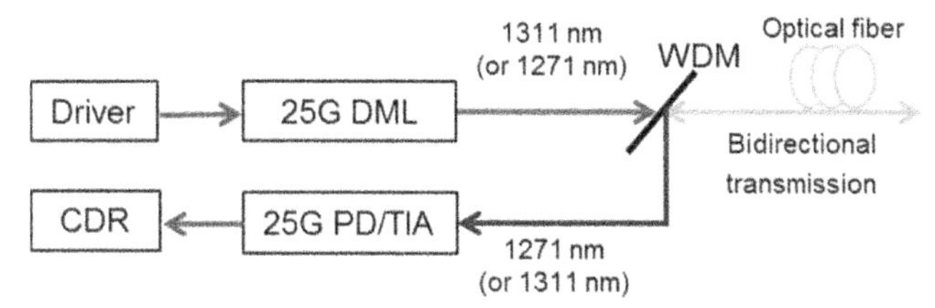

Figure 17.11 Schematic of a 25G bidirectional (BiDi) optical transceiver module that transmit at 1310 nm and receiver at 1270 nm. *CDR*, Clock-data recovery; *DML*, directly modulated laser; *PD*, photo-detector; *TIA*, trans-impedance amplifier; *WDM*, wavelength-division multiplexer.

To reduce the cost of optical transceivers, gray-optics transceivers whose laser wavelengths do not need to be accurately controlled can be used. For typical X-haul reach of about 10 km, O-band optics in the wavelength range between 1260 and 1360 nm is well suited, as the fiber chromatic dispersion is low in the O-band. O-band optical transceivers at 10 Gb/s or below are already quite mature and widely deployed in backhaul/fronthaul optical interconnection for 3G/4G/LTE applications, which can continue to be used in 5G applications where 10 Gb/s or below bit rate per connection is sufficient.

Going beyond 10, 25 Gb/s is the next data rate of interest, for example, for eCPRI fronthaul connections. The Institute of Electrical and Electronics Engineers (IEEE) had completed the P802.3cc project on 25GE in 2016, and the supply chain of 25 Gb/s O-band optical transceivers is becoming mature. Thus, 25-Gb/s eCPRI fronthaul can greatly leverage the 25GE supply chain. Because of the stringent requirement of symmetric latency, single-fiber BiDi optical transceiver is preferred, where two wavelengths are used, one for each direction. Typically, the downstream (DS) wavelength is around 1331 nm and the upstream (US) wavelength is around 1271 nm, as shown in Fig. 17.11.

17.3.2 X-haul via passive wavelength-division multiplexing connection

To achieve higher per fiber data rate, coarse wavelength-division multiplexing (CWDM) could be used. Fig. 17.12 shows the use of four CWDM wavelengths to achieve 100 Gb/s per-fiber data rate.

Preferably, these CWDM wavelengths are in the O-band, for example, 1271, 1291, 1311, and 1331 nm, to minimize the penalty from fiber dispersion.

To further increase the link capacity, wavelength-division multiplexing (WDM) with more wavelength channels can be applied. This can be realized via passive WDM or active WDM. For the passive WDM approach, colored optical modules are deployed at central and remote sites. RRUs occupy their own specific wavelengths and share a pair of fibers for transmission. Passive WDM has the potential benefit of low installation cost. Fig. 17.13 shows the configuration of a passive WDM system for

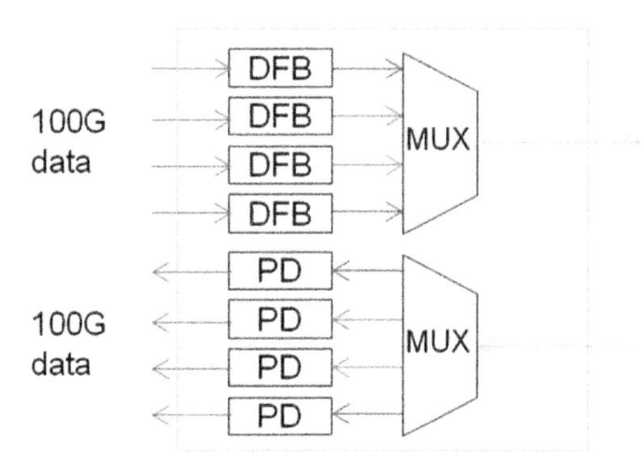

Figure 17.12 The use of coarse wavelength-division multiplexing (CWDM) to achieve 100 Gb/s transmission per direction with four 25-Gb/s wavelengths. *MUX*, CWDM multiplexer or de-multiplexer.

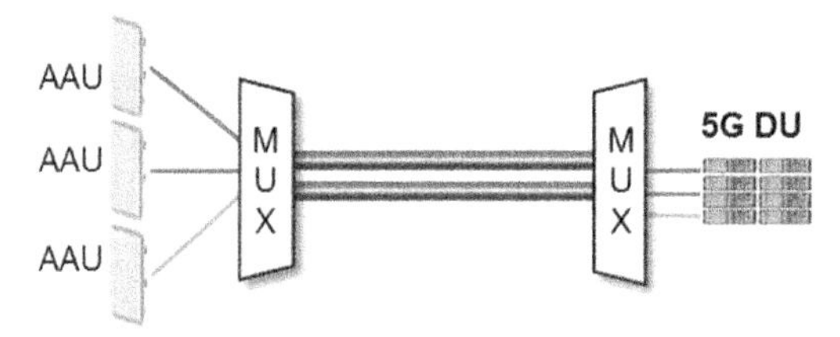

Figure 17.13 Configuration of a passive WDM system for X-haul.

X-haul. Wavelength-division multipliers and de-multiplexers (MUXs) can be used to aggregate WDM channels. BiDi transmission can be realized by using two fibers or one fiber. For BiDi transmission over a single fiber, the DS and US wavelengths can be separated into two bands, for example, the C-band and the L-band with a guard band in between. Alternatively, the DS and US wavelength channels can be interleaved with a given channel spacing, for example, 50 GHz.

Another passive WDM approach is based on wavelength-division multiplexed passive optical network (WDM-PON). WDM-PON offers advantages such as simplified network topology. Fig. 17.14 shows the configuration of a WDM-PON system for X-haul. At the DU site, an optical line terminal (OLT) service line card is added. At each AAU site, an optical network unit (ONU) is added. The OLT is connected to multiple ONUs via an arrayed waveguide grating (AWG) based WDM multiplexer. To manage the WDM-PON system, an auxiliary management and control channel (AMCC) can be introduced into the system. Several methods based on the remodulation have been proposed and tested. There is some sensitivity penalty in the receiver side with remodulation method [15]. The AMCC channel can also be embedded in the forward-error correction (FEC) frame structure, which induces no sensitivity penalty and is compatible to standard passive optical network (PON) interfaces. Due to

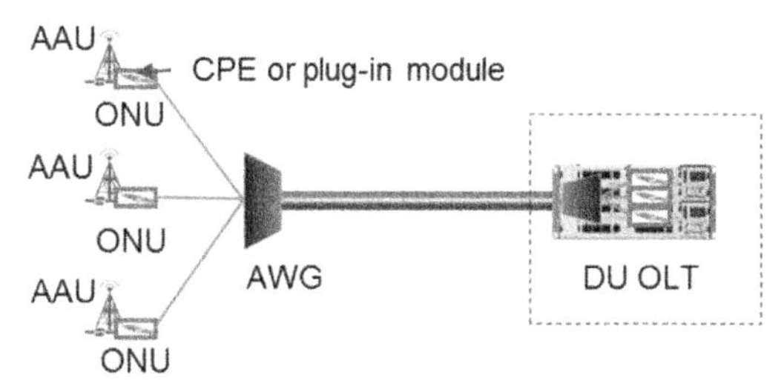

Figure 17.14 Configuration of a wavelength-division multiplexed passive optical network (WDM-PON) system for X-haul.

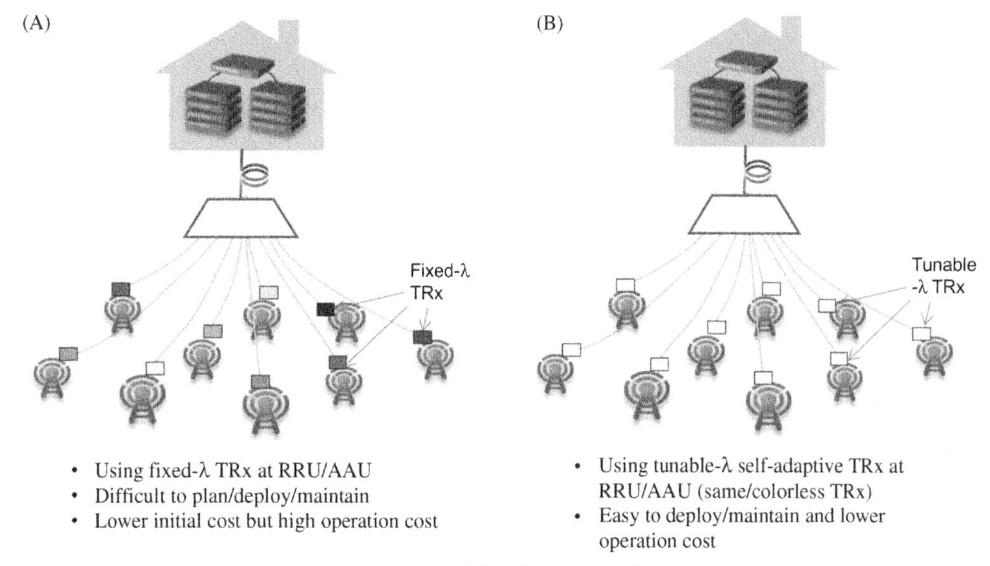

Figure 17.15 Comparison between the use of fixed-wavelength transceivers (A) and tunable transceivers (TRx) (B) for wavelength-division multiplexing (WDM)-based X-haul applications.

the importance of 5G services, PON protection schemes can also be introduced to the system to enhance the system reliability.

In terms of the wavelength band to be used for WDM, the widely used C-band is attractive, as the C-band provides more optical channels than the O-band. Also, one can employ erbium-doped fiber amplifiers (EDFA) to support longer reach. Fixed-wavelength C-band 10G optical transceivers have been used for many years, and 25G optical transceivers are becoming commercially available for WDM deployment, as illustrated in Fig. 17.15A. However, taking the 5G fronthaul application for example, there are huge amount of AAU sites. If we use fixed-wavelength optical transceiver modules, there will be 40 types of modules in case of 40-λ C-band WDM system, which leads to great difficulty to the wavelength planning, transceiver deployment, and WDM system maintenance. Therefore tunable optical transceivers (TRx) are much more preferred in 5G applications, as illustrated in Fig. 17.15B. Moreover, such a tunable transmitter is expected to automatically tune/adapt to the desired wavelength, or be "colorless," in industry jargon.

High-speed transmission in the C-band may be severely limited by fiber chromatic dispersion. For easy and cost-effective deployments, network operators usually prefer no dispersion compensation in the system. In this case 25-Gbaud NRZ signal is found to have a dispersion tolerance of around 15 km, while 25-Gbaud (50-Gb/s) PAM-4 signal has a dispersion tolerance of around 10 km. Such dispersion tolerances are sufficient for most X-haul application scenarios. Thus, 25-Gbaud WDM transceivers are of high importance for X-haul applications.

17.3.3 X-haul via active wavelength-division multiplexing connection

For the active WDM approach, multiple RRUs or AAUs can share one wavelength and the aggregation of wireless signals can be done electronically on a subwavelength basis. Active WDM has the potential benefits of high capacity and scalability. Fig. 17.16 shows the configuration of an active WDM system for X-haul based on OTN equipment. The use OTN equipment allows one to leverage the capabilities of encapsulation, management, and protection naturally provided by OTN. To better address the new requirements for 5G, OTN is being evolved and improved, leading to the so-called mobile-optimized optical transport network (M-OTN) [7]. More details on M-OTN and 5G-oriented optical networks in general will be provided in a following section.

17.3.4 X-haul via bandwidth-efficient modulation formats

It is desirable to increase signal data rate by using bandwidth-efficient modulation formats such as 4-ary pulse amplitude modulation (PAM-4), in which case 50 Gb/s per wavelength can be achieved by using 25-Gbaud modulation. On the other hand,

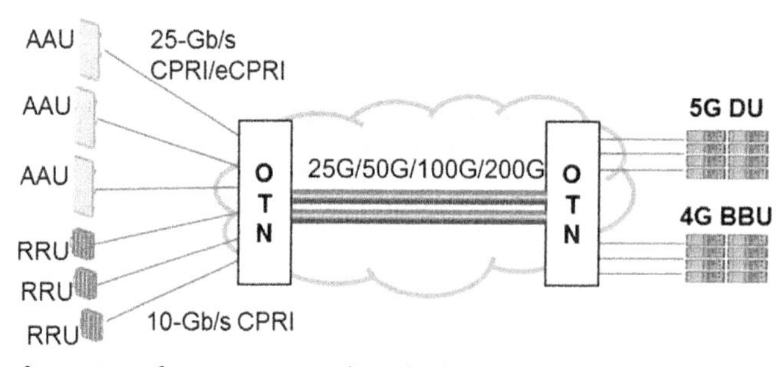

Figure 17.16 Configuration of an active wavelength-division multiplexing (WDM) system for X-haul based on optical transport network (OTN) equipment.

PAM-4 modulation requires linear modulation and detection, which are currently more demanding than the traditional binary modulation and detection used for OOK.

Another attractive modulation format is discrete multitone (DMT). It is an effective approach to leverage low-cost low-bandwidth devices to approach high bit rate using the highest order modulation format in each subcarrier, as shown in Fig. 17.17. For example, using 10-GHz-class devices to achieve 50 Gb/s or even 100 Gb/s have been demonstrated [16,17].

It is regarded as a promising solution to use such DMT system for LTE or 5G fronthaul applications. As described, there are multiple fronthaul signals that need to be transmitted between the cloud BBU and multiple RRUs or AAUs. In many circumstances optical fibers are rare resources and it is required to deliver multiple CPRI/eCPRI/NGFI signals in a single fiber. In this scheme, a small optical network equipment (ONE) is placed together with the BBU equipment. The ONE can accommodate multiple transceivers (TRx) that connect to the BBUs. The data processing circuitry in the ONE aggregates and encapsulates multiple CPRI/eCPRI/NGFI signals into one signal structure, using OTN mapping and framing techniques, for example. Then the aggregated signal is fed into the DMT transceiver to generate an optical DMT signal at the aggregated bit rate and then sent to the RRU/AAU site. For example, a DMT signal at around 100 Gb/s can encapsulate three 25 Gb/s eCPRI signals, or twelve CPRI7 (9.8 Gb/s) signals (with first decoding the 8B/10B coded CPRI7 signal and then mapping to ODUflex). On the other hand, the ONE receives the DMT signal from the RRU/AAU and then extracts the aggregated CPRI/eCPRI/NGFI signals through data processing.

Recently, a CPRI-compatible efficient mobile fronthaul (EMF) technique has been proposed and demonstrated [18–20]. Fig. 17.18 shows the principle of the EMF technique. CPRI input channels are first processed so that the wireless IQ bits are

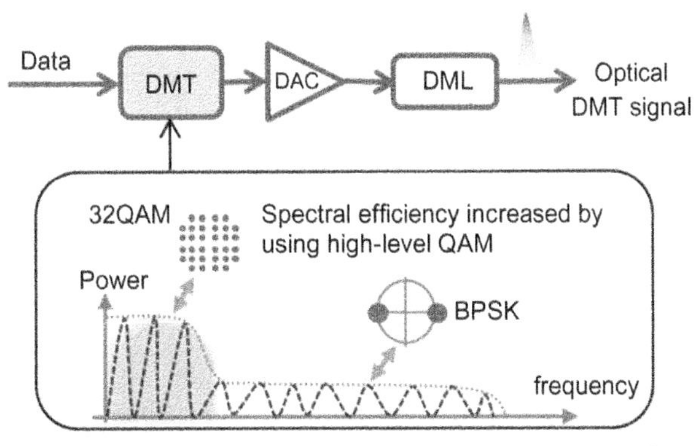

Figure 17.17 Principle of discrete multitone (DMT) modulation to achieve high bandwidth efficiency.

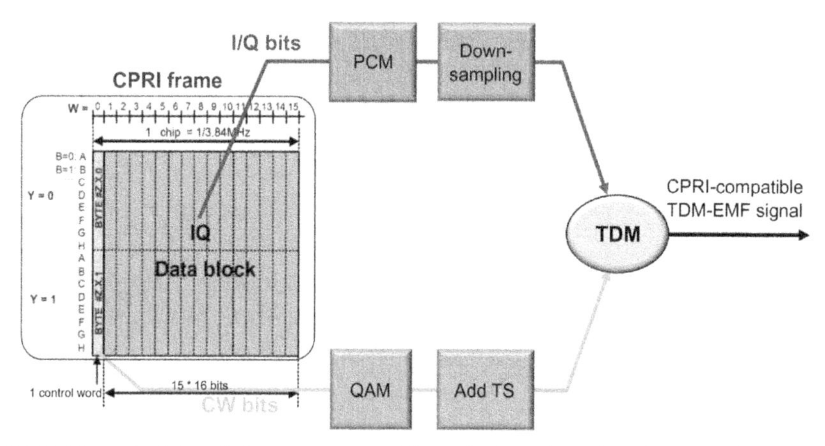

Figure 17.18 The principle of the common public radio interface (CPRI)-compatible time-domain multiplexing-efficient mobile fronthaul (TDM-EMF) scheme.

separated from the CW bits. TDM is then applied to aggregate the IQ bits and CW bits separately. As the CW bits need to be transmitted without errors, we use low-level QAM, for example, 16-QAM, to carry the CW bits. For the IQ bits, we use pulse-code modulation (PCM) in order to achieve high bandwidth efficiency. The PCM signal is quantized to 9 bits for each of the I and Q components of the complex waveform of the signal. Due to the finite effective number of bits (ENOB) of the transmission link, the received PCM signal will have nonzero error-vector magnitude (EVM), which is acceptable as long as it is below a given value, for example, as specified by 3GPP. The PCM signal is downsampled by a factor of ¾ and the QAM signal is inserted with periodic training symbols (TS) for synchronization purpose. The two signal streams are then multiplexed in the time domain, Nyquist pulse shaped, and frequency upconverted to generate a real-valued TDM-EMF signal for IM-DD.

Fig. 17.19A shows the schematic diagram of the transmitter digital signal processing (DSP). Multiple CPRI nput channels are first processed so that IQ bits are separated from the CW bits. Time-division multiple access (TDMA) is then applied to aggregate the IQ bits and CW bits separately, followed by PCM for IQ bits and 16-QAM for CW bits. The PCM signal is downsampled by a factor of ¾ and the QAM signal is inserted with periodic TS for synchronization purpose. The two signal streams are then multiplexed in the time domain, Nyquist pulse shaped, and frequency upconverted to generate a real-valued TDMA RoF/CW signal for IM-DD. Fig. 17.19B shows the receiver DSP. The input signal is downconverted to the baseband, and TS-based synchronization and time-division de-multiplexing are performed to separate the PCM signal stream from the QAM signal stream. The QAM signals are used to train a FIR-based equalizer, which then performs channel equalization (EQ) for both the PCM and the QAM signals. With the use of the equalizer, the signal-to-noise ratio

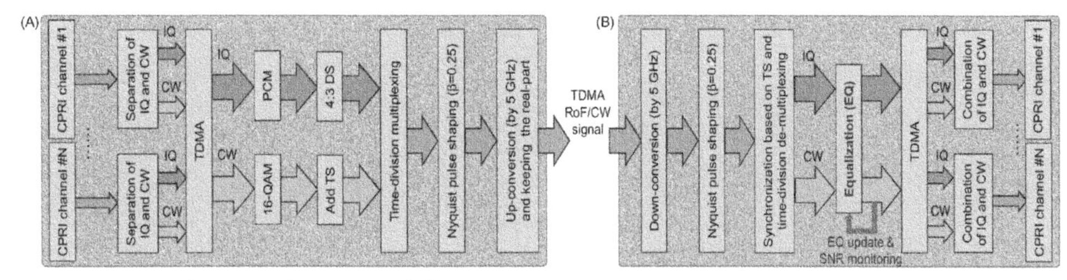

Figure 17.19 Schematic diagrams of the transmitter (A) and receiver (B) digital signal processing (DSP) for common public radio interface (CPRI)-compatible time-domain multiplexing-efficient mobile fronthaul (TDM-EMF).

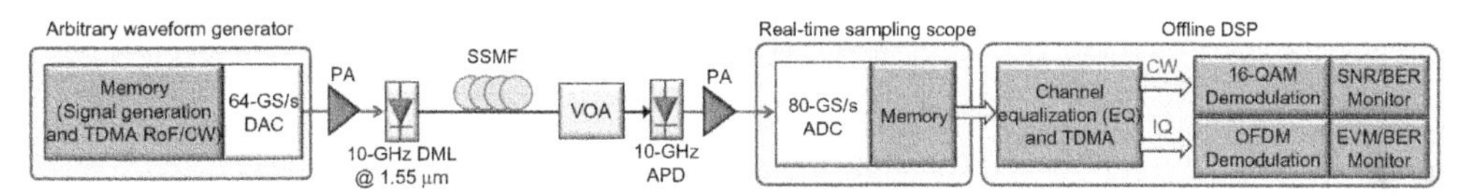

Figure 17.20 Schematic of the time-domain multiplexing-efficient mobile fronthaul (TDMA-EMF) experimental setup. *APD*, Avalanche photodiode; *DML*, directly modulated laser.

(SNR) performance of the fronthaul can be accurately estimated. After EQ, TDMA is applied to de-aggregate the IQ and CW bits for their corresponding CPRI channels. Finally, the IQ and CW bits of each CPRI channel are reconstructed and outputted, and the transmission performances are evaluated.

Fig. 17.20 shows the experimental setup for demonstrating EMF based on TDMA RoF/CW. We first generate an 8-Gbaud TDMA RoF/CW signal, corresponding to a CPRI-equivalent data rate of 256 Gb/s ($=32\times 8$ Gb/s). The signal is stored in an arbitrary waveform generator and outputted by a 64-GS/s DAC. The generated analog signal is then amplified before driving a 1550-nm 10-GHz-bandwidth directly modulated laser (DML). The generated optical signal has a power of 8 dBm and is launched into a 1-km standard single-mode fiber (SSMF). After fiber transmission, a variable optical attenuator is used to vary the optical power (PRX) received by an avalanche photodiode (APD). The detected signal is digitized by an 80-GS/s ADC in a real-time sampling scope. The digitized samples are stored in the scope and later processed by offline DSP.

Fig. 17.21A shows the recovered spectra of a 8-Gbaud TDM-EMF signal, generated by a 10G DML and received by a 10G linear APD in the optical back-to-back (BTB) case ($L=0$ km) and after 1-km SSMF transmission ($L=1$ km). The corresponding CPRI data rate is 256 Gb/s. Fig. 17.21B and C respectively show the SNR of the CW signal and the EVM of the wireless 64-QAM-OFDM signal versus PRX. For the CW signal, BER $<10^{-12}$ is obtained without FEC (or SNR >23.9 dB) at PRX ≥ -4 dBm. For the wireless 64-QAM-OFDM signal, EVM $<8\%$ is achieved

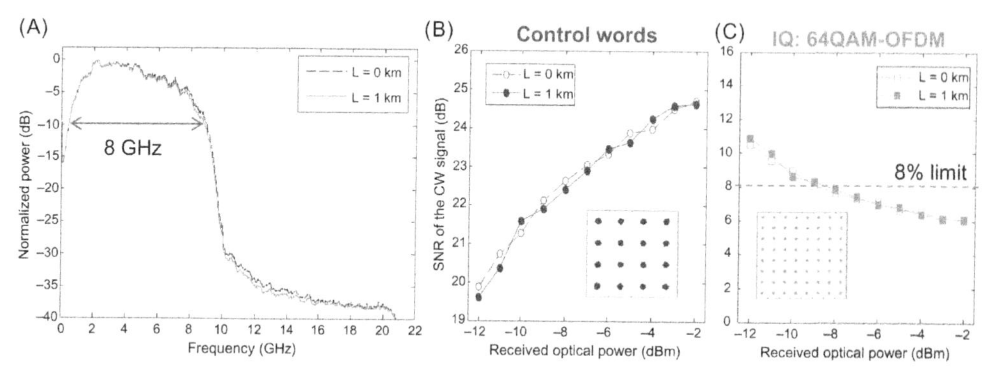

Figure 17.21 Measured radio frequency (RF) spectra of a 8-Gbaud common public radio interface (CPRI)-compatible time-domain multiplexing-efficient mobile fronthaul (TDM-EMF) signal after optical modulation and detection (A), SNR of the CW signal versus received optical power (B), and EVM performance of the 64-QAM-OFDM signal versus received power (C). *Insets*: Typical recovered constellations of the CW signal (with 16-QAM) and the LTE signal (with 64-QAM) at PRX = −10 dBm.

at PRX ≥ −8 dBm. Moreover, negligible fiber transmission penalty is observed by comparing the cases of $L = 0$ and $L = 1$ km. When operating in the O-band, 20-km transmission distance is expected to be supported. It is possible that future PON systems would widely use similar 10 GHz-class linear optics (such as DML and APD), making it possible to use a common optical transceiver platform to cost-effectively support both optical access and mobile fronthaul applications.

17.4 5G-oriented optical networks

Optical networks are supporting diverse communication services including residential services, enterprise services, and mobile services. The upcoming fifth-generation (5G) wireless network brings to optical networking new requirements such as high bandwidth, low latency, accurate synchronization, and the ability to perform network slicing to optimize the resource utilization for any given application. To better address these new requirements for 5G, optical networks need to be evolved and improved, leading to the so-called 5G-oriented optical networking [7–9]. Fig. 17.22 illustrates a 5G-oriented optical communication network with multiple enabling technologies. In this section, we review some of the enabling technologies such as M-OTN [7] for fronthaul and network slicing, advanced coherent transmission for high-performance optical core networks, reconfigurable optical wavelength switching for low-latency optical networks, and low-latency PON for CPRI and eCPRI [21,22].

17.4.1 Mobile-optimized optical transport network for X-haul

C-RAN is playing an important role in mobile networks by improving network performance via coordinated multipoint (CoMP) and increasing network energy

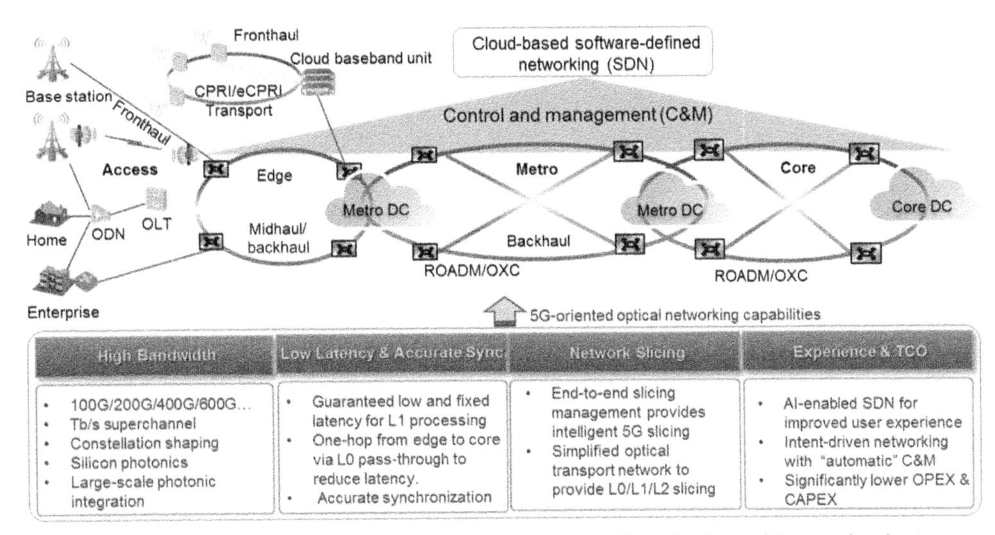

Figure 17.22 An end-to-end 5G-oriented optical network with multiple enabling technologies.

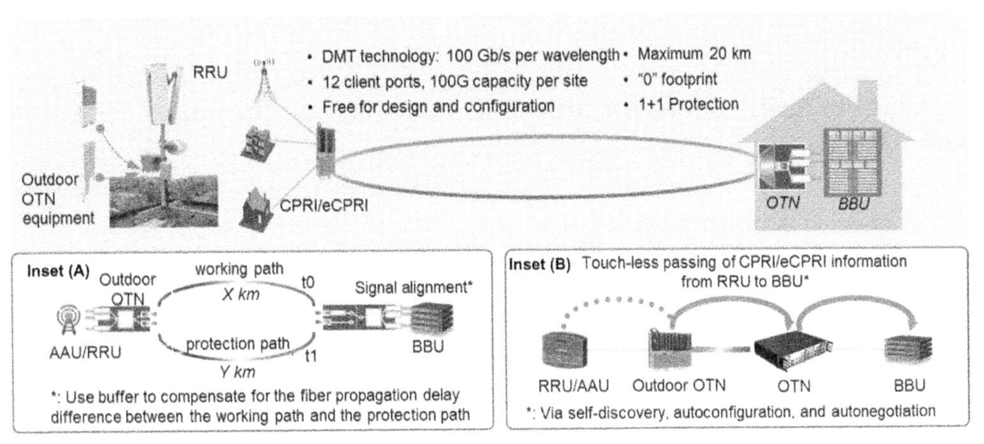

Figure 17.23 A mobile-optimized optical transport network (M-OTN) for fronthaul. Inset (A): Latency compensation between the original and protection paths; Inset (B): Easy operations, administration, and maintenance (OAM) via self-sensing of common public radio interface (CPRI) information and automatic negotiation and configuration.

efficiency via capacity sharing and optimization. Mobile fronthaul is a key network element in the C-RAN architecture, as it connects centralized BBUs with RRUs. The interface for mobile fronthaul is primarily based on the CPRI and the evolved CPRI (eCPRI) via a point-to-point (PTP) architecture. Recently, new OTN equipment has been developed to support the CPRI interface with high bandwidth efficiency, low-cost, high reliability, accurate synchronization among the connected nodes, and easy OAM, as illustrated in Fig. 17.23.

To achieve high bandwidth efficiency at low-cost, intensity modulation and direct detection (IM/DD) with spectrally efficient modulation format such as discrete multi-tone (DMT) with high-order modulation has been applied to achieve 100 Gb/s transmission with low-cost 10 GHz-class optics [16,17]. To achieve high reliability, path protection can be applied. To achieve accurate synchronization among the connected nodes, precision time protocol as defined by the IEEE 1588 standard [23] can be applied. To maintain the accurate synchronization even in the event of path protection, latency compensation needs to be applied to ensure that the protection path and the original path are equal in propagation delay, as illustrated in inset (A). To achieve easy OAM, it is desirable for OTN to be able to sense the CPRI information and conduct configuration and negotiation automatically, as illustrated in inset (B).

At the RRU/AAU site, an outdoor OTN equipment is installed, usually at the same location with the RRU/AAUs. It is important that both the outdoor OTN and the plugged optical modules inside need to satisfy the outdoor environment requirement, including temperature, humidity, etc. The functions in the outdoor OTN equipment include conversion between multiple CPRI/eCPRI/NGFI signal pairs and one optical DMT signal pair, with each signal pair including one for transmitting and the other for receiving.

17.4.2 Advanced coherent transmission for high-performance optical core networks

A key element of optical core network is coherent optical transceiver. Recently, probabilistic shaping (PS) has been introduced to further improve coherent optical transceiver performance to better approach the Shannon capacity limit [24,25]. PS with 64QAM has been demonstrated in field trial environment with offline DSP [25]. More recently, a field trial on the use of PS-programmable real-time 200-Gb/s coherent transceivers in a deploying core optical network has been demonstrated, achieving a twofold increase in reach when the PS is activated [26].

Fig. 17.24A shows the schematic of an intelligent intent-driven optical network with a network cloud engine (NCE) consisting of intent, intelligence, automation, and analytics engines, and an intelligent optical layer consisting of reconfigurable optical add-drop multiplexers (ROADMs), advanced optical monitoring, and PS-programmable optical DSP. Fig. 17.24B illustrates the use of format-and-shaping-programmable coherent transceivers to achieve the trade-off between performance and power consumption for a given application or intent.

The transmission performances of 200G PDM-PS16QAM, PDM-8QAM, and PDM-16QAM were first compared in a lab environment, as shown in Fig. 17.25A. Evidently, PS16QAM offers the best performance, and doubles the reach of 16QAM at a given optical signal-to-noise ratio (OSNR) penalty of 1.5 dB. Then the transmission performances in the field trial were compared [26]. Fig. 17.25B and C show

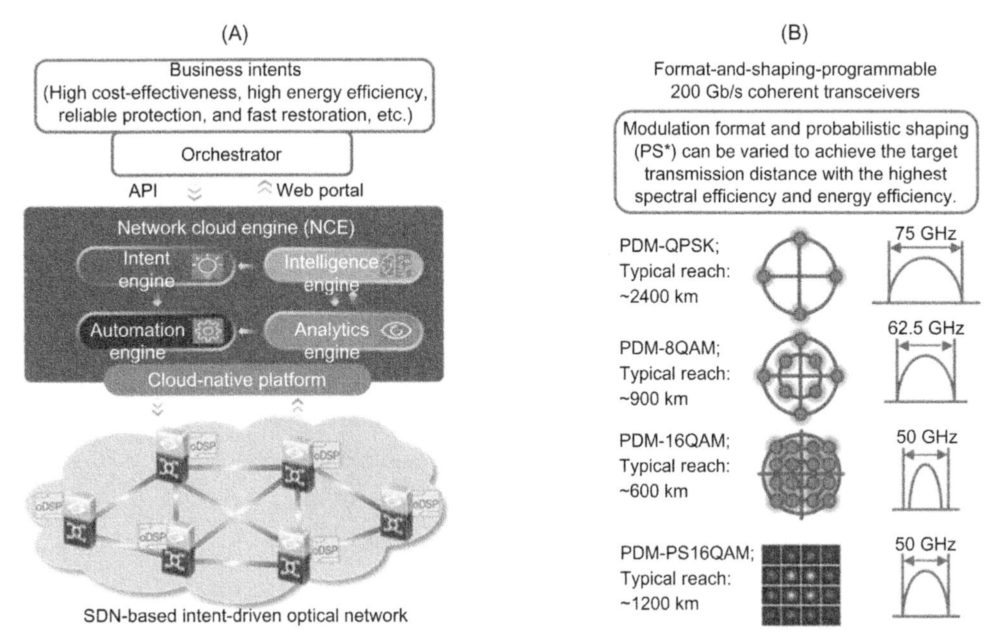

Figure 17.24 (A) Schematic of an intelligent intent-driven optical network with a network cloud engine (NCE) and an intelligent optical layer; (B) illustration of format-and-shaping-programmable coherent transceivers. Source: *After J. Li, et al., Field trial of probabilistic-shaping-programmable real-time 200-Gb/s coherent transceivers in an intelligent core optical network, in: 2018 Asia Communications and Photonics Conference (ACP), PDP Su2C.1, Hangzhou, China, 2018.*

representative recovered constellations, respectively, for 16QAM after 571-km and PS16QAM after 1142-km transmission.

Fig. 17.26A and B show the corresponding BER curves. In the BTB case, the required OSNR values at the FEC threshold of 3×10^{-2} are 18 and 16.5 dB for PDM-16QAM and PDM-PS16QAM, respectively, which represent 1 and 2.5 dB improvements over the previous PDM-16QAM result. At the FEC threshold, PDM-16QAM requires a received OSNR of 20 dB after 571-km transmission, while PS16QAM requires a received OSNR of 17 dB after 1142-km transmission, confirming that PS16QAM offers doubled reach as compared to 16QAM. The measured average OSNR values after 571-km and 1142-km transmissions are 22.3 and 19.3 dB, respectively. Thus, there is an average OSNR margin of 2.3 dB. Ethernet tester report showed error free over 24 hours with over 2×10^{15} bytes received [26].

This real-time field demonstration shows the use of a state-of-the-art probabilistic-shaping-programmable real-time 200-Gb/s coherent transceivers in a deploying intelligent core network to achieve improved performance, energy-efficiency, and protection in case of fiber cuts, showing the benefits of such probabilistic-shaping-programmable transceivers in future high-performance optical core networks.

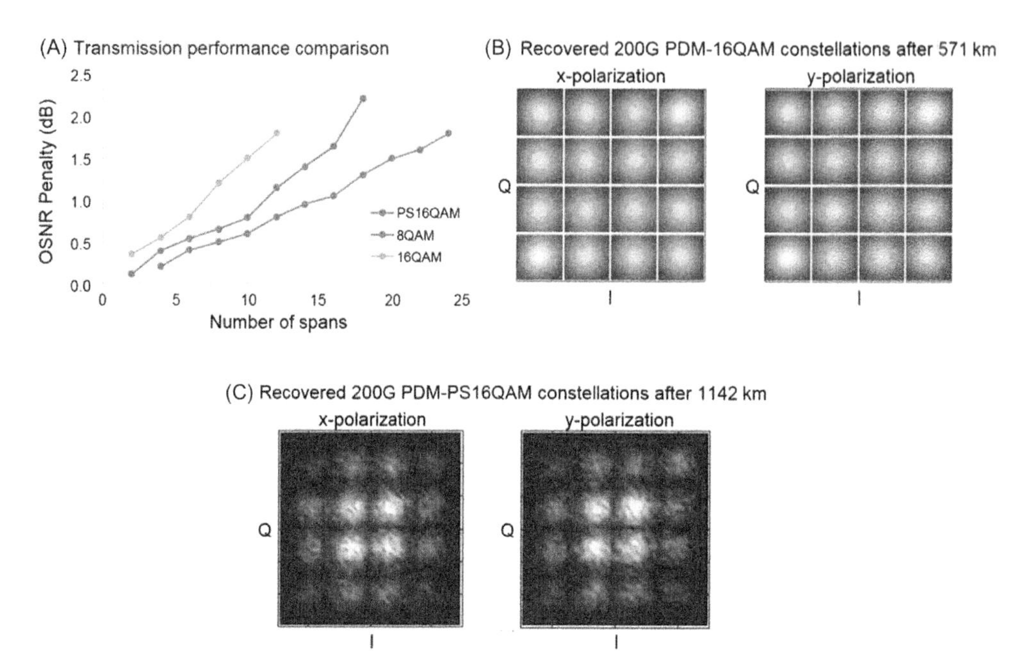

Figure 17.25 (A) Measured transmission performance comparison among 200G PDM-PS16QAM, PDM-8QAM, and PDM-16QAM in a lab environment; (B) Recovered 200 G PDM-16QAM constellations after 571-km transmission (without loop-back) in the field trial; (C) Recovered 200G PDM-PS16QAM constellations after 1142-km transmission (with loop-back) in the field trial. Source: *After J. Li, et al., Field trial of probabilistic-shaping-programmable real-time 200-Gb/s coherent transceivers in an intelligent core optical network, in: 2018 Asia Communications and Photonics Conference (ACP), PDP Su2C.1, Hangzhou, China, 2018.*

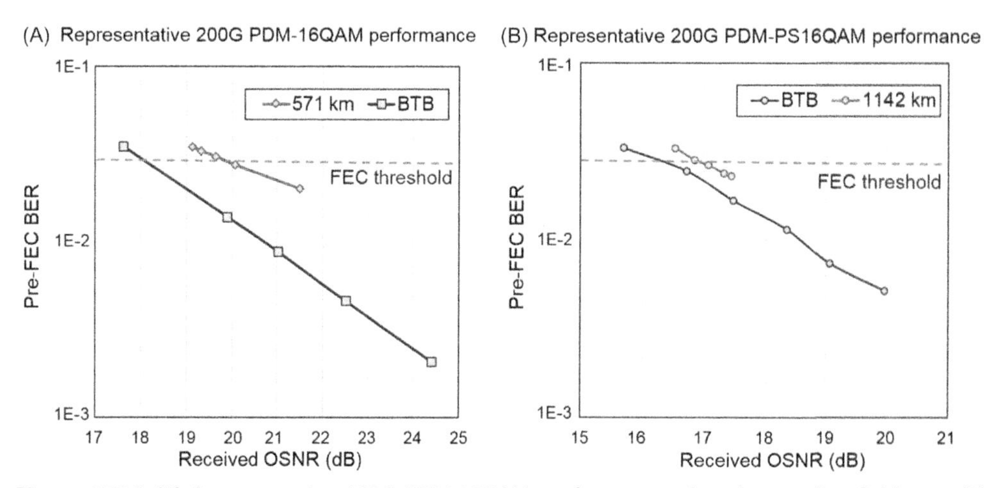

Figure 17.26 (A) Representative 200G PDM-16QAM performance after the 571-km field test; (B) representative 200G PDM-PS16QAM performance after the 1142-km field test. Source: *After J. Li, et al., Field trial of probabilistic-shaping-programmable real-time 200-Gb/s coherent transceivers in an intelligent core optical network, in: 2018 Asia Communications and Photonics Conference (ACP), PDP Su2C.1, Hangzhou, China, 2018.*

17.4.3 Wavelength switching for low-latency optical networks

As mentioned in previous sections, 5G applications impose a stringent requirement on overall network latency. To effectively transport WDM channels with low latency, it is highly preferred to adopt optical wavelength switching as much as possible as shown in Fig. 17.27, so as to achieve the direct wavelength pass-through at the optical layer (L0) over various nodes of the 5G backhaul and backbone networks. Fig. 17.28 illustrates a practical network topology and the adoption of ROADM/optical cross-connects (OXC) for wavelength switching in 5G-oriented metro and backbone optical networks.

Fig. 17.29 shows a feasible evolution path of ROADM/OXC for optical wavelength networking applications [27–31]. The industry has evolved from the conventional optical add drop multiplexing (OADM) era to the multidegree (MD) ROADM era. Currently, much effort has been devoted to enhance the 1 × N wavelength selective switch (WSS) based MD-ROADM for instance, by reducing or eliminating the number of interport fiber connections with an optical backplane, using N × M WSS to improve the reliability, etc. Going forward, it may be desirable to realize compact

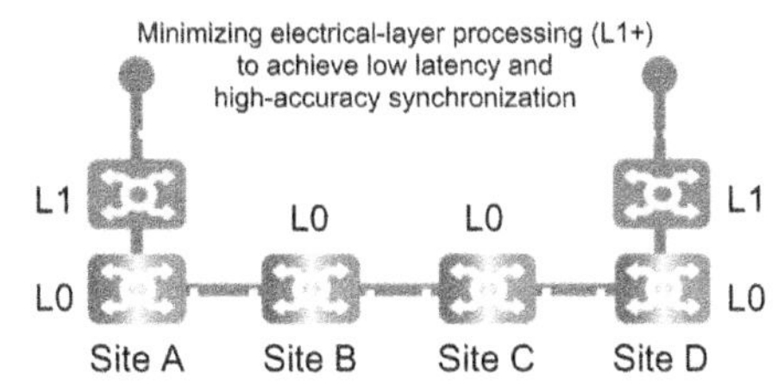

Figure 17.27 The use of optical wavelength switching in L0 to achieve low latency.

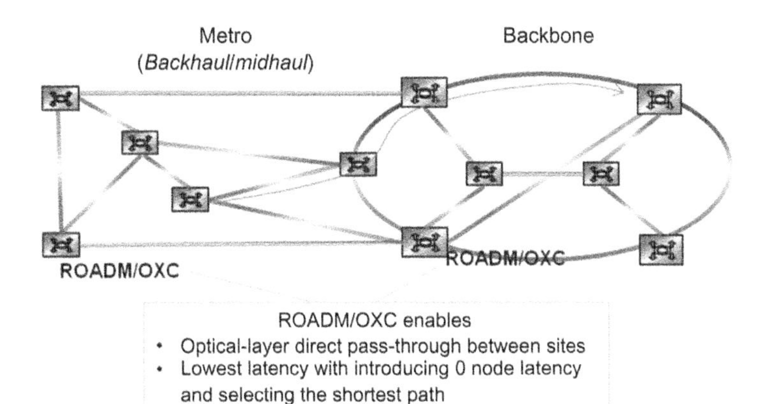

Figure 17.28 The use of reconfigurable optical add/drop multiplexers/optical cross-connects (ROADM/OXC) for wavelength switching in metro and core optical networks to achieve low network latency.

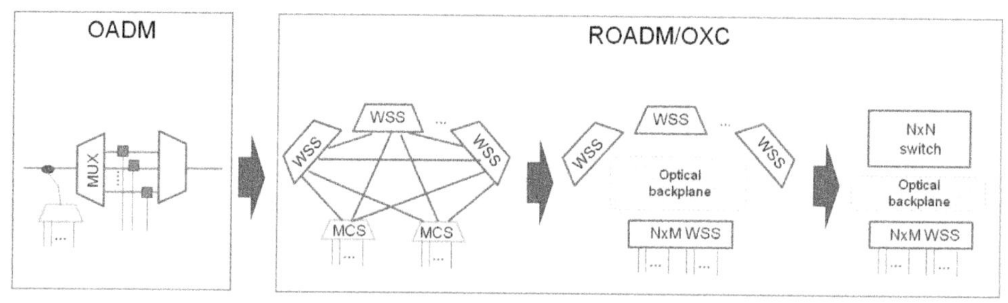

Figure 17.29 A possible evolution path of optical cross connect.

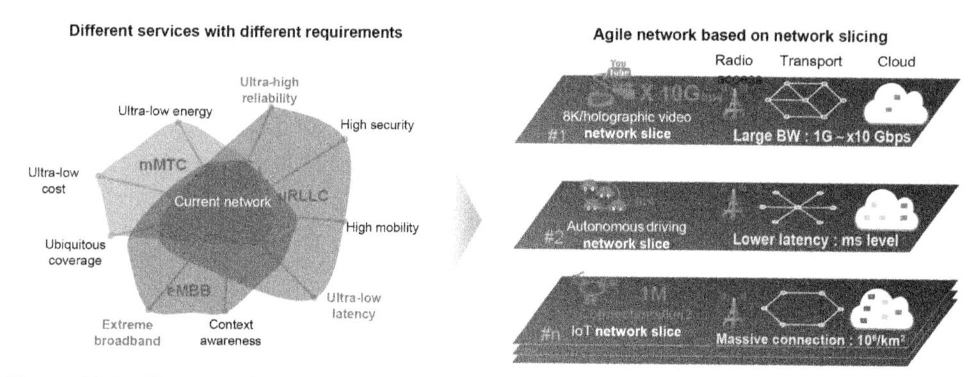

Figure 17.30 The use of network slicing for service-specific optimization in 5G. *eMBB*, Enhanced mobile broadband; *IoT*, internet of things; *mMTC*, massive machine type communications; *uRLLC*, ultra-reliable and low latency communications.

OXC functionality with integrated N × N OXC at wavelength level for line-side switching, integrated N × M WSS for local add/drop and fiber-free connection, or even further integration of all of them.

17.4.4 Mobile-optimized optical transport network for network slicing

Mobile network applications have a diverse set of demands such as ultralow latency, ultrahigh availability, ultralarge bandwidth, and the ability to perform network slicing to achieve service-specific optimization in terms of both quality of service and resource utilization, as shown in Fig. 17.30. With the use of software-defined network (SDN), intelligent optical networks can be built with the network virtualization and slicing capabilities for 5G [32–35]. Fig. 17.31 illustrates a SDN-based intelligent network with slicing capabilities. OTN is currently designed to support multiple businesses and to carry multiservice client signals. To more efficiently carry the upcoming 5G client signals, the "full-stack" OTN can be simplified as well as enhanced to focus on 5G services, as shown in Fig. 17.32. Because of the TDM based multiplexing and

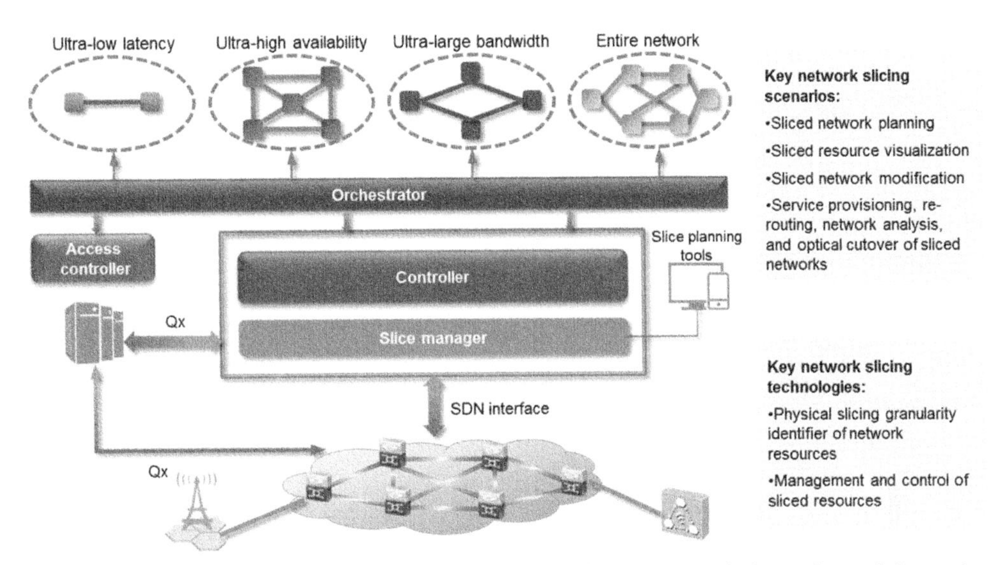

Figure 17.31 Software-defined network (SDN)-based intelligent network slicing for mobile services with diverse demands.

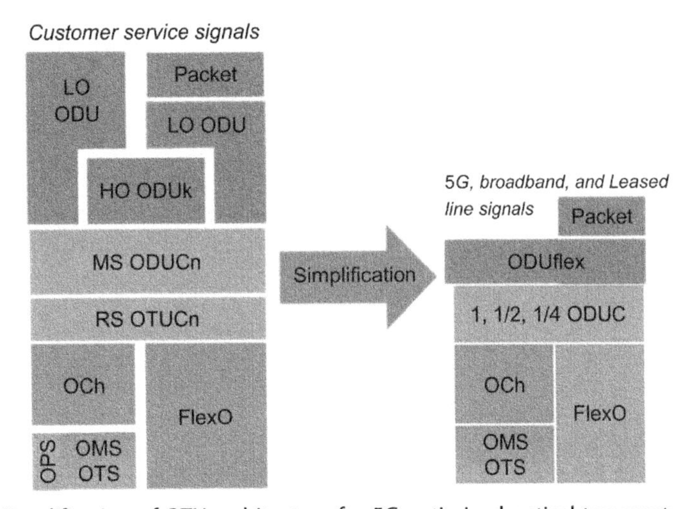

Figure 17.32 Simplification of OTN architecture for 5G-optimized optical transport.

switching in OTN, guaranteed low latency can be achieved to effectively meet the 5G networking latency requirement.

17.4.5 High-speed low-latency passive optical network for common public radio interface/Ethernet-based common public radio interface

To more efficiently support the fronthaul interface, Ethernet-based eCPRI has recently been introduced with a new functional split whose data rate is RB utilization dependent. It allows the sharing of the same optical fronthaul network for multiple RRHs and achieve the statistical multiplexing gain via a point-to-multipoint (PTMP) architecture. Fig. 17.33A shows a PTMP eCPRI architecture supported by TDM-PON, where a single feeder fiber is shared by multiple RRUs and statistical multiplexing gain may be obtained, thus reducing the overall network cost. DSP can be used to support US transmission with fast burst-mode tracking capability, thereby achieving low latency for future wireless applications. To carry eCPRI packets over TDM-PON with low latency, a just-in-time scheme where each TDM-PON burst is transmitted right after the aggregation of multiple US eCPRI packets into this burst is completed, as illustrated in Fig. 17.33B. This can be achieved by coordination between the media access control (MAC) of the RAN and that of the PON.

To support more 5G remote units with a PON, both DS and US speeds of PON need to be much increased over the current 10-Gb/s PON. As a next-generation PON, 50-Gb/s PON is being standardized by ITU-T. Recently, it has been demonstrated that sufficient receiver sensitivity and dispersion tolerance can be achieved by the combined use of low-chirp optical transmitter in the O-band, receiver-side EQ, and advanced FEC [36–38]. For the transmitter, low-chirp electro-absorption-modulated laser and DML are cost-effective options. For receiver-side EQ, DSP based EQ offers substantial performance gain, especially when intersymbol interference is present due to transceiver bandwidth limitation and fiber dispersion. For advanced FEC, hard-decision low-density parity check (LDPC) is been standardized by the IEEE [39], and

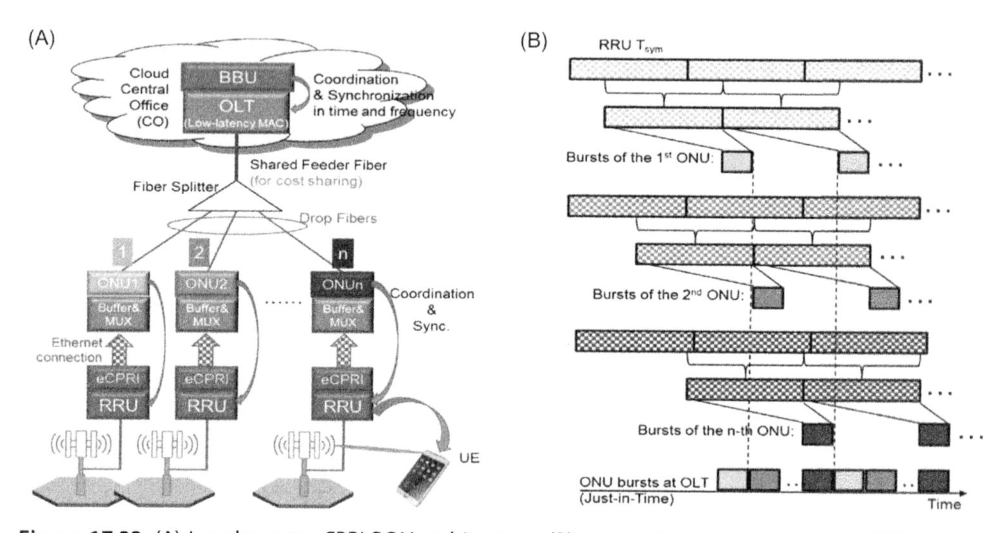

Figure 17.33 (A) Low-latency eCPRI-PON architecture; (B) Just-in-time aggregation of eCPRI packets into TDM-PON bursts in the low-latency eCPRI-PON. Source: *After X. Liu, et al., Enabling technologies for 5G-oriented optical networks, in: Optical Fiber Communication Conference (OFC), Invited paper Tu2B.4, 2019.*

soft-decision LDPC is being considered by the ITU [38]. With 4-ary pulse-amplitude modulation, the DS and US transmission speeds may be doubled to 100 Gb/s. This shows the promise of utilizing the next-generation TDM-PON to support CPRI and eCPRI, especially in cost-sensitive high-density deployment scenarios.

17.5 Industry standards and development for 5G-oriented optical networks

As 5G mobile traffic is expected to be one of the major services and data traffics in communication networks in the next few years, quite a few global standard development organizations are developing the standards and specifications to carry and transport 5G service data in fixed networks. Some exemplary developments include 5G-oriented optical networking aspects in ITU-T Study Group 15 (SG15) [2], time-sensitive networking in IEEE 802.1 Working Group [23], IP networking and optical network control related aspects in IETF, etc. In the following we give more elaboration on the up-to-date 5G-oriented optical networking related standard and industry development.

17.5.1 5G-oriented optical network architecture and signal structure developments

SG15 has developed a Technical Report GSTR-TN5G "Transport network support of IMT-2020/5G" that was published in February 2018 [2]. This Technical Report focuses on requirements on optical networks in order to support 5G networks, particularly at interfaces between 5G wireless network entities and optical networks. It also documents reference models of 5G wireless network and 5G-oriented optical network, as well as a set of deployment scenarios. Aspects of transport network support including network slicing (data plane and control plane), synchronization, and control/management are also documented.

In addition to the completion of the first version of Technical Report GSTR-TN5G, ITU-T SG15 has also initiated two new standard work items. One is G.sup. 5gotn "Application of optical transport network (OTN) to 5G Transport," targeting to generate a Supplement that describes the use of existing OTN technology to address the requirements to support 5G transport in the fronthaul, midhaul, and backhaul networks. OTN technology is a series of technologies and specifications for optical network elements connected by optical fiber links, able to provide functionality of transport, multiplexing, switching, management, supervision, and survivability of optical channels carrying client signals.

The other initiated work item is G.ctn5g "Characteristics of transport networks to support IMT-2020/5G." This work item targets to generate a Recommendation that contains the requirements for and characteristics of a frame format that provides hard isolation between aggregated digital clients. The digital client are the digital streams to/from the 5G entities (RRU, DU, CU, NGC) and other digital clients carried in the access, aggregation, and core networks. The requirements and characteristics will be documented for each of the fronthaul, midhaul and backhaul networks.

Here the hard isolation between aggregated digital clients could be realized, for example, through electrical-layer TDM, in which each digital client is allocated to certain time slot(s), and thus this approach is a "hard" isolation to strictly guarantee the network path, capacity, and latency for each client.

17.5.2 5G-oriented optical interface specification developments

As can be found in the Technical Report GSTR-TN5G [2], the capacity requirement in 5G-oriented optical networks vary from operator to operator, depending on the different network scale and supported number of users. For fronthaul the collected operator's requirements vary from 10 to 825 Gb/s, which means one or multiples of 10 or 25 Gb/s are required. For midhaul and backhaul, one or multiples of 25 or 100 Gb/s (or higher bit rates) would be required.

ITU-T G.698.2 was published a revised version (Edition 3) in November 2018 to include optical interface specifications of 100 Gb/s per λ DWDM applications. These target two typical applications: up to 80–120 km PTP DWDM applications with no intermediate OADM, and up to around 450 km DWDM applications with typical 2–3 OADMs yet not precluding 6–7 OADMs. This standard could be potentially used in 5G-oriented midhaul and backhaul optical networking applications.

ITU-T G.698.4 was published in March 2018 which includes port-agnostic optical interface specifications of 10 Gb/s per λ DWDM applications. The current version targets up to 20 km single-fiber BiDi applications, and the optical interface at the tail end can automatically adapt its wavelength to the OADM or multiplexer it connects to. This standard may be potentially used in 5G-oriented fronthaul optical networking applications.

IEEE 802.3 is continuing to be revised to include new PHY specifications. So far, standards on 25GE up to 40 km reach, 50GE up to 10 km reach, 100GE up to 40 km reach, and 200GE and 400GE up to 10 km reach have been released or almost completed. A new study group has been formed in late 2017 to investigate the feasibility to extend the 50GE, 100GE, 200GE, and 400GE Ethernet interface to achieve even longer reach (40 or 80 km). These may facilitate the operators to have more choices when constructing the 5G midhaul and backhaul networks.

17.5.3 The IEEE Optical Networks 2020 activity

Optical Networks 2020 (ON2020) is an IEEE Industry Connections (IC) activity that drives innovative optical network solutions to better meet the optical networking demands in the decade of the 2020s where emerging applications such as 5G, 8K, and AR/VR are expected to flourish [40]. This IC activity aims to define new optical network requirements and specifications, develop general network technology roadmaps and evolution scenarios, and foster an open and sustainable ecosystem for end users, service providers, and equipment and component vendors to collectively address the optical networking demands in the cloud era. Freed from near-term thinking, competition, and looking beyond existing standardization efforts, the program will help to set longer-term goals and directions that the fiber-optic communications industry should be expected to work toward. Deliverables and outcomes from this Industry Connections activity are expected to include forward-looking technical studies and white papers summarizing the developed vision and overall industry directions. Current activities cover the areas of transport SDN, ROADM/OXC, 5G-oriented optical networks, and next-generation WDM and optical link technologies.

17.6 Conclusions

In this chapter, we provided an up-to-date overview of the emerging optical communication technologies for next-generation wireless networks such as 5G. We presented 5G wireless trends and technologies such as massive MIMO & CoMP, recent advances on CPRI and eCPRI for C-RAN, emerging optical communication technologies such as 100 + Gb/s coherent, low-cost IM/DD transmission, and associated DSP techniques for high-throughput and low-latency wireless fronthaul and backhaul. We also discussed emerging network architectures and design trade-offs among various optical transport and access systems for better converged fiber/wireless networks

In summary, we reviewed enabling technologies in optical transport and access networks to better support the upcoming 5G wireless networks with high bandwidth efficiency, low latency, accurate synchronization, and flexible network slicing. It is expected that future advances in optical transport and access network technologies, enabled by close cooperation and collaboration in the global telecommunication community, will bring unprecedented communication experience for our society in the 5G era.

Acknowledgments

We wish to thank many colleagues at Futurewei Technologies for collaboration. Among them are: Frank Effenberger, Huaiyu Zeng, Sharief Megeed, Naresh Chand, Ning Cheng, Yuanqiu Luo, Andy Shen, and Linlin Li. We also would like to acknowledge valuable discussion with many Huawei colleagues working in both optical and wireless departments. Among them are Guangxiang Fang, Min Zhou, Yin Wang, Lei Zhou, Huafeng Lin, Minghui Tao, Shengping Li, Jiang Qi, Liang Song, and Xiao Sun.

The progresses on optical communication for next-generation wireless networks summarized in this chapter implicitly represent the works of many researchers and engineers around the world. Some of their works are cited in the presentation. Particularly we would like to acknowledge valuable discussion with experts from China Telecom, China Mobile, Ericsson, ETRI, Finisar, Georgia Institute of Technology, New York University, Molex, Nokia, NTT, and Orange Lab.

References

[1] The 3rd Generation Partnership Project (3GPP). Available from: <http://www.3gpp.org/>.
[2] ITU-T Technical Report, Transport network support of IMT-2020/5G. Available from: <https://www.itu.int/pub/T-TUT-HOME-2018>, 2018.
[3] ITU towards "IMT for 2020 and beyond". Available from: <https://www.itu.int/en/ITU-R/study-groups/rsg5/rwp5d/imt-2020/Pages/default.aspx>.
[4] F. Boccardi, R.W. Heath, A. Lozano, T.L. Marzetta, P. Popovski, Five disruptive technology directions for 5G, IEEE Commun. Mag. 52 (2) (2014) 74–80.
[5] E. Larsson, O. Edfors, F. Tufvesson, T. Marzetta, Massive MIMO for next generation wireless systems, IEEE Commun. Mag. 52 (2) (2014) 186–195.
[6] The common public radio interface. Available from: <http://www.cpri.info/>.
[7] M. Vissers, Introduction to Mobile-optimized Multi-service Metro OTN (M-OTN), in: Invited talk in ECOC Workshop WS15—Technology Trends for Optical Networks Towards 2020 and Beyond, Rome, Italy, 2018.
[8] X. Liu, Optical communication technologies for 5G wireless, in: Optical Fiber Communication Conference (OFC), Short Course 444, 2019.
[9] X. Liu, et al., Enabling technologies for 5G-oriented optical networks, in: Optical Fiber Communication Conference (OFC), Invited paper Tu2B.4, 2019.
[10] China Mobile Research Institute, C-RAN: the road towards green RAN, Whitepaper v. 2.6, September 2013.
[11] A. Pizzinat, P. Chanclou, F. Saliou, T. Diallo, Things you should know about fronthaul, J. Lightwave Technol. 33 (5) (2015) 1077–1083.
[12] J. Kani, S. Kuwano, J. Terada, Options for future mobile backhaul and fronthaul, Opt. Fiber Technol. 26 (2015) 42–49.
[13] N. Shibata, T. Tashiro, S. Kuwano, N. Yuki, Y. Fukada, J. Terada, et al., Performance evaluation of mobile front-haul employing Ethernet-based TDM-PON with IQ data compression [Invited], IEEE/OSA J. Opt. Commun. Netw. 7 (11) (2015) B16–B22.
[14] J. Li, Photonics for 5G in China, in: Asia Communications and Photonics Conference, paper S4E.1, 2017.
[15] G. Nakagawa, et al., Experimental investigation of AMCC superimposition impact on CPRI signal transmission in DWDM-PON network, in: 42nd European Conference on Optical Communication (ECOC 2016), Dusseldorf, Germany, 2016.
[16] T. Takahara, et al., Discrete multi-tone for 100 Gb/s optical access networks, in: Proc. OFC, M2I.1, San Francisco, CA, 2014.
[17] W.A. Ling, I. Lyubomirsky, R. Rodes, H.M. Daghighian, C. Kocot, Single-channel 50G and 100G discrete multitone transmission with 25G VCSEL technology, J. Lightwave Technol. 33 (4) (2015) 761–767.
[18] X. Liu, H. Zeng, N. Chand, F. Effenberger, Efficient mobile fronthaul via DSP-based channel aggregation, J. Lightwave Technol. 34 (6) (2016) 1556–1564.
[19] X. Liu, H. Zeng, N. Chand, F. Effenberger, CPRI-compatible efficient mobile fronthaul transmission via equalized TDMA achieving 256 Gb/s CPRI-equivalent data rate in a single 10-GHz-bandwidth IM-DD Channel, in: Proc. OFC, W1H.3, Anaheim, 2016.
[20] H. Zeng, X. Liu, S. Megeed, N. Chand, F. Effenberger, Real-time demonstration of CPRI-compatible efficient mobile fronthaul using FPGA, J. Lightwave Technol. 35 (2017) 1241–1247.
[21] X. Liu, F. Effenberger, Emerging optical access network technologies for 5G wireless [Invited], J. Opt. Commun. Netw. 8 (2016) B70–B79.

[22] H. Zeng, X. Liu, S. Megeed, A. Shen, F. Effenberger, Digital signal processing for high-speed fiber-wireless convergence [Invited], J. Opt. Commun. Netw. 11 (2019) A11–A19.
[23] J.C. Eidson, Measurement, Control and Communication Using IEEE 1588, Springer, 2006.
[24] F. Buchali, et al., Experimental demonstration of capacity increase and rate-adaptation by probabilistically shaped 64-QAM, in: Proc. ECOC PDP3.4, 2015.
[25] J. Cho, et al., Trans-Atlantic field trial using probabilisticallv shaped 64-QAM at high spectral efficiencies and single-carrier real-time 250-Gb/s 16-QAM, in: Proc. OFC, PDP Th5B.3, 2017.
[26] J. Li, et al., Field trial of probabilistic-shaping-programmable real-time 200-Gb/s coherent transceivers in an intelligent core optical network, in: 2018 Asia Communications and Photonics Conference (ACP), PDP Su2C.1, Hangzhou, China, 2018.
[27] ITU-T Recommendation G.672, Characteristics of multi-degree reconfigurable optical add/drop multiplexers, 2012.
[28] W.I. Way, Optimum architecture for M × N multicast switch-based colorless, directionless, contentionless, and flexible-grid ROADM, in: National Fiber Optic Engineers Conference (NFOEC), 2012.
[29] L. Zong, H. Zhao, Z. Feng, Y. Yan, Low-cost, degree-expandable and contention-free ROADM architecture based on M × N WSS, in: Optical Fiber Communication Conference (OFC), paper M3E.3, 2016.
[30] D.M. Marom, et al., Survey of photonic switching architectures and technologies in support of spatially and spectrally flexible optical networking, IEEE/OSA J. Opt. Commun. Netw. 9 (2017) 1–26.
[31] P.D. Colbourne, et al., Contentionless twin 8x24 WSS with low insertion loss, in: Optical Fiber Communication Conference (OFC), post-deadline paper Th4A.1, 2018.
[32] M. Channegowda, et al., Software-defined optical networks technology and infrastructure: enabling software-defined optical network operations [Invited], J. Opt. Commun. Netw 5 (10) (2013) A274.
[33] S. Yan, et al., Field trial of machine-learning-assisted and SDN-based optical network planning with network-scale monitoring database, in: Proc. ECOC, Th.PDP.B.4, 2017.
[34] C. Natalino, et al., Joint optimization of failure management costs, electricity costs, and operator revenue in optical core networks, IEEE Trans. Green Commun. Netw. 2 (1) (2018) 291–304.
[35] R. Vilalta, A.M. López-de-Lerma, R. Muñoz, R. Martínez, R. Casellas, Optical networks virtualization and slicing in the 5G era, in: Proc. OFC, M2A.4, San Diego, CA, 2018.
[36] X. Liu, F. Effenberger, Improved dispersion tolerance for 50G-PON downstream transmission via receiver-side equalization, in: Contribution C-0907, ITU-T Meeting on 9–18 October, Geneva, 2018.
[37] L. Zhou, D. Liu, Symmetric 50G PON using NRZ, in: IEEE 802.3 Plenary, Chicago, IL, March 2018.
[38] L. Li, X. Liu, F. Effenberger, Soft-decision LDPC for 50G-PON, in: D75, ITU-T Q2 Interim Meeting, June 2018.
[39] M. Laubach, Updated Draft text for LDPC, in: IEEE 802.3 Plenary, Chicago, IL, March 2018.
[40] IEEE-SA—Optical Networks 2020. Available from: <https://standards.ieee.org/industry-connections/optical-networks-2020.html> (also see: www.on2020. org).

CHAPTER 18

Optical interconnection networks for high-performance systems

Qixiang Cheng, Madeleine Glick and Keren Bergman
Columbia University in the city of New York, New York, NY, United States

18.1 Introduction

Large-scale high performance computing (HPC) systems in the form of supercomputers and warehouse scale data centers permeate nearly every corner of modern life from applications in scientific research, medical diagnostics, and national security to film and fashion recommendations. Vast volumes of data are being processed at the same time that the relatively long-term progress of Moore's law is slowing advances in transistor density. Data-intensive computations are putting more stress on the interconnection network, especially those feeding massive data sets into machine learning algorithms. High-bandwidth interconnects, essential for maintaining computation performance, are representing an increasing portion of the total energy and cost budgets.

It is widely accepted that new approaches are required to meet these new challenges [1,2]. Photonic interconnection networks are often cited as ways to break through the energy-bandwidth limitations of conventional electrical wires to solve bottlenecks and improve interconnect performance. In this chapter we begin with an overview of the recent trends in HPC and warehouse scale data centers. We briefly review the challenges due to the slowing of Moore's law and the emergence of machine learning, which are both strongly affecting all aspects of HPC. We then focus on the more immediate supercomputer challenges in the race toward exascale of the CPU-to-memory bottleneck and potential architectural solutions using photonics for bandwidth steering. Turning to the data center, we review the requirements for scaling and improved resource utilization and the need for high-bandwidth intradata center links and the move toward disaggregation. We see that there is a current need for high bandwidth density links in both systems into the server and compute node down to the board and chip module level. At the same time to improve energy efficiency and resource utilization, both supercomputers and data centers are exploring new architectures at all levels of the network from the full system to chip modules. Energy efficient, flexible, adaptable networks involve switched fabrics for which silicon photonics is

Optical Fiber Telecommunications V11
DOI: https://doi.org/10.1016/B978-0-12-816502-7.00020-8

ideally suited. We therefore follow with a review of advances in integrated photonics at the device and system level with specific examples of design explorations.

18.2 Trends and challenges in computing architecture

18.2.1 Overview

We are living in an era of major advances in supercomputing and massive accumulation and analysis of "Big Data." The performance of an Apple iPhone 6 or Samsung Galaxy S5 on standard linear algebra benchmarks exceeds that of a Cray-1, which was considered the first successful supercomputer, and has storage capacity rivaling the text-based content of a major research library [3]. Going forward, both next-generation supercomputers and warehouse scale data centers are facing challenges of scale to transmit, store, and compute. The research and development costs to create an exascale computing system have been estimated to exceed one billion US dollars. At the same time, warehouse scale cloud data centers cost more than $500 million to construct [3].

The cost of electricity to power supercomputing systems and large data centers is a substantial portion of the total cost of ownership. This is a significant part of the motivation for the Department of Energy's (DOE's) Exascale Initiative Steering Committee adopting 20 MW as the upper limit for the system design [4,5].

The supercomputer and data center ecosystems, although not identical in structure or purpose share many similar scaling challenges (see Fig. 18.1). Although in many

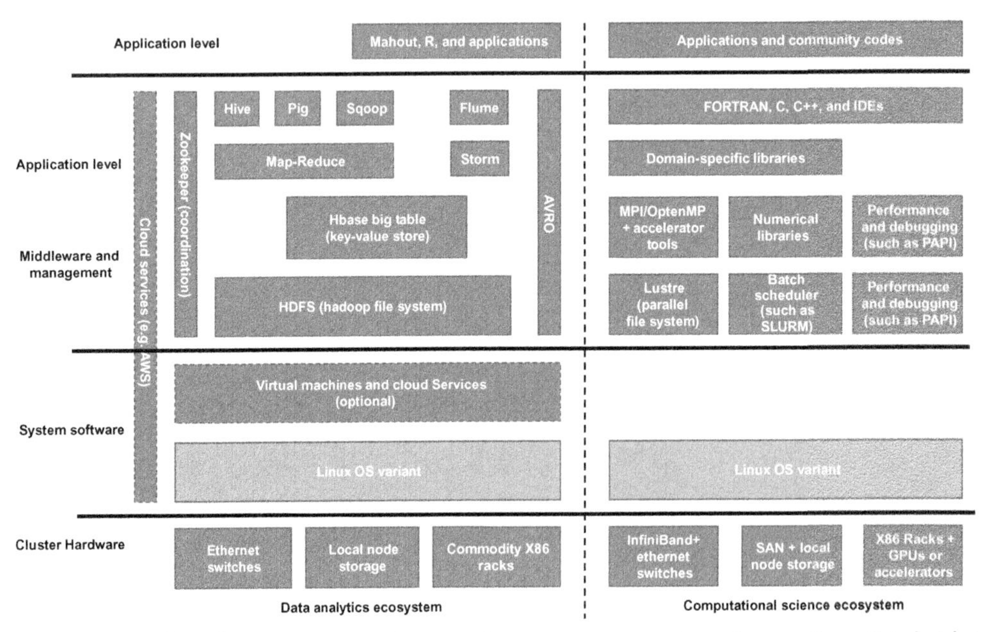

Figure 18.1 High-performance data and computing systems comparison. Source: *From D.A. Reed, J. Dongarra, Exascale computing and big data, Commun. ACM 58 (7) (2015) 56–68 [3].*

ways they are becoming more similar, advanced computing or supercomputing can be defined as those systems computing with multiple petaflops (10^{15} floating operations/second), while cloud data centers can be defined as those with many petabytes of secondary storage. A Cisco study estimates that total data storage capacity will grow from 663 EB in 2016 to 2.6 ZB by 2021 [6].

Supercomputer clusters often make considerable use of accelerators, such as graphical processing units (GPUs) and coprocessors. They use low-latency interconnects (e.g., InfiniBand) and storage area networks. The supercomputer priority is performance rather than minimal cost, while more recently energy efficiency has become a significant metric. Data centers, in contrast to supercomputers, are primarily based on commodity Ethernet servers, often stripped to minimum necessary capabilities, to reduce cost and power consumption.

In the data center, cost and capacity have, until recently, been the primary metrics; however, with new applications, improved performance is also being optimized and accelerators are being used. With increases in scaling, robustness and reliability are becoming higher priorities. Although there are notable differences between the supercomputer platforms and the data/cloud computing centers, significant challenges regarding scaling requirements for power and cost reduction are similar.

Challenges for both the supercomputer and warehouse scale data center are arising from physical hardware limits and burgeoning new applications: from the slowing or ending of Moore's law and the new and almost ubiquitous use of machine learning and data analytics.

18.2.1.1 The end of Moore's law

It has been well known for the last few years that traditional Complementary metal–oxide-semiconductor (CMOS) technology scaling, Moore's law, with the doubling of transistor density every 2 years is ending [7,8]. There is the obvious hard lower limit of the size of molecules in addition to the increasing costs and extreme manufacturing challenges. On-chip power density has reached limitations, and there has been a leveling off of clock frequencies, inhibiting performance increases, which has led to an increased focus on energy efficiency. An exascale supercomputer built with current semiconductor technologies would consume 100s of megawatts of power, an order of magnitude higher than the 20 MW target. This scaling constraint has been met by a turn towards on-chip parallelism and increased use of accelerators. Photonics technology, particularly in the form of wavelength division multiplexing (WDM), is a natural fit for the trends towards parallelism. In addition, CMOS-compatible integrated silicon photonics is enabling high-bandwidth, low-energy interconnects at ever smaller distances, at the board level and possibly eventually on-chip, through monolithic integration and the use of multichip modules (MCMs). In Ref. [10], the authors demonstrate the possibilities of this advanced integration technology with a report on an electronic–photonic system on a single chip integrating over 70 million transistors and 850 photonic components that work together to provide logic, memory, and interconnect functions.

It is worth pointing out that the end of Moore's law or lithographic scaling does not mean the end of performance scaling. There is considerable research being carried out into post-Moore's law technologies, including novel electronic transistors and spintronic technologies, among others [11]. For any of these new technologies, including photonics, to be incorporated into computing systems, metrics, for lowcost and manufacturability, often high volume manufacturability, must be met before the new technology is widely accepted [1,9].

18.2.1.2 Machine learning and data analytics

Recently, data analytics and machine learning applications are driving increasing amounts of both computation and traffic in the network. Applications include voice assistants such as Apple Siri, Google Voice Search, and Amazon Alexa, facial recognition, spam filters, medical imaging, energy efficiency [12], self-driving vehicles, recommendation services such as on Netflix and Amazon, and many, many more.

Machine learning is a form of computational intelligence that provides computers with the ability to learn and adapt without being explicitly programmed. Neural networks have existed as a form of machine learning since the late 1950s. However, neural networks have become truly useful for perceptual problems over the past 10 years. The change is due to three main factors: (1) availability of large data sets (big data), (2) increased computational capabilities of specialized processors including GPUs and TPUs [13], and (3) advances in machine learning algorithms including parallelization. These all allow predictions to be made in more reasonable time scales [14].

The machine learning process is made up of two main steps: the training stage and inference. Very briefly, in the training stage, the neural network learns to set weights or parameters of the model it is training on [15]. During this stage, training sets of sample data are presented to the network in batches and the weights are adjusted step-by-step (often using stochastic gradient descent) until an acceptable, predefined confidence level is achieved. Training is a compute- and data-intensive process. Often the computation does not fit on a single server or GPU. This is partly due to the large sizes of the data sets [16,17]. The training can take hours [18], days [19], or weeks [20] depending on the number of GPUs available. One Baidu Chinese speech recognition model required 4 terabytes of training data, and 20 exaflops of compute across the entire training cycle [21]. High-bandwidth interconnects are required to maintain performance when the computation must be spread over multiple processing units.

The second main step in machine learning is the inference, or predicting, step, which can be done quickly on a single GPU or processing unit. Here the processor is

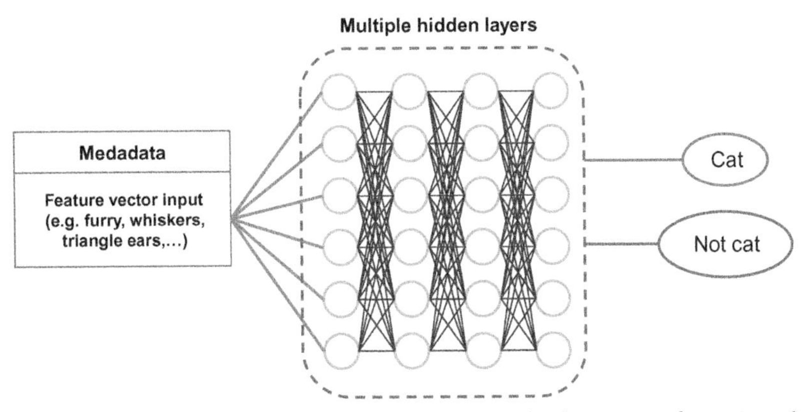

Figure 18.2 Schematic of a machine learning neural network. The input information, the feature vector, gives input values for an appropriate set of features for the problem being studied. The neural network algorithm evaluates the input based on weights on the links defined in the training cycle to determine the prediction. For example, in an image recognition problem, is the picture of a cat?

presented with new information and based on the previously determined weights makes a predictive decision (see Fig. 18.2) [22].

Photonic interconnects have a promising role to play in machine learning in several functions. In particular, as the processing units have significantly differing requirements in the two stages, the architecture may profit from the use of high-bandwidth reconfigurable interconnects. In Ref. [23], the authors use an optical neural network based on a silicon photonic Mach-Zehnder fabric to enhance runtime and energy efficiency.

18.2.2 High performance computing—toward exascale

The next grand challenge for HPC is to reach EFLOPs (10^{18} operations per second), the exascale computer [24,25]. To achieve this in a relatively economical and manufacturably viable manner, the main goal is to design a machine that consumes approximately 20 MW or 50 GFLOPs/W. This goal has been recently made more achievable with major shifts in design that place the memory closer to the GPU [26,27]. Power efficiency has improved in the most recent machines by 2.5 × through the introduction of the new architectures of the Nvidia Tesla P100/Volta V100 and the Zettascaler 2.0 and 2.2. These new architectures including innovative data movement solutions have vastly improved the GFlops/Watt metric [26].

In Figs. 18.3 and 18.4 we can see the trends in the TOP 500 since 2010 [28–30]. The FLOPs/node metric has improved by greater than 50 ×. The byte/FLOP ratio, however, has declined significantly from 0.09 to 0.001. A byte/FLOP ratio below the 0.001 level will result in limitations for programmers, requiring interconnects to scale to far larger bandwidths [27].

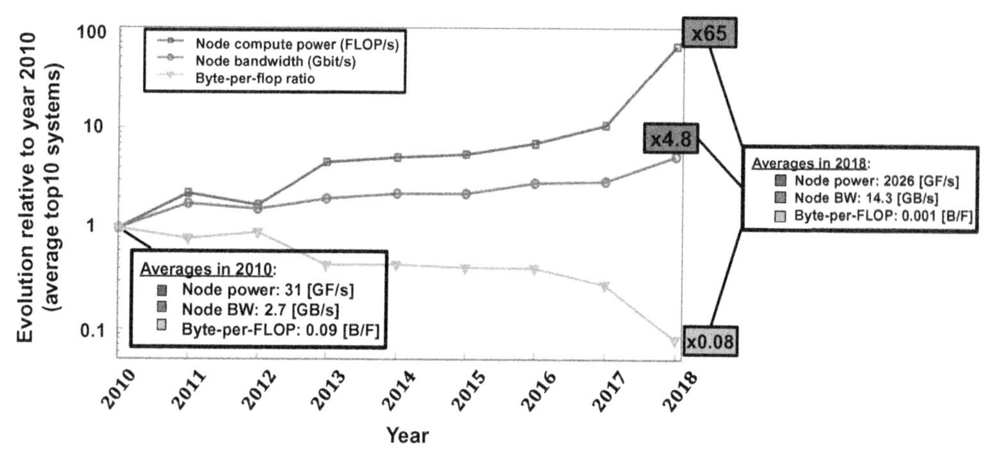

Figure 18.3 Evolution of the average top 10 supercomputers normalized to year 2010.

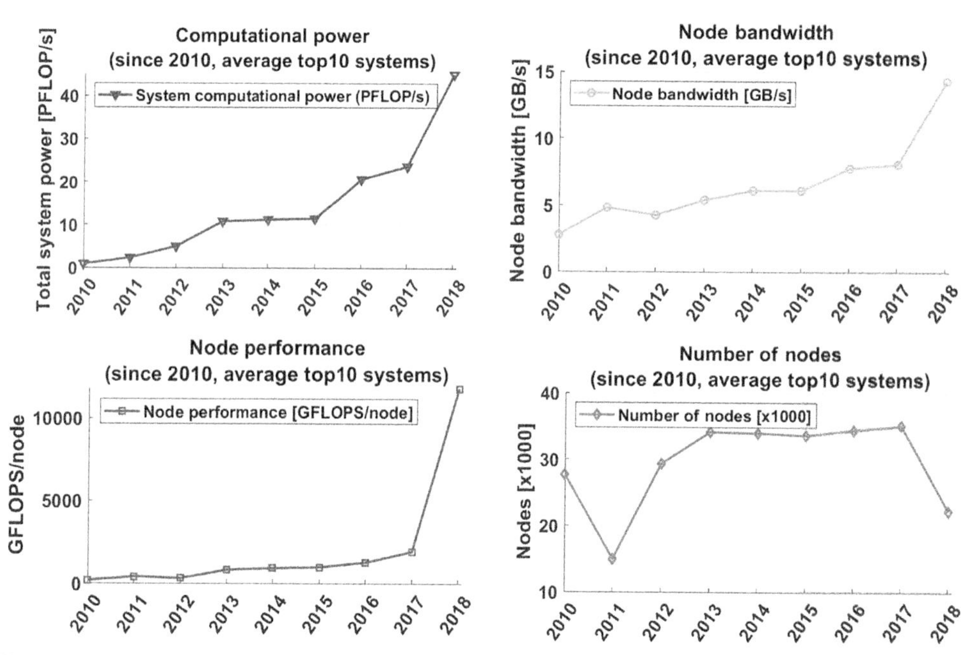

Figure 18.4 Evolution of the average of the top 10 supercomputer systems with respect to computational power, node bandwidth, node performance, and number of nodes.

Photonics is becoming more accepted by the computing community as a technology to provide the required performance. Optical interconnects for supercomputers have been studied since the mid-1980s [31–34]. The use of active optical cables (AOCs) in supercomputers has increased significantly since they were introduced in

2005 [35]. The AOC, however, is a fairly straightforward substitution of an optical link for an electrical cable where longer distance, smaller volume (and larger bend radius), lower weight, and sometimes, even secondarily, higher bandwidth is required. However, photonics can enable further advances in the interconnect, especially with the advent of silicon photonics and photonic integrated circuits (PICs) [36–39].

Current HPC interconnects rely totally on electronics for switching, and still partially for transmission. However, photonics research and development are progressing rapidly in the critical metrics of cost and energy consumption, especially through recent advances of silicon photonics design and manufacturing as described in the next section.

18.2.2.1 The memory bottleneck

For HPC the interconnect has become the bottleneck between CPU and memory. Data movement to other cores is dominating compute power even for short on-chip distances [8,40]. The performance of HPC systems relies heavily on the interconnection network as parallelism increases, resulting in massive data exchange between network endpoints [41–44].

Memory interfaces and communication links on modern computing systems are currently dominated by electrical/copper technology. However, copper wires are reaching limits of bit rate scaling as wire lengths decrease [45,46]. In Refs. [45,46] the author notes that natural bit rate capacity of the wire depends on the aspect ratio, the ratio of the length to the cross-sectional area for a constant input voltage, and does not improve as we shrink the wires down with smaller lithographic processes. As a consequence, power consumption increases proportionally to the bit rate and is highly distance dependent. Photonics technologies have the advantage of having minimal distance dependence and are "transparent" to the signaling rate. Short electronic interconnects are reaching the 1 pJ/bit mark [47–50] but face steep physical limits to get much lower [27,45,46]. Based on these considerations and derived in detail in [4,26,27], energy consumption below 1 pJ/bit has become the target metric for off--chip photonic links.

The memory bandwidth increase is also stressing the pin count limit of the processor package. The pin density of standard chip package cannot scale indefinitely. With pinout for advanced packages already up to 6,000 pins, no opportunity remains to scale performance simply by increasing the pinout [51]. Each SiP waveguide can support terabit/s bandwidth, orders of magnitude higher than what can be achieved with conventional electrical I/O. For example, while an 8-channel (4-layer) high bandwidth memory cube requires a 1024-bit bus for 100 Gb/s, a single SiP waveguide can provide the same bandwidth with 32 wavelengths each at 25 Gb/s [52].

Silicon photonics offers the promise of breaking through the limited bandwidth and packaging constraints of organic carriers using electrical pins, thus solving the challenge of pin-limited bandwidth [8].

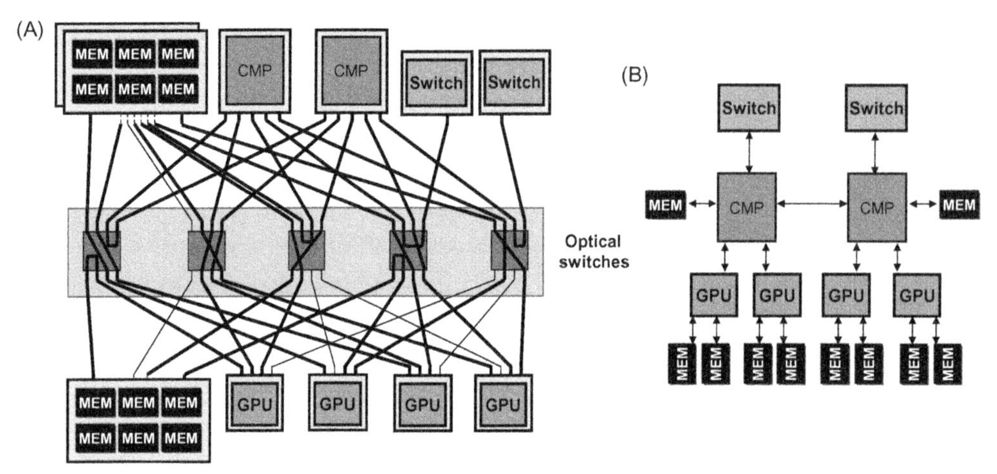

Figure 18.5 An example of beam steering. Photonic switches may be used in (A) to assemble optimized nodes (B) by configuration of the optical switches (shown within the *light blue box*).

18.2.2.2 Bandwidth steering

Applications and network architectures drive the traffic patterns in the computer network. It would enhance the performance of the network architecture if it could match the traffic pattern under consideration. The traffic pattern contains the information of the communication between the nodes in the network. Knowledge of the traffic patterns are therefore critical for optimizing the performance of the architecture. Traffic patterns are usually proprietary. In addition, there can be variations depending on the specific network and the applications running on it [53–55]. Often there is insufficient information on the current and future traffic patterns the architecture should support. A solution to this challenge is the development of flexible, adaptive networks that can take advantage of network resources efficiently and at low cost while meeting bandwidth and latency requirements. The Flexfly network proposed in Ref. [40] uses low to medium radix switches to rewire the interconnect, as required by the application, bandwidth steering, in order to achieve a high-bandwidth low-energy interconnection network with improved resource utilization. More generally, bandwidth steering can be used to change the network configuration dynamically to match the application, as shown schematically in Fig. 18.5 [56]. The original Flexfly network was proposed to modify the Dragonfly architecture, however, the concept of bandwidth steering using silicon photonic switches to improve resource utilization can also be applied to data center networks.

18.2.3 Data centers—scaling and resource utilization

Traffic increases inside the data center are staggering. A Cisco study estimates that the amount of annual global data center traffic in 2016 was 6.8 ZB and will triple to 20.6 ZB per year by 2021 (see Figs. 18.6 and 18.7). This includes traffic within the data center [6]. Total intradata center traffic does not include traffic local to the rack level, which according to the study is approximately twice the size of the within data center volumes shown in the forecast. The inclusion of rack-local traffic would change our traffic distribution to show more than 90% of traffic remaining local to the data center.

With the growing traffic, there are increasing stresses on the network and the hardware. Autonomous vehicles can produce over 500 GB data per vehicle per day [57].

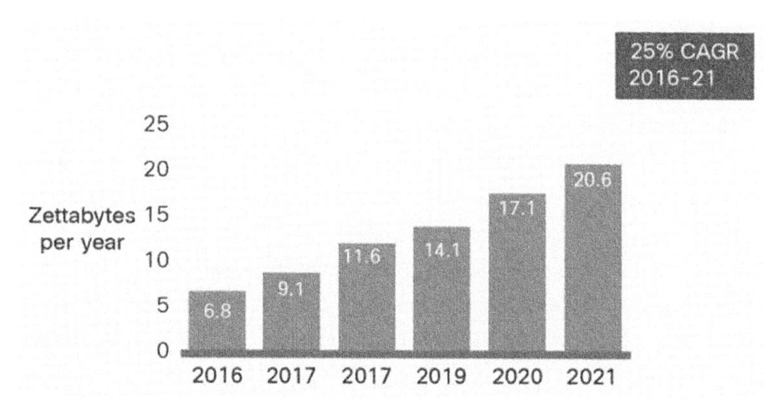

Figure 18.6 Global data center IP traffic growth. Source: *From Cisco Global Cloud Index: Forecast and Methodology, 2016–2021 White Paper. Available from: <https://www.cisco.com/c/en/us/solutions/collateral/service-provider/global-cloud-index-gci/white-paper-c11-738085.html> [6].*

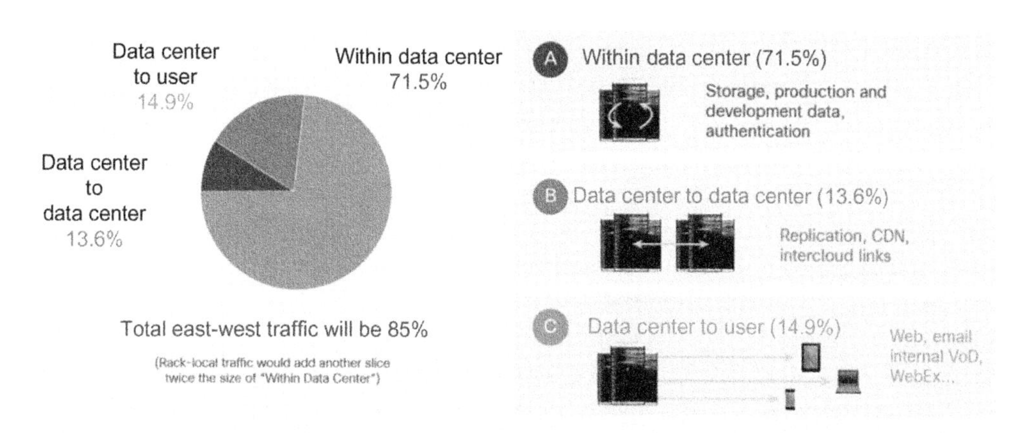

Figure 18.7 Global data center traffic by destination in 2021. Source: *From Cisco Global Cloud Index: Forecast and Methodology, 2016–2021 White Paper. Available from: <https://www.cisco.com/c/en/us/solutions/collateral/service-provider/global-cloud-index-gci/white-paper-c11-738085.html> [6].*

Some machine learning applications, for example, the training of self-driving vehicles, can use 100 TBps TBps of data and are bounded by available resources. Given these constraints on growth, many research and development programs are seeking ways to enhance performance through improvements at all levels of the architecture, software and hardware. Software improvements are often more easily adopted, as they are usually less risky and less costly. However, despite initial increased hardware costs, Facebook, Google, and Microsoft have found it economically justified to move toward multiwavelength links to achieve higher bandwidth transmission, starting with coarse WDM [58–60]. New architectures have been proposed to improve data center performance, many taking advantage of the high bandwidth density of optics and using optical switches [61,62]. The evaluation of the data center network at the system level depends on several metrics beyond those of cost and power consumption of the hardware. Data throughput and job completion time are also prime metrics. These depend on several factors including scheduling packet transmission and congestion control. In this chapter we focus on the performance of the interconnect level hardware as a basis toward improved performance.

Two current trends for improving data center performance are (1) high bandwidth density communication links and (2) improved resource utilization through disaggregation. In both these areas the advantages of photonic interconnects makes photonics an enabling technology.

18.2.3.1 High-bandwidth links in the data center

There have already been considerable advances in high-bandwidth pluggable optical interconnects for the data center. Large-scale data centers adopted optical transmission technologies during the transition from the 1 to 10 Gbps link data rate between 2007 and 2010.

In 2007, Google began using optical interconnects in its data centers with the introduction of 10 Gbps vertical cavity surface-emitting laser (VCSEL) and multimode fiber-based SFP transceivers for link lengths up to 200 m [63]. With the massive increase of traffic from data center servers over the last several years, it was obvious that the transceiver data rate would be increasing as it has from 10 to 40 Gbps, then 40 to 100 Gbps [64,65]. 100 Gbps links have been commercially available since 2014 and are currently installed in production data centers. Increases to higher rates of 400 Gbps are planned [63]. 400 Gbps transceivers are being standardized by the efforts of IEEE 802.3bs 400 Gb/s Task Force on standardizing short-range (500 m to 10 km) intradata center interconnects over standard single-mode fiber [66,67]. Even higher data rates are being studied with the exception of the eventual need for Tpbs transceivers in the near future [27]. Applications involving machine learning are driving a good portion of this increased need. For example, the DGX-1 station from Nvidia, optimized for machine learning, uses 400 Gbps of network bandwidth [58]. In addition to expanded bandwidths, optical equipment with improved energy efficiency [10] is also required. It is widely accepted that to achieve the required bandwidth density for the data center, onboard silicon photonics will be used. This can be accomplished either with 2.5D integration on a MCM (Fig. 18.8B and C) or with more advanced 3D integration using through silicon vias (Fig. 18.8D). 2.5D integration is defined by packages in which chips are placed side by side and interconnected through an interposer or substrate.

Advances in silicon photonics manufacturing capabilities are expected to lead to higher bandwidth and considerable energy savings compared to pluggable optics [68]. QSFP56 based on 50 Gbps signaling should increase the front panel BW to 7.2 Tbps;

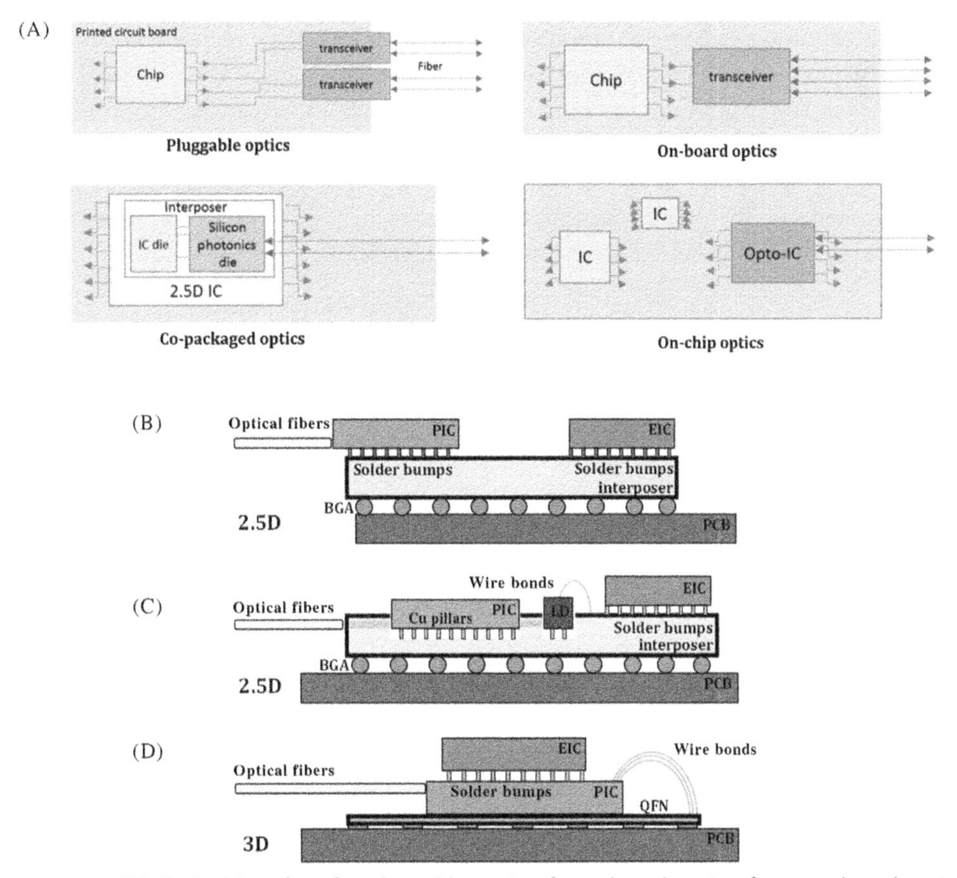

Figure 18.8 (A) Optical interface for pluggable optics, for onboard optics, for copackaged optics and on-chip optics. (B) Schematic of a 2.5D multichip module cointegrating electronics and photonics via an interposer. The interposer only serves as an electrical redistribution layer. (C) Schematic of a 2.5D multichip module. The interposer has both electrical traces and optical waveguides. (D) Schematic of a 3D integrated module.

however, there will eventually be hard limitations to increased bandwidth due to limited area at the front panel and channel impairments on higher data rates [69].

Although the concept of on board optical transceivers is not new, the nearer term data center requirements have provoked vendors to push the technology forward to reduce cost. The Consortium for OnBoard Optics (COBO), led by Microsoft, is defining the standard for optical modules that can be mounted or socketed on a network switch or adapter motherboard. Their initial focus has been on high-density 400 GbE applications [70] with large cloud providers as the early adopters.

Given the requirement for high bandwidth density at low cost and low power consumption, it is not surprising that silicon photonics, fabricated in high volume CMOS-compatible foundries [71,72], is a prime candidate for the interconnection network. Photonic roadmap predictions expect [61] early deployment of 2.5-D integrated-photonic technologies by 2020, and pervasive deployment of WDM interconnects and the beginnings of commercial chip-to-chip intrapackage photonic interconnects by 2025. Roadmapping [73] also sees demand for links to 1 Tbps on boards and 1–4 Tbps within a module by 2020. For very short mm's to cm's distance links on these modules, the energy target is on the order of 0.1 pJ/bit. In the near term the aim is to achieve manufacturable results below 1 pJ/bit.

18.2.3.2 Resource utilization and disaggregation

The traditional data center is built around servers as building blocks. Each server is composed of CPU, memory, one or more network interfaces, specialized hardware such as GPUs, and possibly some storage systems (hard disks or solid state disks). This manner of organizing the hardware is now hitting cost and utilization challenges. Each server element each has its own trends of cost and performance. Upgrading the server to incorporate more recent versions of the CPU or memory requires an entirely new server with new motherboard design [74]. Traditional data centers also suffer from resource fragmentation. Data gathered from data centers show that server memory is unused by as much as 50% or higher [75,76]. This occurs in situations where resources (CPU, memory, storage IO, network IO) are mismatched with workload requirements. For example, a compute-intensive task may not use the full memory capacity or a communication intensive task may not fully use the CPU. These challenges become motivations for disaggregation of the server.

Disaggregation is a concept in which similar resources are pooled and used as required for the application. This enables both the possibility of the resources being independently upgraded and also adaptively configuring the system for optimized performance. The network can be disaggregated at different levels, for example, at the rack or server scale [75,77], as illustrated in Fig. 18.9.

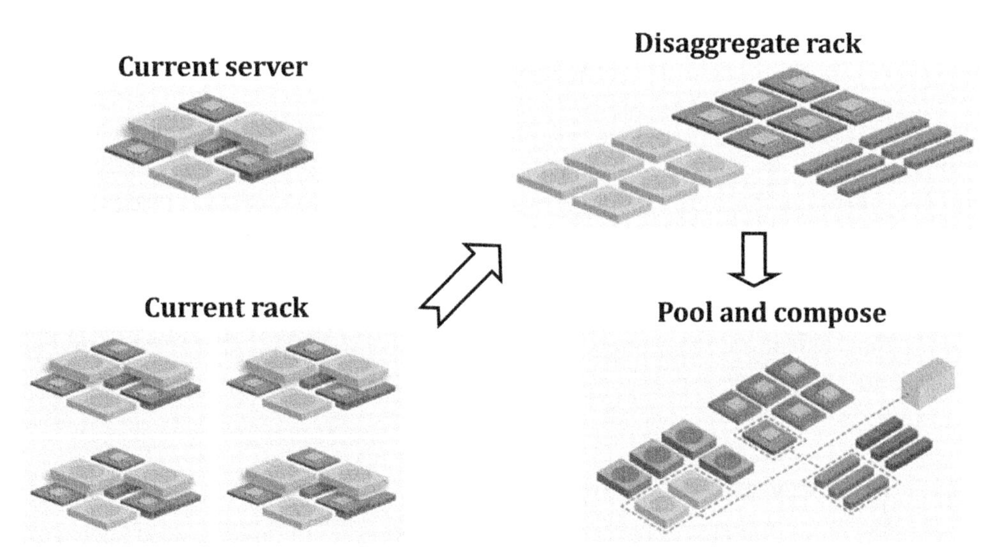

Figure 18.9 A disaggregated rack places resources of different types in different parts of the data center compared to traditional servers and uses networking to pool and compose needed resources together. In the bottom right figure, a logical node is constructed from distant resources.

The disaggregated data center requires a modified interconnection fabric that must carry the additional traffic engendered by the disaggregation and have low latency in order to not only maintain but also improve performance. The network requires a switching fabric to adaptively provision the computing resources. Optical circuit switches are prime candidates for reconfiguration of resources in the disaggregated network. Several reconfigurable data center architectures with optical switch fabrics have been proposed [53,76,78].

In a traditional server, with memory close to the CPU, latency is on the order of 10 seconds of nanoseconds. As a disaggregated network involves additional switched paths, attention must be paid to prevent added latency leading to performance degradation. The cost of the added interconnect components compared to resource savings through improved utilization must also be balanced. Several groups have developed guidelines to achieve these goals [75,77].

Given the requirement for high bandwidth density at low cost and low power consumption, it is not surprising that photonics, and especially silicon photonics, fabricated in high-volume CMOS-compatible foundries [79], is a prime candidate for the disaggregated interconnection network. [75] explores a cost/performance analysis including cost of latency and bandwidth to determine at what point a data center disaggregated memory system would be cost competitive with a conventional direct attached memory system. The authors find that the current cost of an optically switched interconnect should be reduced by approximately a factor of 10 to be an economically viable solution.

18.3 Energy-efficient links

In this section we first review photonic link architectures together with key photonic building blocks. Then we discuss the electrical and optical models used for optical links and present an example of designing the silicon photonic link with performance analysis. The objective of the design is to find the optimal combination of the number of wavelengths, N_λ, and data rate for each channel, r_b, in order to achieve the minimal energy consumption for a given optical aggregation rate. The following work is done through open-source software developed in the Lightwave Research Laboratory, PhoenixSim, which offers a unique and comprehensive modeling platform for efficient design and analysis of the physical layer, link- and system level silicon photonic interconnects.

18.3.1 Anatomy of optical link architectures

As a fundamental building block of optical interconnects, optical transceivers, which consist of the laser light source, modulator, (de)multiplexer and photodetector, are critical for the performance of an optical link. Currently, VCSEL-based transceivers and parallel fibers are the dominant technology in HPC systems. As discussed earlier, the roadmap for ultrahigh-bandwidth, low-energy links requires WDM technology leveraging PICs. Fig. 18.10 schematically shows an anatomy of options for link architectures. Fig. 18.10A shows the transceiver design for a single channel link. Fig. 18.10B shows an approach that is commonly used in telecommunications to combine modulated colored channels using (de)-multiplexers, and in Fig. 18.10C the architecture is equipped with DeMux/Mux stages utilizing broadband modulators, such as electro-absorption modulators and Mach-Zehnder modulators. Another promising architecture, illustrated in Fig. 18.10D, takes advantage of wavelength-selective microring modulators (MRRs) implemented in a cascaded structure, enabling ultrahigh on-chip bandwidth density. To reach the Tbps regime, a large number of wavelengths are required; thus the development of comb lasers with the capability of emitting over 100 individual wavelengths is a promising next step for the progress of transceiver architectures.

18.3.2 Comb laser

The optical frequency comb laser is an appealing alternative to continuous wave (CW) laser arrays as a source for HPC systems in terms of footprint, cost, and energy consumption. A comb laser consists of equally spaced lines in the frequency domain that can be used as separate optical carriers for WDM. Since the comb is generated from a single source and has intrinsically equidistant spacing between its lines, it has the potential to eliminate the energy overhead associated with independently tuning many CW lasers to maintain the desired channel locking. Currently there are two main

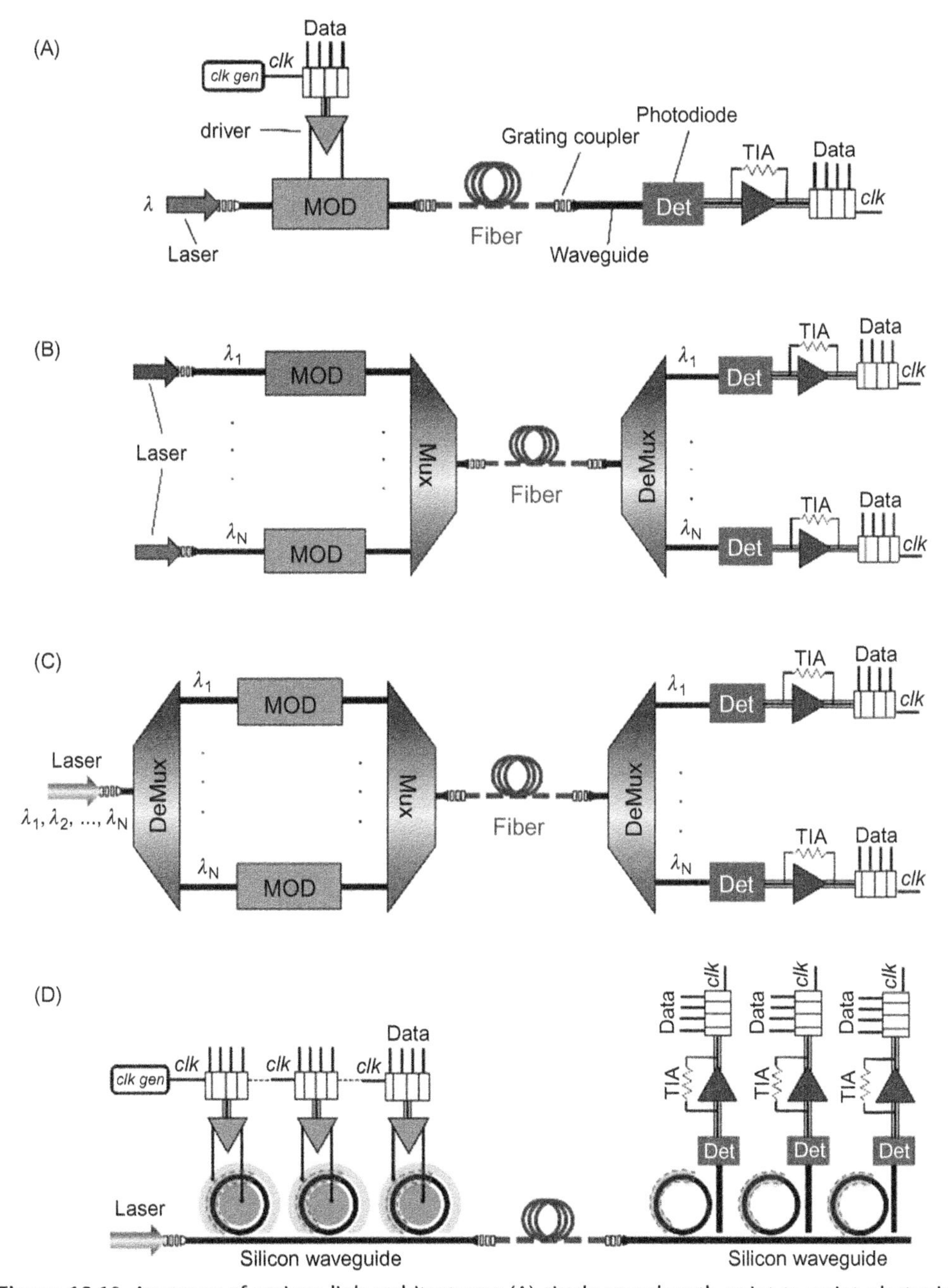

Figure 18.10 Anatomy of various link architectures: (A) single wavelength point to-point photonic link; (B) wavelength division multiplexing (WDM) photonic link based on separate lasers and broadband modulators; (C) photonic link based on parallel broadband modulators and DeMux/Mux. (D) WDM photonic link based on cascaded microring resonators and cascaded drop filters [1].

methods used for generating combs: mode-locking lasers and nonlinear generation using four-wave mixing (FWM) in a microcavity.

Comb generation can occur in a laser by inducing a fixed-phase relation between the longitudinal cavity modes in a Fabry-Perot cavity (mode locking), leading to a stable pulse train in the time domain and therefore a comb with precise spacing in the frequency domain. The channel spacing can be tuned by changing the cavity length. Quantum dot (QD) mode-locked semiconductor lasers, which can be directly grown on silicon [80], are an attractive comb laser source, as the high nonlinear gain saturation of the QD active layer results in low relative intensity noise. Moreover, by controlling the distribution of the sizes of the QDs, intentional inhomogeneous broadening of the gain spectrum, such as a 75 nm broad spectrum of emission, can be achieved [81]. The amplitude and phase noise of such a comb laser source has been greatly reduced through active mode-locking, with reduced optical linewidths of the carriers and increased effective bandwidth that is compatible with coherent systems [82].

Frequency comb generation has also been realized with a CMOS-compatible Si_3N_4 ring resonator through the nonlinear process of FWM in an optical parametric oscillator [83,84], which can be directly integrated in the current silicon photonics platform. In this implementation, numerous equally spaced narrow-linewidth sources can be generated simultaneously using a microresonator with an off-chip CW optical pump, as illustrated in Fig. 18.11A [83]. The pump field undergoes FWM in the resonator and creates signal and idler fields that also satisfy the cavity resonance; these signal and idler fields then seed further FWM, leading to a cascade effect that fills the remaining resonances of the cavity (as illustrated by Fig. 18.11B). This yields many equally spaced optical carriers (with spacing depending on the FSR of the cavity) and has a high pump-to-comb conversion efficiency of up to 31.8% when operating in

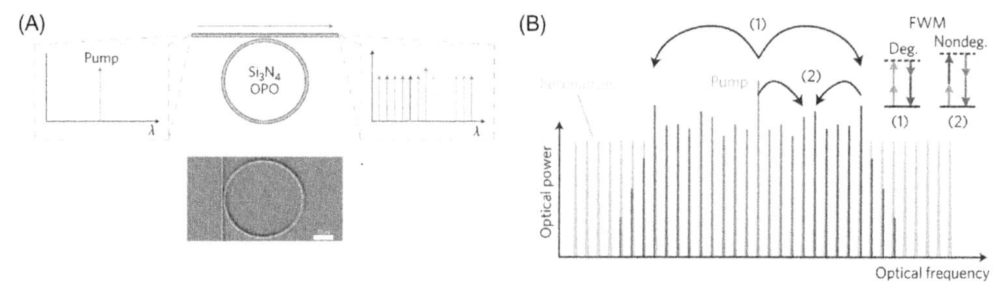

Figure 18.11 (A) On-chip optical comb generator using silicon nitride ring resonator with a single external pump laser. (B) Principle of Kerr comb formation by FWM. Source: *From (A) J.S. Levy, A. Gondarenko, M.A. Foster, A.C. Turner-Foster, A.L. Gaeta, M. Lipson, CMOS-compatible multiple-wavelength oscillator for on-chip optical interconnects, Nat. Photon 4 (2009) 37 [78]; (B) J. Pfeifle, V. Brasch, M. Lauermann, Y. Yu, D. Wegner, T. Herr, et al., Coherent terabit communications with microresonator Kerr frequency combs, Nat. Photon. 8, 375 [84].*

the normal dispersion regime [85]. Recently, a chip-scale comb source was reported [86] using an integrated semiconductor laser pumping an ultrahigh-quality factor (Q) Si_3N_4 ring resonator. This small-footprint device had performance lasting longer than ~200 hours powered from a standard dry cell battery. It is a strong candidate for an energy-efficient optical source for high-performance systems.

Even though the comb laser is a very promising direction of research, in order to be adopted in high-performance systems, several challenges remain. The comb laser must demonstrate advantages over CW laser arrays in terms of energy efficiency, cost, and footprint. To achieve this, the comb lines need to be fully utilized and the comb laser should demonstrate a relatively flat power profile. The optical power per channel needs to be greater than that needed to meet the power budget of the link. Currently most demonstrations have insufficient optical power per comb line and amplification is required to overcome the link power budget. The poor conversion efficiency in the anomalous dispersion regime (~2% pump-to-comb conversion efficiency) poses a major challenge for the wall plug efficiency of the comb source including pump laser.

18.3.3 Microring-based modulators

With its small footprint and wavelength-selective nature, the MRR is a highly promising candidate for realizing high-throughput optical interconnects compatible with comb lasers [87]. Since its introduction in 2005 [88], tremendous improvements have been demonstrated, such as modulation with high speed [89], ZigZag [90], and interdigitated [91] junctions. Advance modulation is also achieved by cascading MRRs along a single bus waveguide, as demonstrated in Refs. [92–94]. Recently the MRR has also been used for higher order amplitude modulation formats such as four-level pulse amplitude modulation (PAM4) at 128 Gb/s [95] and PAM8 at 45 Gb/s [96], achieving an energy consumption of as low as 1fJ/bit [96].

To quantify the performance of an MRR modulator in a cascaded architecture, the power penalty metric associated with the bit error rate (BER) is typically used [97]. The modulated light has a certain optical modulation amplitude (OMA) based on the spectral shift of the resonator. As shown in Fig. 18.12, the spectral response of a PIN-based ring modulator experiences a blue shift with the addition of some excess cavity loss. Such changes are due to the cavity phase shift and round-trip loss, which can be controlled through the driving voltage/current of the modulator [98,99].

Intermodulation crosstalk [97,100] can impact the overall power penalty of modulators in such a cascaded arrangement. A trade-off exists between the spacing of the channels and the shift of the resonance. A larger shift of resonance results in an improved OMA and lower modulator insertion loss but leads to a higher average loss of optical power due to on-off keying (OOK) and higher intermodulation crosstalk. In addition, even though ideally the operation point for the shift of resonance

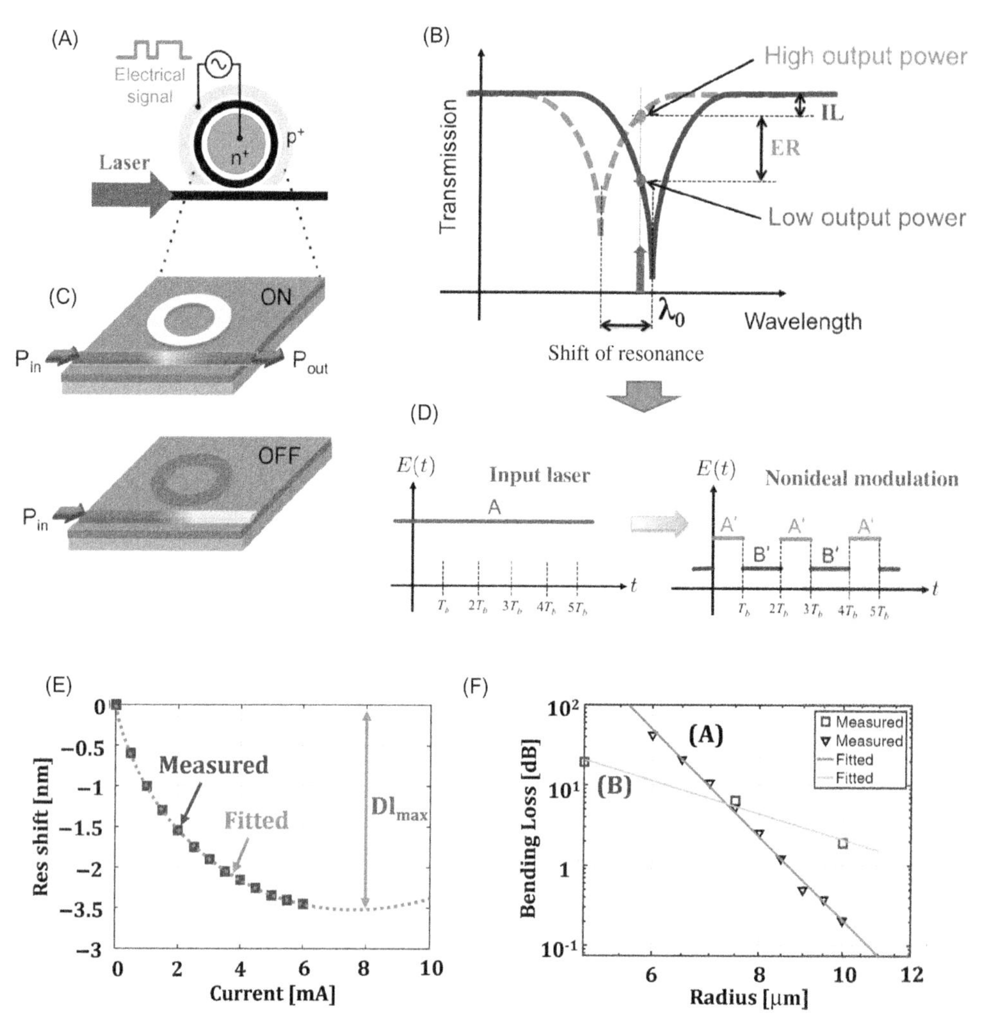

Figure 18.12 Schematic view of a microring modulator (MRR). (A) The high-speed electrical signal is applied to the pn-junction embedded inside the silicon ring. (B) Modulation of the input laser by shifting the resonance of the ring to create high and low levels of optical power at the output. (C) Graphical view of the ON and OFF states of light at the output. (D) Time-domain presentation of a nonideal NRZ OOK modulation. (E) Spectral shift of a PIN-based ring modulator as a function of injected current [94]. (F) Measured bending loss of ring resonators as a function of radius reported in Refs. [103,104] (both horizontal and vertical axes are in log scale). Source: *(A–D) From Q. Cheng, M. Bahadori, M. Glick, S. Rumley, and K. Bergman, "*Recent advances in optical technologies for data centers: a review,*" Optica 5, 1354–1370 (2018) [1].*

should be close to half of the spacing between the channels, PIN-based modulators suffer from Ohmic heating due to the injection of current inside the waveguide. The Ohmic heating limits the blue shift of the spectrum to about 2.5 nm, as shown in Fig. 18.12B. This situation is even worse for PN-based ring modulators due to their relatively low electro-optic modulation efficiency [101]. The choice of PN or PIN design for ring modulators is therefore twofold based on the desired optical penalty and the operation speed. PN-based modulators exhibit higher optical penalty compared to their PIN-based counterparts but benefit from operating at higher speeds [102].

A key step to establishing the design space exploration of MRRs is to relate the spectral parameters (Q-factor, round-trip loss) to the geometrical parameters (radius, coupling gaps) [105]. The bending loss of silicon ring resonators (in dB/cm) is a critical factor. Fig. 18.12F shows two sets of measurements for the bending loss of silicon ring resonators as a function of the radius, as reported in Refs. [103,104]. This leads to a power-law relation between the bending loss α and radius: $\alpha = A^0 \times R^{-B}$. An analytical approach can then be used for estimating the coupling coefficients between the ring and waveguides as a function of radius and coupling gaps [104]. One can then explore the design space of WDM links based on ring parameters [106]. Other design trade-offs also need to be taken into consideration. For instance, a large FSR supports more optical channels in the cascaded WDM configuration but requires a small radius leading to high bending loss [107].

18.3.4 Microring-based drop filters

As a resonance cavity, MRRs can also be employed in the form of add-drop structures. They are capable of performing wavelength de-multiplexing due to their wavelength-selective spectral response. Based on the desired passband and the rejection ratio of the filter, first-order [108,109] or higher order [110] add-drop filters are used. Higher order filters provide a better rejection ratio but suffer from a higher loss or resonance splitting in their passbands.

The power penalty of ring filters can be estimated based on the Lorentzian spectral shape of the filter. As shown in Fig. 18.13C, if the data rate of the OOK channel is much smaller than the 3 dB bandwidth of the ring, the power penalty is simply based on the spectral attenuation of the MRR. However, if the data rate is comparable to the bandwidth of the filter, a correction needs to be introduced to include the data rate impact on the filter power penalty and the crosstalk effects in a cascaded arrangement [109]. An example of the design space of silicon-based add-drop filters under the critical coupling condition is shown in Fig. 18.13D, with preset design metrics of the insertion loss, optical bandwidth, FSR of the filter, and the extinction of resonance.

For each individual channel, an optimization of the add-drop MRR in the cascaded arrangement can be utilized so that the power penalty associated with the entire demultiplexer array is minimized [97]. This optimization depends on the parameters of the ring, as well as data rate, number of channels, and channel spacing. The

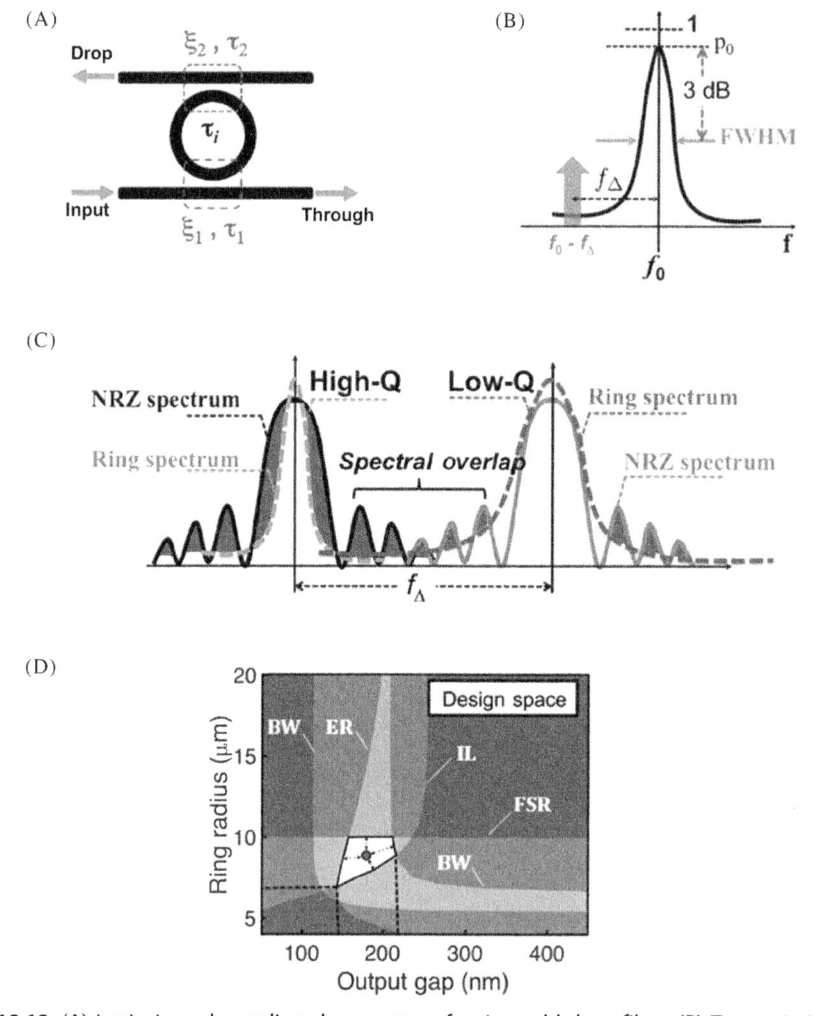

Figure 18.13 (A) Intrinsic and coupling decay rates of a ring add-drop filter. (B) Transmission spectrum of the drop path of a demux ring with the 3-dB bandwidth denoted as FWHM. f_Δ is the possible detuning between the resonance and the channel. (C) Schematic view of two adjacent channels with a fair amount of spectral overlap. A low-Q ring will result in more crosstalk effect but less spectral distortion (highlighted areas on the NRZ spectrum). f_Δ denotes the spacing between channels. (D) Design space of a critically coupled demux add-drop ring. Source: *(A–C) From Q. Cheng, M. Bahadori, M. Glick, S. Rumley, and K. Bergman,* "Recent advances in optical technologies for data centers: a review," *Optica 5, 1354–1370 (2018) [1].*

power penalty imposed on each channel consists of three parts: (1) The insertion loss of the ring—independent of the number of channels and their data rate. (2) The truncation effect—only dependent on the data rate. Strong truncation arises when the 3 dB bandwidth of the MRR is small compared to the signal bandwidth.

(3) Optical crosstalk due to the imperfect suppression of adjacent channels—a function of number of channels and channel spacing. As shown in Fig. 18.13C, the Q-factor of the MRRs is the determining factor in the power penalty space [97]. Increasing the Q will increase the insertion loss of the ring and truncation of the OOK signal, but doing so results in suppression of optical crosstalk. Therefore, an optimized point exists for the minimal penalty.

In addition to the physical properties discussed, other challenges need to be carefully addressed to fully utilize the advantages of MRRs for optical interconnects: (1) Thermal sensitivity: Thermal effects significantly impact the optical response of silicon-based resonant devices due to the strong thermo-optic coefficient of silicon. The resonance of a typical silicon MRR is shifted by $\sim$9 GHz for each degree Kelvin change in the temperature [111]. Such thermal drift in high-Q MRRs can impose more than 1 dB of penalty on high-speed OOK signals [97]. (2) Self-heating: The enhancement of optical power inside the MRR is proportional to the finesse, or Q-factor. Even a slight internal absorption in a high Q MRR can lead to a noticeable thermal drift of resonance. A recent transceiver design has proposed a thermal tuning algorithm based on the statistics of the data stream to counteract this effect [112]. (3) Fabrication variation: The spectral parameters of MRRs such as resonance wavelength, FSR, and the 3 dB optical bandwidth largely depend on their geometrical parameters. It is known that current silicon photonic fabrication imposes variations on the dimensions of the waveguides [113]. This results in deviations of the resonance wavelength from the original design [114] and requires thermal tuning, hence degrading the energy efficiency of the link. Various wavelength locking schemes based on analog and digital feedback [115], bit statistics [112], and pulse width modulation and thermal rectification [111] have been proposed and implemented to overcome the unwanted variations due to the fabrication. (4) Backscattering: In applications where narrow optical linewidths (i.e., $Q>10{,}000$) are required, even a slight roughness on the sidewalls of the MRRs will cause back reflections inside the ring [116]. The effect of backscattering in MRRs is typically observed in the form of a splitting of the resonance in the spectral response [117]. This spectral distortion adds extra complexity to the design of optical links and further narrows the design space of MRRs [118].

18.3.5 Energy-efficient photonic links

In this section we present a design exploration for short-reach silicon photonic links using MRRs and filters. We seek to obtain the maximum achievable aggregate bandwidth. This is achieved by analyzing the impacts due to the induced impairments and translating them into power penalties, the extra optical power required to compensate for the effects of such impairments on the bit error ratio performance of the system

[98,119–122]. We also show how the spectral statistics of modulated light changes as it travels through a ring demultiplexer and how the changes can be become a power penalty.

We concentrate our efforts on MRR-based links, as they offer the highest bandwidth density and most energy-efficient performance among current silicon photonic interconnect devices [123–125]. Due to their small size, multiple microrings can be placed along a single waveguide on chip, facilitating a dense WDM design [126,127]. However, WDM links may suffer from spectral degradation of channels and interchannel crosstalk [128–131]. These impairments eventually set an upper limit on both the number of channels and the modulation speed of each channel, thus placing an upper bound on the aggregate rate to the link [132,133].

Consider a simple chip-to-chip silicon photonic link as shown in Fig. 18.14. Microrings are placed along an on-chip waveguide to modulate the incoming multiwavelength light generated by a comb laser source [87]. The incoming wavelengths, once imprinted with data, are then transmitted through an optical waveguide to a receiver chip. The receiver chip consists of multiple passive microrings with resonances tuned to the channel wavelengths. The total capacity of this link is obtained by multiplying the number of channels N_λ with the modulation bit rate r_b.

Intuitively, it is tempting to maximize the number of wavelengths and/or to choose higher bit rates for each channel. This allows for higher utilization of the available spectrum in the transmission media. However, as the number of wavelengths and/or the bit rate grows, crosstalk between channels and other undesired impairments emerge, which eventually prevent a reliable transmission through the link. Therefore, the total capacity of the link is closely tied to the optical power losses and other unde-

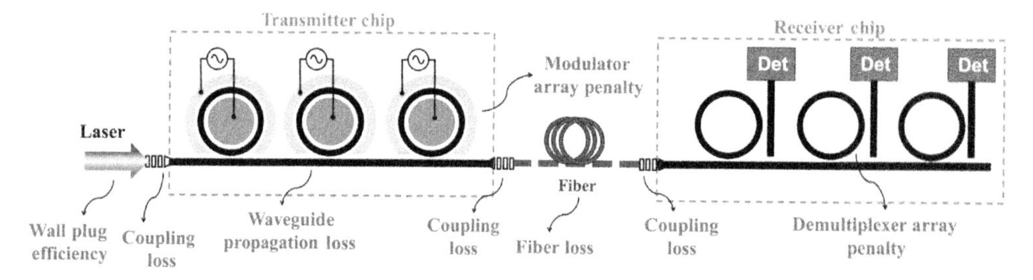

Figure 18.14 Chip-to-chip silicon photonic interconnect with an MRR-based wavelength division multiplexing (WDM) link. The optical interface of the transmitter chip includes MRR modulators that use carrier dispersion for high-speed modulation. The optical interface at the receiver includes demultiplexing filters, photodetectors, and electronic decision circuitry (Det: detector). Wall-plug efficiency corresponds to the electrical to optical power conversion of the laser. Source: *From M. Bahadori, S. Rumley, D. Nikolova, K. Bergman, Comprehensive design space exploration of silicon photonic interconnects, J. Lightw. Technol. 34 (12) (2016) 2975–2987 [97].*

sired optical impairments through the entire link. Summing up all the power penalties of the link, PP^{dB}, for a single channel, the following inequality must hold [134]:

$$\left[P_{\mathrm{laser}}^{\mathrm{dBm}} - 10\log 10(N\lambda)\right] - P_{\mathrm{sensitivity}}^{\mathrm{dBm}} \geq PP^{\mathrm{dB}} \tag{20.1}$$

In general, aggregated optical power P_{laser} (sum over all wavelengths) must stay below the nonlinear threshold of the silicon waveguides at any point of the link [133,135]. On the other hand, the signal powers should stay above the sensitivity of the detectors $P_{\mathrm{sensitivity}}$ (minimum number of photons or equivalently a certain amount of optical power) at the receive side. A typical receiver may have a sensitivity of −12.7 dBm at 8 Gb/s operation [120], while a good receiver may exhibit a sensitivity to −21 dBm at 10 Gb/s [119]. The difference between these higher and lower thresholds can be exploited to find the maximum power budget. This budget accounts for the power penalty, PP^{dB}, per channel over the N_λ channels. We will show that the power impairments induced by the microrings depend on the channel spacing, which is inversely proportional to the number of channels, and on the modulation rate.

Here we present a study on the minimal energy consumption for 200−800 Gbps data rate aggregation on the link, as indicated in Fig. 18.15A, based on the optimization of the physical parameters of the microring. The rings are configured to operate at their critical coupling point. The results in the figure show that the smaller ring radius in the ~7 μm regime leads to the best energy performance of the link. Fig. 18.15B shows the breakdown of the factors contributing to energy consumption. At higher data rates, the laser power consumption becomes critical, but the energy consumed by the static thermal tuning declines. Finally, Fig. 18.15C provides details of the number of channels and the required data rate per channel that lead to the minimal energy consumption for the target aggregation rate.

When the available optical power budget is fully utilized, the study also investigates the maximum aggregation rate based on the product of the number of channels and the optical data rate of each channel. Fig. 18.15D indicates a maximum possible aggregation rate of ~800 Gbps at 15 Gbps data rate per channel. For each data rate, its associated energy efficiency is also plotted. Note that the lowest energy efficiency is not associated with the highest aggregation rate. This further reiterates the fact that designing a silicon photonic link requires a trade-off between energy consumption and high-speed performance. We note that the results depend on the ring resonator parameters and vary if different parameters are used. In addition, the losses in the ring have significant impact on the maximum aggregation rate. The details for the model of the ring resonators associated with these results can be found in Ref. [107].

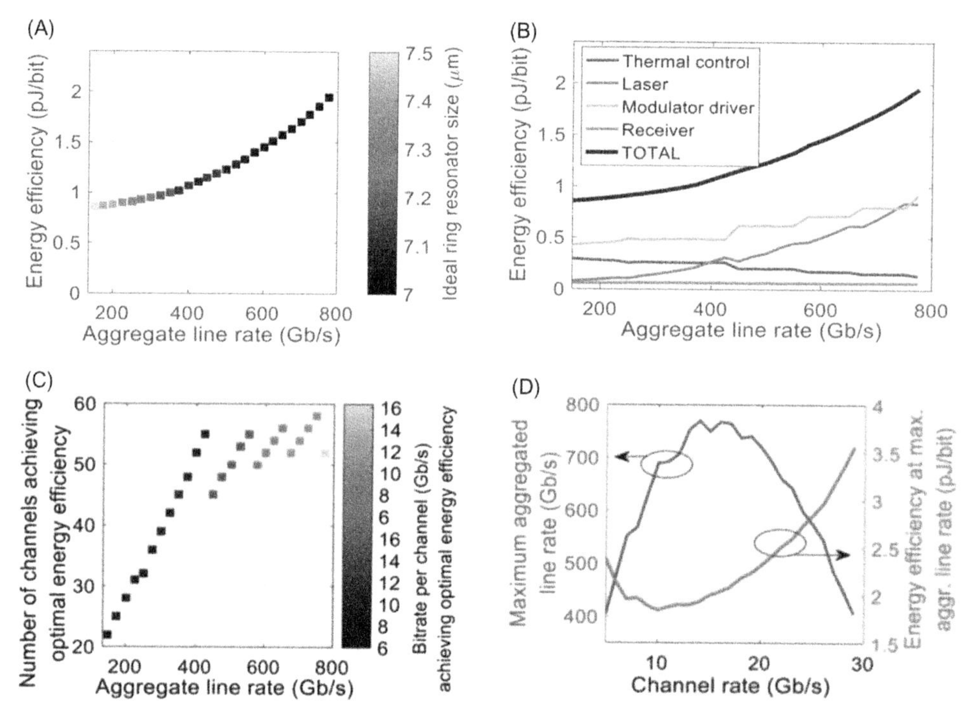

Figure 18.15 (A) Minimum energy consumption of the link for given aggregations based on the optimum value for the ring radius. (B) Breakdown of energy consumption. (C) Breakdown of the number of channels and the required data rate per channel for minimum energy consumption. (D) Evaluation of the maximum supported aggregation and the associated energy consumption for various channel rates Source: *From M. Bahadori, S. Rumley, R. Polster, A. Gazman, M. Traverso, M. Webster, et al., Energy-performance optimized design of silicon photonic interconnection networks for high-performance computing, in: Proceedings of the Conference on Design, Automation & Test in Europe, European Design and Automation Association: Lausanne, Switzerland, 2017, pp. 326–331 [136].*

18.4 Bandwidth steering

In this section, we first briefly survey optical switching technologies, then discuss optimized photonic links with optical switches and introduce the Flexfly architecture—a Dragonfly design that is capable of reconfiguring its bandwidth to match traffic patterns by using low-radix silicon photonics switches [40]. Optical switches are an important component in modern high-speed telecommunications. The advantages they can provide in terms of energy efficiency and high bandwidth density, particularly as cost is reduced through integrated silicon photonics, are an important subject of research and development for computing systems.

18.4.1 Free-space optical switches

Numerous competing optical switching approaches based on free-space technology have been commercially realized, including microelectromechanical systems (MEMS) [137], beam-steering [138], and liquid crystal on silicon [139]. Among these, MEMS-based optical switches are the most common free-space optical switching devices. An electrostatic driver is commonly used because of its low power consumption and ease of control; however, a typical voltage up to 100–150 V is required [61,140]. MEMS spatial switches can be realized in both two-dimensional (2D) and three-dimensional (3D) configurations. The crossbar topology is normally implemented in the 2D configuration with digital operation using a bistable mirror position. 3D MEMS switches, which are assembled using 2D input and output fiber arrays with collimators, have been proposed to support very large-scale optical cross-connect devices [141]. Two stages of independent 2D micromirror arrays with a two-axis tilting structure [137] are used to steer the optical beams in three dimensions.

MEMS switches can support connectivity of hundreds of ports [137,142]; however, the installation and calibration with surface-normal micro optics introduces considerable complexity that is ultimately reflected in the cost per port. This cost remains a challenge for the implementation of MEMS switches in high-performance systems.

18.4.2 Photonic integrated switches

In order to be adopted in high-performance systems, optical switching technologies must demonstrate a path toward high-volume manufacture and ensure low cost per port. This leads to the consideration of lithography-based fabrication and high-level integration. Here we present a brief overview of the switching technologies based on III–V and silicon platforms.

In the integrated devices under consideration, different physical mechanisms have been investigated in order to realize the optical switching process. Physical properties used include phase manipulation through thermal or electrical control in interferometric structures, that is, MZI and MRR, signal amplification/absorption in semiconductor optical amplifiers (SOAs), and MEMS-actuated coupling between different layers of waveguides. In the last decade, photonic integration technologies have quickly matured to realize monolithic integrated circuits of a few thousands of components with increasingly sophisticated functionalities. Notable demonstrations of monolithic switch fabrics are summarized in Fig. 18.16.

InP-based switch fabrics have primarily employed SOA gated elements in the broadcast and select (B&S) topology. B&S networks utilize passive splitters/combiners with each path gated by an SOA element, which can provide chip-level multicast [143]. However, the optical loss due to the various signal splits and recombinations discourage scaling this architecture beyond 4×4 connectivity. As an alternative,

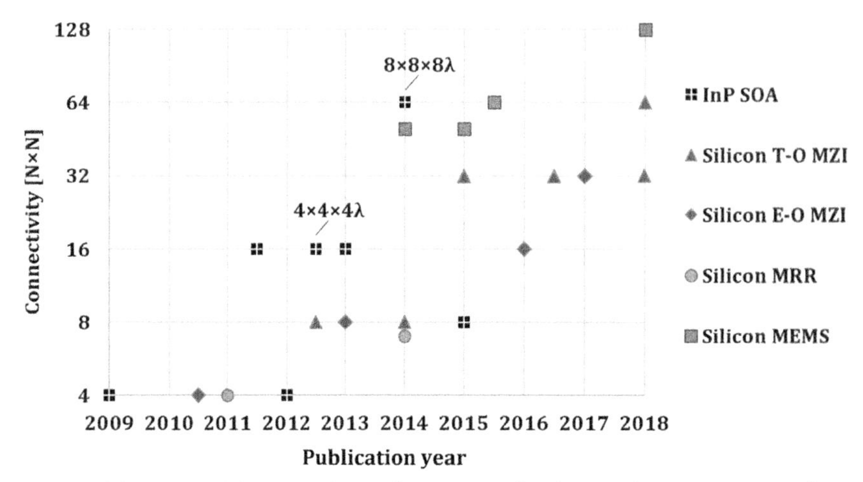

Figure 18.16 High connectivity optical switch matrix technologies shown in terms of input side connectivity. Source: Q. Cheng, M. Bahadori, M. Glick, S. Rumley, and K. Bergman, "*Recent advances in optical technologies for data centers: a review*," Optica 5, 1354–1370 (2018) [1].

multistage architectures using cascaded switching elements have been proposed [144]. 16 × 16 port count SOA-based switches have been demonstrated using both all-active [145] and passive-active [146] integration schemes. Higher on-chip connectivity, scaling up to 64 × 64 connections, has been achieved by combining spatial ports with wavelength channels using co-integrated AWGs [147]. Further scaleup would require a large reduction in component-level excess loss, a more careful design of balancing the summed loss and gain per stage, and a close examination of SOA designs for linear operation [148,149].

In addition to InP based switches, the highly advanced CMOS industry with mature fabrication infrastructures and advances in silicon photonics have stimulated the development of silicon-based optical switches. The current record for a monolithic photonic switch radix is 64 × 64 by a thermo-optic MZI-based Beneš switch [150] and the very recent 240 × 240 by a MEMS-actuated crossbar switch [151]. Other notable advances include a 32 × 32 thermally actuated PILOSS MZI switch with <13.2 dB insertion loss [152] and a 32 × 32 electro-optic MZI-based Beneš switch [153].

To support the scaling of integrated photonic switches for high-performance systems, the switch architecture should be reexamined in terms of crosstalk cancellation, the number of cascaded switch stages, and the total number of switch cells, as signal degradation is introduced from accumulated crosstalk and loss exacerbates with increased cascaded stages. We demonstrate an MRR-based modified switch-and-select (S&S) switching circuit with the concept illustrated in Fig. 18.17A and B, where the $1 \times N$ switch unit is built from MRR add-drop cells assembled in a bus coupled

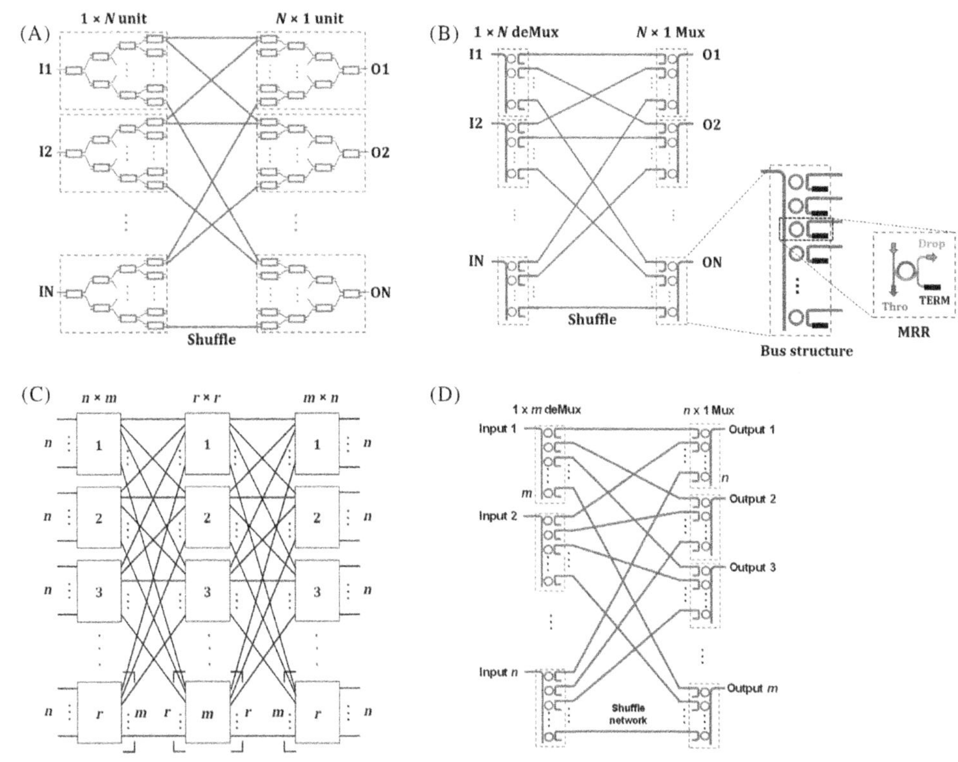

Figure 18.17 (A) Switch-and-select topology with MZI elements arranged in a cascaded structure. (B) Modified switch-and-select topology with MRR based spatial (de)multiplexers. (C) Layout of a generic three-stage Clos network building from r n × m, m r × r, and r m × n blocks. (D) Schematic of an n × m microring-based block in the switch-and-select topology. Source: *From (A-B) Q. Cheng, L. Y. Dai, N. C. Abrams, Y. Hung, P. E. Morrissey, M. Glick, P. O'Brien, and K. Bergman,* "Ultralow-crosstalk, strictly non-blocking microring-based optical switch," *Photon. Res. 7, 155–161 (2019)(C-D) Q. Cheng, M. Bahadori, Y. Hung, Y. Huang, N. Abrams and K. Bergman,* "Scalable Microring-Based Silicon Clos Switch Fabric with Switch-and-Select Stages," *in IEEE Journal of Selected Topics in Quantum Electronics. Doi: 10.1109/JSTQE.2019.2911421 [157]*

structure [154]. Scaling such a structure requires only adding MRRs to the bus waveguide, which effectively reduces the scaling overhead in loss compared to that of the cascaded scheme. The layout of a generic $N \times N$ S&S MRR based switch is depicted in Fig. 18.17B. This configuration has N input spatial $1 \times N$ and N output spatial $N \times 1$ units, maintaining the number of drop microrings at two in any path, therefore the first-order crosstalk is fully blocked. We performed further studies on combining the scalable three-stage Clos network with populated S&S stages [155,156], as shown in Fig. 18.17C and D. The proposed design offers a balance that keeps the number of stages to a modest value while largely reducing the required number of switching elements. The scalability is predicted to be 128×128 [155,156].

Prototyped devices have been fabricated at the American Institute of Manufacturing (AIM photonics) foundry and all designs used the predefined elements in the PDK library to ensure high yield with low cost. The design rules of standard packaging houses, for example, Tyndall National Institute, are employed to achieve low-cost packaging solutions. Fig. 18.18A, B, and C shows the microscope photo of a 4 × 4 Si/SiN dual-layer S&S switch, a 4 × 4 silicon S&S switch, and a 12 × 12 silicon Clos switch, respectively. All devices have been fully packaged with thermo-optical MRRs in use. Small radix switches can be packaged using the QFN-type socket, and directly wire-bonded or flip-chip bonded to a PCB breakout board with a UV-cured fiber array, as shown in Fig. 18.19A, B, and C, respectively. For densely integrated Clos switches, a packaging platform was developed with a silicon interposer (as shown by Fig. 18.18D) as an electrical redistribution layer for an ultra-compact package with low insertion loss. Fig. 18.19D shows the packaged 12 × 12 Clos switch, which was first flip-chip bonded onto a silicon interposer and then wire-bonded to a PCB breakout board. Excellent testing results were achieved for the fully packaged 4 × 4 S&S switch with on-chip loss and crosstalk ratio as low as 1.8 and −50 dB, respectively [154,158].

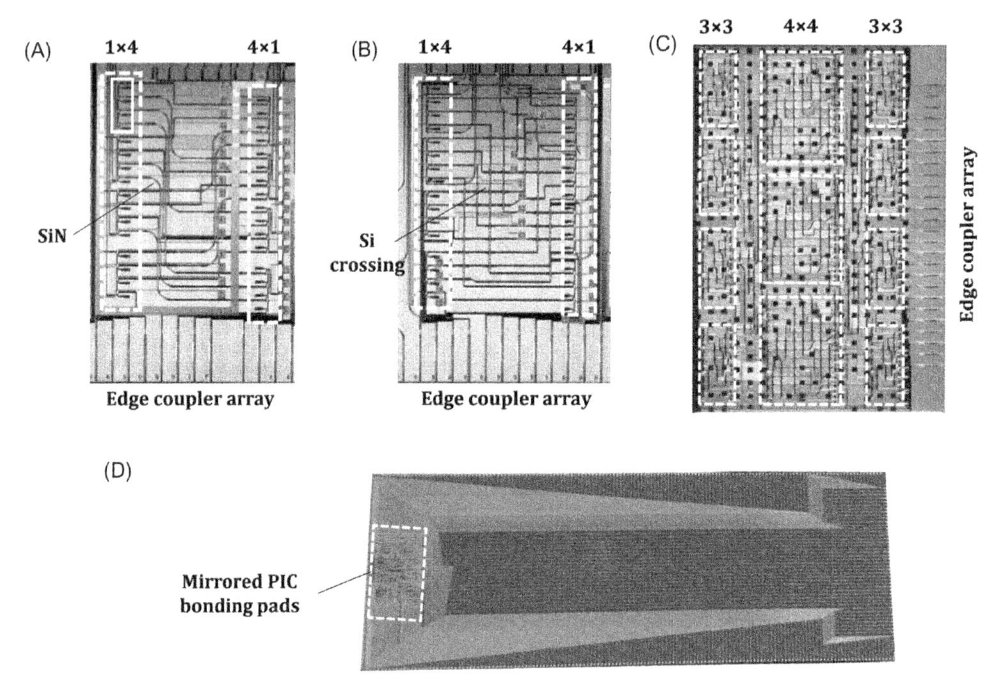

Figure 18.18 Microscope photo of (A) 4 × 4 Si/SiN dual-layered MRR-based S&S switch, (B) 4 × 4 Si MRR-based S&S switch, and (C) 12 × 12 Si MRR-based Clos switch with populated S&S stages. (D) Silicon interposer for the 12 × 12 Clos switch.

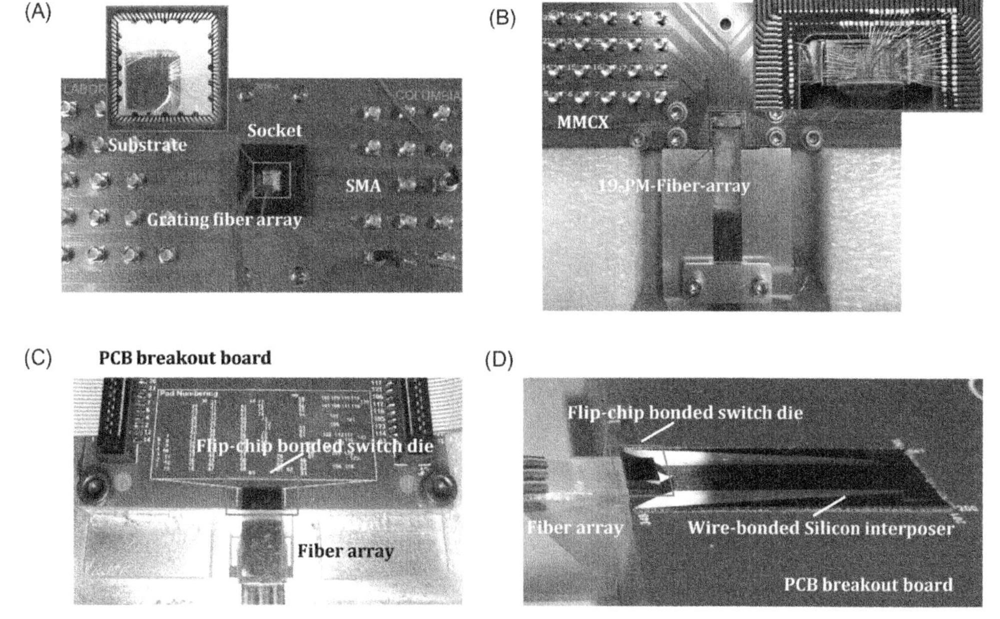

Figure 18.19 (A) QFN type package with socket. (B) Packaged switch device by wire bonding to the PCB breakout board. (C) Packaged switch device by flip-chip bonding on the PCB breakout board. (D) Packaged switch device with silicon interposer. Source: *(B) From Q. Cheng, M. Bahadori, Y. Hung, Y. Huang, N. Abrams and K. Bergman,* "Scalable Microring-Based Silicon Clos Switch Fabric with Switch-and-Select Stages," *in IEEE Journal of Selected Topics in Quantum Electronics. Doi: 10.1109/JSTQE.2019.2911421. (C) From Q. Cheng, L. Y. Dai, N. C. Abrams, Y. Hung, P. E. Morrissey, M. Glick, P. O'Brien, and K. Bergman,* "Ultralow-crosstalk, strictly non-blocking microring-based optical switch", *Photon. Res. 7, 155–161 (2019). (D) From Q. Cheng, M. Bahadori, M. Glick, S. Rumley, and K. Bergman,* "Recent advances in optical technologies for data centers: a review," *Optica 5, 1354–1370 (2018) [158].*

Looking forward, we envision a new class of III–V/Si heterogeneously integrated optical switches leveraging advanced bonding techniques to provide compact, energy-efficient, and low-cost switch fabrics that satisfy the high-performance system metrics [61], where lossless design would be a significant advantage. The implementation can follow the approach demonstrated in the InP MZI-SOA switch fabrics [159–161] or be combined with the S&S topology of MRR add-drop multiplexers. Detailed discussions are found in Ref. [61].

18.4.3 Network performance

In today's computing systems, the high bandwidth densities that optics is capable of are not fully leveraged. Implementing optical interconnects for a system with 10 K computing nodes entirely with optical cables and optical switches remains cost-prohibitive. Currently, network topologies implemented in computing systems use electrical packet switches and optical cables for longer reach links (> 1 m). Supercomputers are often perceived as the first adopters in the market for innovative optical technologies (as occurred, for instance, with AOCs). In the supercomputer, where torus topologies previously dominated, high-radix packet switches have made hierarchical topologies such as the Dragonfly and the fat tree more popular. The Dragonfly topology [41,162] provides high connectivity with all-to-all global links at the intergroup level, and aims at minimizing the number of long distance links to reduce the cost. However, the advantages of high connectivity are diluted by low per link bandwidth. The bandwidth of intergroup (global) links, carrying the traffic between Dragonfly groups, can become the bottleneck for an entire network.

A major reason for this bandwidth bottleneck is due to the highly skewed traffic characteristics of HPC applications. These traffic patterns concentrate traffic on only a small percentage of links, so that only a few links are congested while most others are severely underutilized. Thus, the current, best-for-all approach using static, over-provisioned networks have topologies that are mismatched with the applications that operate over them, which will likely become a bottleneck for the next-generation Exaflop platforms [42–44,163].

Designing networks that properly balance traffic is challenging: over-provisioning the network incurs unnecessary cost and energy [164], while under-provisioning leads to limitations on system performance due to data-starved processors. In this section we present results from a study on Flexfly [40] a photonic architecture that trades global links among dragonfly groups using low-radix silicon photonic switches, allowing the network topology to be dynamically reconfigured to match HPC application traffic. In Flexfly the global links initially defining the all-to-all topology can be taken from their original destination groups and reassigned to traffic-intensive ones. By trading the global links in this way, Flexfly creates additional direct bandwidth for intensively communicating group pairs where and when it is needed. This is illustrated in Fig. 18.20, showing an all-to-all Dragonfly topology being reconfigured to a bandwidth-steered topology that focuses on maximizing bandwidth between neighboring groups. It achieves such reconfigurability through the use of transparent silicon photonic circuit switching. Flexfly is designed to support the use of low-radix optical switches, realizable through low-cost fabrication technologies. Simulations on the Flexfly architecture with applications such as GTC, Nekbone, and LULESH show up to 1.8× speedup over the Dragonfly topology paired with UGAL routing. The hop count and cross-group message latency are also halved compare to the Dragonfly topology.

An experimental 32-node Flexfly prototype was built with four groups connected through a silicon photonic switch [56]. The interconnect reconfiguration time was 820 ns. The network architecture consisting of the control and data planes is shown in Fig. 18.21. The control plane is an Ryu-based SDN controller, and each rack sends a

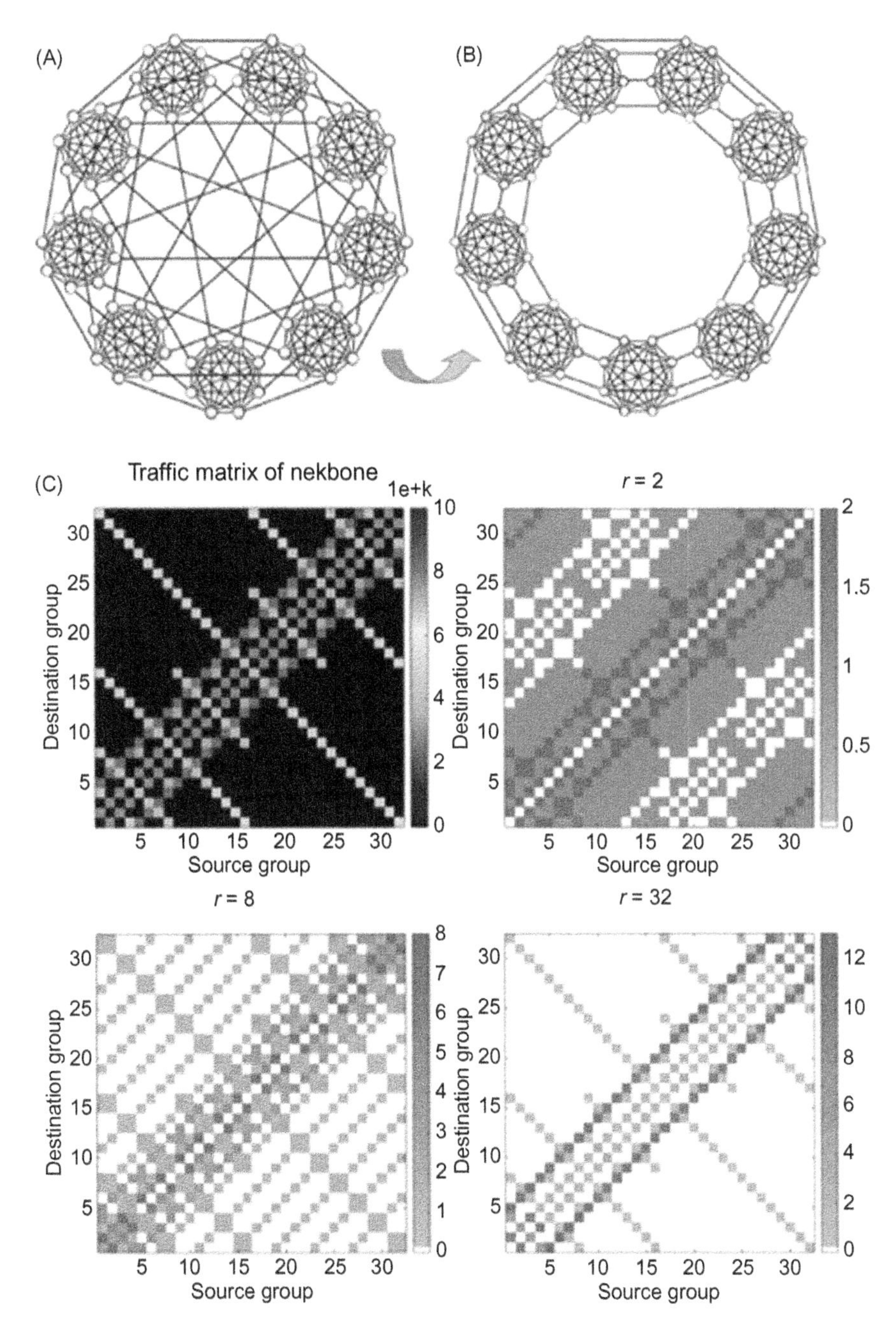

Figure 18.20 (A) Regular Dragonfly topology with all-to-all intergroup links. (B) Reorganized topology after bandwidth steering using optical switching. (C) Top-left matrix shows traffic distribution across pairs of Dragonfly groups during execution of Nekbone workload; other matrices show the network topology for different silicon photonic switch radices. Source: *From J. Wilke, Bringing minimal routing back to HPC through silicon photonics: a study of "flexfly" architectures with the structural simulation toolkit (SST), in: Proceedings of the 2nd International Workshop on Advanced Interconnect Solutions and Technologies for Emerging Computing Systems, ACM, Stockholm, Sweden, 2017, p. 5-5 [165].*

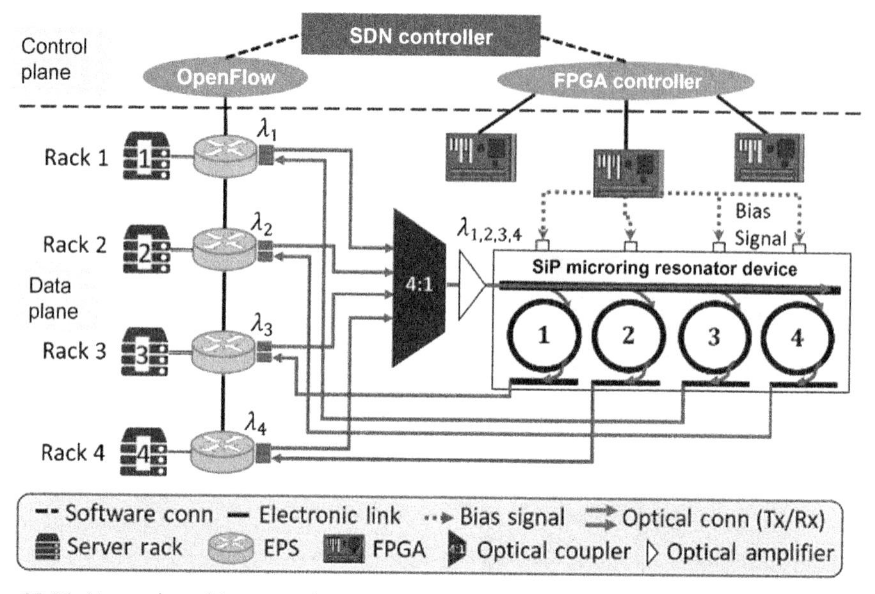

Figure 18.21 Network architecture showing the connections of the servers and top-of-rack electronic packet switches (EPSs) to the SiP MRR device. Source: *From Y. Shen, A. Gazman, Z. Zhu, M.Y. Teh, M. Hattink, S. Rumley, et al., Autonomous dynamic bandwidth steering with silicon photonic-based wavelength and spatial switching for Datacom networks, in: Optical Fiber Communication Conference, Optical Society of America, 2018 [56].*

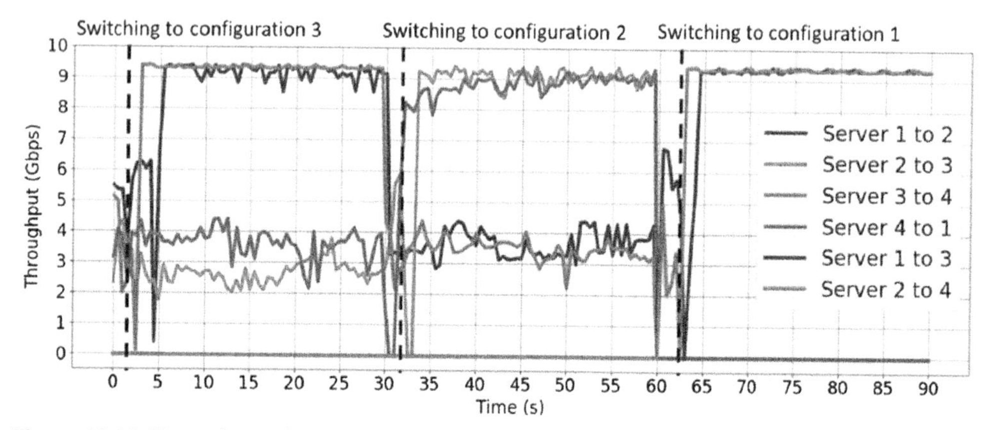

Figure 18.22 Throughput of various intergroup flows over time demonstrating control plane bandwidth steering capabilities. Source: *From Y. Shen, A. Gazman, Z. Zhu, M.Y. Teh, M. Hattink, S. Rumley, et al., Autonomous dynamic bandwidth steering with silicon photonic-based wavelength and spatial switching for Datacom networks, in: Optical Fiber Communication Conference, Optical Society of America, 2018 [56].*

unique wavelength through the electronic packet switches (EPSs). By tuning the rings in different ways, the device allows different servers to be connected bidirectionally and thus act as a wavelength and spatial optical circuit switch.

The results of the bandwidth steering are shown in Fig. 18.22. When traffic is added between servers, the control plane detects the changes and initiates the configuration updates. The configurations provide a direct connection for traffic between particular servers, so that they are able to reach near full link capacity, while the other flows must compete with the background traffic and are therefore limited in throughput. The results demonstrate that Flexfly-based optically switched networks are a promising solution for improved network efficiency and resource allocation.

18.5 Conclusions

Looking forward we see several factors pointing toward excellent opportunities for optical interconnection networks to be increasingly deployed in high-performance systems. Although the details of the architectures of the supercomputer and warehouse scale data center are different, currents trends are leading to similar scaling challenges. The slowing of Moore's law leads to more parallelism and a greater focus on energy efficiency. Photonic solutions are a natural fit for increased parallelism and as data rates get higher and are at an advantage in the energy/bit metric. The need for networks with higher and higher bandwidth is exacerbated by new applications using machine learning algorithms and data analytics. By their nature, these applications require much greater storage capabilities and are communication intensive. Communication-intensive calculations using large amounts of data also favor the photonic interconnects' high bandwidth capabilities. With an increased emphasis on cost savings and energy efficiency, network architects are looking to get the highest utilization out of the equipment in the network. This is leading to a re-architecting of the topologies toward adaptability and reconfiguration to more efficiently match traffic and application requirements. Techniques such as bandwidth steering using low-radix optical switches can efficiently reconfigure the topology to place the bandwidth where required to meet the application. At the same time as the needs are expanding, there are major advances being made to lower the cost of photonic-based circuits through the use of CMOS-compatible silicon photonics and leveraging manufacturing technology used in the electronics industry to reduce the cost of fabrication and packaging. This combination of an avenue toward reduced cost and greater needs suited to photonics capabilities are expected to lead to major advances in research and deployment of photonic interconnection networks in the near future.

References

[1] Q. Cheng, M. Bahadori, M. Glick, S. Rumley, K. Bergman, Recent advances in optical technologies for data centers: a review, Optica 5 (2018) 1354–1370.

[2] Christos A. Thraskias, et al., Survey of photonic and plasmonic interconnect technologies for intra-datacenter and high-performance computing communications, IEEE Communications Surveys & Tutorials 20 (4) (2018) 2758–2783.

[3] D.A. Reed, J. Dongarra, Exascale computing and big data, Commun. ACM 58 (7) (2015) 56–68.

[4] J. Shalf, S. Dosanjh, J. Morrison. Exascale computing technology challenges, in: International Conference on High Performance Computing for Computational Science, Springer, 2010.

[5] R. Meisner, A Platform Strategy for the Advanced Simulation and Computing Program. NA-ASC-113R-07-Vol. 1-Rev. 0, 2007.

[6] Cisco Global Cloud Index: Forecast and Methodology, 2016–2021 White Paper. Available from: <https://www.cisco.com/c/en/us/solutions/collateral/service-provider/global-cloud-index-gci/white-paper-c11-738085.html>.

[7] M.M. Waldrop, The chips are down for Moore's law, Nat. News 530 (7589) (2016) 144.

[8] G. Michelogiannakis, J. Shalf, D. Donofrio, J. Bachan, Continuing the Scaling of Digital Computing Post Moore's Law. 2016.

[9] Ken Giewont, et al., 300mm Monolithic Silicon Photonics Foundry Technology, IEEE Journal of Selected Topics in Quantum Electronics (2019).

[10] C. Sun, M.T. Wade, Y. Lee, J.S. Orcutt, L. Alloatti, M.S. Georgas, et al., Single-chip microprocessor that communicates directly using light, Nature 528 (2015) 534.

[11] D.E. Nikonov, I.A. Young, Benchmarking of beyond-CMOS exploratory devices for logic integrated circuits, IEEE J. Explorat. Solid-State Comput. Dev. Circuits 1 (2015) 3–11.

[12] DeepMind AI Reduces Google Data Centre Cooling Bill by 40%. Available from: <https://deepmind.com/blog/deepmind-ai-reduces-google-data-centre-cooling-bill-40/>.

[13] N.P. Jouppi, C. Young, N. Patil, D. Patterson, G. Agrawal, R. Bajwa, et al., In-datacenter performance analysis of a tensor processing unit, in: Computer Architecture (ISCA), 2017 ACM/IEEE 44th Annual International Symposium on, EEE, 2017.

[14] K. Hazelwood, S. Bird, D. Brooks, S. Chintala, U. Diril, D. Dzhulgakov, et al., Applied machine learning at facebook: a datacenter infrastructure perspective, in: High Performance Computer Architecture (HPCA), 2018 IEEE International Symposium on, IEEE, 2018.

[15] N.M. Nasrabadi, Pattern recognition and machine learning, J. Electr. imaging 16 (4) (2007) 049901.

[16] N. Farrington, A. Andreyev, Facebook's data center network architecture, in: Optical Interconnects Conference, 2013 IEEE. Citeseer, 2013.

[17] S. Kanev, J.P. Darago, K. Hazelwood, P. Ranganathan, T. Moseley, G.-Y. Wei, et al., Profiling a warehouse-scale computer, in: ACM SIGARCH Computer Architecture News, ACM, 2015.

[18] P. Goyal, P. Dollár, R. Girshick, P. Noordhuis, L. Wesolowski, A. Kyrola, et al., Accurate, large minibatch SGD: training imagenet in 1 hour. arXiv preprint arXiv:1706.02677, 2017.

[19] A. Krizhevsky, I. Sutskever, G.E. Hinton, Imagenet classification with deep convolutional neural networks, in: Advances in Neural Information Processing Systems. 2012, pp. 1097–1105.

[20] K. Simonyan, A. Zisserman, Very deep convolutional networks for large-scale image recognition. arXiv preprint arXiv:1409.1556, 2014.

[21] Training Recurrent Neural Networks at Scale. Available from: <https://www.slideshare.net/SessionsEvents/erich-elsen-research-scientist-baidu-research-at-mlconf-nyc-41516>.

[22] Y. Lecun, Y. Bengio, G. Hinton, Deep learning, Nature 521 (7553) (2015) 436.

[23] Y. Shen, N.C. Harris, S. Skirlo, M. Prabhu, T. Baehr-Jones, M. Hochberg, et al., Deep learning with coherent nanophotonic circuits, Nat. Photon. 11 (2017) 441.

[24] J. Hines, Stepping up to Summit, Comput. Sci. Eng. 20 (2) (2018) 78–82.

[25] D. Schneider, US supercomputing strikes back, IEEE Spectrum 55 (1) (2018) 52–53.

[26] K. Bergman, Silicon photonics for high performance interconnection networks, in: Optical Fiber Communication Conference, Optical Society of America, 2018.

[27] S. Rumley, M. Bahadori, R. Polster, S.D. Hammond, D.M. Calhoun, K. Wen, et al., Optical interconnects for extreme scale computing systems, Parallel Comput. 64 (2017) 65–80.
[28] Top 500. Available from: <https://www.top500.org/.>.
[29] The Green 500. Available from: <https://www.top500.org/green500/>.
[30] Graph 500. Available from: <https://graph500.org/>.
[31] B.E. Floren, Optical interconnects in the Touchstone supercomputer program, in: Integrated Optoelectronics for Communication and Processing. International Society for Optics and Photonics, 1992.
[32] D.H. Hartman, Digital high speed interconnects: a study of the optical alternative, Opt. Eng. 25 (10) (1986) 251086.
[33] J.W. Goodman, F.J. Leonberger, S.-Y. Kung, R.A. Athale, Optical interconnections for VLSI systems, Proc. IEEE 72 (7) (1984) 850–866.
[34] A. Louri, H. Sung, An optical multi-mesh hypercube: a scalable optical interconnection network for massively parallel computing, J. Lightw. Technol. 12 (4) (1994) 704–716.
[35] J.A. Kash, Leveraging optical interconnects in future supercomputers and servers, in: High Performance Interconnects, 2008. HOTI'08. 16th IEEE Symposium on, IEEE, 2008.
[36] A.F. Benner, M. Ignatowski, J.A. Kash, D.M. Kuchta, M.B. Ritter, Exploitation of optical interconnects in future server architectures, IBM J. Res. Dev. 49 (4.5) (2005) 755–775.
[37] M.J. Kobrinsky, B.A. Block, J.-F. Zheng, B.C. Barnett, E. Mohammed, M. Reshotko, et al., On-chip optical interconnects, Intel Technol. J. 8 (2) (2004).
[38] C. Gunn, CMOS photonics for high-speed interconnects, IEEE Micro 26 (2) (2006) 58–66.
[39] A. Shacham, K. Bergman, L.P. Carloni, Maximizing GFLOPS-per-Watt: high-bandwidth, low power photonic on-chip networks, in: P = ac2 Conference, Citeseer, 2006.
[40] K. Wen, P. Samadi, S. Rumley, C.P. Chen, Y. Shen, M. Bahadori, et al. Flexfly: enabling a reconfigurable dragonfly through silicon photonics, in: High Performance Computing, Networking, Storage and Analysis, SC16: International Conference for, IEEE, 2016.
[41] J.H. Ahn, N. Binkert, A. Davis, M. Mclaren, R.S. Schreiber, HyperX: topology, routing, and packaging of efficient large-scale networks, in: Proceedings of the Conference on High Performance Computing Networking, Storage and Analysis, ACM, 2009.
[42] B. Arimilli, R. Arimilli, V. Chung, S. Clark, W. Denzel, B. Drerup, et al., The PERCS high-performance interconnect, in: High Performance Interconnects (HOTI), 2010 IEEE 18th Annual Symposium on, IEEE, 2010.
[43] M. Besta, T. Hoefler, Slim fly: a cost effective low-diameter network topology, in: Proceedings of the International Conference for High Performance Computing, Networking, Storage and Analysis, IEEE, 2014.
[44] S. Rumley, D. Nikolova, R. Hendry, Q. Li, D. Calhoun, K. Bergman, Silicon photonics for exascale systems, J. Lightw. Technol. 33 (3) (2015) 547–562.
[45] D. a B. Miller, H. Ozaktas, Limit to the bit-rate capacity of electrical interconnects from the aspect ratio of the system architecture, J. Parallel Distrib. Comput. 41 (1) (1997) 42–52.
[46] D.A. Miller, Rationale and challenges for optical interconnects to electronic chips, Proc. IEEE 88 (6) (2000) 728–749.
[47] T.O. Dickson, Y. Liu, A. Agrawal, J.F. Bulzacchelli, H.A. Ainspan, Z. Toprak-Deniz, et al., A 1.8 pJ/bit $16\times 16\;\text {Gb/s} $ Source-synchronous parallel interface in 32 nm SOI CMOS with receiver redundancy for link recalibration, IEEE J. Solid-State Circ. 51 (8) (2016) 1744–1755.
[48] T.O. Dickson, Y. Liu, S.V. Rylov, A. Agrawal, S. Kim, P.-H. Hsieh, et al., A 1.4 pJ/bit, power-scalable 16 × 12 Gb/s source-synchronous I/O with DFE receiver in 32 nm SOI CMOS technology, IEEE J. Solid-State Circ. 50 (8) (2015) 1917–1931.
[49] J.W. Poulton, W.J. Dally, X. Chen, J.G. Eyles, T.H. Greer, S.G. Tell, et al., A 0.54 pJ/b 20 Gb/s ground-referenced single-ended short-reach serial link in 28 nm CMOS for advanced packaging applications, J. Solid-State Circ. 48 (12) (2013) 3206–3218.
[50] A. Shokrollahi, D. Carnelli, J. Fox, K. Hofstra, B. Holden, A. Hormati, et al. 10.1 A pin-efficient 20.83 Gb/s/wire 0.94 pJ/bit forwarded clock CNRZ-5-coded SerDes up to 12mm for MCM packages in 28nm CMOS, in: Solid-State Circuits Conference (ISSCC), 2016 IEEE International, IEEE, 2016.

[51] K. Bergman, J. Shalf, T. Hausken, Optical interconnects and extreme computing, Optics and Photonics News 27 (4) (2016) 32–39.
[52] K. Wen, S. Rumley, P. Samadi, C.P. Chen, K. Bergman, Silicon photonics in post Moore's Law era: technological and architectural implications, in: Post-Moore's Era Supercomputing (PMES) Workshop, Salt Lake City, IEEE, 2016.
[53] M. Ghobadi, R. Mahajan, A. Phanishayee, N. Devanur, J. Kulkarni, G. Ranade, et al., ProjecToR: agile reconfigurable data center interconnect, in: Proceedings of the 2016 ACM SIGCOMM Conference, ACM, Florianopolis, Brazil, 2016, pp. 216–229.
[54] T. Benson, A. Akella, D.A. Maltz, Network traffic characteristics of data centers in the wild, in: Proceedings of the 10th ACM SIGCOMM conference on Internet measurement, ACM, 2010.
[55] A. Roy, H. Zeng, J. Bagga, G. Porter, A.C. Snoeren, Inside the social network's (datacenter) network, in: ACM SIGCOMM Computer Communication Review, ACM, 2015.
[56] Y. Shen, A. Gazman, Z. Zhu, M.Y. Teh, M. Hattink, S. Rumley, et al., Autonomous dynamic bandwidth steering with silicon photonic-based wavelength and spatial switching for Datacom networks, in: Optical Fiber Communication Conference, Optical Society of America, 2018.
[57] https://cloud.google.com/solutions/designing-connected-vehicle-platform accessed April 16, 2019.
[58] R. Urata, H. Liu, L. Verslegers, C. Johnson, Silicon photonics technologies: Gaps analysis for datacenter interconnects, Silicon Photonics III., Springer, 2016, pp. 473–488.
[59] R. Urata, H. Liu, X. Zhou, A. Vahdat, Datacenter interconnect and networking: from evolution to holistic revolution, in: Optical Fiber Communications Conference and Exhibition (OFC), 2017, IEEE, 2017.
[60] A. Chakravarty, K. Schmidtke, V. Zeng, S. Giridharan, C. Deal, R. Niazmand, 100Gb/s CWDM4 optical interconnect at facebook data centers for bandwidth enhancement, in: Laser Science, Optical Society of America, 2017.
[61] Q. Cheng, S. Rumley, M. Bahadori, K. Bergman, Photonic switching in high performance datacenters [invited], Opt. Express 26 (12) (2018) 16022–16043.
[62] N. Farrington, G. Porter, S. Radhakrishnan, H.H. Bazzaz, V. Subramanya, Y. Fainman, et al., Helios: a hybrid electrical/optical switch architecture for modular data centers, ACM SIGCOMM Comput. Commun. Rev. 40 (4) (2010) 339–350.
[63] X. Zhou, H. Liu, R. Urata, Datacenter optics: requirements, technologies, and trends (Invited Paper), Chin. Opt. Lett. 15 (5) (2017) 120008.
[64] A. Singh, J. Ong, A. Agarwal, G. Anderson, A. Armistead, R. Bannon, et al., Jupiter rising: a decade of clos topologies and centralized control in Google's datacenter network, Commun. ACM 59 (9) (2016) 88–97.
[65] C.F. Lam, Optical network technologies for datacenter networks (invited paper), in: 2010 Conference on Optical Fiber Communication (OFC/NFOEC), collocated National Fiber Optic Engineers Conference, 2010.
[66] Moving data with light. Available from: <https://www.intel.com/content/www/us/en/architecture-and-technology/silicon-photonics/silicon-photonics-overview.html>.
[67] T. Rokkas, I. Neokosmidis, B. Shariati, I. Tomkos, Techno-economic evaluations of 400G optical interconnect implementations for datacenter networks, in: Optical Fiber Communication Conference, Optical Society of America, 2018.
[68] A. Ghiasi, Large data centers interconnect bottlenecks, Opt. Express 23 (3) (2015) 2085–2090.
[69] R.H. Johnson, D.M. Kuchta, 30 Gb/s directly modulated 850 nm datacom VCSELs, in: Conference on Lasers and Electro-Optics/Quantum Electronics and Laser Science Conference and Photonic Applications Systems Technologies, Optical Society of America, San Jose, CA, 2008.
[70] M. Filer, B. Booth, D. Bragg, The role of standards for cloud-scale data centers, in: Optical Fiber Communication Conference, Optical Society of America, San Diego, CA, 2018.
[71] D. Thomson, A. Zilkie, J.E. Bowers, T. Komljenovic, G.T. Reed, L. Vivien, et al., Roadmap on silicon photonics, J. Opt. 18 (7) (2016) 073003.
[72] E.R.H. Fuchs, R.E. Kirchain, S. Liu, The future of silicon photonics: not so fast? Insights from 100G ethernet LAN transceivers, J. Lightw. Technol. 29 (15) (2011) 2319–2326.
[73] M. Glick, L.C. Kimmerling, R.C. Pfahl, A roadmap for integrated photonics, Opt. Photon. News 29 (3) (2018) 36–41.

[74] Disaggregated Servers Drive Data Center Efficiency and Innovation. Available from: <https://www.intel.com/content/www/us/en/it-management/intel-it-best-practices/disaggregated-server-architecture-drives-data-center-efficiency-paper.html>.
[75] B. Abali, R.J. Eickemeyer, H. Franke, C.-S. Li, M.A. Taubenblatt, Disaggregated and optically interconnected memory: when will it be cost effective? arXiv preprint arXiv:1503.01416, 2015.
[76] G. Zervas, H. Yuan, A. Saljoghei, Q. Chen, V. Mishra, Optically disaggregated data centers with minimal remote memory latency: technologies, architectures, and resource allocation, J. Opt. Commun. Netw. 10 (2) (2018) A270−A285.
[77] P.X. Gao, A. Narayan, S. Karandikar, J. Carreira, S. Han, R. Agarwal, et al., Network requirements for resource disaggregation, in: OSDI, 2016.
[78] W.M. Mellette, R. Mcguinness, A. Roy, A. Forencich, G. Papen, A.C. Snoeren, et al., RotorNet: a scalable, low-complexity, optical datacenter network, in: Proceedings of the Conference of the ACM Special Interest Group on Data Communication, ACM, 2017.
[79] K. Hazelwood, S. Bird, D. Brooks, S. Chintala, U. Diril, D. Dzhulgakov, et al., Applied machine learning at facebook: a datacenter infrastructure perspective, in: 2018 IEEE International Symposium on High Performance Computer Architecture (HPCA), 2018.
[80] S. Liu, J.C. Norman, D. Jung, M.J. Kennedy, A.C. Gossard, J.E. Bowers, Monolithic 9 GHz passively mode locked quantum dot lasers directly grown on on-axis (001) Si, Appl. Phys. Lett. 113 (4) (2018) 041108.
[81] A. Kovsh, I. Krestnikov, D. Livshits, S. Mikhrin, J. Weimert, A. Zhukov, Quantum dot laser with 75nm broad spectrum of emission, Opt. Lett. 32 (7) (2007) 793−795.
[82] V. Panapakkam, A.P. Anthur, V. Vujicic, R. Zhou, Q. Gaimard, K. Merghem, et al., Amplitude and phase noise of frequency combs generated by single-section InAs/InP quantum-dash-based passively and actively mode-locked lasers, IEEE J. Quant. Electr. 52 (11) (2016) 1−7.
[83] J.S. Levy, A. Gondarenko, M.A. Foster, A.C. Turner-Foster, A.L. Gaeta, M. Lipson, CMOS-compatible multiple-wavelength oscillator for on-chip optical interconnects, Nat. Photon. 4 (2009) 37.
[84] J. Pfeifle, V. Brasch, M. Lauermann, Y. Yu, D. Wegner, T. Herr, et al., Coherent terabit communications with microresonator Kerr frequency combs, Nat. Photon. 8 (2014) 375.
[85] X. Xue, P.H. Wang, Y. Xuan, M. Qi, A.M. Weiner, High-efficiency WDM sources based on microresonator Kerr frequency combs, in: 2017 Optical Fiber Communications Conference and Exhibition (OFC), 2017.
[86] B. Stern, X. Ji, Y. Okawachi, A.L. Gaeta, M. Lipson, Fully integrated chip platform for electrically pumped frequency comb generation, in: Conference on Lasers and Electro-Optics, Optical Society of America, San Jose, CA, 2018.
[87] C.-H. Chen, M.A. Seyedi, M. Fiorentino, D. Livshits, A. Gubenko, S. Mikhrin, et al., A comb laser-driven DWDM silicon photonic transmitter based on microring modulators, Opt. Express 23 (16) (2015) 21541−21548.
[88] Q. Xu, B. Schmidt, S. Pradhan, M. Lipson, Micrometre-scale silicon electro-optic modulator, Nature 435 (2005) 325.
[89] T. Baba, S. Akiyama, M. Imai, N. Hirayama, H. Takahashi, Y. Noguchi, et al., 50-Gb/s ring-resonator-based silicon modulator, Opt. Express 21 (10) (2013) 11869−11876.
[90] X. Xiao, X. Li, H. Xu, Y. Hu, K. Xiong, Z. Li, et al., 44-Gb/s silicon microring modulators based on zigzag pn junctions, IEEE Photon. Technol. Lett. 24 (19) (2012) 1712−1714.
[91] M. Pantouvaki, H. Yu, M. Rakowski, P. Christie, P. Verheyen, G. Lepage, et al., Comparison of silicon ring modulators with interdigitated and lateral p-n junctions, IEEE J. Select. Topics Quant. Electr. 19 (2) (2013), pp. 7900308−7900308.
[92] Q. Xu, B. Schmidt, J. Shakya, M. Lipson, Cascaded silicon micro-ring modulators for WDM optical interconnection, Opt. Express 14 (20) (2006) 9431−9436.
[93] D. Brunina, X. Zhu, K. Padmaraju, L. Chen, M. Lipson, K. Bergman, 10-Gb/s WDM optically-connected memory system using silicon microring modulators, in: European Conference and Exhibition on Optical Communication, Optical Society of America, Amsterdam, 2012.

[94] J. Li, X. Zheng, A.V. Krishnamoorthy, J.F. Buckwalter, Scaling trends for picojoule-per-bit WDM photonic interconnects in CMOS SOI and FinFET processes, J. Lightw. Technol. 34 (11) (2016) 2730–2742.
[95] J. Sun, M. Sakib, J. Driscoll, R. Kumar, H. Jayatilleka, Y. Chetrit, et al., A 128 Gb/s PAM4 silicon microring modulator, in: 2018 Optical Fiber Communications Conference and Exposition (OFC), 2018.
[96] R. Dubé-Demers, S. Larochelle, W. Shi, Ultrafast pulse-amplitude modulation with a femtojoule silicon photonic modulator, Optica 3 (6) (2016) 622–627.
[97] M. Bahadori, S. Rumley, D. Nikolova, K. Bergman, Comprehensive design space exploration of silicon photonic interconnects, J. Lightw. Technol. 34 (12) (2016) 2975–2987.
[98] R. Wu, C.H. Chen, J.M. Fedeli, M. Fournier, R.G. Beausoleil, K.T. Cheng, Compact modeling and system implications of microring modulators in nanophotonic interconnects, in: 2015 ACM/IEEE International Workshop on System Level Interconnect Prediction (SLIP), 2015.
[99] R. Wu, C.-H. Chen, J.-M. Fedeli, M. Fournier, K.-T. Cheng, R.G. Beausoleil, Compact models for carrier-injection silicon microring modulators, Opt. Express 23 (12) (2015) 15545–15554.
[100] K. Padmaraju, X. Zhu, L. Chen, M. Lipson, K. Bergman, Intermodulation crosstalk characteristics of WDM silicon microring modulators, IEEE Photon. Technol. Lett. 26 (14) (2014) 1478–1481.
[101] O. Dubray, A. Abraham, K. Hassan, S. Olivier, D. Marris-Morini, L. Vivien, et al., Electro-optical ring modulator: an ultracompact model for the comparison and optimization of p-n, p-i-n, and capacitive junction, IEEE J. Select. Topics Quant. Electr. 22 (6) (2016) 89–98.
[102] J.B. Quélène, J.F. Carpentier, Y.L. Guennec, P.L. Maître, Optimization of power coupling coefficient of a carrier depletion silicon ring modulator for WDM optical transmissions, in: 2016 IEEE Optical Interconnects Conference (OI), 2016.
[103] H. Jayatilleka, K. Murray, M. Caverley, N.A.F. Jaeger, L. Chrostowski, S. Shekhar, Crosstalk in SOI microring resonator-based filters, J. Lightw. Technol. 34 (12) (2016) 2886–2896.
[104] M. Bahadori, M. Nikdast, S. Rumley, L.Y. Dai, N. Janosik, T. Van Vaerenbergh, et al., Design space exploration of microring resonators in silicon photonicinterconnects: impact of the ring curvature, J. Lightw. Technol. 36 (13) (2018) 2767–2782.
[105] G. Li, A.V. Krishnamoorthy, I. Shubin, J. Yao, Y. Luo, H. Thacker, et al., Ring resonator modulators in silicon for interchip photonic links, IEEE J. Select. Topics Quant. Electr. 19 (6) (2013) 95–113.
[106] S. Rumley, M. Bahadori, D. Nikolova, K. Bergman, Physical layer compact models for ring resonators based dense WDM optical interconnects, in: ECOC 2016; 42nd European Conference on Optical Communication, 2016.
[107] M.A. Seyedi, R. Wu, C.-H. Chen, M. Fiorentino, R. Beausoleil, 15 Gb/s transmission with wide-FSR carrier injection ring modulator for Tb/s optical links, in: Conference on Lasers and Electro-Optics, Optical Society of America, San Jose, CA, 2016.
[108] M. Bahadori, S. Rumley, H. Jayatilleka, K. Murray, N. a F. Jaeger, L. Chrostowski, et al., Crosstalk penalty in microring-based silicon photonic interconnect systems, J. Lightw. Technol. 34 (17) (2016) 4043–4052.
[109] L. Chen, N. Sherwood-Droz, M. Lipson, Compact bandwidth-tunable microring resonators, Opt. Lett. 32 (22) (2007) 3361–3363.
[110] C.L. Manganelli, P. Pintus, F. Gambini, D. Fowler, M. Fournier, S. Faralli, et al., Large-FSR thermally tunable double-ring filters for WDM applications in silicon photonics, IEEE Photon. J. 9 (1) (2017) 1–10.
[111] M. Bahadori, A. Gazman, N. Janosik, S. Rumley, Z. Zhu, R. Polster, et al., Thermal rectification of integrated microheaters for microring resonators in silicon photonics platform, J. Lightw. Technol. 36 (3) (2018) 773–788.
[112] C. Sun, M. Wade, M. Georgas, S. Lin, L. Alloatti, B. Moss, et al., A 45 nm CMOS-SOI monolithic photonics platform with bit-statistics-based resonant microring thermal tuning, IEEE J. Solid-State Circ. 51 (4) (2016) 893–907.
[113] P.L. Maître, J.F. Carpentier, C. Baudot, N. Vulliet, A. Souhaité, J.B. Quélène, et al., Impact of process variability of active ring resonators in a 300mm silicon photonic platform, in: 2015 European Conference on Optical Communication (ECOC), 2015.

[114] M. Nikdast, G. Nicolescu, J. Trajkovic, O. Liboiron-Ladouceur, Chip-scale silicon photonic interconnects: a formal study on fabrication non-uniformity, J. Lightw. Technol. 34 (16) (2016) 3682–3695.
[115] K. Padmaraju, D.F. Logan, T. Shiraishi, J.J. Ackert, A.P. Knights, K. Bergman, Wavelength locking and thermally stabilizing microring resonators using dithering signals, J. Lightw. Technol. 32 (3) (2014) 505–512.
[116] F. Morichetti, A. Canciamilla, C. Ferrari, M. Torregiani, A. Melloni, M. Martinelli, Roughness induced backscattering in optical silicon waveguides, Phys. Rev. Lett. 104 (3) (2010) 033902.
[117] B.E. Little, J.-P. Laine, S.T. Chu, Surface-roughness-induced contradirectional coupling in ring and disk resonators, Opt. Lett. 22 (1) (1997) 4–6.
[118] M. Bahadori, S. Rumley, Q. Cheng, K. Bergman, Impact of backscattering on microring-based silicon photonic links, in: Optical Interconnects, 2018.
[119] C. Chen, C. Li, R. Bai, K. Yu, J. Fedeli, S. Meassoudene, et al., DWDM silicon photonic transceivers for optical interconnect, in: 2015 IEEE Optical Interconnects Conference (OI), 2015.
[120] A. Biberman, J. Chan, K. Bergman, On-chip optical interconnection network performance evaluation using power penalty metrics from silicon photonic modulators, in: 2010 IEEE International Interconnect Technology Conference, 2010.
[121] Q. Li, D. Nikolova, D.M. Calhoun, Y. Liu, R. Ding, T. Baehr-Jones, et al., Single microring-based 2 × 2 silicon photonic crossbar switches, IEEE Photon. Technol. Lett. 27 (18) (2015) 1981–1984.
[122] R. Ding, Y. Liu, Q. Li, Z. Xuan, Y. Ma, Y. Yang, et al., A compact low-power 320-Gb/s WDM transmitter based on silicon microrings, IEEE Photon. J. 6 (3) (2014) 1–8.
[123] P. Dong, W. Qian, H. Liang, R. Shafiiha, N.-N. Feng, D. Feng, et al., Low power and compact reconfigurable multiplexing devices based on silicon microring resonators, Opt. Express 18 (10) (2010) 9852–9858.
[124] P. Dong, S. Liao, D. Feng, H. Liang, D. Zheng, R. Shafiiha, et al., Low Vpp, ultralow-energy, compact, high-speed silicon electro-optic modulator, Opt. Express 17 (25) (2009) 22484–22490.
[125] W. Bogaerts, R. Baets, P. Dumon, V. Wiaux, S. Beckx, D. Taillaert, et al., Nanophotonic waveguides in silicon-on-insulator fabricated with CMOS technology, J. Lightw. Technol. 23 (1) (2005) 401.
[126] A. Bianco, D. Cuda, R. Gaudino, G. Gavilanes, F. Neri, M. Petracca, Scalability of optical interconnects based on microring resonators, IEEE Photon. Technol. Lett. 22 (15) (2010) 1081–1083.
[127] K. Yu, C.-H. Chen, A. Titriku, A. Shafik, M. Fiorentino, P.Y. Chiang, S. Palermo, 25Gb/s hybrid-integrated silicon photonic receiver with microring wavelength stabilization, in: Optical Fiber Communication Conference, Optical Society of America, 2015.
[128] M. Bahadori, D. Nikolova, S. Rumley, C.P. Chen, K. Bergman, Optimization of microring-based filters for dense WDM silicon photonic interconnects, in: Optical Interconnects Conference (OI), 2015 IEEE, IEEE, 2015
[129] R. Hendry, D. Nikolova, S. Rumley, N. Ophir, K. Bergman, Physical layer analysis and modeling of silicon photonic WDM bus architectures, in: Proc. HiPEAC Workshop, 2014.
[130] B.G. Lee, A. Biberman, P. Dong, M. Lipson, K. Bergman, All-optical comb switch for multiwavelength message routing in silicon photonic networks, IEEE Photon. Technol. Lett. 20 (10) (2008) 767–769.
[131] M. Georgas, J. Leu, B. Moss, C. Sun, V. Stojanović, Addressing link-level design tradeoffs for integrated photonic interconnects, in: Custom Integrated Circuits Conference (CICC), 2011 IEEE, IEEE, 2011.
[132] N. Ophir, C. Mineo, D. Mountain, K. Bergman, Silicon photonic microring links for high-bandwidth-density, low-power chip I/O, IEEE Micro 33 (1) (2013) 54–67.
[133] R. Hendry, D. Nikolova, S. Rumley, K. Bergman, Modeling and evaluation of chip-to-chip scale silicon photonic networks, in: 2014 IEEE 22nd Annual Symposium on High-Performance Interconnects, 2014.
[134] A. Yariv, Universal relations for coupling of optical power between microresonators and dielectric waveguides, Electron. Lett. 36 (4) (2000) 321–322.
[135] A. Biberman, P. Dong, B.G. Lee, J.D. Foster, M. Lipson, K. Bergman, Silicon microring resonator-based broadband comb switch for wavelength-parallel message routing, in: LEOS 2007 - IEEE Lasers and Electro-Optics Society Annual Meeting Conference Proceedings, 2007.

[136] M. Bahadori, S. Rumley, R. Polster, A. Gazman, M. Traverso, M. Webster, et al., Energy-performance optimized design of silicon photonic interconnection networks for high-performance computing, in: Proceedings of the Conference on Design, Automation & Test in Europe, European Design and Automation Association: Lausanne, Switzerland, 2017, pp. 326−331.

[137] J. Kim, C.J. Nuzman, B. Kumar, D.F. Lieuwen, J.S. Kraus, A. Weiss, et al., 1100 x 1100 port MEMS-based optical crossconnect with 4-dB maximum loss, IEEE Photon. Technol. Lett. 15 (11) (2003) 1537−1539.

[138] Polatis technology − Directlight® Beam-Steering All-Optical Switch. Available from: <http://www.polatis.com/polatis-all-optical-switch-technology-lowest-loss-highest-performance-directlight-beam-steering.asp>.

[139] Z. Zhang, Z. You, D. Chu, Fundamentals of phase-only liquid crystal on silicon (LCOS) devices, Light Sci. Appl. 3 (2014). p. e213.

[140] M. Yano, F. Yamagishi, T. Tsuda, Optical MEMS for photonic switching-compact and stable optical crossconnect switches for simple, fast, and flexible wavelength applications in recent photonic networks, IEEE J. Select. Topics Quant. Electr. 11 (2) (2005) 383−394.

[141] R. Ryf, J. Kim, J.P. Hickey, A. Gnauck, D. Carr, F. Pardo, et al. 1296-Port MEMS transparent optical crossconnect with 2.07 petabit/s switch capacity, in: OFC 2001. Optical Fiber Communication Conference and Exhibit. Technical Digest Postconference Edition (IEEE Cat. 01CH37171), 2001.

[142] D.T. Neilson, R. Frahm, P. Kolodner, C.A. Bolle, R. Ryf, J. Kim, et al., 256 × 256 Port optical cross-connect subsystem, J. Lightw. Technol. 22 (6) (2004) 1499.

[143] K. Wang, A. Wonfor, R.V. Penty, I.H. White, Active-passive 4x4 SOA-based switch with integrated power monitoring, in: OFC/NFOEC, 2012.

[144] I. White, E.T. Aw, K. Williams, H. Wang, A. Wonfor, R. Penty, Scalable optical switches for computing applications [Invited], J. Opt. Netw. 8 (2) (2009) 215−224.

[145] A. Wonfor, H. Wang, R.V. Penty, I.H. White, Large port count high-speed optical switch fabric for use within datacenters [invited], J. Opt. Commun. Netw. 3 (8) (2011) A32−A39.

[146] R. Stabile, A. Albores-Mejia, K.A. Williams, Monolithic active-passive 16x16 optoelectronic switch, Opt. Lett. 37 (22) (2012) 4666−4668.

[147] R. Stabile, A. Albores-Mejia, K.A. Williams, Monolithically integrated 8 × 8 space and wavelength selective cross-connect, J. Lightw. Technol. 32 (2) (2014) 201−207.

[148] Q. Cheng, M. Ding, A. Wonfor, J. Wei, R.V. Penty, I.H. White, The feasibility of building a 64x64 port count SOA-based optical switch, in: 2015 International Conference on Photonics in Switching (PS), 2015.

[149] Q. Cheng, A. Wonfor, J.L. Wei, R.V. Penty, I.H. White, Low-energy, high-performance lossless 8 × 8 SOA switch, in: 2015 Optical Fiber Communications Conference and Exhibition (OFC), 2015.

[150] T. Chu, L. Qiao, W. Tang, D. Guo, W. Wu, Fast, high-radix silicon photonic switches, in: 2018 Optical Fiber Communications Conference and Exposition (OFC), 2018.

[151] T.J. Seok, K. Kwon, J. Henriksson, J. Luo, M.C. Wu, "240 × 240 Wafer-Scale Silicon Photonic Switches," in: Optical Fiber Communication Conference (OFC) 2019, OSA Technical Digest (Optical Society of America, 2019), paper Th1E.5.

[152] K. Suzuki, R. Konoike, J. Hasegawa, S. Suda, H. Matsuura, K. Ikeda, et al., Low insertion loss and power efficient 32x32 silicon photonics switch with extremely-high-D PLC connector, in: 2018 Optical Fiber Communications Conference and Exposition (OFC), 2018.

[153] L. Qiao, W. Tang, T. Chu, 32 × 32 silicon electro-optic switch with built-in monitors and balanced-status units, Sci. Rep. 7 (2017) 42306.

[154] Q. Cheng, L.Y. Dai, M. Bahadori, N.C. Abrams, P.E. Morrissey, M. Glick, et al., Si/SiN microring-based optical router in switch-and-select topology, in: European Conference on Optical Communication (ECOC), 2018, p. We1C.3.

[155] Q. Cheng, M. Bahadori, S. Rumley, K. Bergman, Highly-scalable, low-crosstalk architecture for ring-based optical space switch fabrics, in: 2017 IEEE Optical Interconnects Conference (OI), 2017.

[156] Q. Cheng, L.Y. Dai, N.C. Abrams, Y. Hung, P.E. Morrissey, M. Glick, P. O'Brien, K. Bergman, Ultralow-crosstalk, strictly non-blocking microring-based optical switch, Photon. Res 7 (2019) 155–161.

[157] Q. Cheng, M. Bahadori, Y. Hung, Y. Huang, N. Abrams, K. Bergman, Scalable Microring-Based Silicon Clos Switch Fabric with Switch-and-Select Stages, in IEEE Journal of Selected Topics in Quantum Electronics. https://doi.org/10.1109/JSTQE.2019.2911421

[158] Q. Cheng, R. Dai, M. Bahadori, P. Morrissey, R. Polster, S. Rumley, et al., Microring-based Si/SiN dual-layer switch fabric, in: Optical Interconnects, IEEE, Santa Fe, New Mexico, USA, 2018.

[159] M. Ding, A. Wonfor, Q. Cheng, R.V. Penty, I.H. White, Hybrid MZI-SOA InGaAs/InP photonic integrated switches, IEEE J. Select. Topics Quant. Electr. 24 (1) (2018) 1–8.

[160] Q. Cheng, A. Wonfor, J.L. Wei, R.V. Penty, I.H. White, Monolithic MZI-SOA hybrid switch for low-power and low-penalty operation, Opt. Lett. 39 (6) (2014) 1449–1452.

[161] Q. Cheng, A. Wonfor, J.L. Wei, R.V. Penty, I.H. White, Demonstration of the feasibility of large-port-count optical switching using a hybrid Mach-Zehnder interferometer-semiconductor optical amplifier switch module in a recirculating loop, Opt. Lett. 39 (18) (2014) 5244–5247.

[162] A. Bhatele, N. Jain, Y. Livnat, V. Pascucci, P. Bremer, Analyzing network health and congestion in dragonfly-based supercomputers, in: 2016 IEEE International Parallel and Distributed Processing Symposium (IPDPS), 2016.

[163] K. Wen, D. Calhoun, S. Rumley, X. Zhu, Y. Liu, L.W. Luo, et al., Reuse distance based circuit replacement in silicon photonic interconnection networks for HPC, in: 2014 IEEE 22nd Annual Symposium on High-Performance Interconnects, 2014.

[164] J. Kim, W.J. Dally, S. Scott, D. Abts, Cost-efficient dragonfly topology for large-scale systems, in: 2009 Conference on Optical Fiber Communication - Incudes Post Deadline Papers, 2009.

[165] J. Wilke, Bringing minimal routing back to HPC through silicon photonics: a study of "flexfly" architectures with the structural simulation toolkit (SST), in: Proceedings of the 2nd International Workshop on Advanced Interconnect Solutions and Technologies for Emerging Computing Systems, ACM, Stockholm, Sweden, 2017, p. 5-5.

CHAPTER 19

Evolution of fiber access networks

Cedric F. Lam and Shuang Yin
Google Fiber, Mountain View, CA, United States

19.1 Introduction

Optical fiber was originally developed for longhaul transmissions. The idea of using fiber optic for access was initially proposed in the 1980s, way before the internet and broadband access became the norm of our society. After decades of developments, fiber access networks are now mature technologies deployed to hundreds of millions of users around the world. In year 2018, China alone boasted more than 350 million fiber-to-the-home (FTTH) users. Besides directly connecting end customers with optical fibers in FTTH networks, new forms of fiber access networks are indispensable in providing backhaul and fronthaul connectivities to the fourth generation (4G) and the upcoming fifth generation (5G) wireless networks.

Fiber access networks are deployed at the edge of telecommunication networks as end nodes. There are two major challenges in deploying fiber access networks. First, access networks are very cost sensitive, so the equipment cost has to be very low in order for it to be viable as a mass-deployed technology. The ways to achieve this are economy of scale and low-cost optoelectronics packaging techniques. The second major challenge in deploying fiber access networks is labor cost, and speed and ease of deployments. Significant civil engineering cost is incurred especially in developed economies where (1) the labor cost is high and (2) digging and trenching of infrastructure is not easy. Therefore, traditional incumbent carriers in developed nations would like to preserve their legacy copper infrastructure and delay the deployment of fiber in the last mile [from the central office (CO)] as much as possible. In developing economies or greenfield scenarios, there will be fewer architecture constraints, fewer legacy burdens, and more flexibility in technology and architecture choices. But those economies are also very capital cost sensitive and would like to leverage the low cost of existing, mature, and standard-based technologies. These challenges are guiding the design principles of fiber access technologies.

Broadband access was the initial driving force for FTTH fiber access technologies. This was mainly propelled by the booming internet applications, especially over-the-top video streaming applications, which offers any time and any place viewing

Optical Fiber Telecommunications V11
DOI: https://doi.org/10.1016/B978-0-12-816502-7.00022-1

experiences of on-demand contents. Higher resolution videos such as 4 K will demand more bandwidths to end users.

FTTH was mainly provided by passive optical networks (PONs). In fact, PON (especially in the form of time-division multiplexing or TDM implementation) is almost used as a synonym of FTTH, although other FTTH implementations also exist, albeit in much smaller deployments. Most of the deployed residential FTTH networks are based on the IEEE 802.3ah EPON or ITU-T G.984-based G-PON technologies [1], with the latter being the most popular nowadays.

Although carriers have started deploying 10 Gbps based TDM-PON systems around the world, G-PON and EPON networks still offer enough bandwidths to residential homes for at least another 3–5 years. Nevertheless, standard bodies have been busy working on PON technologies beyond 10 Gbps (i.e., 25/50/100 Gbps) [2]. The major driving forces for these higher speed PONs (HSPs) or access technologies are enterprise services and wireless fronthaul applications. Furthermore, besides TDM, WDM (wavelength-division multiplexing) has been introduced to further scale the speed and coverage of future PON networks. These development trends will be discussed in the next sections.

19.2 Evolution of passive optical networks

19.2.1 Mature passive optical network standards

In the broad sense, a PON means that the network between a carrier's CO (where the customer traffic is first terminated) and the end user has no active elements requiring electrical power. Usually there are three ways to achieve this [1]:

1. A point-to-point home-run fiber from the CO to every customer (which makes the medium access control (MAC) protocol and optical transmission the simplest, but requires termination of a large number of fibers and transceivers inside the CO).
2. Using a power splitter in the field as a remote node (RN) to broadcast the signal from a common transceiver at the CO to the multiple end users. The bandwidth at the common transceiver is shared among the users using a TDM protocol. The TDM protocol employs a dynamic bandwidth allocation (DBA) algorithm to efficiently allocate the shared bandwidth of the common transceiver among the end users. This type of TDM-PON is the most common form of commercial deployments today, to which both the ITU-T G-PON and IEEE 802.3 EPON belong.
3. Using a wavelength router in the field as the RN to distribute WDM wavelengths (in the form of virtual fibers) to end users. This kind of WDM-PON has been attracting more and more attentions from wireless carriers for fronthaul connections

because of its protocol transparency and the ability to meet the precise timing requirements needed in wireless networks.

Combinations of WDM and TDM PONs in the form of TWDM PONs have also been implemented in the industry. Examples are the ITU-T G.989 NG-PON2 and Google Fiber's super-PON technology [3].

19.2.1.1 Burst mode operation in time-division multiplexing-passive optical networks

The high-level architecture of a TDM-PON is shown in Fig. 19.1. The equipment in the CO terminating customer traffic is called OLT (optical line terminal), and the customer end modem converting the optical signal from the CO to electrical signal is called ONT (optical network terminal). In a TDM-PON network, the downstream transmission is a point-to-multipoint operation where the OLT broadcast to the same signal to multiple ONTs. An ONT relies on the address field in the downstream frames to determine if the associated data are addressed to itself and discards those not intended for its consumption. The upstream direction, on the other hand, is a multipoint-to-point operation where the ONTs take turns to transmit to the shared OLT receiver. The OLT uses a DBA algorithm to schedule proper time slots for ONTs to transmit in the upstream direction. An ONT not transmitting must shut off its transmitter to avoid collision with signals from other ONTs. The OLT uses a burst mode receiver with fast clock recovery to recover the signal from the ONTs. Each time a different ONT transmits, the OLT has to perform fast clock recovery, which

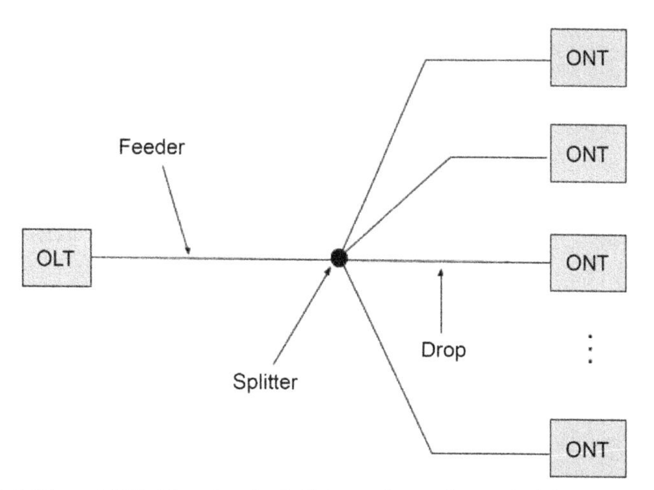

Figure 19.1 1:32, 1:64, or 1:128. The standard distance from the optical line terminal (OLT) to optical network terminals (ONTs) are usually 20 km. Besides power budget, the transmission distance and splitting ratio are also limited by the time-division multiplexing-passive optical network (TDM-PON) protocol such as the size of ranging window and available ONT address space.

adds overhead to upstream transmission. Therefore, guard times are reserved between upstream bursts from different ONTs.

19.2.1.2 Gigabit time-division-multiplexing passive optical network Standards

TDM-PON is commonly specified with 20 km coverage between the CO and the end users with 1:32, 1:64, or 1:128 splitter as the RN. In a PON system. The upstream and downstream signals are separated by wavelengths using a wavelength diplexer inside the OLT and ONT optical transceivers.

The IEEE 802.3ah E-PON is a PON standard that offers 1 Gbps aggregated bandwidth in both upstream and downstream directions, whereas the ITU-T G-PON offers 2.5 Gbps and 1.25 Gbps in the downstream and upstream directions, respectively.

Both G-PON and E-PON have very similar physical layer characteristics. They both use 1310 nm wavelength for upstream transmission and 1490 nm for downstream transmission.

19.2.1.3 10 Gbps time-division-multiplexing passive optical network Standards

In 2009, the IEEE introduced the standard for 10 Gbps capable E-PON or 10GE-PON as IEEE 802.3av. This TDM-PON has proved to be the most successful and mature 10 Gbps PON technology, which is now in commercial deployments in volume.

Symmetric IEEE 802.3av 10GE-PON adopts 1270 nm wavelength for 10 Gbps upstream transmission and 1577 nm wavelength for downstream transmission. These wavelengths are not overlapping with the G-PON and E-PON wavelengths so that one can use wavelength multiplexer to overlay 10GE-PONs to either a legacy G-PON or E-PON network. Fig. 19.2 shows the optical spectrum allocation for existing TDM-PON standards.

Around the time that IEEE developed the 10GE-PON standard, ITU-T has published the standards for XG-PON1 (with 10 Gbps downstream and 2.5 Gbps upstream capacities) and XG-PON2 (with symmetric 10 Gbps in both upstream and downstream directions) [4]. These standards were collectively called NG-PON (next-generation PON) standards. While prototypes of these systems had been built, the very difficult-to-meet requirements for burst-mode timing, coupled with a lack of real demand, delayed commercialization of these standards. Yet the ITU-T SG15 (Study Group 15), which is dominated by service providers, aggressively tried to create a "40 Gbps" TWDM-PON called NG-PON2 (or ITU-T G.989) [5] in order to leapfrog the IEEE 802.3 10GE-PON. This effort, as will be discussed in the next section, was not very successful. In 2015, demands for 10 Gbps PON systems started to emerge. Faced with the competition of the mature IEEE802.3av 10GE-PON technology (which was already in product manufacturing at the time), to quickly come up

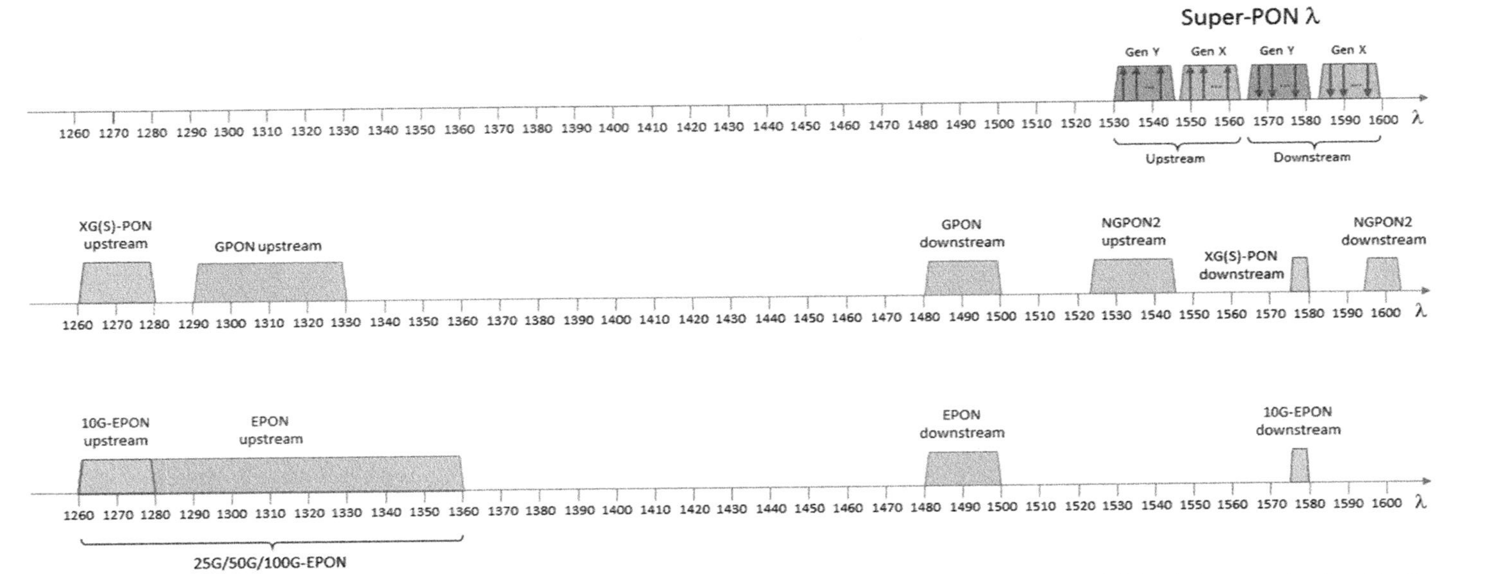

Figure 19.2 Optical spectral allocation for common passive optical network (PON) standards. The super-PON is being standardized in the IEEE 802.3cs Task Force, and it is still in early stage at this writing.

with its own implementable version 10 Gbps PON standard for product offerings, ITU-T neutered the NG-PON2 standard to create an interim "single-wavelength" symmetric 10 Gbps PON, called XGS-PON [6], which will be forward-compatible with future full NG-PON2 standards when the manufacturable technology will be within reach. In fact, the physical layer of XGS-PON almost completely adopted the IEEE 802.3av 10GE-PON physical layer characteristics, except with the modification of burst mode timing requirements. Early implementations of XGS-PON systems even relaxed the timing requirements and completely adopted 10GE-PON transceivers to jump-start the business.

19.2.1.4 TWDM-passive optical network standards

The ITU-T NG-PON2 combines wavelength-division multiplexing and time-division multiplexing together in the same architecture to offer a total capacity of 40 Gbps. It is the first time that dense wavelength-division multiplexing (DWDM) has been adopted in a commercial access network standard. Transmitting a 40 Gbps TDM signal on a single wavelength was not easy at the time the NG-PON2 standard was created. So four (4 ×) 10 Gbps wavelengths between 1524 and 1544 nm are used to transmit the upstream signal and four wavelengths between 1596 and 1603 nm to transmit the downstream signal. The OLT uses an array transceiver with four transmitters and four receivers all multiplexed together with an internal wavelength multiplexer/demultiplexer built into the OLT optical module. To be compatible with conventional PON networks, NG-PON2 inherited the optical power splitter in the field as the RN. Therefore, all the four wavelengths are broadcast to every NG-PON2 ONT. ONTs in NG-PON2 are, however, only specified with a single 10 Gbps transceiver, with both a tunable upstream laser and a tunable receiver. Fig. 19.3 shows the architecture of NG-PON2.

The NG-PON2 design philosophy has several problems, however. For one, the new standard aims at supporting multistage splitting and high splitting ratios (perhaps as great as 1024 users per CO connection), in order to efficiently use the vast bandwidth offered by NG-PON2 and to cut system costs through better sharing of the expensive OLT optics and through reducing the number of fibers required to connect to the CO. But not many of the embedded PON systems currently deployed have such large splitting ratios. Moreover, to support the very high splitting ratio, the OLT burst-mode receiver requires a high dynamic range (greater than 20 dB).

A second problem is the potentially very high power budget needed for NG-PON2 implementation. Having to support very high splitting ratios will further strain the power budget requirements. The NG-PON2 standard specifies a loss budget of up to 35 dB between the OLT and ONT [8,9] (compared to the 28 dB Class B + power budget commonly used in G-PON deployments and 32 dB specified for class C + + of G-PON, the latter of which is already very difficult to produce with high yield, for a transmission of only 2.5 Gbps). This has not accounted for the extra losses from the

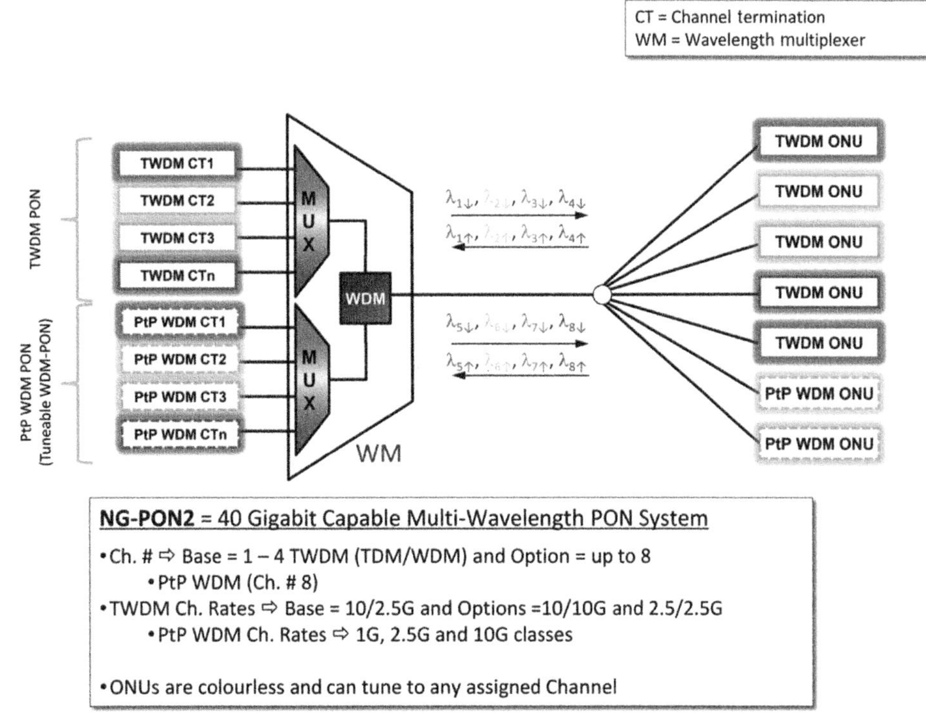

Figure 19.3 Architecture of NG-PON2 [7].

WDM multiplexers inside the OLT and the tunable filter in the ONT optical module. As a result, lasers with very high transmitting power and receivers with high sensitivity will be needed, resulting in potentially significant increase in the system costs—even if the technology is achievable [10].

At the time of writing this article (November 2018, more than 5 years after the proposal of NG-PON2 standard in March 2013), optical-component technologies still have not advanced to the point that NG-PON2 implementation is economically feasible for mass production. The step from G-PON to 10-Gb/s TDM-PON was incremental, as the optical systems are quite similar, apart from the speeds of the optical transceivers. However, in moving from current 10-Gb/s TDM-PONs to NG-PON2, performance requirements of the optical components becomes significantly more stringent, the structures of the optical transceivers are much more complex, and packaging complexity becomes exponentially higher.

Small, low-loss, low-cost, and low-power tunable filters, required for ONTs in an NG-PON2 system, are not easier to manufacture than semiconductor tunable lasers (which are mostly monolithic structures). Innovations in photonic integration circuits are needed to solve these challenges in the long run.

Burst-mode operation of the tunable DWDM laser in NG-PON2 ONT causes transient wavelength drift [11], as the sudden injection of current into the laser heats up the laser structure (unless an external modulator is used, which increases cost and optical losses). The drift increases with the laser bias current and laser output power, and can be as large as 20−30 GHz, with time constants on the order of milliseconds. Such drift not only causes a penalty in OLT receiver sensitivity at the CO but also crosstalk to other wavelength channels in a broadcast-and-select TDM-PON fiber plant. A pure WDM-PON without TDM overlay, on the other hand, does not require burst-mode operation and thus does not suffer from these challenges.

From the MAC protocol perspective, in addition to the usual DBA algorithm used in TDM-PONs, NG-PON2 adds a layer of optical wavelength complexity to manage. The millisecond-scale tuning speeds of the tunable laser and filters used in NG-PON2 will make it economically difficult to do fast wavelength switching on the packet level. Coordinating wavelengths with TDM time slots together complicates the DBA algorithms used in NG-PON2.

In the foreseeable future, for residential FTTH applications, 10-Gb/s PON networks should provide adequate bandwidth to meet existing demands, so there is no immediate justification for dynamically adjustable wavelength allocations, especially with the additional cost and complexity. That fact may underlie ITU-T's decision in 2015 to create the XGS-PON (10-Gb/s symmetric PON with single wavelength) standard as an "initial stage" of NG-PON2.

Despite lack of commercial successes, NG-PON2 started the chapter of using parallel WDM wavelength channels to scale optical access networks. Google Fiber's TWDM Super-PON architecture [12], which will be discussed later, is another example of using WDM to scale FTTH networks, while avoiding some of the implementation challenges of NG-PON2 by sacrificing some flexibilities such as optical tunable receiver and broadcast ODN networks based on power splitting RNs.

19.2.2 Passive optical network standards in the make

Besides the mature PON standards discussed in the previous sections, optical access network is constantly evolving with the discussions of new standards, which is typically driven by the constant growth of subscriber traffic. Among all latest standards under discussion or close to publication, here are a few that generate broad interests among ISPs, system and component vendors, which represent general consensus among those groups in the evolution path toward the next-generation optical access network. First of all, IEEE initiated 802.3ca study group in 2016 to study toward 100G-EPON as the next-generation system after 10G-EPON [2]. Second, FSAN/ITU-T SG15/Q2 started from 2017 to look into HSP systems [13], for example, a multichannel tunable 50 Gbps per channel system. Third, a Super-PON Task Force (IEEE 802.3cs) has been formed by IEEE 802.3 working group in November 2018 to explore physical layer specifications using WDM over an increased-reach PON.

These three major standard efforts in the make will be discussed in the following sections.

19.2.2.1 IEEE 802.3ca

The IEEE 802.3 working group started initial ad hoc discussion on next- generation EPON (NG-EPON) in January 2014, which was expected to serve as the evolution path after 10G-EPON introduced by IEEE 802.3av around 2009. This was mainly driven by the increase of access network bandwidth demand for both business and residential (Table 19.1). Three similar linearly increasing bandwidth drivers lead to exponential growth of the total bandwidth demand [14], that is, growing number of subscribers, increasing number of connected devices per subscriber, proliferating bandwidth demand per device or application. In addition, cellular network backhauling bandwidth also was growing at an unabated pace considering the evolution from 4G to LTE, and eventually 5G cellular networks. Given a typical business access deployment scenario shown in Table 19.2, usually without oversubscription unlike residential

Table 19.1 Guaranteed access bandwidth requirement.

Subscriber type	Guaranteed access bandwidth range (2018–2025) (Gbps)
Small business[a]	0.1–1.0
Large business[b]	1–10
Cellular backhaul[c]	10–25

[a]Google Fiber, Fiber for small business. [Online]. Available from: https://fiber.google.com/smallbusiness/.
[b]AT&T, Dedicated Internet. Exclusively for your business. [Online]. Available from: https://www.business.att.com/products/dedicated-internet/#/.
[c]CSMA, Mobile backhaul options Spectrum analysis and recommendations. [Online]. Available from: https://www.gsma.com/spectrum/wp-content/uploads/2019/04/Mobile-Backhaul-Options.pdf.

Table 19.2 Typical PON capacity requirement.

Typical subscriber combinations on the same PON			Required PON capacity (Gbps)
Small business	Large business	Cellular backhaul	
24	4	–	~35
20	2	1	~40
–	6	1	~51

PON, passive optical network.

deployment, that is, including some business and cellular tower subscribers, guaranteed bandwidth requirement can easily add up to above 30 Gbps, which goes beyond 10G-EPON capacity. On the other hand, residential demand can also break 10G-EPON capacity, especially in dense multidwelling units deployment with split ratio $\geq$ 1:128 [15]. However, due to inevitable high system cost of the NG-EPON system at the time of early market penetration, it is commonly expected that it will be first adopted to serve business customers, and eventually be deployed in residential market as the system cost being continuously driven down with the increase of volume.

In the meantime of NG-EPON ad hoc discussion, FSAN and ITU-T SG15/Q2 was close to finalizing the standardization work on NG-PON2 as the successor for NG-PON standards, that is, XG-PON1 and XG-PON2. Since the beginning of NG-EPON discussion, both system and component vendors in IEEE 802.3 working group were fully aware of the technical difficulties in NG-PON2 system, for example, Tx wavelength drifting in burst mode operation [11], tunable Rx integration [11,16], large power budget requirement [17], etc. In addition, at the same time, bandwidth demand of intra-data center and inter-data center network have been growing exponentially, which led to fast market adoption of multiple variations of 100GE-based transceiver standards for data center networks, mostly 4×25G in slightly different flavors, for example, 100GE CWDM-4 [18], 100GE LAN-WDM-4 [19], 100GE PSM-4 [20], etc. Therefore, leveraging high-volume components, for example, laser diode, laser/modulator driver, photo-detector, etc., used by data center network at 25 Gbps or above per channel/wavelength makes higher serial rate above 10 Gbps a feasible option at the time of developing NG-EPON standard. NG-EPON call for interest (CFI) was completed in July 2015 [14], which moved the ad hoc discussion into an official IEEE 802.3 NG-EPON study group. Project authorization request (PAR), criteria for standards development (CSD), and project objectives were developed in subsequent meetings and reported to and approved by the 802.3 working group to officially form the IEEE 802.3ca 100G-EPON Task Force in November 2015. The initial target of the working group is to standardize 25/50/100 G EPON with the project name IEEE P802.3ca.

The planned timeline for standardization in IEEE of 25/50/100 G EPON is depicted in Fig. 19.4 [21]. It is anticipated that the standard will be finalized by the end of the first quarter of 2020. It was therefore decided to develop a single standard for multiple generations of PON at the same time, which is envisioned as first-generation 25 Gbps system, second generation 50 Gbps system, and third generation 100 Gbps system [22]. All three generations should be able to coexist on the same outside plant (OSP), and network equipments, such as ONTs, should be backward-compatible. Similar to the previous PON system deployment, initial deployment of symmetrical 25 Gbps system is envisioned for business applications, for example, cellular front- and backhaul and business subscribers. On the other hand, asymmetrical

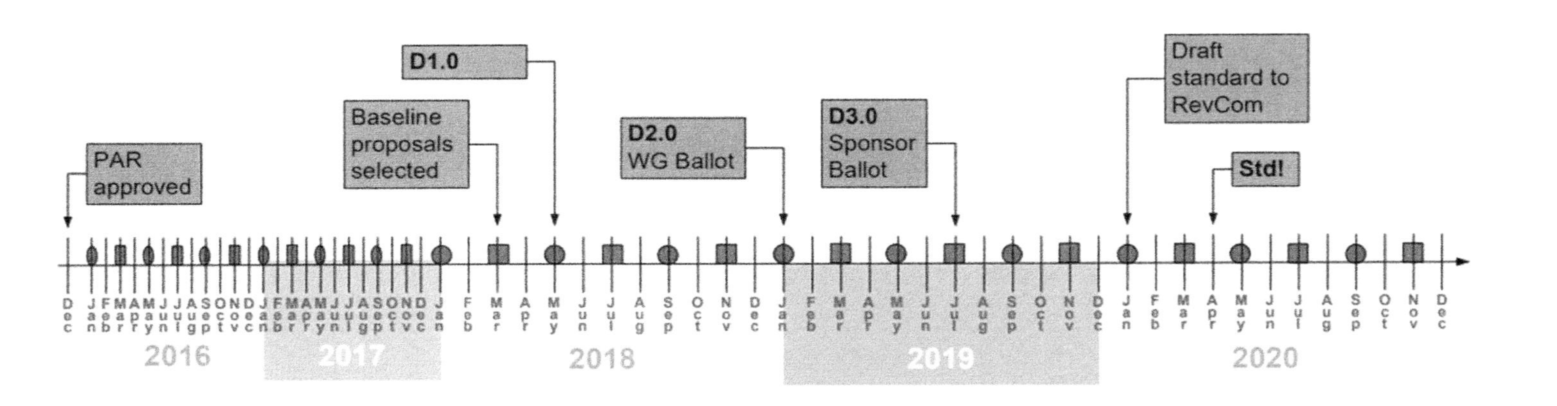

Figure 19.4 25/50/100G-EPON standardization timeline [22].

25/10 Gbps system is considered for residential subscribers, which is driven by headline speed competition.

One of the initially agreed upon key project objectives of the 802.3ca Task Force [23] was to provide specifications for physical layers operating over a single SMF strand and supporting symmetric and/or asymmetric MAC data rates of (1) 25 Gbps in downstream and $\leq$25 Gbps in upstream, (2) 50 Gbps in downstream and $\leq$50 Gbps in upstream, and (3) 100 Gbps in downstream and $\leq$100 Gbps in upstream. However, significant physical layer challenges were observed during the technical discussions in the subsequent meetings, especially in wavelength planning due to narrow low dispersion window and 10G-EPON/XGS-PON coexistence requirements. Additionally, there is also stringent time-to-market requirement for 25 Gbps PON by major US operators. Therefore, the third objective, "100 Gb/s in downstream and less than or equal to 100 Gb/s in upstream," was removed from 802.3ca Task Force objectives in November 2017.

The Task Force eventually settled with the following key specifications for 802.3ca:

- Operate over a single SMF strand
- Support symmetric and/or asymmetric MAC data rates of
  - 25 Gbps in downstream and 10 or 25 Gbps in upstream (25G-EPON)
  - 50 Gbps in downstream and 10, 25, or 50 Gbps in upstream (50G-EPON)
- Have a BER better than or equal to 10^{-12} at the MAC/PLS service interface (or the frame loss ratio equivalent)
- Wavelength plan:
  - Downstream wavelength plan with DS0 (downstream lane 0) at 1358 $\pm$ 2 nm (as in Fig. 19.5) and DS1 (downstream lane 1) at 1342 $\pm$ 2 nm (as in Fig. 19.2). 25 G NG-EPON will use DS0.
  - Upstream wavelength plan with US0-A (upstream lane 0-A): 1300 $\pm$ 10 nm (as in Fig. 19.5), US0-B (upstream lane 0-B): 1270 $\pm$ 10 nm (as in Fig. 19.6)[27], and US1 (upstream lane 1): 1320 $\pm$ 2 nm (as in Fig. 19.2).
  - US0-A is defined to support concurrent operation with 10G-EPON/XG-PON1/XGS-PON (Fig. 19.6).
  - US0-B is defined to support concurrent operation of 25G-EPON and G-PON reduced wavelength set, which employs 1490 $\pm$ 10 nm for downstream and 1310 $\pm$ 20 nm for upstream (Fig. 19.7).

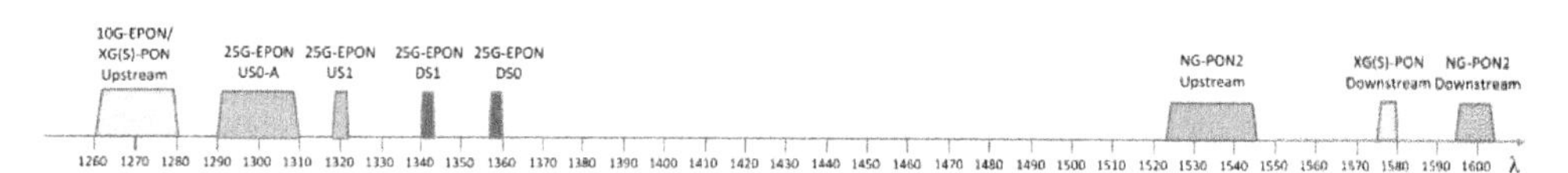

Figure 19.5 Wavelength-division multiplexing (WDM) coexistence w/10 G PON.

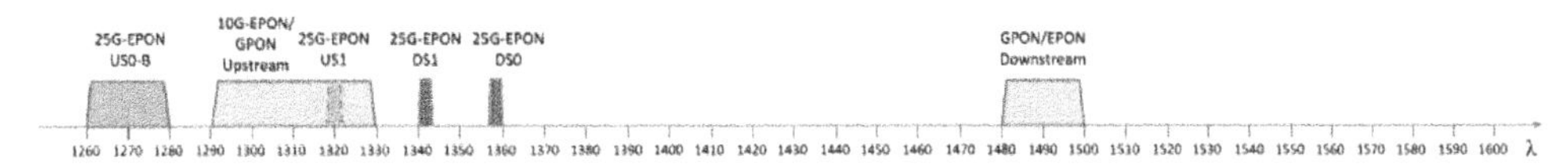

Figure 19.6 Wavelength-division multiplexing (WDM) coexistence w/GPON.

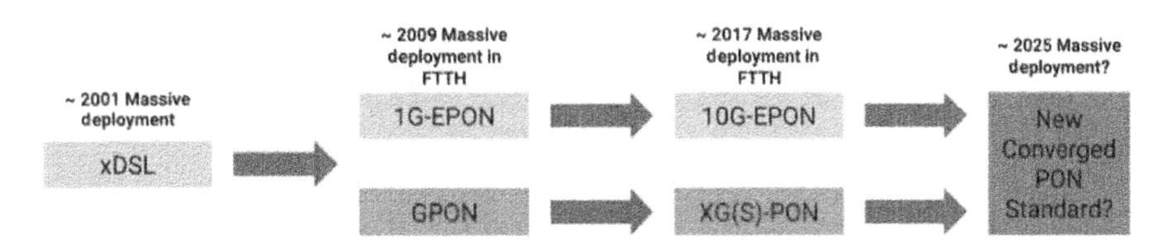

Figure 19.7 Passive optical network (PON) technology evolution in China [28].

- Optical power budgets to accommodate channel insertion losses equivalent to PR20 and PR30. Benchmark PR30 power budget specification [24]:
  - Downstream: 25 Gbps PON OLT transmitter minimum average power (AVPmin) is 4.8 dBm and minimum extinction ratio (ER) is 8 dB, and 25 Gbps PON ONU receiver sensitivity is −25.7 dBm at BER of 10^{-2} and ER of 8 dB.
  - Upstream: 25 G PON ONU transmitter minimum average power is 4 dBm (transmitter dispersion penalty subtracted) and minimum ER is 5 dB, and 25 G PON OLT receiver sensitivity is −25.0 dBm at BER of 10^{-2} and ER of 5 dB.
- LDPC (low-density parity-check) code as FEC (forward error correction) for both directions with optional precoding in the downstream [25]
- Channel bonding is allowed to bond two 25G channels into one 50G channel at the unit of envelope quantum (EQ), which is 8 bytes or 2.56 ns at 25 Gb/s [25,26].

O-band wavelengths are chosen for the IEEE 802.3ca implementation to reduce the transmission penalty in SMF during transmission and to leverage the 25 Gbps optical-component ecosystem developed for data center interconnects. At this writing (November 2018), IEEE 802.3ca Task Force is in the process of finalizing the draft standard, which is expected to be ready by April 2020.

19.2.2.2 FSAN and ITU-T SG15/Q2 next-generation passive optical networks

Although only 1 × 25 and 2 × 25 Gbps physical line rate have been made into IEEE 802.3ca due to technical challenges at higher line rate and stringent time-to-market requirement for 25 Gbps PON, there have been discussions throughout standard meetings on higher line rate at 1 × 50 or 2 × 50 Gbps in comparison to 2 × 25 Gbps.

FSAN/ITU-T SG15/Q2 started call for contributions on a high-speed PON (HS-PON) system from September 2017 within the Next-Generation Passive Optical Network (NG-PON) Task Group. The objective is to study higher speed PON systems, that is ≥ 50 Gbps, as the evolution path after NG-PON2. Currently, 1×50 Gbps physical line rate is generating significant interest among ISPs. This is especially the case for the China market, which just started massive deployment of 10G-EPON or XG(S)-PON in 2017. The interval between the massive deployment of two adjacent generations of broadband access platform has been 7 and 8 years over the last three generation of technologies. Therefore, the expected timeline for the next-generation technology deployment in China would be around 2025. Chinese carriers argue that in terms of bandwidth requirement, $2.5\times$ increase from 10G-EPON or XG(S)-PON to the IEEE 802.3ca PONs is too small of a jump and not future-proof. At least a $5\times$ increase, that is, 50 Gbps per channel/wavelength, should be considered, given that 100 Gbps per channel/wavelength is still quite challenging in the access domain, that is, low-cost system, high power budget, etc.

The general consensus on HS-PON requirements at 50 Gbps per channel/wavelength is as follows:

- TDM architecture with one DS wavelength and one US wavelength
- Support the maximum fiber distance of at least 20 km
- Support the following bit rates
  - Nominally 50 Gbit/s downstream, 10 Gbit/s upstream
  - Nominally 50 Gbit/s downstream, 25 Gbit/s upstream
  - Nominally 50 Gbit/s downstream, 50 Gbit/s upstream
- Common TC (transmission convergence) layer, when employed for 50 G TDM-PON, should support
  - Maximum fiber distance of 60 km
  - Maximum differential fiber distance of up to 40 km
  - Configuring the maximum differential fiber distance with a 20 km step
  - Support 1:256 split ratio
- Support standard single-mode fiber [29] and bending loss insensitive single-mode fiber [30]

At this writing (November 2018), ITU-T SG15/Q2 is in the process of providing a skeleton draft for different key documents, including but not limited to:

- G.hsp.req: Higher Speed Passive Optical Networks: Requirements
- G.HSP.50 GPMD: 50-Gigabit-Capable Passive Optical Networks (50G-PON): Physical Media Dependent (PMD) Layer Specification
- G.HSP.TWDMpmd: Higher Speed TWDM-Passive Optical Networks: Physical Media Dependent (PMD) Layer Specification
- G.hsp.ComTC: Higher Speed Gigabit-Capable Passive Optical Networks: Common Transmission Convergence Layer Specification

At 50 Gbps per channel/wavelength, the physical layer implementation becomes very challenging, that is, higher dispersion penalty and lower receiver sensitivity while moving toward higher serial rate. To cope with deployed OSP and support potential new deployment with high splitting ratio, for example, $\geq 1{:}128$ (inherently 21 dB splitting loss for an ideal splitter already), meaningful power budget [i.e., IEEE PR20 (24 dB), IEEE PR30 (29 dB), ITU-T N1 (29 dB), ITU-T N2 (31 dB), etc.] needs to be achieved. At the transmitter side, the maximum output power is limited by nonlinear distortion, which should be kept $\leq +10$ dBm per wavelength for PON applications [31]. Therefore, receiver sensitivity has to reach below −15 dBm, or preferably −20 dBm, in order to close the link budget. Recently, there have been different transmission experiments for ≥ 50 Gbps per wavelength in intensity modulation and direct detection (IM-DD) system with different modulation formats, for example, NRZ [32−34], EDB (electronic duo-binary) [34−36], PAM-4 (4-level pulse amplitude modulation) [34,37−42], and DMT (discrete multitone) [37], which are surveyed and summarized in Fig. 19.8. Here, the receiver sensitivity at BER of 10^{-3} in optical back-to-back (BtB) case is captured for comparison purposes. Pre-FEC BER of 10^{-3} is enabled via widely available FEC [43], although more advanced FEC can further enhance the results [44]. Fig. 19.8 is not intended to exhaust all up-to-date demonstrations but provide a common performance guidance for ≥ 50 Gbps receiver in optical access network. A few benchmark results are elaborated here for general references:

- 50 Gbps NRZ at O-band has been demonstrated to achieve receiver sensitivity of −25 dBm with 25-Gbps optics and DSP, that is, feedforward equalizer (FFE) and maximum likelihood sequence estimation (MLSE) based equalizer [32]. Negative dispersion penalty, that is, 0.5 dB, was observed after 20 km SMF due to negative chirp and dispersion interactions. Above 30 dB power budget can be achieved in this case without any optical amplification, which is ideal in terms of power consumption and system cost.
- 50 Gbps NRZ at O-band has been shown to reach receiver sensitivity of −25.2 dBm with 25-Gbps optics, optical amplification, that is, semiconductor optical amplifier (SOA), and DSP, that is, FFE [34].
- 50 Gbps EDB at O-band has been proven to fulfill receiver sensitivity of −21.4 dBm with 25-Gbps optics and without DSP [35].
- 50 Gbps EDB at O-band has been illustrated to accomplish sensitivity of −20.2 dBm with 25-Gbps optics, SOA, and without DSP [36]. This is one of the few real time demonstrations using 3-level decoder IC.
- 50 Gbps PAM-4 at C-band has been demonstrated to achieve receiver sensitivity of −25.5 dBm with 10-Gbps optics, optical amplification, and sophisticated DSP, that is, linear preequalization, predistortion for nonlinear impairments, and postequalization [41]. Moderate dispersion penalty, that is, 2.5 dB, was observed after 20

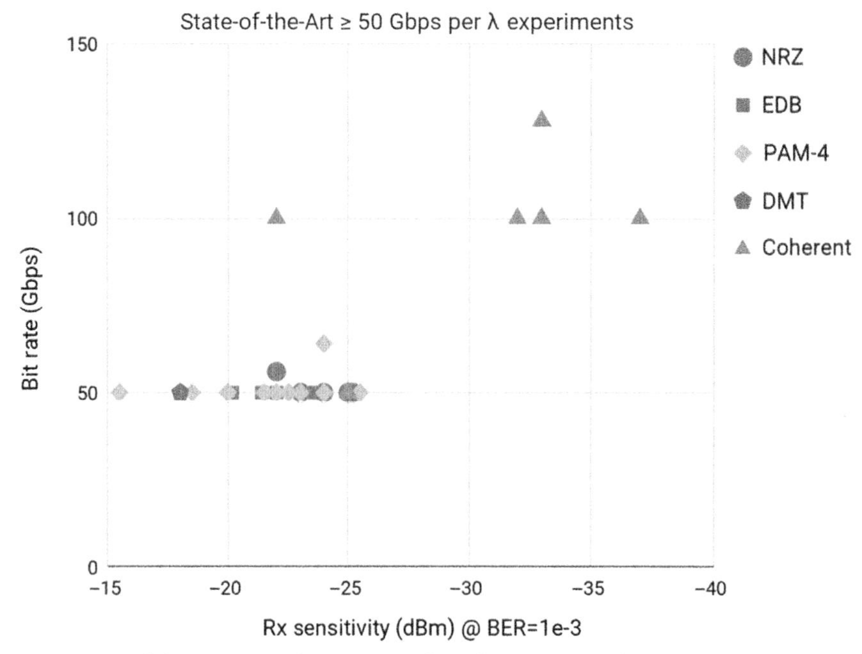

Figure 19.8 State-of-the-art ≥50 Gbps per wavelength experimental demonstrations.

SMF due to high dispersion at C-band, even with predistortion to compensate for chromatic dispersion (CD).

- 50 Gbps PAM-4 at O-band has been shown to achieve receiver sensitivity of −20.0 dBm with 25-Gbps optics and simple DSP, that is, 17-tap FFE, in the receiver side [34]. In addition, optical amplification, that is, SOA, has been proved to provide ~2 dB improvement to the receiver sensitivity.
- 50 Gbps DMT at O-band has been illustrated to reach receiver sensitivity of −18 dBm with 10-Gbps optics [37].

Although some 50 Gbps per wavelength IM-DD transmission systems have been demonstrated, the system performance is significantly limited by receiver sensitivity, dispersion penalty, etc. It is expected that 2× increase in baud rate introduces at least ~5 dB receiver sensitivity degradation due to TIA bandwidth increase, which leads to the decrease in gain and increase of noise density [45]. In addition, dispersion penalty increases proportional to the square of baud rate, which translates into higher power budget degradation. To mitigate the impact of receiver sensitivity degradation in IM-DD system while moving towards higher bit rate, optical amplification, DSP, advanced FEC, and proper dispersion management are widely applied in above experimental demonstrations to close the link budget. These new technologies should be agreed by the industry and standardized for the first time in PON standards.

On the other hand, moving toward even higher bit rate, for example, 100 Gbps per wavelength, coherent detection starts to show significant advantages over direct detection in terms of superior receiver sensitivity and strong dispersion tolerance. For example, 100 Gbps DP-QPSK at C-band has been demonstrated to achieve receiver sensitivity of −33.0 dBm for continuous mode operation and −30.5 dBm for burst mode operation [46]. Furthermore, with coherent detection, dispersion penalty is negligible even at 80 km reach. Standard coherent hardware, that is, integrated tunable Laser assembly and integrated coherent receiver for metro and longhaul applications, were used for the demonstration with simplified coherent DSP. Although superior performance can be achieved with conventional form of coherent detection system, certain aspects have to be specifically tailored for PON applications, for example, optics selection, polarization handling and DSP power consumption. This is especially important for the ONT side, which is cost and power sensitive. Coherent detection and its potential applications in PON will be further discussed in the following section.

As PON moves toward higher data rate, IM-DD or coherent detection has become a hot topic recently for both academia and industry, which led to the organization of multiple workshops in OFC 2018 [47] and ECOC 2018 [48] on the topic. There is general consensus from these discussions that direct detection is feasible in PON up to 50 Gbps per wavelength with the help of certain new technologies, and coherent detection will start to take over at 100 Gbps per wavelength and beyond due to multiple constraints.

19.2.2.3 Super-passive optical network

Besides the conventional and widely deployed IEEE 802.3ah EPON or ITU-T G.984-based G-PON technologies and their successors, super-PON [3,12,49], has been recently proposed by Google Fiber to improve fiber efficiency and reduce the number of COs required to support the network, which is of paramount importance for new ISPs, like Google Fiber, and potential greenfield builds for traditional ISPs. The idea of super-PON, introducing optical amplification and WDM technology in access network to support more subscribers per fiber (from 32 or 64 to ~1000) and longer distance (from 20 to ~100 km), had been proposed in the early 1990s [50−52], which promises lower end-to-end total cost of ownership for access network through bypassing the traditional metropolitan WDM network [52] and simpler network operation and maintenance, hence, lower operational expenses (OpEx), thanks to the significantly fewer number of active sites.

Google Fiber's super-PON architecture, as shown in Fig. 19.9, overlays multiple instances of G-PON networks onto a WDM-PON fiber plant.

- In the downstream direction, 12 fixed wavelength G-PONs and 8 point-to-point (PtP) channels are multiplexed and amplified onto the same feeder fiber in the CO

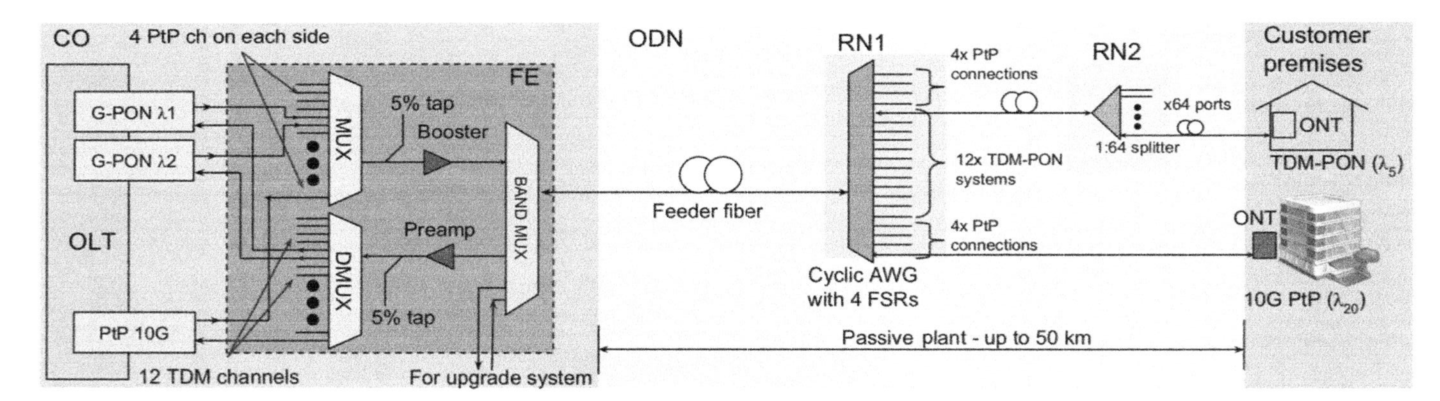

Figure 19.9 Super-passive optical network (PON) implementation in Google Fiber network.

through a fiber expander (FE). In the first remote node (RN1), one CAWG (cyclical arrayed waveguide grating router) is used to demultiplex all 20 wavelength channels. In the second remote node (RN2), a power splitter is used to split each colored TDM channel.

- In the upstream direction, a low-cost wavelength tunable ONT covering 12 channels is deployed at each customer premise. The wavelength channel information is carried in the broadcast PON_ID PLOAMd message from the OLT, detailed in Annex C of Ref. [53]. Each ONT listens for its assigned wavelength in the PON_ID message, and tunes its laser to the corresponding upstream wavelength. The ONT then ranges as per the GPON standard [53]. In RN1, one CAWG multiplexes all wavelengths. The FE in the CO amplifies and demultiplexes the upstream wavelengths.

A thin-film filter based band multiplexer (BM) combines the upstream and downstream signals into a bidirectional signal. Two additional ports on the BM are reserved for the expansion bands as shown in Fig. 19.9, to enable the seamless insertion of a future generation system. This is similar to preinstalling the WDM coexistence filter for upgrading GPON to allow the seamless insertion of a future generation system. This design allows for coexistence of two generation systems, which pave the path for seamless system upgrade.

The CAWG has a free-spectral range (FSR) of 22 channels, with 20 terminated ports, matching the channel plan shown in Fig. 19.10. The first CAWG output port passes the first channel from each of the four subbands, the second port passes the second channel from each of the four subbands, and so on. A 20-port CAWG has significantly lower loss than a 20-port splitter, thereby greatly reducing the required link budget. The wavelengths received by each customer are determined by which CAWG port the customer is attached to. All future system generations on this ODN have to use the wavelength plan shown in Fig. 19.10 and detailed in Table 19.3 with nominally 100-GHz channels.

A 64-way power split per wavelength, for example, enables up to 768 TDM users per fiber strand (12×64), in addition to the eight point-to-point wavelengths. Of course, the allocation of the point-to-point wavelengths and point-to-multiple TDM-PON wavelengths may be flexible. The optical link budget is designed for 50-km reach with a 64-way power split, which allows the use of one or very few centralized OLT locations to serve most metropolitan areas. The ability to support 768 users per fiber strand provides sufficient aggregation to allow very large COs to be connected using thin fiber cables, for example, from traditional 432-strand fiber cable to 12/48-strand fiber cables. This enables low-cost OSP construction method, such as micro-trenching, and simplifies construction process and lowers OSP build cost. In addition, reduced feeder fiber cable promises faster repair speed in case of fiber cut scenarios in operation. Thanks to optical amplification in the CO, longer reach up to 50 km

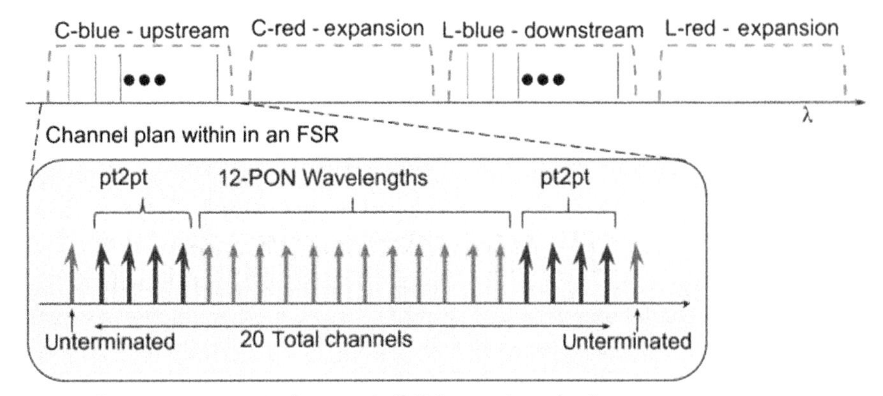

Figure 19.10 Super-passive optical network (PON) wavelength allocation.

Table 19.3 Super-passive optical network (PON) wavelength plan.

Ch #	C-band blue (C-B) upstream		L-band blue (L-B) downstream		C-band red (C-R) upstream		L-band red (L-R) downstream	
	Freq (GHz)	WL (nm)	Freq (GHz)	WL (nm)	Freq (GHz)	WL (nm)	Freq (GHz)	WL (nm)
1	193.991	1545.39	189.609	1581.11	191.800	1563.05	187.418	1599.59
2	194.092	1544.59	189.708	1580.29	191.900	1562.23	187.516	1598.76
3	194.193	1543.78	189.807	1579.46	192.000	1561.42	187.613	1597.93
4	194.294	1542.98	189.906	1578.64	192.100	1560.61	187.711	1597.10
5	194.396	1542.18	190.004	1577.82	192.200	1559.79	187.809	1596.27
6	194.497	1541.38	190.103	1577.00	192.300	1558.98	187.906	1595.44
7	194.598	1540.57	190.202	1576.18	192.400	1558.17	188.004	1594.61
8	194.699	1539.77	190.301	1575.36	192.500	1557.36	188.102	1593.78
9	194.800	1538.97	190.400	1574.54	192.600	1556.56	188.200	1592.95
10	194.901	1538.18	190.499	1573.73	192.700	1555.75	188.297	1592.12
11	195.003	1537.38	190.597	1572.91	192.800	1554.94	188.395	1591.30
12	195.104	1536.58	190.696	1572.09	192.900	1554.13	188.493	1590.47
13	195.205	1535.78	190.795	1571.28	193.000	1553.33	188.590	1589.65
14	195.306	1534.99	190.894	1570.47	193.100	1552.52	188.688	1588.83
15	195.407	1534.19	190.993	1569.65	193.200	1551.72	188.786	1588.00
16	195.508	1533.40	191.092	1568.84	193.300	1550.92	188.883	1587.18
17	195.609	1532.61	191.191	1568.03	193.400	1550.12	188.981	1586.36
18	195.711	1531.82	191.289	1567.22	193.500	1549.32	189.079	1585.54
19	195.812	1531.02	191.388	1566.41	193.600	1548.52	189.176	1584.72
20	195.913	1530.23	191.487	1565.60	193.700	1547.72	189.274	1583.91

enables CO consolidation, which alleviates the pressure of initial CO construction and reduces OpEx during operation.

Google Fiber has successfully demonstrated, trialed, and deployed the first-generation super-PON system in different cities [49] by wavelength stacking conventional G-PONs. While the super-PON system is helpful to Google Fiber's organic deployment, it was also realized that, without a consensus standard from the broadband industry, it will hinder the continual development of the super-PON architecture and the creation a healthy technology ecosystem for the prosperity of the architecture. Therefore, Google Fiber contributed super-PON at the New Ethernet Application ad hoc discussions in IEEE 802.3 working group in January 2018 [3], and its CFI was completed in July 2018 [54], which moved the ad hoc discussion into an official IEEE 802.3 NG-EPON study group. PAR, CSD, and project objectives were developed in subsequent meetings and reported to and approved by the IEEE 802.3 working group to officially form the IEEE 802.3cs task force of *Physical Layers for increased-reach Ethernet optical subscriber access (Super-PON)* in November 2018. Instead of working on the deployed first-generation super-PON at G-PON data rate, the task force has agreed to tackle 10 Gbps downstream and 10/2.5 Gbps upstream system specifications. The objective of IEEE 802.3ca is to provide physical layer specifications that

- Preserve the Ethernet frame format utilizing the Ethernet MAC.
- Support a BER of better than or equal to 10^{-12} at the MAC/PLS service interface (or the frame loss ratio equivalent).
- Support a passive point-to-multipoint ODN with a reach of at least 50 km with at least 1:64 split ratio per wavelength pair.
- Support at least 16 wavelength pairs for point-to-multipoint PON operation.
- Support the MAC data rate of 10 Gb/s downstream.
- Support the MAC data rates of 2.5–10 Gb/s upstream.
- Leverage existing EPON PCS and PMA to support the above MAC data rates.
- Support tunable transmitters.

19.3 Wavelength-division multiplexing and its challenges in access networks

19.3.1 Wavelength-division multiplexing-passive optical network and wireless fronthaul

Fig. 19.11 shows the structure of conventional WDM-PON, which uses a WDM splitter in the form of a CAWG as the RN. Compared to point-to-point homerun fibers from the CO to end users, a WDM-PON network reduces the number of trunk fibers in the CO by overlaying multiple wavelengths for different users on a single strand of trunk fiber. End users are separated at the CO using separate wavelengths

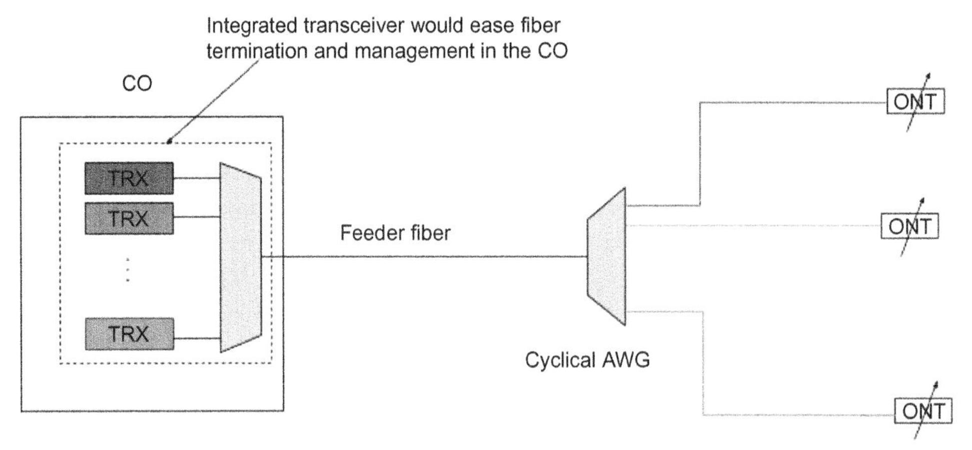

Figure 19.11 Architecture of a conventional wavelength-division multiplexing–passive optical network (WDM-PON).

which serve as virtual fibers. Thus users can easily adopt different protocols and different transmission speeds w/o affecting each other.

Although WDM-PON has been proposed almost around the same time that TDM-PON was proposed, it never went to serious commercial implementation and actual use because of the high costs in packaging WDM transceivers and lack of demand for individual users to be allocated with the bandwidth carried by dedicated wavelengths.

The following considerations are usually taken into account in implementing WDM-PONs [55,56].

1. An integrated array transceiver at the OLT to reduce the number of transceiver terminations in the CO. Advancements in photonic integration in the future will be able to help the implementation of low-cost integrated WDM-PON OLT transceiver arrays.
2. Athermal cyclical AWG in the field to serve as the RNs for wavelength distribution to individual end users.
3. Ultralow-cost tunable lasers at the ONTs in end users to ease the complexity of deployment and operation. Although fixed wavelength transceivers could theoretically be used for ONTs, handling the SKU variations of ONTs, complex inventory, and matching the ONT SKU with the AWG fan-out port during deployment, diagnosis, and troubleshooting time is very undesirable. Therefore, tunable ONTs are usually employed with an algorithm to automatically provision the proper upstream wavelengths which is transparent to the users and installers.

In order to make efficient use of the transmission spectrum inside the fiber, DWDM is usually used for WDM-PON. Another reason for using DWDM is due to the limited tuning range of semiconductor lasers. However, DWDM transceivers require temperature controlled packaging as laser output wavelengths drift with the operating temperature. This makes WDM transceivers more expensive compared to traditional TDM-PON transceivers or short-reach intra-data center optical transceivers [18] which can operate without temperature control (the so-called uncooled operation) as those transceivers operate on widely separated wavelength grids (which will reduce the trunk fiber efficiency in a WDM-PON and demand tunable lasers with unachievable tuning ranges).

The packaging cost of the tunable transceivers could be controlled through the adoption of monolithically tunable laser technologies such as simple tunable DBRs [57], which unfortunately also has a limited tuning range between 8 and 12 nm, using both temperature and current tuning. This limits the number of usable channels to between 12 and 16 channels in the C-band (around 1550 nm). More complex tunable laser technologies (e.g., those using Vernier tuning techniques to widen the tuning range, such as SG-DBR lasers) will be able to support wider tuning range and more channels at the expense of significantly more complex device structure and control circuitry and hence much higher costs.

DBR lasers and TEC (thermoelectric temperature controllers) can be copackaged into a TO-can[1] package to reduce the overall cost and take advantages of the automated packaging systems created for TO-can. Fig. 19.12 shows the diagram of (A) a monolithic BDR laser and (B) the TO package of the laser and TEC controller.

The upstream and downstream channels in WDM-PON are normally segregated by the different FSRs of the CAWG. The cyclical property of the CAWG is also made use of for WDM-PON system upgrades, so that one can put in different services into different orders of FSRs. For example, in the super-PON TWDM system that was created by Google Fiber, four FSRs of the CAWG straddled across C- and L-bands are used for the upstream and downstream transmissions: two for the current generation and two reserved for system upgrades.

WDM-PON received a lot of attentions in 5G wireless network fronthaul applications because of its protocol transparency. A significant advantage of WDM-PON system is the point-to-point wavelength connectivity between the CO and the remote ONTs, which serve as virtual fibers. There is no complex MAC and TDM-PON multiplexing protocol (or DBA) between the OLT and ONTs. This property and the passive nature of WDM-PON made it especially suitable for wireless fronthauls. Fronthaul refers to the mechanism of transmitting the analog waveform of a remote antenna head (RRH) back to a centralized baseband unit (BBU) for processing [58]. In the downstream direction, digitized RF waveform in the form of $I-Q$ modulation

[1] TO-can stands for transistor outline can, which is one of the oldest and most widely used packaging techniques used in the electronic industry.

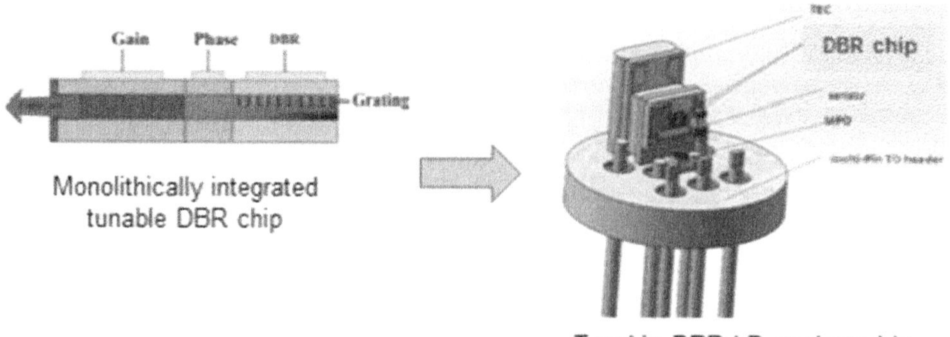

Figure 19.12 (A) Monolithic integrated tunable DBR laser diode (LD) chip. (B) Tunable DBR LD copackaged in TO-CAN with TEC.

amplitudes are sent to the digital-to-analog converters (DAC) to drive the RF (radio frequency) modulator input. In the upstream direction, received quadrature waveforms are sampled and through analog-to-digital converters (ADC) and transmitted to back to BBU.

Obviously, the bandwidths required per antenna increases with the sampling frequency (which is proportional to the RF bandwidths used in the wireless system according to the Nyquist sampling theorem), the RRH dynamic range required (which determines the number of bits per sample). The amount of fronthaul bandwidths required at a cell site is multiplied by the number of antennas at that site. To increase cell capacity and spectral efficiency, each cellular site is usually divided into three to six sectors. Each sector further employs MIMO (multiple-input-multiple-output) antennas to make use of spatial diversity to enhance transmission and increase capacity. 4G LTE systems divides RF spectrum into 20 MHz RF frequency bands and can bundle as many as five 20HMz RF bands for transmission in channel bonding. 5G new radios defined bands as wide as 400 MHz in high frequency radio spectrum and can bond as many as 16 concurrent RF bands.

In a C-RAN [59] (C means centralized as well as cloud, RAN stands for radio access network), remote cell sites are only equipped with RRHs whose analog waveforms are all fronthauled to a central BBU that does joint signal processing of the signals from antenna heads distributed in different cell sites. This architecture had several advantages:

1. It allows antennas from neighboring cell sites to transmit to the same mobile so that the signal from the two cells interferences constructively at the mobile. This smooths out the user experiences of mobiles as they traverse across wireless cell boundaries. It allows smoother transition from one cell to another.
2. It allows interference between neighboring cells to be managed so that a frequency reuse pattern of 1 (as opposed to having to use different carrier frequencies in

neighboring cells to reduce interference in traditional cellular systems) can be used to improve the utilization of the precious RF spectral resources.

In wireless systems, precise timing control is required since all the mobiles share the same radio spectrum through a TDM protocol. As can be imagined, for the joint signal processing at the central BBU to operate correctly, both the amplitudes and phases of the transmitted and sampled waveforms have to be very precisely controlled. 3GPP requires timing precision of 50 ns in 4G LTE fronthaul. The most popular fronthaul framing protocol in 4G LTE implementation is the CPRI (common public radio interface) standard [60]. CPRI defines different levels of transmission speeds. Fronthaul protocols are evolving in more or less an ad hoc manner as capabilities of RF technologies kept evolving, so that the different levels of CPRI transmission standards do not even follow a nicely laid out digital multiplexing hierarchy. Besides, to avoid additional latency and timing complications, CPRI does not even use forward error correction. Transmission of CPRI fronthaul has traditionally been handled with wavelength diplexed BiDi (bidirectional transceiver) using 1470 or 1510 nm CWDM wavelengths for downstream and upstream transmissions. The BBU estimates the signal delays between itself and an RRH through the round-trip signal delay in the fronthaul link. This information is essential in scheduling the transmission from the BBU to the RRH. BiDis allow the use of single strand of fiber to be used for upstream and downstream transmission and removes the uncertainty in the asymmetry between upstream and downstream links in a traditional set up that employs duplex fibers, essential for precision round-trip delay estimation between the BBU and RRH in fronthaul applications.

The traditional fronthaul approach with gray optics on dark fiber requires large number of fiber strands between BBU and cell sites as one adds new RF frequency bands (i.e., new antennas), increases the antenna MIMO sizes (traditional MIMO systems use 2T2R, or 4T4R MIMOs, whereas state-of-the-art massive MIMO systems can be as large as 64T64R). Transparent WDM-PON is an effective way to carry cellular fronthaul signals. It uses single fiber for bidirectional transmission, is passive and easy to deploy, and offers timing and protocol transparency.

Fronthaul transmissions are usually limited by the latency requirements in wireless networks to less than 10 km.

The ITU-T G.698.4 (formerly called G.metro) standard [61] is a WDM-PON standard created to address wireless fronthaul application. Its architecture is show in Fig. 19.13. The definition of this architecture includes a WDM grid with 20–40 wavelengths, intermediate wavelength add/drop capabilities, and cyclical AWGs. The current version ITU-T G.698.4 standard defines transmission speed up to 10 Gbps in C-band (upstream and downstream wavelengths are separated in C + /C− bands).

The upcoming fronthaul transmission standard commands signal baud rate of 25 Gbps, where dispersion penalties can no longer be ignored in the C- and

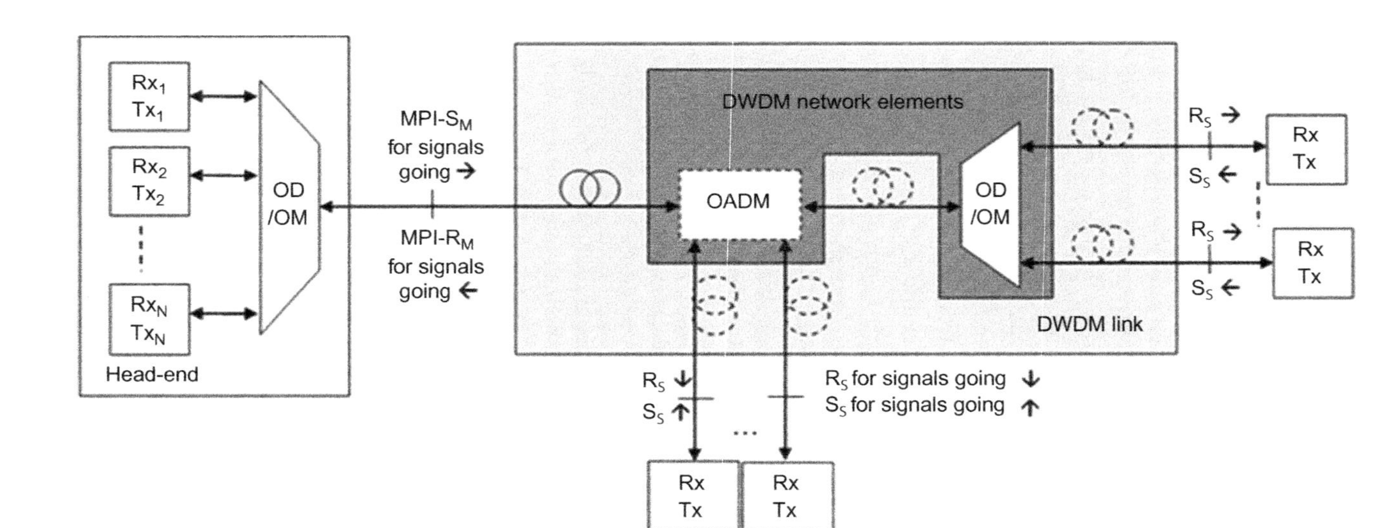

Figure 19.13 Reference architecture of ITU-T G.698.4. Source: *From ITU-T, ITU-T G.698.4 Multichannel bi-directional DWDM applications with port agnostic single-channel optical interfaces, G.698.4, Mar. 2018 [61].*

L-traditional WDM transmission bands. Dispersion managed tunable DBR lasers could be potential solution to solve the dispersion issue [62]. Transmission distances of 60 km at 25 Gbps without dispersion compensation in the 1550 nm band had been demonstrated [63].

19.3.2 TWDM-passive optical networks and their challenges

The two major implementations of TWDM-PONs are represented by the ITU-T G.989 NG-PON2 and Google Fiber's super-PON (and its successors being standardized in the IEEE 802.3cs Task Force). We will simply refer to the later as super-PON in this section although the generic term super-PON has much wider meanings.

The major difference between NG-PON2 and super-PON is in the structure of the OSP. NG-PON2 employees a pure power-splitting-based OSP and broadcast-and-select architecture, whereas super-PON uses a wavelength routed architecture with a wavelength splitter in the field. Both architecture makes use of both TDM and WDM to scale the bandwidths and enable large number of users to share the vast bandwidths.

The pure power-splitting OSP of NG-PON2 offers the most flexibility in optical spectrum usage and compatibility with traditional TDM-PONs. The disadvantage is the need for wavelength selective tunable receiver or WDM-array receiver at individual ONTs, which are both complex and costly. On the other hand, super-PON uses a wavelength router in the field to separate individual downstream wavelengths before they hit the individual ONTs. This sacrifice of flexibility leads to significantly simplified ONTs hardware and much reduced link budget requirements [12].

One of the common issues faced with by both NG-PON2 and super-PON is the burst-mode operation induced wavelength excursions at the ONT transmitter laser [64]. This effect had been measured by passing the transmitter output through the edge of a narrow-band filter (Fig. 19.14, left) whose wavelength is aligned with the nominal output wavelength of the transmitter laser. As the output wavelength drifts during the laser on/off transients, the narrow-band filter slope will translate the wavelength drift into amplitude changes. The amount of wavelength drifts can be inferred by observing the above FM-AM changes. Transient wavelength excursion on tens of microsecond time scale and as big as 30–40 GHz had been observed (Fig. 19.14, right). The magnitude of wavelength excursion increases as the laser output power increases. The broadcast-natured OSP in NG-PON2 demands higher link budget and higher laser output power, which translates into more severe burst-mode induced wavelength excursion and requires more sophisticated compensation. On the other hand, super-PON's reduced link budget and the availability of the preamplifier at the CO both helped to reduce the ONT laser output power requirement and smaller burst-mode induced wavelength excursion. Furthermore, the AWG remote mode in

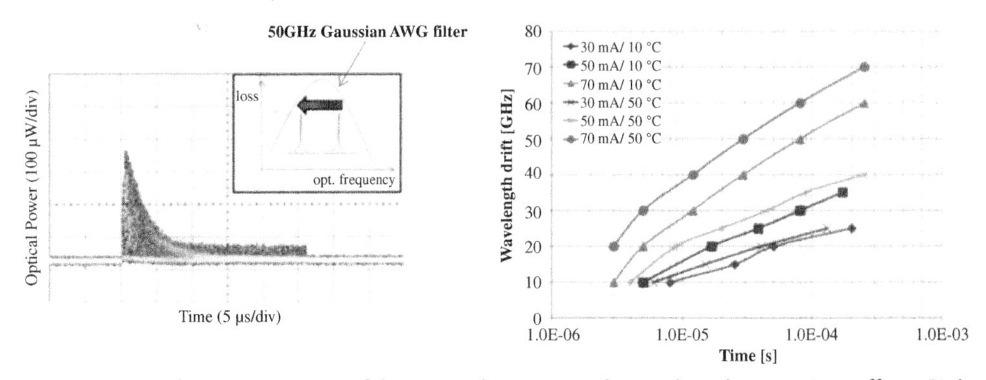

Figure 19.14 Left: Measurement of burst mode generated wavelength excursion effect. Right: Wavelength excursion versus time at different laser bias current (and output power). Source: *From W. Poehlmann, D. van Veen, R. Farah, T. Pfeiffer, P. Vetter, Wavelength drift of burst-mode DML for TWDM-PON [invited], IEEE/OSA J. Opt. Commun. Netw. 7 (1) (2014) A44 [64].*

super-PON also helps to reduce the interchannel crosstalk generated by wavelength excursions.

19.4 Enabling technologies on the horizon

With the rapid increase of the PON data rate, it becomes more and more difficult to meet the system requirement, such as power budget, while still satisfying the aggressive cost target for the PON system. Both academia and industry have started to look into new enabling technologies to help improve the system performance comparing to the previous generation, while not dramatically increasing the system cost. For example, optical amplification has been widely considered as a simple way to increase power budget by boosting the transmitter power or increasing the receiver sensitivity. Once shared by multiple channels in a discrete form factor or integrated with the transmitter/receiver, its extra cost per subscriber can be justified. Among all the novel enabling technologies, digital signal processing (DSP) and coherent detection have great potentials to be adopted in PON-based systems to enable rapid traffic growth, which will be discussed in the following sections.

19.4.1 Digital signal processing

DSP has been widely adopted in optical communication systems from longhaul backbone links to short-reach data center interconnects to compensate transmission impairments and enable the use of lower cost optical components with lower performance. In longhaul and metro coherent detection-based systems, DSP, in the form of digital equalizer, play a key role in adequately modulating and demodulating complex modulation formats and compensating channel impairments, such as CD, polarization mode dispersion (PMD), carrier phase noise, and various channel nonlinearities [65,66]. In addition, digital equalizer is essential to deal with linear and nonlinear distortions arose from limited component frequency response and manufacture tolerances [67], through the use of electrical-to-optical (E/O) and optical-to-electrical (O/E) conversions. FEC has been employed in optical communication systems for a long time to reduce the received bit error rates, which translates into relaxed system requirements, for example, lower optical signal-to-noise ratio and less receiver sensitivity requirements. In this section, we review the application of DSP in optical access networks.

Equalization is a method widely adopted in communication systems to compensate for channel imperfections, for example, bandwidth limitation or channel dispersion induced intersymbol interference (ISI). In general, there are two different major categories of equalizers, linear and nonlinear. The FFE is the most common linear equalizer, which tries to eliminate the impact of both precursor and postcursor symbols from the current symbol [68]. FFE can be applied to both transmitter and receiver. The issue with FFE is that it tries to invert the channel amplitude response to compensate for ISI, which inevitably boosts high frequency noise and limits the final signal quality. Nonlinear equalizer includes the decision-feedback equalizer (DFE), Volterra equalizer, MLSE based equalizer, etc. DFE is the most widely adopted one and is typically applied in the receiver. DFE can only correct for postcursor due to its feedback nature [68]. Comparing to FFE, DFE does not have the noise enhancement issue, but it is subject to error propagation in case of one wrong decision, which will lead to burst errors in detection. In practice, the FIR filter coefficients (both feedforward and feedback filters) are adjusted using an adaptive algorithm, for example, least mean square, and to minimize an estimate of noise plus residual ISI. FFE and DFE can be implemented in either analog or digital fashion. Analog implementation of FFE and DFE does not require digital-to-analog converter and ADC, hence, lower power consumption and implementation complexity. However, this is only valid within limited number FIR filter taps, for example, <10 taps. Analog equalizer is not practically scalable with the increase of filter taps. The following discussions will focus on digital based equalization methods.

Digital equalization provides the following main benefits for high-speed PON systems:

1. It relaxes optical components requirements, such as E/O and O/E bandwidth and linearity. This allows low-cost optics be applied to high-speed transmission, for example, 10-Gbps direct modulated laser (DML) can be used for 25, 40, and even 50 Gbps PON systems with the help of digital equalizer. In Refs. [69,70], a 40-Gbps symmetrical TDM-PON subsystem has been demonstrated with 10-Gbps optics and DSP. In the downstream direction, a 9-GHz electro-absorption modulated laser (EML) was used as the transmitter with a 15-tap FFE for preequalization. A 11-GHz PIN was used at the receiver side with 8-tap FFE and 7-tap DFE were

applied to further equalize the signal. Without pre- and postequalization, it cannot reach pre-FEC threshold at BER of 10^{-3}. However, with both pre- and postequalization, it achieved receiver sensitivity of −13 dBm (without amplification) and −25 dBm (with amplification). In Ref. [37], 50 Gbps PAM-4 downstream transmission has been demonstrated with 10-Gbps optics and DSP, that is, narrow filter compensation (NFC), which includes both FFE and MLSE [71]. Without the NFC, it cannot reach pre-FEC threshold. However, with the NFC, 131-tap FFE and 4-state MLSE-based equalizer, it reached receiver sensitivity of −20 dBm after 20 km in O-band.

2. It compensates for channel impairments, for example, CD, etc. However, CD compensation is less effective in IM-DD systems compare to coherent detection systems, as CD contributes to nonlinear amplitude distortion after square-law photodetection and requires nonlinear equalizer. In Ref. [72], 40 Gbps PAM-4 transmission in C-band had been demonstrated with Volterra algorithm–based nonlinear equalizer at the receiver, and it achieved no dispersion penalty up to 20 km and only 4.5 dB penalty at 40 km. In Ref. [41], 50 Gbps PAM-4 transmission in C-band has been shown with 2.5 dB dispersion penalty after 20 km fiber transmission with frequency domain-based CD precompensation at the transmitter. A dual-drive Mach-Zehnder modulator has to be adopted instead of DML or EML at the transmitter to enable the CD precompensation.
3. It enables high spectral-efficiency modulation formats, for example, orthogonal frequency division multiplexing (OFDM) and DMT, and flexible resource sharing schemes, for example, orthogonal frequency division multiplexing access (OFDMA). In Ref. [73], 40-Gbps DMT transmission in C-band had been demonstrated to achieve receiver sensitivity of −19 dBm after 10 km transmission. Although its performance does not outperform the other single carrier systems [69], it offers the flexibility of bit loading to maximize transmission bit rate based on different channel conditions. In Ref. [74,75], 20-Gbps per wavelength OFDMA-PON has been shown to achieve 20 km fiber transmission and 29 dB equivalent power budget. A centralized laser source in OLT side was transmitted and shared by all the ONTs to achieve source-free ONT operation, which avoided the issue of optical beating noise from different ONT transmitters. In addition, coherent detection was adopted in the upstream direction to eliminate both in- and cross-polarization beating noise. DSP based OFDMA-PON offers the flexibility to dynamically allocate frequency (subcarrier) and time (time slot) domain resources to different applications requirements.

Besides digital equalization, FEC provides additional power budget benefit to PON systems. FEC is a method of encoding the original signal with additional error detection and correction overhead information, for example, parity bytes, so that optical receivers can detect and correct errors that occur in the transmission path. FEC dramatically lowers the required SNR at the receiver, which translates into better receiver sensitivity and higher power budget. In practice, it can be applied to provide extra margin in the system and to relax optical-component specifications, such as launch power and extinction ratio, hence, lower overall system cost. It can also be adopted to offset the reduced receiver sensitivity in higher bit rate PON systems.

FEC has been considered as optional in E-PON and G-PON with Reed-Solomon (255, 239) or its truncated variants, which has 16 parity check symbols per code word and provides 5.6 dB net coding gain. It has the capability of improving pre-FEC BER of 5×10^{-4} to $<1 \times 10^{-12}$. Moving toward 10 Gbps PON standards such as 10GE-PON and XGS-PON, FEC became mandatory with Reed-Solomon (255, 223) or its truncated variants, which provides 7.1 dB net coding gain [76]. It has the capability of improving pre-FEC BER of 1×10^{-3} to $<1 \times 10^{-12}$. Moving toward 25 Gbps and beyond, higher coding gain FEC is required to compensate for the degraded receiver sensitivity. During the standardization process of NG-EPON in IEEE 802.3ca, various FEC have been considered with the requirement of providing 1.0–1.5 dB power budget improvement [22,43], over Reed-Solomon (255, 223) code adopted in 10GE-PON and XGS-PON. LDPC code (18493,15677) has been selected for both downstream and upstream to provide 2.5 dB of extra coding gain and the capability of improving pref-FEC BER of 1×10^{-2} to $<1 \times 10^{-12}$.

More advanced FEC algorithms, such as soft-decision and iterative decoding, can provide further coding gains to >10 dB [77,78], but at the cost of higher power consumption, more implementation complexity, and longer processing latency. Depending on different applications, some parameters may be of higher priority, which needs to be considered at the time of system design.

19.4.2 Coherent detection

The idea of coherent detection in optical communication was first proposed in the 1980s [79,80]. At that time, the major advantages of coherent detection are much higher receiver sensitivity (shot-noise limited receiver) with up to 10–20 dB link budget improvement compare to IM-DD systems. Other advantages include enhanced frequency selectively to separate WDM channels in the electrical domain, and alternative modulation formats, for example, phase-shift keying (PSK). Therefore, it promised to extend the repeaterless transmission beyond 50–100 km, which was of great benefit to any longhaul transmission system in terms of overall system cost. There were a number of research groups working on the topic until the invention of erbium-doped fiber amplifiers (EDFAs) in the late 1980s [81]. This is due to the fact that in a cascaded longhaul system, the overall repeaterless transmission distance is determined by the accumulated amplified spontaneous emission noise rather than the shot noise of the receiver. In a single-span transmission system, the receiver sensitivity can be easily improved by 10–20 dB by adopting low-noise preamplification EDFAs before the typical receiver. Therefore, research and development on coherent detection faded away for about 10 years. However, coherent detection was brought back onto the stage in the early 2000, mainly due to the following three factors. First, self-coherent modulator formats, for example, differential PSK (DPSK), without requiring a local oscillator (LO) at the receiver was introduced in the early 2000 [82]. Second, advancements of high-speed electronics made DSP capable of handling complex demodulation possible for high-speed optical communication, instead of the need for optical phase-lock loop. Furthermore, DSP enables easy mitigation of multiple notorious channel impairments in fiber transmission systems such as CD, polarization mode dispersion, channel nonlinearities, and optical components imperfections comparing to IM-DD systems [83]. Last but not least, nested MZM based optical IQ modulator makes high spectral-efficiency modulation format possible to significantly increase per channel capacity without further pushing the already challenging high modulation baud rate. Shortly after 2000, high-capacity (>100 Gbps per wavelength) commercial coherent detection systems have quickly dominated the longhaul and metro markets [84].

In the PON domain, coherent detection has always been one of many potential technology candidates. It is very beneficial to PON systems in the following aspects. First of all, it significantly increases the receiver sensitivity by >10 dB, which translates into higher splitting ratio and longer reaches. Second, it supports high spectral-efficiency modulation formats, such as QPSK, 16-QAM, etc., which helps to support higher data rate, for example, >50 Gbps per wavelength, at a much lower baud rate (lower speed and lower cost optics and electronics). In fact, there is general consensus among the research community from recent conference workshops [47,48], and discussions that coherent detection will be introduced as PON standards evolving into 100 Gbps per wavelength schemes. Third, it helps to easily compensate for channel impairments such as CD, which is currently limiting the design of high-speed PON systems to the O-band in order to minimize its impact in IM-DD systems. Finally, it conveniently provides tunable optical receiver function, which may avoid the need of band rejection filters in the ONT. This will simplify wavelength planning in the PON standardization process.

Although a coherent detection scheme provides significant advantages to a PON system, current forms of coherent detection were implemented for longhaul and metro system which are higher in cost and greater in complexity. The standard polarization and phase diverse coherent receiver is shown in Fig. 19.15, which includes one LO laser, two polarization beam splitters (PBS), two 90-degree optical hybrids, four balanced photodiodes, and four ADCs. The coherent receiver structure has to be significantly simplified to the level close to a standard IM-DD receiver in order to be meet the cost target of PON systems. In addition, it is preferable to be polarization

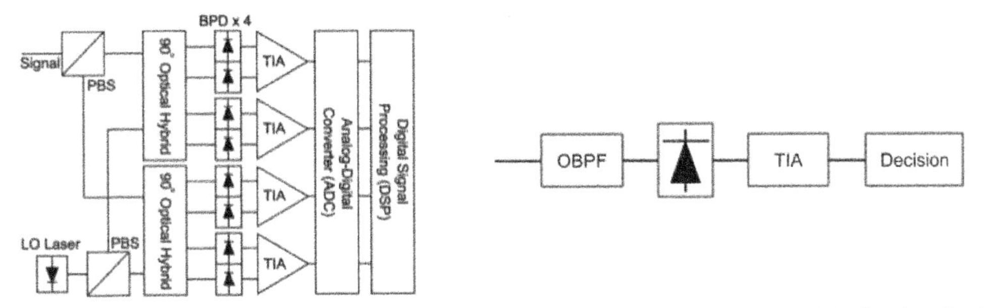

Figure 19.15 Left: Standard polarization and phase diverse coherent receiver [86]. Right: Standard IM-DD receiver.

insensitive. There have been various simplified coherent receiver structures demonstrated for PON applications [85,86]. In Ref. [87], a quasi-coherent receiver has been demonstrated with one LO, one 3-dB coupler, one PBS, and two PDs as shown in Fig. 19.16. It adopted analog heterodyne demodulation scheme with envelope detectors, which completely avoided DSP in the electronic part but was limited to amplitude modulation only. For 10 Gbps NRZ, receiver sensitivity of −31 dBm was achieved with 10.5 dBm of LO power, and no significant dispersion penalty was observed with up to 44 km fiber transmission. However, given no DSP is applied in this scheme, its dispersion tolerance going toward higher bit rate is questionable, unless analog electronic dispersion compensation is implemented at the receiver. This quasi-coherent receiver is polarization insensitive. In Refs. [88,89], a 3 × 3 coupler-based coherent receiver has been illustrated to achieve receiver sensitivity of −49 dBm for 1.25 Gbps DPSK after 50 km transmission. In addition, only 1 dB penalty was observed while varying signal polarization state. In Ref. [90], polarization scrambling at the transmitter using a polarization modulator has been demonstrated to achieve polarization-insensitive coherent detection. Receiver sensitivity of −49 dBm has been achieved for 1.25 Gbps DPSK downstream after 50 km transmission. In Refs. [91−93] Alamouti coding (polarization-time block coding) on the OLT side had been proposed and demonstrated to support one simple coherent receiver design with one LO, one 3-dB coupler, and one balanced photodiode, as shown in Fig. 19.17. Receiver sensitivity of -41.4 dBm has been achieved with 10 Gbps Alamouti-coded OFDM QPSK signal after 80 km. At −34.8 dBm of received power, only 0.6 dB receiver sensitivity degradation was observed while varying signal polarization state. Alamouti coding simplifies ONT coherent receiver and puts additional complexity on the OLT side, which is especially suitable for PON system, as OLTs are shared among all the connected ONTs.

19.4.3 Integrated photonics

Integrated photonics will help the next-generation optical access networks in the following aspects.

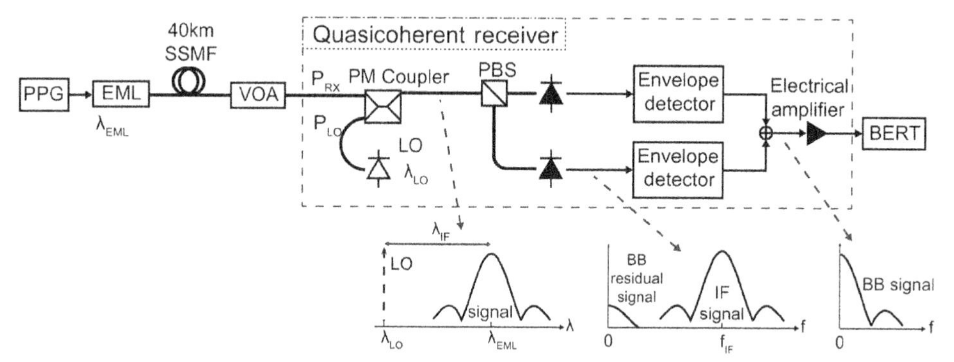

Figure 19.16 Quasi-coherent receiver structure [87].

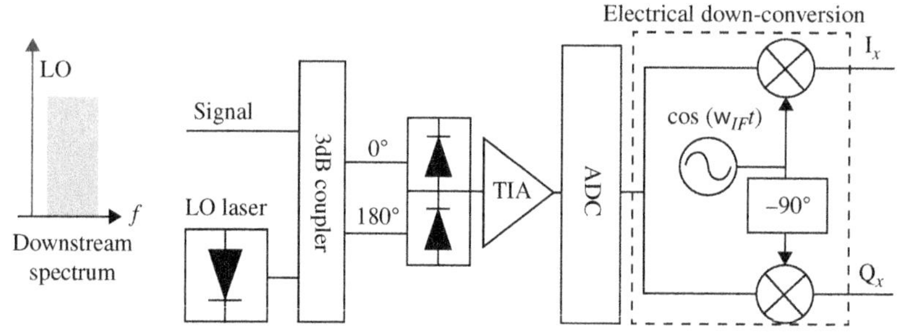

Figure 19.17 Optical line terminal (OLT) Alamouti coding enabled polarization-insensitive coherent receiver [91].

First, although WDM-PON reduces the number of feeder fiber strands required by multiplexing many wavelengths on the same fiber strand, it also increases the number of transceivers required in the CO compared to TDM-PONs. Integrated array transceivers will help to reduce the number of individual transceivers need to be managed and terminated in the CO and significantly simplify WDM-PON operations.

Second, structures of optical transceivers will become more and more complex. NG-PON2 requires the integration of tunable lasers and tunable receivers (or array receivers). IEEE802.3ca 50 Gbps PON requires integration of two 25 Gbps transceivers at two different wavelengths. For backward compatibility reasons, some implementations of 10 Gbps PON systems require transceivers that implement both the previous generation Gigabit PON transceiver and current generation 10 Gbps PON transceivers in the same transceiver package, in the so-called Combo-PON solution [94,95]. Such requirements have been driven by carriers like China Telecom.

Coherent PON systems will require vector modulators and coherent demodulators. Both are complex optical setups that will benefit from integration.

Integrated photonic systems will miniaturize optical components and improve the cost in the long run. Nevertheless, unlike digital electronics, photonic systems are mostly analog components, which poses significant challenges to integration and the yield is often a challenge for moderate-size photonic integrated circuit with a handful of components. Research in photonic integration will eventually benefit access networks, which is a large-volume market.

19.5 Conclusions

In this chapter, we reviewed the latest developments of optical access networks and technologies. Gigabit capable PON technologies will continue to satisfy the bandwidth demand for residential FTTH applications in the next couple years. 10 Gbps PON technologies such as the IEEE 802.3av 10GE-PON and ITU-T XGS-PONs are emerging next-generation commercial PON systems. Standard organizations are now working on PONs with speeds of 25 Gbps, 50 Gbps and beyond. In addition, DWDM technologies are beginning to emerge in optical access networks. WDM helps to scale networks both in speed and coverage. Pure WDM-PON networks offer speed and protocol transparencies and are promising candidates for future wireless fronthaul applications. Instead of FTTH, wireless fronthaul/backhaul and enterprise networks are driving the development of next-generation fiber access networks. TWDM networks helps to scale access networks not only in speed but also in the number of users and network coverage. In particular, super-PONs enable large-scale passive footprint and have the potential of easing network construction and management and improving the access network economy.

Access networks are very cost sensitive. As speed of access networks increases, transmission becomes exponentially harder. Many new techniques are required to enable the implementation of next-generation access networks. First, low-cost tunable lasers is a key enabler for WDM-based PON systems. Such lasers are enabled by novel low-cost packaging techniques. Second, electronic signal processing techniques will be playing more significantly roles in modulation, reducing transmission impairments and allowing lower grade and less expensive optical components (which account for the bulk of the optical access network costs) to be used in implementing next-generation optical access networks. Last, photonic integration technologies will not only help to reduce the cost by miniaturizing component sizes but enable the construction of array transceivers required for WDM-PON systems and vector modulators and coherent receivers required in coherent PONs.

References

[1] C.F. Lam, Passive Optical Networks: Principles and Practice., Elsevier, 2011.

[2] IEEE 802.3ca 50G-EPON Task Force. [Online]. Available from: <http://www.ieee802.org/3/ca/index.shtml>.

[3] C. DeSanti, L. Du, C. Lam, J. Jiang, Super-PON: scale fully passive optical access networks to longer reaches and to a significantly higher number of subscribers, presented at the IEEE 802.3 NEA Meeting, Geneva, CH.

[4] D. Nesset, The PON roadmap, in: Optical Fiber Communication Conference, 2016.

[5] ITU-T, '40-gigabit-capable passive optical network (NG PON2)' ITU-T G.998.x series of recommendations, G.989.

[6] ITU-T, 10-Gigabit-Capable Symmetric Passive Optical Network (XGS-PON). ITU-T Recommendation G.9807.1, G.9807.1, Jun. 2016.

[7] D. Nesset, NG-PON2 technology and standards, J. Lightw. Technol. 33 (5) (2015) 1136–1143.

[8] J.S. Wey, et al., Physical layer aspects of NG-PON2 standards—Part 1: optical link design [invited], IEEE/OSA J. Opt. Commun. Netw. 8 (1) (2015) 33.

[9] ITU-T, 40-Gigabit-capable passive optical networks 2 (NG-PON2): physical media dependent (PMD) layer specification, G.989.2, 2014, G.989.2, 2014.

[10] Y. Luo, et al., Physical layer aspects of NG-PON2 standards—Part 2: system design and technology feasibility [invited], IEEE/OSA J. Opt. Commun. Netw. 8 (1) (2015) 43.

[11] D. Van Veen, W. Pohlmann, B. Farah, T. Pfeiffer, P. Vetter, Measurement and mitigation of wavelength drift due to self-heating of tunable burst-mode DML for TWDM-PON, in: Conference on Optical Fiber Communication, Technical Digest Series, 2014, pp. 1–3.

[12] L.B. Du, et al., Long-reach wavelength-routed TWDM PON: technology and deployment, J. Lightw. Technol. (2018). 1–1.

[13] ITU-T, PON transmission technologies above 10 Gb/s per wavelength, Feb. 2018.

[14] G. Kramer, M. Hajduczenia, D. Remein, E. Harstead, B. Powell, C. Knittle, NG-EPON call for interest, IEEE P802.3ca 50G-EPON Task Force, 2015. [Online]. Available from: <http://www.ieee802.org/3/cfi/0715_1/CFI_01_0715.pdf>.

[15] E. Harstead, R. Sharpe, Forecasting of access network bandwidth demands for aggregated subscribers using Monte Carlo methods, IEEE Commun. Mag. 53 (3) (2015) 199–207.

[16] W. Moench, D. Clague, Challenges in next-gen PON deployment - lightwave. [Online]. Available from: <https://www.lightwaveonline.com/articles/2017/06/challenges-in-next-gen-pon-deployment.html>.

[17] R. Bonk, et al., The underestimated challenges of burst-mode WDM transmission in TWDM-PON, Opt. Fiber Technol. 26 (2015) 59–70.

[18] CWDM4-MSA Group. [Online]. Available from: <http://www.cwdm4-msa.org/>.

[19] 4WDM MSA. [Online]. Available from: <http://4wdm-msa.org/>.

[20] 100G PSM4 MSA. [Online]. Available from: <http://psm4.org/>.

[21] IEEE P802.3ca Timeline, IEEE P802.3ca 50G-EPON Task Force, 2017. [Online]. Available from: <http://www.ieee802.org/3/ca/documents/P802_3ca_timeline.pdf>.

[22] V. Houtsma, D. van Veen, E. Harstead, Recent progress on standardization of next-generation 25, 50, and 100G EPON, J. Lightw. Technol. 35 (6) (2017) 1228–1234.

[23] NG-EPON Objectives, IEEE P802.3ca 50G-EPON Task Force, 2015. [Online]. Available from: <http://www.ieee802.org/3/NGEPONSG/public/2015_11/1511_ngepon_objectives_changes.pdf>.

[24] E. Harstead, S.J. Johnson, 25G EPON PR30 downstream power budget, IEEE P802.3ca 50G-EPON Task Force, 2018. [Online]. Available from: <http://www.ieee802.org/3/ca/public/meeting_archive/2018/01/harstead_3ca_1c_0118.pdf>.

[25] IEEE P802.3ca 50G-EPON task force, approved minutes of the January 2018 meeting, IEEE P802.3ca 50G-EPON Task Force, 2018. [Online]. Available from: <http://www.ieee802.org/3/ca/public/meeting_archive/2018/01/agenda_3ca_1_0118.http://www.ieee802.org/3/ca/public/meeting_archive/2017/11/minutes_unapproved_3ca_1117.pdf>.

[26] G. Kramer, IEEE 802.3ca channel bonding and skew remediation, presented at the Joint IEEE 802 and ITU-T Study Group 15 workshop "Building Tomorrow's Networks," Geneva, Switzerland, 2018.

[27] C. Knittle, IEEE 802.3ca 25Gbps, 50Gbps, and 100Gbps passive optical networking, presented at the FSAN Workshop, Dallas, TX, 2018.

[28] W. Bo, J. Ming, Z. Xiaoxia, S. Yan, W. Lei, Z. Dechao, Proposal to study 50Gb/s TDM PON for higher speed PON and PON convergence, presented at the FSAN NGPON High Speed PON, Paris, France, 2017.
[29] ITU-T, G.652 Characteristics of a single-mode optical fibre and cable, ITU-T, 2016.
[30] ITU-T, G.657 Characteristics of a bending-loss insensitive single-mode optical fibre and cable, ITU-T, 2016.
[31] D.T. Van Veen, V.E. Houtsma, A.H. Gnauck, P. Iannone, Demonstration of 40-Gb/s TDM-PON over 42-km with 31 dB optical power budget using an APD-based receiver, J. Lightw. Technol. 33 (8) (2015) 1675–1680.
[32] L.D.T. Minghui, 50G single wavelength PON analysis and comparison, IEEE P802.3ca 50G-EPON Task Force, 2017. [Online]. Available from: <http://www.ieee802.org/3/ca/public/meeting_archive/2017/11/liu_3ca_2a_1117.pdf>.
[33] M. Nada, Y. Yamada, H. Matsuzaki, Responsivity-bandwidth limit of avalanche photodiodes: toward future ethernet systems, IEEE J. Sel. Top. Quantum Electron. 24 (2) (2018).
[34] J. Zhang, J.S. Wey, X. Huang, Experimental results of single wavelength 50G PON, IEEE P802.3ca 50G-EPON Task Force, 2017. [Online]. Available from: <http://www.ieee802.org/3/ca/public/meeting_archive/2017/11/zhang_junwen_3ca_1_1117.pdf>.
[35] V. Houtsma, D. Van Veen, Bi-directional 25G/50G TDM-PON with extended power budget using 25G APD and coherent detection, J. Lightw. Technol. 36 (1) (2018) 122–127.
[36] X. Yin, et al., An asymmetric high serial rate TDM-PON with single carrier 25 Gb/s upstream and 50 Gb/s downstream, J. Lightw. Technol. 34 (2) (2016) 819–825.
[37] M. Tao, L. Zhou, H. Zeng, S. Li, X. Liu, 50-Gb/s/λ TDM-PON based on 10G DML and 10G APD supporting PR10 link loss budget after 20-km downstream transmission in the O-band, OFC 2017 (2017) 1–3.
[38] H. Zeng, A. Shen, X. Liu, F. Effenberger, Flexible-rate 50G-PON 25 Gb/s date rates, IEEE P802.3ca 50G-EPON Task Force, Chicago, IL, 2018.
[39] K. Zhang et al., Demonstration of 50Gb/s/λ symmetric PAM4 TDM-PON with 10G-class optics and DSP-free ONUs in the O-band, in: OFC 2018, 2018, pp. 1–3.
[40] P.-L.L. Budget et al., Symmetrical 50-Gb/s/λ PAM-4 TDM-PON in O-band with DSP and semiconductor optical amplifier supporting, 1, c, pp. 4–6, 2018.
[41] J. Zhang et al., 64-Gb/s/A downstream transmission for PAM-4 TDM-PON with centralized DSP and 10G low-complexity receiver in C-band, in: European Conference on Optical Communication, ECOC, 2017-Septe, 1, 2017, pp. 1–3.
[42] J. Zhang et al., Experimental demonstration of unequally spaced PAM-4 Signal to improve receiver sensitivity for 50-Gbps PON with power-dependent noise distribution, presented at the Optical Fiber Communication Conference 2018, San Diego, California, 2018, pp. 1–3.
[43] B. Powell, K. Droskiewicz, Latency & complexity for various 25/50/100G EPON FEC code proposals, IEEE P802.3ca 50G-EPON Task Force, 2018. [Online]. Available from: <http://www.ieee802.org/3/ca/public/meeting_archive/2017/09/powell_3ca_2a_0917.pdf>.
[44] I.B. Djordjevic, On advanced FEC and coded modulation for ultra-high-speed optical transmission, IEEE Commun. Surv. Tutor. 18 (3) (2016) 1920–1951.
[45] P. Ossieur, G. Coudyzer, D. Kelly, X. Yin, P.D. Townsend, J. Bauwelinck, ASIC implementation challenges for next generation access networks, in: Advanced Photonics 2018 (BGPP, IPR, NP, NOMA, Sensors, Networks, SPPCom, SOF), 2018, p. SpTh4F.1.
[46] N. Suzuki, H. Miura, K. Matsuda, R. Matsumoto, K. Motoshima, 100 Gb/s to 1 Tb/s based coherent passive optical network technology, J. Lightw. Technol. 36 (8) (2018) 1485–1491.
[47] D. Nesset, N. Suzuki, L. Yi, Ultimate capacity limits for TDM/TDMA PON, in: OFC 2018 Workshops. [Online]. Available from: <https://www.ofcconference.org/en-us/home/about/ofc-blog/2018/march-2018/ofc-daily-wrap-sunday/>.
[48] V. Houtsma, M. Presi, N. Suzuki, D. Van Veen, DSP for next generation optical access, in: ECOC 2018 Workshops. [Online]. Available from: <https://www.ecoc2018.org/programme/workshops/>.
[49] X. Zhao et al., Field trial of long-reach TWDM PON for fixed-line wireless convergence, in: 2017 European Conference on Optical Communication (ECOC), 2017, pp. 1–3.

[50] A.M. Hill, et al., 39.5 million-way WDM broadcast network employing two stages of erbium-doped fibre amplifiers, Electron. Lett. 26 (22) (1990) 1882.
[51] D.J.G. Mestdagh, C.D. Martin, The super-PON concept and its technical challenges, in: Broadband Communications - Global Infrastructure for the Information Age Proceedings of the Internatio, 1996, pp. 333–345.
[52] D.B. Payne, R.P. Davey, The future of fibre access systems? BT Technol. J. 20 (4) (2002) 104–114.
[53] ITU-T, G.984.3: Gigabit-capable passive optical networks (G-PON): Transmission convergence layer specification, 2014.
[54] C. Desanti, Ethernet access PMDs for Central Office Consolidation Call for Interest, IEEE 802.3 Physical Layers for increased-reach Ethernet optical subscriber access (Super-PON) Study Group, 2018. [Online]. Available from: <http://www.ieee802.org/3/cfi/0718_1/CFI_01_0718.pdf>.
[55] R. Urata, C. Lam, H. Liu, C. Johnson, High performance, low cost, colorless ONU for WDM-PON, in: National Fiber Optic Engineers Conference, 2012.
[56] C.F. Lam, FTTH look ahead—technologies & architectures, in: 36th European Conference and Exhibition on Optical Communication, 2010.
[57] L.A. Coldren, G.A. Fish, Y. Akulova, J.S. Barton, L. Johansson, C.W. Coldren, Tunable semiconductor lasers: a tutorial, J. Lightw. Technol. 22 (1) (2004) 193–202.
[58] I.A. Alimi, A.L. Teixeira, P.P. Monteiro, Toward an efficient C-RAN optical fronthaul for the future networks: a tutorial on technologies, requirements, challenges, and solutions, IEEE Commun. Surv. Tutor. 20 (1) (2018) 708–769.
[59] I. Chih-Lin, Quest for 5G - rethink fundamentals, IEEE 5G Summit (2015).
[60] CPRI, Common public radio interface.
[61] ITU-T, ITU-T G.698.4 multichannel bi-directional DWDM applications with port agnostic single-channel optical interfaces, G.698.4, Mar. 2018.
[62] Y. Matsui et al., Transceiver for NG-PON2: wavelength tunablity for burst mode TWDM and point-to-point WDM, in: Optical Fiber Communication Conference, 2016.
[63] Y. Matsui, L. Lin, D. Chen, T. Sudo, Wavelength-tunable DBR laser for burst mode TWDM PON applications, in: IEEE 802.3 Physical Layers for Increased-Reach Ethernet Optical Subscriber Access (Super-PON) Study Group September 2018 Meeting Materials.
[64] W. Poehlmann, D. van Veen, R. Farah, T. Pfeiffer, P. Vetter, Wavelength drift of burst-mode DML for TWDM-PON [invited], IEEE/OSA J. Opt. Commun. Netw. 7 (1) (2014) A44.
[65] S.J. Savory, Digital filters for coherent optical receivers, Opt. Express 16 (2) (2008) 804.
[66] P. Bayvel, C. Behrens, D.S. Millar, Digital signal processing (DSP) and its application in optical communication systems, Optical Fiber Telecommunications VIB: Systems and Networks: Sixth Edition, sixth ed., Elsevier Inc, 2013, pp. 163–219.
[67] C. Fludger, Digital signal processing in optical communications from long-haul to data-centre, in: Optical Fiber Communication Conference, 2016, p. W3G.4.
[68] M. Salehi, J. Proakis, Digital Communications, McGraw-Hill Education, 2007.
[69] S. Yin, D. Van Veen, V. Houtsma, P. Vetter, Investigation of symmetrical optical amplified 40 Gbps PAM-4/duobinary TDM-PON using 10G Optics and DSP, in: Optical Fiber Communication Conference, 2016.
[70] S. Yin, V. Houtsma, D. van Veen, P. Vetter, Optical amplified 40-Gbps symmetrical TDM-PON using 10-Gbps optics and DSP, J. Lightw. Technol. 35 (4) (2017) 1067–1074.
[71] M. Tao et al., 28-Gb/s/λ TDM-PON with narrow filter compensation and enhanced FEC supporting 31.5 dB link loss budget after 20-km downstream transmission in the C-band, in: Optical Fiber Communication Conference, 2016.
[72] X. Tang, et al., 40-Gb/s PAM4 with low-complexity equalizers for next-generation PON systems, Opt. Fiber Technol. 40 (2018) 108–113.
[73] C. Qin, V. Houtsma, D. Van Veen, J. Lee, H. Chow, P. Vetter, 40 Gbps PON with 23 dB power budget using 10 Gbps optics and DMT, in: 2017 Optical Fiber Communications Conference and Exhibition, OFC 2017 - Proceedings, vol. 1, 2017, pp. 23–25.

[74] N. Cvijetic, D. Qian, J. Hu, T. Wang, Orthogonal frequency division multiple access PON (OFDMA-PON) for colorless upstream transmission beyond 10 Gb/s, IEEE J. Sel. Areas Commun. 28 (6) (2010) 781−790.
[75] D. Qian, N. Cvijetic, J. Hu, T. Wang, Optical OFDM transmission in metro/access networks, in: Optical Fiber Communication Conference and National Fiber Optic Engineers Conference, 2009.
[76] F. Effenberger, Enhanced FEC consideration for 100G EPON, IEEE P802.3ca 50G-EPON Task Force, 2016. [Online]. Available from: <http://www.ieee802.org/3/ca/public/meeting_archive/2016/03/effenberger_3ca_2_0316.pdf>.
[77] T. Mizuochi, Y. Miyata, K. Kubo, T. Sugihara, K. Onohara, H. Yoshida, Progress in soft-decision FEC, in: Optical Fiber Communication Conference/National Fiber Optic Engineers Conference 2011, 2011.
[78] T. Mizuochi, Next generation FEC for optical communication, in: OFC/NFOEC 2008 - 2008 Conference on Optical Fiber Communication/National Fiber Optic Engineers Conference, 2008.
[79] T. Okoshi, Recent progress in heterodyne/coherent optical-fiber communications, J. Lightw. Technol. 2 (4) (1984) 341−346.
[80] K. Kikuchi, 3 - Coherent optical communication systems, in: I.P. Kaminow, T. Li, A.E. Willner (Eds.), Optical Fiber Telecommunications V B, fifth ed., Academic Press, Burlington, 2008, pp. 95−129.
[81] P.M. Becker, A.A. Olsson, J.R. Simpson, Erbium-Doped Fiber Amplifiers: Fundamentals and Technology., Elsevier, 1999.
[82] X. Liu, S. Chandrasekhar, A. Leven, Self-coherent optical transport systems, Optical Fiber Telecommunications V B (2008) 131−177.
[83] E. Ip, A.P.T. Lau, D.J.F. Barros, J.M. Kahn, Coherent detection in optical fiber systems, Opt. Express 16 (2) (2008) 753−791.
[84] K. Roberts, D. Beckett, D. Boertjes, J. Berthold, C. Laperle, 100G and beyond with digital coherent signal processing, IEEE Commun. Mag. 48 (7) (2010) 62−69.
[85] M.S. Erkilinc, et al., Comparison of low complexity coherent receivers for UDWDM-PONs (λ-to-the-user), J. Lightw. Technol. (2018). 1−1.
[86] D. Lavery, et al., Opportunities for optical access network transceivers beyond OOK [invited], IEEE/OSA, J. Opt. Commun. Netw. 11 (2) (2019) A186−A195.
[87] J.A. Altabas, et al., Real-Time 10 Gbps polarization independent quasicoherent receiver for NG-PON2 access networks, J. Lightw. Technol. 37 (2) (2019) 651−656.
[88] E. Ciaramella, Polarization-independent receivers for low-cost coherent OOK systems, IEEE Photon. Technol. Lett. 26 (6) (2014) 548−551.
[89] J. Tabares, V. Polo, J. Prat, "Polarization-independent heterodyne DPSK receiver based on 3x3 coupler for cost-effective udWDM-PON, in: Optical Fiber Communication Conference, 2017.
[90] I.N. Cano, A. Lerin, V. Polo, J. Prat, Flexible D(Q)PSK 1.25−5 Gb/s UDWDM-PON with directly modulated DFBs and centralized polarization scrambling, in: 2015 European Conference on Optical Communication (ECOC), 2015.
[91] M.S. Erkilinc, et al., Polarization-insensitive single-balanced photodiode coherent receiver for long-reach WDM-PONs, J. Lightw. Technol. 34 (8) (2016) 2034−2041.
[92] M.S. Erkilinc et al., Polarization-insensitive single balanced photodiode coherent receiver for passive optical networks, in: 2015 European Conference on Optical Communication (ECOC), 2015.
[93] M.S. Erkılınç, et al., Bidirectional wavelength-division multiplexing transmission over installed fibre using a simplified optical coherent access transceiver, Nat. Commun. 8 (1) (Oct. 2017) 1043.
[94] Combo PON Solution, ZTE website. [Online]. Available from: <https://www.zte.com.cn/global/solutions/access/fixed/443954>.
[95] Huawei PON Combo Solutions Enable Smooth Evolution to 10G PON, Lightreading, 20-Sep-2016. [Online]. Available from: <https://www.lightreading.com/ubiquitous-ultra-broadband/huawei-pon-combo-solutions-enable-smooth-evolution-to-10g-pon/d/d-id/726240>.

CHAPTER 20

Information capacity of optical channels

Marco Secondini
Institute of Communication, Information, and Perception Technologies, Scuola Superiore Sant'Anna, Pisa, Italy

20.1 Introduction

The era in which optical fiber systems employed simple transmission techniques such as on-off keying and direct detection, and could waste the huge fiber bandwidth is coming to an end. The current generation of optical fiber systems is designed with the goal of efficiently exploiting the available bandwidth, trying to operate as close as possible to the ultimate limit posed by information theory—the *channel capacity*. To this end, coherent detection and advanced digital signal processing (DSP) are employed to implement the best available coding, modulation, and detection strategies, and to counteract the impairments caused by fiber propagation.

In such circumstances, it sounds perhaps paradoxical to say that the problem of evaluating the capacity of the optical fiber channel is still essentially open [1]. And it is definitely disappointing to start the chapter on the information capacity of optical channels with this very statement. However, despite many years of research on this problem, this is the "elephant in the room" that is largely ignored by the optical community but that we have to face.

Indeed, we are not completely ignorant. For instance, we know that, at low power, when the effect of fiber nonlinearity is negligible, the capacity of the fiber equals that of the additive white Gaussian noise (AWGN) channel that was originally studied by Shannon when he laid the foundations of the information theory [2]. And, at low power, the current generation of optical systems is able to operate quite close to this limit. On the other hand, it seems almost impossible to increase the optical power beyond a certain level without dramatically affecting the performance of the system.

In principle, if the fiber were an AWGN channel, its capacity would grow indefinitely with the available signal power. In this case, to enable a prescribed transmission rate over the channel, it would be sufficient (though possibly

Optical Fiber Telecommunications V11
DOI: https://doi.org/10.1016/B978-0-12-816502-7.00023-3

inefficient) to increase the optical launch power. Unfortunately, if the optical power is increased beyond a certain level, the impact of fiber nonlinearity becomes relevant, so that the system performance begins to deteriorate. In practice all the existing systems operate at an optical power where nonlinear effects have only a moderate impact on performance or are even negligible. The implicit assumption is that, regardless of the available optical power, there is a finite capacity limit that cannot be overcome. In the optical communications community this limit is known as the *nonlinear Shannon limit*, despite the fact that there is no proof that it is a true Shannon limit. Yet it is a practical limitation that plays a fundamental role in the design of optical fiber systems.

From an information-theoretic viewpoint, our level of knowledge is clearly insufficient. But it is rather inconvenient also from a practical viewpoint. In fact, it would be quite disappointing to invest huge resources in the deployment of additional fiber cables or in the replacement of the existing ones only to find out later that it would have been possible, at a much lower cost, to meet the same demand of data traffic by simply improving the transmission techniques. This is one of the most important reasons that keep the research in this field alive—another one being, probably, human curiosity.

This chapter provides an overview of the main results available for the information theoretical analysis of the optical fiber channel. It does not have the ambition to provide a solution to the capacity problem, which is still essentially open, but rather to discuss some relevant capacity bounds that at present can be regarded as practical limitations to the transmission of information through optical fibers. Some simple approaches to compute these bounds will be described. In addition, the theoretical limits of this analysis and the open possibility that the ultimate channel capacity might lay well beyond these lower bounds will be discussed.

The chapter is organized as follows: Section 2 provides an introduction to the notion of channel capacity and to some basic information-theoretic concepts; Section 3 describes the optical fiber channel and some useful channel models; Section 4 studies the capacity of the optical fiber channel, presenting both fundamental bounds and more accurate but complex bounding techniques, and considering different scenarios and link configurations; finally, Section 5 discusses future perspectives and open problems.

Notation: a random variable is denoted by an uppercase letter, e.g., X, its expectation by $\mathrm{E}\{X\}$, and its realization by the corresponding lowercase letter, x. The probability density function (or *distribution*, in short) of the random variable X is simply denoted by $p(x)$, with the argument implicitly defining the specific distribution, so that $p(x)$ and $p(y)$ denote the different distributions of the variables X and Y. A vector is denoted by a boldface letter and one of its elements by the corresponding light letter with a subscript, e.g., $\mathbf{x} = (x_1, x_2, \ldots, x_N)$. Integrals, if not otherwise specified by explicit limits, extend to the whole space in which the integration variable is defined. The conjugate, transpose, and conjugate (Hermitian) transpose are denoted by $(\cdot)^T$, $(\cdot)^*$, and $(\cdot)^H$, respectively.

20.2 Information theory

One of the key results of Claude Shannon's pioneering work is the demonstration that a reliable communication (with arbitrarily low error probability) over a noisy channel is possible, provided that the information rate is less than a characteristic quantity determined by the statistical properties of the channel, which he named *channel capacity*. Conversely, he also showed that this is not possible when this quantity is exceeded. This important result is known as the *noisy-channel coding theorem* [2]. Beyond establishing such a general result and laying the foundations of what is now called *information theory*, Shannon also derived a specific closed-form expression for the capacity of an *additive white Gaussian noise* (AWGN) channel. As we shall see, this expression is widely employed also in the context of optical fibers, though its validity and interpretation in this case need to be carefully discussed. Shannon's work has been since extended and generalized both to account for a broader class of channels and to obtain closed-form capacity expressions for specific channels other than the AWGN one (see Ref. [3a] and references therein).

This chapter deals with the capacity of the optical fiber channel. The composition and characteristics of this kind of channel will be specifically discussed in Section 20.3. In this section, on the other hand, we are mainly interested in the role that this channel plays in a typical digital optical fiber communication system, which can be conveniently represented by the model depicted in Fig. 20.1. A sequence of information bits (the message to be transmitted) is processed by a *forward error correction* (FEC) encoder—which adds some redundant bits (e.g., parity check bits) to be used for error correction—and mapped to a sequence of symbols $\mathbf{x} = (x_1, \ldots, x_K)$ by a digital modulator. Symbols $\mathbf{x}$ are then encoded in a *waveform* $x(t)$ according to some rule,

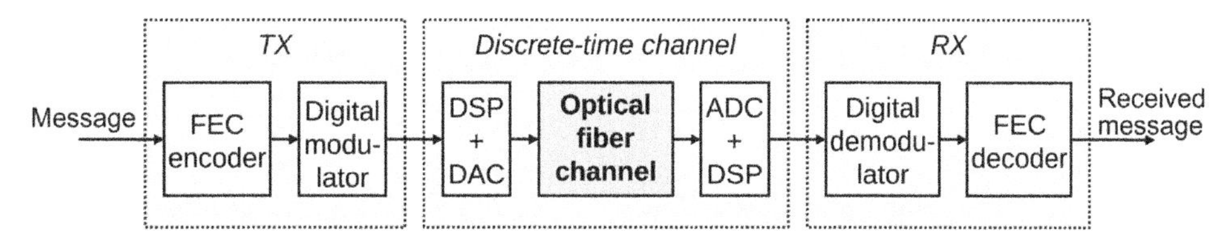

Figure 20.1 Schematic representation of a digital optical fiber communication system, highlighting the main relevant elements needed for an information theoretic analysis. Source: *Adapted from E. Agrell, M. Secondini, Information-Theoretic Tools for Optical Communications Engineers, in: IEEE Photonics Conference (IPC), Reston, VA, USA, 2018 [3b].*

practically implemented by DSP and a digital-to-analog converter (DAC). The input waveform $x(t)$ is coupled to the optical fiber channel, through which it propagates from the transmitter to the receiver. The received output waveform $y(t)$, affected by propagation effects and corrupted by amplifier noise, is filtered, sampled, and processed by an analog-to-digital converter (ADC) and DSP block to obtain the output samples $\mathbf{y} = (y_1, \ldots, y_K)$. The output samples are finally demodulated and decoded to detect the transmitted symbols, correct for possible errors, and eventually extract the message.

From an information-theoretical perspective, the system can be first studied at a higher level by considering the three light blue blocks highlighted in Fig. 20.1: the *discrete-time channel*, comprising the DSP + DAC block, the waveform optical fiber channel, and the ADC + DSP block; the *transmitter*, comprising the FEC encoder and the digital modulator; and the *receiver*, consisting of the digital demodulator and the FEC decoder.[1]

The discrete-time channel is fully characterized by three elements: an input *modulation alphabet*, denoting the set of possible values taken by the input symbols $\mathbf{x}$; an output modulation alphabet, denoting the set of possible values taken by the output samples $\mathbf{y}$; and the channel law, which provides the probability to get some output $\mathbf{y}$ given some input $\mathbf{x}$. The most general case is the one in which no constraints are posed on $\mathbf{x}$ and $\mathbf{y}$, whose elements can take any possible values in the complex plane. This situation entails an infinite resolution of the DAC and ADC and is hence a useful idealization when exploring the ultimate achievable performance. In this case, the channel law is specified by the conditional distribution $p(\mathbf{y}\,|\,\mathbf{x})$, with $\mathbf{x},\mathbf{y} \in \mathbb{C}^K$.

The channel capacity, as defined at the beginning of this section, corresponds to the maximum rate at which information can be transferred through the discrete-time channel, considering *any* possible transmitter and receiver. It is therefore a property of the discrete-time channel alone and is independent of the specific transmitter or receiver implementation. Nonetheless, according to the adopted definition, it depends on the DSP + DAC and ADC + DSP blocks that are used to connect the digital transmitter and receiver to the waveform channel. An interesting question is, hence, if also the optical waveform channel can be characterized by a sort of "capacity"—an intrinsic limit of the waveform channel that is independent of the specific implementation of the DSP + DAC and ADC + DSP blocks. As we shall see, an acceptable answer can be given by posing some reasonable constraints on the system implementation.

The rest of this section is devoted to introduce some useful information theoretical concepts and to briefly review some fundamental results for particular types of

[1] In this representation, the DAC and ADC, as well as some DSP operations aimed at simplifying the channel model, are formally included in the discrete-time channel, though they are usually physically located in the transmitter or in the receiver.

channels. For a more rigorous and in-depth introduction to the problem, we refer to some classical textbooks on information theory [4,5].

20.2.1 Discrete-time memoryless channels

The first kind of channel that we consider is the *discrete-time memoryless channel*. This is a channel whose input and output are sequences of symbols (belonging to discrete or continuous alphabets), such that each output symbol y_k at a given discrete time k depends statistically only on the corresponding input symbol x_k. In this case, the channel can be simply represented as in Fig. 20.2, omitting the time index k for simplicity of notation and considering a single input X, the corresponding output Y, and the conditional distribution $p(Y|X)$.

For a given distribution $p(X)$ of the input variable, the maximum amount of information that can be reliably transmitted over the channel is given by the *mutual information* between the input and output variables

$$I(X;Y) = E\left\{\log_2\frac{p(Y|X)}{p(Y)}\right\} = \iint p(x)p(y|x)\log_2\frac{p(y|x)}{p(y)}dxdy \tag{20.1}$$

where

$$p(y) = \int p(y|x)p(x)dx \tag{20.2}$$

is the output distribution.

The *capacity* of this channel is the supremum of the mutual information (20.1) over all possible input distributions

$$C = \sup_{p(X)} I(X;Y) \tag{20.3}$$

In general, the input and output variables can be real or complex, and can be defined in a monodimensional or multidimensional space, so that the integrals in (20.1) and (20.2) extend to the corresponding spaces. For discrete input and/or output modulation alphabets, the corresponding distributions are replaced by probabilities, and the integrals in (20.1) and (20.2) by summations.

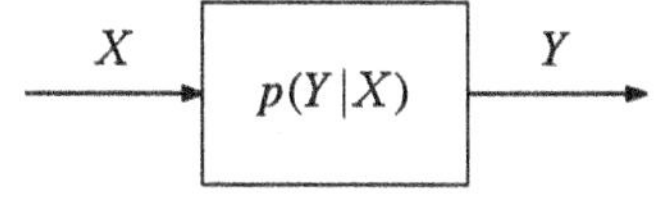

Figure 20.2 Discrete-time memoryless channel.

An efficient numerical computation of the capacity for the case of *finite* input and output alphabets can be obtained by the *Blahut–Arimoto algorithm* [6,7], an iterative algorithm that converges to the solution of the optimization problem (20.3).

In general, only the capacity of some specific channels can be evaluated analytically, the *binary symmetric channel* and the *AWGN channel* being perhaps the two most notable examples.

20.2.1.1 The binary symmetric channel

The binary symmetric channel is illustrated in Fig. 20.3A. The channel has a binary input modulation alphabet, whose elements are conventionally denoted as "0" and "1" and are drawn with probabilities q and $1—q$, respectively. The channel is characterized by an error probability p—that is, the probability that a "0" is received as a "1" or vice versa.

The capacity of this channel can be easily determined by direct evaluation of the discrete-alphabet version of (20.1) and (20.3). The binary nature of the input alphabet restrict the optimization in (20.3) to the single real parameter $0 \leq q \leq 1$. The resulting capacity—reported in Fig. 20.3B as a function of the error probability p—is given by Ref. [5]

$$C = 1 + p \log_2 p + (1-p) \log_2 (1-p) \tag{20.4}$$

and is achieved for $q = 0.5$ (equiprobable input symbols).

For $q = 0.5$, up to $\log_2 2 = 1$ information bit can be encoded on each symbol, corresponding to the *entropy* $H(X)$ of the input alphabet [5]. For either $p = 0$ or $p = 1$, the channel capacity is $C = 1$ bit per symbol. In fact, in this case, since the channel has a deterministic behavior—it is essentially noiseless—all the information can be reliably transmitted (for $p = 1$, errors are systematic and can be easily corrected). On the other hand, for $p = 1/2$, the output symbols become independent of the input ones and are

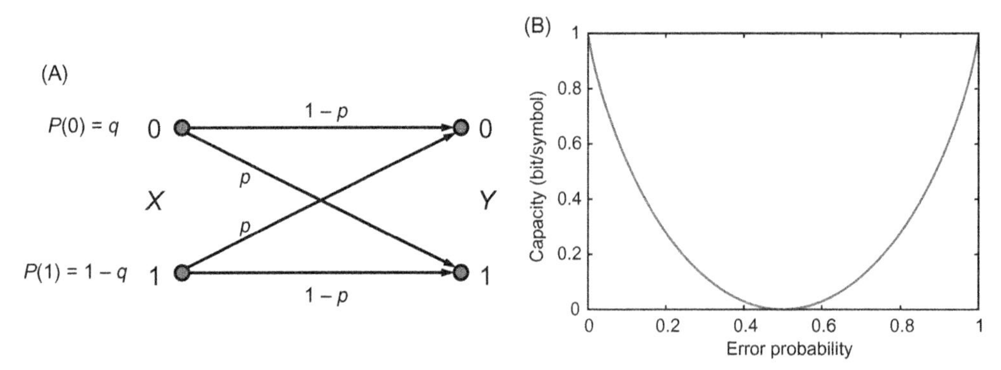

Figure 20.3 (A) The binary symmetric channel and (B) its capacity.

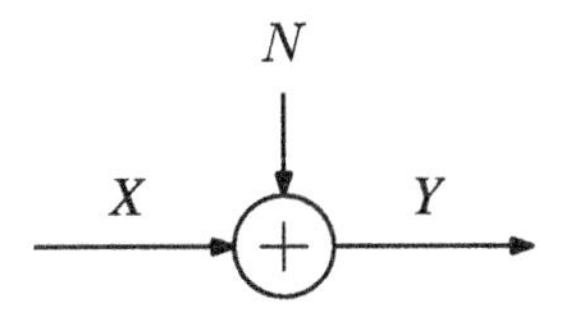

Figure 20.4 The AWGN channel.

determined only by the noisy behavior of the channel. In this case, the channel capacity vanishes and no reliable transmission is possible. In practical cases—that is, for $0 < p < 0.5$ (or, equivalently, $0.5 < p < 1$) —the channel capacity takes some intermediate values in the range $0 < C < 1$. This means that by using a proper error correction code, it is possible to encode and reliably transmit k information bits over n transmitted symbols, provided that the code rate $R = k/n$ is less than the capacity C of the channel. This implies that the remaining $n-k$ symbols are redundant and are employed for error correction.

20.2.1.2 The additive white Gaussian noise channel

A discrete-time AWGN channel is characterized by the input output relation[2]

$$y_k = x_k + n_k \tag{20.5}$$

where the noise samples n_k are realizations of i.i.d. variables with a zero-mean proper complex Gaussian distribution [8]—often referred to as *circularly symmetric complex Gaussian* (CSCG)—with variance $E\{|N_k|^2\} = \sigma_n^2$. Given the absence of memory, the channel can be studied by considering a single pair of input and output variables, as done in (20.1)–(20.3). For a generic time index k, dropped for simplicity of notation, the channel is schematically represented in Fig. 20.4 and is characterized by the conditional distribution

$$p(y|x) = \frac{1}{\pi\sigma_n^2}\exp\left(-\frac{|y-x|^2}{\sigma_n^2}\right) \tag{20.6}$$

In many practical cases, the input symbol X (on which the information bits are encoded) is drawn from a set of M possible values $\chi = \{X_1, X_2, \ldots, X_M\}$ with uniform probability $1/M$. The elements of the set constitutes the M-ary modulation alphabet and, when represented in the complex plane, denote the so-called *input constellation*. Common modulation alphabets are the M-ary *phase-shift keying* (M-PSK) and the M-ary *quadrature amplitude modulation* (M-QAM) [9]. In these cases, a maximum of

[2] A channel characterized by the input-output relation $y_k = ax_k + b + n_k$, with a and b complex constants (known at the receiver), will also be referred to as an AWGN channel, as it can be easily reduced to it by considering the transformed output $y'_k = (y_k - b)/a$.

$\log_2 M$ information bits—corresponding to the *entropy* $H(X)$ of the input constellation [5]—can be encoded on each symbol.

In the special case of $M = 4$, the PSK and QAM alphabets are identical up to a phase rotation of $\pi/4$ of the constellations, and hence equivalent in terms of mutual information over the AWGN channel. The 4-PSK (equivalent to the 4-QAM) modulation is usually denoted as *quadrature phase-shift keying* (QPSK).

For a generic M-ary modulation alphabet with uniform probability, the mutual information (20.1) can be expressed as

$$I(X;Y) = \log_2 M + \frac{1}{M}\sum_{m=1}^{M}\int p(y|X_m)\log_2\frac{p(y|X_m)}{\sum_{\ell=1}^{M} p(y|X_\ell)}dy \tag{20.7}$$

Eq. (20.7) is still not computable in closed form but can be numerically computed, for instance, by a simple Monte Carlo estimation

$$I(X;Y) \approx \log_2 M + \frac{1}{K}\sum_{k=1}^{K}\log_2\frac{p(y_x|x_k)}{\sum_{\ell=1}^{M} p(y_k|X_\ell)} \tag{20.8}$$

where $x_1,\ldots, x_K$ and $n_1,\ldots, n_K$ are K independent input and noise realizations; $y_k = x_k + n_k$ for $k = 1,\ldots, K$ are the corresponding output realizations; and $p(y|x)$ is given by (20.6). Alternatively, a more accurate computation can be obtained by resorting to Gauss-Hermite quadrature [10].

A remarkable result is that for the AWGN channel the capacity (20.3) can be computed analytically. In particular, for an *average power constraint* $E\{|X|^2\} = P$, the capacity is achieved for input symbols with a CSCG distribution and increases logarithmically with the *signal-to-noise ratio* (SNR) P/σ_n^2 according to the well-known expression [2]

$$C = \log_2\left(1 + \frac{P}{\sigma_n^2}\right) \tag{20.9}$$

Sometimes it is useful to express (20.9) in terms of quantities that can be directly measured from the input and output variables

$$C = \log_2\frac{\sigma_x^2\sigma_y^2}{\sigma_x^2\sigma_y^2 - |\sigma_{xy}|^2} = \log_2\frac{1}{1 - |\rho_{xy}|^2} \tag{20.10}$$

where $\sigma_x^2 = P$ is the input variance, σ_y^2 is the output variance, σ_{xy} is the input–output covariance, and $\rho_{xy} = \sigma_{xy}/(\sigma_x\sigma_y)$ the input–output correlation coefficient.

Fig. 20.5 shows the mutual information (20.7) as a function of the SNR for the AWGN channel and different input constellations. The channel capacity, obtained for CSCG input symbols, is also reported. We note that, for each constellation, the

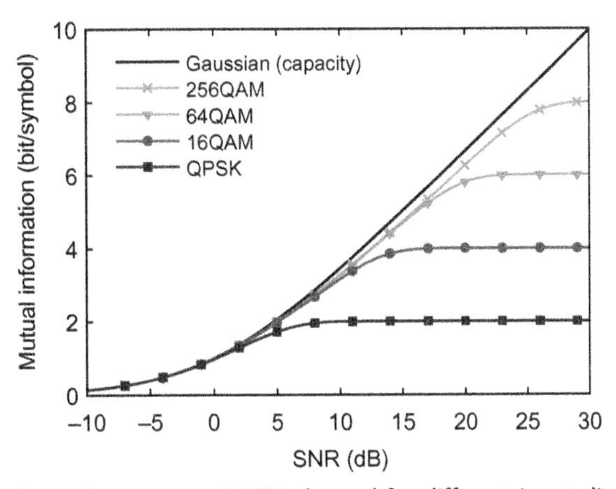

Figure 20.5 Mutual information over an AWGN channel for different input distributions.

mutual information increases with the SNR and saturates to the input entropy. In order to approach the channel capacity, one usually selects the lowest constellation size M that, for the given SNR, has not yet achieved the saturation level. Then, by means of a proper error correcting code and as already discussed in the case of the binary symmetric channel, it is possible to encode and reliably transmit k information bits over n symbols, provided that the code rate $R = k/n$ is less than the mutual information indicated in Fig. 20.5. In principle, a proper combination of code and modulation allows to operate reasonably close to the channel capacity. However, even employing an ideal error correcting code, there is still a gap between the information rate achievable with uniform QAM constellations and the channel capacity, as evident from Fig. 20.5. This gap can be practically closed by employing *constellation shaping* techniques, that is, by properly changing the distribution of the constellation points in the complex plane (*geometric* shaping) and/or their probabilities (*probabilistic* shaping). In particular, probabilistic shaping based on the concatenation of a distribution matcher and a fixed-rate systematic soft-decision forward-error-correction (FEC) code has recently attracted considerable attention in the optical fiber community as an efficient way to both close the gap to the Shannon capacity limit and realize transceivers with a fine rate adaptivity [11].

20.2.2 Discrete-time channels with memory

As opposed to memoryless channels, channels with memory are characterized by the statistical dependence of each output symbol y_k both on the corresponding input symbol x_k and on previous inputs and outputs. This property is generally represented in Fig. 20.6, where the conditional distribution $p(Y_1, Y_2, \ldots | X_1, X_2, \ldots)$ accounts for the dependence of *all* the output symbols on *all* the input ones.

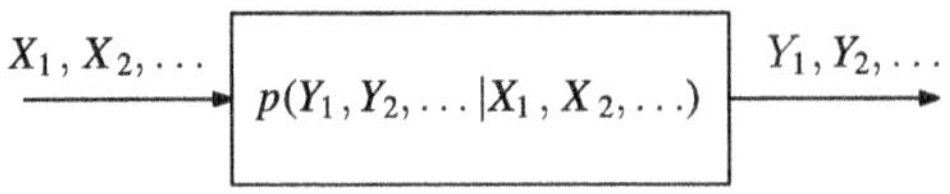

$X_1, X_2, \ldots$ → $p(Y_1, Y_2, \ldots | X_1, X_2, \ldots)$ → $Y_1, Y_2, \ldots$

Figure 20.6 Discrete-time channel with memory.

In the presence of memory, the mutual information (20.1) between a pair of input and output variables at time k—now explicitly denoted by $I(X_k; Y_k)$—corresponds to the information that can be obtained about the symbol X_k from the observation of the symbol Y_k. However, due to memory, some additional information might be gained, for instance, by observing also the symbols received at different times. In this case, the maximum bit rate (in bits per channel use, or bits per symbol) at which information can be transmitted using ideal error-correcting coding at an arbitrarily low error probability is given by the *information rate*[3]

$$I(X; Y) \triangleq \lim_{K \to \infty} \frac{1}{K} E\left\{ \log_2 \frac{p(\mathbf{Y}_K | \mathbf{X}_K)}{p(\mathbf{Y}_K)} \right\} \tag{20.11}$$

With respect to (20.1), the definition (20.11) involves some changes to account for the presence of channel memory. First of all, X and Y represent, respectively, the input and output discrete-time stochastic processes, of which $\mathbf{x}_K$ and $\mathbf{y}_K$ are length-K realizations. The channel is now characterized by a *sequence* of conditional distributions, $p(\mathbf{y}_K | \mathbf{x}_K)$ for $K = 1, 2, \ldots$. Analogously, the input and output processes are characterized by the selected input distribution $p(\mathbf{x}_K)$ and the corresponding output distribution $p(\mathbf{y}_K) = \int p(\mathbf{y}_K | \mathbf{x}_K) p(\mathbf{x}_K) d\mathbf{x}_K$, respectively, for $K = 1, 2, \ldots$. The information rate (20.11) represents the average mutual information per symbol between infinitely long input and output sequences.

The channel capacity C is obtained again by replacing the expectation in (20.11) with the supremum of the same expectation over all input distributions $p(\mathbf{x}_K)$ satisfying a given constraint (usually on the average input power). It is straightforward to verify that, for memoryless channels and i.i.d. inputs, all the distributions in (20.11) factorize into the product of the marginals, the argument of the limit becomes independent of K, and (20.11) reduces to (20.1).

Often, the analytical evaluation of (20.11) is unfeasible. Nevertheless, in many cases—namely, for finite-state sources and channels—an accurate numerical estimate of (20.11) can be efficiently obtained by relying on the *asymptotic equipartition property* [5] and following the procedure described in Ref. [12]:

1. Draw a long input sequence $\mathbf{x}_K = (x_1, \ldots, x_K)$ of samples from the selected input distribution $p(\mathbf{x}_K)$.

[3] We use the same symbol I to denote the mutual information between random variables and the information rate between stochastic processes.

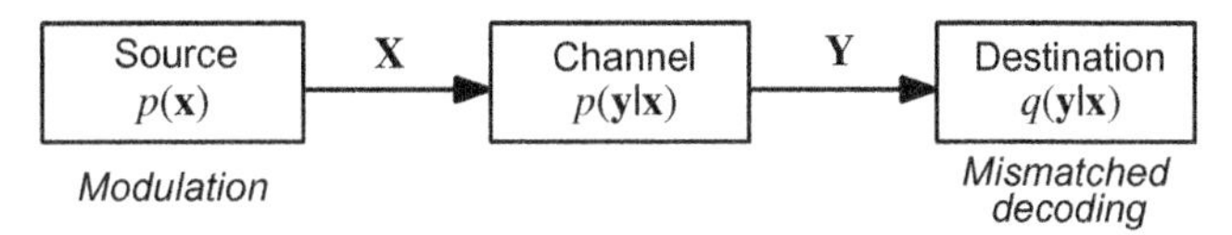

Figure 20.7 Discrete-time channel with mismatched decoding. Source: *Adapted from M. Secondini, E. Forestieri, Scope and limitations of the nonlinear Shannon limit, J. Lightw.Technol. 35 (4) (2017) 893–902.*

2. Compute numerically (or generate experimentally) the corresponding output sequence $\mathbf{y}_K = (x_1, \ldots, x_K)$.
3. Compute $p(\mathbf{x}_K)$, $p(\mathbf{y}_K | \mathbf{x}_K)$, and $p(\mathbf{y}_K)$ by running the *sum-product algorithm* on a suitably defined *factor graph* [13]. This step is equivalent to the forward recursion of the Bahl–Cocke–Jelinek–Raviv (BCJR) algorithm [14].
4. Estimate the information rate as

$$\hat{I}(X; Y) = \frac{1}{K} \log \frac{p(\mathbf{y}_K | \mathbf{x}_K)}{p(\mathbf{y}_K)}$$

The procedure can be extended to the more general case of *continuous-state* sources and channels by resorting to *particle methods* [15].

For the specific case of finite discrete input and output alphabets and a finite-state machine channel, the channel capacity can be computed numerically by an extension of the Blahut-Arimoto algorithm, which maximizes the information rate by iteratively optimizing a finite-state machine source [16].

20.2.3 Mismatched decoding

The computation of the information rate in (20.11) and the implementation of the optimum detector to achieve it require explicit knowledge of the conditional distributions $p(\mathbf{y}_K | \mathbf{x}_K)$, with $K = 1, 2, \ldots$, that characterize the channel. In many practical cases these distributions might be either unavailable or too complicated, so that the detector is designed to make maximum-a-posteriori-probability (MAP) decisions based on a mismatched channel law $q(\mathbf{y}_K | \mathbf{x}_K) \neq p(\mathbf{y}_K | \mathbf{x}_K)$,[4] as schematically shown in Fig. 20.7. As we shall see, this is for instance the actual situation with optical fiber systems in the nonlinear regime.

[4] The detector may be mismatched even when the two conditional distributions are equal up to some finite value K^*, as they might differ for $K > K^*$. This is the case, for instance, when a memoryless detector is employed for a channel with memory, matching the statistics only for $K = 1$. In this case, the mismatched channel law is expressed as $q(\mathbf{y}_K | \mathbf{x}_K) = \prod_{k=1}^{K} p(y_k | x_k)$ even if the output symbols are not conditionally independent.

In these cases, it is useful to define the following quantity [12]

$$I_q(X;Y) \triangleq \lim_{K\to\infty} \frac{1}{K} E\left\{\log_2 \frac{q(\mathbf{Y}_K|\mathbf{X}_K)}{\int p(\mathbf{x}_K)q(\mathbf{Y}_K|\mathbf{x}_K)d\mathbf{x}_K}\right\} \tag{20.12}$$

which we will denote as the achievable information rate (AIR) with mismatched decoding metric $q(\mathbf{y}_K|\mathbf{x}_K)$.[5] With respect to the information rate (20.11), (20.12) is obtained by replacing the actual channel law $p(\mathbf{y}_K|\mathbf{x}_K)$ with an arbitrary mismatched law $q(\mathbf{y}_K|\mathbf{x}_K)$, while the expectation is still taken with respect to the actual distribution $p(\mathbf{y}_K|\mathbf{x}_K)p(\mathbf{x}_K)$ induced by the input distribution and the actual channel law.

The AIR (20.12) has some interesting properties that hold for any channel and mismatched metric and make it suitable for a practical use in optical communications: (1) It is a lower bound to the information rate (20.11) and, therefore, to channel capacity $I_q(X;Y) \leq I(X;Y) \leq C$; (2) its maximization over any possible detection metric, achieved for $q(\mathbf{y}_K|\mathbf{x}_K) = p(\mathbf{y}_K|\mathbf{x}_K)$, yields the actual information rate $I(X;Y)$; (3) its further maximization over the input distribution $p(\mathbf{x}_K)$ yields the channel capacity; (4) it is achievable over the true channel with source probability $p(\mathbf{x}_K)$ and a MAP detector matched to $q(\mathbf{y}_K|\mathbf{x}_K)$; (5) it can be numerically evaluated without an explicit knowledge of $p(\mathbf{y}_K|\mathbf{x}_K)$.

In practice, the system is designed by selecting a modulation format and an approximated (auxiliary) channel model, which determine, respectively, the input distribution $p(\mathbf{x}_K)$ and the mismatched decoding metric $q(\mathbf{y}_K|\mathbf{x}_K)$. The AIR (20.12) for this configuration is then computed through numerical simulations. Possibly, $p(\mathbf{x}_K)$ and/or $q(\mathbf{y}_K|\mathbf{x}_K)$ can be numerically optimized by using the AIR as a performance metric to be maximized.

Whereas an exact analytical evaluation of (20.12) is still unfeasible—as the expectation in (20.12) should be taken with respect to the unknown joint distribution $p(\mathbf{x}_K,\mathbf{y}_K)$—an accurate numerical estimate can be obtained by resorting to the procedure proposed in Ref. [12] and already described at the end of Section 20.2.2 [but replacing $p(\mathbf{y}_K|\mathbf{x}_K)$ with $q(\mathbf{y}_K|\mathbf{x}_K)$ and $p(\mathbf{y}_K)$ with $q(\mathbf{y}_K)$].

Example 1: The AWGN channel

As a simple example, we consider the AWGN channel (20.5) with conditional distribution (20.6) and SNR P/σ_n^2. The capacity of this channel is achieved by a CSCG input distribution and is given by (20.9). On the other hand, considering the same input distribution but a mismatched decoding metric[6]

[5] In Ref. [12], the channel represented by the conditional distribution $q(\mathbf{y}_K|\mathbf{x}_K)$ is referred to as the *auxiliary channel*, and (20.12) as the *auxiliary-channel lower bound* to the information rate.

[6] Being the channel memoryless, we consider only one input and output variable and omit the time index for simplicity.

$$q(y|x) = \frac{1}{\pi\sigma_n'^2}\exp\left(-\frac{|y-x|^2}{\sigma_n'^2}\right) \tag{20.13}$$

that is, a decoder optimized for an AWGN channel with a different SNR $P/\sigma_n'^2$, the corresponding AIR can be obtained, after a few passages, by a direct computation of (20.12)

$$I_q(X;Y) = \log_2(1 + P/\sigma_n'^2) + \frac{1}{\log(2)}\left(\frac{P+\sigma_n^2}{P+\sigma_n'^2} - \frac{\sigma_n^2}{\sigma_n'^2}\right) \tag{20.14}$$

It is simple to verify that (20.14) has a maximum for $\sigma_n'^2 = \sigma_n^2$ (the matched decoder), when it equals the channel capacity.

The mismatch has an impact also on the AIR obtained for an *M*-QAM input alphabet with uniformly distributed symbols $x\in\{X_1,\ldots, X_M\}$. In this case, the AIR can be obtained by a numerical computation of (20.12), considering such a discrete uniform input distribution, the conditional distribution (20.6), and the decoding metric (20.13). For instance, by considering a simple Monte Carlo approach for the estimation of (20.12), we obtain the same expression as in (20.8), but with $q(y|x)$ in (20.13) replacing $p(y|x)$.

Fig. 20.8A shows the AIR as a function of the SNR, obtained for a CSCG or uniform QAM input constellation with either the matched or mismatched decoding metric, where the latter is given by (20.13) with $\sigma_n'^2 = \sigma_n^2/2$, that is, with a 3-dB higher SNR. The corresponding AIR loss, defined as the difference between the AIRs obtained with the matched and mismatched decoder, is reported in Fig. 20.8B. In this case, the mismatch entails only a moderate AIR loss for all the considered input distributions. A maximum AIR loss of ~0.44 bit/symbol is

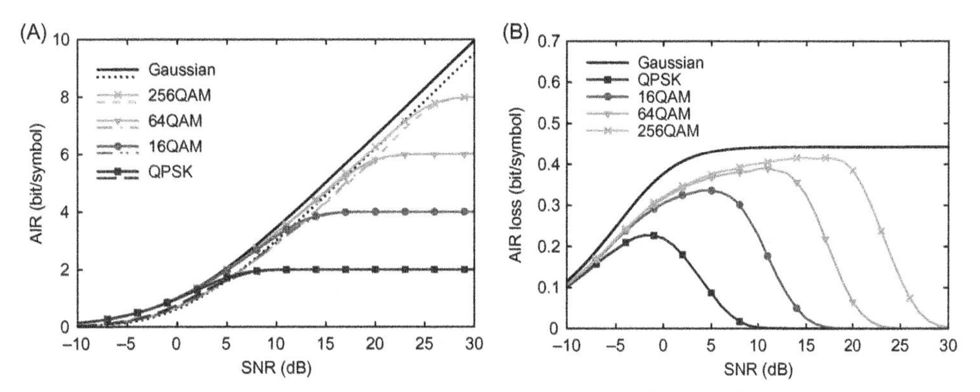

Figure 20.8 (A) AIR with a matched (solid lines) or mismatched (dashed lines) decoding metric over the AWGN channel for various input distributions. The mismatched decoding metric is optimized for a 3-dB higher SNR compared to the actual one. (B) Corresponding AIR loss.

obtained for CSCG input and SNR→ ∞, while for the QAM formats the AIR loss is lower and vanishes at sufficiently higher SNR, when the AIR saturates to the noise-free value of $\log_2(M)$.

Example 2: The phase noise channel

As a second example, we consider a phase noise (PN) channel that, with respect to the AWGN channel (20.5), is characterized by the presence of the additional real PN term θ_k

$$y_k = x_k e^{j\theta_k} + n_k \tag{20.15}$$

The PN term is assumed to evolve as a first-order autoregressive process

$$\theta_{k+1} = \alpha\theta_k + w_k \tag{20.16}$$

with coefficient $\alpha = 0.9$ and i.i.d. Gaussian increments w_k with variance σ_w^2. Because of the presence of the PN term, the channel is no longer memoryless. Fig. 20.9A shows the AIR as a function of the variance σ_w^2 for an SNR of 10 dB, a CSGG input distribution, and three different decoding metrics:

1. The matched metric $q(\mathbf{y}_K|\mathbf{x}_K) = p(\mathbf{y}_K|\mathbf{x}_K)$, corresponding to the true channel law, including memory. This metric can be computed as in Ref. [12], resorting to one of the methods in Ref. [17] (e.g., discretization or parametrization) or to the particle method in Ref. [15] to account for the continuous state space where the PN term is defined.
2. A memoryless PN metric $q(\mathbf{y}_K|\mathbf{x}_K) = \prod_{k=1}^{K} p(y_k|x_k)$, which is obtained by neglecting channel memory (i.e., assuming the conditional independence of the output

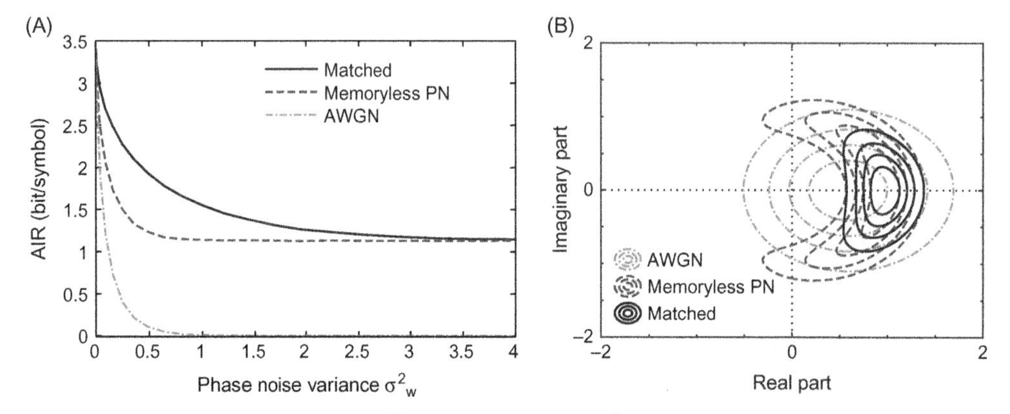

Figure 20.9 (A) AIR versus variance of the PN increments σ_w^2 for the PN channel model (20.15), a CSCG input distribution, and three different decoding metrics; (B) Contour lines of the conditional distributions $q(y_k|x_k = 1, \theta_{k-1} = 0)$ corresponding to the three decoding metrics.

samples given the corresponding input ones) and matching the metric to the true marginal distribution $p(y_k | x_k)$ for the channel (20.15), as explained in Footnote 4.

3. A simple AWGN metric

$$q(y_K|x_K) = \prod_{k=1}^{K} \frac{1}{\pi\sigma_q^2} \exp\left(-\frac{|y_k - x_k e^{-\sigma_\theta^2/2}|^2}{\sigma_q^2} \right) \tag{20.17}$$

where $\sigma_\theta^2 = \sigma_w^2/(1-\alpha^2)$ is the variance of the PN term θ_k and $\sigma_q^2 = \sigma_n^2 + \sigma_x^2(1 - e^{-\sigma_\theta^2})$ the variance of an equivalent AWGN term that accounts for the combined effect of the PN and AWGN terms in (20.15). In practice, the metric in (20.17) neglects channel memory and is optimized for an AWGN channel whose input and output symbols have the same mean and covariance matrix as the symbols x_k and y_k at the input and output of the true channel (20.15).

Fig. 20.9B shows the contour lines of the conditional distributions $q(y_k | x_k = 1, \theta_k = 0)$ corresponding to the three different decoding metrics. With respect to the matched decoding metric, the memoryless PN metric neglects the information that is provided by the knowledge of the previous PN term, resulting in a larger uncertainty about the current PN term. The AWGN metric, on the other hand, is even unaware of the presence of a PN term, resulting in a completely mismatched distribution.

In this example, the behavior of the curves is quite different and the mismatch has more serious consequences compared to the example considered in Fig. 20.8. In fact, when the variance of the PN increments increases, the AIR of the AWGN detector vanishes, while the AIRs of the matched detector and of the memoryless PN detector saturate to a nonzero value. This result shows that it is always possible to transmit at least 1.1 bit/symbol over this channel, regardless of the amount of PN, provided that the decoder is aware of the PN presence. Moreover, if the decoder accounts also for the PN correlation, an additional AIR gain can be obtained. This example suggests that is it not advisable to draw general conclusions about the information rate over a given channel by observing the AIR behavior for a specific mismatched decoder. The important implications of this observation will become apparent in Section 20.4 in the discussion of the optical fiber case.

20.2.4 Waveform channels

The definition of information rate in (20.11) is suitable for discrete-time channels, for which the concept of *channel use* is univocally defined. A message is transmitted through the channel by *using* it a certain number of times, that is, by encoding the message on a sequence of symbols, which are then transmitted through the channel, one per each channel use. The efficiency with which the message is transferred is hence measured by (20.11) in terms of bits of information per channel use (or, equivalently, per symbol).

On the other hand, with regard to waveform channels, the picture gets more complicated. In this case, the sequence of symbols carrying the message is further encoded on a waveform that, when transmitted, occupies the channel for a *certain time*. Thus the efficiency with which the message is transferred is measured in terms of bits per second and depends also on how the symbols are encoded on and decoded from the waveform signal.

A typical example is the use of a linear modulation, in which the input waveform is made of a sequence of pulses, whose amplitudes depend linearly on the symbols to be transmitted. For K input symbols, $\{x_k\}_{k=1}^{K}$, the input waveform takes the following expression [9]:

$$x(t) = \sum_{k=1}^{K} x_k p(t - kT) \tag{20.18}$$

where $p(t)$ is the pulse shape and T the symbol time. At the channel output, the corresponding output samples $\{y_k\}_{k=1}^{K}$ are obtained by filtering the received waveform $y(t)$ by a matched filter (matched to the pulse shape $p(t)$), whose output is sampled at rate $R = 1/T$. The combination of the waveform channel, the input modulator, and the output demodulator constitutes the *equivalent discrete-time channel.*

In this case, while the information rate $I(X;Y)$ through the discrete-time channel is still given by (20.11) and measured in terms of bits per symbol, the information rate through the waveform channel (in bits per second) is given by $I(X;Y)/T$, where the factor $R = 1/T$ accounts for the number of symbols transmitted per unit time.

Physical waveform channels usually have a limited bandwidth. In principle, if the channel bandwidth is wider than the bandwidth of the transmitted waveform, it is possible to exploit the remaining frequency resources to transmit other waveforms at the same time. In this case, the efficiency with which the message is transferred through the waveform channel is more conveniently characterized by the so-called *spectral efficiency* (SE), defined as the amount of information transferred per unit *time* and *bandwidth*. The SE is measured in bits per second per hertz and is related to the information rate (20.11) through the simple expression

$$\text{SE} = \frac{I(X;Y)}{WT} \tag{20.19}$$

where W is a suitable measure of the occupied bandwidth. Analogously, when the AIR in (20.12) is considered to account for mismatched decoding, the corresponding achievable SE is defined as $\text{SE}_q = I_q/(WT)$. Note that properly defining the bandwidth of a signal is a quite subtle point [18]. However, the practical utility of the SE as a performance parameter is mainly restricted to the cases in which many waveforms are multiplexed in the frequency domain and simultaneously transmitted through the same channel. For these cases, a practical and univocal definition of the occupied bandwidth W will be given in Section 20.3.2.

The conditional probability that characterizes the equivalent discrete-time channel depends not only on the underlying waveform channel but also on the employed modulation and demodulation schemes. For instance, depending on the specific choice of $p(t)$ and T in (20.18), as well as on the actual response of the channel, there can be intersymbol interference (ISI) or not. In the latter case, the equivalent discrete-time channel is memoryless, whereas in the former it does have memory.

The maximal information rate through the discrete-time channel is given by its capacity C, from which we can obtain the corresponding information rate C/T (in bit/s) and SE $C/(WT)$ (in bit/s/Hz) for the waveform channel. We might hence be tempted to refer to these quantities as the actual capacity and maximal SE of the waveform channel, respectively. This is, however, incorrect [19]. As already mentioned, a different modulation choice might yield a different discrete-time channel with a different (possibly higher) capacity for the same (or possibly lower) time T or time–bandwidth product WT. Moreover, the symbols $\{y_k\}_{k=1}^{K}$ at the output of the chosen (matched-filter and symbol-time-sampler) demodulator might be not a sufficient statistic for optimal detection, such that a higher information rate might be achieved by changing the demodulator structure. Indeed, the computation of the capacity of the waveform channel would require the extension of the optimization in (20.3) to any possible modulation and demodulation scheme. As shown in the following, this issue can be easily addressed when specific bandwidth constraints are posed on the input and output waveforms.

20.2.4.1 Band-limited channels

Let us consider a waveform channel whose input and output are strictly limited to a bandwidth W.[7] In this case, it is convenient to set $T = 1/W$ and $p(t) = \mathrm{sinc}(t/T)$ in (20.18), so that the input and output symbols correspond to the samples of the input and output waveforms, respectively, taken at the Nyquist rate. This implies, by the sampling theorem [2,9], that the input and output symbols are a sufficient statistic to represent the input and output waveforms, respectively. In other words, any possible coding and decoding strategy for the waveform can be implemented just by working on the input and output symbols of the equivalent discrete-time channel.

[7] By strictly band-limited we mean that both the input and output waveforms would pass unchanged through a rectangular filter of bandwidth W. This condition might be due to an intrinsic behavior of the channel, as if filters of bandwidth W were physically present at its input and output, or to a technological constraint posed on the modulator and demodulator characteristics.

In this case, given the capacity C_{DT} of the equivalent discrete-time channel in bits per channel use, the capacity of the waveform channel in bits per second is simply given by

$$C_{\mathrm{WF}} = \frac{C_{\mathrm{DT}}}{T} = C_{\mathrm{DT}} W \tag{20.20}$$

and its maximal SE in bits per second per hertz by

$$\mathrm{SE}_{\max} = \frac{C_{\mathrm{DT}}}{WT} = C_{\mathrm{DT}} \tag{20.21}$$

20.2.4.2 The band-limited additive white Gaussian noise channel

A special case of band-limited waveform channel is the one characterized by the simple input−output relation

$$y(t) = x(t) + n(t) \tag{20.22}$$

where $x(t)$ and $y(t)$ are band-limited waveforms and $n(t)$ is a band-limited white Gaussian process with power spectral density (PSD) N_0 and independent of $x(t)$ and $y(t)$. Despite its simplicity, in many cases the band-limited AWGN model provides a reasonable approximation and a tractable expression to understand the basic behavior of many physical waveform channels. As we shall see, this is true also for the optical fiber channel.

Considering the same modulation as in (20.18)—with sinc pulses and $T = 1/W$— and the corresponding matched-filter demodulator, the waveform channel (20.22) can be equivalently represented by the discrete-time AWGN channel in (20.5), with noise variance $\sigma_n^2 = N_0 W$, whose capacity is given by (20.9). Eventually, by using (20.20) and (20.21), the capacity in bits per second and the maximal SE in bits per second per hertz of the band-limited AWGN channel (20.22) can be respectively expressed as

$$C_{\mathrm{WF}} = W\log_2\left(1 + \frac{P}{N_0 W}\right) \tag{20.23}$$

and

$$\mathrm{SE}_{\max} = \log_2\left(1 + \frac{P}{N_0 W}\right) \tag{20.24}$$

20.3 The optical fiber channel

Before delving into an information theoretical analysis of the optical fiber channel, we need to provide a definition of the channel itself. Necessarily this definition will entail some degree of simplification and the introduction of specific constraints to ease and

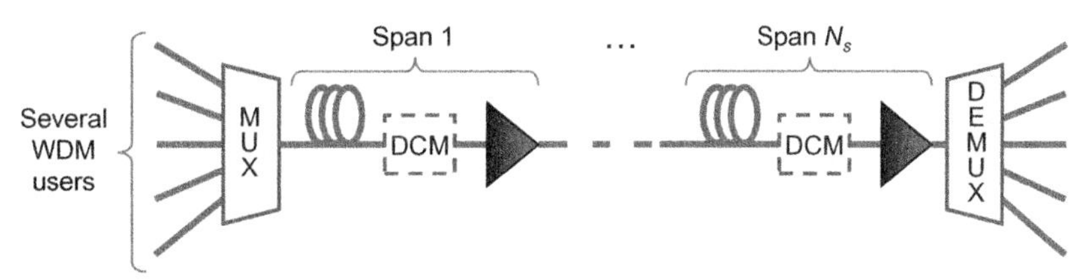

Figure 20.10 Schematic representation of an optical fiber link, shared by several WDM users, with in-line optical amplifiers and optional dispersion compensating modules (DCM).

circumscribe the analysis. For an exhaustive introduction to optical fiber communication systems and their constituent elements and devices we refer to Ref. [20].

The main elements of a typical point-to-point optical fiber link are schematically depicted in Fig. 20.10. Usually, several users access the link simultaneously by adopting a WDM strategy. The optical signals generated by the various users at different wavelengths are combined together by a multiplexer. The resulting WDM signal is then coupled to the input of the optical fiber link. Conversely, at the link output, an optical demultiplexer separates the signals at different wavelengths and sends each of them to the corresponding receiver.

The link consists of several spans of transmission fiber. In the *lumped amplification* (LA) configuration considered in this scheme, each span of fiber is followed by an optical amplifier whose gain exactly compensates for the span loss. Alternatively, we will also consider an *ideal distributed amplification* (IDA) configuration, in which the optical signal is continuously amplified along the transmission fiber, with a distributed gain that exactly equals the fiber attenuation coefficient, so that the signal power remains constant along the whole link. This configuration provides the best performance in terms of SNR and, as we shall see, is also helpful in reducing the impact of nonlinear effects. From a theoretical point of view, the IDA scheme can be seen as the limiting case of an LA scheme with an infinite number of spans of infinitesimal length. In practice, an IDA configuration can be reasonably approximated by means of distributed Raman amplification.

Historically, long-haul systems have deployed in-line dispersion compensation to compensate for the chromatic dispersion of the fiber. In this case, a dispersion compensating module (DCM) is inserted after each span of fiber to compensate, fully or in part, for the dispersion induced by the transmission fiber and obtain the desired *dispersion map*. These kinds of links are referred to as dispersion-managed (DM) links. On the other hand, in the last generation of optical systems, the chromatic dispersion accumulated at the end of the link can be digitally compensated at the receiver, after coherent detection, avoiding the use of DCMs. Links without in-line DCMs are referred to as *dispersion-unmanaged* (DU) links. As we shall see, DU links guarantee a better performance and, of course, a lower cost compared to DM links, so that the former are now preferred over the latter. A DU configuration is not feasible, however, when the optical link is shared with some legacy channels that still employ direct detection and are not equipped with DSP-based dispersion compensation. Moreover, some recent results suggest that the use of DCMs that implement *per-channel* dispersion compensation (compensating dispersion *within* each WDM channel but not *between* different channels, for instance by using fiber Bragg gratings) might ease the mitigation of nonlinear effects and increase the AIR compared to systems operating over DU links [21,22].

The WDM signal propagating through the optical link is subject to many propagation effects. In our analysis, we will consider only attenuation, dispersion, and Kerr nonlinearity, while we will neglect some other effects—usually with a minor impact on performance, though still relevant in some conditions—such as higher order dispersion, polarization-mode dispersion, polarization-dependent loss, and stimulated Raman scattering [20,23]. Moreover, we will consider the ASE noise injected by the optical amplifiers and neglect other sources of noise such as thermal and shot noise, which usually become relevant only in the absence of optical amplification. Despite the simplification, the considered scenario allows us to study and understand the fundamental issues that arise in the optical fiber channel due to the interplay of the considered propagation effects and of ASE noise, and to predict with a reasonable accuracy the performance of optically amplified WDM systems operating over the C-band.

Due to space limitations, we will restrict the analysis to the aforementioned scenario which is, by far, the most relevant for today's optical transport networks. However, it is worth mentioning that a significant research effort is currently focused on designing optical fiber links with higher capacity. This includes the extension of the WDM scenario to wider bandwidths; the in-line insertion of optical phase conjugators [24] or all-optical regenerators [25]; and the replacement of currently deployed single-mode fibers with multimode or multicore fibers [26] and hollow-core fibers [27]. Moreover, in some specific scenarios (mainly for short distances), the highest link capacity might be obtained with no or minimum optical amplification, replacing classical detection strategies with more general approaches based on quantum theory. The ultimate capacity limits are, in this case, determined by the so-called *Holevo bound* [28,29].

In the rest of this section, we provide the equations that describe the propagation of the optical signal through the link under the effect of the aforementioned propagation impairments and ASE noise, we introduce some assumptions that rule the user behavior and the modulation and demodulation of the WDM signals, and we briefly discuss the most popular approaches to model the optical fiber channel.

20.3.1 The equations governing optical fiber propagation

The optical signal propagating through a single-mode fiber is represented by the Jones vector $\mathbf{v}(z,t) = [v_1(z,t),\ v_2(z,t)]^T$, whose components correspond to the signal complex envelopes on two orthogonal states of polarization. The signal power $P(z) = \|\mathbf{v}(z,t)\|^2$ varies along the link due to attenuation and optical amplification, so that it is often more convenient to represent the optical signal through the normalized Jones vector $\mathbf{u}(z,t) = \mathbf{v}(z,t)/\sqrt{a(z)}$, where the function $a(z) = P(z)/P(0)$ represents the power profile along the link. In LA links, the power $P(z)$ decays exponentially within each span of fiber and is restored back to its input value $P(0)$ (the launch power) after each amplifier. In particular, in an LA link with N_s identical spans of length L_s and attenuation coefficient α, the power profile evolves according to

$$a(z) = \exp[-\alpha(z \bmod L_s)], \quad 0 \le z \le N_s L_s \tag{20.25}$$

On the other hand, in IDA links, the power remains constant and $a(z) = 1$ for any z.

In the considered scenario—in particular, neglecting PMD and higher-order dispersion—the propagation of $\mathbf{u}(z,t)$ is governed by the Manakov equation [30]

$$\frac{\partial \mathbf{u}}{\partial z} = j\frac{\beta_2}{2}\frac{\partial^2 \mathbf{u}}{\partial t^2} - ja(z)\gamma\|\mathbf{u}\|^2\mathbf{u} + \mathbf{n}(z,t) \tag{20.26}$$

where β_2 is the group-velocity-dispersion (GVD) parameter, γ the nonlinear coefficient, and $\mathbf{n}(z,t) = [n_1(z,t),\ n_2(z,t)]^T$ a vector collecting the normalized noise components injected by optical amplifiers on the two states of polarization. The first term on the right-hand side of (20.26) accounts for GVD, the second one for Kerr nonlinearity, and the third one for ASE noise. The nonlinear coefficient γ implicitly includes the factor of 8/9 that accounts for rapid mixing of the polarization state on the Poincaré sphere [31]. The fiber parameters α, β_2, and γ are usually constant within each span of fiber but may vary from span to span.

The ASE noise terms injected by each amplifier of an LA link on each polarization state are modeled as independent white Gaussian noise (GN) processes, each with PSD

$$N_{\text{ASE}} = (G-1)h\nu n_{\text{sp}} \tag{20.27}$$

where $G = \exp(\alpha L_s)$ is the gain of the amplifier (equal to the span loss by hypothesis), n_{sp} its spontaneous emission coefficient, h the Plank constant, and ν the optical frequency.

In the special case of a single-polarization signal, the Manakov equation reduces to the nonlinear Schrödinger equation (NLSE) [23]

$$\frac{\partial u}{\partial z} = j\frac{\beta_2}{2}\frac{\partial^2 u}{\partial t^2} - ja(z)\gamma|u|^2 u + n(z,t) \tag{20.28}$$

in which all vectors are replaced by scalar components.

20.3.2 The wavelength division multiplexing scenario

In general, we assume that $2M+1$ users share the fiber channel by transmitting their signals on different portions of the available spectrum, as schematically depicted in Fig. 20.10. In the optical communications community, it is customary to refer to the frequency band allotted to each user as a *WDM channel*. By extension—though with some confusion in the information theory community, which is perhaps not equally favorable to accept metonymy as an equivalence relation—the same term is often used also for the user itself or for the corresponding signal. Each WDM channel is denoted by an index $-M \leq m \leq M$. In our analysis, the central channel ($m=0$) will play the special role of the *channel of interest* (COI), whose properties are under investigation.

The mth user encodes the information on the symbols of a baseband signal[8]

$$x_m(t) = \sum_{k=1}^{K} x_{m,k}\ p(t-kT) \tag{20.29}$$

linearly modulated as in (20.18). The baseband signal modulates an optical carrier at a given frequency $f_m = f_0 + mW$, where W is the channel spacing and f_0 the central frequency of the COI, adopted as a reference frequency for the lowpass equivalent representation of the WDM signal. The $2M+1$ optical signals generated in this way are then multiplexed to obtain the WDM signal at the input of the fiber link, whose complex envelope is[9]

$$u(0,t) = \sum_{m=-M}^{M} x_m(t)e^{j2\pi mWt} \tag{20.30}$$

[8] For the sake of simplicity, we consider here a single-polarization scenario. The same representation can be used also in a dual-polarization scenario, but replacing the input and output symbols with two-element vectors, the corresponding signals with Jones vectors, and considering the Manakov propagation Eq. (20.26).

[9] Here we neglect the random unknown phase and time shifts that are usually present between the various WDM signals. This lack of synchronization has no impact on the performance of practical systems or on the capacity bounds that will be presented in Section 20.4 (with the exception of Section 20.5), but may affect the actual (unknown) capacity.

After propagation through the fiber link according to (20.28), the received optical signal is separated again into its $2M+1$ WDM components, which are sent to a bank of coherent detectors. Each detector demodulates the corresponding optical signal by bringing it back to baseband, filtering it with a matched filter, and sampling it at rate $R=1/T$, where T is the symbol time. The received samples $\{y_{m,k}\}_{k=1}^{K}$ are then processed to extract the transmitted information.

This WDM configuration is quite general. Depending on the employed pulse shape $p(t)$ and spacing W between the channels, it can represent a dense or a coarse WDM system, with different SE. Since we are interested in a theoretical analysis on the maximal achievable SE, we select an ideal Nyquist pulse shape $p(t)=\mathrm{sinc}(t/T)$ as done in Section 20.2.4.1 and a symbol rate equal to the channel spacing, $R=W$. In this way, the symbols $\{x_{m,k}\}_{k=1}^{K}$ correspond to the samples of the baseband signal (20.29) taken at the Nyquist rate $R=1/T$ and univocally represent any arbitrary optical waveform in the bandwidth W allotted to the mth user (WDM channel). Analogously, the output samples $\{y_{m,k}\}_{k=1}^{K}$ are a sufficient statistic to represent the mth received optical waveform in the same frequency band. Therefore without loss of generality, we have provided a discrete-time representation of the waveform channel on the overall WDM bandwidth. The set of complex input and output symbols $\{x_{m,k}\}$ and $\{y_{m,k}\}$ for $m=-M,\ldots, M$ and $k=1,\ldots, K$, with $K\rightarrow\infty$, represent the input and output waveforms, and the corresponding input–output conditional distribution completely characterizes the waveform channel. In particular, capacity results (or bounds) obtained for the discrete-time channel can be directly translated into capacity and SE results (or bounds) for the waveform channel, as discussed in Section 20.2.4.1.

An information theoretical analysis of the scenario described above presents several difficulties. Since the channel is not AWGN, the results in (20.23) and (20.24) are not valid. As dispersion, Kerr nonlinearity, and noise interact during propagation, we have, in general, a *nonlinear channel* with a *long memory* and *non-Gaussian statistics.* Unfortunately, neither analytical solutions nor effective numerical methods are available, in general, to compute the statistics of this kind of channel. In particular, the analysis is complicated by the following effects induced by nonlinearity: deterministic intrachannel nonlinearity, causing a distortion of each WDM signal and a corresponding nonlinear ISI; signal–noise interaction, giving rise to non-GN statistics; interchannel nonlinearity, causing interchannel interference (ICI); and spectral broadening, causing ICI and spectral loss.

The numerical study reported in Ref. [32] shows that, in a typical WDM scenario, deterministic intrachannel nonlinearity can be digitally compensated, for example, by digital backpropagation (DBP) [33,34], while interchannel nonlinearity is the main limiting factor to increase the information rate; on the other hand, spectral broadening and signal–noise interaction are usually negligible. This is, however, not guaranteed if interchannel nonlinearity can be effectively mitigated. For instance, in a single-user scenario ($M = 0$), interchannel nonlinearity is not present, so that signal–noise interaction and spectral broadening become dominant. Studies of achievable rates and channel capacity in the single-user scenario are reported, for instance, in Refs. [35–37].

In the WDM scenario described above, an additional difficulty is that the capacity should be defined and studied from a multiuser perspective [38]. This approach is not commonly considered in optical fiber communications and will not be considered in our study. In fact, the analysis is usually carried out only for the COI, from a single-user perspective, assuming that the other WDM signals are just a source of interference out of the user control and follow a specific *behavioral model* [39]. In particular, the typical behavioral model that will be considered in our study is that of $2M + 1$ independent users with identical input distribution and launch power, referred to as *adaptive interferer distribution* model or behavioral model (c) in Ref. [39]. This behavioral model and the choice of the central channel as the COI generally ensure that the SE achievable by the COI is simultaneously achievable by all the WDM users without any cooperation among them.

20.3.3 Approximated channel models

Under some specific assumptions, commonly met in current WDM systems, the Eqs. (20.26) or (20.28) model optical fiber propagation quite accurately. However, they only provide an implicit description of the waveform channel through a stochastic nonlinear partial differential equation. In this sense, they are not *information theory friendly* [40]. In fact, as discussed in Section 20.2, a good channel model, suitable for capacity analysis, should include an appropriate discrete representation of the input and output waveforms and provide an explicit and mathematically tractable expression of the conditional distribution of the output symbols given the input ones. At the same time, the model should be physically accurate for a wide range of (ideally, for *any*) input distributions, so that the optimization problem involved in the computation of capacity could be solved while remaining within the scope of validity of the model.

Over the years, a number of different approximated models of the optical fiber channel have been developed [41–59], often with a conflicting view about the nature of nonlinear effects. The existing models achieve different trade-offs between the two requirements of physical accuracy and mathematical tractability—being either more accurate but less information-theory friendly, or the other way around [19]—but do not meet both of them sufficiently and simultaneously. As a result, the problems of channel modeling and, in turn, of capacity evaluation remain essentially open [1,60]. Nevertheless, some of these models play an important role in the information theoretic analysis presented later in this chapter and are briefly introduced in this section.

20.3.3.1 The split-step model

The split-step Fourier method (SSFM) is an efficient numerical method to solve the propagation Eqs. (20.26) or (20.28) for specific realizations of the input signal and amplifier noise [23]. The numerical implementation of the method is rather simple and, for a given accuracy, its numerical complexity is significantly lower than that of other known numerical methods [61]. Moreover, the SSFM can be easily extended to account for other propagation effects not included in (20.26) and (20.28), such as PMD, polarization dependent loss, and higher-order dispersion. This characteristics make the SSFM the first choice for the numerical simulation of optical fiber systems.

The key idea behind the SSFM is that of dividing fiber propagation into several short steps, over which the effects of the linear and nonlinear terms of the equations—the first and second term, respectively, on the right-hand side of (20.26) or (20.28)—can be approximately separated. This approximation is particularly convenient, as the separate effects of the two terms can be easily expressed in closed form in frequency and time domain, respectively. The approximation yields the split-step model schematically depicted in Fig. 20.11A.

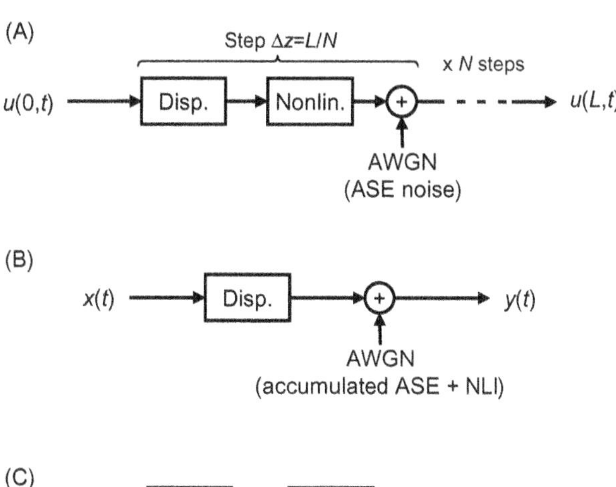

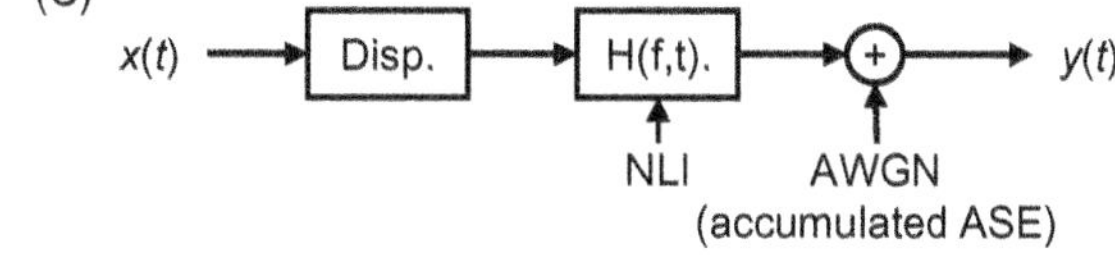

Figure 20.11 Three different kind of models for the nonlinear optical fiber channel: (A) split-step model; (B) GN model; (C) linear time-variant model.

An important characteristic of this model is that its accuracy can be arbitrarily increased by reducing the step-size for the integration along z (i.e., by increasing the number of steps N). Moreover, by considering a discrete-time representation of the propagating signal and using FFTs to go back and forth between time and frequency domains, the model turns into the SSFM—a very efficient algorithm for the numerical simulation of optical fiber systems and for the implementation of DBP.

On the other hand, the use of the model in Fig. 20.11A and of the SSFM for an information theoretical analysis is much more challenging, if not for the mere collection of the input–output statistics. Relevant exceptions are, for instance, the derivation of a capacity upper bound for the optical fiber channel in Ref. [62] and the computation of AIRs based on the stochastic DBP algorithm in [37].

20.3.3.2 The Gaussian noise model

A much simpler description of the channel is provided by the GN model, which assumes that nonlinear interference (NLI) can be modeled as an additive GN, nearly white over the COI bandwidth and independent of the COI signal [52]. In this way the fiber is reduced to a simple linear channel affected by GVD and AWGN, with the peculiar characteristic that the noise PSD depends on the PSD of the transmitted signal. The model is depicted in Fig. 20.11B.

This approach provides a drastic simplification of the problem, as GVD can be easily compensated for by a linear equalizer, while the AWGN channel is one of the best understood kinds of channels, for which efficient coding, modulation, and detection techniques, and closed-form expressions for system performance and channel capacity are available [9]. The GN model allows the computation of the resulting SNR (accounting for both ASE noise and NLI) by means of closed-form approximations [58,63]. The accuracy of the model has been experimentally verified for various system configurations [64] and can be further improved by considering the enhanced version of the model, which accounts for the dependence of NLI on the modulation format [59]. When the WDM approach is extended beyond the C-band of conventional systems, other propagation effects such as third-order dispersion and stimulated Raman scattering become relevant, breaking the symmetry of the system with respect to frequency. Extensions of the GN and enhanced GN models to include these effects have been investigated in Refs. [65–69].

The extreme simplification allowed by the AWGN assumption has an important drawback. In fact, making this assumption entails that we accept a priori the impossibility of mitigating NLI and improving the performance beyond that achievable by conventional systems—this is how we refer to systems that are optimized for the AWGN channel. This implies that the search for improved detectors, performance, and capacity bounds cannot take place within the framework of this hypothesis. Nevertheless, as we will show in Section 20.4.2, the GN model can be used in the context of mismatched decoding to derive a simple capacity lower bound [70,92].

20.3.3.3 Perturbation methods and the linear time-variant model

The characteristics of the two models described above are very different—the split-step model is a very accurate model that is usually employed for the numerical simulation of fiber propagation, while the GN model provides an extremely simplified statistical description of the channel. Indeed, a good channel model for an information theoretical analysis should have the qualities of both models: it should be accurate as the split-step model and mathematically tractable as the GN model; it should allow for a statistical analysis of the channel and also account for the explicit dependence of the statistics on the input signals [40].

The derivation of alternative models for the optical fiber channel has been an active research field during the last four decades. Among many mathematical methods available for the study of nonlinear partial differential equations, perturbation theory has been the most widely used, probably thanks to its simplicity and flexibility. The method is quite general and, depending on the particular approach, yields different kinds of models: a very accurate deterministic model, as the SSFM; a simple statistical model as the GN one; or, yet, a wide range of models with intermediate characteristics.

The main idea is that of starting from a simplified version of the propagation equation, for which a simple exact solution is available. In our case, the simplified equation usually consists in the linear equation obtained by neglecting the nonlinear term in (20.26) or (20.28). The next step is that of considering the neglected term as a small perturbation of the simplified system, such that the solution of the original equation can be written as the solution of the simplified equation, corrected by a small perturbation term. Eventually, by replacing the perturbed solution into the original equation and accounting for the smallness of the perturbation term, it is possible to derive some simple (linearized) equations to compute the perturbation term.

The most classical perturbation approach is the so-called regular perturbation (RP). Formally, considering the NLSE in (20.28), the RP approach consists in expanding the solution in power series in γ as [46,71,72]

$$u(z,t) = \sum_{i=0}^{\infty} \gamma^i u_i(z,t) \tag{20.31}$$

where $u_0(z,t)$ is the linear solution, obtained by setting $\gamma = 0$ in (20.28), $u_1(z,t)$ the first-order perturbation term, $u_2(z,t)$ the second-order perturbation term, and so on. By replacing (20.31) in (20.28) and equating the terms with equal powers of γ, one obtains a system of recursive linear differential equations. Therefore starting from the linear solution u_0, all the perturbation terms u_i's can be recursively computed [72].

In principle, the RP solution can be computed up to an arbitrary order to obtain the desired accuracy. In this sense, the RP method can replace the SSFM as a numerical algorithm to compute the signal propagation through the fiber, though its complexity grows with the perturbation order.

In practice, (20.31) is usually truncated at first order, obtaining a perturbation term $u_1(z,t)$ that depends on all the WDM signals launched into the fiber as well as on the noise $n(z,t)$ generated by all the in-line amplifiers. The various terms that result from the analysis and compose $u_1(z,t)$ correspond to various nonlinear effects that take different names depending on what signal and/or noise components are involved in their generation—namely, intrachannel nonlinearity, interchannel nonlinearity, and signal–noise interaction.

Thanks to its simplicity, the first-order approximation can be used to study the statistics of these nonlinear effects and to eventually obtain the conditional distribution $p(\mathbf{y}_K|\mathbf{x}_K)$ that characterizes the channel. Examples of applications of a first-order RP method can be found in Refs. [44,48,50,53,57,73,74]. Moreover, the Volterra series transfer function method, employed for instance in Refs. [42,75], has been shown to be equivalent to the RP method [46].

The actual channel model that is obtained from the RP method depends on some additional assumptions and approximations that are made to simplify the analysis. For instance, by assuming the Gaussianity of both the propagating WDM signals and of the generated perturbation term $u_1(z,t)$, one obtains the GN model.

The RP approach naturally yields a representation of nonlinear effects through an additive term that affects the received signal as a sort of noise. This is in contrast with the physical nature of Kerr nonlinearity—a change of the refractive index and, hence, of the phase of the propagating signal proportional to the intensity of the optical field but also with the observation of some classical nonadditive effects induced by Kerr nonlinearity in optical fiber systems, such as self- and cross-phase modulation [23] and nonlinear PN [76–78]. This limitation of the RP method has been observed in many different scenarios and has been typically addressed by two different approaches. The first approach consists of using the RP method anyway, deriving the corresponding equations for the perturbation term and then modifying them by recognizing the nonadditive nature of some of its components or, heuristically, to simply increase the numerical accuracy [44,46,57,79,80]. The second approach consists of using a different perturbation approach, such as the logarithmic perturbation, in which nonlinearity is described by a multiplicative rather than an additive perturbation [50,54,56,72,81]. Both approaches have their own merits and, in this context, we are not interested in discussing which one is better—in fact, they often yield similar results.

Indeed, the fundamental message behind this discussion is that describing nonlinear effects through a simple additive Gaussian term, a sort of additional AWGN, might be in some cases an oversimplification of the problem. This is true not only for old systems deploying in-line dispersion compensation and widely spaced low-symbol-rate WDM channels, where the peculiar characteristics of some nonlinear effects can be more easily appreciated, but also in modern DU systems with densely spaced high-symbol-rate WDM channels, where the huge accumulated dispersion seems to turn everything into AWGN [82].

An alternative—perhaps physically more accurate—description of nonlinear effects starts from the observation that a signal propagating through an optical fiber channel changes its properties via the Kerr effect. Such a change affects the propagation of the signal itself and of the other copropagating signals (WDM channels). At first order, this description corresponds to a perturbation of the transfer matrix (or transfer function, in the single-polarization case) of the fiber with respect to the linear case. As the perturbation depends on the propagating WDM signals, it is time variant. The corresponding *linear time-variant* channel model is depicted in Fig. 20.11C, where $H(f,t)$ is the time-variant transfer function of the channel that represents the effect of the other WDM channels on the propagating signal.

The most apparent effects of this perturbation, at least on a narrow bandwidth signal, is a time-varying rotation of the phase and polarization of the received signals—a sort of phase and polarization noise (PPN). The importance of PN and polarization rotation noise induced by NLI in DU links have been pointed out in Refs. [54,57,83,84] and in Ref. [85], respectively, showing that their temporal correlation properties can be exploited for the mitigation of interchannel nonlinearity. Such a PPN term is fundamentally different from the AWGN term in the model of Fig. 20.11B, which cannot be mitigated by any signal processing techniques. This difference will play an important role in the next section, where we shall use these models to lower bound the channel capacity.

20.4 The capacity of the optical fiber channel

In this section, we study the capacity of the optical fiber channel defined in the previous section by using the information theoretical tools introduced in Section 20.2 and the approximated channel models described in Section 20.3.3. We will mainly focus on the case of coherent detection, which provides a full demodulation of the modulated optical signal without any loss of information. Studies on channel capacity with direct detection receivers can be found, for instance, in Refs. [86–89].[10]

[10] In particular, we mention the recent result in Ref. [89] which proves the following relation between C_d, the capacity with direct detection, and C_c, the capacity with coherent detection

$$C_c - 1 \leq C_d \leq C_c$$

where the capacity is measured in bits per (complex) symbol. The same relation holds also between the corresponding SEs in bits per second per hertz.

20.4.1 The linear regime

For low enough optical launch power, the nonlinear term in (20.26) becomes negligible. In such a *linear regime*, the solution of (20.26) can be expressed in the frequency domain as

$$\mathbf{U}(L,f) = \mathrm{H}(f)\mathbf{U}(0,f) + \mathbf{N}(f) \tag{20.32}$$

where $\mathbf{U}(z,f)$ is the Fourier transform of $\mathbf{u}(z,t)$; $\mathrm{H}(f)$ is a unitary fiber transfer matrix, accounting for the accumulated GVD and, possibly, for other linear propagation effects, such as polarization rotations and polarization mode dispersion, which are not included in (20.26); and $\mathbf{N}(f)$ is the Fourier transform of $\mathbf{n}(t)$, a vector of two independent AWGN processes, each with PSD N_0, representing the accumulated ASE noise on the two polarizations, i.e., the integral of the term $\mathbf{n}(z,t)$ in (20.26).

The PSD of the ASE noise can be readily calculated for any specific amplifier configuration starting from (20.27). For instance, in an LA system with N_s identical spans of length L_s, each followed by an optical amplifier that exactly compensates for the span loss, the PSD on each polarization is

$$N_0 = N_s N_{\mathrm{ASE}} = N_s(e^{\alpha L_s} - 1)h\nu n_{\mathrm{sp}} \tag{20.33}$$

On the other hand, an IDA system of length L can be seen as the limiting case of an LA system for $N_s \to \infty$ and $L_s = L/N_s \to 0$. In this case, (20.33) converges to

$$N_0 = \alpha L h\nu n_{\mathrm{sp}} \tag{20.34}$$

The transfer matrix $\mathrm{H}(f)$ due to GVD depends on the accumulated GVD at the end of the link. In a generic DM link of total length L, where the GVD parameter β_2 may change along the link, the transfer matrix can be expressed as

$$\mathrm{H}(f) = \exp\left[-j2\pi^2 f^2 \int_0^L \beta_2(z)dz\right]\mathrm{I} \tag{20.35}$$

where I is the 2×2 identity matrix. The integral expression in (20.35) represents the residual GVD at the end of the link, which can be zero if the dispersion map guarantees an exact compensation of GVD. On the other hand, in DU links, the residual GVD can be simply expressed as $\beta_2 L$.

The propagation equation (20.26) does not include polarization rotations and PMD, which are neglected. However, even when explicitly including these propagation effects, the channel model (20.32) remains valid. In this case, the transfer matrix $\mathrm{H}(f)$ has not the simple expression in (20.35) and is no longer diagonal. However, it remains unitary, so that the following analysis of system capacity is still valid. Moreover, polarization rotations and PMD are stochastic effects which may change over time. Since these changes are very slow compared to the symbol rate, the analysis is done by assuming that the transfer matrix is independent of time and exactly known at the receiver.

In the linear regime described by (20.32), the propagation effects can be exactly compensated for by including, either at the receiver or at the transmitter end, a linear block with transfer matrix $\mathrm{H}^H(f)$ which, being unitary, does not affect the PSD of the noise. In practice, this is equivalently but more conveniently done in the digital domain by implementing a digital filter in one of the DSP blocks of Fig. 20.1. The filter may include an adaptive equalizer to account for an arbitrary and slowly changing channel transfer matrix. According to the data processing inequality [5, Theorem 2.8.1], such a processing does not change the information rate through the channel. However, it basically reduces the channel to a simple AWGN model, allowing for a simple implementation of the optimum detector and a closed-form computation of channel capacity. In particular, in such a linear regime, there is no interaction between the WDM channels, so that each channel can be studied independently. Moreover, thanks to the processing operated by $\mathrm{H}^H(f)$, each WDM channel can be further decomposed into two independent scalar AWGN channels, each modeled as in (20.22). Therefore in the linear regime, the capacity and maximal SE per each polarization of the optical fiber channel are given by (20.23) and (20.24), respectively.

For convenience, we report here the expression for the maximal SE

$$\mathrm{SE}_{\max} = \log_2\left(1 + \frac{P}{N_0 W}\right) \tag{20.36}$$

where N_0 is the noise PSD in (20.33) or (20.34) for LA or IDA links, respectively. This expression holds for the ideal case in which both the amplitude response of the channel and the noise PSD are flat over the whole bandwidth, as we have assumed in our analysis. In realistic cases, the gain (and ASE noise PSD) of the optical amplifiers varies over the transmission bandwidth, and some optical filters may be present along the link. A generalization of (20.23) to account for a frequency-dependent noise PSD and channel response can be found in Ref. [4] and is based on the so-called *water-filling* approach.

When the system is not truly band-limited, that is, if we have the freedom to increase the available bandwidth W, the most effective way (from a capacity viewpoint) of allocating some additional optical power is that of spreading it over a wide bandwidth. In practice, this means increasing the launch power P and the bandwidth W proportionally, keeping the signal PSD P/W nearly constant. According to (20.36), this approach allows the SE to remain unchanged and, hence, the total capacity available for the COI user to increase linearly with P (and W). In alternative, it is also possible to use the additional power to increase the number of WDM users $2M+1$. This would leave the capacity per each user unchanged, with the total capacity of the link still increasing proportionally to P (and M).

In practice, either increasing the bandwidth W per user or the number of users $2M+1$ entails increasing the total WDM bandwidth, for instance by extending the WDM approach from the C-band (1530–1565 nm), primarily used by today's systems, to the L-band (1565–1625 nm). In perspective, the same benefits in terms of channel capacity can be achieved also by a space-division multiplexing strategy, deploying fibers with multiple modes or cores, or even bundles of fibers [1]. These approaches have the additional benefit that, by keeping the signal intensity low in each mode/core/fiber, nonlinear effects remain negligible and do not affect system performance, allowing a *linear growth of the capacity* with the exploited bandwidth and number of modes/cores/fibers. However, all these approaches require an upgrade of the network and the deployment of new technologies.

On the other hand, when the system is band-limited (that is, if we assume that such an upgrade is not possible or convenient), the only way to increase the capacity of the channel is that of increasing the optical launch power P. In this case, according to (20.36), the SE (and the capacity) increases only logarithmically with the available power. In practice, the optical power needs to be nearly doubled to increase the SE by 1 bit/s/Hz. Moreover, as soon as the optical power increases beyond a certain value, the assumption of a linear regime breaks down and nonlinear effects become relevant, with a detrimental effect on system performance.

20.4.2 The Gaussian achievable information rate and the nonlinear Shannon limit

In the linear regime the capacity problem has a simple solution, as shown in the previous section. The situation is totally different when nonlinear effects come into play. From a practical viewpoint, the presence of nonlinear effects makes apparently impossible to increase the SE beyond a certain level, as if we were approaching some capacity limit. From a theoretical viewpoint, on the other hand, the presence of nonlinear effects makes the "linear regime" analysis presented in Section 20.4.1 no longer valid, meaning that a different analysis is required to determine the actual channel capacity.

20.4.2.1 The Gaussian achievable information rate

Given the unavailability of an exact channel model, a possible approach to obtain a lower bound of the capacity of the channel is that of resorting to the mismatched decoding framework introduced in Section 20.2.3. Taking inspiration from the linear regime analysis of Section 20.4.1, a simple idea is that of optimizing the input distribution and the detection metric as if the channel were AWGN, exactly as done in the third metric considered in the Example 2 of Section 20.2.3. The result is that of a memoryless auxiliary channel with jointly Gaussian input and output samples. Moreover, to minimize the mismatch with respect to the true channel, we select the auxiliary channel to have the same input-output covariance matrix as with the true channel.[11] In this case, the AIR (20.12) reduces to the simple expression [70]

$$I_G = \log_2 \frac{1}{1 - |\rho_{xy}|^2} \quad \text{(bit/symbol)} \tag{20.37}$$

denoted as *Gaussian AIR* in the following. Remarkably, this expression holds for *any* channel (linear or nonlinear, with or without memory) and depends only on a quantity which can be directly measured from the channel input and output—the input-output correlation coefficient ρ_{xy}.

We cannot fail to notice that (20.37) equals the expression in (20.10) for the capacity of the AWGN channel. In fact, (20.37) can be used for any channel but with some caveats about its actual meaning: it gives the true channel capacity for AWGN channels, whereas it provides only a capacity lower bound for generic channels—indeed, a particular one, achievable by a conventional system that is optimized for the AWGN channel. As we shall see in Section 20.4.2.2, this particular lower bound is widely used in the optical communications community.

According to the scheme of Fig. 20.1, some DSP operations may be included at the input and output of the channel. While these operations cannot increase the capacity of the channel (which, by definition, is a property of the channel itself and includes any possible input and output processing), they can increase the Gaussian AIR by modifying the input distribution and/or by reducing the mismatch between the channel and the detection metrics. This is the case, for instance, of the GVD compensation discussed in Section 20.4.1. In the linear regime, GVD compensation completely removes the effect of GVD, eliminating such a mismatch and making the Gaussian AIR equal the channel capacity.

In the nonlinear regime, a similar approach can be followed to mitigate the impact of both linear and nonlinear effects, for instance by employing DBP. As discussed in Section 20.3.2, DBP removes only deterministic nonlinear effects that take place within the receiver bandwidth (intrachannel nonlinearity), while leaving signal−noise interaction and interchannel nonlinearity substantially unaffected. The latter, in particular, is usually the dominant effect which makes the channel deviate from the AWGN behavior at high power, reducing the Gaussian AIR [32].

[11] This is the same approach followed in Ref. [90] to determine a capacity lower bound for the optical fiber channel. Here, this approach is recast in the mismatched decoding framework, ensuring the achievability of the bound with an AWGN detection metric.

With respect to the evaluation of the actual channel capacity, the evaluation of the Gaussian AIR (20.37) is much simpler as it consists in the mere computation of the input–output correlation coefficient ρ_{xy}. This can be done either numerically, by estimating the empirical correlation coefficient from a long sequence of input and output samples, or analytically, for example, by employing one of the approximate channel models reviewed in Section 20.3.3.

The numerical estimation is based on a straightforward procedure. Given a sequence of i.i.d. CSCG input symbols $x_1,\ldots, x_K$, and the corresponding output sequence $y_1,\ldots, y_K$—obtained from an accurate numerical simulation of the system or from an experimental setup—the correlation coefficient can be estimated as

$$\rho_{xy} \approx \frac{\sum_{k=1}^{K} y_k x_k^*}{\sqrt{\sum_{k=1}^{K} |y_k|^2 \sum_{k=1}^{K} |x_k|^2}} \tag{20.38}$$

where it is assumed that the output samples have zero mean. The Gaussian AIR is then obtained by replacing (20.38) in (20.37).

The correlation coefficient (or some closely related quantity) can be evaluated also analytically. In this case the procedure is more involved and the result depends on which approximated model is adopted for the computation. For instance, the Feynman's path integral approach was employed in Refs. [45,90,91], the GN model in Ref. [41,92], an RP time-domain model in Ref. [53], a Volterra series expansion in Refs. [93,94], and the FRLP model in Refs. [54,56]. The actual results found in the aforementioned papers cannot be directly compared, as they often refer to different scenarios and configurations. However, many of them have been checked against the numerical approach (20.38), showing a good accuracy.

As a practical example, we report the analytical expression derived in Ref. [56] through the FRLP model and valid for the central channel of a single-polarization Nyquist-WDM system with $2M+1$ identical channels, N_s identical spans of length L_s, LA, and in-line GVD compensation that leaves an uncompensated length L_u per span. In this case, the correlation coefficient can be expressed as

$$|\rho_{xy}|^2 \simeq \frac{P}{P+N_0 W} e^{-\sigma_\theta^2} \tag{20.39}$$

where P, N_0, and W have the same meaning as in (20.36) and

$$\sigma_\theta^2 = \left(\frac{P}{W}\right)^2 \left(\frac{4\gamma^2 L_e}{\pi|\beta_2|}\right)\left[\log(2M+1) + \log(4/3)\right] \tag{20.40}$$

is the effective variance of XPM over the signal bandwidth. In (20.40), L_e plays the role of an equivalent length and can be expressed as

$$L_e = \sum_{n=-n_0}^{n_0} (N_s - |n|) \frac{e^{-\alpha L_s} \sinh[\alpha(L_s - |n| L_u)]}{\alpha} \tag{20.41}$$

where $n_0 = \min(N_s - 1, \lfloor L_s/L_u \rfloor)$ and $\lfloor x \rfloor$ denotes the largest integer not greater than x. The DU case is recovered by setting $L_u = 0$ in (20.41); the IDA case by setting L_e equal to the total link length in (20.40).

In (20.39), the factor $P/(P + N_0 W)$ accounts for the impact of ASE noise, while the factor $\exp(-\sigma_\theta^2)$ accounts for the impact of nonlinear effects. At low power, when σ_θ^2 is very small, the correlation is essentially determined by the first factor and vanishes for $P \to 0$. On the other hand, at high power, the first factor tends to one and the correlation is determined by the second factor, which vanishes for $P \to \infty$. Consequently, the correlation (20.39) and, hence, the Gaussian AIR (20.37) reach a maximum at some optimum power level P_{opt}, which provides the best trade-off between the impacts of ASE noise and nonlinear effects. Alternative expressions for the correlation coefficient can be derived also from the GN model. In particular, we refer to Refs. [58,63,64] for closed-form approximations valid also for dual-polarization systems and for more general configurations.

20.4.2.2 Relation to the nonlinear Shannon limit

In a generic optical fiber system, the Gaussian AIR (20.37) depends on the link and system parameters. However, its qualitative dependence on the optical launch power is always the one reported in Fig. 20.12: initially, it increases with power as in the

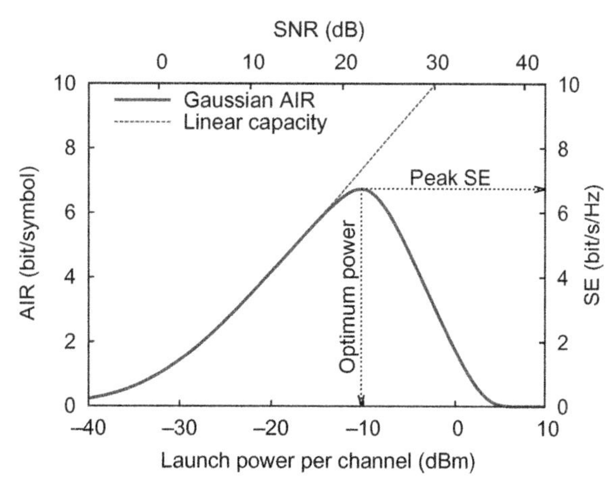

Figure 20.12 Dependence of the Gaussian AIR on the optical launch power with and without Kerr nonlinearity. The peak AIR in the nonlinear case is often denoted, with an abuse of terminology, as nonlinear Shannon limit. The corresponding SNR and SE, respectively given by $P/(N_0W)$ and (20.19) with $WT = 1$, are also reported.

linear regime; then, after reaching a peak value at some optimum power P_{opt}, it decreases again. This can be easily verified, for instance, by analyzing the asymptotic behavior of (20.39) for $P \to 0$ and for $P \to \infty$. This is in apparent contrast with the monotonically increasing behavior typically observed in linear channels. The peak AIR, often referred to as *nonlinear Shannon limit* [95] in the optical communications community, has been sometimes over-interpreted as a true capacity limit. Some possible reasons for such an over-interpretation will be discussed later in this section. However, it is worth recalling at this point that (20.37) is just an AIR with a fixed input distribution and a mismatched decoder optimized for an AWGN channel. It can be therefore used to *lower-bound* channel capacity, but it does not provide any indications on a possible *upper bound*. In fact, as it will be shown in Section 20.4.3, better choices of the input distribution and detection metric might yield higher AIRs.

To the best of our knowledge, a finite capacity limit for a fiber-optic channel was predicted for the first time in Ref. [41].[12] This work anticipated all others by several years, deriving a capacity curve (a lower bound, in fact, as discussed before) as in Fig. 20.12 based on a sort of GN model. Maybe for being ahead of time and not easily accessible, this paper went seemingly unnoticed for almost two decades. A similar result was independently rediscovered in Refs. [45,90,91,93], exploiting different nonlinear fiber models. The name nonlinear Shannon limit was introduced several years later in Ref. [95] with reference to the formula presented in Ref. [90]. Even though based on different approximate models, all of these studies share two important hypotheses on the fiber channel: the absence of memory and the joint Gaussianity of input and output samples. These hypotheses are explicitly mentioned in Refs. [45,90,91] and implied in the Pinsker formalism used in Ref. [93] and in the Shannon formula for the AWGN channel used in Ref. [41].

A different approach to estimate the capacity of the optical fiber channel was followed in Ref. [32], considering an optimized ring constellation as the input distribution and using the SSFM algorithm to accurately emulate fiber propagation and estimate the conditional distribution $p(y_k|x_k)$ by fitting a bivariate Gaussian distribution to a large set of realizations. The capacity was then numerically estimated from (20.1) by assuming a memoryless channel model. This approach ensures that all relevant fiber effects are accounted for and can be selectively switched on or off, so that their relative importance can also be estimated. The dominant effect turns out to be the interchannel nonlinearity (also considered in Refs. [41,45,90,91,93]). Other relevant papers on the channel capacity are [92] and [53]. The former deepens and makes explicit the GN model concept used in Ref. [41], extend the analysis to polarization-multiplexed systems, and draw the consequent conclusions in terms of channel capacity. The latter

[12] This work was developed in mathematical detail in the doctoral thesis (in German) available at https://doi.org/10.14279/depositonce-5080.

develops a detailed analytical model of the NLI based on a regular perturbation method, producing results in agreement with [32]. Similar results were also derived in Ref. [54] by using i.i.d. Gaussian input samples, neglecting channel memory, and computing (20.37) through the FRLP model.

A remarkable fact is that the aforementioned results, though obtained by following different approaches, are all in substantial agreement. Such a convergence of results is probably the cause of the widespread but unproven belief that the actual channel capacity is well approximated by those results—or that, at least, it follows the same trend—despite in many of those papers it was clearly stated that the adopted procedure yields only a *capacity lower bound* [32,54,90]. However, the observed convergence of results is rather due to the fact—not always explicitly stated—that they are all related to the computation of the same quantity (20.12), with different approaches but similar assumptions (usually i.i.d. input symbols with CSCG or uniform-ring distribution and detection metrics optimized for a Gaussian memoryless channel). Regardless of whether it is tight or not, the capacity lower bound in (20.37), shown in Fig. 20.12, has, so far, played an important role in the optical communications community: It is easy to compute, even analytically, and represents a practical limit for conventional systems. In fact, it can be reasonably assumed that all currently deployed systems operate within this limit. For this reasons, in the next Section we analyze more in details the dependence of this capacity lower bound on the link parameters and system configuration.

20.4.2.3 Dependence on link parameters and configuration

The AIR over the optical fiber channel depends on the system configuration and link parameters. In this section we analyze such a dependence, focusing on those parameters that may easily change from system to system and that are commonly object of optimization—for example, launch power, amplifier spacing and characteristics, transmission distance, in-line dispersion compensation—whereas, unless otherwise stated, the other parameters are kept constant as reported in Table 20.1. Fig. 20.13 shows the achievable SE as a function of the launch power per channel for a single-polarization fully loaded (81 channels) WDM system with IDA and various transmission distances. All the curves reach a maximum at the optimum power level P_{opt}, which, for DU-IDA links, is nearly independent of the link length—about -10 dBm per channel in this case. Doubling the distance, the SE decreases of about 1 bit/s/Hz.

At the optimum launch power P_{opt}, we obtain the peak SE. As mentioned, this can be considered as a practical limit for conventional systems. It is therefore interesting to see how this limit depends on the actual system configuration and link parameters. A first important link characteristic to consider is the amplifier configuration and spacing. Whereas the IDA link considered in Fig. 20.13 corresponds to an

Table 20.1 System parameters.

Parameter	Symbol	Value
Channel spacing	W	50 GHz
Number of subcarriers	N	variable
Symbol time	T	$T = N/W$
Pulse shape	$p(t)$	$\mathrm{sinc}(t/T)$
Attenuation	α	0.2 dB/km
Dispersion	β_2	$-21.7\ \mathrm{ps}^2/\mathrm{km}$
Nonlinear coefficient	γ	$1.27\ \mathrm{W}^{-1}\ \mathrm{km}^{-1}$
Spontaneous emission coefficient	n_{sp}	1 (IDA) or 1.6 (LA)

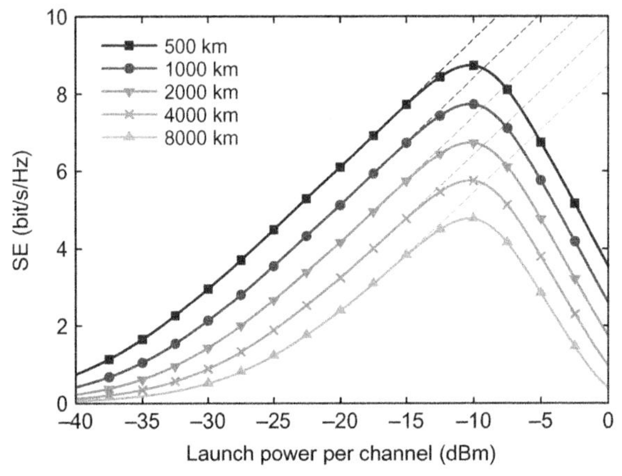

Figure 20.13 Achievable SE (based on the Gaussian AIR lower bound) versus launch power for different transmission distances. Single-polarization fully-loaded Nyquist-WDM system with 50 GHz spacing over a DU-IDA link. Dashed lines refer to the linear regime.

ideal case, hardly achievable in practical systems, a more realistic and widely deployed configuration is the LA one. Fig. 20.14 shows the peak SE as a function of the total link length for different amplifier spacings. As expected, given the amplifier spacing, the peak SE decreases with the total link length, that is, with the number of spans, due to the increasing amount of accumulated ASE noise. However, given the total link length, the peak SE increases for shorter amplifier spacing, that is, when more amplifiers are employed for the same link, as in this case the increasing number of amplifiers is more than compensated by the lower gain (and, hence, noisiness) of each amplifier.

As explained in Section 20.3, when the amplifier spacing goes to zero (and, hence, the number of amplifiers goes to infinity), the LA configuration tends to the IDA configuration. This is confirmed also by the trend of the SE curves. In practice, this limit is nearly achieved for an amplifier spacing shorter than 20 km. A further idealization is

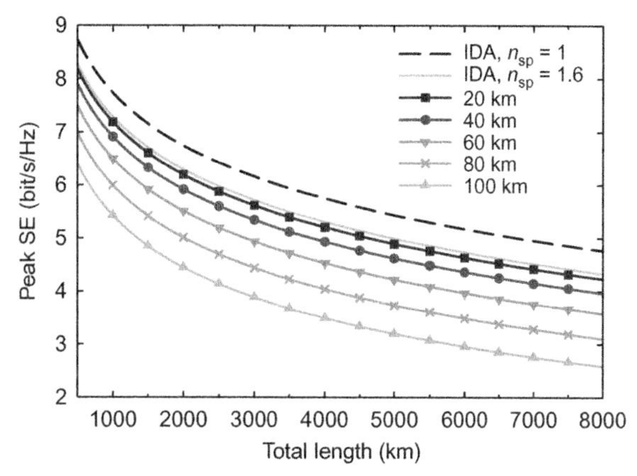

Figure 20.14 Peak achievable SE (based on the Gaussian AIR lower bound) versus distance for different amplifier spacings. Single-polarization fully-loaded Nyquist-WDM system with 50 GHz spacing over a DU-LA link. The DU-IDA links with $n_{sp} = 1.6$ or $n_{sp} = 1$ are also reported as a reference.

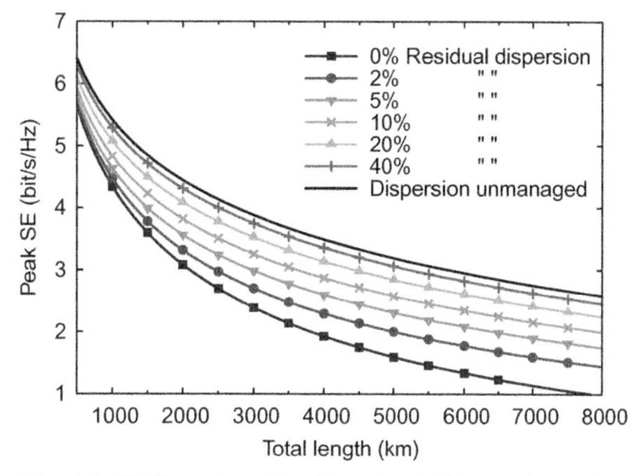

Figure 20.15 Peak achievable SE (based on the Gaussian AIR lower bound) versus distance for different dispersion maps. Single-polarization fully-loaded Nyquist-WDM system with 50 GHz spacing over a DM-LA link. The DU-LA link is also reported as a reference.

obtained when reducing the ASE spontaneous emission factor to its minimum possible value, $n_{sp} = 1$, which is the case considered in Fig. 20.13.

The current trend is that of avoiding in-line dispersion compensation. In fact, with coherent detection and DSP, DU links provide a better performance at a lower cost. It is therefore interesting to verify if the theoretical analysis on the achievable SE supports this observation. Moreover, some links with in-line dispersion compensation still exist, for example, to support legacy intensity-modulated direct-detection channels. Fig. 20.15 shows the peak SE as a function of the total link length for a DM-LA link with different values of the residual dispersion per span, ranging from full in-line compensation (0% residual dispersion) to the DU case. It is apparent that even a little amount of residual dispersion per span can significantly improve the achievable SE. This is due to the beneficial effect of the (residual) walk-off between WDM channels, which avoids a coherent accumulation of the NLI. For a sufficiently long walk-off time (corresponding to about 40% of residual dispersion in this case), the NLI contributions generated at each span become independent, so that NLI accumulates incoherently. This gives nearly the same peak SE as the DU case, and no significant improvements are obtained by further increasing residual dispersion.

We remark here that the achievable SE reported in Figs. 20.13–20.15 is obtained for a CSCG input distribution and an AWGN detection metric. The peak SE values and even their dependence on the system parameters might change if different input distributions or detection metrics are considered. In fact, *higher* peak SE values are obtained by employing better detection metrics, as it will be shown in Section 20.4.3.

20.4.3 Improved lower bounds

As discussed at the end of Section 20.4.2.2, the peak SE values in Figs. 20.13–20.15 can be easily obtained analytically and have a great value as a practical benchmark for conventional systems. However, at the current state of knowledge, they are by no means to be considered as true capacity limits. For instance, an improved capacity lower bound obtained by accounting for the presence of phase noise in single-polarization DU systems was presented in Ref. [83].

In this section it is shown that higher SE values are achievable by using subcarrier modulation (SCM) and improved detection metrics. These results can be interpreted as the tightest available (to the best of the author's knowledge) capacity lower bounds for single- and dual-polarization systems in the WDM scenario described in Section 20.3. The improved lower bounds obtained in this way still have the typical behavior of Fig. 20.12, achieving a peak value at some optimum power and decreasing after it. Again, there is no proof that the actual capacity of the nonlinear optical fiber channel (and the corresponding maximal SE) follows the same behavior rather than increasing unboundedly with power as in linear channels. This issue will be specifically discussed in Section 20.5.

According to the definition of AIR in (20.12), based on the mismatched decoding concept, there are essentially two ways to increase the AIR and get closer to channel capacity: the optimization of the detection metric $q(\mathbf{y}_K|\mathbf{x}_K)$ and the optimization of the input distribution $p(\mathbf{x}_K)$. The first approach focuses on the receiver end and, from a practical viewpoint, it requires the implementation of optimized DSP algorithms for nonlinearity mitigation and signal detection. The second approach, on the other hand, focuses on the transmitter end and involves the optimization of coding and modulation.

In this section we present some improved capacity lower bounds obtained through the combination of the two approaches mentioned above. For what concerns the detection metric, a key observation that has been made in Section 20.3.3.3 is that interchannel NLI is not truly equivalent to AWGN. For instance, as indicated by some channel models, a relevant portion of NLI can be more accurately modeled as a *linear time-variant distortion* that is significantly correlated over time and causes PPN and ISI [54,56,57,85,96,97]. This corresponds to the channel model represented in Fig. 20.11C and means that NLI can be, in part, mitigated by some classical algorithms that are commonly employed to counteract the linear version of this kind of impairments—caused, for instance by laser PN, chromatic dispersion, and polarization mode dispersion. Some examples in this sense are reported in Refs. [80,84,98,99].

The main challenge with this approach is that channel variations induced by nonlinear impairments are typically much faster than those observed in linear channels, requiring algorithms with a higher adaptation speed. Here we are not concerned with practical implementation issues, but with the AIR theoretically achievable by including this kind of mitigation strategies. To this end, limiting the analysis to PPN compensation, we consider an auxiliary channel affected by PPN and, assuming a certain statistical model for the temporal evolution of PPN, we compute the corresponding conditional distribution $q(\mathbf{y}_K|\mathbf{x}_K)$ to be replaced in (20.12). In particular, we assume a simple random walk model for both the phase and the polarization—over the circle for the former, over the Poincaré sphere for the latter—and resort to the procedure introduced in Ref. [12] and described at the end of Section 20.2.2 for the computation of $q(\mathbf{y}_K|\mathbf{x}_K)$, using a particle method to extend it to the continuous state space defined by the PPN variables [15]. The whole procedure can be found with more details in Ref. [100].

For what concerns the input distribution $p(\mathbf{x}_K)$, some improvement is achieved by considering SCM and optimizing the number of subcarriers N (or, equivalently, the symbol rate $R = 1/T = W/N$ of each subcarrier) [97,100–102], still modulating each subcarrier with i.i.d. CSCG symbols.

This is a quite subtle point since, in this case, the overall input distribution $p(\mathbf{x}_K)$ for each WDM user remains CSCG and is independent of the number of subcarriers. This means that the actual information rate (20.11) through the channel—which depends only on $p(\mathbf{x}_K)$ and on the channel law $p(\mathbf{y}_K|\mathbf{x}_K)$—remains unchanged. However, by dividing each WDM channel into many subcarriers and adopting a PPN auxiliary channel model for each of them, we are in fact modifying the auxiliary channel model $q(\mathbf{y}_K|\mathbf{x}_K)$ used for the overall COI signal. In particular, since the bandwidth of each subcarrier is only a fraction of the bandwidth of the overall COI signal, the transfer matrix $\mathrm{H}(f,t)$ of the linear time-variant model in Fig. 20.11C can be considered nearly frequency-independent over each subcarrier bandwidth. This makes the effect of $\mathrm{H}(f,t)$ much closer to a true PPN, reducing the mismatch between the true and auxiliary channels. As a consequence, the AIR in (20.12) increases and get closer to the actual information rate (20.11). However, when increasing the number of subcarriers N, also their symbol time T increases proportionally, making the time variations of $H(f,t)$ harder to track and compensate. The best performance is therefore obtained for some optimal value of N which provides the best trade-off between the impact of the time and the frequency variations of $H(f,t)$—in other words, the value of N for which the bandwidth W/N and the symbol time $T=N/W$ of each subcarrier are better matched to the coherence bandwidth and time of the channel [103,104].

Since no accurate closed-form expressions are available to estimate such improved bounds, we will resort to numerical simulations to compute (20.12) as explained above. This will also ensure that no approximations are involved in the computation of these bounds (but for a small statistical uncertainty that is inherent in the Monte Carlo averaging process). In order to provide a fair and accurate comparison with the Gaussian AIR lower bounds provided in the previous section, also the latter will be reevaluated by using the same numerical procedure. Due to computational complexity constraints, simulations will be performed by considering only $2M+1=5$ WDM channels rather than a fully loaded WDM system. Nonetheless, the gains with respect to the Gaussian AIR lower bound are expected to increase when considering more channels, as shown for instance in Ref. [84].

First, we consider a 1000 km DU-IDA link. The link parameters are the same as in Fig. 20.13 and are reported in Table 20.1. In SCM systems, each subcarrier may have a different AIR, as the "external" ones are closer to the other WDM channels and, hence, more seriously impaired by their NLI. For a fair comparison among systems with different number of subcarriers, we define the average achievable SE over the COI

$$\overline{\mathrm{SE}}_q = \frac{\sum_{n=1}^{N} I_q^{(n)}}{WT} \tag{20.42}$$

where $I_q^{(n)}$ is the AIR over the nth subcarrier. Fig. 20.16 shows the average achievable SE as a function of the launch power for (A) single-polarization and (B) polarization-multiplexed transmission, considering different numbers of subcarriers N and an improved detection metric [PN in (A) and PPN in (B)] [97,100]. As a comparison, the SE obtained with the AWGN detector is also reported. To ease the comparison between single- and dual-polarization cases, here and in the following figures, the launch power and average SE per polarization are reported.

As done in the previous section, we now proceed to analyze the dependence of the peak SE on the link parameters for the improved detection metrics considered in this section, comparing it to the peak SE achievable with the AWGN metric.

Fig. 20.17 shows the peak SE achievable over a DU-IDA link as a function of the link length for an optimized number of subcarriers and either the AWGN or the improved

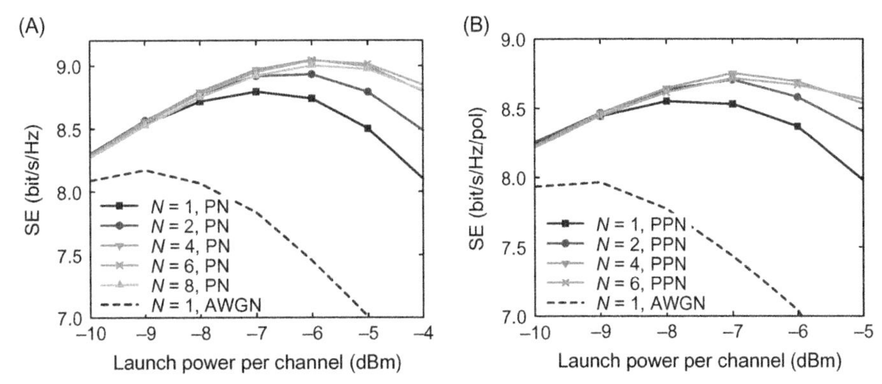

Figure 20.16 Average achievable SE per polarization based on PN or PPN detection metric and optimization of the number of subcarriers N for the COI of a 5-channel Nyquist-WDM system on a 1000 km DU-IDA link: (A) single-polarization with PN detection metric; (B) dual-polarization with PPN detection metric. *Adapted from M. Secondini, E. Agrell, E. Forestieri, D. Marsella, Fiber nonlinearity mitigation in WDM systems: strategies and achievable rates, in: Proc. Eur. Conf. Opt. Commun. (ECOC), 2017. and M. Secondini, E. Agrell, E. Forestieri, D. Marsella, M.R. Camara, Nonlinearity Mitigation in WDM Systems: Models, Strategies, and Achievable Rates, J. Lightw. Technol. 37 (10) (2019) 2270–2283, respectively.*

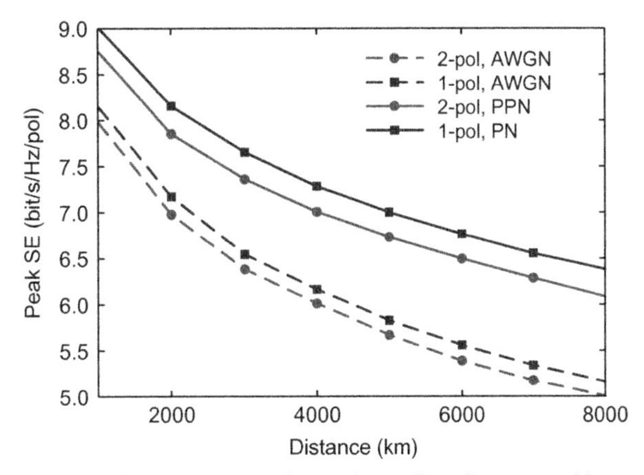

Figure 20.17 Peak SE versus distance: AWGN lower bound and improved lower bound.

(PN or PPN) detection metric. For the sake of comparison, both single-polarization and polarization-multiplexed transmissions are considered. Relevant gains (difference between the peak SE values) are obtained in both cases by using the improved detection metric. For instance, at 4000 km, the SE gain is about 1.1 bit/s/Hz in the single-polarization case, and 1 bit/s/Hz/pol in the polarization-multiplexed case. The gains slightly increase for longer distances, as the coherence times of the PN and PPN processes increase with the accumulated dispersion [54,57,105]. The gains appear even more relevant if measured in terms of reach: for a target SE of 7 bit/s/Hz/pol, the reach is doubled in the 2-pol case and more than doubled in the 1-pol case.

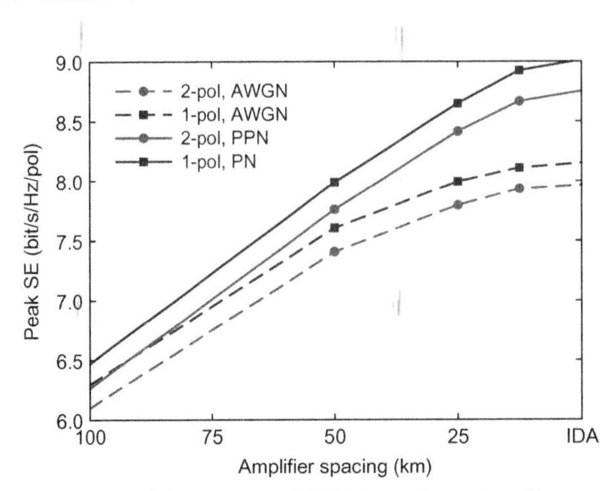

Figure 20.18 Peak SE versus amplifier spacing: AWGN lower bound and improved lower bound.

Unfortunately, the gains achievable in practical LA links are significantly lower than those achievable in IDA links. As an example, Fig. 20.18 reports the peak SE as a function of the amplifier spacing for a 1000 km DU-LA link. In order to get a continuity between the IDA case (actually corresponding to the zero-spacing limit) and the LA case, the same spontaneous emission factor $n_{sp} = 1$ is considered in both cases (rather than the value $n_{sp} = 1.6$ indicated in Table 20.1 for LA links). In the polarization-multiplexed case, the SE gain gradually decreases from 0.8 to 0.2 bit/s/Hz/pol when increasing the amplifier spacing from the IDA configuration to 100 km. A similar behavior is observed in the 1-pol case. The reason of this behavior is to be ascribed, again, to the decreased coherence times of the PN and PPN processes in LA links (where nonlinear interaction takes place only in the first portion of each span and, hence, only with shorter portions of the other WDM channels) compared to IDA links (where nonlinear interaction is evenly distributed along the whole link) [84,105].

20.5 Future perspectives and the quest for an infinite capacity

A key theoretical issue concerning the capacity of the optical fiber channel is its asymptotic behavior at high signal power. Fig. 20.19 shows three different hypothetical behaviors: Behavior A, which reaches a finite maximum and then vanishes at infinite power; behavior B, which saturates to some finite capacity value; and behavior C, which increases unbounded with power.

Posed in this way, the issue sounds purely theoretical. What is the practical value of an "infinite capacity" at infinite power, as in behavior C, given that the fiber will melt at some finite power? Or, yet, what is the practical difference between behaviors A and B?

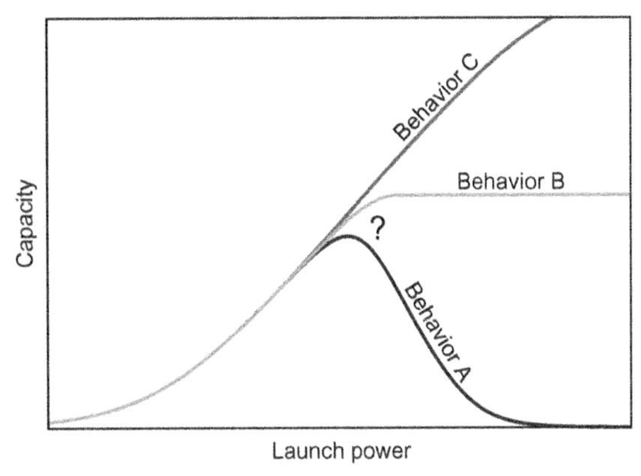

Figure 20.19 What is the right asymptotic behavior of channel capacity among the three possible ones depicted in this figure?

Indeed, this issue does have also a practical value. If A is the right answer—as all the available lower bounds and the widely spread notion of a nonlinear Shannon limit suggest—then we know what is the ultimate limit and the right operating power. We can understand how far we actually are from the limit, if it is worth or not trying to improve the performance, and when the fiber capacity will likely be saturated.

On the other hand, if C is the answer—as in many linear channels—then Kerr nonlinearity is not the main limitation to channel capacity. This means that it might be worth focusing on practical approaches for nonlinearity mitigation, before giving up to new generations of optical fibers. And that the ultimate limitation will come from another source, which is not included in the fiber propagation model but that definitely should be.

And what about the third option, behavior B? Does it make any difference with respect to behavior A? Apparently not—why should one ever consider using more power than the one needed to achieve the maximum capacity value? Indeed, the capacity for some given average input power P^* may even be defined by considering P^* as an *available* power rather than as a *fixed* input power. Formally, this means defining the capacity with the inequality constraint $P \leq P^*$ rather than the equality constraint $P = P^*$ on the input power. With this kind of constraint, channel capacity cannot decrease with the available power, so that behavior A is impossible.

Yet understanding what is the right behavior when the equality constraint is used might be more than a mathematical curiosity. This very issue was studied in Ref. [106], where it is shown that the capacity of a discrete-time channel, even when defined through an equality constraint on the input power, cannot decrease with power. This entails that behavior A, observed in all the lower bounds seen so far, is not a characteristic of the true channel capacity and should be more properly ascribed to the use of nonoptimized input distributions and detection metrics. For instance, the lower bounds in Fig. 20.13 or 20.16, which are obtained with Gaussian input distributions, are very tight at low power but still follow behavior A and vanish at high power. On the other hand, the concept of *satellite distributions*—distributions that, for instance, are obtained from some initial distribution by adding a symbol with very high power and low probability, a "satellite," such that the average power is increased arbitrarily without affecting the information rate obtained with the original distribution—can be used to obtain a nondecreasing capacity lower bound, changing the behavior from A to B [106]. This example does not provide a practical way to go beyond the peak AIR of behavior A, but clearly shows that the optimal distribution depends on power and that, at high power, might be quite different from the Gaussian one, that is optimal in the linear regime. Indeed, it should be considered as a recommendation not to underestimate the importance of optimizing the input distribution by considering even nonconventional distributions and modulation formats.

We have seen many examples of AIRs following behavior A, and one possible way to modify the input distribution in order to get an AIR following behavior B. Are there any examples of an ever increasing AIR, following behavior C? To the best of the author's knowledge, not for a realistic optical fiber channel. However, many authors have demonstrated an infinite asymptotic capacity (behavior C) for some optically related channels—channels that have some of the peculiar characteristics of the nonlinear optical fiber channel, but that are simple enough to allow for an analytical study of their capacity.

The first example is the zero-dispersion sampled fiber channel in the single-user scenario considered in Ref. [107], which is obtained by setting $\beta_2 = 0$ in (20.28), considering a single user in (20.30), and a sampling demodulator (without matched filtering) at the output. This simplified channel avoids the complication of memory and allows for a relatively simple closed-form description of the input-output conditional distribution, still preserving the effect of signal-noise interaction induced by Kerr nonlinearity. When computing the AIR with an AWGN detection metric, even considering a single-channel transmission (hence removing interchannel nonlinearity), this effect is enough to make the AIR follow behavior A and vanish at high power. On the other hand, by accounting for the non-Gaussian statistics of the channel induced by signal-noise interaction, it is possible to obtain the following capacity lower bound

$$C \geq \frac{1}{2}\log_2\left(\frac{P}{N}\right) + O(1) \tag{20.43}$$

where P and N are the signal and noise power, respectively. According to (20.43), the capacity grows unbounded as $P \rightarrow \infty$, as in behavior C, at about half the rate of the corresponding linear channel [107]. An intuitive way to understand this result is the following. Considering the amplitude and phase of the transmitted symbols as two different degrees of freedom to encode information, one degree of freedom (the phase) is completely lost at high power due to the random phase modulation induced by signal–noise interaction, while the other (the amplitude) is fully preserved at any power, being unaffected by signal–noise interaction. Some improved capacity lower and upper bounds for this channel have been proposed in Refs. [108] and [109], respectively, confirming the asymptotic capacity growth rate shown in Ref. [107]. The main limitations of this analysis come from the absence of memory and by the use of a sampling receiver—a receiver that has no band limitations. Without a band limitation it is not possible to translate capacity values into SE values, as done in (20.21). In fact, at zero dispersion, any amplitude modulation of the propagating signal (or even small amplitude fluctuation caused by noise) is directly converted into a phase modulation, with a modulation depth that grows proportionally with the launch power. As a result, asymptotically, also the signal bandwidth grows linearly with power and the SE vanishes [108]. Moreover, it should be noted that the unbounded increase of the information rate is demonstrated for a *fixed noise power* N. However, when the signal bandwidth increases, the demodulator bandwidth should also increase, letting also the noise power N increase proportionally. In this case, if the proportional increase of noise and signal power is accounted for, both the SNR and the information rate asymptotically saturates to a finite value, as in behavior B.[13]

Another interesting example is the memoryless four-wave mixing channel studied in Ref. [39], which still avoids the complication of memory and neglects spectral broadening issues, but introduces the ICI induced by Kerr nonlinearity in a multiuser scenario, considering different possible behavioral models for the interfering users with respect to the COI user. In particular, in the adaptive interferer distribution model considered so far in our analysis and introduced in Section 20.3.2, the use of a fixed Gaussian input distribution makes the mutual information vanish at high power. On the other hand, the use of an optimized discrete input distribution allows to obtain a capacity lower bound that increases unbounded with power (behavior C).

A third example, which we analyze here more in detail, is the RP channel studied in Ref. [70]. The considered channel model is based on the first-order RP model developed in Refs. [53,57] and accounts both for the ICI induced by Kerr nonlinearity and for the memory induced by dispersion. For two WDM channels (with the same modulation format and power) and a memory of N symbols, the channel is described by the following expression

[13] These are not rigorous demonstrations that the SE vanishes or that the capacity saturates to a finite value, but just indications of the limits of the simplified analysis based on the sampling receiver.

$$y_k = x_k + \sum_{\ell,m,n=0}^{N-1} c_{\ell,m,n} z_{k-m} z_{k-n}^* x_{k-\ell} + n_k \tag{20.44}$$

where z_k is the symbol transmitted on the interfering channel at time k and $\{c_{\ell,m,n}\}$, with $\ell, m, n = 0, \ldots, N-1$ are the channel coefficients.

The key characteristic of this model is that it provides an accurate description of the main nonlinear effects which are relevant to determine the Gaussian AIR lower bound presented in Section 20.4.2 and widely accepted as the one determining the so-called nonlinear Shannon limit. Indeed, by considering a sufficient number of channels and the right values of the channel coefficients, (20.44) can be used to obtain an accurate estimate of this lower bound [53,57]. The model neglects other nonlinear effects, such as intrachannel nonlinearity, four wave mixing, signal–noise interaction, and spectral broadening, which have, however, a negligible impact on the lower bound.

Fig. 20.20 compares the AIRs as a function of the SNR for: (A) CSCG input symbols and a mismatched detector optimized for an AWGN channel (the Gaussian AIR); (B) various M-PSK input alphabets and a symbol-by-symbol matched detector—that is, matched to the marginal distribution $p(y_k | x_k)$, still assuming conditional independence of the output samples. In the figure, the latter AIR is referred to as symbol-by-symbol mutual information (S × S MI). In this example, the noise has unitary variance, the coefficients $\{c_{\ell mn}\}$ are randomly and independently drawn from a Gaussian distribution with variance 0.01, and the SNR is varied by varying the mean energy of the input symbols.

Fig. 20.20 shows that, while the Gaussian AIR has a maximum and then vanishes at high SNR, channel capacity grows unbounded with power. In particular, for any PSK alphabet order M and at sufficiently high SNR, the S × S MI saturates to $\log_2 M$ bits/symbol. This is rigorously demonstrated in Ref. [70].

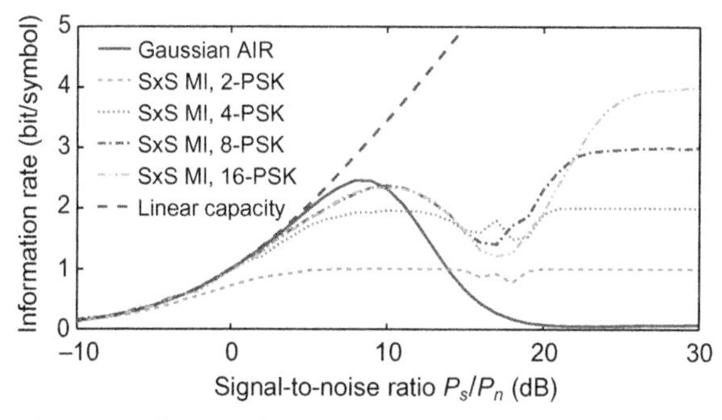

Figure 20.20 AIR as a function of power for the RP channel (20.44) and different input modulations and detection metrics. Source: *Adapted from M. Secondini, E. Forestieri, Scope and limitations of the nonlinear Shannon limit, J. Lightw.Technol. 35 (4) (2017) 893–902.*

A fourth example is the single-channel perturbative model reported in Ref. [110], which accounts for the memory induced by dispersion and where both deterministic intrachannel nonlinearity and signal–noise interaction are included. For this channel model, it is shown that by optimizing the input distribution within a family of particular distributions—referred to as ripple distribution, where the distribution of the amplitude is given by a weighted sum of Rice distributions and the phase is uniformly distributed—a monotonically increasing capacity lower bound can be derived, at least up to the limit of validity of the model.

In all the aforementioned examples, the Gaussian AIR has a finite maximum, but the (per symbol) capacity grows unbounded with power. This is still very far from a practical scheme that can guarantee such an unlimited capacity over a realistic fiber channel, or even to demonstrate that its capacity actually follows the behavior C in Fig. 20.19. In fact, in all the examples, the considered channel models entail some approximation or simplification with respect to the original channel, so that their validity is confined to some finite power range and/or idealized scenario. Moreover, none of the previous results account for the spectral broadening induced by fiber nonlinearity, such that a finite SE limit should still be expected. However, these results are sufficient to cast some reasonable doubts on the existence of a finite nonlinear Shannon limit or, at least, to hope that the lower bounds available so far are not tight at high power and could be significantly improved.

In information theory, when the capacity of a specific channel cannot be exactly calculated, the problem is usually faced by trying to confine the capacity between a *lower bound* and an *upper bound*. If the bounds are close to each other, they provide a reasonable (though not exact) answer to the problem. So far, in our analysis, we have faced the capacity problem only from the lower bound perspective, without providing any upper bound. Clearly, even a couple of loose lower and upper bounds could be sufficient to solve the problem highlighted in Fig. 20.19, if the bounds have the same asymptotic behavior. For instance, both a lower and an upper bound are provided in Ref. [109] for the zero-dispersion fiber with the sampling receiver. In this case, both the bounds follow behavior C and show that, asymptotically, the capacity grows with power at about half the rate compared to the AWGN channel.

To the best of the author's knowledge, the tightest available upper bound for a realistic optical fiber channel (based on a split-step model of the fiber) is the one derived in Ref. [62]. The work uses two basic information theoretical tools—the entropy power inequality [5] and the maximum entropy theorem [8]—and the fact that both dispersion and nonlinearity induce unitary transformations on the signal, to demonstrate that the capacity of the nonlinear optical fiber channel is upper bounded by the capacity of the corresponding AWGN channel. In other words, the work demonstrates that fiber nonlinearity does not increase the capacity of the channel. The extension of the result to the SE is, however, not straightforward.

In conclusion, the distance between the tightest available lower and upper bounds for a realistic optical fiber channel increases with power. While all the available lower bounds follow behavior A, the upper bound follow behavior C, leaving the capacity problem at high power completely open.

Acknowledgments

The author thanks Menelaos Ralli and Enrico Forestieri from Scuola Superiore Sant'Anna, Erik Agrell from Chalmers University of Technology, and Domenico Marsella from Nokia for useful and stimulating discussions and for their contributions to some of the results discussed in this chapter.

References

[1] E. Agrell, M. Karlsson, A.R. Chraplyvy, D.J. Richardson, P.M. Krummrich, P. Winzer, et al., Roadmap of optical communications, J. Optics 18 (6) (2016) 063002.
[2] C.E. Shannon, A mathematical theory of communication, Bell Syst. Tech. J. 27 (3/4) (1948). 379–423/623–656.
[3a] S. Verdú, Fifty years of Shannon theory, IEEE Trans. Inform. Theory 44 (6) (1998) 2057–2078.
[3b] E. Agrell, M. Secondini, Information-Theoretic Tools for Optical Communications Engineers, in: IEEE Photonics Conference (IPC), Reston, VA, USA, 2018.
[4] R.G. Gallager, Information Theory and Reliable Communication, Wiley, New York, 1968.
[5] T.M. Cover, J.A. Thomas, Elements of Information Theory, second ed., Wiley, Hoboken, NJ, 2006.
[6] S. Arimoto, An algorithm for computing the capacity of arbitrary discrete memoryless channels, IEEE Trans. Inf. Theory 18 (1) (1972) 14–20.
[7] R. Blahut, Computation of channel capacity and rate-distortion functions, IEEE Trans. Inf. Theory 18 (4) (1972) 460–473.
[8] F.D. Neeser, J.L. Massey, Proper complex random processes with applications to information theory, IEEE Trans. Inform. Theory 39 (4) (1993) 1293–1302.
[9] J.G. Proakis, Digital Communications, fourth ed., McGraw Hill, 2001.
[10] A. Alvarado, T. Fehenberger, B. Chen, F.M.J. Willems, Achievable information rates for fiber optics: applications and computations, J. Lightw. Technol. 36 (2) (2018) 424–439.
[11] F. Buchali, F. Steiner, G. Böcherer, L. Schmalen, P. Schulte, W. Idler, Rate adaptation and reach increase by probabilistically shaped 64-QAM: An experimental demonstration, J. Lightw. Technol. 34 (7) (2016) 1599–1609.
[12] D.M. Arnold, H.-A. Loeliger, P.O. Vontobel, A. Kavvcic, W. Zeng, Simulation-based computation of information rates for channels with memory, IEEE Trans. Inform. Theory 52 (8) (2006) 3498–3508.
[13] F.R. Kschischang, B.J. Frey, H.-A. Loeliger, Factor graphs and the sum-product algorithm, IEEE Trans. Inf. Theory 47 (2) (2001) 498–519.
[14] L. Bahl, J. Cocke, F. Jelinek, J. Raviv, Optimal decoding of linear codes for minimizing symbol error rate, IEEE Trans. Inf. Theory 20 (2) (1974) 284–287.
[15] J. Dauwels, H.-A. Loeliger, Computation of information rates by particle methods, IEEE Trans. Inf. Theory 54 (1) (2008) 406–409.
[16] P.O. Vontobel, A. Kavčić, D.M. Arnold, H. Loeliger, A generalization of the Blahut–Arimoto algorithm to finite-state channels, IEEE Trans. Inf. Theory 54 (5) (2008) 1887–1918.
[17] G. Colavolpe, A. Barbieri, G. Caire, Algorithms for iterative decoding in the presence of strong phase noise, IEEE Journal on selected areas in communications 23 (9) (2005) 1748–1757.
[18] D. Slepian, On bandwidth, Proc. IEEE 64 (1976) 292–300.

[19] E. Agrell, Capacity bounds in optical communications, in: Proc. Eur. Conf. Opt. Commun. (ECOC), 2017.
[20] G.P. Agrawal, Fiber-Optic Communications Systems, third ed., Wiley, 2002.
[21] L.B. Du, A.J. Lowery, Channelized chromatic dispersion compensation for XPM suppression and simplified digital SPM compensation, in: Proc. Optical Fiber Communication Conf. (OFC), San Francisco, CA, USA, 2014.
[22] K. Keykhosravi, M. Secondini, G. Durisi, E. Agrell, How to Increase the Achievable Information Rate by Per-Channel Dispersion Compensation, J. Lightw. Technol. 37 (10) (2019) 2443–2451.
[23] G.P. Agrawal, Nonlinear Fiber Optics, third ed., Academic Press, San Diego, CA, 2001.
[24] A. Ellis, M. McCarthy, M. Al-Khateeb, S. Sygletos, Capacity limits of systems employing multiple optical phase conjugators, Opt. Exp. 23 (16) (2015) 20381–20393.
[25] M. Sorokina, S. Turitsyn, Regeneration limit of classical shannon capacity, Nature Communications 5 (2014) 3861.
[26] C. Antonelli, A. Mecozzi, M. Shtaif, Scaling of inter-channel nonlinear interference noise and capacity with the number of strongly coupled modes in SDM systems, in: Optical Fiber Communication Conference, 2016, p. W4I.2.47.
[27] F. Poletti, N. Wheeler, M. Petrovich, N. Baddela, E.N. Fokoua, J. Hayes, D. Gray, Z. Li, R. Slavík, D. Richardson, Towards high-capacity fibre-optic communications at the speed of light in vacuum, Nature Photonics 7 (4) (2013) 279.
[28] A.S. Holevo, The capacity of the quantum channel with general signal states, IEEE Trans. Inf. Theory 44 (1) (1998) 269–273.
[29] V. Giovannetti, R. Garcia-Patron, N.J. Cerf, A.S. Holevo, Ultimate classical communication rates of quantum optical channels, Nature Photonics 8 (10) (2014) 796.
[30] E. Wang, C.R. Menyuk, Polarization evolution due to the Kerr nonlinearity and chromatic dispersion, J. Lightw. Technol. 17 (12) (1999) 2520–2529.
[31] P.K.A. Wai, C.R. Menyuk, H.H. Chen, Stability of solitons in randomly varying birefringent fibers, Opt. Lett. 16 (16) (1991) 1231–1233.
[32] R.-J. Essiambre, G. Kramer, P.J. Winzer, G.J. Foschini, B. Goebel, Capacity limits of optical fiber networks, J. Lightw. Technol. 28 (4) (2010) 662–701.
[33] R.J. Essiambre, P.J. Winzer, Fibre nonlinearities in electronically pre-distorted transmission, in: Proc. Eur. Conf. Opt. Commun. (ECOC), vol. 2, 2005, pp. 191–192.
[34] E. Ip, J.M. Kahn, Compensation of dispersion and nonlinear impairments using digital backpropagation, J. Lightw. Technol. 26 (20) (2008) 3416–3425.
[35] E.E. Narimanov, P. Mitra, The channel capacity of a fiber optics communication system: Perturbation theory, J. Lightwave Technol. 20 (3) (2002) 530–537.
[36] E. Forestieri, M. Secondini, The nonlinear fiber-optic channel: Modeling and achievable information rate, Proc. of Progress In Electromagnetics Research Symposium (PIERS) (2015) 1276–1283.
[37] N.V. Irukulapati, M. Secondini, E. Agrell, P. Johannisson, H. Wymeersch, Improved lower bounds on mutual information accounting for nonlinear signal-noise interaction, J. Lightw. Technol. 36 (22) (2018) 5152–5159.
[38] M.H. Taghavi, G.C. Papen, P.H. Siegel, On the multiuser capacity of WDM in a nonlinear optical fiber: coherent communication, IEEE Trans. Inform. Theory 52 (11) (2006) 5008–5022.
[39] E. Agrell, M. Karlsson, Influence of behavioral models on multiuser channel capacity, J. Lightw. Technol. 33 (17) (2015) 3507–3515.
[40] E. Agrell, G. Durisi, P. Johannisson, Information-theory-friendly models for fiber-optic channels: a primer, in: IEEE Information Theory Workshop (ITW), 2015.
[41] A. Splett, C. Kurtzke, K. Petermann, Ultimate transmission capacity of amplified optical fiber communication systems taking into account fiber nonlinearities, in: Proc. Eur. Conf. Opt. Commun. (ECOC), vol. 2, 1993, pp. 41–44.
[42] K.V. Peddanarappagari, M. Brandt-Pearce, Volterra series transfer function of single-mode fibers, J. Lightw. Technol. 15 (1997) 2232–2241.
[43] A. Cartaxo, Cross-phase modulation in intensity modulation-direct detection WDM systems with multiple optical amplifiers and dispersion compensators, J. Lightw. Technol. 17 (2) (1999) 178–190.

[44] R. Holzlöhner, V.S. Grigoryan, C.R. Menyuk, W.L. Kath, Accurate calculation of eye diagrams and bit error rates in optical transmission systems using linearization, J. Lightw. Technol. 20 (3) (2002) 389–400.
[45] A.G. Green, P.B. Littlewood, P.P. Mitra, L.G.L. Wegener, Schrödinger equation with a spatially and temporally random potential: effects of cross-phase modulation in optical communication, Phys. Rev. E 66 (4) (2002). 046627.
[46] A. Vannucci, P. Serena, A. Bononi, The RP method: a new tool for the iterative solution of the nonlinear Schrödinger equation, J. Lightw. Technol. 20 (7) (2002) 1102–1112.
[47] K.-P. Ho, Error probability of DPSK signals with cross-phase modulation induced nonlinear phase noise, J. Sel. Topics Quantum Electron. 10 (2) (2004) 421–427.
[48] S. Kumar, D. Yang, Second-order theory for self-phase modulation and cross-phase modulation in optical fibers, J. Lightw. Technol. 23 (6) (2005) 2073–2080.
[49] P. Serena, A. Bononi, J.-C. Antona, S. Bigo, Parametric gain in the strongly nonlinear regime and its impact on 10-Gb/s NRZ systems with forward-error correction, J. Lightw. Technol. 23 (8) (2005) 2352–2363.
[50] M. Secondini, E. Forestieri, C.R. Menyuk, A combined regular-logarithmic perturbation method for signal-noise interaction in amplified optical systems, J. Lightw. Technol. 27 (16) (2009) 3358–3369.
[51] M. Winter, C.A. Bunge, D. Setti, K. Petermann, A statistical treatment of cross-polarization modulation in DWDM systems, J. Lightw. Technol. 27 (17) (2009) 3739–3751.
[52] P. Poggiolini, The GN model of non-linear propagation in uncompensated coherent optical systems, J. Lightw. Technol. 30 (24) (2012) 3857–3879.
[53] A. Mecozzi, R.-J. Essiambre, Nonlinear Shannon limit in pseudolinear coherent systems, J. Lightw. Technol. 30 (12) (2012) 2011–2024.
[54] M. Secondini, E. Forestieri, Analytical fiber-optic channel model in the presence of cross-phase modulation, IEEE Photon. Technol. Lett. 24 (22) (2012) 2016–2019.
[55] L. Beygi, E. Agrell, P. Johannisson, M. Karlsson, H. Wymeersch, A discrete-time model for uncompensated single-channel fiber-optical links, IEEE Trans. Commun. 60 (11) (2012) 3440–3450.
[56] M. Secondini, E. Forestieri, G. Prati, Achievable information rate in nonlinear WDM fiber-optic systems with arbitrary modulation formats and dispersion maps, J. Lightw. Technol. 31 (23) (2013) 3839–3852.
[57] R. Dar, M. Feder, A. Mecozzi, M. Shtaif, Properties of nonlinear noise in long, dispersion-uncompensated fiber links, Opt. Exp. 21 (22) (2013) 25685–25699.
[58] P. Poggiolini, G. Bosco, A. Carena, V. Curri, Y. Jiang, F. Forghieri, The GN-model of fiber nonlinear propagation and its applications, J. Lightw. Technol. 32 (4) (2014) 694–721.
[59] A. Carena, G. Bosco, V. Curri, Y. Jiang, P. Poggiolini, F. Forghieri, EGN model of non-linear fiber propagation, Opt. Exp. 22 (13) (2014) 16335–16362.
[60] E. Agrell, A. Alvarado, F.R. Kschischang, Implications of information theory in optical fibre communications, Philos. Trans. R. Soc. A 374 (2016). Art. no. 20140438.
[61] T.R. Taha, M.J. Ablowitz, Analytical and numerical aspects of certain nonlinear evolution equations. II. Numerical, nonlinear Schrödinger equation, J. Comput. Phys. 55 (1984) 203–230.
[62] G. Kramer, M.I. Yousefi, F.R. Kschischang, Upper bound on the capacity of a cascade of nonlinear and noisy channels, in: IEEE Information Theory Workshop (ITW), 2015.
[63] P. Poggiolini, A generalized GN-model closed-form formula, arXiv preprint arXiv:1810.06545.
[64] P. Poggiolini, M.R. Zefreh, G. Bosco, F. Forghieri, S. Piciaccia, Accurate nonlinearity fully-closed-form formula based on the GN/EGN model and large-data-set fitting, in: Proc. Opt. Fiber Commun. Conf. (OFC), 2019, paper M1I-4.
[65] D. Semrau, R.I. Killey, P. Bayvel, The Gaussian noise model in the presence of interchannel stimulated Raman scattering, J. Lightw. Technol. 36 (14) (2018) 3046–3055.
[66] I. Roberts, J.M. Kahn, J. Harley, D.W. Boertjes, Channel power optimization of WDM systems following Gaussian noise nonlinearity model in presence of stimulated Raman scattering, J. Lightw. Technol. 35 (23) (2017) 5237–5249.

[67] M. Cantono, D. Pilori, A. Ferrari, C. Catanese, J. Thouras, J.-L. Augé, V. Curri, On the interplay of nonlinear interference generation with stimulated Raman scattering for QoT estimation, J. Lightw. Technol. 36 (15) (2018) 3131–3141.
[68] D. Semrau, R.I. Killey, P. Bayvel, A closed-form approximation of the Gaussian noise model in the presence of inter-channel stimulated Raman scattering, J. Lightw. Technol. 37 (9) (2019) 1924–1936.
[69] D. Semrau, E. Sillekens, R.I. Killey, P. Bayvel, A modulation format correction formula for the Gaussian noise model in the presence of inter-channel stimulated Raman scattering, arXiv preprint arXiv:1903.02506.
[70] M. Secondini, E. Forestieri, Scope and limitations of the nonlinear Shannon limit, J. Lightw. Technol. 35 (4) (2017) 893–902.
[71] D. Zwillinger, Handbook of Differential Equations, third ed., Academic Press, 1998.
[72] E. Forestieri, M. Secondini, Solving the nonlinear Schrödinger equation, in: E. Forestieri (Ed.), Optical Communication Theory and Techniques, Springer, New York, 2005, pp. 3–11.
[73] M. Karlsson, Modulational instability in lossy optical fibers, J. Opt. Soc. Am. B 12 (11) (1995) 2071–2077.
[74] A. Carena, V. Curri, R. Gaudino, P. Poggiolini, S. Benedetto, New analytical results on fiber parametric gain and its effects on ASE noise, IEEE Photon. Technol. Lett. 9 (4) (1997) 535–537.
[75] B. Xu, M. Brandt-Pearce, Comparison of FWM- and XPM-induced crosstalk using the Volterra series transfer function method, J. Lightw. Technol. 21 (1) (2003) 40–53.
[76] J.P. Gordon, L.F. Mollenauer, Phase noise in photonic communications systems using linear amplifiers, Opt. Lett. 15 (23) (1990) 1351–1353.
[77] A. Mecozzi, Limits to long-haul coherent transmission set by the Kerr nonlinearity and noise of the in-line amplifiers, J. Lightw. Technol. 12 (1994) 1993–2000.
[78] K.-P. Ho, Phase-Modulated Optical Communication Systems, Springer, 2005.
[79] B. Xu, M. Brandt-Pearce, Modified Volterra series transfer function method, IEEE Photon. Technol. Lett. 14 (1) (2002) 47–49.
[80] R. Dar, M. Feder, A. Mecozzi, M. Shtaif, Inter-channel nonlinear interference noise in WDM systems: modeling and mitigation, J. Lightw. Technol. 33 (5) (2015) 1044–1053.
[81] E. Ciaramella, E. Forestieri, Analytical approximation of nonlinear distortions, IEEE Photon. Technol. Lett. 17 (1) (2005) 91–93.
[82] A. Carena, V. Curri, G. Bosco, P. Poggiolini, F. Forghieri, Modeling of the impact of nonlinear propagation effects in uncompensated optical coherent transmission links, J. Lightw. Technol. 30 (10) (2012) 1524–1539.
[83] R. Dar, M. Shtaif, M. Feder, New bounds on the capacity of the nonlinear fiber-optic channel, Opt. Lett. 39 (2) (2014) 398–401.
[84] M. Secondini, E. Forestieri, On XPM mitigation in WDM fiber-optic systems, IEEE Photon. Technol. Lett. 26 (22) (2014) 2252 2255.
[85] R. Dar, M. Feder, A. Mecozzi, M. Shtaif, Pulse collision picture of inter-channel non-linear interference in fiber-optic communications, J. Lightw. Technol. 34 (2) (2016) 593–607.
[86] A. Mecozzi, M. Shtaif, On the capacity of intensity modulated systems using optical amplifiers, IEEE Photon. Technol. Lett. 13 (9) (2001) 1029–1031.
[87] A. Lapidoth, On phase noise channels at high SNR, in: Proc. of Information Theory Workshop, 2002.
[88] S. Hranilovic, F.R. Kschischang, Capacity bounds for power-and band-limited optical intensity channels corrupted by Gaussian noise, IEEE Trans. Inf. Theory 50 (5) (2004) 784–795.
[89] A. Mecozzi, M. Shtaif, Information capacity of direct detection optical transmission systems, J. Lightw. Technol. 36 (3) (2018) 689–694.
[90] P.P. Mitra, J.B. Stark, Nonlinear limits to the information capacity of optical fiber communications, Nature 411 (6841) (2001) 1027–1030.
[91] L.G.L. Wegener, M.L. Povinelli, A.G. Green, P.P. Mitra, J.B. Stark, P.B. Littlewood, The effect of propagation nonlinearities on the information capacity of WDM optical fiber systems: cross-phase modulation and four-wave mixing, Phys. D Nonlinear Phenom. 189 (1–2) (2004) 81–99.

[92] G. Bosco, P. Poggiolini, A. Carena, V. Curri, F. Forghieri, Analytical results on channel capacity in uncompensated optical links with coherent detection, Opt. Exp. 19 (26) (2011) B438−B449.
[93] J. Tang, The channel capacity of a multispan DWDM system employing dispersive nonlinear optical fibers and an ideal coherent optical receiver, J. Lightw. Technol. 20 (7) (2002) 1095−1101.
[94] J. Tang, A comparison study of the Shannon channel capacity of various nonlinear optical fibers, J. Lightw. Technol. 24 (5) (2006) 2070−2075.
[95] A.D. Ellis, Z. Jian, D. Cotter, Approaching the non-linear Shannon limit, J. Lightw. Technol. 28 (4) (2010) 423−433.
[96] R. Dar, M. Feder, A. Mecozzi, M. Shtaif, Time varying ISI model for nonlinear interference noise, in: Proc. Opt. Fiber Commun. Conf. (OFC), 2014.
[97] M. Secondini, E. Agrell, E. Forestieri, D. Marsella, Fiber nonlinearity mitigation in WDM systems: strategies and achievable rates, in: Proc. Eur. Conf. Opt. Commun. (ECOC), 2017.
[98] P. Serena, A. Ghazisaeidi, A. Bononi, A new fast and blind cross-polarization modulation digital compensator, in: Proc. Eur. Conf. Opt. Commun. (ECOC), 2012.
[99] M.P. Yankov, T. Fehenberger, L. Barletta, N. Hanik, Low-complexity tracking of laser and nonlinear phase noise in WDM optical fiber systems, J. Lightw. Technol. 33 (23) (2015) 4975−4984.
[100] M. Secondini, E. Agrell, E. Forestieri, D. Marsella, M.R. Camara, Nonlinearity Mitigation in WDM Systems: Models, Strategies, and Achievable Rates, J. Lightw. Technol. 37 (10) (2019) 2270−2283.
[101] D. Marsella, M. Secondini, E. Agrell, E. Forestieri, A simple strategy for mitigating XPM in nonlinear WDM optical systems, in: Proc. Opt. Fiber Commun. Conf. (OFC), 2015.
[102] R. Dar, P.J. Winzer, Nonlinear interference mitigation: methods and potential gain, J. Lightw. Technol. 35 (4) (2017) 903−930.
[103] P. Bello, Characterization of randomly time-variant linear channels, IEEE Trans. on Commun. Syst. 11 (4) (1963) 360−393.
[104] K. Liu, T. Kadous, A.M. Sayeed, Orthogonal time-frequency signaling over doubly dispersive channels, IEEE Trans. Inform. Theory 50 (11) (2004) 2583−2603.
[105] O. Golani, R. Dar, M. Feder, A. Mecozzi, M. Shtaif, Modeling the bit-error-rate performance of nonlinear fiber-optic systems, J. Lightw. Technol. 34 (15) (2016) 3482−3489.
[106] E. Agrell, Conditions for a monotonic channel capacity, IEEE Trans. Commun. 63 (3) (2015) 738−748.
[107] K.S. Turitsyn, S.A. Derevyanko, I.V. Yurkevich, S.K. Turitsyn, Information capacity of optical fiber channels with zero average dispersion, Phys. Rev. Lett. 91 (20) (2003). 203901.
[108] M.I. Yousefi, F.R. Kschischang, On the per-sample capacity of nondispersive optical fibers, IEEE Trans. Inf. Theory 57 (11) (2011) 7522−7541.
[109] K. Keykhosravi, G. Durisi, E. Agrell, A tighter upper bound on the capacity of the nondispersive optical fiber channel, in: Proc. Eur. Conf. Opt. Commun. (ECOC), 2017.
[110] M. Sorokina, S. Sygletos, S. Turitsyn, Ripple distribution for nonlinear fiber-optic channels, Opt. Exp. 25 (3) (2017) 2228−2238.

CHAPTER 21

Machine learning methods for optical communication systems and networks

Faisal Nadeem Khan[1], Qirui Fan[1], Chao Lu[2] and Alan Pak Tao Lau[1]
[1]Photonics Research Centre, Department of Electrical Engineering, The Hong Kong Polytechnic University, Hong Kong
[2]Photonics Research Centre, Department of Electronic and Information Engineering, The Hong Kong Polytechnic University, Hong Kong

21.1 Introduction

Artificial intelligence (AI) makes use of computers/machines to perform cognitive tasks, that is, the ones requiring knowledge, perception, learning, reasoning, understanding, and other similar cognitive abilities. An AI system is expected to do three things: (1) store knowledge, (2) apply the stored knowledge to solve problems, and (3) acquire new knowledge via experience. The three key components of an AI system include knowledge representation, machine learning (ML), and automated reasoning. ML is a branch of AI that is based on the idea that patterns and trends in a given data set can be learned automatically through algorithms. The learned patterns and structures can then be used to make decisions or predictions on some other data in the system of interest [1].

ML is not a new field as ML-related algorithms exist at least since the 1970s. However, tremendous increase in computational power over the last decade, recent groundbreaking developments in theory and algorithms surrounding ML, and easy access to an overabundance of all types of data worldwide (thanks to three decades of Internet growth) have all contributed to the advent of modern deep learning (DL) technology, a class of advanced ML approaches that displays superior performance in an ever-expanding range of domains. In the near future, ML is expected to power numerous aspects of modern society such as web searches, computer translation, content filtering on social media networks, healthcare, finance, and laws [2].

ML is an interdisciplinary field that shares common threads with the fields of statistics, optimization, information theory, and game theory. Most ML algorithms perform one of two types of pattern recognition tasks, as shown in Fig. 21.1. In the first type, the algorithm tries to find some functional description of given data with the aim of predicting values for new inputs, that is, *regression problem*. The second type attempts to find suitable decision boundaries to distinguish different data classes, that is, *classification problem* [3], which is more commonly referred to as *clustering problem* in ML literature.

Optical Fiber Telecommunications V11
DOI: https://doi.org/10.1016/B978-0-12-816502-7.00029-4

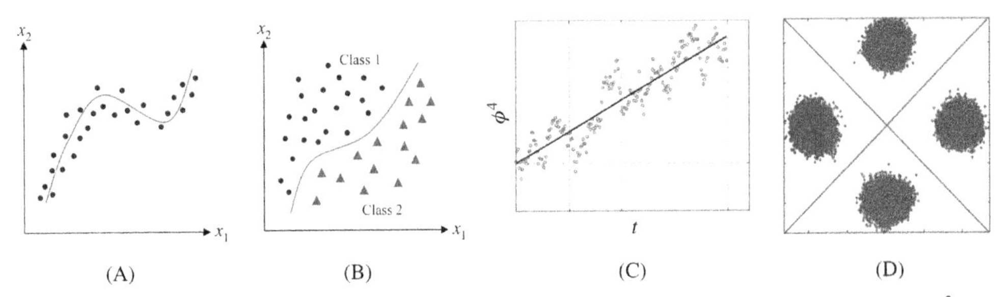

Figure 21.1 Given a data set, machine learning (ML) attempts to solve two main types of problems: (A) functional description of given data and (B) classification of data by deriving appropriate decision boundaries. (C) Laser frequency offset and phase estimation for quadrature phase-shift keying (QPSK) systems by raising the signal phase ϕ to the fourth power and performing regression to estimate the slope and intercept. (D) Decision boundaries for a received QPSK signal distribution.

ML techniques are well known for performing exceptionally well in scenarios in which it is too hard to explicitly describe the problem's underlying physics and mathematics.

Optical communication researchers are no strangers to regressions and classifications. Over the last decade, coherent detection and digital signal processing (DSP) techniques have been the cornerstone of optical transceivers in fiber-optic communication systems. Advanced modulation formats such as 16 quadrature amplitude modulation (16-QAM) and above together with DSP-based estimation and compensation of various transmission impairments such as laser phase noise have become the key drivers of innovation. In this context, parameter estimation and symbol detection are naturally regression and classification problems, respectively, as demonstrated by examples in Fig. 21.1C and D. Currently, most of these parameter estimation and decision rules are derived from probability theory and adequate understanding of the problem's underlying physics. As high-capacity optical transmission links are increasingly being limited by transmission impairments such as fiber nonlinearity, explicit statistical characterizations of inputs/outputs become difficult. An example of 16-QAM multispan dispersion-free transmissions in the presence of fiber nonlinearity and inline amplifier noise is shown in Fig. 21.2A. The maximum likelihood decision boundaries in this case are curved and virtually impossible to derive analytically. Consequently, there has been an increasing amount of research on the application of ML techniques for fiber nonlinearity compensation (NLC). Another related area where ML flourishes is short-reach direct detection systems that are affected by chromatic dispersion (CD), laser chirp, and other transceiver components imperfections, which render the overall communication system hard to analyze.

Optical performance monitoring (OPM) is another area with an increasing amount of ML-related research. OPM is the acquisition of real-time information about

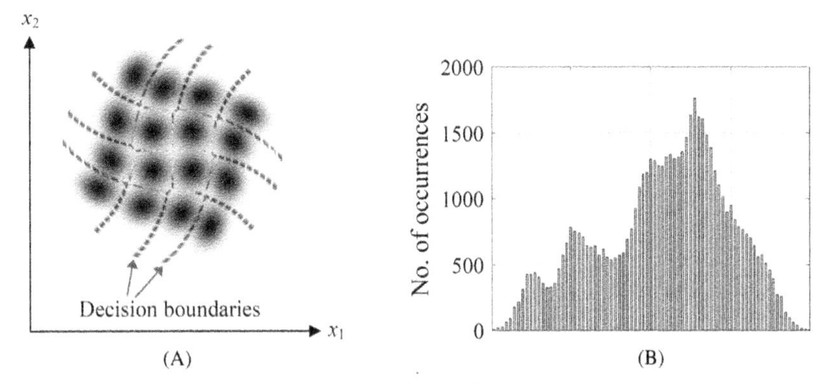

Figure 21.2 (A) Probability distribution and corresponding optimal decision boundaries for received 16 quadrature amplitude modulation (16-QAM) symbols in the presence of fiber nonlinearity are hard to characterize analytically. (B) Probability distribution of received 64-QAM signal amplitudes. The distribution can be used to monitor optical signal-to-noise ratio (OSNR) and identify modulation format. However, this task will be extremely difficult if one relies on analytical modeling.

different channel impairments ubiquitously across the network to ensure reliable network operation and/or improve network capacity. Often OPM is cost-limited so that one can only employ simple hardware components and obtain partial signal features to monitor different channel parameters such as optical signal-to-noise ratio (OSNR), optical power, CD, etc. [4]. In this case, the mapping between input and output parameters is intractable from underlying physics/mathematics, which in turn warrants ML. An example of OSNR monitoring using received signal amplitudes distribution is shown in Fig. 21.2B.

Besides physical layer-related developments, optical network architectures and operations are also undergoing major paradigm shifts under the software-defined networking framework and are increasingly becoming complex, transparent, and dynamic in nature [5]. One of the key features of SDNs is that they can assemble large amounts of data and perform so-called big data analysis to estimate the network states as shown in Fig. 21.3. This in turn can enable (1) adaptive provisioning of resources such as wavelength, modulation format, routing path, etc., according to dynamic traffic patterns and (2) advance discovery of potential components faults so that preventative maintenance can be performed to avoid major network disruptions. The data accumulated in SDNs can span from physical layer (e.g., OSNR of a certain channel) to network layer (e.g., client-side speed demand) and obviously have no underlying physics to explain their interrelationships. Extracting patterns from such cross-layer parameters naturally demands the use of data-driven algorithms such as ML.

This chapter is intended for the researchers in optical communications with a basic background in probability theory, communication theory, and standard DSP techniques used in fiber-optic communications such as matched filters, maximum

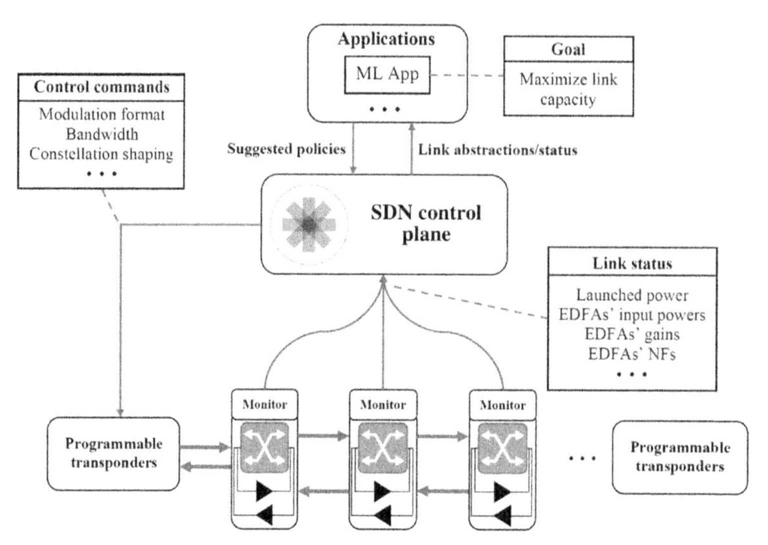

Figure 21.3 Dynamic network resources allocation and link capacity maximization via cross-layer optimization in software-defined networks (SDNs).

likelihood/maximum a posteriori detection, equalization, adaptive filtering, etc. In this regard, a large class of ML techniques such as Kalman filtering, Bayesian learning, hidden Markov models , etc. are actually standard statistical signal processing methods and hence will not be covered here. We will first introduce artificial neural networks (ANNs) and support vector machines (SVMs) from communication theory and signal processing perspectives. This will be followed by other popular ML techniques like K-means clustering, expectation-maximization (EM) algorithm, principal component analysis (PCA), and independent component analysis (ICA), as well as more recent DL approaches such as deep neural networks (DNNs), convolutional neural networks (CNNs), and recurrent neural networks (RNNs). The analytical derivations presented in this chapter are slightly different from those in standard introductory ML text to better align with the fields of communications and signal processing. We will then provide an overview of applications of ML techniques in various aspects of optical communications and networking. By discussing ML through the language of communications and DSP, we hope to provide a more intuitive understanding of ML, its relation to optical communications and networking, and why/where/how it can play a unique role in specific areas of optical communications and networking.

The rest of the chapter is organized as follows. In Section 21.2 we will illustrate the fundamental conditions that warrant the use of a neural network and discuss the technical details of ANN and SVM. Section 21.3 will describe a range of basic unsupervised ML techniques and briefly discuss reinforcement learning (RL). Section 21.4

will be devoted to more recent ML algorithms. Section 21.5 will provide an overview of existing ML applications in optical communications and networking, while Section 21.6 will discuss their future role. Links for online resources and codes for standard ML algorithms will be provided in Section 21.7. Section 21.8 will conclude the chapter.

21.2 Artificial neural network and support vector machine

What are the conditions that need ML for classification? Fig. 21.4 shows three scenarios with two-dimensional (2D) data $\mathbf{x} = [x_1\ x_2]^T$ and their respective class labels depicted as "o" and " × " in the figure. In the first case, classifying the data is straightforward: the decision rule is to see whether $\sigma(x_1 - c)$ or $\sigma(x_2 - c)$ is greater or less than 0 where $\sigma(\cdot)$ is the decision function as shown. The second case is slightly more complicated as the decision boundary is a slanted straight line. However, a simple rotation and shifting of the input, that is, $\mathbf{Wx} + \mathbf{b}$ will map one class of data to below zero and the other class above. Here the rotation and shifting are described by matrix $\mathbf{W}$ and vector $\mathbf{b}$, respectively. This is followed by the decision function $\sigma(\mathbf{Wx} + \mathbf{b})$. The third case is even more complicated. The region for the "green" class depends on the outputs of the "red" and "blue" decision boundaries. Therefore one will need to implement an extra decision step to label the "green" region. The graphical representation of this "decision of decisions" algorithm is the simplest form of an ANN [7]. The intermediate decision output units are known as hidden neurons, and they form the hidden layer.

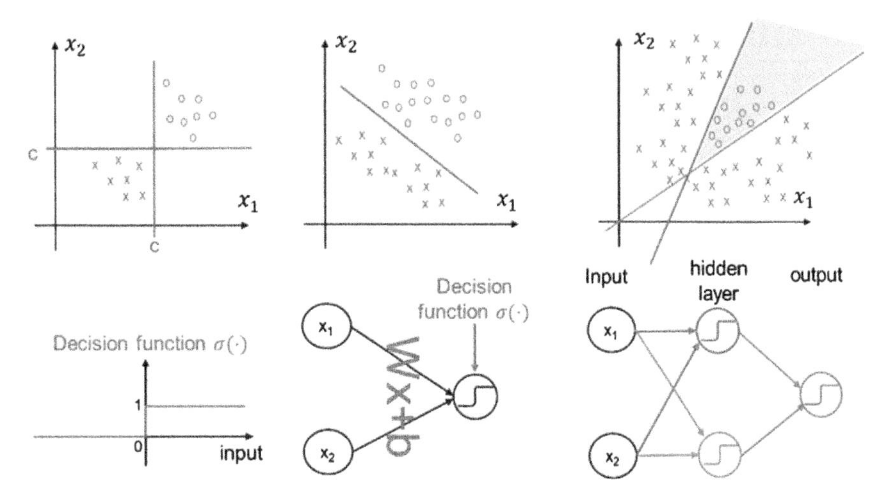

Figure 21.4 The complexity of classification problems depends on how the different classes of data are distributed across the variable space [6].

21.2.1 Artificial neural networks

Let $\{(\mathbf{x}(1), \mathbf{y}(1)), (\mathbf{x}(2), \mathbf{y}(2)), \ldots (\mathbf{x}(L), \mathbf{y}(L))\}$ be a set of L input–output pairs of M and K dimensional column vectors. ANNs are information processing systems comprising an input layer, one or more hidden layers, and an output layer. The structure of a single hidden layer ANN with M input, H hidden, and K output neurons is shown in Fig. 21.5. Neurons in two adjacent layers are interconnected where each connection has a variable weight assigned. Such ANN architecture is the simplest and most commonly used one [7]. The number of neurons M in the input layer is determined by the dimension of the input data vectors $\mathbf{x}(l)$. The hidden layer enables the modeling of complex relationships between the input and output parameters of an ANN. There are no fixed rules for choosing the optimum number of neurons for a given hidden layer and the optimum number of hidden layers in an ANN. Typically the selection is made via experimentation, experience, and other prior knowledge of the problem. These are known as the *hyperparameters* of an ANN. For regression problems, the dimension K of the vectors $\mathbf{y}(l)$ depends on the actual problem nature. For classification problems, K typically equals to the number of class labels such that if a data point $\mathbf{x}(l)$ belongs to class k, $\mathbf{y}(l) = [0\ 0\ \cdots\ 0\ 1\ 0\ \cdots\ 0\ 0]^{\mathrm{T}}$ where the "1" is located at the kth position. This is called *one-hot encoding*. The ANN output $\mathbf{o}(l)$ will naturally have the same dimension as $\mathbf{y}(l)$, and the mapping between input $\mathbf{x}(l)$ and $\mathbf{o}(l)$ can be expressed as

$$\begin{aligned}\mathbf{o}(l) &= \sigma_2(\mathbf{r}(l)) \\ &= \sigma_2(\mathbf{W}_2\mathbf{u}(l) + \mathbf{b}_2) \\ &= \sigma_2(\mathbf{W}_2\sigma_1(\mathbf{q}(l)) + \mathbf{b}_2) \\ &= \sigma_2(\mathbf{W}_2\sigma_1(\mathbf{W}_1\mathbf{x}(l) + \mathbf{b}_1) + \mathbf{b}_2)\end{aligned} \tag{21.1}$$

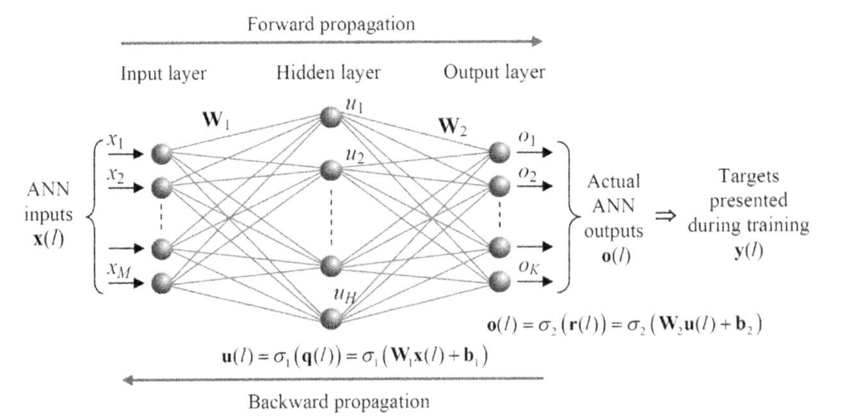

Figure 21.5 Structure of a single hidden layer artificial neural network (ANN) with input vector $\mathbf{x}(l)$, target vector $\mathbf{y}(l)$, and actual output vector $\mathbf{o}(l)$.

where $\sigma_{1(2)}(\cdot)$ are the *activation functions* for the hidden and output layer neurons, respectively. $\mathbf{W}_1$ and $\mathbf{W}_2$ are matrices containing the weights of connections between the input and hidden layer neurons and between the hidden and output layer neurons, respectively, while $\mathbf{b}_1$ and $\mathbf{b}_2$ are the bias vectors for the hidden and output layer neurons, respectively. For a vector $\mathbf{z} = [z_1 z_2 \cdots z_K]$ of length K, $\sigma_1(\cdot)$ is typically an element-wise nonlinear function such as the sigmoid function

$$\sigma_1(\mathbf{z}) = \left[\frac{1}{1+e^{-z_1}} \frac{1}{1+e^{-z_2}} \cdots \frac{1}{1+e^{-z_K}} \right]. \quad (21.2)$$

As for the output layer neurons, $\sigma_2(\cdot)$ is typically chosen to be a linear function for regression problems. In classification problems, one will normalize the output vector $\mathbf{o}(l)$ using the *softmax* function, that is,

$$\mathbf{o}(l) = softmax(\mathbf{W}_2 \mathbf{u}(l) + \mathbf{b}_2), \quad (21.3)$$

where

$$softmax(\mathbf{z}) = \frac{1}{\sum_{k=1}^{K} e^{z_k}} [e^{z_1} e^{z_2} \cdots e^{z_K}]. \quad (21.4)$$

The softmax operation ensures that the ANN outputs conform to a probability distribution for reasons we will discuss below.

To train the ANN is to optimize all the parameters $\theta = \mathbf{W}_1, \mathbf{W}_2, \mathbf{b}_1, \mathbf{b}_2$ such that the difference between the actual ANN outputs $\mathbf{o}$ and the target outputs $\mathbf{y}$ is minimized. One commonly used objective function (also called *loss function* in ML literature) to optimize is the mean square error (MSE)

$$E = \frac{1}{L} \sum_{l=1}^{L} E(l) = \frac{1}{L} \sum_{l=1}^{L} \left\| \mathbf{o}(l) - \mathbf{y}(l) \right\|^2. \quad (21.5)$$

Like most optimization procedures in practice, gradient descent is used instead of full analytical optimization. In this case, the parameter estimates for $n+1$th iteration are given by

$$\theta^{(n+1)} = \theta^{(n)} - \alpha \frac{\partial E}{\partial \theta} \bigg|_{\theta^{(n)}}, \quad (21.6)$$

where the step size α is known as the *learning rate*. Note that for computational efficiency, one can use a single input–output pair instead of all the L pairs for each iteration in Eq. (21.6). This is known as stochastic gradient descent (SGD), which is the standard optimization method used in common adaptive DSP such as constant modulus algorithm (CMA) and least mean squares algorithm. As a trade-off between computational efficiency and accuracy, one can use a *mini-batch* of data

$(\mathbf{x}(nP+1), \mathbf{y}(nP+1)), (\mathbf{x}(nP+2), \mathbf{y}(nP+2)) \ldots (\mathbf{x}(nP+P), \mathbf{y}(nP+P))$ of size P for the nth iteration instead. This can reduce the stochastic nature of SGD and improve accuracy. When all the data set has been used, the update algorithm will have completed one *epoch*. However, it is often the case that one epoch equivalent of updates is not enough for all the parameters to converge to their optimal values. Therefore one can reuse the data set and the algorithm goes through the second epoch for further parameter updates. There is no fixed rule to determine the number of epochs required for convergence [8].

The update algorithm is comprised of following main steps: (1) *Model initialization*: All the ANN weights and biases are randomly initialized, for example, by drawing random numbers from a normal distribution with zero mean and unit variance; (2) *Forward propagation:* In this step, the inputs $\mathbf{x}$ are passed through the network to generate the outputs $\mathbf{o}$ using Eq. (21.1). The input can be a single data point, a mini-batch, or the complete set of L inputs. This step is named so because the computation flow is in the natural forward direction, that is, starting from the input, passing through the network, and going to the output; (3) *Backward propagation and weights/biases update:* For simplicity, let us assume SGD using 1 input–output pair $(\mathbf{x}(n), \mathbf{y}(n))$ for the $n+1$th iteration, sigmoid activation function for the hidden layer neurons, and linear activation function for the output layer neurons such that $\mathbf{o}(n) = \mathbf{W}_2\mathbf{u}(n) + \mathbf{b}_2$. The parameters $\mathbf{W}_2, \mathbf{b}_2$ will be updated first followed by $\mathbf{W}_1, \mathbf{b}_1$. Since $E(n) = \left\| \mathbf{o}(n) - \mathbf{y}(n) \right\|^2$ and $\frac{\partial E(n)}{\partial \mathbf{o}(n)} = 2(\mathbf{o}(n) - \mathbf{y}(n))$, the corresponding update equations are

$$\begin{aligned} \mathbf{W}_2^{(n+1)} &= \mathbf{W}_2^{(n)} - 2\alpha \sum_{k=1}^{K} \frac{\partial o_k(n)}{\partial \mathbf{W}_2} (o_k(n) - y_k(n)) \\ \mathbf{b}_2^{(n+1)} &= \mathbf{b}_2^{(n)} - 2\alpha \frac{\partial \mathbf{o}(n)}{\partial \mathbf{b}_2} (\mathbf{o}(n) - \mathbf{y}(n)) \end{aligned} \tag{21.7}$$

where $o_k(n)$ and $y_k(n)$ denote the kth element of vectors $\mathbf{o}(n)$ and $\mathbf{y}(n)$, respectively. In this case, $\partial \mathbf{o}(n)/\partial \mathbf{b}_2$ is the Jacobian matrix in which the jth row and mth column is the derivative of the mth element of $\mathbf{o}(n)$ with respect to the jth element of $\mathbf{b}_2$. Also, the jth row and mth column of the matrix $\partial o_k(n)/\partial \mathbf{W}_2$ denotes the derivative of $o_k(n)$ with respect to the jth row and mth column of $\mathbf{W}_2$. Interested readers are referred to Ref. [9] for an overview of matrix calculus. Since $\mathbf{o}(n) = \mathbf{W}_2\mathbf{u}(n) + \mathbf{b}_2$, $\frac{\partial \mathbf{o}(n)}{\partial \mathbf{b}_2}$ is simply the identity matrix. For $\partial o_k(n)/\partial \mathbf{W}_2$, its kth row is equal to $\mathbf{u}(n)^{\mathrm{T}}$ (where $(\cdot)^{\mathrm{T}}$ denotes transpose) and is zero otherwise. Eq. (21.7) can be simplified as

$$\begin{aligned} \mathbf{W}_2^{(n+1)} &= \mathbf{W}_2^{(n)} - 2\alpha(\mathbf{o}(n) - \mathbf{y}(n))\mathbf{u}(n)^{\mathrm{T}} \\ \mathbf{b}_2^{(n+1)} &= \mathbf{b}_2^{(n)} - 2\alpha(\mathbf{o}(n) - \mathbf{y}(n)). \end{aligned} \tag{21.8}$$

With the updated $\mathbf{W}_2^{(n+1)}$ and $\mathbf{b}_2^{(n+1)}$, one can calculate

$$\begin{aligned}\mathbf{W}_1^{(n+1)} &= \mathbf{W}_1^{(n)} - 2\alpha\sum_{k=1}^{K}\frac{\partial o_k(n)}{\partial \mathbf{W}_1}(o_k(n) - y_k(n))\\ \mathbf{b}_1^{(n+1)} &= \mathbf{b}_1^{(n)} - 2\alpha\frac{\partial \mathbf{o}(n)}{\partial \mathbf{b}_1}(\mathbf{o}(n) - \mathbf{y}(n)).\end{aligned} \tag{21.9}$$

Since the derivative of the sigmoid function is given by $\sigma_1'(\mathbf{z}) = \sigma_1(\mathbf{z})\circ(\mathbf{1} - \sigma_1(\mathbf{z}))$ where $\circ$ denotes element-wise multiplication and $\mathbf{1}$ denotes a column vector of 1's with the same length as $\mathbf{z}$,

$$\begin{aligned}\frac{\partial \mathbf{o}(n)}{\partial \mathbf{b}_1} &= \frac{\partial \mathbf{q}(n)}{\partial \mathbf{b}_1}\frac{\partial \mathbf{u}(n)}{\partial \mathbf{q}(n)}\frac{\partial \mathbf{o}(n)}{\partial \mathbf{u}(n)}\\ &= \mathrm{diag}\left\{\mathbf{u}(n)\circ(\mathit{1} - \mathbf{u}(n))\right\}\cdot\left(\mathbf{W}_2^{(n+1)}\right)^{\mathrm{T}}\end{aligned} \tag{21.10}$$

where diag{$\mathbf{z}$} denotes a diagonal matrix with diagonal vector $\mathbf{z}$. Next,

$$\begin{aligned}\frac{\partial o_k(n)}{\partial \mathbf{W}_1} &= \sum_j \frac{\partial o_k(n)}{\partial u_j(n)}\frac{\partial u_j(n)}{\partial q_j(n)}\frac{\partial q_j(n)}{\partial \mathbf{W}_1}\\ &= \sum_j w_{2,k,j}^{(n+1)} u_j(n)(1 - u_j(n))\frac{\partial q_j(n)}{\partial \mathbf{W}_1}\end{aligned} \tag{21.11}$$

where $w_{2,k,j}^{(n+1)}$ is the kth row and jth column entry of $\mathbf{W}_2^{(n+1)}$. For $\partial q_j(n)/\partial \mathbf{W}_1$, its jth row is $\mathbf{x}(n)^{\mathrm{T}}$ and is zero otherwise. Eq. (21.11) can be simplified as

$$\frac{\partial o_k(n)}{\partial \mathbf{W}_1} = \left(\left(\mathbf{w}_{2,k}^{(n+1)}\right)^{\mathrm{T}}\circ\mathbf{u}(n)\circ(\mathit{1} - \mathbf{u}(n))\right)\mathbf{x}(n)^{\mathrm{T}}, \tag{21.12}$$

where $\mathbf{w}_{2,k}^{(n+1)}$ is the kth row of $\mathbf{W}_2^{(n+1)}$. Since the parameters are updated group by group starting from the output layer back to the input layer, this algorithm is called back-propagation (BP) algorithm [not to be confused with the digital back-propagation (DBP) algorithm for fiber NLC]. The weights and biases are continuously updated until convergence.

For the learning and performance evaluation of an ANN, the data sets are typically divided into three groups: training, validation, and testing. The training data set is used to train the ANN. Clearly, a larger training data set is better since the more data an ANN sees, the more likely it is that it has encountered examples of all possible types of input. However, the learning time also increases with the training data size. There is no fixed rule for determining the minimum amount of training data needed since it

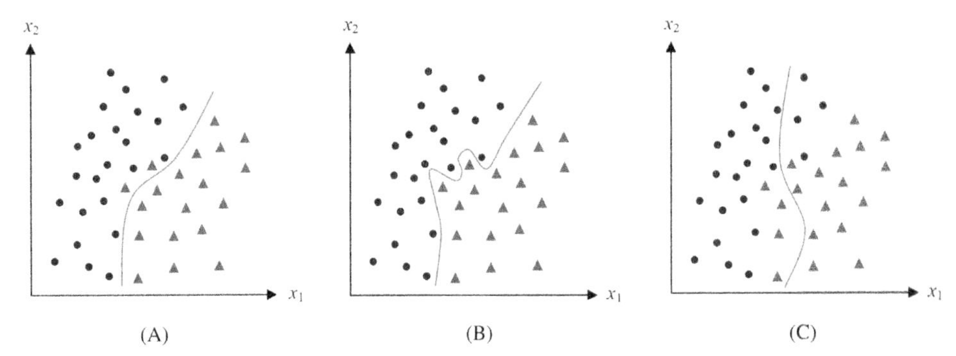

Figure 21.6 Example illustrating artificial neural network (ANN) learning processes with (A) no overfitting or underfitting, (B) overfitting, and (C) underfitting.

often depends on the given problem. A rule of thumb typically used is that the size of the training data should be at least 10 times the total number of weights [1]. The purpose of the validation data set is to keep a check on how well the ANN is doing as it learns, as during training there is an inherent danger of *overfitting* (or *overtraining*). In this case, instead of finding the underlying general decision boundaries as shown in Fig. 21.6A, the ANN tends to perfectly fit the training data (including any noise components of them) as shown in Fig. 21.6B. This in turn makes the ANN customized for a few data points and reduces its generalization capability, that is, its ability to make predictions about new inputs that it has never seen before. The overfitting problem can be avoided by constantly examining ANN's error performance during the course of training against an independent validation data set and enforcing an early termination of the training process if the validation data set gives large errors. Typically, the size of the validation data set is just a fraction ($\sim 1/3$) of that of the training data set. Finally, the testing data set evaluates the performance of the trained ANN. Note that an ANN may also be subjected to an *underfitting* problem, which occurs when it is undertrained and thus unable to perform at an acceptable level, as shown in Fig. 21.6C. Underfitting can again lead to poor ANN generalization. The reasons for underfitting include insufficient training time or number of iterations, inappropriate choice of activation functions, and/or insufficient number of hidden neurons used.

It should be noted that given an adequate number of hidden neurons, proper nonlinearities, and appropriate training, an ANN with one hidden layer has great expressive power and can approximate any continuous function in principle. This is called the *universal approximation theorem* [10]. One can intuitively appreciate this characteristic by considering the classification problem in Fig. 21.7. Since each hidden neuron can be represented as a straight-line decision boundary, any arbitrary curved boundary can be approximated by a collection of hidden neurons in a single hidden layer ANN. This important property of an ANN enables it to be applied in many diverse applications.

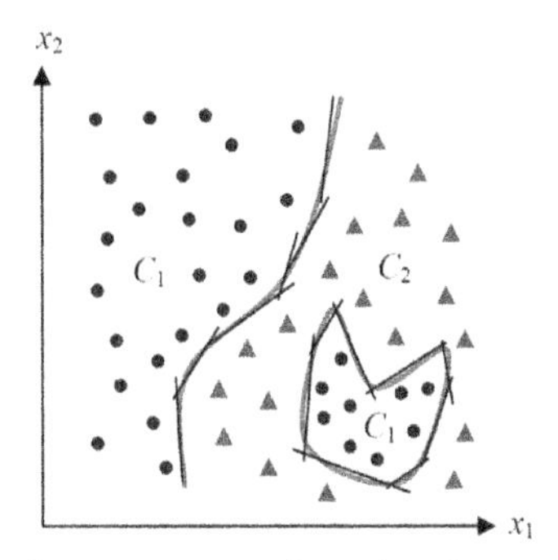

Figure 21.7 Decision boundaries for appropriate data classification obtained using an artificial neural network (ANN).

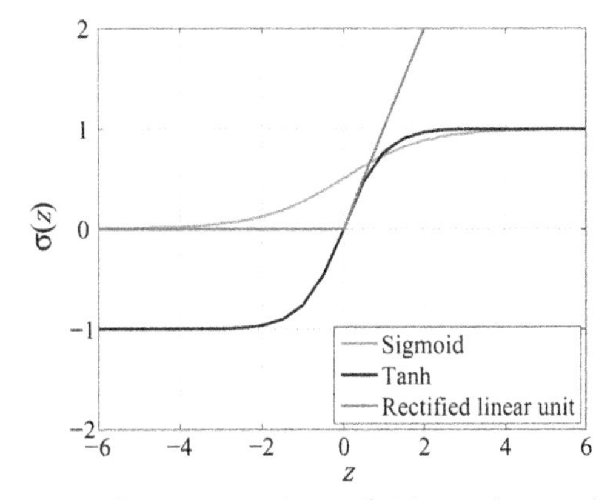

Figure 21.8 Common activation functions used in artificial neural networks (ANNs).

21.2.2 Choice of activation functions

The choice of activation functions has a significant effect on the training dynamics and final ANN performance. Historically, sigmoid and hyperbolic tangent have been the most commonly used nonlinear activation functions for hidden layer neurons. However, the rectified linear unit (ReLU) activation function has become the default choice among ML community in recent years. The above-mentioned three functions are given by

$$\begin{aligned} &\text{Sigmoid: } \sigma(z) = \frac{1}{1+e^{-z}} \\ &\text{Hyperbolic tangent: } \sigma(z) = \frac{e^{z}-e^{-z}}{e^{z}+e^{-z}} \\ &\text{Rectified linear unit: } \sigma(z) = \max(0, z) \end{aligned} \tag{21.13}$$

and their plots are shown in Fig. 21.8. Sigmoid and hyperbolic tangent are both differentiable. However, a major problem with these functions is that their gradients tend to zero as $|z|$ becomes large and thus the activation output gets saturated. In this case the weights and biases updates for a certain layer will be minimal, which in turn will slow down the weights and biases updates for all the preceding layers. This is known as *vanishing gradient problem* and is particularly an issue when training ANNs with a large number of hidden layers. To circumvent this problem, ReLU was proposed since its gradient does not vanish as z increases. Note that although ReLU is not differentiable at $z = 0$, it is not a problem in practice because the probability of having an entry exactly equal to 0 is generally very low. Also, as the ReLU function and its derivative are 0 for $z < 0$, around 50% of hidden neurons' outputs will be 0, that is, only half of total neurons will be active when the ANN weights and biases are randomly initialized. It has been found that such sparsity of activation not only reduces computational complexity (and thus training time) but also leads to better ANN performance [11]. Note that while using ReLU activation function, the ANN weights and biases are often initialized using the method proposed by He et al. [12]. On the other hand, Xavier initialization technique [13] is more commonly employed for the hyperbolic tangent activation function. These heuristics-based approaches initialize the weights and biases by drawing random numbers from a truncated normal distribution (instead of standard normal distribution) with variance that depends on the size of the previous ANN layer.

21.2.3 Choice of loss functions

The choice of loss function E has a considerable effect on the performance of an ANN. The MSE is a common choice in adaptive signal processing and other DSP in telecommunications. For regression problems, MSE works well in general and is also easy to compute. On the other hand, for classification problems, cross-entropy loss function defined as

$$E = -\frac{1}{L}\sum_{l=1}^{L}\sum_{k=1}^{K} y_k(l)\log(o_k(l)) \tag{21.14}$$

is often used instead of MSE [10]. The cross-entropy function can be interpreted by viewing the softmax output $\mathbf{o}(l)$ and the class label with one-hot encoding $\mathbf{y}(l)$ as probability distributions. In this case, $\mathbf{y}(l)$ has zero entropy and one can subtract the zero-entropy term from Eq. (21.14) to obtain

$$\begin{aligned} E &= -\frac{1}{L}\sum_{l=1}^{L}\sum_{k=1}^{K} y_k(l)\log(o_k(l)) + \underbrace{\frac{1}{L}\sum_{l=1}^{L}\sum_{k=1}^{K} y_k(l)\log(y_k(l))}_{=0} \\ &= \frac{1}{L}\sum_{l=1}^{L}\sum_{k=1}^{K} y_k(l)\log\left(\frac{y_k(l)}{o_k(l)}\right), \end{aligned} \tag{21.15}$$

which is simply the Kullback–Leibler (KL) divergence between the distributions $\mathbf{o}(l)$ and $\mathbf{y}(l)$ averaged over all input–output pairs. Therefore the cross-entropy is in fact a measure of the similarity between ANN outputs and the class labels. The cross-entropy function also leads to simple gradient updates as the logarithm cancels out the exponential operation inherent in the softmax calculation, thus leading to faster ANN training. The appendix in this chapter shows the derivation of BP algorithm for the single hidden layer ANN in Fig. 21.5 with cross-entropy loss function and softmax activation function for the output layer neurons.

In many applications a common approach to prevent overfitting is to reduce the magnitude of the weights, as large weights produce high curvatures that make the decision boundaries overly complicated. This can be achieved by including an extra regularization term in the loss function, that is,

$$E' = E + \lambda \|\mathbf{W}\|^2, \tag{21.16}$$

where $\|\mathbf{W}\|^2$ is the sum of squared element-wise weights. The parameter λ, called regularization coefficient, defines the relative importance of the training error E and the regularization term. The regularization term thus discourages weights from reaching large values, and this often results in significant improvement in ANN's generalization ability [14].

21.2.4 Support vector machines

In many classification tasks it often happens that the two data categories are not easily separable with straight lines or planes in the original variable space. SVM is an ML technique that preprocesses the input data $\mathbf{x}(i)$ and transforms it into (sometimes) a higher dimensional space $\mathbf{v}(i) = \varphi(\mathbf{x}(i))$, called *feature space*, where the data belonging to two different classes can be separated easily by a simple straight plane decision boundary or *hyperplane* [15]. An example is shown in Fig. 21.9 where one class of data

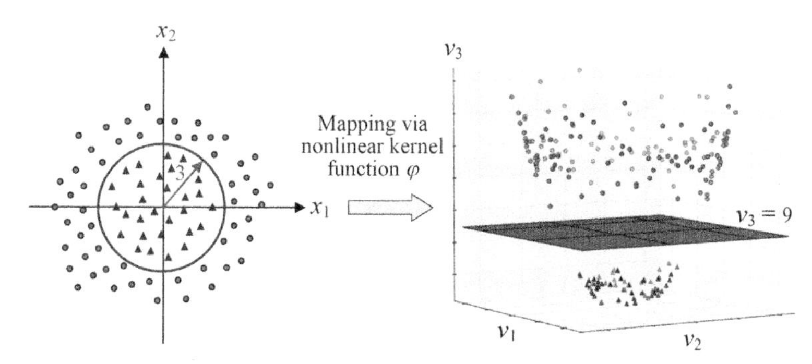

Figure 21.9 Example showing how a linearly inseparable problem [in the original two-dimensional (2D) data space] can undergo a nonlinear transformation and become a linearly separable one in the three-dimensional (3D) feature space.

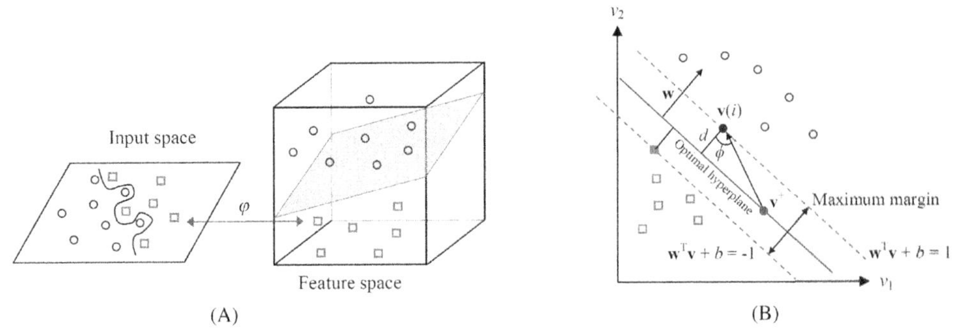

Figure 21.10 (A) Mapping from input space to a higher dimensional feature space using a nonlinear kernel function φ. (B) Separation of two data classes in the feature space through an optimal hyperplane.

lies within a circle of radius 3 and the other class lies outside. When transformed into the feature space $\mathbf{v} = (v_1, v_2, v_3) = (x_1, x_2, x_1^2 + x_2^2)$, the two data classes can be separated simply by the hyperplane $v_3 = 9$.

Let us first focus on finding the right decision hyperplane after the transformation into feature space as shown in Fig. 21.10A. The right hyperplane should have the largest (and also equal) distance from the borderline points of the two data classes. This is graphically illustrated in Fig. 21.10B. Had the data points been generated from two probability density functions (PDFs), finding a hyperplane with maximal margin from the borderline points is conceptually analogous to finding a maximum likelihood decision boundary. The borderline points, represented as solid dot and square in Fig. 21.10B, are referred to as *support vectors* and are often most informative for the classification task.

More technically, in the feature space, a general hyperplane is defined as $\mathbf{w}^{\mathrm{T}}\mathbf{v} + b = 0$. If it classifies all the data points correctly, all the blue points will lie in the region $\mathbf{w}^{\mathrm{T}}\mathbf{v} + b > 0$ and the red points will lie in the region $\mathbf{w}^{\mathrm{T}}\mathbf{v} + b < 0$. We seek to find a hyperplane $\mathbf{w}^{\mathrm{T}}\mathbf{v} + b = 0$ that maximizes the margin d as shown in Fig. 21.10B. Without loss of generality, let the point $\mathbf{v}(i)$ reside on the hyperplane $\mathbf{w}^{\mathrm{T}}\mathbf{v} + b = 1$ and is closest to the hyperplane $\mathbf{w}^{\mathrm{T}}\mathbf{v} + b = 0$ on which $\mathbf{v}^+$ resides. Since the vectors $\mathbf{v}(i) - \mathbf{v}^+, \mathbf{w}$ and the angle ϕ are related by $\cos\phi = \mathbf{w}^{\mathrm{T}}\left(\mathbf{v}(i) - \mathbf{v}^+\right) / \left(\|\mathbf{w}\|\,\|\mathbf{v}(i) - \mathbf{v}^+\|\right)$, the margin d is given as

$$
\begin{aligned}
d &= \|\mathbf{v}(i) - \mathbf{v}^+\| \cos\phi \\
&= \|\mathbf{v}(i) - \mathbf{v}^+\| \cdot \frac{\mathbf{w}^{\mathrm{T}}\left(\mathbf{v}(i) - \mathbf{v}^+\right)}{\|\mathbf{w}\|\,\|\mathbf{v}(i) - \mathbf{v}^+\|} \\
&= \frac{\mathbf{w}^{\mathrm{T}}\left(\mathbf{v}(i) - \mathbf{v}^+\right)}{\|\mathbf{w}\|} = \frac{\mathbf{w}^{\mathrm{T}}\mathbf{v}(i) - \mathbf{w}^{\mathrm{T}}\mathbf{v}^+}{\|\mathbf{w}\|} \\
&= \frac{\mathbf{w}^{\mathrm{T}}\mathbf{v}(i) + b}{\|\mathbf{w}\|} = \frac{1}{\|\mathbf{w}\|}.
\end{aligned}
\tag{21.17}
$$

Therefore we seek to find $\mathbf{w}, b$ that maximize $1/\|\mathbf{w}\|$ subject to the fact that all the data points are classified correctly. To characterize the constraints more mathematically, one can first assign the blue class label to 1 and red class label to -1. In this case, if we have correct decisions for all the data points, the product $y(i)\left(\mathbf{w}^{\mathrm{T}}\mathbf{v}(i)+b\right)$ will always be greater than 1 for all i. The optimization problem then becomes

$$\begin{gathered}\operatorname{argmin}_{\mathbf{w},b}\frac{1}{\|\mathbf{w}\|}\\ \text{subject to } y(l)\left(\mathbf{w}^{\mathrm{T}}\mathbf{v}(l)+b\right)\geq 1, l=1,2,\ldots,L\end{gathered} \tag{21.18}$$

and thus standard convex programming software packages such as CVXOPT [16] can be used to solve Eq. (21.18).

Let us come back to the task of choosing the nonlinear function $\varphi(\cdot)$ that maps the original input space $\mathbf{x}$ to feature space $\mathbf{v}$. For SVM, one would instead find a kernel function $K(\mathbf{x}(i),\mathbf{x}(j))=\varphi(\mathbf{x}(i))\cdot\varphi(\mathbf{x}(j))=\mathbf{v}(i)^{\mathrm{T}}\mathbf{v}(j)$ that maps to the inner product. Typical kernel functions include the following:

- *Polynomials*: $K(\mathbf{x}(i),\mathbf{x}(j))=\left(\mathbf{x}(i)^{\mathrm{T}}\mathbf{x}(j)+a\right)^{b}$ for some scalars a, b;
- *Gaussian radial basis function*: $K(\mathbf{x}(i),\mathbf{x}(j))=\exp(-a\|\mathbf{x}(i)-\mathbf{x}(j)\|^{2})$ for some scalar a;
- *Hyperbolic tangent*: $K(\mathbf{x}(i),\mathbf{x}(j))=\tanh(a\mathbf{x}(i)^{\mathrm{T}}\mathbf{x}(j)+b)$ for some scalars a, b.

The choice of a kernel function is often determined by the designer's knowledge of the problem domain [3].

Note that a larger separation margin typically results in better generalization of the SVM classifier. SVMs often demonstrate better generalization performance than conventional ANNs in various pattern recognition applications. Furthermore, multiple SVMs can be applied to the same data set to realize nonbinary classifications such as detecting 16-QAM signals [17,18,19] (to be discussed in more detail in Section 21.5).

It should be noted that ANNs and SVMs can be seen as two complementary approaches for solving classification problems. While an ANN derives curved decision boundaries in the input variable space, the SVM performs nonlinear transformations of the input variables followed by determining a simple decision boundary or hyperplane as shown in Fig. 21.11.

21.3 Unsupervised and reinforcement learning

The ANN and SVM are examples of *supervised learning* approach in which the class labels $\mathbf{y}$ of the training data are known. Based on this data, the ML algorithm generalizes to react accurately to new data to the best possible extent. Supervised learning can be considered as a closed-loop feedback system, as the error between the ML algorithm's actual outputs and the targets is used as a feedback signal to guide the learning process.

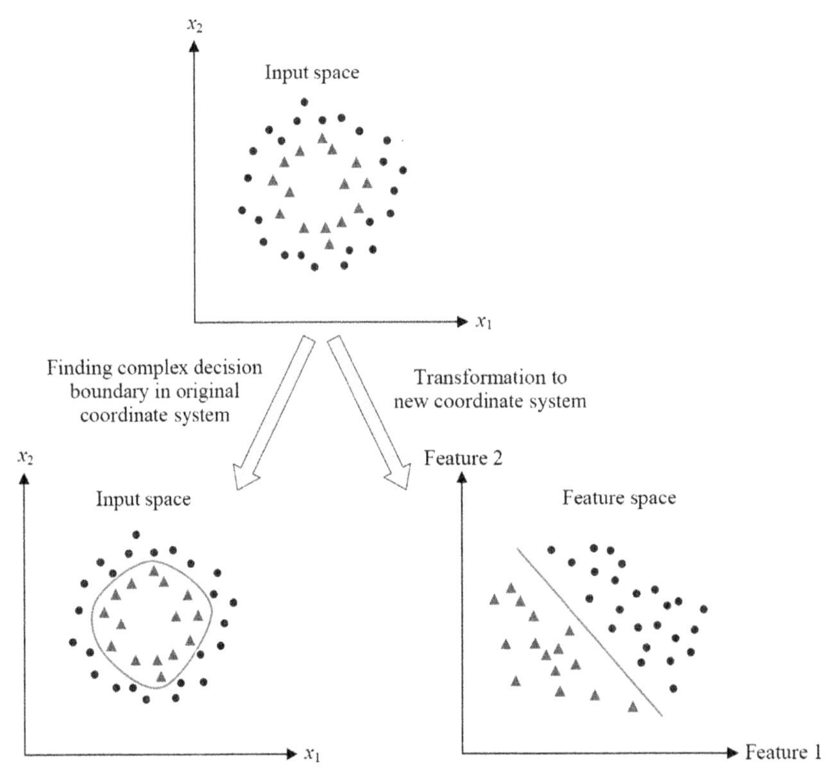

Figure 21.11 Example showing how an artificial neural network (ANN) determines a curved decision boundary in the original input space while a support vector machine (SVM) obtains a simple decision boundary in the transformed feature space.

In *unsupervised learning* the ML algorithm is not provided with correct labels of the training data. Rather, it learns to identify similarities between various inputs with the aim either to categorize together those inputs that have something in common or to determine some better representation/description of the original input data. It is referred to as "unsupervised" because the ML algorithm is not told what the output should be; rather, it has to come up with it itself [20]. One example of unsupervised learning is data clustering, as shown in Fig. 21.12.

Unsupervised learning is becoming more and more important because in many real circumstances, it is practically not possible to obtain labeled training data. In such scenarios an unsupervised learning algorithm can be applied to discover some similarities between different inputs for itself. Unsupervised learning is typically used in tasks such as clustering, vector quantization, dimensionality reduction, and features extraction. It is also often employed as a preprocessing tool for extracting useful (in some particular context) features of the raw data before supervised learning algorithms can be applied. We hereby provide a review of few key unsupervised learning techniques.

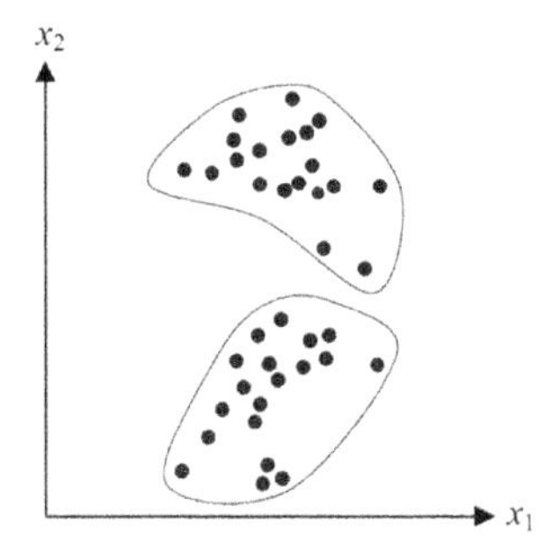

Figure 21.12 Data clustering based on unsupervised learning.

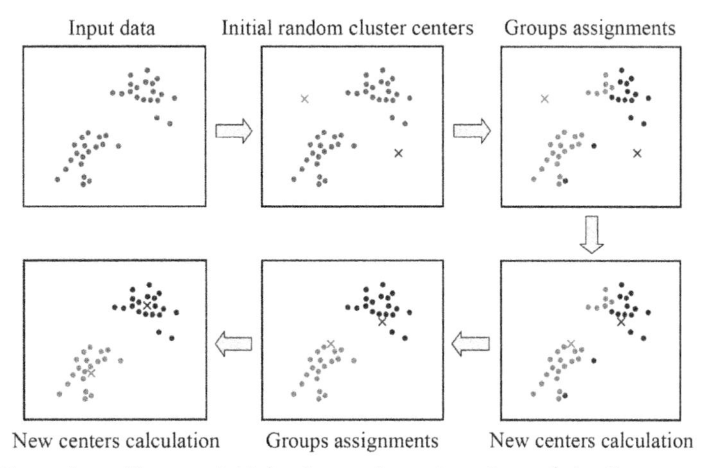

Figure 21.13 Example to illustrate initialization and two iterations of the *K*-means algorithm. The data points are shown as dots, and cluster centers are depicted as crosses.

21.3.1 *K*-means clustering

Let $\{\mathbf{x}(1), \mathbf{x}(2), \ldots \mathbf{x}(L)\}$ be the set of data points that is to be split into K clusters $C_1, C_2, \ldots C_K$. K-means clustering is an iterative unsupervised learning algorithm that aims to partition L observations into K clusters such that the sum of squared errors for data points within a group is minimized [14]. An example of this algorithm is graphically shown in Fig. 21.13. The algorithm initializes by randomly picking K locations $\boldsymbol{\mu}(j)$, $j = 1, 2, \ldots, K$ as cluster centers. This is followed by two iterative steps. In the first step each data point $\mathbf{x}(i)$ is assigned to the cluster C_k with the minimum Euclidean distance, that is,

$$C_k = \left\{\mathbf{x}(i): \|\mathbf{x}(i) - \boldsymbol{\mu}(k)\| < \|\mathbf{x}(i) - \boldsymbol{\mu}(j)\| \, \forall \mathrm{j} \in \{1, 2, \ldots, K\} \backslash \{k\}\right\}. \tag{21.19}$$

In the second step the new center of each cluster C_k is calculated by averaging out the locations of data points that are assigned to cluster C_k, that is,

$$\mu(k) = \sum_{\mathbf{x}(i) \in C_k} \mathbf{x}(i). \tag{21.20}$$

The two steps are repeated iteratively until the cluster centers converge. Several variants of the K-means algorithm have been proposed over the years to improve its computational efficiency as well as to achieve smaller errors. These include fuzzy K-means, hierarchical K-means, K-means++, K-medians, and K-medoids, among others.

21.3.2 Expectation-maximization algorithm

One drawback of the K-means algorithm is that it requires the use of hard decision boundaries whereby a data point can only be assigned to one cluster even though it might lie somewhere midway between two or more clusters. The EM algorithm is an improved clustering technique that assigns a probability to the data point belonging to each cluster rather than forcing it to belong to one particular cluster during each iteration [20]. The algorithm assumes that a given data distribution can be modeled as a superposition of K jointly Gaussian probability distributions with distinct means and covariance matrices $\mu(k), \Sigma(k)$ (also referred to as *Gaussian mixture models*). The EM algorithm is a two-step iterative procedure comprising expectation (E) and maximization (M) steps [3]. The E step computes the a posteriori probability of the class label given each data point using the current means and covariance matrices of the Gaussians, that is,

$$\begin{aligned} p_{ij} &= p\left(C_j | \mathbf{x}(i)\right) \\ &= \frac{p\left(\mathbf{x}(i) | C_j\right) p\left(C_j\right)}{\sum_{k=1}^{K} p(\mathbf{x}(i) | C_k) p(C_k)} \\ &= \frac{N(\mathbf{x}(i) | \mu(k), \Sigma(k))}{\sum_{k=1}^{K} N(\mathbf{x}(i) | \mu(k), \Sigma(k))} \end{aligned} \tag{21.21}$$

where $N(\mathbf{x}(i) | \mu(k), \Sigma(k))$ is the Gaussian PDF with mean and covariance matrix $\mu(k), \Sigma(k)$. Note that we have inherently assumed equal probability $p(C_j)$ of each class, which is a valid assumption for most communication signals. In scenarios where this assumption is not valid, for example, the one involving probabilistic constellation shaping , the actual nonuniform probabilities $p\left(C_j\right)$ of individual symbols shall instead be used in Eq. (21.21). The M step attempts to update the means and covariance matrices according to the updated soft-labeling of the data points, that is,

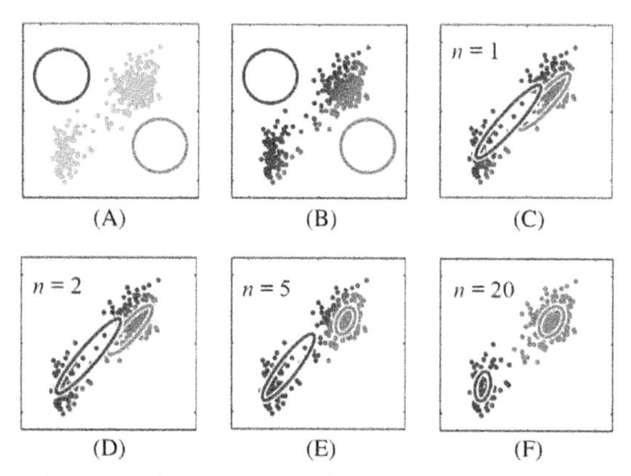

Figure 21.14 Example showing the concept of expectation-maximization (EM) algorithm. (A) Original data points and initialization. Results after (B) first E step; (C) first M step; (D) two complete EM iterations; (E) five complete EM iteratons; and (F) 20 complete EM iterations [3].

$$\begin{aligned} \boldsymbol{\mu}(j) &= \frac{\sum_{i=1}^{L} p_{ij}\mathbf{x}(i)}{\sum_{i=1}^{L} p_{ij}} \\ \boldsymbol{\Sigma}(k) &= \sum_{i=1}^{L} p_{ij}(\mathbf{x}(i) - \boldsymbol{\mu}(j))(\mathbf{x}(i) - \boldsymbol{\mu}(j))^{\mathrm{T}} \end{aligned} \tag{21.22}$$

A graphical illustration of the EM algorithm and its convergence process is shown in Fig. 21.14. Fig. 21.14A shows the original data points in green that are to be split into two clusters by applying the EM algorithm. The two Gaussian probability distributions are initialized with random means and unit covariance matrices and are depicted using red and blue circles. The results after the first E step are shown in Fig. 21.14B where the posterior probabilities in Eq. (21.21) are expressed by the proportion of red and blue colors for each data point. Fig. 21.14C depicts the results after first M step where the means and covariance matrices of the red and blue Gaussian distributions are updated using Eq. (21.22), which in turn uses the posterior probabilities computed by Eq. (21.21). This completes the first iteration of the EM algorithm. Fig. 21.14D–F show the results after 2, 5, and 20 complete EM iterations, respectively, where the convergence of the algorithm and consequently effective splitting of the data points into two clusters can be clearly observed.

21.3.3 Principal component analysis

PCA is an unsupervised learning technique for features extraction and data representation [21,22]. It is often used as a preprocessing tool in many pattern recognition applications for the extraction of limited but most critical data features. The central idea behind PCA is to project the original high-dimensional data onto a lower dimensional

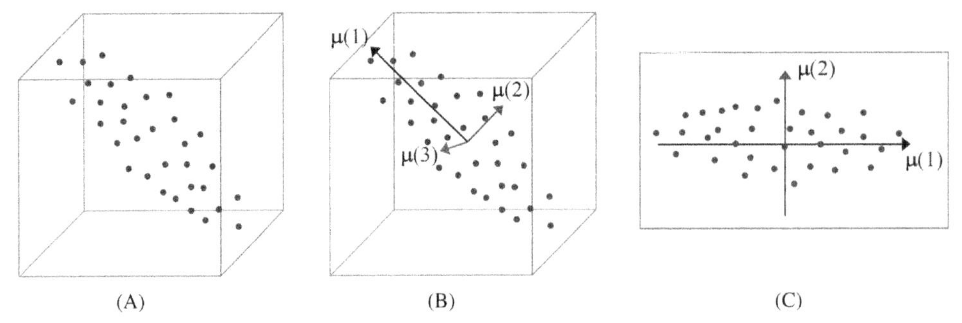

Figure 21.15 Example to illustrate the concept of principle component analysis (PCA). (A) Data points in the original 3D data space. (B) Three principal components (PCs) ordered according to the variability in original data. (C) Projection of data points onto a plane defined by the first two PCs while discarding the third one.

feature space that retains most of the information in the original data, as shown in Fig. 21.15. The reduced dimensionality feature space is spanned by a small (but most significant) set of orthonormal eigenvectors, called principal components (PCs). The first PC points in the direction along which the original data has the greatest variability and each successive PC in turn accounts for as much of the remaining variability as possible. Geometrically, we can think of PCA as a rotation of the axes of the original coordinate system to a new set of orthogonal axes that are ordered based on the amount of variation of the original data they account for, thus achieving dimensionality reduction.

More technically, consider a data set $\{\mathbf{x}(1), \mathbf{x}(2), \ldots \mathbf{x}(L)\}$ with L data vectors of M dimensions. We will first compute the mean vector $\overline{\mathbf{x}} = (1/L)\sum_{i=1}^{L}\mathbf{x}(i)$, and the covariance matrix $\mathbf{\Sigma}$ can then be estimated as

$$\mathbf{\Sigma} \approx \frac{1}{L}\sum_{i=1}^{L}(\mathbf{x}(i) - \overline{\mathbf{x}})(\mathbf{x}(i) - \overline{\mathbf{x}})^{\mathrm{T}}, \tag{21.23}$$

where $\mathbf{\Sigma}$ can have up to M eigenvectors $\boldsymbol{\mu}(i)$ and corresponding eigenvalues λ_i. We then sort the eigenvalues in terms of their magnitude from large to small and choose the first S (where $S << M$) corresponding eigenvectors such that

$$\frac{\sum_{i=1}^{S}\lambda_i}{\sum_{i=1}^{M}\lambda_i} > R, \tag{21.24}$$

where R is typically above 0.9 [22]. Note that as compared to the original M-dimensional data space, the chosen eigenvectors span only an S-dimensional subspace that in a way captures most of the data information. One can understand such a procedure intuitively by noting that for a covariance matrix, finding the eigenvectors with large eigenvalues corresponds to finding linear combinations or particular directions of the input space that give large variances, which is exactly what we want to capture. A data vector $\mathbf{x}$ can then be approximated as a weighted sum of the chosen eigenvectors in this subspace, that is,

$$\mathbf{x} \approx \sum_{i=1}^{S} w_i \boldsymbol{\mu}(i), \tag{21.25}$$

where $\boldsymbol{\mu}(i), i = 1, 2, \ldots, S$ are the chosen orthogonal eigenvectors such that

$$\boldsymbol{\mu}^{\mathrm{T}}(m)\boldsymbol{\mu}(l) = \left\{ \begin{array}{l} 1 \ if \ l = m \\ 0 \ if \ l \neq m \end{array} \right\}. \tag{21.26}$$

Multiplying both sides of Eq. (21.25) with $\boldsymbol{\mu}^{\mathrm{T}}(k)$ and then using Eq. (21.26), we get

$$w_k = \boldsymbol{\mu}^{\mathrm{T}}(k)\mathbf{x}, \ k = 1, 2, \ldots, S. \tag{21.27}$$

The vector $\mathbf{w} = [w_1 \ w_2 \ldots w_S]^{\mathrm{T}}$ of weights describing the contribution of each chosen eigenvector $\boldsymbol{\mu}(k)$ in representing $\mathbf{x}$ can then be considered as a feature vector of $\mathbf{x}$.

21.3.4 Independent component analysis

Another interesting technique for features extraction and data representation is ICA. Unlike PCA, which uses orthogonal and uncorrelated components, the components in ICA are instead required to be statistically independent [1]. In other words, ICA seeks those directions in the feature space that are most independent from each other. Fig. 21.16 illustrates the conceptual difference between PCA and ICA. Finding the

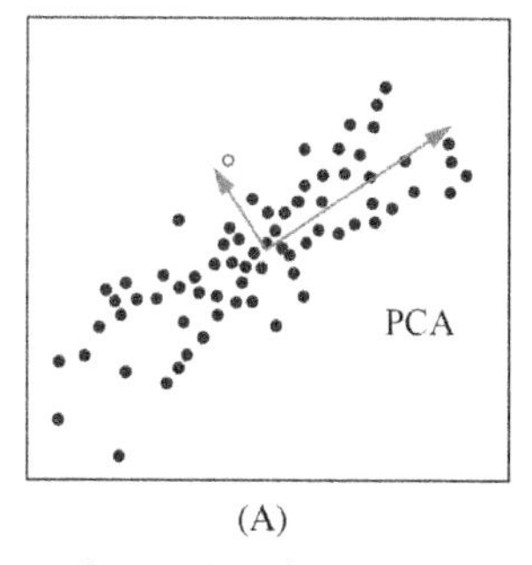

(A)

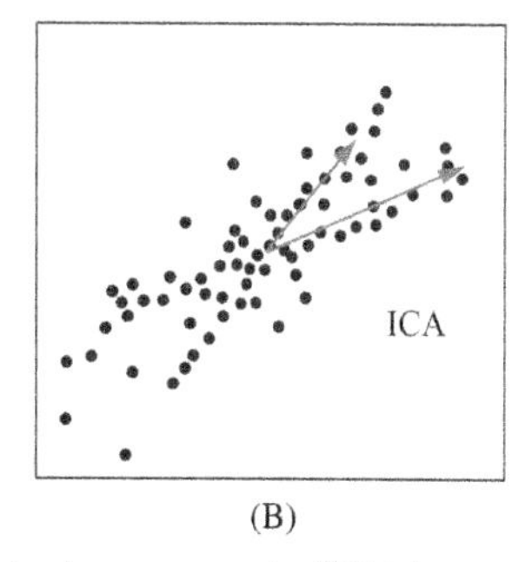

(B)

Figure 21.16 Example 2D data fitted using (A) principal components (PCs) bases and (B) independent components (ICs) bases. As shown, the orthogonal basis vectors in principle component analysis (PCA) may not be efficient while representing nonorthogonal density distributions. In contrast, independent component analysis (ICA) does not necessitate orthogonal basis vectors and can thus represent general types of densities more effectively.

independent components (ICs) of the observed data can be useful in scenarios where we need to separate mutually independent but unknown source signals from their linear mixtures with no information about the mixing coefficients. An example is the task of polarization demultiplexing at the receiver using DSP. For a data set $\{\mathbf{x}(1), \mathbf{x}(2), \ldots, \mathbf{x}(L)\}$, one seeks to identify a collection of basis vectors $\mathbf{v}(1), \mathbf{v}(2), \ldots \mathbf{v}(S)$ so that $\mathbf{x} \approx \sum_{k=1}^{S} w_k \mathbf{v}(k)$ and the empirical distributions of w_k, $k = 1, 2, \ldots, S$ across all the data $\mathbf{x}$ are statistically independent. This can be achieved by minimizing the mutual information between different w_k.

ICA is used as a preprocessing tool for extracting data features in many pattern recognition applications and is shown to outperform conventional PCA in many cases [23]. This is expected because unlike PCA, which is derived from second-order statistics (i.e., covariance matrix) of the input data, ICA takes into account high-order statistics of the data as it considers complete probability distribution.

We would like to highlight here that the dimensionality of the transformed space in ML techniques can be higher or lower than the original input space depending upon the nature of the problem at hand. If the objective of the transformation is to simply reduce the input data dimensionality (e.g., for decreasing the computational complexity of the learning system), then the dimensionality of the transformed space should be lower than that of original one. On the other hand, a transformation to a higher dimensional space may be desirable if the data classes can be separated more easily by a classifier in the new space.

21.3.5 Reinforcement learning

In this learning type, the input of the ML model (called observation) is associated with a reward or reinforcement signal. The output (called action) determines the value of the next observation and hence the reward through the predefined action-observation relationship of a particular problem. The objective here is to learn a sequence of actions that optimizes the final reward. However, unlike supervised learning, the model is not optimized through SGD-like approaches. Rather, the model tries different actions until it finds a set of parameters that lead to better rewards. In RL the model is rewarded for its good output result and punished for the bad one. In this way it can learn to choose actions that can maximize the expected reward [24]. Like supervised learning, RL can also be regarded as a closed-loop feedback system, since the RL model's actions will influence its later inputs. RL is particularly useful in solving interactive problems in which it is often impossible to attain examples of desired behavior that are not only correct but are also representative of all the possible situations in which the model may have to act ultimately. In an uncharted territory, an RL model should be able to learn from its own experiences instead of getting trained by an external supervisor with a training data set of labeled examples.

Due to their inherent self-learning and adaptability characteristics, RL algorithms have been considered for various tasks in optical networks including network self-configuration, adaptive resource allocation, etc. In these applications, the actions performed by the RL algorithms may include choosing spectrum or modulation format, rerouting data traffic, etc., while the reward may be the maximization of network throughput, minimization of latency or packet loss rate, etc. (to be discussed in more detail in Section 21.5). Currently there are limited applications of RL in the physical layer of optical communication systems. This is because in most cases the reward (objective function) can be explicitly expressed as a continuous and differentiable function of the actions. An example is the CMA algorithm where the actions are the filter tap weights and the objective is to produce output signals with a desired amplitude. For such optimization procedures, we simply refer to them as adaptive signal processing instead of RL.

21.4 Deep learning techniques

21.4.1 Deep learning versus conventional machine learning

The recent emergence of DL technologies has taken ML research to a whole new level. DL algorithms have demonstrated comparable or better performance than humans in a lot of important tasks including image recognition, speech recognition, natural language processing, information retrieval, etc. [2,25]. Loosely speaking, DL systems consist of multiple layers of nonlinear processing units (thus deeper architectures) and may even contain complex structures such as feedback and memory. DL then refers to learning the parameters of these architectures for performing various pattern recognition tasks.

One way to interpret DL algorithms is that they automatically learn and extract higher level features of data from lower level ones as the input propagates through various layers of nonlinear processing units, resulting in a hierarchical representation of data. For example, while performing a complex human face recognition task using a DL-based multilayer ANN (called DNN), the first layer might learn to detect edges and the second layer can learn to recognize more complex shapes such as circles or squares that are built from the edges. The third layer may then recognize even more complex combinations and arrangements of shapes such as the location of two ovals and a triangle in between, which in turn starts to resemble parts of a human face with two eyes and a nose. Such an ability to automatically discover and learn features at increasingly high levels of abstraction empowers DL systems to learn complex relationships between inputs and outputs directly from the data instead of using human-crafted features.

As an example of this notion of hierarchical learning, Fig. 21.17 shows the use of a DNN as well as a conventional ANN on a signal's eye diagrams to monitor OSNR.

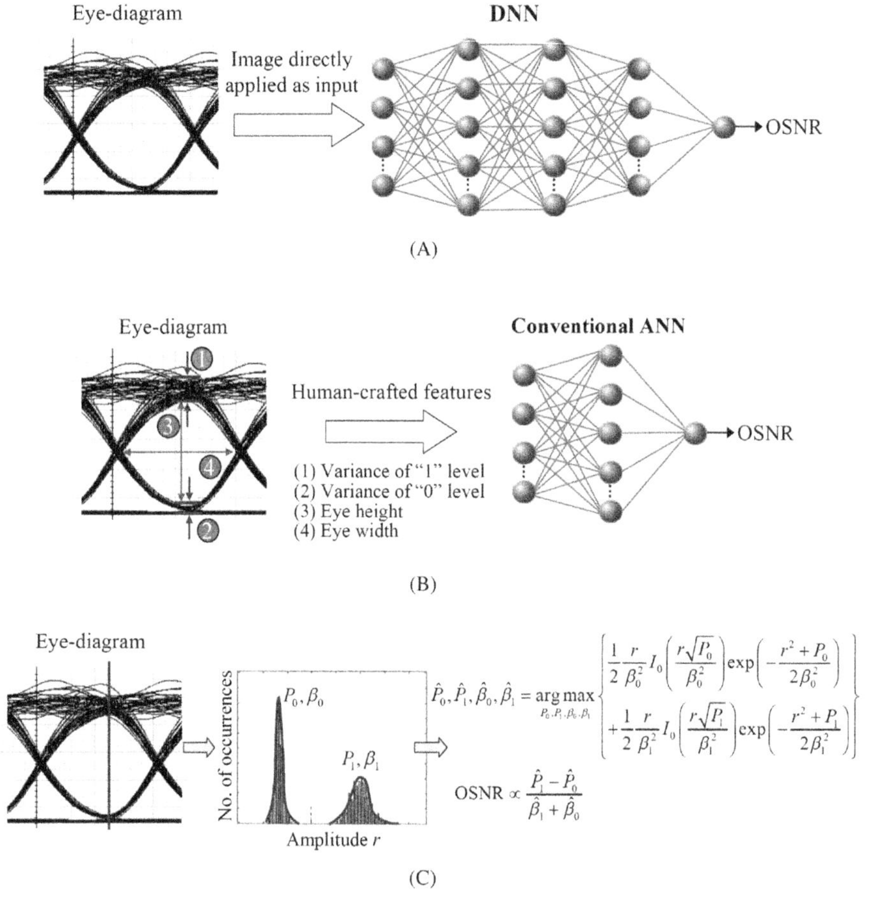

Figure 21.17 Example illustrating optical signal-to-noise ratio (OSNR) monitoring using eyediagrams' features by applying (A) deep neural network (DNN), (B) conventional artificial neural network (ANN), and (C) analytical modeling and parameters fitting.

In the first approach, the eye diagrams are directly applied as images at the input of the DNN, as shown in Fig. 21.17A, and it is made to automatically learn and discover OSNR-sensitive features without any human intervention. The extracted features are subsequently exploited by DNN for OSNR monitoring. In contrast, with conventional ANNs, prior knowledge in optical communications is utilized in choosing suitable features for the task, for example, the variances of "1" and "0" levels and eye opening can be indicative of OSNR. Therefore these useful features are manually extracted from the eye diagrams and are then used as inputs to an ANN for the estimation of OSNR as shown in Fig. 21.17B. For completeness, Fig. 21.17C shows an analytical and non-ML approach to determine OSNR by finding the powers and noise variances that best fit the noise distributions of "1" and "0" levels knowing that

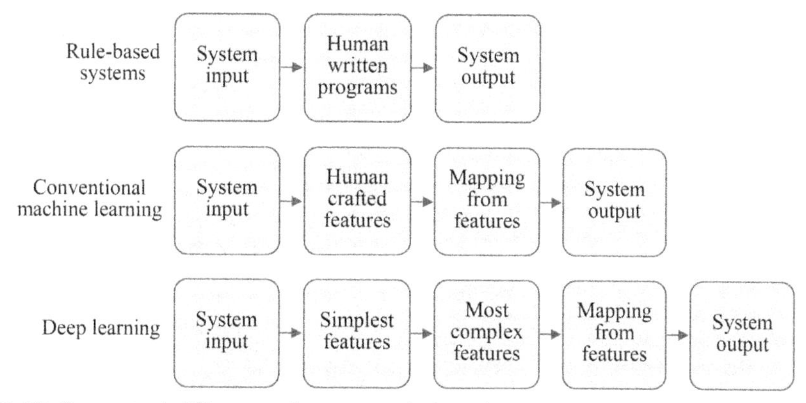

Figure 21.18 Conceptual differences between rule-based systems, conventional machine learning (ML), and deep learning (DL) approaches for pattern recognition.

they follow Rician distribution. In this case, a specific mathematical formula or computational instruction is precoded into the program and there is nothing to learn from the input data. Fig. 21.18 compares the underpinning philosophies of the three different approaches discussed. Note that, in principle, there is no hard rule on how many layers are needed for an ML model in a given problem. In practice, it is generally accepted that when more underlying physics/mathematics of the problem is used to identify and extract the suitable data features as inputs, the ML model tends to be simpler.

It should be noted that deep architectures are more efficient or more *expressive* than their shallow counterparts [26]. For example, it has been observed empirically that compared to a shallow neural network, a DNN requires much fewer number of neurons and weights (i.e., around 10 times fewer connections in speech recognition problems [27]) to achieve the same performance.

A major technical challenge in DL is that the conventional BP algorithm and gradient-based learning methods used for training shallow networks are inherently not effective for training networks with multiple layers due to the *vanishing gradient problem* [2]. In this case, different layers in the network learn at significantly different speeds during the training process, that is, when the layers close to the output are learning well, the layers close to the input often get stuck. In the worst case, this may completely stop the network from further learning. Several solutions have been proposed to address the vanishing gradient problem in DL systems. These include (1) choosing specific activation functions such as ReLU [11], as discussed earlier; (2) pre-training of network one layer at a time in a greedy way and then fine-tuning the entire network through BP algorithm [28]; (3) using some special architectures such as long short-term memory (LSTM) networks [29]; and (4) applying network optimization approaches that avoid gradients (e.g., global search methods such as genetic algorithm). The choice of a given solution typically depends on the type of DL model being trained and the degree of computational complexity involved.

21.4.2 Deep neural networks

Unlike shallow ANNs, DNNs contain multiple hidden layers between input and output layers. The structure of a simple three hidden layers DNN is shown in Fig. 21.19 (top). DNNs can be trained effectively using the BP algorithm. To avoid vanishing gradient problem during training of DNNs, two approaches are typically adopted. In the first method, the ReLU activation function is simply used for the hidden layers neurons due to its nonsaturating nature. In the second approach, a DNN is first pretrained one layer at a time and then the training process is fine-tuned using the BP algorithm [28]. For pretraining of hidden layers of the DNNs, *autoencoders* are typically employed which are essentially feed-forward neural networks. Fig. 21.19 (bottom) shows two simple autoencoders used for the unsupervised pretraining of first two hidden layers of the DNN. First, hidden layer-1 of the DNN is pretrained in isolation using autoencoder-1 as shown in the figure. The first part of autoencoder-1 (called encoder) maps input vectors $\mathbf{x}$ to a hidden representation $\mathbf{f}_1$, while the second part (called the decoder) reverses this mapping in order to synthesize the initial inputs $\mathbf{x}$. Once autoencoder-1 learns these mappings successfully, hidden layer-1 is considered to be pretrained. The original input vectors $\mathbf{x}$ are then passed through the encoder of autoencoder-1 and the corresponding representations $\mathbf{f}_1$ (also called feature vectors) at the output of pretrained hidden layer-1 are obtained. Next, vectors $\mathbf{f}_1$ are utilized as

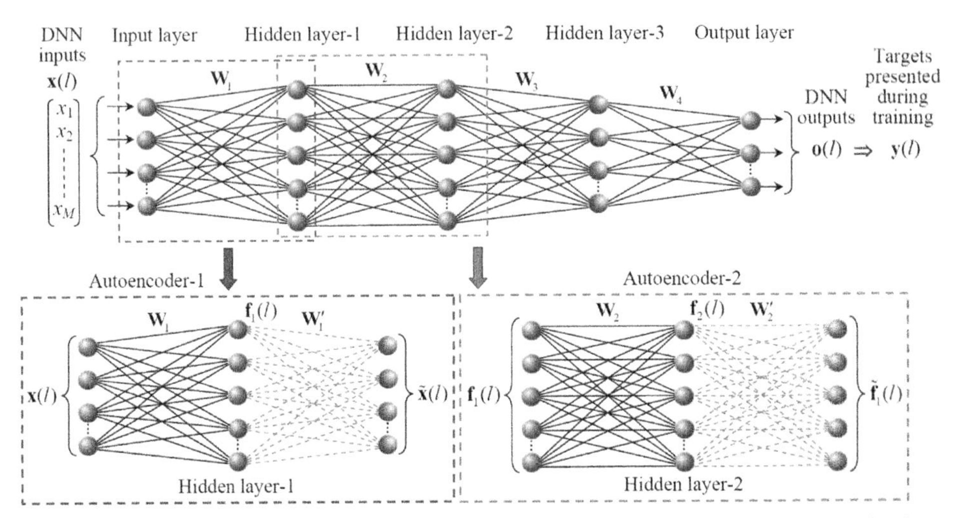

Figure 21.19 Schematic diagram of a three hidden layers deep neural network (DNN) (*top*). Two autoencoders used for the pretraining of first two hidden layers of the DNN (*bottom*). The decoder parts in both autoencoders are shown in gray color *with dashed weight lines*.

inputs for the unsupervised pretraining of hidden layer-2 using autoencoder-2, as depicted in the figure. This procedure is repeated for the pretraining of hidden layer-3, and the corresponding feature vectors $\mathbf{f}_3$ are then used for the supervised pretraining of final output layer by setting the desired outputs $\mathbf{y}$ as targets. After isolated pretraining of hidden and output layers, the complete DNN is trained (i.e., fine-tuned) using BP algorithm with $\mathbf{x}$ and $\mathbf{y}$ as inputs and targets, respectively. By adopting this autoencoders-based hierarchical learning approach, the vanishing gradient problem can be successfully bypassed in DNNs.

21.4.3 Convolutional neural networks

CNNs are a type of neural network primarily used for pattern recognition within images, though they have also been applied in a variety of other areas such as speech recognition, natural language processing, video analysis, etc. The structure of a typical CNN is shown in Fig. 21.20A comprising a few alternating *convolutional* and *pooling* layers followed by an ANN-like structure toward the end of the network. The convolutional layer consists of neurons whose outputs only depend on the neighboring

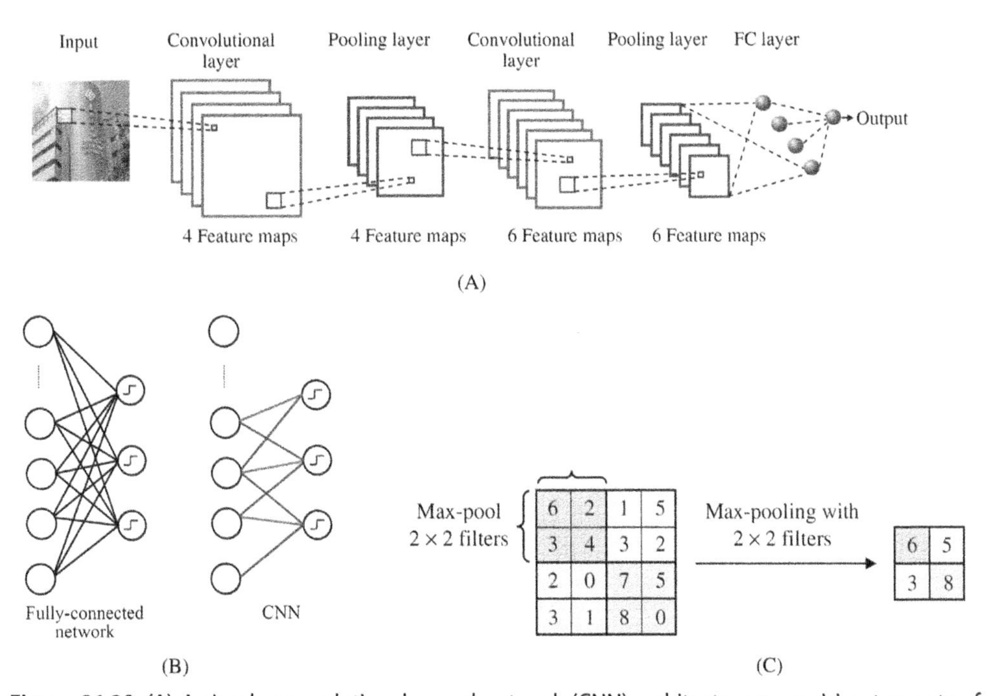

Figure 21.20 (A) A simple convolutional neural network (CNN) architecture comprising two sets of convolutional and pooling layers followed by a fully connected (FC) layer on top. (B) In a CNN, a node in the next layer is connected to a small subset of nodes in the previous layer. The weights (indicated by *colors of the edges*) are also shared among the nodes. (C) Nonlinear downsampling of feature maps via a max-pooling layer.

pixels of the input as opposed to fully connected (FC) layers in typical ANNs as shown in Fig. 21.20B. That is why it is called a *local network*, *local connected network*, or *local receptive field* in ML literature. The weights are also shared across the neurons in the same layer; that is, each neuron undergoes the same computation $\mathbf{w}^{\mathrm{T}}(\cdot) + b$ but the input is a different part of the original image. This is followed by a decision-like nonlinear activation function, and the output is called a *feature map* or *activation map*. For the same input image/layer, one can build multiple feature maps, where the features are learned via a training process. A parameter called *stride* defines how many pixels we slide the $\mathbf{w}^{\mathrm{T}}(\cdot) + b$ filter across the input image horizontally/vertically per output. The stride value determines the size of a feature map. Next, a *max-pooling* or *subsampling* layer operates over the feature maps by picking the largest value out of four neighboring neurons as shown in Fig. 21.20C. Max-pooling is essentially nonlinear downsampling with the objective of retaining the largest identified features while reducing the dimensionality of the feature maps.

The $\mathbf{w}^{\mathrm{T}}(\cdot) + b$ operation essentially multiplies part of the input image with a 2D function $g(s_x, s_y)$ and sums the results as shown in Fig. 21.21. The sliding of $g(s_x, s_y)$ over all spatial locations is the same as *convolving* the input image with $g(-s_x, -s_y)$ (hence the name convolutional neural networks). Alternatively, one can also view the $\mathbf{w}^{\mathrm{T}}(\cdot) + b$ operation as *cross-correlating* $g(s_x, s_y)$ with the input image. Therefore a high value will result if that part of the input image resembles $g(s_x, s_y)$. Together with the decision-like nonlinear activation function, the overall feature map indicates which location in the original image best resembles $g(s_x, s_y)$, which essentially tries to identify and locate a certain feature in the input image. With this insight, the interleaving convolutional and subsampling layers can be intuitively understood as identifying higher level and more complex features of the input image.

The training of a CNN is performed using a modified BP algorithm that updates convolutional filters' weights and also takes the subsampling layers into account.

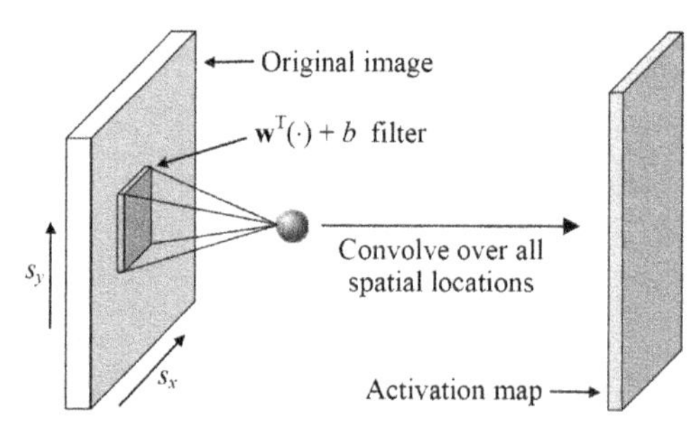

Figure 21.21 Convolution followed by an activation function in a convolutional neural network (CNN). Viewing the $\mathbf{w}^{\mathrm{T}}(\cdot) + b$ operation as *cross-correlating* a 2D function $g(s_x, s_y)$ with the input image, the overall feature map indicates which location in the original image best resembles $g(s_x, s_y)$.

Since a lot of weights are supposedly identical, as the network is essentially performing the convolution operation, one will update those weights using the average of the corresponding gradients.

21.4.4 Recurrent neural networks

In our discussion up to this point, different input–output pairs $(\mathbf{x}(i), \mathbf{y}(i))$ and $(\mathbf{x}(j), \mathbf{y}(j))$ in a data set are assumed to have no relation with each other. However, in a lot of real-world applications such as speech recognition, handwriting recognition, stock market performance prediction, inter-symbol interference (ISI) cancellation in communications, etc., the sequential data has important spatial/temporal dependence to be learned. An RNN is a type of neural network that performs pattern recognition for data sets with memory. RNNs have feedback connections, as shown in Fig. 21.22, and thus enable the information to be temporarily memorized in the networks [30]. This property allows RNNs to analyze sequential data by making use of their inherent memory.

Consider an RNN as shown in Fig. 21.22 with an input $\mathbf{x}(t)$, an output $\mathbf{o}(t)$ and a hidden state $\mathbf{h}(t)$ representing the memory of the network, where the subscript t denotes time. The model parameters $\mathbf{W}_1$, $\mathbf{W}_2$, and $\mathbf{W}_r$ are input, output, and recurrent weight matrices, respectively. An RNN can be unfolded in time into a multilayer network [31], as shown in Fig. 21.22. Note that unlike a feed-forward ANN which employs different parameters for each layer, the same parameters $\mathbf{W}_1$, $\mathbf{W}_2$, $\mathbf{W}_r$ are shared across all steps, which reflects the fact that essentially same task is being performed at each step but with different inputs. This significantly reduces the number of parameters to be learned. The hidden state $\mathbf{h}(t)$ and output $\mathbf{o}(t)$ at time step t can be computed as

$$\mathbf{h}(t) = \sigma_1(\mathbf{W}_1\mathbf{x}(t) + \mathbf{W}_r\mathbf{h}(t-1) + \mathbf{b}_1) \tag{21.28}$$

$$\mathbf{o}(t) = \sigma_2(\mathbf{W}_2\mathbf{h}(t) + \mathbf{b}_2), \tag{21.29}$$

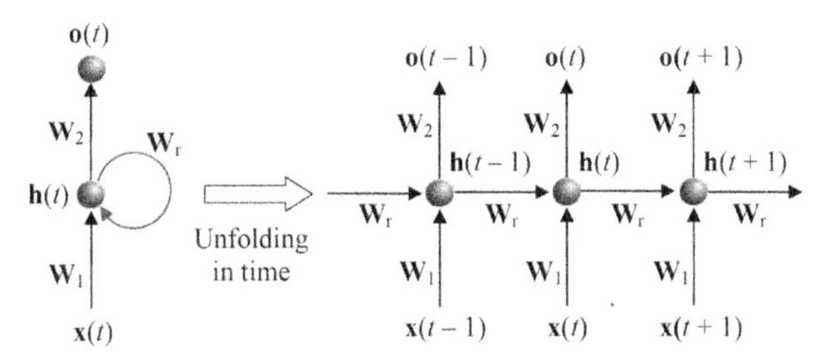

Figure 21.22 Schematic diagram of a recurrent neural network (RNN) and the unfolding in time.

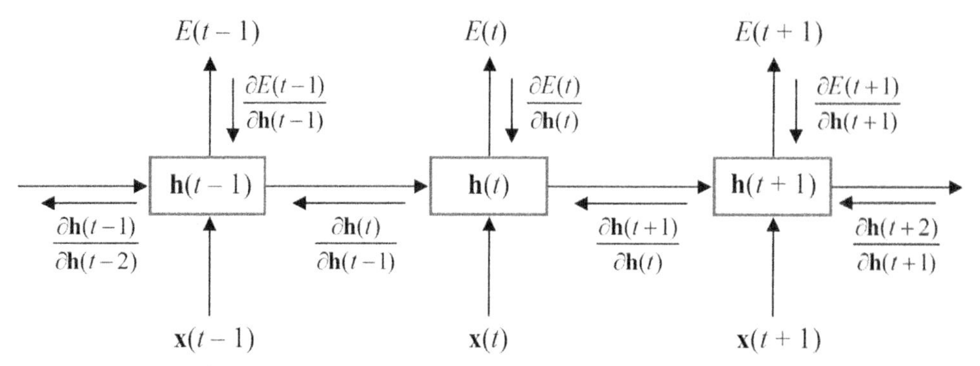

Figure 21.23 Flow of gradient signals in a recurrent neural network (RNN).

where $\mathbf{b}_1$ and $\mathbf{b}_2$ are the bias vectors while $\sigma_1(\cdot)$ and $\sigma_2(\cdot)$ are the activation functions for the hidden and output layer neurons, respectively. Given a data set $\{(\mathbf{x}(1), \mathbf{y}(1)), (\mathbf{x}(2), \mathbf{y}(2)), \ldots (\mathbf{x}(L), \mathbf{y}(L))\}$ of input–output pairs, the RNN is first unfolded in time to represent it as a multilayer network and then BP algorithm is applied on this graph, as shown in Fig. 21.23, to compute all the necessary matrix derivatives $\{(\partial E/\partial \mathbf{W}_1), (\partial E/\partial \mathbf{W}_2), (\partial E/\partial \mathbf{W}_{\mathrm{r}}), (\partial E/\partial \mathbf{b}_1), (\partial E/\partial \mathbf{b}_2)\}$. The loss function can be cross-entropy or MSE. The matrix derivative $\partial E/\partial \mathbf{W}_{\mathrm{r}}$ is a bit more complicated to calculate, since $\mathbf{W}_{\mathrm{r}}$ is shared across all hidden layers. In this case,

$$\frac{\partial E}{\partial \mathbf{W}_{\mathrm{r}}} = \sum_{t=1}^{L} \frac{\partial E(t)}{\partial \mathbf{W}_{\mathrm{r}}} = \sum_{t=1}^{L} \frac{\partial E(t)}{\partial \mathbf{h}(t)} \frac{\partial \mathbf{h}(t)}{\partial \mathbf{W}_{\mathrm{r}}} = \sum_{t=1}^{L} \frac{\partial E(t)}{\partial \mathbf{h}(t)} \sum_{l=1}^{t} \frac{\partial \mathbf{h}(t)}{\partial \mathbf{h}(l)} \frac{\partial \mathbf{h}(l)}{\partial \mathbf{W}_{\mathrm{r}}}, \tag{21.30}$$

where most of the derivatives in Eq. (21.30) can be easily computed using Eqs. (21.28) and (21.29). The Jacobian $\partial \mathbf{h}(t)/\partial \mathbf{h}(l)$ is further decomposed into $\frac{\partial \mathbf{h}(t)}{\partial \mathbf{h}(t-1)} \frac{\partial \mathbf{h}(t-1)}{\partial \mathbf{h}(t-2)} \cdots \frac{\partial \mathbf{h}(l+1)}{\partial \mathbf{h}(l)}$ so that efficient updates naturally involve the flow of matrix derivatives from the last data point $(\mathbf{x}(L), \mathbf{y}(L))$ back to the first $(\mathbf{x}(1), \mathbf{y}(1))$. This algorithm is called BP through time (BPTT) [32].

In the special case when the nonlinear activation function is absent, the RNN structure resembles a linear multiple-input multiple-output channel with memory 1 in communication systems. Optimizing the RNN parameters will thus be equivalent to estimating the channel memory given input and output signal waveforms followed by maximum likelihood sequence detection of additional received signals. Consequently, an RNN may be used as a suitable tool for channel characterization and data detection in *nonlinear* channels with memory such as long-haul transmission links with fiber Kerr nonlinearity or direct detection systems with CD, chirp, or other component nonlinearities. Network traffic prediction may be another area where RNNs can play a useful role.

One major limitation of conventional RNNs in many practical applications is that they are not able to learn long-term dependencies in data (i.e., dependencies between events that are far apart) due to the so-called *exploding* and *vanishing gradient problems* encountered during their training. To overcome this issue, a special type of RNN architecture called LSTM network is designed that can model and learn temporal sequences and their long-range dependencies more accurately through better storing and accessing of information [29]. An LSTM network makes decision on whether to forget/delete or store the information based on the importance that it assigns to the information. The assigning of importance takes place through weights that are determined via a learning process. Simply put, an LSTM network learns over time which information is important and which is not. This allows LSTM network's short-term memory to last for longer periods of time as compared to conventional RNNs, which in turn leads to improved sequence learning performance.

21.5 Applications of machine learning techniques in optical communications and networking

Fig. 21.24 shows some significant research works related to the use of ML techniques in fiber-optic communications. A brief discussion on these works follows.

21.5.1 Optical performance monitoring

Optical communication networks are becoming increasingly complex, transparent, and dynamic. Reliable operation and efficient management of these complex fiber-optic networks require incessant and real-time information of various channel impairments ubiquitously across the network, also known as OPM [33]. OPM is widely regarded as a key enabling technology for SDNs. Through OPM, SDNs can become aware of the real-time network conditions and subsequently adjust different transceiver/ network elements parameters such as launched powers, data rates, modulation formats,

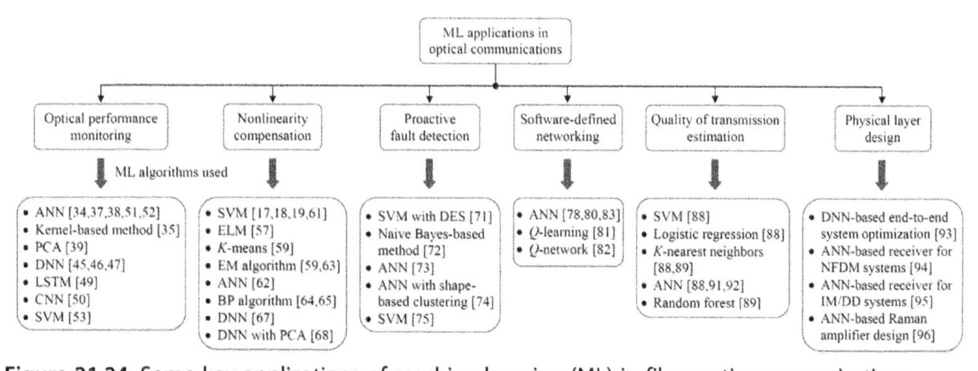

Figure 21.24 Some key applications of machine learning (ML) in fiber-optic communications.

spectrum assignments, etc., for optimized transmission performance [4]. Unfortunately, conventional OPM techniques have shown limited success in simultaneous and independent monitoring of multiple transmission impairments, since the effects of different impairments are often difficult to separate analytically. Another crucial OPM requirement is low complexity since the OPM devices need to be deployed ubiquitously across optical networks. ML techniques are proposed as an enabler for realizing low-complexity (and hence low cost) multiimpairment monitoring in optical networks and have already shown tremendous potential.

Most existing ML-based OPM techniques adopt a supervised learning approach utilizing training data sets of labeled examples during the offline learning process of selected ML models. The training data may, for example, consist of signal representations like eye diagrams, asynchronous delay-tap plots (ADTPs), amplitude histograms (AHs), etc., and their corresponding known impairments values such as CD, differential group delay (DGD), OSNR, etc., serving as data labels, as shown in Fig. 21.25. During the training phase, the inputs to an ML model are the impairments-indicative

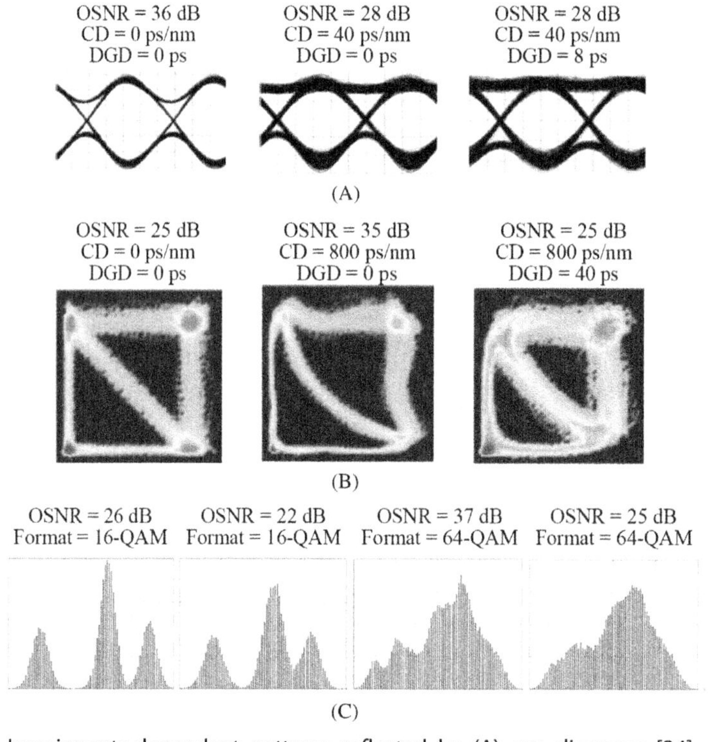

Figure 21.25 Impairments-dependent patterns reflected by (A) eye diagrams [34], (B) asynchronous delay-tap plots (ADTPs) [35], and (C) amplitude histograms (AHs), and their corresponding known impairments values, which serve as data labels during the training process.

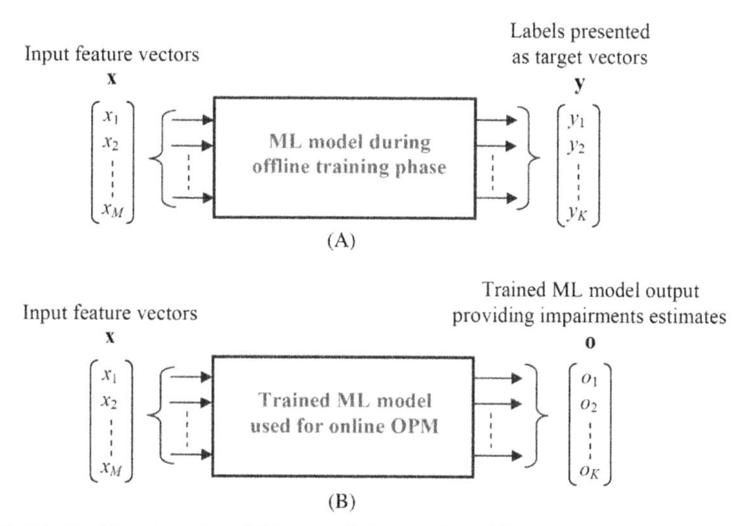

Figure 21.26 (A) Machine learning (ML) model during the offline training phase with feature vectors **x** as inputs and the labels **y** as targets. (B) Trained ML model used for online optical performance monitoring (OPM) with feature vectors **x** as inputs and the impairments estimates **o** as outputs.

feature vectors **x** of eye diagrams/ADTPs/AHs, while their corresponding labels **y** are used as the targets, as shown in Fig. 21.26A. The ML model then learns the mapping between input features and the labels. Note that in case of eye diagrams, the features can be parameters like eye closure, Q-factor, root-mean-square jitter, crossing amplitude, etc. [34]. On the other hand, for AHs/ADTPs, the empirical one-dimensional (1D)/2D histograms can be treated as features [35,36]. Once the offline training process is completed, the ML model can be used for real-time monitoring (as the computations involved are relatively simple) in deployed networks as shown in Fig. 21.26B.

ML algorithms have been applied successfully for cost-effective multiimpairment monitoring in optical networks. Wu et al. [34] exploited impairments-sensitive features of eye diagrams using an ANN for simultaneous monitoring of OSNR, CD, and DGD. Similarly, Anderson et al. [35] demonstrated joint estimation of CD and DGD by applying kernel-based ridge regression on ADTPs. In Ref. [37], we showed that the raw empirical moments of asynchronously sampled signal amplitudes are sensitive to OSNR, CD, and DGD. The first five empirical moments of received signal amplitudes are thus used as input features to an ANN for multiimpairment monitoring. Unlike Refs. [34,35], which can only determine the magnitude of the CD, this technique enables monitoring of both magnitude and sign of accumulated CD. In Ref. [38], the low-frequency part of the received signal's radio frequency spectrum is used as input to an ANN for OSNR monitoring in the presence of large inline uncompensated CD. Apart from supervised learning, unsupervised ML techniques have also been employed for OPM. In Ref. [39], PCA and statistical distance measurement based pattern recognition is applied on ADTPs for joint OSNR, CD, and DGD monitoring, as well as for identification of bit-rate and modulation format of the received signal.

The emergence of SDNs imposes new requirements on OPM devices deployed at the intermediate network nodes. As a lot of OSNR/CD/polarization-mode dispersion (PMD) monitoring techniques are modulation format dependent, the OPM devices are desired to have modulation format identification (MFI) capabilities in order to select the most suitable monitoring technique. Although the modulation format information of a signal can be obtained from upper layer protocols in principle, it is practically not available for OPM task at the intermediate network nodes because the OPM units are often stand-alone devices and can only afford limited complexity [4]. Note that MFI may also be beneficial for digital coherent receivers in elastic optical networks (EONs), since it can enable fast switching between format-dependent carrier recovery modules, as conventional supervisory channels may not be able to provide modulation format information that quickly [40]. Reported ML-based MFI techniques in the literature include the *K*-means algorithm [41], ANNs [36,42], variational Bayesian expectation-maximization [43], and DNN [44] based methods.

Recently, DL algorithms have also been applied for OPM. In Ref. [45], we demonstrated joint OSNR monitoring and MFI in digital coherent receivers, as shown in Fig. 21.27A, using DNNs in combination with AHs depicted in Fig. 21.25C. The DNNs automatically extracted OSNR and modulation format sensitive features of AHs and exploited them for the joint estimation of these parameters. The OSNR monitoring results for one signal type are shown in Fig. 21.27B, and it is clear from the figure that OSNR estimates are quite accurate. The confusion table/matrix in Fig. 21.27C summarizes MFI results (in the absence of fiber nonlinear effects) for 57 test cases used for evaluation. The upper element in each cell of this table represents the number of instances of correct/incorrect identifications for a given actual modulation format, while the bottom element shows percentage of correct/incorrect identifications for a given actual modulation format. It is evident from the table that no errors are encountered in the identification of all three modulation formats under consideration. The performance of this technique in the presence of fiber nonlinearity is shown in Fig. 21.27D, and it is clear from the figure that identification accuracies decrease slightly in this case [44]. However, they still remain higher than 99%, thus showing the resilience of this technique against fiber nonlinear effects. Since this technique uses DL algorithms inside a standard digital coherent receiver, it avoids extra hardware costs. Tanimura et al. [46] applied DNN on asynchronously sampled raw data for OSNR monitoring in a coherent receiver. Using a deep 5-layers architecture and a large training data set of 400,000 samples, the DNN is shown to learn and extract useful OSNR-sensitive features of incoming signals without involving any

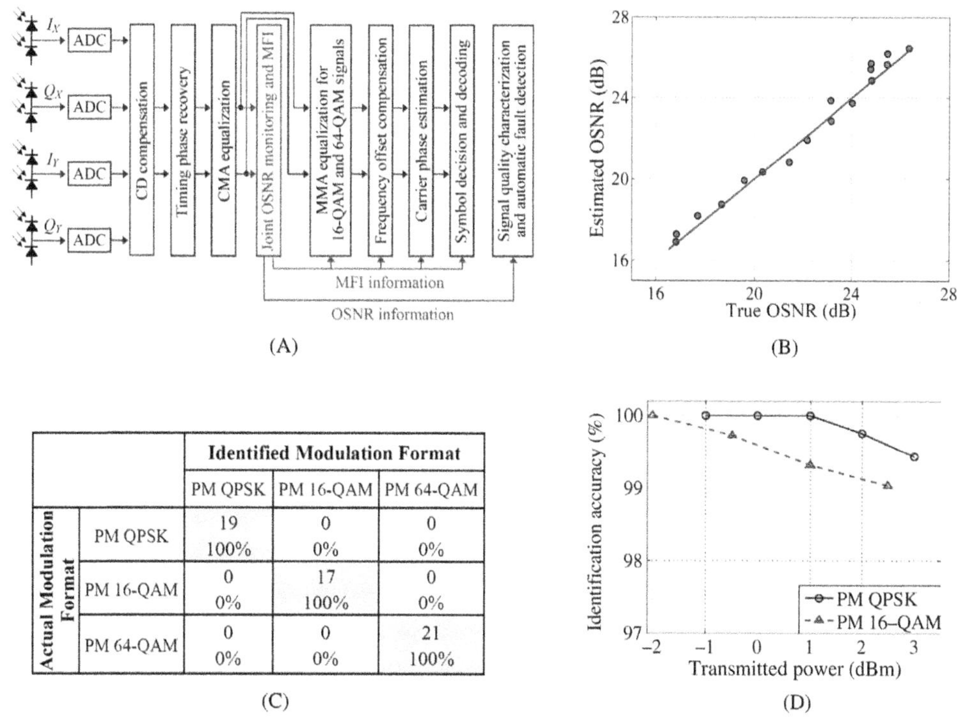

		Identified Modulation Format		
		PM QPSK	PM 16-QAM	PM 64-QAM
Actual Modulation Format	PM QPSK	19 100%	0 0%	0 0%
	PM 16-QAM	0 0%	17 100%	0 0%
	PM 64-QAM	0 0%	0 0%	21 100%

(C)

Figure 21.27 (A) Receiver digital signal processing (DSP) configuration with the deep neural network (DNN)-based optical signal-to-noise ratio (OSNR) monitoring and modulation format identification (MFI) stage shown in *red*. (B) True versus estimated OSNRs for 112 Gb/s polarization-multiplexed (PM) 16 quadrature amplitude modulation (16-QAM) signals. (C) MFI accuracies (in number of instances and percertange of correct identifications) for different modulation formats in the absence of fiber nonlinear effects [45]. (D) Effect of fiber nonlinearity on the MFI accuracies [44].

manual feature engineering. An extension of this method is presented in Ref. [47], where the DNN-based monitor is enhanced using the dropout technique [48] at the inference time (unlike typical approach of using dropout during training) so that multiple "thinned" DNNs with slightly different configurations provide multiple OSNR estimates. This in turn enables them to compute confidence intervals of the OSNR estimates as an auxiliary output. Similarly, Wang et al. [49] applied an LSTM network on temporal sequences of signal samples in a digital coherent receiver for simultaneous monitoring of OSNR and CD. In Ref. [50], raw eye diagrams are treated as images (composed of various pixels) and are processed using a CNN for automatic extraction of features, which are then used for joint OSNR monitoring and MFI. This technique exhibits better performance than conventional ML approaches.

Open issues: While ML-based OPM has received significant attention over the last few years, certain issues still need to be addressed. For example, accurate OSNR monitoring in long-haul transmission systems in the presence of fiber nonlinearity is still a challenging task, as nonlinear distortions are incorrectly treated as noise by most OSNR monitoring methods. Developing ML-based techniques to estimate actual OSNR irrespective of other transmission impairments, channel power, and wavelength-division multiplexing (WDM) effects is highly desirable in future optical networks. Recently there have been some initial attempts in this regard exploiting amplitude noise covariance (ANC) of received symbols [51], and features of nonlinear phase noise along with time correlation properties of fiber nonlinearities [52]. Zhang et al. [53] used multiple ANCs corresponding to different numbers of symbol delays along with accumulated CD as independent input features to an SVM for accurate nonlinear noise-to-signal ratio monitoring in the absence of channel state information (i.e., number of WDM channels transmitted). Another open issue is the development of ML-based monitoring techniques using only low-bandwidth components, as this can reduce the computational complexity and cost of OPM devices installed at the intermediate network nodes. Alternatively, instead of physical deployment of OPM units across the network, capturing of various physical layer data (such as launched powers, parameters of various optical amplifiers and fibers, etc.) via network management and using the ML algorithms to uncover complex relationships between these parameters and the actual link OSNRs can also be investigated.

21.5.2 Fiber nonlinearity compensation

The optical fiber communication channel is nonlinear due to the Kerr effect, and thus classical linear systems equalization/detection theory is suboptimal in this context. Signal propagation in optical fibers in the presence of Kerr nonlinearity together with distributed copropagating amplified spontaneous emission (ASE) noise $n(t, z)$ can be described by the stochastic nonlinear Schrödinger equation (NLSE)

$$\frac{\partial}{\partial z}u(t,z) + j\frac{\beta_2}{2}\frac{\partial^2}{\partial t^2}u(t,z) = j\gamma\left|u(t,z)\right|^2 u(t,z) + n(t,z), \tag{21.31}$$

where $u(t, z)$ is the electric field while β_2 and γ are group velocity dispersion parameter and fiber nonlinear coefficient, respectively. Although the NLSE can be numerically evaluated using the split-step Fourier method (SSFM) to simulate the waveforms evolution during transmission, the interplay between signal, noise, nonlinearity, and dispersion complicates the analysis. This is also essentially the limiting factor of the DBP technique [54]. At present, stochastic characteristics of nonlinearity- induced noise depend in a complex manner on dispersion, modulation format, transmission distance, amplifier, and type of optical fiber [55]. As an example, two received signal distributions after linear compensation of various transceiver and transmission impairments are shown in Fig. 21.28 for long-haul systems with and without inline dispersion

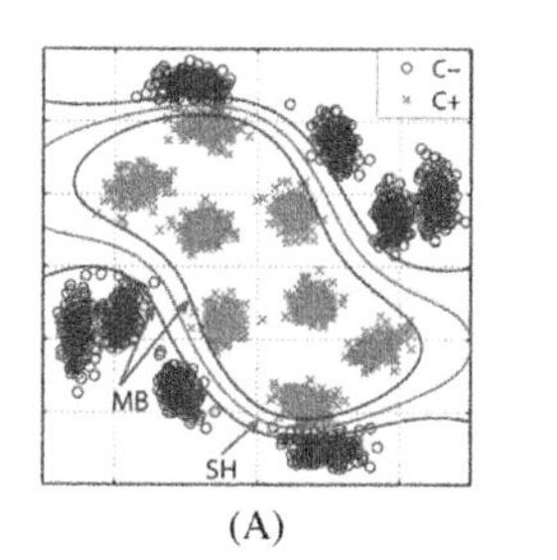

(A)

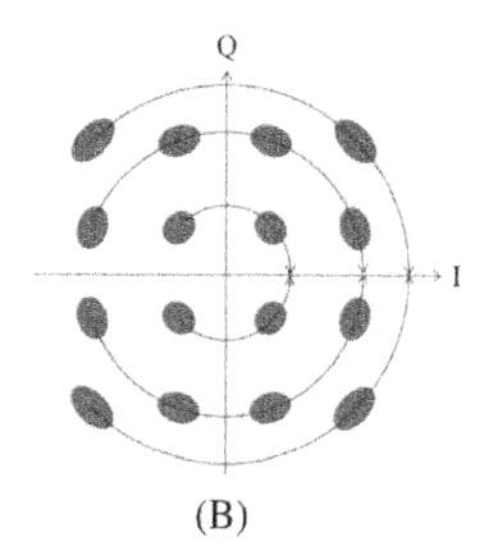

(B)

Figure 21.28 Received signal distributions after linear equalization for long-haul (A) dispersion-managed [17] and (B) dispersion-unmanaged [55] transmissions.

compensation. From Fig. 21.28A, it is obvious that the decision boundaries are nonlinear, which naturally calls for the use of ML techniques. In contrast, the nonlinear noise is more Gaussian-like in dispersion-unmanaged transmissions [56], as shown in Fig. 21.28B. However, the noise is correlated in time and hard to model analytically.

More specifically, consider a transmission link with inline distributed erbium-doped fiber amplifiers (EDFAs). Let $\mathbf{y} = [y_1 y_2 \ldots]$ be a symbol sequence with overall transmitted signal $q(t) = \sum_j y_j s(t - jT)$ where $s(t)$ is the pulse shape and T is the symbol period. The received signal is given by

$$x(t) = f_{\mathrm{NLSE}}(q(t), n(t, z)), \tag{21.32}$$

where $f_{\mathrm{NLSE}}(\cdot)$ is the input–output mapping characterized by the NLSE. It should be noted that electronic shot noise and quantization noise also act as additional noise sources but are omitted here, as ASE noise and its interaction with Kerr nonlinearity and dispersion are the dominant noise sources in long-haul transmissions. Other WDM effects are also omitted here for simplicity. The received signal is sampled to obtain $\mathbf{x} = [x_1 x_2 \ldots]$. While using ML for NLC, one seeks to develop a neural network or generally a mapping function $g(\cdot)$ so that the output vector

$$\mathbf{o} = g(\mathbf{x}) \tag{21.33}$$

of ML model is as close as possible to $\mathbf{y}$.

There are a few classes of approaches to learn the best mapping $g(\cdot)$. One direction is to completely ignore the intricate interactions between nonlinearity, CD, and noise in NLSE and treat the nonlinear fiber as a black- box system. To this end, in Ref. [57], we proposed the use of an ANN after CD compensation in a digital coherent receiver and applied a training technique called extreme learning machine (ELM) that avoids SGD-like iterative weights and biases updates [58]. Fig. 21.29 shows the simulated Q-factor for 27.59 GBd/s return-to-zero (RZ) quadrature phase-shift keying (QPSK) transmissions over 2000 km standard single-mode fiber (SSMF). The proposed ELM-based approach provides comparable performance to DBP but is computationally much simpler.

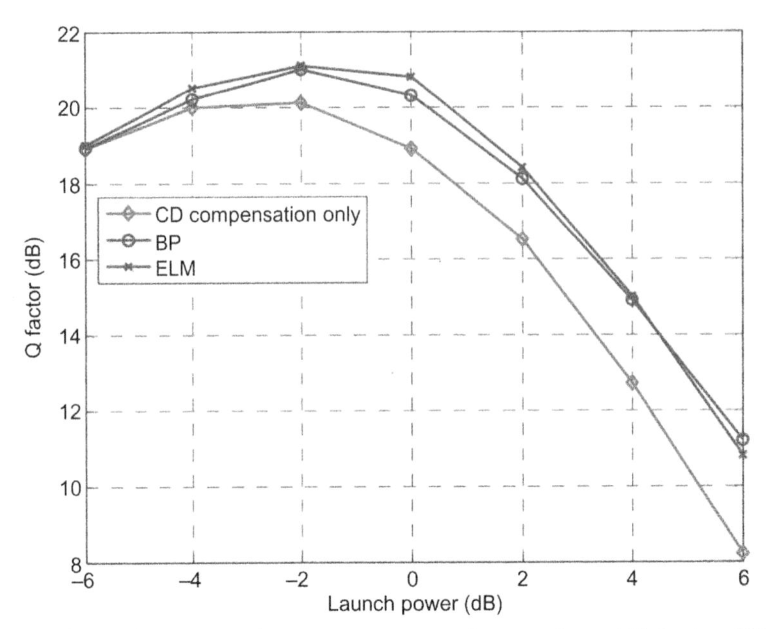

Figure 21.29 *Q*-factor for 27.59 GBd/s return-to-zeroquadrature phase-shift keying (RZ-QPSK) signals after transmission over 2000 km standard single-mode fiber (SSMF) [57].

For dispersion-managed systems, Zibar et al. [59] investigated the use of EM and *K*-means algorithms to derive nonlinear decision boundaries for optimized transmission performance. By applying the EM algorithm after regular DSP, the distribution of the combined fiber nonlinear distortions, laser phase noise and transceiver imperfections can be estimated. EM algorithm assumes that each cluster is Gaussian distributed with different mean and covariance matrix (the outer clusters in Fig. 21.28A are expected to have larger variances) so that the received signal distribution is a Gaussian mixture model. EM algorithm optimizes the mean and covariance matrix parameters iteratively to best fit the observed signal distribution in a maximum likelihood sense. The converged mixture model is then used for symbol detection. In contrast, the *K*-means algorithm provides less improvement because it assumes that the clusters share the same covariance matrix [60]. Nevertheless, for dispersion-unmanaged systems, both EM and *K*-means algorithms provide minimal performance gain as optimal decision boundaries in this case are nearly straight lines as shown in Ref. [59]. Li et al. [17] applied SVMs for fiber NLC. However, since basic SVM is only a binary classifier, classifying *M*-QAM signals would require $\log_2 M$ binary SVMs in this case. Other SVM-related works with different complexities and performance concerns are reported in Refs. [18,19,61] with a 0.5−2 dB gain in *Q*-factor compared to linear equalization methods. In Ref. [62], the use of an ANN per subcarrier to equalize nonlinear distortions in coherent optical orthogonal frequency-division multiplexing (CO-OFDM) systems is studied.

Another class of ML techniques incorporates limited amount of underlying physics to develop better fiber NLC methods. Pan et al. [63] noted that the received phase distortions of a 16-QAM signal are correlated across symbols due to nonlinearity in addition to the effect of slowly varying laser phase noise. They proposed to interleave the soft forward error correction decoding with a modified EM algorithm. In particular, the EM algorithm takes the output of the soft-decision low-density parity-check (LDPC) decoder as input and provides phase noise estimates $\hat{\phi}_k$, which are then used to compensate the symbols and fed back to the LDPC decoder, thus forming an adaptive feedback equalizer. An additional regularization term is added inside the EM algorithm to prevent large phase jumps.

Finally, ML approaches can also augment well-developed analytical models and signal processing strategies for NLC. Häger et al. [64] consider the standard DBP algorithm

$$\mathbf{W}^{-1}\sigma^{-1}(\mathbf{W}^{-1}\sigma^{-1}(\ldots.)), \tag{21.34}$$

where $\mathbf{W}$ is a matrix (representing linear CD operation) and $\sigma(z) = e^{j\gamma|z|^2}$ is the nonlinear phase rotation. However, when viewed from an ML perspective, this sequence of interleaved linear and nonlinear operations resembles a standard neural network structure and hence, all the parameters in $\mathbf{W}$ as well as the nonlinear function parameters can be learned. A dedicated complex-valued DNN-like model following time-domain DBP procedures but generalizing the filter of the linear step and the scaling factor of the nonlinear phase rotation is developed. The loss function is differentiable with respect to the real and imaginary parts of the parameters and thus can be learned using standard BP techniques. The learned DBP algorithm performs similar to conventional DBP but is computationally simpler. For high baud rate systems, a subband division solution to reduce the required linear filter size was recently proposed [65]. In another approach, perturbation analysis of fiber nonlinearity is used to analyze the received signal $x(t) = x_{\text{lin}}(t) + \Delta x(t)$, where $x_{\text{lin}}(t)$ is the received signal if the system is linear while the intra-channel four-wave mixing (IFWM) perturbations $\Delta x(t)$ are given by

$$\Delta x(t) = \sum_{m,n} P^{3/2} C_{m,n} x_{\text{lin}}(t - mT) x^*_{\text{lin}}(t - (m+n)T) x_{\text{lin}}(t - nT), \tag{21.35}$$

where T is the symbol period, P is the signal power, and $C_{m,n}$ is determined by the fiber parameters [66]. From an ML perspective, the $x_{\text{lin}}(t - mT)x^*_{\text{lin}}(t - (m+n)T)x_{\text{lin}}(t - nT)$ triplets and $x(t)$ can serve as inputs to an ANN to estimate $C_{m,n}$ and $\Delta x(t)$, which can then be used to predistort the transmitted signal and obtain better results.

The proposed technique is demonstrated in an 11,000 km subsea cable transmission, and it outperforms transmitter side perturbation-based predistortion methods by 0.3 dB in both single-channel and WDM systems [67]. Gao et al. [68] applied PCA for extracting limited but most significant features of intra-channel cross-phase modulation and IFWM perturbation triplets, which are then used as inputs to a DNN for fiber NLC. The PCA-assisted decrease in number of inputs is shown to reduce the DNN computational complexity by 90% with only 0.1 dB Q-factor penalty incurred.

Open issues: Table 21.1 shows some key techniques using ML for fiber NLC. Most of these works incorporate ML as an extra DSP module placed either at transmitter or receiver. While effective to a certain extent, it is not clear what is the best sequence of conventional signal processing and ML blocks in such a hybrid DSP configuration. One factor driving the choice of sequence is the dynamic effects such as carrier frequency offset, laser phase noise, PMD, etc. that are hard to be captured in the learning process of an ML algorithm. In this case, one can perform ML-based NLC after linear compensations so as to avoid tackling these time-varying dynamics in ML. In the other extreme, RNN structures can embrace all the time-varying dynamics in principle, but it may be an overkill since we do know their underlying physics and it should be exploited in the overall DSP design. Also, in case of hybrid configurations, the accuracy of conventional DSP algorithms such as CMA or carrier phase estimation (CPE) plays a major role in the quality of the data sets which ML fundamentally relies on. Therefore there are strong dependencies between ML and conventional DSP blocks and the right balance is still an open area of research. Finally, to the best of our knowledge, an ML-based single-channel processing technique that outperforms DBP in practical WDM settings has yet to be developed.

Numerous studies are conducted to also address the computational complexity issues of conventional and ML techniques for fiber NLC. For conventional NLC algorithms, we direct the readers to the survey paper [69]. On the other hand, the computational complexity of ML algorithms for NLC varies significantly with the architecture and the training process used, which make comparison with the conventional techniques difficult. Generally, the training processes are too complex to be performed online, as they require a lot of iterations and potentially massive training data. For the inference phase (i.e., using the trained model for real-time data detection), most ML algorithms proposed involve relatively simple computations, leading to the perception that ML techniques are generally simple to implement, since offline training processes are typically not counted toward the computational complexity. However, in reality, the training does take up a lot of computational resources and time, which should not be completely disregarded while one is evaluating the complexity of ML approaches for NLC.

Table 21.1 Some key ML-based fiber NLC techniques [6].

Reference, year	ML algorithm used	Data rate	Transmission link	Modulation format	Polarization multiplexing	WDM	Experimental demonstration
[57], 2011	ANN	27.59 GBd/s	2000 km SSMF	RZ-QPSK	No	No	No
[59], 2012	EM	14 GBd/s	$<$ 800 km SSMF/DCF	16-QAM	Yes	No	Yes
[62], 2015	ANN	40 Gb/s	200 km SSMF	16-QAM CO-OFDM	No	No	Yes
[64], 2018	Learned DBP	20 GBd/s	3200 km SSMF	16-QAM	No	No	No
[67], 2018	Learned PPD using DNN	4×12.25 GBd/s	11,000 km Trans-Pacific cable	DSM-PS-8/16/64-QAM	Yes	Yes	Yes

PPD, Pre/post-distortion; *DCF*, dispersion-compensating fiber; *DSM*, digital subcarrier modulation; *PS*, probabilistic shaped.

21.5.3 Proactive fault detection

Reliable network operations are essential for the carriers to provide service guarantees, called service-level agreements (SLAs), to their customers regarding the system's availability and promised quality levels. Violation of these guarantees may result in severe penalties. It is thus highly desirable to have an early warning and proactive protection mechanism incorporated into the network. This can empower network operators to know when the network components are beginning to deteriorate, and preventive measures can then be taken to avoid serious disruptions [33].

Conventional fault detection and management tools in optical networks adopt a rigid approach where some fixed threshold limits are set by the system engineers and alarms are triggered to alert malfunctions if those limits are surpassed. Such traditional network protection approaches have following main drawbacks: (1) These methods protect a network in a passive manner, that is, they are unable to forecast the risks and tend to reduce the damages only after a failure occurs. This approach may result in the loss of immense amounts of data during network recovery process once a failure happens. (2) The inability to accurately forecast the faults leads to ultraconservative network designs involving large operating margins and protection switching paths, which in turn result in an underutilization of the system resources. (3) They are unable to determine the root cause of faults. (4) Apart from hard failures (i.e., the ones causing major signal disruptions), several kinds of soft failures (i.e., the ones degrading system performance slowly and slightly) may also occur in optical networks which cannot be easily detected using conventional methods.

ML-enabled proactive fault management has recently been conceived as a powerful means to ensure reliable network operation [70]. Instead of using traditional fixed pre-engineered solutions, this new mechanism relies on dynamic data-driven operations, leveraging immense amounts of operational data retrieved through network monitors (e.g., using simple network management protocol). The data repository may include network components' parameters such as optical power levels at different network nodes, EDFAs' gains, current drawn and power consumption of various devices, shelf temperature, temperatures of various critical devices, etc. ML-based fault prediction tools are able to learn historical fault patterns in networks and uncover hidden correlations between various entities and events through effective data analytics. Such unique and powerful capabilities are extremely beneficial in realizing proactive fault discovery and preventive maintenance mechanisms in optical networks. Fig. 21.30 illustrates various fault management tasks powered by the ML-based data analytics in optical networks including proactive fault detection, fault classification, fault localization, fault identification, and fault recovery.

Recently a few ML-based techniques have been developed for advanced failure prediction in networks. Wang et al. [71] demonstrated an ML-based network equipment failure prediction method in software-defined metropolitan area networks using

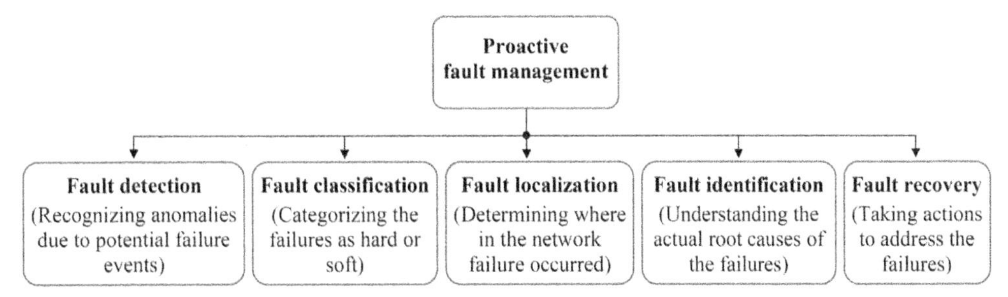

Figure 21.30 Fault management tasks enabled by machine learning (ML)-based approaches.

a combination of double exponential smoothing (DES) and an SVM. Their approach involves constant monitoring of various physical parameters of the boards used in WDM nodes. The set of parameters includes the boards' power consumption, laser bias current, laser temperature offset, and environment temperature. However, to realize proactive failure detection, DES, which is basically a time-series prediction algorithm, is used to predict the future values of these parameters. Next, an SVM-based classifier is used to learn the relationship between forecasted values of boards' parameters and the occurrence of failure events. This method is shown to predict boards' failures with an average accuracy of 95%.

Similarly, in Ref. [72], proactive detection of fiber damages is demonstrated using ML-based pattern recognition. In their work, the state-of-polarization rotation speed is constantly monitored in a digital coherent receiver, and if it exceeds a certain predefined limit, the system considers it as an indication of some fiber stress event (leading to certain fiber damages) and a flag is raised. Next, Stokes parameters' traces are recorded which are shown to exhibit unique patterns for various mechanical stress events on the fiber such as bending, shaking, etc. These patterns are exploited using a naive Bayes classifier for their recognition. This technique is shown to predict various fiber stress events (and thus fiber breaks before their actual occurrence) with 95% accuracy.

In Ref. [73], a cognitive fault detection architecture is proposed for intelligent network assurance. In their work, an ANN is used to learn historical fault patterns in networks for proactive fault detection. The ANN is trained to learn how the monitored optical power levels evolve over time under normal or abnormal network operation (i.e., recognize power level abnormalities due to occurrence of certain faults). The trained ANN is then shown to detect significant network faults with better detection accuracies and proactive reaction times as compared to conventional threshold-based fault detection approaches, as shown in Fig. 21.31. An extension of this method is presented in Ref. [74] which makes use of an ANN and shape-based clustering algorithm to not only proactively detect and localize faults but also determine their likely root causes. The two-stage fault detection and diagnosis framework proposed in their work involves monitoring optical power levels across various network nodes as well as

Fault Label	Description
I	Point abnormalities due to random flash event and may lead to abrupt device damage
II	Local abnormalities indicating potential flaws with potential long-term impact on service performance
III	Steady abnormalities due to preceding system configuration changes, and may lead to damage and/or consistent performance loss
IV	Ramp abnormalities representing gradual system and/or service distortion possibilities

(A)

		Abnormality Detection Rate [%]	Proactive Reaction Time [hours]	Detected / True Faults
Condition-based	Label I	100	0	1
Data-driven		100	0	1
Condition-based	Label II	0	0	0
Data-driven		~57	>100*	0.57
Condition-based	Label III	~45	0	0.45
Data-driven		100	10	1
Condition-based	Label IV	25	0	.25
Data-driven		~93	96	1.25

(B)

Figure 21.31 (A) Fault types typically encountered in commercial fiber-optic networks. (B) Comparison of fault detection rates and proactive reaction times of data-driven and condition-based methods for the fault types given in (A) [73]

nodes' local features such as temperature, amplifier gain, current draw profiles, etc. In the first stage, an ANN is applied to detect faults by identifying optical power level abnormalities across various network nodes. The faulty node is then localized using network topology information. In the second stage, the faulty node's local features (which are also interdependent) are further analyzed using a clustering technique to identify potential root causes.

In Ref. [75], an SVM-based anomaly detection method is proposed to predict soft failures in optical links. The soft failures considered in their work include laser malfunctioning (i.e., increase in laser linewidth), interchannel interference, reconfigurable optical add-drop multiplexer abnormalities (i.e., reduction in wavelength-selective switch bandwidth or shifting of its center frequency), and OSNR degradation (caused due to increase in EDFA's noise). The SVM is trained to recognize the patterns exhibited by the adaptive filter weights in receiver DSP under normal link operation. The detection of any abnormal filter weights pattern by the trained SVM is then considered as an indication of the potential soft failures. This technique is shown to detect (but not localize) the above-mentioned four soft failures with $>96\%$ accuracy.

Open issues: Realization of ML-based proactive fault management in optical networks is still at its nascent stage. While few techniques for detecting and localizing hard failures have been proposed and deployed, the development of effective automated solutions for soft failures is still a relatively unexplored area. Furthermore, most of the existing works focus on the detection/localization of faults, while the development of mechanisms that can uncover actual root causes of these faults as well as facilitate efficient fault recovery process is an open area for research. Another major problem faced while implementing ML-based fault detection/prevention is the unavailability of extensive data sets corresponding to different faulty operational conditions. This is mainly because current network operators adopt ultraconservative designs with large operating margins in order to reduce the fault occurrence probability in their networks. This, however, limits the chances to collect sufficient examples of various network failure scenarios. In this context, the development of ML algorithms that could predict network faults accurately despite using minimal training data sets is an interesting area for research.

21.5.4 Software-defined networking

The software-defined networking approach centralizes network management by decoupling the data and control planes. software-defined networking technologies enable the network infrastructure to be centrally controlled/configured in an intelligent way by using various software applications. Data-driven ML techniques naturally fit in SDNs where abundant data can be captured by accessing the monitors spanning the whole network. Many studies have demonstrated the applications of ML in solving particular problems in SDNs such as network traffic prediction, fault detection, quality of transmission (QoT) estimation, etc. We refer the readers to two recent survey papers [76,77] for comprehensive reviews on these topics. In contrast, systematic integration of those ML applications into an SDN framework for cross-layer optimization is less reported, which is what we will focus on here. Morales et al. [78] performed ANN-based data analytics for robust and adaptive network traffic modeling. Based on the predicted traffic volume and direction, the virtual network topology (VNT) is adaptively reconfigured to ensure that the required grade of service is supported. Compared to static VNT design approaches, this predictive method decreases the required number of transponders to be installed at the routers by 8% − 42%, as shown in Fig. 21.32, thus reducing energy consumption

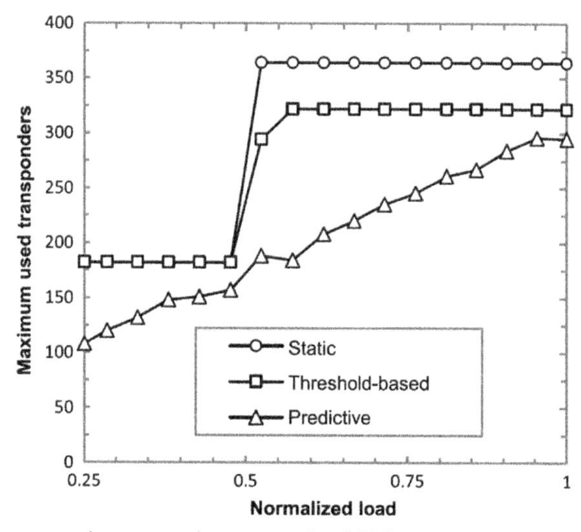

Figure 21.32 Maximum used transponders versus load [79].

and costs. Similarly, Alvizu et al. [80] used ML to predict tidal traffic variations in a software-defined mobile metro-core network. In their work, ANNs are employed to forecast traffic at different locations of an optical network and the predicted traffic demands are then exploited to optimize the online routing and wavelength assignments using a combination of analytical derivations and heuristics. Energy savings of $\sim$31% are observed as compared to traditional static methods used in mobile metro-core networks.

In Ref. [81], an RL technique called Q-learning is used to solve the path and wavelength selection problem in optical burst switched (OBS) networks. Initially, for each burst to be transmitted between a given source-destination pair, the algorithm picks a path and wavelength from the given sets of paths and wavelengths, respectively, as action and then a reward that depends on the success or failure of that burst transmission is determined. In this way, the algorithm learns over time how to select optimal paths and wavelengths that can minimize burst loss probability (BLP) for each source-destination pair. It has been shown that the Q-learning algorithm reduces BLP significantly as compared to other adaptive schemes proposed in the literature. Similarly, Chen et al. [82] applied an RL algorithm, called Q-network, for joint routing, modulation, and spectrum assignment (RMSA) provisioning in SDNs. In their work, the Q-network self-learns the best RMSA policies under different network states and time-varying demands based on the feedback obtained from the network for the RMSA actions taken in those conditions. Compared to shortest-path (SP) routing and the first-fit (FF) spectrum assignment approach, 4 times reduction in request blocking probability is reported using this method.

In Ref. [83], we demonstrated an ML-assisted optical network planning framework for SDNs. In this work, the network configuration as well as the real-time information about different link/signal parameters such as launched power, EDFAs' input and output powers, EDFAs' gains, EDFAs' noise figures (NFs), etc. is stored in a network configuration and monitoring database as shown in Fig. 21.33A. Next, an ANN is trained using this information where vectors $\mathbf{x}$ comprising the above-mentioned link/signal parameters are applied at the input of the ANN while actual known OSNR values y corresponding to those links are used as targets, as depicted in Fig. 21.33B. The ANN is then made to learn the relationship between these two sets of data by optimizing its various parameters. After training, the ANN is able to predict the performance (in terms of OSNR) of various unestablished lightpaths in the network, as shown in Fig. 21.33C, for optimum network planning. We demonstrated that the ML-based performance prediction mechanism can be used to increase the transmission capacity in an SDN framework by adaptively configuring a probabilistic shaping-based spectral efficiency tunable transmitter. Similarly, Bouda et al. [84] proposed field measurements and ML based Q-factor estimation for real-time

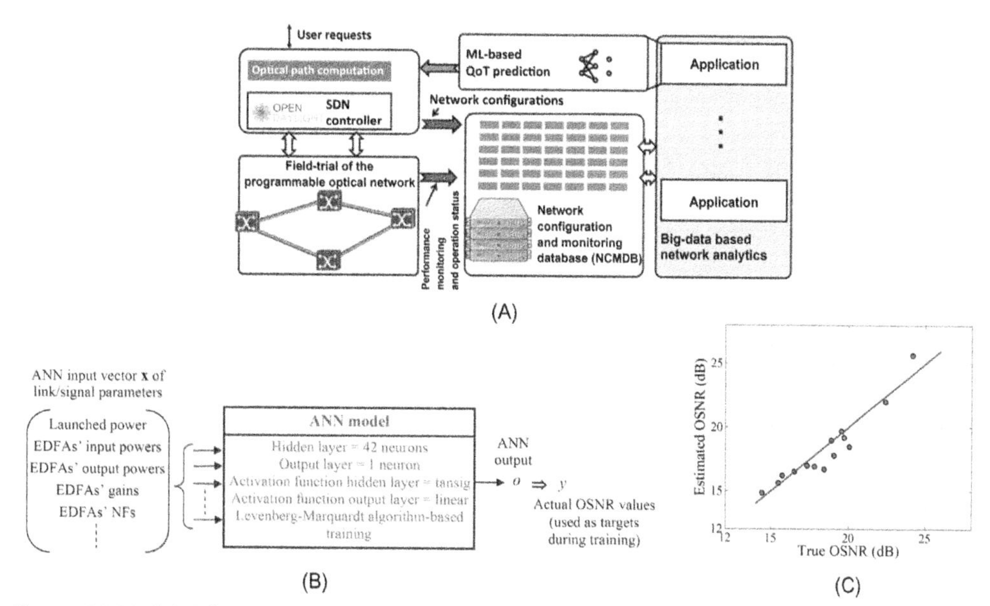

Figure 21.33 (A) Schematic diagram of machine learning (ML)-assisted optical network planning framework for software-defined networks (SDNs). (B) Artificial neural network (ANN) model with link/signal parameters as inputs and estimated optical signal-to-noise ratios (OSNRs) as outputs. (C) True versus estimated OSNRs using the ANN model [6].

network planning and design in SDNs. In their work, Q-factor for each lightpath in the network is continuously estimated and based on the predicted Q-factor values suitable modulation formats are adaptively provisioned to increase the network capacity and utilization.

Open issues: We already observe some benefits of ML-assisted network state prediction and decision making in SDNs. However, some practical concerns need to be addressed when applying ML in SDNs. First, real networks still require worst-case performance guarantees, which in turn necessitates a full understanding of the robustness of the chosen ML algorithms. Second, network characteristics can vary significantly in different network scenarios. An ML model trained using one particular data set may not be able to generalize to all network scenarios, and thus the scalability of such a method becomes questionable.

A number of concerns also need to be addressed to realize more active use of RL in SDNs. First, it must be shown that RL algorithms are scalable to handle large and more complex networks. Second, the RL algorithms must demonstrate fast convergence in real network conditions so as to limit the impact of nonoptimal actions taken during the early learning phase of these algorithms.

The interpretability of ML methods is another issue, as it is not desirable in practice to adopt an algorithm without really understanding how and why it works.

Consequently, much needs to be done in understanding the fundamental properties of ML algorithms and how to properly incorporate them into SDN framework.

21.5.5 Quality of transmission estimation

QoT estimation is considered vital for minimizing link margins and network optimization. Existing QoT prediction methods in optical networks can be broadly categorized into three main types: (1) Techniques based on sophisticated analytical models, such as SSFM [85], that exploit the information about various physical layer impairments to predict the bit-error rate (BER) of a given lightpath with good accuracy. These approaches typically involve high computational complexity (and hence require more computation time) thus limiting their application in large and dynamic optical networks. (2) Techniques based on simple approximated formulas [86] that are fast but relatively less accurate thus resulting in higher link margins [87]. (3) ML-based approaches [88–92] that learn the relationship between monitored field data and the QoT of already deployed lightpaths to predict the QoT of unestablished lightpaths. ML-based QoT estimation has gained significant attention recently due to the fact that ML algorithms can effectively learn and discover the hidden correlations in big data typically encountered in large SDNs [70].

Morais et al. [88] compared the performance of four ML algorithms namely ANN, SVM, *K*-nearest neighbors (KNN) and logistic regression for estimating the QoT of unestablished lightpaths. In their work, synthetic data for four reference network topologies with different characteristics (i.e., number of nodes, number of paths, fiber types used, etc.) is analyzed. The set of lightpaths features used for the training of four ML models includes number of hops, number of spans, average and maximum link lengths, attenuation and CD values of the links, and modulation format type. The ML models are trained in a supervised manner to learn the relationship between aforementioned features and the feasibility of the lightpaths. The results in Fig. 21.34 show that all four ML models could correctly predict the QoT of more than 90% lightpaths with ANN demonstrating the best accuracy, that is, $>99\%$. Similarly, in Ref. [89], KNN and random forest algorithms are applied to predict whether the BERs of candidate lightpaths meet the required system threshold by using the network parameters such as traffic volume, modulation format, total length of the lightpath, length of the longest link, and the number of links in the lightpath as input features. The performance of both algorithms is analyzed for different network topologies and for various combinations of the input features. It has been shown that the use of three specific features, that is, total length of the lightpath, traffic volume, and modulation format leads to the best QoT prediction accuracy.

Sartzetakis et al. [90] proposed two ML-based QoT estimation methods and studied their suitability under various parameter uncertainty scenarios. The first approach

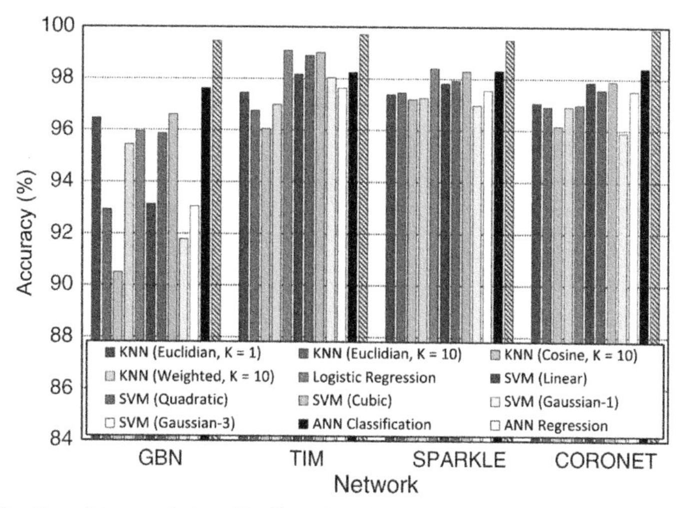

Figure 21.34 Quality of transmission (QoT) estimation accuracies using various machine learning (ML) algorithms for four different network topologies [88].

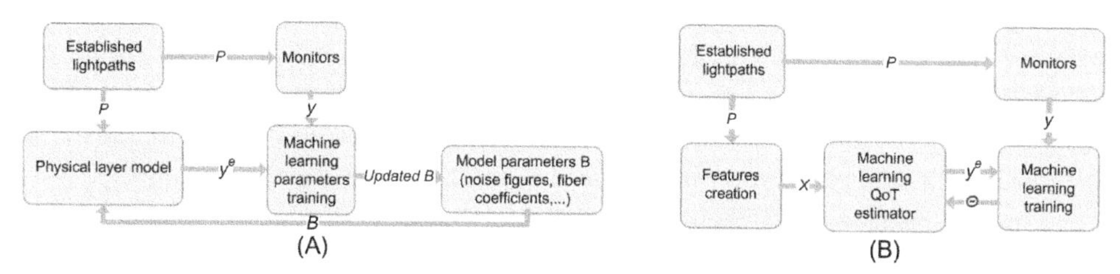

Figure 21.35 Block diagrams of (A) analytical physical layer model (APLM) and (B) machine learning model (MLM) used for quality of transmission (QoT) estimation [90].

used an analytical physical layer model (APLM) (i.e., Gaussian noise (GN) model) $Q(P,B)$ whose inputs are the parameters P of the established lightpaths (such as the routes, launched power, central frequency, modulation format, baud rate, etc.) as well as the physical layer parameters B of each span (such as the length, EDFAs' gains and NFs, fiber attenuation, CD and nonlinear coefficients, etc.) and whose outputs are the QoT estimates y^e of the lightpaths, as shown in Fig. 21.35A. In this approach, the monitored QoT values y and the APLM estimates y^e are exploited using ML to obtain better estimates of the input parameters B and thus improve the accuracy of the APLM. The second approach used a machine learning model (MLM) that is trained using the features X of the established lightpaths (such as the lightpath length, baud rate, number of crossed EDFAs, etc.) and their monitored QoT values y, as shown in Fig. 21.35B. The trained MLM is then employed for predicting the QoT of the unestablished lightpaths. The performance of both methods is evaluated (using GN model estimates as ground truths) under various parameter uncertainty scenarios where the fiber attenuation, and CD and nonlinear coefficients of the spans are varied randomly. Both approaches demonstrate good QoT estimation accuracies, with MLM performing better than APLM under high CD and nonlinear coefficients uncertainties.

Li et al. [91] investigated the effect of various chosen lightpaths features on the accuracy of ANN-based QoT estimation approaches. For this purpose, they identified five commonly used features: the number of hops, total length of the lightpath, length of the longest link of the lightpath, and the number of 15 dB and 22 dB EDFAs traversed in the lightpath. In their work, an ANN is first trained and evaluated using all five features. Next, the values of the features are set equal to zero, one at a time, and the prediction performance of ANN is determined based on the remaining nonzero features. In this way, the least important feature among the five chosen features is determined, which is then discarded. This process continues until only one feature, that is, the most important one, is left. It has been found that among the five features, the total length of the lightpath most impacts the QoT prediction accuracy.

Typically, the ML-based QoT estimation techniques developed for one system configuration are not quite suitable for predicting QoT in other configurations, thus requiring time-consuming training data recollection and the ML model retraining in such scenarios. In this context, Mo et al. [92] proposed an ANN-based transfer learning approach to accurately predict the QoT in various system configurations with different data rates, modulation formats, and fiber types. In their work, the weights of an ANN, which is initially trained in a source domain system, are tuned/adjusted via a simple training process in a different target domain system. The benefit of such transfer learning approach is that only a small amount of new data from the target domain system is needed for the training of pretrained weights, thus speeding up the training process. This approach demonstrates accurate QoT prediction for three different target domain systems with much faster training speeds as compared to the case where the ANN models are retrained from the scratch.

Open issues: A large number of existing ML-based QoT estimation methods predict mainly whether the QoT of the lightpath under investigation is above or below a certain threshold. However, such binary classification of the lightpaths does not give any information about the residual margin (i.e., the difference from the acceptable threshold), which is useful for the operators to avoid network overdimensioning. Therefore for realizing low-margin network operations, ML-based approaches that could quantify the QoT of lightpaths in terms of actual OSNR, Q-factor, BER, etc. shall be developed. Another issue with current ML-based QoT prediction techniques is that they typically rely on information such as network configuration, network topology, and transmitter parameters, which is insufficient, as the operational behavior of the network devices may also significantly affect the lightpaths' QoT. Therefore for robust and accurate QoT estimation, the dynamics of various network elements such as EDFAs, WDM nodes, etc. should also be considered.

21.5.6 Physical layer design

ML techniques offer the opportunity to optimize the design of individual physical components as well as complete end-to-end fiber-optic communication systems. Recently we have seen some noticeable research works in this regard with quite encouraging results.

In Ref. [93], a complete optical communication system including transmitter, receiver and nonlinear channel is modeled as an end-to-end fully-connected DNN. This approach enables the optimization of transceivers in a single end-to-end process where the transmitter learns waveform representations that are robust to channel impairments while the receiver learns to equalize channel distortions. The results for 42 Gb/s intensity modulation/direct detection (IM/DD) systems show that the DL-based optimization outperforms the solutions based on two- and four-level pulse amplitude modulation (PAM2/PAM4) and conventional receiver equalization, for a range of transmission distances.

Jones et al. [94] proposed an ANN-based receiver for nonlinear frequency-division multiplexing (NFDM) optical communication systems. Unlike standard nonlinear Fourier transform (NFT) based receivers which are vulnerable to losses and noise in NFDM systems, the ANN-based receiver tackles these impairments by learning the distortion characteristics of previously transmitted pulses and applying them for inference for future decisions. The results demonstrate improved BER performance as compared to conventional NFT-based receivers for practical link configurations. In Ref. [95], an ANN is used in receiver DSP for mitigating linear and nonlinear impairments in IM/DD systems. The ANN infers linear and nonlinear channel responses simultaneously, which are then exploited for increasing the demodulation reliability beyond the capability of linear equalization techniques. Using an ANN along with standard feed-forward equalization (FFE), up to 10 times BER improvement over FFE-only configurations is demonstrated for 84 GBd/s PAM4 transmission over 1.5 km SSMF.

Zibar et al. [96] applied an ANN for obtaining specific Raman gain profiles. The ANN is trained to learn the mapping between Raman gain profiles and the corresponding required pump powers and wavelengths settings. The trained ANN is then used to accurately predict pump powers and wavelengths for a targeted Raman gain profile. This approach is shown to attain Raman gain profiles with maximum error of 0.6 dB as compared to the targeted gain profiles, as shown in Fig. 21.36.

21.6 Future role of machine learning in optical communications

The emergence of SDNs with their inherent programmability and access to enormous amount of network-related monitored data provides unprecedented opportunities for

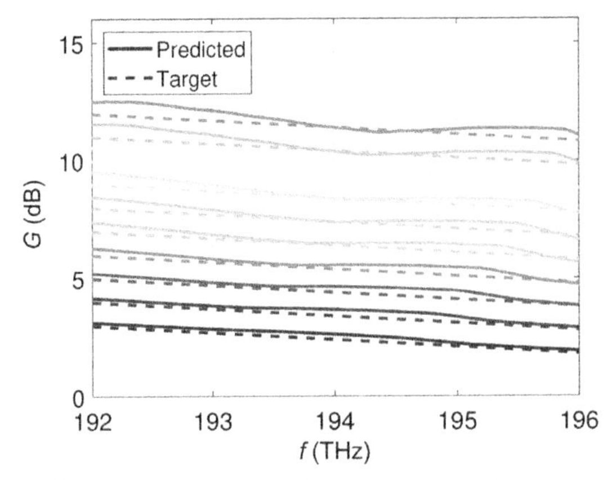

Figure 21.36 Comparison between the predicted and target tilted gain profiles [96].

the application of ML methods in these networks. The vision of future intelligent optical networks integrates the programmability/automation functionalities of SDNs with data analytics capabilities of ML technologies to realize self-aware, self-managing, and self-healing network infrastructures. Over the past few years we have seen an increasing amount of research on the application of ML techniques in various aspects of optical communications and networking. As ML is gradually becoming a common knowledge to the photonics community, we can envisage some potential significant developments in optical networks in the near future ushered in by the application of emerging ML technologies.

Looking to the future, we can foresee a vital role played by ML-based mechanisms across several diverse functional areas in optical networks, for example, network planning and performance prediction, network maintenance and fault prevention, network resources allocation and management, etc. ML can also aid cross-layer optimization in future optical networks requiring big data analytics, since it can inherently learn and uncover hidden patterns and unknown correlations in big data, which can be extremely beneficial in solving complex network optimization problems. The ultimate objective of ML-driven next-generation optical networks will be to provide infrastructures that can monitor themselves, diagnose and resolve their problems, and provide intelligent and efficient services to the end users.

21.7 Online resources for machine learning algorithms

Standard ML algorithms' codes and examples are readily available online and one seldom needs to write their own codes from the very beginning. There are several off-the-shelf powerful frameworks available under open-source licenses such as TensorFlow, Pytorch, Caffe, etc. Matlab, which is widely used in optical communications researches is not the most popular programming language among the ML community. Instead, Python is the preferred language for ML research partly because it is freely available, multiplatform, and relatively easy to use/read, and it has a huge number of libraries/modules available for a wide variety of tasks. We hereby report some useful resources including example Python codes using TensorFlow library to help interested readers get started with applying simple ML algorithms to their problems. More intuitive understanding of ANNs can be found at this visual playground [97]. The Python codes for most of the standard neural network architectures discussed in this chapter can be found in these Github repositories [98,99] with examples. For nonstandard model design, TensorFlow also provides low-level programming interfaces for more custom and complex operations based on its symbolic building blocks, which are documented in detail in Ref. [100].

21.8 Conclusions

In this chapter we discussed how the rich body of ML techniques can be applied as a unique and powerful set of signal processing tools in fiber-optic communication systems. As optical networks become faster, more dynamic, and more software-defined, we will see an increasing number of applications of ML and big data analytics in future networks to solve certain critical problems that cannot be easily tackled using conventional approaches. A basic knowledge and skills in ML will thus become necessary and beneficial for researchers in the field of optical communications and networking.

Acknowledgments

This work was supported by the Hong Kong Government General Research Fund under Project number PolyU 152757/16E.

References

[1] S. Marsland, Machine Learning: An Algorithmic Perspective, second ed., CRC Press, Boca Raton, USA, 2015.
[2] Y. Bengio, A. Courville, P. Vincent, Representation learning: a review and new perspectives, IEEE Trans. Pattern Anal. Mach. Intell. 35 (8) (2013) 1798–1828.
[3] C.M. Bishop, Pattern Recognition and Machine Learning, Springer-Verlag, New York, 2006.
[4] Z. Dong, F.N. Khan, Q. Sui, K. Zhong, C. Lu, A.P.T. Lau, Optical performance monitoring: a review of current and future technologies, J. Lightwave Technol. 34 (2) (2016) 525–543.
[5] A.S. Thyagaturu, A. Mercian, M.P. McGarry, M. Reisslein, W. Kellerer, Software defined optical networks (SDONs): a comprehensive survey, IEEE Commun. Surveys Tuts 18 (2) (2016) 2738–2786.
[6] F.N. Khan, Q. Fan, C. Lu, A.P.T. Lau, An optical communication's perspective on machine learning and its applications, J. Lightwave Technol 37 (2) (2019) 493–516.
[7] R.A. Dunne, A Statistical Approach to Neural Networks for Pattern Recognition, John Wiley & Sons, Hoboken, 2007.

[8] I. Kaastra, M. Boyd, Designing a neural network for forecasting financial and economic time series, Neurocomputing 10 (3) (1996) 215–236.
[9] C.D. Meyer, Matrix Analysis and Applied Linear Algebra, Society for Industrial and Applied Mathematics, Philadelphia, 2000.
[10] R.O. Duda, P.E. Hart, D.G. Stork, Pattern Classification, second ed., John Wiley & Sons, New York, 2007.
[11] X. Glorot, A. Bordes, Y. Bengio, Deep sparse rectifier neural networks, in: Proc. AISTATS, Fort Lauderdale, FL, USA, 2011, vol. 15, pp. 315–323.
[12] K. He, X. Zhang, S. Ren, J. Sun, Delving deep into rectifiers: surpassing human-level performance on ImageNet classification, in: Proc. ICCV, Santiago, Chile, 2015, pp. 1026–1034.
[13] X. Glorot, Y. Bengio, Understanding the difficulty of training deep feedforward neural networks, in: Proc. AISTATS, Chia Laguna Resort, Sardinia, Italy, 2010, pp. 249–256.
[14] A. Webb, Statistical Pattern Recognition, second ed., John Wiley & Sons, Chicester, 2002.
[15] A. Statnikov, C.F. Aliferis, D.P. Hardin, I. Guyon, A Gentle Introduction to Support Vector Machines in Biomedicine, World Scientific, Singapore, 2011.
[16] M.S. Andersen, J. Dahl, L. Vandenberghe, CVXOPT: Python software for convex optimization. Available from: <https://cvxopt.org>.
[17] M. Li, S. Yu, J. Yang, Z. Chen, Y. Han, W. Gu, Nonparameter nonlinear phase noise mitigation by using M-ary support vector machine for coherent optical systems, IEEE Photonics J. 5 (2013) 6. Art. no. 7800312.
[18] D. Wang, M. Zhang, Z. Li, Y. Cui, J. Liu, Y. Yang, et al., Nonlinear decision boundary created by a machine learning-based classifier to mitigate nonlinear phase noise, in: Proc. ECOC, Valencia, Spain, 2015, Paper P.3.16.
[19] T. Nguyen, S. Mhatli, E. Giacoumidis, L.V. Compernolle, M. Wuilpart, P. Mégret, Fiber nonlinearity equalizer based on support vector classification for coherent optical OFDM, IEEE Photonics J. 8 (2016) 2. Art. no. 7802009.
[20] M. Kirk, Thoughtful Machine Learning with Python, O'Reilly Media, Sebastopol, 2017.
[21] I.T. Jolliffe, Principal Component Analysis, second ed., Springer-Verlag, New York, 2002.
[22] J.E. Jackson, A User's Guide to Principal Components, John Wiley & Sons, Hoboken, 2003.
[23] L.J. Cao, K.S. Chua, W.K. Chong, H.P. Lee, Q.M. Gu, A comparison of PCA, KPCA and ICA for dimensionality reduction in support vector machine, Neurocomputing 55 (1–2) (2003) 321–336.
[24] R.S. Sutton, A.G. Barto, Reinforcement Learning: An Introduction, second ed., MIT Press, Cambridge, 2018.
[25] Y. Bengio, Learning deep architectures for AI, Found. Trends Mach. Learn. 2 (1) (2009) 1–127.
[26] Y. Bengio, O. Delalleau, On the expressive power of deep architectures, in: J. Kivinen, C. Szepesvári, E. Ukkonen, T. Zeugmann (Eds.), Algorithmic Learning Theory, Springer-Verlag, Heidelberg, 2011, pp. 18–36.
[27] L.J. Ba, R. Caurana, Do deep nets really need to be deep? in: Proc. NIPS, Montreal, Canada, 2014, pp. 2654–2662.
[28] H. Larochelle, Y. Bengio, J. Louradour, P. Lamblin, Exploring strategies for training deep neural networks, J. Mach. Learn. Res. 10 (2009) 1–40.
[29] I. Goodfellow, Y. Bengio, A. Courville, Deep Learning, MIT Press, Cambridge, MA, 2016.
[30] D.P. Mandic, J. Chambers, Recurrent Neural Networks for Prediction: Learning Algorithms, Architectures and Stability, John Wiley & Sons, Chicester, 2001.
[31] R. Pascanu, C. Gulcehre, K. Cho, Y. Bengio, How to construct deep recurrent neural networks, in: Proc. ICLR, Banff, Canada, 2014.
[32] R. Pascanu, T. Mikolov, Y. Bengio, On the difficulty of training recurrent neural networks, in: Proc. ICML, Atlanta, GA, USA, 2013, pp. 1310–1318.
[33] F.N. Khan, Z. Dong, C. Lu, A.P.T. Lau, Optical performance monitoring for fiber-optic communication networks, in: X. Zhou, C. Xie (Eds.), Enabling Technologies for High Spectral-Efficiency Coherent Optical Communication Networks, John Wiley & Sons, Hoboken, 2016. Chapter 14.
[34] X. Wu, J.A. Jargon, R.A. Skoog, L. Paraschis, A.E. Willner, Applications of artificial neural networks in optical performance monitoring, J. Lightwave Technol. 27 (16) (2009) 3580–3589.

[35] T.B. Anderson, A. Kowalczyk, K. Clarke, S.D. Dods, D. Hewitt, J.C. Li, Multi impairment monitoring for optical networks, J. Lightwave Technol. 27 (16) (2009) 3729–3736.
[36] F.N. Khan, Y. Zhou, A.P.T. Lau, C. Lu, Modulation format identification in heterogeneous fiber-optic networks using artificial neural networks, Opt. Express 20 (11) (2012) 12422–12431.
[37] F.N. Khan, T.S.R. Shen, Y. Zhou, A.P.T. Lau, C. Lu, Optical performance monitoring using artificial neural networks trained with empirical moments of asynchronously sampled signal amplitudes, IEEE Photonics Technol. Lett. 24 (12) (2012) 982–984.
[38] T.S.R. Shen, Q. Sui, A.P.T. Lau, OSNR monitoring for PM-QPSK systems with large inline chromatic dispersion using artificial neural network technique, IEEE Photonics Technol. Lett. 24 (17) (2012) 1564–1567.
[39] M.C. Tan, F.N. Khan, W.H. Al-Arashi, Y. Zhou, A.P.T. Lau, Simultaneous optical performance monitoring and modulation format/bit-rate identification using principal component analysis, J. Opt. Commun.Netw. 6 (5) (2014) 441–448.
[40] P. Isautier, K. Mehta, A.J. Stark, S.E. Ralph, Robust architecture for autonomous coherent optical receivers, J. Opt. Commun. Netw. 7 (9) (2015) 864–874.
[41] N.G. Gonzalez, D. Zibar, I.T. Monroy, Cognitive digital receiver for burst mode phase modulated radio over fiber links, in: Proc. ECOC, Torino, Italy, 2010, Paper P6.11.
[42] F.N. Khan, Y. Zhou, Q. Sui, A.P.T. Lau, Non-data-aided joint bit-rate and modulation format identification for next-generation heterogeneous optical networks, Opt. Fiber Technol 20 (2) (2014) 68–74.
[43] R. Borkowski, D. Zibar, A. Caballero, V. Arlunno, I.T. Monroy, Stokes space-based optical modulation format recognition in digital coherent receivers, IEEE Photonics Technol. Lett 25 (21) (2013) 2129–2132.
[44] F.N. Khan, K. Zhong, W.H. Al-Arashi, C. Yu, C. Lu, A.P.T. Lau, Modulation format identification in coherent receivers using deep machine learning, IEEE Photonics Technol. Lett 28 (17) (2016) 1886–1889.
[45] F.N. Khan, K. Zhong, X. Zhou, W.H. Al-Arashi, C. Yu, C. Lu, et al., Joint OSNR monitoring and modulation format identification in digital coherent receivers using deep neural networks, Opt. Express 25 (15) (2017) 17767–17776.
[46] T. Tanimura, T. Hoshida, J.C. Rasmussen, M. Suzuki, H. Morikawa, OSNR monitoring by deep neural networks trained with asynchronously sampled data, in: Proc. OECC, Niigata, Japan, 2016, Paper TuB3-5.
[47] T. Tanimura, T. Kato, S. Watanabe, T. Hoshida, Deep neural network based optical monitor providing self-confidence as auxiliary output, in: Proc. ECOC, Rome, Italy, 2018, Paper We1D.5.
[48] N. Srivastava, G. Hinton, A. Krizhevsky, I. Sutskever, R. Salakhutdinov, Dropout: a simple way to prevent neural networks from overfitting, J. Mach. Learn. Res. 15 (1) (2014) 1929–1958.
[49] C. Wang, S. Fu, M. Tang, L. Xia, D. Liu, Deep learning enabled simultaneous OSNR and CD monitoring for coherent transmission system, in: Proc. OFC, San Diego, CA, USA, 2019, Paper Th2A.44.
[50] D. Wang, M. Zhang, Z. Li, J. Li, M. Fu, Y. Cui, et al., Modulation format recognition and OSNR estimation using CNN-based deep learning, IEEE Photonics Technol. Lett. 29 (19) (2017) 1667–1670.
[51] A.S. Kashi, Q. Zhuge, J.C. Cartledge, A. Borowiec, D. Charlton, C. Laperle, et al., Fiber nonlinear noise-to-signal ratio monitoring using artificial neural networks, in: Proc. ECOC, Gothenburg, Sweden, 2017, Paper M.2.F.2.
[52] F.J.V. Caballero, D.J. Ives, C. Laperle, D. Charlton, Q. Zhuge, M. O'Sullivan, et al., Machine learning based linear and nonlinear noise estimation, J. Opt. Commun. Netw. 10 (10) (2018) D42–D51.
[53] K. Zhang, Y. Fan, T. Ye, Z. Tao, S. Oda, T. Tanimura, et al., Fiber nonlinear noise-to-signal ratio estimation by machine learning, in: Proc. OFC, San Diego, CA, USA, 2019, Paper Th2A.45.
[54] E. Ip, Nonlinear compensation using backpropagation for polarization-multiplexed transmission, J. Lightwave Technol. 28 (6) (2010) 939–951.

[55] P. Poggiolini, Y. Jiang, Recent advances in the modeling of the impact of nonlinear fiber propagation effects on uncompensated coherent transmission systems, J. Lightwave Technol. 35 (3) (2017) 458–480.

[56] A. Carena, G. Bosco, V. Curri, Y. Jiang, P. Poggiolini, F. Forghieri, EGN model of non-linear fiber propagation, Opt. Express 22 (13) (2014) 16335–16362.

[57] T.S.R. Shen, A.P.T. Lau, Fiber nonlinearity compensation using extreme learning machine for DSP-based coherent communication systems, in: Proc. OECC, Kaohsiung, Taiwan, 2011, pp. 816–817.

[58] G.-B. Huang, Q.-Y. Zhu, C.-K. Siew, Extreme learning machine: theory and applications, Neurocomputing 70 (1–3) (2006) 489–501.

[59] D. Zibar, O. Winther, N. Franceschi, R. Borkowski, A. Caballero, V. Arlunno, et al., Nonlinear impairment compensation using expectation maximization for dispersion managed and unmanaged PDM 16-QAM transmission, Opt. Express 20 (26) (2012) B181–B196.

[60] E. Alpaydin, Introduction to Machine Learning, second ed., MIT Press, Cambridge, 2010.

[61] E. Giacoumidis, S. Mhatli, M.F.C. Stephens, A. Tsokanos, J. Wei, M.E. McCarthy, et al., Reduction of nonlinear intersubcarrier intermixing in coherent optical OFDM by a fast Newton-based support vector machine nonlinear equalizer, J. Lightwave Technol. 35 (12) (2017) 2391–2397.

[62] E. Giacoumidis, S.T. Le, M. Ghanbarisabagh, M. McCarthy, I. Aldaya, S. Mhatli, et al., Fiber nonlinearity-induced penalty reduction in CO-OFDM by ANN-based nonlinear equalization, Opt. Lett. 40 (21) (2015) 5113–5116.

[63] C. Pan, H. Bülow, W. Idler, L. Schmalen, F.R. Kschischang, Optical nonlinear phase noise compensation for 9 × 32-Gbaud PolDM-16QAM transmission using a code-aided expectation-maximization algorithm, J. Lightwave Technol. 33 (17) (2015) 3679–3686.

[64] C. Häger, H.D. Pfister, Nonlinear interference mitigation via deep neural networks, in: Proc. OFC, San Diego, CA, USA, 2018, Paper W3A.4.

[65] C. Häger, H.D. Pfister, Wideband time-domain digital backpropagation via subband processing and deep learning, in: Proc. ECOC, Rome, Italy, 2018, Paper Tu4F.4.

[66] Z. Tao, L. Dou, W. Yan, L. Li, T. Hoshida, J.C. Rasmussen, Multiplier-free intrachannel nonlinearity compensating algorithm operating at symbol rate, J. Lightwave Technol. 29 (17) (2011) 2570–2576.

[67] V. Kamalov, L. Jovanovski, V. Vusirikala, S. Zhang, F. Yaman, K. Nakamura, et al., Evolution from 8QAM live traffic to PS 64-QAM with neural-network based nonlinearity compensation on 11000 km open subsea cable, in: Proc. OFC, San Diego, CA, USA, 2018, Paper Th4D.5.

[68] Y. Gao, Z.A. El-Sahn, A. Awadalla, D. Yao, H. Sun, P. Mertz, et al., Reduced complexity nonlinearity compensation via principal component analysis and deep neural networks, in: Proc. OFC, San Diego, CA, USA, 2019, Paper Th2A.49.

[69] R. Dar, P.J. Winzer, Nonlinear interference mitigation: methods and potential gain, J. Lightwave Technol. 35 (4) (2017) 903–930.

[70] F.N. Khan, C. Lu, A.P.T. Lau, Optical performance monitoring in fiber-optic networks enabled by machine learning techniques, in: Proc. OFC, San Diego, CA, USA, 2018, Paper M2F.3.

[71] Z. Wang, M. Zhang, D. Wang, C. Song, M. Liu, J. Li, et al., Failure prediction using machine learning and time series in optical network, Opt. Express 25 (16) (2017) 18553–18565.

[72] F. Boitier, V. Lemaire, J. Pesic, L. Chavarria, P. Layec, S. Bigo, et al., Proactive fiber damage detection in real-time coherent receiver, in: Proc. ECOC, Gothenburg, Sweden, 2017, Paper Th.2.F.1.

[73] D. Rafique, T. Szyrkowiec, H. Griesser, A. Autenrieth, J.-P. Elbers, Cognitive assurance architecture for optical network fault management, J. Lightwave Technol. 36 (7) (2018) 1443–1450.

[74] D. Rafique, T. Szyrkowiec, A. Autenrieth, J.-P. Elbers, Analytics-driven fault discovery and diagnosis for cognitive root cause analysis, in: Proc. OFC, San Diego, CA, USA, 2018, Paper W4F.6.

[75] S. Varughese, D. Lippiatt, T. Richter, S. Tibuleac, S.E. Ralph, Identification of soft failures in optical links using low complexity anomaly detection, in: Proc. OFC, San Diego, CA, USA, 2019, Paper W2A.46.

[76] J. Mata, I. de Miguel, R.J. Durán, N. Merayo, S.K. Singh, A. Jukan, et al., Artificial intelligence (AI) methods in optical networks: a comprehensive survey, Opt. Switch. Netw. 28 (2018) 43–57.

[77] F. Musumeci, C. Rottondi, A. Nag, I. Macaluso, D. Zibar, M. Ruffini, et al., An overview on application of machine learning techniques in optical networks, IEEE Commun. Surveys Tuts. to be published. doi: 10.1109/COMST.2018.2880039.
[78] F. Morales, M. Ruiz, L. Gifre, L.M. Contreras, V. López, L. Velasco, Virtual network topology adaptability based on data analytics for traffic prediction, J. Opt. Commun.Netw. 9 (1) (2017) A35–A45.
[79] D. Rafique, L. Velasco, Machine learning for network automation: overview, architecture, and applications, J. Opt. Commun. Netw. 10 (10) (2018) D126–D143.
[80] R. Alvizu, S. Troia, G. Maier, A. Pattavina, Matheuristic with machine-learning-based prediction for software-defined mobile metro-core networks, J. Opt. Commun. Netw. 9 (9) (2017) D19–D30.
[81] Y.V. Kiran, T. Venkatesh, C.S. Murthy, A reinforcement learning framework for path selection and wavelength selection in optical burst switched networks, IEEE J. Sel. Areas Commun. 25 (9) (2007) 18–26.
[82] X. Chen, J. Guo, Z. Zhu, R. Proietti, A. Castro, S.J.B. Yoo, Deep-RMSA: a deep-reinforcement-learning routing, modulation and spectrum assignment agent for elastic optical networks, in: Proc. OFC, San Diego, CA, USA, 2018, Paper W4F.2.
[83] S. Yan, F.N. Khan, A. Mavromatis, D. Gkounis, Q. Fan, F. Ntavou, et al., Field trial of machine-learning-assisted and SDN-based optical network planning with network-scale monitoring database, in: Proc. ECOC, Gothenburg, Sweden, 2017, Paper Th. PDP.B.4.
[84] M. Bouda, S. Oda, Y. Akiyama, D. Paunovic, T. Hoshida, P. Palacharla, et al., Demonstration of continuous improvement in open optical network design by QoT prediction using machine learning, in: Proc. OFC, San Diego, CA, USA, 2019, Paper M3Z.2.
[85] J. Shao, X. Liang, S. Kumar, Comparison of split-step Fourier schemes for simulating fiber optic communication systems, IEEE Photonics J. 6 (2014) 4. Art. no. 7200515.
[86] P. Poggiolini, G. Bosco, A. Carena, V. Curri, Y. Jiang, F. Forghieri, The GN-model of fiber non-linear propagation and its applications, J. Lightwave Technol. 32 (4) (2014) 694–721.
[87] Y. Pointurier, Design of low-margin optical networks, J. Opt. Commun. Netw. 9 (1) (2017) A9–A17.
[88] R.M. Morais, J. Pedro, Machine learning models for estimating quality of transmission in DWDM networks, J. Opt. Commun. Netw. 10 (10) (2018) D84–D99.
[89] C. Rottondi, L. Barletta, A. Giusti, M. Tornatore, Machine-learning method for quality of transmission prediction of unestablished lightpaths, J. Opt. Commun. Netw. 10 (2) (2018) A286–A297.
[90] I. Sartzetakis, K. Christodoulopoulos, E. Varvarigos, Formulating QoT estimation with machine learning, in: Proc. ECOC, Rome, Italy, 2018, Paper We1D.3.
[91] L. Li, G. Shen, X. Fu, S. Jie, S.K. Bose, Which features most impact: prediction of ANN-based lightpath quality of transmission? in: Proc. ACP, Hangzhou, China, 2018, Paper Su3E.4.
[92] W. Mo, Y.-K. Huang, S. Zhang, E. Ip, D.C. Kilper, Y. Aono, et al., ANN-based transfer learning for QoT prediction in real-time mixed line-rate systems, in: Proc. OFC, San Diego, CA, USA, 2018, Paper W4F.3.
[93] B. Karanov, M. Chagnon, F. Thouin, T.A. Eriksson, H. Bulow, D. Lavery, et al., End-to-end deep learning of optical fiber communications, J. Lightwave Technol. 36 (20) (2018) 4843–4855.
[94] R.T. Jones, S. Gaiarin, M.P. Yankov, D. Zibar, Time-domain neural network receiver for nonlinear frequency division multiplexed systems, IEEE Photonics Technol. Lett 30 (12) (2018) 1079–1082.
[95] J. Estaran, R.R. Müller, M.A. Mestre, F. Jorge, H. Mardoyan, A. Konczykowska, et al., Artificial neural networks for linear and non-linear impairment mitigation in high-baudrate IM/DD systems, in: Proc. ECOC, Düsseldorf, Germany, 2016, Paper M.2.B.2.
[96] D. Zibar, A. Ferrari, V. Curri, A. Carena, Machine learning-based Raman amplifier design, in: Proc. OFC, San Diego, CA, USA, 2019, Paper M1J.1.
[97] D. Smilkov, S. Carter, TensorFlow—A neural network playground. Available from: <http://playground.tensorflow.org>.
[98] A. Damien, GitHub repository—TensorFlow tutorial and examples for beginners with latest APIs. Available from: <https://github.com/aymericdamien/TensorFlow-Examples>.

[99] M. Zhou, GitHub repository—TensorFlow tutorial from basic to hard. Available from: <https://github.com/MorvanZhou/Tensorflow-Tutorial>.
[100] TensorFlow, Guide for programming with the low-level TensorFlow APIs. Available from: <https://www.tensorflow.org/programmers_guide/low_level_intro>.

Appendix

For cross-entropy loss function defined in Eq. (21.14), the derivative with respect to the output is given by

$$\frac{\partial E(n)}{\partial o_j(n)} = -\frac{y_j(n)}{o_j(n)}. \tag{21.36}$$

With softmax activation function for the output neurons,

$$\begin{aligned}\frac{\partial o_j(n)}{\partial r_k(n)} &= \frac{\left(\sum_{m=1}^{K} e^{r_m(n)}\right) e^{r_j(n)} \delta_{j,k} - e^{r_j(n)} \cdot e^{r_k(n)}}{\left(\sum_{m=1}^{K} e^{r_m(n)}\right)^2} \\ &= \frac{\left(\sum_{m=1}^{K} e^{r_m(n)}\right) e^{r_j(n)} \delta_{j,k} - e^{r_j(n)} \cdot e^{r_k(n)}}{\left(\sum_{k=1}^{K} e^{r_k(n)}\right)^2} \\ &= o_j(n)\delta_{j,k} - o_j(n)o_k(n),\end{aligned} \tag{21.37}$$

where $\delta_{j,k} = 1$ when $j = k$ and 0 otherwise. Consequently,

$$\begin{aligned}\frac{\partial E(n)}{\partial r_k(n)} &= \sum_{j=1}^{K} \frac{\partial E(n)}{\partial o_j(n)} \frac{\partial o_j(n)}{\partial r_k(n)} \\ &= \sum_{j=1}^{K} -\frac{y_j(n)}{o_j(n)} \left(o_j(n)\delta_{j,k} - o_j(n)o_k(n)\right) \\ &= \sum_{j=1}^{K} -y_j(n)\left(\delta_{j,k} - o_k(n)\right) = o_k(n) - y_k(n)\end{aligned} \tag{21.38}$$

as $\sum_{j=1}^{K} y_j(n) = 1$. Therefore

$$\frac{\partial E(n)}{\partial \mathbf{r}(n)} = \mathbf{o}(n) - \mathbf{y}(n).$$

Now, since $(\partial \mathbf{r}(n)/\partial \mathbf{b}_2)$, $(\partial \mathbf{r}(n)/\partial \mathbf{b}_1)$, $(\partial r_k(n)/\partial \mathbf{W}_2)$, $(\partial r_k(n)/\partial \mathbf{W}_1)$ are the same as $(\partial \mathbf{o}(n)/\partial \mathbf{b}_2)$,$(\partial \mathbf{o}(n)/\partial \mathbf{b}_1)$,$(\partial o_k(n)/\partial \mathbf{W}_2)$,$(\partial o_k(n)/\partial \mathbf{W}_1)$ for MSE loss function and linear activation function for the output neurons (as $\mathbf{o}(n) = \mathbf{r}(n)$ for that case), it follows that the update equations Eqs. (21.8)−(21.12) also hold for the ANNs with cross-entropy loss function and softmax activation function for the output neurons.

CHAPTER 22

Broadband radio-over-fiber technologies for next-generation wireless systems

Jianjun Yu[1], Xinying Li[2] and Xiaodan Pang[3]
[1]ZTE TX Inc., Morristown, NJ, United States
[2]Georgia Institute of Technology, Atlanta, GA, United States
[3]KTH Royal Institute of Technology, Stockholm, Sweden

22.1 Introduction on radio-over-fiber

With the growth of the number of internet users, the demand for data communication has been increasing rapidly in the past decade. This demand for bandwidth puts significant pressure on the communication network infrastructure. Hence, the development of the next generation network (NGN) has attracted great concern in the research community to meet the capacity requirements [1]. The NGN consists of two parts: the next generation core network and the next generation access network [2]. The transmission data rate of the core network has been increasing rapidly due to the development of photonic technology and the worldwide deployment of optical fibers. Nevertheless, for the access network between the users and core networks, the cost becomes huge for operators by means of massive deployment of new optical fiber infrastructure to cover a wide range of end users. Besides, the upgrade of the future optical fiber access network can be predicted to be rather complicated and time-consuming [3,4].

By contrast, for wireless access network, less infrastructure is required and its deployment is time-saving, which can easily be upgraded over time [4]. An illustration for current wireless standards aimed at different scenarios based on the relationship between throughput and carrier frequency is shown in Fig. 22.1, such as the long-term evolution, the wireless wide area network (WWAN), the WiFi network (802.11), and the wireless personal area network (WPAN). From the consumers' perspective, the number of laptops, multifunctional smart phones and tablet computers has been exponentially growing for the past few years, indicating that mobility is an ideal feature. Moreover, for many applications, the connection by wireless media is superior to that by fixed wired way, under the same condition of connection speed. In this context, a combination of fiber-optic network and wireless network can be

Optical Fiber Telecommunications V11
DOI: https://doi.org/10.1016/B978-0-12-816502-7.00030-0

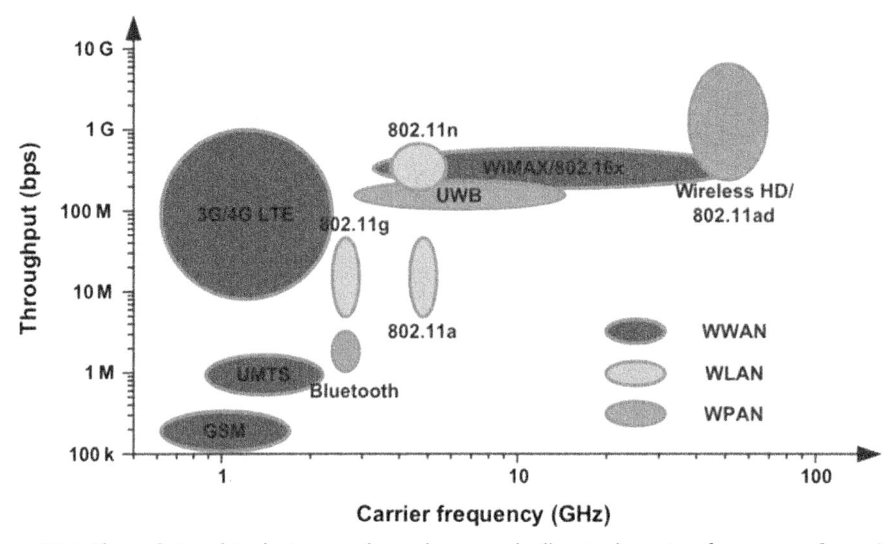

Figure 22.1 The relationship between throughput and allocated carrier frequency for selected wireless standards. *GSM*, Global System for Mobile Communications; *UMTS*, Universal Mobile Telecommunications System; *LTE*, long-term evolution; *UWB*, ultrawide band; *WiMAX*, Worldwide Interoperability for Microwave Access; *WWAN*, wireless wide area network; *WLAN*, wireless local area network; *WPAN*, wireless personal area network.

used to create a fiber-wireless hybrid access network architecture. The optical network can provide high-capacity backhaul for wireless access points or base stations (BSs), while the wireless network can realize the mobility and flexibility of broadband services for end users. It is likely for such hybrid architectures to solve the "last mile" access network problem. For that, radio-over-fiber (RoF) technology is regarded as a rather promising scheme for signal delivery and BS simplification [5].

RoF networks are regarded as a new and promising communication scheme for delivering broadband wireless signal services and fronthaul at microwave or millimeter-wave (mm-wave), relying on the synergy between fixed optical and mm-wave technologies. RoF technology enables radio-frequency (RF) signals to be transported over fiber up to 100 km and can be engineered for unity gain RF links. Hence, it is thought that it could do a lot to ease spectrum constraints, and one can replace multiple coax cables with a single optical fiber cable. RoF requires light to be modulated with radio data for optical transmission. It offers a huge bandwidth increase over existing solutions and requires no digital-to-analog conversion (DAC), resulting in a low-latency solution.

Fig. 22.2 shows the architecture of RoF. For uplink path, the RF signal is optically transported from the BS to the Central office (CO). For the downlink, we generate optical mm-wave at the CO before the mm-wave optical signals are transported to the BS [6,7].

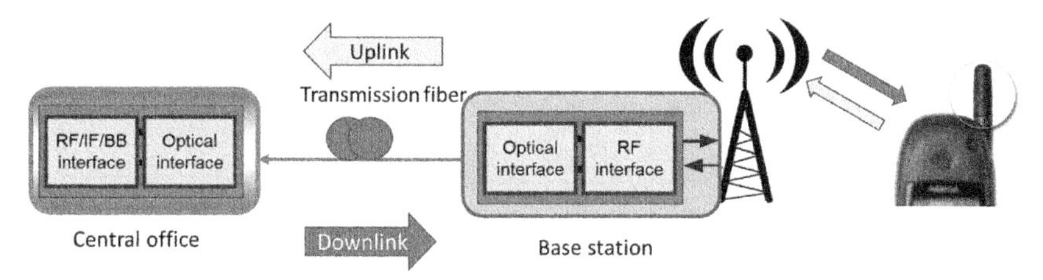

Figure 22.2 Architecture of radio-over-fiber (RoF). *RF*, Radio frequency; *IF*, immediate frequency; *BB*, baseband.

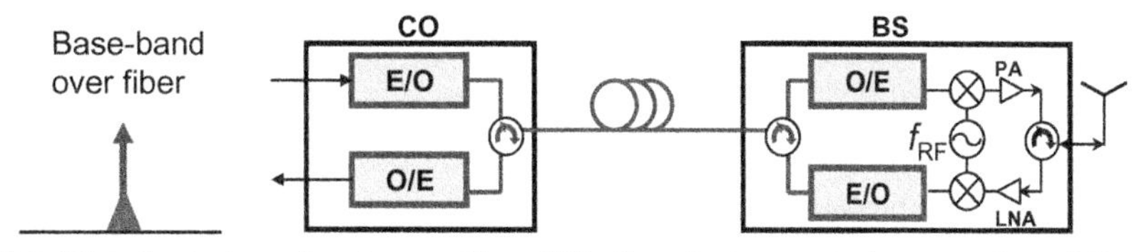

Figure 22.3 Digital wireless signal over fiber. *E/O*, Electrical-to-optical conversion; *O/E*, optical-to-electrical conversion; *RF*, radio frequency.

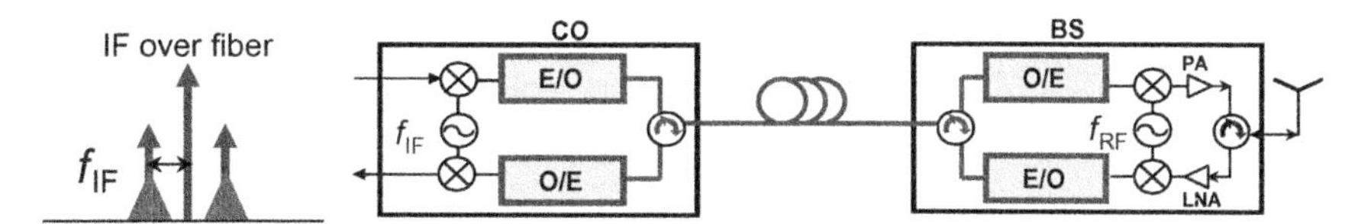

Figure 22.4 IF analog wireless signal over fiber. *IF*, Intermediate frequency.

Fig. 22.3 shows the baseband signal over fiber link between the CO and the BS, in order to compare with the RoF link. The baseband digital electrical signal is converted to optical signal. After fiber transmission, the optical signal is converted to electrical signal. After RF up-conversion, the RF electrical signal is launched by the antenna. The advantages of baseband over fiber are that low-cost optical-to-electrical (O/E) interface is needed at the CO and there is no dispersion limitation. However, the frequency conversion at the BS is complex and complicated multiband and multiservice schemes are required.

Fig. 22.4 shows the principle of intermediate-frequency (IF) analog wireless signal transmission over fiber. The baseband signal is firstly IF up-converted. Then the IF electrical signal is converted to optical signal. After fiber transmission, the optical signal is converted to electrical signal. After RF up-conversion, the electrical signal is launched by the antenna. This IF-over-fiber scheme has mature IF hardware and high dispersion tolerance. However, complicated frequency conversion is needed at the BS.

The process of RoF is given by Fig. 22.5. After RF up-conversion, the electrical signal is converted to optical signal. After fiber transmission, the optical signal is

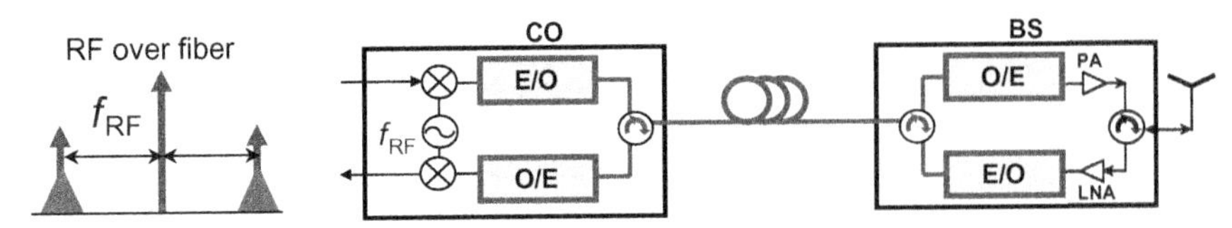

Figure 22.5 Radio-frequency (RF) analog wireless signal over fiber.

converted to electrical signal. Afterwards, the electrical signal is launched by the antenna. This RF-over-fiber scheme requires simple BS design without complicated frequency conversion. Centralized control and remote monitoring can be offered. However, it will suffer dispersion impact and linearity of O/E devices. RoF is an analog solution to realize full fiber-wireless centralization and has even a higher degree of centralization than the digital solutions. In order to overcome dispersion impact and nonlinear of O/E conversion, some advanced digital signal processing (DSP) is needed in the CO or the user units.

Some typical application scenarios of hybrid optical fiber-wireless systems in access and in-building networks are shown in Fig. 22.6, including high-definition television, video conferencing, interactive online gaming, e-learning, e-health services and so on. Additionally, high-speed wireless links, can be used as backup links for optical fibers to protect and recover data transmission when the fiber-optic traffic is down. Furthermore, high-speed wireless can be used to connect separated segments of optical fiber networks for metropolitan or rural area users, which in many cases cannot be realized by only optical fiber deployment.

The straightforward approach to move wireless signal carrier to mm-wave range is proposed to address current frequency-band congestion and to develop broadband wireless communication. The mm-wave range is defined to be a series of electromagnetic waves with the frequency between 30 and 300 GHz or, in other words, it can be described as those with the wavelength between 1 and 10 mm. At present, there are limited bands below 60 GHz which can be utilized without authorization. At 60 GHz and even more specifically, at V-band (50–75 GHz), regulators assign up to 7-GHz bandwidth for unauthorized use to North America, 7 GHz (57–64 GHz) to Korea, 7 GHz (59–66 GHz) to Japan and up to 9 GHz (57–66 GHz) to the European Union [8]. The communication at 60-GHz frequency band is standardized by several workgroups, such as the wireless HD, the ECMA-387, the IEEE802.15.3c employed in the WPAN scenario, and the HDMI as well as the cable replacement for the short-range wireless application [8–10]. In addition, there are more proposals which apply the 60-GHz technology to the WPAN [11], the WLAN [12] and the data center interconnection [13].

To further increase the wireless communication speed to, for example, over 100 Gbit/s requires a wider bandwidth. Therefore the underutilized higher frequency

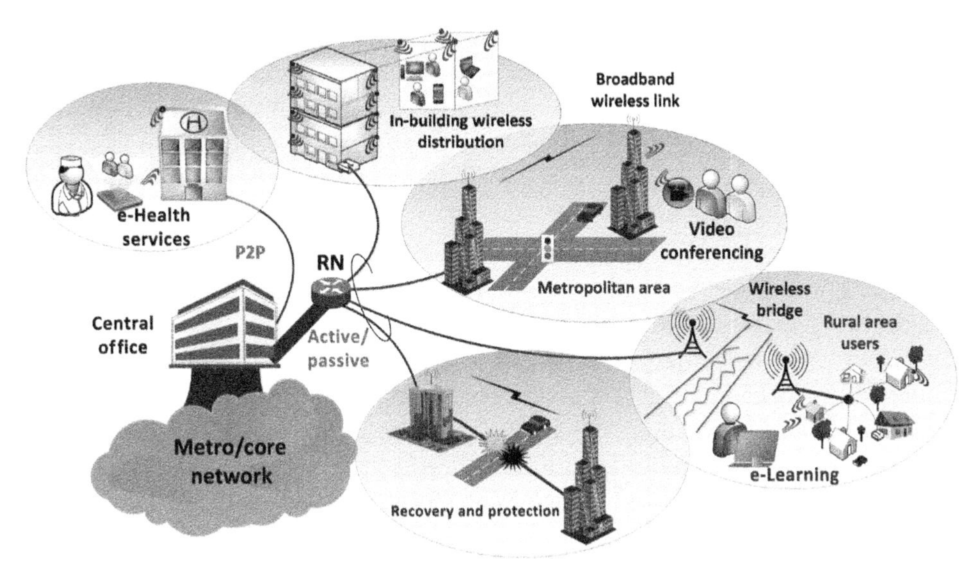

Figure 22.6 Application scenarios of next generation hybrid optical fiber-wireless systems in access and in-building networks. *RN*, Remote node; *P2P*, point-to-point.

range of 100 GHz and above is becoming a timely research topic due to its wider available bandwidth. Recently, W-band (75–110 GHz) has attracted more and more attention due to its potential to provide higher capacity [14]. In the United States, the Federal Communications Commission has started commercial applications of frequency bands of 71–75.5, 81–86, 92–100, and 102–109.5 GHz, which will be recommended for high-speed wireless communications [15]. All these facts have promoted industrial considerations for the incorporation of mm-wave communication links into next-generation fiber-optic wireless hybrid networks. However, there still remains technical challenges in terms of broadband mm-wave generation and long-distance transport, DSP for broadband RoF systems, and RoF networks.

22.2 Broadband optical millimeter-wave generation

22.2.1 Basic photonic up-conversion schemes

Due to the high free-space loss and atmospheric attenuation of mm-wave signals, the coverage of each radio transmitter is reduced to tens to hundreds of meters, which means that a large number of BSs are required to provide a wide geographic coverage [16]. Therefore a low-cost simplified BS design is critical to the commercialization of the fiber-optic wireless hybrid networks with mm-wave links. To carry high-speed wireless data, it is necessary to generate and transport mm-wave signal with high transmission power, high bandwidth and high phase noise tolerance. High frequency RF sources, mixers, or cascaded frequency multiplexers [17,18] are commonly required to generate mm-wave signals using an electrical up-conversion method. This method can usually satisfy the requirements of phase noise and power performance of the generated signal. However, given the trade-off between further expansion of signal bandwidth and increased system complexity, this approach is not considered the best solution in the long run.

On the other hand, using photonic technology to generate wireless signals has the advantage of broad bandwidth that can be achieved by opto-electronic devices. The principle of this method is photonic heterodyne mixing [16,19,20]. Various techniques for generating mm-wave signals have been proposed, and they can usually be divided into two basic categories. The first uses coherent laser sources for heterodyne mixing, which can be achieved by using Mach–Zehnder modulator (MZM) [21], dual-mode distributed feedback laser [22], subharmonic mode-locked laser [23,24] or optical frequency comb. Fig. 22.7 shows three typical coherent up-conversion schemes using single wavelength lasers to realize photonic wireless communication links, among many others.

Scheme A is similar to a traditional double-sided band (DSB) RoF system in which data is first modulated onto an RF carrier before being fed to optical modulator. The difference is that the carrier suppression-DSB (CS-DSB) scheme is used to eliminate periodic RF power fading caused by dispersion in optical fibers and reduces half of the RF source demand [25]. Similarly, CS-DSB modulation technology is employed in scheme B. In this scheme, instead of modulating the data to the RF carrier, the data is directly modulated to the lightwave to form an optical baseband signal, which is then fed to the dual-tone generator for frequency conversion. By doing so, the bandwidth limitations of RF devices such as mixers can be eliminated. The frequency of optical mm-wave signals is doubled by the CS-DSB scheme, so we can reduce the bandwidth requirement for optical and electrical devices [25]. Both schemes A and B support amplitude modulation formats, such as on-off keying and pulse amplitude modulation signals. However, since both the upper and lower bands contain modulated signals, heterodyne mixing with common square-law detection will result in the loss of phase information. Therefore they are not suitable for complex signal formats containing phase modulation, such as phase shift keying, orthogonal amplitude modulation (QAM), and complex valued orthogonal frequency division multiplexing (OFDM) signals. Researchers have made several efforts to adapt to systems that transmit complex signals. In scheme A, by converting the baseband signal to IF signal before mixed with RF signal, after heterodyne jitter, the converted complex signal can be removed from the frequency component generated by the sideband which contains data, or so-called "jitter noise", so that the intensity and phase information can be reserved [26]. Different from scheme A and B, scheme C separates the upper and lower bands after the two-tone generator, as shown in Fig. 22.7. The data is modulated to one sideband

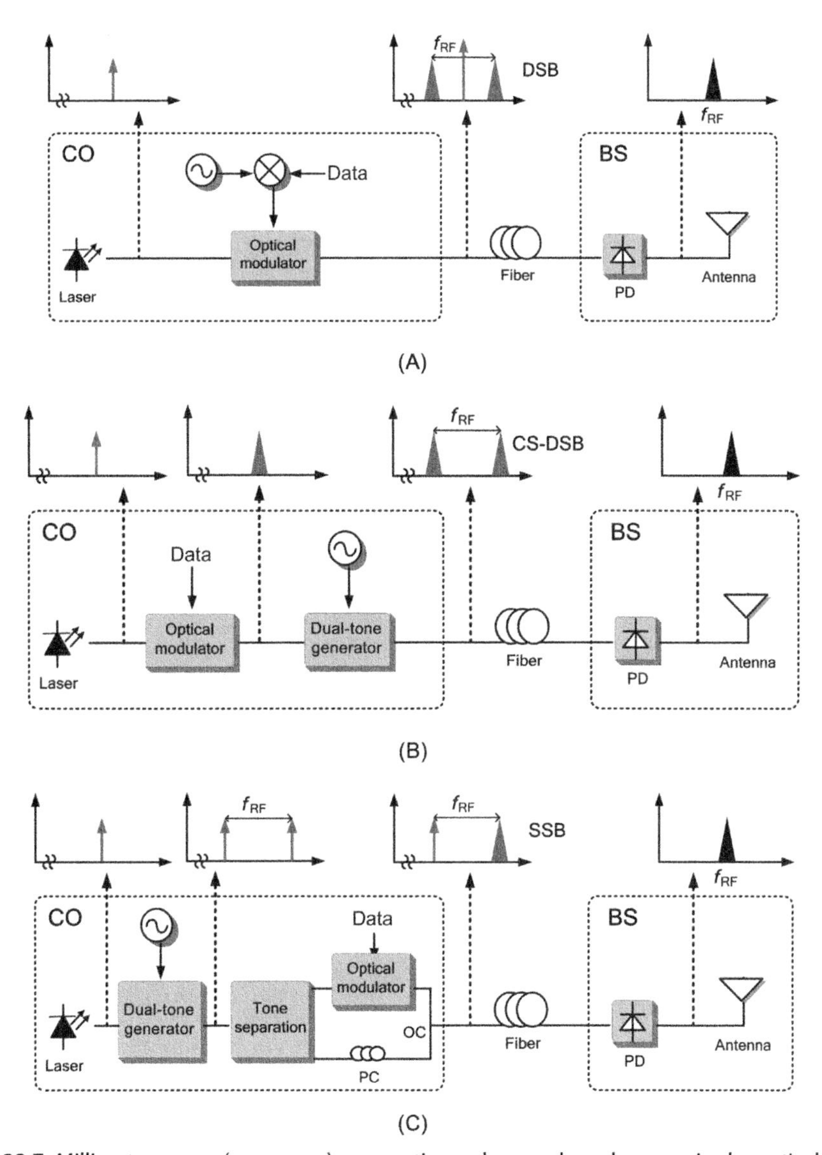

Figure 22.7 Millimeter-wave (mm-wave) generation schemes based on a single optical source. (A) Scheme A, (B) scheme B, and (C) scheme C. *CO*, Central office; *BS*, base station; *CS-DSB*, carrier suppressed-double sideband; *SSB*, single sideband; *PC*, polarization controller; *OC*, optical coupler.

and the other sideband is used as the carrier signal. In this scheme, the signal and carrier are combined together to form a single sideband (SSB) RoF signal before transmission over the fiber link, thus it can preserve the complex signal format after heterodyne mixing and eliminate the RF power fading effect in the fiber. Since the modulated signal by the optical modulator is a baseband signal and the optical

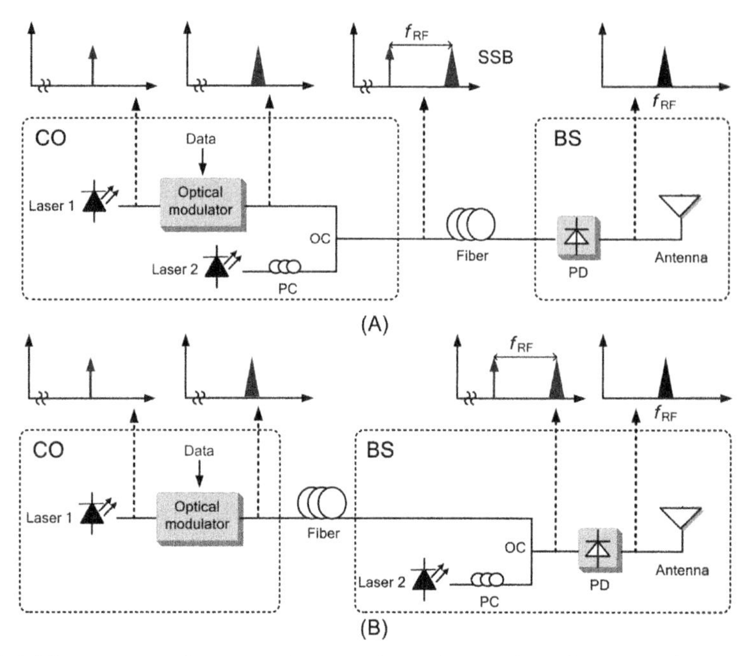

Figure 22.8 Millimeter-wave (mm-wave) generation schemes based on two free-running lasers. (A) Scheme D and (B) scheme E.

modulator has a bandwidth up to 40 GHz, we can generate very large bandwidth RoF signals by this scheme. The mm-wave signals generated by the schemes A, B and C based on a single external modulator have a stable frequency [27–32].

Fig. 22.8 shows another two schemes (schemes D and E) [27–30,33], where the two independent lasers run freely without frequency locking. By changing frequency spacing between the two lasers, the mm-wave signals at any frequency after heterodyne beating in a photodiode (PD) can be generated. But the generated mm-wave signals have a frequency instability problem since the two independent lasers used for heterodyne beating are not frequency-locked [33,34]. In scheme D, the continuous-wave (CW) lightwave and the modulated lightwave are generated at the CO, so the BS has no laser source, which makes it simple. In scheme E, the CW laser locates in the BS. This scheme can be used to realize polarization multiplexing mm-wave signal generation [33], and it is employed to demonstrate large-capacity mm-wave signal generation with multidimensional multiplexing. Moreover, the optical signal generated at the CO is a baseband signal, and we can easily realize over 100-km fiber transmission from the CO to the BS.

As each of the aforementioned five basic photonic up-conversion schemes has its own limitations, there is an improved scheme to generate optical mm-wave signal based on the external modulator [27,28]. This scheme simplifies the system structure and reduces the cost because only one laser source and one modulator are used. In addition, mm-wave signals can carry different types of modulation formats, including single carrier and multicarrier OFDM. Since OFDM usually has arbitrary amplitude and phase, the processing for OFDM vector signal generation will be different from that for single-carrier vector signal generation. We will discuss separately how to generate vector QAM mm-wave signal based on the two different modulation formats in the following section.

22.2.2 Simplified architecture for millimeter -wave generation

In scheme B of Fig. 22.7, the need for two modulators to generate the optical-carrier-suppression (OCS) optical mm-wave signal is considered too complicated. Also, as explained earlier, generation of vector signal is not directly supported with such scheme. For example, if a quadrature-phase-shift-keying (QPSK) signal is modulated, after beating in a PD, both the frequency and phase of this signal will be doubled. Therefore the regular QPSK signal is irretrievable after the PD detection [27]. Recent research shows that we can largely simplify this architecture if we generate the vector signal by a single optical modulator [27]. Fig. 22.9 shows the schematic diagram of the photonic vector mm-wave signal generation employing a single intensity modulator (IM) based on OCS. With the aid of MATLAB programming, a pseudo-random binary sequence (PRBS) with a certain length is first constellation mapped, then precoded and low-pass filtered, and finally digitally up-converted to generate the

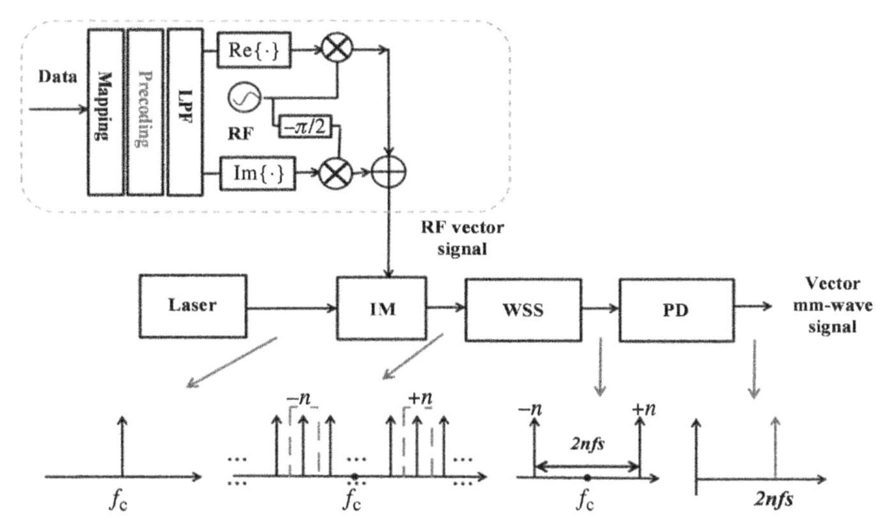

Figure 22.9 Schematic diagram of photonic vector mm-wave signal generation employing a single IM based on optical-carrier-suppression (OCS). *IM*, Intensity modulator; *WSS*, wavelength selective switch; *PD*, photodiode.

precoded vector RF signal at frequency f_s, as shown in the dotted box of Fig. 22.9 [29,35–37]. The vector RF signal is used to drive a single IM which operates at OCS mode, that is, even- or odd-order subcarrier suppression mode. A CW lightwave at frequency f_c from a laser is modulated by the IM. Then, we use a wavelength selective switch (WSS) or other types of optical filters to select the two symmetrical $\pm n$-order optical subcarriers with a frequency spacing of $2nf_s$, which are heterodyne beat in a single-ended PD to obtain the vector signal at the frequency of $2nf_s$.

Assume that the CW lightwave at f_c has constant amplitude E_1 and can be formulated as

$$E_{\text{in}}(t) = E_1 \exp(j2\pi f_c t). \tag{22.1}$$

Assume that the driving vector RF signal at f_s can be formulated as

$$V_{\text{driver}}(t) = V_{\text{RF}} A \cos(2\pi f_s t + \varphi), \tag{22.2}$$

where V_{RF} is the driving RF voltage on the IM. A and φ denote the precoded amplitude and phase of the driving vector RF signal at f_s, respectively. The output from the IM can thus be formulated as

$$\begin{aligned} E_{\text{out}}(t) &= \frac{1}{2} E_{\text{in}}(t) \left[\exp\left(\text{j}\pi \frac{V_{\text{driver}} + V_{\text{DC}}}{V_\pi} \right) + \exp\left(-\text{j}\pi \frac{V_{\text{driver}} + V_{\text{DC}}}{V_\pi} \right) \right] \\ &= E_1 \cos\gamma \sum_{m=-\infty}^{\infty} (-1)^m J_{2m}(\beta A) \exp(\text{j}2\pi f_c t + \text{j}2m \cdot 2\pi f_s t + \text{j}2m\varphi) \\ &\quad - E_1 \sin\gamma \sum_{m=-\infty}^{\infty} (-1)^m J_{2m+1}(\beta A) \exp\left[\text{j}2\pi f_c t + \text{j}(2m+1)2\pi f_s t + \text{j}(2m+1)\varphi\right], \end{aligned} \tag{22.3}$$

where V_π and V_{DC} is the half-wave voltage and DC-bias voltage of the IM, respectively. $\beta = \pi V_{\text{RF}}/V_\pi$ is the modulation index of the IM, and $\gamma = \pi V_{\text{DC}}/V_\pi$ is the initial phase caused by V_{DC}. J_n is the first-kind Bessel function of order n. When V_{DC} is set to zero, that is, the IM is biased at its maximum transmission point, the even-order subcarriers, including the central one, are remained, while the odd-order ones are suppressed. The desired $\pm 2m$-order subcarriers can be selected via a WSS and expressed as

$$\begin{aligned} E_{\text{filter}}^{\text{even}}(t) = &-(-1)^m E_1 J_{2m}(\beta A) \exp\left[j2\pi f_c t + j(2m)2\pi f_s t + j(2m)\varphi\right] \\ &-(-1)^{-m-1} E_1 J_{-2m}(\beta A) \exp\left[j2\pi f_c t - j(2m) f_s t - j(2m)\varphi\right]. \end{aligned} \tag{22.4}$$

Similarly, for the case of even-order subcarrier suppression, the two desired subcarriers can be selected by a WSS from the output of the MZM biased at $V_{DC} = V_\pi/2$ and formulated as

$$\begin{aligned} E_{\text{filter}}^{\text{odd}}(t) = & -(-1)^m E_1 J_{2m+1}(\beta A) \exp\left[j2\pi f_c t + j(2m+1)2\pi f_s t + j(2m+1)\varphi\right] \\ & -(-1)^{-m-1} E_1 J_{-2m-1}(\beta A) \exp\left[j2\pi f_c t - j(2m+1) f_s t - j(2m+1)\varphi\right]. \end{aligned} \tag{22.5}$$

According to the square-law direct-detection rule of the PD, the output current is generated by the heterodyne beating of two $\pm 2m$-order or $\pm(2m+1)$-order optical subcarriers, and can be expressed as

$$i_{\text{PD}}^{\text{even}}(t) = 2RE_1^2 J_{2m}^2(\beta A) + 2RE_1^2 J_{2m}^2(\beta A) \cos\left[2(2m)\cdot 2\pi f_s t + 2(2m)\varphi\right] \tag{22.6}$$

$$i_{\text{PD}}^{\text{odd}}(t) = 2RE_1^2 J_{2m+1}^2(\beta A) + 2RE_1^2 J_{2m+1}^2(\beta A) \cos\left[2(2m+1)\cdot 2\pi f_s t + 2(2m+1)\varphi\right], \tag{22.7}$$

where R is the conversion efficiency of the PD. Eqs. (22.6) and (22.7) clearly demonstrate that the frequency of the output current of the PD increases to $2Nf_s$ for two selected $\pm N$-order ($N = 1, 2, 3, 4 \ldots$) subcarriers, and therefore photonic frequency multiplication can be realized based on the scheme shown in Fig. 22.9. Correspondingly, the current phase is linearly transformed into $2N\varphi$ and the current amplitude is nonlinearly expanded by $2RE_1{}^2 J_N{}^2(\beta A)$ times. Therefore in order to obtain a vector mm-wave signal displaying standard vector modulation, such as QPSK, 8 quadrature amplitude modulation (8QAM), and 16 quadrature amplitude modulation (16QAM), the transmitter standard vector RF signal needs to be processed according to the phase and amplitude precoding rules as follows:

$$A_p = J_N{}^{-1}(A_s \beta), \tag{22.8}$$

$$\varphi_p = \varphi_s / 2N, \tag{22.9}$$

where A_s and φ_s denote the amplitude and phase of the standard vector signal, respectively, A_p and φ_p are the amplitude and phase of the precoded vector signal at transmitter side, respectively. J_N^{-1} is the inverse function of J_N.

As one example, we show QPSK vector signal generation with photonic frequency doubling based on a single MZM. The experimental setup and main results are shown in Fig. 22.10 [29,36]. To generate a standard QPSK signal, we first need to precode the driving QPSK RF signal. Due to the constant amplitude characteristic of the QPSK constellation unaffected by the MZM and PD, only QPSK phase information should be precoded as the half of its original value according to Eq. (22.9).

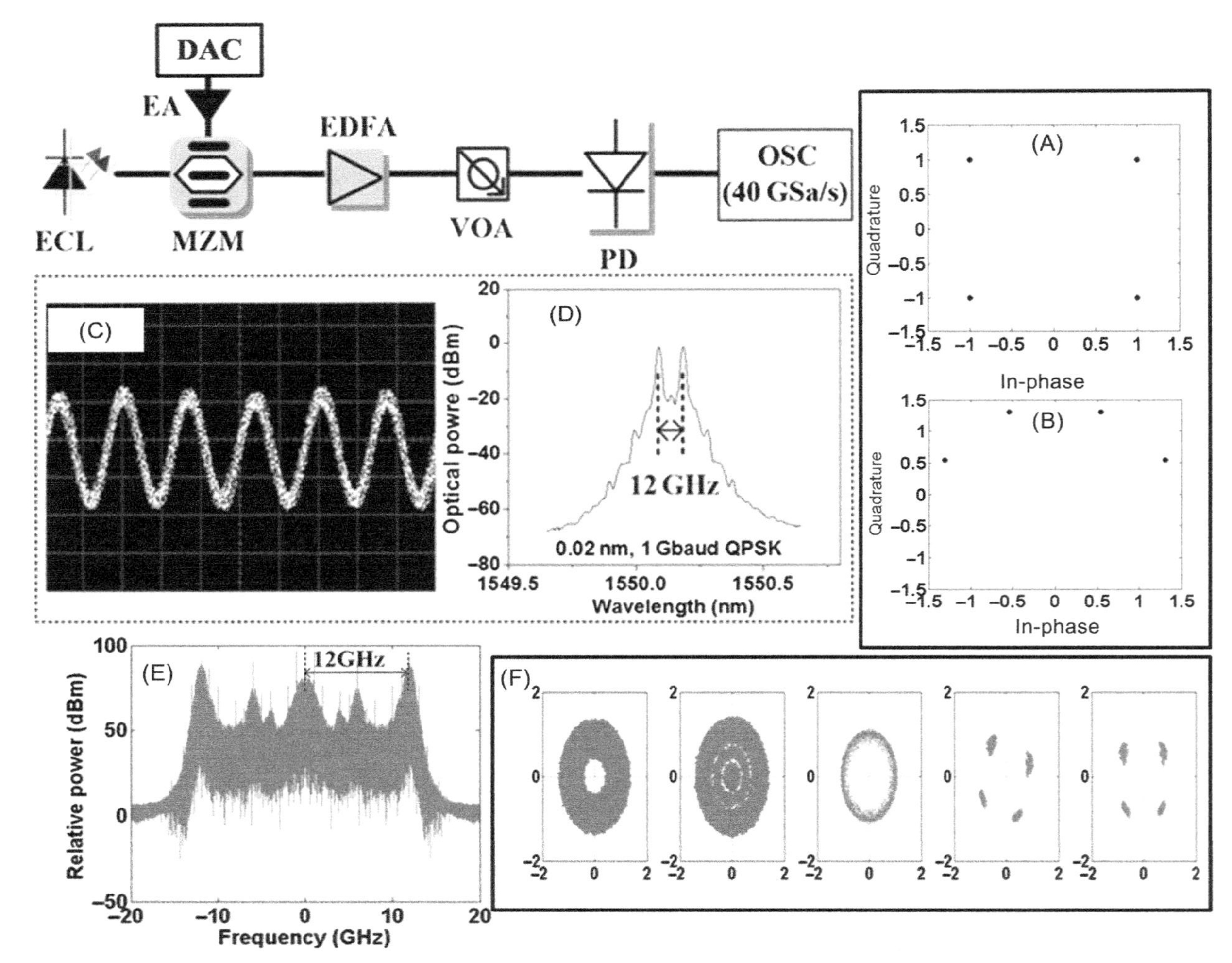

Figure 22.10 Experimental setup for quadrature-phase-shift-keying (QPSK) vector signal generation employing photonic frequency doubling. *ECL*, External cavity laser; *EDFA*, erbium-doped fiber amplifier; *EA*, electrical amplifier; *PD*, photodiode; *VOA*, variable optical attenuator; *OSC*, oscilloscope. (A) Standard QPSK constellation. (B) Precoded QPSK constellation. (C) Measured electrical spectrum after digital-to-analog conversion (DAC). (D) Measured optical spectrum after Mach–Zehnder modulator (MZM). (E) Received electrical spectrum after OSC. (F) Received QPSK constellations.

Fig. 22.10A and B give the QPSK constellations before and after phase precoding, respectively. A 1-Gbaud precoded QPSK RF signal at 6 GHz can be generated offline and then digital-to-analog converted by a DAC with 64-GSa/s sampling rate. The time-domain waveform of the RF electrical signal is illustrated in Fig. 22.10C. As expressed in Eq. (22.5), the MZM is DC-biased at the even-order subcarrier suppression mode, and the driving power of the RF signal is adjusted to ensure that only symmetrical ± 1-order subcarriers are remained with 12-GHz frequency spacing, as shown in Fig. 22.10D.

Fig. 22.10E shows the electrical spectrum. Fig. 22.10F give the recovered QPSK constellations by the DSP processing, including down-conversion, clock extraction, constant-modulus-algorithm (CMA) equalization, frequency offset estimation, and carrier phase estimation.

Similarly, vector mm-wave generation at Q-band [38] and W-band [30–33,35,39,40] employing the photonic quadrupling [41] and octupling [30,31,40] schemes can be achieved. Also higher frequency mm-wave bands offer a much wider bandwidth. Moreover, 8QAM and 16QAM [29,39,40,42] vector mm-wave signal can also be generated to further improve the spectrum efficiency. However, for high-order modulation formats, the filtering effect caused by the insufficient DAC bandwidth results in a relatively large phase jitter, which limits the transmission baud rate. Thus, there is a trade-off between the modulation order and transmission speed.

The single IM shown in Fig. 22.10 can be replaced by a single phase modulator (PM) to realize single-carrier vector mm-wave signal generation, since the PM, compared to the IM, has the advantages of less insert loss, larger optical signal-to-noise ratio, and higher stability without DC bias [34,43–46]. Base on a single PM, we experimentally demonstrated W-band QPSK, 8QAM, and 16QAM vector mm-wave signal generation employing photonic frequency quadrupling [34], and also experimentally demonstrated W-band QPSK vector mm-wave signal generation employing random and asymmetrical photonic frequency tripling [45]. The laser and the single IM in Fig. 22.10 can be also replaced by a single directly-modulated laser (DML) or electro-absorption modulated laser (EML) to realize single-carrier vector mm-wave signal generation, since the DML or EML is cheaper and more compact compared to the IM and PM [47]. Based on a single DML, we experimentally demonstrated Q-band QPSK vector mm-wave signal generation employing photonic frequency doubling [47]. The principle of the PM and DML/EML schemes are quite similar to that of the IM scheme, and therefore here we do not describe it in detail.

The high-frequency vector mm-wave signal will be degraded by the DAC filtering effect at transmitter end, which leads to the interaction of some constellation points and the system bit-error ratio (BER) degradation. To solve this problem, Ref. [48] proposed an improved balanced precoding scheme, in which the transmitter standard vector RF signal needs to be processed according to the phase precoding rule as follows:

$$\varphi_{\mathrm{p}} = \frac{\varphi_{\mathrm{s}} + 2k\pi}{2N}(k = 0, 1 \ldots 2N - 1), \tag{22.10}$$

where φ_{s} and φ_{p} denote the phase of the standard vector signal and the precoded vector signal, respectively. N denotes the order of the two selected optical subcarriers. For example, to generate a frequency-doubling QPSK vector mm-wave signal, the phase of the driving QPSK vector RF signal after balanced precoding is

$$\varphi_{\mathrm{p}} = \frac{\varphi_{\mathrm{QPSK}} + 2k\pi}{2}\left(k = 0, 1;\ \varphi_{\mathrm{QPSK}} = \frac{\pi}{4}, \frac{3\pi}{4}, \frac{5\pi}{4}, \frac{7\pi}{4}\right). \tag{22.11}$$

Figs. 22.11A–C show the standard QPSK constellation, the constellation after imbalanced precoding as given by Eq. (22.9), and the constellation after balanced

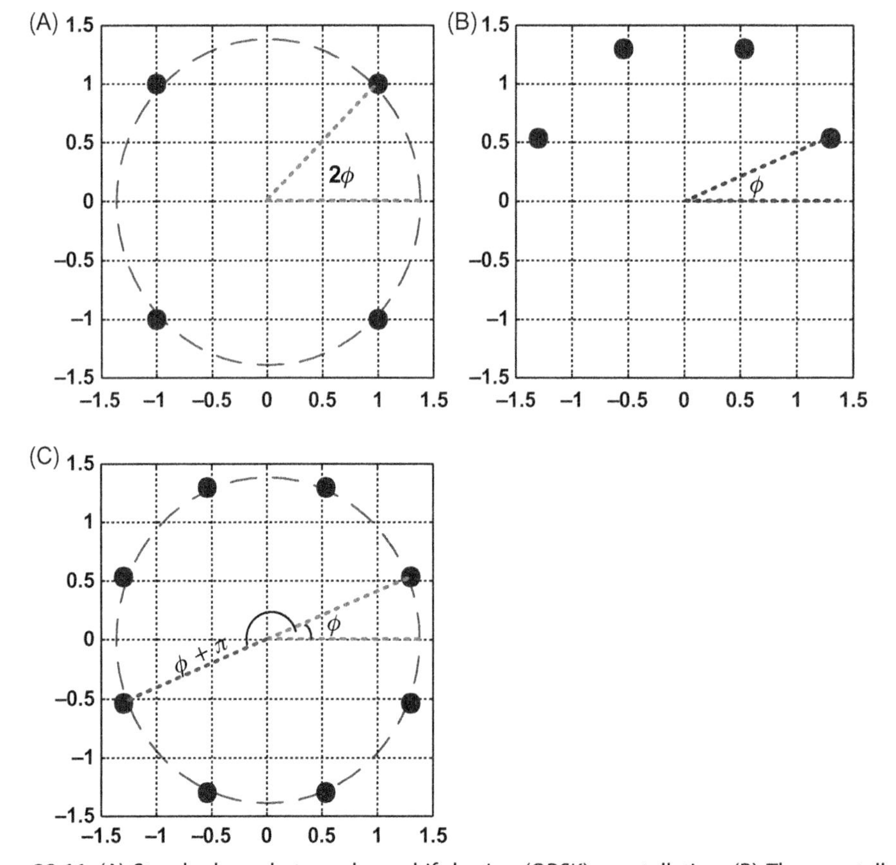

Figure 22.11 (A) Standard quadrature-phase-shift-keying (QPSK) constellation. (B) The constellation after imbalanced precoding. (C) The constellation after balanced precoding.

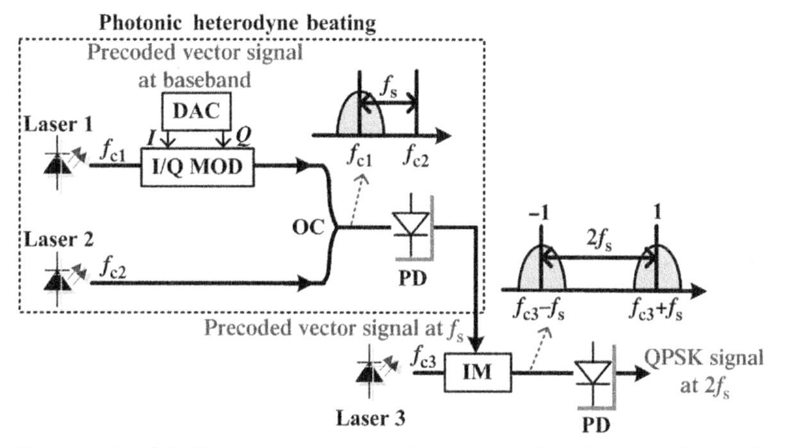

Figure 22.12 Principle of QPSK vector mm-wave signal generation based on heterodyne beating, phase precoding and OCS. *QPSK*, Quadrature-phase-shift-keying; *mm*, millimeter; *OCS*, optical-carrier-suppression.

precoding as given by Eq. (22.11), respectively. Based on the balanced precoding scheme, we experimentally demonstrated 8-Gbaud 16-GHz QPSK vector signal generation, with a 2.5-dB receiver sensitivity improvement and more symmetrical receiver constellation distribution compared to the imbalanced precoding scheme [49].

Also, we can introduce the technique of heterodyne beating into the IM scheme given in Fig. 22.12, to avoid the filtering effect caused by the DAC insufficient bandwidth. Fig. 22.12 gives the principle of the QPSK vector mm-wave signal generation based on heterodyne beating, phase precoding, and OCS [50]. The heterodyne beating based on two free-running lasers (laser 1 and laser 2) are used to generate the precoded vector signal at frequency f_s, which is then used to drive the IM biased at its even-order carrier suppression mode to realize frequency-doubling photonic vector mm-wave signal generation. Here, the CW lightwave generated from laser 1 is modulated by a precoded vector signal at baseband via an in-phase/quadrature (I/Q) modulator, while the CW lightwave generated from laser 2 is unmodulated.

Compared to the scheme shown in Fig. 22.10, though a more complicated RF signal generation architecture, based on I/Q modulation and PD detection, is added, the DAC can be realized at baseband instead of RF band. It indicates that a vector mm-wave signal can be generated without the constraint of the limited DAC bandwidth. Based on the scheme shown in Fig. 22.12, a 17.6-GHz QPSK vector signal generation with a BER lower than 3.8×10^{-3} was demonstrated in Ref. [50].

In addition, a photonic vector signal generation scheme employing single-sideband (SSB) I/Q modulation without precoding is proposed, which is used to further simply the system, as shown in Fig. 22.13 [51]. At transmitter side, based on offline DSP, a

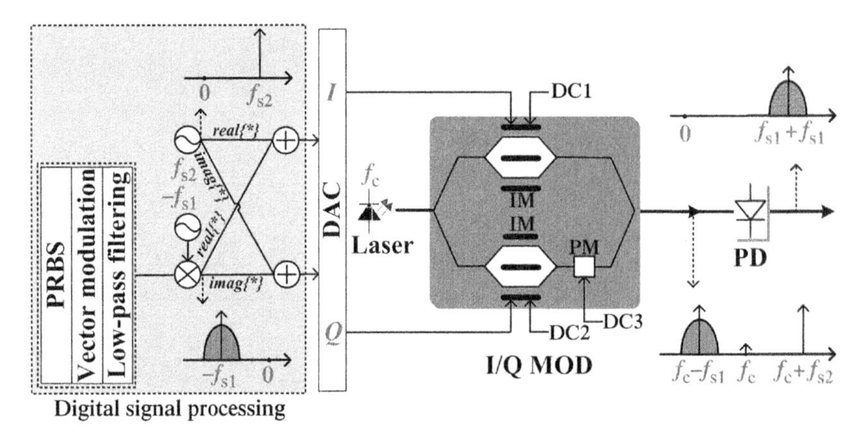

Figure 22.13 Principle for photonic vector signal generation employing (SSB) I/Q modulation without precoding. *SSB*, Single sideband; *I/Q*, in-phase/quadrature.

vector-modulated digital SSB at frequency—f_{s1} and an unmodulated digital SSB at frequency f_{s2} are generated, and then, the real and imaginary parts of these two digital SSBs are added, respectively. After analog-to-digital conversion, the real and imaginary summation of these two SSBs are used to drive an I/Q modulator. The electrical-to-optical conversion via the I/Q modulator is linear [51], and therefore the output of the I/Q modulator is a vector-modulated optical SSB at frequency ($f_c - f_{s1}$) and an unmodulated optical SSB at frequency ($f_c + f_{s2}$) with a suppressed central optical carrier at frequency f_c. After further PD detection, we can obtain an electrical vector-modulated mm-wave signal at frequency ($f_{s1} + f_{s2}$).

Due to the advantage of fiber dispersion mitigation, the approach can be also combined with multiple-input multiple-output (MIMO) technology to improve the transmission capacity. Employing SSB I/Q modulation, a 2×2 MIMO mm-wave transmission system based on different modulation formats, transmission rates and mm-wave frequencies is proposed, which can be realized without optical polarization multiplexers [52].

OFDM vector signal not only can tolerate fiber chromatic dispersion and polarization mode dispersion (PMD), but also can provide a good robustness to wireless multipath effects and wireless channel impairments [53–56]. As shown in Fig. 22.14, the scheme of the OFDM photonic vector signal generation employing a single MZM based on OCS is proposed in Ref. [36]. Firstly, the input data is converted into parallel data stream, and then mapped. The digital time domain signal is obtained by using inverse Fast Fourier Transformation (IFFT). To avoid intersymbol interference (ISI) caused by channel dispersion, cyclic prefix is inserted. Next the OFDM baseband signal is precoded firstly based on phase information and then amplitude information. Then the OFDM precoded baseband signal is up-converted into RF band. The whole process is based on offline DSP. Then, the photonic vector signal can be generated by a MZM driven by the OFDM RF signal based on OCS.

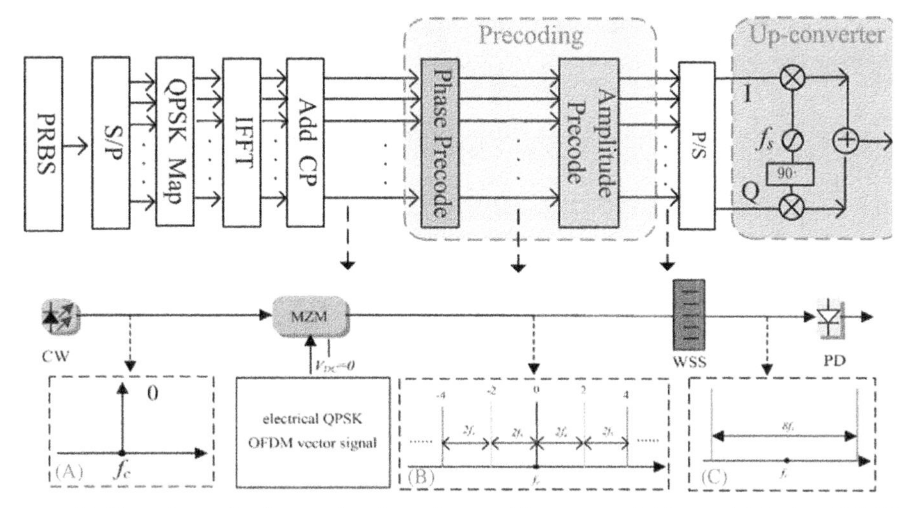

Figure 22.14 The schematic diagram for the orthogonal-frequency-division-multiplexing (OFDM) vector signal generation based on precoding. (A), (B), and (C) are optical spectra at different points.

The CW lightwave generated from a laser is described

$$E_{\text{in}}(t) = E_1 \exp(2\pi f_c t), \tag{22.12}$$

where E_1 and f_c are the amplitude and frequency of the optical carrier, respectively. The baseband electrical OFDM vector signal can be expressed as

$$S_{\text{BB\_OFDM}}(t) = \sum_{n=0}^{N-1} d_n \exp\left(j2\pi \frac{n}{T} t\right), \tag{22.13}$$

where t is the discrete time index, N is the number of the subcarriers, and T is the symbol duration. d_n is the data symbol modulated onto the nth subcarrier. The baseband electrical OFDM vector signal can be simplified as

$$S_{\text{BB\_OFDM}}(t) = s(t) \exp\left(j\varphi_t\right), \tag{22.14}$$

where $s(t)$ and φ_t is the amplitude and phase of the OFDM signal, respectively. Therefore the OFDM vector signal at frequency f_s can be expressed as

$$S_{\text{OFDM\_drive}}(t) = s(t) \cos\left(2\pi f_s t + \varphi_t\right). \tag{22.15}$$

The generated optical OFDM vector signal by the MZM is written as

$$E_{\text{out}}(t) = \frac{1}{2} E_{\text{in}}(t) \left[\exp\left(j\pi \frac{V_{\text{OFDM\_driver}} + V_{\text{DC}}}{V_\pi} \right) + \exp\left(-j\pi \frac{V_{\text{OFDM\_driver}} + V_{\text{DC}}}{V_\pi} \right) \right], \tag{22.16}$$

where V_{DC} is the DC-bias voltage of the MZM, and V_π is the half-wave voltage of the MZM. The optical vector signal can be further expanded into

$$E_{out}(t) = E_1 \cos\gamma \sum_{m=-\infty}^{\infty} (-1)^m J_{2m}[\beta s(t)] \cdot \exp\left(j2\pi f_c t + j2m \cdot 2\pi f_s t + j2m\varphi_t\right)$$
$$- E_1 \sin\gamma \sum_{m=-\infty}^{\infty} (-1)^m J_{2m+1}[\beta s(t)] \cdot \exp\left[j2\pi f_c t + j(2m+1)2\pi f_s t + j(2m+1)\varphi_t\right], \tag{22.17}$$

where $\beta = \pi V_{RF}/V_\pi$ is the modulation index of the MZM, and $\gamma = \pi V_{DC}/V_\pi$ is the initial phase caused by V_{DC}. According to Eq. (22.17), we can conclude that the baseband OFDM signal is modulated onto the optical subcarriers with different orders. The frequency spacing between adjacent subcarriers is the RF carrier frequency f_s.

When $V_{DC} = V_\pi/2$, the MZM operates at even-order subcarrier suppression mode, and its output can be written as

$$E_{out}(t) = -E_1 \sum_{m=-\infty}^{\infty} (-1)^m J_{2m+1}[\beta s(t)] \times \exp\left[j2\pi f_c t + j(2m+1)2\pi f_s t + j(2m+1)\varphi_t\right]. \tag{22.18}$$

Then the two selected optical subcarriers, with the order of $\pm(2m+1)$, by a WSS, can be written as

$$E_{filter}(t) = (-1)^{m-1} E_1 J_{2m+1}[\beta s(t)]\left\{\exp\left[j2\pi f_c t + j(2m+1)2\pi f_s t + j(2m+1)\varphi_t\right] + \exp\left[j2\pi f_c t - j(2m+1)2\pi f_s t - j(2m+1)\varphi_t\right]\right\}. \tag{22.19}$$

Then the output current of the square-law PD can be expressed as

$$i_{PD}(t) = 2RE_1^2 J_{2m+1}^2[\beta s(t)] + 2RE_1^2 J_{2m+1}^2[\beta s(t)] \cdot \cos\left[2(2m+1) \cdot 2\pi f_s t + 2(2m+1)\varphi_t\right], \tag{22.20}$$

where R is the conversion efficiency of the PD. As shown in Eq. (22.20), the amplitude and phase of the OFDM signal on each subcarrier after the PD is $2RE_1^2 J_{2m+1}^2[\beta s(t)]$ and $2(2m+1)$ times of that of the driving OFDM vector signal on corresponding subcarrier, respectively. The phase of the vector signal is a linear transformation, and its amplitude is a nonlinear transformation. Therefore the driving OFDM vector signal after IFFT requires amplitude and phase precoding according to the rules as follows:

$$p(t) = \frac{J_{2m+1}^{-1}\left(\sqrt{s(t)}\right)}{\beta}, \tag{22.21}$$

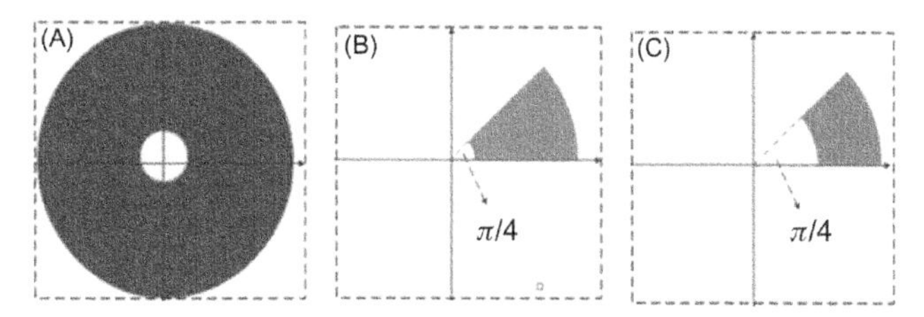

Figure 22.15 (A) Constellation of standard orthogonal-frequency-division-multiplexing quadrature-phase-shift-keying (QPSK OFDM) signal after inverse fast Fourier transformation (IFFT). (B) Constellation of QPSK OFDM signal after phase precoding. (C) Constellation of QPSK OFDM signal after both phase and amplitude precoding.

$$\varphi_p = \frac{\varphi_t}{2(2m+1)}, \tag{22.22}$$

where $p(t)$ and φ_p are the amplitude and phase of the precoded OFDM vector signal at transmitter side, respectively. When $V_{DC} = 0$, the MZM operates at odd-order subcarrier suppression mode. The case of odd-order subcarrier suppression has a theoretical derivation quite similar to the case of even-order subcarrier suppression, and therefore we do not give a detailed description in this paper.

Fig. 22.15A–C show the standard QPSK OFDM signal constellation after IFFT, the OFDM constellation after phase precoding, and the OFDM constellation after both phase and amplitude precoding, respectively. Based on the scheme shown in Fig. 22.14, 14-GHz QPSK OFDM vector signal generation employing photonic frequency doubling is demonstrated in Ref. [36], and also W-band OPSK OFDM vector signal generation employing photonic frequency octupling is also experimentally demonstrated [32].

In addition, quite similar to the scheme shown in Fig. 22.13, OFDM vector signal can also be generated employing SSB I/Q modulation without precoding. The SSB OFDM vector signal can tolerate fiber dispersion better and has a higher spectral efficiency than the DSB OFDM vector signal [37,57–63]. 38-GHz SSB OFDM vector signal generation employing SSB I/Q modulation is demonstrated in Ref. [64], and its software-defined characteristic is compatible with 5G wireless communication [48,65–68].

22.3 Broadband millimeter-wave detection in the radio-over-fiber system

Mm-wave signals can be detected by direct or heterodyne detection. Direct detection is commonly used in the mm-wave communication [69–72]. The structure of the direct detection scheme is shown in Fig. 22.16. The wireless mm-wave is usually received by the antenna at the receiver and amplified by the electrical amplifier. Then

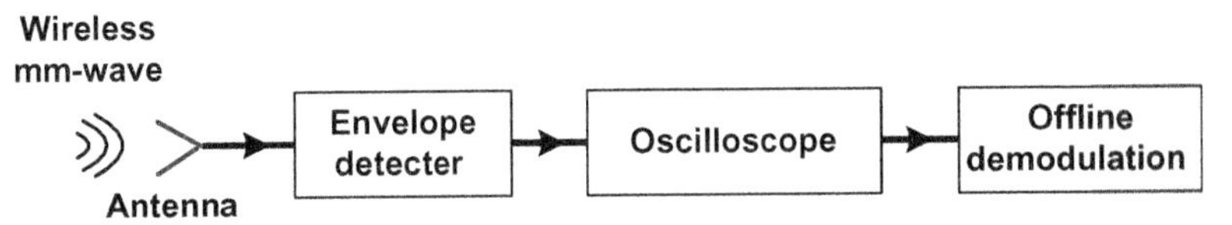

Figure 22.16 The structure of the direct detection scheme.

high-frequency mm-wave signal is down-converted to the baseband signal by an envelope detector. The direct detection scheme is simple, low-cost and phase-noise insensitive [70]. But this scheme cannot detect the vector signal carried by the mm-wave, and only intensity modulation signal can be detected in the direct detection scheme. The theoretical derivation of the square-law detection in the direct detection scheme is as follows. The mm-wave signal received at the receiver can be expressed as

$$R(t) = [A + ms(t)]\cos(2\pi f_c t), \tag{22.23}$$

where A represents the DC-bias in intensity modulation, $s(t)$ represents the amplitude modulation signal, m represents the modulation factor, and f_c represents the frequency of mm-wave. After the square-law detection, the output current of the envelope detector can be described as

$$\begin{aligned} I(t) &= \left|[A+ms(t)]\cos(\omega_c t)\right|^2 \\ &= \frac{1}{2}A^2 + \frac{1}{2}Ams(t) + \frac{1}{2}m^2s^2(t) + \frac{1}{2}[A+ms(t)]^2\cos(4\pi f_c t). \end{aligned} \tag{22.24}$$

From Eq. (22.24), we can know that the first item is the direct current (DC) component, the second item represents the amplitude modulation signal component, the third item is the nonlinear noise, and the fourth item is the high-frequency component that needs to be filtered out. We can retrieve the amplitude modulation signal linearly. The detected signal will be distorted and suffer from the nonlinear noise. The power of the output of the envelope detector may be too small, so a baseband amplifier is needed. The direct detection system is relatively wasteful of power and bandwidth, but it is easy to implement and economical.

Heterodyne detection is widely used in the mm-wave communication [73–75]. In addition to the mm-wave communication, heterodyne technique is also commonly used in radar [76,77] and optical communication. In contrast to the only one frequency in the homodyne detection, heterodyne means two or more frequency beating at the receiver. After the beating between two signals at frequencies f_1 and f_2, the new frequencies called heterodynes are created. One of the heterodynes is at the frequency $f_1 + f_2$, and the other is at the frequency $f_1 - f_2$. In order to realize the mm-wave down-conversion, the signal at low-frequency $f_1 - f_2$ is reserved and the signal at high-frequency $f_1 + f_2$ is filtered out.

The structure of the heterodyne detection scheme in the mm-wave communication is shown in Fig. 22.17. The scheme mainly consists of a local oscillator (LO) and

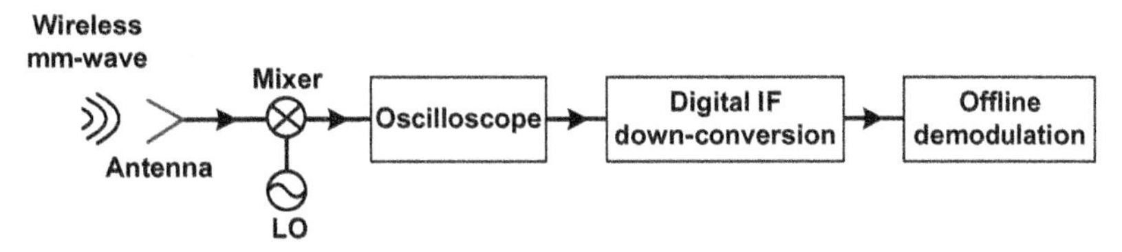

Figure 22.17 The structure of the heterodyne detection scheme.

a mixer. The wireless mm-wave is usually received by the antenna at the receiver and amplified by the electrical amplifier. Then the mm-wave combines with the local RF signal in the mixer to achieve the mm-wave down-conversion. The down-converted IF signal is captured by a digital storage oscilloscope and digitally down-converted to the baseband signal. The heterodyne detection scheme requires additional RF source and mixer resulting in high cost, but the system has a high receiving sensitivity. This detection scheme can be used to detect both intensity modulation signal and vector signal, and the offline DSP algorithm can be used to compensate for noise in the received signal. The theoretical derivation of the heterodyne detection scheme is as follows. The mm-wave vector signal received at the receiver can be expressed as

$$S(t) = A\left[I(t)\sin(2\pi f_c t + \theta_c(t)) + Q(t)\cos(2\pi f_c t + \theta_c(t))\right], \tag{22.25}$$

where A represents the amplitude index of the mm-wave, $I(t)$ and $Q(t)$ represent the vector signal carried by the mm-wave, f_c represents the frequency of the mm-wave and $\theta_c(t)$ represents the phase of mm-wave.

The local RF signal can be expressed as

$$S_{LO}(t) = A_{LO}\cos(2\pi f_{LO} t + \theta_{LO}(t)), \tag{22.26}$$

where A_{LO} represents the amplitude of the RF signal, f_{LO} represents the frequency of RF signal, and $\theta_{LO}(t)$ represents the phase of RF signal. The down-converted IF signal after the mixer can be described as

$$S_{IF}(t) = A_{LO}A\begin{bmatrix} I(t)\sin(2\pi f_c t - 2\pi f_{LO} t + \theta_c(t) - \theta_{LO}(t)) \\ + Q(t)\cos(2\pi f_c t - 2\pi f_{LO} t + \theta_c(t) - \theta_{LO}(t)) \end{bmatrix}. \tag{22.27}$$

The equation can be simplified as

$$\begin{aligned} S_{IF}(t) &= A_{LO}A\left[I(t)\sin(2\pi f_{IF} t + \theta_{IF}(t)) + Q(t)\cos(2\pi f_{IF} t + \theta_{IF}(t))\right] \\ &= -jA_{LO}A\left[I(t) + jQ(t)\right]\exp(j2\pi f_{IF} t + j\theta_{IF}(t)), \end{aligned} \tag{22.28}$$

where f_{IF} and $\theta_{IF}(t)$ represent the frequency and the phase of the down-converted IF signal, respectively. From Eq. (21.28), it can be seen that the transmitted vector signal $I(t) + jQ(t)$ will be recovered after the digital down-conversion and the phase correction in offline DSP.

The heterodyne detection is more complex and expensive compared with the direct detection, while the heterodyne detection based on DSP can effectively recover the distorted signal and show high receiving sensitivity. The heterodyne detection is more suitable for the ultrabroadband mm-wave communication. In the next section we will discuss the DSP in detail.

22.4 Digital signal processing for radio-over-fiber systems

The lasers run without frequency locking in the schemes as shown in Fig. 22.8, and therefore we have to use DSP to compensate for the frequency drifting [33]. Also, the frequency response of the optical or electrical devices at mm-wave band is usually not flat, which will lead to the signal distortion [47]. The broadband mm-wave optical signal after transmission over fiber will be suffered from fiber dispersion, which leads to sever ISI [48]. In fact, the nonlinear effects in the optical or electrical devices or optical fiber will also degrade the mm-wave signal performance. If we use high-order QAM, these nonlinear effects become much more severe. All these impairments can be compensated by advanced DSP [38,39]. The receiver side for broadband mm-wave detection usually use heterodyne coherent detection, which is widely used in radio systems. Coherent detection combined with advanced DSP has many advantages, such as high receiver sensitivity, long transmission distance, and large transmission capacity [78–96]. Coherent detection can be divided into homodyne detection and heterodyne detection [97]. Homodyne detection is mainly used for coherent optical communication systems, while heterodyne detection is mainly used for wireless communication systems [98–102]. Fig. 22.18 shows a schematic coherent receiver based on DSP, including two polarization beam splitters (PBSs), two optical 90° hybrids, four balanced PDs (BPDs), and four analog-to-digital converters (ADCs). For homodyne detection, the optical LO and the optical signal has identical carrier frequency, and therefore an electrical baseband signal is generated after BPD detection. For heterodyne detection, the

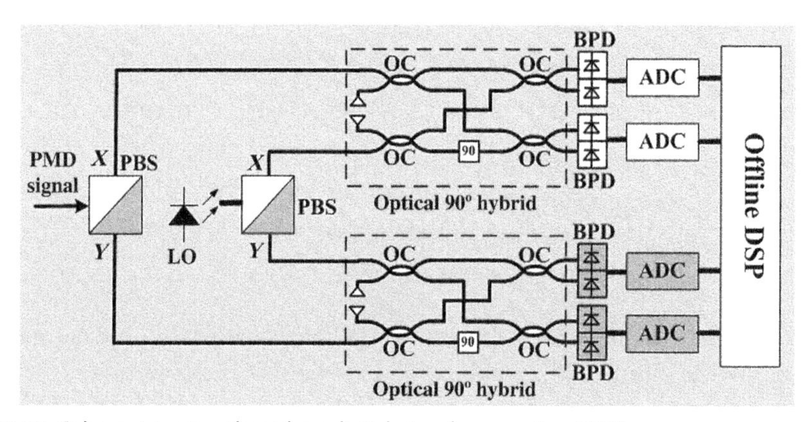

Figure 22.18 Coherent receiver based on digital signal processing (DSP).

optical LO and the optical signal has different carrier frequencies, and therefore an electrical IF signal is generated after BPD detection. The subsequent DSP can further down-convert the electrical IF signal to baseband in the digital domain.

22.4.1 Principle of simplified heterodyne coherent detection based on digital intermediate-frequency down-conversion

Fig. 22.19 shows the schematic simplified heterodyne coherent receiver [103]. Compared to the coherent receiver shown in Fig. 22.18, the simplified heterodyne coherent receiver only needs two PBSs and two optical couplers (OCs) to do optical polarization diversity for the received optical signal and the optical LO, while the number of BPDs and ADCs is also reduced to one half. Moreover, with the aid of high-speed large-bandwidth BPDs and ADCs, the traditional analog down-conversion based on sinusoidal RF signals and electrical mixers in heterodyne coherent receivers can be replaced by DSP-based down-conversion in digital frequency domain.

Fig. 22.20 presents the principle of IF down-conversion in digital frequency domain. The I and Q components centering on the IF are simultaneously received.

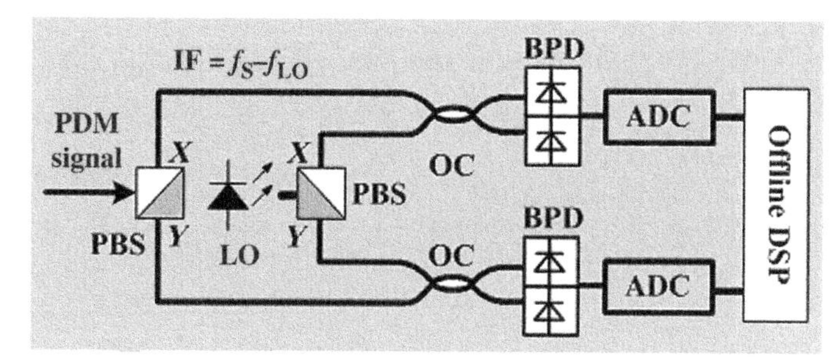

Figure 22.19 Schematic simplified heterodyne coherent receiver.

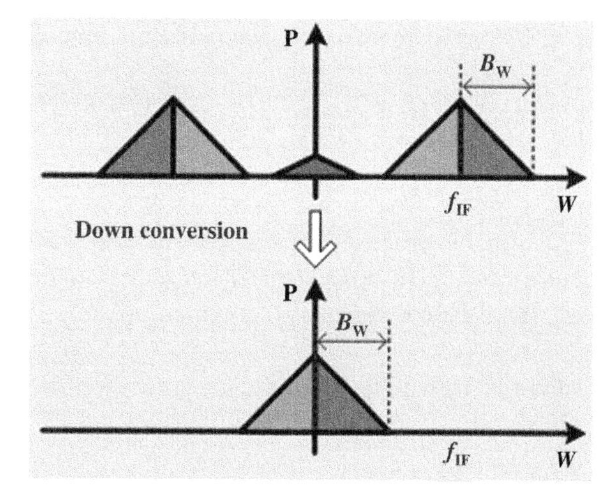

Figure 22.20 Principle of intermediate-frequency (IF) down-conversion in digital frequency domain.

Here f_{IF} is the IF, B_W is the bandwidth of the I or Q component. To separate I and Q components without crosstalk, $f_{IF} \geq B_W$ should be satisfied. Then the received IF signal can be down-converted to baseband in the digital frequency domain by simultaneously multiplying the cosine and sinusoidal signals generated by a digital LO.

After optical polarization diversity, the optical signal at X-polarization can be presented as

$$E_S(t) = \sqrt{P_S}\exp[j2\pi f_S t + \theta_S(t)], \tag{22.29}$$

where P_S, f_S, and θ_S denote optical power, carrier frequency, and phase at X-polarization, respectively.

Assuming that the bandwidth of the PD and the ADC is large enough, the output photocurrent of the BPD at X-polarization can be expressed as

$$\begin{aligned} I_{BPD}(t) &= 4R\sqrt{P_S P_{LO}}\exp\{j[2\pi f_{IF} t + \theta_{IF}(t)]\}; \\ f_{IF} &= f_S - f_{LO}; \\ \theta_{IF}(t) &= \theta_S(t) - \theta_{LO}(t), \end{aligned} \tag{22.30}$$

where R is the conversion efficiency of the PD, while f_{IF} and θ_{IF} are frequency and phase of the IF. The output photocurrent of the BPD at Y-polarization has an expression similar to Eq. (22.30).

After the ADC, a digital LO with a frequency of f_{IF} can provide synchronous cosine and sinusoidal signals to realize DSP-based IF down-conversion, and it can be expressed as

$$E_{LO\_d}(t) = \sqrt{P_{LO\_d}}\exp\{-j[2\pi f_{IF} t + \theta_{LO\_d}(t)]\}, \tag{22.31}$$

where P_{LO_d} and θ_{LO_d} are the power and phase of the digital LO. After multiplying Eq. (22.30) with Eq. (22.31), the down-converted signal can be expressed as

$$\begin{aligned} E_{IF}(t) &= K\sqrt{P_S}\exp\{j[\theta_{IF}(t) - \theta_{LO\_d}(t)]\} = K[\mathrm{I}(t) + j\mathrm{Q}(t)]\exp\{j[\theta_{IF}(t) - \theta_{LO\_d}(t)]\}; \\ K &= 4R\sqrt{P_{LO}P_{LO\_d}}, \end{aligned} \tag{22.32}$$

where $[\theta_{IF}(t) - \theta_{LO\_d}(t)]$ can be processed by subsequent DSP-based carrier recovery algorithms. Compared to analog IF down-conversion based on sinusoidal RF signals and electrical mixers, DSP-based digital IF down-conversion is more hardware-effective.

Similar to the analysis in Ref. [80], the signal-to-noise ratio (SNR) in heterodyne detection can be expressed as

$$SNR_{\mathrm{He}} = \frac{S_{\mathrm{He}}}{N_{\mathrm{He}}} = \frac{2 \times 0.5 I^2{}_{\mathrm{BPD}}}{2B_{\mathrm{W}}\zeta} = 2P_{\mathrm{S}}P_{\mathrm{LO}}R^2/B_{\mathrm{W}}\zeta, \tag{22.33}$$

where ζ represents the noise density of the signal. However, for homodyne detection satisfying $f_{\mathrm{IF}} = 0$, a classical dual-hybrid structure is needed to realize the separation of I and Q components. In this case, the photocurrent after the BPD can be expressed as

$$I_{\mathrm{BPD\_i/q}} = 2R\sqrt{P_{\mathrm{S}}P_{\mathrm{LO}}}\cos\left[\theta_{\mathrm{S}}(t) - \theta_{\mathrm{LO}}(t)\right]. \tag{22.34}$$

And therefore the SNR for homodyne detection is

$$SNR_{\mathrm{Ho}} = \frac{S_{\mathrm{Ho}}}{N_{\mathrm{Ho}}} = \frac{I^2{}_{\mathrm{BPD\_i/q}}}{B_{\mathrm{W}}\zeta} = 4P_{\mathrm{S}}P_{\mathrm{LO}}R^2/B_{\mathrm{W}}\zeta. \tag{22.35}$$

It can be seen that heterodyne detection has a 3-dB SNR penalty compared to homodyne detection.

Overall, compared to homodyne detection, heterodyne detection has a much simpler structure. In addition, heterodyne detection, with the generation of IF signal, can be used for mm-wave generation and high-speed optical wireless transmission and has a great application prospect [104—108].

22.4.2 Equalization algorithm of heterodyne coherent detection

The baseband signal after digital IF down-conversion needs further DSP. Fig. 22.21 shows the algorithm flow diagram of heterodyne coherent detection. First, the received IF signal is digitally down-converted to baseband before I/Q imbalance compensation based on the Gram-Schmidt orthogonal process algorithm [109]. Then the dispersion and nonlinearity [110,111] impairments is compensated. Then the signal clock is recovered to get the best sample point [112]. Then polarization, antenna

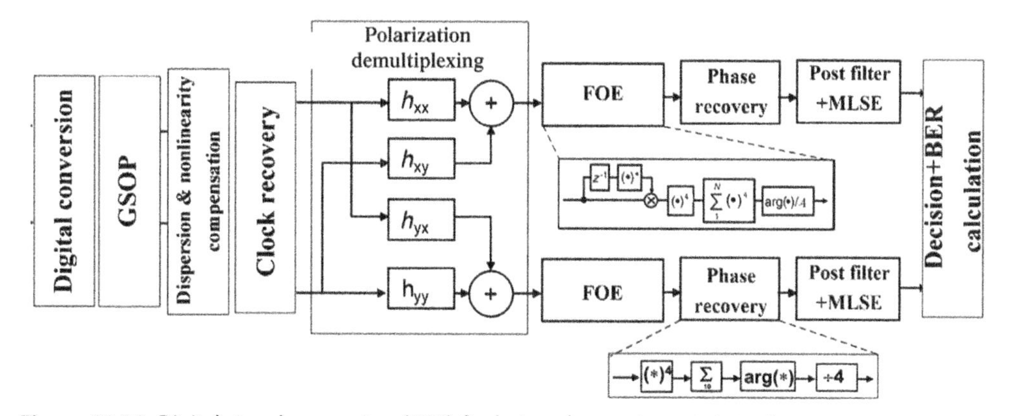

Figure 22.21 Digital signal processing (DSP) for heterodyne coherent detection.

polarization, and PMD are recovered by the CMA algorithm [109–113]. Carrier recovery is divided into two steps. The 4th-power algorithm is used to estimate the frequency offset, and then the Viterbi–Viterbi phase estimation algorithm is used to eliminate the phase noise caused by the laser linewidth. The algorithms for carrier recovery can tolerate larger laser linewidth and frequency jitter. Finally, a digital post filter combined with maximum likelihood sequence estimation is used to eliminate the ISI [92,113] before the decision.

22.4.2.1 Fiber chromatic dispersion compensation

When the optical signal is transmitted in the fiber, the group velocity of different frequency components is not uniform, which can widen the signal in time domain and therefore deteriorate the signal quality. In the practical system, the dispersion will lead to the blur of eye diagram, the disappearance of clock component, the ISI, and so on, which may cause the signal distortion. As a kind of static linear impairment [109,112,114,115], the dispersion is constant for optical signals with a certain baud rate. Therefore dispersion compensation is usually placed in the second step of DSP equalization. DSP-based dispersion compensation can be easily achieved by using time- or frequency-domain transfer function.

The principle of the electrical dispersion compensation in the time domain and frequency domain is shown in Fig. 22.22A and B, respectively [111–113]. According to Ref. [109], the finite-impulse-response (FIR) tap coefficient a_k in the time-domain dispersion compensation is determined by the following equation:

$$a_k = \sqrt{\frac{jcT^2}{D\lambda^2 z}} \exp\left(-j\frac{\pi c T^2}{D\lambda^2 z}k^2\right), \quad -\left\lfloor\frac{N}{2}\right\rfloor dkd \left\lfloor\frac{N}{2}\right\rfloor, \quad N = 2\times\left\lfloor\frac{|D|\lambda^2 z}{2cT^2}\right\rfloor + 1. \tag{22.36}$$

where D is the dispersion coefficient, λ is the signal wavelength, z is the length of the fiber, c is the velocity of light, T is the signal symbol duration, k is the tap number, and N is the tap order. For a certain wavelength and a certain dispersion coefficient, the size of N is determined by the fiber length and symbol duration. Therefore for

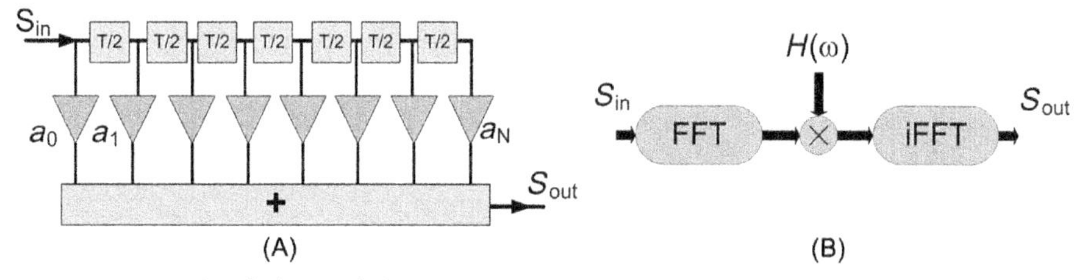

Figure 22.22 Principle of electrical dispersion compensation in (A) time domain and (B) frequency domain.

high-speed or long-distance optical signals, the value of N is larger, and the taps of the FIR filter and the calculation complexity is also increased.

Similarly, according to Ref. [109], the principle of frequency-domain dispersion compensation is relatively simple. According to the dispersion transfer function of the fiber channel, the dispersion can be directly compensated by multiplying the inverse transfer function, which can be expressed as

$$H(w) = \exp\left(\mathrm{j}\frac{D\lambda^2 z}{4\pi c}\omega^2\right). \tag{22.37}$$

22.4.2.2 Clock recovery

As shown in Fig. 22.19, the electrical signal after BPD detection can be converted to a digital signal by the ADC. However, in the practical system, since the local sampling clock is not synchronized with the transmitter signal clock, the sampling points of the ADC are usually not the best ones of the signal [116–119]. Furthermore, the instability of the local clock source may cause the sampling error of the system, which includes both the sampling phase error and the sampling frequency error. On the one hand, the clock error of the sampling points will introduce the ISI due to the imperfect sampling points. On the other hand, the sampling clock jitter can also cause signal performance fluctuations. Therefore in order to realize the best digital signal recovery, the actual system needs to use the clock recovery module to eliminate the impact of the clock error. Considering that the dispersion will cause the disappearance of the clock component, the clock recovery module is usually placed after the dispersion compensation module, or cooperates with the dispersion compensation module to form a uniform equalization feedback module.

Actually, the clock recovery algorithm is a common requirement of communication systems and has been widely used in early wired and wireless communication systems [117–119]. In the optical communication systems, some classical clock recovery algorithms are also widely used, which proves the universality of these algorithms. Nowadays, in the coherent optical communication systems, the main clock recovery algorithms include feedforward and feedback ones, and the typical algorithms are presented as follows.

The square timing recovery algorithm adopts a feedforward clock synchronization structure to realize the clock recovery. This algorithm redefines the clock and finds the best interpolation points by extracting the timing error phase of the signal asynchronous sampling sequence. This algorithm requires a resampling rate at four times of the signal baud rate and the calculation complexity is high.

The Gardner time-domain clock recovery algorithm uses a feedback clock synchronization structure, which estimates the phase of the digital clock source by calculating timing error. The estimation of the timing error can track the frequency jitter of the signal, and therefore this algorithm can achieve dynamic clock recovery. Moreover, the Gardner clock recovery algorithm requires a resampling rate only at twice of the signal baud rate, and the algorithm complexity is low. This algorithm is widely used in the DSP modules of coherent optical communication systems.

22.4.2.3 Polarization demultiplexing and channel dynamic equalization

Since the fiber channel has a birefringence effect, in the coherent optical communication system with polarization multiplexing, polarization rotation occurs due to the randomness of the polarization state. Moreover, since the optical fiber is not perfect, the propagation constants of the two orthogonal polarization modes are different, which directly leads to a time delay between the two polarization modes, that is, the PMD. The PMD is ever-changing in the practical fiber channel, and therefore we need dynamically equalize the channel and complete the polarization demultiplexing at the same time [119–123]. The polarization multiplexing system is actually a 2 × 2 MIMO structure, and therefore its polarization demultiplexing can use classical channel equalization algorithms, such as CMA and decision-directed least mean square (DD-LMS) algorithm.

The CMA algorithm was proposed by Godard et al. in 1980 [119]. As a blind equalization algorithm for QPSK signals, this algorithm can also be applied to polarization multiplexing coherent optical transmission systems. Ref. [122] has given a butterfly adaptive equalization filter structure for coherent optical transmission system with polarization multiplexing, which is showed in Fig. 22.23. This structure contains four adaptive filters to do channel estimation and signal equalization. h_{xx} and h_{yy} denote the equalization for single polarization itself. And the introduction of h_{xy} and h_{yx} takes the effects of polarization selection, polarization coupling, and PMD into consideration. The performance of the CMA algorithm depends on the number of taps and the convergence coefficients. The optimal number of taps and accurate convergence

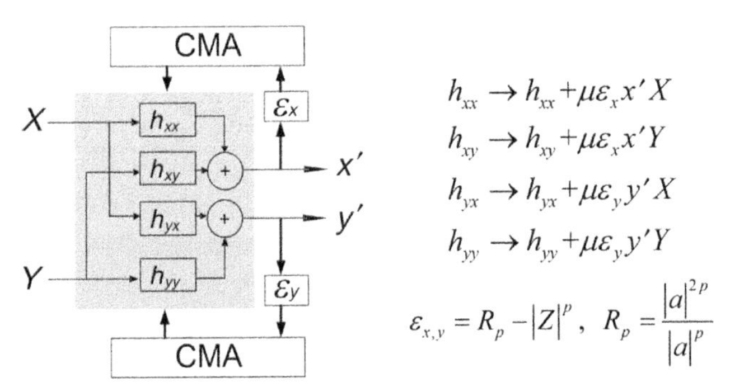

Figure 22.23 The principle of constant-modulus-algorithm (CMA) algorithm and its update functions.

coefficients can be obtained by experimental investigation to reduce the algorithm complexity.

For high-order modulation formats, such as 16QAM, 32QAM, and even higher level QAM, many references also present dynamic equalization algorithms for these modulation formats. It is worth noting that, since the update of the error function of the CMA aims at the single mode of QPSK, the error function cannot converge to zero for higher-level modulation formats, such as 16QAM, and therefore there is a larger error in the channel equalization. X. Zhou et al. proposed a cascaded multimodulus algorithm (CMMA) for higher level QAM, such as 8QAM and 16QAM [109]. This algorithm chooses the best decision value based on the different modulus radius of high-level QAM, which improves the accuracy of error function and makes it converge to zero. Winzer et al. also experimentally verified the application of the radius-directed equalization (RDE) algorithm for 16QAM and higher level QAM [109]. This algorithm carries out a continuous decision of modulus radius when calculating the error function, and finally chooses the modulus radius after decision as the calculation value of the error function. In addition, the commonly used polarization equalization algorithms also include independent component analysis algorithm and so on [121]. DD-LMS can be also used for channel equalization, which needs to integrate carrier frequency and phase recovery into the feedback loop [123]. Since it is more accurate with a slower convergence rate, DD-LMS is usually placed after the first-stage CMA or CMMA preconvergence, as reported in Ref. [123].

22.4.2.4 Carrier recovery

The carrier recovery usually includes frequency offset estimation and phase recovery [114,115,124–127]. In a practical communication system, the LO frequency is not locked with the signal frequency, which may lead to a certain frequency offset from a few MHz to several hundred MHz or even GHz. Moreover, the signal light and the LO light will have a frequency drift effect due to the change of the environmental conditions including temperature. This frequency offset will lead to a large phase rotation of the signal light until covering the phase information of the signal itself. In addition, phase noise is introduced due to the laser linewidth, and it randomly changes at a certain rate, which can cause the tailing, extension, and aliasing of the constellation points. These two kinds of impairments will result in the deterioration of the signal quality.

In general, the carrier and phase recovery algorithms are divided into feedforward and feedback ones. Phase-locking loop is a kind of classical feedback loop [114], which can track the slow change of signal frequency and phase. However, with the increase of the optical transmission rate, the feedback carrier recovery algorithm sometimes cannot quickly track the signal change. The Viterbi–Viterbi algorithm [114] is a classic feedforward carrier phase recovery algorithm. According to the signal modulation format, this algorithm can eliminate the phase information after the Nth power of the signal, leaving only the frequency offset and phase noise. The frequency offset can be extracted and compensated after the correlation operation between neighboring symbols. The average of multiple symbols can reduce the effect of the amplified spontaneous emission noise and improve the accuracy of phase noise estimation.

At present, the Viterbi–Viterbi algorithm has been widely used in carrier and phase recovery for QPSK and mPSK signals. However, for 16QAM and higher-level QAM, the simple Nth power operation cannot completely eliminate the phase modulation of the signal. At the same time, due to the different constellation radius, Nth power operation makes the subsequent direct phase extraction difficult. A constellation partitioning algorithm was proposed to solve this problem [125]. This algorithm divides higher-level QAM constellations according to the radius of the constellation points, and therefore the constellation points with the same radius can be treated as mPSK signals. Therefore the Viterbi–Viterbi algorithm can be used for carrier recovery in this scenario.

For mQAM signal, Pfau et al. proposed a simple and effective feedforward phase recovery algorithm, that is, blind phase search algorithm [124]. This algorithm premultiplies the signal by a recovery phase to do blind phase recovery. Then this algorithm calculates the Euclidean distance between the prerecovery signal and the standard signal, and determines the best recovery phase by searching the smallest Euclidean distance. This algorithm only needs to change the decision function according to different constellation points, and therefore it is suitable for all constellation structures and has been widely used in high-level modulation formats, such as 16QAM, 64QAM, and 1024QAM [115,125–127].

22.4.3 Digital signal processing for orthogonal-frequency-division-multiplexing millimeter -wave signal detection

OFDM, with the advantages of high spectral efficiency and robustness to a variety of dispersions, has recently received much attention on the application into photonics-assisted mm-wave systems [128–131]. However, OFDM has a major disadvantage of high peak-to-average ratio (PAPR). Too high PAPR will affect the normal operation of a variety of components in the photonics-assisted mm-wave systems, and it can be overcome with the aid of advanced DSP [132].

22.4.3.1 Discrete-Fourier-transform spread and intra-symbol frequency-domain averaging

DFT-S can be used to effectively reduce the PAPR, and makes the OFDM modulation display the characteristics of the single-carrier signal [133,134]. The time-domain waveform of the OFDM signal in a symbol duration can be expressed as

$$s(t) = \sum_{k=1}^{N} a_k e^{j2\pi\frac{k-1}{T_s}t}, t \in [0, T_s], \tag{22.38}$$

where N is the number of subcarriers, T_s is the symbol duration, and a_k is the data modulated onto each subcarrier. The PAPR of the OFDM signal is defined as

$$\text{PAPR} = \frac{\max\left\{|s(t)|^2\right\}}{\text{E}\left\{|s(t)|^2\right\}}, t \in [0, T_s]. \tag{22.39}$$

Assume that the power is normalized as follows:

$$\text{E}\left\{|s(t)|^2\right\} = N, t \in [0, T_s] \tag{22.40}$$

$$\begin{aligned} |s(t)|^2 &= \sum_{n=1}^{N}\sum_{k=1}^{N} a_k a_n{}^* e^{j2\pi\frac{k-n}{T_s}t} \\ &= N + 2\text{Re}\left\{\sum_{n=1}^{N-1}\sum_{k=n+1}^{N} a_k a_n{}^* e^{j2\pi\frac{k-n}{T_s}t}\right\} \\ &= N + 2\text{Re}\left\{\sum_{k=1}^{N-1} e^{j2\pi\frac{k-n}{T_s}t}\sum_{n=1}^{N-k} a_{k+n} a_n{}^*\right\}. \end{aligned} \tag{22.41}$$

When z is a complex number, $\text{Re}(z) \le |z|$ and $\left|\sum z_n\right| \le \sum |z_n|$ [135,136]. The PAPR can be written as

$$\begin{aligned} \text{PAPR} &= \frac{N + 2\text{Re}\left\{\sum_{k=1}^{N-1} e^{j2\pi\frac{k-n}{T_s}t}\sum_{n=1}^{N-k} a_{k+n} a_n{}^*\right\}}{N}, \\ &\le 1 + \frac{2}{N}\sum_{k=1}^{N-1} |\rho(k)|, \end{aligned} \tag{22.42}$$

where $\rho(k)$ is the aperiodic autocorrelation coefficient of the IFFT signal, and can be expressed as

$$\rho(k) = \sum_{n=1}^{N-k} a_{k+n} a_n{}^*, k = 1, 2, \ldots, N. \tag{22.43}$$

According to Eq. (22.43), when $\rho(k)$ is smaller, the PAPR of the OFDM signal is smaller. With the aid of the DFT-S matrix, the autocorrelation of the data sequence can be reduced, and therefore the PAPR of the OFDM signal can be reduced.

In the photonics-assisted OFDM mm-wave transmission system, the presence of random noise can reduce the accuracy of channel estimation. To eliminate the influence of random noise, several training sequences (TSs) are usually inserted into the transmission data, and then at the receiver, the channel response of each TS is calculated before the averaging operation. However, since this method uses quite a few of known assisted data, the system spectral efficiency will be decreased. To eliminate the random noise without increasing the amount of assisted data, the technique of ISFA is proposed to improve the channel estimation accuracy [137–139].

In the traditional OFDM systems, the frequency response H_k of the kth subcarrier is usually estimated according to the TS. After introducing the technique of ISFA, the optimized frequency response H_{ISFA} of the kth subcarrier is the average of the frequency response of $(2m + 1)$ subcarriers, which include m subcarriers before the kth subcarrier, the kth subcarrier, and m subcarriers after the kth subcarrier. That is,

$$H_{ISFA} = \frac{\sum_{k'=k-m}^{k+m} H_{k'}}{\min\left(k'_{\max}, k+m\right) - \max\left(k'_{\min}, k-m\right) + 1}, \tag{22.44}$$

where $k'_{\max}$ and $k'_{\min}$ is the maximum and minimum subcarrier index within one OFDM symbol duration, respectively. If k' is larger than $k'_{\max}$ or smaller than $k'_{\min}$, the value of $H_{k'}$ is defaulted as zero. A larger ISFA tap number is more beneficial to the suppression of random noise, but it will decrease the subcarrier correlation and therefore decrease the accuracy of the channel estimation. Thus, the choice of the ISFA tap number should balance the noise suppression and the subcarrier correlation.

With the aid of the advanced DSP-based DFT-S and ISFA techniques, we experimentally demonstrated an OFDM mm-wave system adopting optical polarization-division multiplexing (PDM) combined with MIMO reception, which can deliver 30.67-Gb/s PDM OFDM signal through 40-km fiber and 2-m 2×2 MIMO wireless link at 100 GHz. The experimental results verified that, both the DFT-S technique with the reduction of the PAPR and the ISFA technique with the optimization of the channel estimation, can improve the system transmission performance [140].

22.4.3.2 Volterra equalizer in direct detection of orthogonal-frequency-division-multiplexing millimeter -wave signal

The OFDM mm-wave system suffers from linear and nonlinear effects. The nonlinear effects can severely degrade system performance compared to linear effects. The sources of nonlinear effects mainly include the nonlinearity of the optoelectronic

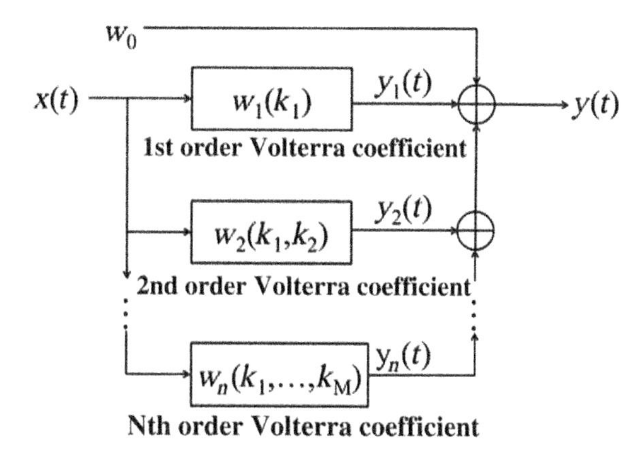

Figure 22.24 Nonlinear equalizer based on the Volterra series.

devices, the nonlinear effects of the transmission fiber, and the nonlinearity generated by the square-law PD detection. These three kinds of nonlinear effects all will cause the system to produce the square term, even the third, quadratic and other higher order terms, of the signal data. The nonlinear equalizer can well suppress the nonlinear effects of the OFDM mm-wave system. The nonlinear equalizer based on the Volterra series is a hot spot with great potential for development. The Volterra series is a generic function similar to the Taylor series, and it is used in the nonlinear field. The output of the system based on the Taylor series only depends on the current input. However, the Volterra series have an ability of memory, and the output of the system based on the Volterra series depends on not only the input of the current time, but also the input of other time.

The structure of the adaptive nonlinear equalizer based on the Volterra series is shown in Fig. 22.24, and it can be expressed as

$$
\begin{aligned}
y(t) = w_0 &+ \sum_{k=0}^{M-1} w_1(k)x(t-k) + \sum_{k_1=0}^{M-1}\sum_{k_2=k_1}^{M-1} w_2(k_1,k_2)x(t-k_1)x(t-k_2) \\
&+ \cdots + \sum_{k_1=0}^{M-1}\cdots\sum_{k_n=k_{n-1}}^{M-1} w_n(k_1,\ldots,k_n)x(t-k_1)\ldots x(t-k_n),
\end{aligned}
\tag{22.45}
$$

where $y(t)$ represents the equalizer output, $x(t)$ represents the equalizer input, M is the memory length, and $(w_0, w_1, w_2, \cdots w_n)$ is the equalizer parameters. In order to balance the output accuracy of the adaptive equalizer and the computational complexity of the system, we usually keep the terms of the Volterra series expansion until the second-power, with higher-power terms discarded.

The received discrete multitone (DMT) signal after square-law PD detection can be expressed as

$$\begin{aligned} R_i &= \left|\sqrt{S_i + P_0} \otimes h\right|^2 \\ &\approx \sum_{k=-N}^{N} |h_k|^2 S_{i-k} + \sum_{l=-N}^{N} \sum_{k=-N}^{l-1} \mathrm{Re}(h_k h_l^*)(S_{i-l} + S_{i-k}) \\ &- \frac{1}{4P_0} \sum_{l=-N}^{N} \sum_{k=-N}^{l-1} \mathrm{Re}(h_k h_l^*)(S_{i-l} - S_{i-k})^2, \end{aligned} \tag{22.46}$$

where R_i represents the received signal, S_i is the transmitted DMT signal, P_0 is the average power determined by the bias voltage, h_k is the channel response, and N is one half of symbol length. i, k, and l represent the subcarrier index. Eq. (22.46) keeps only the first- and second-power terms. The first-power term represents the linear noise, and the second-power term represents the nonlinear noise.

According to Eq. (22.46), Ref. [141] further proposed a nonlinear equalizer for DMT signal to compensate for linear and nonlinear noise in the system. This equalizer includes two adaptive FIR filters. One is used to compensate for the linear noise and the other is used to compensate for the nonlinear noise caused by subcarrier-subcarrier beating interference (SSBI). The expression of this Volterra equalizer is

$$y_i = \sum_{k=-N}^{N} c_k x_{i-k} + \sum_{l=-N}^{N} \sum_{k=-N}^{l-1} h_{l,k}(x_{i-l} - x_{i-k})^2, \tag{22.47}$$

where x and y respectively represent the input and output signals of the equalizer. N is the tap number. The memory length is $(2N+1)$. c_k and $h_{l,k}$ represent the weight parameters of the linear filter and the nonlinear filter, respectively. DMT modulation ensures that the transmitted signal is real. So x, y, c_k, and $h_{l,k}$ are also all real.

Fig. 22.25 shows the structure of the Volterra equalizer when $N=1$. The weight parameters of the linear and nonlinear filters are both 3. However, when the weight

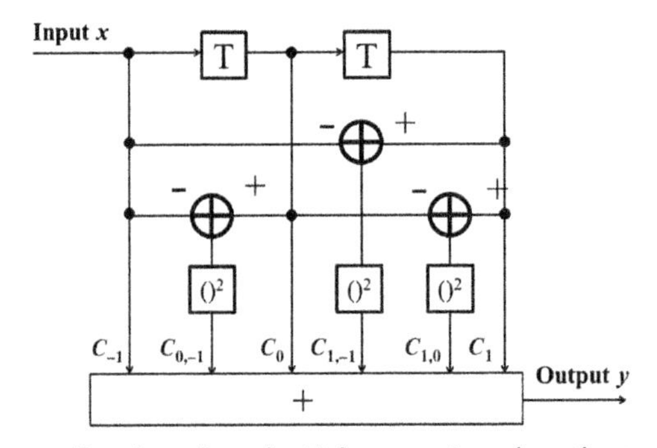

Figure 22.25 Nonlinear equalizer based on the Volterra series when the tap number is 1.

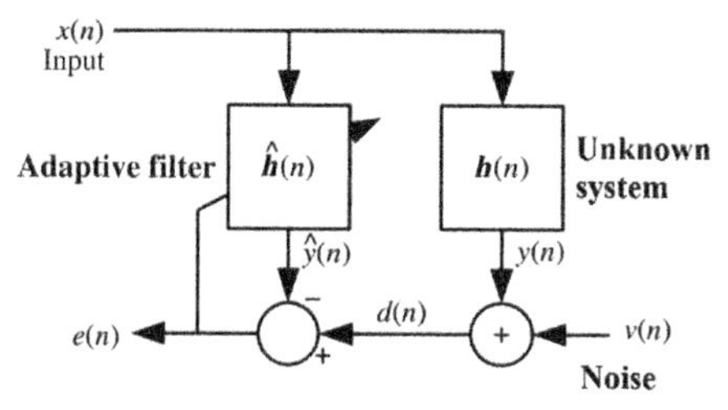

Figure 22.26 Principle of least mean square (LMS) filter.

parameter of the linear filter is $2N+1$, the weight parameter of the nonlinear filter is $N(2N+1)$. Therefore when N is large, the computation complexity will depend on $O(N^2)$. The nonlinear parameter will be huge and the equalizer structure will become more complicated. So, the size of N needs to be controlled in the practical implementation.

The linear-term and nonlinear-term coefficient of the adaptive equalizer can be obtained by the least-mean-square (LMS) algorithm. The LMS algorithm is a classical algorithm for solving the coefficients of adaptive filters. The criterion of this algorithm is to ensure that the error between the theoretical and actual values has the smallest root-mean-square value. The cost function is the mean-square error. This algorithm utilizes the principle of stochastic gradient descent to adjust the weight coefficients.

Fig. 22.26 shows the principle of the LMS filter. Here, x is the input, y is the output, h is the channel response, and v is the noise. $\hat{y}$ and $\hat{h}$ represent the estimated output values and channel responses, respectively. The system can be expressed as

$$x(n) = \left[x_0(n), x_1(n), \ldots, x_{p-1}(n-p+1)\right]^T; \tag{22.48}$$

$$h(n) = \left[h_0(n), h_1(n), \ldots, h_{p-1}(n)\right]^T; \tag{22.49}$$

$$y(n) = h^H(n) \cdot x(n); \tag{22.50}$$

$$d(n) = y(n) + v(n); \tag{22.51}$$

$$e(n) = d(n) - \hat{y}(n) = d(n) - \hat{h}^H(n) \cdot x(n), \tag{22.52}$$

where n is the nth moment and p is the tap number. $\{\cdot\}^H$ represents the conjugate transpose operation and e is the error between the theoretical value and the actual value.

The instant cost function of the system is

$$C(n) = E\left\{|e(n)|^2\right\}, \tag{22.53}$$

where $E\{\cdot\}$ represents the expectation operation. The partial derivative of this function can be expressed as

$$\frac{\partial C(n)}{\partial \hat{h}^H} = \frac{\partial E\{e(n)e*(n)\}}{\partial \hat{h}^H} = 2E\left\{\frac{\partial e(n)}{\partial \hat{h}^H} e*(n)\right\}. \tag{22.54}$$

After putting Eq. (22.52) into Eq. (22.54), we can obtain

$$\frac{\partial C(n)}{\partial \hat{h}^H} = 2E\left\{\frac{\partial\left\{d(n) - \hat{h}^H(n)\cdot x(n)\right\}}{\partial \hat{h}^H} e*(n)\right\} = -2E\{x(n)e*(n)\}. \tag{22.55}$$

The estimated channel response coefficient is adjusted as

$$\hat{h}(n+1) = \hat{h}(n) - \frac{\mu}{2}\frac{\partial C(n)}{\partial \hat{h}^H} = \hat{h}(n) + \mu E\{x(n)e*(n)\}, \tag{22.56}$$

where μ represents the size of adjustment step.

For most systems, the expectation function needs to take an approximate estimation as

$$\hat{E}\{x(n)e*(n)\} = \frac{1}{N}\sum_{i=0}^{N-1} x(n-i)e*(n-i), \tag{22.57}$$

where N represents the number of samples for estimation. Since the LMS only utilizes the current sample to adjust the weight coefficients, $N=1$ in Eq. (22.57) can be used to obtain the estimated channel response

$$\hat{h}(n+1) = \hat{h}(n) + \mu x(n)e*(n). \tag{22.58}$$

Therefore the calculation steps of the LMS adaptive algorithm can be summarized as follows: Set the tap length to p. Set the size of adjustment step to μ. Set $\hat{h}(0)$ as a vector with a length of p and values of 0. Then for the different values of n, we cycle the following calculation process from the minimum value of n until the maximum value of n

$$x(n) = \left[x_0(n), x_1(n), \ldots, x_{p-1}(n-p+1)\right]^T; \tag{22.59}$$

$$e(n) = d(n) - \hat{h}^H(n)\cdot x(n); \tag{22.60}$$

$$\hat{h}(n+1) = \hat{h}(n) + \mu x(n)e*(n). \tag{22.61}$$

When applying the LMS algorithm into the aforementioned nonlinear equalizer for the DMT signal to compensate for the linear and nonlinear noises in the system, the iterative formula for the LMS algorithm is as follows:

$$y_i = \sum_{k=-N}^{N} c_k x_{i-k} + \sum_{l=-N}^{N} \sum_{k=-N}^{l-1} h_{l,k}(x_{i-l} - x_{i-k})^2; \tag{22.62}$$

$$\varepsilon_i = d_i - y_i; \tag{22.63}$$

$$c_k = c_k + \mu_c \varepsilon_i x_i \quad k = -N, -N+1, \ldots, N; \tag{22.64}$$

$$h_{l,k} = h_{l,k} + \mu_h \varepsilon_i (x_{i-l} - x_{i-k})^2 \;\; k = -N, -N+1, \ldots, l-1; \;\; l = -N, -N+1, \ldots, N, \tag{22.65}$$

where x and y represent the input and output signals of the equalizer, respectively. N is the tap number. c_k and $h_{l,k}$ represent the weight coefficients of the linear and nonlinear filters, respectively. ε_i is the error between the theoretical and actual values. μ_c and μ_h represent the adjustment step size of linear and nonlinear equalizers, respectively. Eqs. (22.64)–(22.66) represents the process of minimizing the root-mean-square error of the cost function, by adjusting the weight coefficients of linear and nonlinear filters based on the LMS algorithm.

Employing the aforementioned Volterra nonlinear equalizer, we experimentally demonstrated a 2-GHz 16QAM-OFDM RoF system with 4-m wireless link [142]. The experimental results show that the Volterra nonlinear equalizer can improve the system performance by about 1-dB receiver sensitivity.

In addition, the preemphasis technique [143] and the companding transform technique [130] can also be applied into the OFDM mm-wave signal system to improve the system performance. The preemphasis technique can overcome the subcarrier-subcarrier mixing interference and frequency fading caused by dispersion [143], while the companding transform technique can reduce the PAPR of the OFDM signal [130].

22.5 Broadband millimeter -wave delivery

Large-capacity (>100-Gb/s) wireless transmission system can be effectively realized by using multidimensional multiplexing techniques [33,131,136,144–148], including optical PDM combined with MIMO reception [33], antenna polarization multiplexing [149], advanced multilevel modulation, optical multicarrier modulation, and multiband multiplexing [131,136,144–149]. These techniques can effectively overcome the limitation of bandwidth-insufficient optical and electrical devices and simultaneously reduce the signal baud rate. Table 22.1 summarizes the experimentally demonstrated

Table 22.1 Experimentally demonstrated large-capacity (> 100 Gb/s) wireless mm-wave signal transmission.

Wireless capacity (Gb/s)	Wireless distance (m)	Mm-wave band	Carrier frequency (GHz)	Modulation format	BER threshold	Applied techniques and references
100	2	W-band	95	PDM-QPSK	3.8×10^{-3}	T1, T2 [150,151]
108	1	W-band	100	PDM-QPSK	3.8×10^{-3}	T1, T2 [33]
112	0.5	Q-band	35	PDM-16QAM	3.8×10^{-3}	T1, T2 [108]
112	2	Q-band	37.5	PDM-QPSK	3.8×10^{-3}	T1, T2, and T3 [149]
120	1	Q-band	37.5	PDM-64QAM	3.8×10^{-3}	T1, T2 [152]
120	1	V-band	57.2	PDM-64QAM	3.8×10^{-3}	T1, T2 [153]
120	2	W-band	79.5, 92, and 104.5	PDM-QPSK	3.8×10^{-3}	T1, T2, and T4 [154]
128	1	W-band	100	PDM-16QAM	2×10^{-2}	T1, T2 [38]
128	0.1	D-band	137.5	PDM-16QAM	2×10^{-2}	T1, T2 [155]
144	2	W-band	100	PDM-QPSK	3.8×10^{-3}	T1, T2, and T3 [73,156]
184	0.6	D-band	137.5	PDM-QPSK	2×10^{-2}	T1, T2 [155]
216	0.7	W-band	100	PDM-QPSK	3.8×10^{-3}	T1, T2, T3, and T4, [131]
224	1.5	Q-band	37.5	PDM-16QAM	3.8×10^{-3}	T1, T2, T3, and T4, [131]
352	3.1	D-band	140	PDM-64QAM-PS5.5	4×10^{-2}	T1, T2 [157]
412	0.6	Q-, V-, W-band	62.5, 97.5, 137.5	PDM-QPSK	2×10^{-2}	T1, T2, T3, and T5 [158]
432	2	W-band	94	PDM-16QAM	3.8×10^{-3}	T1, T2, and T3 [39]
440	1	Q-, W-band	37.5, 100	PDM-QPSK PDM-16QAM	3.8×10^{-3}	T1, T2, T3, and T5 [131]
1056	3.1	D-band	124.5, 150.5	PDM-64QAM-PS5.5	4×10^{-2}	T1, T2, T3, and T4 [157,159]

Note: *T1*, MIMO; *T2*, optical polarization multiplexing; *T3*, antenna polarization multiplexing; *T4*, optical multicarrier modulation; *T5*, multiband multiplexing.

large-capacity (> 100-Gb/s) wireless mm-wave signal transmission based on the aforementioned techniques.

22.5.1 Multiple-input multiple-output for millimeter-wave signal delivery

MIMO wireless transmission technique, adopting multiple transmitting and receiving antennas, can effectively increase wireless transmission capacity on the premise that no additional spectrum resource is required. Also, due to the diversity gain caused by MIMO, relatively small transmitter power is needed for each transmitting antenna, thus avoiding the use of expensive wide-range electrical amplifier. The MIMO technique, combined with space-division multiplexing [149,157,160], PDM [33,131], and wavelength-division multiplexing (WDM), can further improve the system performance and achieve the mm-wave signal transmission with high spectral efficiency.

Fig. 22.27 shows the architecture of mm-wave signal transmission over fiber-wireless 2 × 2 MIMO system using optical PDM. The optical PDM signal is received by optical heterodyne up-converter, composed of a LO laser, two PBSs, two OCs, and two PDs. The frequency offset between the PDM signal and LO laser is set within the mm-wave band to generate the mm-wave signal. Two PBSs and two OCs are used to implement polarization diversity of the received signal and LO laser in optical domain. It is worth noting that each output port of the PDs contains both the *X*- and *Y*- polarization components of the PDM signal. In the following part, we define the two outputs of the PBS as *X*- and *Y*-components just for simplification. After heterodyne beating, the PDM mm-wave signal is injected into a 2 × 2 MIMO link that includes two pairs of transmitting and receiving HAs. Note that the relative location of HAs can be adjusted based on different conditions. In this case, both the transmitting HAs and the receiving HAs are deployed very close at each side, and thus full crosstalk exists in the 2 × 2 MIMO link. At the wireless receiver, two-stage down conversion is implemented for the *X*- and *Y*-polarization components of PDM signal. In the first stage, the mm-wave signal is down-converted to a lower IF based on

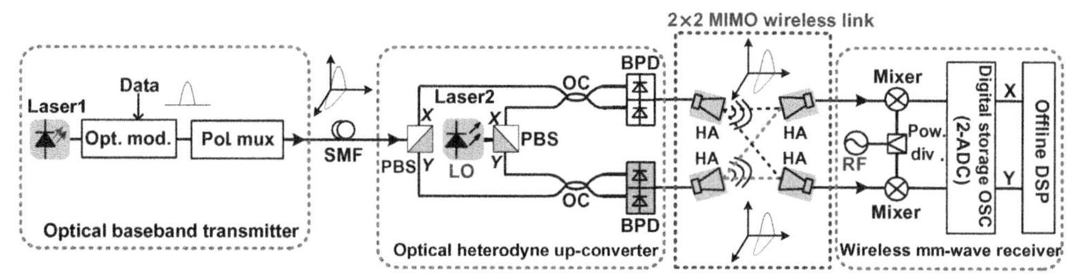

Figure 22.27 Architecture of wireless millimeter (mm)-wave signal transmission over fiber-wireless 2 × 2 multiple-input multiple-output (MIMO) system using optical polarization-division multiplexing (PDM).

balanced mixer and sinusoidal RF signal. Then the second-stage down-conversion and final data recovery is conducted by offline DSP.

In RoF system, the optical polarization demultiplexing and 2 × 2 MIMO demultiplexing can be simultaneously solved by CMA [109]. In other words, the PDM signal can be taken as another MIMO system because it has the similar formula with 2 × 2 MIMO. To clarify this point, we define E_x and E_y as *X*- and *Y*-polarization components of the PDM signal. After fiber transmission and heterodyne detection, the signal is expressed as

$$\begin{pmatrix} E^1{}_x \\ E^1{}_y \end{pmatrix} = \begin{pmatrix} J_{xx} & J_{yx} \\ J_{xy} & J_{yy} \end{pmatrix} \begin{pmatrix} E_x \\ E_y \end{pmatrix} \cos \omega_W t = T \begin{pmatrix} E_x \\ E_y \end{pmatrix} \cos \omega_W t, \tag{22.66}$$

where T is the Jones matrix, $J_{x/y}$ is the element of the matrix T, and ω_W is the mm-wave frequency. After the 2 × 2 MIMO link, the received signal can be formulated as

$$\begin{pmatrix} E^2{}_x \\ E^2{}_y \end{pmatrix} = W \begin{pmatrix} E^1{}_x \\ E^1{}_y \end{pmatrix} = \begin{pmatrix} W_{11} & W_{21} \\ W_{12} & W_{22} \end{pmatrix} \begin{pmatrix} J_{xx} & J_{yx} \\ J_{xy} & J_{yy} \end{pmatrix} \begin{pmatrix} E_x \\ E_y \end{pmatrix} \cos \omega_W t = H \begin{pmatrix} E_x \\ E_y \end{pmatrix} \cos \omega_W t, \tag{22.67}$$

where W is 2 × 2 MIMO transfer function matrix and $W_{x/y}$ is the element of the matrix W. As we can see, the procedure of fiber transmission and 2 × 2 MIMO wireless transmission can be simplified as a matrix H

$$H = WT = \begin{pmatrix} W_{11}J_{xx} + W_{21}J_{xy} & W_{11}J_{yx} + W_{21}J_{yy} \\ W_{12}J_{xx} + W_{22}J_{xy} & W_{12}J_{yx} + W_{22}J_{yy} \end{pmatrix} = \begin{pmatrix} h_{11} & h_{21} \\ h_{12} & h_{22} \end{pmatrix}, \tag{22.68}$$

where $h_{1/2}$ is the elements of 2 × 2 fiber-wireless MIMO transfer function matrix H.

Therefore in order to obtain E_x and E_y, we can estimate the 2 × 2 fiber-wireless MIMO transfer function matrix H by adopting CMA, a four-butterfly adaptive digital equalizer used for polarization and wireless MIMO demultiplexing. Based on the architecture given in Fig. 22.27, 108-Gb/s PDM-QPSK data is delivered over 80-km fiber and 1-m 2 × 2 MIMO wireless link at 100 GHz [33].

The MIMO techniques are also potentially compatible with antenna horizontal (H-) and vertical polarization (V-polarization) multiplexing [149]. The antenna H- and V-polarization are orthogonal polarization states that have large isolation. As shown in Fig. 22.28, the near location of the transmitting HAs and receiving HAs at each side gives rise to full crosstalk over the 2 × 2 MIMO link, which makes signal recovery more difficult. However, the full crosstalk can be largely eliminated if antenna polarization multiplexing is taken into consideration.

Fig. 22.28 depicts the architectures of single antenna polarization and antenna polarization diversity over a 2 × 2 MIMO wireless link. In Fig. 22.28A, each receiving

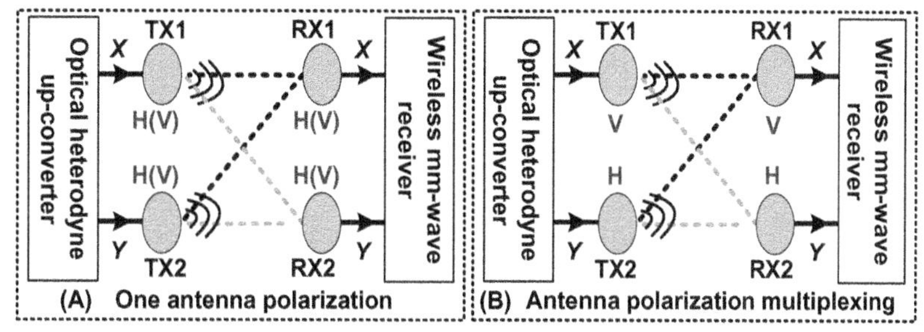

Figure 22.28 2 × 2 MIMO wireless link: (A) single antenna polarization and (B) antenna polarization diversity. *MIMO*, Multiple-input multiple-output.

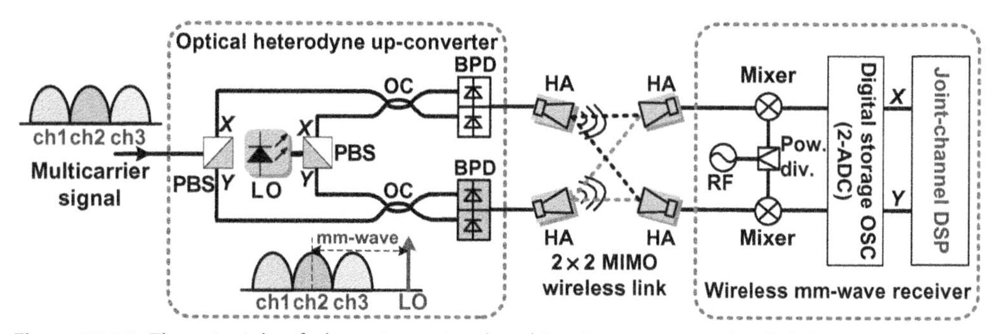

Figure 22.29 The principle of photonics-assisted multicarrier mm-wave signal delivery.

HA is able to receive the signal power from both transmitting HAs, thus leading to wireless crosstalk. Moreover, with the increase of wireless transmission distance, wireless crosstalk becomes increasingly serious and it is more difficult to deploy the receiving HAs for accurate signal recovery. By contrast, in Fig. 22.28B, each receiving HA can only detect the signal from the transmitting HA with the same antenna polarization state, thus avoiding the existence of wireless crosstalk. Based on the architecture given in Fig. 22.28B, Ref. [156] demonstrates 146-Gb/s PDM-QPSK data transmission over 80-km fiber and 2-m 2 × 2 MIMO wireless link at 100 GHz.

22.5.2 Multicarrier millimeter -wave signal delivery

Multicarrier technique is another popular approach to provide multigigabit mobile data transmission for mm-wave signal delivery. Adopting multicarrier technique can reduce the signal baud rate and bandwidth requirement for optical and wireless devices. Moreover, multicarrier technique has a better tolerance to the spectrum fading effect, and its introduction into the mm-wave signal delivery can further improve the performance and increase the data rate.

Fig. 22.29 gives the principle of the photonics-assisted multicarrier mm-wave signal delivery. Here, we take a dense WDM (DWDM) signal with three channels, that is, ch1, ch2, and ch3, as an example. We define the frequency spacing between LO and ch2 as the wireless mm-wave carrier frequency. The optical heterodyne up-conversion and wireless mm-wave reception for the multicarrier case is quite similar to that for the aforementioned single-carrier case. The generated digital signal with full channel information after analog-to-digital conversion is processed by joint-channel DSP [154]. Based on the architecture given in Fig. 22.29, three-channel 3×40-Gb/s data is delivered over 80-km fiber and 2-m 2×2 MIMO wireless link at 92 GHz [154].

22.5.2.1 Multiband millimeter-wave signal delivery

The large-capacity wireless signal can be transmitted at K-, Q-, V-, W-, and D-band [150,152–155,158,161–164]. Multiband multiplexing can also effectively reduce the signal baud rate and performance requirements for optical and wireless devices at the cost of more antennas and devices. In addition, advanced multilevel modulation, such as 16QAM and 64QAM, can make each symbol carry more bits and further reduce the signal baud rate with higher receiver sensitivity. Table 22.2 shows the optical wireless transmission systems at different mm-wave bands with different multilevel modulation formats and advanced DSP algorithms [150,152–155,158,161–164]. In Ref. [163] it was experimentally demonstrated that 32-Gb/s PDM-16QAM data is transmitted through 100-km fiber and $>$1-km 2×2 MIMO wireless link at 23-GHz K-band, and the BER is less than 3.8×10^{-3}. In Ref. [152], a large-capacity mm-wave transmission system at Q-band was experimentally demonstrated, and up to 10-Gbaud (120-Gb/s) PDM-64QAM signal at 37.5 GHz can be generated and delivered over a 1-m wireless link, with a BER less than 3.8×10^{-3}. Within V-band, it was experimentally demonstrated that 120-Gb/s PDM-64QAM mm-wave signal at 57.2 GHz can be generated with a BER less than 3.8×10^{-3} [153]. Using lower frequency mm-wave bands, such as K-band, Q-band, and V-band, can avoid analog down conversion, but at the expense of the reduction of available signal bandwidth. Within D-band, it was experimentally demonstrated that up to 184-Gb/s

Table 22.2 Optical wireless transmission within different mm-wave bands and different multilevel modulation formats.

Mm-wave band	Frequency range (GHz)	Carrier frequency (GHz)	PDM-QPSK (Gb/s)	PDM-16QAM (Gb/s)	PDM-64QAM (Gb/s)
K	18–27	23	–	32	–
Q	33–50	37.5	–	–	120
V	50–75	60	–	–	120
W	75–110	94	–	432	–
D	110–170	137.5	184	–	128

Note: The transmission rates refer to the HD-FEC threshold of 3.8×10^{-3}.

PDM-QPSK signal is transmitted over 80-km fiber and 0.6-m 2 × 2 MIMO wireless link at D-band and up to 128-Gb/s PDM-16QAM signal is transmitted over 10-cm 2 × 2 MIMO wireless link at D-band [164].

22.5.3 Advanced multilevel modulation

Wireless transmission distance and spectral efficiency cannot be met at the same time. To use high-order QAM, we can get high spectral efficiency because each baud can carry more bits in high-order QAM signals. However, high-order QAM mm-wave signals need more input power or SNR at the receiver side because it has smaller Euclidean distance between adjacent constellation points. If we need longer wireless delivery, we should use low-order QAM such as QPSK or 8QAM [163].

By adopting multidimensional multiplexing techniques, an optical wireless transmission system was experimentally demonstrated, and it can simultaneously deliver 2 × 112-Gb/s two-channel PDM-16QAM wireless signal at 37.5 GHz and 2 × 108-Gb/s two-channel PDM-QPSK wireless signal at 100 GHz through 1.5-m and 0.7-m 4 × 4 MIMO wireless link, respectively [131]. As shown in Fig. 22.30, both Q-band and W-band, two orthogonal antenna polarizations, and advanced DSP algorithms are employed to realize multidimensional multiplexing large-capacity transmission. The wireless multipath interference can be removed completely if the CMA taps based on DSP are appropriately selected.

In Ref. [37], a 4 × 4 MIMO was employed to realize up to 432-Gb/s PDM-16QAM modulated W-band wireless signal delivery. The experimental setup is shown in Fig. 22.31. 4 × 4 W-band horn antenna (HA) array based on antenna polarization multiplexing is used to deliver mm-wave signals. The wireless distance between the transmitter and receiver is 2 m.

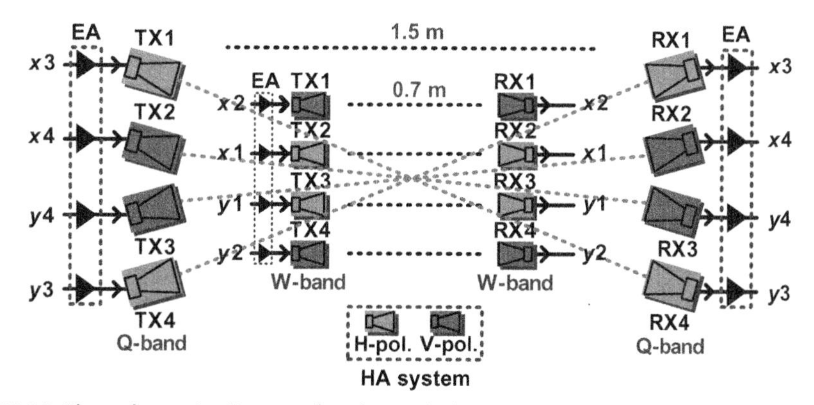

Figure 22.30 The schematic diagram for the multidimensional multiplexing horn antenna (HA) transmission system

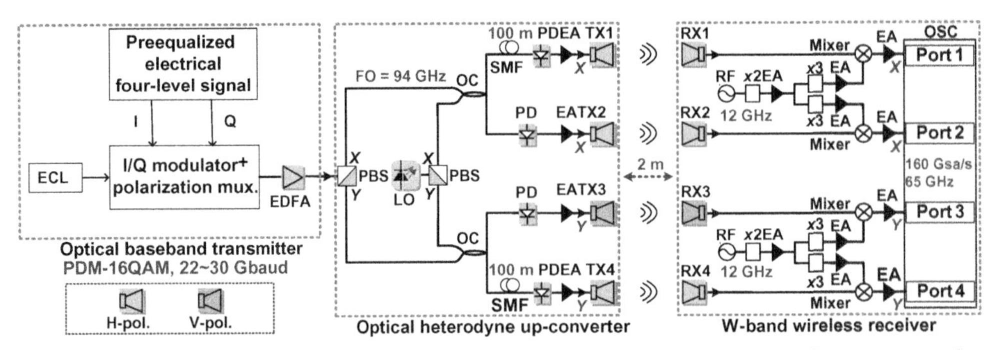

Figure 22.31 Experimental setup for up to 432-Gb/s polarization-division multiplexing-16 quadrature amplitude modulation (PDM-16QAM) modulated wireless signal delivery at W-band.

The frequency range of D-band is from 110 to 170 GHz [155]. It has a wider bandwidth relative to W-band. In Ref. [157], over 1-Tb/s vector signal delivery over the fiber-wireless integration transmission link at D-band has been reported. The experimental setup is shown in Fig. 22.32. The modulation format is Probabilistic shaping (PS) 64QAM [157], the fiber transmission distance is 10 km, and the wireless transmission distance is 3.1 m. It simultaneously employs two different D-band mm-wave carrier frequencies, that is, 124.5 and 150.5 GHz. The employment of advanced DSP techniques, including PS, Nyquist shaping, and look-up-table algorithm, significantly improves the transmission capacity and system performance. Two dual-subcarrier PDM-64QAM-PS5.5 modulated mm-wave signals, with a total baud rate of $2 \times 2 \times 24 = 96$ Gbaud and a total bit rate of $96 \times 5.5 \times 2 = 1.056$ Tb/s, can be delivered over 3.1-m wireless distance with a BER under the soft-decision forward-error-correction (SD-FEC) threshold of 4×10^{-2}. After removing the SD-FEC and PS overhead, the corresponding net bit rate is 762.2 Gb/s.

22.6 Long-distance millimeter-wave transmission in the radio-over-fiber system

In addition to large-capacity wireless mm-wave transmission, long-distance wireless mm-wave transmission is also required to meet the increasing mobile communication demands. We have investigated long-distance mm-wave transmission systems at both K-band and W-band based on advanced devices and enabling technologies, mainly including high-gain small-beamwidth Cassegrain antennas (CAs), large-gain/high-power W-band electrical amplifiers, wideband optical/electrical components, optical PDM combined with MIMO reception, antenna polarization diversity, and advanced DSP algorithms [149,150,152–154,156–158,160–164]. The CMA equalization based on DSP can be used at the wireless receiver to simultaneously implement PDM signal polarization demultiplexing and wireless crosstalk suppression. Table 22.3 summarizes

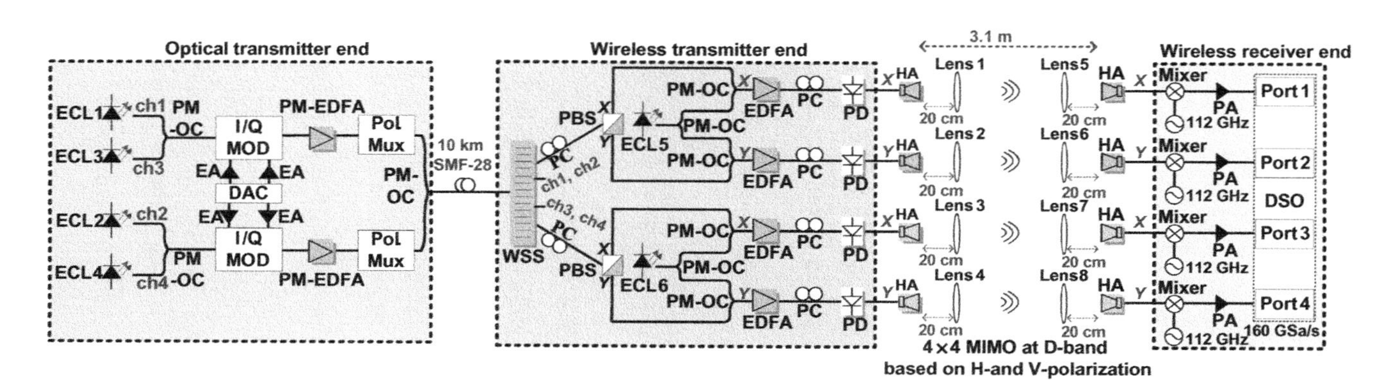

Figure 22.32 Experimental setup of > 1-Tb/s vector signal delivery over the fiber wireless-integration transmission link at D-band.

Table 22.3 Experimentally demonstrated long-distance (>100-m) wireless millimeter (mm)-wave signal transmission.

Wireless capacity (Gb/s)	Wireless distance (km)	Product of capacity and distance (Gb/s km)	Mm-wave band	Carrier frequency (GHz)	Modulation format	BER threshold and references
40	0.16	6.4	W-band	85.5	PDM-QPSK	3.8×10^{-3} [165]
100	0.1	10	W-band	96	PDM-QPSK	2×10^{-2} [166]
80	0.3	24	W-band	85.5, 95.5	PDM-QPSK	3.8×10^{-3} [164]
32	1	32	K-band	23	PDM-16QAM	3.8×10^{-3} [163]
20	1.7	34	W-band	85.5	PDM-QPSK	3.8×10^{-3} [74,162]
32	2.5	80	K-band	23	PDM-16QAM	3.8×10^{-3} [161,167]
54	2.5	135	W-band	94	PDM-8QAM	3.8×10^{-3} [161,167]

the experimentally demonstrated long-distance (>100-m) wireless mm-wave signal transmission based on the aforementioned techniques.

In mm-wave communication system, the type and parameter of antennas plays an important role in communication performance. Mm-wave antenna usually has a relatively high gain. In practical applications, the antenna gain is usually the gain in the direction of the maximum radiation. The antenna gain can be expressed as [168]

$$G=\frac{4\pi A_e}{\lambda^2}, \tag{22.69}$$

where A_e is the antenna effective area, that is, the ratio of the received power to the power density of the incident field, and $\lambda = c/f$ is the signal wavelength (c is the velocity of light, and f is the frequency of the signal). From Eq. (22.69), we can know that higher frequency causes shorter wavelength, which enables higher gain at the same antenna effective area. As a result, in the mm-wave band, it is not difficult to design high gain antenna.

According to the Friis equation [169], the received power P_R can be expressed as

$$(P_R)dB=(P_T)dB+(G_T)dB+(G_R)dB-(L_P)dB, \tag{22.70}$$

where P_T denotes transmitted power, G_T and G_R are the gain of transmitting antenna and receiving antenna, respectively, and L_P is the path loss and given as follows:

$$(L_P)dB=20\log(4\pi d/\lambda), \tag{22.71}$$

where d is the wireless transmission distance. The shorter the signal wavelength, the larger the path loss and the smaller the received power. However, in the mm-wave

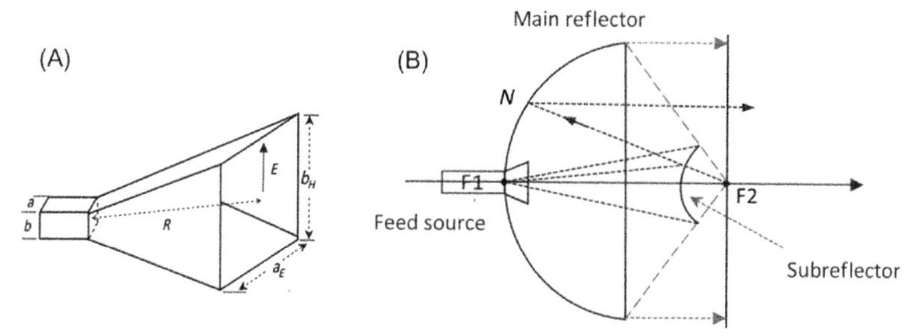

Figure 22.33 (A) The schematic of horn antenna (HA); (B) the schematic of Cassegrain antenna (CA).

band, the loss of received power caused by high carrier frequency can be compensated by high gain antenna, as shown in Eq. 22.70.

For the antennas with the same size, the higher the frequency, the narrower the beamwidth, so mm-wave antenna has better performance in terms of antiintercept and antijamming. Because of the narrow beamwidth, mm-wave antenna is more suitable for point-to-point communication. Among many kinds of antennas, horn antennas, parabolic antennas, and CAs are commonly used as mm-wave antennas [170–172]. Horn antennas is often used for short-range communications, while parabolic antennas and CAs are used in long-distance communications due to their larger gain. Fig. 22.33 shows the schematic of the rectangular horn antenna and the CA, respectively.

Rectangular horn antennas, also known as the pyramidal horn antennas, are made up of the rectangular waveguide E-plane and H-plane. As shown in Fig. 22.33A, a and b represent the width and height of the rectangular waveguide, respectively; a_E and b_H denote the side length of the aperture on the E- and H-plane, respectively; R is the distance between the upper and lower horn aperture. The CA consists of feed source, parabolic main reflector and hyperbolic subreflector. As shown in Fig. 22.33B, F1 and F2 are two focal points of hyperboloid and F2 is also the focal point of paraboloid. First, the electromagnetic wave is emitted from the feed source, and then reflected through the hyperboloid subreflector to the paraboloid main reflector. Finally, a plane beam with the same direction is radiated by the main reflector. The CA is evolved from a parabolic antenna with a higher aperture efficiency than a standard parabolic antenna and can be equivalent to a long focal length parabolic antenna, and thus it has a higher antenna gain [168].

As shown in Table 22.4, we give out the comparison of some key parameters of a typical horn antenna and a typical CA [162]. Compared to the horn antenna, the CA has a larger gain of 45dBi and a smaller half-power beamwidth of 0.8° at the cost of a larger size and weight.

Table 22.4 The comparison of some key parameters of a typical HA and a typical CA.

Parameters	CA	HA
Frequency range (GHz)	75–110	75–110
Gain (dBi)	45	25
Beamwidth	0.8°	12°
Polarization	H- or V-polarization	H- or V-polarization
XPD (dB)	>35	>33
Diameter (cm)	30.5	~1.8
Weight (kg)	8.8	~0.1

HA, Horn antenna; *CA*, cassegrain antenna.

Based on the aforementioned enabling technologies and advanced devices, long-distance multiband bidirectional wireline and wireless link delivery was further experimentally demonstrated [167]. Fig. 22.34 shows the schematic diagram for the long-distance multiband bidirectional wireline and wireless link delivery. MIMO and advanced DSP algorithms are employed in order to increase the transmission capacity and extend the transmission distance. In the field-trial demonstration, bidirectional delivery of 54-Gb/s PDM-8QAM signal at W-band and 32-Gb/s PDM-16QAM signal at K-band was realized over 20-km fiber and 2500-m MIMO wireless distance with a BER under the HD-FEC threshold of 3.8×10^{-3}. In order to overcome the power saturation effect of the wireless transmitter, the PDM-8QAM driving electrical baseband signal is predistorted. This demonstrated mm-wave transmission system may have the potential to be used for building-to-building broadband connection, disaster emergency communications, and the aforementioned other application scenarios.

22.7 Radio-frequency-transparent photonic demodulation technique applied for radio-over-fiber networks

To match the capacity of high-speed fiber-optic communication systems, there is an urgent demand for multigigabit-per-second wireless systems, which can offer high-speed mobile backhaul between wireless macro stations as well as emergency services when large-capacity long-haul optical cables are cut during natural disasters (e.g., earthquake and tsunami) [173–177]. Mm-wave band and terahertz-wave band have sufficient bandwidth and may be the location of future multigigabit-per-second wireless communication systems, which has been intensively studied in the research community. If the generated wireless mm-wave signal is demodulated in the electrical domain, at such high frequencies, the RF cable transmission distance is limited. Furthermore, electrical mm-wave demodulation will become more complicated with the increase of transmission bit rate and mm-wave carrier frequency. In Ref. [178], a RF-transparent photonic mm-wave demodulation technique was proposed based on

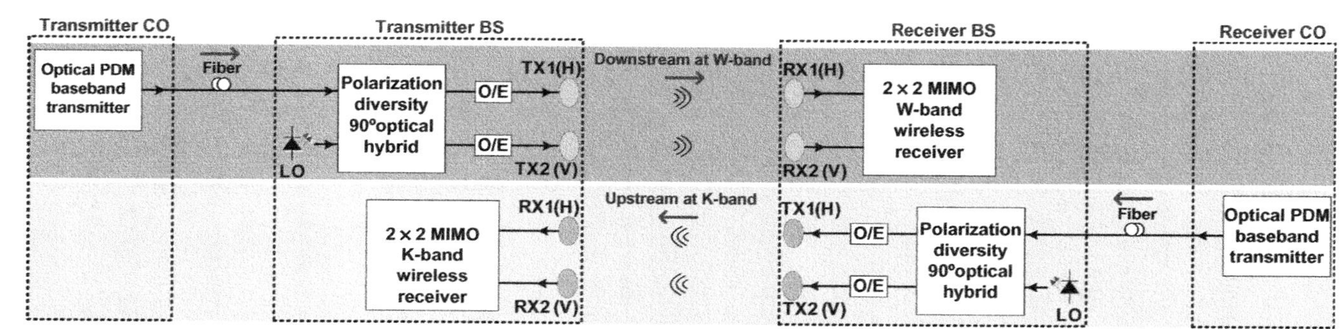

Figure 22.34 The schematic diagram for multiband bidirectional wireline and wireless link delivery.

coherent detection and baseband DSP, which has an advantage of converting the wireless mm-wave signal into the optical baseband signal. The converted optical baseband signal can be directly transmitted in a fiber-optic network. In Ref. [151], a fiber-wireless-fiber integration system at W-band was demonstrated enabled by photonic mm-wave generation and demodulation techniques. Analog down conversion enabled by sinusoidal RF signal and balanced mixer is first performed for the received W-band mm-wave signal before photonic demodulation to overcome the insufficient bandwidth of the MZM. The proposed system has throughput comparable to that of fiber-optic communication and is suitable for emergency backup communications.

Fig. 22.35 shows the principle of photonic demodulation based on push-pull MZM for the PDM-QPSK wireless mm-wave signal. The PDM-QPSK wireless mm-wave signal is generated by remote heterodyne beating technology. At the transmitter CO, a CW of wavelength λ_1 is first externally modulated and then polarization multiplexed to generate the PDM-QPSK optical baseband signal.

At the transmitter BS, the optical PDM-QPSK baseband signal is beat with the CW lightwave with a wavelength of λ_2, and up-converted to wireless mm-wave signal. The generated wireless mm-wave signal, with a frequency of $f_{RF} = c|1/\lambda_1 - 1/\lambda_2|$, is then transmitted by a 2 × 2 MIMO wireless link.

At the receiver BS, when the received PDM-QPSK wireless mm-wave signal is carried by a high frequency (e.g., W-band), due to the limited modulation bandwidth of the MZM, the analog down-conversion based on electrical mixer and sinusoidal RF signal is adopted to down-convert the wireless mm-wave signal to a mm-wave signal with a lower carrier frequency. A CW lightwave with a wavelength of λ_3 is first

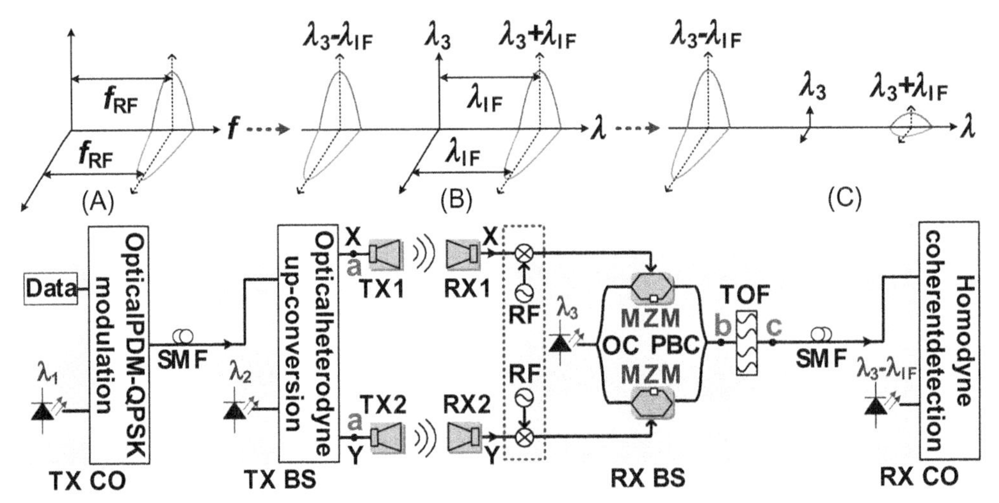

Figure 22.35 The principle of photonic demodulation based on push–pull Mach–Zehnder modulator (MZM).

divided into two tributaries by a polarization maintaining OC, and then each tributary is modulated by the X- or Y-component of the down-converted wireless mm-wave signal in a MZM. A polarization beam combiner (PBC) is used to reconstruct the two modulated tributaries.

The CW lightwave with a wavelength of λ_3 can be expressed as

$$E_{\text{in}}(t) = E_c \cos(2\pi f_c t), \tag{22.72}$$

where E_c and f_c represent the amplitude and the frequency of the CW lightwave, respectively. The IF signal at the input of the push-pull MZM can be expressed as

$$S_{\text{IF}}(t) = V_{\text{IF}} s(t) \cos(2\pi f_{\text{IF}} t), \tag{22.73}$$

where $s(t)$ represents the transmitted signal, while V_{IF} represents the amplitude of the mm-wave carrier after the analog down-conversion and its frequency is f_{IF}. Thus, the output of the push–pull MZM operating at the point of OCS can be expressed as

$$\begin{aligned} E_{\text{MZM}}(t) &= E_c \cos(2\pi f_c t) \cos\left[\frac{\pi}{2} + \beta_{\text{MZM}} s(t) \cos(2\pi f_{\text{IF}} t)\right] \\ &= -2E_c s(t) \cos(2\pi f_c t)\left\{\sum_{n=0}^{+\infty} J_{2n+1}(\beta_{\text{MZM}}) \sin[(2n+1)(2\pi f_{\text{IF}} t)]\right\}, \end{aligned} \tag{22.74}$$

where $\beta_{\text{MZM}} = \pi(V_{\text{IF}}/V_\pi)$ (V_π is the half-wave voltage of MZM), and it presents the modulation index of the MZM. When β_{MZM} is small enough, the high-order components ($n \geq 1$) in the MZM output can be ignored, and Eq. (22.74) can be approximated as

$$\begin{aligned} E_{\text{MZM}}(t) &\approx 2E_c s(t) J_1(\beta_{\text{MZM}}) \cos(2\pi f_c t) \cos\left(2\pi f_{\text{IF}} t + \frac{\pi}{2}\right) \\ &= E_c s(t)\left\{J_1(\beta_{\text{MZM}}) \cos\left[2\pi(f_c - f_{\text{IF}})t - \frac{\pi}{2}\right] + J_1(\beta_{\text{MZM}}) \cos\left[2\pi(f_c + f_{\text{IF}})t + \frac{\pi}{2}\right]\right\}. \end{aligned} \tag{22.75}$$

It can be seen from Eq. (22.75) that the output of the MZM only theoretically contains two first-order components, with carrier frequencies of $f_3 \pm f_{\text{IF}}$, carrying the transmitted data in amplitude. In practical application, the OCS signal generated after PBC usually contains two PDM-QPSK modulated sidebands, with carrier wavelengths of $\lambda_3 \pm \lambda_{\text{IF}}$ ($\lambda_{\text{IF}} = (\lambda_3^2 f_{\text{IF}})/c$), and a small optical carrier with a center wavelength of λ_3, since the MZM cannot completely suppress the central optical carrier due to its limited extinction ratio. Subsequently, the upper sideband and the central optical carrier are suppressed by a tunable optical filter (TOF), so that a PDM-QPSK baseband signal at the carrier frequency of $\lambda_3-\lambda_{IF}$ is sent to the receiver CO. Fig. 22.35B and C show the optical spectra after PBC and TOF, respectively.

At the receiver CO, the transmitted data is recovered from the PDM-QPSK baseband signal via homodyne coherent detection and baseband DSP. The operating wavelength of LO in the receiver CO is $\lambda_3-\lambda_{IF}$. It is worth noting that the TOF in the receiver BS can also filter the lower sideband and the central optical carrier with the operating wavelength of $\lambda_3-\lambda_{IF}$. In this case, the carrier wavelength of the LO in receiver and the generated optical baseband signal after the TOF is both $\lambda_3+\lambda_{IF}$.

22.8 Conclusions

We introduced the different RoF schemes to meet different requirements in the future optical wireless networks. A series of ultrahigh-speed large-capacity (>100 Gb/s) RoF systems have been realized, employing mm-wave generation based on photonics up-conversion technology, advanced multilevel modulation, DSP technology for single-carrier and OFDM detection, multicarrier technology, antenna polarization multiplexing, multiband multiplexing and so on. These approaches can effectively improve the transmission capacity and distance, as well as reduce the required bandwidth for optical and electrical devices. To keep data transmission efficiently in the future optical wireless networks, we can use the fiber-wireless-fiber integration based on photonics up-conversion and down-conversion schemes.

Acknowledgments

We'd like to thank Dr. Junwen Zhang, Dr. Fan Li, Dr. Ze Dong, Dr. Hung-Chang Chien, Dr. Wen Zhou, Kaihui Wang, Li Zhao, Zihang Zhu, Dr. Jiangnan Xiao, Can Wang, Mingming Zhao, Bowen Zhu, Kaibo Fang for their help to write this chapter. One of the authors, Dr. Jianjun Yu, likes to thank many colleagues at Georgia Institute of Technology, NEC Labs America and ZTE TX Inc. for their collaboration.

References

[1] J.E. Berthold, L.Y. Ong, Next-generation optical network architecture and multidomain issues, Proc. IEEE 100 (5) (2012) 1130–1139.

[2] E. Wong, Next-generation broadband access networks and technologies, J. Lightwave Technol. 30 (4) (2012) 597–608.

[3] J.-i Kani, F. Bourgart, A. Cui, A. Rafel, M. Campbell, R. Davey, et al., Next-generation PON—Part I: technology roadmap and general requirements, IEEE Commun. Mag. 47 (11) (2009) 43–49.

[4] L. Kazovsky, W.-T. Shaw, D. Gutierrez, N. Cheng, S.-W. Wong, Next-generation optical access networks, J. Lightwave Technol. 25 (11) (2007) 3428–3442.

[5] K. Prince, Converged wireline and wireless signal distribution in optical fiber access networks (Ph.D. thesis), DTU Fotonik, 2010. ISBN 87-92062-43-1.

[6] J. Capmany, D. Novak, Microwave photonics combines two worlds, Nat. Photonics 1 (2007) 319–332.

[7] D. Wake, A. Nkansah, N. Gomes, Radio over fiber link design for next generation wireless systems, J. Lightwave Technol. 28 (16) (2010) 2456–2464.
[8] Wireless HD Specification Version 1.1 Overview, Wireless HD, White paper, 2010. (Online). Available from: <www.wirelesshd.org>.
[9] R. Fisher, 60 GHz WPAN Standardization within IEEE 802.15.3c, in: International Symposium on Signals, Systems and Electronics, ISSSE'07, 2007, pp. 103–105.
[10] ECMA-387 Standard, High Rate 60GHz PHY, MAC and HDMI PAL, 2008 December (Online). Available from: <www.ecmainternational.org>.
[11] Defining the future of multi-gigabit wireless communication, WiGig white paper, 2010. (Online). Available from: <wirelessgigabitalliance.org>.
[12] Gigabit wireless applications using 60 GHz radios, Bridgewave white paper, 2006. (Online). Available from: <www.bridgewave.com>.
[13] X. Pang, High-Capacity Hybrid Optical Fiber-Wireless Communications Links in Access Networks (PhD thesis).,Technical University of Denmark (DTU), Lyngby, Denmark, 2013.
[14] D. Zibar, A. Caballero Jambrina, X. Yu, X. Pang, A.K. Dogadaev, I. Tafur Monroy, Hybrid optical fibre-wireless links at the 75–110 GHz band supporting 100 Gbps transmission capacities, in: International Topical Meeting on Microwave Photonics Conference, MWP'11, Singapore, November 2011, pp. 445–449.
[15] FCC online table of frequency allocations (Online). Available from: <www.fcc.gov>.
[16] J. Yao, Microwave photonics, J. Lightwave Technol 27 (2009) 314–335.
[17] C. Jastrow, S. Priebe, B. Spitschan, J. Hartmann, M. Jacob, T. Kurner, et al., Wireless digital data transmission at 300 GHz, Electron. Lett. 46 (9) (2010) 661–663.
[18] A. Hirata, R. Yamaguchi, T. Kosugi, H. Takahashi, K. Murata, T. Nagatsuma, et al., 10-Gbit/s wirelesslink using InP HEMT MMICs for generating 120-GHz-band millimeter-wave signal, IEEE Trans. Microw. Theory Tech 57 (5) (2009) 1102–1109.
[19] H.-J. Song, K. Ajito, A. Hirata, A. Wakatsuki, T. Furuta, N. Kukutsu, et al., Multi-gigabit wireless data transmission at over 200-Ghz, in: Proc. 34th International Conference on Infrared, Millimeter and Terahertz Waves, IRMMW-THz'09, 2009, pp. 1–2.
[20] A. Stohr, S. Babiel, P. Cannard, B. Charbonnier, F. van Dijk, S. Fedderwitz, et al., Millimeter-wave photonic components for broadband wireless systems, IEEE Trans. Microw. Theory Tech. 58 (11) (2010) 3071–3082.
[21] J.J. O'Reilly, P.M. Lane, R. Heidemann, R. Hofstetter, Optical generation of very narrow line-width millimetre wave signals, Electron. Lett. 28 (25) (1992) 2309–2311.
[22] D. Wake, C.R. Lima, P.A. Davies, Optical generation of millimeter-wave signals for fiber-radio systems using a dual-mod DFB semiconductor laser, IEEE Trans. Microw. Theory Tech. 43 (9) (1995) 2270–2276.
[23] R.-P. Braun, G. Grosskopf, D. Rohde, F. Schmidt, Low-phase noise millimeter-wave generation at 64 GHz and data transmission using optical sideband injection locking, IEEE Photonics Technol. Lett 10 (5) (1998) 728–730.
[24] A. Hirata, M. Harada, T. Nagatsuma, 120-GHz wireless link using photonic techniques for generation, modulation, and emission of millimeter-wave signals, J. Lightwave Technol. 21 (10) (2003) 2145–2153.
[25] J. Yu, Z. Jia, L. Yi, Y. Su, G.K. Chang, T. Wang, Optical millimeter-wave generation or up-conversion using external modulators, IEEE Photonics Technol. Lett 18 (1) (2006) 265, 267.
[26] W. Jiang, C.T. Lin, A. Ng'oma, P.T. Shih, J. Chen, M. Sauer, et al., Simple 14-Gb/s short-range radio-over- fiber system employing a single-electrode MZM for 60-GHz wireless applications, J. Lightwave Technol. 28 (16) (2010) 2238–2246.
[27] C.T. Lin, P.T. Shih, W.J. Jiang, E.Z. Wong, J. Chen, S. Chi, Photonic vector signal generation at microwave/millimeter-wave bands employing optical frequency quadrupling scheme, Opt. Lett. 34 (14) (2009) 2171–2173.
[28] A. Kanno, K. Inagaki, I. Morohashi, T. Sakamoto, T. Kuri, I. Hosako, et al., 40 Gb/s W-band (75–110 GHz) 16-QAM radio-over-fiber signal generation and its wireless transmission, Opt. Express 19 (26) (2011) B56–B63.

[29] X. Li, J. Yu, J. Zhang, J. Xiao, Z. Zhang, Y. Xu, et al., QAM vector signal generation by optical carrier suppression and precoding techniques, IEEE Photonics Technol. Lett. 27 (18) (2015) 1977–1980.
[30] X. Li, J. Yu, J. Xiao, N. Chi, Y. Xu, W-band PDM-QPSK vector signal generation by MZM-based photonic frequency octupling and precoding, IEEE Photonics J. 7 (4) (2015).
[31] X. Li, J. Yu, Z. Zhang, J. Xiao, G.K. Chang, Photonic vector signal generation at W-band employing an optical frequency octupling scheme enabled by a single MZM, Opt. Commun. 349 (2015) 6–10.
[32] J. Xiao, X. Li, Y. Xu, Z. Zhang, L. Chen, J. Yu, W-band OFDM photonic vector signal generation employing a single Mach–Zehnder modulator and precoding, Opt. Express 23 (18) (2015) 24029–24034.
[33] X. Li, Z. Dong, J. Yu, N. Chi, Y. Shao, G.K. Chang, Fiber wireless transmission system of 108-Gb/s data over 80-km fiber and 2 × 2 MIMO wireless links at 100GHz W-Band frequency, Opt. Lett. 37 (24) (2012) 5106–5108.
[34] X. Li, J. Xiao, J. Yu, W-band vector millimeter-wave signal generation based on phase modulator with photonic frequency quadrupling and precoding, J. Lightwave Technol. 35 (13) (2017) 2548–2558.
[35] J. Yu, Photonics-assisted millimeter-wave wireless communication., IEEE J. Quantum Electron. 53 (6) (2017) 8000517.
[36] X. Li, J. Xiao, Y. Xu, J. Yu, QPSK vector signal generation based on photonic heterodyne beating and optical carrier suppression, IEEE Photonics J 7 (5) (2015).
[37] A.S. Mirfananda, M. Suryanegara, 5G spectrum candidates beyond 6 GHz: a simulation of Jakarta environment, in: Region 10 Symposium (TENSYMP), 2016, pp. 30–35.
[38] X. Li, J. Yu, J. Xiao, Y. Xu, Fiber-wireless-fiber link for 128-Gb/s PDM-16QAM signal transmission at W-band, IEEE Photonics Technol. Lett 26 (19) (2014) 1948–1951.
[39] J. Yu, X. Li, J. Zhang, and J. Xiao, 432-Gb/s PDM-16QAM signal wireless delivery at w-band using optical and antenna polarization multiplexing, in: ECOC 2014: We.3.6.6.
[40] X. Li, J. Zhang, J. Xiao, Z. Zhang, Y. Xu, J. Yu, W-band 8QAM vector signal generation by MZM-based photonic frequency octupling, IEEE Photonics Technol. Lett 27 (12) (2015) 1257–1260.
[41] Z. Dong, 64QAM vector radio-frequency signal generation based on phase precoding and optical carrier suppression modulation, IEEE Photonics J. 8 (6) (2016) 1–7.
[42] X. Li, J. Yu, G.K. Chang, Frequency-quadrupling vector mm-wave signal generation by only one single-drive MZM, IEEE Photonics Technol. Lett 28 (12) (2016) 1302–1305.
[43] J. Yu, Z. Jia, T. Wang, G.K. Chang, A novel radio-over-fiber configuration using optical phase modulator to generate an optical mm-wave and centralized lightwave for uplink connection, IEEE Photonics Technol. Lett 19 (3) (2007) 140–142.
[44] J. Xiao, Z. Zhang, X. Li, Y. Xu, L. Chen, J. Yu, High-frequency photonic vector signal generation employing a single phase modulator, IEEE Photonics J. 7 (2) (2015).
[45] X. Li, Y. Xu, J. Xiao, J. Yu, W-band millimeter-wave vector signal generation based on precoding-assisted random photonic frequency tripling scheme enabled by phase modulator, IEEE Photonics J. 8 (2) (2016) 1–10.
[46] L. Zhao, J. Yu, L. Chen, P. Min, J. Li, R. Wang, 16QAM vector millimeter-wave signal generation based on phase modulator with photonic frequency doubling and precoding, IEEE Photonics J. 8 (2) (2016) 1–8.
[47] J. Yu, X. Li, W. Zhou, Tutorial: Broadband fiber-wireless integration for 5G + communication, in: APL Photonics 3 (11), 111101.
[48] J. Ma, J. Yu, C. Yu, X. Xin, J. Zeng, L. Chen, Fiber dispersion influence on transmission of the optical millimeter-waves generated using LN-MZM intensity modulation, IEEE/OSA J. Lightwave Technol. 25 (11) (2007) 3244–3256.
[49] Y. Wang, Y. Xu, X. Li, J. Yu, N. Chi, Balanced precoding technique for vector signal generation based on OCS, IEEE Photonics Technol. Lett 27 (13) (2015) 2469–2472.
[50] H.-C. Chien, Z. Jia, J. Zhang, Z. Dong, J. Yu, Optical independent-sideband modulation for bandwidth-economic coherent transmission, Opt. Exp. 22 (8) (2014) 9465–9470.

[51] C.T. Lin, J. Chen, P.T. Shih, W.J. Jiang, S. Chi, Ultra-high data-rate 60 GHz radio-over-fiber systems employing optical frequency multiplication and OFDM formats, J. Lightwave Technol. 28 (16) (2010) 2296–2306.
[52] C.T. Lin, Y.M. Lin, J. Chen, S.P. Dai, P.T. Shih, P.C. Peng, et al., Optical direct-detection OFDM signal generation for radio-over-fiber link using frequency doubling scheme with carrier suppression, Opt. Exp 16 (9) (2008) 6056–6063.
[53] L. Zhang, T. Zuo, Q. Zhang, E. Zhou, G. Liu, X. Xu, Transmission of 112-Gb/s + DMT over 80-km SMF enabled by twin-SSB technique at 1550 nm, Eur. Conf. Opt. Commun. (2015).
[54] Z. Xu, R. Hui, M. O'Sullivan, Dual-band OOFDM system based on tandem single-sideband modulation transmitter, Opt. Expr. 17 (16) (2009) 13479–13486.
[55] L. Zhang, E. Zhou, Q. Zhang, X. Xu, G.N. Liu, T. Zuo, C-band single wavelength 100-Gb/s IM-DD transmission over 80-km SMF without CD compensation using SSB-DMT, Opt. Fiber Commun. Conf. Exp (2015).
[56] J. Yu, L. Chen, Phase factor optimization for QPSK signals generated from MZM based on optical carrier suppression, IEEE Photonics J. 9 (2) (2017) 1–7.
[57] X. Li, Y. Xu, J. Yu, Single-sideband W-band photonic vector millimeter-wave signal generation by one single I\Q modulator, Opt. Lett. 41 (18) (2016) 4162–4165.
[58] X. Li, J. Yu, 2 × 2 multiple-input multiple-output optical–wireless integration system based on optical independent-sideband modulation enabled by an in-phase/quadrature modulator, Opt. Lett. 41 (13) (2016) 3138–3141.
[59] X. Li, Optimization of pre-coding phase distribution for frequency-multiplication vector signal generation, IEEE Photonics J. (2017). pp (99): 1-1.
[60] A. Gudipati, SoftRAN: software defined radio access network, in: Proc. ACM SIGCOMM Wksp. Hot Topics in Software Defined Networking, 2013, . 25–30.
[61] H.-H. Cho, C.-F. Lai, T.K. Shih, H.-C. Chao, Integration of SDR and SDN for 5G, IEEE Access 2 (2014) 1196–1204.
[62] K.-I. Kitayama, A. Maruta, Y. Yoshida, Digital coherent technology for optical fiber and radio-over-fiber transmission systems, J. Lightwave Technol. 32 (20) (2014) 3411–3420.
[63] S.E. Alavi, M.R.K. Soltanian, I.S. Amiri, M. Khalily, A.S.M. Supa'at, H. Ahmad, Towards 5G: a photonic based millimeter wave signal generation for applying in 5G access fronthaul, Sci. Rep. 6 (19891) (2016) 1–11.
[64] X. Zhang, M. Jia, L. Chen, J. Ma, J. Qiu, Filtered-OFDM—enabler for flexible waveform in the 5th generation cellular networks, in: Proc. IEEE Conf. Global. Commun., 2015, pp. 1–6.
[65] D. Qian, S.H. Fan, N. Cvijetic, J. Hu, T. Wang, 64/32/16 QAM-OFDM using direct-detection for 40G-OFDMA-PON downstream, in: Opt. Fiber Commun. Conf., 2011.
[66] J. Armstrong, OFDM for optical communications, J. Ligthwave Technol. 27 (3) (2009) 189–204.
[67] C. Lim, A. Nirmalathas, M. Bakaul, P. Gamage, K.L. Lee, Y. Yang, et al., Fiber-wireless networks and subsystem technologies, J. Lightwave Technol. 28 (4) (2010) 390–405.
[68] J. Yu, J. Hu, D. Qian, Z. Jia, G.K. Chang, T. Wang, 16 Gbit/s super broadband OFDM-radio-over-fibre system, Electron. Lett. 44 (6) (2008) 450–452.
[69] M. Chen, X. Xiao, J. Yu, X. Li, F. Li, Real-time generation and reception of OFDM signals for X-band RoF uplink with heterodyne detection, Photonics Technol. Lett 29 (1) (2017) 51–54.
[70] C.-H. Li, M.-F. Wu, C.-H. Lin, C.-T. Lin, W-band OFDM RoF system with simple envelope detector down-conversion, in: Proc. OFC 2015, Los Angeles, California, W4G. 6.
[71] S. Mikroulis, M.P. Thakur, J.E. Mitchell, Investigation of a robust remote heterodyne envelope detector scheme for cost-efficient E-PON/60 GHz wireless integration, in: 2014 16th International Conference on Transparent Optical Networks (ICTON), Graz, 2014, pp. 1–4.
[72] D. Zibar, R. Sambaraju, A. Caballero, J. Herrera, U. Westergren, A. Walber, et al., High-capacity wireless signal generation and demodulation in 75- to 110-GHz band employing all-optical OFDM, Photonics Technol. Lett 23 (12) (2011) 810–812.
[73] X. Li, J. Yu, N. Chi, J. Xiao, Antenna polarization diversity for high-speed polarization multiplexing wireless signal delivery at W-band, Opt. Lett. 39 (5) (2014) 1169–1172.

[74] J. Xiao, J. Yu, X. Li, Y. Xu, Z. Zhang, L. Chen, 20-Gb/s PDM-QPSK signal delivery over 1.7-km wireless distance at W-band, in: Proc. OFC 2015, Los Angeles, CA, W4G.4.
[75] S. Inudo, Y. Yoshida, A. Kanno, P. Tien Dat, T. Kawanishi, K.-I. Kitayama, On the MIMO channel rank deficiency in W-band MIMO RoF transmissions, in: Proc. OFC 2015, Los Angeles, CA, W4G. 5.
[76] M.T. Dao, D.H. Shin, Y.T. Im, S.O. Park, A two sweeping VCO source for heterodyne FMCW radar, IEEE Trans. Instrum. Meas. 62 (1) (2013) 230–239.
[77] C. Gu, C. Li, J. Lin, J. Long, J. Huangfu, L. Ran, Instrument-based noncontact Doppler radar vital sign detection system using heterodyne digital quadrature demodulation architecture, IEEE Trans. Instrum. Meas. 59 (6) (2010) 1580–1588.
[78] K. Kikuchi, Digital coherent optical communication systems: fundamentals and future prospect, IEICE Electron. Exp. 8 (20) (2011) 1642–1662.
[79] E. Ip, J. Kahn, Compensation of dispersion and nonlinear impairments using digital backpropagation, J. Lightwave Technol. 26 (20) (2008) 3416–3425.
[80] E. Ip, J. Kahn, Digital equalization of chromatic dispersion and polarization mode dispersion, J. Lightwave Technol. 25 (8) (2007) 2033–2043.
[81] Y. Han, G. Li, Coherent optical communication using polarization multiple-input-multiple-output, Opt. Express 13 (19) (2005) 7527–7534.
[82] X. Zhou, J. Yu, M. Huang, Y. Shao, T. Wang, L. Nelson, et al., 64-Tb/s 8 b/s/Hz PDM-36QAM transmission over 320 km using both pre- and post-transmission digital signal processing, J. Lightwave Technol. 29 (4) (2011) 571–577.
[83] G. Bosco, V. Curri, A. Carena, P. Poggiolini, F. Forghieri, On the performance of Nyquist-WDM terabit superchannels based on PM-BPSK PM-QPSK PM-8QAM or PM-16QAM subcarriers, J. Lightwave Technol. 29 (1) (2011) 53–61.
[84] J. Yu, Z. Dong, H.-C. Chien, Y. Shao, Nan Chi, 7-Tb/s (7 × 1.284 Tb/s/ch) signal transmission over 320 km using PDM-64QAM modulation, IEEE Photonics Technol. Lett 24 (4) (2012) 264–266.
[85] J. Zhang, J. Yu, Z. Jia, H.-C. Chien, 400 G transmission of super-Nyquist-filtered signal based on single-carrier 110-GBaud PDM QPSKwith 100-GHz grid, J. Lightwave Technol. 32 (19) (2014) 3239–3246.
[86] J. Zhang, Z. Dong, H. Chien, Z. Jia, Y. Xia, Y. Chen, Transmission of 20 × 440-Gb/s super-Nyquist-filtered signals over 3600 km based on single-carrier 110-GBaud PDM QPSK with 100-GHz Grid, in: Opt. Fiber Commun. Conf., San Francisco, CA, USA, 2014.
[87] J. Yu, J. Zhang, Z. Dong, Z. Jia, H.-C. Chien, Y. Cai, et al., Transmission of 8 × 480-Gb/s super-Nyquist-filtering 9-QAM-like signal at 100 GHz-grid over 5000-km SMF-28 and twenty-five 100 GHz-grid ROADMs, Opt. Express 21 (2013) 15686–15691.
[88] G. Gavioli, E. Torrengo, G. Bosco, A. Carena, S. Savory, F. Forghieri, et al., Ultra-narrow-spacing 10-channel 1.12 Tb/s D-WDM long-haul transmission over uncompensated SMF and NZDSF, IEEE Photonics Technol. Lett 22 (19) (2010) 1419–1421.
[89] J. Yu, Z. Dong, N. Chi, 1.96 Tb/s (21x100 Gb/s) OFDM optical signal generation and transmission over 3200-km fiber, IEEE Photonics Technol. Lett 23 (15) (2011) 1061–1063.
[90] Y. Ma, Q. Yang, Y. Tang, S. Chen, W. Shieh, 1-Tb/s single-channel coherent optical OFDM transmission over 600-km SSMF fiber with subwavelength bandwidth access, Opt. Express 17 (11) (2009) 9421–9427.
[91] J.-X. Cai, H. Zhang, H.G. Batshon, M. Mazurczyk, O.V. Sinkin, Y. Sun, et al., Transmission over 9100 km with capacity of 49.3 Tb/s using variable spectral efficiency 16 QAM based coded modulation, in: Opt. Fiber Commun. Conf., San Francisco, CA, USA, 2014.
[92] Z. Jia, J. Yu, H.-C. Chien, Z. Dong, D. Di Huo, Field transmission of 100G and beyond: Multiple baud rates and mixed line rates using Nyquist-WDM technology, J. Lightwave Technol. 30 (24) (2012) 3793–3804.
[93] H.-C. Chien, J. Yu, Z. Jia, Z. Dong, X. Xiao, Performance assessment of noise-suppressed Nyquist-WDM for Terabit superchannel transmission, J. Lightwave Technol. 30 (24) (2012) 3965–3971.

[94] B. Huang, J. Zhang, J. Yu, Z. Dong, X. Li, H. Ou, et al., Robust 9-QAM digital recovery for spectrum shaped coherent QPSK signal, Opt. Express 21 (2013) 7216–7221.

[95] J. Zhang, B. Huang, X. Li, Improved quadrature duobinary system performance using multi-modulus equalization, IEEE Photonics Technol. Lett 25 (16) (2013) 1630–1633.

[96] J. Zhang, J. Yu, N. Chi, Z. Dong, J. Yu, X. Li, et al., Multi-modulus blind equalizations for coherent quadrature duobinary spectrum shaped PM-QPSK digital signal processing, J. Lightwave Technol. 31 (7) (2013) 1073–1078.

[97] L.G. Kazovsky, Optical heterodyning versus optical homodyning: a comparison, J. Opt. Commun 6 (1) (1985) 18–24.

[98] E. Coersmeier, E. Zielinski, Adaptive pre-equalization in analog heterodyne architectures for wireless LAN, in: Radio and Wireless Conference, Rawcon, 2002, pp. 107–110.

[99] E. Coersmeier, E. Zielinski, Comparison between different adaptive pre-equalization approaches for wireless LAN, in: IEEE PIMRC, vol. 3, 2002, pp. 1136–1140.

[100] S.U.H. Qureshi, Adaptive equalization, Proc. IEEE 73 (9) (1985) 1349–1387.

[101] C.R. Johnson, P. Schniter, T.J. Endres, J.D. Behm, D. Brown, R.A. Casas, Blind equalization using the constant modulus criterion: a review, Proc. IEEE 86 (1998) 1927–1950.

[102] J.G. Proakis, Digital Communications, McGraw-Hill, 2001, p. 1002.

[103] J. Zhang, Z. Dong, J. Yu, N. Chi, L. Tao, X. Li, et al., Simplified coherent receiver with heterodyne detection of eight-channel 50 Gb/s PDM-QPSK WDM signal after 1040 km SMF-28 transmission, Opt. Lett. 37 (19) (2012) 4050–4052.

[104] C. Liu, W. Jian, H.C. Chien, A. Chowdhury, G.K. Chang, Experimental analyses and optimization of equalization techniques for 60-GHz OFDM radio-over-fiber system, in: Opt. Fiber Commun. Conf., San Diego, California, 2010.

[105] N.-W. Chen, H.-J. Tsai, F.-M. Kuo, J.-W. Shi, High-speed W-band integrated photonic transmitter for radio-over-fiber applications, IEEE Trans. Microw. Theory Tech. 59 (2011).

[106] G.L. Stüber (Stuber), Broadband MIMO-OFDM wireless communications, Proc. IEEE 92 (2) (2004) 271–294.

[107] A. Kanno, P.T. Dat, T. Kuri, I. Hosako, T. Kawanishi, Y. Yoshida, et al., Coherent radio-over-fiber and millimeter-wave radio seamless transmission system for resilient access networks, IEEE Photonics J. 4 (6) (2012) 2196–2204.

[108] Z. Dong, J. Yu, X. Li, G.-K. Chang, L. Chen, Z. Cao, Integration of 112-Gb/s PDM-16QAM wireline and wireless data delivery in millimeter wave RoF system, in: Proc. Opt. Fiber Commun. Conf. Exposit. Nat. Fiber Opt. Eng. Conf. (OFC/NFOEC), Mar. 2013, pp. 1–3.

[109] X. Zhou, J. Yu, Multi-level, multi-dimensional coding for high-speed and high-spectral-efficiency optical transmission, J. Lightwave Technol. 27 (16) (2009) 3641–3653.

[110] K. Kikuchi, Electronic post-compensation for nonlinear phase fluctuations in a 1000-km 20-Gbit/s optical quadrature phase-shift keying transmission system using the digital coherent receiver, Opt. Express 16 (2) (2008) 889–896.

[111] J. Zhang, X. Li, Z. Dong, Digital nonlinear compensation based on the modified logarithmic step size, J. Lightwave Technol. 31 (22) (2013) 3546–3555.

[112] E. Ip, J.M. Kahn, Feedforward carrier recovery for coherent optical communications, J. Lightwave Technol. 25 (9) (2007) 2675–2692.

[113] J. Li, E. Tipsuwannakul, T. Eriksson, M. Karlsson, P.A. Andrekson, Approaching Nyquist limit in WDM systems by low-complexity receiver-side duobinary shaping, J. Lightwave Technol. 30 (11) (2012) 1664–1676.

[114] A.J. Viterbi, Error bounds for convolutional codes and an asymtotically optimum decoding algorithm, IEEE Trans. Inf. Theory 13 (2) (1967) 260–269.

[115] I. Fatadin, D. Ives, S.J. Savory, Blind equalization and carrier phase recovery in a 16-QAM optical coherent system, J. Lightwave Technol. 27 (15) (2009) 3042–3049.

[116] K. Koizumi, K. Toyoda, T. Omiya, et al., 512 QAM transmission over 240 km using frequency-domain equalization in a digital coherent receiver, Opt. Express 20 (21) (2012) 23383–23389.

[117] D. Godard, Passband timing recovery in an all-digital modem receiver, IEEE Trans. Commun. 26 (5) (1978) 517–523.

[118] K. Mueller, M. Muller, Timing recovery in digital synchronous data receivers, IEEE Trans. Commun. 24 (5) (1976) 516–531.
[119] D. Godard, Self-recovering equalization and carrier tracking in two-dimensional data communication systems, IEEE Trans. Commun. 28 (11) (1980) 1867–1875.
[120] F. Gardner, A BPSK/QPSK timing-error detector for sampled receivers, IEEE Trans. Commun. 34 (5) (1986) 423–429.
[121] A. Nafta, P. Johannisson, M. Shtaif, Blind equalization in optical communications using independent component analysis, J. Lightwave Technol. 31 (12) (2013) 2043–2049.
[122] S. Savory, Digital filters for coherent optical receivers, Opt. Express 16 (2) (2008) 804–817.
[123] P. Winzer, High-spectral-efficiency optical modulation formats, J. Lightwave Technol. 30 (24) (2012) 3824–3835.
[124] T. Pfau, S. Hoffmann, R. Noe, Hardware-efficient coherent digital receiver concept with feedforward carrier recovery for M-QAM constellations, J. Lightwave Technol. 27 (8) (2009) 989–999.
[125] X. Zhou, An improved feed-forward carrier recovery algorithm for coherent receivers with M-QAM modulation format, IEEE Photonics Technol. Lett 22 (14) (2010) 1051–1053.
[126] K. Zhong, J. Ke, Y. Gao, J. Cartledge, Linewidth-tolerant and low-complexity two-stage carrier phase estimation based on modified QPSK partitioning for dual-polarization 16-QAM systems, J. Lightwave Technol. 31 (1) (2013) 50–57.
[127] X. Zhou, C. Lu, A.P.T. Lau, K. Long, Low-complexity carrier phase recovery for square M-QAM based on S-BPS algorithm, IEEE Photonics Technol. Lett 26 (18) (2014) 1863–1866.
[128] J. Yu, M.F. Huang, D. Qian, L. Chen, G.K. Chang, Centralized lightwave WDM-PON employing 16-QAM intensity modulated OFDM downstream and OOK modulated upstream signals, IEEE Photonics Technol. Lett 20 (18) (2008) 1545.
[129] W.-R. Peng, X. Wu, V.R. Arbab, K.-M. Feng, B. Shamee, L.C. Christen, et al., Theoretical and experimental investigations of direct-detected RF-tone assisted optical OFDM systems, J. Lightwave Technol. 27 (10) (2009) 1332–1339.
[130] F. Li, et al., Reducing the peak-to-average power ratio with companding transform coding in 60 GHz OFDM-ROF systems, IEEE/OSA J. Opt. Commun. Netw 4 (3) (2012) 202–209.
[131] X. Li, J. Yu, J. Zhang, Z. Dong, F. Li, N. Chi, A 400G optical wireless integration delivery system, Opt. Express 21 (16) (2013) 18812–18819.
[132] L. Tao, J. Yu, Y. Fang, J. Zhang, Y. Shao, N. Chi, Analysis of noise spread in optical DFT-S OFDM systems, J. Lightwave Technol. 30 (20) (2012) 3219–3225.
[133] F. Li, X. Li, J. Yu, L. Chen, Optimization of training sequence for DFT-spread DMT signal in optical access network with direct detection utilizing DML, Opt. Express 22 (19) (2014) 22962–22967.
[134] C. Tellambura, Upper bound on peak factor of N-multiple carriers, Electron. Lett. 33 (19) (1997) 1608–1609.
[135] J. Zhao, H. Shams, Fast dispersion estimation in coherent optical 16QAM fast OFDM systems, Opt. Express 21 (2) (2013) 2500–2505.
[136] W. Yan, B. Liu, L. Li, Z. Tao, T. Takahara, J.C. Rasmussen, Nonlinear distortion and DSP-based Compensation in metro and access networks using discrete multi-tone, in: European Conference and Exhibition on Optical Communication, Amsterdam, 2012, p. Mo.1.B.2.
[137] W.-R. Peng, K. Takeshima, I. Morita, H. Takahashi, H. Tanaka, Scattered pilot channel tracking method for PDM-CO-OFDM transmissions using polar-based intra-symbol frequency-domain average, in: Presented at the Opt. Fiber Commun. Conf., San Francisco, CA, USA, 2014, Paper OWE6.
[138] X. Liu, F. Buchali, Intra-symbol frequency-domain averaging based channel estimation for coherent optical OFDM, Opt. Express 16 (26) (2008) 21944–21957.
[139] L. Chen, J. Xiao, J. Yu, Application of volterra nonlinear compensation in 75-GHz mm-wave fiber-wireless system, IEEE Photonics J. 9 (1) (2017) 1–8.
[140] F. Li, Z. Cao, X. Li, Z. Dong, L. Chen, Fiber-wireless transmission system of PDM-MIMO-OFDM at 100 GHz frequency, J. Lightwave Technol. 31 (14) (2013) 2394–2399.

[141] W. Yan, B. Liu, L. Li, Z. Tao, T. Takahara, J.C. Rasmussen, Nonlinear distortion and DSP-based compensation in metro and access networks using discrete multi-tone, in: European Conference and Exhibition on Optical Communication, Amsterdam, 2012, p. Mo.1.B.2.
[142] Y. Wang, J. Yu, N. Chi, Demonstration of 4 128-Gb/s DFT-S OFDM signal transmission over 320-km SMF With IM/DD, IEEE Photonics J. 8 (2) (2016) 1–9.
[143] F. Li, J. Yu, J. Xiao, Z. Cao, L. Chen, Reduction of frequency fading and imperfect frequency response with pre-emphasis technique in OFDM-ROF systems, Opt. Commun. 284 (19) (2011) 4699–4705.
[144] Y. Xu, Z. Zhang, X. Li, J. Xiao, J. Yu, Demonstration of 60 Gb/s W-band optical mm-wave signal full-duplex transmission over fiber-wireless-fiber network, IEEE Commun. Lett. 18 (12) (2014) 2105–2108.
[145] X. Li, et al., Performance improvement by pre-equalization in W-band (75-110 GHz) RoF system, in: Proc. Opt. Fiber Commun. Conf., Anaheim, CA, USA, 2010, Paper OW1D.3.
[146] M. Zhu, L. Zhang, J. Wang, L. Cheng, C. Liu, G.-K. Chang, Radio-over-fiber access architecture for integrated broadband wireless services, J. Lightwave Technol. 31 (23) (2013) 3614–3620.
[147] Z. Cao, et al., Direct-detection optical OFDM transmission system without frequency guard band, IEEE Photonics Technol. Lett. 22 (11) (2010) 736–738.
[148] Z. Cao, et al., Reduction of intersubcarrier interference and frequency-selective fading in OFDM-ROF systems, J. Lightwave Technol. 28 (16) (2010) 2423–2429.
[149] X. Li, J. Yu, J. Zhang, Z. Dong, N. Chi, Doubling transmission capacity in optical wireless system by antenna horizontal-and vertical-polarization multiplexing, Opt. Lett. 38 (12) (2013) 2125–2127.
[150] X. Li, J. Yu, J. Zhang, F. Li, Fiber-wireless-fiber link for 100-Gb/s PDM-QPSK signal transmission at W-band, IEEE Photonics Technol. Lett 26 (18) (2014) 1825–1828.
[151] X. Li, J. Yu, Z. Cao, J. Zhang, F. Li, G.K. Chang, Ultra-high-speed fiber-wireless-fiber link for emergency communication system, in: OFC 2014: M3D.6, San Francisco, California, USA.
[152] X. Li, J. Yu, Generation and heterodyne detection of > 100-Gb/s Q -Band PDM-64QAM mm-wave signal, IEEE Photonics Technol. Lett 29 (1) (2017) 27–30.
[153] X. Li, Y. Xu, J. Yu, Over 100-Gb/s V-band single-carrier PDM-64QAM fiber-wireless-integration system, IEEE Photonics J. 8 (5) (2016) 1–7.
[154] J. Zhang, J. Yu, N. Chi, Z. Dong, X. Li, G.-K. Chang, Multichannel 120-Gb/s data transmission over 2 × 2 MIMO fiber-wireless link at Wband, IEEE Photonics Technol. Lett 25 (8) (2013) 780–783.
[155] X. Li, J. Yu, Over 100 Gb/s ultrabroadband MIMO wireless signal delivery system at the D-band, IEEE Photonics J. 8 (5) (2016) 1–10.
[156] X. Li, J. Yu, J. Zhang, F. Li, J. Xiao, Antenna polarization diversity for 146 Gb/s polarization multiplexing QPSK wireless signal delivery at W-band, in: OFC 2014, Paper M3D.7.
[157] X. Li, J. Yu, L. Zhao, K. Wang, W. Zhou, J. Xiao, 1-Tb/s photonics-aided vector millimeter-wave signal wireless delivery at D-Band, in: Optical Fiber Communication Conference Postdeadline Papers, OSA Technical Digest (online) (Optical Society of America, 2018), paper Th4D.1.
[158] X. Li, J. Yu, J. Xiao, Y. Xu, L. Chen, Photonics-aided over 100-Gbaud all-band (D-, W- and V-band) wireless delivery, in: Proc. ECOC 2016, Düsseldorf, Germany, 2016, pp. 303–305.
[159] X. Li, J. Yu, L. Zhao, K. Wang, C. Wang, M. Zhao, et al., 1-Tb/s millimeter-wave signal wireless delivery at D-band, IEEE/OSA J. Lightwave Technol. 37 (1) (2019) 196–204.
[160] J. Yu, X. Li, N. Chi, Faster than fiber: over 100-Gb/s signal delivery in fiber wireless integration system, Opt. Express 21 (19) (2013) 22885–22904.
[161] X. Li, J. Yu, K. Wang, Y. Xu, L. Chen, L. Zhao, et al., Bidirectional Delivery of 54-Gbps 8QAM W-Band Signal and 32-Gbps 16QAM K-Band Signal over 20-km SMF-28 and 2500-m Wireless Distance, in: presented at the pt. Fiber Commun. Conf. Exhib., Los Angeles, CA, USA, 2017, Paper Th5A.7.
[162] X. Li, J. Xiao, J. Yu, Long-distance wireless mm-wave signal delivery at W-band, J. Lightwave Technol. 34 (2) (2016) 661–668.

[163] X. Li, J. Yu, Photonics-assisted 32-Gb/s wireless signal transmission over 1km at K-band, IEEE Photonics Technol. Lett 29 (13) (2017) 1120–1123.
[164] X. Li, J. Yu, J. Xiao, Demonstration of ultra-capacity wireless signal delivery at W-band, J. Lightwave Technol. 34 (1) (2016) 180–187.
[165] J. Xiao, X. Li, J. Yu, Y. Xu, Z. Zhang, L. Chen, 40-Gb/s PDM-QPSK signal transmission over 160-m wireless distance at W-band, Opt. Lett 40 (6) (2015) 998–1001.
[166] X. Li, J. Yu, and J. Xiao, 100^3 (100Gb/s × 100m × 100GHz) optical wireless system, in: ECOC 2015, Valencia, Spain, ID: 0638.
[167] X. Li, J. Yu, K. Wang, Y. Xu, L. Chen, L. Zhao, and et al., Bidirectional delivery of 54-Gbps 8QAM W-band signal and 32-Gbps 16QAM K-band signal over 20-km SMF-28 and 2500-m wireless distance, in: Presented at the Pt. Fiber Commun. Conf. Exhib., Los Angeles, CA, USA, 2017, Paper Th5A.7.
[168] C.A. Balanis, Antenna Theory: Analysis and Design, John Wiley & Sons, 2016.
[169] H.T. Friis, T. Harald, A note on a simple transmission formula, Proc. Inst. Radio Eng. 34 (5) (1946) 254–256.
[170] M. Schlechtweg, A. Tessmann, From 100 GHz to terahertz electronics—activities in Europe, in: Proc. CSIC 2006, San Antonio, TX, pp. 8–11.
[171] V. Dyadyuk, J.D. Bunton, Y.J. Guo, Study on high rate long range wireless communications in the 71–76 and 81–86 GHz bands, in: Proc. EuMC 2009, pp. 1315–1318.
[172] T.P. McKenna, A.N. Jeffrey, R.C. Thomas, Experimental demonstration of photonic millimeter-wave system for high capacity point-to-point wireless communications, J. Lightwave Technol. 32 (20) (2014) 3588–3594.
[173] The NTT Group's Response to the Great East Japan Earthquake, NTT Group Tokyo, Japan [Online]. Available from: <http://www.ntt.co.jp/csr_e/2011report/>.
[174] C. Wang, C. Lin, Q. Chen, B. Lu, X. Deng, J. Zhang, A 10-Gbit/s wireless communication link using 16-QAM modulation in 140-GHz band, IEEE Trans. Microw. Theory Techn. 61 (7) (2013) 2737–2745.
[175] A. Hirata, et al., 120-GHz-band wireless link technologies for outdoor 10-Gbit/s data transmission, IEEE Trans. Microw. Theory Tech. 60 (3) (2012) 881–895.
[176] H.-J. Song, K. Ajito, Y. Muramoto, A. Wakatsuki, T. Nagatsuma, N. Kukutsu, 24 Gbit/s data transmission in 300 GHz band for future terahertz communications, Electron. Lett. 48 (15) (2012) 953–954.
[177] S. Koenig, et al., 100 Gbit/s wireless link with mm-wave photonics, in: Proc. Opt. Fiber Commun. Conf. Exposit. Nat. Fiber Opt. Eng. Conf. (OFC/NFOEC), March 2013, pp. 1–3, paper PDP5B.4.
[178] R. Sambaraju, D. Zibar, R. Alemany, A. Caballero, J. Herrera, I.T. Monroy, Radio-frequency transparent demodulation for broadband hybrid wireless-optical links, IEEE Photonics Technol. Lett. 22 (11) (2010) 784–786.

Further reading

K. Ramachadran, R. Kokku, R. Mahindra, S. Rangarajan, 60 GHz Data-Center Networking Wireless ⇒ Worry less? NEC Technical Report, 2008.

Index

Note: Page numbers followed by '*f*' and '*t*' refer to figures and tables, respectively.

C

D

E

G

H

I

J

K

M

N

O

R

S

T

CPI Antony Rowe
Eastbourne, UK
November 15, 2022